The
ENCYCLOPEDIA
of
PALEONTOLOGY

ENCYCLOPEDIA OF EARTH SCIENCE SERIES

The
ENCYCLOPEDIA
of
PALEONTOLOGY

EDITED BY

Rhodes W. Fairbridge
Columbia University

David Jablonski
Yale University

Dowden, Hutchinson & Ross, Inc.
Stroudsburg   Pennsylvania

Copyright © 1979 by **Dowden, Hutchinson & Ross, Inc.**
Library of Congress Catalog Card Number: 79-11468
ISBN: 0-87933-185-2

83 82 81 80 79   5 4 3 2 1
Manufactured in the United States of America.

Library of Congress Cataloging in Publication Data
Main entry under title:
Encyclopedia of paleontology.
  (Encyclopedia of earth sciences series; v. 7)
  Includes index.
  1. Paleontology. I. Fairbridge, Rhodes Whitmore, 1914–   II. Jablonski,
David, 1953–   III. Series.
QE703.E52   560'.3   79-11468
ISBN 0-87933-185-2

Distributed world wide by Academic Press,
a subsidiary of Harcourt Brace Jovanovich,
Publishers.

PREFACE

Paleontology is a cognate science, one contributing to and deriving much information from such diverse disciplines as geochemistry, population biology, histology, botany, anthropology, and geophysics. Consequently, the study of ancient life will reveal the wide-ranging nature of the subject matter. We have attempted herein to provide a survey of many of the objects and concepts encompassed by this multifaceted field within the limits of a single volume.

The entries are written at several conceptual levels, from the most basic and inclusive articles such as *Invertebrate Paleontology* and *Evolution*, to specialized, technical treatments such as *Morphology, Constructional* and *Taxonomy, Numerical*. They are the contributions of many different people whose diverse backgrounds and approaches reflect the interdisciplinary nature of the field. In the more specialized articles, emphasis often has been on the most topical or interesting aspects of the subject. For example, the history of paleontology entries are highly individualistic in their style and content. Martin Rudwick's pre-Darwinian section is a broad overview, covering centuries of gradual discovery and development. We found, not unexpectedly, that no single authority was able to perform a similar synthesis of the tumultuous period that gave rise to the modern discipline of paleontology. We include Peter Bretsky's more intensive study of late nineteenth- and early twentieth-century paleontological thought in the hopes of stimulating interpretation, discussion, and eventual synthesis of this yet uncharted interval in the history of the science. The final article on the history of paleontology, by Campbell and Valentine, reports on the current trends of the present post-plate tectonics era of paleontology.

The systematic entries dealing with groups of organisms also range from the general to the specific. The larger groups such as *Vertebrata* and *Arthropoda* are given general coverage, while their corresponding smaller categories such as *Mammals, Mesozoic* and *Trilobita* are discussed in greater detail.

But whether the entry is general or specialized, the extensive reference lists and cross-references at the end of each are designed to lead the reader deeper into the subject. These items and the index will also enable the interested reader to contrast the different viewpoints generated by disagreement among the authors on certain points—disagreement due to the fact that paleontology is a rapidly developing science and many of our authors were anxious to include the newest material possible and the most modern concepts available.

How to Use This Encyclopedia

Key words, taxonomic names such as *Reptilia* or conceptual or theoretical terms such as *Extinction*, appear alphabetically. If the reader fails to find a sought-after term, the comprehensive Subject Index may be of help.

Throughout this encyclopedia there is extensive cross-referencing from one article to another. (The abbreviation q.v., which is used to denote cross-references in the text of the articles, is from the Latin *quod vide,* "which see.") For example, if interested in the ecology of ancient environments, the reader should start with *Paleoecology.* Checking the cross-references therein, the reader desiring further conceptual background may consult such listed entries as *Diversity* and *Population Dynamics.* It was not possible to cross-reference *Paleoecology* to every article that contained paleoecologic information; the Subject Index will be of help, and for further reading on a particular environment, the reader should consult the taxonomic entries of the organisms which inhabit it. As another example, useful entries for shallow marine environments would include *Brachiopoda* and *Mollusca* and the cross-references therein. Higher taxa having particularly important subgroups carry cross-references covering further entries: thus, *Mollusca* will lead to *Bivalvia, Cephalopoda, Gastropoda,* and a number of others.

Although it is customary in the prefaces of these encyclopedias to list standard references in the field, paleontology with its naturally-imposed divisions of study precludes this. We refer the reader to our entries on *Paleontology, Invertebrate Paleontology, Vertebrate Paleontology, Paleobotany, Micropaleontology,* and *History of Paleontology: Post-Plate Tectonics* for introductory references.

For relevant material outside the immediate scope of this volume but of interest to paleontologists, there are the other volumes in the *Encyclopedia of Earth Sciences* series. Regional physical and chemical data relating to marine ecology will be found in *Oceanography;* climatic material and extraterrestrial life appear in *Atmospheric Sciences and Astrogeology* (which also contains geophysical conversion tables);

Phanerozoic Geologic Time ("Absolute" ages are in millions of years).

Time term	Epoch	Period	Era
Rock term	Series	System	
0.01	Holocene	Quaternary	Cenozoic
1.8	Pleistocene		
5.2	Pliocene	Tertiary	
22.0	Miocene		
35.0	Oligocene		
49.0	Eocene		
	Paleocene		
64			
135	Cretaceous		Mesozoic
192	Jurassic		
	Triassic		
225			
280	Permian		Paleozoic
320	Carboniferous	Pennsylvanian	
345		Mississippian	
395	Devonian		
430	Silurian		
500	Ordovician		
	Cambrian		
570			
	Precambrian		

Conversion Table for Metric to English Units

1 micron (μ) = 0.001 millimeter (mm) = 0.00004 inch (in.)
1 mm = 0.1 centimeter (cm) = 0.03937 in.
1000 mm = 100 cm = 1 meter (m) = 39.37 in. = 3.2808 foot (ft)
1 inch = 2.54 cm
12 in. = 1 ft = 0.3048 m = 30.480 cm
1 cm = 0.39370 in. = 0.032808 ft
1 km = 10^5 cm = 0.62137 mile
1 fathom = 6 ft = 1.8288 m
a degree C = 1.8°F
a degree F = 0.5556°C
0°F = -18°C
32°F = 0°C
212°F = 100°C
°F = 9/5 (°C + 32)
°C = 5/9 (°F - 32)

land surface descriptions and details of Quaternary and Holocene earth history are in *Geomorphology* (which also contains physical statistics of continents and oceans); chemical information, including descriptions of all the elements and the major biochemical cycles appear in *Geochemistry*. Regional geologies of different countries including many paleontological notes are found in the three parts of *World Regional Geology*. The deformation of fossils will be treated in *Structural Geology*. . . and so on.

Acknowledgments

The execution of a work of this size and scope is dependent upon the generous assistance of very many persons. First we must thank Tamar Gordon, our editorial assistant, whose energy, care, and creativity made this encyclopedia a much better volume. Susan M. Kidwell contributed much paleontological and editorial aid in the final stages of this book. Thanks also to Edith and Edward Jablonski. Joanne Bourgeois lent expertise accrued while coediting her own volume. We would like to thank the many publishers and insitutions who permitted us to reproduce their copyrighted material; The Geological Society of American and The University of Kansas Press in particular allowed us to use many fine illustrations. Original drafting was done by Patricia Manley and Susan Lundstedt.

We are also most grateful to a veritable army of scientists who aided our endeavors in various ways: by critical reading, by recommending authors, by sending reprints and illustrations. To each and all, including any we may have omitted inadvertently, we express our appreciation: W. H. Adey, D. I. Axelrod, S. Barghoorn, R. L. Batten, A. K. Behrensmeyer, R. H. Benson, R. S. Boardman, O. M. B. Bulman, L. H. Burckle, R. L. Carroll, W. C. Cornell, J. H. Dickson, R. E. Dodge, C. Downie, G. F. Elliot, W. K. Emerson, G. Engelman, J. O. Farlow, T. D. Fouch, M. Fracasso, D. G. Frey, P. D. Gingerich, M. F. Glaessner, S. J. Gould, R. E. Grant, T. Hansen, W. W. Hay, R. Herrin, D. Hill, H. G. Hofmann, E. G. Kauffman, S. M. Kidwell, Z. Kielan-Jaworowska, J. A. W. Kirsch, M. LaBarbera, D. W. Larson, D. R. Lawrence, E. B. Leopold, J. A. Lillegraven, R. M. Linsley, A. Logan, K. B. Macdonald, M. C. McKenna, D. K. Meinke, W. Mintz, C. Nelson, R. G. Osgood, Jr., A. L. Panchen, R. E. Peck, B. F. Perkins, T. L. Phillips, R. Plotnik, H. Polz, D. M. Raup, F. H. T. Rhodes, W. R. Riedel, J. K. Rigby, J. M. Schopf, A. Seilacher, B. J. Stahl, F. M. Swain, R. D. K. Thomas, K. S. Thomson, K. M. Towe, M. A. Turner, J. W.

Valentine, M. R. Voorhies, H. E. Wright, Jr., K. M. Waage, S. Walker, S. D. Webb, and E. L. Yochelson. With much regret we must record the deaths of seven of our contributors; they will surely be missed: Maurice Black, G. Colom, Paul Edwards, Louis T. Grambast, Walter Häntzschel, A. G. Smith, and P. C. Sylvester-Bradley.

Finally, a word of appreciation for our farsighted, patient, and friendly publishers, Charles Hutchinson and James Ross, without whose constant enthusiasm, faith, and encouragement the *Encyclopedia of Earth Sciences* series may well have fallen by the wayside. We are also grateful to Shirley End, Production Manager, and Bernice Wisniewski, Production Editor, for the care and diligence that they put into the preparation of this book.

DAVID JABLONSKI
RHODES W. FAIRBRIDGE

CONTRIBUTORS

DEREK V. AGER, Dept. of Geology, University College of Swansea, Singleton Park, Swansea, W. Glam. SA2 8PP, United Kingdom. *Paleoecology.*

FRANCISCO J. AYALA, Dept. of Genetics, University of California, Davis, California, 95616. *Genetics.*

BASIL E. BALME, Dept. of Geology, University of Western Australia, Nedlands, Western Australia. *Palynology, Paleozoic and Mesozoic.*

HARLAN P. BANKS, Dept. of Botany, Cornell University, Ithaca, New York, 14850. *Pteridophyta.*

STEVEN BARGHOORN, Dept. of Vertebrate Paleontology, American Museum of Natural History, Central Park West and 79th Street, New York, New York, 10024. *Systematic Philosophies.*

K. WERNER BARTHEL, Institut für Geologie und Paläontologie, Technische Universität Berlin, Hardenbergstrasse 42, D – 1 Berlin 12, Germany. *Solnhofen Formation.*

R. L. BATTEN, Dept. of Fossil and Living Invertebrates, American Museum of Natural History, Central Park West and 79th Street, New York, New York, 10024. *Mollusca.*

MAURICE BLACK, deceased. *Coccoliths.*

J. PLATT BRADBURY, U.S. Geological Survey, Mail Stop 919, P&S Branch, Box 25046, Federal Center, Denver, Colorado, 80225. *Paleoecology, Inland Aquatic Environments.*

PETER W. BRETSKY, Dept. of Earth and Planetary Sciences, State University of New York, Stony Brook, New York, 11794. *History of Paleontology, Post-Darwinian.*

DAVID BUKRY, U.S. Geological Survey (A-015), Scripps Institute of Oceanography, La Jolla, California, 92093. *Coccoliths; Discoasters.*

LLOYD H. BURCKLE, Lamont-Doherty Geological Observatory, Palisades, New York, 10964. *Diatoms.*

JOHN G. BYRNES, Geological and Mining Museum, 36 George Street, Sydney 2000, Australia. *Receptaculitoids.*

CATHRYN A. CAMPBELL, Dept. of Genetics, University of California, Davis, California, 95616. *History of Paleontology, Post-Plate Tectonics.*

ROBERT L. CARROLL, Redpath Museum, McGill University, Montreal, Quebec H3C 2G1, Canada. *Amphibia; Reptilia.*

KEITH CHAVE, Dept. of Oceanography, University of Hawaii, Honolulu, Hawaii, 96822. *Biomineralization.*

CHARLES S. CHURCHER, Dept. of Zoology, University of Toronto, Toronto, Ontario M5S 2E1, Canada. *Marsupials.*

J. L. CISNE, Dept. of Geological Sciences, Cornell University, Ithaca, New York, 14850. *Arthropoda; Population Dynamics.*

GEORGE R. CLARK, II, Dept. of Geology, Kansas State University, Manhattan, Kansas, 66506. *Growth Lines.*

J. S. H. COLLINS, 63 Oakhurst Grove, London S.E. 22, England. *Cirripedia.*

G. COLOM, deceased. *Tintinnids.*

G. A. COOPER, Dept. of Paleobiology, Smithsonian Institution, E-205 Natural History Museum, Washington, D.C., 20560. *Brachiopoda.*

RICHARD COWEN, Dept. of Geology, University of California, Davis, California, 95616. *Morphology, Functional.*

JOEL CRACRAFT, Dept. of Anatomy. University of Illinois at the Medical Center, Chicago, Illinois, 60680. *Paleobiogeography of Vertebrates.*

T. DELEVORYAS, Dept. of Botany, University of Texas at Austin, Austin, Texas, 78712. *Gymnosperms.*

ROBERT H. DENISON, Todd Pond Road, Lincoln, Massachusetts, 01773. *Agnatha; Pisces.*

J. ROBERT DODD, Dept. of Geology, Indiana University, Bloomington, Indiana, 47401. *Biogeochemistry.*

ROBERT G. DOUGLAS, Dept. of Geological Sciences, University of Southern California, Los Angeles, California, 90007. *Foraminifera, Planktic.*

PAUL EDWARDS, deceased. *Coprolite.*

HEINRICH K. ERBEN, Institut für Paläontologie, Rheinische Freidrich-Wilhelms Universität, Nussallee 8, 5300 Bonn 1, West Germany. *Biomineralization.*

JAMES O. FARLOW, Dept. of Geology and Geophysics, Yale University, New Haven, Connecticut, 06520; Dept. of Geology, Hope College, Holland, Michigan, 49423. *Paleoecology, Terrestrial.*

ROBERT M. FINKS, Dept. of Earth and Environmental Sciences, Queens College, Flushing, New York, 11367. *Fossils and Fossilization.*

KARL W. FLESSA, Dept. of Geosciences, University of Arizona, Tucson, Arizona, 85721. *Extinction.*

R. L. FOLK, Dept. of Geological Sciences, University of Texas at Austin, Austin, Texas, 78712. *Coprolite.*

RICHARD C. FOX, Dept. of Geology, University of Alberta, Edmonton, Alberta T6G 2E1, Canada. *Mammals, Mesozoic.*

ROBERT W. FREY, Dept. of Geology, University of Georgia, Athens, Georgia, 30601. *Trace Fossils.*

R. A. GANGLOFF, Dept. of Geology, Merritt College, Oakland, California, 94619. *Archaeocyatha.*

BRIAN F. GLENISTER, Dept. of Geology, University of Iowa, Iowa City, Iowa, 52242. *Cephalopoda.*

ROBERT M. GOLL, Duke University Marine Laboratory, Beaufort, North Carolina, 28516. *Radiolaria.*

TAMAR GORDON, Dept. of Anthropology, University of California, Berkeley, California, 94720. *Living Fossil.*

ALAN GRAHAM, Dept. of Biological Sciences, Kent State University, Kent, Ohio, 44240. *Angiospermae; Paleobotany; Paleophytogeography.*

LOUIS T. GRAMBAST, deceased. *Charophyta.*

JOSEPH T. GREGORY, Dept. of Paleontology, University of California, Berkeley, California, 94720. *Chordata; Vertebrata; Vertebrate Paleontology.*

RAYMOND C. GUTSCHICK, Dept. of Earth Sciences, University of Notre Dame, Notre Dame, Indiana, 46556. *Fossil Record; Sclerites, Holothurian.*

ALLAN M. GUTSTADT, Dept. of Geology, California State University, Northridge, California, 91330. *Pseudofossil.*

WALTER HÄNTZSCHEL, deceased. *Actualistic Paleontology; Trace Fossils.*

JOHN H. HANLEY, U.S. Geological Survey, Mail Stop 919, P&S Branch, Box 25046, Federal Center, Denver, Colorado, 80225. *Paleoecology, Inland Aquatic Environments.*

THOR HANSEN, Dept. of Geological Sciences, University of Texas at Austin, Austin, Texas, 78712. *Larvae of Marine Invertebrates: Paleontological Signifiance.*

PHILIP H. HECKEL, Dept. of Geology, University of Iowa, Iowa City, Iowa, 52240. *Reefs and Other Carbonate Buildups.*

CALVIN J. HEUSSER, Dept. of Biology, New York University, New York, New York, 10003. *Palynology.*

SEWELL H. HOPKINS, Route 3, Box 232, Gloucester, Virginia, 23061. *Oysters.*

HILDEGARDE HOWARD, 2045-Q Via Mariposa East, Laguna Hills, California, 92653. *Aves.*

DAVID JABLONSKI, Dept. of Geology and Geophysics, Yale University, New Haven, Connecticut, 06520. *Larvae of Marine Invertebrates: Paleontological Significance; Living Fossil; Reefs and Other Carbonate Buildups; Solnhofen Formation.*

R. P. S. JEFFERIES, British Museum (Natural History), Cromwell Road, London SW7 5BD, England. *Calcichordates.*

W. A. M. JENKINS, Petro-Canada Exploration, Inc., 650 Guinness House, 7th Avenue S. W., Calgary, Alberta TP2 0Z6, Canada. *Chitinozoa.*

ROGER L. KAESLER, Dept. of Geology, University of Kansas, Lawrence, Kansas, 66045. *Computer Applications in Paleontology: Taxonomy, Numerical.*

ERLE G. KAUFFMAN, Dept. of Paleobiology, Smithsonian Institution, E-307 Natural History Museum, Washington, D.C., 20560. *Bivalvia; Rudists.*

ERIK N. KJELLESVIG-WAERING, Box 699, Marco Island, Florida, 33937. *Eurypterida.*

LOUIS S. KORNICKER, Smithsonian Institution, U.S. National Museum, Washington, D.C., 20560. *Ostracoda.*

WILLIAM S. LACEY, School of Plant Biology, University College of North Wales, Banhor, Gwynedd LL57 2UW, Wales, United Kingdom. *Bryophyta.*

G. P. LARWOOD, Dept. of Geological Sciences, University of Durham, Durham DH1 3LE, England. *Bryozoa.*

DAVID R. LAWRENCE, Dept. of Geology, University of South Carolina, Columbia, South Carolina, 29208. *Biostratinomy; Diagenesis of Fossils–Fossildiagenese; Taphonomy.*

U. LEHMAN, Geologische-Paläontologisches Institüt, Universität Hamburg, 2, Hamburg 13, West Germany. *Aptychus, Anaptychus.*

E. B. LEOPOLD, Quarternary Research Center, University of Washington, Seattle, Washington, 98195. *Palynology, Late Tertiary.*

JERE H. LIPPS, Bodega Marine Laboratory, P. O. Box 247, Bodega Bay, California, 94923. *Ebridians; Silicoflagellates.*

L. G. LOVE, Dept. of Geology, University of Sheffield S1 3JD, England. *Pyritospheres.*

J. R. MACDONALD, 1128 Reinclaud Court, Sunnyvale, California, 94087. *Paleopathology.*

DEBORAH K. MEINKE, Dept. of Geology and Geophysics, Yale University, New Haven, Connecticut, 06520. *Bones and Teeth.*

D. MELLOR, JR., Dept. of Geology and Geophysics, Yale University, New Haven, Connecticut, 06520. *Ediacara Fauna.*

E. GEORGES MERINFELD, Dept. of Oceanography, Dalhousie University, Halifax, Nova Scotia, Canada. *Radiolaria.*

FRANK A. MIDDLEMISS, Dept. of Geology, Queen Mary College, University of London, London E1 4NS, England. *Paleobiogeography.*

B. A. MASTERS, Amoco Production Company, 4502 East 41st Street, P. O. Box 591, Tulsa, Oklahoma, 74102. *Calcispheres and Nannoconnids.*

CLAUDE MONTY, Centre d'analyses Paléoecologiques, Université de Liège, B-4000 Liège, Belgium. *Stromatolites.*

T. C. MOORE, JR., Graduate School of Oceanography, University of Rhode Island, Kingston, Rhode Island, 02881. *Micropaleontology.*

SIMON CONWAY MORRIS, Dept. of Earth Sciences, The Open University, Walton Hall, Milton Keynes, MK7 6AA, England. *Burgess Shale.*

K. J. MÜLLER, Institut für Paläontologie, Rheinische Friedrich-Wilhelms Universität, Nussallee 8, 5300 Bonn 1, West Germany. *Conodonts; Silicification of Fossils.*

JOHN W. MURRAY, Dept. of Geology, Exeter University, Exeter EX4 4PS, England. *Foraminifera, Benthic.*

D. NICHOLS, Dept. of Biological Sciences, Exeter University, Exeter EX4 4PS, England. *Echinodermata.*

MICHAEL J. NOVACEK, Dept. of Zoology, San Diego State University, San Diego, California, 92812. *Mammals, Placental.*

WILLIAM A. OLIVER, JR., U.S. Geological

Survey, E-305 Natural History Building, Washington, D.C., 20560. *Coelenterata; Coloniality; Conulata; Corals.*

JOHN H. OSTROM, Peabody Museum, Yale University, New Haven, Connecticut, 06520. *Amphibia; Reptilia.*

C. P. PALMER, Dept. of Paleontology, British Museum (Natural History), Cromwell Road, London SW7 5BD, England. *Scaphopoda.*

ROBERT H. PARKER, Coastal Ecosystems Management Inc., 3600 Hulen Street, Fort Worth, Texas, 76107. *Paleotemperature and Depth Indicators.*

J. S. PENNY, Dept. of Biology, La Salle College, Philadelphia, Pennsylvania, 19141. *Palynology, Early Tertiary.*

JOHN POJETA, JR., U.S. Geological Survey, E-302 Natural History Building, Washington, D.C., 20560. *Rostroconchia.*

STAN RACHOOTIN, Dept. of Biology, Yale University, New Haven, Connecticut, 06520. *Zooxanthellae.*

R. E. H. REID, Dept. of Geology, Queen's University, Belfast, Ireland. *Porifera.*

CHARLES R. REMINGTON, Dept. of Biology, Yale University, New Haven, Connecticut, 06520. *Insecta.*

R. A. REYMENT, Paleontologiska Institutionen, Uppsala Universitet, Box 558, S-751 22 Uppsala, Sweden. *Biometrics in Paleontology.*

F. H. T. RHODES, Dept. of Geology and Minerology, University of Michigan, Ann Arbor, Michigan, 48104. *Invertebrate Paleontology; Paleontology.*

R. PETER RICHARDS, Dept. of Geology, Oberlin College, Oberlin, Ohio, 44074. *Cornulitidae.*

R. B. RICKARDS, Dept. of Geology, University of Cambridge, Cambridge CB2 3EQ, England. *Graptolithina; Hemichordata.*

ROBERT RIDING, University College, P. O. Box 78, Cardiff CF1 1XL, Wales, Great Britain. *Algae.*

CHARLES A. ROSS, Dept. of Geology, Western Washington State University, Bellingham, Washington, 98225. *Fusulinacea.*

MARTIN J. S. RUDWICK, Vrije Universiteit, de Boelelaan 1088, Postbus 7161, 1007 MC Amsterdam, Netherlands. *History of Paleontology, Before Darwin.*

BRUCE RUNNEGAR, Dept. of Geology, University of New England, Armidale, N.S.W. 2351, Australia. *Rostroconchia.*

WILLIAM A. S. SARJEANT, Room 108.3 (Geological Sciences), General Purpose Building, University of Saskatchewan, Saskatoon, Saskatchewan S7N 0W0, Canada. *Acritarchs; Anellotubulates; Dinoflagellates; Hystrichospheres; Melanosclerites; Tasmanitids.*

J. WILLIAM SCHOPF, Dept. of Geology, Paleobiological Laboratory, University of California, 405 Hilgard Avenue, Los Angeles, California, 90024. *Precambrian Life.*

F. R. SCHRAM, Dept. of Paleontology, San Diego Natural History Museum, P. O. Box 1390, San Diego, California, 92112. *Crustacea.*

R. W. SCOTT, Amoco Production Company, 4502 East 41st Street, P. O. Box 591, Tulsa, Oklahoma, 74102. *Calcispheres and Nannoconnids.*

R. E. SLOAN, Dept. of Geology and Geophysics, University of Minnesota, Minneapolis, Minnesota, 55455. *Multituberculata.*

A. G. SMITH, Dept. of Geology, University of Cambridge, Cambridge CB2 3EQ, England. *Paleogeographic Maps.*

ALLYN G. SMITH, deceased. *Polyplacophora.*

NORMAN F. SOHL, U.S. Geological Survey, E-309 Natural History Building, Washington, D.C., 20560. *Gastropoda; Rudists.*

STEVEN M. STANLEY, Dept. of Earth and Planetary Sciences, The Johns Hopkins University, Baltimore, Maryland, 21218. *Evolution.*

ROBERT J. STANTON, JR., Dept. of Geology, Texas A & M University, College Station, Texas, 77843. *Diversity.*

COLIN W. STEARN, Dept. of Geological Sciences, McGill, University, Montreal, Quebec, Canada. *Stromatoporoids.*

GRAEME R. STEVENS, New Zealand Geological Survey, Lower Hutt, New Zealand. *Belemnitida.*

W. E. SWINTON, Massey College, University of Toronto, Toronto, Ontario M5S 2E1, Canada. *Dinosaurs.*

P. C. SYLVESTER-BRADLEY, deceased. *Biostratigraphy.*

HELEN TAPPAN, Dept. of Earth and Space Sciences, University of California, Los Angeles, California, 90024. *Plankton; Protista.*

PAUL TASCH, Dept. of Geology and Geography, Wichita State University, Wichita, Kansas, 67208. *Branchipoda; Chitin; Dwarf Faunas; Scolecodonts.*

R. D. K. THOMAS, Dept. of Geology, Franklin and Marshall College, Lancaster, Pennsylvania, 17604. *Morphology, Constructional.*

IDA THOMPSON, Dept. of Geological and Geophysical Sciences, Princeton University, Princeton, New Jersey, 08540. *Annelida; Mazon Creek.*

E. R. TRUEMAN, Dept. of Zoology, The Uni-versity of Manchester, Manchester M13 9PL, England. *Species Concept.*

JAMES W. VALENTINE, Dept. of Geological Science, University of California, Santa Barbara, California, 93106. *History of Paleontology, Post-Plate Tectonics; Plate Tectonics and the Biosphere.*

KENNETH R. WALKER, Dept. of Geology, University of Tennessee, Knoxville, Tennessee, 37916. *Communities, Ancient; Substratum; Trophic Groups.*

JOHN E. WARME, Dept. of Geology, Rice University, Houston, Texas, 77001. *Actualistic Paleontology.*

G. E. G. WESTERMANN, Dept. of Geology, McMaster University, Hamilton 76, Ontario, Canada. *Sexual Dimorphism.*

H. B. WHITTINGTON, Sedgwick Museum, Downing Street, Cambridge CB2 3EQ, England. *Trilobita.*

ELLIS L. YOCHELSON, U.S. Geological Survey, E-501 Natural History Building, Washington, D.C., 20560. *Hyolitha; Mattheva; Monoplacophora; Stenthecoida; Tentaculita.*

KEITH YOUNG, Dept. of Geological Sciences, University of Texas at Austin, Austin, Texas, 78712. *Ammonoidea.*

The
ENCYCLOPEDIA
of
PALEONTOLOGY

ACRITARCHS

Acritarchs are microorganisms of uncertain affinity, having a hollow shell of highly varied shape (spherical, ellipsoidal, discoidal, elongate, or polygonal) composed of an organic substance or substances. Their size is between 7 and 1000 μm, most often less than 150 μm. The shell may be pitted, granular, or entirely smooth; it may be unornamented or may bear spines or other processes, raised ridges (*crests*), pits, or granules. The distribution of the ornament may be quite random or it may show a consistent positional relationship—e.g., confinement to, or arrangement around, the poles of an ellipsoidal shell. Sometimes the shell is multiple, consisting of two separate, roughly concentric membranes of comparable thickness; sometimes the main shell is partially or entirely surrounded by a much more tenuous, often incomplete outer membrane; sometimes the shell bears a median flange or wing. (Fig. 1.)

Individual acritarchs may exhibit rupture or splitting. This may occur along a predetermined line (a *cryptosuture*), opening progressively until one entire pole of the shell may be lost; alternatively, the opening may be in the form of a more regularly shaped crescentic slit (an *epityche*) or by loss of a circular or subpolygonal section of the shell wall, forming a *pylome.* Apparently intact shells are also common, some species consistently lacking observable openings. The individual shells are usually quite separate, but they occasionally show a loose linkage into chains or association into monospecific masses, possibly suggesting that they were originally surrounded by some sort of enclosing membrane, such as an egg sac or spore case.

History of Study

In 1961, Evitt demonstrated that a number of microfossil genera previously classed into an order, the Hystrichosphaeridea, of uncertain systematic reference, were in fact the reproductive cysts of dinoflagellates (q.v.) and should be classified as such. But since these misclassified genera included the type genus, *Hystrichos-*

phaera, the order had to be abandoned and the term "hystrichosphere" assumed a more restricted usage. Of the genera that provided no evidence of dinoflagellate affinity, some, such as *Tasmanites,* were subsequently shown by Wall (1962) to be referable to a living group of unicellular plants, the Prasinophyceae; these are spherical to ellipsoidal forms, with thick shell walls penetrated by systems of pores.

However, a number of genera remained, of very varied form but of undetermined affinity, for which there was now no group name nor any taxonomic niche. For these, Evitt (1963) proposed the name "acritarchs" (Greek: "uncertain origin"). For convenience they were, like the dinoflagellates, to be dealt with under the botanical rules of nomenclature; however, as genera defined on morphology only, without implication as to relationships, it was undesirable that they should be classed into the Linnaean hierarchy of orders and families. It was clearly preferable that any classification adopted should give no false impression of any natural relationship and should facilitate the easy transfer away of genera when their affinities came to be determined.

The classification now in most widespread use is usually designated the "D-E-S classification," since it was proposed by Downie, Evitt, and Sarjeant (1963). This is recognizedly an entirely artificial system in which the groupings are based exclusively on morphology, without implication of expressing natural relationships, and do not have specified type genera. The basic unit is the Group Acritarcha, which is divided into fourteen Subgroups, e.g., Acanthomorphitae (comprising spherical to ellipsoidal acritarchs bearing isolated processes—most often simple or branched spines), Sphaeromorphitae (spherical to ellipsoidal forms without processes), etc. Since this is a non-Linnaean classification, it has no recognized status in taxonomy; and a number of modifications or alternative schemes have since been proposed, from mere transformation of names from Group to Paraorder and from Subgroup to Parafamily, proposed by G. and M. Deflandre (1964), to the formulation of a new hierarchy of names, exactly similar in scope but with the suffix "-morphyda," proposed by Timofeyev (1965).

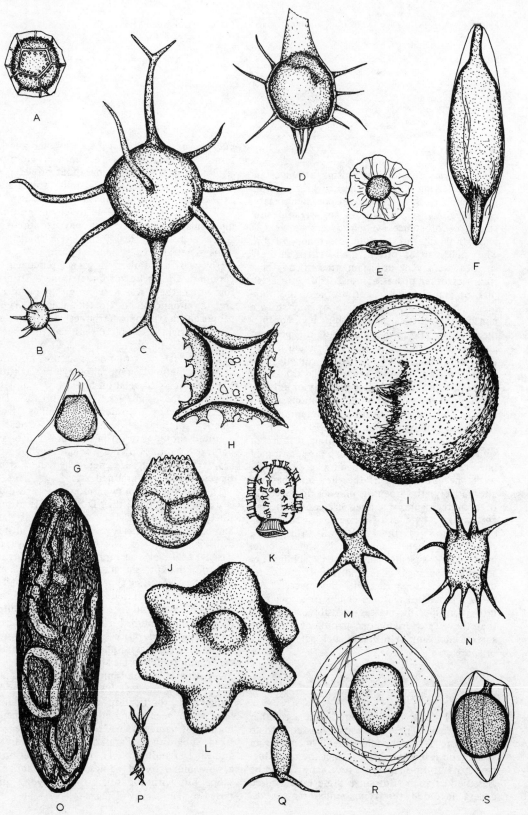

Geological Distribution

The earliest acritarchs are among the oldest known fossils: acritarchs of a very simple type, with spherical or ellipsoidal shells ornamented only by granules or pits (belonging to the Subgroup Sphaeromorphitae), range far back into the Precambrian. They are the only group to have been successfully used for purposes of stratigraphic correlation in Precambrian deposits, the same broad subdivisions emerging in sediments of Scotland and of Russia (Naumova and Pavlovsky, 1961). Throughout the lower Paleozoic, the acritarchs are tremendously abundant and varied, dominating fossil microplankton assemblages and acting as satisfactory stratigraphic indices. In the upper Paleozoic, they remain abundant but are significantly less varied in morphology: though their distribution is clearly subject to environmental control, recent studies suggest that they are still useful for correlation in the Permian and Triassic. However, from the middle Mesozoic onward, dinoflagellate cysts become the most striking constituents of assemblages: acritarchs with spherical tests bearing simple processes, of the genera *Micrhystridium* and *Solisphaeridium* (Subgroup Acanthomorphitae), remain intermittently abundant until the Lower Cretaceous but afford no apparent potential for correlation: other acritarchs are relatively infrequent and even more erratic in distribution. There are few Tertiary records and, though acritarchs have been recorded from present-day plankton and from marine sediments, they have been studied very little.

Acritarchs are consistently asssociated with water-deposited sediments. They are known from riverine and lacustrine sediments (see Harland and Sarjeant, 1970) but are most characteristic of marine sediments. They have been recorded from all types of marine deposits but are perhaps most abundant in shales and clays, totals of 10,000 individuals per gram being in no way exceptional. (For references see Downie and Sarjeant, 1964).

Affinities

The acritarchs are quite certainly plankton, as seen from their distribution; but otherwise, their relationships remain doubtful. Since they are of simple morphological character, it is highly probably that they are polyphyletic. It is possible that they may include spores of higher plants or the eggs of organisms such as copepods. However, it is much more likely that the majority represent stages in the life cycle of unicellular algae—cysts, spore cysts, vegetative stages, etc. The style of opening in some genera suggests that they may prove to be dinoflagellate cysts (see *Dinoflagellates*); other algal groups may also be represented; and Downie (1973) believes that some lineages may be referable to the Prasinophyceae. However, the incompleteness or ambiguity of evidence means that the true affinity of some groups may never be conclusively demonstrated.

WILLIAM A. S. SARJEANT

FIGURE 1. Fossil acritarchs. A, Subgroup Herkomorphitae, *Cymatiosphaera imitata* (Upper Cretaceous). B–D, Subgroup Acanthomorphitae; B, *Micrhystridium fragile* (Middle Jurassic–Lower Cretaceous); C, *Baltisphaeridium longispinosum* (Ordovician-Silurian); D, *Aremoricanium rigaudae* (Silurian). E, Subgroup Pteromorphitae; *Pterospermopsis australiensis* (Lower Cretaceous). F, Subgroup Dinetromorphitae; *Diplofusa gearlensis* (Upper Cretaceous). G, Subgroup Platymorphitae; *Trigonopyxidia ginella* (Upper Cretaceous). H, Subgroup Prismatomorphitae; *Polyedryxium diabolicum* (Devonian). I, Subgroup Sphaeromorphitae; *Leiosphaeridia granulata* (Silurian). J, Subgroup Oomorphitae; *Oodium sablincaense* (Upper Cambrian). K, Subgroup Stephanomorphitae; *Stephanelytron redcliffense* (Upper Jurassic). L–M, Subgroup Polygonomorphitae; L, *Pulvinosphaeridium pulvinellum* (Silurian); M, *Veryhachium stelligerum* (Devonian). N, Subgroup Diacromorphitae; *Acanthodiacrodium petrovi* (Upper Cambrian–Lower Ordovician). O–Q, Subgroup Netromorphitae; O, *Leiofusa navicula* (Middle Ordovician); P, *Anthatractus insolitus* (Devonian); Q, *Domasia elongata* (Silurian). R–S, Subgroup Disphaeromorphitae; R, *Disphaeria macropyla* (Upper Cretaceous). S, *Pterocystidiopsis stephaniana* (Upper Cretaceous). Magnification ca. X500.

References

Deflandre, G., and M., 1964. Notes sur les Acritarches, *Rev. Micropaléont.*, 7, 111–114.

Downie C., 1973. Observations on the nature of the acritarchs, *Palaeontology*, 16, 239–260.

Downie, C., and Sarjeant, W. A. S., 1964. Bibliography and index of fossil dinoflagellates and acritarchs, *Geol. Soc. Amer. Mem.* 94, 180p.

Downie, C.; Evitt, W. R.; and Sarjeant, W. A. S., 1963. Dinoflagellates, hystrichospheres and the classification of the acritarchs, *Stanford Univ. Publ., Geol. Sci.*, 7(3), 1–16.

Downie, C.; Jardine, S.; and Visscher, eds., 1974. Acritarchs, *Rev. Palaeobot. Palynol.*, 18, 1–186.

Evitt, W. R., 1963. A discussion and proposals concerning fossil dinoflagellates, hystrichospheres and acritarchs, *Proc. Nat. Acad. Sci.*, 49, 158–164, 298–302.

Harland, R., and Sarjeant, W. A. S., 1970. Fossil freshwater plankton (dinoflagellates and acritarchs) from Flandrian (Holocene) sediments of Victoria and Western Australia, *Proc. Roy. Soc. Victoria*, 13, 211–234.

Naumova, S. N., and Pavlovsky, E. V., 1961. The discovery of plant remains (spores) in the Torridonian shales of Scotland (in Russian), *Doklady Akad. Nauk SSSR,* **141,** 181–182.

Norris, G., and Sarjeant, W. A. S., 1965. A descriptive index of genera of fossil Dinophyceae and Acritarcha, *New Zealand Geol. Surv. Palaeont. Bull.* **40,** 72p.

Sarjeant, W. A. S., 1965. The Xanthidia; the solving of a palaeontological problem, *Endeavour,* **24**(91), 33–9.

Sarjeant, W. A. S., 1970. Xanthidia, palinospheres and "Hystrix." A review of the study of fossil microplankton with organic cell walls, *Microscopy; J. Quekett Microsc. Club,* **31,** 221–253.

Timofeyev, B. V., 1965. Phytoplankton of the late Proterozoic and early Palaeozoic seas (in Russian), *Tez. Doklad. K Perv. Vses. Paleoalgo. Sov., Novosibirsk,* **1965,** 112–114. (Translation available in Russian Translating Programme R.T.S. 4006. Boston Spa, England: National Lending Library.)

Wall, D., 1962. Evidence from recent plankton regarding the biological affinities of *Tasmanites* Newton, 1975, and *Leiosphaeridia* Eisenack, 1958, *Geol. Mag.,* **99,** 36–62.

Cross-references: *Algae; Dinoflagellates; Hystrichospheres; Plankton; Protista.*

ACTUALISTIC PALEONTOLOGY

Aktuo-Paläontologie

The German term *Aktuo-paläontologie* (Actualistic Paleontology) was proposed by Richter (1928), and defined by him as the science of the origin and present-day mode of formation of future fossils, in the widest sense. It is simply the application of the uniformitarian principle ("the present is the key to the past") to paleontological problems. The term is very similar in meaning to. *neontology,* as opposed to paleontology, and may be regarded as the "paleontology" of living animals, including the "paleoecology" of modern environments. Herein, the term is used synonymously with neontology.

Some authors regarded the term Aktuo-paläontologie as inappropriate because *actual* (= present; current) and *paleo* are contradictory and ought to be avoided in a single term, and because it seemed impossible to speak of paleontology of the Recent. Most early actual-paleontological studies were independent of fossil application. Today, however, focus comes from paleontological problems and vice versa; modern workers compare extant depositional settings with ancient strata in order to ask the proper paleontological questions of the former and to better interpret the latter. Thus, actual-paleontology now broadly overlaps with other topics such as biostratinomy, taphonomy, and thanatology, and applies as well to studies of functional morphology, paleoecology, community ecology, substrate ecology, trace fossils, and other topics also covered in this volume.

Most modern biologists have little appreciation for the task confronting paleontologists in reconstructing prefossilization events in ancient environments. For this reason, paleontologists find themselves observing organisms in modern environments and studying them for their interrelationships with the substrate, associations with other organisms, and other paleontologically important aspects of their life habits, especially their tracks, trails, burrows, or borings. With its paleontological emphasis, such neontology also involves the problems of organism death, transportation, physical and mechanical processes of decomposition and disintegration, and the ways in which it, or its hard parts, are buried in the sediment. Reliable data on these processes help to provide an understanding of how organisms lived in their respective ancient sedimentary environments, and hence give a deeper meaning to the occurrences of fossils. Finally, neontologists carefully attempt to judge how faithfully species and communities are preserved for the fossil record, in terms of abundances and characteristics of component species, and the relative proportions of these species that lived together. Thus, empty shells, rotting carcasses, and scattered bones hold great fascination for the neontologist.

Early Work

Actualistic paleontological investigations were initiated and developed notably by Rudolf Richter and Johannes Weigelt. Richter's work was begun about 1920 in the North Sea and has been continued by the members of the "Forschungsanstalt fur Meeresgeologie 'Senckenberg'" in Wilhelmshaven (see summary of work and list of early papers of the Senckenberg Institute in Bucher, 1938; Häntzschel, 1956; Schäfer, 1962, 1972; and Reineck and Singh, 1973). Besides similar investigations in the North and Baltic seas, Weigelt studied the mode of death and burial (defined as *biostratonomy,* later altered to *biostratinomy*) of vertebrates, especially their mass death owing to severe weather.

An excellent book dealing with actualistic paleontology was written by Schäfer (1962) in German with short English summaries, and later translated into English (1972). For many years, Schäfer studied the fauna and sediments of the tidal flats and shallow-water regions of the southern North Sea. He did this as a zoologist and ecologist of marine invertebrates, but with the orientation of a paleontologist. His book not only summarizes present knowledge, but also describes and illustrates many new observations in superb detail, some of which are used as examples below.

Most neontological work has been in shallow marine environments of temperate-climate coasts. Large regions such as the ocean depths, the continental shelves, the shallow seas of the tropics, and even some continental habitats were long largely ignored, but are now also under study using the methods of actual-paleontology. However, there still remains much to be done in all biotopes of the world.

Studies in Actualistic Paleontology

Actualistic paleontology may be divided into several broadly overlapping areas of study, many of which command separate sections in this volume. To illustrate the work of neontologists, results of several actual-paleontological investigations are outlined briefly below.

Ichnology. No systematic studies on ichnology by zoologists are known. Recent tracks, trails, and other *Lebensspuren* (life traces) have mostly been studied by neontologists, especially in order to compare fossil forms with recent ones (e.g., Howard, 1971). In every regard paleoecology has received inspiration from neoichnological investigations in both the identity of the trace producers and their environmental, ecological, and ethological (behavioral) interpretation. Surface trails, observable everywhere (e.g., Heezen and Hollister, 1971), have little chance for fossilization. Much better is the probability that burrows will be preserved, as made by animals within the sediment, as well as borings made in hard objects such as rocks and shells.

Observations of recent tracks and trails remind us to be critical and cautious while explaining fossil surface traces. Studies on recent lebensspuren show great variety in their size and geometry. Animals of very different phylogenetic affinities may produce similar or identical forms of traces; and, conversely, a single organism can respond differently to sediment plasticity, water content, and grain size, as well as change its behavior for feeding, locomotion, burrowing, etc.

Analysis of trace fossils is enhanced by consideration and use of all data drawn from studies on present-day biogenic sedimentary structures. As an example, Boyer (1974; Boyer and Warme, 1975) described the occurrence of a trace fossil tentatively designated as *Phycodes* (see *Trace Fossils*) from Eocene shallow-water facies of California (Fig. 1A). This structure can be interepreted variously, e.g. as a mud-filled cavity of a burrower, as fecal-stuffed excursions of a deposit-feeder, as a resting trace of a

medusoid coelenterate, or as a mud-lined and -filled burrow of a sediment-dwelling anemone or other tentaculate creature. Another attractive hypothesis is that the *Phycodes* trace was constructed by an ophiuroid (brittle star), feeding and respiring with two or three arms upward and the remaining arms anchored downward. The animal's activity, presumably using its arms for filtering of the overlying water and perhaps scavenging along the sediment surface, resulted in a concentration of mud beneath the appendages; such a condition has been described by Hertwick (1972) for the modern ophiuroid *Hemipholas elongata* (Fig. 1B), as shown in the hypothetical reconstruction of Fig. 1C. Such reconstructions depend largely upon our knowledge of the ecology and natural history of modern animals, and much research remains to be done in this regard.

For reviews on trace fossils and ichnological research see the references in Häntzschel (1975), and the books edited by Crimes and Harper (1970, 1977) and Frey (1975).

Taphonomy. Another important division of actualistic paleontology is the study of the death and burial of organisms. In 1928, Richter distinguished several sections of this field of work, such as thanatology (causes of death and its direct consequences), biostratinomy (mode of burial), and necrology (alterations of the

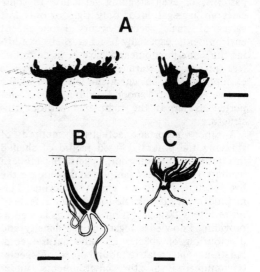

FIGURE 1. **A:** Two examples of a trace fossil from the Eocene of California, constructed of mud in a sand matrix (after Boyer and Warme, 1975). **B:** Life position and sediments associated with the brittle star *Hemipholas elongata* (after Hertwick, 1972). **C:** Hypothetical reconstruction of the trace fossil showing the multiple arm positions of the brittle star and the implied origin of the trace (after Boyer, 1974). Bar scales = 3 cm.

dead bodies before rock diagenesis); the entire postmortem history has been termed *taphonomy* by Efremov (1940).

Some early work was done by Hecht (1933), who undertook systematic, especially chemical, investigations on the decomposition of dead marine animals buried artificially in sediments of the North Sea. Similarly, Schäfer (1962; 1972) studied processes in the mechanical disintegration of dead marine invertebrates (e.g., skeletons of crustaceans after their death) and especially the steps in fossilization of jellyfishes and of fishes, birds, and sea mammals. The now classic work of Weigelt (1927) describes the various postmortum processes on corpses of vertebrates, on the basis of the biological consequences of blizzards in the southern United States. He later (1935) applied his observations to the Eocene vertebrate assemblages found in the lignites near Halle, Germany.

All processes of biostratinomy were summarized and described in full detail by Müller (1951). An example of biostratinomic observations to be made in the present and applied to the past is the various kinds of burial of pelecypod valves. Normally they lie convex upwards, this being the most stable position in fluid currents; they may be accumulated into great masses, giving rise to "coquinas" or "lumachelles," where the shells are deposited in large numbers forming a pavement, or in imbricated patterns, or even standing vertically. In some cases an aggregation of only right or only left valves of single species occurs. From recent environments, especially beaches, shells exhibiting the same phenomenon have been described (see review by Warme, 1971). An obvious example is *Mya*, in which the left valve of the bivalve, with its "spoon," will not be transported as far as the right one without this structure.

Another important actualistic method of studying the potential fossil record of shelled invertebrates or microorganisms in modern habitats is to collect samples and compare the live specimens with the empty skeletons. This method makes it possible to learn the effects of postmortem transportation of shells, as well as providing knowledge regarding the density and distribution of species and the natural combinations and relative abundances of species (communities) in the environments under study. Results of most such studies are optimistic, suggesting that the ecological distributions of skeletonized invertebrates are faithfully preserved in the sedimentary environments in which they lived (Warme et al., 1976). However, more data are needed on postdepositional changes that effect skeletons of various sizes, surface areas, and mineralogies, within the spectrum of diagenetic conditions in natural environments.

Functional Morphology. Usually only skeletal hard parts can be used to investigate the mode of life of fossil organisms. In the fossil record, only calcareous, siliceous, or chitinous components are commonly preserved, and these only selectively and accidentally. *Functional morphology* interprets the hard parts in terms of their biologic function, and of the organism's mode of life.

Adaptive morphology, a term often used in English papers, is not an exact synonym of functional morphology. The former deals with adaptations evolved by a species for its role in an environment, the latter concerns the functional connections within a single organism.

Usually only very few ecological interpretations can be made from the morphology of a skeleton or its parts. Indeed, such conclusions require detailed observations of the nearest living representatives of the fossil animals or plants under investigation, or a thorough and imaginative grasp of the ecological problems confronting fossil forms, which helped mold their adaptive history. Functional morphology is a somewhat neglected field in modern zoology.

Trueman (1964) discussed the functional anatomy of the hinge ligament of several living bivalved mollusks (clams). In this case, the detailed anatomy and function of the ligament must be understood, because it is often difficult to reconstruct the ligament structure in fossil valves and to recognize certain adaptive features of the ligament merely from its base of attachment. Trueman found that the rocking of the valves during deep burrowing is reflected by strong reduction of the outer layer of the ligament and the functioning of only a part of the inner layer. However, broad generalizations linking ligament strength with mode of life were not always valid (as with many biologic studies), and single observations should not be extrapolated to a general case without further testing.

Functional morphological investigations in some living irregular echinoids (e.g., *Echinocardium*) enabled Nichols (1959) to explain the evolution of individual characters in *Micraster* tests of the British Upper Cretaceous Chalk. The depth of burrowing of the animals is reflected by various features of the test such as the width of the subanal fasciole, the existence of a sanitary tube, and the degree of development of heavily ciliated regions for respiratory purposes. These skeletal parts, used for taxonomic classification within the genus *Micraster,* were thus also interpreted functionally.

Schäfer (1954) studied the functional structure of brachyuran crustaceans (crabs), which

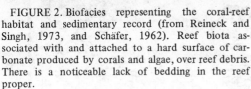

FIGURE 2. Biofacies representing the coral-reef habitat and sedimentary record (from Reineck and Singh, 1973, and Schäfer, 1962). Reef biota associated with and attached to a hard surface of carbonate produced by corals and algae, over reef debris. There is a noticeable lack of bedding in the reef proper.

FIGURE 3. Biofacies representing shallow shelf conditions with abundant live invertebrates, erosion surfaces and shell layers associated with reworked sands and muds, and burrows and escape structures of sediment-dwelling animals (from Schäfer, 1962).

may be divided ecologically into four groups: running, burrowing, climbing, and swimming. The claws of brachyurans have different forms according to their different main function (e.g., for nutrition, defense, or threatening). Schäfer's results have not yet been applied to fossil forms, though this seems possible if done with caution and based upon sufficiently well-preserved specimens.

Biofacies. The study of Recent biofacies is an important element of actualistic paleontology. In every habitat, special combinations of species represent the result of adaptations to environmental conditions. Typical organisms of each habitat (the German *Lebensform-Typen*) are useful for the ecological interpretation of fossil faunas and their environments. Thus, biofacies analysis, in the combined study of organisms and their related sedimentary environment, is the neontological equivalent, in many respects, of the ecosystem analyses of ecologists, wherein the biota and the environment are investigated for their mutual relationships.

Coral reefs are one of the most dramatic examples of a biofacies (Fig. 2). They are characterized by a permanent and very typical community and a rarity or absence of stratification.

Schäfer (1962, 1972) outlined four major biofacies present in clastic environments. These are distributed worldwide, reoccurring under similar conditions of hydroclimate and depth. However, each is generally separated by a transitional zone, and thus seldom can be sharply differentiated one from another. One biofacies is characteristic of the littoral region of strong tidal currents and much surf, an environment of frequent erosion and redeposition of sediments and shells. Thus, the sedimentary record of this biofacies is dominated by laminated and crossbedded sand and current-concentrated shell beds (Fig. 3). Schäfer's other major biofacies are further combinations of sedimentary processes, bedding characteristics, and organism abundances.

A simple example of biofacies reconstruction in an ancient setting is provided by outcrops of Eocene Delmar Formation in southern California (Boyer, 1974; Boyer and Warme, 1975). Three kinds of facies exist in the Delmar, all characteristic of shallow-water, brackish to marine conditions (Fig. 4): (1) biostromic concentrations of cemented oyster shells and complementary molluscan fauna, representing oyster "reefs"; (2) thin-bedded, ripple-marked, carbon-rich interbeds of sand and mud, lacking macrofossils and ascribed to tidal flats and saltmarsh deposits; and (3) thick-bedded (ca. 1 m) sequences, grading upward from a scoured erosion surface at the base, with an immediately overlying shell and pebble conglomerate, through sandy horizontal or inclined cross-beds, and into heavily bioturbated or homogeneous mud, representing migrating tidal channels. The three facies are laterally equivalent, and together epitomize habitats comprised of a shallow bay or lagoon with adjoining oyster reefs and tidal flats.

The third facies tells a clear story, pieced together using both sedimentological and faunal evidence. As tidal channels meander laterally they erode and undermine adjacent habitats, producing mud clasts such as are found at the base of the bed (Fig. 5). Some clasts exhibit holes, traces of excavating mollusks or crustaceans that burrowed or bored into the semiconsolidated muds of the adjoining flats, marshes, or creek banks before they were undermined. Shells at the base of the bed represent animals that lived in the channel, or that were washed into it from the adjoining flats as the creek migrated. Above, the laminated sands represent channel fill and point-bar sequences, usually rapidly deposited and thus containing only a few burrows and even fewer shells. The mud cap on each sequence represents finer-grained sediments deposited on the newly reconstructed tidal flat as the channel migrated away. This mud is commonly intensely bioturbated—devoid of physical sedimentary structures—and is mixed with the underlying sand in a graded sequence owing to the biologic reworking of the substrate by burrowers. Thus, physical processes dominate the record in the lower part of the sequence, exemplified by an erosion surface, a lag deposit of coarse clasts, clean sand, intact sediment laminations, and evidence of rapid deposition. In contrast, biologic energy had time to leave its stamp on the upper part of the sequence

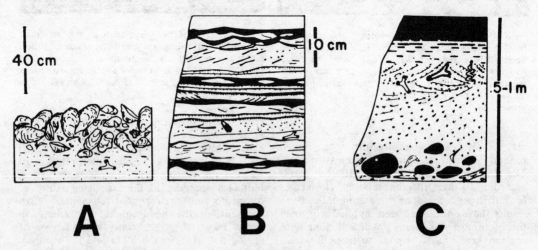

A   **B**   **C**

FIGURE 4. Three biofacies from the Eocene Delmar Formation, California, being all laterally contiguous (from Boyer and Warme, 1975). **A**: Oyster beds with associated other molluscs as well as burrows. **B**: Tidal flats with interbedded sands and muds and small-scale cross-bedding and ripples. **C**: Coarse-to-fine-grained upward sequences representing meandering tidal creeks; see Fig. 5.

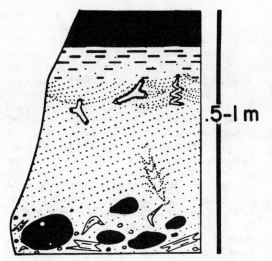

.5–1 m

FIGURE 5. Detail of the coarse-to-fine-grained sequence in Fig. 4. The base is characterized by an erosion-scour surface immediately overlain by coarse clasts of clay as well as of shells and other transported debris mostly derived from the adjoining channel banks and tidal flats. The central portion of the sequence is cross-bedded clean sand with a few animal-escape structures or large burrows. The upper part is medium-to-small-scale crossbedding with increasing proportions of mud and burrows upward; and, finally, a cap of bioturbated mud ends the sequence. See text for discussion.

where finer-grained sediment was deposited, probably more slowly, and was homogenized and mixed by an active fauna of burrowers over a longer period of time. By such analyses, sediments can be used to determine physical ecological conditions in the ancient environment, and the faunas, both body fossils and trace fossils, can be employed as both paleoecological indicators and as sedimentological tools to interpet episodes in earth history.

JOHN E. WARME
WALTER HÄNTZSCHEL*

References

Boyer, J. E., 1974. Sedimentary facies and trace fossils in the Eocene Delmar Formation and Torrey Sandstone, California. M.A. Thesis, Rice University, Houston, Texas, 176p.

Boyer, J. E., and Warme, J. E., 1975. Sedimentary facies and trace fossils in the Eocene Delmar Formation and Torrey Sandstone, California, in *Future Energy Horizons of the Pacific Coast*. Long Beach, Calif.: Ann. Meetings Pac. Sect. AAPG-SEPM-SEG, 65–98.

Bucher, W. H., 1938. Key to papers published by an

*Deceased.

institute for the study of modern sediments in shallow seas, *J. Geol.*, **46**, 726–755.

Crimes, T. P., and Harper, J. C., 1970. *Trace Fossils*. Liverpool: Seel House Press, 547p.

Crimes, T. P., and Harper, J. C., eds., 1977. *Trace Fossils 2*. Liverpool: Seel House Press, 351p.

Efremov, I. A., 1940. Taphonomy—A new branch of paleontology (in Russian), *Akad. Nauk SSSR Biul., Biol. Ser.*, **3**, 405–413. Transl. in *Pan-Amer. Geol.*, **74**, 181–193.

Frey, R. W., ed., 1975. *The Study of Trace Fossils*. New York: Springer-Verlag, 592p.

Häntzschel, W., 1956. Rückschau auf die paläontologischen und neontologischen Ergebnisse der Forschungsanstalt "Senckenberg am Meer," *Senckenbergiana Lethaea*, **37**, 319–330.

Häntzschel, W., 1975. Trace fossils and problematica, in C. Teichert, ed., *Treatise on Invertebrate Paleontology*, pt. W, Miscellanea, Supp. 1. Lawrence, Kansas: Geol Soc. Amer. and Univer. Kansas Press, 269p.

Hecht, F., 1933. Der Verbleib der organischen Substanz der Tiere bei meerischer Einbettung, *Senckenbergiana*, **15**, 165–249.

Heezen, B. C., and Hollister, C. D., 1971. *The Face of the Deep*. New York: Oxford Univ. Press, 659p.

Hertwick, G., 1972. Georgia coast region, Sapelo Island, U.S.A., sedimentology and biology. V. Distribution and environmental significance of lebensspuren and in-situ skeletal remains, *Senckenbergiana Maritima*, **4**, 125–167.

Howard, J. D., 1971. Comparison of the beach-to-offshore sequence in modern and ancient sediments, in *Recent Advances in Paleoecology and Ichnology*. Washington, D.C.: Amer. Geol. Inst., 148–183.

Müller, A. H., 1951. Grundlagen der Biostratonomie, *Abh. Deutsch. Akad. Wiss. Berlin Kl. Math. Allg. Naturwiss.*, **1950**(3), 147p.

Nichols, D., 1959. Changes in the chalk heart urchin *Micraster* interpreted in relation to living forms, *Phil. Trans. Roy. Soc. London, Ser. B*, **242**, 347–437.

Reineck, H. E., and Singh, I. B., 1973. *Depositional Sedimentary Environments*. New York: Springer-Verlag, 439p.

Richter, R., 1928. Aktuopaläontologie und Paläobiologie, eine Abgrenzung, *Senckenbergiana*, **10**, 285–292.

Schäfer, W., 1954. Form und Funktion der Brachyuren-Schere, *Abh. Senckenberg. Naturforsch. Gesell.*, **489**, 1-65.

Schäfer, W., 1962. *Aktuo-paläontologie nach Studien in der Nordsee*. Frankfurt: Waldemar Kramer, 688p.

Schäfer, W., 1972. *Ecology and Paleoecology of Marine Environments* (transl. of Schäfer, 1962). Chicago: Univ. Chicago Press, 568p.

Trueman, E. R., 1964. Adaptive morphology in paleoecological interpretation, in J. Imbrie and N. Newell, eds., *Approaches to Paleoecology*. New York: Wiley, 45–74.

Warme, J. E., 1971. The fidelity of the fossil record and paleoecology, in *Recent Advances in Paleoecology and Ichnology*. Washington, D.C.: Am. Geol. Inst., 11–28.

Warme, J. E.; Ekdale, A. A.; Ekdale, S. F.; and Peterson, C. H., 1976. The raw material of the fossil record, in R. W. Scott and R. R. West, eds., *Struc-*

ture and Classification of Paleocommunities.
Stroudsburg, Pa.: Dowden, Hutchinson & Ross,
143–169.

Weigelt, J., 1927. *Rezente Wirbeltierleichen und ihre
paläobiologische Bedeutung.* Leipzig: Max Weg,
227p.

Weigelt, J., 1935. Was bezwecken die Hallenser Uni-
versitätsgrabungen in der Braunkohle des Beiseltales,
Natur und Volk, **65,** 347–356.

Cross-references: *Diversity; Growth Lines; Larvae
of Marine Invertebrates–Paleontological Signifi-
cance; Morphology, Constructional; Morphology,
Functional; Paleoecology; Paleotemperature and
Depth Indicators; Population Dynamics; Taphon-
omy; Trace Fossils.* Vol. I: *Abyssal Zone; Bathyal
Zone; Benthonic Zonation; Hadal Zone; Littoral
Zone; Neritic Zone.*

AGNATHA

Agnatha, a class of primitive aquatic vertebrates, are distinguished by the fact that none
of their gill arches has been modified to form
jaws. Living ones, the eel-like cyclostomes, including hagfishes (*Myxine*) and lampreys
(*Petromyzon*), lack a mineralized skeleton and
so are very rare as fossils; the only one that has
been discovered is *Mayomyzon,* a lamprey from
the Pennsylvanian of Illinois. Other fossil Agnatha (Stensiö, 1964; Romer, 1966; Obruchev,
1967; Moy-Thomas and Miles, 1971; Halstead
and Turner, 1973), sometimes called ostracoderms, occur in Ordovician, Silurian, and Devonian rocks in North America, Europe, Asia,
and Australia. Agnatha are classified in two subclasses: (1) Monorhina with a single, median
nostril, including cyclostomes and the fossil
Osteostraci and Anaspida; and (2) Diplorhina
with paired nostrils, known only from fossil
Heterostraci and Thelodonti.

Heterostraci

Order Heterostraci (Middle Ordovician to Upper Devonian) includes the earliest and some of
the most primitive known vertebrates (Fig. 1).
Usually only their dermal skeleton is mineralized and consists of an acellular bone-like substance, aspidine, surmounted by ridges or

tubercles of dentine. Two Ordovician genera,
Astraspis and *Eriptychius,* are North American
forms with a shield of small polygonal plates
and a body covered by thick, overlapping
scales. *Arandaspis* and *Porophoraspis* from the
Middle Ordovician of Australia have a non-
tesserate shield. Recently reported occurrences
in the Lower Ordovician and Cambrian are questionable. Silurian and Devonian Heterostraci are
grouped in a number of families characterized
by the number of plates in the shield. With the
exception of the tail, they had no fins. Their
eyes were small and placed typically at the sides
of the head, and their paired nostrils opened
just inside the mouth. They had seven pairs of
gills with a single external opening on each side.
Early Heterostraci were marine, but many
moved into fresh waters late in the Silurian.
Most were small, but one attained a length of
5 ft. (1.5 m). *Anglaspis* typifies an early
family, the Cyathaspidae, in which the dorsal
shield was formed in one piece. They gave rise
to the Traquairaspidae of the Silurian and
Lower Devonian with a dorsal shield of eight
plates, and the Pteraspidae with ten plates
in the dorsal shield; the latter, especially
Pteraspis, have proved to be good stratigraphic guides in the Lower Devonian. The
mostly European Psammosteidae (Lower to
Upper Devonian) were flat-bodied, bottom-
dwelling derivatives of Pteraspidae; *Drepanaspis*
and *Psammosteus* are well-known representatives. The North American Lower Devonian
Cardipeltidae paralleled the Psammosteidae in
some respects. The Siberian Amphiaspidae are
pecularly specialized Lower and Middle Devonian forms in which the shield of adults is
without sutures. The Lower Devonian *Poly-
branchiaspis* from China is possibly a relative of
Heterostraci, though its mouth opens dorsally
through the dorsal shield and each gill opens
separately to the exterior; perhaps it represents
a distinct order of Diplorhina.

Thelodonti

Order Thelodonti (Coelolepida; Lower Silurian
to Lower or perhaps Middle Devonian) had the
skin set with small, simple scales or denticles of
dentine (Fig. 2). Since Thelodonti lacked any

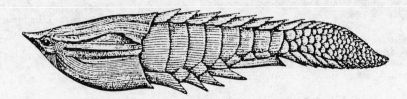

FIGURE 1. Agnatha, Heterostraci, *Anglaspis heintzi* (From Stahl, 1974).

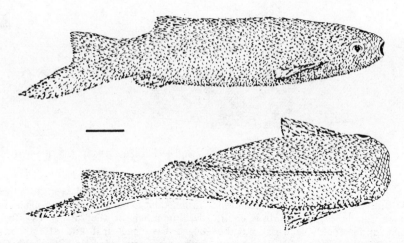

FIGURE 2. Agnatha, Thelodonti, *Phlebolepis elegans,* lateral and dorsal views (from Stahl, 1974, after Ritchie, 1968). Scale bar is 10 mm.

mineralized skeleton, little is known of their structure and relationships. Their body was depressed, their eyes lateral, the mouth subterminal, and the eight pairs of gills opened separately to the exterior. In some, the tail was hypocercal as in Anaspida, and projecting lateral angles of the head may have served as stabilizing fins. *Thelodus,* a well-known genus, occurs in Europe and North America. Ordovician records of Thelodonti are questionable.

Osteostraci

Order Osteostraci (Middle Silurian to Upper Devonian) had the head and anterior part of the body encased in an armor of true bone, commonly covered by or bearing tubercles of a special kind of dentine (mesodentine). The dorsally placed eyes, ventral mouth, and flattened head and body suggest that they were bottom dwellers (Fig. 3). Most are known from fresh- or brackish-water deposits. Characteristic features of the head shield are the paired and median sensory fields covered by small platelets. An internal skeleton of perichondral bone

has permitted detailed study of the internal structure (Stensiö, 1964), and indicates that the head contained a large mouth and gill chamber used not only for respiration, but also for feeding by straining out food particles. Structural details, especially the single, dorsally placed nostril and hypophysial opening, indicate a relationship to Anaspida and more distantly to living cyclostomes. *Tremataspis* and *Oeselaspis* are primitive Upper Silurian genera without paired fins, while *Cephalaspis* is a well-known Devonian genus with a pair of pectoral fins protected on each side by a projecting spine or cornu. The Chinese Lower Devonian *Galeaspis* is an aberrant osteostracan without sensory fields.

Anaspida

Order Anaspida (Middle Silurian to Upper Devonian) were small Agnatha with a fusiform body, lacking a mineralized internal skeleton, and usually covered externally by small scales and plates of acellular bone. Very elongate paired fins are preserved in some genera, and

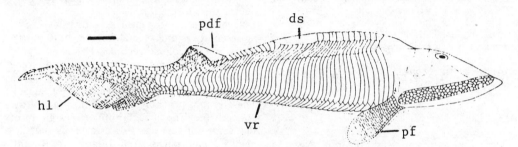

FIGURE 3. Agnatha, Osteostraci, *Hemicyclaspis murchisoni* (from Stahl, 1974). ds = dorsal scales (= anterior dorsal fin); hl = hypochordal lobe; pdf = posterior dorsal fin; pf = pectoral fin; vr = ventrolateral ridge. Scale bar is 10 mm.

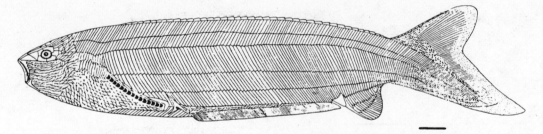

FIGURE 4. Agnatha, Anaspida, *Pharyngolepis oblongus* (from Ritchie, 1964). Scale bar is 10 mm.

the tail is remarkable in being hypocercal, that is with a downturned muscular lobe surmounted by a large epichordal lobe. The mouth is small, round, terminal, and probably suctorial; but the nature of their food is uncertain. The single nostril on top of the head indicates a relationship to modern lampreys and to Osteostraci. Anaspida occur in fresh- and brackish-water deposits. One of the best-known genera is the Upper Silurian *Pharyngolepis* from Norway (Fig. 4).

ROBERT H. DENISON

References

Halstead, L. B. and Turner, S., 1973. Silurian and Devonian ostracoderms, in A. Hallam, ed., *Atlas of Palaeobiogeography.* Amsterdam: Elsevier, 67–79.

Moy-Thomas, J. A., and Miles, R. S., 1971. *Palaeozoic Fishes,* 2nd ed. Philadelphia: Saunders, 259p.

Obruchev, D. B., 1967. *Fundamentals of Paleontology,* vol. 11, Agnatha, Pisces (transl. from Russian). Jerusalem: Israel Program for Scientific Translations, 36–167.

Ritchie, A., 1964. New light on the morphology of the Norwegian Anaspida, *Skr. Norske Videnskaps-Akad., Oslo, I, Mat.-Nat. Kl.,* n.s., **14,** 1–35.

Ritchie, A., 1968. *Phlebolepis elegans* Pander, an Upper Silurian thelodont from Oesel, with remarks on the morphology of thelodonts, in T. Ørvig, ed., *Current Problems of Lower Vertebrate Phylogeny.* New York: Wiley-Interscience, 81–88.

Romer, A. S., 1966. *Vertebrate Paleontology,* 3rd ed. Chicago: Univ. Chicago Press, 15–23.

Stahl, B. J., 1974. *Vertebrate History: Problems in Evolution.* N.Y.: McGraw-Hill, 594.

Stensiö, E. A., 1964. Les cyclostomes fossiles ou ostracodermes, in J. Piveteau, ed., *Traité de Paléontologie,* vol. 1. Paris: Masson et Cié, 96–382.

Cross-references: *Chordata; Pisces; Vertebrata; Vertebrate Paleontology.*

ALGAE

Algae were among the earliest forms of life on earth and are very abundant in aqueous environments at the present day. They are an extremely diverse group of lower plants characterized by relatively unspecialized reproductive organs rather than by general morphological simplicity. During much of the history of life on earth, algae have had profound effects on the biosphere as sources of both atmospheric oxygen and food for higher organisms.

Ten or more divisions of algae can be recognized, each being equivalent in rank to an animal phylum. They exhibit a wide range of vegetative structure, often with marked parallelism between groups. Consequently, the primary classification of algae is based on more fundamental details such as cell structure and composition, and the nature of food resources and photosynthetic pigments. As a result, the classification of fossil algae is difficult, and must be based primarily on gross morphological features. An additional problem is that relatively few algae have hard parts. The only algae with a good fossil record are those that construct skeletons or cysts of calcium carbonate, silica, or tough organic material. Soft algae, which represent the great majority, have a very low preservation potential and only poor carbonaceous remains of them are occasionally found unless, exceptionally, they are silicified prior to decay and preserved in cherts. Stromatolites, laminated sediments representing algal mats (commonly of blue-green algae in carbonate environments) have a long geologic record, but they are organo-sedimentary structures rather than true fossils and usually contain little direct information regarding the algae responsible for their formation.

Nevertheless, most algal divisions have fossil representatives and several families have particularly good geologic records. The majority of calcareous algae are benthic. They include some Rhodophyta (red algae), Chlorophyta (green algae), and Cyanophyta (blue-green algae). These groups probably all occur throughout the Phanerozoic. The coccolithophores (a group of the Chrysophyta) are also calcareous, but are planktic and range from Jurassic to Recent. Other planktic algae with good fossil records, the siliceous diatoms (Bacillariophyta) and organic-walled dinoflagellate cysts (Pyrrophyta), are also essentially Mesozoic and Cenozoic in age although acritarchs, which probably include some dinoflagellate cysts, range from Precambrian to Quaternary. Stromatolites range from >3000 m yr to the Holocene. Soft algae preserved in silica are rare but important microfossils, especially in the Precambrian where they are mainly represented by blue-green and green algae. Soft benthic Phaeophyta (brown algae) are known as carbonaceous fossils from various Phanerozoic periods.

Fossil planktic algae are microscopic, ranging in size from 10–500 μm. Fossil benthic algae are often macroscopic, commonly being in the range .1–10 cm. Stromatolites are usually much larger, commonly with dimensions in decimeters, and examples even up to hundreds of meters across have been reported. Stromatolites are volumetrically important algal deposits in the geologic record, but the majority of fossil algae are inconspicuous when compared with invertebrate fossils.

Cyanophyta

Blue-green algae are simple organisms which, together with bacteria, comprise the Procaryota (organisms lacking nuclei and organelles such as mitochondria and plastids in their cells). They were among the dominant organisms on earth during much of the Precambrian and inhabit a wide range of environments at the present day, including soil, snow, and hot springs, as well as fresh water and the oceans. Most blue-greens are filamentous but unicellular coccoid forms, usually aggregated into palmelloid clusters, and colonial forms also occur. The cells are usually surrounded by a mucilaginous sheath that can become mineralized. Calcified sheaths of filamentous forms produce tubiform fossils (Porostromata), which are relatively common throughout the Paleozoic and Mesozoic and are also known from the Recent. They include *Girvanella* and *Ortonella.*

Wholesale silicification of blue-green algae has occasionally occurred and is particularly significant in the Precambrian. These microfloras can show details of the cellular structure of the algae and have an important bearing on our knowledge of the early development of life on earth (Schopf, 1970). They indicate that both coccoid and filamentous blue-green algae were in existence 2300 m yr ago.

Stromatolites (Fig. 1) are usually assumed to have been formed primarily by blue-green algae trapping sediment grains on their mucilaginous sheaths and binding them into the mat by upward growth. The algae were soft, and decomposed, leaving the sediment arranged in irregular fine laminae containing small irregular

voids (fenestrae or bird's-eyes). These deposits with a clotted, fenestral microfabric have been termed Spongiostromata to distinguish them from (Porostromata) deposits containing minute tubes (Pia, 1927). Stromatolites normally have a spongiostrome microfabric; but if the algal filaments have been calcified, they have a porostromate microfabric. Stromatolites formed by coccoid blue-green algae have less distinct lamination than do those formed by filamentous blue-greens and have been called thrombolites (a reference to their clotted microfabric). Calcification of the sheath produces microspherulites in the case of coccoid forms and fine porostromate tubes in the case of filamentous forms. These *skeletal* stromatolites (Riding, 1977a) are locally common in the Phanerozoic

Stromatolites (q.v.) are known from marine, lacustrine, hypersaline, and hot-spring environments. They exhibit a range of shapes—flat, undulose, domical, conical, columnar, digitate, and spherical—depending upon the environment and the type of alga. Spherical or ovoid forms develop around a loose grain and are termed oncolites. Columnar, digitate, and spherical stromatolites preferentially form in subaqueous environments, while partially subaerial situations such as lake margins and intertidal and supratidal environments are characterized by flat stromatolites. Stromatolites occurred as early as 3000 m yr ago in southern Africa and Canada and were abundant and diverse in late Precambrian carbonate shelf environments. They have been used for biostratigraphic purposes in the late Precambrian (Cloud and Semikhatov, 1969), especially by workers in the USSR; but their utility for long-distance correlation appears to be limited. Stromatolites are also common in the early Paleozoic but decline in both abundance and diversity thereafter, probably due to competition from higher organisms for space and to grazing of algal mats by gastropods and other herbivores (Garrett, 1970). Flat stromatolites continue to be common around lakes and in intertidal and supratidal environments but are uncommon in subtidal marine environments at the present day; columnar stromatolites are very rare, the best examples occurring in hypersaline lagoons at Shark Bay, Western Australia.

Stromatolites present descriptive and classificatory problems, since important features such as external form, internal arrangement of laminae, and microfabrics show gradations that can be difficult to characterize (Hofmann, 1969). Early attempts to apply a Linnean-type nomenclature to stromatolites tended later to be superseded by a simple geometric classificatory scheme emphasizing lateral linkage and

FIGURE 1. Stromatolite from the Cambrian of Colorado.

vertical stacking of hemispheroidal growths (Logan et al., 1961). More recently, however, a binomial system using a group name and form name (e.g., *Conophyton basalticum*) has proved to be more appropriate for the complex digitate forms characteristic of the late Precambrian, although the geometric descriptive scheme is often applied to simpler Phanerozoic stromatolites. Groups (e.g., *Baicalia*, *Gymnosolen*) are based on column morphology and lamina shape; form is defined by finer morphological and microstructural details (Walter, 1972). Bacteria are probably the most important microorganisms after blue-green algae in stromatolite formation. Deep-sea manganese nodules have been reported to be bacterial stromatolites (Monty, 1973).

Chlorophyta

Three families of green algae include calcareous forms. The Codiaceae and Dasycladaceae belong to the Chlorophyta. The Charophyceae may be included with them or placed in a separate division (Charophyta; q.v.). All the Holocene calcified members of these groups are macroscopic benthic plants in which the calcium carbonate is an intimate part of the plant body but is, nevertheless, extracellular. They all fragment upon decomposition, and the resulting debris can be a significant addition to carbonate sediment in both marine (codiaceans and dasycladaceans) and freshwater (charophytes) environments.

Holocene codiaceans such as *Penicillus* and *Halimeda* are common in some tropical and subtropical marine environments. *Penicillus* disintegrates into tiny needle-shaped aragonite crystals which form lime mud (Stockman et al., 1967), and this process prevents the plant from being preserved as a recognizable fossil. *Halimeda* also shows spontaneous post mortem disintegration following decomposition of the soft tissue, but it breaks up into segments, usually several mm across, which show a well-defined internal tube system (Figs. 2, 3). Fragments of *Halimeda*-like algae are known from the Ordovician (*Dimorphosiphon*) and several subsequent periods. *Halimeda* itself ranges from Cretaceous to Holocene.

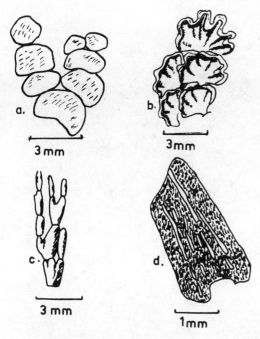

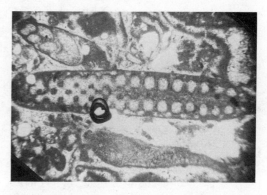

FIGURE 4. *Cylindroporella* (Dasycladaceae) Cretaceous, Texas; X17.

FIGURE 2. *Halimeda* (Codiaceae). a, b, c, Parts of plants showing form of segments in different species (after Barton, 1901); d, sketch of internal structure of segment.

Many dasycladaceans are calcified and the group has a good fossil record from the lower Paleozoic onwards (Pia, 1927). The plant typically consists of an unbranched main stem or axis from which numerous lateral branches radiate. Calcification is variable. Commonly the outer part of the stem is calcified and the lateral branches leave rounded holes in this layer where they pass through it (Fig. 4). The stem and branches may be segmented and are often fragile. Segments and fragments are usually circular in cross-section and those of the stem have pores representing the sites of the lateral branches. Dasycladaceans were important contributors of loose carbonate sediment during the Jurassic to Eocene and also in several earlier periods.

In most Paleozoic genera the laterals are irregularly arranged and usually unbranched, e.g., *Rhabdoporella,* although some do show tufts of secondary and even tertiary branches, e.g., *Coelospheridium* and *Primicorallina* respectively. Considerable diversification of the group occurred during the Mesozoic and early Cenozoic and evolutionary trends become apparent (Pia, 1927; Elliott, 1968). By the early Mesozoic, lateral branches were often arranged in regular whorls on the main axis (Fig. 5). Reproductive cysts, unknown in Paleozoic forms but presumably formed in the axis (endosporic), were produced on the lateral branches (cladosporic). Eventually, special

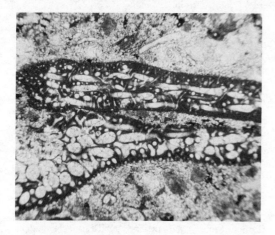

FIGURE 3. Sections of *Halimeda* (Codiaceae) Miocene, Phillippine Islands; X17.

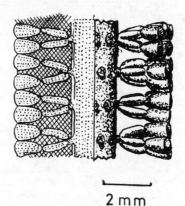

FIGURE 5. *Diplopora* (Dasycladaceae) Triassic (after Pia, 1920).

15

gametangia were developed (choristosporic). In the Holocene genus *Acetabularia* these are arranged in whorls and contain numerous cysts. In *Chalmasia* the cysts are calcified and resemble calcispheres. Middle Cenozoic extinctions considerably reduced the number of dasycladacean taxa and there are only ten Holocene genera. These characteristically occur in shallow tropical and subtropical seas but are also known in brackish and hypersaline environments.

The Cyclocrineae and Receptaculitaceae are large dasycladacean-like Paleozoic fossils (Rietschel, 1969; Nitecki, 1972). *Cyclocrinus* and its relatives (which include *Coelospheridium*) have generally been regarded as a dasycladacean group, but the Receptaculitaceae (see *Receptaculitids*) have also been regarded as sponges. These fossils are globular, ovoid, or irregular in form and have a smooth surface consisting of plates formed, in the Cyclocrineae, at the ends of primary branches radiating from a central stem.

The Characeae is a distinctive group of algae with relatively complex reproductive organs. Extant genera include large (up to 1 m) bushy plants, such as *Nitella* and *Chara*, normally inhabiting clear quiet fresh water, principally lakes and ponds, although some species can live in brackish water. Short laterals are borne in whorls on branching stems (Fig. 6). The surface of the plant is moderately calcified, but the heavily calcified oogonia (gyrogonites) are the parts most commonly found as fossils. These female reproductive bodies are small (250–1000 µm) subspherical or pyriform objects with a spiral ornament (Fig. 6). In some cases, the external covering (utricle) is also present. It is difficult to formulate a "natural" classification based on fossil oogonia alone or to use them to reconstruct the evolution of the group in any detail, but they indicate that characeans have existed since the Silurian and they have been applied to the biostratigraphic subdivision and correlation of nonmarine deposits (Peck, 1953; see *Charophyta*).

Rhodophyta

Extant calcareous red algae include the Corallinaceae (coralline algae) and Squamariaceae. Coralline algae have heavily calcified cell walls and consequently have a good fossil record. There are two major subgroups: articulated corallines and crustose corallines. Articulated corallines are erect branched segmented plants (Fig. 7). They first appear in the Cretaceous and include the Holocene genera *Corallina* and *Jania*. Upon death, they fragment into individual segments. Crustose forms are essentially prostrate encrusting plants although some of them give rise to erect unsegmented branches.

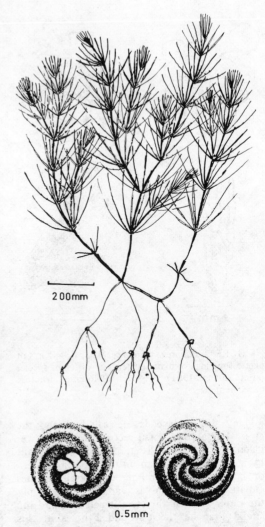

FIGURE 6. Above, *Chara fragifera* (Characeae), Holocene; below, oogonia of *Chara oehlerti* (Characeae) (from Johnson, 1961).

Since the thallus is entire and attached, crustose corallines readily remain in place after death and are important components of reef structures, especially in the Cenozoic. They first appear in the Jurassic, but possible ancestral forms such as *Archaeolithophyllum* are known in the Pennsylvanian. Free nodular growths of crustose corallines are termed maerl or rhodoliths. Holocene genera include *Lithothamnium* (Fig. 8) *Lithophyllum* (Fig. 9) and *Goniolithon*.

Coralline algae are normally differentiated into large-celled internal (hypothallic) and small-celled external (perithallic) tissue. Sporangia occur in spherical or lenticular cavities (conceptacles) within the perithallus (Fig. 9). The form of the asexual conceptacles is important in classifying both living and fossil speci-

FIGURE 7. *Calliarthron antiquum* (Corallinaceae), Miocene, Saipan (from Johnson, 1961). Three segments with nodes between and a conceptacle on the lower right, X50.

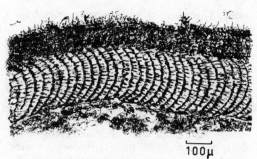

FIGURE 9. *Lithophyllum prelichenoides* (Corallinaceae), Miocene, Spain (from Johnson, 1961).

mens. Coralline algae are widely distributed in present-day marine environments from the tropics to high latitudes (Adey and Macintyre, 1973).

Like crustose corallines, Squamariaceae, e.g., *Peyssonnelia,* also form crusts with a cellular internal structure (Fritsch, 1945, vol. 2), but they are usually more weakly calcified and consequently are less well preserved as fossils. Calcified forms are most abundant in tropical and subtropical marine environments. They range from Cretaceous to Holocene, but some Pennsylvanian leaf-shaped (phylloid) algae such as *Ivanovia* resemble squamariaceans.

The Solenoporaceae (*Solenopora, Parachaetetes*) are a long-ranging (Cambrian to Miocene) extinct group usually regarded as being related to the Corallinaceae. They form nodular or bulbous growths often several cm across with an internal reticulate cellular structure (Fig. 10), which is coarser than that in coralline algae and is also undifferentiated. Conceptacle-like cavities occur only very rarely. In the Paleozoic, solenoporaceans sometimes occur in reef structures; but representatives of the group are most abundant in the Jurassic and Cretaceous.

The Gymnocodiaceae is commonly included in the Rhodophyta (Elliott, 1955) although it has also been regarded as a green algal group. These are small but macroscopic fossils preserved as irregular, sometimes branched, tubes with perforate walls. Only two genera are included in the group: *Gymnocodium* (Permian) and *Permocalculus* (Permian to Cretaceous).

Chrysophyta

A variety of unicellular planktic algae referred to the Chrysophyta produce mineralized scales, skeletons, and cysts which provide a fossil record during the Mesozoic and Cenozoic. Chrysophytes that produce organic-walled cysts sometimes impregnated by silica are usually termed "chrysomonads." They range from Cretaceous to Holocene, as do the silicoflagellates, which may be related to the Chrysophyta and have internal skeletons of silica. But the major group of fossil chrysophytes is the haptophycean family Coccolithophoridaceae which stud their cells with tiny calcite plates (coccoliths; Black, 1965). Most Holocene coccolithophores are marine flagellates, but brackish and freshwater forms also

FIGURE 8. *Lithothamnium* (Corallinaceae) showing crustose form, hypothallic tissue (below), and perithallic tissue with conceptacle (above). Thin-section, Eocene, Guam, X40.

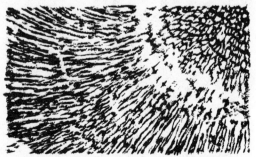

FIGURE 10. *Solenopora* (Solenoporaceae), Silurian, Estonia (after Maslov, from Johnson, 1961).

occur. The majority of oceanic species have a life cycle that alternates between free-swimming forms and floating resting cysts. Neritic species tend to show alternation of free-swimming and sessile forms. The motile and cyst stages commonly bear coccoliths, whereas the sessile forms are uncalcified.

Coccoliths (q.v.) are usually elliptical or circular plates or rings, often only 1–3 µm across, which sometimes bear stalk-like projections. Most are composed or overlapping rhombohedral calcite crystals. The coccoliths appear to be formed inside the cell prior to being moved to their external position. Commonly, sediments contain individual coccoliths or fragments of them; but complete coccospheres, cysts or motile forms 10–100 µm in diameter covered by overlapping coccoliths (Fig. 11A), also occur. Despite the small size of coccoliths, and the fact that individual coccospheres may be composed of two different types, the classification of both living and fossil coccolithophores is largely based on coccolith morphology.

Coccoliths range from Early Jurassic to Holocene. They show rapid evolution, and this together with their minute size makes them very suitable for biostratigraphic work where small samples are involved. Important extinctions occurred in the Late Cretaceous and late Tertiary followed by rapid recovery each time. Coccolithophores are the most abundant Holocene *calcareous* phytoplankton group, but they are nevertheless subordinate to the dinoflagellates and diatoms. Coccoliths are comparable with other haptophycean scales that are composed of organic material. Selective calcification of these might account for some of the sudden appearances of new coccolith types that characterize the history of the group (Black, 1968). No pre-Jurassic coccoliths are known; it is possible that diagenesis may have obscured or destroyed them in older rocks. In the Mesozoic, they are important rock-forming fossils, constituting up to 95% of some limestones and being prominent components of chalk. Coccoliths are also important in Holocene deep-sea sediments. In the deep Atlantic 5–20% of the sediment is composed of coccoliths.

Discoasters (q.v.) are extinct microfossils that show similarities to coccoliths, although they have also been compared with dinoflagellates. They commonly consist of a group of calcite crystals arranged in a stellate snowflake-like pattern with 3 to 10 rays (Black, 1968). They range from Upper Cretaceous to Pliocene.

Bacillariophyta

Diatoms (q.v.) (Bacillariophyta) are unicellular, sometimes colonial, algae with a silicified boxlike cell wall (frustule) consisting of an outer lid-like epitheca and an inner hypotheca. They are commonly 40 to 300 µm in size. There are two orders: Centrales, which typically have a circular, wheel-shaped appearance although they may also be triangular or irregular; and Pennales, which are usually elongate, commonly being oblong or spindle shaped. The degree of silicification is variable. The frustule of pennate forms tends to be thicker and tougher than that of centric forms although some Pennales are only weakly silicified. Centric forms are mainly marine and planktic (Fritsch, 1935). They are nonmotile, and the flat disc-like form they commonly exhibit is an adaptation to a free-floating existence. Some also form ribbon-like colonies for the same purpose (Fig. 11B). Pennate forms also occur in the sea but many inhabit fresh water and commonly occur in mud, soil, or as epiphytes. Many pennate forms are capable of a gliding movement. Diatom-like organisms have been reported from the Jurassic, and even from the Cambrian, but the earliest established occur-

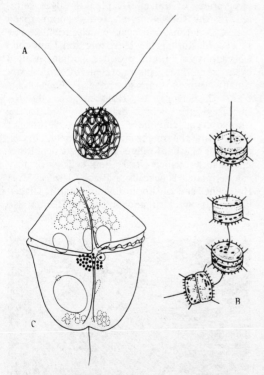

FIGURE 11.A, *Syracosphaera* (Coccolithophoridaceae), coccosphere covered by numerous coccoliths; Holocene. B, *Thalassiosira* (Bacillariophyceae), centric diatoms connected by mucilage; Holocene. C, *Gymnodinium* (Dinophyceae), unarmored motile form; Holocene.

rences are from the Cretaceous. These first diatoms are centric forms. Pennate forms appear in the Cenozoic. At the present day, diatoms are prominent components of both marine and freshwater phytoplankton assemblages.

Pyrrophyta

Dinoflagellates (q.v.) are motile unicellular planktic algae usually less than 200 μm in size, which inhabit marine and freshwater tropical to polar environments. They are dinophyceans, members of the Pyrrophyta. Together with diatoms and coccolithophores they dominate Holocene marine phytoplankton assemblages. The motile phase itself is not readily preserved. Unarmored "naked" forms (Gymnodiniales) have only a thin pellicle enclosing the cell (Fig. 11C) and even the rigid cellulosic wall (theca) of "armored" dinoflagellates (e.g., Peridiniales) is prone to disintegration and decay. Only a few uncertain records exist of fossil motile stages. Instead, the geological history of the dinoflagellates is recorded by their cysts (resting spores) formed during reproduction or in response to adverse conditions such as winter. The cysts are important microfossils in the Mesozoic and Cenozoic, exhibiting rapid evolution from the Late Triassic onward (Sarjeant, 1974). Formerly, many fossil dinoflagellate cysts were included with the hystrichospheres (see below).

Dinoflagellate cysts are generally 60–120 μm in size and have a basically spherical, ovoid, or polyhedral form often with spinose outgrowths. Most are composed of a highly resistant noncellulosic organic material although "calciodinellids" have a calcareous outer layer and siliceous cysts also occur. Today, encystment is a feature of temperate neritic species; those in stable oceanic environments tend not to form cysts. The difficulties of relating cysts to motile stages in fossil dinoflagellates has necessitated a classification primarily based on cyst morphology. There are four major groups: proximate cysts, similar in shape to the theca of the motile stage; chorate cysts, formed well inside the theca and linked by processes to its inner surfaces; proximochorate cysts, intermediate between the first two; and cavate cysts, with two separated wall layers—the outer one resembling the shape of the theca, the inner one separated from it by cavities. The alga emerges from the cyst through an opening (archaeopyle) in the wall.

Proximate cysts have been reported from the Silurian and Permian but the continuous record of dinoflagellate cysts begins in the Upper Triassic, again with proximate forms. Chorate, proximochorate, and cavate cysts all appear in

the Lower and Middle Jurassic and the biostratigraphic value of dinoflagellates rivals that of ammonites in the Upper Jurassic. Calcareous forms appear in the Upper Jurassic and siliceous forms in the Tertiary. All four major cyst groups are probably extant although modern cavate forms have still to be found.

Acritarchs

Hystrichospheres (q.v.) not recognizable as dinoflagellate cysts have been classed as acritarchs (Greek: uncertain origin). These hollow organic-walled microfossils, mostly 10–50 μm in size, are the most common Proterozoic fossils and were also the dominant marine plankton during the Paleozoic (Downie, 1967). The earliest acritarchs (q.v.) are simple globular bodies, termed *sphaeromorphs,* which are probably a diverse group of algal spores. One genus, *Huroniospora,* has been recorded from the Gunflint Chert (2000 m yr) of Ontario. In the Cambrian, more complex spinose forms arose and probably represent both dinoflagellate and prasinophycean (a small but distinct algal group of uncertain affinity) cysts. Dinoflagellate-like cysts include the long-ranging genera *Micrhystridium* and *Veryhachium.* Prasinophycean and similar forms include *Baltisphaeridium, Cymatiosphaera, Dictyotidium, Tasmanites,* and *Pachysphaera.* Acritarchs declined rapidly in the Carboniferous and dinoflagellate cysts dominate organic-walled phytoplankton assemblages in post-Paleozoic rocks. Freshwater acritarchs first appear in the Quaternary. Holocene acritarchs have not been identified with certainty.

Soft Algae

Carbonaceous remains and impressions of plants that may be algae have been described from several periods, particularly Ordovician-Devonian. They are mostly poorly preserved macroscopic fossils whose precise affinities are uncertain but which possibly represent unmineralized red, green, or brown algae. *Drydenia, Enfieldia,* and *Hungerfordia* are flattened fronds borne on stems, from the marine Devonian of New York State. *Buthotrephis* has narrower fronds similar to those of laminarian brown algae. *Chaetocladus* and *Inopinatella* resemble dasycladaceans, and *Parka* and *Powysia* may also be green algae. *Prototaxites* and *Nematothallus* are possibly freshwater algae.

Discussion

While some fossil algae are readily recognizable as such, others present serious problems of identity. Because many algae have a relatively simple construction they sometimes show superficial morphological similarities to lower invertebrates as well as exhibiting marked parallelism among themselves. In the past, many trace fossils were regarded as algal remains and termed "fucoids." Difficulties continue in placing groups such as acritarchs, Solenoporaceae, Receptaculitaceae, Gymnocodiaceae, and others; and in the Paleozoic there is probably considerable overlap between "algae" and small calcareous invertebrates such as foraminifers (Riding, 1977b).

Fossil algae provide little check on phyletic relationships elucidated by neobotany although they do indicate the sequence of appearance of major groups, and also supply detailed information concerning the evolution of groups such as the Dasycladaceae. Blue-greens were the first algae to appear; acritarchs and chlorophytes also originated during the Precambrian. The Pyrrophyta, Rhodophyta, Phaeophyta, and characeans probably all appeared during the early Paleozoic. Other important events in algal history include the early Paleozoic decline of stromatolites, the late Paleozoic reduction of the acritarchs, and the Mesozoic mineralization of phytoplankton groups (diatoms, coccolithophores, calciodinellids).

Algal photosynthesis during the Precambrian gradually raised the level of free oxygen in the atmosphere, which in turn gave rise to ozone, thus providing a shield against ultraviolet solar radiation. Both these effects were probably of crucial importance to the evolution of metazoans in the late Precambrian and their rapid diversification in the early Paleozoic (Cloud, 1968). Phytoplankton have been an important food source for suspension feeders and zooplankton since the late Precambrian and their temporary decline in the Carboniferous has been suggested as a factor in the widespread late Paleozoic extinctions of animal life (Tappan, 1968). Some algae also influence animals by being symbiotic with them. Blue-greens, greens, pyrrophytes, and diatoms all have forms that exist within the tissue of a variety of invertebrates. Their presence is not recorded directly in fossils but their geologic importance is reflected in the success of certain groups such as reef-forming scleractinian corals, which can grow and calcify very rapidly in shallow water due to the presence of symbiotic unicellular dinophycean and bacillariophyte algae termed zooxanthellae (q.v.).

Most fossil algae are recorded by mineralized skeletons; and groups with few mineralized forms (Cryptophyta, Euglenophyta, Phaeophyta, Xanthophyta) have a very poor fossil record unless, like the dinoflagellates, they form tough organic-walled cysts. The processes

and materials utilized in mineralization vary (Johnson, 1961; Milliman, 1974). Calcification in red algae and coccolithophores involves mainly calcite or high-magnesian calcite which is produced within the cells by the golgi apparatus. Silica production in diatoms appears to rely on a similar internal process. In blue-green and green algae, calcification is extracellular and relatively unsophisticated; the calcium carbonate is in the form of calcite or aragonite. The abundance of calcified phytoplankton in the late Mesozoic has been suggested to have altered the calcium/magnesium ratio in sea water by extracting calcium. Algae can also be preserved by being coated or replaced by minerals. Calcareous crusts form on algae in freshwater tufa deposits and in environments of subsea cementation. Similar siliceous coatings can form in hot springs. Complete impregnation of algae by silica is known in chert deposits.

Algal mineralization and trapping of loose grains has important sedimentary effects. In situ biostromes and bioherms have been produced or contributed to, by stromatolites, solenoporaceans, receptaculitaceans, squamariaceans and crustose corallines (Wray, 1971). The latter are especially important in association with scleractinian corals in Cenozoic reefs and locally dominate encrusting communities as in Holocene cup reefs and algal ridges around oceanic islands. Green algae, dasycladaceans during the Mesozoic–early Cenozoic and codiaceans more recently, have been major sources of loose carbonate sediment in shallow tropical and subtropical marine environments. Characeans have produced carbonate sediment in freshwater lacustrine environments. Planktic coccolithophores and diatoms have made significant contributions to deep-sea sediment certainly since the late Mesozoic. In very deep water, however, dissolution of skeletons also takes place, first of carbonate shells and then of siliceous ones, unless deposition and burial are relatively rapid.

The importance, interest and uses of fossil algae vary according to the group. They can be applied to environmental and biostratigraphic analysis; are sedimentologically important; have significant bearing upon the history of life; and in their morphology, secretion of hard parts, evolution, and modes of life present a diversity of interest that belies their apparent simplicity.

ROBERT RIDING

References

Adey, W. H., and Macintyre, I. G., 1973. Crustose coralline algae; a re-evaluation in the geological sciences, *Geol. Soc. Amer. Bull.*, **84**, 883–903.

Barton, E. S. 1901. The genus *Halimeda, Siboga Expedition Monogr.* **60**, 1–32.

Black, M. 1965. Coccoliths, *Endeavour,* **24,** 131–137.

Black, M. 1968. Taxonomic problems in the study of coccoliths, *Palaeontology,* **11,** 793–813.

Cloud, P. E., Jr., 1968, Pre-metazoan evolution and the origins of the Metazoa, in E. T. Drake, ed., *Evolution and Environment.* New Haven: Yale Univ. Press, 1–72.

Cloud, P. E., Jr., and Semikhatov, M. A., 1969. Proterozoic stromatolite zonation, *Amer. J. Sci.,* **267,** 1017–1061.

Downie, C., 1967. The geological history of the microplankton, *Rev. Palaeobot. Palynol.,* **1,** 269–281.

Elliott, G. F., 1955. The Permian calcareous alga Gymnocodium, *Micropaleontology,* 1, 83–90.

Elliott, G. F., 1968. Permian to Palaeocene calcareous algae (Dasycladaceae) of the Middle East, *Bull. British Mus. (Nat. Hist.) Geology,* Suppl. 4, 111p.

Fritsch, F. E., 1935, 1945. *The Structure and Reproduction of the Algae.* vols. 1, 2. London: Cambridge Univ. Press, 791p. 939p.

Flügel, E., ed., 1977. *Fossil Algae.* Berlin: Springer, 375p.

Garrett, P., 1970. Phanerozoic stromatolites: Noncompetitive ecologic restriction by grazing and burrowing animals, *Science,* **169,** 171–173.

Hofmann, H. J., 1969. Attributes of stromatolites, *Geol. Surv. Canada Paper 69-39,* 58p.

Johnson, J. H., 1961. *Limestone-building Algae and Algal Limestones.* Golden, Co.: Colorado School of Mines, 187p.

Logan, B. W., 1961. *Cryptozoon* and associate stromatolites from the Recent, Shark Bay, Western Australia, *J. Geol.,* **69,** 517–533.

Logan, B. W.; Rezak, R.; and Ginsburg, R. N., 1964. Classification and environmental significance of algal stromatolites, *J. Geol.,* **72,** 68–83.

Milliman, J. D., 1974. *Marine Carbonates.* Berlin: Springer, 375p.

Monty, C. L. V., 1973. Les nodules de manganèse sont des stromatolithes océaniques, *Comptes Rendus Acad. Sci.,* Ser. D, **276,** 3285–3288.

Nitecki, M. H., 1972. North American Silurian receptaculitid algae, *Fieldiana, Geol.,* **28,** 108p.

Peck, R. E., 1953. Fossil charophytes, *Bot. Rev.,* **19,** 209–227.

Pia, J. 1920. Die Siphoneae Verticillatae vom Karbon bis zur Kreide, *Verh. Zool.-Bot. Gesell. Wien,* **11**(2), 1–263.

Pia, J., 1927. Thallophyta, in M. Hirmer, ed. *Handbuch der Paläobotanik.* München: Oldenbourg, 31–136.

Riding, R., 1977a. Skeletal stromatolites, in E. Flügel, ed., *Recent Research on Fossil Algae.* Berlin: Springer, 57–60.

Riding, R. 1977b. Problems of affinity in Palaeozoic calcareous algae, in E. Flügel, ed., *Recent Research on Fossil Algae.* Berlin: Springer, 202–211.

Rietschel, S., 1969. Die Receptaculiten, *Senckenbergiana Lethaea,* **50,** 465–517.

Sarjeant, W. A. S., 1974. *Fossil and Living Dinoflagellates.* London: Academic Press, 182p.

Schopf, J. W., 1970. Precambrian micro-organisms and evolutionary events prior to the origin of vascular plants, *Biol. Rev. Cambridge Phil. Soc.,* **45,** 319–352.

Stockman, K. W.; Ginsburg, R. N.; and Shinn, E. A.,

1967. The production of lime-mud by algae in south Florida, *J. Sed. Petrology,* **37,** 633–648.

Tappan, H., 1968. Primary production, isotopes, extinctions and the atmosphere, *Palaeogeography, Palaeoclimatology, Palaeoecology,* **4,** 187–210.

Walter, M. R., 1972. Stromatolites and the biostratigraphy of the Australian Precambrian and Cambrian, *Spec. Paper Palaeontol.,* **11,** 190p.

Wray, J. L., 1971. Algae in reefs through time, *Proc. N. Amer. Paleontological Conv.,* J, 1358–1373.

Wray, J. L. 1977. *Calcareous Algae.* New York: Elsevier, 185p.

Cross-references: *Acritarchs; Calcispheres and Nannoconids; Charophyta; Coccoliths; Diatoms; Dinoflagellates; Plankton; Precambrian Life; Protista; Reefs and Other Carbonate Buildups; Stromatolites; Tasmanitids; Zooxanthelle.* Vol. I: *Phytoplankton.* Vol. III: *Algal Reefs; Algal Rims, Terraces and Ledges; Fringing Reef.* Vol. VI: *Algal Reef Sedimentology; Stromatolites.*

AMMONOIDEA

The Ammonoidea (ammonoids) are an order of the class Cephalopoda (q.v.), phylum Mollusca. The nearest living relative of the ammonoids is the pearly *Nautilus* of which Oliver Wendell Holmes wrote *"This is the ship of pearl, which, poets feign, /Sails the unshadowed main."*

The hollow shell of most ammonoids is essentially a cone, coiled in a plane, regularly partitioned into chambers. The center of the outer margin of the coil is the *venter;* the corresponding part of the inner margin is the *dorsum* (Fig. 1). The essential difference between the nautiloid and ammonoid shell is in the partitions that separate the chambers; the nautiloid partitions are undulating and of "free-form" appearance, whereas those of the ammonoid are much more highly crenulated and therefore more complex.

Ammonoids became extinct near the end of the Cretaceous, and are therefore known only from fossil shells. The soft parts of the body must be inferred: one may imagine an ammonoid as a coiled shell with a protruding head similar to that of a squid (Fig. 2).

Most ammonoids have the spirally coiled shells described above, but a few groups do not and are collectively called heteromorphs (Fig. 3). The heteromorphs can generally be classified, nonphylogenetically, by form into (1) those with helical shells, either high- or low-spired and with whorls in contact (as in snail shells), or loosely coiled with whorls not in contact; (2) those that are straight, but with a very small initial coil; (3) those with straight shafts connected by one or more hairpin (180°) curves; (4) those with a spiral coil ending in a

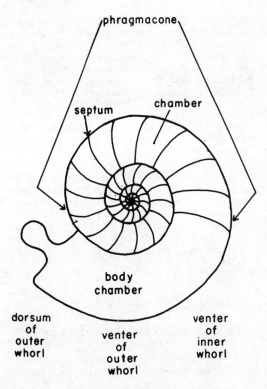

FIGURE 1. The ammonoid shell.

straight shaft that turns back toward the coil (scaphitoid); and (5) a few rare forms that defy simple description.

The ammonoid shell can normally be divided into the chambered part, called the *phragmacone,* and the *body chamber,* which was the last chamber occupied by the living animal. As the animal grew, the shell was enlarged at the

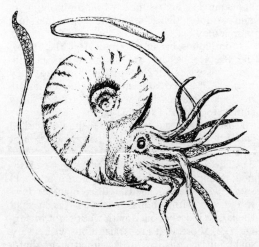

FIGURE 2. Hypothetical reconstruction of a living ammonoid.

open end of the body chamber, and a new partition or *septum* was added behind the body of the animal to enlarge the phragmacone. The intersection of the septum with the outer wall can be observed if the wall is removed. This line of intersection is termed the *suture,* of which part is bent orad (toward the open end of the shell) and part aborad (away from the open end of the shell). Areas extending orad are called *saddles* and areas extending aborad are called *lobes.* On some ammonoids in which the lobes are relatively simple and undivided, the suture is said to be goniatitic; on others the lobes are divided into second-order lobes and saddles (ceratitic); and in a third group both lobes and saddles contain second-, third-, or even fourth-order subdivisions (ammonitic). The latter forms are said to have highly diverticulate sutures. These groupings are purely descriptive and not phylogenetic.

Ammonite shells range from smooth to highly ornate. Common ornamentation includes (1) radial ridges from the umbilicus (the umbilicus is that area of lateral shell surface enclosed by the outer whorl) outward, which if fine are called *lirae* (sing.: lira) and if coarse and strong are called *ribs* or *costae* (sing.: costa); and (2) bumps in various arrangements, usually on the costae if the shell is ribbed, termed *nodes* if small and round, or *tubercles* if high and hollow. In some species the nodes are elongate spirally, or less often radially, and in others the tubercles develop long spines. Spiral ridges and/or grooves are also sometimes present. Modifications occurring at the opening of the body chamber (the *aperture*) may restrict the size and shape of the opening. Apertures are seldom preserved, so these ornaments are not always useful in classification. Some forms have a calcareous or horny (chitinous) plate, the *aptychus* (q.v.) closing the aperture.

Classification of the ammonoids is based partly on ornamentation and septum configuration and partly on concepts of phylogeny, which are inferred from morphological trends observed in stratigraphic sequence.

In spite of many hypotheses the exact origin of the ammonoidea is not known. This mystery results in large part from collection failure, since the specimens of Silurian and Devonian nautiloids that might have given rise to the Ammonoidea are rare, and early Ammonoidea are equally scarce. Two hypotheses (both strictly conjectural) are prevalent. The first assumes nautiloid-to-ammonoid transition in uncoiled shells, ammonoids later becoming coiled. The second assumes coiled ammonoid ancestry in coiled nautiloids, with uncoiling a later specialization in some ammonoids. This writer usually assumes the second hypothesis only be-

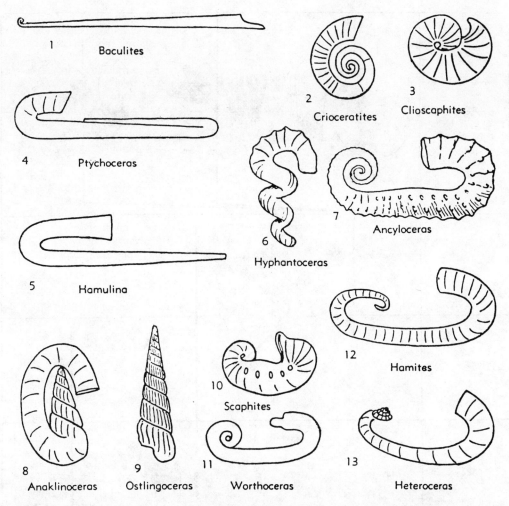

FIGURE 3. Types of coiling in ammonoid shells; heteromorphs (from Arkell et al., 1957; courtesy of The Geological Society of America and University of Kansas Press).

cause it seems to be slightly favored by the age relationships of known possible ancestors and descendents.

The transition of nautiloid to ammonoid is largely marked by the increased crenulation of the slightly curved suture of a coiled nautiloid to the greater or multiple arcs of the primitive ammonoid. This development occurs in the earliest of the eight orders of Ammonoidea, the Anarcestina, which is restricted to the Devonian. The anarcestines range from a species very similar to the presumed ancestral nautiloid to forms that have highly complex goniatitic sutures. Although later nautiloid offshoots produce goniatitic-like sutures, they are never as complex as most anarcestines. The suture is the most important character in the taxonomy of this suborder, but some lineages also developed quite strong ornamentation, which is useful in

classification. The first rapid expansive adaptive radiation of ammonoids occurred largely within the Anarcestina, and the other three dominantly Paleozoic ammonoid suborders (Prolecanitina, Goniatitina, and Clymeniina) are descended independently from the anarcestines (Fig. 4).

The suborder Clymeniina constitutes a small, apparently monopyletic group of ammonoids descended from an anarcestine with simple sutures. The sutures never become as highly digitate (with many secondary folioles) as in the major Paleozoic suborders, the Prolecanitina and the Goniatitina. Some clymeniines depart from the normal coil, having a triangular coiling, and in other species the chambers become globular, reminiscent of the foraminifer, *Globigerina*. Although many forms are smooth, radial ribbing is characteristic of this suborder.

23

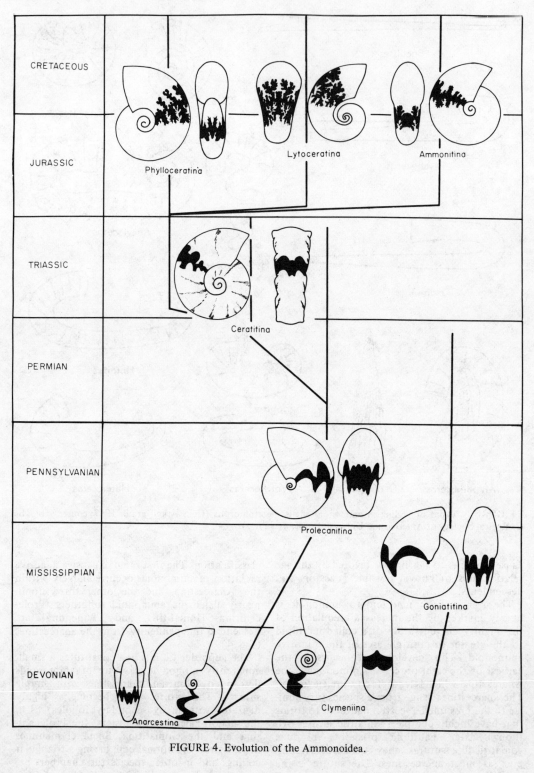

FIGURE 4. Evolution of the Ammonoidea.

The Clymeniina are restricted to the Upper Devonian.

A third suborder, the Goniatitina, is descended from more primitive Anarcestina, but rapidly develops a suture with 8 lobes, characteristic of all its members except primitive transitional forms. Lobes are often subdivided and this group has species that have goniatitic sutures, a few species in which the sutures might be termed ceratitic, and a good number of species with ammonitic sutures. In other words, the evolution of the suture in the Goniatitina is toward increased complexity, with the ceratitic stage of the evolution rushed through rather rapidly. Although goniatitines are generally without marked ornamentation, a few species bear fine sigmoid or curved lirae, and an exceptional species may be ribbed. The Goniatitina range from the Devonian to Upper Permian.

The suborder Prolecanitina is descended from less primitive anarcestines. The conchs of Prolecanitina are characteristically discoidal or thinly lenticular with broad flanks, resulting in a large number of lobes and saddles, decreasing in size somewhat regularly from venter to umbilicus. Sutures may be goniatitic, but are characteristically ceratitic, paralleling the development of the goniatitine pattern. The ammonitic stage was reached only in the Triassic by the prolecanitine-derived suborder Ceratitina. The Prolecanitina range from the Upper Devonian to the Upper Triassic.

A lineage representing the beginnings of the suborder Ceratitina branches off of the major Prolecanitina stock in the Permian, and only this lineage plus the Prolecanitina proper cross the Paleozoic/Mesozoic boundary. During the Triassic, the Prolecanitina remain a conservative group; but in the earliest Triassic the Ceratitina udergo a tremendous and rapid radiation, becoming the dominant suborder of the Triassic. As in the Anarcestina and Prolecanitina, the sutures of the Ceratitina pass from goniatitic through ceratitic to ammonitic stages. Furthermore, the transition from ceratitic to ammonitic sutures occurs independently in several different lineages of Ceratitina. Many Ceratitina became much more highly ornamented, with nodes, ribs, and other rugose features, than did the Paleozoic ammonoids. The Ceratitina range from Permian through the Triassic.

In the Early Triassic a lineage of ammonoids branched from the rapidly radiating Ceratitina. This new group, the Phylloceratina, persisted to the end of the Cretaceous as a rather conservative stock from which all post-Triassic ammonoids were derived. The Pylloceratina are generally less ornate than associated species of other suborders. The secondary and tertiary folioles of the saddles are phylloid; the lobes are usually tripartite. The Phylloceratina range from Lower Triassic to the end of the Cretaceous.

Following the extinction of the Ceratitina and Prolecanitina at the end of the Jurassic, there was again a radiation of new groups of ammonoids.

Very early in the Jurassic the Lytoceratina were derived from the Phylloceratina. Coiled Lytoceratina, like the Phylloceratina, are usually less ornate than other associated ammonoid groups. They are usually round in whorl section, more evolute (more loosely coiled) than the Phylloceratina, and with sutures usually not phylloid. In addition to the coiled species, most species of Jurassic and Cretaceous hetermorphs belong to the Lytoceratina. Although frequently stated to be typical of the late Cretaceous, heteromorphs date from at least late Bajocian (Middle Jurassic) and appear quite extensively in the Lower Cretaceous. Lycoceratina range from the base of the Jurassic to the top of the Cretaceous.

The evolution of Jurassic and Cretaceous ammonoids is not as well understood as is the evolution of their predecessors. Consequently, in most ammonoid studies the last group of Ammonoidea is the polyphyletic suborder, the Ammonitina. This suborder exists out of necessity; no leading ammonitologist has as yet succeeded in dividing this group into its proper monophyletic lineages to the satisfaction of his colleagues. Thus, polyphyletism results partly from disagreement and ignorance concerning the true affinities and partly from the reluctance of some authorities to abandon or modify a well-established taxon.

The suborder Ammonitina comprises those coiled ornate ammonoids plus less ornate transitional forms to either Lytoceratina or Phylloceratina. There are modifications from dwarfs to giants, and suture patterns range from near-ceratitic to extremely complex. The more ornate and exotic Mesozoic ammonoids are retained in this suborder because their affinities (with Lytoceratina or Phylloceratina) are not yet clear. The suborder thus includes species from early Jurassic through Cretaceous.

Since the Ammonoidea died out about 70 m yr ago, their paleoecology cannot be approached by direct comparison with modern animals, and has necessarily been derived empirically. Studies of containing sediments indicate that ammonites were either absent or extremely rare in reef environments but were widely adapted to other oceanic and near-shore habitats.

From more theoretical considerations it has been shown that the body chamber is so related to the phragmacone that the phragmacone probably served as a float, providing the buoyancy that would suspend in water the heavier body chamber containing the animal. This is not only true of regularly coiled forms, but of heteromorphs as well, and thus nearly all ammonites were probably primarily nektonic or planktonic organisms.

Ammonoidea have long been considered of greatest utility for detailed stratigraphic correlations. For some ages other animals are as well suited as ammonoids, and for special purposes and economic purposes replace them. In most of the middle Paleozoic strata ammonoids are rare, and are of no aid to biostratigraphic studies. They become more common in many upper Paleozoic rocks, although they are not abundant in many environments. The suture pattern is the most useful character for dating Paleozoic rocks, although the specialist always uses all the available information.

There are no fossils that surpass the ammonoids for dating Mesozoic rocks, and the Triassic and Jurassic zonal sequences are particularly detailed. In Jurassic and Cretaceous rocks the more ornate species and genera are most useful for zonation, and suture patterns take on a secondary importance.

Ammonoidea have long been used to demonstrate mechanics and principles of evolution, and some basic generalities can be made concerning evolution in this group. This evolution is best studied by ontogenetic series from successional species, the data for which must be substantiated by superposition of those species in the rocks.

Parallel evolution is common in ammonoids and homoemorphy is not uncommon. This is because the total possible combinations of shape and ornamentation of a coiled ammonoid shell is somewhat limited. It is inevitable that at different times similar ornament and shape appear in entirely unrelated stocks (homoeomorphy). Furthermore, different stocks of similar genetic inheritance (as the Prolecanitina and Goniatitina, both descended from Anarcestina) met similar environments with the selection of similar structural modifications (parallel evolution).

The Ammonoidea through the upper Paleozoic and the Triassic develop several lineages. The evolution of the suture in these lineages is parallel in morphology, but not always parallel in time. In other words, a true program is absent (in program evolution the parallel events in different lineages occur at the same time).

There are two prevalent ideas concerning the evolution of Jurassic and Cretaceous Ammonoidea: (1) there was iterative replenishment of the polyphyletic Ammonitina from the conservative Phylloceratina and Lytoceratina lineages, and (2) the ammonitines were more persistent

through the Jurassic and/or Cretaceous and included less periodic offshoots from the conservative Phylloceratina and Lytoceratina.

Two other antithetical concepts concerning the evolution of the Ammonoidea have been actively debated: recapitulation and cenogenesis. Many lineages of Ammonoidea were set up in the last century on the assumption that the juvenile or young forms were similar in suture and/or shell morphology to the adult ancestors. In other words, the young stages of the descendent recapitulate the adult stage of the progenitor. The second idea, that of cenogenesis, suggests that new characters appear in the young stage as a result of special adaptations to the environment occupied by the young, and that these special characters, if of survival value to adults, gradually spread through time to the adult stage.

Neither recapitulation nor cenogenesis is the entire answer. In some stocks the first seems to be dominant, and in others the second. In many stocks it is evident that both processes are involved, depending on the particular character, and in a single species some characters involve cenogenesis and others involve recapitulation. Neither hypothesis can be validated unless proper transitional forms are found in place in the correct stratigraphic sequence.

A further complication, long known, but recently restudied in considerable detail, is the effect of sexual dimorphism (q.v.) in the Ammonoidea on extant taxonomy, nomenclature, and evolutionary concepts (e.g., Makowski, 1962; Holder, 1975; Kennedy and Cobban, 1976).

KEITH YOUNG

References

Arkell, W. J., 1950. A classification of the Jurassic ammonites, *J. Paleontology,* **24**, 354–364.

Arkell, W. J., et al., 1957. Cephalopoda: Ammonoidea, in R. C. Moore, ed., *Treatise on Invertebrate Paleontology,* pt. L, Lawrence Kansas: Geol. Soc. Amer. and Univ. Kansas Press, 490p.

Bather, F. A., 1894. Cephalopod beginnings, *Nature Sci.,* **5**, 421–436.

Donovan, D. T., 1964. Cephalopod phylogeny and classification, *Biol. Rev. Cambridge Phil. Soc.,* **39**, 259–287.

Donovan, D. T., 1973. The influence of theoretical ideas on ammonite classification from Hyatt to Trueman, *Univ. Kansas Paleontol. Contrib. Paper,* **62**, 16p.

Erben, H. K., 1964. Die evolution der ältesten Ammonoidea (Lieferung 1), *Neues Jahrb. Geol. Paläontol., Abh.,* **120**, 107–212.

Gordon, W. A., 1976. Ammonoid provinciality in space and time, *J. Paleontology,* **50**, 521–535.

Holder, H., 1975. Forchungsbericht uben Ammoniten, *Paläont. Zeitschr.,* **49**, 443–511.

Kennedy, W. J., 1977. Ammonite evolution, in A. Hallam, ed., *Patterns of Evolution*. Amsterdam: Elsevier, 251–304.

Kennedy, W. J., and Cobban, W. A., 1976. Aspects of ammonite biology, biogeography and biostratigraphy, *Spec. Paper Palaeontology,* **17**, 94p.

Kullmann, J., and Weidmann, J., 1970. Significance of sutures in the phylogeny of Ammonoidea, *Univ. Kansas Paleontol. Contrib. Paper*, **47**, 32p.

Kummel, B., 1952. A classification of the Triassic ammonoids, *J. Paleontology,* **26**, 847–853.

Kummel, B., 1954. Status of invertebrate paleontology, 1953. V. Mollusca: Cephalopoda, *Bull. Mus. Comp. Zool. Harvard Univ.,* **112**, 181–192.

Makowski, H. K., 1962. Problem of sexual dimorphism in ammonites, *Paleontologia Polonica,* **12**, 89p.

Miller, A. K., and Furnish, W. M., 1954. The classification of the Paleozoic ammonoids, *J. Paleontology,* **28**, 685–692.

Nelson, C. M., 1968. Ammonites: Ammon's horns into cephalopods, *J. Soc. Bibliography Nat. Hist.,* **5**, 1–18.

Ruzhencev, V. E., 1960. Ammonoid classification problems, *J. Paleontology,* **34**, 609–619.

Schindewolf, O. H., 1968. Homologie and Taxonomie —Morphologische Grundlegung und phylogenetische Auslegung, *Acta Biotheoretica,* **18**, 235–283.

Spath, L. F., 1933. The evolution of the Cephalopoda, *Biol. Rev. Cambridge Phil. Soc.,* **8**, 418–462.

Stenzel, H. B., 1964. Living *Nautilus,* in R. C. Moore, ed., *Treatise on Invertebrate Paleontology,* pt. K, Mollusca 3 Lawrence, Kansas: Geol. Soc. Amer., and Univ. Kansas Press, K59–K93.

Trueman, A. E., 1941. The ammonite body-chamber, with special reference to the buoyancy and mode of life of the living ammonite, *Quart. J. Geol. Soc. London,* **96**, 339–383.

Westermann, G. E. G., 1971. Form, structure and function of the shell and siphuncle in coiled Mesozoic ammonites, *Roy. Ontario Mus. Life Sci. Contrib.,* **78**, 39p.

Wright, C. W., 1952. A classification of Cretaceous ammonites, *J. Paleontology,* **26**, 213–222.

Cross-references: *Cephalopoda; Mollusca; Sexual Dimorphism.*

AMPHIBIA

The Amphibia is a class of backboned animals classified in an intermediate position between fish and reptiles. Included are living frogs, salamanders, and limbless caecilians together with numerous extinct types. As a class, the amphibians are terrestrial as adults; but most living species are dependent on the water for reproduction. The amphibian egg, like that of fish, lacks the amniotic and allantoic membranes which in reptiles and birds protect the embryo and contain blood vessels for the transport of nutrients, respiratory gases, and metabolic wastes. Without these membranes, the eggs of amphibians must be small (<10 mm in diameter) to provide a large surface-to-volume ratio for the passive diffusion of these substances. The eggs are typically deposited in the water, where the hatchlings depend on the continued support of the fluid medium. Living amphibians typically have a distinct larval stage which feeds on small particles suspended in the water and respires with external gills. Fossils of larvae, grossly similar to those of living salamanders, show that a similar developmental pattern was present in some Paleozoic amphibians, as was probably the case for their fish ancestors. Fertilization of the egg is external and occurs in the water in primitive modern amphibians. This must have been the typical practice for their Paleozoic ancestors, in common with most fish.

Living amphibians are cold-blooded or ectothermic (the body temperature is dependent on that of the environment). The reliance of modern amphibians on the general body surface for exchange of respiratory gases requires that the skin be moist, and the body size small. The larger size of most Paleozoic amphibians, together with the presence of a heavy covering of bony dermal scales, indicates that cutaneous respiration would not have been practical. The presence of well-developed ribs indicates that they pumped air into the lungs in a manner similar to that of reptiles and mammals. All living amphibians are carnivorous, with the exception of the anuran larval stage. It is probable that most, if not all, Paleozoic amphibians were also.

Classification

Three subclasses of amphibians are currently recognized: Labyrinthodontia, Lepospondyli, and Lissamphibia (Romer, 1966). The labyrinthodonts are known from the Upper Devonian into the Lower Jurassic. They are generally large forms (up to 3 m in length), heavily scaled, and characterized by having the vertebral centra ossified in more than a single unit. The lepospondyls are known from the Lower Carboniferous to the end of the Lower Permian. Like the labyrinthodonts, they are heavily scaled; but generally of smaller size and typically have a single central ossification. The term Lissamphibia is applied to the three living orders—Urodela, Anura, and Apoda (Gymnophiona, Caecilia)—whose fossil record begins, respectively, in the Middle Jurassic, Lower Jurassic, and Paleocene. These forms are generally small, and always have only a single central ossification. All three practice cutaneous respiration. A striking similarity of all living frogs, salamanders, and apodans is the presence of pedicellate teeth in which the base and crown are separated by a band of fibrous tissue. Anurans and salamanders lack body scales, but these are present in some apodans. The interrelationships among the three subclasses have not been established. There is not convincing evidence as to whether or not either lepospondyls or the Lissamphibia are natural groups.

Because of the continuing question as to the nature of relationships between major groups, any classification of the Amphibia is somewhat arbitrary. A history of the classification is given in tabular form by Olson (1971). The following scheme (much simplified) is based on Romer (1966) with modifications to incorporate more recent work:

CLASS AMPHIBIA
  Subclass Labyrinthodontia
    Order Ichthyostegalia
    Order Temnospondyli
    Order Anthracosauria
      Suborder Herpetospondyli
      Suborder Embolomeri
      Suborder Seymouriamorpha
  Subclass Lepospondyli
    Order Aistopoda
    Order Nectridea
    Order Microsauria
    Families without ordinal designation
      Adelogyrinidae
      Lysorophidae
      Acherontiscidae
  Subclass Lissamphibia
    Order Apoda
    Order Urodela
    Order Proanura
    Order Anura

Labyrinthodontia

Labyrinthodonts were clearly derived from rhipidistian fish, with which they share a similar pattern of the bones of the skull, labyrinthine infolding of the dentine of the teeth, and presence and pattern of large palatal tusks. The pattern of the girdles and limbs of amphibians can be readily derived from that of the rhipidistians. The closest affinities of amphibians appear to be with the rhizodontid rhipidistians on the basis of skull proportions and details of limb and girdle structure, but no rhipidistians are known that can be specifically cited as labyrinthodont ancestors. On the basis of the degree of differentiation of Upper Devonian and Lower Mississippian amphibians, it has been assumed that divergence must have occurred by the Middle Devonian. The major groups of Carboniferous labyrinthodonts are sufficiently similar that all could have evolved from a single group of rhipidistians. Romer (1966) recognized three orders of labyrinthodonts: Ichthyostegalia, Temnospondyli, and Anthracosauria.

Ichthyostegalia. The Ichthyostegalia (Fig. 1A) are only definitely known from the Upper Devonian of East Greenland. They were large, with limbs and girdles fully developed for a terrestrial way of life. The ribs were massive to support the body against the force of gravity. The ichthyostegids are clearly the most primitive of labyrinthodonts, retaining the break between the two parts of the braincase characteristic of rhipidistians, as well as two elements of the opercular series between the cheek and the pectoral girdle, and a vertebral construction of essentially rhipidistian grade. Most of the support for the body was supplied by the notochord. The centra are made up of an anterior ventral element in the shape of a crescent, compared with the intercentra of higher tetrapods, and posterior paired dorsal elements, termed pleurocentra. Dorsally, the neural arches are joined to one another by special articulating surfaces, the zygopophyses. Despite their generally primitive anatomy, the pattern of the bones of the skull roof in ichthyostegids makes them unlikely ancestors of either temnospondyls or anthracosaurs, and they are considered an early specialization among labyrinthodonts that left no descendents.

Because ichthyostegids are unquestionably labyrinthodonts and yet not ancestral to either of the more advanced orders, one must assume that at least one other group of labyrinthodonts was present in the Upper Devonian, but left no fossil record. The presence of Devonian footprints in Australia indicates a wide geographic distribution as well.

Temnospondyli. Temnospondyls were the dominant amphibians from the Mississippian to the end of the Triassic. They are clearly differentiated from anthracosaurs on the basis of the pattern of the dermal bones of the skull roof as well as that of the vertebral centra (Romer, 1947). The pleurocentra (the true centra of advanced tetrapods) are dominant in the anthracosaurs, and the intercentra (an anterior, usually crescentic element) in the temnospondyls. It is clear from the known Pennsylvanian fossils that the record of temnospondyl evolution in the Mississippian is very incomplete. The better known early temnospondyls have much less well-developed limbs than the ichthyostegids and a long vertebral column, indicating adaptation to an aquatic way of life quite unlike that of the rhipidistians, and one that must be assumed to be secondary for amphibians. A single fossil of a more terrestrial temnospondyl from the Late Mississippian (Holmes and Carroll, 1977) suggests that the primary radiation of this group was toward a terrestrial, or at least semiterrestrial way of life during the Carboniferous. The fossil record im-

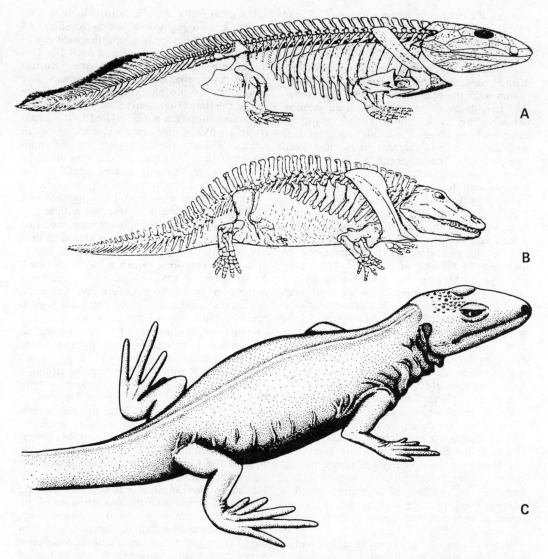

FIGURE 1. Examples of extinct amphibians. A, *Ichthyostega* (from Romer, 1966, copyright © 1966 by The University of Chicago Press). B, *Eryops* (from Romer, 1966, copyright © 1966 by The University of Chicago Press). C, *Gephyrostegus* (from Carroll, Kuhn, and Tatarinov, 1972).

proves in the Pennsylvanian and documents an extensive radiation including fully terrestrial genera, as well as a spectrum of semiterrestrial, semiaquatic, secondarily aquatic, and even neotenic forms. The Lower Permian genus *Eryops* (Fig. 1B) is an extremely common representative of this stage of evolution.

During the Carboniferous, temnospondyls were the dominant terrestrial vertebrates. With the radiation of reptiles in the Late Pennsylvanian, and their dominance in the terrestrial environment by the end of the Lower Permian, the terrestrial temnospondyl groups gradually became extinct. The Late Permian and Triassic,

however, evidence a renewed radiation, with a preponderance of aquatic forms. In general, the Late Permian and Triassic temnospondyls were large animals, with skulls up to a meter in length. The limbs are reduced and the bodies flattened. The skull in many groups is also greatly flattened, with double condyles, large palatal fossae, and an expanded junction between the braincase and the palate. Many groups elaborated the intercentrum until it became the primary vertebral element. These forms have been termed *stereospondyls* and placed in a distinct suborder, implying a common origin. Although the specific ancestry of most of the subgroups remains unspecified, it is probable that they evolved from several different groups of primitive temnospondyls, and are more realistically classified within the same larger taxa. The aquatic temnospondyls remained numerous and had a world wide distribution until the end of the Triassic. One group, the trematosaurs, are known from marine deposits; this is the only known group of amphibians to invade a saltwater environment. Another particularly aberrant group, the plagiosaurs, of Late Triassic age, had a very flat but short and broad skull; the body was flat and covered with dorsal armour. Some genera were neotenic. Their vertebrae had elongate centra, homologized with the pleurocentra of other tetrapods. Their specific affinities are unknown.

With the extensive radiation of aquatic reptiles in the Triassic and the dominance of dinosaurs by the end of that period, the temnospondyls became extinct. No published accounts document their presence later than the Lower Jurassic. They left no descendants.

Anthracosauria. The anthracosaurs were never as numerous or diverse as the temnospondyls. They presumably originated from a common ancestor in the Upper Devonian or Lower Mississippian. Panchen (1975) has recently coined the term *Herpetospondyli* for the primitive anthracosaurian stock. These forms have well-developed limbs and girdles, approximately 32 presacral vertebrae, and must have been capable of an active life on land, although traces of lateral line canals remain in the adult. The vertebrae have large, nearly cylindrical pleurocentra and small crescentic intercentra. This is essentially the pattern of amniotes, and it is probable that the line leading to reptiles diverged at only a slightly earlier stage in anthracosaur evolution.

Forms similar to the known herpetospondyls could have given rise to all the subsequent anthracosaur lines, although no intermediate forms are known. In the Carboniferous, several divergent groups can be recognized. The embolomeres represent a major aquatic adaptation based on elongation of the trunk region, with approximately 40 presacral vertebrae. The individual vertebrae have both intercentra and pleurocentra elaborated as subequal cylinders, presumably allowing a great deal of flexibility in the column associated with an undulatory swimming mode. The limbs, however, retain a pattern similar to that of their more terrestrial antecedents. Some genera reevolved a dorsal caudal fin.

The remaining anthracosaurs were apparently more terrestrial in their habits. The Pennsylvanian gephyrostegids were relatively small forms, with only 24 trunk vertebrae and rela-

tively great limb-to-trunk ratio (Fig. 1C). The earliest known genus, *Bruktererpeton* (Boy and Bandel, 1973) has body proportions approaching those of primitive reptiles. Forms allied to the known gephyrostegids may have given rise to the seymouriamorphs of the Permian. The dominance of the pleurocentra and the large size of the limbs suggested to early workers that seymouriamorphs might be related to the ancestry of reptiles. Gilled larval stages are known in some genera, and the late appearance and primitive cranial anatomy indicate that they could be no more than distantly related relicts of the group that gave rise to reptiles.

Other anthracosaurs, from the Upper Permian of Russia, are difficult to associate with any of the previous groups. They have been classified among the seymouriamorphs because they may have diverged from the same general stock. Some have very long narrow skulls (rather similar to those of some embolomeres) and large plates of dermal armour. Others have very wide, low skulls, fenestrate in the temporal region. None of these Upper Permian groups show any evidence of continuing into the Triassic, although beds of this age in Russia have abundant remains of temnospondyls.

A number of other families may be affiliated with the Anthracosauria but their specific position is difficult to assess. In the Upper Carboniferous and Lower Permian are several groups that approach the reptilian grade of development in some skeletal features, but retain others that are typically amphibian. Members of the Limnoscelidae, Solenodonsauridae, and Tseajaiidae may not be particularly closely related to one another, but all retain an otic notch and have a primitive configuration of the occiput and temporal region that bar them from close relationship to unquestioned reptiles. All are relatively large, and their skeleton suggests that they were primarily terrestrial in habits, at least as adults. The diadectids are another group of Permian-Carboniferous forms that have been classified as both reptiles and amphibians. Their skeletal anatomy is basically reptilian, however.

Lepospondyli

The term lepospondyl is used in reference to a variety of small Paleozoic amphibians that differ from labyrinthodonts in lacking the distinctive infolding of the dentine, and in the absence of an otic notch and palatine fang pairs. The vertebral centra are typically ossified as a single cylindrical element, presumably homologous with the pleurocentra of labyrinthodonts and amniotes. Three orders are typically recognized: Microsauria, Nectridea, and

Aistopoda. Their fossil record is restricted to North America and Europe.

Aistopoda. Aistopods are snake-like in having a very long body with up to 230 vertebrae. Limbs and girdles are completely absent. The braincase, unlike that of other Paleozoic amphibians, is highly ossified and surrounded by a loose framework of dermal bones. A single specimen from the early Lower Carboniferous has been referred to this order, but is not adequately known. Its affinities with the better-known genera from the Pennsylvanian and Lower Permian have not been established.

Nectridea. Nectrideans are typically aquatic forms; the caudal vertebrae are specialized to form a swimming organ; both neural and haemal spines are expanded and laterally compressed. Limbs may be small, but are always present. The group appears in the Early Pennsylvanian, is known from diverse forms in the later Pennsylvanian and Early Permian, but disappears from the fossil record in the Late Permian. It has been suggested that nectrideans are related to anthracosaurs (Thomson and Bossy, 1970), but there is no convincing evidence. The ancestral forms were apparently terrestrial in habit, with large and well-ossified limbs. The terminal members of the group—*Diploceraspis* and especially *Diplocaulus*—were large forms, with the skulls greatly flattened and the tabular region expanded posteriorly in the form of "horns." *Diplocaulus* is common in the later Lower Permian redbeds of Texas. They may be said to parallel the general habitus of the aquatic temnospondyl amphibians of the Triassic.

Microsauria. Fossils of microsaurs, known from the Lower Pennsylvanian through the Lower Permian, are rare, but demonstrate radiation into a wide range of habitats. Some microsaurs were obligatorily aquatic, with external gills as adults, others had proportions similar to those of living lizards and were probably fully terrestrial, at least as adults. Several families have a pointed skull and long vertebral column suggestive of burrowing habits. All have limbs, but they may be relatively small. Among typical microsaurs, the number of presacral vertebrae ranges from 19 to 44. Several families have intercentra in the trunk region in addition to the normal spool-shaped centra. Such elements have never been reported in nectrideans and aistopods. Usually allied with microsaurs are two additional families—the lysorophids and adelogyrinids. In both, the pattern of the skull appears to preclude close affinities with typical microsaurs. Lysorophids have more than 70 presacral vertebrae, but retain small limbs. Functionally, the skull resembles that of living salamanders in being open in the cheek region

(in contrast with typical microsaurs) and having the jaw suspension far forward.

Acherontiscidae. A final lepospondyl family, Acherontiscidae (Carroll, 1969) is represented by a single specimen from the Mississippian of Scotland. The centra of the very long column are formed from two subequal cylinders, grossly resembling the pleurocentra and intercentra of embolomeres. The skull is poorly known but is solidly roofed and has lateral line canal grooves. The limbs are poorly developed. Lysorophids, adelogyrinids, and acherontiscids appear like the isolated orders of Mesozoic mammals. They have no obvious time or phylogenetic position of origin. They hint at an even greater diversity of small amphibians in the Paleozoic.

History. Unlike the labyrinthodonts, the lepospondyls show no obvious affinities with the rhipidistians. The characteristics by which they differ from both these groups may all be related to their small size, and do not necessitate close relationship. It may be suggested that the various lepospondyl groups arose during the Late Devonian and Early Carboniferous from various stocks of labyrinthodonts. It is conceivable that one or more might have evolved directly from rhipidistian fish.

In addition to the uncertainty of the origin of various lepospondyl groups, there is also a question as to the time of their extinction. Their absence in the Upper Permian is difficult to explain. There are, in fact, almost no small amphibians of any sort between the Middle Permian and the Upper Jurassic. It would be rather surprising to have what would appear to be a fairly wide adaptive zone completely empty until the differentiation of Lissamphibia in the late Mesozoic. It seems more probable that preservational or collecting bias has left these elements of the amphibian fauna unsampled.

Lissamphibia

One of the most profound gaps in all of vertebrate phylogeny separates the Paleozoic and Triassic labyrinthodonts and lepospondyls from the modern amphibians orders. Not only are the structure and way of life of the modern orders totally distinct from those of the earlier forms, but no intermediates are known and no specific interrelationships have been established. Despite the tendency of herpetologists to accept the inclusion of the three living amphibian orders within a single group, the Lissamphibia Fig. 2), it is necessary to discuss the structure and general biology of frogs, salamanders, and apodans separately, for each group differs markedly from the others throughout their known history.

31

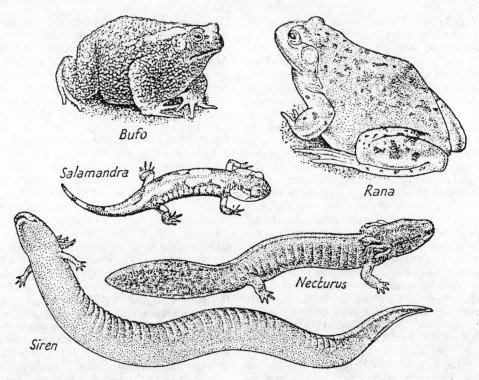

FIGURE 2. Examples of living amphibians—Subclass Lissamphibia (from Young, 1950). (Copyright © 1962 by Oxford University and reprinted by permission.)

Anura. Frogs all adhere to a single skeletal pattern that is among the most specialized of any vertebrate order. The vertebral column is greatly reduced, with only 5 to 9 trunk vertebrae and no tail in the adult stage. The immediately postsacral region is modified to form a longitudinal rod, the urostyle. The rear limbs are emphasized, with both terrestrial and aquatic locomotion typically being produced by their symmetrical movement. The skull is typically, and presumably primitively, specialized in being very open, in contrast with the labyrinthodont condition, with the main support being provided by the braincase which forms a stout longitudinal bar. In most genera, there is a large tympanum supported by a tympanic annulus set into a notch in the squamosal. This is almost certainly a primitive condition for frogs and is almost universally accepted as being directly derived from the pattern considered typical for labyrinthodonts.

Among the most striking specialization of frogs is the evolution of a larval stage that is radically different in structure and biology from the adult. In contrast with the high degree of specialization of the larval stage, anurans remain primitive in other aspects of reproduction for nearly all species practice external fertilization.

Urodela. Salamanders may be considered the least specialized of living amphibians in retaining conservative body proportions, with limbs present in most forms and with most of the power for locomotion resulting from waves of contraction of the axial musculature. Locomotor specialization within the group includes limb reduction and loss of the pelvic limbs in one superfamily, and elongation of the trunk. The skull of salamanders is modified in a manner analogous and possible homologous with that of frogs, with the main support provided by the braincase. Salamanders lack a tympanum. A short stapes is frequently present and sound is said to be transmitted from the substrate through the lower jaw to the inner ear by bone conduction. Both frogs and many salamanders have a second system for transmission of external vibrations to the inner ear, including a second ear ossicle, the operculum. This is an element forming part of the wall of the otic capsule that is attached via a muscle to the shoulder girdle. The complexity of the peculiar hearing system is a strong argument for the close relationship or common ancestry of frogs and salamanders.

The reproductive system of primitive salamanders presumably resembles that of Paleozoic amphibians, for external fertilization is the rule and the larvae are similar to the adult in general anatomy, except for the presence of external gills. In advanced forms, both aquatic and terrestrial, fertilization is internal. In the family Plethodontidae, many genera lay their eggs on land and development is direct, with the young miniature replicas of the adults, with no specifically larval features.

Apoda. Apodans are restricted to the damp tropics throughout the world. Most are aquatic or burrowing in habit. They are without trace of girdles or limbs; and although the trunk region may have more than 200 vertebrae, the tail is very short or absent. In contrast with frogs and salamanders, there are well-developed ribs. The skull roof is solidly ossified as in Paleozoic amphibians. The orbital opening is small, and may be completely covered with bone. A specialized tactile or chemosensory organ, the tentacle, extends out from the skull anterior to the orbit. Apodans all practice in-

ternal fertilization; in contrast with frogs and salamanders, the males possess a copulatory organ.

History. Fossils of frogs essentially like modern genera are known as early as the Lower Jurassic (Estes and Reig, 1973) and remains of fairly specialized salamanders from the Upper Jurassic (Hecht and Estes, 1960). The only fossil apodan is a single vertebra from the Paleocene (Estes and Wake, 1972) indicating an essentially modern pattern. The only fossil to suggest an earlier appearance of any of the modern orders is a single skeleton of *Triadobatrachus* from the Lower Triassic of Madagascar (Estes and Reig, 1973). The skull is frog-like, with a squamosal suggesting support of a tympanum. There are 14 trunk vertebrae and 6 caudals. The ilium is specialized in a manner like that of frogs, but to a lesser degree. This fossil is a plausible ancestor to frogs, but shows no specific affinities with any of the advanced families. Unfortunately,

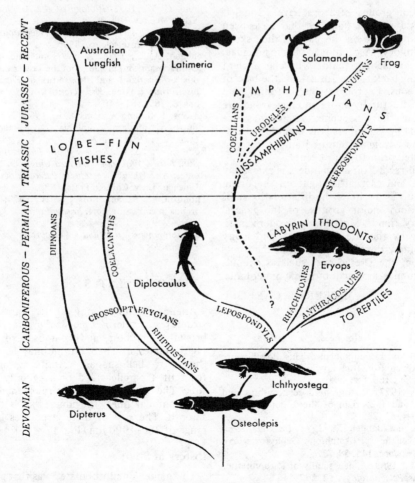

FIGURE 3. Evolution of the amphibians and the lobe-finned fishes (from Colbert, 1969; artwork prepared by Lois M. Darling).

Triadobatrachus provides no clues of affinities with any of the Paleozoic amphibian groups, or with apodans or salamanders.

The living amphibian orders must have been distinct from one another since at least the late Paleozoic (Fig. 3). In the absence of any known common ancestor, it is at least possible that the ancestors of the various orders are to be found among different Paleozoic groups. The specialized nature of the skeleton of salamanders and frogs separates them markedly from all Paleozoic groups, and it is reasonable to think that the definitive pattern was not established until at least the end of the Paleozoic. If apodans share a common ancestry with frogs and urodeles, the divergence of each order is presumably late in the Paleozoic. If the apodans (as suggested most recently by Carroll and Currie, 1975) have evolved separately from these groups, the Late Devonian or Early Carboniferous may mark the time of phylogenetic divergence.

Most major groups of frogs have a long, if spotty, fossil record. Families that have been considered primitive on the basis of their skeletal morphology—Ascaphidae, Discoglossidae, Pipidae, and Palaeobatrachidae—have an ancient record, going back to the Jurassic or the base of the Cretaceous. Leptodactylids, bufonids, and hylids, all structurally more modern frogs, are known from the Paleocene, where they are represented by members of modern species groups. The advanced family Ranidae has not been reported prior to the Eocene.

The fossil record of salamanders is extremely incomplete and casts little light on the origin of the group as a whole, the interrelationships of the various major groups, or the general evolutionary trends. Most of the modern families appear by the late Cretaceous, and many living genera by the Miocene. Most of the forms known from the Cretaceous, such as sirenids and amphiumids, are paedomorphic or neotenic.

ROBERT L. CARROLL

References

Bolt, J. R., 1969. Lissamphibian origins: Possible Protolissamphibian from the Lower Permian of Oklahoma, *Science,* **166,** 888–891.

Bolt, J. R., 1977. Dissorophoid relationships and ontogeny, and the origin of the Lissamphibia, *J. Paleontology,* **51,** 235–249.

Boy, J. A., and Bandel, K., 1973. *Bruktererpeton fiebigi* n. gen. n. sp. (Amphibia: Gephyrostegida), *Palaeontographica,* **145,** 39–77.

Carroll, R. L., 1969. A new family of Carboniferous amphibians, *Palaeontology,* **12,** 537–548.

Carroll, R. L., 1977. Patterns of amphibian evolution: An extended example of the incompleteness of the fossil record, in A. Hallam, ed., *Patterns of Evolution.* Amsterdam: Elsevier, 405–437.

Carroll, R. L., and Currie, P. J., 1975. Microsaurs as possible apodan ancestors, *J. Linnean Soc. London, Zool.,* **57,** 229–247.

Carroll, R. L., and Gaskill, P., 1978. The Order Microsauria, *Amer. Phil. Soc. Mem.,* **126,** 1–211.

Carroll, R. L.; Kuhn, O.; and Tatarinov, L., 1972, in O. Kuhn, ed., *Handbuch der Paläoherpetologie,* vol. 5B. Jena: Gustav Fischer Verlag, 81p.

Colbert, E. H., 1969. *Evolution of the Vertebrates.* N.Y.: Wiley, 535p.

Estes, R., and Reig, O., 1973. The early fossil record of frogs: A review of the evidence, in James L. Vail, *Evolutionary Biology of the Anurans.* Columbia: Univ. Missouri Press, 11–63.

Estes, R., and Wake, M., 1972. The first fossil record of caecilian amphibians, *Nature,* **239,** 228–231.

Hecht, M., and Estes, R., 1960. Fossil amphibians from Quarry Nine, *Peabody Mus. Nat. Hist., Yale Univ., Postilla,* **46,** 19p.

Holmes, R., and Carroll, R. L., 1977. A temnospondyl from the Mississippian of Scotland, *Bull. Mus. Comp. Zool. Harvard Univ.,* **147,** 489–511.

Noble, G. K., 1931. *Biology of the Amphibia.* New York: McGraw-Hill, 577p.

Olson, E. C., 1971. *Vertebrate Paleozoology.* New York: Wiley-Interscience, 839p.

Panchen, A. L., 1975. A new genus and species of anthracosaur amphibian from the Lower Carboniferous of Scotland and the status of *Pholidogaster pisciformis* Huxley, *Phil. Trans. Roy. Soc. London, Ser. B,* **269,** 581–640.

Parsons, T. S., and Williams, E. E., 1963. The relationships of modern Amphibia: A re-examination, *Quart. Rev. Biol.,* **38,** 26–53.

Romer, A. S., 1947. *Review of the Labyrinthodontia, Bull. Mus. Comp. Zool. Harvard Univ.,* **99,** 1–368.

Romer, A. S., 1966. *Vertebrate Paleontology,* 3rd ed. Chicago: Univ. Chicago Press, 468p.

Thomson, K. S., and Bossy, K. H., 1970. Adaptive trends in early amphibia, *Forma Functio,* **3,** 7–31.

Cross-references: *Reptilia; Vertebrata; Vertebrate Paleontology.*

ANELLOTUBULATES

Anellotubulates, "ring-tubers," are tubular structures of microscopic size, creamy white to brownish in color, and composed of a phosphatic material. Their outline may be trumpet shaped to bell shaped, vermiform, or in the form of a parallel-sided or irregularly swollen tube. The outer walls may be roughened or may show annular lines; the inner wall is typically smooth. The size of these structures is in the range of 200–1000 μm (Fig. 1).

History of Study

The name "anellotubulate" was first proposed in 1959 by Wetzel, in a paper published only in title. Two years later he mentioned the name

FIGURE 1. An anellotubulate ("*Mikrokalyx pullulans*" of Wetzel) obtained by chemical treatment of German Lower Jurassic sediments; X800 (after Wetzel, 1967).

incidentally in a second paper, in which a "worm-like tubular fragment, Form C" was illustrated. However, the first formal publication of this name and the first description of these microstructures did not appear until 1967 (Wetzel, 1967). Wetzel then proposed their classing as a Group Anellotubulata of undetermined affinity, comprising a single family, the Mikrocalycidae. All described types were considered as forms of a single genus and species, *Mikrocalyx pullulans*.[1]

Wetzel had recorded anellotublates only from marine sediments of the Lower Jurassic and Upper Cretaceous. However, McLachlan (1973) reported them also from nonmarine, even freshwater sediments of the Karroo Beds; and similar structures were subsequently obtained in preparations of other sediments of many different types, including Carboniferous nonmarine mudstones and contemporary deep-sea sediments.

This apparently wide stratigraphical range and lack of ecological discrimination by a group of supposed "microorganisms" so lately discovered caused disquiet among a number of micropaleontologists. In particular, it was noted that these structures, albeit of such fragile character, were well preserved in preparations in which more robust microfossils were poorly preserved; that, unlike other microfossils from

[1] The author has found no evidence of publication of this name in 1957, the date indicated in the title of Wetzel's 1967 paper.

the same samples, they were never found filled by sediment particles or secondary minerals; and that they occurred encrusting fragments of the rock and other microfossils rather than within the rock fragments themselves. Most disquieting of all was the fact that they appeared exclusively in samples prepared with hydrogen peroxide and were not found when other chemicals had been employed.

In a series of experiments conducted quite independently and almost simultaneously, three British and two Australian geologists discovered that "anellotubulates" were produced artificially by treating pyrite-rich sediments with commercial-grade hydrogen peroxide (Richardson et al., 1973; Picket and Scheibnerova, 1974). The British scientists observed the following sequence of events:

Bubbles accumulated on the shale, originating on and below the surface, eventually leading to streams of bubbles as the reaction proceeded. After a period varying from a few minutes to several hours a transparent to white gel-like deposit appeared around the edges of streaming bubbles. Where bubble streaming was persistent, we observed that the deposit formed a tubular growth around and parallel to the bubble stream, the internal diameter of the tubes being directly related to that of the emergent bubbles, and the original gel-like deposit appeared to harden around the bubble stream, enabling the deposit to build up into a tube-like growth. As the tube hardened at its base, constricting the diameter of the bubbles, the softer, upper portion of the tube expanded, corresponding to the increasing circumference of the bubbles. Trumpet-shaped tubes, formed by this process, were common. Vermiform tubes of an even diameter were also produced. . . . Variations in the quantity of gas produced, and in bubble diameter, resulted in tubes of variable diameter. The hardening of the tubes, while still immersed in the reagent, was progressive and after several hours they were quite brittle. (Richardson et al., 1973, p. 347)

It was found that hydrogen peroxide of analytical grade produced no such structures; they developed only when phosphates—often used as stabilizing agents in this reagent—were present, the Fe^{+++} ions being precipitated as iron phosphate. Indeed, sodium hexametaphosphate, added to hydrogen peroxide, produced abundant tube growth.

Since both Wetzel and McLachlan had specifically mentioned their use of commercial-grade hydrogen peroxide in sample preparation (the single apparent exception in McLachlan's report may be discounted; see Picket and Scheibnerova, 1974, p. 100), it is clear that these supposed "microfossils" are in fact inorganic artefacts produced during sample preparation. Thus, while the name "anellotubulate" is possibly of some utility for designating such inorganic artefacts, the supposed "Group Anellotubulata" and its constituent genus and species must be abandoned.

WILLIAM A. S. SARJEANT

References

McLachlan, I. R., 1973. Problematic microfossils from the Lower Karroo Beds in South Africa, *Palaeontol. Africana,* **15**, 1–21.

Picket, J., and Scheibnerova, V., 1974. The inorganic origin of "anellotubulates," *Micropaleontology,* **20**, 97–102.

Richardson, G., Gregory, D., and Pollard, J., 1973. Anellotubulates are manufactured "microfossils," *Nature,* **246**, 347–348.

Wetzel, O., 1967. Rätselhafte Mikrofossilien des Oberlias (ϵ); Neue Fünde von "Anellotubulaten" O. We. 1957, *Neues Jahrb. Geol. Paläontol., Abh.,* **128**, 341–352.

Cross-reference: *Pseudofossil.*

ANGIOSPERMAE

The angiosperms are plants bearing reproductive structures organized as flowers. They constitute the major element of the modern flora, with some 220,000 species distributed among 12,000 genera and 330 families. The characteristic feature of the angiosperms is the flower which, phylogenetically, represents a vegetative shoot modified for reproduction (Fig. 1).

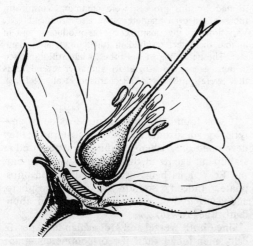

FIGURE 1. Drawing of angiosperm flower. The pistil in the center of the flower is the female structure in angiosperm reproduction and consists of the terminal stigma, the slender style, and the basal ovary in which the ovules are borne. The stamens surround the pistil and consist of the anther bearing the pollen grains, and the elongated filament. The petals collectively constitute the corolla and the sepals form the calyx.

According to this view, proposed by the German philosopher Goethe in 1790, the sepals, petals, stamens, and pistil are modified leaves borne on an axis with shortened internodes. The product of reproduction in the angiosperms, as in the gymnosperms, is the seed; but in the flowering plants, this structure is borne within a fruit, as opposed to the gymnosperm seed, which is exposed on the surface of the cone scale.

Anatomically the angiosperms are characterized by vessel elements in the xylem; sieve-tube elements in the phloem; and fibers, which may occur throughout the vascular system of the plant. The vessel elements are cells with open or perforated end walls arranged end to end to form long tubes (vessels), which function in the upward transport of water and mineral nutrients from the soil to the site of photosynthesis in the leaves. The sieve-tube elements are elongated cells with minute thin areas on the end walls clustered into structures called sieve plates. The sieve plates of adjacent cells overlap and the connected elements form sieve tubes, which serve in the downward transport of manufactured photosynthetic products from the leaves to storage areas in the stem and root. Fibers are elongated thick-walled cells that function in support of the plant. Although vessels, sieve tubes, and fibers are not unique to the angiosperms, they are characteristic and most highly developed in the flowering plants. For further discussion of angiosperm anatomy and morphology, the treatments of Esau (1965), Eames (1961), and Sinnott (1960) may be consulted.

Reproduction

The essential parts of the angiosperm flower are the stamens and pistil. The stamen is the male element in angiosperm reproduction and typically consists of a filament bearing a terminal anther. The anther is divided into two halves, each with two cavities (microsporangia) that contain diploid (2n) microspore mother cells. These cells undergo meiosis to produce pollen grains with 2 haploid (1n) sperm nuclei. The pistil is the female part of the flower and consists of a stigma, style, and ovary. The ovules are borne within the ovary and each contains an egg cell (1n) resulting from meiosis of the 2n megaspore mother cell.

For fertilization to occur, the sperm-bearing pollen must land on the stigma of the pistil. In some species pollen may germinate on the stigma of the same flower (self-compatible species), while in others (self-incompatible species) the pollen must be transported to another flower (cross pollination). The transporting vector may be wind, water, insects, birds, or bats and the pollen and flower are often modified to facilitate pollination by a particular vector (for discussion of pollination mechanisms see Faegri and van der Pijl, 1966; Grant and Grant, 1965). The fact that pollination attracts insects, some of which may also feed on the vegetative parts of the plants, has resulted in the evolution of complex defense mechanisms in the form of secondary chemical compounds (e.g., phenols, terpenoids, alkaloids, nitriles, mustard oils, ketones). In turn, some insects have developed enzyme systems whereby these compounds may be digested and used as a source of energy, thus converting a plant defense mechanism into an attractant. Such interactions between plants and animals are termed coevolution and have recently been recognized as important in the speciation of both groups (Ehrlich and Raven, 1964; Macior, 1971; Janzen, 1966, 1974).

Once the pollen has landed on the stigma of a compatible species, a pollen tube containing the sperm nuclei grows through the stigma to the ovules. The sperm are released and one fertilizes the egg to form the diploid (2n) zygote, which develops into the embryo, and the other unites with two nuclei of the ovule to form the triploid endosperm nucleus. The latter nucleus often develops into the endosperm, a nutrient tissue for the embryo during early stages of seed germination. This process of "double fertilization" is a common feature of the angiosperms. The wall (integument) of the ovule hardens to form the seed coat; the ovary enlarges to form the fruit; and the seed then usually enters a period of dormancy. An angiosperm seed may thus be defined as an embryo surrounded by a nutritive layer (the endosperm) and a protective layer (the seed coat), enclosed in a fruit, and in a dormant condition. The fruits of many species are provided with spines or sticky coverings that attach to passing animals; others may be impervious to marine waters or possess fleshy outer coverings that may be eaten by birds and other animals, the seeds passing unharmed through the digestive tract of the animal. These morphological features provide an effective dispersal potential for the propagules (van der Pijl, 1972) and are an important factor in plant migration and evolution (e.g., see Mayr, 1963, for Founder Principle). The success of the angiosperms in the modern flora is due at least in part to greater genetic variability resulting from cross pollination and the evolutionary development of the seed and fruit.

Classification

Several systems have been proposed for the classification of plants. According to one, the presence of vascular tissue places the angiosperms in the Division Tracheophyta. Within the Tracheophyta there are four subdivisions: Psilopsida, Lycopsida, Sphenopsida, and Pteropsida. The last subdivision includes plants having leaf gaps and megaphyllous leaves. A leaf gap is an area of the vascular cylinder devoid of conducting tissue as a result of departure of strands of xylem and phloem to the leaf. In a megaphyllous leaf this strand ramifies after it enters the leaf to form the vascular pattern (veins) of the blade, in contrast to the microphyllous leaf which lacks leaf gaps and has a single unbranched vein. Leaves of the Psilopsida, Lycopsida, and Sphenopsida are microphyllous, and those of the Pteropsida are megaphyllous. Within the Pteropsida are the Classes Filicineae (ferns), which lack seeds; the Gymnospermae (conifers, evergreens, cone-bearing plants), which bear seeds exposed on the surface of cone scales; and the Angiospermae (flowering plants), in which the seeds are enclosed in the fruit.

Further classification of the angiosperms into Orders and Families is currently an unsettled field of active research. With the publication in 1940 of Huxley's, *The New Systematics,* the multidisciplinary approach to angiosperm classification and phylogeny became standard procedure. Results of these various studies have been published for the fields of numerical taxonomy (Sneath and Sokal, 1973) and related computer techniques (cf. *Taxon,* 23(1), 1974), biochemical systematics (Alston and Turner, 1963; Gibbs, 1974), pollen morphology and plant taxonomy (Erdtman, 1952), electron microscopy with reference to systematics (Heywood, 1971), phytogeography and taxonomy (Valentine, 1972), and cytology (Stebbins, 1971). These modern approaches are summarized by Heywood (1968; cf. also *Taxon,* 20(1), 1971).

This new information has been incorporated into four recent attempts to classify flowering plants according to natural or evolutionary relationships, by Cronquist (1968), Hutchinson (1959), Takhtajan (1969), and Thorne (1968). According to most modern authors, the "woody Ranales" (Subclass Magnoliidae of Cronquist, 1968) constitute the ancestral group for dicotyledonous angiosperms. In earlier classification systems, the Amentiferae (plants with inflorescences of lax, cone-like structures, as oaks, birch) were considered primitive because of the simple structure of the flowers. Recent studies have shown this group to be simple through reduction in response to wind pollination, rather than primitive, and they no longer occupy a basal position in angiosperm phylogeny (see *Brittonia* 25(4),(1973).

Collectively, members of the Magnoliidae illustrate certain primitive features, as numerous, separate floral parts spirally arranged, woody habit, absence of vessels, monocolpate pollen without well-defined columellae in the exine, petaloid stamens, broad stigmatic surface, and conduplicate ("unsealed") carpels. Examples of primitive angiosperm families are Magnoliaceae (e.g., *Magnolia*), Austrobaileyaceae, Winteraceae, Degeneriaceae, Annonaceae, and Illiciaceae. The Compositae (represented by *Aster,* sunflower) is regarded as the most advanced dicot family. Among the monocots (Class Liliatae of Cronquist, 1968) the Subclass Alismatidae is considered primitive and the Subclass Lillidae (e.g., lilies, orchids) most advanced. Evidence presently available favors early derivation of the monocots from Magnoliidae dicots.

Distribution

In form and structure, ecological amplitude, and distribution the angiosperms are the most diverse of the plant groups. They occur in marine and fresh waters, and on land surfaces ranging in altitude from below sea level to over 4500 m (15,000 ft), where temperatures in some areas may be as low as –45°C (–50°F) and in others as high as 52°C (125°F), and under a rainfall regimen of 12.5–1525 cm (5–600 in.) per year. The group includes the duckweeds (*Lemna* sp.), < 6 mm (.25 in.) in length, as well as the *Eucalyptus* of the Australian coast, which are over 90 m (300 ft) in height, and *Rafflesia,* a genus of Malaysian plants with flowers up to .9 m (3 ft) in diameter. With the exception of a few sites dominated locally by conifers, lichens, or bryophytes, the angiosperms are of cosmopolitan distribution (see *Paleophytogeography*).

Geologic History

The angiosperms first appear in beds of late Early Cretaceous age, and by Middle to Late Cretaceous times had become the dominant land plants. This rapid increase in numbers and diversity suggests the angiosperms had a long pre-Cretaceous history. There are numerous reports of late Paleozoic and early Mesozoic plant structures (wood, pollen, seeds, leaves) thought to represent angiosperms, but Scott et al. (1960, 1972) have reviewed these reports and conclude there is little or no direct paleobotanical evidence for the presence of angiosperms in pre-Cretaceous sediments.

A large amount of material has been published on the origin of angiosperms (for recent views, see Hughes, 1976; Beck 1976), and almost every plant group has been proposed at one

time or another as a possible ancestor for flowering plants. The cycadeoid inflorescence, in its reconstruction from fossil remains, shows superficial resemblance; but recent studies have shown it to be basically different in its structure and organization. Similarly, the Caytoniales, Triassic representatives of the Paleozoic seed ferns, were thought to be ancestors of the angiosperms; but later investigations have shown this to be unlikely. Paleobotanical investigations alone will probably not reveal the ancestral stock of flowering plants, but must be supplemented by data from the taxonomy, morphology, genetics, and distribution of living plants. In the meantime, the problem remains as described by Darwin in a letter to Sir Joseph Dalton Hooker in 1879: "The rapid development, so far as we can judge, of all the higher plants within Recent geological time is an abominable mystery. I would like to see the whole problem solved."

ALAN GRAHAM

References

Alston, R. E., and Turner, B. L., 1963. *Biochemical Systematics.* Englewood Cliffs, N.J.: Prentice-Hall, 404p.

Beck, C. B., ed., 1976. *Origin and Early Evolution of Angiosperms.* N.Y.: Columbia Univ. Press, 341p.

Cronquist, A., 1968. *The Evolution and Classification of Flowering Plants.* Boston: Houghton Mifflin, 396p.

Doyle, J. A., 1977. Patterns of evolution in early angiosperms, in A. Hallam, ed., *Patterns of Evolution.* Amsterdam: Elsevier, 501–546.

Eames, A., 1961. *Morphology of the Angiosperms.* New York: McGraw-Hill, 518p.

Ehrlich, P. R., and Raven, P. H., 1964. Butterflies and plants: A study in co-evolution, *Evolution,* **18,** 586–608.

Erdtman, G., 1952. *Pollen Morphology and Plant Taxonomy—Angiosperms.* Stockholm: Almqvist & Wiksell, 539p.

Esau, K., 1965. *Plant Anatomy.* New York: Wiley, 767p.

Faegri, K., and van der Pijl, L., 1966. *The Principles of Pollination Ecology.* Oxford: Pergamon Press, 248p.

Gibbs, R. D., 1974. *Chemotaxonomy of Flowering Plants.* 4 vols. Montreal: McGill-Queen's Univ. Press.

Grant, V., and Grant, K. A., 1965. *Flower Pollination in the Phlox Family.* New York: Columbia Univ. Press, 180p.

Heywood, V. H., ed., 1968. *Modern Methods in Plant Taxonomy.* New York: Academic Press, 312p.

Heywood, V. H., ed., 1971. *Scanning Electron Microscopy—Systematic and Evolutionary Applications.* New York: Academic Press, 331p.

Hughes, N. F., 1976. *Palaeobiology of Angiosperm Origins.* Cambridge: Cambridge Univ. Press, 216p.

Hutchinson, J., 1959. *The Families of Flowering Plants.* 2 Vols. London: Oxford Univ. Press, 792p.

Huxley, J., ed., 1940. *The New Systematics.* London: Oxford Univ. Press, 583p.

Janzen, D. H., 1966. Coevolution of mutualism between ants and acacias in Central America, *Evolution,* **20,** 249–275.

Janzen, D. H., 1974. Swollen-Thorn Acacias of Central America, *Smithsonian Contrib. Bot.,* **13,** 131p.

Macior, L. W., 1971. Co-evolution of plants and animals—Systematic insights from plant-insect interactions, *Taxon,* **20,** 17–28.

Mayr, E., 1963. *Animal Species and Evolution.* Cambridge, Mass.: Harvard Univ. Press, 797p.

Scott, R. A., Barghoorn, E. S.; and Leopold, E. B., 1960. How old are the angiosperms? *Amer. J. Sci.,* **258,** 284–299.

Scott, R. A.; Williams, P. L.; Craig, L. C.; Barghoorn, E. S.; Hickey, L. J.; and MacGintie, H. D., 1972. "Pre-Cretaceous" angiosperms from Utah: Evidence for Tertiary age of the palm woods and roots, *Amer. J. Bot.,* **59,** 886–896.

Sinnott, E. W., 1960. *Plant Morphogenesis.* New York: McGraw-Hill, 550p.

Sneath, P., and Sokal, R. R., 1973. *Numerical Taxonomy.* San Francisco: Freeman, 573p.

Stebbins, G. L., 1971. Chromosomal evolution in higher plants. London: Edward Arnold, 216p.

Takhtajan, A., 1969. *Flowering Plants, Origin and Dispersal.* Washington, D.C.: Smithsonian Inst. Press, 310p.

Thorne, R. F., 1968. Synopsis of a putatively phylogenetic classification of the flowering plants, *Aliso,* **6,** 57–66.

Valentine, D. H., ed., 1972. *Taxonomy, Phytogeography and Evolution.* New York: Academic Press, 446p.

Van der Pijl, L., 1972. *Principles of Dispersal in Higher Plants.* Berlin: Springer-Verlag, 162p.

Walker, J. W., et al., 1975. The bases of angiosperm phylogeny, *Ann. Missouri Bot. Garden,* **62,** 515–834.

Cross-references: *Paleobotany; Paleophytogeography; Palynology, Tertiary.*

ANNELIDA

The annelids are the most diverse and abundant members of almost all soft-bottom marine communities. The main representatives of this phylum of segmented worms are earthworms, polychaetes, and leeches. As earthworms tunnel in the soil, most polychaetes burrow in marine mud and leeches seek a living host. The fossil record is almost barren of earthworms and leeches. They leave no hard parts and almost always live in freshwater or terrestrial environments, where the chances of fossilization are poorer than in marine conditions.

Polychaetes have the best fossil record and therefore are of most interest to paleontologists. Many species of these worms have hard parts, either chitinous jaws or mineralized tubes, which are readily fossilized. In addition, they live in areas of near-shore sedimentation, where eventual recovery of fossils is most likely. But most polychaete fossils are jaws, and these are easily overlooked. They are dark

and tiny, usually <1 mm long; special techniques are necessary to concentrate them from limestones and silicates. Only recently have paleontologists come to recognize the potential of fossil polychaete jaws as aids to stratigraphic correlation (see *Scolecodonts*).

As paleontologists have overlooked polychaetes in the fossil record, neontologists have similarly ignored polychaetes in Holocene marine communities, for the animals themselves, like their fossils, are easily not seen. Most are tiny and timid; not surprising, since they are meat to many predators—fish, crustaceans, other polychaetes, etc. Their main defense has been inconspicuousness; almost all are infaunal. Collectors of polychaetes go armed with shovel and screen for sieving their prey from the sand and mud in which they live.

Classification

There are two minor classes of annelids and three major classes. The minor classes are the Archiannelida, a group of tiny marine worms found between sand grains and in mud and splash pools; and the Myzostomia, a small group of highly specialized parasites on crinoids and brittle-stars. The three major classes are the Polychaetea; the Oligochaeta, which includes the earthworms; and the Hirudinea, or leeches. The polychaetes are probably the ancestors of the freshwater oligochaetes. The oligochaetes, in turn, gave rise to the leeches, the most specialized class.

Polychaetes are the most diverse class with over 5000 species that range in size from <2 mm to 3 m. The "ideal" polychaete has the most primitive features among the major annelid classes: a prostomium, or cephalic lobe, bearing tentacles, eyes, and a pair of palps; a mouth lying between the prostomium and the first body segment; an extensible proboscis armed with jaws; and a series of identical body segments, each with a pair of fleshy parapodia stiffened with aciculae and bearing two bundles of setae as well as cirri and gills. There is a terminal lobe, the pygidium, which bears the anus and one or two anal cirri.

The free-living, predaceous polychaetes have evolved little from this primitive plan. The sessile polychaetes, dwelling in tubes and permanent burrows, have undergone the greatest evolution. In most cases they have lost internal segmentation and all but traces of parapodia, and have increased specialization of body regions.

Traditionally, the polychaetes have been divided into two subclasses or orders: the free-living polychaetes were placed in the Errantia and the sessile polychaetes in the Sedentaria.

This is a convenient but artificial classification, not reflecting phylogeny. A recent classification (Dales, 1962) groups polychaetes into orders on the basis of elaboration of the stomodeum—an ectodermal evagination that forms the anterior part of the digestive tract.

The Oligochaeta contain about 3000 species. Although earthworms are most familiar, more species of oligochaetes live in fresh water than in soil, and a few are marine. They evolved from the polychaetes by the loss of parapodia and all but a few setae.

There are only about 300 species of leeches, although they are often very abundant in fresh water. A few species are adapted to moist ter-restrial conditions, estuaries, or the sea. Leeches are epiparasites. They have lost all setae as well as internal septa and developed a rasping mouth in an anterior sucker. A posterior sucker attaches to the host and aids locomotion.

Polychaete Ecology

Polychaetes may be classified by feeding type: predaceous, omnivorous, suspension feeders filtering food particles from the water, nonselective deposit (sediment) feeders, or selective deposit feeders. They may also be classified by habitat as planktic, epifaunal, or infaunal (Fig. 1).

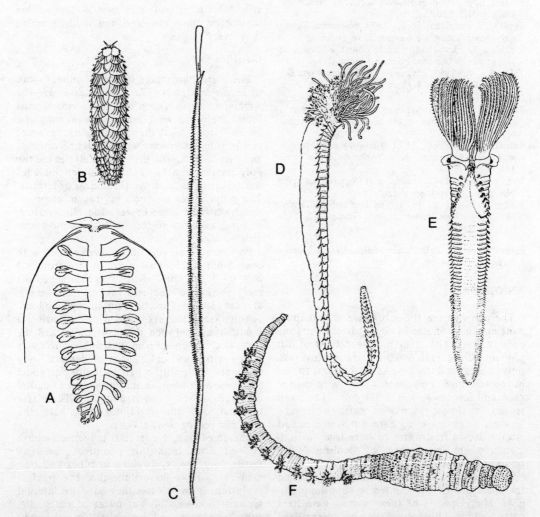

FIGURE 1. Recent polychaetes of various feeding types (from Day, 1967; reproduced with permission of The British Museum [Natural History]). A, *Tomopteris elegans,* a planktonic predator or omnivore; B, *Pontogenia chrysocoma,* an epifaunal omnivore; C, *Goniada emerita,* an infaunal predator drawn with the proboscis everted; D, *Nicolea macrobranchia,* a tubiculous selective deposit feeder that attaches its tube to rocks; E, *Branchiomma violacea,* a tubiculous epifaunal suspension feeder; F, *Abarenicola affinis,* an infaunal nonselective deposit feeder. (Not drawn to the same scale.)

Planktic species are always predaceous or omnivorous. The epifaunal polychaetes are either free-living, in which case they will be predaceous or omnivorous, or tube-dwelling, in a permanent tube attached to a firm substrate. Tubed epifaunal polychaetes are usually suspension feeders, collecting small part̶ ̶s of food from the water column, but s̶ ̶cies of predaceous polychaetes also ̶ ̶ ̶ ̶e ̶ ̶ ̶ ̶nal tubes and snatch prey wi̶t̶ ̶ ̶ ̶ ̶ ̶ ̶ ̶ and eversible probosces (Ju̶ ̶ ̶ ̶ ̶ ̶ ̶ ̶r̶ ̶auchald, 1977).

Some infaunal polychaetes live in permanent tubes or burrows and filter water or use tentacles to select sediment particles from the surface. In ecosystems where sedimentation rate is high, permanent living structures would be disadvantageous, so species here tend to move freely through the sediment or have temporary tubes or burrows. They may be predaceous or selective or nonselective deposit feeders. In most soft-bottom ecosystems, sedimentation rate is high and nonselective deposit feeders dominate (Sanders, 1956). The process of thousands or millions of worms eating the sediment and digesting out organic material profoundly changes the physical structure and chemistry of the sediments and influences the geologic record (Rhoads and Young, 1970).

Nothing is known about the food consumed by many species of polychaetes and even by entire families (see "Biological Notes" for polychaete families in Day, 1967). In addition, little is known about the functional morphology of this group. This lack of information about living polychaetes hinders understanding of polychaete fossils and restricts their usefulness in paleoecology. Increased research in feeding behavior and functional morphology of living polychaetes would not only aid paleontologists but increase under-standing of present marine communities where polychaetes play a dominant role.

The Fossil Record of Annelids

Annelids may be preserved as whole-body fossils under unusual conditions which prevent decay of soft tissue. A few oligochaetes and leeches have been preserved in this way (Howell, 1962). Whole-body fossils are more common for polychaetes, due to their longer history and greater abundance in marine ecosystems (Figs. 2, 3). In addition, many polychaetes secrete mineralized tubes or have chitinous jaws and these hard parts are less rare in the fossil record (Figs. 4, 5).

Jaws, or scolecodonts (q.v.), are the most common polychaete fossils. They range in size from 50 μm to several mm. They are similar to conodonts (it has been suggested that conodonts are the jaws of annelids) but can be distinguished by the presence of a myocoele, or pulp cavity. Scolecodonts are found on bedding planes or can be concentrated from limestones and silicates by dissolving the matrix in acid.

The first polychaete fossils are probably tracks and burrows in the Precambrian, but the first unambiguous polychaetes are whole-body fossils from the Ediacara fauna of Australia, 680 m yr old (Glaessner, 1961). The Ediacara polychaetes (Fig. 2) resemble some Holocene free-living forms and were probably predaceous or omnivorous—suggesting this is the primitive ecological role of polychaetes.

Except for calcified tubes (Glaessner, 1976), there is a gap in the annelid record until the Middle Cambrian, when rapid burial preserved soft-bodied animals in the Burgess Shale of Canada. Several species of polychaetes have been described but some of these probably be-

FIGURE 2. Polychaetes from the Ediacara fauna of South Australia. Left, *Dickinsonia costata,* 6 cm long; right, *Spriggina floundersi,* 3.5 cm long.

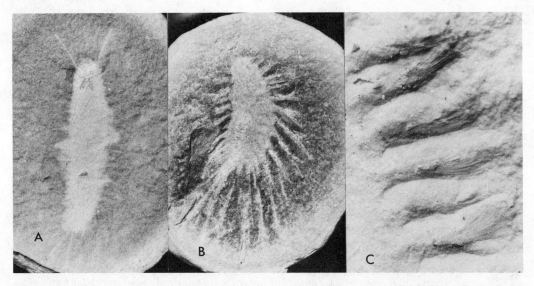

FIGURE 3. Polychaetes from the Middle Pennsylvanian of Essex, Illinois. A, *Fossundecima konecnyi,* 26 mm long, with tentacles, anal cirri, and a pair of large jaws; B, 18 mm long, *Rhaphidiophorus hystrix* with very long, heavy setae; C, 4.5 mm long, Parapodia of *Rutellifrons wolfforum* showing ventral setal bundles; parapodia, and setae.

long to other phyla (Walcott, 1911). The Burgess Shale polychaetes are distinguished by elaborate setae modified as spines and scales (Fig. 6). The setae may have been protection from predation for the polychaetes, which were probably epifaunal predators and omnivores.

Beginning in the Ordovician, scolecodonts

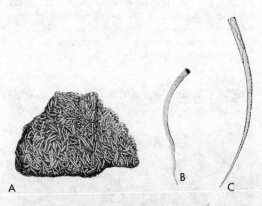

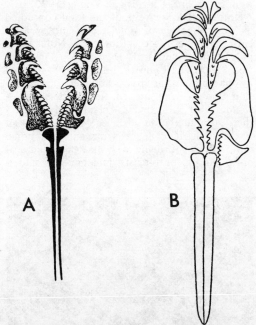

FIGURE 4. Fossil polychaete tubes (from Howell, 1962, courtesy of The Geological Society of America and University of Kansas Press). A, *Serpulites coacervatus,* Mesozoic of Germany—the slab is 3 cm high; B, *Tubulelloides gracilis,* Ordovician, New York, 4.5 cm long; C, *Tubulella flagellum* Cambrian, British Columbia, 6.5 cm long.

FIGURE 5. Jaw apparatuses. A, *Aglaurides fulgida,* Recent of South Africa (from Day, 1967; reproduced with permission of The British Museum [Natural History]). B, *Atraktoprion cornutus,* Ordovician of Poland (from Kielan-Jaworowska, 1966).

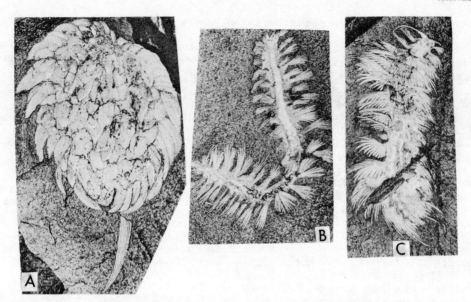

FIGURE 6. Polychaetes from the Burgess Shale, Middle Cambrian, British Columbia (from Walcott, 1911). A, *Wiwaxia corrugata*, 2.5 cm long; B, *Canadia setigera*, 2 cm long; C, *C. spinosa*, 2.5 cm long.

appear and are abundant. Most of these probably belong to the superfamily Eunicea, a group of free-living polychaetes possessing elaborate jaw apparatuses with many separate elements. These jaws must represent a specialization for predation (Fig. 5).

Few whole-body fossils are available until the Middle Pennsylvanian Essex fauna of Illinois. The rapid formation of siderite concretions around polychaetes appears to have halted decay and such delicate features as tentacles and gills are preserved. There are 15–20 species; some are remarkably similar to living genera; others have been placed in separate families (Thompson and Johnson, 1976; Thompson, 1976). The Essex polychaetes (Fig. 3) are remarkably abundant. Thousands of specimens have been collected from coal strip-mining areas.

This abundance and diversity has permitted strong inferences about the role of polychaetes in soft-bottomed communities near the end of the Paleozoic, which demonstrates how useful polychaetes can be in understanding geologic history. The Essex polychaetes were dominantly epifaunal or shallowly burrowing predators or omnivores and lived on and in a soft, muddy bottom. In a similar ecosystem today, polychaetes would be dominantly infaunal deposit feeders. The evolutionary reasons for this difference are unclear. Deposit feeders had evolved in Pennsylvanian time, but were very rare. This may indicate a stable food supply in

late Paleozoic ecosystems, which would have allowed polychaetes to ignore sedimentary organic material in favor of the more concentrated food source of whole plants and animals. Perhaps only with ecosystem change at the end of the Paleozoic did deposit feeding become advantageous.

Another hypothesis to explain the predominance of predation in late Paleozoic polychaetes is a scarcity of fish predators. The fish in the Essex fauna were generally smaller than the polychaetes. With the rise of teleosts in the Mesozoic and their specializations for bottom feeding and exploitation of polychaetes, the polychaetes may have been forced to change their role from predators to infaunal deposit eaters. The Mesozoic does show a drop in the abundance of scolecodonts; this may mean a drop in the abundance of predatory polychaetes and possibly an increase in the proportion of polychaete suspension and deposit feeders.

The Mesozoic yields a few whole-body polychaete fossils in the Triassic Buntsandstein of France (Gall and Grauvogel, 1967) and the Jurassic Solenhofen of Germany (Ehlers, 1868). Polychaete fossils continue to be rare in the Tertiary.

Problems and Possibilities

Annelid whole-body fossils will probably always show a patchy distribution in both space and time. There may not exist another such locality as Essex, Illinois, where polychaetes are both diverse and abundant. However, scolecodonts are ubiquitous and abundant in the Paleozoic, at least from the Ordovician through the Devonian. The dearth of fossils from other periods may reflect a collection bias rather than the true state of distribution through time. Yet scolecodonts have not been used for stratigraphic correlation or for environmental reconstruction. Scolecodonts differ from most other microfossils in that they are not destroyed by heat, acid, or recrystallization of the matrix and can be readily concentrated by dissolving the matrix in acid. Thus scolecodonts are of particular potential value to the paleontologist.

Jansonius and Craig (1971) think that the evolution of jaw apparatuses was rapid enough to permit scolecodont use in stratigraphic correlation. But first a ubiquitous taxonomic problem must be solved for the polychaetes. Paleontologists are often forced to resort to a dual taxonomic system for lack of evidence, assigning a specimen to one genus and what may be its track of coprolite to a different genus. Similarly, sometimes a polychaete's jaw apparatus is assigned to one genus and an isolated scolecodont to another—even though the actual biological species may be the same. A larger set of descriptions of complete polychaete jaw apparatuses will permit the gradual discard of elements of the dual taxonomic system for polychaetes.

The prospects for using scolecodonts for correlation improve with the continuing recognition that scolecodonts are present in many strata where they were unexpected (Charletta and Boyer, 1974), and with collecting techniques that allow for the recovery of whole apparatuses. And because of increasing interest in the structure and function of modern marine ecosystems, more is being learned about living polychaetes. This knowledge can be applied to fossil polychaetes and aid in the understanding of their evolution and ecology.

IDA THOMPSON

References

Boyer, P. S., 1975. Polychaete jaw apparatus from the Devonian of central Ohio, *Acta Palaeontol. Polonica*, 20,(3), 184–205.

Charletta, A. C., and Boyer, P. S., 1974. Scolecodonts from Cretaceous greensand of the New Jersey coastal plain, *Micropaleontology*, 20, 354–366.

Dales, R. P., 1962. The polychaete stomodeum and the inter-relationships of the families of Polychaeta, *Proc. Zool. Soc. London*, 139, 389–428.

Dales, R. P., 1967. *Annelids*. London: Hutchinson, 200p.

Day, J. H., 1967. *A monograph on the Polychaeta of Southern Africa*. London: British Museum (Natural History), 878p.

Ehlers, E., 1868. Ueber fossile Würner aus dem lithographischen Schiefer in Bayern, *Palaeontographica*, 17, 145–175.

Eisenack, A., 1975. Beiträge zur Anneliden-Forschung, I, *Neues Jahrb. Geol. Paläontol. Abh*, 150, 227–252.

Gall, J. and Grauvogel, L., 1967. Faune du Buntsandstein III. Quelques annélides du Grès à Voltzia des Vosges, *Ann. Paleontol. Invertebr.*, 53(2), 105–110.

Glaessner, M. F., 1961. Pre-Cambrian animals, *Sci. American*, 204, 72–78.

Glaessner, M. F., 1976. Early Phanerozoic annelid worms and their geological and biological significance, *J. Geol. Soc. London*, 132, 259–275.

Glaessner, M. F., and Wade, M., 1966. The late Precambrian fossils from Ediacara, South Australia, *Palaeontology*, 9, 599–628.

Hartmann-Schröder, G., 1967. Feinbau und Funktion des Kieferapparates der Euniciden am Beispiel von *Eunice (Palola) siciliensis* Grube (Polychaeta), *Mitt. Hamburg. Zool. Mus. Inst.*, 64, 5–27.

Howell, B. F., 1957. Vermes, in H. S. Ladd, ed., Treatise on marine ecology and paleoecology vol. 1, Paleoecology, *Geol. Soc. Amer. Mem.*, 67, vol. 2, 805–816.

Howell, B. F., 1962. Worms, in R. C. Moore, ed., *Treatise of Invertebrate Paleontology*. Pt. W, Miscellanea, Lawrence, Kansas: Geol. Soc. Amer. and Univ. Kansas Press, W144–W177.

Jansonius, J., and Craig, J. H., 1971. Scolecodonts: I. Descriptive terminology and revision of systematic nomenclature; II. Lectotypes, hew names for homonyms, index of species, *Bull. Canadian Petroleum Geol.*, 19, 251–302.

Jumars, P. A., and Fauchald, K., 1977. Between-community contrasts in successful polychaete feeding strategies, in B. C. Coull, ed., *Ecology of Marine Benthos. Belle Baruch Libr. Mar. Sci. 6.* Columbia: Univ. South Carolina Press, 1–20.

Kielan-Jaworowska, Z., 1966. Polychaete jaw apparatuses from the Ordovician and Silurian of Poland and a comparison with modern forms, *Palaeontol. Polonica*, 16, 1–152.

Kielan-Jaworowska, Z., 1968. Scolecodonts versus jaw apparatuses, *Lethaia*, 1, 39–49.

Kozur, H., 1970. Zur Klassifikation und phylogenetischen Entwicklung der fossilen Phyllodocida und Eunicida (Polychaeta), *Freiberger Forschungshefte*, C260 *Paläontologie*, 35–81.

Kozur, H., 1971. Die Eunicida und Phyllodocida des Mesozoikums, *Freiberger Forschungshefte*, C267, *Paläontologie*, p. 73–111.

Rhoads, D. C., and Young, D. K., 1970. The influence of deposit-feeding organisms on sediment stability and community trophic structure, *J. Marine Research*, 28, 150–178.

Sanders, H. L., 1956. Oceanography of Long Island Sound, 1952–1954. X. The biology of marine bottom communities. *Bull. Bingham Oceanogr. Collect.* 15, 345–414.

Schwab, K. W., 1966. Microstructure of some fossil and Recent scolecodonts, *J. Paleontology*, 40, 416–423.

Szaniawski, H., 1974. Some Mesozoic scolecodonts congeneric with Recent forms, *Acta Palaeontol. Polonica*, 19, 179–200.

Thompson, I., 1976. Nine new species of whole-body fossil polychaetes from Essex Illinois, *Fieldiana, Geol.,* **26,** 146–183.

Thompson, I., and Johnson, R. G., 1976. New fossil polychaete from Essex Illinois, *Fieldiana, Geol.,* **26,** 122–146.

Walcott, C. D., 1911. Cambrian geology and paleontology, II, no. 5, Middle Cambrian annelids, *Smithsonian Misc. Collect.,* **57,** 109–144.

Cross-references: *Burgess Shale; Ediacara Fauna; Mazon Creek; Scolecodonts.*

APTYCHUS, ANAPTYCHUS

The term *aptychus* is from the Greek meaning a body apparently folded into two parts. Aptychi are symmetrical pairs of calcareous plates which in vivo were in contact along their hinge-like straight line of symmetry. They are often found close to or inside the shells of ammonoid cephalopods, and sometimes even closing the aperture of the shell.

Superficially, aptychi resemble bivalve shells, in that their concave surface is smooth and the convex surface is ornamented in various ways: granulated, ribbed, furrowed, or punctate (Fig. 1). Between the outer and inner layers, a middle layer with a cellular or tubular structure may be found. If aptychi are completely preserved, a thin dark layer of horny or chitinous material covers the innermost calcareous layer. In section, aptychi consist of an outer calcareous layer; a middle layer with cellular or tubular structure; an inner calcareous layer; and, if completely preserved, a thin dark innermost layer of horny or chitinous material.

Aptychi occur in Jurassic and Cretaceous rocks, disappearing, with the ammonites, at the end of the Cretaceous. In some beds, aptychi are abundant, even if no ammonites are preserved in them. This may be explained in two ways: (1) While the calcareous material of the aptychi is calcitic, that of the ammonite shells is aragonitic, and thus more easily dissolved. (2) When the animals died, the bodies decayed, and the aptychi fell to the bottom, while the empty shells floated away.

In contrast to aptychi, *anaptychi* (from the Greek meaning unfolded) are single chitinous plates. Their outline, however, is also rather similar to that of the ammonoid aperture, and they are also found close to or within the aperture of ammonoids (Fig. 1). The earliest anaptychi occur in the Devonian together with the first ammonoids, and their peak is in late Paleozoic to early Jurassic time.

The majority of anaptychi and aptychi may be associated with particular ammonoid genera. Nevertheless, aptychi have been treated as

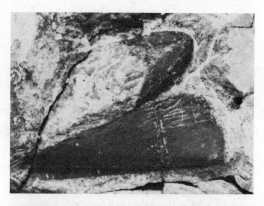

FIGURE 1. Aptychus and anaptychus (from Lehmann, 1971). *Above,* Aptychus (=lower jaw) and upper jaw in the living chamber of the ammonite *Eleganticeras;* lower Toarcian, Hamburg, Germany; X2. *Below,* Anaptychus (= lower jaw) found in the Upper Triassic Pedata Limestone, Austria; X2.

zoological entities of their own and classified in "form genera," such as *Cornaptychus, Lamellaptychus, Praestriaptychus, Laevaptychus,* mainly by Trauth (1927–1936) and others.

For a long time, the true nature of these structures was open to controversy. At first they were sometimes considered as shells of animals that had been eaten by the ammonoids or, alternatively, had eaten the ammonoids. Most of the early paleontologists regarded them as the shells of bivalve mollusks, while others supposed them to be palatal bones of fishes, plates of stalked cirripides or some additional element of the ammonite organism. The last opinion was strengthened by the fact that aptychi are quite frequently found in a particular position in the front part of the ammonoid living chamber close to the external side.

In a series of papers, Trauth advocated that those structures should be interpreted as the opercula of ammonoids, by which such animals could close their shells when in danger. His interpretation was almost universally accepted, especially as many aptychi fit fairly well into the aperture of the ammonoid shells with which they are associated. In some cases, they were actually found in the position of an operculum in front of the aperture.

As early as 1864, Meek and Hayden suggested that aptychi are the jaws of ammonites. This interpretation did not find acceptance until 1970, when Lehmann happened to find an upper jaw within the anaptychus of an ammonite specimen (*Amaltheus*) and subsequently sectioned a number of anaptychi and aptychi found within the living chambers of ammonites. In several cases, upper jaws were found between anaptychi and aptychi in this position (Fig. 2).

Investigations are still going on, but at the moment, the following conclusions seem reasonable: Jurassic anaptychi were actually the lower jaws of ammonites. Phylogenetically, they developed into aptychi when they were covered with a layer of calcite which parted along the midline. Their primary function was that of a jaw, but in some cases it widened to that of a shovel. Secondarily, use as an operculum became possible and increasingly important in some lineages. The upper jaw always retained its characteristic shape and was never covered by calcite. Upper jaws of ammonoids are very rarely found as fossils.

As to Paleozoic anaptychi, no recent investigations have been published.

U. LEHMANN

References

Kaiser, P., and Lehmann, U., 1971. Vergleichende Studien zur Evolution des Kieferapparates rezenter und fossiler Cephalopoden, *Paläont. Zeitschr.*, **45**, 18–32.

Lehmann, U., 1970. Lias-Anaptychen als Kieferelemente (Ammonoidea), *Paläont. Zeitschr.*, **44**, 25–31.

Lehmann, U., 1971. New aspects in ammonite biology, *Proc. N. Amer. Paleontological Conv.*, pt. I, 1251–1269.

Lehmann, U., 1972. Aptychen als Kieferelemente der Ammoniten, *Paläont. Zeitschr.*, **46**, 34–48.

Schindewolf, O. H., 1958. Über Aptychen (Ammonoidea), *Palaeontographica*, A, **111**, 1–46.

Trauth, F., 1927–1936. Aptychenstudien, I–VIII, *Ann. Naturhist. Mus. Wien*, 41–47.

Cross-reference: *Cephalopoda.*

ARCHAEOCYATHA

The Archaeocyatha are an extinct group of bottom-dwelling organisms, most of which inhabited warm shallow seas during the Lower and Middle Cambrian. They are represented in the fossil record only by their complex conical to discoidal skeletons of calcium carbonate, which range from a few mm to 60 cm in diameter, with most averaging 10–20 mm. Archaeocyatha have been positively identified in collections from all continents and subcontinents except South America.

Affinities

The biologic or systematic position of the archaeocyathids has always been unclear and highly debated by paleontologists. They have at various times been identified as protozoans, sponges, calcareous algae, or coelenterates. Most workers have considered them to be either representative of a distinct phylum or a major subdivision of the phylum Porifera. French and Soviet paleontologists who have been involved in this problem recently have concluded that the Archaeocyatha are an independent phylum of multicellular organisms with a position somewhere between the Protozoa and Porifera or between the Porifera and the Coelenterata.

History of Study

The first descriptions of archaeocyathids were based upon specimens collected by E. Billings in 1861 near Anse au Loup, Labrador. Subse-

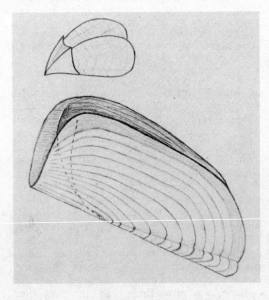

FIGURE 2. Lower jaw (= the paired aptychi) and upper jaw of the ammonite *Hildoceras*, lower Toarcian. Enlarged restoration, X1.5.

quent finds were reported from New York and Nevada during the 1800s. It wasn't until the 1890s and early 1900s with discoveries in the western Mediterranean and Siberia that the focus shifted outside North America. The 1930s saw new discoveries in Siberia, Mongolia, and China. The next forty years saw numerous discoveries and a great deal of published data

come from the work of Soviet geologists such as A. G. Vologdin, I. T. Zhuravleva, and A. Yu. Rozanov in eastern Siberia. The 1950s and 1960s were also a period of intense and significant work by F. Debrenne (1964) on new collections from Morocco and France, as well as restudy of material from Sardinia and Australia. The late sixties and early seventies have wit-

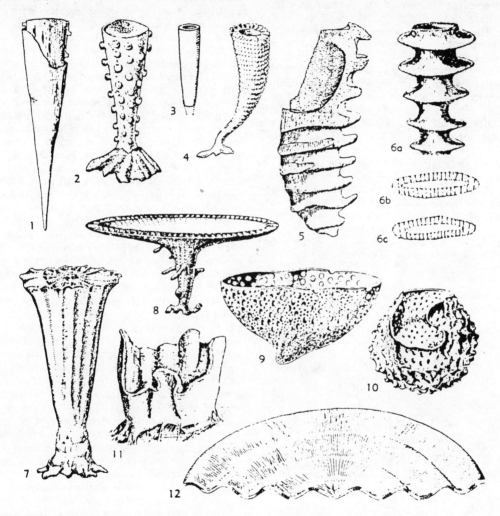

FIGURE 1. Representative archaeocyathans (from Hill, 1972; courtesy of The Geological Society of America and University of Kansas Press). 1, Slenderly conical and erect form of *Dokidocyathus simplicissimus* Bedford and Bedford, X0.3; 2, slenderly conical and erect form with basal holdfasts shown by *Tumuliolynthus karakolensis* Zhuravleva, X1; 3, cylindrical form of *Sigmacoscinus sigma* Bedford and Bedford, X0.7; 4, curved conical form of *Kotuyicyathus kotuyikensis* Zhuravleva, X1; 5, transversely annulated, annulation not involving inner wall, as seen in *Pycnoidocyathus synapticulosus* Bedford and Bedford, X0.3; 6, transversely annulated type, annulation involving inner wall, as shown by *Orbicyathus mongolicus* Vologdin (6a, external view X2; 6b, c tang. sec. of 2 rings X4); 7, longitudinally ribbed and fluted externally with basal holdfasts exemplified by *Beltanacyathus ionicus* Bedford and Bedford X0.3; 8, conical suddenly expanding form of *Paranacyathus subartus* Zhuravleva X0.7; 9, broadly conical form of *Cryptoporocyathus junicanensis* Zhuravleva X5; 10, subspherical form of *Fransuasaecyathus subtumulatus* Zhuravleva X5; 11, bowl-shaped form, with irregular longitudinal folds affecting both walls, shown by *Cosinoptycta convoluta* (Taylor), X0.7; 12, discoid form with concentric waves (cut in half diametrically) seen in *Okulitchicyathus discoformis* (Zhuravleva), X0.16.

nessed new discoveries and a rekindled interest in North American archaeocyathids with new discoveries from northern Mexico, California, Nevada, British Columbia, and the Yukon.

General Morphology

Colonial and solitary skeletons of archaeocyathids are known. The Archaeocyatha are most often represented by a variety of solitary skeletal forms (Fig. 1). Although the typical archaeocyathan cup (Fig. 2) is comprised of two porous walls separated by radial or near radial partitions (septa), the space between the walls (intervallum) may also be filled with a system of irregular and wavy porous plates or a network of flattened rods and/or bars. (Fig. 3) The inner wall usually surrounds a central cavity which may extend to a position just short of the apex of the cup. In some forms the septa are linked by round bars (synapticulae). The intervallum may be spanned by horizontal to subhorizontal porous plates (tabulae) or contain numerous imperforate cystose plates, rods, or laminar skeletal material. Most archaeocyathan cups possess a skeleton in which most components are porous. A great deal of study has focused upon the variety and complexities of both the inner and outer walls. The walls can be perforated by rows of simple round to oval pores, rows of inclined tubes or canals, or systems of branching interconnected tubes. Some archaeocyathids possess walls that are combinations of porous plates faced with thin annuli (rings) of various shapes. The walls of archaeocyathids, in toto, represent an impressive array of structural variations on a number of basic plans. It is often the reconstruction of these complex and intricate walls from a series of thin sections that has excited and challenged the paleontologist.

Little is known about the soft-part anatomy of archaeocyathids. To date, there are no positively identified impressions of archaeocyathan soft parts in the rocks containing their skeletons. A few workers have cited the presence of calcified "organs" within the central cavity and intervallum. Among Soviet specialists on archaeocyathids, including A. G. Vologdin, I. T. Zhuraveleva, A. B. Maslov, and V. D. Fonin, there are two interpretations as to the complexity and distribution of archaeocyathan soft parts. There are those who have concluded that the Archaeocyatha were multicellular organisms that reached a tissue and organ grade of organization. The principal organs and life processes, they further conclude, were contained in the central cavity and/or the intervallum space between the two walls. Others point to the high degree of porosity of all major skeletal components, forms with no central cavity, and some types of skeletal repair as being indicative of a continuous sheath-like agglomeration of undifferentiated cells. All fundamental vital processes would therefore be carried on intracellularly as in sponges and some protozoans.

A final solution to the problem of the exact nature of archaeocyathan soft parts is directly related to a better understanding of the hydrodynamics of circulation within the test or cup. Recent studies (Balsam and Vogel, 1973) are focusing in on this aspect and preliminary results indicate that the potential for both passive and active current movement through archaeocyathan cups is present. A review of the current knowledge regarding the archaeocyathan skeleton clearly indicates that if the Archaeocyatha had reached an organ and specialized multicellular grade of organizations, these organs must have occupied only a small portion of either the intervallum or central cavity and that there was continuous tissue connection throughout most of the cup.

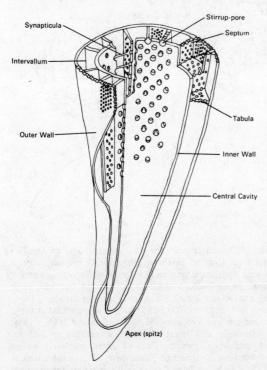

FIGURE 2. A generalized archaeocyathan (from Handfield, 1971; by permission of the Geological Survey of Canada).

Classification

Most workers on archaeocyathids recognize two classes, the Regulares and Irregulares Vologdin. The Regulares are often subdivided

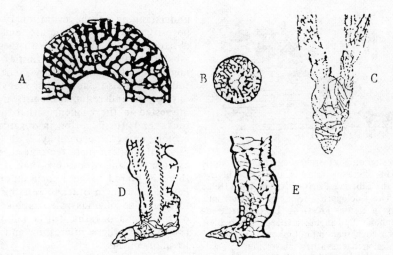

FIGURE 3. Representative (Class Irregulares) Archaeocyathans (from Hill, 1965). A,B,C,E, Family Archaeocyathidae; A, *Archaeocyathus atlanticus* Billings, transverse section, X2; B, *Retecyathus laqueus* Vologdin, transverse section, X2.5; C, *Spirocyathella kyzlartauensis* Vologdin, median longitudinal section, X2.5; E, *Potekhinocyathus bateniensis* Vologdin, median longitudinal section, X3.5; D, Family Protocyocyathidae; *Protocyclocyathus irregularis* Vologdin, median longitudinal section, X3.5.

into the orders Monocyathida Okulitch, Putapacyathida Vologdin, and Ajacicyathida R. and J. Bedford. The Irregulares contain the orders Thalassocyathida Vologdin, Archaeocyathida Okulitch, and the Syringocnematidae Okulitch. Often three or four suborders are recognized and as many as 20 superfamilies as well as over 80 families. There are over 370 published genera and more than 600 species. The classification of the Regulares is based upon the following:

Classes are defined by rate of ontogenetic changes, regularity of wall pore patterns, degree of wall complexity, and regularity of radial intervallar elements.

Orders are defined by the type of skeletal elements that fill the intervallum and apex of the cup and the presence or absence of an inner wall.

Suborders are defined by the elaboration of the different types of intervallum structures and the presence or absence of structures such as tabulae.

Superfamilies are defined by the differentiation and type of outer wall.

Families are defined by the differentiation and type of inner wall.

Genera are defined by the porosity of intervallum structures, details of form and complexity of inner and outer walls.

Species are defined by statistical differentiation based upon skeletal coefficients.

F. Debrenne (1970) has proposed a similar scheme for the classification of the Irregulares.

Paleoecology

Archaeocyatha are predominantly sessile benthic marine organisms that lived in shallow (< 200 m) shelf-sea waters which had a low to moderate turbidity and were rich in calcium carbonate. Conditions such as these are presently found within the tropics and subtropics. It is firmly established that the archaeocyathids often constructed carbonate buildups or bioherms. However, allusion to archaeocyathan "reefs" by many early workers is not fully justified in the light of present knowledge since this term often implies that they built structures closely similar to modern coralline-algal reefs. Although some form of carbonate buildup has been reported from most archaeocyathan localities, the most extensive geographic and stratigraphic examples are to be found in the southeastern part of the Siberian Platform along the Lena and Aldan Rivers (Fig. 4). It is doubtful that the Archaeocyatha built structures truly comparable in size and complexity to modern coralline-algal reefs. However, it is interesting to note that they did establish consistent symbiotic relationships with a variety of calcareous algae such as *Epiphyton* and *Renalcis*. Therefore, it is probably justified to conclude that the Archaeocyatha represent the first evolutionary step which eventually led to the complex biogenic features that are such important and beautiful parts of the tropical and subtropical seas today.

The archaeocyathids are often closely associated with a wide variety of other organisms. Specimens of trilobites, brachiopods, monoplacophorans, cephalopods, protogastropods, eocrinoids, helicoplacoids, and sponges are found in samples that contain archaeocyathids.

49

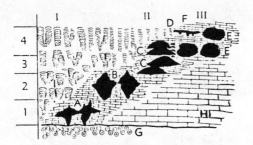

FIGURE 4. Migration of archaeocyathan and algal-archaeocyathan bioherms in time and space in the Aldanian of the Siberian Platform (from Hill, 1972; courtesy of The Geological Society of America and the University of Kansas Press). A, Littoral archaeocyathan dilophoid; B, fringing archaeocyathan dilophoid; C, archaeocyathan monolophoid; D, pseudostromatolic algal-archaeocyathan biostrome; E, algal-archaeocyathan oncoidal dilophoid; F, bedded algal-archaeocyathan biostrome; G, oncolites; H, limestone without bioherms. I, stromatolites; and I, shallow, littoral zones; II, depth over 10 m; III, depth some tens of meters.

In addition, a number of unidentified and enigmatic forms are also found in archaeocyathan-rich rocks.

Conclusions regarding the temperature, depth, and salinity of the waters in which archaeocyathids lived are highly speculative and are based upon a limited amount of chemical and physical data. The conclusions reached thus far are based upon the uniformitarian application of modern studies of carbonate deposition, associated floral and faunal analogs, and the analysis of primary rock structures and textures.

Direction of Future Research

Despite the enormous wealth of data that has been accumulated over the last forty years regarding the morphology and paleoecology of the Archaeocyatha, knowledge of this group is still far from adequate. Most specialists on archaeocyathids are still in the process of establishing the basic temporal and geographic distribution of these fossils. The great bulk of the work, outside of the Soviet Union, has been directed toward the gathering of raw data. However, the last ten years has seen an ever increasing focus on such aspects as ontogeny, phylogeny, paleogeography, statistical analysis of morphologic variation, functional morphology, and paleoecology.

The advent of automatic thin-sectioning equipment, the electron probe, electron microscope, and computers will greatly help to speed up both the preparation of data on archaeocyathids and the analysis of it. The next ten years should see the establishment of a better understanding of the evolutionary pattern of porosity, basic microstructure, and the hydrodynamics of the skeleton. With these advances, a better understanding of affinities, and refinement of the classificatory scheme should follow. It is also evident that many more paleontologists and sedimentologists are becoming interested in the challenges that the Cambrian rocks and fossils offer in regards to paleoecology and paleogeography. With the development of modern plate tectonic theory and the excitement and questions that it has raised, there is bound to be a focus upon the lower Paleozoic soon in order to test this hypothesis further. The obvious paleographic significance of the Archaeocyatha will certainly accelerate research on these interesting and challenging fossil organisms.

R. A. GANGLOFF

References

Balsam, W. L., and Vogel, S., 1973. Water movement in Archaeocyathids: Evidence and implications of passive flow in models, *J. Paleontology*, **47**, 979–984.

Beerbower, J. R., 1968. *Search for the Past: An Introduction to Paleontology*. 2nd ed. New Jersey: Prentice-Hall, 512p.

Debrenne, F., 1964. Archaeocyatha: Contribution à l'étude des faunes Cambriennes du Maroc, de Sardaigne et de France, *Serv. Mines Carte Geol. Maroc, Notes et Mem.*, 1 & 2, 265p.

Debrenne, F., 1969. Lower Cambrian Archaeocyatha from the Ajax Mine, Beltana, South Australia, *Bull. British Mus. (Nat. Hist.), Geol.*, **17**, 299–376.

Debrenne, F., 1970. A revision of Australian genera of Archaeocyatha, *Trans. Roy. Soc. S. Australia*, **94**, 21–48.

Handfield, R., 1971. Archaeocyatha from the Mackenzie and Cassiar Mountains, Northwest Territories, Yukon Territory and British Columbia, *Geol. Surv. Canada, Bull.*, **201**, 119p.

Hill, D., 1965. Archaeocyatha from Antarctica and a review of the phylum, *Trans-Antarct. Exped., Sci. Repts.*, **10**, 151p.

Hill, D., 1972. Archaeocyatha, in C. Teichert, ed., *Treatise on Invertebrate Paleontology*, pt. E (rev.). Lawrence, Kansas: Geol. Soc. Amer. and Univ. Kansas Press, I, 158p.

Rozanov, A. Yu., 1973. Zakonomernosti morfologicheskoi evolyutsii arkheotsiat i voprosy yarusnogo raschieneniya nizhnego kembriya, *Trudy, Geol. Inst. Akad. Nauk SSSR*, **241**, 164p.

Rozanov, A. Yu., and Debrenne, F., 1974. Age of archaeocyathid assemblages, *Amer. J. Sci.*, **274**, 833–848.

Zhuravleva, I. T., 1960. *Arkheotsiaty Sibirskoy Platformy*. Moscow: Adad. Nauk SSSR, 344p.

Zhuravleva, I. T., 1970. Marine faunas and Lower Cambrian stratigraphy, *Amer. J. Sci.*, **269**, 417–445.

Cross-references: *Reefs and Other Carbonate Build-ups.*

ARTHROPODA

Represented by over 1,000,000 modern species, the Arthropoda comprise by far the most diverse animal phylum. Modern arthropods fall into several quite distinct groups (Fig. 1). The three major ones are Crustacea (40,000+ species, mostly marine) such as shrimp and crabs; Chelicerata (60,000+ species, almost all terrestrial) such as terrestrial spiders, terrestrial and aquatic mites, and marine horseshoe crabs; and Uniramia (900,000+ species, virtually all nonmarine) such as centipedes, millipedes, and insects (see Table 1). Pantopoda, sea spiders (500 species) are allied to the Chelicerata. Tardigrada, water bears (350 species) and Pentastomida, mite-like parasites on mammals (65 species), are little-known groups classed as arthropods. A variety of anthropod types, including representatives of all three major modern groups, are known through-

out the fossil record from the Cambrian onward. The Trilobitomorpha, a heterogeneous collection of primitive marine forms restricted to the Paleozoic, comprise a fourth major group. Best known of these are trilobites (15,000+ species, Cambrian-Permian), the first appearance of which is traditionally taken to mark the base of the Cambrian.

The name "Arthropoda" ("jointed foot") refers to the characteristic development of the cuticle, which encases the body, as a jointed exoskeleton of more or less rigid sclerites articulated by flexible arthrodial membranes. Such development of the cuticle permits the body to attain relatively great strength at the small sizes that are typical of adult arthropods. But at the same time, arthropodization imposes limitations on maximum body size in several ways. Adult arthropods are typically no more than a few mm long. There have, nevertheless, been arthropod "dinosaurs" that have evidently

---

TABLE 1. Classification of Arthropoda

Phylum ARTHROPODA
  Subphylum TRILOBITOMORPHA
    Class TRILOBITA—Cambrian-Permian
    Class TRILOBITOIDEA—Cambrian-Devonian
  Subphylum CRUSTACEA
    Class CEPHALOCARIDA—Holocene
    Class BRANCHIPODA, clam shrimp, brine shrimp, water fleas, and allies—Devonian-Holocene
    Class MYSTACOCARIDA—Holocene
    Class OSTRACODA—Cambrian-Holocene
    Class EUTHYCARCINOIDEA—Carboniferous-Triassic
    Class COPEPODA—Cretaceous-Holocene
    Class BRANCHIURA—Holocene
    Class CIRRIPEDIA, barnacles—Silurian-Holocene
    Class MALACOSTRACA, true shrimp, including edible shrimp and lobsters—Cambrian-Holocene
  Subphylum CHELICERATA
    Class MEROSTOMATA, eurypterids and horseshoe crabs—Cambrian-Holocene
    Class ARACHNIDA, scorpions, spiders, mites, and allies—Silurian-Holocene
  Subphylum PYCNOGONIDA
    Class PALEOPANTOPODA—Devonian
    Class PANTOPODA, sea spiders—Holocene
  Subphylum UNIRAMIA
   Superclass ONYCHOPHORA
    Class PROTONYCHOPHORA—Cambrian
    Class EUONYCHOPHORA, walking worms—? Carboniferous-Holocene
   Superclass MYRIAPODA
    Class ARCHIPOLYPODA—Silurian-Carboniferous
    Class ARTHROPLEURIDA—Carboniferous
    Class DIPLOPODA, millipedes—Carboniferous-Holocene
    Class PAUROPODA—Holocene
    Class CHILOPODA, centipedes—Carboniferous-Holocene
    Class SYMPHYLA, garden centipedes—Tertiary-Holocene
   Superclass HEXAPODA
    Class PROTURA—Holocene
    Class DIPLURA—Tertiary-Holocene
    Class COLLEMBOLA, spring-tails—Devonian-Holocene
    Class THYSANURA, bristle-tails and silverfish—Triassic-Holocene
    Class MONURA—Carboniferous-Permian
    Class INSECTA—Carboniferous-Holocene

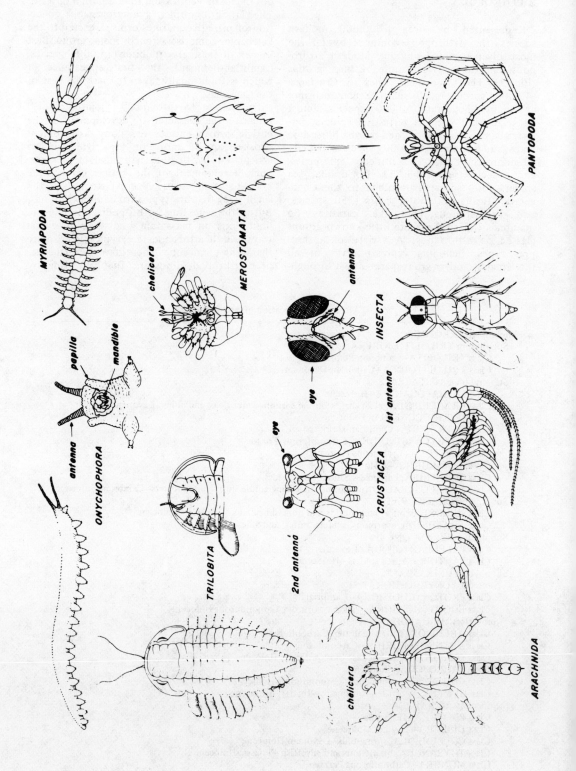

MYRIAPODA

MEROSTOMATA

chelicera

PANTOPODA

antenna

INSECTA

eye

ONYCHOPHORA

antenna

papilla

mandible

CRUSTACEA

1st antenna

eye

2nd antenna

TRILOBITA

ARACHNIDA

chelicera

overcome problems of body mechanics, respiration, and circulation necessary for the attainment of much larger size. These include Silurian eurypterids (sea scorpions) and Carboniferous millipede relatives (arthropleurids) that were 2 m long, and Carboniferous dragon-flies that had a wingspan of > 1 m. Among the largest living arthropods are a horshoe crab nearly a meter long and beetles about 30 cm long.

Arthropods are found throughout the biosphere, in the sea, on land, and in the air. They are an ecologically diverse group of grazers, scavengers, predators, and parasites. The former three feeding types are represented among Cambrian arthropods. In terms of individuals, minute copepod crustaceans are among the most abundant multicellular animals, and are of great importance in aquatic communities as scavengers, as grazers on planktonic algae, and as food for larger crustaceans and fish. Arthropods have attained by far their greatest diversity on land. The majority of chelicerate species are terrestrial predators (e.g., spiders) and parasites (e.g., many mites). The great majority of arthropods are uniramians, and the great majority of these are winged hexapods, the insects (see *Insecta*). Most uniramians are associated in one way or another with land plants, as herbivores, as decomposers, or as predators on herbivores or decomposers. Beetles, by far the most diverse arthropod group with over 400,000 modern species, typically feed on fungi and other decomposers.

Definition of Arthropoda

Arthropoda are characterized by (1) metameric segmentation; (2) specialization of

FIGURE 1. Representative anthropods. (Drawings by Kelly Dempster.) Subphylum Uniramia, Superclass Onychophora, *Peripatopsis,* view of right side, X1.5, and ventral view of head and first trunk segment, X5 (after Manton); Subphylum Uniramia, Superclass Myriapoda, the centipede *Otocryptops,* dorsal view, X1.5 (after Snodgrass); Subphylum Trilobitomorpha, Class Trilobita, *Triarthrus,* dorsal view, X.75, and ventral view of head and first trunk segment, X1; Subphylum Chelicerata, Class Merostomata, the horseshoe crab *Limulus,* ventral view of the central region of the body showing the appendages and dorsal view of the entire body, X0.25 (after Snodgrass); Subphylum Chelicerata, Class Arachnida, the scorpion *Centuriodes,* dorsal view, X1.5 (after Snodgrass); Subphylum Crustacea, Class Malacostraca, the shrimp *Anaspides,* anterior view of the head, X4, and view of right side of body X2.5 (after Snodgrass); Subphylum Uniramia, Class Insecta, the fly *Drosophila,* anterior view of head, X25, and dorsal view of body, X12 (after Papp); Subphylum Pycnogonida, Class Pantopoda, the sea spider *Nymphon,* dorsal view, X4 (after Barns).

segments at the anterior end as a head; (3) development of paired segmental limbs; (4) an open circulatory system that includes a heart bearing segmental openings (ostia) through which blood (hemolymph) enters it, and a hemocoel that forms the body cavity; (5) reduction of the coelom in the adult; and (6) a cuticle composed largely of chitin (q.v.) and sclerotized protein that is typically developed as a jointed exoskeleton.

Recent research (Tiegs and Manton, 1958; Manton [papers summarized in Manton, 1977]; Anderson, 1973; Cisne, 1974a) has pointed out the likelihood that arthropodization of the body mechanical system evolved independently in Unirama (defined as a phylum by Manton, 1972), the primitively terrestrial arthropods, and in the primitively marine groups, which include Trilobitomorpha, Crustacea, and Chelicerata. It perhaps evolved more than once among these latter groups. If one takes the exoskeleton as the single most important character defining Arthropoda, as the very name suggests, one is forced to conclude that arthropods fall into as least two phyla, Uniramia, and at least one other group of primitively marine forms.

Anatomy and Morphological Mechanisms

Segmentation and Tagmatization. Excepting the anterior presegmental region (acron) and the posterior postsegmental region (telson)—which are not always present—the arthropod body is made up of a chain-like series of segments. What ultimately defines a segment is the presence of a pair of mesoderm somites in embryonic development. A typical segment bears a complement of segmental organs that includes a pair of nerve ganglia, a set of muscles, and a pair of limbs. Segments are commonly set off by grooves or articulations in the exoskeleton that to varying degrees coincide with boundaries between segmental sets of internal organs. Adjacent segments may become fused in the course of evolution, that is, their segmental sclerites, which were separately articulated in an ancestral form, may fuse to form a single skeletal element. Segmentation appears to be initiated in development by segmentation of the mesoderm. In the course of development, whole segments that are clearly set off in the embryonic mesoderm may become vanishingly reduced in the adult. In the course of evolutionary specialization of segments, different sets of segmental organs may be enhanced in size and complexity or reduced to the point of vanishing.

The number of segments in the adult body varies from <10 to >100 from group to group.

53

Except in the most primitive forms, this number is fixed for adults of a given species and sex. The body is in various ways divided into distinct regions (*tagmata*), each defined by a group of segments that are developed together as an anatomical unit. The head is one tagma. Its varying structure and segmental composition are the bases for classification of the major arthropod groups. The trunk, an internally undifferentiated region in the most primitive forms, is variously subdivided in more advanced forms. It is divided into a thorax and abdomen in crustaceans and hexapods. In some malacostracan crustaceans, a cephalothorax is developed through fusion of segments. Chelicerates have an anterior prosoma, which is vaguely similar to the cephalothorax in some crustaceans.

A general trend in the evolution of free-living arthropods has been the building of tagmata through progressive specialization of segments. In general, this process has proceeded from front to back, commencing with the building of the head region, and continuing with subdivision of the trunk. It corresponds to a progressive division of labor among segments. The phenomenon, progressive tagmosis, affords a convenient scale for judging degree of phyletic specialization and degree of complexity of limb mechanisms (Lankester, 1904; Cisne, 1974b). Sessile and parasitic forms have followed opposite evolutionary trends toward reduction in expressions of segmentation and reduction of in cephalization. Certain isopod crustaceans are among the most degenerate parasites known, the adult female consisting of little more than ripe gonads.

Body Mechanical System. The exoskeleton forms a system of more or less rigid levers, the sclerites, operated by internal muscles that extend across joints. To varying degrees, the fluid-filled body cavity acts as a hydrostatic skeleton in opposing the action of muscles on sclerites.

While its precise structure and the terminology used to describe it varies from major group to major group, the cuticle basically consists of a thin, nonchitinous layer (epicuticle) underlain by a thick, lamellar layer (procuticle). The procuticle is largely made up of chitin, a polysaccharide, and arthropodin, a protein. It is divided into an outer layer (exocuticle) that is shed at the next molt, and an inner layer (endocuticle) that is resorbed and replaced by new cuticle prior to the next molt. Reinforcement of the cuticle in sclerite formation takes place through sclerotization (tanning) of the protein component and, particularly in some crustaceans and millipedes, deposition of calcium carbonate and calcium phosphate in the exocuticle. Reinforcing mate-

rial is commonly resorbed from the cuticle prior to molting and stored in the body.

In all its detail, the cuticle has a very complex surficial anatomy. Elaborate hinge structures are developed that control movement across joints. Setae, hair-like projections that are involved in various tacticle and prehensile operations, can be present in great numbers over the body and limbs. Areas of the cuticle that bear muscle attachments are commonly strengthened and modified, and often have surface expressions as protuberances, depressions, or furrows.

The skeletal muscles, formed of striated muscle in typical arthropods, can be divided into three sets: body muscles that flex sclerites of the body proper with respect to one another, extrinsic limb muscles that move the basal parts of each limb with respect to the body, and intrinsic limb muscles that move the parts of the limb with respect to one another. A ventral endoskeleton of tendons or of invaginated cuticular material is sometimes developed. Onychophorans, which lack an exoskeleton, have the musculature formed of smooth muscle, and striated muscle is developed only in the jaw musculature. Body form is controlled by a body-wall musculature (sheets of fibers that surround the body cavity just beneath the thin cuticle) that is typically absent in arthropodized forms.

Segmental Limbs and Limb Mechanisms. The jointed segmental limbs are developed in a great variety of configurations in accordance with their many functions in locomotion; feeding; respiration; sensing the environment; mating; defense; and, in some instances, care of the young, communication, and dwelling construction. A simple limb developed as a walking leg is a basically conical structure that is usually made up of several more or less cylindrical limb segments and a small, conical segment at its tip. Joints are developed between limb segments and between the basal limb segment and the body. Trunk limbs in many crustaceans and the most primitive chelicerates have a second branch (ramus, pl. rami), often of feather-like or whip-like structure, that lies above a branch of much the same structure as the simple limb described, and is attached on the basal part of the entire limb. Being comprised of two rami, these limbs are said to have a biramous structure. Limbs of primitive crustaceans may include additional rami. Limbs in uniramians consist only of a single ramus that has basically the same structure of the simple limb first described, hence the name "Uniramia."

In general, specialization of limb mechanisms through division of labor among limb pairs has taken place through reduction in the parts of primitively complex limbs. In the most primitive marine arthropods, for example, the trunk limbs, all having basically the same structure, take on manifold functions in morphological mechanisms. As specialization has taken place, the parts of the primitive, complex limb have been reduced in limbs of some segments and amplified in limbs of others in accordance with division of labor and maintenance of morphological integration among mechanisms. There are numerous instances of convergent evolution of limb structure and limb mechanisms. The most notable are the convergences of head structure, head limbs, and mandibular mechanisms, and also of thoracic limb structure and ambulatory mechanisms between hexapods and certain malacostracan crustaceans (see Manton, 1964, 1977). These similarities were often taken as the basis for placing Crustacea and Uniramia together in a Subphylum Mandibulata.

Feeding and the Digestive System. Feeding is performed by several sorts of limb mechanisms. While there are many variations on the theme, uniramians and some more advanced crustaceans directly seize and masticate food with the mandibles and adjacent mouthparts. The mouth region is accordingly anteriorly or ventrally directed. More primitive crustaceans have various mechanisms in which trunk limbs, beating to a metachronal rhythm, at once create a feeding current, collect suspended material from it, and channel the collected material forward toward the mouth along the midline on the underside of the body. Accordingly, the mouth is posteriorly directed. While animals with this general sort of trunk limb mechanism may feed on detritus deposited on the bottom or on material suspended in the water column above, or may even feed on both, food being taken from a suspension created by the animal itself. More advanced malacostracan crustaceans have mechanisms in which head limbs set up feeding currents and collect material suspended in them. In one such mechanism found in many malacostracans, the maxillae, the third of three pairs of head limbs behind the mouth, set up a respiratory and feeding current and filter suspended material from it. Chelicerates seize and tear apart their food, often other organisms, with a pair of small, clawed limbs in front of the mouth, the chelicerae. Many predaceous forms first subdue their prey with a neurotoxic sting. Claws developed on one or more pairs of limbs immediately behind the mouth in many chelicerates assist in food manipulation. Some chelicerates masticate food between the spiny basal segments of one or more pairs of these limbs.

A sucking action of the muscular region of the gut adjacent to the mouth region is normally involved in the actual ingestion of food. Ingestion is commonly aided by lubricating glandular secretions and, particularly in arachnids, by enzyme-bearing secretions that reduce food to a more or less liquid form prior to ingestion. Suctorial ingestion is developed in many specialized ways in parasitic forms such as mosquitoes and biting lice, and in insects that feed on plant fluids.

The digestive system consists of the gut and associated salivary and digestive glands. The gut, a tube that extends from the head region through the last segment at the posterior end of the body, is divided into an anterior foregut that bears a cuticular lining, a typically unlined midgut (intestine), and a hindgut that also bears a cuticular lining. A portion of the foregut is commonly developed as a crop, an enlarged region surrounded by muscle and lined interiorly with cuticular spines, in which ingested food is further broken down for digestion. Crustaceans, chelicerates, and trilobitomorphs have digestive glands and diverticula that connect to the gut in the head region. Uniramians, however, have salivary glands that empty near the mouth region, and lack the often extensive proliferation of glands found in the other groups.

Cardiovascular System. The arthropod open circulatory system includes a heart and dorsal vessel, accessory arteries and veins, and the body cavity, the hemocoel. Internal organs are bathed in the flow of blood (hemolymph) that passes through a loose network of connective tissue around them. In simplest form, the circulatory organs consist of a tube that extends along the midline near the top of the body cavity. A section of several segments' length at variable positions in the body is enlarged and modified as the heart. The dorsal vessel, a continuation of the tube, extends anteriorly into the head and posteriorly along the trunk. Lateral vessels may be developed, particularly in larger arthropods. Hemolymph enters the heart from the pericardium, a partially walled space surrounding it, through segmental perforations in its own wall, the ostia. It leaves through the dorsal vessel and other vessels, if developed.

Very small arthropods, which have a very large surface area relative to their volume, respire across the general body surface and commonly lack specialized respiratory organs, and may even lack respiratory pigments. Larger marine arthropods generally have the upper branches of certain trunk limbs specialized as gills, and may have sections of the body surface specially modified for gas exchange. Uniramians have very ramose invaginations of thin cuticle (tracheae) that serve in gas exchange in air. Certain spiders and terrestrial crustaceans have similarly formed respiratory structures. Tracheal structures, once taken as the prime character of a Subphylum Tracheata, are now known to have evolved independently in onychophorans, in myriapods and hexapods, in arachnids, and in crustaceans. Arachnids typically have differently formed invaginated structures forming a book lung, which is thought to be an adaptation of the book gills of segmental limbs found in marine horseshoe crabs. Crustaceans often have more than one set of respiratory structures, and those spiders with tracheae also have book lungs. Large marine arthropods with elaborate respiratory mechanisms also have a more elaborate set of blood vessels, in accordance with problems of circulation of hemolymph to and from the respiratory structures and through the body. They generally have respiratory pigments—usually hemoglobin or hemocyanin—but there is no broad pattern to their occurrence among the major modern groups. Insects, having a relatively very large surface area exposed in tracheae, typically lack respiratory pigments.

Urogenital System. The organs of excretion and ionic regulation, and of reproduction, take on many forms in Arthropoda. The gonads, which are associated with the segment on which the genitial openings are located, and certain excretory glands, which are associated with one or more segments, are derivatives of embryonic coelomic sacs. True nephridia are absent, and cilia are found only in the genital ducts of onychophorans. Except in onychophorans, in which a series of small excretory glands occur on trunk segments, organs derived from the coelom disappear in all but a few segments. The coelom is said to be reduced in arthropods in this sense, and in the sense that coelomic organs never closely and extensively surround the gut.

Sexes are generally separate, though hermaphroditism and parthenogenesis occur in some groups. The position of the genital opening, which is often marked by modifications of limbs in that area of the body, is variable from group to group. In onychophorans, centipedes, and hexapods, it occurs near the posterior end of the body; but in millipedes and other myriapods, it occurs just behind the head. In crustaceans, it is generally located at the back of the thorax on segments that vary from group to group and between sexes. In chelicerates, it is always located in the first segment of the opisthosoma.

The structure and function of an arthropod's excretory system is related to its adaptations for life on land or in an aqueous medium, and in particular to adaptations for solving problems of water and salt regulation and of nitrogen excretion in one or the other of the two quite different situations. Aquatic forms secrete nitrogen from nutrient metabolism as a dilute aqueous solution of ammonia. Life on dry land generally demands a mechanism for water retention, and hence demands a higher concentration of nitrogenous wastes in the excreta, which in turn demands the reduction of nitrogenous wastes to form less toxic than ammonia. Fully terrestrial arthropods have excretory organs that are different in structure and embryological derivation from those in marine arthropods. All crustaceans and chelicerates have one or two segmental pairs of excretory organs of coelomic derivation. Terrestrial chelicerates have in addition the Malpighian tubules, an organ involved in reduction of wastes to guanine crystals. These crystals are passed through the intestine and out of the body. The excretory organs in uniramians typically include Malpighian tubules and a modified section of the gut. Forms living in dry habitats typically secrete nitrogenous wastes as crystals of uric acid and urates. Malpighian tubules have evolved independently in the two major groups, as shown by the fact that they develop from ectodermal tissue in uniramians but from endodermal tissue in arachnids.

While semiaquatic, semiterrestrial representatives of all three major modern groups occur in intertidal environments, the transition from marine to fully terrestrial life has been made only a few times in arthropod evolution—perhaps only once in the early evolution of uniramians, probably once in the early evolution of arachnids, and once in the evolution of crustaceans. The reverse transition has also taken place only a few times. With one exception, certain free-living mites are the only marine arachnids. The only insects found in truly marine environments are a group of water striders. The one most profound division in relationships among arthropod groups is that between the basically terrestrial Uniramia and the other major groups, which are primitively marine. That the transition between aquatic and terrestrial environments has been hindered by physiological hurdles in the evolution of necessary adaptations is suggested by the case of oniscoid isopods, the only fully terrestrial crustaceans. As with the marine isopods from which they are descended, the structure of the head and of walking appendages is similar to that in certain insects. They are the only crustaceans in which trachea-like respiratory structures are developed. But their excretory system is of typically crustacean form, and they secrete nitrogenous waste as ammonia. The toxicity problem has been solved in their case by severe limitation of nitrogen metabolism.

Nervous System. The nervous system is organized around a pair of nerve cords that extend from the segments to along the ventral part of the body cavity. Each segment is enervated by nerves that branch off from a swelling or ganglion on each nerve cord. In the head and in at least the anterior part of the trunk region, the pair of ganglia are connected transversely by at least one commissure. The first few ganglia in the head are enlarged and fused to form the brain.

The only part of the nervous system at all commonly reflected in fossils is the eyes, which are often developed as compound eyes made up of tens to hundreds of separate visual units. The cuticular covering of the eye and, particularly in trilobites (see *Trilobita*), the individual lenses of the visual components, are often preserved. One of the most profound evolutionary convergences in the biological world is the independent development of compound eyes of virtually identical structure in certain insects and crustaceans. Not all arthropods have compound eyes. Sometimes visual units are distributed around the head region singly or in small groups. A median eye is functional at least in sensing light in crustacean nauplius larvae, in some adult chelicerates, and in certain neotenous crustacean groups in which it is retained in the adult. Many burrowing marine crustaceans, and a majority of those living much below continental shelf depths, lack functional eyes.

Growth and Development

While specific patterns of embryonic development are quite variable among arthropods, each major modern group appears to have its own basic developmental pattern (see Anderson, 1973, for review).

Development is commonly marked by one or more metamorphoses between larval and juvenile and/or juvenile and adult stages. The most primitive marine forms (trilobites and very primitive crustaceans) hatch as tiny larvae that include around three or four well-formed segments and follow a sequence of gradual development in which segments are added at the posterior end, one or two at each molt until the adult form is reached (see Whittington, 1957, on trilobites and Sanders, 1963, on crustaceans). The early-hatching larvae are usually developed as dispersal stages in the life cycle. Metamorphosis is more pronounced among more advanced crustaceans, and often corresponds to a developmental break between an antenna feeding nauplius larva and a trunk-

limb feeding j
been an evolution
degree of metamorph
an evolutionary trend tow
larval stages to the egg such
emerge more or less as miniature a
gence of the young in essentially th
form (direct development) is common an
more advanced crustaceans and is the rule
among chelicerates. The only early-hatching larvae known in Arthropoda are the trilobite protaspis, the crustacean nauplius, and the pycnogonid protonymphon.

The young of onychophorans and myriapods emerge in essentially the adult form, though the full complement of trunk segments may not be present. There has been an evolutionary trend toward increasing degree of metamorphosis among hexapods. Primitive, wingless hexapods hatch in more or less the adult form. Metamorphosis is marked primarily by the appearance of functional sexual organs. More primitive insects, more advanced than wingless hexapods, hatch as nymphs, and undergo greater changes at metamorphosis. Often the change is marked by the appearance of wings, as in dragonflies and grasshoppers. The very advanced insects undergo profound change in morphology and ecology in changing from larva and adult through a transitional pupal stage. The development of a caterpillar into a butterfly is a familiar example.

Growth takes place with molting. The actual molting of one skeleton and the formation of a new one, which may take place over hours or a few days, is only a small segment of the complex cycle. Different arthropods have different mechanisms of escape from the shell appropriate to differing morphologies; all involve splitting of the old exocuticle along certain suture lines. The molted skin (exuvium) is often consumed after it is shed. Molting is ordinarily a time of increased likelihood of mortality, owing to increased physiological stress and difficulties of escaping from the old shell, as well as to increased vulnerability to predators and parasites. In general, the rate of molting decreases with age. At maturity, the molt cycle is often coupled with the reproductive cycle.

Fossil Record

The arthropod fossil record consists primarily of the most resistant parts of exoskeletons, particularly those sclerites that have been reinforced with mineral matter. Commonly preserved parts used in biostratigraphy include the heavily mineralized dorsal exoskeletons of trilobites and the tiny but heavily calcified carapace valves of ostracod crustaceans. The most fre-

d crustacean larvae, lice, mites, and other
forms no more than a millimeter long, as well
as representatives of most modern insect orders
and numerous extinct ones.

Classification

The classification of Arthropoda is complex.
The major groups are characterized by head
structure as follows:

1. *The number of preoral segments in the adult.* The
preoral segments are those anterior to the interseg-
mental mouth. Onychophora are characterized by one
preoral segment, Chelicerata by two, and Myriapoda,
Hexapoda, and Crustacea by three. However, evidence
on the number of embryonic mesoderm somites that
define these particular segments is not entirely clear
for chelicerates and some crustaceans (see Anderson,
1973). In development, segments take on a preoral
position by growing forward and around the mouth
region. In Crustacea, the second antennal segment is
postoral in the nauplius larva but preoral in later stages.

2. *The number and construction of preoral limbs.*
Uniramia are characterized by a single pair of antennae, though it is developed on noncorresponding segments in Onychophora and in Myriapoda and Hexapoda. A single pair of chelicerae are the only preoral
limbs in the adult in Chelicerata and the little-known
Pycnogonida. Two pairs of antennae are typically
present in Crustacea.

3. *The construction of the feeding mechanism.* Uniramia and Crustacea have a single pair of jaws immediately behind the mouth. In Uniramia, the jaw is
developed from a whole limb, but in Crustacea, the
jaw is developed from only the basal segment of the
limb. Onychopora have one pair of segmental limbs
behind the jaws that are involved in feeding. Other
Uniramia have two pairs of segmental limbs behind the
jaws incorporated in the feeding mechanism. Crustacea
likewise have two pairs of segmental limbs behind the
jaws that participate in the mechanism; and in more
advanced forms additional pairs of trunk limbs are
incorporated in the feeding apparatus. Chelicerates
and pycnogonids lack jaws of the sort found in crustaceans and uniramians. In chelicerates, the bases of at
least the first pair of postoral limbs are usually developed for mastication of food. Chelicerates are also
characterized by the prosoma, which always includes
six postoral segments, the limbs of at least the middle
four of which are typically developed as locomotory
appendages. Pycnogonids, which have a strong sucking
organ developed around the mouth, lack masticatory
appendages. The construction of the anterior end of
the body resembles that of the chelicerate prosoma,
and the abdomen is almost vanishingly reduced.

Trilobitomorpha represent a heterogeneous
group of primitive marine forms that includes
the Trilobita (q.v.), a distinctive group, and
various crustacean-like and chelicerate-like
forms as well as arthropods of altogether distinctive appearance that are placed in the
Trilobitoidea. Recent research has failed to
confirm the established idea that limb structure
unites Trilobitomorpha. While head segmenta-

e
oft
ugh
from
reserv-
natomy
thropods
om tracks,
na trails of
af mines in
plant, out the paleo-
botanical in some of the
earliest land p s, personal communication). Certa ic barnacles are
first known from their racteristic cysts in
fossil crinoids and borings in fossil shells.

Because a mature arthropod may have gone
through a few tens of molts prior to death,
there is reason to believe that exuvia may be
more abundant than actual skeletons. Owing to
differences in the physiology and behavior
associated with molting, as well as to factors
of fossil assemblage formation and preservation, the proportion of parts of actual skeletons
to parts of exuvia is probably quite variable
among arthropod groups and among fossil assemblages. While parts of skeletons and exuvia
should be theoretically distinguishable on the
basis of cuticle structure and evidence of escape,
preservation often does not make it possible to
clearly distinguish them. Because material is
commonly resorbed from the exocuticle prior
to molting, exuvia may be intrinsically less
preservable than skeletons, and because exuvia
are often consumed by newly molted individuals
or other organisms, they may be less frequently
preserved relative to the total number produced.

The arthropod fossil record is strongly biased
toward faunas of shallow marine environments,
and to coastal terrestrial forms that are often
swept into these environments by winds and
floods. Knowledge of the anatomical details of
extinct arthropods comes primarily from relatively few occurrences of exceptionally preserved fossils, such as the Middle Cambrian
Burgess Shale (see *Burgess Shale*) of British
Columbia (marine), the Middle Devonian
Rhynie Chert of Scotland (freshwater and
terrestrial), the Carboniferous Francis Creek
Shale of Illinois (marine, freshwater, terrestrial),
and the Oligocene Baltic Amber (terrestrial).
Tens of orders are known exclusively from such
local occurrences. The variety of arthropods
represented under these exceptional circumstances is truly amazing, and includes trilobite

tion is difficult if not impossible to ascertain from fossil arthropods in general, it does appear that a variety of head structures are found among trilobitomorphs. Trilobita appear to be characterized by a head that includes four limb-bearing segments: a preoral segment that bears an antenna, and three postoral segments that bear biramous limbs that have basically the same structure as trunk limbs and have their basal segment developed for handling food (Cisne, 1975).

The higher classification of arthropods has changed substantially in the last two decades as the external anatomical characteristics on which older classifications have been based have been revealed to have evolved convergently in different groups. Evolutionarily more conservative characters of development and morphological mechanisms have been discovered and applied to constructing a classification that more accurately reflects evolutionary history. The fundamental problem in continuing development of this classification is whether or not the common ancestor of all arthropods (which is unknown and must be reconstructed) was itself an arthropod. If it was not, Arthropoda would comprise at least two distinct phyla. The fact that the major modern groups are so distinct from one another in the characteristics of their modern representatives is consistent with the idea that the arthropod condition evolved in Uniramia independent of its evolution in other groups, and that it evolved at least once in these other groups. However, the apparent distinctness of the major groups as they are represented today is at least in part an artifact of the extinction of primitive, intermediate forms, some of which are known in the fossil record. While the dichotomy between primitively marine and primitively terrestrial forms is known to extend virtually to the origin of arthropods, very much remains to be done in resolving problems of evolutionary relationships over geologic time in order that classification may appropriately reflect them.

The classification presented in Table 1 represents one among several current, conflicting views on the taxonomic rank of the major groups. Excluded from the classification are two little-known and exclusively modern groups: Pentastomida, which have often been included in Arthropoda on the basis of resemblance to certain specialized mites, and Tardigrada, which, quite unlike arthropods, are said to be entercoelous and may very well represent a distinct phylum.

Geologic History

Arthropods originated in the series of evolutionary radiations of marine invertebrates

that took place near the beginning of the Cambrian. The major lineages, and numerous minor ones that have not survived to the present, diverged very early (Fig. 2). Trilobitomorpha, Crustacea, and Chelicerata are known from the Early Cambrian. The earliest known Uniramia, Onychophora, are known from the Middle Cambrian, where these early forms, protonychphorans, were evidently marine organisms. Onychphorans subsequently made the transition to terrestrial life, but are unknown in the post-Cambrian fossil record. Onychophorans, probably Euonychophora, have recently been discovered from the Carboniferous Francis Creek Shale in the Mazon Creek area. Trilobites, crustaceans, chelicerates, and related forms may have diverged rapidly from a trilobite-like common ancestor that was adapted for trunk-limb feeding on particulate detritus (Hessler and Newman, 1975; Cisne, 1975). While adaptations for this mode of feeding were retained in trilobites and primitive crustaceans, chelicerates diverged very early in becoming adapted for predaceous feeding.

The next major event in the marine realm was the replacement, in the later Devonian, of

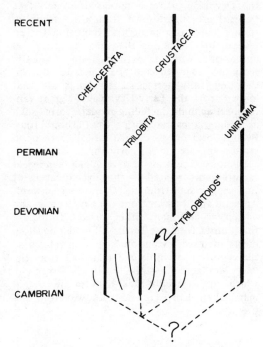

FIGURE 2. A phylogenetic tree showing the geologic ranges and inferred relationships among the major arthropod groups. (from Cisne, 1974a; copyright © 1974 by the American Association for the Advancement of Science). Was the common ancestor of arthropods itself an arthropod? If not, arthropods would fall into more than one phylum.

trilobites by crustaceans as the most diverse group of larger arthropods. This transition is marked by the appearance and diversification of more advanced malacostracan crustaceans (Eumalacostraca) in the later Devonian, and by the extinction of all but three trilobite families prior to the Carboniferous. The eumalacostracan radiation can be related to the appearance of a maxillary filter-feeding mechanism, a device that freed thoracic limbs from their roles in the primitive feeding mechanism and in this way made possible the evolution of many new and different sorts of thoracic limb mechanisms (Cannon, 1927; Cisne, 1974b). While copepods are recorded as early as the Cretaceous, virtually nothing is known of their ecological history. The radiation of true crabs, the most diverse group of modern malacostracans, is paleontologically well documented from its beginnings in the Jurassic.

The evolution of terrestrial arthropods has been closely related to the evolution of land plants. Middle Silurian archipolypodan myriapods are the first known terrestrial arthropods, and their appearance in the record is more or less coincident with the first occurrence of vascular plants. Hexapods are first known from the Devonian. Diverse faunas of insects (see *Insecta*) are first known from the Carboniferous. Chelicerates made the transition to terrestrial life at about the same time as did uniramians. While an extinct group of aquatic scorpions is known as early as the later Silurian, the first fully terrestrial arachnids are first known from the Early Devonian. A great variety of arachnids, including modern forms such as spiders, appeared during the Carboniferous. Beetles, the most diverse of all arthropod groups, appeared in the Permian.

The most recent major developments in insect evolution are related to the diversification of flowering plants from the Cretaceous onward. Ants are first known from the Cretaceous, and butterflies and many other modern forms are first known from the Tertiary. Further developments in insect evolution have been related to the evolution of birds and mammals from the later Mesozoic onwards. Fleas, for example, which are specially adapted for life amid mammalian hair, are first known from the Cretaceous.

J. L. CISNE

References

Anderson, D. T., 1973. *Embryology and Phylogeny in Annelids and Arthropods.* Oxford: Pergamon Press, 495p.

Cannon, H. G., 1927. On the feeding mechanism of *Nebalia bipes, Trans. Roy. Soc. Edinburgh,* 55, 355–369.

Cisne, J. L., 1974a. Trilobites and the origins of arthropods, *Science,* 186, 13–18.

Cisne, J. L., 1974b. Evolution of the world fauna of aquatic free-living arthropods, *Evolution,* 28, 337–366.

Cisne, J. L., 1975. Anatomy of *Triarthrus* and the relationships of the Trilobita, *Fossils and Strata,* 4, 45–63.

Hessler, R. R., and Newman, W. A., 1975. Origin of Crustacea, *Fossils and Strata,* 4, 437–459.

Lankester, E. R., 1904. The structure and classification of Arthropoda, *Quart. J. Micros. Sci.,* 47, 523–582.

Manton, S. M., 1964. Mandibular mechanisms and the evolution of arthropods, *Phil. Trans. Roy. Soc. London, Ser. B,* 247, 1–183.

Manton, S. M., 1965. The evolution of arthropod locomotory mechanisms, pt. 8, *J. Linnean Soc. London, Zool.,* 46, 251–483.

Manton, S. M., 1972. The evolution of arthropod locomotory mechanisms, pt. 10, *J. Linnean Soc. London, Zool.,* 51, 203–400.

Manton, S. M., 1977, *The Arthropoda: Habits, Functional Morphology, and Evolution.* Oxford: Clarendon Press, 527p.

Sanders, H. L., 1963. The Cephalocarida: comparative external anatomy, functional morphology, and larval development, *Mem. Conn. Acad. Arts Sci.,* 15, 1–80.

Tiegs, O. W., and Manton, S. M., 1958. The evolution of the Arthropoda, *Biol. Rev. Cambridge Phil. Soc.,* 33, 255–337.

Whittington, H. B., 1957. The ontogeny of trilobites, *Biol. Rev. Cambridge Phil. Soc.,* 32, 421–469.

Cross-references: *Branchiopoda; Burgess Shale; Chitin; Cirripedia; Crustacea; Eurypterida; Insecta; Mazon Creek; Population Dynamics; Solnhofen Formation; Trilobita.*

AVES

Approximately 985 extinct species of birds (Class Aves) have been described from all fossil deposits in the world. A nearly equal number of still-existing species is found in the geologic record. Their combined total, however, is less than one-fourth the number of avian species living in the world today. Since the record of birds spans a period of nearly 140 m yr, it is obvious that there is still much to learn about the birds of the past.

The fragility of avian material complicates the problem of fossil preservation in this group. Fossil egg shells, footprints, and imprints of feathers are recorded but contribute little toward classification of the birds responsible for them. Identification of fossil avian species is based largely on skeletal fragments, particularly limb bones.

The study of fossil birds, therefore, depends upon a detailed knowledge of the separate skeletal elements of the various orders, families, and genera of living birds. Although ornithologists have recognized the importance of certain parts of the skeleton in classifying avian groups, these lie largely in the palatal area of the skull that is rarely preserved in fossils. Modern keys for classification are, for the most part, based on external characters. The paleornithologist must develop a separate system of identification based on osteology, and must attempt to relate it to the accepted taxonomy of living birds. A good comparative collection of modern bird skeletons is a prime requisite for the student of fossil birds. A concise description of the avian skeleton, with bibliographic references, is given by George and Berger (1966, 5–21).

Systematics

Including all extinct forms, birds have been classified in 31–35 orders, the arrangement and subdivisions varying somewhat depending on the authority. The classification given below is essentially that of Wetmore (1960), but includes some later revisions by Storer (1971) and Brodkorb (1963–1971). Many authorities do not recognize the separation of the ratites (Palaeognathae) from the carinates (Neognathae). And among those who do, there is difference of opinion regarding the placement of the Tinamous.

Two extinct orders, known only from incomplete jaw parts, are omitted in the list for want of clear determination of their relationship to birds or to reptiles. Both orders, Caenagnathiformes and Gobipterygiformes, are from the Cretaceous.

The earliest geologic record is given below for each order or its subdivisions. The terminal record is added for extinct groups that extended over more than one geologic period or epoch. Extinct entries are marked with an asterisk.

CLASS AVES
  *Subclass Archaeornithes, ancestral birds
    *Order Archaeopterygiformes—Jurassic (only)
  Subclass Neornithes, true birds
    *Superorder Odontognathae, toothed birds
      *Order Hesperornithiformes—Cretaceous (only)
      *Order Ichthyornithiformes—Cretaceous (only)
    Superorder Palaeognathae, ratites
      Order Struthioniformes, ostriches—Eocene
      Order Rheiformes, rheas—Pliocene
      Order Casuariiformes, cassowaries, emus—Miocene
      *Order Aepyornithiformes, elephant birds—Eocene-Subrecent
      *Order Dinornithiformes, moas—Miocene-Subrecent
      Order Apterygiformes, kiwis—Pleistocene
      Order Tinamiformes, tinamous—Pliocene

Superorder Neognathae, carinates
  Order Gaviiformes, loons—Cretaceous
  Order Podicipediformes, grebes—(Cretaceous?)-Miocene
  Order Sphenisciformes, penguins—Eocene
  Order Procellariiformes, albatrosses, shearwaters—Eocene
  Order Pelecaniformes
    *Suborder Odontopterygia, bony-toothed birds—Eocene-Pliocene
    *Suborder Cladornithes—Oligocene-Miocene
    Suborder Phaethontes, tropic birds—Eocene
    Suborder Pelecani
      Superfamily Pelecanoidea, pelicans—Miocene
      Superfamily Suloidea, boobies, cormorants—Cretaceous
    Suborder Fregatae, frigatebirds—Eocene
  Order Ciconiiformes
    Suborder Ardeae, herons, bitterns—Eocene
    Suborder Balaenicipites, whaleheaded stork—Miocene?
    Suborder Ciconiae
      Superfamily Scopoidea, hammerhead—no fossils
      Superfamily Threskiornithidae, ibises—Cretaceous
      Superfamily Ciconioidea, storks—Eocene
  Order Phoenicopteriformes, flamingoes—Cretaceous
  Order Anseriformes
    Suborder Anhimae, screamers—Pleistocene
    Suborder Anseres, ducks, geese, swans—Eocene
  Order Falconiformes
    Suborder Cathartae
      *Superfamily Neocathartoidea—Eocene (only)
      Superfamily Cathartoidea, New World vultures—Eocene
    Suborder Accipitres
      Superfamily Sagittarioidea, secretary bird—Miocene
      Superfamily Accipitroidea, hawks, Old World vultures—Eocene
      Superfamily Falconoidea, falcons—Miocene
  Order Galliformes
    Suborder Galli
      Superfamily Cracoidea, curassows—Eocene
      Superfamily Phasianoidea, pheasants, quail, turkeys—Eocene
    Suborder Opisthocomi, hoatzins—Miocene
  Order Gruiformes
    Suborder Grues, cranes, limpkins, trumpeters
      Superfamily Gruoidea, cranes—Eocene
      Superfamily Ralloidea, rails, gallinules—Eocene
    Suborder Mesitornithes, roatelos—no fossils
    Suborder Turnices, buttonquail—Pleistocene
    Suborder Heliornithes, finfoots, sungrebes—no fossils
    Suborder Rhyncoceti, kagu—no fossils
    Suborder Eurypygae, sun-bitterns—no fossils
    *Suborder Gastornithes, *Diatryma* etc.—Paleocene-Eocene
    Suborder Cariamae
      *Superfamily Phororhacoidea—Oligocene-Pleistocene
      Superfamily Cariamoidea—Oligocene
    Suborder Otides, bustards—Eocene

Order Charadriiformes
  Suborder Charadrii
    Superfamily Jacanoidea, jacanas—Pleistocene
    Superfamily Charadrioidea, shorebirds—Cretaceous
    Superfamily Chionidoidea, seedsnipes—no fossils
  Suborder Lari, gulls, terns—Oligocene
  Suborder Alcae, auks, murres, puffins—Oligocene
Order Columbiformes
  Suborder Pterocletes, sandgrouse—Eocene
  Suborder Columbae, pigeons, dodo, solitaire—Miocene
Order Psittaciformes, parrots, macaws—Miocene
Order Cuculiformes
  Suborder Musophagi, plantain-eaters—Eocene
  Suborder Cuculi, cuckoos, roadrunners—Eocene
Order Strigiformes, owls—Cretaceous
Order Caprimulgiformes
  Suborder Steatornithes, oil-birds—no fossils
  Suborder Caprimulgi, goatsuckers—Eocene
Order Apodiformes
  Suborder Apodi, swifts—Eocene
  Suborder Trochili, hummingbirds—Pleistocene
Order Coliiformes, colies—Miocene
Order Trogoniformes, trogons—Eocene? or Oligocene
Order Coraciiformes
  Suborder Alcedines
    Superfamily Alcedinoidea, kingfishers—Oligocene
    Superfamily Todoidea, todies—Pleistocene
    Superfamily Momotoidea, motmots—Pleistocene
  Suborder Meropes, bee-eaters—Pleistocene
  Suborder Coracii, rollers, hoopoes—Eocene?-Oligocene
  Suborder Bucerotes, hornbills—Eocene
Order Piciformes
  Suborder Galbulae
    Superfamily Galbuloidea, puffbirds—Eocene
    Superfamily Capitonoidea, toucans, barbets—Miocene
  Suborder Pici, woodpeckers—Pliocene
Order Passeriformes
  Suborder Eurylaimi, broadbills—Miocene
  Suborder Furnarii, wood hewers—Pleistocene
  Suborder Tyrani, cotingas—Pleistocene
  Suborder Menurae, lyrebirds—no fossils
  Suborder Passeres, songbirds—Eocene

The Fossil Record

Avian fossils are now recorded on all continents. The greatest number have come from North America and Europe, but the occurrence is undoubtedly less a matter of actual distribution than of scientific exploration. In Australia, for example, until the 1960s, records of fossil birds, except for penguins, were largely confined to the Pleistocene. Owing to persistent paleontologic efforts, a promising Tertiary avifauna has now come to light (Rich, 1976).

Rich (1974) also reviewed the African fossil record, noting that on this continent, also, new discoveries are at hand and are in process of study by several authors.

Mesozoic Birds. The most famous avian Mesozoic locality in the world is that of the lithographic limestone deposits in Solnhofen, Bavaria. Here, in Late Jurassic time (approximately 140 m yr ago) very fine sediments accumulated along the coast of a shallow sea or lagoon. Remains of ammonites and other marine creatures are mingled with those of land-dwelling forms—twigs of trees; flying insects; an occasional flying reptile (pterosaur); and the primitive bird, *Archaeopteryx.*

There are six known skeletal specimens of *Archaeopteryx,* in addition to a single feather impression that was the first evidence of birds to be discovered in the Solnhofen area. Shortly after the feather discovery, in 1861, a nearly complete skeleton was found, with the impressions of feathers along wings, body, and tail. The imagination could not conceive a more perfect example of the combination of reptilian and avian characters than is found in *Archaeopteryx.* The jaws are toothed. The vertebrae have simple, concave articular facets, more reptilian than avian; the tail is long, consisting of many vertebrae, in contrast to the shortened, bony tail of modern carinate birds in which the feathers are borne fanwise on a large terminal pygostyle. The hand bones, though reduced to three, are not fused as in modern birds; the fingers have sharp claws. Had the feather impressions been lacking, the creature might well have been considered reptilian. Although a sternum is present, it is a simple, flat structure that could not have borne strong musculature such as is necessary for sustained flight. The rib basket, also, lacks the rigidity of that of modern flying birds. *Archaeopteryx,* apparently, was capable of only the simplest feats of flight.

The first skeleton of *Archaeopteryx,* described as *Archaeopteryx lithographica* (Meyer, 1861), is housed in the British Museum (Natural History). Another excellent specimen, found in 1877 (see Fig. 1), is in the Berlin Museum. Three others have been retrieved in recent years. Still another was actually found in 1857, before any of the others, but was described as a species of pterosaur. Its true identity was not recognized for over a hundred years (Ostrom, 1970).

Authorities have differed regarding the naming of the British and Berlin museum specimens. The first name used was *Archaeopteryx lithographica.* Under the belief that this name applied only to the single feather, the designation *Archaeopteryx macrura* was proposed for the

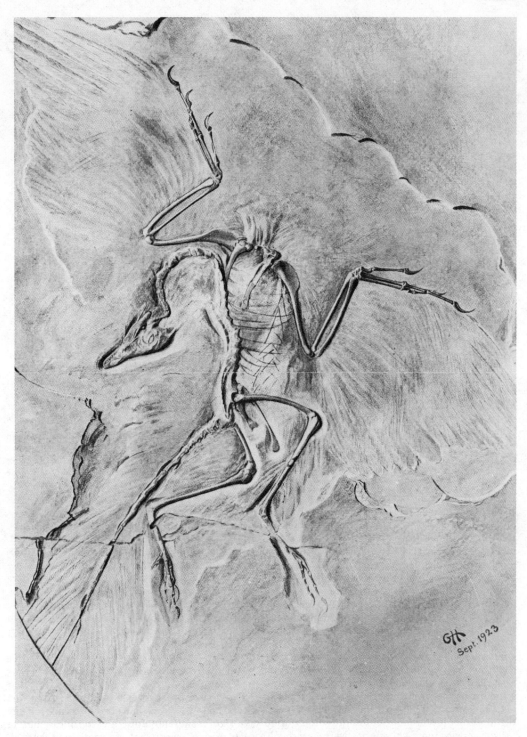

FIGURE 1. *Archaeopteryx siemensii,* primitive bird of the Jurassic (from Heilmann, 1927; drawing based on specimen in the Berlin Museum).

skeleton in the British Museum. Later, the Berlin Museum bird was assigned to a separate genus and species, *Archaeornis siemensii.* Most recent studies place all specimens in *Archaeopteryx,* but recognize two species, *A litho-graphica* and *A. siemensii* (Brodkorb, 1963). A detailed account of the original dating of the report on the feather and the first skeleton is given by de Beer (1954).

Only one other Jurassic occurrence of a bird is recorded. This is a poorly preserved cranial section found in the United States (Wyoming). The specimen (described as *Laopteryx prisca*) has been subjected to critical study since its description in 1881. Reviewing the evidence, Brodkorb (1971b) concludes "the probability seems strong that *Laopteryx prisca* is a reptile, not a bird."

In the 50 to 60 m yr that elapsed between the time when *Archaeopteryx* practiced its first attempts at flight, and the later Mesozoic records, the avian type was firmly established. Thirty-six extinct species are recorded from the Cretaceous of the United States, Canada, South America, Europe, and Mongolia.

The best preserved and most numerous specimens of Cretaceous birds are found in the United States in the Kansas chalk deposits. They represent the extinct orders Hesperornithiformes and Ichthyornithiformes. Skeletal and partial skull parts of the typical genus of each order are preserved. Marsh, who monographed these birds (1880), described the typical genera of both orders as being toothed, though possessing recognizably avian characters of the skeleton.

Three genera and five species are referred to the Hesperornithiformes, all but one of which are from Kansas. A nearly complete skeleton of the typical genus, *Hesperornis,* was found. This bird was highly specialized for diving and swimming, with strong legs more markedly adapted for this purpose than in present-day divers such as loons and grebes. The clavicles were not united to form a furcula; the sternum lacked a keel; and the wing bones were so diminished in size that the bird was completely flightless. Although *Hesperornis* had the avian type of saddle-shaped articular facets of the vertebrae, and the tail was shortened, the vertebral column lacked a terminal pygostyle. Most outstanding of the several species was *Hesperornis regalis,* some 6 feet (1.8 m) in length (see Fig. 2).

Two genera and eight species were described for the Ichthyornithiformes, all but one of which came from Kansas. The several partial skeletons of the type genus, *Ichthyornis,* show these birds to be distinctly different from *Hesperornis.* Gull-like in size, they may have

FIGURE 2. *Hesperornis regalis,* flightless, toothed diving bird of the Cretaceous. Artist's restoration in display at Natural History Museum of Los Angeles County, based on skeleton in Peabody Museum, Yale University. (Photo by Natural History Museum.)

been gull-like in habits, as well. The wings, furcula, and breast bone were well developed, indicating good capacity for flight. The vertebrae were more primitive than those of *Hesperornis,* having biconcave articular facets. There was, however, indication of a terminal pygostyle. For a time it was believed that the toothed jaws assigned to *Ichthyornis* were, in reality, reptilian (mosasaur) jaws (Gregory, 1952). Later studies by Walker (1967) and Gingerich (1972) have confirmed Marsh's belief that the jaws are in fact avian and belong to *Ichthyornis.*

Only two other orders recorded from the Cretaceous (both extinct) are known from the skull or mandible. Both are toothless, but in each case the specimens are so distinct from all other known forms that there is question as to whether they are truly avian. *Caenognathus* (Order Caenagnathiformes) was described from a lower mandible found in Alberta, Canada (Sternberg, 1940). Cracraft (1971), who described a second species of the genus from a fragmentary specimen, notes its reptilian features, but believes the preponderance of evidence favors avian affinities. He suggests possible relationship with the Galliformes and Anseriformes. The other order, Gobipterygiformes, was described from the Gobi Desert region of Mongolia. It is based on a small partial skull with mandible. The single species, *Gobipteryx minuta* (Elzanowski, 1974) is tentatively placed with the ratites.

In addition to the species representing the four extinct orders, another twenty Cretaceous species are referred to modern orders of birds. None, however, is represented by a skull, or even partially complete skeleton. Whether or not the allocations are precise, they do indicate the occurrence before the end of the Mesozoic (70 m yr ago) of birds suggestive of ancestral loons, grebes (?), cormorants, pelicans, ibises, flamingos, owls, and shorebirds.

Cenozoic Birds. Most of the carinate orders are known to have been in existence before the end of the Eocene. There are, however, no early Tertiary records of parrots, colies, or woodpeckers; and of the ratites, only the ostriches and elephant birds are recorded. Although many orders are represented by families now extinct, at least 38 existing families are represented in the Eocene or Oligocene.

The finest early Tertiary avian fossils have come from Montmartre and the phosphorites of Quercy in France. They were monographed in detail in a classic work by Milne-Edwards (1867–1871). The London clays, and the fresh water deposits of Wyoming, U.S.A., have also yielded important bird remains. Scattered Paleocene and Eocene records are known from other North American and European areas and from Africa, South America, Asia, Australia, and New Zealand.

By the middle Tertiary (Miocene), many avian fossils so closely resemble existing forms that they are recorded under modern genera. In North America, the mid-Tertiary marine localities in southern California are noteworthy for the unusual preservation of avian fossils. Partial skeletons, some nearly complete, are found impressed in fine-grained shales. Twelve species have been described from six localities, the most notable of which is near the town of Lompoc, California (Miller, 1925b). See Fig. 3.

A few existing species have been identified from the late Tertiary (Pliocene), but there is some doubt as to the authenticity of these identifications as they are based on incomplete skeletal material. In the Pleistocene, however, hundreds of existing species are recorded. In fact, it is suggested that all living species had their origin in the Pleistocene.

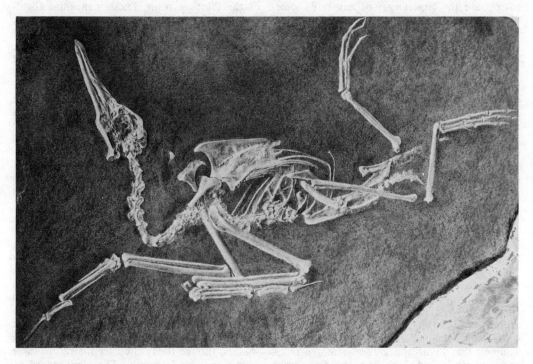

FIGURE 3. Fossil imprint of the skeleton of Miocene shearwater from the diatomaceous shales of Lompoc, California (from Miller, 1925b).

The richest Pleistocene bird-bearing localities are found in the United States in the asphalt deposits of southern California, particularly those at Rancho La Brea, in Los Angeles. Over 100,000 bones of birds have been recovered from the latter site where, during the late Pleistocene (14,000 to 40,000 yr ago), sticky tar accumulated, entrapping many forms of life. Pioneer work in this field was accomplished by Miller (1925a). A census of the birds (Howard, 1962) shows a preponderance of flesh-eaters. Other rich Pleistocene deposits yielding avian remains are found in Florida, Oregon, Ecuador, Peru, England, Italy, Hungary, Soviet Union, Israel, and Africa.

There is nothing in the avian fossil record to equal the sequences of genera that mark the evolution of many groups of mammals. Nevertheless, the avian record does present a number of significant facts regarding relationship, abundance and distribution of some groups of birds. It also reveals the past occurrence of unique and spectacular forms.

Ratites. Bones and eggshell fragments, presumably representing the elephant birds (Order Aepyornithiformes) are found in the Eocene and Oligocene of Egypt and the Eocene of the Middle East. Except for these early Tertiary records, members of this order are known only from the prehistoric Holocene of Madagascar. These huge, ostrich-like birds are credited with having laid the largest eggs of any bird, some being of 2-gallon capacity.

The most diverse of all flightless terrestrial birds were the moas (Dinornithiformes). Seven extinct genera are known, the birds ranging in size from that of a turkey to giants of 3–3.5 m (10–12 ft) in height, the largest of all known birds (*Dinornis maximus*). All are found in New Zealand, and, except for one mid-Tertiary representative, all have been found in deposits of Quaternary (prehistoric Holocene) age; see Oliver (1955).

Two primitive genera related to the ostriches appear in the Eocene of Europe. The modern genus, *Struthio,* is first recorded from the late Miocene of Moldavia. Pliocene and Pleistocene records are known from China, Mongolia, Europe, and Africa. Fossils of the present-day species, *Struthio camelus,* are cited from the Pleistocene of Mongolia, Arabia, and Algeria.

All fossil records of rheas and cassowaries are limited to the Pliocene and Pleistocene, the former from Argentina, the latter from Australia. Emus and an extinct family of the Casuariiformes appear in the Miocene of Australia.

Carinates. Among diving birds, it was long assumed that loons and grebes stemmed from common stock, and *Colymoboides* of the early Tertiary was cited as proof. Storer's (1956) study of over a hundred separate bones of *Colymboides* has now demonstrated that the skeletal characters of this early diver, while more primitive than those of living loons, fall strictly within this order and do not indicate relationship to the grebes.

The diving adaptation occurred independently in several orders of birds, involving a high degree of specialization of either the hind or fore limb. The loons, grebes, and the Cretaceous *Hesperornis* are examples of the former type of adaptation. The penguins and flightless auks show a flattening of the wing to form a flipper.

Fossil records of penguins date back to the Eocene and include 32 extinct genera. Simpson (1946) and Marples (1952) demonstrate that even the earliest forms known had already lost aerial flight and had wings adapted for swimming; only slight differences in proportions of the bones are evident. The entire fossil record of penguins is confined to the southern hemisphere, within the limits of their present distribution.

Flightlessness developed twice within the auk family. In neither was the adaptation of the forelimb as highly developed for swimming as in the penguins, although it was closely approached in the Pliocene *Mancalla* (Miller and Howard, 1949). See Fig. 4. *Praemancalla,* of the late Miocene, is viewed as a possible ancestor of the Pliocene form. These mancalline flightless auks are known only from the west coast of California and Mexico. The Great Auk, which became extinct within historic time, represents a separate evolutionary line toward flightlessness. Pleistocene fossils of the Great Auk are known as far south as Florida.

A remarkable group of marine birds, the Odontopterygia (Order Pelecaniformes) appears in Tertiary records. The jaws of these birds bore tooth-like projections, but a section made through jaw and "tooth" reveals that the projections are not true teeth but extensions of the jaw bone. The most outstanding specimen of this group, described as *Osteodontornis orri,* was found in the Miocene of California. It is a partial skeleton, with feather imprints, embedded in a slab of shale (see Fig. 5). The skull is approximately 40 cm (16 in) long, the wing span estimated at 5 m (16 ft). This group was widespread and is known from the Eocene of England, Miocene of Europe and both eastern and western United States, and the late Miocene or Pliocene of New Zealand.

The fossil record of flamingos indicates that they were once a dominant group among waders, though today they are limited to three genera in a single family. The fossil record includes five extinct families and at least eleven

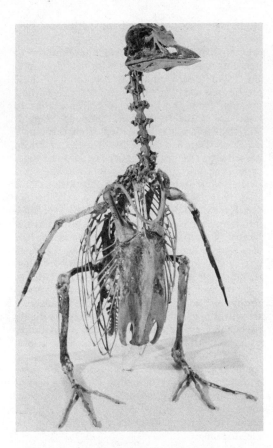

FIGURE 4. *Mancalla*, flightless auk of the Pliocene of the west coast of North America. Composite skeleton in Natural History Museum of Los Angeles County. (Photo by Natural History Museum.)

extinct genera, found in Europe, Asia, North and South America, and Australia. In addition to three Cretaceous records, the group is represented in every epoch of the Tertiary as well as the Pleistocene, but with greatest abundance in the Oligocene and Miocene. The genus *Palaelodus* of Europe is the best known of the primitive Tertiary flamingos that were shorter-legged and straighter-billed than are living forms. A related genus, *Megapaloelodus,* is known from the Tertiary of North America.

The fossil record of the Order Anseriformes, comprising the ducks, geese, and swans, was reviewed by Howard (1973). More than a hundred extinct species and 25 extinct genera are now listed from Eocene through Pleistocene. Two extinct subfamilies, whose relationship to living ducks or geese is ambiguous, are recorded in the Eocene of France and the United States (Utah). True ducks and swans are recognizable from the Oligocene, and geese of the existing genus *Anser* occur in the Miocene.

Vultures and hawks (Order Falconiformes) are first recorded from the Eocene. The cathartid vultures, now limited to the New World, have a fossil record in Europe as well (Cracraft and Rich, 1972). On the other hand, the present-day Old World vultures, the Suborder Accipitres, had North American representation in the Tertiary and Quaternary (Brodkorb, 1964). One of the most spectacular members of the Order Falconiformes was the giant *Teratornis* of the Pleistocene. The bird was undoubtedly vulture-like in habits; the feet were not adapted for grasping live prey and the lower jaw was weak. But the high, hooked

FIGURE 5. Skull of *Osteodeotornis orri,* a strange "bony-toothed" bird of the Miocene (from Howard, 1957). Skeleton in Santa Barbara Museum of Natural History.

beak must certainly have been used for tearing flesh. The best known of the teratorns is *Teratornis merriami,* with a wingspread of probably 3.5 m, represented in the Rancho La Brea asphalt deposits of California by many bones and found elsewhere throughout the United States and into Mexico in Pleistocene deposits.

In addition to one Late Cretaceous record from Romania (Harrison and Walker, 1975), the owls (Order Strigiformes) are well represented in the early Tertiary of France and North America.

Approximately 90 genera of the Order Gruiformes are recorded from Tertiary and Quaternary deposits throughout the world. Five existing families, the cranes, rails, limpkins, bustards, and seriamas, are represented in the early epochs of the Tertiary. Several extinct gruiform families also occurred in Tertiary time. Among these are the spectacular large flightless terrestrial birds that appear to have been formidable predators. These are best known by *Gastornis* and *Diatryma* (Fig. 6) of the Paleocene and Eocene of the United States and Europe (Suborder Gastornithes), and the members of the Superfamily Phororhacoidea (Suborder Cariama) of the Oligocene through the Pliocene of Argentina, with a Pleistocene record from Florida. Some of these great birds stood as much as 2 m (7 ft) in height, on long,

stout legs, and were equipped with a massive, hooked beak.

Less than half of the existing families of perching birds (Order Passeriformes) appear in the fossil record, and only one-third of these are represented in the Tertiary. The earliest records, from the Eocene of Europe, refer to relatives of the titmice and starlings. Wagtails and shrikes are first recorded from the Oligocene of Europe. The first North American passeriform record also occurs in the Oligocene. The specimen, found in Colorado, is a partial skeleton with feather impressions in a slab of shale. It has been assigned to a distinct family, Palaeospizidae, near the larks. A similarly preserved specimen from another extinct family, Palaeoscinidae, is recorded from the Miocene of California. The bird had a thrush-like beak, but other skeletal characters align it more closely with waxwings and bulbuls. This type of preservation is rare among passeriforms. As the members of this order are largely tree-dwelling in habit, their remains are likely to become scattered before there is possibility of preservation by entombment. As a consequence, the number of fossil perching birds is small in proportion to the multitude of these delightful birds that compose so large a part of our present-day avifauna. There are, however, more than 40 extinct species recorded, and at least 275 existing species are known from the Pleistocene.

Origin of Birds

There is no dispute among scientists that birds have evolved from reptiles. *Archaeopteryx,* the earliest avian fossil, is distinctly reptilian in many features, and even modern birds reflect this relationship. Opinions differ, however, regarding the immediate ancestors and the origin of flight.

Most paleontologists are agreed that the root stock from which birds emerged is to be found in the Triassic archosaurian reptiles of the Order Thecodontia (see *Reptilia*), which also gave rise to dinosaurs, pterosaurs, and crocodiles. The basic archosaurian structure is exemplified in representatives of the Triassic suborder Pseudosuchia, such as *Euparkeria* from the Lower Triassic (approximately 230 m yr. ago) of South Africa. This little reptile was about 1 m long with a long, bony tail. Many of the bones were hollow as in birds. The skull was delicately structured, and the jaws were toothed. The hind legs were long, the forelimbs shortened, the fingers free. A paired row of slightly elongate, ribbed scales occured along the back. It is suggested that such scales may have been forerunners of feathers.

FIGURE 6. *Diatryma,* a 2-m tall, gruiform bird of the early Tertiary. (from Matthew and Granger, 1917). Artist's conception based on a nearly complete skeleton in the American Museum of Natural History.

But what ancestral forms are postulated to fill the 100 m yr interval between these Triassic archosaurians and the Jurassic primitive bird, *Archaeopteryx?* Many paleontologists suggest a tree-dwelling form, jumping from branch to branch in the manner of flying squirrels or flying lizards. Proponents of this theory point to the clawed, free fingers and reverse hind toe of *Archaeopteryx* as evidence of prior adaptation to tree climbing and perching. Others believe that the immediate ancestors were bipedal, running forms, either paralleling in development the smaller cursorial dinosaurs (Order Saurischia; see *Dinosaurs*) or actually related to them. In line with the latter theory, some contend that feathers originated independently of flight as a body covering minimizing heat loss. It is even suggested that the ratites represent the descendants of early cursorial dinosaurs and are not degenerate fliers (Lowe, 1935; Jensen, 1969). Proponents of the theory of a cursorial ancestry of birds account for elongation of the feathers of the forelimb as associated with attaining speed in running, or for assault or snaring of prey.

Detailed descriptions of *Archaeopteryx* with discussions of its ancestry are presented in the works of Heilmann (1927) and de Beer (1954). More recently, Ostrom (1976) has reviewed the subject. He cites numerous similarities between *Archaeopteryx* and the coelurosaurian dinosaurs (Suborder Theropoda) in substantiation of his belief that not only were birds descended from these predaceous, fleet-footed theropods, but, like them, were predaceous in habit.

History of Paleornithological Research

Compared with the number of researchers in mammalian paleontology, workers in the field of fossil birds have been few. Among the earliest studies to have far-reaching effect are those of George Cuvier [1769-1832]. Cuvier's work on the fossils of the Paris Basin formed the basis of the species descriptions by Paul Gervais [1816-1879] and Alphonse Milne-Edwards [1845-1900]. The latter's monograph published over a period from 1867 to 1871 is one of the classics of paleornithology. In Germany, Meyer published the first description of *Archaeopteryx* in 1861. In England, Sir Richard Owen [1804-1892] and Richard Lydekker [1849-1915] were important pioneers, with Owen's work extending to the New Zealand moas. Interest in paleornithology in the United States arose in the 1870s with the descriptions by Othniel Charles Marsh [1831-1899] of the Cretaceous birds of Kansas and New Jersey, followed soon after by the work of Edward Drinker Cope [1840-1897] on the Pleistocene

lake beds of Oregon. Toward the end of the 19th century, the brothers Ameghino (Carlos and Florentino) had begun their work on the Miocene of Argentina, including the first discoveries of the giant predaceous terrestrial bird, *Phororhacos.* Although Cope had opened up interest in fossil birds in Oregon in the 1880s, extensive paleornithological studies in the western United States began in the early 20th century with the work of Loye Holmes Miller [1874-1970] on the newly discovered Rancho La Brea asphalt pits of southern California. The first comprehensive world-wide review of the field of paleornithology was compiled in a monumental work by Kalman Lambrecht in 1933. Useful later works include Pierce Brodkorb's Catalogue of Fossil Birds (1963-1971) and the following general reviews; Brodkorb (1971b), Fisher (1967a,b), Kuhn (1971), Kurotchkin (1971) and Swinton (1958).

Paleornithology has grown since its infancy a little over a hundred years ago. A section on the subject was included in the International Ornithological Congress of 1970 and 1974. A symposium was held in the Soviet Union in 1971. A recent volume celebrating the 90th birthday of the outstanding American paleornithologist, Alexander Wetmore, was issued while this volume was in preparation; it includes the latest works of 18 avian paleontologists concerning fossil birds from all continents (Olson, 1976). The number of persons studying fossil birds today probably totals twice this number. This generation is not only seeking new areas of discovery, but is attempting analysis of individual orders and families of birds based on a restudy of previously published scattered records.

HILDEGARDE HOWARD

References

*Brodkorb, P., 1963. Catalogue of fossil birds: Part 1, Archaeopterigiformes through Ardeiformes, *Bull. Florida State Mus. Biol. Sci.,* 7, 179-293.

*Brodkorb, P., 1964. Catalogue of fossil birds: Part 2, Anseriformes through Galliformes, *Bull. Florida State Mus. Biol. Sci.,* 8, 195-335.

*Brodkorb, P., 1967. Catalogue of fossil birds: Part 3, Ralliformes, Ichthyornithiformes, Charadriiformes, *Bull. Florida State Mus. Biol. Sci.,* 11, 99-220.

*Brodkorb, P., 1971a. Catalogue of fossil birds: Part 4, Columbiformes through Piciformes, *Bull. Florida State Mus. Biol. Sci.,* 15, 163-266.

*Brodkorb, P., 1971b. Origin and evolution of birds, in Farner and King, 1971, vol. 1, 19-55.

*Brodkorb, P., 1978. Catalogue of fossil birds (Passeriformes), *Bull. Florida State Mus.,* 23, 139-228.

Cracraft, J., 1971. Caenagnathiformes: Cretaceous birds convergent in jaw mechanism to dicynodont reptiles, *J. Paleontology,* 45, 805-809.

Cracraft, J., and Rich, P. V., 1972. The systematics

and evolution of the Cathartidae in the Old World Tertiary, *Condor,* 74, 272-283.

*de Beer, G., 1954. Archaeopteryx lithographica. London: British Mus. (Nat. Hist.), 68p.

Elzanowski, A., 1974. Results of the Polish-Mongolian palaeontological expeditions: Part V, Preliminary note on the palaeognathous bird from the Upper Cretaceous of Mongolia, *Palaeontol. Polonica,* 30, 103-109.

Farner, D. S., and King, J. R., eds., 1971. *Avian Biology.* New York: Academic Press, 586p.

Fisher, J., 1967a. Fossil birds and their adaptive radiation, in Harland et al., 1967, 133-154.

Fisher, J., 1967b. Aves, in Harland et al., 1967, 733-762.

*George, J. C., and Berger, A. J., 1966. *Avian Myology.* New York: Academic Press, 500p.

Gingerich, P. D., 1972. A new partial mandible of *Ichthyornis, Condor,* 74, 471-473.

Gregory, J. T., 1952. The jaws of the Cretaceous toothed birds, Ichthyornis and Hesperornis, *Condor,* 54, 73-88.

Harland, W. B., et al., eds., 1967. *The Fossil Record.* London: Geol. Soc. London, 827p.

Harrison, C. J., and Walker, C. A., 1975. The Bradycnemidae, a new family of owls from the Upper Cretaceous of Romania, *Palaeontology,* 18, 563-570.

*Heilmann, G., 1927. *The Origin of Birds.* New York: Appleton, 210p.

Howard, H., 1957. A gigantic "toothed" marine bird from the Miocene of California, *Santa Barbara Mus. Nat. Hist. Geol. Bull.,* 1, 1-23.

Howard, H., 1962. A comparison of avian assemblages from individual pits at Rancho La Brea, California, *Los Angeles Co. Mus. Contrib. Sci.,* 58, 1-24.

*Howard, H., 1973. Fossil Anseriformes, in J. Delacour, ed., *Waterfowl of the World,* Vol 4. London: Country Life, 233-326, 371-378.

Jensen, J. A., 1969. Fossil eggs from Utah and a concept of surviving feathered reptiles, *Utah Acad. Proc.,* 46, 125-133.

Kuhn, O., 1971. Die vorzeitlichen Vögel, *Neues Brehm-Bücherei,* 435, 72p.

Kurotchkin, E. N., 1971. Basic questions in the study of fossil birds (in Russian). *Zoologia pozvonochnykh voprosy ornitologii. Itogi Nauki VINITI T-11795,* 116-151.

*Lambrecht, K., 1933. *Handbuch der Palaeornithologie.* Berlin: Gebrüder Borntraeger, 1024p.

Lowe, P. R., 1935. On the relationship of the Struthiones to the dinosaurs and the rest of the avian class with especial reference to the position of *Archaeopteryx, Ibis,* 1935, 398-432.

Marples, B. J., 1952. Early Tertiary penguins of New Zealand, *New Zealand Geol. Surv., Paleontol. Bull.,* 20, 1-66.

Marsh, O. C., 1880. Odontornithes: A monograph of the extinct toothed birds of North America, *Rept. Geol. Explor. 40th Parallel,* 7, 201p.

Matthew, W. D., and Granger, W., 1917. The skeleton of *Diatryma, Amer. Mus. Nat. Hist. Bull.,* 37, 307-326.

Meyer, H. Von., 1861. *Archaeopteryx lithographica* (Vogel-Feder) und *Pterodactylus* von Solenhofen, *Neues Jahrb. Mineral., Geol. Palaeontol.,* 1861, 678-679.

Miller, L., 1925a. The birds of Rancho La Brea, *Carnegie Inst. Washington Publ.,* 349, 63-106.

Miller, L., 1925b. Avian remains from the Miocene of Lompoc, California, *Carnegie Inst. Washington Publ.,* 349, 107-117.

Miller, L., and Howard, H., 1949. The flightless Pliocene bird *Mancalla, Carnegie Inst. Washington Publ.,* 584, 201-228.

Milne-Edwards, A., 1867-1871. *Recherches Anatomiques et Paléontologiques pour Servir à l'Histoire des Oiseaux Fossiles de la France,* 2 vols. and atlas. Paris: Masson, 474p., 632p.

Oliver, W. R. B., 1955. *New Zealand Birds,* 2nd ed. Wellington: H. H. and A. W. Rand, 661p.

*Olson, S. L., ed., 1976. Collected papers in avian paleontology honoring the ninetieth birthday of Alexander Wetmore, *Smithsonian Contrib. Paleobiol.,* 27, 211p.

Ostrom, J. H., 1970. Archaeopteryx: Notice of a "new" specimen, *Science,* 170, 537-538.

Ostrom, J. H., 1976. *Archaeopteryx* and the origin of birds, *Biol. J. Linnean Soc.,* 8, 91-182.

*Rich, P. V., 1974. Significance of the Tertiary avifaunas from Africa (with emphasis on the Mid to Late Miocene avifauna from southern Tunisia), *Ann. Geol. Surv. Egypt,* 4, 167-210.

Rich, P. V., 1976. The history of birds on the island continent Australia, *Proc. 16th International Ornithological Congress, 1974,* 53-65.

Simpson, G. G., 1946. Fossil penguins, *Bull. Amer. Mus. Nat. Hist.,* 87, 1-100.

Sternberg, R. M., 1940. A toothless bird from the Cretaceous of Alberta, *J. Paleontology,* 14, 81-85.

Storer, R. W., 1956. The fossil loon Colymboides minutus, *Condor,* 58, 413-426.

Storer, R. W., 1971. Classification of birds, in Farner and King, 1971, vol. 1. 1-18.

Swinton, W. E., 1958. *Fossil Birds.* London: British Mus. (Nat. Hist.), 63p.

Walker, M. V., 1967. Revival of interest in the toothed birds of Kansas, *Trans. Kansas Acad. Sci.,* 70, 60-66.

Wetmore, A., 1960. A classification for the birds of the world, *Smithsonian Misc. Collect.,* 139(11), 1-37.

*Extensive references

Cross-references: *Bones and Teeth; Reptilia; Vertebrata; Vertebrate Paleontology.*

B

BELEMNITIDA

The Order Belemnitida comprises fossil coleoid (= dibranchiate) cephalopods that were the Mesozoic equivalents (but not ancestors, according to Jeletzky, 1966; see Fig. 1) of the modern squids (Order Teuthida) and cuttlefish (Order Sepiida).

Jeletzky (1965, 1966) has recognized three suborders in the Belemnitida:

(i) Belemnitina, which includes families Belemnitidae (Lower and Middle Jurassic), Hastitidae (Lower Jurassic to lower part of Upper Jurassic), Cylindroteuthididae (Middle Jurassic–Lower Cretaceous), Oxyteuthididae (Lower Cretaceous), Bayanoteuthididae (? Upper Cretaceous-Eocene), Belemnoteuthididae (Middle Jurassic–mid-Upper Jurassic) and Chondroteuthididae (upper part of the Lower Jurassic).

(ii) Belemnopsina, with the families Belemnopsidae (Middle Jurassic–lower part of the Upper Cretaceous), Duvaliidae (upper part of the Lower Jurassic–Lower Cretaceous), Belemnitellidae (Upper Cretaceous) and Dimitobelidae (Lower and Upper Cretaceous).

(iii) Diplobelina, with the single family Diplobelidae (upper part of the Upper Jurassic–lower part of the Upper Cretaceous).

At the end of the Cretaceous, Belemnitida became extinct except for a single family, Bayanoteuthididae, rare representatives of which occur in the Upper Eocene of France, Italy, and southern Germany (e.g., Roger, 1952, p. 726).

All the belemnites have an internal calcareous shell that typically comprises an elongate guard (or rostrum), composed of radiating

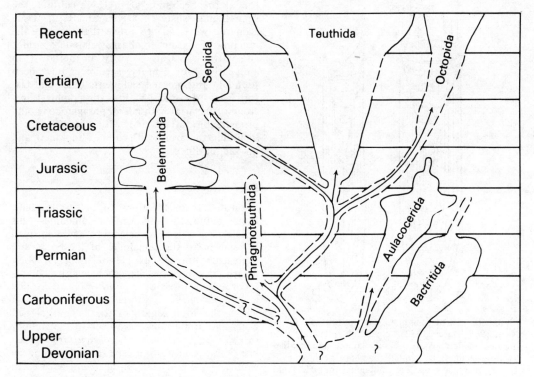

FIGURE 1. Phylogenetic relationships of orders of Coleoidea (from, Jeletzky, 1966. Courtesy of the University of Kansas).

calcite crystals; a chambered phragmocone enclosed by the anterior end of the guard, with the protoconch (or initial chamber) at its apex and the individual chambers linked by a siphuncle; and the proostracum, a very delicate plate-like extension of the dorsum of the phragmocone (Fig. 2). The guard is the only part of the belemnite shell that is commonly preserved; and, although the phragmocone may sometimes be found still attached, it has quite frequently become separated, leaving a conical depression (alveolus) in the anterior of the guard. The proostracum is rarely preserved.

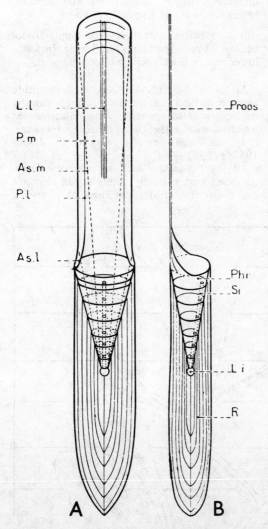

FIGURE 2. The complete belemnite shell (after Naef, 1922; Roger, 1952). A, ventral; B, lateral, with dorsum to left. As.l., As.m., lateral and median asymptotic lines; L.i., initial chamber or protoconch; L.l., longitudinal lines; Phr., phragmocone; P.l., P.m., lateral and median plates; Proos., proostracum; R., rostrum or guard; Si., siphuncle.

Isolated tentacle hooklets (described as *Onychites*) may sometimes be found.

Belemnite guards occur in great abundance in certain formations of the Jurassic and Cretaceous and were well known to ancient peoples. The word "belemnite" is a modern scientific term, coined by the 18th-century naturalists, from Greek words meaning "arrowstone." The ancient Greeks themselves, however, called them by the name "lynkourion," and the Romans called them "lapis lyncis"— both expressions meaning "lynx-stone," and based on the strange superstition that they are stones secreted by wild cats. The New Zealand Maori children used to gather belemnites to play with them; and called them "Rokekanae," or excrement of the mullet, which they thought was in the habit of leaping out of the water and leaving them behind on the shore. In medieval times, belemnites were believed to be the toenails and fingernails of devils, presumably of devils drowned in Noah's Flood. Later and more plausible interpretations were that belemnites are stalactites produced underground, or thunderbolts fallen from the sky ("thunderstones"). In the 17th century, the Sicilian naturalist and painter Agostino Scilla concluded that belemnites must be some kind of shell formed by an unknown type of mollusk. His inference we now know to have been correct for, at the beginning of the 19th century, the French naturalists G. Cuvier and J. B. M. de Lamarck were able to prove that they are the remains of extinct mollusks related to squids and cuttlefish. More recently, in the late 19th and early 20th century, specimens have been found in fine-grained sediments (Lias and Oxford Clay of Great Britain, Lithographic Limestone of Bavaria) with guard, phragmocone, and proostracum intact and with the soft parts, including the arms (with rows of hooklets), and ink sac, outlined.

Reconstructions

Various reconstructions of belemnites have been published and one accompanies this article (Fig. 3). The belemnite guard is small in relation to the remainder of the body, and represents only 1/7 or 1/8 of the total body length. Among the largest belemnite guards recorded are those of *Hibolithes ingens* Stolley and *Megateuthis giganteus* (Schlotheim), in which the total length of the belemnite (i.e., body and arms) must have been in the vicinity of 3–5 m. By comparison, *Architeuthis longimanus* Kirk, one of the largest of the modern squids, has a body length of 1.8 m, body girth of 1.5 m, and a total length of about 17 m. Other giant squids, *Architeuthis verrilli* Kirk

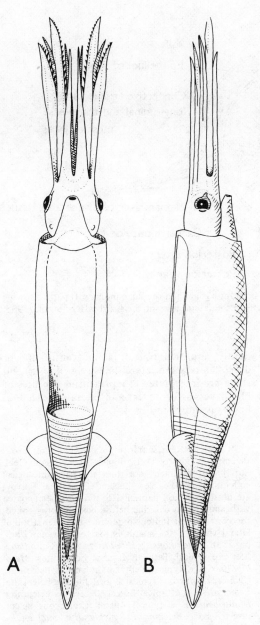

Lesueur, *L. opalescens* Berry) inhabiting the coastal shelves of the present-day oceans.

Habitat and Life History

It is thought that the belemnites and modern squids had similarities in both habitat and life history. Belemnites are commonly found in sediments of shelf facies and only rarely in those of geosynclinal or abyssal facies. The migration and paleogeographic distribution of belemnites can best be explained by assuming a shelf habitat, that is, regions of deep water would have been effective barriers to migration. These two points suggest that the belemnites, though free swimming, were probably confined to the shelves as are many of the modern common squids—though there are less well-known deep-water forms, including the giant squids. Belemnites, like present-day squids, were probably carnivorous, so their habitat was probably largely determined by that of their prey. If their prey was confined to the shelves, the belemnites in turn presumably frequented the shelves.

Like present-day squids, belemnites probably preyed upon small fish and crustacea and in turn were preyed upon by larger fish. Cuts and scratches on the surface of belemnite guards have been attributed to attacks by predators, and it has been suggested that the principal enemies of belemnites were fish (especially sharks) and ichthyosaurs. Over 250 belemnite guards were contained in the body of a shark found in the Lias of S Germany.

The common squids of the present day become sexually mature late in their second year or in their third year; and usually migrate to inshore waters for the breeding season and form great swarms on communal spawning grounds. Spawning is the climax of the lives of at least the smaller species and most die on the spawning grounds. The belemnites are thought to have had a similar life cycle. Three or four growth stages may be distinguished in many well-preserved adult belemnite guards; and it is likely that these are annual stages, as oxygen isotope analyses of samples taken across the diameter of belemnite guards have recorded temperature fluctuations interpreted as seasonal and consistent with a life cycle of 3–4 years.

Belemnites frequently occur as concentrations of guards ("belemnite battlefields"), apparently all of nearly the same stage of development. It is thought that these probably represent individuals that have spawned and died in the vicinity of the spawning grounds. After hatching, young squids remain in swarms in the shallow inshore waters. If the young

FIGURE 3. A reconstruction of the soft parts of a belemnite (*Megateuthis gigantus* Schlotheim), juvenile form (after Naef, 1922, Fig. 67d,e). A, ventral; b, lateral, with dorsum to left. Approximately $\frac{1}{5}$ natural size.

and *A. stockii* (Kirk) have body lengths of 2.7 m and 3.3 m respectively, and body girths of 2.7 m and 2.1 m. Belemnite guards, however, normally range in length from 38 mm to 127 mm and the entire body probably attained a length of between 0.3–1.2 m, comparable in size to the common squids (e.g., *Loligo forbesii* Steenstrup, *L. vulgaris* Lamarck, *L. pealeii*

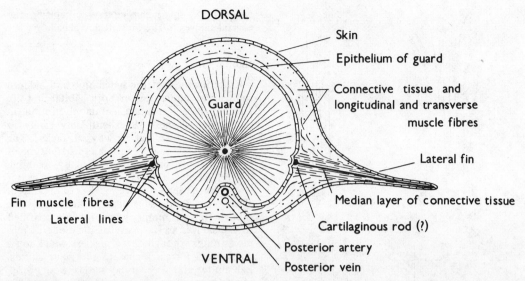

DORSAL

Skin

Epithelium of guard

Connective tissue and
longitudinal and transverse
muscle fibres

Guard

Lateral fin

Lateral fin

Fin muscle fibres

Lateral lines

Median layer of connective tissue

Cartilaginous rod (?)

Posterior artery

VENTRAL

Posterior vein

FIGURE 4. Hypothetical transverse section through a belemnite guard and the surrounding soft parts, showing relationship of the lateral fins and associated musculature to the lateral lines of the guard (after Stevens, 1965, Fig. 18).

belemnites had the same habits a swarm might die, for example, by poisoning or stranding, which could explain the concentrations of juvenile guards that are sometimes found.

Functional Significance of the Guard

Earlier writers envisaged the guard as functioning as a plough or digging point. Belemnites, however, probably propelled themselves in much the same way as do the modern squids (i.e., by slow rhythmic expulsion of water from the mantle cavity and/or undulation of the lateral fins for leisurely movement and rapid expulsion of the water from the mantle cavity—"jet propulsion"—for rapid movement). As these movements are mainly in the horizontal plane, it is possible that the distribution of soft parts in the belemnite required a counterweight (i.e., the guard) in the posterior part of the body to achieve stability for horizontal propulsion (Fig. 4). The chambered phragmocone probably contained liquid as well as gas so that the belemnite could change its density and posture by varying the amounts of liquid in the chambers, in much the same way as the modern cuttlefish vary amounts of liquid in the porous cuttlebone.

In some belemnite genera (e.g., *Belemnitella, Belemnella, Actinocamax;* See Fig. 5, M–O), the guard shows vascular markings, anastomosing and divaricating over its surface and their presence attest to the correctness of the assumption that the guard was internal. Other genera have grooves, median (ventral more commonly, and/or dorsal) or lateral (ventro- and/or dorso-lateral) and lateral lines (e.g., Fig. 5) and these are interpreted as representing the sites of blood vessels (grooves) and lateral fins (lateral lines), as shown in Fig. 4.

FIGURE 5. Representative Jurassic and Cretaceous Belemnites. (Photo: D. L. Homer, New Zealand Geol. Surv.) All specimens are from the palaeontological collections of the New Zealand Geological Survey, are illustrated 0.60 natural size, and have been coated with ammonium chloride before being photographed; the convention adopted to describe the orientation of lateral views of the guards is that of Stevens (1965, p. 233). *Belemnopsis, Hibolithes,* and *Conodicoelites* are typical Tethyan Jurassic belemnites; *Mesohibolites* is a typical Tethyan Cretaceous belemnite; *Cylindroteuthis* is a typical Boreal Jurassic belemnite; *Belemnitella* is a typical Boreal Cretaceous belemnite; *Dimitobelus* is a typical Austral Cretaceous belemnite. A,B,C, *Belemnopsis aucklandica aucklandica* (Hochstetter), lower Tithonian, New Zealand (A, ventral; B, dorsal; C, right lateral); D,E,F, *Hibolithes marwicki marwicki* Stevens, lower Tithonian, New Zealand (D, ventral; E, dorsal; F, left lateral); G,H, I, *Cylindroteuthis (Cylindroteuthis) puzosiana* (d'Orbigny), upper Callovian, Sutherland, Northern Scotland (G, ventral; H, dorsal; I, right lateral); J,K, L, *Mesohibolites uhligi* (Schwetzoff), upper Barremian, Bulgaria (J, ventral; K, dorsal; L, left lateral); M,N,O, *Belemnitella americana* (Morton), lower Maastrichtian, New Jersey, U.S.A. (M, ventral; N, dorsal; O, right lateral); P,Q,R, *Dimitobelus lindsayi* (Hector), Campanian, New Zealand (P, ventral; Q, dorsal; R, left lateral); S,T,U, *Conodicoelites flemingi* Stevens, Kimmeridgian, New Zealand (S, ventral; T, dorsal; U, right lateral).

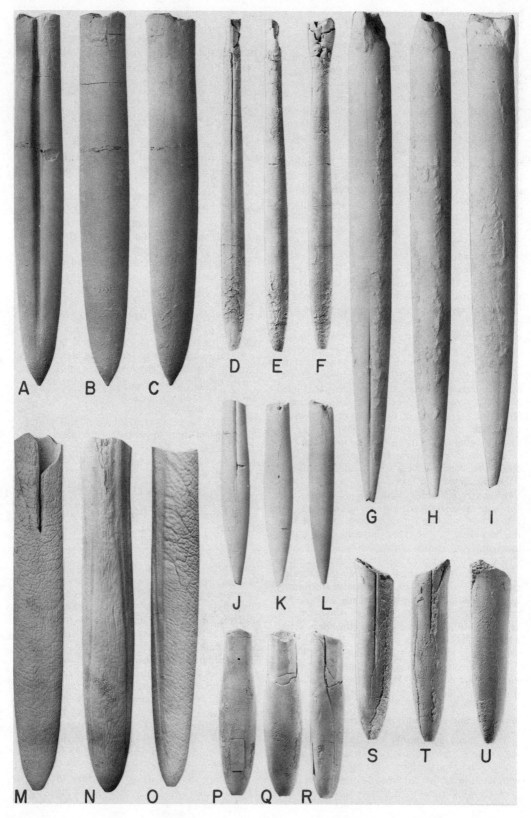

Evolution and Development of the Belemnite Shell

The belemnite shell may be thought of as descending from that of an archetypal mollusk, with a simple cone-shaped external shell. Heightening of the cone and cutting-off of the apex by septa gave rise to early Paleozoic primitive nautiloids such as *Endoceras, Piloceras,* and *Orthoceras* (see *Cephalopoda*). In *Orthoceras,* the shell takes the form of a straight, simple phragmocone, external in position and with a central siphuncle. In the belemnites, the symmetry of the orthocone is lost: the siphuncle is ventral; a guard has been developed posterior to the phragmocone and partly enclosing it; and the dorsal margin of the phragmocone is extended anteriorly as the proostracum. The septa of the phragmocone are more closely spaced than in *Orthoceras* and the entire shell is internal. In the early coleoid *Aulacoceras* (Upper Triassic; Order Aulacocerida) the guard is very small, enclosing only the apex of the phragmocone, and the phragmocone resembles that of *Orthoceras* more closely than those of the Jurassic and Cretaceous belemnites, as it increases only gradually in width and has the septa more widely spaced. Many authors (e.g., Naef, 1922; Donovan, 1964) have maintained that the modern cuttlefish (Sepiida), squid (Teuthida), and octopus (Octopida) evolved via the belemnites through a process of shell reduction, mainly by loss of the guard or its modification. Jeletzky (1966) maintains, however, that the modern coleoid orders evolved independently of the Belemnitida (cf. Fig. 1).

In some primitive Teuthids (*Plesioteuthis,* Jurassic; *Styloteuthis,* Upper Cretaceous) development of the proostracum occurred at the expense of the phragmocone and guard, until in the living squid *Ommastrephes* the guard has disappeared and the phragmocone is represented by a small hollow structureless cone at the posterior end of an extraordinarily long proostracum. In the living *Loligo,* the common squid, the proostracum alone remains, as a long thin plate of conchiolin shaped like a lance (the "pen" of a squid).

The tendency toward reduction in shell went a stage further in the Order Octopida (which includes the common Octopus) where total loss of shell resulted.

Migration and Distribution of Belemnites

It has been commonly assumed that belemnites were free-swimming (nektic) animals and as such, capable of wide distribution in a manner similar to that of the ammonites. However, to judge from their distributional patterns this does not appear to have been true, at least for most of the Jurassic and Cretaceous genera. Free-floating forms may have been present, but the observed distribution patterns can best be explained by proposing that the belemnites were essentially confined to the shelves, as are many of the present-day common squids (see above).

Good seaway connections in the Early Jurassic, probably around the margin of Pangaea, enabled the belemnites of that time to become widely distributed, but not, apparently, to the SW Pacific, probably because of the presence of deep water.

After the Lower Jurassic, belemnite faunas become differentiated into two distinct faunal realms: a Boreal Realm mainly restricted to northern regions of the Northern Hemisphere, and a Tethyan Realm extending at various times over most of the remainder of the world. This faunal differentiation is thought to reflect the start of the fragmentation of Pangaea. In this, the movement of landmasses disrupted existing oceanic current systems, leading to cooling of the poles and rotation of some regions polewards. Thus the Boreal belemnites are interpreted as cold-temperate stenothermal animals populating the northern regions of the Northern Hemisphere, grouped around the North Pole (Fig. 6).

The boundary between Boreal and Tethyan belemnites is a gradational one, and isotopic data suggest that it is the boundary between cold-temperate (Boreal) and warm-temperate/tropical (Tethyan) faunas (Stevens, 1971, 1973a; Stevens and Clayton, 1971).

In the Middle and Late Jurassic, Tethyan belemnites were able to spread to most regions apart from those in the northern areas of the Northern Hemisphere, populated by Boreal belemnites. These widespread migrations have been interpreted as movement of warm-temperate and tropical animals along the Tethyan seaway, formed as Pangaea split into Laurasia and Gondwanaland (Stevens, 1973a). Orogenic movements that accompanied the splitting influenced belemnite migration and differentiation. Thus, diastrophism in the Alpine-Balkan-Caucasian region in Oxfordian-Kimmeridgian times isolated the Mediterranean belemnites from those in the Indo-Pacific.

Distinctive belemnite faunas, grouped as an Ethiopian province, appeared at the same time in E Africa, Arabia, and India; and these have been interpreted as Tethyan derivatives populating the seaway, the ancestral Indian Ocean, that opened as Antarctica and India moved away from Africa (Stevens, 1973a). Orogenic movements extending over Middle and Upper Jurassic, and probably relating to the move-

In the Kimmeridgian-Tithonian, distinctive belemnite faunas appeared in Madagascar and Patagonia and these have been interpreted as the forerunners of a cold-temperate Austral Realm (Stevens, 1973a). If this interpretation is correct, it suggests that the continuing dispersal of the Gondwana lands had brought Madagascar and Patagonia into the cold-temperate climatic zone by the end of the Jurassic.

FIGURE 6. The relation of Kimmeridgian belemnite provinces to assemblies of Gondwanaland and Laurasia (after Stevens, 1971, Fig. 3). Grouping of the Laurasia landmasses around the North Pole at this time is reflected in the differentiation of Boreal belemnites, adapted to cold-temperate waters. On the other hand, grouping of the Gondwana countries away from the Jurassic South Pole provided Tethyan belemnites with tropical and warm-temperate dispersal routes. Intermittent shallow-water routes were available between Gondwanaland and Laurasia, probably via island arc systems. Just before the Kimmeridgian, India had split away from E. Africa, opening a seaway and allowing belemnites populating the southern shore of the Tethys to spread and differentiate.

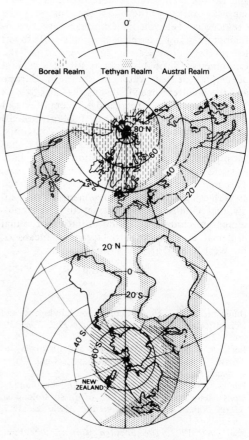

ment of New Zealand away from Australia, ended the isolation of the SW Pacific by establishing shallow-water routes along which moved a flood of Tethyan immigrants.

Tethyan belemnites were able to migrate from 40-50°N Lat. southward to 75°S Lat. (Stevens, 1967). This pattern of distribution, together with the absence of an anti-Boreal belemnite fauna for most of Jurassic time, has been interpreted to mean that none of the Southern Hemisphere belemnite occurrences were further south than about 45°S Lat., as Boreal belemnites dominated the Northern Hemisphere north of 45°N Lat. Thus the belemnite evidence suggests that although some splitting of Gondwanaland was occurring in the Middle and Late Jurassic, the Gondwana lands were still grouped some distance away from the Jurassic South Pole (Fig. 6).

FIGURE 7. The relation of Aptian-Albian belemnite provinces to assemblies of Gondwanaland and Laurasia (after Stevens, 1973b, Fig. 5). As in the Jurassic (Fig. 6), Boreal belemnites, adapted to life in cold-temperate seas, populated the northern regions of the Northern Hemisphere, grouped around the Cretaceous North Pole. Fragmentation of Gondwanaland had been in progress since Middle Jurassic time; and by the Aptian-Albian, South America and Africa had drifted apart to form the South Atlantic Ocean. Tethyan belemnites, adapted to life in tropical and warm-termperate seas, migrated along the dispersal routes that became available at this time. Movement of Australasia, South America, and Antarctica had brought them closer to the South Pole (cf. Fig. 6) and these countries were populated by Austral belemnites adapted to cold-temperate seas.

In the Cretaceous, Boreal faunas continued to populate the northern regions of the Northern Hemisphere and like their Jurassic equivalents are interpreted as cold-temperate populations living peripheral to the Cretaceous North Pole (Stevens, 1973b).

Tethyan belemnites populated central and southern Europe and spread at various times to South America, India, Africa, and Asia.

The relationship of these faunas to the assumed paleogeography of the time (Fig. 7), and isotopic temperature data (Stevens, 1971), suggest that like those of the Jurassic, the Cretaceous Tethyan belemnites were warm-temperate and tropical animals. Tethyan belemnites disappeared however after the Cenomanian, probably as a result of a general deepening of sea water in southern Europe and the Mediterranean (Stevens, 1973b).

Throughout the Cretaceous, belemnites and other Mollusca of Austral (= anti-Boreal) affinities populated Madagascar, Patagonia, W Antarctica, and Australasia; countries that throughout most of the Middle and Late Jurassic had been populated by Tethyan faunas. This marked change in affinities is interpreted as reflecting southward movement of the individual Gondwana landmasses into the cold-temperate climatic zone and initiation of faunal dispersal in the Southern Ocean by means of the West Wind Drift, as in modern times.

GRAEME R. STEVENS

References

Donovan, D. T., 1964. Cephalopod phylogeny and classification, *Biol. Rev. Cambridge Phil. Soc.*, 39, 259–287.

Jeletzky, J. A., 1965. Taxonomy and phylogeny of fossil Coleoidea (= Dibranchiata), *Geol. Surv. Canada Paper* 65-2, 72–76.

Jeletzky, J. A., 1966. Comparative morphology, phylogeny and classification of fossil Coleoidea, *Univ. Kansas Paleontol. Contrib.: Mollusca*, art 7, 62p.

Naef, A., 1922. *Die Fossilen Tintenfische*. Jena: Fischer, 322p.

Roger, J., 1952. Sous-classe des Dibranchiata, in J. Piveteau, ed., *Traite de Paleontologie*, vol. 2. Paris: Masson, 689–755.

Stevens, G. R., 1965. The Jurassic and Cretaceous Belemnites of New Zealand and a review of the Jurassic and Cretaceous Belemnites of the Indo-Pacific region, *New Zealand Geol. Surv. Palaeontol. Bull.*, 36, 283p.

Stevens, G. R., 1967. Upper Jurassic fossils from Ellsworth Land, West Antarctica and notes on Upper Jurassic biogeography of the South Pacific region, *New Zealand J. Geol. Geophys.* 10, 345–393.

Stevens, G. R., 1971. Relationship of isotopic temperatures and faunal realms to Jurassic-Cretaceous palaeogeography, particularly of the Southwest Pacific, *J. Roy. Soc. New Zealand*, 1, 145–158.

Stevens, G. R., 1973b. Cretaceous Belemnites, *in* A. Hallam, ed., *Atlas of Palaeobiogeography*. Amsterdam: Elsevier, 385–401.

Stevens, G. R., 1973b. Cretaceous Belemnites, *Atlas of Palaeobiogeography*. Amsterdam: Elsevier, 385–401.

Stevens, G. R., and Clayton, R. A., 1971. Oxygen Isotope studies on Jurassic and Cretaceous Belemnites from New Zealand and their biogeographic significance, *New Zealand J. Geol. Geophys.*, 14, 829–897.

Cross-reference: *Cephalopoda*.

BIOGEOCHEMISTRY

Biogeochemistry can be defined as the study of the chemical processes and products related to the activity of ancient organisms. Biogeochemists are especially interested in fossilizable products produced by ancient organisms and in the conditions under which they were produced. One branch of biogeochemistry is concerned with organic compounds in sediments and fossils, and particularly in using these compounds as chemical fossils or evidence of ancient life (Eglington and Murphy, 1969). Of special interest is the search for biologically produced compounds in very old rocks as evidence of some of the earliest forms of life (see *Precambrian Life*). Another branch of biogeochemistry is concerned with the chemistry of mineralized products of ancient organisms. The three chemical attributes of mineralized fossils that have received the most intensive study are mineral composition, trace chemistry, and isotopic composition.

The factors controlling these three properties may often be a complex mixture of physical chemical, genetic, and environmental parameters (Dodd and Schopf, 1972). In some cases, the chemical properties of fossils can be explained by the same physical chemical processes that affect the properties of inorganic precipitates. In other cases, the chemical properties vary between different taxonomic groups of organisms and thus are under genetic control. Finally, some chemical properties vary with conditions in the environment in which the fossil organisms lived. One of the goals of biogeochemistry is to sort out these factors in order to better interpret the chemical characteristics of the fossil and the conditions under which it formed. The above factors control the chemical characteristics of the mineralized skeleton of the living organism before it becomes a fossil. After death of the organism another factor, diagenesis (q.v.), may begin to alter the chemistry of the fossil. Obviously, extensive diagenesis will prevent determination of the conditions under which the fossilized organism originally lived.

Many uses have been made of data on the chemistry of fossils, including (1) study of the history of the chemistry of sea water; (2) measurement of past environmental conditions—especially temperature and salinity of sea water; (3) study of the chemistry and diagenesis of limestones, which often have fossils as major constituents; (4) study of geochemical cycles; (5) study of the taxonomy of organisms based on skeletal chemistry; and (6) study of the skeletal formation process (see *Biomineralization*).

Mineralogy

A wide variety of minerals is found in invertebrate skeletons. By far the most important minerals among the invertebrates are the two crystalline forms of calcium carbonate ($CaCO_3$), calcite, and aragonite; the opaline silicates and phosphates also make major contributions (Lowenstam, 1963). In addition to skeletons that are pure calcite or pure aragonite, many invertebrates have skeletons consisting of a combination of these two polymorphs. The minerals are always present in separate microarchitectural units rather than in intimate intermixtures. Crystal sizes, shapes, orientations, and interrelationships within these microarchitectural units may be quite varied and complex giving rise to another interesting area of study: the variation in these skeletal structure units and the factors responsible for that variation.

Several lines of evidence indicate that aragonite is not a stable mineral under surface temperature-pressure conditions. Aragonite is, however, an extremely common mineral in invertebrate skeletons and in modern marine sediments, and is probably the most common carbonate mineral in the shallow-water environment. For years biogeochemists and sedimentologists have been attempting to explain this paradox. A related question is why both calcite and aragonite are precipitated by marine organisms, in some cases both within the same organism.

Laboratory precipitation experiments and field studies by many different workers have shown that the concentration of certain ions has a strong influence over whether calcite or aragonite is formed. The most effective ion in controlling mineralogy seems to be Mg^{++}. At low Mg^{++} concentrations, calcite usually forms when the solution exceeds the $CaCO_3$ solubility product; at greater Mg^{++} concentrations, aragonite usually forms. The Mg^{++} concentration in sea water (1330 mg/liter) is considerably in excess of the amount needed to prevent for-

mation of calcite, hence inorganic precipitation from sea water most commonly produces aragonite. Variation in the Mg^{++} concentration at the calcification site in organisms may have a similar effect in determining whether calcite or aragonite forms. Mg^{++} may act as a "crystal poison," i.e., the Mg^{++} ions occupy Ca^{++} sites in the growing calcite lattice and block or greatly slow the further growth of the crystal. Due to its small size, the Mg^{++} ion is not readily accommodated in the aragonite lattice so that it is able to grow much more rapidly than calcite. Other ions may have a similar crystal poisoning effect.

Each crystal of a molluscan skeleton is surrounded by a thin, proteinaceous sheaf or matrix. The organic matrix is obviously intimately interrelated with the calcification process, especially in serving as the site upon which crystal growth is nucleated. The matrix consists largely of protein molecules, the amino acid composition of which varies between aragonitic and calcitic skeletons and between aragonitic and calcitic portions of the same skeleton. The amino acids may be arranged into protein molecules in such a way that side chains act as the sites upon which the Ca^{++} and CO_3^- ions are positioned when $CaCO_3$ crystals are formed. In some proteins these side chains are arranged to conform to the aragonite lattice and in others to the calcite lattice.

The distribution of aragonite and calcite is apparently controlled in part by environmental factors, most notably temperature but perhaps also salinity and other factors. Three expressions of the temperature effect have been noted (Fig. 1): (1) Some groups of organisms having aragonitic skeletons are far more abundant in the tropics than in the higher latitudes. The scleractinian corals are the best example of this effect. (2) Certain groups of organisms having aragonitic skeletons are found only in the tropics and semitropics. The best examples of this effect are the calcareous green algae. (3) In groups having skeletons composed of a combination of aragonite and calcite, the proportion of aragonite increases as temperature increases.

Data on skeletal mineralogy have been extensively used to explain a variety of features observed in the fields of sedimentology and paleontology. For example, variation in the skeletal mineralogy of carbonate sediment has been explained as being due to variation in the mineralogy of the biota that contribute the skeletal debris producing the sediment. Differences in skeletal mineralogy may also explain differential preservation of different fossil groups. Because they are slightly more soluble,

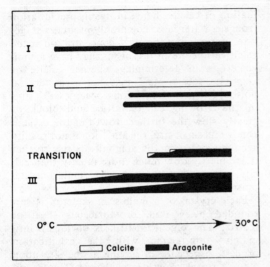

FIGURE 1. Schematic representation of three types of temperature effect on skeletal mineralogy. (from Lowenstam, 1954). I, Taxa with aragonitic skeletons that are much more abundant in warm than cool water; II, aragonite taxa restricted to warm waters whereas calcitic groups cover the entire temperature range; "transition" refers to aragonitic taxa that add small amounts of calcite at the low-temperature end of their range; III, taxa having skeletons of combined calcite and aragonite but with relatively more aragonite in warmer water forms.

aragonitic skeletons will tend to dissolve before calcitic ones. If a buried aragonitic skeleton dissolves before the matrix surrounding it is lithified, the resulting void will collapse and all evidence of its former existence may be lost, or at best only an imprint will be left. If the calcite skeletons are not dissolved, the resulting fossil assemblages will be highly biased, consisting only of calcite shells.

A rather obvious potential use of skeletal mineralogy is for the quantitative determination of paleotemperatures based on the relative proportions of aragonite and calcite within a shell (Bowen, 1966). Such a study would be very difficult because of the complexity of the temperature-mineralogy relationship and its lack of consistency between different groups of organisms. The study of skeletal mineralogy can be of value in identifying general temperature trends. The presence of one of the aragonitic groups that is not found outside of the tropics or the abundance of one of the aragonitic groups that is not normally abundant outside of the tropics could certainly be used for identifying such general trends. An increase in the proportion of aragonite in skeletons with mixed mineralogies could also be used to identify temperature trends.

Most biogeochemical studies of skeletal mineralogy have emphasized the invertebrates. Some work has been done on the aragonitic otoliths of fish and on the organic compounds in the phosphatic bones and teeth. One of the potentially most useful results of vertebrate biogeochemical studies is the development of a possible method of dating fossil bones by the extent of alteration of the amino acids in their organic matrix (Bada, 1972).

Trace Chemistry

Though little work has been done on the trace chemistry of noncarbonate skeletal material, many different ions have been reported from $CaCO_3$ skeletons. Of the trace elements found in carbonate skeletons, Mg and Sr are by far the most abundant. This is due to two factors: The ionic radii and charge of Mg^{++} and Sr^{++} allow them to substitute readily for Ca^{++} in either the calcite or aragonite lattice. Both ions are relatively abundant in natural waters, especially in sea water.

Several factors appear to affect the trace chemistry of carbonate skeletons. Other factors being equal, the Mg/Ca and Sr/Ca ratios in a carbonate skeleton are proportional to those ratios in the water in which the carbonates formed. In other words, the higher the concentration of a trace element relative to Ca in the water in which a shell grows, the higher will be the concentration of that element in the shell. This relationship can be expressed mathematically:

$$(M/Ca)_{skeleton} = K(M/Ca)_{water}$$

in which M represents the molar concentration of either Mg or Sr (or any other cation) and Ca the molar concentration of calcium. K is a proportionality constant commonly called a distribution or partition coefficient.

Other factors being equal, Mg concentration will be greater in calcitic skeletons and Sr concentration will be greater in aragonitic skeletons. The small Mg^{++} ion (radius 0.66Å) substitutes for Ca^{++} (radius 0.99Å) more readily in the calcite lattice, which is isostructural with magnesite ($MgCO_3$), than it does in the aragonite lattice. Likewise the larger Sr^{++} ion (radius 1.12Å) substitutes more readily in the aragonite lattice, which is isostructural with strontianite ($SrCO_3$).

No clear model has emerged to explain the obvious fact that different groups of organisms are characterized by different ranges of concentrations of trace and minor elements. The differences must in some way be produced by differences in the biochemistry of the skeletal formation process between different groups. Although the Sr and Mg concentrations in the skeletal carbonates of the various groups of organisms show no well-defined trends, a generalized pattern toward lower Mg and Sr concentrations in more highly organized groups has been recognized (Fig. 2).

On the basis of their Mg concentration, skeletal calcites tend to fall into two groups: a low Mg-calcite group with less than 5–6 mol % MgCO$_3$ (including the mollusca, brachiopods, and planktonic genera of forams) and a high Mg-calcite group with higher concentrations of MgCO$_3$ (including the calcareous algae, echinoderms, and several other groups). Skeletal aragonites also tend to fall into a high Sr group, typified by the corals and green algae, and a low Sr group, consisting especially of the mollusks.

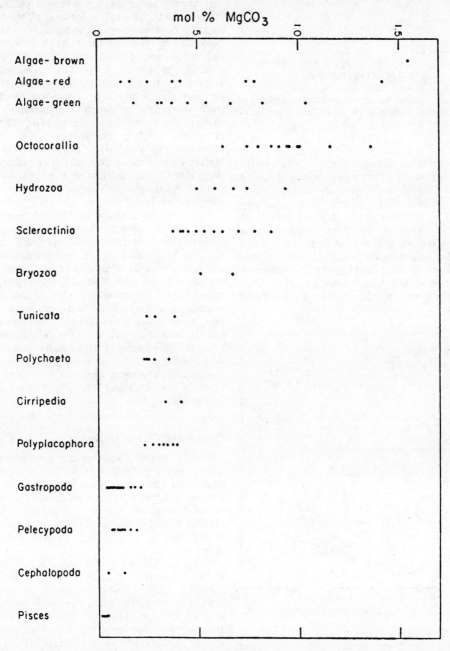

FIGURE 2. Mol % MgCO$_3$ in aragonite skeletons of various taxa arranged in order of increasing organic complexity (from Lowenstam, 1963). All samples from Bermuda.

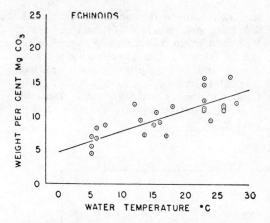

FIGURE 3. Relation between water temperature and magnesium content in the echinoids (from Chave, 1954).

Environmental factors (other than water chemistry) may be considered indirect because they act through the physiology of the calcification process. Mg concentration increases regularly with temperature in many groups of organisms and is especially apparent in forms characterized by a generally high Mg content (Fig. 3). Temperature also affects the Sr concentration in a few taxa (Fig. 4). In some cases the Sr concentration correlates positively with temperature and in other cases the correlation

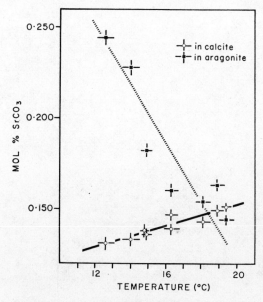

FIGURE 4. Variation with temperature of mol % $SrCO_3$ in the last formed portion of the calcite outer prismatic layer (open circles) and aragonite nacreous layer (solid squares) of *Mytilus* (from Dodd, 1965).

is negative. In several studies, salinity has also been shown to affect the trace chemistry of invertebrate skeletons.

Data on the trace chemistry of skeletal carbonates has been used in the study of a number of different types of geologic problems. Interpretation of the trace chemistry of carbonate sediments depends on a knowledge of the trace chemistry of the organically derived carbonates that usually comprise the principal portion of the sediments. The major features of the trace-element composition of the sediments can be explained in terms of the relative proportions and taxonomic position of the major biotic contributors to the sediment. Studies of the diagenesis of carbonate sediments and rocks are also greatly aided by a knowledge of the trace chemistry of the skeletal materials that usually are their major constituent. The approximate starting chemistry of an altered carbonate can usually be predicted from the biotic composition of the rock. Knowing the original composition allows one to propose explanations of the observed chemical composition.

Of particular interest to the paleoecologist is the use of skeletal chemical data to make environmental interpretations. As noted above, Mg and Sr concentrations are correlated with temperature in many groups of organisms. The Mg and Sr concentration in fossil representatives of these groups should then be usable for making paleotemperature interpretations. However, three major problems must be considered and/or overcome before trace chemically determined paleotemperatures can be used: diagenesis; lack of phylogenetic change in the temperature-chemistry relationship; and stability of the Sr/Ca and Mg/Ca ratio in sea water through time.

Oxygen and Carbon Isotopes

Oxygen isotopic analysis has been the most widely used geochemical technique in the study of skeletal carbonates. This is probably because oxygen isotopic composition of skeletal material is biologically simple and readily explainable in physical-chemical terms although it is instrumentally a complex determination to make.

Oxygen has three stable isotopes, O^{16}, O^{17}, and O^{18}, occurring in air in the ratio of 99.759:-0.0374:0.2039. These isotopes are in fact not always found in the same relative proportions: the O^{18}/O^{16} ratio may vary by as much as 10%, the highest value being found in air and the lowest in Antarctic snow (Epstein, 1959). The correlation of the O^{18}/O^{16} ratio in the carbonate of skeletal materials with growth temperature is the basis of the well-known oxygen paleothermometer.

The oxygen isotopic paleothermometer is based on separation or fractionation of the different isotopes of oxygen during chemical reactions. The extent of this fractionation is proportional to temperature, being greater at lower than at higher temperatures. Thus the O^{18}/O^{16} ratio in the carbonate skeleton will become progressively more different from the O^{18}/O^{16} ratio in the water in which it is forming as the temperature decreases. When grown in open, well-circulated sea water of constant oxygen isotopic composition, the O^{18}/O^{16} ratio in the carbonate skeletons of most groups of organisms decreases as temperature increases (Fig. 5). For unknown reasons some groups of organisms do not precipitate carbonate in oxygen isotopic equilibrium with the water in which they are growing.

The O^{18}/O^{16} ratio of a shell may also vary with the salinity of the water in which it forms, due to variation in the O^{18}/O^{16} ratio of the water. Because of its slightly lower vapor pressure, H_2O^{16} tends to evaporate more readily than H_2O^{18}, enriching the sea water in H_2O^{18}, which is left behind in the evporation process. The H_2O^{16} concentrates slightly in the vapor and in the rain that results from its condensation. Brackish water which represents a mixture of sea water and fresh water has intermediate values. If the temperature at which the shells are growing is constant, the O^{18}/O^{16} ratios of

the shells will reflect the composition of the water in which they are growing and thus the salinity.

The carbon isotopic composition of shells and fossils has also been extensively studied. Carbon has two stable isotopes, C^{12} and C^{13}, which are fractionated in natural materials in a manner similar to the isotopes of oxygen. The isotopic composition of carbon in bicarbonate dissolved in sea water (the source of carbon in shells) is too variable to allow determination of paleotemperatures. Carbon isotopic composition has been used to distinguish freshwater shells (characterized by low C^{13}/C^{12} ratios) from marine shells (characterized by high C^{13}/C^{12} ratios).

J. ROBERT DODD

References

Bada, J. L., 1972. The dating of fossil bones using the racemization of isoleucine, *Earth Planetary Sci. Letters*, **15**, 223–231.

Bowen, R., 1966. *Paleotemperature Analysis.* Amsterdam: Elsevier, 265p.

Chave, K. E., 1954. Aspects of the biogeochemistry of magnesium. I. Calcareous marine organisms, *J. Geol.*, **62**, 266–283.

Dodd, J. R., 1965. Environmental control of strontium and magnesium in *Mytilus, Geochim. Cosmochim. Acta*, **29**, 385–398.

Dodd, J. R., and Schopf, T. J. M., 1972. Approaches to biogeochemistry, in T. J. M. Schopf, ed., *Models in Paleobiology.* San Francisco: Freeman, Cooper, 46–60.

Eglington, G., and Murphy, M. T., eds., 1969. *Organic Geochemistry, Methods and Results.* New York: Springer-Verlag, 828p.

Epstein, S., 1959. The variation of the O^{18}/O^{16} ratio in nature and some geologic implications, in P. H. Abelson, ed., *Researches in Geochemistry.* New York: Wiley, 217–240.

Epstein, S; Buchsbaum, R.; Lowenstam, H. A.; and Urey, H. C., 1953. Revised carbonate-water isotopic temperature scale, *Geol. Soc. Amer. Bull.*, **64**, 1315–1326.

Lowenstam, H. A., 1954. Factors affecting the aragonite: calcite ratios in carbonate-secreting marine organisms, *J. Geol.*, **62**, 284–322.

Lowenstam, H. A., 1963. Biologic problems relating to the composition and diagenesis of sediments, in T. W. Donnelly, ed., *The Earth Sciences: Problems and Progress in Current Research.* Chicago: Univ. Chicago Press, 137–195.

Cross-references: *Biomineralization; Paleotemperature and Depth Indicators.* Vol. I: *Pelagic Biogeochemistry.* Vol. IVA: *Biochemicals; Biogeochemistry; Calcium Carbonate: Geochemistry; Carbon Isotope Fractionation; Ecology; Geochemistry of Sediments (Ancient); Isotope Fractionation; Organic Geochemistry; Oxygen Isotope Geochemistry; Paleosalinity; Paleotemperatures–Isotopic Determinations.*

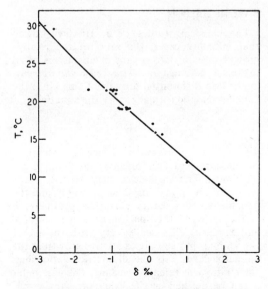

FIGURE 5. Variation with growth temperature of the O^{18}/O^{16} ratio (expressed as δO^{18} relative to the PDB standard) of molluscan carbonates grown at known temperatures (from Epstein et al., 1953).

BIOMETRICS IN PALEONTOLOGY

The statistical analysis of metrical characters of fossils is a useful aid to paleontologists in the study of species and infraspecific categories. The description of a type specimen, backed up by observations on a few paratypes, can often be profitably supplemented by the statistical study of an association consisting of many specimens. The quantitative assessment of such material will then be a helpful complement to the qualitative diagnosis. Such topics as the study of growth patterns, phenotypic response to environmental stimuli, predation, etc. may also be assisted by quantitative methods.

Species and Populations

A zoological species consists of an interbreeding natural population (see *Species Concept*). Obviously, it is not possible to make direct inferences about the breeding habits of fossil organisms; the paleontologist is therefore obliged to reconstruct as complete a picture as possible from the fragmentary evidence available, using morphology; stratigraphic location; geographic distribution; and, where pertinent, relationship to the enclosing sediment. Clearly, such a reconstruction cannot be expected to yield an infallible representation of the true relationships that once existed and in some cases valid species may be lumped together whereas variants will be split. There is also the difficulty of knowing where to draw a limit between morphotypes showing, apparently, a gradual shift from the one to the other over time. A correctly designed statistical analysis of sequential data of this kind can often provide a useful guideline.

There has been much discussion concerning the formal definition of what should constitute a paleontological species—some people have defended the typological concept vigorously while others have put forward arguments for basing the concept on associations (cf. Sylvester-Bradley, 1958).

A collection of fossils selected for study is referred to statistically as a sample. The sample is analyzed so that inferences may be drawn concerning the statistical universe (the alternative designation for a statistical population, i.e., "universe," is employed here to avoid confusion) from which it was taken. The statistical universe from which the sample was drawn comprises, theoretically, all the individuals of that species that ever existed. Obviously, the paleontologist has no chance at all of ever being able to collect all individuals of the species in which he is interested, in contrast with the situation for, say, the population of all cars of a certain make, for which a feeble chance could exist of getting hold of all of them.

Frequently, two or more samples have to be compared. These may derive from two or more localities, either from the same, or from different, stratigraphical levels. Another problem to be resolved is whether a collection from a single locality is homogeneous and, if not, can it be subdivided into the original components?

Univariate Analysis

This area of statistics is concerned with the analysis of one character at a time. The mean and the standard deviation of variables such as length, height, and breadth are calculated and, where necessary, differences in means are tested for statistical significance by standard methods (see standard works on statistics and, for example, the biologically oriented Simpson et al., 1960; Sokal and Rohlf, 1969).

Where a morphological character is correlated with the age of the organism, as for example with the length of the body in humans, univariate comparisons of means must be misleading, if made with samples of different age compositions. In some cases, e.g., arthropods which have discrete growth stages and organisms reaching a terminal growth condition, univariate comparisons may be enlightening. The basic technique for pairwise univariate comparisons of means of normally distributed data is the "Student" *t* test. The extension of this to more than two samples is known as the analysis of variance (in its simplest form).

Bivariate Analysis

The classical studies of J. Huxley showed that in many cases, the ratio of the relative growth rates of two organs or characters of an organism is constant. The term *allometry* is applied to differential growth of this kind; the concept may be formalized by the equation

$$y = ax^b$$

where x and y are two variables (size variables or measurements on organs) and a and b are constants. The variable x may be the whole length of an animal, or some part of it, and the variable y, some other part. The parameter b is the ratio of the specific growth rates of y and x; these will usually vary with the age of the animal, but so long as relative growth in the animal accords with the allometric relationship, their ratio will remain constant. The parameter a can be defined as the value of y for $x = 1$. When the exponent $b = 1$, the growth rates do not change with time and the growth is said to be *isometric* (cf. Gould, 1966). The subject of allometric growth in fossils has been studied particularly by Gould (1966, 1972). Various methods have been proposed for ameliorating allometric studies, Jolicoeur (1963) has developed a method of study that provides a good solution.

Apart from allometric studies, standard methods of regression analysis are often useful for comparing samples of the same taxonomic entity from different environments. Providing care is taken to minimize the effects of age–size differences (such as by the use of a suitable transformation viz. the logarithmic transformation; cf., allometry), much valuable information can be obtained by bivariate regression analysis, for it is backed up by a great volume of statistical theory and methodology, allowing a great range of hypotheses to be tested, something which cannot be said for more special methods.

Correlation coefficients can be computed readily between pairs of variables and these can then be used for interpreting bivariate relationships in the material. Correlation analysis can be expanded to include a study of the correlational structure between many variables by finding partial correlation coefficients (cf. Simpson et al., 1960; Reyment, 1971).

Multivariate Analysis

Most problems in paleontology tend to be highly multivariate and, in fact, some of the central theory of multivariate statistics was invented in order to answer problems of interest to the paleontologist and the biostratigrapher. Multivariate analysis has shown itself to be of great utility in unravelling intricate relationships between morphological variables, e.g., growth studies, patterns of variation, and paleoecological analyses.

Normally, the first kind of question that will be of interest will concern the homogeneity of a multivariate sample of some organism and the relationships among the several variables measured on it. In this respect, heterogeneity usually derives from mixing of the material (postmortem transport, mixture of growth stages, deformation).

The step from bivariate to multivariate analysis is not great theoretically but quite considerable from the point of view of the increase in complexity of the computations.

Analysis of a Single Multivariate Sample. In analyzing a single multivariate sample, interest will be directed toward ascertaining, firstly, whether the sample comes from a multivariate normally distributed population. If significant deviations occur, many of the standard methods of multivariate analysis may be invalidated.

Determination of the statistical properties of the sample can provide useful information on size and growth relationships in the organism under study. Examples are given in Blackith and Reyment (1971), ch. 12.

Techniques for analyzing a multivariate sample may be grouped under the general concept of *factor analysis* (Jöreskog et al., 1976). Factor analysis comprises a number of methods used for analyzing interrelationships within a set of variables (R-mode) or objects (Q-mode). Although the various techniques differ greatly in their objectives, and in the mathematical models underlying them, they all have one feature in common, to wit, the construction of a few hypothetical variables (or objects), called factors, that are supposed to contain the essential information in a larger set of observed variables or objects. The factors are constructed in a manner such that the overall complexity of the data is reduced by capitalizing on inherent interdependencies between variables. As a consequence of this, a small number of factors will usually account for a large proportion of the information contained in the larger set of original variables.

Factor analysis can be in the R-mode, in which relationships between variables are analyzed, or the Q-mode, in which the interrelationships between objects are portrayed. Factor analysis has been widely applied in paleontology. Some references are: R-mode analysis was used by Gould (1967) in analyzing pelycosaurs; Birks (1974) studied Pleistocene palynological problems; and Reyment (1966) investigated morphological forms of the foraminiferal life cycle. The use of Q-mode factor analysis has been greatly popularized by Imbrie and his coworkers. Imbrie and Kipp (1971) investigated the paleoecology of foraminifers by Q-mode factor analysis.

Providing the correct form of factor analysis is used, valuable results can often be obtained. Q-mode factor analysis usually succeeds in displaying natural groupings of organisms, whereby similar individuals are located near to each other. R-mode factor analysis can, for example, provide a useful representation of multivariate allometric relationships (Jolicoeur, 1963).

Study of Two or More Multivariate Samples. The methods in this section are all designed for multivariate normally distributed variables. Normally, the necessary testing for this requirement will have been done as an introductory part of the analysis. Some of the more common methods are mentioned briefly below.

The analogue of the well-known *t* test of univariate statistics is the T^2 test, in which, instead of two single means, one is concerned with testing the difference between two vectors of means. The T^2 test is useful for studies on sexual dimorphism (q.v.) and for picking up slight morphological differences. Here, and elsewhere, the confusing element of size differences can be eliminated by a suitable growth-invariant transformation (Burnaby, 1966). T^2 is the test equivalent of the Mahanolobis' *generalized statistical distance,* originally developed for expressing morphological differences between human populations.

The group of procedures for multivariate analysis of variance and canonical variates forms a partial generalization of the *analysis of variance* of univariate statistics. And here, too, growth-invariant versions can be computed if desired. Connected to the analysis of differences in several multivariate means, we have the method of *discriminant functions,* a biological account of which is to be found in Reyment (1973). Expressed simply, the discriminant function attempts to allocate a specimen on which certain characters have been measured into one of two or more groups, on the assumption that the specimen really derives from one of these groups. For example, this method can be used in micropaleontology for assigning an ostracod specimen to its correct growth stage, for identifying a level in a borehole by its fossil content, using a mean vector of morphological variables on some diagnostic species, and for expanding studies in sexual dimorphism. For testing differences in several multivariate means, the generalized analysis of variance provides the appropriate technique. Useful graphical displays are obtained from the *canonical variate* means computed for the multivariate samples.

Related mathematically to the discriminant function, the *generalized statistical distance* is a useful way of computing values, growth-invariant if necessary, that can be employed in graphical displays, in order to illustrate multivariate differences between taxonomical entities. Many examples of the techniques referred to here are discussed in Blackith and Reyment (1971, chs. 8,19,20, and 29).

Relationships Between Organism and Environment

An important question arising in quantitative studies is whether it is possible to identify variables in the environment that have influenced the morphology of the organisms contained in the sediment. A suitable technique for analyzing this problem is that of *canonical correlation,* a method devised to assess the strength of the association between two sets of variables (more sets can be included if needed). Other methods that can be applied in conjunction with canonical correlations are partial correlations, multivariate regression, and multiple correlation. All of these are very sensitive to departures from multivariate normality. Examples are cited in Blackith and Reyment (1971).

Time Series Analysis

Borehole sequences, for example, constitute natural sets of data ordered in time series. Univariate means can be studied by standard methods of time series analysis, such as serial correlation and the periodogram. Graphical representations can be produced from such analyses and used to provide a kind of paleontological log for sequences of fossils.

Illustrative Examples

Variation in the Cretaceous Echinoid *Micraster*. Kermack (1954) made a bivariate statistical analysis of species of *Micraster* from the Late Cretaceous of England. Nichols (1959) continued this line of investigation, supplementing his work with an ethological study of living echinoids. Kermack separated *M. coranguinum* Leske from *M. senonensis* Lambert by measurements of the subanal fasciole, which, when plotted against the length of the specimen, showed diffuse clustering with individuals of *senonensis* tending to concentrate in the middle of the scatter diagram.

Reyment and Ramdén (1970) analyzed Nichols' measurements (length, height, and breadth of the test, and number of tube feet) on eight samples of species of *Micraster,* using R- and Q-mode techniques. The Q-mode analytical method of principal coordinates analysis (Blackith and Reyment, 1971, ch. 12) shows three main clusters to occur in the material, one for each of the species *Micraster coranguinum, M. cortestudinarum,* and *M. corbovis,* as well as the presence of forms intermediate between these groups (Fig. 1). The canonical variate analysis arranges the multivariate means of the eight samples equally over the plot with only one tendency to cluster (Fig. 2). The analysis suggests that the supposed intermediary forms between *coranguinum* and *senonensis,* on the one hand, and *glyphus* and *stolleyi,* on the other, are all actually more like *glyphus,* for the four variables considered, than they resemble any of the other species.

Geographical Variation in the ostracod *Brachycythere sapucariensis* Krömmelbein. The early Turonian (Cretaceous) ostracod species *Brachycythere sapucariensis* occurs in Gabon, Nigeria, and northeastern Brazil (Sergipe). If the present positions of South America and

Africa had been the same in the middle Cretaceous as they are today, one would expect the Nigerian and Gabonese individuals to be more alike than either of them with the South American individuals, assuming all other things (facies, local population variation) to be equal. Using eight measurements on the carapace, Reyment and Neufville (1974) showed by means of R-mode analysis that the Gabonese and Brazilian individuals are more alike than

either are to the Nigerian specimens. Such a paleobiogeographical situation could have arisen if, for example, the continents had been close to each other in the early Turonian, as is postulated in continental-drift interpretations of the origin of the South Atlantic Ocean. The relationship between the ostracods is shown in Fig. 3 on a plot using the first two axes of the principal components analysis of the data (cf. Jöreskog et al., 1976).

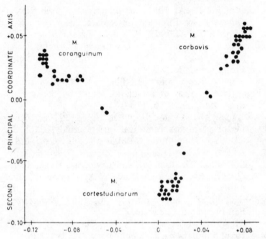

FIGURE 1. Principal coordinates analysis of species of the Cretaceous echinoid genus *Micraster,* showing the evolution of two species from an ancestral one (from Blackith and Reyment, 1971; data from Nichols, 1959).

FIGURE 2. Canonical variates analysis of species of *Micraster,* based on three measurements on the test and one on the frequency of tube feet (from Blackith and Reyment, 1971; data from Nichols, 1959).

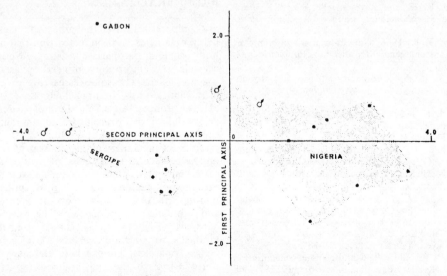

FIGURE 3. Plot of transformed variable scores of principal components, an R-mode factor-analytical variant, for *Brachycythere sapucariensis* Krömmelbein (from Reyment and Neufville, 1974). The diagram displays the geographical differentiation in the material.

Nonparametric Methods

The statistical methods discussed in the foregoing sections are all designed for normally distributed variables, such as size dimensions of an organism. Such variables are termed *continuous variables*. A great number of the variables encountered in paleontology are, however, *discontinuous* or *dichotomous* (presence-absence data) and not always amenable for treatment by the methods available for continuously distributed variables. Nonparametric statistics are available for such data and, in fact, their history is as old as modern statistics itself, to which witnesses the chi-squared distribution, a procedure specifically designed for testing assumptions about such observations as differences in frequencies.

Nonparametric methods are particularly valuable in the study of orientatory data (cf. Reyment, 1971), in cases where there is no evidence that the data accord with the circular or spherical normal distribution. Another area of application concerns variations in the frequencies of spines on organisms and variation in ornamental categories in, e.g., crustaceans. Of recent years, statisticians have developed interesting techniques for the analysis of multivariate categorical data. Although such methods are still uncommon in paleontological work, they are gaining ground.

R. A. REYMENT

References

Birks, J. J. B., 1974. Numerical zonations of Flandrian pollen data, *New. Phytologist.,* **73**, 351–358.

Blackith, R. E., and Reyment, R. A., 1971. *Multivariate Morphometrics.* London: Academic Press, 412p.

Burnaby, T. P., 1966. Growth-invariant discriminant functions and generalized distances, *Biometrics,* **22**, 96–100.

Gould, S. J., 1966. Allometry and size in ontogeny and phylogeny, *Biol. Rev. Cambridge Phil. Soc.,* **41**, 587–640.

Gould, S. J., 1967. Evolutionary patterns in pelycosaurian reptiles: A factor-analytical study, *Evolution,* **21**, 385–401.

Gould, S. J., 1972. Allometric fallacies and the evolution of *Gryphaea:* A new interpretation based on White's criterion of geometric similarity, *Evol. Biol.,* **6**, 91–118.

Imbrie, J., and Kipp, N. G., 1971. A new micropaleontological method for quantitative micropaleontology: Application to a late Pleistocene Carribean core, in K. Turekian, ed. *The Late Cenozoic Glacial Ages.* New Haven: Yale Univ. Press, 71–181.

Jolicoeur, P., 1963. The multivariate generalization of the allometry equation, *Biometrics,* **19**, 497–499.

Jöreskog, K. G.; Klovan, J. E.; and Reyment, R. A., 1976. *Geological Factor Analysis.* Amsterdam: Elsevier, 180p.

Kermack, K. A., 1954. A biometrical study of *Micraster coranguinum* and *M. (Isomicraster) senonensis, Phil. Trans. Roy. Soc. London, Ser. B,* **237**, 375–428.

Kermack, K. A., and Haldane, J. B. S., 1950. Organic correlation and allometry, *Biometrika,* **37**, 30–41.

Nichols, F., 1959. Changes in the chalk heart-urchin *Micraster* interpreted in relation to living forms, *Phil. Trans. Roy. Soc., London, Ser. B,* **242**, 347–437.

Reyment, R. A., 1966. *Afrobolivina africana* (Graham, deKlasz, Rérat): quantitative Untersuchung der Variabilität einer palaeozänen Foraminifera, *Eclogae Geol. Helvetia,* **59**, 319–338.

Reyment, R. A. 1971. *Introduction to Quantitative Paleoecology.* Elsevier, Amsterdam, 226p.

Reyment, R. A. 1973. The discriminant function in systematic biology, in *Discriminant Analysis and Applications.* New York and London: Academic Press, 311–335.

Reyment, R. A., and Neufville, E. M. H., 1974. Multivariate analysis of populations split by continental drift, *Math. Geol.,* **6**, 173–181.

Reyment, R. A., and Ramdén, H. Å., 1970. Fortran IV program for canonical variates analysis for the CDC 3600 computer, *Computer Contr. Geol. Surv. Kansas,* **47**, 39p.

Simpson, G. G.; Roe, A.; and Lewontin, R. C., 1960. *Quantitative Zoology.* New York: Harcourt Brace, 440p.

Sokal, R. R., and Rohlf, F. J., 1969. *Biometrics.* San Francisco: Freeman, 776p.

Sylvester-Bradley, P. C., 1958. The description of fossil populations, *J. Paleontology,* **32**, 214–235.

Cross-references: *Computer Applications in Paleontology; Population Dynamics; Sexual Dimorphism; Species Concept; Taxonomy, Numerical.*

BIOMINERALIZATION

Biomineralization is the process that enables living organisms to produce, within or on their body, skeletal hardparts consisting of inorganic crystals or amorphous mineral phases and organic matrix. In some cases this process is intracellular (e.g., in unicellular algae, higher plants, sponges, etc.), but in most cases it is extracellular. In intracellular biomineralization the hardpart is formed in vacuoles in the cytoplasm. Extracellular biomineralization occurs on the surface of single cells, or on the surface of ectodermal or mesenchymal tissue which presumably secretes physiological fluids containing the basic constituents of the organic and inorganic components of the prospective hardpart (Fig. 1). These fluids originate, in mollusks for instance, in the vesicles of the Golgi apparatus, from where they are transported through the cell, and released beyond the surface of the mantle epithelium to become part of the colloidal extrapalleal fluid. Enamel and dentine of vertebrate teeth are produced

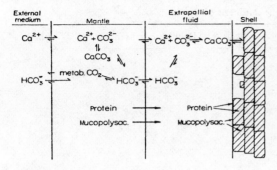

FIGURE 1. Diagram showing ions and organic compounds concerned in the biomineralization of a molluscan shell (from Wilbur, 1972).

by ameloblasts (ectodermal cells) and odontoblasts (mesenchymal cells), whereas certain avian and reptilian eggshells are produced by uterine glands. Biomineralization can also be pathogenic (see *Paleopathology*).

A wide variety of minerals are found as hardparts of living organisms. Calcium minerals predominate. Table 1 lists the minerals known to occur in the hardparts of marine and terrestrial organisms. Further investigations will undoubtedly document more.

Ultimately, the source of the materials for mineral formation is from the environment of the living organism. Varying degrees of biological concentration of elements are required to produce various mineral phases. In the marine environment, for instance, significant bioconcentration of phosphorous and silica is required to form skeletal dahllite (Chave, unpublished), or opal (Hurd, 1972). On the other hand, the formation of calcium carbonate hardparts in the surface waters requires no bioconcentration because these waters are supersaturated with respect to both calcite and aragonite (Lyakhin, 1968). Little is known about problems of bioconcentration for the more exotic minerals.

The most significant feature of Table 1 is that organisms have evolved the potential of forming hardparts from a broad spectrum of mineral substances having a wide variety of physical and chemical properties. However, extrapolation of the data in Table 1 into the geologic past could be very risky because in only a few cases do we know with any certainty when specific organisms evolved the ability to produce these minerals. In general, however, it is assumed that the ability to produce the first mineralized hardpart evolved during the Early Cambrian (Glaessner, 1972).

The organic matrix is formed by condensation or polymerization of amino acids and sugars as collagen, conchiolin, "ovokeratin," etc., from the body fluids in the shape of fibrils, which may form organic membranes. The process of organization and ordering of this matrix is not yet fully understood, but in many cases it may be comparable to the formation of ionotropic gels (Thiele, 1967).

The inorganic crystals are formed by nucleation and growth, and they, too, in some cases are derivatives of the body fluids. Their specific origin, however, is still problematical. In certain cases (e.g., initial stage of bivalve shell regeneration) tiny crystals (crystallites) and their spherolitic aggregates seem to be formed without any chemical control exerted by organized organic matrices. In many other cases, however, there is a distinct dependence of the crystallites on such an organic matrix. This dependence is at least topological and perhaps epitaxic, but it is assumed by a number of authors that there is also a causal relation.

While the compartment hypothesis assumes that biocrystals are formed in, and shaped by, preexisting compartments consisting of organic material, the template hypothesis suggests that specific crystal nucleation is induced by active sites of the underlying organic matrix. In such cases, a certain analogy in lattice dimensions of the active site and the crystalline solid phase would selectively control at least the mineral that is formed. (In the model case of molluscan nacre, an appraisal of both working hypotheses is contained in the papers of Bevelander and Nakahara, 1969; Towe, 1972; Erben, 1972 and 1974; as well as Erben and Watabe, 1974).

Two enzymes commonly found at sites of biomineralization are carbonic anhydrase and alkaline phosphatase. Carbonic anhydrase catalyzes reactions in the system CO_2–H_2O, and alkaline phosphatase catalyzes reactions between inorganic and organophosphates.

In the past, certain objections have been raised against aspects of the matrix models pointed out above. For instance, identical or nearly identical organic matrices are found in nonmineralizing organisms and in nonmineralized parts of mineralizing organisms. The same is true of the enzymes mentioned. In fact, Pocklington (personal communication) has observed higher concentrations of carbonic anhydrase in uncalcified cnidarians (sea anemones) than in calcified forms (corals).

On the other hand, several facts actually seem to suggest the existence of active sites: (a) the general correlations of collagen with calcium phosphate and of conchiolin with calcium carbonate, but never the reverse; (b) some indications of a mineral specifity of the organic matrix from calcite- or aragonite-producing molluscan species (Watabe and Wilbur, 1960);

TABLE 1. Mineralogy of the Hardparts of Organisms

Organism	Calcite	Aragonite	Mg-Calcite	Opal	Dahllite	Francolite	Other
PLANTS							
Coccolithophorids	ES						
Diatoms				ES			
Chlorophyta		ES					
Phaeophyta		ES (?)					
Rhodophyta		CS	CS				
Angiosperms		OS		CS			WS
ANIMALS							
Foraminifera							
planktic	ES						
benthic		OS	CS				
Radiolaria				CS			CS
Porifera							
calcarea		ES					
hexactinellida				ES			
demospongiae				ES			
sclerospongia		CM		CM			
Cnidaria							
hydrozoa		ES					
anthzoa		CS	CS				
		OM	OM				
Bryozoa		CS					
		CM	CM				
Brachiopoda							
inarticulata						ES	
articulata	OS		CS				
Mollusca							
polyplacophora		CS					
		OM					
		OM					LM (R)
		OM					MaM (R)
		OM			OM (R)		AM (R)
scaphopoda		ES					
gastropoda		CS					
	CM	CM					
		OM					FM (G)
		OM					MaM (G)
		OM					WM (G)
		OM					GM (R)
		OM		OM (R)			
bivalvia	CM	CM					
	CM	OM			OM		
cephalopoda		CS	OS (E)				
Echinodermata							
asteroidea			ES				
ophuroidea			ES				
echinoidea			ES				
holothuroidea			ES				
crinoidea			ES				
Annelida							
polychaeta		CS	CS				
		CM	CM				
Arthropoda							
ostracoda			ES				
cirripedia	OS		CS				
malacostraca			CM		CM (?)		AM (?)
Chordata							
tunicata		CS (?)					V
pisces	CM (O)				CM		
					OM		MoM (O)
					OM		AM (O)
					OM	OM	
	OM (O)				OM		

TABLE 1. continued

Organism	Calcite	Aragonite	Mg-Calcite	Opal	Dahllite	Francolite	Other
amphibia					ES		
reptilia					CS		
		OM (E)			OM		
aves	EM (E)	EM (O)			EM		
mammalia		EM(O)			EM		

Explanation of symbols

Abundance*

E - Exclusively
C - Commonly
O - Occasionally
*Refers only to mineralized forms;
  taxonomic abundance rather than biomass.

Occurrence

S - Single mineral
M - Mixed minerals

Other Minerals

C - Celestite
F - Fluorite
Mo - Monohydrocalcite
W - Weddellite
L - Lepidocrocite
Ma - Magnetite
A - Amorphous phosphates
G - Goethite
V - Vaterite

Non-skeletal Parts ( )

(O) - Otoliths
(E) - Egg cases
(S) - Statocysts
(R) - Radular teeth
(G) - Gizzard plates

Note: It would be impossible to give references to each entry in Table 1. Most of the mineralogic determinations can be found in Meigen (1903), Bøggild (1930) and various papers by H. A. Lowenstam, D. McConnell, D. Carlstrom, N. Watabe, K. E. Chave, and H. K. Erben.

(c) in some cases, taxon-specific regular distribution of crystal nucleation on organic matrices (e.g., on the collagen fibers in mammalian bone on the matrix in gastropod and cephalopod nacre); (d) the invariable production of specific minerals in mollusk species and other taxa (e.g., aragonite only, calcite only, or both in layers and the production of constant microarchitectural units). All of these observations strongly suggest a genetic control within each biological species, a control that could hardly be achieved without selective induction by the organic matrix.

The exact nature of such active sites is still under investigation. Several authors assume that such sites are represented by acid mucopolysaccharides. Others suggest that such sites are represented by free side-chains of certain amino acids contained in the proteins involved, for example the free COOH chain of the aspartic acid in conchiolin (Matheja and Degens, 1968). G. Krampitz produced evidence pointing toward a calcium-binding glycoprotein, which is contained, as a fraction, in three independent, organic, mineralizable matrices (personal communication). Further research may lead to additional clarifications. It would not be surprising if the production of the corresponding mineral could be induced, in different cases of biomineralization, by more than just one type of organic compound within the organic matrices.

The tiny crystallites produced by biomineralization are usually of irregular shape, but, according to the specific conditions of crystal growth, they sometimes become euhedral. Another occasional phenomenon is that of crystal twinning. Spherolitic and dendritic growth are very frequent. In some cases these crystallites are arranged at random, in many others the orientation of the crystallographical axes depends on the orientation of the biological tissues. (This is most clearly shown in the echinoderms and mollusks.) In many invertebrates the biocrystallites, arranged in aggregates and layers form a specific and strictly regular pattern ("ultrastructure" of the skeletal hardpart), which does not vary significantly within each species. In such cases even the ontogenetic changes of this pattern are constant at the species level. This seems to indicate that the ultrastructure, too, is genetically controlled.

Although further research is required, several observations have shown that such patterns have changed during biological evolution, following certain trends: The crossed lamellar type of ultrastructure originated apparently as the result of parallel development in polyplacophorans, gastropods, and bivalves (Erben, unpublished). The nacre in the shells of gastropods and cephalopods is characterized by a "stack-of-coins pattern," while most bivalves develop a "step-like pattern" (Wise, 1970; Erben, 1972). Another general trend was the reduction and loss of the nacreous layer during the evolution of bivalves (Newell, 1969, Fig. 101), gastropods, and cephalopods (Erben, unpublished), or the trend in the calcified eggshells of certain reptiles to increase the calcitic spongy layer at the expense of the aragonitic mamillary layer (Erben, 1970).

Some of the most important types of ultrastructural patterns are shown in Figs. 2–7 and described below (see also Rhoads and Lutz, 1979).

Acicular ("prismatic")—Needle-shaped crystallites arranged in different or parallel orientation; very frequent (e.g., some algae, octocorals).

Columnar ("prismatic")—Vertical polygonal columns, composed of rod-like or acicular crystallites sometimes arranged in the shape of a spherolite sector; occurring occasionally (e.g., *Neopilina*, a monoplacophoran; several bivalves; enamel rods of vertebrate teeth).

Spherolitic aggregates—Globular aggregates of radiating acicular, rod-like or blade-shaped crystallites (Fig. 2); very frequent (e.g., turtle eggshells, *Argonauta*).

Spherolite sectors—Unilaterally extended cone-shaped sectors of spherolites (Fig. 3); very frequent (e.g., some algae, many mollusks, gastropod opercula, some tubicole worms).

Crossed lamellar—Complicated pattern consisting of vertical 1st-order lamellae composed of 2nd-order lamellae containing, as 3rd-order elements, acicular or blade-like crystallites (Figs. 4,5): in successive 1st-order lamellae, the 2nd- and 3rd-order elements are arranged in opposite direction (typical in mollusks).

Nacreous—Superimposed lamellae, each one formed by lateral coalescence of pseudohexagonal or rounded tabular aragonite crystals (occurring in mollusks only; Figs. 6,7).

FIGURE 3. Spherolite sectors, aragonite. Vertical fracture of the shell in the gastropod *Gibbula adriatica*. Picture shows a detail from the outer "prismatic" layer. Scanning electron micrograph, X575.

FIGURE 2. Spherolitical structure, aragonite. View of the inner surface of eggshell in the turtle *Testudo pardalis*. Note the resorption craters in the center of the spherolites; they have been caused by calcium extraction exerted by the embryo during the calcification of its skeletal bones. Scanning electron micrograph, X60.

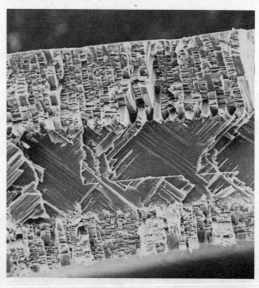

FIGURE 4. Crossed lamellar structure, aragonite. Vertical fracture of the shell in the gastropod *Rostellaria fusus. Upper part;* outer crossed lamellar sublayer with transversal section of first order lamellae; *central part;* middle crossed lamellar sublayer with longitudinal section subparallel to first order lamellae; *lower part;* inner crossed lamellar sublayer with transversal section of first order lamellae. Scanning electron micrograph, X60.

FIGURE 5. Crossed lamellar structure, aragonite. Detail (close-up) from vertical fraction of the shell in the gastropod *Volva volva*. Picture shows four first-order lamellae in which the second-order lamellae composed of blade-like third-order elements are inclined in opposite directions. Scanning electron micrograph, X1100.

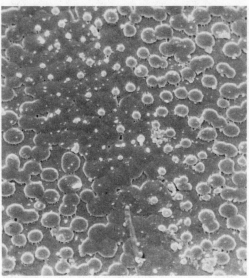

FIGURE 7. Nacreous structure, represented by circular to oval tabular crystals (fast crystal growth). Vertical view of growing surface of the nacreous layer in the pearl oyster, *Pinctada martensi*. Scanning electron micrograph, X900.

Foliate—Thin superimposed lamellae, each one formed by lath-like crystals (e.g., calcitostracum in mollusks).
Grained—Irregular to polygonal grains (different animal groups).
Homogeneous—Containing no observable pattern (different animal groups).

Biomineralized hardparts grow by the addition of crystallites to their flat surface (increase in thickness) and particularly by the addition of increments at their growing edges (increase in length). On the surface, these accretions can be identified sometimes by growth lines, while, in the internal ultrastructure, they may become evident from incremental boundaries. Such growth lines or increments occur in mollusks, corals, brachiopods, echinoderms, annelid worms, barnacles, and in the scales, otoliths, and teeth of vertebrates (Clark, 1975); and they may reflect periodical changes in environmental conditions. Knutson et al. (1972) have shown that reef corals from Enewetak, where nuclear tests have occurred periodically providing radioactive markers in the carbonate, have growth bands that are definitely yearly. In some bivalves, periodicities of different orders seem to reflect daily growth, tides, spawning seasons, lunar cycles, etc. (Pannella and MacClintock, 1968). As shown by Wada (1972), seasonal growth in mollusk nacre is expressed by a change from euhedral crystals

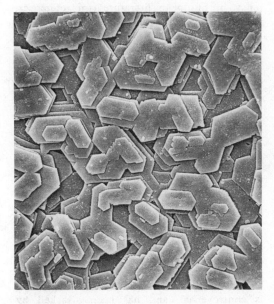

FIGURE 6. Nacreous structure, represented by euhedral tabular crystals (slow crystal growth). Vertical view of adoral face in a septum of the cephalopod *Nautilus pompillius*. Scanning electron micrograph, X675.

(slow growth in winter) to rounded tablets without crystal faces (fast growth in summer time). Additional evidence for seasonal periodicity is furnished by partial dissolution of the uppermost nacreous crystal layers in marine bivalves (Wada, 1972) as well as in at least one freshwater species (Watabe and Erben, unpublished) during late autumn and winter.

KEITH E. CHAVE
HEINRICH K. ERBEN

References

Bevelander, G., and Nakahara, H., 1969. An electron microscope study of the formation of the nacreous layer in the shell of certain bivalve molluscs, *Calc. Tissue Research,* 3, 84–92.

Bøggild, O. B., 1930. The shell structure of mollusks, *Danske Vidensk. Selsk. Skr.,* 9, 235–326.

Clark, G. R., 1975. Growth lines in invertebrate skeletons, *Ann. Rev. Earth Planetary Sci.,* 2, 77–99.

Erben, H. K., 1970. Ultrastruckturen und Mineralisation rezenter und fossiler Eischalen bei Vögeln und Reptilien, *Biominer. Research Repts.,* 1, 1–66.

Erben, H. K., 1972. Uber die Bildung und das Wachstum von Perlmutt, *Biominer. Research Repts.,* 4, 15–46.

Erben, H. K., 1974. On the structure and growth of the nacreous tablets in gastropods, *Biominer. Research Repts.,* 7, 14–27.

Erben, H. D., and Watabe, N., 1974. Crystal formation and growth in bivalve nacre, *Nature,* 248, 128–130.

Glaessner, M. F., 1972. Precambrian palaeozoology, *Univ. Adelaide Centre Precambr. Research, Spec. Paper,* 1, 43–52.

Hurd, D. C., 1972. Factors affecting solution rate of biogenic opal in seawater, *Earth Planetary Sci. Letters,* 15, 411–417.

Knutson, D. W.; Buddemeier, R. W.; and Smith, S. V., 1972. Coral chronometers: Seasonal growth bands in reef corals, *Science,* 177, 270–272.

Lyakhin, Y. I., 1968. Calcium carbonate saturation of Pacific Water, *Oceanology,* 8, 44–53.

Matheja, J. and Degens, E. T., 1968. Molekulare Entwicklung mineralisationsfähiger organischer Matrizen, *Neues Jahrb. Geol. Paläontol. Mh.,* 1968, 215–229.

Meigen, W., 1903. Beitrage zur Kenntnis des kohlensauren Kalk, *Naturwiss. Gesel. Freiburg Ber.,* 13, 1–55.

Newell, N. D., 1969. Classification of Bivalvia, in R. C. Moore, ed., *Treatise on Invertebrate Paleontology,* pt. N, Mollusca 6, Bivalvia. Lawrence, Kansas: Geol. Soc. Amer. and Univ. Kansas Press, 1, N205–N224.

Pannella, G., and MacClintock, C., 1968. Biological and environmental rhythms reflected in molluscan shell growth, *J. Paleontol.,* 42(5, Suppl.), 64–80.

Rhoads, D. C., and Lutz, R. A., eds., 1979. *Skeletal Growth: Biological Records of Environmental Change.* New York: Plenum.

Thiele, H., 1967. Geordnete Kristallisation, Nucleation und Mineralisation, *J. Biomed. Mater. Research.,* 1, 213–238.

Towe, K. M., 1972. Invertebrate shell structure and the organic matrix concept, *Biominer. Research Repts.,* 4, 1–14.

Wada, K., 1972. Nucleation and growth of aragonitic crystals in the nacre of some bivalve molluscs, *Biominer. Research Repts.,* 6, 141–159.

Watabe, N., and Wilbur, K. M., 1960. Influence of the organic matrix on crystal type in molluscs, *Nature,* 188, 334.

Wilbur, K. M., 1972. Shell formation in mollusks, in M. Florkin and B. T. Scheer, eds., *Chemical Zoology,* vol. VII, Mollusca. New York: Academic Press, 103–145.

Wise, S. W., Jr., 1970. Microarchitecture and formation of nacre (mother-of-pearl) in pelecypods, gastropods and cephalopods, *Eclogae Geol. Helvetiae,* 63, 775–797.

Cross-references: *Biogeochemistry; Bones and Teeth; Chitin; Growth Lines; Paleopathology.* Vol. IVA; *Calcium Carbonate; Geochemistry.* Vol. IVB: *Invertebrate Mineralogy.*

BIOSTRATIGRAPHY

Biostratigraphy is the study of the stratigraphical distribution of fossils. Its principles were laid down in 1816 by William Smith when he published his *Strata Identified by Organized Fossils.* The term was first introduced by L. Dollo in 1904. R. Wedekind, C. Diener, and K. Andrée have used it in the restricted sense of "paleontological methods applied to stratigraphy"; and it has gained considerable currency in European literature (for references, see Teichert, 1958; Hupé, 1960; and Schindewolf, 1960). It is usual to distinguish the investigation of facies-controlled faunas (e.g., Imbrie's study of the Florena Shale of Kansas, 1955) as a part of paleoecology rather than biostratigraphy. The two disciplines are closely allied. Thus the term *comparative biostratigraphy* is used in Russia to denote the comparison of faunas of the same age but showing provincial or environmental differences; and the Termiers in France include in biostratigraphy both ecology and "ethology" (meaning, after G. St. Hilaire and L. Dollo, the study of adaptive modifications—not the study of animal behavior, which is a better-known modern usage of the word ethology). The term *biostratigraphy* has only recently come into general use in Britain, but has been much at the fore in American literature dealing with stratigraphic classification and nomenclature. Its use in opposition to "chronostratigraphy" is controversial, and has been attacked by Arkell and Schindewolf (see references in Schindewolf, 1960).

It is noteworthy that the scope and content of the discipline has been formulated by a long line of European geologists (William Smith,

A. d'Orbigny, A. Oppel. S. S. Buckman, Brink-mann, Arkell, Schindewolf), all of whom have been specialists in Mesozoic stratigraphy in general and Jurassic ammonites in particular. The principles thus evolved have been adopted by other stratigraphers and applied to other systems and other fossils, particularly in America, where the Jurassic system is ill represented. As the discipline itself is still actively developing, it is not surprising to find that principles first applied to Jurassic rocks, may now be employed to classify or correlate (for example) the Cambrian or Mississippian, even though they have been superseded in the classification of the Jurassic by different methods embodying new principles.

The principles of biostratigraphy stem from the fundamental precept that William Smith claimed to be a general law: "The same strata are found always in the same order of superposition and contain the same peculiar fossils." The subject can be considered under four headings: (1) biostratigraphic correlation; (2) biostratigraphic classification; (3) biostratigraphic nomenclature; and (4) applied biostratigraphy.

Biostratigraphic Correlation

A unit in biostratigraphy is composed of a stratum (or of strata) containing a peculiar assemblage of fossils. It can be distinguished from other units in a sequence by differences in the fossil assemblages they contain. Units can be correlated with each other if they contain the same assemblage of peculiar fossils, and if they occur in the same order of superposition. (see Fig. 1). These two processes of *distinction* (or diagnosis) and of *correlation*

presuppose that it is possible to draw boundaries between successive units. Commonly, any two successive units are separated by a bedding plane that marks a change of lithology and that therefore denotes a change of environment, and a break in the continuity of sedimentation. When such breaks in the succession are chosen as boundaries between biostratigraphic units, it is clear that successive units are separated by planes that represent an unknown lapse of time during which, in other places, fossiliferous strata may have been deposited. The first task of a biostratigrapher is to construct a geological column in which the succession of units in one area is correlated with the succession in other areas. In this way, nonsequences in a succession can be detected, and a hypothetical model constructed representing a complete succession with all nonsequences made good. Hypothetical columns can be constructed on varying scales. The most detailed so far attempted was that by S. S. Buckman who postulated 367 biostratigraphic units for the Jurassic System. He called his units "hemerae" (Greek for "days"). His system became known as the "polyhemeral system," and is now almost completely discredited. The rejection of the polyhemeral system marked an important turning-point in the development of biostratigraphic thought. It almost coincided with the death of Buckman (1929) and the publication of Arkell's first paper in 1926.

The reasons for abandoning Buckman's hemerae were two; one of them was sound, the other less so. First, J. A. Douglas and W. J. Arkell showed that many of the hemerae proposed by Buckman for the "Cornbrash" were not based on any known section any-

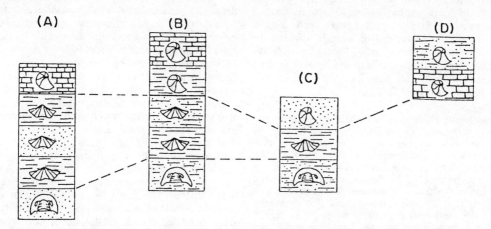

FIGURE 1. Correlation of local sequences (from Beerbower, 1968, by permission of Prentice-Hall, Inc.). A similar sequence of fossils occurs in each of four different local sections. The occurrence of a particular type or association of fossils thus provides the basis of matching local sections.

where. They were based on the hypothetical phylogeny of a series of ammonites whose exact location in the section was unknown. Other parts of Buckman's hemeral scheme were equally hypothetical. Secondly, L. F. Spath showed that a succession of hemerae known from part of the Lias in England occurred (according to French authors) in a different sequence in the Jura Mountains. It is now suspected that Buckman in this case may have been right, and the succession in France was read incorrectly (Arkell, quoted by Teichert, 1958). Indeed, Buckman was one of the most remarkable intuitive paleontologists of his time; and some of his stratigraphical ideas, apparently unsupported by facts, and at first attacked by his successors, have later turned out to be correct. But intuition is unfortunately not communicable, and biostratigraphy has had to develop along more orthodox lines. The rejection of the polyhemeral scheme focussed attention on the unsatisfactory nature of using a nonsequence, a disconformity, or an unconformity as the boundary of a biostratigraphic unit. If an arbitrary plane could be chosen in a continuous sequence of uniform sediment, and if the fossils above and below such a plane could be distinguished, the boundary chosen would represent a precise instant of geological time. But such a situation can only occur when two rare circumstances coincide, continuous sedimentation and rapid and continuous evolution in a fossil lineage. Nevertheless, the two kinds of boundary emphasize two contrasting methods of diagnosis employed in distinguishing succeeding biostratigraphic units, i.e., correlation by faunal replacement; and by evolutionary succession.

Correlation by Faunal Replacement. The early stratigraphers distinguished the units in their biostratigraphic systems entirely on the grounds of faunal replacement, and indeed d'Orbigny and Oppel had already divided up the Jurassic System into Stages and Zones before the publication, in 1859, of Darwin's exposition of evolution. The method used is well described by d'Orbigny:

Geologists in their classifications allow themselves to be influenced by the lithology of the beds, while I take for my starting point . . . the annihilation of an assemblage of life-forms and its replacement by another. I proceed solely according to the identity in the composition of the faunas, or the extinction of genera or families. . . . Since my first observations on the rocks of France, I have realized that in crossing the successive beds from the older to the younger, I have everywhere met with the same sequence of fossil faunas, restricted within the same vertical limits in the geological succession, whatever might be the lithological composition of the beds containing them.

After having obtained in all parts of France—north, south, east and west, in Provence as in Normandy, in the Ardennes as well as in the Vendée,—the same results, and having found nothing but one confirmation after another over a period of fifteen years, without encountering a single contradictory fact, I at last became convinced that the Jurassic rocks were divisible into ten zones or stages, demarcated as well by the different faunas they contain as by stratigraphical boundaries which re-appeared again and again at every point. I have followed them one after another around the basins in France and outside; I have ascertained that in no single locality do they become confused, and that they represent as many distinct geological epochs succeeding one another in constant and regular order. . . . I was therefore bound to adopt them for the double reason that there is nothing arbitrary about them and that they are, on the contrary, the expression of the divisions which nature has delineated with bold strokes across the whole earth. (Translation by Arkell, 1933)

Although the increase of knowledge with time has somewhat blurred the boldness of nature's strokes, d'Orbigny's precepts of "annihilation" and "replacement" remain the principles by which far the greatest number of biostratigraphic units are discriminated. But it is not the whole "assemblage of life forms" that can be used in this way. Only some taxa are restricted to the units. Others are "facies fossils" or "zone-breakers," and are not restricted so closely by time, but more by environment. In every system, the various groups of fossils present have varying ranges. Those with the widest geographic distribution and most restricted stratigraphic range are selected as the main indicators (Fig. 2). Thus, in the Cambrian, we have the trilobites; in the Silurian, graptolites; in the Mesozoic, ammonites. But these main indicators can be supplemented by others. Commonly, macrofossils are partnered by microfossils, by foraminifera, ostracodes, or others. Sometimes the main indicators are restricted by facies, and their place is taken by quite another group. Freshwater facies of the Jurassic are zoned by ostracodes; terrestrial facies in the Tertiary by mammals. But, irrespecitve of the size of the assemblage that is considered as characteristic of the Zone, the special feature of this system of biostratigraphic diagnosis is the phenomenon of replacement. The species taking the place of the previous indicators are not their descendants. They are often quite unrelated. The range of the indicators has been assessed in three different ways, listed as follows by Arkell (1933): (a) acme (i.e., where most abundant); (b) absolute duration (i.e., as established over the whole area of distribution); (c) local duration (i.e., as established in a single section). These diversities of procedure have given rise to

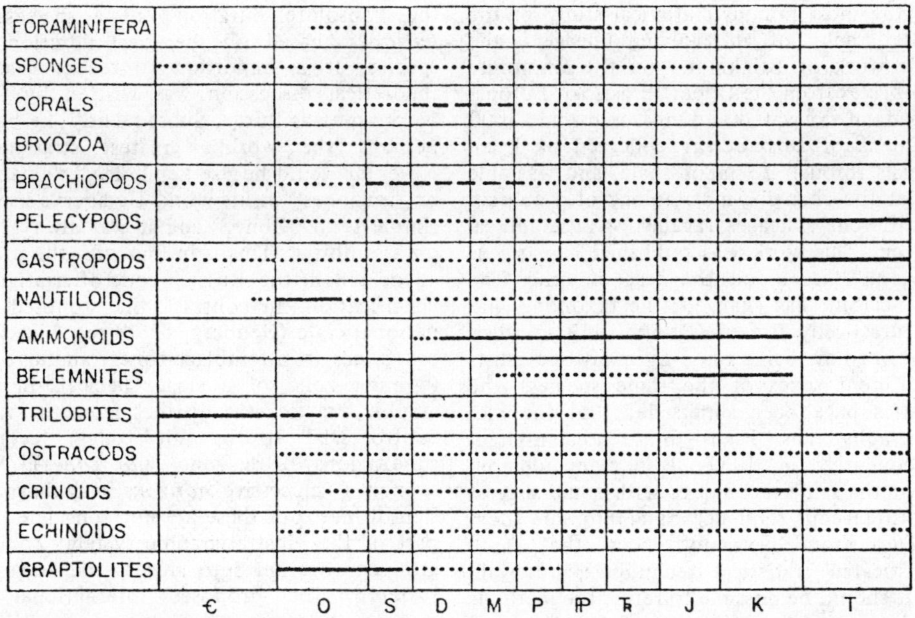

FIGURE 2. Stratigraphic significance of fossil groups (after Teichert, 1958; from Beerbower, 1968, by permission of Prentice-Hall, Inc.). The solid line indicates the group is important for world-wide zoning and correlation; the dashed line means that it was important in regional correlation; the dotted line signifies only occasional or rare use as zonal fossils.

voluminous theoretical discussions, not in practice important, for the units diagnosed by faunal replacement are not based on single species but on assemblages (a fact realized well by Arkell, but often overlooked by others). The boundary (despite d'Orbigny) is chosen arbitrarily in a particular section. Some of the indicating assemblage will cross it, others not, for their ranges will, of course, vary. Provided we know the range of each species, it is possible to correlate one section with another with an accuracy that can be quantified, with statistically determined confidence limits (Shaw, 1964). To achieve this, it is necessary to know the actual ranges (in thicknesses of strata) of all species common to the two sections. If there are enough species with limited vertical ranges, it is possible to calculate an equation of correlation, and from this to recognize time equivalents in the two sections. Alternatively, the data (fossil ranges) can be plotted graphically. The statistical significance of this method increases with the number of species whose ranges are not exactly coincident. However, if the ranges of a group of species do not overlap but are exactly coincident, it is necessary to consider an alternative hypothesis in which the correlation is based not on time-equivalence, but on the migration of faunas with changing facies.

Correlation by Evolutionary Succession. Less sophisticated, but perhaps even more valuable, the method of correlation by evolutionary succession is presented whenever evolution proceeds at a rate fast enough, to allow changing morphology to be plotted against stratigraphic position. Several classic biostratigraphic studies have been based on such changes. Often these occur in strata with uniform lithology over a considerable thickness. The amateur H. T. Rowe studied the changes in the echinoid *Micraster* in the Chalk of England. The main morphological change in the evolution of *Micraster* is the development of a trend toward an elaboration of the anteal ambulacral areas, but this trend is accompanied by others, such as changing proportions, movement of mouth, development of labrum. The various trends are not exactly correlated, so that any particular specimen may have a primitive labrum, advanced ambulacra, intermediate shape, etc.; and in this way contains within itself a statistical sample. Thus a skilled biostratigrapher can place a single specimen in its proper place in the stratigraphical column with considerable accuracy. Although such skill is in part intuitive, the changes observed can be quantified and treated with considerable statistical rigor (Kermack, 1954); see also *Biometrics in Paleontology*. These pioneer studies have been followed by a great host of other examples describing phylogenetic successions, especially in micropaleontology. However, the most famous statistical study of the biostratigraphy of an evolving lineage is the classic work of Brinkmann on the ammonite *Kosmoceras* from the Upper Jurassic of Europe. Brinkmann excavated and measured some 3000 specimens of *Dosmoceras* collected at 1 cm intervals through 13 m of shale, and was able to plot the changing morphology of a number of different lineages against stratigraphical position. This work was published as early as 1929, and has stood the test of time. The nomenclature has changed; the taxonomy has been drastically revised (Brinkmann's so-called subgenera now being rated by many specialists as different sexes of the same species); the statistics have been superseded; but the biostratigraphy has been fully substantiated. Subsequently, in 1937, Brinkmann demonstrated the same techniques with equal success in a Cretaceous sequence of ammonites (*Leymeriella*) from Germany. Now that more sophisticated statistical techniques are available it should be possible to refine the methods of Brinkmann. Quantification of morphological characters has been developed in the system of "numerical taxonomy" (q.v.) devised by P. H. A. Sneath and R. R Sokal; such systems could be applied to biostratigraphy if they could be combined with stratigraphic data in a manner analogous to that adopted by Shaw when dealing with assemblages. But so far, although computerized techniques have been successfully applied to the study of phylogenetic trends, they have not yet been linked with quantitative stratigraphic data of the kind employed so successfully by Brinkmann. When this is achieved, it seems likely that the improvement in rigor of correlation by phylogenetic succession will be even more startling than the improvement Shaw's (1964) techniques have brought to correlation by assemblage.

For discussion of a wide variety of modern approaches to biostratigraphy, see the volumes edited by Tedford (1970) and Kauffman and Hazel (1977).

Biostratigraphic Classification

That "a biostratigraphic unit is a body of rock strata which is unified with respect to adjacent strata by certain elements of its fossil content" (Hedberg, 1961; 1976) is now generally agreed (though it has been disputed). In 1933, Arkell summarized biostratigraphic theory by recognizing two kinds of unit, those based on assemblages (which he called "faunizones"), and those based on single species. The latter he divided into three: *epiboles*, based on the acme of a single species; *biozones*, based on

the "absolute duration" of a species; and *teilzones*, based on the local duration of a species. It is important to realize that this theoretical discussion was written by Arkell before he was thirty. Subsequently, he became perhaps the world's greatest stratigrapher. Never once did he use "epiboles," "biozones," or "teilzones" in his work. He always used the simple term "Zone," and it was always based on assemblage. This is in line with the practice advocated in the Russian code of stratigraphic classification. In contrast, the proposed international code (Hedberg, 1961) recognizes both *cenozones* (= assemblage-zones) and *acrozones* (= range-zone of a singe species). Biostratigraphic practice (as distinct from the various controversial stratigraphic codes) recognizes subdivisions of a zone (*sub zones*). Within subzones, successive *horizons* may have local significance, but these are not considered to be part of the biostratigraphic system. Zones are grouped together into *stages*, and stages into *systems*. The proposed international code regards the Stage and System as "chronostratigraphic" rather than biostratigraphic, but this distinction seems hardly possible to maintain in practice. Arkell in 1933 emphasized that the biostratigraphic hierarchy was one that referred to bodies of rock, not to periods of time. Separate terms were recognized as the chronological equivalents of the biostratigraphic classes, *secule* or *moment* for the faunizone, *age* for the stage, but again, this was theoretical. He did not later use these terms in stratigraphic practice. One can more easily speak of "Parkinsoni Zone time," "Bajocian time," or "Jurassic time" to make one's meaning clear, and this incidentally is the procedure recommended by the present Russian Code.

Biostratigraphic Nomenclature

The system of nomenclature in current use names stages after place names, often Latinized, with the adjectival suffix -ian (French -ien) (e.g., Toarcian after the French town Thouars, "Toarcium" in the Latin). Zones are named after the specific name of an index species of the zonal assemblage (e.g., Bifrons Zone after the ammonite *Hildoceras bifrons*). The zone is still a "faunizone," i.e., based on an asemblage, and may be recognized in the absence of the index species. Ages (time equivalents of stages) are called after the generic name of a characteristic fossil, to which is added the suffix -an (e.g., Hildocertan Age after the ammonite genus *Hildoceras*). Although ages were first introduced by Buckman as time equivalents of stages, he later refined them to such an extent that they became in his hands the time equivalents of zones. They are seldom used now, though they have been proposed for certain systems, and are recognized by most national (and international) codes of stratigraphic nomenclature.

In order to attain a degree of objectivity in biostratigraphic nomenclature it has been proposed that the International Geological Congress should authorize a set of Stratigraphic Rules analogous to the Codes of Zoological and Botanical Nomenclature. The two most important suggestions are for a "Law of Priority" and for the establishment of "Stratotypes."

Rules for these concepts have not yet been devised, but various subcommissions have produced some interesting reports. The issues are controversial. For example, it has been suggested that a Stage can only be characterized by its component zones, and that the type of a Stage is not a stratum, but a Zone. By analogy with genera and species in Zoological nomenclature, a zone has a stratum which can be chosen arbitrarily as the "stratotype." This must (for comple objectivity) be a single stratum in a single locality. It has been suggested that the *base* of the Zone in its type-locality should be the legal "stratotype." "Synstratotypes" and "Lectostratotypes" could be recognized by analogy with syntypes and lectotypes in biological nomenclature.

Applied Biostratigraphy

Biostratigraphy is an evolving discipline, and recent increase in activity promises greatly improved techniques of correlation in the future. Consequently, existing compendia of "index fossils" are of limited value, even when revised editions have attempted to keep pace with current research. For all detailed work, individual monographs for each group of fossils are required. However, for general purposes inclusive compendia have some value; among the better known are: Shimer and Shrock (1944), Neaverson (1955), and Termier and Termier (1960).

P. C. SYLVESTER-BRADLEY*

References

Arkell, W. J., 1933. *The Jurassic System in Great Britain.* Oxford: Clarendon Press, 681p.
Beerbower, J. R., 1968. *Search For the Past,* 2nd ed. Englewood Cliffs, N.J.: Prentice-Hall, 512p.
Brinkmann, R., 1929. Stratistisch-biostratiographische Untersuchungen an Mitteljurassischen Ammoniten Ulber Artbegriff und Stammesentwicklung, *Abh.*

*Deceased

Gesell. Wissensch. zu Gottingen. M. P. Klasse, N. F., 13, 1–250.
Hedberg, H. D., 1961. Stratigraphic classification and terminology, *21st Internat. Geol. Cong. Rept.,* 25, 38p.
Hedberg, H. D., ed., 1976. *International Stratigraphic Guide.* New York: Wiley-Interscience, 200p. (Biostratigraphic units, pp. 45–65.)
Hupé, P., 1960. Les zones stratigraphiques, *BRGM, Serv. Inform. Geol., Bull, Trimest.,* 12(49), 1–20.
Imbrie, I., 1955. Quantitative lithofacies and biofacies study of Florena Shale (Permian) of Kansas, *Bull. Am. Assoc. Petrol. Geologists,* 39, 649–670.
Kauffman, E. G., and Hazel, J. E., eds., 1977. *Concepts and Methods in Biostratigraphy.* Stroudsburg, Pa.: Dowden, Hutchinson & Ross, 658p.
Kermack, K. A., 1954. A biometrical study of *Micraster coranguinum* and *M. (Isomicraster) senonensis, Phil. Trans. Roy. Soc. London, Ser. B,* 237, 375–428.
Neaverson, E., 1955. *Stratigraphical Palaeontology,* 2nd ed. Oxford: Clarendon Press, 806p.
Schindewolf, O. H., 1960. Stratigraphische Methodik und Terminologie, *Geol. Rundschau,* 49, 1–35.
Shaw, A. B., 1964. *Time in Stratigraphy.* New York: McGraw-Hill, 365p.
Shimer, H. W., and Shrock, R. R., 1944. *Index Fossils of North America.* Cambridge, Ma.: M.I.T. Press, 837p.
Tedford, R. H., ed., 1970. Correlation by fossils, *Proc. N. Amer. Paleontological Conv.,* pt. F, 533–703.
Teichert, C., 1958. Some biostratigraphical concepts, *Geol. Soc. Amer. Bull.,* 69, 99–120.
Termier, H. and Termier, G., 1960. *Paléontologie Stratigraphique,* pt. 1–4. Paris: Masson et Cie.

Cross-references: *Computer Applications in Paleontology; Evolution; Fossil Record; Invertebrate Paleontology; Larvae of Marine Invertebrates— Paleontological Significance; Micropaleontology; Palynology.*

BIOSTRATINOMY

The term biostratinomy (originally biostratonomy; Weigelt, 1919) is now defined as the study of the environmental factors that affect organic remains between organisms' death and the final burial of the remains (Müller, 1963). Biostratinomy is thus a very important part of work in taphonomy—the study of the entire postmortem history of organic remains. Studies in present-day settings do provide valuable comparative data for the biostratinomy of fossils; these studies have been given the rather enigmatic name of actuopaleontology (Schäfer, 1962, 1972).

In terms of reconstructing once-living communities of organisms, the most important biostratinomic events involve the dissipation of soft tissue parts, normally by microbial activity.

Johnson (1964) reviewed work on the preservability of marine level-bottom invertebrates and concluded that an average of 70% of the individuals and species in these settings lacked resistent hard parts and would normally be lost to the fossil record. Decay of vertebrate soft parts leads to disarticulation of the skeletal remains (Weigelt, 1927); subsequent dispersal of the bones can make the original populations and communities very difficult to unravel (Shotwell, 1955; Voorhies, 1969). The lack of hard tissues in plants, and the common degradation of even thick-walled plant structures (Barghoorn and Scott, 1958) are major causes of the relative scarcity of plants as fossils.

In regard to organisms with hard tissues, the processes most distinctive of the early postmortem period result in the transport and/or reorientation of fossilizable particles, and the vertical (stratigraphic) displacement of remains before or during final burial. It has been argued (e.g., Lawrence, 1968) that skeletal transport is a very important factor influencing the accumulation of the main part of the fossil deposits found in the geologic record. Recent studies have suggested, however, that postmortem transport is not a decisive factor in the abundant, widespread, and invertebrate-dominated communities of ancient epeiric seas (Johnson, 1972). However, transport can be an important postmortem event in many coastal-marine and some nonmarine settings. For a review of important studies comparing living and dead assemblages of mollusks, see Stanton (1976).

Studies of skeleton transport in the coastal zone have emphasized the bivalved mollusks. The most critical recent studies on shell transport have been done by a Dutch group working on the sandy beaches of the North Sea (e.g., Lever and Thijssen, 1968). With thoughtful work, the Dutch group has been able to show that left-right valve asymmetries and architectural differences, the size of valves, the presence and diameter and location of valve perforations, and the specific weights of valves do all affect the upbeach and longshore mobility of bivalve shells (Fig. 1). These findings have applicability to a wide spectrum of fossil skeletons, and the experimental studies of Voorhies (1969) have pointed the way toward a similar increased knowledge of the dispersal of vertebrate skeletal parts.

That asymmetric skeletons can be reoriented by waves and currents is a well-known phenomenon; oriented groupings such as the "belemnite battlefields" are often cited in the literature (for example, Ager, 1963, pp. 78–80). The work on mollusks by Nagle (1967) has spurred a rebirth of interest in the nature and

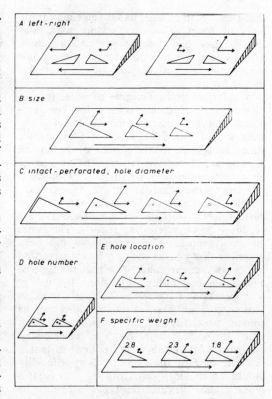

FIGURE 1. Summary of various parameters' effects upon transport of bivalves on sandy beaches (from Lever and Thijssen, 1968). Horizontal arrows show direction of longshore current; right-angled arrows show relative upbeach and longshore mobility of valves.

causes of these reorientation patterns in the coastal zone. Nagle found that current- and wave-dominated regimes did produce different reorientation patterns in asymmetric, elongate skeletons (Fig. 2). Vectorial long axes in cur-

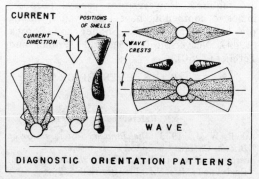

FIGURE 2. Diagnostic reorientation patterns of mollusks in wave- vs. current-dominated depositional regimes, shown as vectorial rose diagrams of skeletal long axes (from Nagle, 1967; copyright © 1967 by the Society of Economic Paleontologists and Mineralogists).

rent regimes showed one maximum, oriented parallel to the direction of current flow; wave-regime long axes showed two maxima oriented along, or slightly inclined to, the direction of advancing wave fronts. As in skeletal transport, these theoretical reorientation patterns are affected by the density, architecture, and other characteristics of the skeletons, as well as by the lithic nature and bedforms of the surface of deposition. With an understanding of these factors, Nagle was able to differentiate wave- and current-dominated deposition in fossil-iferous Devonian rocks from Pennsylvania, USA.

The original stratigraphic and sequential positions of fossils can also be changed before or during final burial. Striking displacements of these types have been documented for ammonites from the Mediterranean Jurassic (Wendt, 1970; 1971). In this setting, regions of submarine rises experienced little or no sedimentation for extensive periods of time. One result of this was the marked condensation of the normal Jurassic stratigraphic sequences.

Reworking of the condensed surficial deposits led to the mixing and, in some cases, inversion of key Zonal ammonites. Also, the formation of Fe/Mn crusts and coatings often accompanied submarine lithification in this nondepositional setting; the reworking history of individual ammonites can be deciphered by tracing the episodes of encrusting and chamber infilling that are preserved in the shells (Fig. 3). A third distinctive feature of this setting was the forma-tion of submarine fissures consequent to sea-floor lithification, with the fissures opening both parallel to and diagonal to bedding. In-filling of the fissures by surficial fossils did result in vertical displacements of up to several hundred meters, and the recognition of these displacements proved to be very important in the interpretation of local geologic histories (Wendt, 1969). The oftimes enhanced fossil preservation in the fissure deposits makes them noteworthy to paleontologists. Indeed, the richest invertebrate biota known from the Mediterranean Jurassic comes from fissures at Rocca Busambra, Sicily (Wendt, 1971).

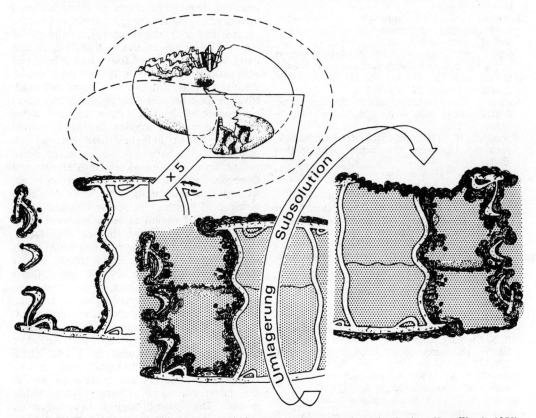

FIGURE 3. Early postmortem history of a typical ammonite from a condensation horizon (from Wendt, 1970). Essential stages include: more or less complete loss of body chamber and breaching of the phragmacone (*top*); encrustation of exposed surfaces with limonite *(left);* sediment infilling of chambers, with interruptions reflected by additional encrustations (*center*); overturning of shell, dissolution of newly exposed shell and part of infilling, additional encrustation with limonite (*right*); and final epizoan overgrowth.

Additional features of similar nondepositional regimes are described by Bathurst (1971); other, detailed reviews of biostratinomic events, processes, and products include those of Müller (1951; 1963), Rolfe and Brett (1969), Schäfer (1962; 1972), and Seilacher (1973).

DAVID R. LAWRENCE

References

Ager, D. V., 1963. *Principles of Paleoecology*. New York: McGraw-Hill, 371p.

Barghoorn, E. S., and Scott, R. A., 1958. Degradation of the plant cell wall and its relation to certain tracheary features of the Lepidodendrales, *Amer. J. Bot.*, 45, 222-227.

Bathurst, R. G. C., 1971. *Carbonate Sediments and their Diagenesis*. Amsterdam: Elsevier, 620p.

Johnson, R. G., 1964. The community approach to paleoecology, in J. Imbrie, and N. Newell, eds., *Approaches to Paleoecology*, New York: Wiley, 107-134.

Johnson, R. G., 1972. Conceptual models of benthic marine communities, in T. J. M. Schopf, ed., *Models in Paleobiology*. San Francisco: Freeman, Cooper, 148-159.

Lawrence, D. R., 1968. Taphonomy and information losses in fossil communities, *Geol. Soc. Amer. Bull.*, 79, 1315-1330.

Lever, J., and Thijssen, R., 1968. Sorting phenomena during the transport of shell valves on sandy beaches studied with the use of artificial valves, *Symp. Zool. Soc. London*, 22, 259-271.

Müller, A. H., 1951. Grundlagen der Biostratonomie, *Deutsche Akad. Wiss. Berlin Abh.*, Jahrg. 1950(3), 147p.

Müller, A. H., 1963. *Lehrbuch de Paläozoologie*, Band 1. *Allgemeine Grundlagen*. Jena: Gustav Fischer Verlag, 387p.

Nagle, J. S., 1967. Wave and current orientation of shells, *J. Sed. Petrology*, 37, 1124-1138.

Rolfe, W. D. I., and Brett, D. W., 1969. Fossilization processes in G. Eglinton, and M. T. J. Murphy, eds., *Organic Geochemistry: Methods and Results*. Berlin and New York: Springer-Verlag, 213-244.

Schäfer, W., 1962. Aktuo-Paläontologie nach Studien in der Nordsee. Frankfurt am Main: Verlag Waldemar Kramer, 666 p.

Schäfer, W., 1972. *Ecology and Palaeoecology of Marine Environments*. Edinburgh: Oliver and Boyd, 568p.

Seilacher, A., 1973. Biostratinomy: The sedimentology of biologically standardized particles, in R. N. Ginsburg, ed., *Evolving Concepts in Sedimentology*. Baltimore: Johns Hopkins Univ. Press, 159-177.

Shotwell, J. A., 1955. An approach to the paleoecology of mammals. *Ecology*, 36, 327-337.

Stanton, R. J., 1976. Relationship of fossil communities to original communities of living organisms, in R. W. Scott and R. R. West, eds., *Structure and Classification of Paleocommunities*. Stroudsburg, Pa.: Dowden Hutchinson & Ross, 107-142.

Voorhies, M. R., 1969. Taphonomy and population dynamics of an early Pliocene vertebrate fauna, Knox County, Nebraska, *Wyoming Contrib. Geol., Spec. Paper 1*, 69p.

Weigelt, J., 1919. Geologie und Nordseefauna, *Der Steinbruch*, 14, 228-231, 244-246.

Weigelt, J., 1927. Über Biostratonomie, *Der Geologe*, 42, 1069-1076.

Wendt, J., 1969. Stratigraphie und Paläogeographie des Roten Jurakalks im Sonnwendgebirge (Tirol, Osterreich), *Neues Jahrb. Geol. Paläontol., Abh.*, 132, 219-238.

Wendt, J., 1970. Stratigraphische Kondensation in triadischen und jurassischen Cephalopodenkalken der Tethys, *Neues Jahrb. Geol. Paläontol.*, 1970(7), 443-448.

Wendt, J., 1971. Genese und Fauna submariner sedimentarer Spaltenfullungen im mediterranen Jura, *Palaeontographica, Abt. A*, 136, 121-192.

Cross-references: *Diagenesis of Fossils—Fossildiadenese; Fossils and Fossilization; Taphonomy*, Vol VI: *Biostratinomy*.

BIVALVIA

The molluscan Class Bivalvia (Class Pelecypoda or Lemellibranchiata of many authors) includes the clams, oysters, scallops, and similar groups, and is characterized by laterally paired external calcareous shells which enclose the soft parts, a highly modified foot which is large and adopted for burrowing in many forms, paired sheath-like gills, and lack of a functional head. Bivalves have become an increasingly important component of benthic marine to fresh water faunas since their origins in the Cambrian (Stanley, 1968) and their hard external calcareous shells have left an excellent fossil record since the Ordovician. In Mesozoic and Cenozoic aquatic biotas, bivalves are consistently a common to numerically dominant element of benthic "paleocommunities"; they are equally abundant at more scattered levels in the Paleozoic, mainly in nearshore facies. From the Mesozoic onward, their broad range of adaptations and ecological strategies insures their presence in virtually all major facies and aquatic environments, from fresh and intertidal niches to abyssal habitats. Between 6,000 and 15,000 living species are recognized, approximately 20,000 Cenozoic forms, 15,000 Mesozoic species, and over 7,000 Paleozoic species (Pojeta and Runnegar, 1974). These factors make the bivalves important in biostratigraphic zonation and correlation, and in the interpretation of diverse paleonenvironments. The paleobiological importance of Bivalvia is enhanced by the fact that well-studied, living, taxonomic, or adaptive counterparts exist for most fossil groups, and also because the Bivalvia have been utilized as a principal fossil invertebrate group in the study of evolution, marine ecology, and paleobio-geography. We possibly know more about Bivalvia and their evolution than we do about any other fossil invertebrates; few groups are so useful in geological and biological inter-pretation. Cox et al. (1969) provides the most up-to-date classification of the Bivalvia found in one publication.

Life Habit

The Bivalvia are almost wholly aquatic benthos; rare freshwater Pisidiidae can live in wet leaf mold. No living adult bivalves are permanently planktic or nektic even though most bivalve larvae are planktic for a few days to a few weeks. The claim that certain Mesozoic bivalves were pelagic during their adult life (e.g., *Posidonia, Monotis, Halobia,* etc.; Jeffries and Minton, 1965) is not sup-ported by good functional morphological evidence; their wide biogeographic spread does not exceed that which can be explained through the normal current drift of long-lived planktotrophic larvae characteristic of many bivalve groups (Kauffman, 1975).

The Bivalvia are predominantly marine to brackish water organisms, although a generi-cally diverse freshwater fauna is also known, representing few families (predominantly Union-idae and Pisidiidae). Freshwater bivalves have widely converged on marine adaptive plans at a relatively late stage in their evolution. Within the marine realm, bivalves occupy every major benthic environment except for those that are too anaerobic to support metazoan life; bivalves are known at all depths, from the littoral zone to the abyss. A few additional forms have special adaptations that allow them to live predominantly in or on floating or rooted vegetation (algae, wood) and other materials. But these epibenthic forms represent a very small percentage of marine species. Most bivalves found attached to such objects are rare chance occurrences of taxa which more normally live as bottom benthos. Floating wood, etc. is not an important mechanism in biogeographic dispersal as suggested by Hallam (1967) and Westermann (1973); Kauffman (1975) discusses this problem. Rare bivalves are commensal on echinoderms and crustaceans (Cyamiacea, Erycinacea), or on sponges (*Vulsella*); *Entovalva* is a parasite in the esoph-agus of holothurians.

In the marine benthos, the Bivalvia have exploited and partitioned diverse habitats through a broad series of adaptive strategies. These include epibenthic forms (Fig.-1) that are free living and which swim periodically as an escape mechanism (e.g., many Pectinidae), sessile, mainly unattached epibenthic forms (e.g., the tridacnid, *Hippopus*), weakly to firmly attached byssate epibenthic forms (Pteriidae and some Arcidae, respectively), cemented epibenthos like the Ostreidae, and those that inhabit crevices and fissures, even modifying them with the shell to a better fit (e.g., some Arcidae). Other groups are semi-infaunal (e.g., Pinnidae, some Modiolinae and Arcidae; Figs. 1, 2), partially or almost wholly buried in the substrate but with the feeding margins and sensory mantle edges exposed. Wholly infaunal burrowing bivalves (Fig. 2) range from habitats just below the sediment-water interface to burial at depths several times the length of the shell. Similarly, boring bivalves excavate holes through physical and chemical processes in a variety of substrates and at a variety of penetration depths. Inas-much as most bivalves feed on suspended material in the water column, and do not seem to compete for this resource, these varying adaptations are primarily concerned with protection and the partitioning of benthic living space.

Bivalves also have a variety of trophic strat-egies that allow partitioning of available re-sources in the benthic zone. Most are suspension or filter feeders, utilizing phytoplankton as a primary food source. Among suspension-feeding bivalves, however, there are varying degrees of selectivity. Some taxa, e.g., many lucinaceans, do not greatly sort suspended food as to size, type, or quality and have only crude sorting mechanisms on the gills, adductor muscles, and mantle areas (Allen, 1958; Kauff-man, 1967). These are well adapted for difficult environments with limited resources, including aphotic deep-water habitats. Other bivalves show varying degrees of selectivity in suspended food resources; some are highly specialized. Detritus feeders are common at all depths but dominate along with nonselective suspen-sion feeders in many deep-water communities, below the zone of phytoplankton production and its dominance in the water column. Bivalve carnivores (septibranchs, or Poromyacea), which feed on small to microscopic animals (arthropods, worms, zooplankton), also are an important element of deep-water communities below the photic zone.

Bivalve diversity increases from intertidal and shallow-water marine habitats, across the shelf zone, into the deep sea (Sanders, 1969), but this is a deceptive statement in that diversity at the higher taxonomic levels, and in terms of adaptive strategies, is greatest in shallow shelf habitats, whereas diversity at the generic and specific levels is highest in deep-water, time-stable habitats among particular groups specifi-

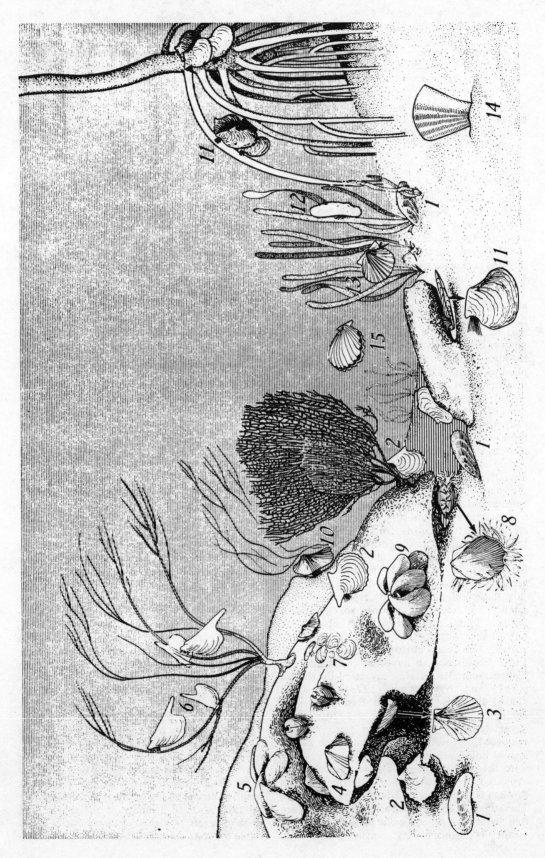

FIGURE 1 (at left). Diversity in habitat and form among epifaunal, crevice-dwelling, and very shallow infaunal bivalves, generalized (slightly modified from Kauffman, 1969), as follows: 1, *Modiolus (Geukensia)*, semi-infaunal, normally intertidal to shallow subtidal, attached to rooted vegetation and buried to umbonal ridge, may be epifaunal nestled in vegetal mats, etc.; 2, *Pinctada*, byssally firmly attached flat against hard substrates; 3, *Chlamys*, byssally loosely attached in fissures, under ledges; 4, *Barbatia*, firmly attached to hard substrates and in crevices in both light-positive and light-negative zones; 5, *Mytilus*, moderately firm attachment to light-positive hard substrates, intertidal to shallow subtidal, gregarious; 6, *Pteria*, firmly (erect wing, normal) to loosely attached (uncommon) to elevated hard substrates by byssal threads; 7, *Anomia*, moderately to firmly attached by byssus to firm substrates, usually light positive; 8, *Lima*, loosely attached to byssal nest-building in protected, usually light-negative habitats; 9, *Brachidontes*, brackish to marine, moderately firm byssal attachment to hard substrates, singly or clustered, usually light positive; 10, *Arca*, crevice nestler, with ability to modify shape of crevice with shell, firmly byssally attached to hard substrates, light-positive and light-negative habitats; 11, *Isognomon*, separate species loosely attached, pendant on elevated substrates, or crevice and hard-surface dwellers with firm byssal attachment; 12, *Amygdalum*, byssally attached to firm vegetal surfaces; 13, *Leptopecten*, firm byssal attachment to algal surfaces; 14, *Pinna*, semi-infaunal form, half buried, byssally attached to subsurface grains; 15, *Pecten*, free-living epibenthic swimmer.

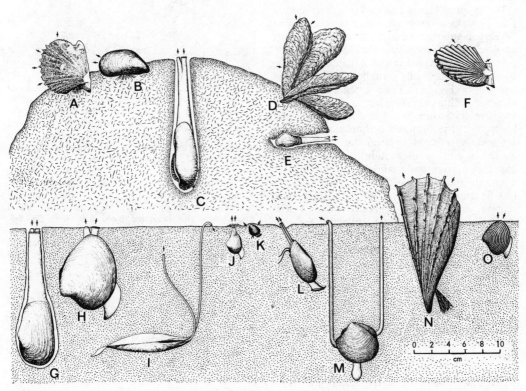

FIGURE 2. Diversity and mode of life habits among infaunal and selective epifaunal bivalves, all to same scale, with water currents over incurrent and excurrent areas, apertures, or siphons marked with arrows (modified after Stanley, 1968). A,B, byssally closely attached epifaunal suspension feeders—*Pinctada* (Pteriacea) and *Mytilus* (Mytilacea), respectively; C,E, rock-boring, siphonate, infaunal suspension feeders—*Pholas* (Pholadacea) and *Hiatella* (Hiatellacea), respectively; D, cemented epifaunal suspension feeder—*Crassostrea* (Ostreacea); F, predominantly free-living epifaunal suspension feeder with swimming capabilities—*Pecten* (Pectinacea); G, deep infaunal, siphonate suspension feeder—*Mya* (Myacea); H, shallow infaunal, siphonate, suspension feeder—*Mercenaria* (Veneracea); I, Infaunal, siphonate, detrital and plankton suspension feeder—*Tellina* (Tellinacea); J, shallow infaunal, siphonate carnivor—*Cuspidaria* (Poromyacea); K, shallow, sedentary infaunal, labial palp deposit (detritus) feeder—*Nucula* (Nuculacea); L, mobile, moderately deep to moderately shallow infaunal, labial palp-deposit feeder—*Yoldia* (Nuculacea); M, moderately deep infaunal, mucus-sediment tube suspension feeder, and possibly (with mucus covered tip of vermiform foot) surface-detritus feeder—*Phacoides* (Lucinacea); N, semi-infaunal, byssate, nonsiphonate suspension feeder—*Atrina* (Pteriacea); O, shallow infaunal, nonsiphonate, suspension feeder—*Astarte* (Astartacea).

cally adapted for this environment (e.g., Nuculacea, Lucinacea, septibranchs).

The Bivalve Animal

The bivalve animal is characteristically bilaterally symmetrical, anteriorly-posteriorly elongated, and enclosed completely, or nearly so, within paired (left and right) calcium carbonate valves which are hinged along the dorsal margin by teeth and/or ligamental material (conchiolin, with or without calcification). In some rock and wood boring taxa (e.g., Teredinidae), the shell is greatly reduced and modified as a boring instrument, and the soft parts are naked, covered by a thick integument, or enclosed within a secondarily deposited calcium carbonate tube.

The "typical" bivalve animal (e.g., *Mercenaria;* Fig. 3) lacks a head, radula or jaws, and well-developed anterior sensors, and is seemingly not as specialized as many other Mollusca. Soft parts may be divided into 5 categories in most Bivalvia: mantle, gills, muscles, visceral mass, and foot (Fig. 3). Each category is highly variable within the Bivalvia, reflecting broad adaptive radiation into diverse habitats and life styles. The foot and certain major muscles are greatly reduced or secondarily eliminated in the evolution of certain bivalve lineages, like the oysters and scallops. Extensive treatments of bivalve soft parts are found in Morton (1958) and Cox et al. (1969), and references therein; these are only generally treated below, especially where they are reflected in shell morphology and can be interpreted from fossils.

Mantle. The three-layered mantle is a thin sheet of tissue that encloses all other soft parts in a mantle cavity, and is responsible for secreting the organic periostracum and ligament and all parts of the shell. Different mantle surfaces and layers have different functions. The periostracum, including that over the ligament, is secreted in the groove between the outer and middle mantle layers, the outer shell layer by the free edge of the outer mantle layer, and the inner shell layer and ligament by the outer surface of the mantle where it lies within the pallial line (Fig. 4). Shell and conchiolin secretion on various parts of the mantle is intricately programmed to a combination of: (a) nature of the mantle cells; (b) nature of the organic template for crystal growth; and (c) chemistry of the pallial fluid. The middle mantle layer bears diverse marginal sensory organs and tentacles in various bivalves, including light, chemical, pressure, and heat sensors, and tentacles used in primary sorting of particles and sediment screening. This is the outer

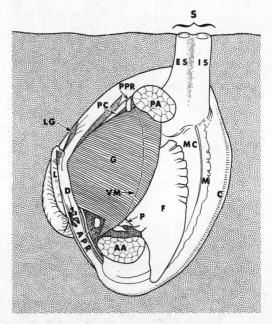

FIGURE 3. Schematic view of the shallow infaunal bivalve *Mercenaria mercenaria* (Linné) in life position, showing major features of internal morphology as follows: AA, anterior adductor muscle (catch and quick portions not shown); APR, anterior pedal retractor muscle; F, foot; G, gills or ctenidia; L, ligament, external; M, mantle, MC, mantle cavity; PA, posterior adductor muscle; PC, pericardial cavity; P, labial palps; PPR, posterior pedal retractor muscle; S, siphon, divided into the inhalent siphon (IS) and the exhalent siphon (ES); VM, outline of visceral mass below gills indicated by dashed line. The following interior shell features are visible: C, commissure, set here with marginal denticles or crenulations; D, dentition, consisting of alternating teeth and sockets; LG, ligamental groove.

defense line for the bivalve, since it lacks a well-defined head with its usual battery of sensors, and is primarily used in detection of predators and the onset of deleterious environments, and to trigger escape responses in the animal. Bivalves with more exposed epifaunal to semi-infaunal habitats (Fig. 1) tend to have more highly developed mantle sensors than those of infaunal habitats; and in many cases these sensory organs are protected by marginal folds, flanges, or spinose extensions of the valve itself which are preserved in fossils. Permanent small shell gapes along the commissure allow constant sensing in many bivalves, even when the shell is closed. The inner mantle layer bears small radial muscles which attach the mantle to the shell—primarily in an arc around the mantle cavity known as the pallial line but also within this line, with clustering of small mantle suspender muscles commonly developed dorsally

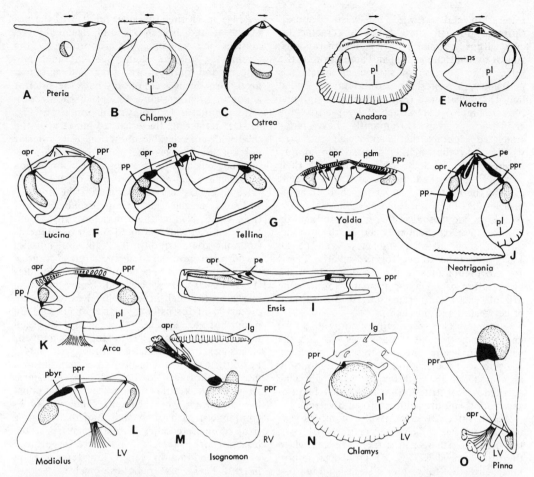

FIGURE 4. Interior views (generalized), examples of typical bivalve musculature in diverse dimyarian, hetero-myarian, and monomyarian forms. (modified after Cox et al., 1969, Figs. 31–35). Pedal and pedalbyssal muscles in black, main pallial muscles shown as line, adductor muscle in stippled pattern. Where the adductor is shown as being divided into component muscles, the "quick" or striated muscle is unshaded, the "catch" or smooth muscle is stippled. Taxa and habitats as follows: *Top row,* relative size of catch and quick adductor muscle segments in A, epifaunal, pendent to closely attached, byssate *Pteria;* B, epifaunal, free-living to weakly byssate *Chlamys* with swimming ability (note large "quick" muscle); C, cemented epifaunal *Ostrea;* D, slow burrowing, shallow infaunal *Anadara;* and E, moderately rapid-burrowing, moderately deep infaunal *Mactra. Middle row,* size and shape of foot, comparative position and size of pedal protractor and retractor muscles, and adductors, in diverse infaunal bivalves—I, slow-burrowing, moderately deep infaunal *Lucina;* G, rapid-burrowing, moderately deep infaunal *Tellina;* H, rapid-burrowing, shallow to moderately deep, mobile infaunal detritus feeder, *Yolida;* I, rapid-burrowing, deep infaunal *Ensis;* J, strong but moderately slow-burrowing, shallow infaunal *Neotrigonia,* with most inflated shell of group. *Bottom row,* epifaunal to semi-infaunal, mostly sedentary suspension feeders— K, strongly byssate crevice and hard-surface dweller, moving shell to modify hard surface by grinding in some cases, *Arca;* L, semi-infaunal to (less commonly) epifaunal byssate *Modiolus* with moderately strong attach-ment; M, pendant, moderately to loosely attached, byssate *Isognomon;* N, free-living to weakly byssate *Chlamys,* with swimming capabilities; O, semi-infaunal, byssate *Pinna.* Note relationship of muscle size, number and placement to habitat and behavior throughout. *Key to letter symbols:* apr, anterior pedal retractor; lg, ligament; LV, left valve; pbyr, posterior byssal retractor; pdm, dorsomedian pedal muscles; pe, pedal elevator; pl, pallial line; pp, pedal protractor; ppr, posterior pedal retractor; ps, pallial sinus; RV, right valve.

(Fig. 4). The inner mantle surface may also perform a respiratory function, and contain cilia used in primary sorting of food and other particles and the creation of currents in the mantle cavity.

The mantle is slightly to extensively fused posteriorly to posteroventrally in most bivalves to form distinct inhalent and exhalent apertures or extendable tubes (siphons; Figs. 2,3), which direct water currents inward for feeding and respiration and outward for waste disposal. Currents through these tubes are created by cilia within the mantle cavity and on the gills, and/or by alternate gaping and adduction of the valves. The position of siphons in bivalves is marked by a posterior reentrant in the pallial line—the pallial sinus (Fig. 4E)—where siphonal retractor muscles are attached. Deep-burrowing and boring bivalves frequently have large, elongated siphons that cannot be retracted wholly into the shell. These are commonly covered with a thick integument and their presence is marked by a large, permanent gape in the posterior part of the shell. In a few bivalves, like the tropical reef-dwelling *Tridacna*, symbiotic zooxanthellae live in exposed mantle tissue. These biochemically aid in the production of shell material and may even be an emergency food source for the bivalve. Pearls are also predominantly formed in mantle tissue by secretion of calcium carbonate around some irritant (e.g., a sand grain) lodged in the tissue, or between it and the shell.

Mantle muscles (especially along the pallial line), principal mantle blood vessels, and the pallial sinus—reflecting the presence, size, and position of the siphons—are all commonly impressed on the interior of shells, allowing detailed reconstruction of mantle features and from them, life habits in fossil bivalves.

Musculature. Bivalvia have a complex system of muscles which operate the soft parts, tie them together, and attach the animal to the protective shell (Fig. 4). Of special importance are the following systems: adductors; pedal and/or pedalbyssal muscles; pallial (including siphonal) muscles; and visceral and gill suspender muscles. All systems have numerous muscles attaching directly to a special shell layer (myostracum), and their insertion areas or "muscle scars" are commonly preserved in fossils from throughout the Phanerozoic (see Cox et al., 1969, pp. N30–N39, figs. 30–39). By understanding the relationship between muscle number, size, strength of insertion, placement, and the nature of the structures they serve in living bivalves, the paleontologist can use preserved muscle "scars" for detailed interpretation of soft-part morphology, behavior, and habitat in fossil Bivalvia (Kauffman, 1969).

The largest muscles are the adductors (Fig. 4), which contract to close the shell against the force of the elastic ligament. In most primitive bivalves, and in virtually all modern infaunal groups, the adductors are large, strongly attached, paired, and of equal or near-equal size (dimyarian, isomyarian condition; Fig. 4D–K). They are placed medially or dorsolaterally just inside the shell margin, insuring optimal

leverage in closing the shell relative to the position of the hinge axis and ligament. The muscles are unequal in proportion to the asymmetry of the shell. Semi-infaunal pinnaform and modioliform taxa and epifaunal mytiliform bivalves commonly have the anterior adductor greatly reduced relative to the posterior adductor (anisomyarian condition; Fig. 4L–O), reflecting the great asymmetry of the shell with considerable reduction of its anterior portion. In bivalves that have taken on a cemented, byssate, or free-living life habit which is basically recumbent on one or the other valve (e.g., Ostreidae, Pectinidae) the anterior adductor muscle has been secondarily lost (Fig. 4B, C, N) and these taxa have only an enlarged, subcentrally situated posterior adductor (monomyarian condition). Adductor muscles are divided into striated or "quick" muscle bundles, and smooth or "catch" muscle bundles (Fig. 4A–E). These leave different insertion marks on the shell in many bivalves, which are preserved in fossils. Large striated muscles are important to bivalves that close the shell rapidly in swimming, burrowing, and frequent cleansing; striated muscles contract very rapidly but fatigue easily under constant strain. Smooth muscles contract slowly but have long-term strength for keeping the valves closed against the pressure of the ligament. The two types compliment each other well, and their relative size, often detectable from the nature of the insertion area, depicts the relative importance of rapid adduction versus long-term closure in the behavior patterns of the bivalve.

Pedal and pedalbyssal muscles (a) project the foot in probing and locomotion, and/or the byssal gland for attachment to stable surfaces, by contraction of the extensors or protractors, and then (b) pull the shell down onto the byssus or the foot once they are anchored, employing the retractor muscles. These may be single or multiple muscles, and their size and location closely define the forces involved in these two activities, and the direction in which they are applied. A pedal retractor (or retractors), for example, operates most efficiently when placed directly opposite the axis of elongation of the foot in burrowing or when this direction is the summation of all independent vector forces from multiple retractors. In fossil material, therefore, analysis of size and placement of pedal muscles allows reconstruction of the behavior of the foot, its direction of projection, and the alignment of the shell relative to the foot in burrowing. Pedal and pedalbyssal muscles are usually concentrated dorsally just inside dimyarian or anisomyarian adductors, under the umbonal septum, and below the hinge line and dorsal margin (Fig. 4). In addition, a large retractor may be located posteroventrally adjacent to the posterior adductor muscle, as in *Isognomon* (Fig. 4M). Pedal and byssal muscles, though smaller than the adductors, usually have well incised and commonly buttressed muscle platforms.

Pallial muscles, including those that depict the presence and size of siphons and degree of mantle fusion, were dealt with in the section on the "mantle." These, and various small mantle, gill, and visceral suspenders usually situated dorsally in the shell below the hinge axis, are commonly preserved in fossil bivalves and allow detailed interpretation of major features of the soft-parts.

Gills. The gills or ctenidia are predominantly flat, anteriorly-posteriorly elongated, thin organic sheets that extend from the sides of the visceral mass and/or the base of the foot, downward into the mantle cavity on either side of the foot (Fig. 3) in all bivalves except the carnivorous Poromyacea (septibranchs). The gills are attached along their dorsal margin by a series of small muscles and membranes to the mantle-shell (outer demibranch) and to the foot (inner demibranch); the outer line of muscle insertions is marked dorsally on the inner shell surface by a small line or series of muscle pits in some cases, and/or by the position of the umbonal fold or sulcus on the shells of many bivalves, allowing the paleobiologist to reconstruct the position of the gills in fossil shells (e.g., as in *Thyasira;* Kauffman, 1969). The gills consist of one or, more commonly, two filamentous sheets of tissue on either side of the mantle cavity (the inner and outer demibranchs).

The gills serve two principal functions; respiration and feeding. Respiration is the primary function in most detritus-feeding bivalves, such as the nuculaceans. But in most suspension-feeding bivalves, the gills serve an equally or more important role in obtaining and sorting food, and in carrying it to the mouth. This is done through the action of extremely abundant, minute cilia situated on the surfaces of gill lamellae, the mantle, and rarely the adductor muscle surfaces, which create complex current systems. These currents draw nutrient and oxygen-bearing water in through the inhalent aperture and across the gills where food particles are strained by the filamentous demibranches, entangled in mucous, and transported to the mouth via the labial palps. The palps are fleshy leaflike to tentaculate extensions around the mouth, related to the gills structurally and physiologically; they consist of inner and outer lamellae between which food particles are passed along ciliated surfaces, where final

sorting takes place, and on to the mouth. In many detritus-feeding bivalves (e.g., nuculaceans), the palps are extended into tentaculate proboscides which collect detritus and pass it to the mouth.

Cox et al. (1969, p. N21) note one instance where Trigoniacian demibranchs have been fossilized, but otherwise the paleontologist must depend upon gill musculature, the position and extent of the line of gill suspender muscle insertion (where marked by muscle insertion areas, ridges, or folds), and a knowledge of the life habits and phylogenetic relationships of fossil material to be able to reconstruct much about gill size, shape, orientation, and structure.

Foot and Byssus. The foot of the bivalve is a muscular mass of variable size and shape extending ventrally or anteroventrally from the visceral mass into the mantle cavity (Figs. 3,4). It contains numerous blood vessels which, when filled, cause the foot to expand in size. The foot is the principal organ of locomotion and burrowing among semi-infaunal and infaunal bivalves. Burrowing rate and efficiency is partially dependent upon the size and shape of the foot. Strong burrowers commonly have a large foot with an anchor-like projection at the base when inflated (see Cox et al., 1969, figs. 12, 31, 33, 34; Fig. 4). Burrowing takes place in a series of rhythmic movements which involve: (a) downward projection of the foot into the substrate by pedal extensor or protractor muscles; (b) expansion and anchorage of the tip of the foot in the sediment by pumping blood into its large vessels; (c) downward movement of the shell into the substrate as the animal contracts on the foot using the pedal retractor muscles; in most cases this is accompanied by a rocking motion caused by alternate contraction of anterior and posterior retractors; (d) resting period as blood is pumped back out of the foot; (e) downward probing of the now diminished foot, etc. The size, strength, and direction of projection of the foot in fossil bivalves can be indirectly interpreted from the size, distribution, number, depth of impression, and buttressing of the pedal protractor and retractor muscles (Kauffman, 1969; Fig. 4).

In deep-burrowing and boring taxa with an exceptionally large foot, the shell has a permanent gape anteriorly to accomodate it.

The foot bears a gland in many bivalves which secretes the byssus, bundles of conchiolin threads for attachment to the substrate, especially in the juvenile stage. The byssus and its gland become greatly reduced or atrophied in adults of most bivalves with active locomotion, but become enlarged in bivalves that have a more sessile or semi-active free living epifaunal habitat, great reduction or virtual loss of the foot, and that depend upon a large byssus for permanent or periodic attachment (e.g., Mytilidae, many Pectinidae). Cemented bivalves, like the oysters, have a greatly reduced foot and, in the adult, a reduced or atrophied byssal gland as well. Reduction or loss of the foot is reflected by decrease in size and number of pedal muscles and shell insertion areas (Fig. 4).

Visceral Mass. The main body of organs and life systems in the bivalve is contained in the visceral mass, a complex and concentrated ovate mass situated dorsocentrally between the most convex portion of the valves (Fig. 3). The visceral mass contains primarily the digestive and excretory systems, the main part of the circulatory and nervous systems, the reproductive system, and complex musculature. The visceral mass leaves little evidence in fossils of its morphology or the complexity and arrangement of organs, and thus will be treated only briefly (see Morton, 1958, for details). The size and position of the mass can be estimated from the size and inflation of the umbonal and early postumbonal portions of the shell—in many cases the only part of the shell with significant convexity. To a lesser extent, small visceral suspender muscles are inserted dorsally on the inner shell surface (Fig. 4), leaving a trace in some fossil bivalves, and the size, number, and spacing of these suspender muscles indirectly suggest both the position and size of the visceral mass.

The digestive system consists of a mouth without jaws and radula, a short esophagus, a relatively large rounded stomach served by digestive glands and containing a rodlike structure of hyaline mucroprotein, and crystalline style—a unique feature of the Bivalvia. Fecal material in mucous strings passes from the stomach through a simply looped and loosely coiled intestine and is passed eventually through the anus into the mantle cavity, where it is carried out the exhalent aperture or siphon by cilia-driven currents and/or through rapid cleansing of the valves by abrupt contraction of the adductor muscles. Fossil casts of the intestine of *Nuculana* sp. have been reported from the Jurassic of England (Cox et al., 1969, p. N24).

The circulatory system of the bivalve is a simple, closed system with a dorsocentral chambered heart situated in the pericardial cavity; veins and arteries contain a clear blood which assists in transport and assimilation of food, in excretion, in conveying calcium carbonate for shell formation, and also carries oxygen in solution to the cells; Cox, Nuttall, and Trueman show characteristic examples (1969, fig. 24). Main arteries serve the foot, the stomach-intestine, the gills, the rectum, and the mantle. The ability of the bivalve animal to rapidly fill and drain the pedal vessels with blood is critical to burrowing in infaunal taxa inasmuch as it determines the shape of the foot at any time during burrowing and allows alternating probing and anchorage functions for the foot. The nervous system of the bivalve consists of three pairs of ganglia (Cox et al., 1969, p. N26, fig. 25): a cerebropleural pair of ganglia on either side of the esophagus, a visceral pair of ganglia just below the rectum, and a pedal pair of ganglia set in the middle of the foot. These are interconnected by main nerves, and serve a moderately complex set of secondary, tertiary, etc. nerves.

Reproduction. Most bivalves are dioecious and have a simple reproductive system consisting of a pair of gonads symmetrically placed on either side of the visceral mass. In most taxa, the sexual products are released and fertilized in the mantle cavity or in the open water. About 4% of living bivalves are not dioecious, but rather ambisexual (functional hermaphrodites): protandric (consecutive) dimorphs undergoing a single sex change in most individuals (e.g., *Astarte*); alternating dimorphs with yearly sex changes (e.g., *Crassostrea*); and rhythmic consecutive dimorphs (e.g., *Ostrea edulis*) with a series of alternating male-female changes, commonly producing one male and one female cycle each year in the individual, with part of the population undergoing male to female change, and the other part having a reverse trend to maintain a balance of sexes. These changes can be seen in shell features of some ambisexual groups.

Sexual dimorphism is expressed in Holocene and fossil bivalve shells mainly in those relatively few groups that incubate the larvae on or between the gill lamallae—in which case the female shell is larger, more inflated, and commonly has a larger umbonal ridge or fold. Female dimorphs may also have coarser concentric ornamentation (rugae, folds, growth lines) resulting from longer and more frequent growth rests, during which time metabolic energy is mainly directed to producing eggs and incubating larvae. Female members of the Astartidae and perhaps other families as well, more commonly have pronounced crenulations ("denticles") along the inner valve margin than do males (see Kauffman and Buddenhagen, 1969). Heaslip (1969) reports not only greater size and valve inflation, but also greater umbonal curvature and more complex radial ornamentation in female shells of *Venericardia* and related forms than in male shells. Some taxa actually have shelly brood chambers along the ventral shell margins of incubatory females (e.g., *Thecalia, Milneria* and among Carditadae). Thus, numerous morphologic features preservable in fossil shells express sexual dimorphism in the Bivalvia.

Ontogeny and Growth. All bivalves have a trochophore larvae; and a majority continue development into a planktic, swimming veliger stage. Planktotrophic larvae feed mainly on phytoplankton in the upper water layers and may spend from a few hours to a few months drifting in the plankton before undergoing metamorphosis and settling on the bottom (average planktonic duration 2–3 weeks). The number of bivalve taxa having long-lived planktotrophic larvae, and the frequency of breeding and larval production increases significantly from cold-water environments into warm-temperate to tropical environments. A long planktotrophic stage, and active, far-reaching oceanic currents combined in the past, as today, to disperse bivalve species widely and rapidly, ensuring broad biogeographic spread and high biostratigraphic potential for long-range correlation among ancient bivalves (Kauffman, 1975). Some groups of bivalves, especially in colder-water environments, have direct development of the larvae in a large-yolked egg, without a pelagic stage. As an intermediate plan (lecithotrophic larval development), the larvae undergo a short pelagic stage of a few hours to a few days but draw their nourishment from a large-yolked egg.

In planktotrophic development, the first shell (prodissoconch) is secreted by the shell gland of the veliger larva and is an extremely thin, delicate calcareous structure sparsely preserved in most fossil and many living bivalve shells. Prodissoconchs have ornament, shape, and dentition that are morphologically distinct from adult shells of the species. When shell secretion is assumed by the mantle in later development, the larval shell features begin to change with marginal growth (Prodissoconch II stage), and become transitory to the normal adult morphology. Other developmental plans among bivalves have distinct larval shells: lecithotrophic larval shells lack the Prodissoconch II stage; directly developed larvae have Prodissoconch I and II stages which are distinct from equivalent stages in planktotrophic bivalves. Larval history, and thus dispersal and biostratigraphic potential, can thus be determined by the study of the prodissoconch of fossil bivalves. Freshwater unionacean bivalves have a distinct larval history, with a parasitic larval stage (glochidium) that initially clings to passing fish by means of hooks or spines on the larval shell and ultimately becomes embedded in the host's tissue, where the larvae metamorphose and eventually break through the host's tissue to drop to the bottom and begin benthic life.

Settling bivalve larvae normally first attach by a byssus, which eventually disappears in the adults of most taxa as they take on a different lifestyle. Young bivalves may go through two or more distinct living habits during ontogeny; and in both living and fossil shells, this is commonly marked by changes in shell morphology. Attainment of reproductive maturity may also introduce changes in shell morphology. At each stage of ontogeny, therefore, the life habits of the bivalve are imprinted on the adaptive morphology of the shell and successively preserved as ontogeny progresses, allowing the paleontologist to interpret the entire life history of fossil bivalve species (see *Larvae of Marine Invertebrates, Paleontological Significance*).

The Bivalve Shell

The bivalve shell consists of three or more main layers (Fig. 5): all bivalves have a thin outer uncalcified periostracum which is composed of conchiolin. The calcified part of the shell (ostracum) usually consists of two principal layers composed of calcite and/or aragonite, interlaced with thin, mainly aragonitic layers of myostracum, a special structure which underlies muscle attachment platforms.[1] The calcified layers possess highly varied structure in different groups; a tremendous literature describes this (e.g., Bøggild, 1930; Taylor et al., 1969; Cox et al., 1969, and references therein). Principal types are (Fig. 5):

a. *nacreous structure,* mainly confined to the inner shell layers and consisting of very thin (1 μm or less) sheets of aragonite basically parallel to the inner shell surface; mother of pearl is composed of nacre, which is among the strongest of shell structures;

b. *prismatic structure,* mainly composed of calcite, less commonly aragonite, and predominantly found in the outer shell layers; it consists of small, regular to subregular, crowded prisms which are perpendicular or at a high angle to the shell surface;

c. *simple or complex cross-lamellar structure,* primarily of aragonite and potentially in all parts of the shell; it consists of rows of moderately inclined rectangular lathes or plates, alternate rows of which dip in opposite directions to the shell surface; this is another high strength structure;

d. *foliated or cross foliated structure* composed of calcite in parallel sheets which are thicker than those

---

[1] The commonly cited "three-layer" model of bivalve shell structure (periostracum, outer ostracum, inner ostracum) does not apply to many bivalves and is misleading.

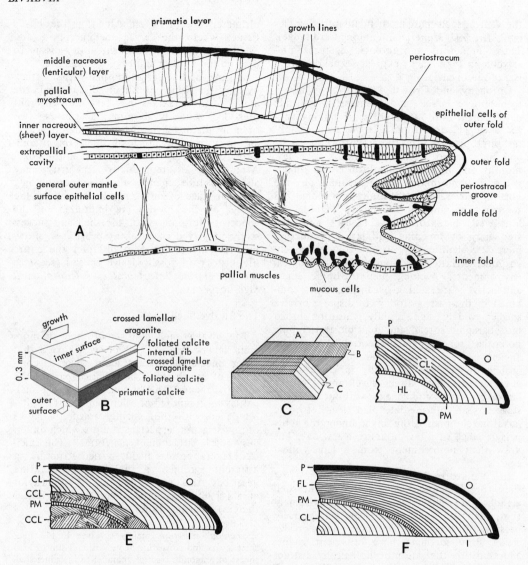

FIGURE 5. Principal shell structure types among Bivalvia, and dominant combinations within the class; highly generalized. **A:** Cross section through typical "three layered" prismatonacreous shell and mantle edge of the freshwater bivalve *Anodonta cygnea* showing relationship between mantle lobes and various shell layers, and between muscles and special muscle platform structure, the myostracum (after Taylor, et al., 1969). **B:** Schematic block diagram through complexly layered shell of *Propeamusium* (periostracum not shown) showing distribution of mineralogy and microstructure (after Waller, 1972). **C:** Three-dimensional diagram of cross-lamellar structure (after Wise, 1971). D–E are schematic cross sections through three additional common bivalve shell structures. **D:** An outer aragonitic cross lamellar layer (CCL), and inner aragonitic homogeneous layer (HL), separated by aragonitic myostracum for attachment of pallial muscles (PM), as in *Venus subimbricata* Sowerby. **E:** An outer cross lamellar aragonitic layer, an inner complex cross lamellar aragonitic layer (CCL), with aragonitic pallial myostracum shown, as in *Anadara notabilis* (Röding). **F:** An outer foliated calcite layer (FL), and an inner aragonitic cross lamellar layer separated by aragonite pallial myostracum, as in *Chlamys squamata* (Gmelin). (A–C show first-, second-, and third-order lamels, respectively; D–F only show side views) O, outer shell surface; I, inner shell surface; P, periostracum. Figures not to relative scale.

of nacre and which either parallel the shell surface or dip at low angles to it in alternate directions, as in cross-lamellar structure;

e. *homogeneous shell layers* where patterned microstructure is not evident.

These different structural plans have distinct tensile and compressive strength qualities, and they are adaptive to the main life habit of the bivalve groups they characterize (Taylor and Layman, 1972; Currey and Taylor, 1974, and references therein). Thus, most infaunal bivalves have very strong shells of cross-lamellar and homogeneous aragonite, whereas many pendant byssate epifaunal bivalves have thinner and more flexible prismatonacreous shells.

The periostracum covers the outer shell surface and ligament; it protects the shell from solution and may discourage encrustation by its texture or biochemical qualities. It commonly bears the shell coloring, a product of specialized metabolic waste disposal, and is preserved on many fossils with original shell material. The elastic ligament basically consists of an outer periostracum, middle lamellar layer, and inner fibrous layer; it takes on many forms (Trueman, 1969, p. N58). The ligament may be internal and/or external on the shell and serves to open the valves, as well as hinge them in many cases; it aids in preventing independent rotation of the valves in forms with weak or nonexistent dentition. The internal ligament (resilium) may be single or multiple; it is compressed when the valves are closed, and opens the valves by expansion when the adductor muscles relax. External ligaments are stretched when the valves are closed, and open the shell by contraction. Ligaments are variously calcified and commonly preserved in fossils. The position and nature of the ligament is expressed in fossil shells by the size, shape, and position of the ligamental groove(s), pits, resilifer(s), etc. on various bivalves. Ligament structure, size, and placement are directly related to life habits in the Bivalvia.

The bivalve shell is highly variable in size, shape, structure, thickness, and morphologic complexity. Adult size ranges from <1 mm to ca. 2.5 m (fossil Inoceramidae), shell thickness from translucent shells only microns thick to nearly 30 cm thick (Cretaceous rudists), weight from milligrams to >45 kg (Tridacnidae), and shape from almost flat and symmetrical to highly convex or complexly coiled, and grossly asymmetrical (Raup, 1966, models the possible range of bivalve form).

The bivalve shell serves many functions, some primary, others secondary or tertiary in nature; and the class shows a broad range of structural adaptations in the shell, which reflect the lift habit and the nature of the preferred environment for each taxon (see Kauffman, 1969; Stanley, 1968, 1970, 1972; Cox et al., 1969; and references in each). Primarily, the shell protects the soft parts inside of it and acts as a rigid surface for muscle and other tissue attachment necessary to support the organ and tissue systems, and to allow them to function. The structural qualities of the shell that reflect these basic functions are its size; shape; convexity; thickness; mineralogy; crystalline structure; and the presence or absence of additional structural supports like radial costae, plicae, folds, concentric rugae, or raised ridges, etc. For example, the thickest, strongest, most inflated, and in some cases the most ornate shells belong to dimyarian bivalves with epifaunal to shallow infaunal habitats in high-energy shallow-water environments (e.g., rudists, tridacnids, certain venerids). Thin, flexible, narrow, smooth anisomyarian shells with hydrodynamically streamlined shape, belonging to byssate epifaunal groups (e.g., Pteriidae, Mytilidae, Isognomonidae) represent yet another adaptive strategy for current swept environments. Among infaunal taxa—protected to varying degrees from currents and predators by the surrounding sediment shell convexity, thickness, and ornamentation in general decrease with increasing depth of burial (Kauffman, 1969; Stanley, 1970); these features are no longer necessary to provide adequate protection for the soft parts. Most deep infaunal bivalves have smooth, relatively thin, and only moderately inflated valves.

Shape is another highly adaptive feature of both epifaunal and infaunal Bivalvia (Kauffman, 1969; and Stanley, 1968, 1970, 1972, discuss this in depth). Virtually all bivalves have inequilateral valves with the plane of symmetry between them, and most are inclined anteriorly (prosocline) with anteriorly curved beaks (prosogyre); a lesser number are erect (orthocline, orthogyre) or posteriorly curved (opisthocline, opisthogyre). For example, swimming bivalves predominantly have nearly symmetrical shells of moderate to low convexity with a cross section resembling an airfoil; the best swimmers (e.g., *Placopecten*) have the smoothest, least inflated, most symmetrical shells, and the poorest swimmers have the most asymmetrical, inflated, inequivalve, and ornamented shells among Pectinidae. Kauffman (1969, fig. 89) has suggested that generally the degree of streamlining by posterior and posteroventral elongation of pendent, byssally attached epifaunal bivalve shells is correlative with the degree of exposure to currents in the normal habitat; the more elongated shells act as rudders in the face of strong currents to orient the shell in the direction of least resistance when it is not pulled down against the substrate on the byssus.

Shape is equally important for infaunal bivalves (Kauffman, 1969; Stanley, 1968, 1970); it is highly correlative with the efficiency and speed of burrowing, and more generally to the depth of burrowing. Efficient, rapid burrowing allows escape from predation and re-establishment of normal life position following chance exhumation by current and wave scour. The efficiency of burrowing is related to: the size of the burrow necessary to accomodate the entire shell; the friction created by the shell as it is drawn into the sediment; the degree to which the shell and its ornament serve to buttress the shell against the sediment during probing of the foot; and, naturally, the size, shape, and activity of the foot. It follows that for any given foot plan, the less convex, less ornamented, and more elongated the shell is in the direction of burrowing, the more rapidly it can be drawn into the sediment. Fig. 6 suggests therefore that burrowing rate, and to some degree burrowing depth, increases as shell convexity decreases and elongation along the burrowing axis increases; Stanley (1970) has shown that this generalization may be modified considerably if the bivalve possesses a large active foot to offset the effects of high valve convexity, and therefore that some more inflated forms are capable of relatively deep burial.

Shell shape can be highly modified, and tremendously variable within species populations of gregarious byssate bivalves and/or cemented forms such as the Mytilidae, the oysters, and the extinct Cretaceous rudists (q.v.). This plasticity, and in cemented forms an upward growth habit, are essential for gregarious living where individuals compete strongly for space in the primary habitat. Evolution of rudist bivalve shells toward the conical "cup" plan of corals and ricthofenid brachiopods for life in reefoid frameworks is a classic example of convergence on a morphological strategy which allows maximum packing of gregarious individuals into a small space within optimal habitats.

Bivalve ornamentation consists of both radial elements (lirae, costellae, costae, plicae, folds, and sulcae) and concentric elements such as growth lines, lamellae, ridges, rugae, and large flexures. All of these elements have examples among Bivalvia where they are further modified into nodes, squamae, and spines. The shell ornament is reviewed by Cox et al. (1969). Bivalve ornament has complex functions; in some cases a single type of structure may serve several functions during on-

togeny, a primary one as it is formed at the mantle margin, and secondary or tertiary functions as growth proceeds beyond it's point of origin. In most cases, radial elements are formed at the mantle-shell margin to strengthen the shell with structural "struts" in this vulnerable and most preyed-upon zone. Folds in the mantle margin produce costae, plicae, etc. on the shell edge; and these may also serve to cover and protect sensory organs or mantle extensions, as in some Pectinidae and (where spinose) in *Spondylus* (Kauffman, 1969). Folds and plicae may also act as a screen against infiltration of course sediment particles by increasing the surface area of the growing mantle margin and therefore reducing the necessary shell gape for adequate water intake (Rudwick, 1964). With growth, the shell continues to be strengthened by both radial and concentric elements of the ornamentation. In addition, these structures may serve secondarily to discourage epizoan encrustation, to create anchorage for shallow infaunal and crevice-dwelling epifaunal bivalves, and to break up water currents flowing over the shell into a turbulent layer, thus reducing frictional drag on shallow-water epifaunal bivalves (Kauffman, 1969). Stanley (1970, and work in progress) has shown that certain ornamentation in infaunal bivalves, for example, the divaricate asymmetrical ribs of *Divaricella*, or the heavy curved ridges of most Trigoniidae, are functional both as "skids" in guiding the shell into the sediment during burrowing, and as buttresses against the sediment during downward projection of the foot in burrowing. Many other functions are known for bivalve shell sculpture; only the primary ones are discussed above. Shell form and sculpture preserved on fossils are an important clue to the life habits in extinct bivalve taxa.

In addition to muscle "scars," previously discussed, the interior morphology of the shell consists normally of internal ligamental grooves or pits, and the dentition—interlocking teeth and sockets along the dorsal margin, which are set onto a thickened plate and serve primarily to prevent independent rotation of the valves past one another when differential torque is placed on the valves during burrowing or predation by gastropods and echinoderms. Most bivalves have some sort of dentition, well preserved in fossils, which can be classified into the following main groups:

a. *taxodont* dentition, a primitive type consisting of numerous, subequal teeth spread along the entire dorsal margin (e.g., Nuculacea);

b. *actinodont* dentition, also primitive, with narrow, elongate teeth radiating downward from the beak—the so-called schizodont dentition of Trigoniidae are an advanced form of this type;

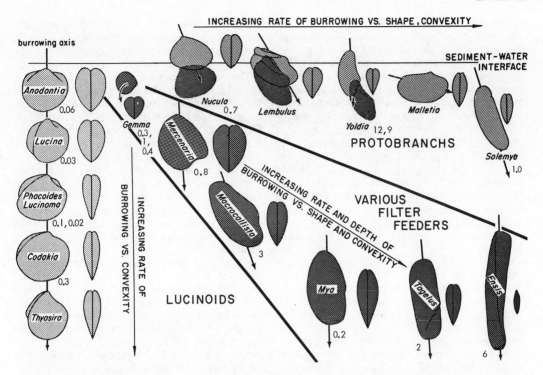

FIGURE 6. General relationships between shell shape (outline, convexity), burrowing rate, and (to a lesser degree) depth of burrowing among selected infaunal suspension and detritus feeding bivalves (after Kauffman, 1969, with new data from Stanley, 1970). Dark arrows indicate main axis of burrowing; all shell outlines with dense cross-hatched pattern shown in approximate living position. Depth of burrowing relative to other taxa or to sediment-water interface, and relative size of shells, are not to scale. Mean burrowing rate indices from living examples (Stanley, 1970) are inserted next to specimens where known. For protobranchs, depth distribution is approximately to scale except that some species of *Yoldia* and *Solemya* prefer markedly deeper infaunal burial. Slow burrowing index for *Solemya* seems out of phase and is for *S. velum* Say as reported in Stanley (1970, p. 119); C. M. Yonge reports more rapid burrowing in this genus. Dual images of *Nucula, Lembulus,* and *Yolida* represent ultimate depth and position attained in burrowing (dense cross hatch), and progress toward these depths attained during the time it would take *Malletia* and *Solemya* to reach normal burrowing depths. Mobile protobranchs are in general morphologically more specialized detritus feeders than are more sedentary forms. Among siphonate, infaunal, selective suspension (filter) feeders (middle group), relative burial-depth relationships are generally as shown and match well the burrowing rates where known excpet for *Mya*, a slow but deep burrower with a large foot. Among nonselective suspension-detritus (?) feeding Lucinaceans, burrowing rate versus shape relationships are approximately as shown, but are generally inversely related to depth of burial which is, from shallow to deep, *Codakia, Lucina, Lucinoma, Phacoides, Anodontia,* and some *Thyasira* (shallower in some species). Thus, effects of shell shape on burrowing rate and depth are modified by the relative strength and size of the foot, effectiveness of the rocking motion in burrowing, number of burrowing sequences, etc.

c. *heterodont* dentition, in which the teeth are divided into two types, the more massive ones below the beak (cardinal teeth), and the narrow, elongate ones along the dorsolateral margins (lateral teeth);

d. *pachydont* dentition, consisting of one or a few heavy blunt amorphous teeth below the beaks (e.g., Megalodontidae); and

e. *isodont* dentition, consisting of two equal teeth on each valve placed symmetrically on either side of a central resilifer which contains the ligament, as in certain Pectinidae.

Many groups, like the oysters, have secondarily lost their dentition.

Inasmuch as bivalve dentition functions primarily to keep the valves aligned under stress, it is not surprising to find that generally the type and extent of the dentition is correlative with the degree of torsion placed on the valves during burrowing and possibly through predation. Bivalves living epifaunally, especially those elevated above the bottom on vegetation, etc., tend to have weak dentition or none at all (e.g., Mytilidae, Pteriidae, Isognomonidae, Ostreidae, Pectinidae, Anomiidae). Notable exceptions among epifaunal bivalves are crevice nestlers that use the valve margins

115

to modify the substrate so that it better fits the shell (e.g., some Arcidae, Tridacnidae), or certain thick-shelled, cemented forms that are commonly predated or preferentially live crowded in very shallow water (e.g., Chamidae, Spondylidae, Cretaceous rudists). Shallow infaunal bivalves commonly have strong interlocking dentition related not only to torque placed on the shell in burrowing (and high predation levels?) but also to possession, in many, of a large foot which requires broad gaping of the valves (and consequently lateral parting of teeth and sockets along the hinge line) during burrowing. Deeper burrowing bivalves, and especially those with permanent shell gapes to accommodate the foot during burrowing, have weaker, mainly heterodont dentition.

Other interior shell features common to bivalves include central spoon- or trough-shaped depressions (chondrophore, resilifer, respectively) for the internal ligament, or alternatively, multiple ligament pits or a single internal groove situated posteriorly just inside the dorsal margin to accommodate the ligament. Special structures for muscle attachment include low raised or depressed platforms; internal folds and ridges; platforms raised on elongated "stalks" that originate beneath the beaks (e.g., the apophysis on Pholadidae); or a small, triangular, and commonly notched plate below the umbo (umbonal septum). Development of specialized muscle insertion areas usually denotes some special function, unusual stress, or special positioning for the adductor and pedal-byssal muscles, allowing identification of this function or behavioral trait in fossils. More bizarre internal shell features of the Bivalvia include: internal septa, dissepiments, and tubules, as in the Megalodontidae and their Late Jurassic–Cretaceous descendents the rudists, for support of viscera and increasing mantle area; shelly pouches or chambers along the ventral shell margin for brooding larvae; internal hollow tubes over which muscles attach (also brood pouches?); and various solid internal ribs which serve to partition the shell-mantle cavity into chambers for as yet unknown purposes.

Evolution of Bivalvia

The origin of the Bivalvia is a matter of debate, but the hypothesis of Pojeta and Runnegar (1974) that they arose from laterally compressed Rostroconchia (Mollusca) like *Heraultia* seems most reasonable at present. *Heraultia* is an anteriorly-posteriorly elongated univalve with prominent anterior, ventral, and posterior shell gapes, and a rigid dorsal valve area. Pojeta and Runnegar postulate that muta-

tion of this type of rostroconch, producing a more flexible, uncalcified, membranous organic hinge dorsally, would have given rise to the basic bivalve condition.

The earliest known bivalve is *Fordilla troyensis* (Fig. 7) from the late Early Cambrian of New York (540–570 m yr B.P.). *Fordilla* also occurs in the Early Cambrian of Newfoundland, Greenland, Denmark, Siberia, and possibly England and Portugal (Pojeta et al., 1973). *Fordilla* is small, laterally compressed, and displays the basic musculature of the Class Bivalvia (Fig. 7); its shape indicates a shallow infaunal habitat; it was probably a suspension feeder. The grade of evolution, abundance, and geographic spread of *Fordilla* suggests an even earlier evolutionary development of true bivalves.

There is a 40 m yr gap between the occurrence of *Fordilla* and the next known bivalve, the paleotaxodont *Ctenodonta? iruyensis* (Harrington) of Tremadocian (Early Ordovician) age. But presumably considerable evolution occurred in this interval because the Ordovician was a major period of bivalve radiation (with strong Early and Middle Ordovician pulses) in which nearly all major modes of life characteristic of later Paleozoic through Holocene bivalves—exclusive of swimming, cementation, and boring life habits—were developed (Fig. 8).

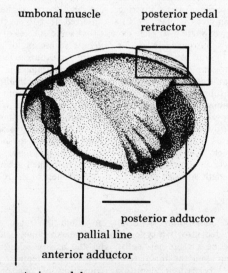

umbonal muscle   posterior pedal retractor

posterior adductor

pallial line

anterior adductor

anterior pedal retractor

FIGURE 7. Composite reconstruction of right valve of *Fordilla troyensis* Barrande, late Early Cambrian, New York, the oldest fossil bivalve (from Pojeta and Runnegar, 1974). Interior view with muscle scars (labeled) shown by dark stippled areas. Portions of hinge within rectangles have been observed and are known to lack teeth, but main part of hinge is not preserved on known specimens. Bar equals 1 mm.

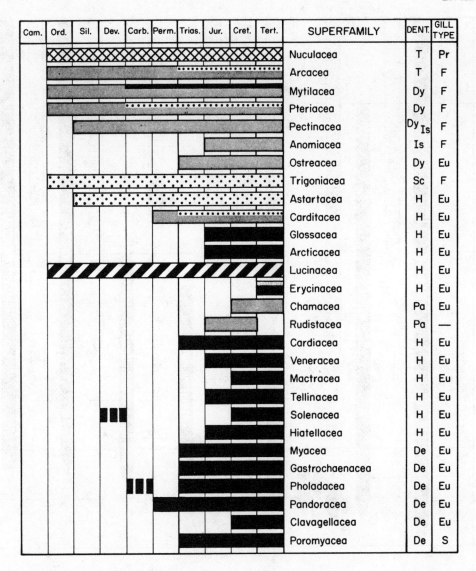

FEEDING GROUPS

⊠ Labial palp deposit feeders

▨ Epifaunal suspension feeders

⠿ Infaunal, non-siphonate suspension feeders

■ Infaunal siphon feeders

▨ Infaunal mucus tube feeders

DENTITION

T   Taxodont
Dy Dysodont
Is Isodont
Sc Schizodont
H   Heterodont
Pa Pachyodont
De Desmodont

GILL TYPE

Pr Protobranch
F   Filibranch
Eu Eulamellibranch
S   Septibranch

FIGURE 8. Distribution of feeding groups among marine bivalve superfamilies since the Ordovician (after Stanley, 1968, who gives references for data on superfamilies). Dentition and gill type given for each superfamily in right columns. Note that superfamilies characterized by common boring and cementing taxa (e.g., Ostreacea, Rudistacea, Pholadacea, Gastrochaenacea, etc.) did not arise until latest Paleozoic and Mesozoic time. Swimming did not develop in Pectinacea until late Paleozoic.

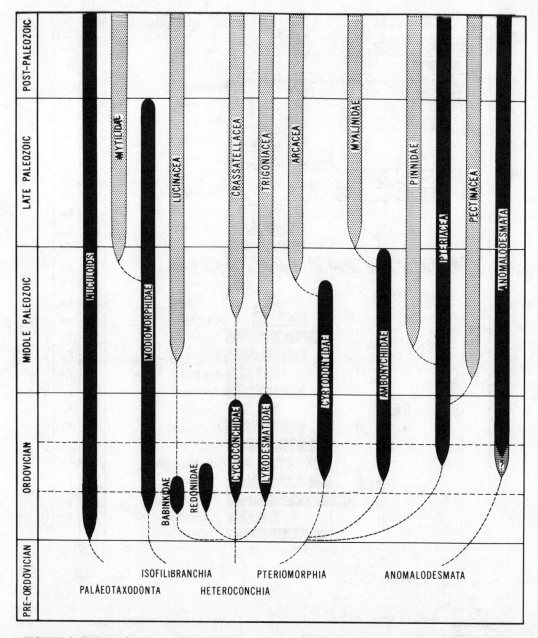

FIGURE 9. Outline of phylogenetic relationships of Paleozoic bivalves demonstrating evolution of all major subclasses by the Middle Ordovician during the first major radiation of the Bivalvia, and their replacement by more advanced adaptive and ecological counterparts, as well as the origin of new families, superfamilies, and subclasses during the Middle Palezoic radiation (after Pojeta and Runnegar, 1974).

All major subclasses of bivalves or their immediate ancestors had evolved by Middle Ordovician time (Pojeta, 1971, 1978; Pojeta and Runnegar, 1974, fig. 1; see Fig. 9).

Seven major Early Paleozoic groups had Early Ordovician origins. Presumably this radiation was based on a moderately diverse suite of as yet undiscovered Cambrian bivalve lineages. Detritus-feeding paleotaxodont (nuculoid) bivalves appeared first in the earliest Ordovician (Tremadocian), but were soon followed by diverse epifaunal to shallow infaunal suspension-feeding groups; the semi-infaunal Modiomorphidae, and the infaunal Babinkidae, Redondiidae,

Cycloconchidae, and Lyrodesmatidae (Fig. 9). *Fordilla* is probably directly ancestral only to the Cycloconchidae (Pojeta, 1975, p. 373); it is reasonable to assume other major Ordovician groups were either indirectly derived from *Fordilla,* or came from as yet unknown lineages of Cambrian bivalves.

A second, Middle Ordovician radiation produced an additional large infaunal to semi-infaunal suspension-feeding group, the Cyrtodontidae, but was mainly concerned with radiation into semi-infaunal and epifaunal habitats (Ambonychiidae, Pteriacea, Anomalodesmata; Fig. 9). Middle and Late Ordovician time also marked the first major bivalve extinction, with the elimination of four more primitive families (Babinkidae, Redondiidae, Cycloconchidae, Lyrodesmatidae).

A third major radiation of the Bivalvia occurred in the Middle Paleozoic (Silurian, Devonian Periods; 345–440 m yr B.P.) and again was largely concerned with diversification and niche partitioning among new groups of semi-infaunal to epifaunal suspension-feeding bivalves—the Pectinacea (epifaunal), Pinnidae (semi-infaunal), Arcacea (epifaunal to shallow infaunal), and the true mussels (Mytilidae, mainly endobyssate, semi-infaunal forms; Fig. 9). New, predominantly shallow infaunal suspension-feeding taxa (Lucinacea, Crassatellacea, Trigonacea) were merely adaptive replacements for more archaic Ordovician groups (Fig. 9).

The late Paleozoic was a relatively quiet period in bivalve radiation, but two important new adaptations arose. Swimming in the Pectinaceans, an escape mechanism attached to free-living epibenthos may have evolved as early as the Devonian in aviculopectinids, but was definitely an adaptation of late Paleozoic *Pernopecten* and related taxa (Waller, pers. comm., January 1976). Cementation to hard substrates, providing stability in shallow high-energy environments, also first evolved among late Paleozoic Pseudomonotidae, Terquemiidae, and Anomiidae (Newell and Boyd, 1970).

Although widespread genus and species-level extinction in the Bivalvia occurred at the Permian-Triassic boundary, a point of "catastrophic" extinction for many Paleozoic groups, only the Modiomorphidae became extinct among major subclasses, superfamilies, and families; and the bivalves survived this crisis with no large-scale change to radiate into many new ecological niches in the Mesozoic, giving rise to most modern bivalve groups and life habits (Fig. 10).

Thus, Paleozoic bivalve evolution initially involved diversification among primitive shallow, infaunal suspension and detritus feeders, and, beginning in the Middle Ordovician, was pri-

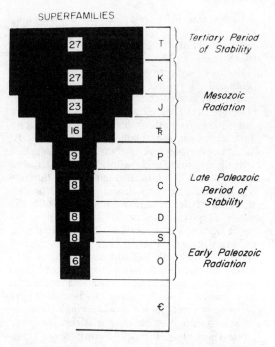

FIGURE 10. Evolution of diversity in marine Bivalvia through the Phanerozoic, as measured by the number of superfamilies existing for each geological period, showing major periods of radiation (after Stanley, 1968).

marily concerned with adaptation to semi-infaunal and, eventually, epifaunal habitats. Stanley (1972) argues that evolution of epifaunal life habits was gradual during the Paleozoic and early Mesozoic; in their evolution, many groups went through a prolonged endobyssate, semi-infaunal stage (modioliform shells) before mytiliform epibyssate adaptive types arose in the late Paleozoic and Mesozoic. Endobyssate species comprised more than 40% of Ordovician bivalve faunas, decreasing to only 15% in the Permian-Carboniferous, and to even lower percentages in the Mesozoic and Cenozoic (Stanley, 1972).

Adaptive experimentation in swimming, cementation, and possibly primitive boring modes of life occurred in the late Paleozoic. With few exception, Paleozoic infaunal suspension-feeding bivalves lacked siphons and were limited to shallow burrowing. Paleozoic bivalves were common only in facies shunned by diverse brachiopod assemblages, in particular shallow-water, nearshore, turbid clastic facies. Eurytopic adaptations allowed them to survive higher stress environments than those typically occupied by brachiopods. This suggests that bivalves, especially epifaunal forms, may have been competitive with brachiopods to some degree, and that the great Mesozoic bivalve radiation into offshore epibenthic habitats may have reflected the great decline of competing brachiopods at the Permian-Triassic boundary (see Stanley, 1972).

Mesozoic evolution of the Bivalvia, which produced the rootstocks of the modern bivalve fauna, was largely concerned with three trends: (a) widespread development of mantle fusion and siphon formation, desmodont and heterodont dentition and eulamellibranch gills in 13 diverse new superfamilies of infaunal suspension-feeding bivalves (Fig. 8), allowing occupation of new infaunal habitats through deeper burrowing, increased protection, and greater niche partitioning of the substrate (Stanley, 1968); the major adaptive radiation of the Mesozoic Bivalvia occurred as a result of these related evolutionary events (Fig. 8); (b) great ecological spread of epifaunal and infaunal bivalves into nearly all benthic habitats, from intertidal to abyssal; among existing epifaunal adaptations, this involved widespread radiation at lower taxonomic levels; and (c) widespread development of three highly specialized adaptive strategies among Bivalvia—boring; free, unattached surface-dwelling, and swimming; and cementation to firm substrates. All had sparse roots in the Paleozoic.

Boring apparently arose through certain Paleozoic Modiomorphidae and Mytilidae with crevice-nestling life habits, to better fit and protect the shell; but boring evolved polyphyletically in diverse Mesozoic lineages, including the Mytilidae, the Teredilidae (shipworms, in wood), and the Pholadidae, as a means of gaining protection in shallow, photic, highly active marine environments. Swimming evolved widely in weakly byssate to free-living Mesozoic surface-dwelling bivalves, in particular the Pectinidae, Entoliidae, Limidae, and possibly various Monotidae.

Cementation became widespread as an adaptive strategy in four new Mesozoic superfamilies and gave rise to the most bizarre adaptations among Bivalvia—the coral-like, reef- or framework-forming rudists (Hippuritacea; see *Rudists*) with their large, cemented, cup-like lower valve and typically smaller, lid-like free (upper) valve. Cementation allowed a major radiation of Mesozoic and Cenozoic bivalves into previously unoccupied hard substrate niches, including those of very shallow to intertidal, high-energy environments, and again appears to have arisen polyphyletically in such diverse groups as the oysters, pseudomonotids, pectinids, rudists, chamids, spondylids, anomiids, and other less common forms.

The bivalves of the Triassic were largely cosmopolitan and much like those of the late Paleozoic. However, the Jurassic and Cretaceous saw a tremendous radiation linked in part to plate-tectonic movements and isolation, especially in tropical-subtropical bivalves, at the family and all lower taxonomic levels. Tropical biotas were dominated by rudists and diverse oysters at the peak of their evolution, whereas the ubiquitous Permian-Mesozoic Inoceramidae were the most abundant temperate bivalves in nearly all facies. Bivalves took on a very modern aspect during this time, however, and evolved complex ecological relationships. The great Cretaceous extinction had a profound effect on the dominant Mesozoic Bivalvia, in particular those of the tropical and subtropical realm, and on the cosmopolitan Inoceramidae. Rudists, trigonaceans, and a tremendous number of less common tropical bivalves became greatly reduced or died out at the peak of their Cretaceous radiation, possibly due to the effect of deleterious oceanic surface environments on their largely planktonic larvae, and to climatic cooling and widespread marine regression at the end of the Cretaceous. Among superfamilies, only the Rudistacea became extinct. Nevertheless, many major groups of bivalves, especially those of temperate or cooler environments, crossed over the Cretaceous-Paleogene boundary with little change to give rise to the Cenozoic and Holocene faunas.

Cenozoic evolution of bivalves was low key by comparison to the Mesozoic and Paleozoic, and mainly concerned with diversification and niche partitioning at lower taxonomic levels. One new superfamily, the Erycinacea, evolved in the lower Cenozoic and all major groups carried over from the Mesozoic are still extant; many of them continued their radiation throughout the Cenozoic and are now at the known peak of their diversity.

Evolutionary Rates, Biogeography, and Biostratigraphic Potential. Despite the claims of many authors (e.g., Simpson, 1944, 1953; Stanely, 1973) that bivalves generally evolve more slowly than do more complex animals such as cephalopods and mammals, actual measurements of species duration and rates of speciation within fossil lineages, utilizing a detailed radiometric scale, show a broad range of evolutionary rates, including some of the most rapid yet recorded in animals (0.08 m yr/species, averaging 0.3 m yr for selected lineages during the Cretaceous; Kauffman, 1970; 1972; 1977, 1978). Evolutionary rates in bivalves vary tremendously between lineages subjected to the same geological environments, and within lineages subjected to regional environmental changes through time. Evolutionary rates seem to be most closely tied to ecological factors in bivalves, to their degree of tolerance for environmental stress (degree of eurytopy), to the magnitude of this stress in the form of unpredictable environmental perturbations, and to the rate at which stress is imposed on species populations. In the Cretaceous, Kauffman (1977, and references therein) has shown that environmentally highly tolerant taxa of very shallow marine, intertidal, brackish, and freshwater habitats have relatively very slow evolutionary rates, as do deep-water taxa protected for long periods of time from unpredictable environmental perturbations. More stenotopic marine-shelf organisms have relatively faster evolutionary rates. Among these, more protected deep infaunal bivalves seem to have evolved more slowly than shallow infaunal bivalves, and much more slowly than epifaunal bivalves. Trophic generalists (e.g. detritus feeders) have evolved more slowly than nonselective suspension feeders, and these more slowly than selective suspension feeders, which are more likely to be stressed by unpredictable variations in food supply. The rate at which stress is applied to evolving bivalve populations (e.g., rate of a major global marine regression) is directly related to evolutionary rates among various bivalve groups. Other factors relating to evolutionary rates in bivalves, and rate data for the above generalizations are given in Kauffman (1977).

Further, many but not all groups with broad, rapid biogeographic dispersal seem to have slower evolutionary rates than do geographically more isolated taxa (e.g., Jackson, 1973, 1974). This was probably truer of times of high climatic variability in the past (as now) than during times of warm, globally equable climates when there was higher production of long-lived, widely dispersed planktotrophic larvae among a greater percentage of the bivalve fauna, and fewer environmental barriers for widespread migration. Warm climatic periods during the past greatly enhanced their dispersal, as did closer proximity of marine shelf areas tied to plate tectonic movements. Bivalves of all types, including those with the potential for rapid evolution, could have been widely and rapidly dispersed during a large portion of geological time. The combination of these characters—rapid evolution and wide rapid biogeographic dispersal—added to their good fossil record (especially in the Mesozoic and Cenozoic) give Bivalvia a prominent role in regional biostratigraphy. This has been clearly demonstrated, for example with Cretaceous Buchiidae, Inoceramidae, and Ostreidae (Kauffman, 1975, and references therein).

Certainly not all bivalves are so blessed with this biostratigraphic potential. The Bivalvia are so diverse, especially in the Mesozoic and Cenozoic, that they are important elements of most benthonic communities. Many common taxa were stenotopic to varying degrees and sensitive to specific benthic environments to the point that they had restricted biogeographic and environmental ranges despite the dispersal potential of their dominantly planktotrophic larvae. A smaller number of bivalve groups lack this dispersal mechanism in their life histories. These biogeographically restricted groups are common and ecologically important in their preferred areas and habitats, and with other benthic mollusks, arthropods, etc., form the basis today, and in the past, for the detailed definition of marine biogeographic units of the benthos (Hall, 1964; Kauffman, 1975).

ERLE G. KAUFFMAN

References

Allen, J. A., 1958. On the basic form and adaptations to habitat in the Lucinacea (Eulamellibranchia), *Phil. Trans. Roy. Soc. London, Ser. B.,* **241,** 421–481.

Bøggild, O. B., 1930. The shell structure of the mollusks, *Kgl. Danske Vidensk. Selsk., Skrifter,* ser. 9, **2,** 231–325.

Cox, L. R.; Nuttall, C. P.; and Trueman, E. R., 1969. General features of Bivalvia, in Moore, 1969, vol. 1, N2–N129.

Currey, J. D., and Taylor, J. D., 1974. The mechanical behavior of some molluscan hard tissues. *J. Zool.,* **173,** 395–406.

Hall, C. A., 1964. Shallow water marine climates and molluscan provinces, *Ecology,* **54,** 226–234.

Hallam, A., 1967. The bearing of certain palaeozoogeographic data on continental drift. *Palaeogeography, Palaeoclimatology, Palaeoecology,* **3,** 201–241.

Heaslip, W. G., 1969. Sexual dimorphism in bivalves, in G. Westermann, ed., Sexual dimorphism in fossil Metazoa and taxonomic implications, *Internat. Union Geol. Sci.,* Ser. A, **1,** 60–75.

Jackson, J. B. C., 1973. The ecology of molluscs of *Thalassia* communities, Jamaica, West Indies, I. Distribution, environmental physiology, and ecology of common shallow water species, *Bull. Marine Sci.,* **23,** 313–350.

Jackson, J. B. C., 1974. Biogeographic consequences of eurytopy and stenotopy among marine bivalves and their evolutionary significance. *Amer. Naturalist,* **108,** 541–560.

Jeffries, R. P. S., and Minton, P., 1965. The mode of life of two Jurassic species of "*Posidonia*" (Bivalvia), *Palaeontology,* **8,** 156–185.

Kauffman, E. G., 1967. Cretaceous *Thyasira* from the Western Interior of North America, *Smithsonian Misc. Collect.,* **152**(1), 159p.

Kauffman, E. G., 1969. Form, function, and evolution, in Moore, 1969, vol. 1, N129–N205.

Kauffman, E. G., 1970. Population systematics, radiometrics, and zonation—a new biostratigraphy, *Proc. N. Amer. Paleontological Conv.,* pt. F, 612–666.

Kauffman, E. G., 1972. Evolutionary rates and patterns of North American Cretaceous Mollusca, *24th Internat. Geol. Cong., Proc.,* **7**, 174–189.

Kauffman, E. G., 1975. Dispersal and biostratigraphic potential of Cretaceous benthonic Bivalvia in the Western Interior, *Geol. Assoc. Canada, Spec. Paper 13,* 163–194.

Kauffman, E. G., 1977. Evolutionary rates and biostratigraphy, in E. G. Kauffman and J. E. Hazel, ed., *Concepts and Methods of Biostratigraphy.* Stroudsburg, Pa.: Dowden, Hutchinson & Ross, 109–141.

Kauffman, E. G., 1978. Evolutionary rates and patterns among Cretaceous Bivalvia, *Phil. Trans. Roy. Soc. London,* **284B**, 277–304.

Kauffman, E. G., and Buddenhagen, C. H., 1969. ·Protandric sexual dimorphism in Paleocene *Astarte* of Maryland, in G. Westermann, ed., Sexual dimorphism in fossil Metazoa and taxonomic implications, *Internat. Union Geol. Sci.,* ser. A, **1**, 76–93.

Moore, R. C., ed., 1969. *Treatise on Invertebrate Paleontology,* pt. N, Mollusca 6, Bivalvia, 3 vols. Lawrence, Kansas: Geol. Soc. Amer. and Univ. Kansas Press, 1224p.

Morton, J. E., 1958. *Molluscs.* London: Hutchinson Univ. Library, 229p.

Newell, N. D., and Boyd, D. W., 1970. Oyster-like Permian Bivalvia, *Amer. Mus. Nat. Hist. Bull.,* **143** (4), 221–281.

Pojeta, J., Jr., 1971. Review of Ordovician pelecypods, *U.S. Geol. Surv. Prof. Paper 695,* 46p.

Pojeta, J., Jr., 1975. *Fordilla troyensis* Barrande and early pelecypod phylogeny, *Bull. Amer. Paleontol.,* 67(287), 363–384.

Pojeta, J., Jr., 1978. The origin and early taxonomic diversification of pelecypods, *Phil. Trans. Roy. Soc. London,* **284B**, 225–246.

Pojeta, J., Jr., and Runnegar, B., 1974. *Fordilla troyensis* and the early history of the pelecypod mollusks, *Amer. Scientist,* **62**, 706–711.

Pojeta, J., Jr.; Runnegar, B.; and Kříž, J., 1973. *Fordilla troyensis* Barrande: The oldest known pelecypod, *Science,* **180**, 866–868.

Raup, D. M., 1966. Geometric analysis of shell coiling: General problems, *J. Paleontology,* **40**, 1178–1190.

Rudwick, M. J. S., 1964. The function of zigzag deflections in the commissures of fossil brachiopods, *Palaeontology,* **7**, 135–171.

Sanders, H. L., 1969. Benthic marine diversity and the stability-time hypothesis, *Brookhaven Symp. Biol.,* **22**, 71–81.

Simpson, G. G., 1944. *Tempo and Mode in Evolution.* New York: Columbia Univ. Press, 237p.

Simpson, G. G., 1953. *The Major Features of Evolution.* New York: Columbia Univ. Press, 434p.

Stanley, S. M., 1968. Post-Paleozoic adaptive radiation of infaunal bivalve molluscs–a consequence of mantle fusion and siphon formation, *J. Paleontology,* **42**, 214–229.

Stanley, S. M., 1970. Relation of shell form to life habits of the Bivalvia (Mollusca), *Geol. Soc. Amer., Mem. 125,* 496p.

Stanley, S. M., 1972. Functional morphology and evolution of bysally attached bivalve mollusks, *J. Paleontology,* **46**, 165–212.

Stanley, S. M., 1973. Effects of competition on rates of evolution, with special reference to the bivalve mollusks and mammals, *Syst. Zool.,* **22**, 486–506.

Taylor, J. D.; Kennedy, W. J.; and Hall, A., 1969. The shell structure and mineralogy of the Bivalvia, Introduction, Nuculacea-Trigonacea, *British Mus. (Nat. Hist.) Bull. Zool., Suppl.,* **3**, 125p.

Taylor, J. D., and Layman, M., 1972. The mechanical properties of bivalve (Mollusca) shell structure, *Palaeontology,* **15**, 73–87.

Trueman, E. R., 1969. Ligament, in Moore, 1969, vol. 1, N58–N64.

Waller, T. R., 1972. The functional significance of some shell microstructures in the Pectinacea (Mollusca: Bivalvia), *24th Internat. Geol. Cong., Proc.,* **7**, 48–56.

Westermann, G. E. G., 1973. The Late Triassic bivalve, *Monotis,* in A. Hallam, ed., *Atlas of Palaeobiogeography.* Amsterdam: Elsevier, 251–257.

Wise, S. W., Jr., 1971. Shell ultrastructure of the taxodont pelecypod *Anadara nobilis* (Röding). *Ecolgae Geol. Helvetiae,* **64**, 1–12.

Cross-references: *Biomineralization; Growth Lines; Larvae of Marine Invertebrates–Paleontological Significance; Mollusca; Oysters; Paleotemperature and Depth Indicators; Rostroconchia; Rudists.*

BONES AND TEETH

The fossil record is composed almost entirely of the preserved hard parts of organisms. Shells comprise most of the invertebrate record, while vertebrate remains consist primarily of bones and teeth. Despite the lack of soft parts (those most commonly studied in recent organisms), a reasonably complete story of vertebrate evolution has been reconstructed. Interest in bones and teeth has been generated on both the macroscopic and microscopic levels, each of which yields important information about the biology of recent and, by inference, fossil organisms.

Although the ability to form mineralized tissues is widespread in the animal kingdom, bones and teeth are unique to the vertebrates. In mammals (where both bones and teeth acquired their original definitions), the two are very distinct; however, in primitive vertebrates, the boundaries separating them are often not very sharp. This has led many workers to the conclusion that bones and teeth are part of an evolutionary continuum of tissues that are governed by similar morphogenetic processes. With the fundamental interrelatedness of bones and teeth in mind, then, the following discussion may appear to artificially separate them. However, this is necessary in order to simplify and shorten this discussion.

Function and Evolution of Bone

The term *bone* can refer either to the gross element or to the tissue. The evolution of bones as architectural elements has allowed great advances in feeding and locomotion, beginning with the early fishes and culminating, thus far, in the mammals and birds. Their changes in form have been the paleontologist's clues to the life habits of extinct organisms. These changes are most obviously related to the function of bones as supporting elements.

Bone, the tissue, is a remarkable product. It is an unusual structural material that can withstand great stress from both tensile and compressive forces, and under normal conditions it will not fail (in the engineering sense; Halstead, 1974). Other properties of bone that affect its suitability as a structural material are its capacity for growth and its ability to respond to forces acting upon it. As is well known, bone will repair itself amazingly quickly after a fracture. Also, when a bone is stressed, it tends to build up in areas where the stress is greatest. This ability seems to be related to bone's piezoelectric properties. Study of these indicates that mechanical deformation results in the generation of an electric potential, which may stimulate reorganization of the bone tissue (Bassett, as cited by Halstead, 1974). In other words, the shape of a bone will change according to demands being made upon it. Conversely, as has been found in the case of astronauts subjected to long periods of weightlessness, bone begins to deteriorate if it is not used (Hancox, 1972).

Bone, therefore, is not an inert tissue. Most of it is remodelled throughout the life of the organism. However, this capacity for turnover is not exclusively related to its role as a structural material. Bone also plays a major part in the homeostatic mechanisms of the body and, in fact, forms what Urist (1962) terms a continuum with the body fluids. This metabolic role of bone assists in maintaining ionic or chemical equilibrium within the body.

Romer (1963) speculated that the development of bony armor in early vertebrates gave them a selective advantage over unarmored forms in defending themselves against the large, predaceous eurypterids. Despite the conspicuous function of bone as a supportive and protective device, however, there is good reason to believe that its original evolution was more closely connected to its metabolic function. Some workers (Berrill, 1955; Smith, 1939) tied the development of a skeleton to the evolution of vertebrates in fresh water. They conceived of it as an osmoregulatory device, protecting the body fluids from excessive influx of water, or as an excretory by-product to rid the body of excess phosphate (Berrill, 1955), or alternatively, calcium (Westoll, as cited in Halstead, 1974). However, almost all workers now agree that the earliest fossil vertebrates lived in a marine environment; and therefore, it is assumed that they evolved there (Denison, 1956). Pautard (1961) suggested that bone first evolved as a phosphate store to guarantee constant protection from the seasonal fluctuations of the marine phosphate cycle. Halstead (1974) maintains that, relative to the body, there is an excess of calcium ions in the sea, which must continually be pumped out to preserve equilibrium in the body fluids. The combination of the two above theories accounts for both major components of bone salt and seems to be the most plausible explanation at the present time.

Composition of Bone. The strength of bone seems to be due to the fact that bone is a compound material. The inorganic fraction, about 70% of the bone, is composed of a variety of the mineral apatite. It is usually hydroxyapatite, a calcium phosphate with the general chemical formula $Ca_{10}(PO_4)_6(OH)_2$ (Hancox, 1972). The presence of a carbonate apatite has also been postulated; however, the actual mechanism of substitution of the carbonate ion for the phosphate is still not known. There may be a separate phase of calcium carbonate instead (Armstrong and Singer, 1956; Brown, 1966). Small amounts of elements such as strontium, barium, and lead can substitute for the calcium. Fluorine freely replaces the hydroxyl group while chlorine is a rarer substitute (Palache et al., 1951).

The remaining portion of bone is organic matter. Most of this matrix consists of the protein collagen, which forms fibers that have a very characteristic appearance under the microscope, exhibiting periodic banding every 640 Å. The amino acid composition of vertebrate collagen is unique and constitutes another identifying character. One third of the amino acids are glycine, while proline and hydroxyproline form another quarter. While apatite is highly resistant to compression, collagen (being a fibrous protein) is able to withstand considerable tension (Hildebrand, 1974). The other part of the organic matrix is an amorphous ground substance composed of a noncollagenous protein and a mixture of mucopolysaccharides. It may function to maintain the order of the collagen fibers or to store water (Hancox, 1972).

Types of Bone. There are two sets of terms given to types of bone. The first refers to the two ontogenetic modes of formation of bone, although the final products are histologically identical. The simpler of the two modes involves the direct formation of bone from mesenchyme cells. The bone that forms in this manner is called *dermal* (or *membrane*) *bone*. Most of the bones formed in this way are associated with the dermis, close to the body surface. For example, many of the superficial bones of the skull form by direct ossification. Some dermal bones have secondarily sunk deeper into the mesoderm, but nevertheless retain the name because they still ossify directly (Hildebrand, 1974).

The other method of production results in *endochondral* (or *replacement*) *bone.* It has acquired this name because the bone is preformed in cartilage, which is replaced at some later point in development by bone. It is important to note that cartilage is not transformed into bone but is replaced by it. The long bones of the body are endochondral, as are the braincases of most vertebrates (Hildebrand, 1974).

The second set of terms applied to bone does reflect structural differences in the products. The following is a brief synopsis of Hancox's (1972) discussion and applies primarily to man and other higher mammals. It should be noted here that there is a great deal of variation in vertebrate bone tissue. For a more complete treatment of this variation than is possible here, the reader is referred to Enlow and Brown (1956–1958).

The first morphological type is woven bone. It is characterized by coarse collagen fibers that are oriented randomly within the ground substance. The matrix itself forms trabeculae that are irregular in form and enclose sac-like spaces containing blood vessels (Fig. 1). The cells present in the bone are also oriented randomly. Woven bone is a transient phenomenon. Almost all bone is initially deposited as woven bone, in embryos or fracture calluses. Later it is resorbed and replaced by the more highly organized lamellar bone.

In lamellar bone, the collagen fibers are finer and they are arranged in layers called lamellae. These lamellae may be stacked flat upon one another or arranged concentrically around a central vascular canal, forming an Haversian system (see Fig. 1). Haversian bone is not very common among most vertebrate groups and its functional significance is not very well understood (Enlow and Brown, 1956–1958). Depending on the amount of vascularity within the tissue, both woven and lamellar bone can be either cancellous (spongy) or dense (compact).

A third structural type of bone is the acellular bone of teleosts and the extinct heterostracan fishes. This bone contains no osteocytes. However, in the teleost fishes, it is now known to have been derived from bone with cells, and therefore, in an evolutionary sense, it is a secondary product (Moss, 1963).

Osteogenesis. The cells that deposit the bone matrix are called osteoblasts. These cells secrete the ground substance and extrude the components of the collagen fibers, which are then assembled extracellularly (Halstead, 1974). When active, these cells are columnar with the nucleus positioned away from the developing bone surface and the mitochondria and endoplasmic reticulum adjacent to this surface. Osteoblasts have numerous cell processes in the

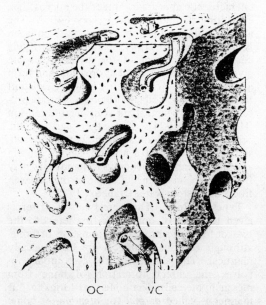 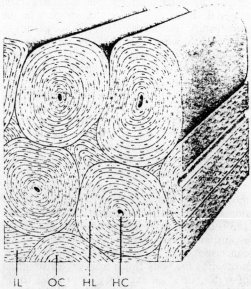

OC   VC   IL   OC   HL   HC

FIGURE 1. Diagrammatic representation of woven and lamellar bone (from Hancox, 1972, courtesy of Cambridge University Press). *Left,* woven bone; *right* lamellar bone; OC, osteocytes; VC, vascular channels; HL, Haversian lamellae; IL, interstitial lamellae; HC, Haversian canal (containing blood vessel).

matrix and Halstead (1974) states that these processes may play a role in organizing the collagen fibers as well as in ossification. The continuation of bone formation causes the osteoblasts to become trapped in the matrix, at which point they become osteocytes. Osteocytes closely resemble osteoblasts and retain the cytoplasmic processes, but the cellular organelles are not as well developed (Halstead, 1974).

Ossification of the bone matrix occurs by a process of crystallization. For a long time it was not understood how bone could form at physiological concentrations of calcium and phosphate when precipitation *in vitro* required three times as much. Crystallization, however, can be effected at much lower concentrations because the growth of a crystal lattice (in this case, hydroxyapatite) involves the gradual addition of ions by epitaxy. For this process, the collagen fibers act as templates, sites for nucleation (Neuman and Neuman, 1968).

Collagen is the most widespread protein in the body, however. Workers wondered why mineralization did not occur everywhere. It is now known that specific enzymes called pyrophosphatases destroy crystal poisons (substances that prevent crystals from forming) and allow mineralization to proceed in certain areas. The initial deposits of mineral are associated with the collagen, the apatite crystals aligning themselves with their long axes parallel to those of the fibers. There is still some question regarding the initial form of the apatite, however, and some authors believe that it is first an amorphous deposit, becoming crystalline somewhat later (Halstead, 1974). Others contend that octacalcium phosphate forms first and is then exchanged for hydroxyapatite (Brown, 1966).

Remodelling of bone is effected by osteoclasts, multinucleated cells that resorb bone. These giant cells form an irregular border with the bone surface, showing their active role in resorption. Within cell vacuoles, fragments of collagen fibers and apatite crystals can sometimes be found (Halstead, 1974). Evidence of remodelling can be seen in some mammalian bone where the remains of Haversian systems are visible as fragments between the complete ones that have replaced them (see Fig. 1).

Function and Evolution of Teeth

The primary function of teeth as masticatory devices is obvious. However, teeth (or their precursors) have not always been associated specifically with the mouth (Miles and Poole, 1967). The earliest vertebrates known possessed tissues in their dermal armor that seem closely comparable to those that now belong almost exclusively to the teeth. The exact function of the protuberances, variously called denticles or tubercles, on this armor is not known but it is assumed that the tissues functioned in ways similar to bone, to which they were intimately connected in these early forms. When jaws evolved, some dermal bones became associated with them (with the presumed exception of the cartilaginous fishes) and the denticles they bore evolved into teeth (Miles and Poole, 1967).

The basic tooth form is a sharp cone. However, this type has been considerably modified even in some of the fishes. In many sharks, for example, this cone is flattened in a parasagittal plane and serrated, giving it a blade-like appearance (Peyer, 1968). In the rays and lungfishes, on the other hand, the teeth have become very low and blunt, often fusing into tooth plates that are suitable for crushing food (Miles and Poole, 1967; Denison, 1974).

The teeth of amphibians remain fairly simple, although some modern adults have bicuspid teeth (Peyer, 1968). Typical reptiles have conical teeth as well but there are numerous exceptions to this rule in the fossil record. Many therapsids had complex crown patterns (Miles and Poole, 1967) whereas some dinosaurs have been identified as herbivores from a complex tooth morphology (Ostrom, 1964) analogous to that of mammalian herbivores (see below).

In fishes, amphibians, and most of the reptiles, the teeth are without sockets and are simply ankylosed to the jawbone in different ways. Teeth attached to the inner side of the jawbone are termed *pleurodont* while those fixed to the crest of the bone are *acrodont*. Teeth with roots housed in sockets appeared in the thecodont reptiles and are present today in their descendants, the crocodiles. This thecodont condition is almost universal in the mammals and has also been found in their ancestors, the therapsids. In mammals and some reptiles, the teeth are restricted to the jaw margins, whereas in more primitive vertebrates they are found on other bones in the mouth, or elsewhere, as on the gill arches in fishes.

The evolution of mammalian teeth has long been a subject of great interest to paleontologists. The morphology of the teeth has been the major criterion for distinguishing groups of fossil mammals. The development of the mammalian heterodont dentition is evidence of a preoccupation with food preparation but it also serves other functions such as grooming or sexual display (Crompton and Hiiemae, 1969).

The incisors are small teeth at the front of the jaws and are used for securing food and for grooming in some groups. They become large and blade-like in gnawing creatures. Canines are single rooted, conical teeth, excellent for slashing meat and holding prey. The premolars are the first group of cheek teeth and are important in chewing, especially if they become molarized. The molars have complex crowns and at least two roots. They are the only members of the mammalian dentition that are not replaced once during life. The primitive dental formula for placental mammals is

$$\frac{3 \cdot 1 \cdot 4 \cdot 3}{3 \cdot 1 \cdot 4 \cdot 3} \times 2 = 44$$

These numbers represent, respectively, numbers of incisors, canines, premolars and molars on each side of the upper and lower jaws (Romer, 1966). Most later mammals have reduced this number somewhat.

It is the changes in the molars that have formed the basis for mammalian phylogeny. As stated above, most reptilian teeth are simple and conical. However, many primitive placentals have tribosphenic molars that are triangular in shape with three major cusps (Romer, 1966). Understanding the nature of the transition from one to the other condition proved to be a major problem in vertebrate paleontology for a number of years.

In the tribosphenic molar, the apical cusp or protocone is on the inside. It was given this name because it was believed to represent the original reptilian cone (Romer, 1966). External to the protocone lie the paracone and metacone (Fig. 2). The lower molar forms a reversed triangle with the apical cusp, the protoconid, on the outside. Behind the main triangle there is a talonid (heel) that has two, and occasionally three, cusps (Fig. 2). The upper teeth are positioned slightly behind the corresponding lower ones so that they shear past each other during chewing.

The evolution of the tribosphenic molar was originally explained by rotation of the three linearly arranged cusps of triconodont teeth (belonging to the oldest known order of mammals). These were considered to be one step away from the reptilian condition. As Romer (1966) points out, to accept this theory, one would expect the protocone to be the first to develop embryologically and that it would line up with the main cusp of the premolars. This is true of the protoconid in the lower teeth but in the upper molars it is the paracone that develops first and lines up with the premolars.

Crompton (1971) proposed a more likely explanation for the evolution of the tribosphenic molar after studying the wear facets on the teeth in a series of early fossil mammals. The results of his study show that evolution has pro-

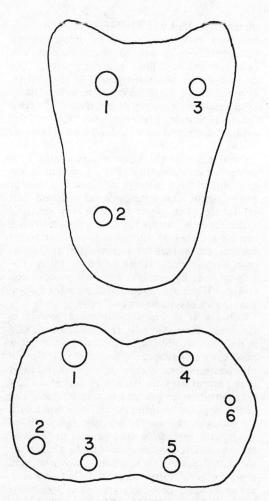

FIGURE 2. Diagrammatic representation of the mammalian tribosphenic molar. (after Hershkovitz, 1971). Upper molar: 1, paracone; 2, protocone; 3, metacone. Lower molar: 1, protoconid; 2, paraconid; 3, metaconid; 4, hypoconid; 5, entoconid; 6, hypoconulid.

ceeded by addition of cusps rather than rotation. The trends that Crompton demonstrates are (1) increase in the amount of transverse jaw movement, (2) increased size and efficiency of the shearing surfaces, and (3) addition of successive shearing surfaces to accommodate upward and medial movement of the lower molars.

Evolution of the mammalian molar has proceeded far beyond the triangular, however. There are two main morphological trends away from the puncturing and shearing actions possible with this primitive molar. One leads to an actively carnivorous mode of life. The small shearing surfaces have been modified into fewer large ones, culminating in the evolution of carnassial teeth. These teeth developed from the fourth upper premolar and first lower molar in the true carnivores (Order Carnivora), whereas they arose from the first upper and second lower molars in the creodonts (not closely related to the former group; Stahl, 1974). Much of the remaining cheek dentition is reduced or absent. Because the teeth are used mostly for shearing meat, the jaws are restricted to largely vertical movement with only a minor transverse component (Crompton and Hiiemae, 1969).

The herbivorous mode of life requires different adaptations of the teeth and jaws. In most herbivores, either ridges have developed between the cusps yielding a lophodont pattern, or the cusps have expanded into crescents, the selenodont condition. The teeth become much squarer to form a continuous grinding surface and the premolars become molarized to a greater or lesser degree. The incisors may be modified for nipping or gnawing and may be evergrowing in the latter case. In grazing mammals, the molars have developed high crowns with thick layers of enamel and cementum to counteract the constant wear on the teeth due to the highly siliceous diet. Because grinding is a major activity, the jaws are very mobile in the transverse direction.

Composition of Teeth. Mammalian teeth are composed of several types of tissues (see Fig. 3). The bulk of the tooth consists of enamel and dentine surrounding a pulp cavity. A minor component of the tooth is cementum. These tissues are easily distinguished in mammalian teeth; but there is much greater variation among them in other vertebrates, particularly the fishes.

Dentine has approximately the same amount of mineral as bone, about 70%. The apatite crystallites are very small, about 20–35 Å thick and up to 1000 Å long (Johansen, 1967). Dentine also contains collagen fibers and mucopolysaccharide ground substance. In thin sections it appears permeated by a series of tubules that radiate from the pulp cavity. In some types of dentine, these tubules are highly branched and variable in size (Bradford, 1967).

The dentine in mammalian teeth is called *orthodentine* because it is penetrated by straight, parallel dentinal tubules. There are many other varieties that have been given different names because they differ histologically from orthodentine (Ørvig, 1967). Many of these have been identified in the scales of extinct fishes (differing considerably from the scales of most living fishes) or the dermal armor of the earliest vertebrates. Both scales and dermal armor are related to teeth (Stahl, 1974). Ørvig (1967) has described many of these forms of dentine and attempted to derive a phylogenetic scheme in which to fit them.

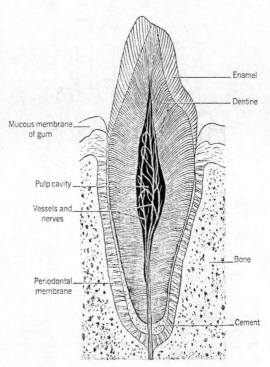

FIGURE 3. Longitudinal section of a mammalian tooth showing tissue layers (from Hildebrand, 1974, courtesy of John Wiley & Sons).

The term *dentine* also indicates a specific place and time of formation. It is derived from the mesoderm and precedes the deposition of the enamel layer (Peyer, 1968).

Like dentine, enamel was first described from mammalian teeth. It is highly mineralized and, when mature, retains only 3% organic matter by weight (Allan, 1967). The mineral is again apatite but the crystals are much larger (Trautz, 1967). The organic matrix protein in enamel is noncollagenous but it has not yet been fully characterized. There is also a striking difference in the proteins of mature and immature enamel (Smillie, 1973).

Mammalian enamel is organized into prisms enclosed by a prism sheath. The prismatic structure is not known in more primitive organisms and Halstead (1974) attributes its development to a change in the behavior of the ameloblasts, the enamel-forming cells. The prisms are usually distributed in zones that differ in direction (Boyde, 1971). Cross-sectional shape of the enamel prisms also differs from group to group (Fig. 4). Both of the above characteristics have proven useful for taxonomic studies (Boyde, 1971). Other microscopic features of enamel include incremental growth lines, enamel tufts (which represent softer areas), and spindles

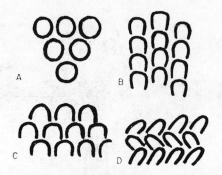

FIGURE 4. Cross-sectional appearance of enamel prisms in different groups of mammals (from Poole, 1967). A, cetaceans, insectivores, bats, and sirenians; B, ungulates, marsupials, and some primates; C, primates, proboscideans, and carnivores; D, murine rodent.

(continuations of dentinal tubules; Gustafson and Gustafson, 1967).

Enamel is characterized by the above aspects of structure; but, like dentine, it is also defined by developmental properties. Enamel forms external to the basement membrane that separates mesoderm from ectoderm and is, therefore, an ectodermal product. Secondly, it begins to form after the beginning of dentine deposition. Finally, while dentine grows toward the pulp cavity, enamel is laid down so that the last-formed layer is outermost (Peyer, 1968).

There is a third type of tissue present in mammalian teeth, usually forming a covering around the roots. This is called *cementum*, although it is really a type of bone, being similar in composition and mode of formation (Halstead, 1974). Part of the cementum, that which is close to the crown, is acellular while the portion adjacent to the root has enclosed cells, cementocytes.

A final tissue to be discussed has been given many names, reflecting the confusion that has surrounded its classification. Poole (1967) has termed it enameloid; however, it has also been called mesodermal enamel (Kvam, 1946), modified dentine (Peyer, 1968), durodentine (Schmidt and Keil, 1971), vitrodentine (Röse, 1898), and hypermineralized dentine (Bradford, 1967). This substance, present in fishes, resembles mammalian enamel in its shiny appearance, its relative lack of structure, and its high degree of mineralization. Some authors believe that it is homologous to the enamel of reptiles and mammals (e.g., Peyer, 1968).

However, studies of the development of this tissue seem to have shown that it is formed below the basement membrane and thus is a mesodermal derivative (Kerr, 1955, 1960; Grady, 1970a). It also seems likely that the matrix is collagenous, at least in part (Shellis and Miles, 1974). These results indicate that this tissue does not fit the definition of enamel as given by Peyer (1968). Nevertheless, the question has not been entirely resolved because several workers claim to have identified ectodermal components in the tooth, which would then be true enamel (Grady, 1970b; Peyer, 1968). Regardless of the outcome of this debate, the evolutionary significance of this tissue is interesting and merits further study.

Formation of Teeth. The ontogeny of tooth formation has been rather well studied in a number of organisms. Mammalian tooth development will be described below (see also Fig. 5). In the embryonic mouth region, the ectoderm penetrates into the mesoderm and forms a double-layered wall called the dental lamina. At intervals, invaginations of this lamina occur, forming structures with the appearance of inverted goblets. The cells inducing these invaginations are of neural crest origin and form the dental, or dermal, papilla. Next, the outermost cell layer of the dermal papilla differentiates into odontoblasts under induction by the enamel organ (ameloblasts).

When the odontoblasts have been formed, they immediately begin to lay down the organic matrix of dentine. When a layer 10–20 μm thick has been produced, mineralization begins. Dentine formation proceeds inward, the odontoblasts retreating toward the pulp cavity. They are not surrounded by the matrix as bone cells are and leave only cell processes encased (the dentinal tubules; Symons, 1967).

The ameloblasts, which are ectodermal derivatives, begin to produce enamel matrix soon after dentine begins to form. This begins earliest in the crown region and then spreads down the sides of the tooth (Reith and Butcher, 1967). The ameloblasts retreat centrifugally (Peyer, 1968). Mineralization occurs at the expense of the organic matrix, most of which is removed during maturation. Eruption occurs after the crown of the tooth is formed although the tooth gradually moved toward the surface during development (Hildebrand, 1974).

Fossilization

Since it is by virtue of fossilization that vertebrate paleontologists have the opportunity to study past life, a few words about the process are in order. As stated previously, soft parts are rarely preserved. After decay of the organic matter or due to scavengers, the bones themselves are often scattered and broken. Mechanical destruction by wind or water further reduces the number of elements available (Raup and Stanley, 1971). Finally, potential vertebrate

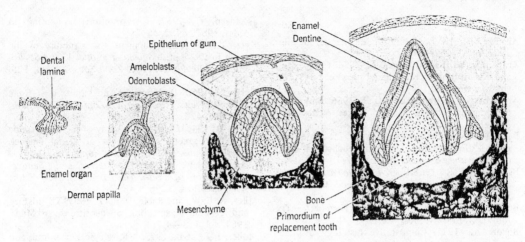

FIGURE 5. Stages in mammalian tooth development (from Hildebrand, 1974, courtesy of John Wiley & Sons).

fossils may be worn away by chemical action, although they are not as prone to this as are invertebrate remains composed of calcium carbonate (Beerbower, 1968).

These agents serve to obscure the macroscopic morphological features of bones and teeth. Unless they are well preserved, these elements may be able to tell us little about the once-living animal. Fortunately, there is a good enough record for many organisms of intact bones and dentitions to reconstruct them realistically.

Often the microscopic structure is also obliterated, usually by water circulating through the cavities in bones and teeth. Replacement with different minerals or the same ones, usually with larger crystals and different in orientation, may occur (Beerbower, 1968). This process can take place soon after death or it may not occur until millions of years after burial. Nevertheless, it is surprising how many of the vertebrate fossils that have been recovered and observed under the microscope are not damaged. The earliest vertebrates would surely not have been recognized as members of the subphylum if they had not been found to possess tissues that bear a remarkable resemblance to bone and dental tissues as we know them in higher vertebrates. Preservation, then, can be good enough to preserve delicate tubules in the dermal armor of heterostracans that lived 450 m yr ago (Denison, 1967). Recent studies using the electron microscope have shown that the orientation and structure of the tiny apatite crystallites themselves are little altered in many cases (Doberenz and Wyckoff, 1967).

Tests for trace elements such as strontium, lead, arsenic, and yttrium, which are incorporated into bones, have led to interesting speculations on aspects of vertebrate paleoecology (Wyckoff, 1971). On the other hand, evidence exists for postdepositional replacement of hydroxyl ions with fluorine (e.g., Rhodes and Bloxam, 1971), so care must be taken in using trace-element data for paleoecological reconstruction.

Although soft parts are quickly decomposed, is this true for the organic matter within bones and teeth? In general, the answer is yes; however, collagen fibers have been found intact in samples of Pleistocene bone (Wyckoff, 1971). Also, degradation products of collagen (free amino acids) have been found in fossils as old as Mesozoic dinosaurs (Wyckoff, 1971) and Devonian ostracoderms (Isaacs et al., 1963). Caution must be exercised in testing for these substances to avoid contamination with extra organic matter. Analysis of the sediment surrounding the fossil is also advised to determine whether amino acids have diffused in or out of the fossil (Armstrong and Tarlo, 1966). The preservation of amino acids makes possible the dating of fossils by a process that makes use of the slow racemization of amino acids through time (Bada and Schroeder, 1975).

In summary, the macroscopic and microscopic aspects of bone and teeth have been discussed as an introduction to some of the work of vertebrate paleontology. The forms that bones and teeth take are keys to their functions as well as to taxonomic problems. With the information gained from fossilized skeletal and dental elements, vertebrate paleontologists have sought, with much success, to reconstruct the appearances and habits of extinct organisms.

DEBORAH K. MEINKE

References

Allan, J. H., 1967. Maturation of enamel, in Miles, 1967, vol. 1, 467–494.

Armstrong, W. D., and Singer, L., 1956. Composition and constitution of the mineral phase of bone, *Clin. Orthopaedics,* 38, 179–190.

Armstrong, W. G., and Tarlo, L. B., 1966. Amino acid components in fossil calcified tissues, *Nature,* 210, 481–482.

Bada, J. L., and Schroeder, R. A., 1975. Amino acid racemization reactions and their geochemical implications, *Naturwissenschaften,* 62, 71–79.

Beerbower, J. R., 1968. *Search for the Past: An Introduction to Paleontology.* Englewood Cliffs, N.J.: Prentice-Hall, 512p.

Berrill, N. J., 1955. *The Origin of Vertebrates.* Oxford: Oxford Univ. Press, 257p.

Boyde, A., 1971. Comparative histology of mammalian teeth, in A. A. Dahlberg, ed., *Dental Morphology and Evolution.* Chicago: Univ. Chicago Press, 81–94.

Bradford, E. W., 1967. Microanatomy and histochemistry of dentine, in Miles, 1967, vol. 2, 3–34.

Brown, W. E., 1966. Crystal growth of bone mineral, *Clin. Orthopaedics,* 44, 205–220.

Crompton, A. W., 1971. The origin of the tribosphenic molar, in D. M. Kermack and K. A. Kermack, eds., *Early Mammals.* London: Academic Press, 65–88.

Crompton, A. W., and Hiiemae, K., 1969. How mammalian molar teeth work, *Discovery,* 5, 23–34.

Denison, R. H., 1956. A review of the habitat of the earliest vertebrates, *Fieldiana, Geol.,* 11(8), 361–457.

Denison, R. H., 1967. Ordovician vertebrates from the western United States, *Fieldiana, Geol.,* 16(6), 131–192.

Denison, R. H., 1974. The structure and evolution of teeth in lungfishes, *Fieldiana, Geol.,* 33(3), 31–58.

Doberenz, A. R., and Wyckoff, R. W. G., 1967. The microstructure of fossil teeth, *J. Ultrastruct. Research,* 18, 166–175.

Enlow, D. H., and Brown, S. O., 1956–1958. A comparative histological study of fossil and recent bone tissue, pts. I–III, *Texas J. Sci.,* 8, 405–443; 9, 186–214; 10, 187–230.

Grady, J. E., 1970a. Tooth development in sharks, *Arch. Oral Biol.,* 15(7), 613–619.

Grady, J. E., 1970b. Tooth development in *Latimeria chalumnae, J. Morph.,* 132(4), 377–387.

Gustafson, G., and Gustafson, A. -G., 1967. Microanatomy and histochemistry of enamel, in Miles, 1967, vol. 2, 76–134.

Halstead, L. B., 1974. *Vertebrate Hard Tissues.* New York: Springer-Verlag, 179p.

Hancox, N. M., 1972. *Biology of Bone.* Cambridge: Cambridge Univ. Press, 199p.

Hershkovitz, P., 1971. Besic crown patterns and cusp homologies of mammalian teeth, in A. A. Dahlberg, ed., *Dental Morphology and Evolution.* Chicago: Univ. Chicago Press, 95–150.

Hildebrand, M., 1974. *Analysis of Vertebrate Structure.* New York: Wiley, 710p.

Isaacs, W. A., et al. 1963. Collagen and a cellulose-like substance in fossil dentine and bone, *Nature,* 197, 192.

Johansen, E., 1967. Ultrastructure of dentine, in Miles, 1967, vol. 2, 35–75.

Kerr, T., 1955. Development and structure of the teeth in the dogfish *Squalus acanthias* L. and *Scyliorhinus caniculus* L., *Proc. Zool. Soc. London,* 125, 95–144.

Kerr, T., 1960. Development and structure of some actinopterygian teeth, *Proc. Zool. Soc. London,* 133, 401–422.

Kvam, T., 1946. Comparative study of the ontogenetic and phylogenetic development of dental enamel, *Norske Tannlaegeforen. Tidsskr. Suppl.,* 56, 1–198.

Miles, A. E. W., ed., 1967. *Structural and Chemical Organization of Teeth,* vols. 1 and 2. New York: Academic Press, 525p. + 489p.

Miles, A. E. W., and Poole, D. F. G., 1967. The history and general organization of dentitions, in Miles, 1967, vol. 1, 3–44.

Møller, I. J., et al., 1975. A histological, chemical, and x-ray diffraction study on contemporary (*Carcharias glaucus*) and fossilized (*Macrota odontaspis*) shark teeth. *Arch. Oral Biol.,* 20, 797–802.

Moss, M. L., 1963. The biology of acellular teleost bone, *Ann. N.Y. Acad. Sci.,* 109, 337–350.

Neuman, W., and Neuman, G., 1968. *Chemical Dynamics of Bone Mineral.* Chicago: Univ. Chicago Press, 209p.

Olson, E. C., 1971. *Vertebrate Paleozoology.* New York: Wiley, 839p.

Ørvig, T., 1967. Phylogeny of tooth tissues: Evolution of some calcified tissues in early vertebrates, in Miles, 1967, vol. 1, 45–110.

Ostrom, J. H., 1964. A reconsideration of the paleoecology of hadrosaurian dinosaurs, *Amer. J. Sci.,* 262, 975–997.

Palache, C.; Berman, H.; and Frondel, C., 1951. Apatite series, in *The System of Mineralogy,* vol. 2. New York: Wiley, 878–889.

Pautard, F., 1961. Calcium, phosphorus, and the origin of backbones. *New Scientist,* 260, 364–366.

Peyer, B., 1968. *Comparative Odontology.* Chicago: Univ. Chicago Press, 347p.

Poole, D. F. G., 1967. Phylogeny of tooth tissues: Enameloid and enamel in Recent vertebrates, with a note on the history of cementum, in Miles, 1967, vol. 1, 111–150.

Raup, D. M., and Stanley, S. M., 1971. *Principles of Paleontology.* San Francisco: Freeman, 388p.

Reith, E. J., and Butcher, E. O., 1967. Microanatomy and histochemistry of amelogenesis, in Miles, 1967, vol. 1, 371–397.

Rhodes, F. H. T., and Bloxam, T. W., 1971. Phosphatic organisms in the Paleozoic and their evolutionary significance, *Proc. N. Amer. Paleontol. Conv.,* pt. K, 1485–1513.

Röse, C., 1898. Über die verschidenen Abänderungen der Hartgewebe bei niederen Wirbeltieren, *Anat. Anz.,* 14, 21–31, 33–69.

Romer, A. S., 1963. The ancient history of bone, *Ann. N.Y. Acad. Sci.,* 109, 168–176.

Romer, A. S., 1966. *Vertebrate Paleontology.* Chicago: Univ. Chicago Press, 468p.

Schmidt, W. J., and Keil, A., 1971. *Polarizing Microscopy of Dental Tissues.* Oxford: Pergamon Press, 584p.

Shellis, R. P., and Miles, A. E. W., 1974. Autoradiographic study of the formation of enameloid and dentine matrices in teleost fishes using tritiated amino acids, *Proc. Roy. Soc. London, Ser. B,* **185,** 51–72.

Smillie, A. C., 1973. The chemistry of the organic phase of teeth, in I. Zipkin, ed., *Biological Mineralization.* New York: Wiley, 139–164.

Smith, H. W., 1939. *Physiology of the Kidney.* Lawrence, Kansas: Univ. Kansas Press, 106p.

Stahl, B. J., 1974. *Vertebrate History: Problems in Evolution.* New York: McGraw-Hill, 594p.

Symons, N. B. B., 1967. The microanatomy and histochemistry of dentinogenesis, in Miles, 1967, vol. 1, 285–324.

Trautz, O., 1967. Crystalline organization of dental mineral, in Miles, 1967, vol. 2, 165–200.

Urist, M., 1962. The bone-body fluid continuum: Calcium and phosphorus in the skeleton and blood of extinct and living vertebrates, *Perspect. Biol. Med.,* **6,** 75–115.

Wyckoff, R. W. G., 1971. Trace elements and organic constituents in fossil bones and teeth, *Proc. N. Amer. Paleontol. Conv.,* pt. K, 1514–1524.

Cross-references: *Amphibia; Aves; Biomineralization; Dinosaurs; Fossils and Fossilization; Mammals, Mesozoic; Mammals, Placental; Marsupials; Morphology, Constructional; Morphology, Functional; Multituberculata; Paleoecology, Terrestrial; Paleopathology; Pisces; Reptilia; Sexual Dimorphism; Taphonomy; Vertebrata; Vertebrate Paleontology.*

BRACHIOPODA

The Brachiopoda are a phylum of small sessile marine animals having their bodies enclosed in two shells ventral and dorsal in position, unequal in size, and bilaterally symmetrical. At present they are scattered in all of the seas from pole to pole. Modern species are a mere remnant of the brachiopod hosts that dominated the seas of the past.

Classification

The major categories dividing the brachiopods into classes are stated below. Most taxonomists agree on the 41 superfamilies now recognized, but arranging these into orders produces considerable disagreement (Table 1).

Two classes of brachiopods are recognized; they differ in the form and composition of their shells, their hingement, their musculature, presence or absence of an anus, and their embryology. The less abundant and less diversified group at present and in the past is the Class Inarticulata, having shells held together by a complicated system of muscles rather than a hinge, presence of an anus, the shell usually composed of chitin and calcium phosphate, and the caudal segment of the embryo becoming

TABLE 1. Brachiopod Superfamilies and Their Ranges

Superfamily	Range
CLASS INARTICULATA	
Rustellacea	Lower Cambrian
Lingulacea	Ordovician?–Recent
Trimerellacea	Ordovician–Silurian
Craniacea	Paleozoic?–Recent
Acrotretacea	Lower Cambrian–Devonian
Discinacea	Ordovician–Recent
Obolellacea	Lower–Middle Cambrian
Siphonotretacea	Upper Cambrian–Middle Ordovician
Paterinacea	Lower Cambrian–Middle Ordovician
Kutorginacea	Lower–Middle Cambrian
CLASS ARTICULATA	
Orthacea	Lower Cambrian–Upper Devonian
Billingsellacea	Lower Cambrian–Lower Ordovician
Clitambonitacea	Lower–Upper Ordovician
Triplesiacea	Middle Ordovician–Silurian
Porambonitacea	Middle Cambrian–Lower Devonian
Camerellacea	Lower Ordovician–Lower Devonian
Pentameracea	Upper Ordovician–Upper Devonian
Eichwaldiacea	Middle Ordovician–Permian
Plectambonitacea	Lower Ordovician–Middle Devonian
Strophomenacea	Middle Ordovician–Mississippian
Davidsoniacea	Middle Ordovician–Triassic
Chonetacea	Upper Ordovician–Permian
Cadomellacea	Jurassic
Productacea	Lower Devonian–Permian
Strophalosiacea	Lower Devonian–Permian
Richthofeniacea	Permian
Lyttoniacea	Pennsylvanian–Permian
Rhynchonellacea	Middle Ordovician–Recent
Stenoscismatacea	Lower Devonian–Permian
Rhynchoporacea	Mississippian–Permian
Atrypacea	Middle Ordovician–Upper Devonian
Retziacea	Silurian–Triassic
Athyridacea	Middle Ordovician–Triassic
Koninckinacea	Triassic–Jurassic
Spiriferacea	Middle Ordovician–Permian
Rhipidomellacea	Ordovician–Permian
Enteletacea	Lower Ordovician–Permian
Stringocephalacea	Silurian–Permian
Terebratulacea	Silurian–Recent
Cancellothyridacea	Jurassic–Recent
Zeilleriacea	Triassic–Cretaceous
Terebratellacea	Triassic–Recent
Thecidiacea	Triassic–Recent
Spiriferinacea	Devonian?–Jurassic

the pedicle. Leading modern examples are: the tongue-shaped *Lingula* (Fig. 1D), biconical *Discinisca* (Fig. 1A,B), and the cemented, conical *Crania* (Fig. 1I).

The class Articulata have hinged shells of calcium carbonate (calcite), simple musculature, a blind intestine; and, in their development, the pedicle is formed by an outgrowth of the ventral mantle. Leading common modern articulates are the lenticular *Terebratella* (Fig. 2D) and *Magellania* (Fig. 2F), the triangular *Hemithiris* (Fig. 2A-C), and small cemented *Lacazella* (See page 134).

Both brachiopod valves are rostrate with the beaks posterior; the ventral or pedicle valve is usually the larger and bears the pedicle that attaches the shell to the sea floor. The beak of the opposite valve, dorsal or brachial valve, is usually not strongly rostrate. The length of the

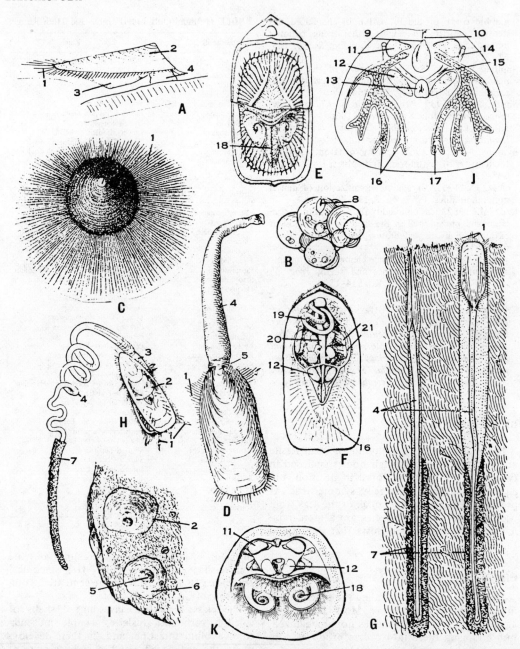

FIGURE 1. Modern examples of Brachiopoda, Class Inarticulata (A,C,D,G–J, from Hyman, 1959, and B, from Shrock and Twenhofel, 1953, copyrights © 1959 and 1953 by McGraw Hill and used with permission of McGraw-Hill Book Co.: E,F, from Dall, 1870; and K from Davidson, 1851). A–B, *Disciniscs lamellosa* (A, in side view showing both valves and pedicle; B, cluster of specimens, X.25); C, *Pelagodiscus atlanticus,* an abyssal species. D–G, *Lingula,* a tongue-shaped brachiopod (E, ventral and F, dorsal, valves showing details of the interior and organization of fleshy parts; G, in its burrow); H, *Glottidia,* another linguloid showing agglutinated sand capsule in which the pedicle is attached; I–K, cap-shaped *Crania* (I, attached to a pebble, seen from the dorsal side; J, ventral mantle, showing canals and gonads lodged in them; K, interior, showing lophophore). 1, setae; 2, dorsal or brachial valve; 3, pedicle or ventral valve; 4, pedicle; 5, apex; 6, lines of shell growth; 7, agglutinated sand capsule; 8, pedicle opening or foramen; 9, anus; 10, levator muscle of anus; 11, posterior adductor; 12, anterior adductor; 13, knob of ventral valve for muscle attachment; 14, superior oblique muscle; 15, inferior oblique muscle; 16, mantle canals or pallial sinuses; 17, gonads in mantle canals; 18, lophophore in mantle cavity; 19, intestine in coelom; 20, stomach in coelom; 21, oblique muscles.

shell is the line from the posterior to anterior margins; the width is the widest part at right angles to the length; and the thickness is the maximum distance between the valves at right angles to the length-width plane.

Inarticulata. The known genera of Inarticulata are usually small, some ancient ones are minute, but a few elongated shells attain a length of 7.6 cm. They are usually brown, yellow, or black, but mottled with rich green in some linguloids. *Lingula* (Fig. 1D), the commonest and best known of the Inarticulata, is unusual among brachiopods in its capability of limited movement because of its long wormlike pedicle.

Inside, the inarticulate shell is divided into a posterior body cavity, or coelom, separated by the body wall from an anterior mantle cavity. The body is enclosed by the mantle and an extension of the coelom extends into a dorsal and ventral mantle lobe, which line the shell to form the mantle cavity. The shell is secreted by the margin of the mantle, which is lined with setae (Fig. 1D,H). The setae group to form a central outgoing and two lateral incoming tubes for feeding, the central one being for the elimination of waste. The mouth, an opening in the body wall of the mantle cavity, leads into a gullet, stomach, and convoluted intestine ending in an anus (Fig. 1F). A paired digestive gland surrounds the stomach. The gonads are in ventral and dorsal pairs. Waste is eliminated by two nephridia, opening into the mantle cavity, which also serve as gonoducts. The nervous system is primitive. Blood is circulated mainly by ciliated vessels, although a weakly beating heart is present (see Fig. 2Q). The shells are held together by several sets of muscles which slide the valves laterally (Fig. 1E,F).

The mantle cavity contains the chief food-gathering organ, the lophophore (Fig. 1E), a ciliated, coiled filamentous extension of the body wall. The lophophore creates water currents containing food. These enter the mantle cavity from the sides and waste is expelled at the front. Microscopic food particles caught by the filaments are moved to the mouth by cilia.

Other types of Inarticulata are strongly inequivalve, such as *Discinisca*, which has a highly conical dorsal valve and a flatly conical ventral valve that bears a posteromedian slit for the emergence of the pedicle (Fig. 1B). It is attached by its pedicle to the sea bottom (Fig. 1A). *Crania* is another conical inarticulate differing from *Discinisca*, having a calcareous shell, the ventral valve of which is cemented to the substrate (Fig. 1I).

Articulata. The class Articulata is characterized by shells having three layers: an outer, probably chitinous periostracum; a thin median layer of lamellar calcite; and a thicker inner layer of fibrous calcite. The articulate shell has a great variety of forms, all characterized by a hinge consisting of two teeth in the ventral valve and sockets in the opposite valve, located just anterior to the beaks on each side of the pedicle opening. The ventral valve generally has a triangular opening, the delthyrium, under the beak; and the dorsal valve has a similar one, the notothyrium. In primitive, early brachiopods, the delthyrium was the pedicle opening; but in later forms this was modified by cover plates which restrict the pedicle opening to a round or oval foramen of varying size or, if the pedicle is atrophied, cover it completely. The deltidium is a cover composed of two laterally growing deltidial plates (Fig. 2D), and is characteristic of some superfamilies. The pseudodeltidium is a single plate growing anteriorly from the beak and characterizes several other ancient superfamilies.

The articulate shell of some fossil and living superfamilies is pierced by minute channels called punctae that contain caecae of the mantle. Another group of superfamilies contains small rods of calcite, taleolae, that leave coarse pits on exfoliation of the shell. This type is spoken of as *pseudopunctate* and the former as *punctate*. Still other superfamilies lack these structures and are called *impunctate*. The value to classification of these shell characters is not well understood, but several superfamilies are grouped by their shell structure.

Articulate valves have varying degrees of convexity, color, and size. Shells vary from biconvex to plano-convex and concavo-convex, either valve being affected. The exterior varies from smooth to radially capillate to costate to plicate and from concentrically filate to lamellose and frilled. Some genera have elaborate spines resulting from aberrations of the radii and concentric lamellae, but others (superfamilies Productacea and Strophalosiacea) have a variety of hollow spines that serve for attachment (Fig. 2N), support (Fig. 2M), or protection as well as adornment. Some modern brachiopods are gaily colored by curving or straight red bands on a matrix of pink or straw. Most brachiopods, however, have monotones ranging from white through straw-yellow, salmon, brown, and black. Minute shells as well as giants are known among the articulates. Modern *Gwynia* and fossil *Cardiarina* measure a few mm in length, but a modern species of *Magellania* is 7.6 cm long, Silurian *Conchidium* measures 12.7 cm, and Permian *Scacchinella* is 17.8 cm long; the Carboniferous *Gigantoproductus* and *Titanaria* are respectively 30.5 and 38.1 cm in width.

The internal organization of the articulate animal (Fig. 2Q) is similar to that of the Inarticulata but significant differences occur. The coelom contains the vital organs and is extended into mantle flaps lining the anterior two-thirds of the valves to form the mantle cavity. The digestive system ends in a blind intestine. All genera have a pair of nephridea but some rhynchonelloids have two pairs. The musculature of the articulate (Fig. 2P) is entirely different from that of the Inarticulata. Large diductor muscles extend obliquely from the ventral-valve posterior to attach under the dorsal beak. When contracted, these pull the dorsal beak down and the fulcrum formed by the teeth and sockets thus raises the anterior and opens the valves. Two adductors appear between the diductors in the ventral valve. These divide dorsally in passing between the valves and close the shells by contraction. Additional

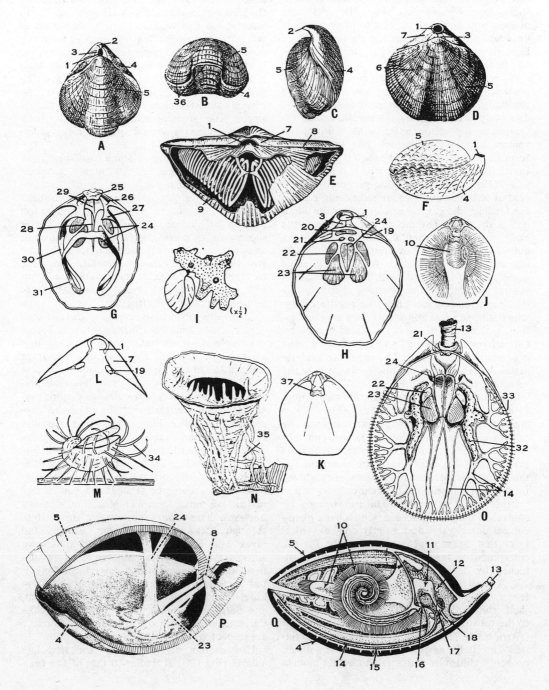

pairs of muscles attached to the pedicle adjust the shell on its stalk. The various muscles leave scars, helpful in classification, on the inside shell surface.

The beak region of both valves is usually complicated by diagnostic septa and struts of various sorts. The teeth of the ventral valve are often buttressed by dental plates, separate and subparallel in some genera but convergent and united with a median septum in others. The posterior structures of the brachial valve (Fig. 2G), the cardinalia, include a hinge plate for attachment of pedicle muscles, a central boss or septum under the beak, the cardinal process, for attachment of the diductor muscles, and crural bases that support the brachidium and lophophore (Fig. 2J). The hinge plate is often divided but in some genera is complete forming a flat or concave plate. The hinge plate varies from free to supported by a septum, but some genera have a small Y-shaped chamber, the cruralium or septalium depending on its origin.

Various types of brachidia attach to the dorsal hingeplate (Fig. 2E,G,K). These are calcified supports of the lophophore and are diagnostic of groups of brachiopods. The simplest type is a short loop (Fig. 2K) of two descending branches and a transverse ribbon (Terebratulacea); a more complicated loop consists of two long descending branches which reverse to form an ascending closed branch (Fig. 2G). This is the end member of three different processes of loop formation, in two of which the loop is initially tied to the shell medially by a septum (Terebratellacea). Another type of brachidium is the spire (Spiriferacea), which takes three forms: one coiling posterolaterally like a pulled-out watch spring (Fig. 2E); another, similar to the preceding but coiling from both sides of the shell toward the center of the dorsal valve (Atrypacea); and the third in which the spire

ends coil medially toward each other. In a large number of early brachiopods and some Holocene ones, the lophophore was not strengthened by calcification (Rhynchonellacea), but simple extensions from the hinge-plate, the cura, determine its position.

Coelomic extensions into the mantle, the pallial sinuses, or mantle canals, contain prolongations of the gonads and often leave complicated and diagnostic patterns on the inner shell.

Distribution

Geographic Distribution of Modern Brachiopods. Inarticulate brachiopods are most common in the warmer regions. *Lingula* is confined mainly to southern waters but its counterpart *Glottidia* occurs on the SE and W coasts of the United States. Groups of brachiopods and some genera are restricted geographically. One northern realm with *Hemithiris* is circumpolar. Another association occurs in the Mediterranean and off the W coasts of Spain and Africa and has some relatives in the West Indies. Some West Indian types occur off the coast of Florida and in the Gulf of Mexico. An assemblage characterized by Terebratellidae is found in abundance off New Zealand and southern Australia, and related forms are found in the Antarctic seas and off South America. Some peculiar species are limited to the waters off South Africa. In the Pacific, brachiopods occur along the W coast of the United States, Canada, and Alaska, and some similar forms occur off the Hawaiian Islands. The seas about Japan have a variety of beautiful species, and other brachiopods are found around the Philippines and East Indies. Recent studies have shown diminutive brachiopods, such as *Argyrotheca* and *Thecidellina*, to be consistent inhabitants of shaded, protected niches in coral reefs of all oceans, where they

FIGURE 2. Modern examples of Brachiopoda, Class Articulata (A-H, L,O from Hyman, 1959, and I,M, N,Q from Shrock and Twenhofel, 1953, copyrights © 1959 and 1953 by McGraw-Hill and used with permission of McGraw-Hill Book Co.; P, from Fenton, 1958, © Doubleday & Co.; and J,K from Davidson, 1851). A-C, *Hemithiris psittacea,* a rhynchonelloid (A, dorsal, B, anterior, and C, side view); D, dorsal view of *Terebratella,* a terebratuloid; E, *Spirifer,* an ancient spire-bearer; F-H, *Magellania,* a terebratelloid (F, side view; G, inner view of the dorsal valve showing loop; H, inner view of ventral valve); I. Terebratuloid attached to pebble by pedicle; J-K, inside of dorsal valve of terebratuloid *Gryphus vitreus* (J, showing lophophore; K, with lophophore removed to show short loop); L, posterior part of ventral valve of *Macandrevia* showing hinge teeth; M, productoid brachiopod showing spines which keep it on mud surface; N, *Hercosia,* having the form of a cup-coral and anchored to the substrate by spines; O, ventral mantle of *Macandrevia* showing mantle canals containing gonads; P, section through a brachiopod showing important muscles—diductors (23) for opening the shell, and adductors (24) for closing it; Q, cross section of terebratelloid *Magellania* showing soft anatomy; 1, foramen; 2, beak; 3, deltidial plates; 4, ventral or pedicle valve; 5, dorsal or brachial valve; 6, lines of shell growth; 7, palintrope; 8, hinge line; 9, spiralium, the spiral brachidium; 10, lophophore; 11, stomach; 12, heart; 13, pedicle; 14, mantle canal; 15, mantle; 16, mouth; 17, nephridium; 18, intestine; 19, tooth; 20, scar of pedicle attachment; 21, accessory diductor muscles; 22, adjustor muscles; 23, diductor muscles; 24, adductor muscles; 25, cardinal process to which the diductors are attached; 26, socket; 27, crural process; 28, median septum; 29; hinge plates; 30, descending lamella of the long loop; 31, ascending lamella of the long loop; 32, gonad; 33, setae; 34, spines for support; 35, spines for attachment; 36, marginal sinuosities producing a fold and sulcus; 37, short loop.

form part of a characteristic brachiopod-coralline sponge association (Jackson et al., 1971).

The bathymetric range of the Holocene brachiopods is very great; some articulates around New Zealand and off the coasts of Oregon and British Columbia are exposed by the tides. *Lingula* is generally confined to water less than 23 (42 m) fathoms deep but also lives in often exposed tidal falts. Others, such as *Abyssothyris* and *Pelagodiscus,* have been taken at depths slightly in excess of 6100 m (20,000 ft). About 35% of modern brachiopod species dwell in shallow water (tide zone to 200 m). About 36% are confined to the deeps but the remainder have variable ranges. *Frieleia* off the California coast ranges from 21 fathoms (38 m) to 1059 fathoms (1938 m).

Geological Distribution. Both classes of brachiopods appear simultaneously in the oldest Cambrian sediments as fairly complicated forms, indicating a long prior evolution in the Precambrian. The Cambrian linguloids are like modern *Lingula* in organization and living habits (see *Living Fossil*). The Cambrian is the heyday of the Inarticulata; they were somewhat less abundant and varied in the succeeding Ordovician period. They declined in the Silurian and after Devonian time reached the position subordinate to the articulates that they now occupy. Today, only seven genera of Inarticulata are known.

By Late Cambrian time, the Articulata became abundant and in the succeeding Ordovician period gained ascendancy over the inarticulates and remained the dominant class throughout the rest of geological time. Peaks of brachiopod diversity (see Fig. 3), according to present information (e.g., Williams and Hurst, 1977), occurred in the Ordovician (393 genera), the Devonian (480 genera), the Permian (308 genera), and the Jurassic (193 genera). The Ordovician was characterized by many simplified types without other lophophore supports than straight, rod-like plates, brachiophores, which served the dual purpose of forming socket ridges as well as defining the position of the lophophore. The spire-bearing brachiopods originated at this time. The loop-bearing brachiopods originated in the Silurian and proliferated from then to now. Later Paleozoic time saw a great diversity of brachiopods in the Mississippian. A decline ensued in the Pennsylvanian, but the Permian was an age of bizarre developments.

The end of the Permian witnessed a great extinction of many animal groups. The brachiopods were seriously depleted by the loss of three superfamilies (Rhipidomellacea, Productacea, and Strophalosiacea). By Jurassic time, all residual Paleozoic hangovers, such as the Spiriferacea, Strophomenacea, and Cadomellacea, disappeared and essentially modern types became dominant; i.e., the short-looped Terebratulacea, the long-looped Terebratellacea, and the triangular crura-bearing Rhynchonellacea. The latter superfamily, persistent and vigorous since the Ordovician, declined from the Jurassic, which was its golden age, into modern times and is now represented by only a meager 14 genera.

The three leftover groups at the end of the Mesozoic could not compete, except locally, with the vigorous and ascendant Mollusca. They have therefore steadily declined from the Cretaceous to the present. Modern brachiopods number about 300 species distributed among 74 genera. Approximately 2300 genera (including 148 genera of Inarticulata) and 30,000 spe-

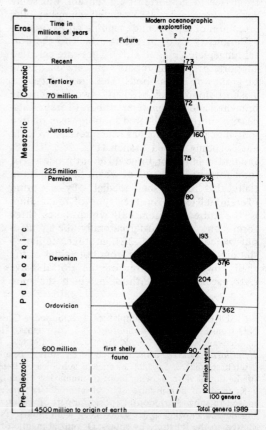

FIGURE 3. Spindle diagram showing the range of brachiopods through geological time. Note bulges of great development and reentrants showing decline, or more probably insufficient research on the fossils of these times. Data to about 1970. The numbers of genera given are an approximation because all paleontologists do not agree on all of the genera; some recognize genera that others regard as synonyms. (Courtesy of the Smithsonian Institution)

cies of all brachiopods have been described (see Fig. 3).

Biology

Habits. Many inarticulates attach to the substrate by a pedicle (Fig. 1A) as do the articulates (Fig. 2I), but *Lingula* and its relatives live in burrows (Fig. 1G) in mud or sand with the tip of the long pedicle anchored by agglutination of debris (Fig. 1H). Here the projecting setae form pseudo-siphons for water inhalation and exhalation.

The pedicle of many articulates atrophied causing some of them to lie loose on the sea bottom, weighted down by the heavy ventral beak region, or anchored by frills, corrugations, or spines. Many Paleozoic brachiopods were cemented by the ventral beak or a large part of the ventral valve (Orthotetacea). Two large superfamilies, the Productacea and Strophalosiacea, were provided with elaborate spines, best developed on the pedicle valve, which anchored the shell to the substrate by ramifying, root-like spines, or helped to keep it on the mud surface (Fig. 2M) by widely spreading spines. The brachiopods are filter feeders and bring food to themselves by their ciliated lophophores (Steele-Petrovic, 1976, reviews the feeding processes).

Adaptations. Modern brachiopods are a standardized lot, but in the past, especially about the Permian reefs of W Texas and Palermo Province, Sicily, a number of bizarre forms lived (Rudwick, 1970; Cooper and Grant, 1972–1977). The Richthofeniacea were conical, like corals, with the dorsal valve a lid deep within the cup (Fig. 2N). These were anchored by spines to form small patch reefs or clusters. *Scacchinella*, a type found in Sicily and W Texas, was also conical, but most of the cone was internally strengthened by cystose shell and the animal occupied a small chamber at the top of the cone capped by the dorsal valve, in contrast to the richthofenias. This development foreshadowed the ponderous rudistid Bivalvia of the Mesozoic (see *Rudists*). In contrast to these conical forms, the Oldhaminacea developed an ostreiform habit, the dorsal valve becoming a pinnate series of lobes that fitted over corresponding lateral troughs in the ventral valve.

Adaptations to facilitate movement of feeding currents resulted in anterior lobation of the shell (Fig. 2A,B), usually in the form of a major median plication of either valve, commonly the dorsal one. Lobation was carried to extremes by some forms in which folds opposed folds to produce an elaborately scalloped anterior. In the key-hole brachiopod (*Pygites*), two lobes grew laterally but united anteriorly to leave a large posteromedian hole.

Reproduction. All brachiopods except *Argyrotheca* are dioecious. Different species spawn at definite times while others spawn at intervals during the year. In some species, the eggs and sperm are discharged into the water and fertilization takes place externally; in others, the eggs are discharged into the mantle cavity where fertilization takes place. A few brachiopods have brood pouches in which the young develop. The free—swimming stage of the larvae is short, usually longer among the Inarticulata, the brevity of this stage being a limiting factor in brachiopod distribution.

Enemies. Since Ordovician time, brachiopods have been found with neatly bored shells, the boring presumably wrought by carnivorous gastropods. It is probable that fish were prime enemies since the Silurian. Man can also be numbered as an enemy of these defenseless creatures because *Lingula* is used as food in some places where it is common.

Pathology. Inasmuch as brachiopods live closely crowded in clusters, the progeny often weight the parent to suffocation, and the shells are commonly distorted. The larvae frequently settle in unfavorable positions which do not permit normal growth, consequently, malformations are numerous. Aberrations of the flesh occasionally leave traces of unsymmetrical muscle marks. In the Permian and other periods, the brachiopods were plagued by boring organisms, such as sponges and barnacles, with the result that shells are often riddled by holes or have internal blisters.

G. A. COOPER

References

Ager, D. V., 1967. Brachiopod palaeoecology, *Earth-Sci. Rev.,* **3,** 157–179.

Brunton, H., 1975. Some lines of brachiopod research in the last decade, *Paläont. Zeitschr.,* **49,** 512–529.

Cooper, G. A., and Grant, R. E., 1972–1977. Permian brachiopods of West Texas, I–VI *Smithsonian Contrib. Paleobiol.,* **14,** 231p.; **15,** 233–793; **19,** 795–1921; **21,** 1925–1607; **24,** 2609–3159; **32,** 3163–3370.

Dall, W. H., 1870. A revision of the Terebratulidae and Lingulidae, with remarks on and descriptions of some recent forms, *Amer. J. Conchol.,* **6,** 88–168.

Davidson, T., 1851. *British Fossil Brachiopoda.* London: Palaeontographical Society, 136p.

Fenton, C. L., and Fenton, M. A., 1958. *The Fossil Book.* Garden City, N.Y.: Doubleday, 482p.

Foster, M. W., 1974. Recent Antarctic and subantarctic brachiopods. *Antarctic Research Ser.,* **21,** 189p.

Fürsich, F. T., and Hurst, J. M., 1974. Environmental factors determining the distribution of brachiopods, *Palaeontology,* **17,** 879–900.

Hyman, L. H., 1959. *The Invertebrates,* vol. 5, Smaller

Coelomate Groups. New York: McGraw-Hill, 766p. (Brachiopoda, 516–609).

Jackson, J. B. C.; Goreau, T. F.; and Hartman, W. D., 1971, Recent brachiopod-coralline sponge communities and their paleoecological significance, *Science,* **173,** 623–625.

Moore, R. C., ed., 1965. *Treatise on Invertebrate Paleontology.* pt. H, Brachiopoda, 2 vols. Lawrence, Kansas: Geol. Soc. Amer. and Univ. Kansas Press, 927p.

Muir-Wood, H. M., 1955. *A History of the Classification of the Phylum Brachiopoda.* London: British Mus. (Nat. Hist.), 124p.

Rudwick, M. J. S., 1970. *Living and Fossil Brachiopods.* London: Hutchinson Univ. Library, 199p.

Shrock, R. R., and Twenhofel, W. H., 1953. *Principles of Invertebrate Paleontology,* 2nd ed. New York: McGraw-Hill, 816p. (Phylum Brachiopoda, 260–349.)

Steele-Petrovic, M., 1976. Brachiopod food and feeding processes. *Palaeontology,* **19,** 417–436.

Thomson, J. A., 1927. Brachiopod morphology and genera (Recent and Tertiary). *New Zealand Board Sci. and Art Manual,* **7,** 338p.

Williams, A., and Hurst, J. M., 1977. Brachiopod evolution, in A. Hallam, ed., *Patterns of Evolution.* Amsterdam: Elsevier, 79–121.

Cross-references: *Living Fossils; Rudists.*

BRANCHIOPODA

Fairy shrimp (Anostraca), clam shrimp (Conchostraca), tadpole shrimp (Notostraca), and water fleas (Cladocera) are members of this cosmopolitan crustacean class today. Common components of the nonmarine biota, they inhabit temporary ponds, pools, and shallow margins of lakes. Equivalents of one or more of these orders and several extinct orders occupied similar habitats in the geologic past, from Devonian on, with few exceptions (marine cladocerans, marine conchostracans).

Classification

A chitinous carapace (shield), univalve (bent shield), or bivalve (halved shield) may or may not have been present in branchiopods, most of which are small to minute and have distinctive swimming appendages, variable in number, that are leaflike, flattened, and lobate (see Figs. 1–3). There are seven orders (three extinct) in the class Branchiopoda, which may be grouped under three subclasses.

Subclass Calmanostraca—all the forms have a dorsal shield (cephalothorax) and/or a telson, and a supra-anal plate
Order Acercostraca (Devonian)—dorsal shield only (Fig. 1A; Lehmann, 1955)
Order Notostraca (Upper Carboniferous–Holocene)—with shield, telson, and supra-anal plate (Fig. 2A).

Order Kazacharthra (Lower Jurassic)–like notostracans but have fewer appendages and more variable telson (Fig. 2B; Novojilov, 1957)
Subclass Diplostraca
Order Cladocera (Permian*; Oligocene–Holocene)–a chitinous ornamented, folded univalve (Fig. 1B; Tasch, 1969; *Smirnov, 1970)
Order Conchostraca (Lower Devonian*–Holocene)–a smooth or ribbed and ornamented bivalve (Fig. 3A-C; *Kobayashi, 1973, would start this order in Cambrian time–a questionable assignment)
Subclass Sarostraca–both orders lack a carapace and have an elongate body.
Order Anostraca (Upper Carboniferous(?); Cretaceous–Holocene)–with body ending in caudal furca; posterior somites without legs, i.e., apodous (Fig. 3D)
Order Lipostraca (Middle Devonian)–like anostracans but body terminates in caudal segment having double furca, and trunk appendages are modified into two series, i.e., branchiopod and copepod (Fig. 3E; Scourfield, 1926)

Fossil Record

Notostracan and Related Forms (Calmanostraca). Modern notostracans (*Triops, Lepiduras*) are burrowers in soft substrates and detritus feeders as well as carnivores. Constant foraging for food in the substrates leaves distinctive feeding trails. Most commonly fossilized is the supra-anal plate and/or cercopods, or dorsal shield.

Triops species are known from both the Upper Carboniferous, Lower and Upper Triassic of Germany as well as the Permian of the United States. Fossils of *Lepidurus* occur in the Triassic of South Africa, Lower Cretaceous of Turkestan, and the Pleistocene of England (references, in Tasch 1969).

The Hunsrück shale (Lower Devonian of Germany) contains the only acerostracan genus, *Vachonisia.* Similarly, another apparently restricted order is the kazacharthrans of Kazakhstan, USSR.

A common pre-Devonian ancestor is thought to have given rise to the three Calmanostraca orders (Tasch, 1973).

Conchostracans and Cladocerans (Diplostraca). The habitat of modern bivalved, equivalved conchostracans is any seasonally filled depression ranging from puddles to ponds, in fields, on floodplains, or more enduring margins of lakes. They will eat vegetal debris, microorganisms, or, among other forms, minute ostracods or copepods. Conchostracans are soft substrate burrowers and leave trails on the substrate that may be fossilized. Chief fossilized parts are the chitinous valves, which are often carbonized.

Like other crustaceans, conchostracans were originally marine (Devonian, Germany; USA [Williston Basin]) and fossilized with brachio-

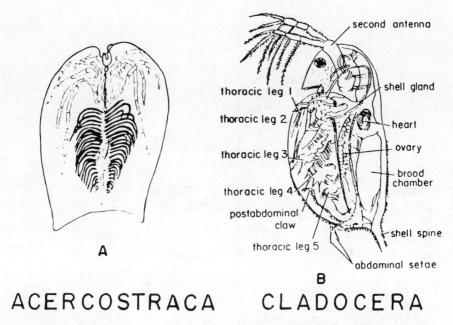

ACERCOSTRACA   CLADOCERA

FIGURE 1. Branchiopod orders (from Tasch, 1973). A, Order Acercostraca, *Vachonisia rogeri,* Lower Devonian, Germany, ventral side shown by x-ray, X9; B, Order Cladocera, *Daphnia pulex,* Holocene, male, schematic.

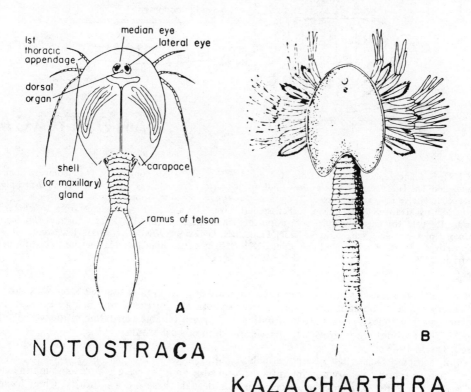

NOTOSTRACA

KAZACHARTHRA

FIGURE 2. Branchiopod orders (from Tasch, 1973). A, Order Notostraca, *Triops cancriformis.* Carboniferous-Holocene, schematic; B, Order Kazacharthra, *Jeanrogerium sornoyi,* Lower Jurassic, USSR, reconstruction, X9.

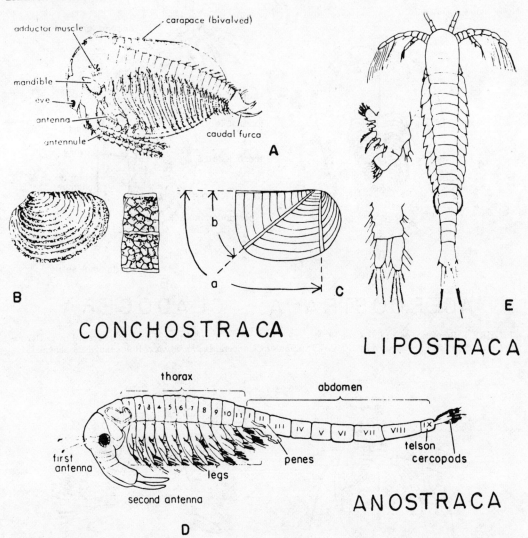

FIGURE 3. Branchiopod orders (A, from Moore and McCormick, 1969, courtesy of The Geological Society of America and The University of Kansas Press; B–E from Tasch, 1973). Order Conchostraca, A, *Cyzicus obliquus,* Holocene, left valve removed, enlarged; B, *Cyzicus (Euestheria) minuta,* Upper Triassic, Gt. Brit., ca X5: *Leaia tricarinata,* Coal Measures (Westphalian) N. France, left valve, X10; alpha (a) angle extends from dorsal margin to anterior rib; beta (b) angle extends from dorsal margin to posterior rib. D, Order Anostraca, *Branchinecta paludosa,* Holocene, male, schematic. E, Order Lipostraca, *Lepidocaris rhyniensis,* dorsal view, adult male showing appendages, Middle Devonian, Rhynie Chert, X25 (assignment to Branchiopoda questioned by Stürmer and Bergström, 1976).

pods and trilobites among others. Transition to the freshwater environment transpired across estuaries, i.e., in brackish water (e.g., *Limnestheria ardra,* Carboniferous, Ireland), in several pulses (Tasch, 1963).

In the restricted facies they occupy, conchostracan fossils, found on every continent including Antarctica, are excellent paleoecological indicators. They also have stratigraphic utility in the European Carboniferous, the American Permian, the Asian-Russian Mesozoic and are valuable in Gondwana Paleozoic/Mesozoic correlations. Furthermore, being seasonal forms they are reliable for dating sequences of paleolimnological events.

All types of fresh, brackish, and slightly alkaline water may be inhabited by modern cladocerans. There are also some marine genera although these are a minority. Chief fossilized cladoceran parts are: postabdominal segments, the ephippia bearing one or two eggs (which in life separate from rest of the shell—an apparent survival mechanism), and all or part of the bent univalve that is often distinctively ornamented.

Fossilized daphnid ephippia until recently have been known only from the Tertiary (Oligocene, Germany [Braunkohle]; Miocene, USA [Humboldt Formation]), and the Pleistocene-Holocene. Smirnov (1970), however, reported daphnid-like ephippia in the Permian of Kazakhstan. Since cladocerans evolved from conchostracans, and that separation was formerly thought to have occurred in post-Cretaceous time (Tasch, 1963), it must now be put back to the Permian.

Fossilized remains of cladocerans, including appendages, are especially valuable in deciphering the history of glacial and postglacial lakes (Frey, 1958).

Anostracans and Lipostracans (Sarsostraca). Fossils of these two orders are their soft parts, cephalon, segmented body, and appendages; in females, the ovisac may be fossilized.

Modern anostracans live in small ponds that are alkaline and temporary. Plants and insects are commonly fossilized with them. Fossils have been recorded from the Upper Carboniferous(?) and Lower Cretaceous, Victoria, Australia (Koonomara Formation), and the Tertiary, USA, (Miocene, Mohave Desert, California; Rolfe ftn. [p. R183] and other references in Tasch 1969; personal communication, Edgar Riek; and examination of site and fossils [Cretaceous, Australia] by Tasch).

Lipostracans are known only from the Rhynie Chert (Middle Devonian, Scotland) where they were fossilized in silicified peats (Tasch, 1969). Microscopic anatomical details can be observed in flakes of this chert when covered with a fine oil layer.

PAUL TASCH

References

Frey, D. G., 1958. The late glacial cladoceran fauna of a small lake, *Arch Hydrobiol.,* **54,** 14–270.

Kobayashi, T., 1973. On the classification of the fossil Conchostraca, etc. *Geology and Paleontology of Southeast Asia, Vol. 13.* Univ. Tokyo Press, 14–72.

Lehmann, W. H., 1955. *Vachonia rogeri,* n. gen., n. sp., ein Branchiopod aus dem unterdevonischen Hunrückschiefer, *Paläont. Zeitschr.,* **29,** 126–130.

Moore, R. C., and McCormick, L., 1969. *Treatise of Invertebrate Paleontology,* pt. R, Arthropoda 4 (1). Lawrence, Kansas: Geol. Soc. Amer. and Univ. Kansas Press, R57–R119.

Novojilov, N., 1957. Un nouvel order d'arthropodes particuliers: Kazacharthra du Lias des monts Ketmen. *Soc. Géol. France, Bull.,* sér. 6, **7,** 171–184.

Scourfield, D. J., 1926. On a new type of crustacean from the Old Red Sandstone (Rhynie Chert Bed, Aberdeenshire)–*Lepidocaris rhyniensis* gen. et. sp. nov., *Phil. Trans. Roy. Soc. London, Ser. B,* **214,** 153–187.

Smirnov, H. H., 1970. Cladocera (Crustacea) from the Permian deposits of eastern Kazahkstan, *Palaeontol. Zhur.,* **4**(3), 95–100.

Stürmer, W., and Bergström, J., 1976. The arthropods *Mimetaster* and *Vachonisia* from the Devonian Hunsrück Shale, *Paläont. Zeitschr.,* **50,** 78–111.

Tasch, P., 1963. Evolution of the Branchiopoda, in H. B. Whittington and W. D. I. Rolfe, eds., Phylogeny and Evolution of Crustacea, *Mus. Comp. Zool. Harvard Univ. Spec. Pub.,* 145–159.

Tasch, P., 1969. Branchiopoda, in Moore and McCormick, 1969, R128–185.

Tasch, P., 1973. *Paleobiology of the Invertebrates.* New York: Wiley, 923p.

Cross-references: *Arthropoda; Crustacea.*

BRYOPHYTA

Bryophytes (liverworts and mosses) are, for the most part, relatively small plants usually not more than a few centimeters in length. They are characterized by a unique life cycle, having two alternating generations of about equal importance but differing markedly in size and morphology, one almost completely parasitic upon the other. The generation that is usually the more conspicuous (commonly called the "plant body") is generally a green, perennial, free-living haploid gametophyte with the ability to reproduce vegetatively and also sexually by the fusion of male and female gametes produced in organs known respectively as antheridia and archegonia. The sexual process results in the formation of a diploid spore-producing generation, the sporophyte, of limited growth and life span, attached to and parasitic upon the gametophyte. The plant body of a liverwort may have a flattened, freely branched, ribbon-like form (thalloid liverworts) or a system of stems with simple leaf-like appendages (leafy liverworts); the plant body of a moss shows more differentiation and more nearly resembles on a small scale, the stem and leaves of a flowering plant. The sporophyte generation of all bryophytes usually differentiates into three parts (not always all fully developed); a basal foot, which attaches it to the gametophyte (plant body); a stalk-like seta; and a distal spore-producing part or capsule. In liverworts, the seta extends rapidly after the capsule has matured, the latter typically containing hair-like elaters which assist in the dispersal of the spores. In mosses, the seta grows slowly over a period of several months; and the capsule, which is the last part of the sporophyte to mature, contains no elaters but spore dispersal is normally facilitated by the opening of a characteristic ring of teeth, known as the peristome. Both liverworts and mosses are devoid of true vascular tissues (xylem and phloem), although internal conducting strands are present in some species. They lack extensive mechanical tissue and have little or no cuticle covering their exposed surfaces.

Their delicate nature has often been quoted (e.g., Parihar, 1961) as the explanation for the apparent scarcity of fossil remains. Yet patient paleobotanical studies during the last 50 years have shown that fossil bryophytes are, in fact, widespread both geographically and stratigraphically and that their remains are often very well preserved, even in the oldest deposits.

It is clear that the chances of preservation do not always depend necessarily on the presence of resistant structures, but on the occurrence of the appropriate kind of sedimentation in the right situation at the right time. Except for the special case of preservation in Baltic Amber, all the examples of well-preserved fossil bryophytes have one feature in common—effective preservation has depended on the inclusion of the plants in fine sediments, probably accumulated in fresh water under anaerobic conditions. The impervious matrix thus produced ensured that little of the original carbon content of the plant is lost. Since the organic remains so preserved have also been found to withstand quite drastic maceration in the laboratory, it is very tempting to believe that the chemical changes that take place during the fossilization process may result in the formation of resistant membranes or, indeed, that in some cases at least a cuticle-like layer was present in the first place.

It is rather curious that bryophytes have not been recognized in the extensive studies on coal-balls that have been carried out in Britain, Holland, Belgium, the USA and the USSR, for a solitary record due to Lignier (1914) shows that petrified bryophytes can occur. A possible explanation may be found in the fact that coal-ball floras consist for the most part of the accumulated and sometimes drifted debris while the bryophytes found in Carboniferous shales may represent mud-dwelling species preserved in the original place of growth.

The absence of bryophytes in coal-balls seems to indicate that the forest trees of the Carboniferous coal swamps were not clothed with the thick festoons of epiphytic species that characterize both temperate and tropical rain forests at the present day. Although the trunks of arborescent lycopsids and ferns, such as *Lepidodendron* and *Psaronius,* would seem to have provided suitable lodgement for them, no epiphytic bryophytes have been described.

The best hope for finding further examples of early fossil bryophytes probably lies in the systematic searching of very fine-grained de-

TABLE 1. Classification of the Bryophyta

Division BRYOPHYTA
  Class HEPATICOPSIDA
    Subclass Jungermanniae
      Order Metzgeriales–Devonian (Frasnian)–Holocene
      Order Jungermanniales–Eocene–Holocene
      (Orders Takakiales and Calobryales–no known fossils)
    Subclass Marchantiae
      Order Marchantiales–Middle Triassic–Holocene
      Order Sphaerocarpales–Triassic (Rhaetian)–Holocene
      (Order Monocleales–no known fossils)
  Class ANTHOCEROTOPSIDA
    Order Anthocerotales–Upper Oligocene–Holocene
  Class BRYOPSIDA *sensu lato*
    Subclass Sphagnidae
      Order Protosphagnales–Lower Permian–Upper Permian
      Order Sphagnales–Jurassic (Liassic)–Holocene
    Subclass Andreaeidae
      Order Andreaeales–Pleistocene–Holocene
    Subclass Bryidae *sensu lato*
      Order Incertae Sedis–Upper Carboniferous (Stephanian)–Holocene
      Order Eubryales: Lower Permian–Holocene
      (many other extant orders–Tertiary–Holocene

posits—clays, soapstones, and shales—of known freshwater origin, by the bulk maceration methods first described by Harris (1926).

The scheme of classification used here is modified from that used by Jovet-Ast (1967) and recognizes only three classes (Table 1), namely Hepaticopsida and Anthocerotopsida (together constituting the Hepaticae or liverworts of some authors) and the Bryopsida *sensu lato* (Musci or mosses of some authors).

Class Hepaticopsida

Order Metzgeriales. The order Metzgeriales (Jungermanniales Anacrogynae of some authors) are simple thalloid liverworts, but with a tendency to "leafiness" in some genera. The earliest stratigraphic record is *Hepaticites devonicus* described by Hueber (1961) from the lowermost Upper Devonian (Frasnian) of New York State. This plant consisted of a thalloid part and a "rhizomatous" part. No reproductive organs are known. This species is referred to Metzgeriales on the basis of vegetative morphology. Hueber compares his material with the extant genera *Pallavicinia* and *Metzgeria.*

The recent recognition by Oschurkova (1967) of *Hepaticites metzgerioides* Walton in various localities in the Upper Carboniferous of the Karanganda Basin, USSR along with the original description by Walton (1925, 1928) of many species in the Coal Measures of England, indicates quite clearly that thalloid hepatics were widespread in Carboniferous times.

Jovet-Ast (1967) lists many other species of *Hepaticites, Thallites,* and *Metzgeriites* from Jurassic and Cretaceous deposits. Most of these are incompletely known, but some seem to be clearly referable to the Metzgeriales, others to Marchantiales.

Order Jungermanniales. The order Jungermanniales (Jungermanniales Acrogynae of some authors) are leafy liverworts with a strong tendency toward dorsiventrality of the leafy shoot. They do not appear with certainty before the early Tertiary. Jovet-Ast (1967) lists some 50 species in 20 genera from Tertiary and Quaternary deposits. Most of the Tertiary species are placed in the two pre-Quaternary genera *Jungermannites* as emended by Steere (1946), and *Plagiochilites* Straus (1952), but a few of the Tertiary species and all the Quaternary ones are referred to extant genera in the families Trichocoleaceae, Lophoziaceae, Jungermanniaceae, Lophocoleaceae, Plagiochilaceae, Lejuneaceae, and Frullaniaceae. Some of the most beautifully preserved examples occur in the Oligocene Baltic Amber deposits, the flora of which has been reviewed by Savicz-Ljubitzkaja and Abramov (1959) and by Czeczott (1959). Steere (1946) has reviewed the occurrence of leafy liverworts in the Tertiary of North America.

Order Marchantiales. The thalloid liverworts of the order Marchantiales are characterized by the differentiation of specialized photosynthetic tissues, either in the form of closely packed filaments on the dorsal surface of the plant body (Ricciaceae) or contained in air chambers in its upper part (Marchantiaceae, Targioniaceae, and allied families). The Marchantiales are not known with certainty from the Paleozoic. *Thallites willsi* Walton (1949) from the coal measures of Staffordshire, England, and *T. lichenoides* from the coal measures of New Brunswick, Canada (Lundblad, 1954) have some similarity to the extant genus *Riccia*, while *Marchantites lorea* Zalessky (1937) from the Middle Permian of Bardinsky, Urals, USSR is said to resemble the living species *Marchantia polymorpha* in some respects. But these three taxa do not seem to be sufficiently well characterized to constitute reliable evidence of the Marchantiales.

The earliest acceptable record is *Hepaticites cyathodoides*, described by Townrow (1959) from the Middle Triassic shales in Natal, South Africa. Townrow compares *H. cyathodoides* with the extant genus *Cyathodium*, formerly placed in the Targioniaceae, but now in a family of its own. There are similarities in habit, rhizoids, ventral scales, and midrib structure; but the pores of the airchambers in the upper part of the plant body in *H. cyathodoides* differ from those in the genus *Cyathodium*.

Later Mesozoic records of plants of undoubted marchantialean affinity include species in the genera *Ricciopsis* (resembling *Riccia*) and *Marchantiolites* (close to Marchantiaceae), both established by Lundblad (1954 for material from the Rhaetian-Liassic (Triassic-Jurassic) of Skromberg, Scania, Sweden; also several species of *Hepaticites* resembling Marchantiaceae described by Harris (1961) from the Bajocian (Jurassic) of Yorkshire, England; and a *Marchantites* (probably belonging to Marchantiaceae) from the Lower Cretaceous of Patagonia (Lundblad, 1955).

From Tertiary and Quaternary deposits, about sixteen species have been described (Jovet-Ast, 1967). These consist of five species each in the genera *Marchantites* and *Marchantia* and two species of *Riccia* in the Tertiary; and from the Quaternary, the extant species *Riccia fluitans, Marchantia polymorpha, Clevea hyalina,* and *Conocephalum conicum.* Nagy (1968) has described species of spores of Ricciaceae from the Neogene of Hungary.

Apart from the Quaternary species, the most completely known fossil representative of the Marchantiaceae is *Marchantites sézannensis* from the Eocene Travertine of Sézanne, Marne, France. The large dichotomous thallus shows not only the air pores characteristic of the family but also gemma cups (asexual propagating organs) on the thallus surface, ventral scales, and stalked structures similar to the antheridium-bearing branches of *Preissia* and *Marchantia*.

Order Sphaerocarpales. In the Cotham Beds of the English Rhaetic there occurs locally an abundance of fragments of the minute leafy plant *Naiadita lanceolata.* The plant has been known for more than 100 yr and has had a very checkered history, having been regarded in turn as a Monocotyledon resembling *Naias,* a moss close to *Fontinalis,* a lycopod, and finally a liverwort. However, Harris (1938) has proved conclusively that this plant is a bryophyte. Stems, leaves, rhizoids, gemma cups and gemmae, archegonia, sporophytes and spores are all described. Only the antheridia are lacking. *Naiadita lanceolata* remains to this day the most fully known fossil bryophyte. Harris considers it to be a submerged aquatic liverwort and assigns it tentatively to the Riellaceae (Sphaerocarpales).

Class Anthocerotopsida

Order Anthocerotales. The thalloid liverworts that constitute the order Anthocerotales are characterized by the possession of an unusual sporophyte. This has a massive foot, no seta, complex internal organization, considerable photosynthetic capability, and a capsule that is both actually long and long-lived, produced from a basal intercalary meristem.

Very little is known of the early history of the Anthocerotales. Neither gametophyte thalli nor sporophytes have been found as fossils and the earliest occurrences of Anthocerotales are based on certain kinds of Tertiary spores. Thiergart (1942) and later Krutzsch (1963) record *Anthoceros* spores from the Upper Oligocene, Miocene, and Pliocene at various localities in Eastern Germany. Stuchlik (1964) has described from the Miocene of Poland a spore known as *Rudolphisporis rudolphi,* which he refers to the extant genus *Anthoceros.* Nagy (1968) has recorded anthocerotalean spores from the Neogene of Hungary.

Class Bryopsida *sensu lato* (or Musci)

The earliest geological records of what would seem to be undoubted mosses are *Muscites polytrichaceus* Renault and Zeiller (1888) and *M. bertrandi* Lignier (1914) from the Stephanian (Upper Carboniferous) of France, but their affinities are obscure. Paleozoic mosses that can be classified more satisfactorily are the undoubted representatives of the Sphagnidae and Bryidae described by Neuburg (1960) from the Permian of the USSR.

Subclass Sphagnidae

Order Protosphagnales. One of the most outstanding advances in the knowledge of early fossil bryophytes during the last 30 years results from the extensive researches of the late Professor Maria F. Neuburg (1960). This author established 14 species of mosses distributed in 10 genera (of which 12 species and 9 genera were new to science) from the Lower and Upper Permian deposits of the Petchora, Kuznetsk, and Tunguska Basins, Angarida, USSR. The plants were embedded in shales and were so beautifully preserved as to permit the preparation of cellulose peel and balsam transfers.

Three of the species (*Junjagia glottophylla, Vorcutannularia plicata,* and *Protosphagnum nervatum;* Fig. 1 [1–4]) have leaves resembling those of the extant genus *Sphagnum* in possessing two kinds of cells. They differed from modern *Sphagnum* in not showing a marked differentiation into narrow photosynthetic cells and large hyaline waterholding cells. In addition they had a midrib and sometimes a suggestion of lateral nerves as well. Neuburg created the order Protosphagnales for them.

Her work is being continued by S. V. Meyen (1963, 1966) who has extended the geographical range of the Protosphagnales with the discovery of a probable second species of *Protosphagnum* in the Upper Permian of Southern Priuralia.

Order Sphagnales. Bog mosses are characterized by the occurrence of two-kinds of cells in the gametophyte leaves (narrow, green, photosynthetic cells forming a network round large empty hyaline cells) and a sporophyte consisting of little more than a globular capsule carried up on a seta-like extension (pseudopodium) of the apex of the gametophyte; there is a capsule opening by an apical pore. Undoubted *Sphagna* have been recorded from Mesozoic, Tertiary, and Quaternary deposits. Reissinger (1950), for example, has described authentic leaves and spores of a *Sphagnum* fom the Liassic (Lower Jurassic) of Nuremberg in Bavaria, while Arnold (1947) claimed to have *Sphagnum* leaves from the Upper Cretaceous of Disko Island, Western Greenland. A little caution is necessary in accepting this latter record, for Steere (1946) suggested that it could have been due to the contamination of the Cretaceous material by recent *Sphagnum* fragments. Jovet-Ast (1967) lists some 12 Mesozoic species of *Sphagnum* based on spores. The same author records about 12 Tertiary species of *Sphagnum* and about 30, many of which are modern, from Quaternary deposits.

Subclass Andreaeidae

Order Andreaeales. The small, dark-colored mosses of the order Andreaeales are characterized (like Sphagnales) by having the capsule carried on a pseudopodium, and are further distinguished by spore dispersal effected through four longitudinal splits. The fossil record for Andreaeales, like that for the Anthocerotales, is almost blank. Nothing is known until the Quaternary, when the three extant species *Andreaea huntii, A. petrophila,* and *A. rothii* have been found in glacial, interglacial or postglacial deposits in Bavaria, Poland, and Scotland (Szafran, 1952; Jovet-Ast, 1967).

Subclass Bryidae

The name Bryidae is here used in a broad sense to include all the mosses other than Sphagnidae and Andreaeidae, from which they differ chiefly in the form of the sporophyte and especially of the capsule, which usually possesses a peristome. The earliest true moss floras known so far are those described by Neuburg (1960) from the Permian of various localities in Angarida, USSR. No less than 11 species, referred to 7 genera, occur variously distributed throughout both Lower and Upper Permian strata.

Some of the mosses described by Neuburg, especially four species included in the genus *Intia* Fig. 1 [3]) have striking similarities to the extant genera *Mnium* and *Bryum;* others are quite unlike any living mosses. Although all the species are undoubted members of the Bryidae,

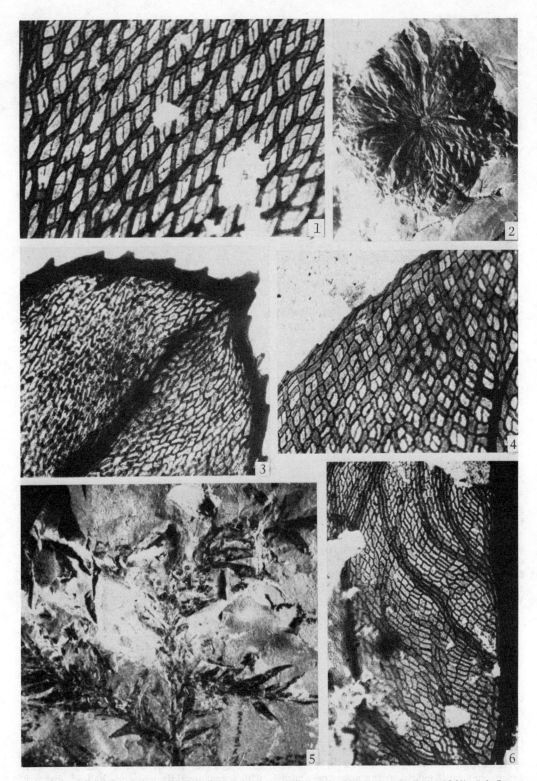

FIGURE 1. Permian mosses from Angaraland (from K. Mägdefrau, 1961, after Neuberg, 1960). 1,4, *Protosphagnum nervatum* Neuburg, ca. X110; 2, *Vorcutannularia plicata* Neuburg, ca. X1.75; 3, *Inita vermicularis* Neuburg, ca. X110; 5, *Uskatia conferta* Neuburg, ca. X4; 6, *Polyssaievia spinulifolia* Neuburg, ca. X110.

the absence of sporophytes and capsules makes further classification difficult. Jovet-Ast (1967) places all except *Muscites uniforme* in a "Famille incertaine" within the order Eubryales. *M. uniforme* is placed in Bryophyta Incertae Sedis.

Mesozoic records of mosses are scarce. However, *Muscites guescelini* described by Townrow (1959) from the Middle Triassic of Natal, South Africa, is of some interest as it is sufficiently well preserved to permit comparison with extant members of the family Leucodontaceae.

By Tertiary times mosses are abundant and widespread and many of them can be referred to living genera (Savicz-Ljubitzkaja and Abramov, 1959; Jovet-Ast, 1967). The numbers increased steadily throughout the Tertiary and Quaternary and many hundreds are known, mostly referable to extant species. Jovet-Ast (1967) for example, describes examples in at least 40 families referable to Fleischer's orders Polytrichales, Fissidentales, Dicranales, Pottiales, Grimmiales, Funariales, Tetraphidales, Eubrayles, Isobryales, Hookeriales, and Hypnobryales. Dickson (1967) states that macroscopic remains of more than 100 species of mosses are known from British deposits of the last glaciation alone. Nagy (1968) has recorded the familes Encalyptaceae and Ephemeraceae, based on spores, from the Neogene of Hungary.

Many of these more recent records of mosses relate to beautifully preserved material, as, for example, the Upper Miocene Moss Flora of Arjuzanx in Landes, France, recently described by Jovet-Ast and Huard (1966) (Fig. 2). Dickson (1973) has provided a very full account of British Pleistocene Bryophyta. Despite good preservation, the majority of Tertiary and Quaternary mosses are known only from vegetative remains. Fertile mosses worthy of note are *Muscites florissanti* (Knowlton) Steere (1946) from the Upper Miocene of Florissant, Clorado,; *M. yallournensis* Clifford and Cookson (1953) from the Oligocene of Yallourn, Victoria, Australia; and *Plagiopodopsis cockerelliae* (Britton and Hollick) Steere (1946) from the Oligocene of Florissant, Colorado.

Bryophyte Evolution

Many well-known bryologists (for example, D. H. Campbell, G. M. Smith, and J. Proskauer, to mention but three) have suggested at various times that the Anthocerotales occupy an important ancestral position in the evolution of the Hepaticae. More recently, Jeffrey (1962) suggests that *Anthoceros* "of all bryophytes approaches most nearly to the ancestral condition." Features considered to be ancestral include (1) the single large plastid that normally

FIGURE 2. *Thomnites marginatus* Jovet-Ast and Huard, a Tertiary moss of the family Neckeraceae, from the Neogene lignites of Arjuzanx, Landes, France X30 (from Jovet-Ast and Huard, 1966).

occurs in each photosynthetic cell of the gametophyte and suggests a link with the green algae; (b) the absence of mycotrophic associations, shown by many other bryophytes and presumably developed at a later stage in their evolution; and (c) the long green cylindrical sporophyte, indicating a basic similarity of gametophyte and sporophyte generation, the latter being seen as an "archaic survival" in terms of an homologous theory of bryophyte origin. It would be fine if the fossil evidence furnished additional support for these views based on comparative morphology, anatomy, and cytology of modern species. Unfortunately it does not, for so far no Anthocerotales are known before Tertiary times.

Other bryologists have suggested a "downgrade" line of bryophyte evolution; that is, the gametophyte was at first a "leafy" structure which became flattened and thalloid, either by the loss of the leaves and enlargement of the axis (A. W. Evans) or by the fusion of the leaves

(P. N. Mehra, B. R. Vashist). If the Paleozoic bryophytes do not yet establish the Anthocerotales as an early ancestral group, do they throw any light in general terms on the question of "up-grade" as against "down-grade" theories of bryophyte evolution? The earliest known example, *Hepaticites devonicus* from the Upper Devonian of the USA, is thalloid and tips the scales slightly in favor of the "up-grade" view, but the occurrence of a wide range of foliose or near-foliose forms relatively soon afterward in the Upper Carboniferous and Permian must cast doubt on the wisdom of basing arguments on a single record from the Devonian. Putting this another way, although thalloid liverworts appear from the fossil record to have arisen before foliose ones, and liverworts in general to have arisen before mosses, the complexity and diversity of the moss floras from the Permian of Russia, for example, must surely imply an evolutionary origin considerably earlier in the late Paleozoic and probably contempoary with that of the liverworts.

Within the liverworts, on present fossil evidence, the Metzgeriales seem to have predated the Marchantiales and these in turn predated the Sphaerocarpales; but if the Lower Carboniferous *Tetrapterites* (Sullivan and Hibbert, 1964; Lacey, 1969) is ever shown to have sphaerocarpalean affinities, this order would be largely reversed. Jungermanniales, Anthocerotales, and, within the Musci, the Andreaeales all appear, so far as fossil evidence goes, to be of relatively recent origin.

On a lower taxonomic level, a similar situation exists within orders. In the Marchantiales, for example, the Ricciaceae with dorsal assimilatory filaments are sometimes placed in an ancestral position, but sometimes the reverse view is held. The late Paleozoic and early Mesozoic marchantialean fossils give no unequivocal answer. The Middle Triassic South African *Hepaticites cyathodoides* with its dorsal assimilatory chambers and pores predating by a few million years the Rhaetian-Liassic *Ricciopsis scanica* from Sweden, might be taken to support a Marchantiaceae → Ricciaceae "down-grade" line of evolution; but should the *Riccia*-like cf. *Hepaticites* sp. from the Upper Carboniferous of Scotland (Walton, 1949) prove to be a *Riccia*, the position would be reversed.

Early fossil bryophytes, then, give no clear answers, so far, to most phylogenetic questions. They permit perhaps one firm conclusion and one reasonable conjecture, namely (a) that, as Watson (1971) has pointed out, some groups of both liverworts and mosses already existed in a highly differentiated form before the end of the Paleozoic, and (b) that a polyphyletic origin of the Bryophyta seems highly probable. To speculate beyond this without much more fossil evidence than is presently available is an exercise that is hardly worth while.

WILLIAM S. LACEY

References

Arnold, C. A., 1947. *An Introduction to Paleobotany.* New York: McGraw-Hill, 433p.

Clifford, H. T., and Cockson, I. C., 1953. *Muscites yallournensis,* a fossil moss capsule from Yallourn, Victoria, *Bryologist,* **56,** 53–55.

Czeczott, H., 1959. The flora of the Baltic Amber and its age, 4, *Pr. Muz. Ziemi,* **4,** 119–131.

Dickson, J. H., 1967. The British moss flora of the Weichselian Glacial, *Rev. Palaeobot. Palynol.,* **2,** 245–253.

Dickson, J. H., 1973. *Bryophytes of the Pleistocene.* London: Cambridge Univ. Press, 256p.

Harris, T. M., 1926. Note on a new method for the investigation of fossil plants, *New Phytologist,* **25,** 58–60.

Harris, T. M., 1938, *The British Rhaetic flora.* London: British Mus. (Nat. Hist.) 84p.

Harris, T. M., 1961. *The Yorkshire Jurassic Flora. I, Thallophyta, Pteridophyta.* London: British Mus. (Nat. Hist.), 212p.

Hueber, F. M., 1961. *Hepaticites devonicus,* a new fossil liverwort from the Devonian of New York, *Ann. Mo. Bot. Gdn.,* **48,** 125–132.

Jeffrey, C., 1962. The origin and differentiation of the Archegoniate land-plants, *Bot. Notiser.,* **115,** 446–454.

Jovet-Ast, S., 1967. Bryophyta, in E. Boureau, ed., *Traité de Paléobotanique,* vol. 2, Paris: Masson et Cie, 19–186.

Jovet-Ast, S., and Huard, J., 1966. Mousses de la flore néogène d'Arjuzanx (Landes), *Rev. Bryol. Lichenol.,* **34,** 807–815.

Krutzsch, W., 1963. *Atlas der mittel-und jungtertiaren dispersen Sporen- und Pollen-sowie der Mikroplankton formen des nordlichen Mitteleuropas.* 2. Die Sporen der Anthocerotaceae und der Lycopodiaceae. Berlin: Deutscher Verlag der Wissenschaften, 141p.

Lacey, W. S., 1969. Fossil bryophytes, *Biol. Rev. Cambridge Phil. Soc.,* **44,** 189–205.

Lignier, O., 1914. Sur une Mousse houillère à structure conservée. *Bull. Soc. Linneenne Normandie,* **7,** 128–131.

Lundblad, B., 1954. Contributions to the geological history of the Hepaticae. Fossil Marchantiales from the Rhaetic-Lissic coal mines of Skromberga (Province of Scania), Sweden, *Svensk Bot. Tidskrift* **48,** 381–417.

Lundblad, B., 1955. Contributions to the geological history of the Hepaticae. 2. On a fossil member of the Marchantineae from the Mesozoic plant-bearing deposits near Lago San Martin, Patagonia (Lower Cretaceous), *Bot. Notiser.,* **108,** 22–39.

Mägdefrau, K., 1961. Paläobotanik, *Fortschr. Bot.,* **23**(6), 106–128.

Meyen, S. V., 1963. Mosses from the Palaeozoic of Angarida (in Russian), *Piroda (Moscow),* **5,** 73–76.

Meyen, S. V., 1966. On the occurrence of true Mosses in the Upper Permian deposits of Southern Priuralia (in Russian), *Akad. Nauk. SSSR Doklady*, **166**, 924–927.

Nagy, E., 1968. Moss spores in Hungarian Neogene strata. *Acta Bot. Acad. Sci. Hungaricae*, **14**, 113–132.

Neuburg, M. F., 1960. Leafy mosses from the Permian deposits of Angarida (in Russian), *Akad. Nauk S.S.S.R. Trudy Geol. Inst. Leningrad*, **19**, 1–104.

Oschurkova, M. V., 1967. Palaeophytological stratigraphy of the Karanganda Basin (in Russian), *Akad. Nauk. U.S.S.R. Geol. Inst. (V.S.E.G.E.I.)*, 1–152.

Parihar, N. S., 1961. *An Introduction to Embryophyta*, vol. 1, Bryophyta, 4th rev. ed. Allahabad: Central Book Depot, 338p.

Reissinger, A., 1950. Die "Pollenanalyse" ausgedehnt auf alle Sedimentgesteine der geologischen Vergangenheit, *Palaeontographica*, B, **90**, 99–126.

Renault, B., and Zeiller, R., 1888. Flore fossile de Commentry, *Bull. Soc. Indus. Minerale St.-Etienne*, **2**, 366p.

Savicz-Ljubitzkaja, L. I., and Abramov, I. I., 1959. The geological annals of Bryophyta, *Rev. Bryol. Lichenol.*, **28**, 330–342.

Steere, W. C., 1946. Cenozoic and Mesozoic Brophytes of North America, *Amer. Midland Naturalist*, **36**, 298–324.

Straus, A., 1952. Beiträge zur Pliocänflora von Willerhausen. 3. Die niederen Pflanzengruppen bis zu den Gymnospermen, *Palaeontographica*, B, **93**, 1–44.

Stuchlik, L., 1964. Pollen analysis of the Miocene deposits at Rypin, *Acta Palaeobot., Cracow*, **5**, 1–111.

Sullivan, H. J., and Hibbert, F. A., 1964. *Tetrapterites visensis*–A new spore-bearing structure from the Lower Carboniferous, *Palaeontology*, **7**, 64–71.

Szafran, B., 1952. Miocene mosses from Poland and the adjacent eastern territories, *Biul. Pánstw. Inst. Geol. Warszawa*, **68**, 5–38.

Thiergart, F., 1942. *Anthoceros* Sporen aus jüngeren Braunkohlen, *Beih. Bot. Zentralbl.*, **61B**, 619–621.

Townrow, J. A., 1959. Two Triassic bryophytes from South Africa, *J. South African Bot.*, **25**, 1–22.

Walton, J., 1925. Carboniferous Bryophyta. I. Hepaticae, *Ann. Bot. (London)*, **39**, 563–572.

Walton, J., 1928. Carboniferous Bryophyta. 2. Hepaticae and Musci, *Ann. Bot. (London)*, **42**, 707–716.

Walton, J., 1949. A thalloid plant (cf. *Hepaticites* sp.) showing evidence of growth *in situ* from the Coal Measures at Dollar, Clackmannanshire, *Trans. Geol. Soc. Glasgow*, **21**, 278–280.

Watson, E. V., 1971. *The Structure and Life of Bryophytes*, 3rd ed. London: Hutchinson Univ. Library, 211p.

Zalessky, M. D., 1937. Sur la distinction de l'étage Bardien dans le Permien de l'Oural et sur sa flore fossile, *Problemy Paleontol.*, **2–3**, 37–101.

Cross-reference: *Paleobotany*.

BRYOZOA

Bryozoa (also known as Polyzoa and Ectoprocta) are aquatic sessile coelomate invertebrates forming colonies of very varied shape and inhabiting marine and freshwater environments. They are common today and occur frequently as fossils in marine sediments from the Ordovician onward. General accounts of many important aspects of the Bryozoa are to be found in those works marked by an asterisk (*) in the list of references. In his book *Bryozoans*, Ryland (1970), dealing with both living and fossil forms, gives an account of the structure, evolution, physiology, and ecology of the phylum.

Morphology

In living Bryozoa, each colony comprises a number of minute individuals (*zooids*) each with a crown of tentacles protrusible through an orifice and a body that is permanently attached within the exoskeleton. The skeletal structure of the entire colony is termed the *zoarium* (pl. zoaria) and this is composed of *zooecia* (sing. zooecium), the exoskeletons secreted by individual zooids. Zooecia are seldom greater than 0.50 mm in diameter or width, and, in marine Bryozoa, consist of chitin and calcium-carbonate. Only the calcareous parts are normally preserved as fossils. Bryozoan zoaria increase by asexual zooidal budding, secreting zooecia which are normally contiguous. Mature zooids also reproduce sexually. The resulting free-swimming larvae attach to substrata and metamorphose into the first individuals (*ancestrulae*) of new colonies. Different types of zooids may differentiate in one colony; and, in fossil Bryozoa, such variation is reflected in zooecia of different structure.

Zoaria vary in size from a few mm to massive accumulations as much as 1 m across. Theoretically, the size of the colony may be limited only by the extent of an available and suitable substratum. The great majority are sessile, attached to the sea floor, or encrust hard substrata such as the shells of larger organisms. In the past, as today, many zoaria encrusted seaweeds or other perishable substrata. Apparently, unattached zoaria of fossil Bryozoa, occurring free in sediments, may have encrusted such perishable substrata. Many fossil Bryozoa have delicate frond-like or branched zoaria, which were broken up before or during their burial in sediments. Actual encrusting zoaria apart, their broken remains are commonly widely scattered and the numerous fragments derived from one colony may give a false impression of abundance of individual species at a particular horizon. The calcareous skeletons may be replaced by various minerals during the process of fossilization; but the original calcareous skeleton is frequently well preserved, especially in finer grained calcareous mudstones and shales or in limestones. Bryozoan remains occur, however, in all types of marine sediments, particularly those accumulated in shallow-water environments. But in general, in the past Bryozoa appear to have inhabited as great a range of depth as they do today; from the shoreline to over 5,000 m.

Taxonomy

Study of living Bryozoa shows that three classes may be recognized: the Stenolaemata, the Gymnolaemata, and the Phylactolaemata (Table 1). The latter are exclusively freshwater forms in which the zooids have horseshoe-shaped crowns of tentacles and secrete chitinous or gelatinous zooecia forming delicate zoaria. Examples of these have not been preserved as fossils, but the Phylactolaemata may be presumed to have a very long geological history. A doubtful fossil from the Cretaceous of Bohemia has been tentatively assigned to the Phylactolaemata. In the class Stenolaemata, the zooids secrete cylindrical zooecia usually with calcified walls, some with interzooecial communication pores. New zooids are produced in a common bud region of the colony, consisting of confluent coeloms, which is ultimately divided into zooids by newly grown walls. The extant Cyclostomata serve as a model for interpretation of the other, now extinct, orders of this class—the Trepostomata, the Cystoporata, and the Cryptostomata.

Cyclostomata. One of the most important orders of Bryozoa surviving to the present day is the Cyclostomata. The order has a very long geological history, ranging from the Ordovician to the Holocene and includes nearly 300 genera. Structurally, cyclostomatous Bryozoa are comparatively simple. They consist of bundles or expansions of distally open calcareous tubes, which may be unpartitioned or have dividing calcareous diaphragms and which are budded off one another and connected by open pores. Each zooecium has a circular orifice, not closed by an operculum, at its distal extremity and

TABLE 1. Classification of Bryozoa

Phylum BRYOZOA
  Class STENOLAEMATA
    Order Cyclostomata—Ordovician–Holocene
    Order Trepostomata—Ordovician–Permian, ?Triassic
    Order Cystoporata—Ordovician–Permian
    Order Cryptostomata—Ordovician–Permian
  Class GYMNOLAEMATA
    Order Cetnostomata—Ordovician–Holocene
    Order Cheilostomata—Upper Jurassic–Holocene
      Suborder Anasca—Upper Jurassic–Holocene
      Suborder Ascophora—Upper Cretaceous–Holocene
      Suborder Cribrimorpha—Upper Cretaceous–Holocene
  Class PHYLACTOLAEMATA

houses a tentacled zooid which protrudes from the zoarial surface. Zoarial form is extremely varied: they may be encrusting, flat, unilaminar or aborescent, cyclindrical or ribbon-like, branched or massive. Orindary zooecial tubes may be modified as gonozooecia, i.e., enlarged zooecia which function as brood chambers for developing larvae (see Fig. 1), or neighboring zooecia may be incorporated in a swollen gonocyst. Although there is great variety of form in the whole colony the structural pattern of individual zooecia is simple. Numerically, the Cyclostomata were subordinate to the Trepostomata, Cryptostomata, and Cystoporata in the Paleozoic. In the Mesozoic the Cyclostomata were the only surviving order of Bryozoa—apart, presumably, from a small number of Ctenostomata—until the Cretaceous. In that period, the Cheilostomata appeared and rapidly surpassed the Cyclostomata in both numbers and variety of form. Although subordinate to the Cheilostomata from the later Mesozoic onward, the Cyclostomata continue with great variety of zoarial form (e.g., Brood, 1972) and in large numbers to the present. During the Jurassic and the earlier part of the Cretaceous, the Cyclostomata reached their acme of development and their remains accumulated in shallow-water marine sediments in many parts of the world.

Trepostomata. Of the extinct orders, the Trepostomata are usually the more massive, forming heavily calcified zoaria (see Fig. 2), often with a lamellar or coarsely branched form. In branched zoaria, the zooecia are calcareous tubes budded off in the axial endozone where they are thin walled and subparallel to the branch axis. In the surrounding exozone, the tubular zooecia are thicker walled and bend outward to meet the branch surface almost at right angles. The zooecia are divided by complete transverse partitions (diaphragms), which are more frequent in the exozone, and lack interconnecting pores. Between the large zooecia (autozooecia), smaller modified zooecia occur; these may be narrower tubes crowded with transverse partitions (mesopores). Autozooecia and mesopores are accompanied by small strengthening rod-like features known as acanthopores. These project above the zoarial surface as short spines. Some forms of Trepostomata have an irregular zoarial surface with protruberances (monticules) consisting mainly or large, thicker-walled zooecia surrounded by normal zooecial tubes. Earlier workers attempted to separate the Trepostomata into two large groups distinguished by supposed differences in zooecial wall structure. More recently, it has been shown that at least five distinctive wall microstructures occur in which the fine

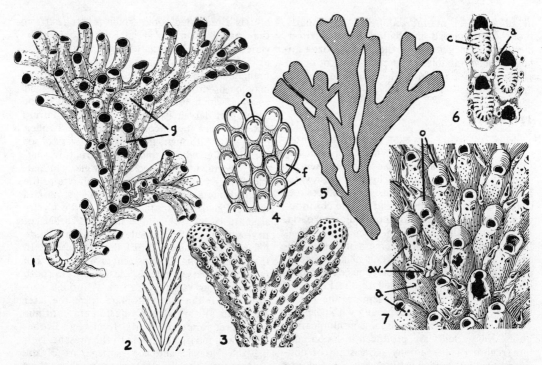

FIGURE 3. 1-3, Cyclostomata; 1, *Proboscina boryi*–Holocene (zoarium with gonozooecia, g, X10); 2,3, *Cardioecia neocomiensis*–Lower Cretaceous (longitudinal section and distal part of zoarium, respectively, X10); 4-7, Cheilostomata; 4,5, *Flustra foliacea*–Holocene (4, surface of part of zoarium showing chitinous front walls, f, of zooecia and opercula, o, X10; 5, form of zoarium, X1); 6, *Sandalopora soccata*–Upper Cretaceous (zooecia, showing costate secondary front wall, c, and avicularia, a, X25); 7, *Diplotresis sparsiporosa*–Eocene (zooecia, showing paired ascopores, a, avicularia, av, and ovicells, o. X25).

calcareous laminae that compose the walls are arranged in different ways. As with other orders of Bryozoa, investigations of this kind demonstrate the difficulty of supra-generic classification in the phylum. Over 200 trepostome genera are recognized, ranging from the Ordovician to the Permian and possibly surviving into the Triassic.

Cystoporata. The extinct order of the Cystoporata was established in 1964 to include certain Bryozoa formerly placed in the Trepostomata or in the Cyclostomata. Members of the Cystoporata form massive calcareous zoaria. These may be encrusting or branched or bifoliate and are composed of tubular zooecia divided by diaphragms. The order Cystoporata includes the Fistuliporidae, a family in which vesicular tissue is abundantly developed between the tubular zooecia. The Cystoporata comprise more than 80 genera, the earliest occurring in the Ordovician and the latest ranging to the end of the Permian.

Cryptostomata. In the extinct order of the Cryptostomata the zoaria are cylindrical or ribbon-shaped and branched, or, commonly, they form reticulate expansions with zooecia

opening on one side of the zoarium only (see Fig. 2). The zooecial tubes are short, thin walled nearer the origin of growth, thicker walled near the zoarial surface, and lacking mural pores. In many species, each tube is partly closed by a transverse calcareous partition some distance below the surface of the zoarium. This partition extends about half-way across each zooecium and is termed a hemiseptum, marking the position of the orifice in life. The space between the hemiseptum and the opening of the zooecium at the zoarial surface has been termed the vestibule. In this outer region, the vestibules of adjacent zooecia are commonly separated by solid or vesicular calcium carbonate skeletal material which, as in the Trepostomata, may contain acanthopore-like structures. In the Cryptostomata, some of these may have been points of attachment for colonial cuticular layers in life (Tavener-Smith, 1969). Like the Trepostomata, the Cryptostomata range from the Ordovician to the Permian, and include about 130 genera. The Cryptostomata were most numerous and varied in Devonian and Mississippian marine deposits. At these horizons particularly, many reticulate

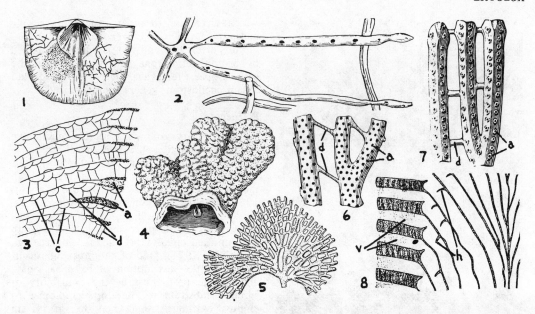

FIGURE 2. 1,2, Ctenostomata, *Vinella repens*—Middle Ordovician (1, zoarial excavation in brachiopod valve, X1; 2, stolon-like zooecia, X25); 3,4, Trepostomata, *Monticulipora mammulata*—Ordovician (3, longitudinal section near surface of zoarium showing diaphragms, d, and cystiphragms, c, in larger normal zooecia with granulose acanthopores, a, between, X25; 4, massive zoarium, X1); 5–8, Cryptostomata; 5,6, *Polypora dendroides*—Lower Carboniferous (5, obverse of zoarium, X1; 6, obverse of part of zoarium showing zooecial apertures, a, and connections, d, X5); 7, *Fenestella antiqua*—Silurian (obverse of part of three branches showing zooecial apertures, 1, nonporous tubular connections, d, X10); 8, *Nemataxis fibrosus*—Middle Devonian (longitudinal section showing hemisepta, h, and dense skeletal material between vestibules, v, X20).

frond-like forms of fenestellid Cryptostomata occur. The zoaria of these are reticulate or net-like expansions consisting of short zooecial tubes opening alternately in two rows on the obverse of each branch. The reverse of each branch lacks zooecial openings and adjacent branches are connected, approximately at right angles, by cross tubes lacking openings (Tavener-Smith, 1969).

The third class, the Gymnolaemata, comprises extant Bryozoa of two orders, the Ctenostomata and the Cheilostomata. The class is almost exclusively marine, the zooids having a circular crown of tentacles which are extruded by muscular flexing of part of the body wall. Zooids are markedly polymorphic, forming colonies of extremely varied shape with adjacent zooecia connected by tissue-filled pores.

Ctenostomata. The orifices of ctenostome zooids are closed by a crenulate collar. The zooecia of the 40 or so bryozoan genera belonging to this order are not calcified, but some have the habit of excavating cavities or hollows in the shells of larger organisms leaving delicate stolon-like traces which are preserved in fossil shells and other substrates from the Ordovician onwards (see Fig. 2).

Cheilostomata. The Cheilostomata, which have been dominant since the later Mesozoic, have calcareous and chitinous zoaria of delicate structure (see Fig. 1). The zoaria are commonly unilaminar encrustations or uni- or bilaminar foliaceous expansions or branched stem-like growths; some zoaria are unattached (Lagaaij, 1963). The zooecia are generally small, box-like and contiguous, being budded off one another in rows (Fig. 3). Each has a distal orifice closed by a hinged chitinous operculum and a chitinous front wall, which may become calcified in some groups. Polymorphism occurs and is reflected by structurally specialized zooecia. In addition to normal zooecia containing tentacled zooids, ovicells are common. These are specialized globular calcareous structures developed at the distal extremity of mature zooecia and house developing larvae. Other modified zooecia occur, the most common of which are avicularia polymorphs of distinctive shape without tentacles but possessing a musculature that activates a snapping mandible-like operculum. This device prevents the larvae of other organisms from encrusting the growing bryozoan colony. Vibraculae, smaller structures with long whip-like mandibles, function in the same way.

The Cheilostomata probably have greater

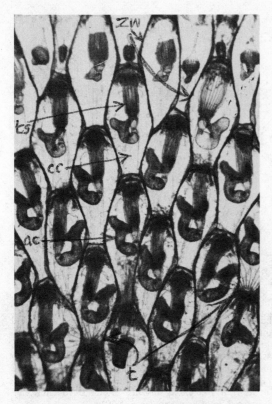

FIGURE 3. Part of a multiserial colony or zoarium of a Cheilostome Bryozoan, *Carbasea carbasea* (Ellis and Solander), X25. ac, looped, alimentary tract from the mouth surrounded by tentacles to the anus lying near the mouth but outside the lophophore supporting the tentacles; cc, coelomic cavity; t, expanded tentacles; cc, coelomic cavity; t, expanded tentacles; ts, tentacles retracted into sheath; zw, zooecial wall.

variety of form than other bryozoan orders. Three main types are found as fossils and at the present day. The Anasca have a flexible chitinous front wall and depression of this by muscles extrudes the tentacle crown and gut through the orifice, enabling the individual to filter food. In the Ascophora, the front wall is calcified and therefore rigid. A flexible chitinous compensation sac lies beneath the rigid front wall of the Ascophora and connects to the exterior by a pore which allows sea water to enter the sac, thus extruding the polypide through the orifice. The third group of Cheilostomata is the Cribrimorpha, in which calcareous marginal spines overarch and fuse medially to produce a costate secondary front wall above the primary chitinous front wall (Larwood, 1962). These forms extrude the polypide as in the Anasca.

The oldest cheilostomes occur in sediments of Portlandian age in the youngest Jurassic rocks of Europe. It is a simple anascan form

which gives rise to other structurally simple anascans during the Lower Cretaceous. These rapidly diversified and were soon accompanied by both Cribrimorpha and Ascophora in the early Upper Cretaceous. Over 1000 fossil and extant cheilostome genera are recognized.

Paleoecology

Since the Bryozoa, which secrete preservable calcareous skeletons, are tolerant of a remarkably wide ecological range (Schopf, 1969), their remains occur in a great variety of marine sediments throughout the fossil record (Schopf, 1969). Ecological studies of fossil Bryozoa depend on an understanding of present-day forms, and studies of the paleoecology of this group are confined chiefly to Cenozoic faunas (e.g., Cheetham, 1971). One of the most significant factors in the distribution of Bryozoa today, and in the past, must be the availability of suitable substrata for developing colonies to encrust. Although there may be, at certain horizons, a distinct correlation between the presence of hard shells of other organisms, affording a suitable base for colony growth, and the frequency and variety of byrozoan genera and species, the bryozoan free-larval stage, the adult encrustation of perishable floating substrata, and the delicacy of many branched Bryozoa ensures their widespread occurrence and more even distribution at many horizons. Thus, stratigraphically, their remains are well distributed and distinctive assemblages of Bryozoa are most useful for correlation and for the precise determination of horizons.

Bryozoans, phoronids, and brachiopods comprise the lophophorate invertebrate phyla. These have, as common features, a body trunk; a circumoral lophophore; and, with the exception of the gymnolaemate bryozoa, a preoral epistome. Zimmer (1973) has argued that the phoronids are most similar to the ancestral lophophorates and that they share with the brachiopods and bryozoans some common larval and, particularly, adult characters indicating general deuterostome affinities. Farmer et al. (1973) also discuss some of these relationships in their assessment of the ectoproct ground plan.

Evolution

It is probable that the Bryozoa are polyphyletic. The freshwater forms may reasonably be presumed to have a long geological history—their remains not being preserved because of the perishable nature of their skeletons. Evidence of the excavations of ctenostomatous Bryozoa occurs sparsely from the Ordovician onward. Cryptostomatous and trepostomatous

Bryozoa and Cystoporata occur in the Ordovician and become abundant in subsequent Paleozoic horizons. Cyclostomata also occur, although in comparatively small numbers, in the Ordovician. Apart from the Ctenostomata, they are the only marine order of Bryozoa to survive into the Mesozoic, where they become extremely abundant in the Jurassic and continue to evolve many new genera and species until the end of the Cretaceous. In the Late Cretaceous, they are rapidly outnumbered by the great variety of the Cheilostomata, and continue subordinate to that order through the Cenozoic to the present. Thus, all orders of Bryozoa, apart from the Cheilostomata, were distinct from the early Paleozoic. It has been suggested that the Cheilostomata were derived from the Cryptostomata at the end of the Paleozoic. Attention has been drawn to the short zooecial tubes of the Cryptostomata as most closely resembling the box-like form of cheilostome zooecia. Some Cyclostomata in the Jurassic and Early Cretaceous have short zooecial tubes and many have structurally modified zooecia including gonozooecia. No Cheilostomata occur in the Mesozoic before the Portlandian Stage of the Jurassic and no Cryptostomata are recorded from post-Permian horizons. The Ctenostomata, known from the Ordovician onward, may well have given rise to the early cheilostome bryozoans, with which they have some structural affinity.

G. P. LARWOOD

References

*Annoscia, E., ed., 1968. Proceedings of the First International Conference on Bryozoa, *Atti Soc. Ital. Sci. Nat.*, 108, 1–377.

*Bassler, R. S., 1953. *Treatise on Invertebrate Paleontology*, pt. G, Bryozoa. Lawrence, Kansas: Geol. Soc. Amer. and Univ. Kansas Press, 253p.

Boardman, R. S., 1960. Trepostomatous Bryozoa of the Hamilton Group of New York State, *U.S. Geol. Surv. Prof. Paper 340*, 87p.

*Boardman, R. S., and Cheetham, A. H., 1969. Skeletal growth, intracolony variation, and evolution in Bryozoa, *J. Paleontology*, 43, 205–233.

Boardman, R. S., and Cheetham, A. H., 1973. Degrees of colony dominance in stenolaemate and gymnolaemate Bryozoa, in R. S. Boardman, A. H. Cheetham, and W. A. Oliver, Jr., eds., *Animal Colonies*. Stroudsburg, Pa.: Dowden, Hutchinson & Ross, 121–220.

*Boardman, R. S.; Cheetham, A. H.; and Cook, P. L., 1970. Intracolony variation and the genus concept in Bryozoa, *Proc. N. Amer. Paleontological Conv.*, pt. C, 294–320.

Brood, K., 1972. Cyclostomatous Bryozoa from the Upper Cretaceous and Danian in Scandinavia, *Stockholm Contrib. Geol.*, 26, 1–464.

Cheetham, A. H., 1971. Functional morphology and biofacies distribution of cheilostome Bryozoa in the

Danian Stage (Paleocene) of southern Scandinavia, *Smithsonian Contrib. Paleobiol.*, 6, 87p.

Farmer, J. D.; Valentine, J. W.; and Cowen, R., 1973. Adaptive strategies leading to the ectoproct ground plan, *Systematic Zool.*, 22, 233–239.

*Hyman, L. H., 1959. The Lophophorate Coelomates—Phylum Ectoprocta, in L. H. Hyman, *The Invertebrates: Smaller Coelomate Groups*. New York: McGraw-Hill, 275–515.

Lagaaij, R., 1963. *Cupuladria canariensis* (Busk)—Portrait of a bryozoan, *Palaeontology*, 6, 172–217.

Larwood, G. P., 1962. The morphology and systematics of some Cretaceous cribrimorph Polyzoa (Pelmatoporinae), *Bull. British Mus. (Nat. Hist.), Geol.*, 6, 1–285.

*Larwood, G. P., ed., 1973. *Living and Fossil Bryozoa. Recent Advances in Research*. London and New York: Academic Press, 634p.

*Ryland, J. S., 1970. *Bryozoans*. London: Hutchinson Univ. Library, 175p.

Schopf, T. J. M., 1969. Paleoecology of ectoprocts (bryozoans), *J. Paleontology*, 43, 234–244.

Schopf, T. J. M. 1977. Patterns and themes of evolution among the Bryozoa, in A. Hallam, ed., *Patterns of Evolution*. Amsterdam: Elsevier, 159–207.

Tavener-Smith, R., 1969. Skeletal structure and growth in the Fenestellidae (Bryozoa), *Palaeontology*, 12, 281–309.

Tavener-Smith, R., and Williams, A., 1972. The secretion and structure of the skeleton of living and fossil Bryozoa, *Phil. Trans. Roy. Soc. London Ser. B*, 264, 97–159.

Zimmer, R. L., 1973. Morhphological and developmental affinities of the lophophorates, in Larwood, 1973, 593–599.

Cross-reference: *Coloniality.*

BURGESS SHALE

The soft-bodied fossils from the Burgess Shale (Middle Cambrian, *Bathyuriscus-Elrathina* Zone) are among the most exquisitely preserved in the fossil record. Recognizable muscles, gut, and nerve cord are preserved in some specimens. The Shale was discovered by C. D. Walcott, Secretary of the Smithsonian Institution, in 1910 on the W side of the ridge connecting Mt Field and Wapta Mtn, southern British Columbia (Fig. 1), following the discovery the previous year of a detached block containing soft-bodied fossils. Exploitation by quarrying, over a number of years, of a stratum near the base of the Shale, the 7′7″ (2.31 m) thick Phyllopod bed, as well as another unit 65′ (19.8 m) higher now known as Raymond's quarry, yielded over 50,000 fossils. The collection is in the National Museum of Natural History, formerly the U.S. National Museum, Washington, D.C., from which Walcott issued a series of preliminary publications on much of the biota. In 1966 and 1967 a Geological Survey of Canada team led

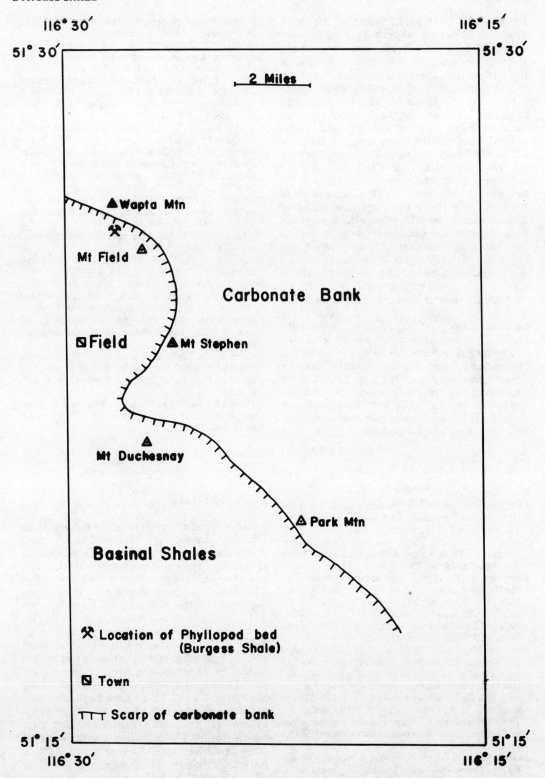

FIGURE 1. Map of area around Field, British Columbia (after McIlreath, 1974).

by Dr. J. D. Aitken reopened the two quarries with the cooperation of the authorities of the Yoho National Park and the Parks Canada, Dept. of Indian and Northern Affairs, Ottawa. The history of research on the Burgess Shale and its fossils was broadly reviewed by Whittington (1971a).

Preservation

It has long been known that the Burgess Shale is a predominantly shale lentil within a basinal calcareous shale and limestone succession, the "thick" Stephen Formation (Fig. 2). The Burgess Shale is immediately adjacent to a massive carbonate bank consisting of thick bedded dolomitic limestones, the "thick" Cathedral Formation. Limestone tongues derived from the bank form a veneer over the bank front. One of these tongues, the Boundary Limestone unit, projected far into the basin and acts as an invaluable chronological and stratigraphic marker bed. Fritz (1971) demonstrated by using trilobite zonation that the bank formed a submarine scarp about 530 ft (162 m) high at the time of deposition of the Shale. This scarp trends NNW and has been traced by McIlreath (1974) over several miles. The Shale

is located within a prominent embayment of the scarp (McIlreath, 1974) (see Fig. 1).

The numerous laminated beds consisting of calcareous siltstones grading up into mudstone that extend unbroken across the outcrop (Piper, 1972), the variable orientation of fossils such as the trilobitoid *Marrella splendens* (Fig. 3) with the appendages separated by sediment (Whittington, 1971a), and the twisting of compressed worms like *Pikaia gracilens* (Fig. 4) constitute the chief evidence for a turbiditic origin of the Phyllopod bed. It is possible to distinguish between a preslide environment, where the benthic fauna and flora lived, and a postslide environment to which the fauna and flora were transported and fossilized. Despite the number of fossils preserved in the Phyllopod bed, the absence of bioturbation and scavenging strongly suggests that the postslide environment was lifeless, anoxic and probably poisoned by H_2S extending above the sediment-water interface. All levels of the Phyllopod bed are undisturbed, and there is no evidence of periodic renewal of oxygenated conditions. The position of the preslide environment with respect to the carbonate bank and off-bank basin is uncertain. It is unlikely that the biota was swept over the bank edge, as there is no evi-

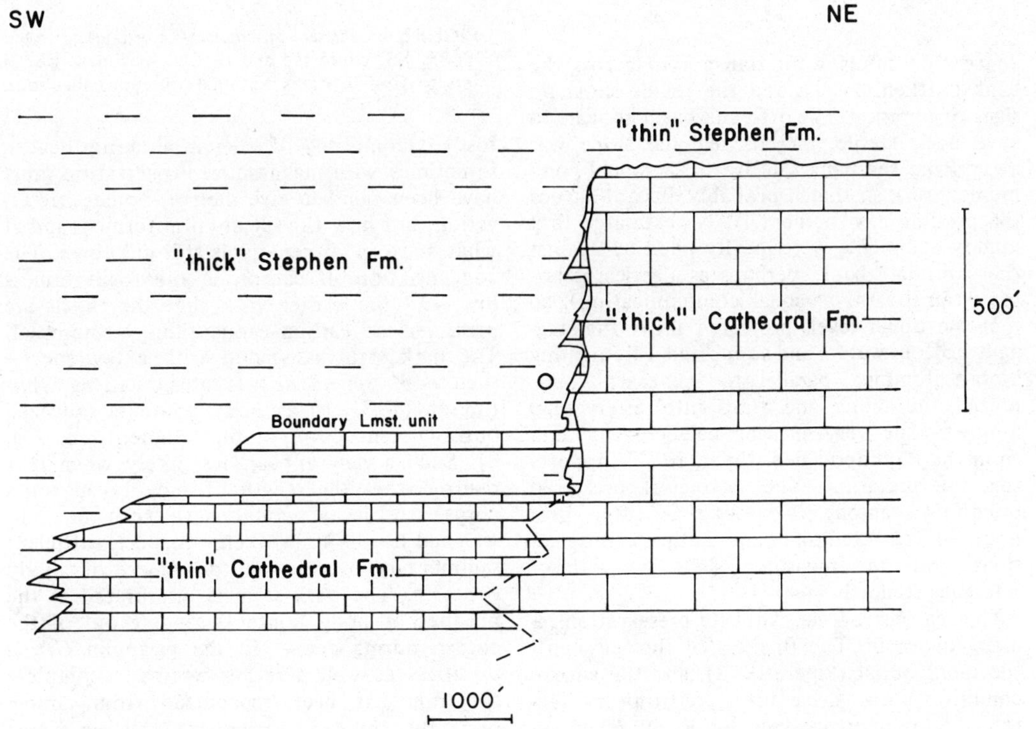

SW   NE

FIGURE 2. Schematic cross section through the Cathedral and Stephen Formations (after Fritz, 1971). The position of the Burgess Shale is marked by the circle above the Boundary Limestone unit.

FIGURE 3. *Marrella splendens* (Trilobitoidea), GSC 26592 (counterpart), X4, ventral view with prominent dark stain at posterior. This and the following photographs were taken in ultra-violet light; GSC refers to Geological Survey of Canada collections.

FIGURE 4. *Pikaia gracilens* (Chordata), USNM 198683, X5, contorted and twisted specimen. USNM refers to United States National Museum collections.

dence that muds were transported across the bank (Aitken, 1971), and the fossils show no signs of abrasion. The off-bank basin appears to have been sterile, and its regional slope was away from the bank (McIlreath, personal communication), so that it probably did not harbor the preslide environment. It is postulated that muddy sediments were locally piled up against the carbonate bank, perhaps as a series of terraces (McIlreath, personal communication), so that the upper levels projected above the stagnant bottom waters and supported a flourishing biota. Slumping parallel to the scarp transported the fauna and flora into intervening hollows. The very meager paleocurrent data from the Phyllopod bed (Piper, 1972) supports this interpretation. The degree of preferred orientation among the fossils is often pronounced (R. Lindholm and author's observations), but the measured slabs are without reference to north.

The reasons for the exquisite preservation remain uncertain. The fineness of the sediment, the rapid burial (Piper, 1972), and the anoxic conditions must have been contributory factors. Preliminary analyses by R. A. Chappell (National Physical Laboratory, Teddington) indicate that the reflective film that forms the

fossil is composed of calcium aluminosilicates, sometimes with magnesium. Fossilization must have been rapid to give the soft-bodied preservation, but how the silicate film formed, and at what stage in diagenesis, is still unknown. The concentration of carbon in the fossil film is low, and the earlier view that the fossils are preserved as carbonaceous films is incorrect. The dark stain associated with a few species such as *M. splendens* was interpreted by Whittington (1971a,b) as body contents squeezed out by compaction of the sediment (see Fig. 3). Such a view appears less likely when it is realized that the length of time taken for a corpse to decay would have been far outweighed by the time taken to deposit sufficient sedimentary overburden to squeeze out body contents. The dark stain is interpreted as the product of body contents seeping out of the corpse during decay. In the priapulid *Ottoia prolifica*, as well as other worms, a complete spectrum has been recognized from almost perfectly preserved specimens with an intact body wall (Fig. 5) to very decayed specimens (Fig. 6).

FIGURE 5. *Ottoia prolifica* (Priapulida), GSC 40972, X2, well-preserved specimen.

FIGURE 6. *Ottoia prolifica* (Priapulida), USNM 198572, X1–5 decayed and twisted specimen.

Fauna

The composition of the fauna is given in Table 1. The fauna includes numerous arthropods, other than trilobites, which are yielding valuable morphological details of the limbs and other appendages. Many of the trilobites consist of the exoskeleton only, but some specimens of *Olenoides serratus* have the appendages preserved (Whittington, 1975a). Some species previously interpreted as arthropods, such as *Opabinia regalis,* are now known to be only distantly, if at all, related to this group (Whittington, 1975b). The worms embrace many phyla and include priapulids; polychaetous annelids; ?aplacophoran mollusks; possible protochordates and hemichordates; and a lophophorate, *Odontogriphus omalus,* which

TABLE 1. Percentage Composition by Genera of the Phyllopod bed

Arthropods, excluding trilobites	23.5%
Trilobites	10.0
Arthropod-like invertebrates	3.5
Echinoderms	3.5
Mollusks	2.5
Polychaetes	4.0
Priapulids	5.8
Hemichordates and chordate-like worms	3.5
Lophophorates, excluding brachiopods	1.5
Brachiopods	4.0
Miscellaneous worms	9.2
Coelenterates and ?ctenophores	4.0
Sponges	25.0

may represent a conodontophorid (or "cono-dont animal"; Conway Morris, 1976). Rare echinoderms are represented by eocrinoids and primitive crinoids (Sprinkle, 1973), although the assignation of *Eldonia ludwigi* (Fig. 7) to the holothurians (Durham, 1974) is doubtful. The coelenterates include hydroid and medusoid forms. A possible ctenophore has also been recognized. At least 10 worm-like gen-era, however, have so far defied taxonomic assignment.

The paleontological significance of the fauna is twofold: (1) A number of genera can be clearly related to other fossil and recent groups. These genera sometimes give important phylo-genetic clues, and in the priapulids from the Phyllopod bed the recognition of primitive fea-tures has been greatly facilitated and allows a tentative phylogeny to be proposed. More fre-quently, however, they serve to emphasize that many phyla in the Cambrian were far more diverse and specialized then previously realized. (2) Other genera cannot be related to any modern phyla, and are sometimes of bizarre appearance. They seem to represent early "experiments" in metazoan evolution.

Evidence of two biological communities is preserved in the Phyllopod bed. The benthic one is termed the *Marrella-Ottoia* community after the numerically predominant epi- and infaunal genera respectively. The epifauna was composed of sessile forms such as sponges and brachiopods, and vagrant arthropods, poly-chaetes, and mollusks. Most of the sessile forms were unattached, although there were a few ex-ceptions (see Fig. 8). The infauna was domi-nated by priapulids. Details of population structure are under study. The areal extent of the *Marrella-Ottoia* community probably was not large, but it must have lain within the photic zone judging by the presence of many different types of algae. The persistence of species such as *M. splendens* and *O. prolifica* through the entire Phyllopod bed suggests that only one benthic community was being eroded to give the slumps. Other species with a more restricted vertical distribution, such as the arthropod *Yohoia tenuis* (Whittington, 1974), indicate that they had a more patchy distribu-tion over the seafloor. Comparison with similar modern benthic communities is showing how communities evolved, and demonstrating how different groups have played the same ecologi-cal role through geological time. The benthic

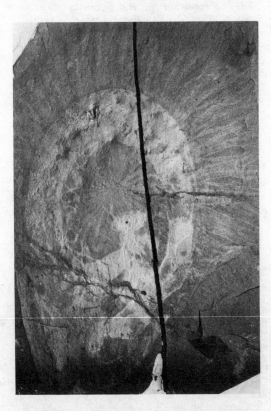

FIGURE 7. *Eldonia ludwigi* (?Holothuroida), USNM 57540, X1.

FIGURE 8. Empty tube of *Selkirkia columbia* (Pri-apalida) USNM 198623, X4, with attached brachiopod.

FIGURE 9. *Amiskwia sagittiformis (incertae sedis),*
USNM 57644, X4.

fauna recovered from the stratigraphically
higher Raymond's quarry lacks many genera
found in the Phyllopod bed, and is greatly
enriched in the arthropod *Leanchoilia superlata*
and the worm *Banffia constricta,* and so represents a somewhat different benthic community.
The pelagic *Amiskwia-Odontogriphus* community, which is preserved with the *Marrella-Ottoia*
community, is very poorly represented because
of the unlikelihood of its members being involved in the benthic turbidites. The elongate
worm *Amiskwia sagittiformis* (Fig. 9) and
possible conodontophorid *Odontogriphus
omalus* are both dorsoventrally flattened gelatinous animals. More active animals, which
occupied the same ecological niche as chaetognaths do today, are also known.

The possibility of constructing a food web for
the *Marrella-Ottoia* community from a study of
gut contents and the type of mouth parts is
frustrated by the difficulty of identifying gut
contents, although the hyolithids in the gut of
some specimens of *O. prolifica* are a notable
exception (Fig. 10).

Research on the biota and ecology of the
Phyllopod bed is still incomplete, but the next
decade will see much clarification on this the
most celebrated fossil locality in the world.

SIMON CONWAY MORRIS

References

Aitken, J. D., 1971. Control of Lower Palaeozoic sedi-
mentary facies by the Kicking Horse Rim, southern
Rocky Mountains, Canada, *Bull. Canadian Petro-
leum Geol.,* **19**, 557–569.
Conway Morris, S., 1976. A new Cambrian lopho-
phorate from from the Burgess Shale, British
Columbia, *Palaeontology,* **19**, 199–222.
Durham, J. Wyatt, 1974. Systematic position of
Eldonia ludwigi Walcott, *J. Paleontology,* **48**, 750–
755.
Fritz, W. H., 1971. Geological setting of the Burgess
Shale, *Proc. N. Amer. Paleontological Conv.,* pt. I,
1155–1170.
Hughes, C. P., 1975. Redescription of *Burgessia bella*
from the Middle Cambrian Burgess Shale, British
Columbia, *Fossils and Strata,* **4**, 415–435.
McIlreath, I. A., 1974. Stratigraphic relationships at
the western edge of the Middle Cambrian carbonate
facies belt, Field, British Columbia, *Geol. Surv.
Canada Paper* **74-1**, 333–334.
McIlreath, I. A., 1975. Stratigraphic relationships at
the western edge of the Middle Cambrian carbonate
facies belt, Field, British Columbia, *Geol. Surv.
Canada Paper* **75-1**, 557–558.
Piper, D. J. W., 1972. Sediments of the Middle Cam-
brian Burgess Shale, Canada, *Lethaia,* **5**, 169–175.
Sprinkle, J., 1973. Morphology and evolution of
blastozoan echinoderms, *Spec. Pub. Mus. Comp.
Zool. Harvard Univ.,* 283p.
Whittington, H. B., 1971a. The Burgess Shale: History

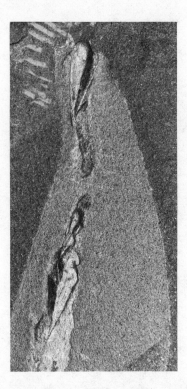

FIGURE 10. *Ottoia prolifica* (Priapulida), USNM
188604, X6, hyolithids in the posterior gut.

of research and preservation of fossils, *Proc. N. Amer. Paleontological Conv.,* pt. I, 1170–1201.

Whittington, H. B., 1971b. Redescription of *Marrella splendens* (Trilobitoidea) from the Burgess Shale, Middle Cambrian, British Columbia, *Geol. Surv. Canada Bull.,* **209,** 24p.

Whittington, H. B., 1974. *Yohoia* Walcott and *Plenocaris* n. gen., arthropods from the Burgess Shale, Middle Cambrian, British Columbia, *Geol. Surv. Canada Bull.,* **231,** 27p.

Whittington, H. B., 1975a. Trilobites with appendages from the Middle Cambrian, Burgess Shale, British Columbia. *Fossils and Strata,* 4, 97–136.

Whittington, H. B., 1975b. The enigmatic animal *Opabinia regalis,* Middle Cambrian, Burgess Shale, British Columbia. *Phil. Trans. Roy. Soc. London, Ser. B,* **271,** 1–43.

Cross-references: *Annelida; Fossil Record.*

C

CALCICHORDATES

The calcichordates are a strange group of lower Paleozoic and Devonian fossils that are traditionally placed among the carpoid echinoderms or Homalozoa (see Ubaghs 1968, 1971). They show undoubted echinoderm affinities, but detailed study of their anatomy indicates that they were chordates, ancestral to all three living subphyla of chordates, i.e. amphioxus and its allies, tunicates, and vertebrates. The author of this article previously regarded them as a subphylum of the Chordata, but as a group they are not logically comparable with the living subphyla of chordates (see Jefferies and Lewis, 1978). The calcichordate group is coextensive with the class Stylophora Gill and Caster (1960). The calcichordates owe their name to the echinoderm-like skeleton in which each plate is a single crystal of calcite.

The first author to suggest that calcichordates were allied to chordates was Matsumoto (1929). The case was argued in detail by Gislén (1930) and many of Gislén's major results are correct, though most of his embryological supporting arguments seem to be mistaken. The author of the present article has published since 1967 a series of papers arguing for the chordate nature of the calcichordates (Jefferies 1967, 1968a,b, 1969, 1971, 1973, 1975; Jefferies and Lewis, 1978; Jefferies and Prokop, 1972). Authors broadly sympathetic to his conclusions include Eaton (1970) and Bone (1972). Authors emphatically out of sympathy include Parsley and Caster (1975), Ubaghs (1971, 1975), and Halstead-Tarlo (discussion in Jefferies 1967, 1968b).

All calcichordates consist of a massive part and an appendage. There are various names for these two parts but, by homology with vertebrates, they are best referred to as head and tail (previously called body and tail, or by probably mistaken homology with crinoids, theca and stem). Unlike vertebrates, but like tunicate tadpoles, the calcichordates had no trunk region. The nonpharyngeal gut and other viscera were not ventral to the anterior part of the notochord and foremost muscle blocks, but anterior to them, inside the head. The head and tail of calcichordates correspond to the visceral and somatic regions of the hypothetical primitive vertebrate imagined by Romer (1972). By homology with vertebrates, the calcichordate head is defined as anterior and the tail as posterior; the habitually lower surface of the head is ventral and the upper surface dorsal; and this also defines right and left. It is likely, however, that the habitual direction of crawling in calcichordates was backwards, pulled by the tail.

The calcichordates are divided into two groups, conventionally ranked as orders of the class Stylophora. These groups are the cornutes, which are the more primitive, and the mitrates, which evolved from cornates. The living chordate subphyla probably evolved from mitrates.

A representative cornute is *Cothurnocystis* (Figs. 1 and 2) known from Ordovician marine rocks of France and Scotland. The head was bootshaped (*cothurnos* = boot). The head skeleton consisted of a frame of big marginal plates to which were attached dorsal and ventral flexible integuments. In the left part of the dorsal integument were up to 16 branchial slits with the structure of outlet valves. (Larval amphioxus likewise has left branchial slits without right ones). The gonoporeanus was just left of the tail, behind the most median of the gill slits, so that feces and gametes could be washed away by the branchial current.

Inside the head of *Cothurnocystis* (Fig. 3A) there is positive evidence of four soft chambers and indirect evidence for a fifth chamber. The *buccal cavity* was posterior to the mouth and filled the "ankle" part of the "boot." The *posterior coelom* was just anterior to the tail; it was homologous with the chamber known as the left epicardium of living tunicates. The *pharynx* filled most of the rest of the head, with the gill slits opening through its roof. The *right anterior coelom* lay on the floor in the right posterior angle of the head beneath the pharynx, in the "heel" part of the "boot." The *left anterior coelom,* which would have had no cavity and for which there is only indirect evidence, probably overlay the dorsal surface of the pharynx.

Inside the right anterior coelom, there would have been situated the gonads, heart, nonpharyngeal gut, and other viscera. A large and a small groove in the skeleton run out of it un-

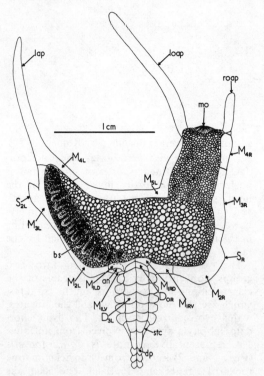

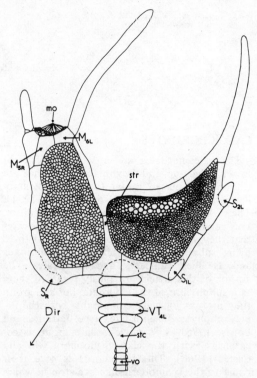

FIGURE 1. *Cothurnocystis elizae* Bather, Upper Ordovician, Girvan, Scotland, reconstruction of dorsal aspect, most of hind tail omitted (from Jefferies, 1968a; reproduced with permission of the British Museum [Natural History] and Zoological Society of London). an, Gonopore-anus; bs, branchial slit; dp, dorsal plate of hind tail; lap, left head appendage; loap, left oral appendage; mo, mouth; stc. stylocone; roap, right oral appendage; D, dorsal plates of fore tail; M, marginal plates; S, ventral spikes of head. Details of plate notations in this and other figures are explained in Jefferies (1973, 423).

FIGURE 2. *Cothurnocystis elizae* Bather, reconstruction of ventral aspect (from Jefferies, 1968a; reproduced with permission of the British Museum [Natural History] and Zoological Society of London). mo, mouth; stc, stylocone; str, strut; vo, ventral ossicle of hind tail, Dir, probable direction of movement in life; M, marginal plates; S, ventral spikes; VT_{4L}, 4th left ventral plate of fore tail.

der the posterior coelom to the gonopore-anus just left of the tail. The large groove probably carried the rectum and the small groove the gonoduct.

The tail of *Cothurnocystis* (Figs. 1 and 2) was divided into fore, mid, and hind parts. The hind tail contained a series of hemicylindrical ventral ossicles, roofed over by paired dorsal plates. A groove in the dorsal surface of the ventral ossicles probably carried a posterior extension of the notochord. The fore tail was broad with a large lumen and rather loose skeleton. It presumably contained muscles; and a noncompressional axial notochord would have existed inside it to prevent disarticulation or telescoping when the muscles contracted. The mid tail contained a massive ventral element, the stylocone. This served as a socket by which the muscles of the fore tail

could wag the mid and hind tail as a unit from side to side. Spikes on the ventral surface of the head would have prevented *Cothurnocystis* from sliding forward, but would have helped it to slide backward over the sea bottom, pulled by the tail. The tip of the hind tail was probably thrust into the sea bottom to secure purchase. A totally different interpretation of the tail has been given by Ubaghs (1975) and earlier papers).

The brain of *Cothurnocystis* was at the anterior end of the tail, in a large depression in the skeleton. It was at the anterior end of the notochord as in other chordates. Its position, strategically situated to help liaison between the tail and head, is analogous, but probably not homologous, to that of the aboral nerve center between the theca and stem of crinoids.

A more primitive cornute is *Ceratocystis*, in which the gonopore and anus opened directly out of the right anterior coelom to the right of the tail. Nearby was an echinoderm-like hydro-

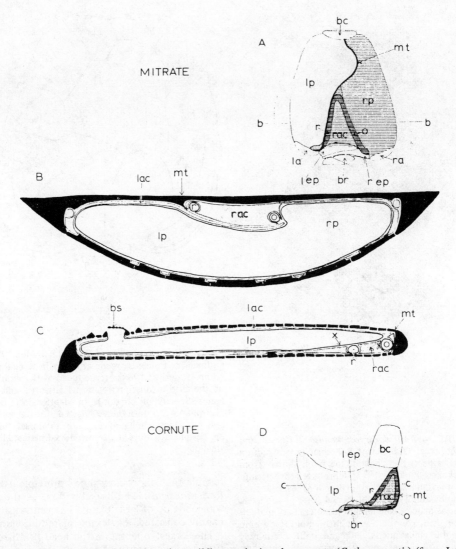

FIGURE 3. The head chambers of a mitrate (*Mitrocystites*) and a cornute (*Cothurnocystis*) (from Jefferies, 1975; reproduced by permission of the Zoological Society of London). B is a transverse section of A through b–b; C is a transverse section of D through c–c. The mitrate condition can be derived from the cornute condition by the growth of a right pharynx (rp) out of the left pharynx (lp) at the point marked x so as to lift up the cavity and contents of the right anterior coelom (rac), squash them against the roof of the head, and force them in a medial direction. Other features: bc, buccal cavity; br, brain; bs, branchial slit; la, left atrium; lac, left anterior coelom (purely virtual); l ep, left epicardium (= posterior coelom of cornutes, part of posterior coelom of mitrates); mt, mesenteric trace, between the left and right anterior coeloms (oblique ridge of mitrates); o, oesophagus; r, rectum; ra, right atrium; r ep, right epicardium (left and right epicardia = posterior coelom of mitrates).

pore. The head openings of *Ceratocystis* can be identified by direct comparison with primitive echinoderms. The head of *Ceratocystis* was covered with big plates, but lines of slight flexibility between the dorsal plates correspond to the edges of the dorsal integument of *Cothurnocystis*. *Ceratocystis* can be directly compared with the modern pterobranch hemichordate *Cephalodiscus* resting on its right side. Such a

pterobranch resting on its right side may have been the latest common ancestor both of chordates and echinoderms. Both *Ceratocystis* and *Cothurnocystis* lived mainly on fine sand, which was probably the primitive environment of the cornutes.

By contrast, the advanced cornute *Reticulocarpos* (Fig. 4), like mitrates, lived on mud, resting on the soft surface and relying only

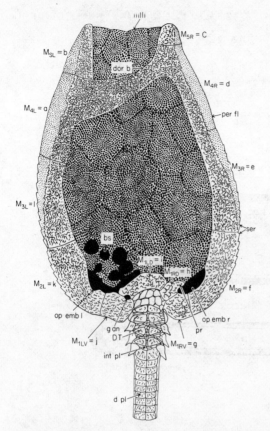

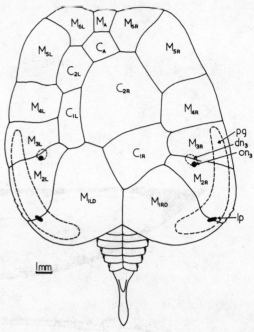

FIGURE 4. *Reticulocarpos hanusi* Jefferies and Prokop, Lower Ordovician, Bohemia, reconstruction of dorsal aspect—the whole of the tail is shown (from Jefferies and Prokop, 1972; reproduced by permission of the Linnean Society of London). bs, branchial slit; dor b, dorsal bar; d pl, dorsal plate of hind tail, DT, dorsal plate of fore tail; int pl, intercalary plate of fore tail; gan, gonopore-anus; mth, mouth; op emb l, r, optic embayments, left and right, containing transpharyngeal eyes; per fl, peripheral flange; pr, prong anterior to right transpharyngeal eye; ser, serration of peripheral flange; M, marginal plates (other letters refer to plate notations).

FIGURE 5. *Mitrocystites mitra* Barrande, Lower Ordovician, Bohemia, reconstruction in dorsal aspect (from Jefferies, 1968a; reproduced with permission of the British Museum [National History] and Zoological Society of London). dn, depression at end of transpharyngeal optic nerve on$_3$, i.e., for transpharyngeal eye; lp, opening of $n_{4\&5}$ onto dorsal surface; pg, peripheral groove; C, centro-dorsal plates; M, marginal plates.

on the weak strength of the mud to support its weight. To this end it was small and flat and had a very light skeleton. *Reticulocarpos* represents the group of cornutes from which mitrates evolved. Its head was bilaterally symmetrical in outline, as in mitrates, but the branchial slits were still restricted to the left side only.

A representative mitrate is *Mitrocystites* (Figs. 5 and 6), known from Ordovician rocks of Czechoslovakia and France. The head was externally almost bilaterally symmetrical with a flat dorsal surface and a convex ventral surface. It would partly have sunk into the soft mud of the sea bottom and thus stayed up partly by displacement ("buoyancy"). This was more

reliable than the "snow-shoe" principle used by *Reticulocarpos*. The dorsal surface of the head was made up of large plates. The ventral surface mainly consisted of a flexible plated integument.

The largest opening in the head of *Mitrocystites* was the mouth at the anterior end. There were no external gill slits, but there is evidence of right and left gill openings near the posterior right and left angles of the head. A groove just right of the tail probably represents the beginnings of the lateral-line system.

The inside of the head of *Mitrocystites* was less symmetrical than the outside (Fig. 3B). The same chambers can be discerned as in cornutes, but with additions and modifications. A *buccal cavity* existed just behind the mouth. *Right and left atria,* whose anterior walls were presumably penetrated by gill slits, were situated just anterior to the gill openings. There was a *posterior coelom* just anterior to the tail, as in cornutes; but there is evidence in mitrates that this was a double chamber. (The posterior coelom of cornutes was equivalent to only the left epicardium of tunicates, while the posterior coelom of mitrates was equivalent to both right

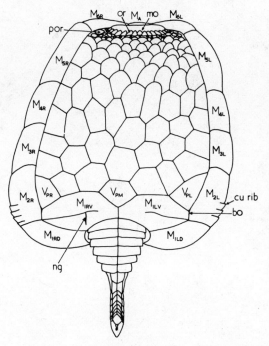

FIGURE 6. *Mitrocystites mitra* Barrande, reconstruction in ventral aspect (from Jefferies, 1968a; reproduced with permission of the British Museum [Natural History] and Zoological Society of London). por, postoral plate; or, oral plate; mo, mouth; M, marginal plates; V, ventral plates; ng, narrow groove, i.e., lateral line; cu rib, cuesta-shaped rib; bo, branchial openings (paired).

and left epicardia.) The *left pharynx* was strictly homologous with the pharynx of cornutes and had the same oblique disposition, running from anterior right to posterior left. The *right pharynx* was a new chamber which pouched out of the left pharynx to reach the right atrium. Both at the phylogenetic origin of the mitrates from cornutes, and in the ontogeny of each mitrate, the right pharynx pouched out beneath the *right anterior coelom* at X in Fig. 3C). The right anterior coelom was consequently lifted up and squashed against the roof of the head, while its cavity and contents were forced toward the mid line. As in cornutes, the *left anterior coelom* would have no cavity and would cover the left pharynx. The arguments for its existence are purely comparative.

The functioning of the pharynx of *Mitrocystites,* and mitrates in general, can be understood by comparison with tunicates. These feed, as do other primitive living chordates, by means of a mucus trap inside the pharynx. The mucus is secreted in tunicates by a mid-ventral endostyle and carried dorsally by right and left peripharyngeal bands which run dorsal-

ward from the front end of the endostyle. When the mucus reaches the dorsal mid line, it is rolled up into a mucous rope, together with entrapped food particles, by a fold of tissue called the dorsal lamina or finger-like dorsal languets. The mucus rope is pulled backward into the oesophagus, which opens posteriorly between right and left pharynxes.

In mitrates it is possible to identify the positions of the endostyle, the right and left peripharyngeal bands, and the oesophageal opening. It is also possible to locate the structures known as the retropharyngeal band, the opening of the neural gland or hypophysis, and connections between the right and left pharynxes and right and left epicardia (= posterior coelom) (Jefferies 1975, pp. 297ff; Jefferies and Lewis, 1978, pp. 249ff). The rectum of mitrates opened into the left atrium as in a tunicate tadpole.

The tail of *Mitrocystites* consisted of fore, mid, and hind portions; but these are homologous only with the fore tail of cornutes. In the early evolution of mitrates, the old cornute mid and hind portions of the tail seem to have dropped off. (In all calcichordates the end of the tail is abrupt, indicating that some such process of autotomy was a normal part of ontogeny). The hind tail contains big dorsal ossicles and paired ventral plates. The sculpture of the ventral surface of the ossicles reflects the presence of the notochord, with a dorsal nerve cord overlying it and paired segmental ganglia. The mid tail contained a massive dorsal styloid which acted as a socket for the muscles of the fore tail by which the mid and hind tail could be moved as a unit.

The brain and cranial nerves of *Mitrocystites* (Fig. 7) were complicated and fish-like. The brain lay between the tail and the head and was divided into prosencephalon and rhombencephalon like the primordial brain of a vertebrate embryo (Starck, 1975, p. 369). (Formerly the brain was seen as divided into anterior, medial and posterior parts, but the so-called anterior part probably represents a concentration of olfactory fibres before these connected with the brain proper.) The cranial nerves include an olfactory system, trigeminal complex, optic nerves, and an acustico-lateralis system.

The latest common ancestor of the three living chordate subphyla was probably a mitrate rather than a cornute, since all living adult chordates have right branchial slits as well as left ones, and right branchial slits are likely only to have evolved once. The stem group of vertebrates among mitrates probably included *Mitrocystites* which, unlike more primitive mitrates, possessed a lateral line. The stem group of amphioxus and its allies probably included

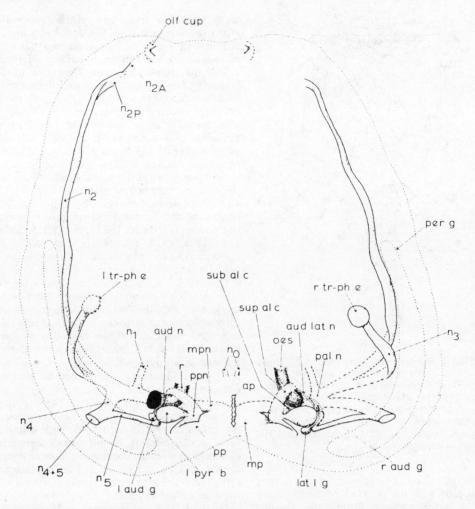

FIGURE 7. *Mitrocystites,* brain and cranial nerves in dorsal aspect (from Jefferies, 1975; reproduced by permission of the Zoological Society of London). ap, "anterior part of brain" = concentration of olfactory fibres; aud lat n, common auditory and lateral-line nerve of right side; and n, auditory nerve; lat 1 g, lateral-line ganglion; l aud g, left auditory ganglion; l pyr b, left pyriform body (left trigeminal ganglion); l tr-ph e, left transpharyngeal eye; mp, "medial" part of brain = prosencephalon; mpn, medial part nerves (bases of optic nerves); n_0, nerves n_0, (endostylar nerves); n_1, 1st nerves of palmar complex (mandibular trigeminal); n_2, 2nd nerves of palmar complex (maxillary trigeminal) divided anteriorly, on the left, into n_{2A} and n_{2P}; n_3, 3rd nerves of palmar complex (transpharyngeal optic nerves); n_4, $n_{4.5}$, n_5, 4th and 5th nerves of palmar complex to dorsal surface; oes, oesophagus; olf cup, olfactory cup, in buccal cavity; pal n, palmar nerve; per g, peripheral groove; pp, "posterior" part of brain = rhombencephalon; ppn, "posterior" part nerves = nerves from rhombencephalon; r, rectum; r aud g, right auditory ganglion; sub al c, subalimentary component of palmar nerve, going beneath rectum or oesophagus; sup al c, supraalimentary component of palmar nerve, going over rectum or oesophagus; r tr-ph e, right transpharyngeal eye.

the genus *Lagynocystis,* which, like amphioxus, possessed a median ventral atrium opening to the outside by way of a median ventral atriopore (Jefferies, 1973). The stem group of tunicates probably included the genus *Balanocystites,* which had dorsal gill openings like tunicates.

Vertebrate morphologists are divided into two schools. The "antisegmentationist" school, represented by A. Froriep, B. F. Kingsbury, A. S. Romer, and D. Starck, emphasizes the distinctness of the head from the tail, and of gill-slit segmentation from muscle-block segmentation. The "segmentationist" school, by contrast, regards the gill slits as having opened between the anterior muscle blocks of primitive vertebrates. Representatives of this school include F. M. Balfour, C. Gegenbaur, J. W. van

Wijhe, E. S. Goodrich (1930, ch. 5), G. R. de Beer, and H. Damas. At first sight, the existence of mitrates confirms the antisegmentationist view.

However, both schools have in common the belief, strongly supported by the condition in mitrates, that muscle blocks primitively extended on either side of the notochord up to its anterior end. Moreover, in all living vertebrates the visceral clefts are ventral to the notochord, and therefore to the usually vestigial, anterior somites, whereas in all mitrates the clefts were anterior both to somites and notochord. It is therefore possible that, in migrating ventral to the anterior somites of primitive vertebrates, each gill slit came to emerge just ventral to an intersomitic boundary. In that case, both the main schools of vertebrate morphology would be partly right. The hypothetical protovertebrate imagined by Froriep and Romer, which was very like a mitrate, would be ancestral to the hypothetical protovertebrate imagined by Goodrich, in which gill-slit segmentation had partly come to coincide with muscle-block segmentation. All living vertebrates would in that case be descended from Goodrich's animal. And, in fact, the classical head somites of vertebrate comparative anatomy can be identified in the mitrates (Jefferies and Lewis, 1978, pp. 267ff.).

R. P. S. JEFFERIES

References

Bone, Q., 1972. *The origin of chordates,* Oxford Biology Readers. London: Oxford Univ. Press.

Eaton, T. H., 1970. The stem-tail problem and the ancestry of chordates, *J. Paleontology,* 44, 969–979.

Gill, E. D., and Caster, K. E., 1960. Carpoid echinoderms from the Silurion and Devonian of Australia, *Bull. Amer. Paleontol.,* 41, 1–71.

Gislén, T., 1930. Affinities between the echinodermata, enteropneusta and chordonia, *Zool. Bijdr. Upps.,* 12, 199–304.

Goodrich, E. S., 1930. *Studies on the Structure and Development of Vertebrates.* London: Constable, 837p. (Reissued New York: Dover, 1958, 2 vols.)

Jefferies, R. P. S., 1967. Some fossil chordates with echinoderm affinities, *Symp. Zool. Soc. London,* 20, 163–208.

Jefferies, R. P. S., 1968a. The subphylum Calcichordata Jefferies 1967– Primitive fossil chordates with echinoderm affinities, *Bull. British Mus. (Nat. Hist.), Geol.,* 16, 243–339.

Jefferies, R. P. S., 1968b. Fossil chordates with echinoderm affinities, *Proc. Geol. Soc. London,* 1649, 128–140.

Jefferies, R. P. S., 1969. *Ceratocystis perneri* Jaekel– A Middle Cambrian chordate with echinoderm affinities, *Palaeontology,* 12, 494–535.

Jefferies, R. P. S., 1971. Some comments on the origin of chordates, *J. Paleontology,* 45, 910–912.

Jefferies, R. P. S., 1973. The Ordovician fossil *Lagynocystis pyramidalis* and the ancestry of amphioxus, *Phil. Trans. Roy. Soc. London, Ser. B,* 265, 409–469.

Jefferies, R. P. S., 1975. Fossil evidence concerning the origin of the chordates, *Symp. Zool. Soc. London,* 36, 253–318.

Jefferies, R. P. S., and Lewis, D. N., 1978. The English Silurian fossil *Placocystites forbesianus* and the ancestry of the vertebrates, *Phil. Trans. Roy. Soc. London, Ser. B,* 282, 205–323.

Jefferies, R. P. S., and Prokop, R. J., 1972. A new calcichordate from the Ordovician of Bohemia and its anatomy, adaptations and relationships, *Biol. J. Linnean Soc.,* 4, 69–115.

Matsumoto, H., 1929. Outline of a classification of the Echinodermata, *Sci. Rept. Tohoku. Univ. (Geol.),* 13, 27–33.

Parsley, R. L., and Caster, K. E., 1975. Zoological affinities and functional morphology of the Mitrata (Echinodermata), *Geol. Soc. Amer. Abstr. Prog.,* 7, 1225–1226.

Romer, A., 1972. The vertebrate as a dual animal–Somatic and visceral, *Evol. Biol.,* 6, 121–156.

Starck, D., 1975. *Embryologie.* Stuttgart: Thieme, 704p.

Ubaghs, G., 1968. Stylophora, in R. C. Moore, ed., *Treatise on Invertebrate Paleontology,* pt. 5, Echinodermata 1. Lawrence, Kansas: Geol. Soc. Amer. and Univ. Kansas Press, 3–60.

Ubaghs, G., 1971. Diversité et spécialisation des plus anciennes echinodermes que l'on connaisse, *Biol. Rev.,* 46, 157–200.

Ubaghs, G., 1975. Early Paleozoic echinoderms, *Ann. Rev. Earth Planetary Sci.,* 3, 79–98.

Cross-references: *Chordata; Echinodermata; Vertebrata; Vertebrate Paleontology.*

CALCISPHERES AND NANNOCONIDS

Calcispheres are a heterogeneous assemblage of organisms 400 μm or less in diameter, while nannoconids are an order of magnitude smaller. Both are usually examined in thin sections. The biologic relationship of these two groups of microfossils is unknown; thus, they are classed as Incertae Sedis. The nannoconids and the Mesozoic representatives of the calcispheres commonly occur together associated with planktic fossils and in such numbers as to suggest that they, too, were planktic. On the other hand, evidence does not support such a habitat for the Paleozoic calcispheres. Calcisphere-like objects range discontinuously from Devonian to Cretaceous and possibly to the Holocene. Nannoconids are stratigraphically more restricted, being found in rocks of Late Jurassic to Late Cretaceous age. The Mesozoic calcispheres and the nannoconids have proven biostratigraphic value primarily within the Tethyan realm. The Paleozoic-aged calcispheres appear to be geographically more widespread. Both types of organisms are found in a variety of fine-grained rock types.

Calcispheres

Calcispheres are spherical to ellipsoidal, single chambered, calcite bodies. The diameter ranges from 30 to 400 μm, and wall thickness ranges from 3 to 170 μm (Fig. 1A). The diameter of the chamber ranges from 20 to 200 μm. Although apparently quite simple, calcispheres are separated into several groups by wall structure, presence or absence of an aperture or canals, and cross-sectional outline (e.g., Villain, 1975; Masters and Scott, 1978).

Paleozoic calcispheres are quite different from Mesozoic forms and probably are not related. The exterior of Paleozoic calcispheres is either smooth or spiny. Wall microstructure of the smooth forms is a single micrite layer or consists of micrite layers alternating with layers of radial spar microlites. The wall of spiny forms consists of broad, symmetrically terminated calcite prisms (Fig. 1B). In some spiny species, narrow canals extend the length of the spines. Smooth species are either perforate or imperforate, but none are known to have an aperture. Most Paleozoic calcispheres have a circular outline.

Mesozoic calcispheres are without spines, and the calcite wall consists of radial or lamellar prisms, or granular microspar. In some specimens, a secondary micritic layer lines the chamber and a sparry layer is exterior (Fig. 1A). Scanning electron micrographs of the exterior surface show a reticulate or rhomboidal pattern of the wall, imparted by the ends of the calcite prisms. Many species possess a single aperture 5-20 μm wide. The aperture is either flush with the exterior surface or it is countersunk (Fig. 1C). Known Mesozoic species are imperforate. Test shape and outline are constant within a species. Some are spherical, others are ellipsoidal, and still others are heart shaped.

The biologic affinities of calcispheres are unknown. Several calcisphere species have subsequently been removed and identified as charophytes and foraminifers. Before Williamson (1880) created the genus *Calcisphaera*, workers thought the specimens to be radiolarians replaced by calcite. Currently, most

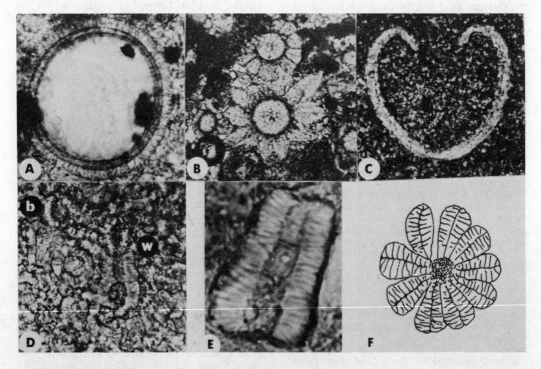

FIGURE 1. Calcispheres and Nannoconids. A, *Stomiosphaera* sp., X880, James Limestone, Aptian, Texas subsurface; B, radiosphaerid calcispheres, X100, Leduc Formation, Upper Devonian, Alberta (from Stanton, 1967); C, *Bonetocardiella* sp., X380, Buda Limestone, Cenomanian, Texas subsurface; D, *Nannoconus bucheri* Brönnimann, (b), and *N. wassalli* Brönnimann, (w), X865, Aptian limestone, Cuba; E. *Nannoconus planus* Stradner, X1700. James Limestone, Aptian, Texas subsurface; F, *Nannoconus steinmanni* Kamptner, ≈X900, Cretaceous Limestone, San Vincente, Mexico (from Trejo, 1960).

workers consider calcispheres to be algae or protists. The algal hypothesis is based upon analogy with spherical reproductive cysts of dasycladacean algae and upon the occurrence of calcispheres in lagoonal carbonates (Rupp, 1966). The ciliate protist hypothesis is based upon a morphologic analysis of apertures and exterior ornamentation (Banner, 1972). Other workers have classed calcispheres with Foraminiferida, Flagellata, and Porifera, or as calcified gas bubbles. The consistent prismatic microstructure, aperture, and external ornamentation all support the organic origin of calcispheres. Approximately 10 genera and 3 families have been defined, some of which may be synonyms. The systematic framework is far from complete, and several genera have not yet been placed within a family taxon.

Paleozoic calcispheres range from Devonian to Permian. Rocks older than Devonian contain poorly preserved calcisphere-like objects, which may ultimately extend the range of this group. Mesozoic forms range from Lower Jurassic to Upper Cretaceous. Several calcisphere zones aid the correlation of the Cretaceous (e.g., Bonet, 1956). The range of calcispheres would extend to the Holocene if living algal cysts were classed with these forms.

The common depositional environment of Paleozoic calcispheres is the nearshore shallow, carbonate shelf lagoon. They are most abundant in fossiliferous, pelletal, and laminated limestone; associated fossils are foraminifers, brachiopods, gastropods, pelmatozoans, bryozoans, and girvanellid and dasyclad algae. However, calcispheres may range into open-shelf and back-reef deposits. They are not yet reported from basinal deposits (Stanton, 1967). Cretaceous calcispheres are common in shallow-water, back-reef carbonates, and range into shallow, epicontinental basins on the one hand, and into deep-water, shelf-margin to slope deposits associated with planktic foraminifera and radiolaria, and benthic echinoids, bivalves, and ostracodes on the other hand.

Nannoconids

Nannoconus are ultramicroscopic organisms ranging from 3 to 55 μm in length and 3 to 18.5 μm in width. The external longitudinal outline varies from rectangular to subcircular and triangular to pear shaped, while the cross-sectional outline is nearly always circular (Fig. 1D). The shape of the axial cavity generally conforms to the test outline (Fig. 1E). This cavity forms an aperture at each pole. The test wall is composed of calcite microlites, measuring 1–2 μm in length and 0.5–1 μm in diameter, that may be arranged spirally around

the axis and are normal to the test wall. Although nannoconids typically occur as isolated specimens, several individuals may be grouped into rosettes (Fig. 1F).

The first published illustrations of *Nannoconus* were by J. de Lapparant in 1925. He regarded them as an embryonic growth stage of the unilocular foraminifer *Lagena*. Subsequently, these organisms have been variously regarded as inorganic or as some type of alga, perhaps related to the coccoliths or to the charaphytes. None of these suggested affinities is satisfactory. The organic origin of *Nannoconus* is no longer questioned, but the lack of living descendants prevents accurate biological assignment.

The Family Nannoconidae Deflandre, 1959, is represented by the single genus, *Nannoconus* Kamptner, 1931. To date, 18 species have been described. These are based upon the external outline and the shape of the axial canal of isolated individuals.

Species of *Nannoconus* have been recognized in rocks dating from Late Jurassic (Tithonian) to Late Cretaceous (Campanian). Most species appear to have a relatively short stratigraphic range. This, no doubt, is partially due to insufficient study. Brönnimann (1955) was first to propose a zonal scheme using three assemblages of nannoconids to subdivide Lower Cretaceous rocks. Subsequently, Trejo (1960) and others have devised zones using *Nannoconus*. Those species found in the Jurassic and Lower Cretaceous are particularly useful because, during this time, the planktonic foraminifers are least diverse.

Nannoconus frequently occur in limestones and calcareous shales. They can, in fact, be present in such numbers as to be the basic building component of these rocks, to which the adjective "nannoconid" could be appropriately applied (Fig. 1D).

The genus is found primarily in nearshore and shelf environments, but may be associated with deeper water faunal elements. The wide Tethyan distribution, high numbers of individuals, and common association with planktic foraminifers and tintinnids suggest that *Nannoconus* is also planktic over at least a portion of its life history.

B. A. MASTERS
R. W. SCOTT

References

Banner, F. T., 1972. *Pithonella ovalis* from the early Cenomanian of England, *Micropaleontology*, 18, 278–284.

Bonet, F., 1956. Zonificación microfaunística de las calizas cretácicas del este de México, *Bol. Asoc. Mexicana Geol. Petróleros*, 8, 389–488.

Brönnimann, P., 1955. Microfossils incertae sedis from the Upper Jurassic and Lower Cretaceous of Cuba, *Micropaleontology,* **1,** 28–51.

Masters, B.A., and Scott, R. W., 1978. Microstructure, affinities and systematics of cretaceous calcispheres, *Micropaleontology,* **13,** 210–221.

Rupp, A. W., 1966. Origin, structure, and environmental significance of Recent and fossil calcispheres, *Geol. Soc. Amer. Abstr. Prog.,* **186.**

Stanton, R. J., 1967. Radiosphaerid calcispheres in North America and remarks on calcisphere classification, *Micropaleontology,* **13,** 465–472.

Trejo, M., 1960. La Familia Nannoconidae y su alcance estratigrafico en America. (Protozoa, Incertae Saedis), *Bol. Asoc. Mexicana Geol. Petróleros,* **12,** 259–314.

Villain, J.-M., 1975. "Calcisphaerulidae" (incertae sedis) du Crétacé Supérieur du Limbourg (Pays-Bays), et d'autres regions, *Palaeontographica,* A, 149, 193–242.

Villain, J.-M., 1977. Les Calcisphaerulidae: Arcitectures, calcification de la paroi et phylogenese. *Palaeontographica,* A, 159, 137–179.

Williamson, W. C., 1880. On the organization of the Fossil Plants of the Coal-Measures, pt. 10, Including an Examination of the supposed Radiolarians of the Carboniferous rocks, *Phil. Trans. Roy. Soc. London,* 171(2), 493–539.

Cross-references: *Algae; Protista.*

CEPHALOPODA

Mollusca of the class Cephalopoda are diverse, predacious marine carnivores that are judged to have achieved degrees of structural complexity and metabolic efficiency unmatched by any other group of unsegmented invertebrates. They are generally characterized by bilateral symmetry, a well-defined head, an undivided mantle, and by differentiation of the molluscan foot into both a cluster of tentacles surrounding the mouth and a muscular funnel beneath. Rhythmic ejection of water from the mantle cavity through the funnel, or hyponome, powers backward locomotion by jet propulsion, making most cephalopods agile rapid swimmers. Living representatives include the squid, cuttlefish, octopus, and the pearly *Nautilus*—the latter a solitary relic of the host of externally shelled cephalopods that dominated Phanerozoic seas. Cephalopods have been among the largest and most abundant invertebrates for one-half billion years, from the Early Ordovician to the present. Appearing first in the Late Cambrian (Trempealeauan), the Cephalopoda diversified spectacularly in the Ordovician, where their gradually tapered conical shells approached a length of 10 m. Despite several periods of crisis during which major subgroups disappeared and the class approached extinction, the cephalopods remain common to the present with about 650 living species. Modern representatives, like their Ordovician ancestors, include the largest living invertebrate, the giant squid *Architeuthis* with a length up to 17 m. Many cephalopods secreted robust mineralized skeletons that were readily fossilized. and details of shell morphology permit confident interpretation of evolutionary lineages. The paleontological record is voluminous, especially in Paleozoic and Mesozoic strata, and cephalopods constitute one of the most valuable references for relative age determination. In total there have been in excess of 10,000 species, now assigned to 400 families and 25 orders.

Taxonomy

Living cephalopods are separable into two numerically unequal groups. One includes only the single genus *Nautilus* (Fig. 1), which is characterized by two pairs of gills (ctenidia), numerous sheathed cirrate tentacles, and a chambered external shell. By contrast, all remaining living cephalopods (e.g., Fig. 2) have a single pair of gills, eight or ten tentacles, and a shell which is internal and commonly vestigial or absent. These differences were first noted by Richard Owen in 1832 who, in reference to the gills, proposed Tetrabranchiata and Dibranchiata as the two major subdivisions of the class. It has become customary to include all externally shelled fossil cephalopods with *Nautilus* in the Tetrabranchiata, generally with subdivision into Nautiloidea and Ammonoidea. Similarly, those fossil forms with an internal skeleton have been classified as Dibranchiata. However attractive this subdivision may appear for living cephalopods, assignment of fossils into one of these categories must be based on tenuous assumptions. The terms Ectocochlia and Endocochlia were introduced by W. E. Leach in 1894 as direct synonyms for Tetrabranchiata and Dibranchiata. They refer to the location of the shell with respect to soft parts, but fail to reflect the biological groupings considered most plausible by modern students of the Cephalopoda. Recently, additional major taxonomic categories have been differentiated, and most students now accept some variation of the arrangement that follows. However, the older names are still useful for purely descriptive purposes. The term Coleoidea is synonymous with Dibranchiata-Endocochia, and the remaining six subclasses in Table 1 refer to Tetrabranchiata-Ectocochlia.

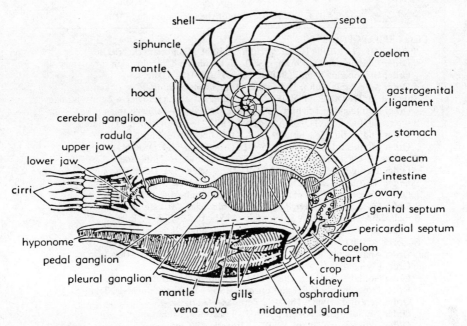

FIGURE 1. Median longitudinal section of *Nautilus* showing anatomical features of an ectocochliate cephalopod, X0.5 (from Sweet *in* Moore, 1964; after Naef, 1928; courtesy of The Geological Society of America and University of Kansas Press).

Morphology

Principal anatomical features of the two main groups of living cephalopods are portrayed by Figs. 1 and 2. The body is bilaterally symmetrical, with a well-developed anterior head, U-shaped digestive tract, and fleshy mantle that surrounds the viscera and defines a ventral mantle cavity in which the gills are suspended. *Nautilus* has a robust chambered external shell, whereas the shell of *Sepia* is chalky and internal. The vast majority of cephalopod fossils resemble *Nautilus* in possession of a chambered external shell. However, internal shells are sporadically abundant from strata as old as the Late Mississippian Chesteran Stage (Flower and Gordon, 1959) and most steps in the progressive modification of the ancestral external shell to form the varied internal structures of the modern coleoids are preserved in the geological record. The shell has become vestigial or absent in the Octopoda, although the female of *Argonauta* secretes a calcareous unchambered external egg case that is similar to but not

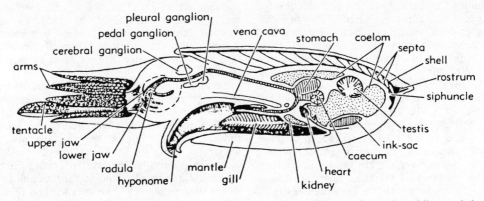

FIGURE 2. Median longitudinal section of *Sepia* showing anatomical features of an endocochliate cephalopod, X0.5 (from Sweet *in* Moore, 1964; after Naef, 1928; courtesy of The Geological Society of America and University of Kansas Press).

171

TABLE 1. Classification of the Cephalopoda

Class CEPHALOPODA
  Subclass ORTHOCERATOIDEA Kuhn, 1940–U. Cambrian–U. Triassic
    Order Ellesmerocerida Flower, 1950–U. Cambrian–U. Ordovician
    Order Orthocerida Kuhn, 1940–L. Ordovician–U. Triassic
    Order Ascocerida Kuhn, 1949–Mid. Ordovician–U. Silurian
  Subclass ACTINOCERATOIDEA Teichert, 1933–Mid. Ordovician–Pennsylvanian
    Order Actinocerida Teichert, 1933–Mid. Ordovician–Pennsylvanian
  Subclass ENDOCERATOIDEA Teichert, 1933–L. Ordovician–Mid. Silurian
    Order Endocerida Teichert, 1933–L. Ordovician–Mid. Silurian
    Order Intejocerida Balashov, 1960–L. Ordovician–Mid. Ordovician
  Subclass NAUTILOIDEA Agassiz, 1847–L. Ordovician–Holocene
    Order Tarphycerida Flower, 1950–L. Ordovician–Mid. Devonian
    Order Nautilida Agassiz, 1847–L. Devonian–Holocene
    Order Discosorida Flower, 1950–Mid. Ordovician–Mid. Devonian
    Order Oncocerida Flower, 1950–Mid. Ordovician–Mississippian
  Subclass BACTRITOIDEA Shimanskiy, 1951–U. Silurian–U. Triassic
    Order Bactritida Shimanskiy, 1951–U. Silurian–U. Triassic
  Subclass AMMONOIDEA Agassiz, 1847–L. Devonian–U. Cretaceous
    Order Anarcestida Miller and Furnish, 1954–L. Devonian–U. Devonian
    Order Clymeniida Hyatt, 1884–U. Devonian
    Order Goniatitida Hyatt, 1884–Mid. Devonian–U. Permian
    Order Prolecanitida Miller and Furnish, 1954–L. Mississippian–U. Triassic
    Order Ceratitida Hyatt, 1884–L. Permian–U. Triassic
    Order Phylloceratida Arkell, 1950–L. Triassic–U. Cretaceous
    Order Lytoceratida Hyatt, 1889–L. Jurassic–U. Cretaceous
    Order Ammonitida Zittel, 1884–L. Jurassic–U. Cretaceous
  Subclass COLEOIDEA Bather, 1888–Mississippian–Holocene
    Order Aulacocerida Jeletzky, 1965–Mississippian–U. Jurassic
    Order Belemnitida Zittel, 1885–L. Jurassic–Eocene
    Order Phragmoteuthida Jeletzky, 1965–U. Permian–U. Triassic
    Order Teuthida Naef, 1916–L. Jurassic–Holocene
    Order Sepiida Naef, 1916–U. Cretaceous–Holocene
    Order Octopoda Leach, 1818–?Triassic–Holocene

homologous with the shell of the ectocochliates.

The head bears two conspicuous eyes and a mouth surrounded by eight or more tentacles that commonly are contractile, and in all but *Nautilus* are provided with suckers. The number of tentacles is of high taxonomic value for living cephalopods: all Octopoda have eight, remaining Coleoidea have ten, and *Nautilus* has ten times that number. Coleoid eyes are highly complex and similar in both structure and development to those of the vertebrates. However, the eye of *Nautilus* is a simple water-filled pit which has a retina but lacks eyelids, iris, cornea, and lens. In *Nautilus,* the dorsal surface of the head is covered by a tough but flexible hood which can be positioned to close the shell opening. Comparable structures occurred in some ammonoids for which single or paired calcareous or horny plates, termed aptychi (see *Aptychus*), are known to conform to the contours of the shell aperture. Recent studies (Lehmann, 1970) have demonstrated that at least some of the unpaired horny plates (anaptychi) interpreted previously as operculae are in fact flattened beaks.

The digestive tract consists of a two-jawed parrot-like beak (rhyncholite) followed by a multitoothed radula into which two pairs of salivary glands discharge; a long esophagus; and a large stomach and associated caecum, intestine, and anus. Food is seized by the tentacles, and large pieces are bitten off by the beak and pushed down the esophagus by a rasping action of the radula. Extracellular digestion is accomplished by enzymes discharged into the caecum. Indigestible refuse passes down the intestine, through the anus into the mantle cavity, and is ejected by way of the hyponome.

An undivided mantle, internal in *Nautilus* but external in other living cephalopods, envelops all but the head region. In the living coleoids, it has thick layers of both circular and longitudinal muscles; but in *Nautilus,* the mantle lies inside the rigid shell and is only weakly muscular. Ventrally, the mantle forms a cavity which lodges the gills, four in *Nautilus* and two in the living coleoids. The anterior end of the mantle cavity is partially blocked by the muscular funnel-shaped hyponome. Sea water is drawn into the mantle cavity either through the hyponome or around its sides and base, by action of the muscles in the hyponome or those of the mantle. Discharge of water is through the hyponome and results from foreceful rhythmic contraction of the circular mantle muscles. This jet propulsion provides the main means of locomotion, although mantle fins or tentacles may serve a subsidiary role in the coleoids. Maneuverability is afforded by flexibility of the hyponome, and the ability for violent expulsion of sea water places the squids among the fastest of marine organisms.

The circulatory system of coleoids is of the closed type, but in *Nautilus* it is partially lacunar as in most other Mollusca. All blood contains a blue respiratory pigment, haemocyanin. The heart has a median ventricle and as many auricles as there are gills. Blood is pumped by the ventricle through aortas that branch to form several arterial complexes. In the coleoids, it returns to the ventricle through a closed venous system; from a branchial heart at the base of each gill, it enters the gill capillaries before being sucked into one of the auricles. In *Nautilus,* returning blood is not confined to closed vessels but trickles back to the haemocoel through interconnected sinuses then passes through the vena cava to the gills.

The nervous system of living cephalopods is exceptionally well developed, being centralized in the head to form a massive circumesophageal brain within a cartilaginous cephalic capsule. Squids have the largest of all known nerve fibers and thus serve significantly in medical research.

Coleoids are extraordinary in the rapidity with which they are able to achieve spectacular color changes. This ability to assume effective protective coloration even extends to the faithful replication of sharp geometric patterns. Color changes are effected by contraction or expansion of pigment-charged sacs, chromatophores, which overlie the pearly iridocysts.

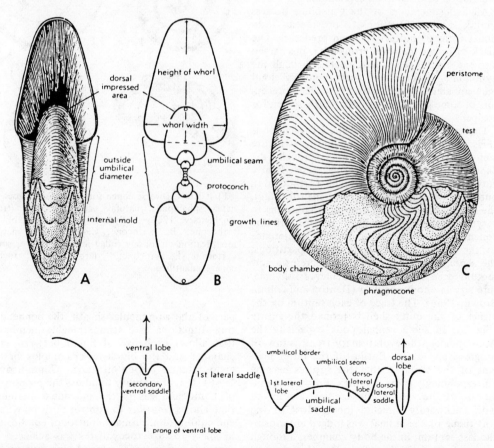

FIGURE 3. Views of a characteristic Devonian ammonoid, *Manticoceras,* X1 (from Miller and Furnish *in* Moore, 1957; courtesy of The Geological Society of America and University of Kansas Press). A, ventral view; B, cross-sectional view; C, lateral view, D, enlarged representation of a suture of the same. The upper parts of A and C portray the exterior of the shell and show the growth increments, whereas the lower parts represent the internal mold with the sutures.

The chromatophores expand to display their individual colors, and are nearly invisible when contracted. *Nautilus* lacks the ability to change color, but displays permanent protective camouflage. The hood is brown (burnt orange) with while mottling, while the lateral and dorsal margins of the shell are characterized by prominent transverse brown bands against a white background. Only the mature body chamber is uniformly white, without pattern. There are minor interspecific differences in coloration, but patterns are consistent within each species. Viewed from above, the hood and shell of *Nautilus* tend to blend with the tropical sea floor, whereas the mature body chamber merges with the ocean surface if viewed from below. Traces of color banding in fossil ectocochliates are preserved sporadically as bands of light and darker grey back to the Ordovician (Foerste, 1930). Concentration of color banding on the upper surface may provide a clue to life-orientation of these fossil cephalopods.

All living members of the Coleoidea possess an ink sac that functions to provide an additional form of protection from predators. The black or brown melaniniferous ink is discharged into the mantle cavity and ejected through the hyponome. It spreads to form a murky cloud that apparently anesthetizes the chemoreceptors of some predators and provides a screen for escape.

The conch of the ectocochliate cephalopods (Figs. 1,3,4) is a robust aragonitic structure comprising a variously modified gradually expanded cone. Ancestral forms were slightly curved (cyrtoconic) or straight (orthoconic); but most descendants are coiled (nautiliconic), the outer coils covering preceding volutions to varying degrees. The apical portion of the conch is termed the phragmocone in reference to the numerous shelly transverse partitions, the septa. Apparently, *Nautilus* secretes a new septum each fortnight and achieves fully mature size in about one year (Denton and Gilpin-Brown, 1966). The trace of each septum on the inside of the outer shell is termed the suture (Fig. 3). In some cephalopods, especially the Ammonoidea, the evolutionary trend is toward progressively greater fluting of the outer margins of the septa so that the sutures became correspondingly more complex. Forward crenulations of the suture, the saddles, are linked with backwardly directed lobes. Most of the soft tissue of the animal was lodged anterior to the last septum, in the body chamber, although a specialized derivative of the mantle extended back through each successive septum to the initial chamber or protoconch (Fig. 5). This backward prolongation and the sheath that housed it comprise the siphuncle. In *Nautilus,*

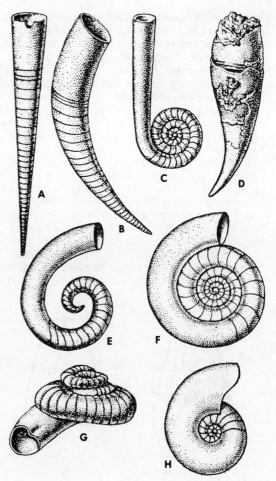

FIGURE 4. Diverse conch forms of fossil ectocochliate cephalopods, natural size or reduced (from Ruzhentsev, 1962; partly after Flower, 1964b). A, orthocone; B, cyrtocone; C, orthoconic mature modification on loosely coiled juvenile; D, breviconic cyrtocone; E, gyrocone; F, tarphycone; G, trochocone; H, nautilicone.

part of the siphuncular sheath, the connecting ring, functions as a semipermeable membrane that allows exchange of liquid between each chamber and the blood-vessel complex of the endosiphuncle (Denton and Gilpin-Brown, 1966). It is this ability to adjust the proportion of liquid and gas in the individual chambers that allows *Nautilus* to control both buoyancy and equilibrium. Primary control of equilibrium in many fossil ectocochliates was achieved by secretion of layered mineral deposits in the chambers and in the siphuncle (Figs. 6,7), although even these forms certainly made short-term adjustments by varying the fluid content of the chambers. The aragonitic cameral and

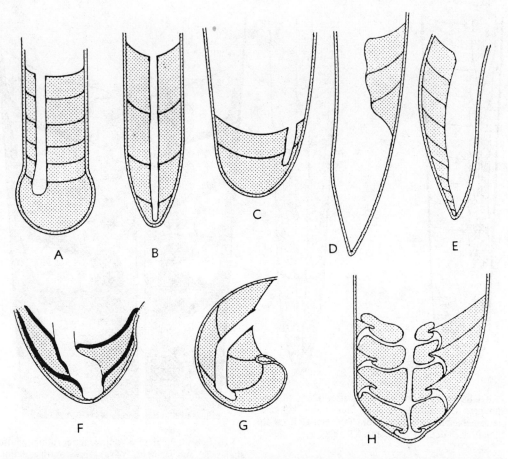

FIGURE 5. Initial parts of the conch (camerae and siphuncular deposits stippled) in several ectocochliate cephalopod orders, not to scale (from Teichert *in* Moore, 1957; courtesy of The Geological Society of America and University of Kansas Press). A,B, Orthocerida; C, Oncocerida; D,E, Endocerida; F, Nautilida; G, Tarphycerida; H, Actinocerida.

siphuncular deposits were secreted in a regular ontogenetic progression, with the most voluminous masses appearing apically and ventrally. Commonly, the anterior chambers and siphuncle are devoid of primary deposits; but progressively greater volumes appear adapically until the apical chambers and adjacent siphuncle are filled completely. It was largely this regular progression that led to general acceptance of the hypothesis that these mineral deposits are primary, secreted by the living animal (Teichert, 1933). Current practice is to credit the form of both the cameral and siphuncular deposits with major taxonomic significance.

Most living coleoids are denser than sea water and must maintain themselves by active swimming. However, *Sepia* and *Spirula* resemble *Nautilus* in possession of rigid shell structures for storage of gas, and like it they are approximately weightless in water (Denton and Gilpin-Brown, 1973; Denton, 1974). *Sepia* maintains effective control of buoyancy by pumping liquid into and from the chambers of a highly modified ectocochliate phragmocone (Denton and Gilpin-Brown, 1961). The phragmocone of *Spirula* (Fig. 8A-D) more closely resembles those of ancestral ectocochliates, and buoyancy control is basically similar to that in *Nautilus* (Denton and Gilpin-Brown, 1971). All three genera probably adjust buoyancy to aid in vertically extensive diurnal migration.

The hard parts of Coleoidea are internal (Figs. 2,8). They represent varying degrees of modification of the ectocochliate shell, having evolved from the Bactritida. The Aulacocerida and Belemnitida arose independently (Jeletzky, 1966), but shells of both display a phragmocone enveloped posteriorly by a solid calcareous guard (Fig. 8E-G).

The sexes are separate in living cephalopods,

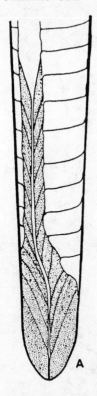 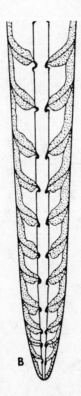

FIGURE 6. Progressive ontogenetic filling of the siphuncle (A, endocerid; after Flower, 1964b) and the camerae (B, orthocerid; from Teichert, 1933) with primary shell deposits stippled), X1.

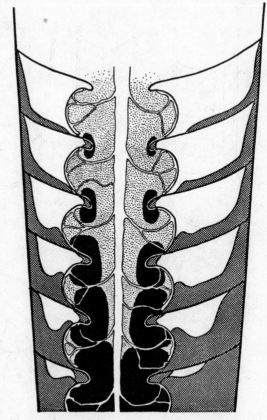

FIGURE 7. Progressive ontogenetic filling of the siphuncle and camerae of a representative actinocerid cephalopod with primary shell material (stippled), X.08 (from Teichert, 1933).

and genitalia are highly developed. Sexual dimorphism may be conspicuous, and in fossil ectocochliates it is commonly reflected as differences in size and general form. In living cephalopods, the males usually possess arms modified for transfer of spermatophores to the mantle cavity or suboral seminal receptacles of females. These copulatory organs are permanent in *Nautilus,* but they undergo seasonal modification in the coleoids. Eggs are either released free in the water or are attached to some solid object. In some cases, such as the *Octopus,* eggs are brooded by the female. Ontogenetic development differs from that of other mollusks, proceeding directly to the adult without the typical intermediate molluscan trochophore or veliger larval stages. Ectocochliates generally retain all growth stages of the shell, permitting detailed reconstruction of the full ontogeny. Many undergo marked modification of the conch as maturity is approached and cease growth at a finite size. Mature modifications include change in curvature (Fig. 4C), whorl section, shell thickness, ornamentation and contours of the aperture.

Ecology and Diet

Endocochliates are ubiquitous in modern oceans where they are known to occur in sporadic abundance. They form a significant component of human diet, especially in the Mediterranean and E Asia, and represent an important link in the oceanic food chain. The overall ecology of the group is poorly known, mainly because most representatives live at bathyal depths during the day. Extensive daily vertical migration is common, as is seasonal geographic migration. Vast shoals occur in depths exceeding 1000 m; and similar abundance is recorded in shallow waters, especially at night. Individuals feed voraciously upon fish and crustaceans, and are ferocious cannibals under some conditions. There is no compelling evidence that ancient coleoids differed significantly from their living descendants in ecological provenance.

Living *Nautilus* is largely restricted to the

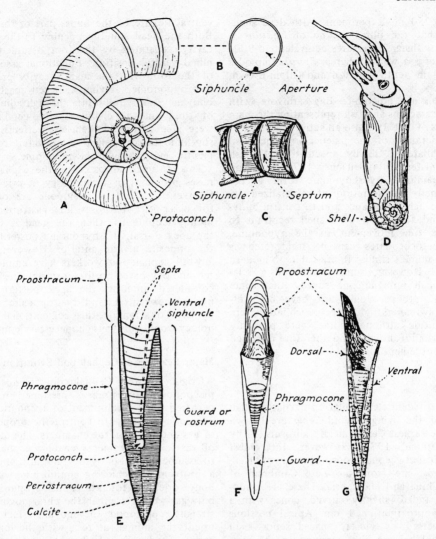

FIGURE 8. Representative living and fossil endocochliate cephalopods (from Shrock and Twenhofel, 1953; copyright © 1953 by McGraw-Hill and used by permission of McGraw-Hill Book Co.). A–C enlarged, D–G X1. A–D, *Spirula,* a living sepiid; E–G, representative Jurrassic belemnitids.

SW Pacific, although rarer occurrences are known from parts of the adjacent Indian Ocean and around the entire coast of Australia. Determination of the limits of distribution is complicated by extensive postmortem dispersal of empty shells. Drift specimens are found sporadically from S Africa to Japan, indicating current transportation approaching 10,000 km following putrefaction and loss of soft tissues. Comparable dispersal may be anticipated for many fossil ectocochliates (Reyment, 1958), especially forms such as the Ammonoidea and those Nautiloidea that lacked voluminous mineral deposits in the chambers and siphuncle. However, geological occurrence of even some of these named groups suggests only limited dispersal after death.

Live juvenile *Nautilus* occurs, although rarely observed, in water depths of a few meters (Davis and Mohorter, 1973); but when larger specimens are captured from water less than 50 m deep they are usually wounded or diseased. During daylight hours, healthy large specimens are at depths greater than 100 m; but they are actively diurnal, rising to as little as 50 m at night (Haven, 1972) or perhaps shallower (Denton and Gilpin-Brown, 1966). Baited traps are most productive if set between 80 and 180 m; *Nautilus* certainly occurs at greater depths, but logistic difficulties preclude fishing beyond 250 m. Experiments with dead shells reveal that the phragmocone of *Nautilus* is able to withstand pressures equivalent to 700–900 m of sea water, whereas *Spirula* implodes at 1700 m depth (Denton and Gilpin-Brown, 1971).

Nautilus is a bottom-feeding carnivore, with small crustaceans and fish representing the main elements of diet. It is also an active scavenger.

Consideration of the paleoecology of fossil ectocochliates is highly speculative (Furnish and Glenister, 1964; Seilacher, 1975). Based on the streamlined hydrodynamic form of the outer shell, the secretion of mineralized deposits to achieve balance and to control buoyancy, and formation of a shell reentrant to facilitate free movement of the hyponome, many ectocochliates can be interpreted as facile swimmers similar in life habits to modern coleoids. However, squat form and a relatively small phragmocone suggest that some Paleozoic groups would not have been effective swimmers. Most ectocochliates were nektobenthic although some were probably only partially buoyant and therefore confined to the motile benthos.

Ancestry

Bands of denticles similar to but larger than those of the living coleoid *Sepia* are known from the earliest Cambrian of California (Firby and Durham, 1974). However, the oldest authenticated cephalopods, species of *Plectronoceras,* are from the Late Cambrian of North China and Manchuria. These are gradually expanded, slightly curved, cones with a length approximating 1 cm. Apical portions are traversed by closely spaced septa, shell partitions that are also common in the molluscan classes Gastropoda and Monoplacophora. However, *Plectronoceras* and all other ancestral cephalopods are unique in possession of a tube, the siphuncle, that penetrates each successive septum to permit communication between the apical septate portion of the shell and the basally situated body chamber. Development of this siphuncle forms the point of evolutionary divergence of the Cephalopoda, probably from a high-spired septate monoplacophoran (Yochelson et al., 1973). The siphuncle functioned to adjust the proportions of gas and liquid in the chambers, allowing the cephalopod to control both buoyancy and equilibrium. Facile swimming ability followed, opening an entirely new adaptive realm for spectacular Early Ordovician increase in individual size, diversification, and proliferation.

Elongation of the body chamber of the ancestral cephalopods necessitated pronounced ventral flexure of the upper part of the body. Both the head and the opening to the mantle cavity, separated by the foot, assumed a terminal ventral position. Functional association of the foot with the mantle cavity produced the hyponome. Having achieved mastery of buoyancy and equilibrium through adjustment of fluids in the chambers, the cephalopods were able to propel themselves effectively by rhythmic forceful ejection of water from the mantle cavity through the hyponome.

This assumed ancestry of the cephalopods raises problems in terminology of orientation (Mutvei, 1957). If cephalopods are oriented for descriptive purposes like gastropods and monoplacophorans, then the head is ventral, the apex dorsal, the hyponome posterior, and the opposite margin anterior. However, virtually all cephalopod workers have employed a terminology based on life position. The head is considered anterior, the apex posterior, the hyponome ventral, and the opposite margin dorsal. To scramble this conventional terminology would only cause unnecessary confusion.

Major Features of Cephalopod Evolution

Origin of the Cephalopoda from monoplacophoran ancestors is presumed to have occurred with development of a siphuncle and consequent ability to control the proportions of gas and liquid in the chambers. The presence of gas in these camerae enabled cephalopods to assume neutral buoyancy, but also presented a problem in achieving equilibrium with the body oriented for horizontal propulsion. If only gas were present in the phragmocone of a straight cephalopod, the buoyant effect would orient the animal, at rest, with the long axis vertical and part of the phragmocone projecting above water. The evolution of the cephalopods may be viewed, in broad terms, as a series of responses to this problem of equilibrium. Many fundamentally different strategies were developed to achieve hydrostatic equilibrium with the anterior part of the body chamber, and hence the axis of the hyponome, in approximately horizontal orientation (Furnish and Glenister, 1964). These include: coiling of the entire conch or a major part of it, secretion of various types of mineralized cameral or siphuncular deposits, development of a heavy guard on the outside of the phragmocone, ontogenetic shell truncation, formation of a squat shell, and differential secretion of fluids in the chambers. The morphology and evolution of each of the main groups of cephalopods is reviewed beneath in terms of these various responses to the need for hydrostatic and hydrodynamic equilibrium.

Subclass Orthoceratoidea. Members of the subclass Orthoceratoidea range widely in gross form of the conch, from squat (breviconic) to elongate (longiconic) and curved (cyrtoconic) to straight (orthoconic). Most are less than a few tens of centimeters long, and all possess a siphuncle that is narrow in proportion to the corresponding conch diameter. Ancestral orthoceratoids are referred to the Order Ellesmerocerida (Upper Cambrian–Upper Ordovician). Characteristically, they are brevicones with closely spaced septa, and a body chamber that is large in relation to the phragmocone. Mineralized deposits are generally absent from the chambers and siphuncle. Most representatives are poorly streamlined, and overall conch morphology suggests that they may not have been fully buoyant. They are best considered as benthic or nektobenthic in habit. Most other cephalopod orders evolved from the Ellesmerocerida during the Early Ordovician period of eruptive diversification (Teichert, 1967).

Representatives of the Order Orthocerida (Lower Ordovician–Upper Triassic) are moderately large orthoconic longicones with voluminous cameral and siphuncular deposits (Fig. 5A,B, 6B). Concentration of these deposits in ventral and posterior positions suggests hydrostatic equilibrium with horizontal conch orientation. Streamlined hydrodynamic form confirms that orthocerids were active swimmers.

The Ascocerida (Middle Ordovician–Upper Silurian) are small, bizarre, rare orthoceratoids that underwent periodic truncation of the apical portion of the conch as a normal part of growth (Fig. 9). The shell is thin, and deposits are absent from the chambers. In the mature conch (Fig. 9C), the chambers were concentrated dorsally, and the aperture was modified to provide reentrants for the eyes and the hyponome. All features indicate that mature ascocerids were facile swimmers.

Subclass Actinoceratoidea. The single order, Actinocerida (Middle Ordovician–Pennsylvanian), comprises medium to large orthocones in which the siphuncle may occupy virtually all of the initial few chambers of the phragmocone (Fig. 5H) and commonly reaches one-third the corresponding conch diameter in succeeding chambers (Fig. 7). Voluminous layered mineral

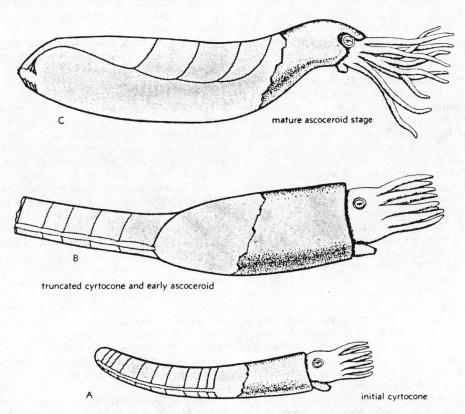

FIGURE 9. Ontogenetic series of a Silurian ascocerid, *Glossoceras*, showing the shell truncation and mature modifications that characterize the order, X2.5 (from Furnish and Glenister, 1964; courtesy of The Geological Society of America and University of Kansas Press). Apical portions are shown as in median section; soft parts are schematic.

179

deposits formed successively from the apex in both the camerae and the siphuncle. They may fill the phragmocone posteriorly, with the exception of a system of central and radial canals which is presumed to have housed a complex of blood vessels. Distribution of deposits within the phragmocone indicates achievement of hydrostatic equilibrium with the conch in horizontal orientation and suggests that most actinocerids were nektic.

Subclass Endoceratoidea. For practical purposes the subclass Endoceratoidea comprises the abundant large Ordovician orthocones of the Order Endocerida. However, a few representatives of that order occur in the Middle Silurian, and rare Ordovician endoceratoids with radial blades in the siphuncle are referred to the Order Intejocerida. A distinctive feature of the Endocerida is the possession of an exceptionally large ventral marginal siphuncle that may occupy virtually all of the phragmocone apically and commonly is half the corresponding conch diameter in subsequent growth stages (Figs. 5D,E, 6A). The order evolved from the Ellesmerocerida by progressive enlargement of the siphuncle and by lengthening of the septal necks, the backward prolongations of the septa where they meet the siphuncle. Size of the siphuncle is such that it must have housed substantial parts of the viscera, not just the blood-vessel complex found in *Nautilus* and assumed for most other cephalopods. The siphuncle was filled progressively with distinctive conical sheets of aragonite, the endocones. These provided balance which would have enhanced swimming ability. No deposits occur in the chambers; the bulk of the endocones rendered them unnecessary.

Subclass Nautiloidea. Morphologically diverse coiled cephalopods (nautilicones), brevicones, and cyrtocones ranging from Lower Ordovician to Holocene comprise the Nautiloidea. Most are of intermediate size, have a proportionally small siphuncle, and with some notable exceptions lack voluminous cameral deposits. The ancestral representatives, Order Tarphycerida (Lower Ordovician–Middle Devonian), evolved from the Ellesmerocerida by progressive coiling of the shell (Fig. 5G). Cameral deposits are rare in the Tarphycerida, and for the majority of its species buoyancy control must have been primarily through differential secretion of liquid in the chambers. The common presence of a hyponomic sinus confirms that they were swimmers, although agility and speed probably varied greatly. Members of the Order Nautilida resemble the tarphycerids but display tighter coiling and achieved larger size (Figs. 1,5F). General similarity to *Nautilus* permits the assumption of a comparable mode of life for fossil forms. Both the Discosorida (Middle Ordovician–Middle Devonian) and the Oncocerida (Middle Ordovician–Mississippian) are medium sized brevicones (Fig. 5C) and cyrtocones that generally lack voluminous deposits in the camerae. The gross conch form of most representatives of both orders would have been poorly suited for rapid propulsion, and many members were probably not fully buoyant. Hence, most are best interpreted as sluggish nektobenthos.

Subclass Bactritoidea. Members of the single component order, Bactritida, are small generalized poorly-known cephalopods that occur rarely in the Silurian and represent inconspicuous elements of faunas ranging from the Lower Devonian through the Upper Triassic. Derived from an orthocerid stock, they assume great significance as the rootstock for the subclasses Ammonoidea and Coleoidea. Bactritids are characterized by a bulbous protoconch, gradually expanded straight conch and a small ventral marginal siphuncle. Neither cameral nor siphuncular deposits occur, so that balance must have been achieved by differential control of fluids in the chambers. Streamlined hydrodynamic form of most representatives probably indicates a nektic mode of life.

Subclass Ammonoidea (q.v.) The ancestral ammonoid order, Anarcestida (Lower Devonian–Upper Devonian), was derived from the Bactritida by progressively tighter curvature and subsequent coiling of the conch (Erben, 1964, 1965). The siphuncle is small and marginal, generally ventral, and is associated with a backward flexure (lobe) in the suture (Fig. 3). Tightly coiled ammonoids of the Middle Devonian initiated a trend, repeated in successive stocks, toward crenulation of the septal margins to produce successively more complex sutures. It is largely by reference to the sutural contours that ontogenetic and evolutionary progressions can be discerned (Figs. 10, 11). Ammonoids are sporadically abundant from the Devonian through the Cretaceous, and throughout that entire span they are accepted as one of the most valuable groups for relative dating. Periods of evolutionary crisis for the ammonoids occurred at the end of the Devonian, Permian, and Triassic periods, before final extinction in the Late Cretaceous. Ammonoids lack any mineralized deposits in the chambers. They controlled buoyancy by fluid exchange through the siphuncle, and most would have been effective swimmers.

Subclass Coleoidea. Representatives of the Coleoidea are distinguished by an internal shell, sometimes vestigial (Jeletsky, 1966). The geological record is adequate for only those groups in which the modified ectocochliate conch is

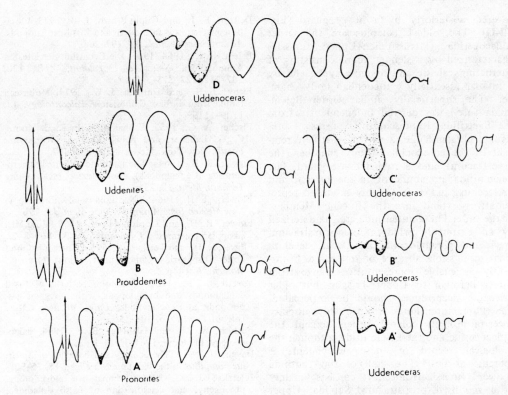

FIGURE 10. Phylogenetic and ontogenetic development of sutures in *Uddenoceras* (from Miller and Furnish, in Moore, 1957; courtesy of The Geological Society of America and University of Kansas Press). Evolutionary sequence of mature sutures (A–D) is paralleled closely by growth stages (A'–C',D) of the most advanced representative, *Uddenoceras,* exemplifying the principle of recapitulation; all specimens Pennsylvanian, SW United States.

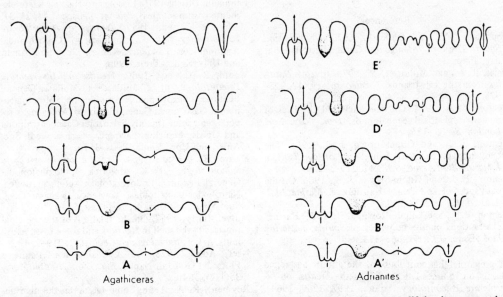

FIGURE 11. Ontogenetic development of sutures in two Permian ammonoids, exemplifying homeomorphy (from Miller and Furnish, in Moore 1957, courtesy of The Geological Society of America and University of Kansas Press). External sutures are superficially similar at maturity, but the taxa belong in separate superfamilies.

covered posteriorly by a heavy guard (Fig. 8E–G). The oldest coleoids are the Order Aulacocerida (Mississippian–Upper Jurassic) characterized by a slender guard consisting of alternating calcitic and organic layers, and by a tubular essentially ectocochlian body chamber. The superficially similar Order Belemnitida (q.v.) was derived independently from the Bactritida. Representatives possess a breviconic phragmocone, a narrow spatulate remnant of the ectocochliate body chamber (the proostracum), and a robust dense guard. The Belemnitida are abundant in the Jurassic and Cretaceous, and a few highly modified representatives extend into the Eocene. Members of the order Phragmoteuthida are characterized by an extremely wide fanlike proostracum, breviconic phragmocone, and weak development or possible absence of the guard. Confidently assignable representatives range from Upper Permian to Upper Triassic, but older Paleozoic occurrences should be anticipated. The Phragmoteuthida are probably directly ancestral to all coleoid orders except the Aulacocerida and the Belemnitida, although the geological record of these descendants is sporadic at best. The shell of the Teuthida (Lower Jurassic–Holocene) consists mainly of a modified proostracum. Sepiida (Upper Cretaceous–Holocene) comprise a group in which the various features of the conch are modified drastically in diverse manners (Fig. 8A–D). Octopoda are essentially devoid of any shell and are virtually unknown as fossils.

BRIAN F. GLENISTER

References

Cousteau, J.-Y., and Diolé, P., 1973. *Octopus and Squid: The Soft Intelligence.* Garden City: Doubleday, 304p.

Davis, R. A., and Mohorter, W., 1973. Juvenile *Nautilus* from the Fiji Islands, *J. Paleontology,* 47, 925–928.

Denton, E. J., 1974. On buoyancy and the lives of modern and fossil cephalopods, *Proc. Roy. Soc. London, Ser. B,* 185, 273–299.

Denton, E. J., and Gilpin-Brown, J. B., 1961. The buoyancy of the cuttlefish, *Sepia officinalis* (L.), *J. Marine Biol. Assoc. U.K.,* 41, 319–342.

Denton, E. J., and Gilpin-Brown, J. B., 1966. On the buoyancy of the Pearly Nautilus, *J. Marine Biol. Assoc. U.K.,* 46, 723–759.

Denton, E. J., and Gilpin-Brown, J. B., 1971. Further observations on the buoyancy of *Spirula, J. Marine Biol. Assoc. U.K.,* 51, 363–373.

In preparing this contribution, the author has relied heavily on volumes I, K, and L of the *Treatise on Invertebrate Paleontology* (Moore, 1957, 1960, 1964). Particularly extensive use has been made of summary statements in these volumes by Walter C. Sweet and C. M. Yonge.

Denton, E. J., and Gilpin-Brown, J. B., 1973. Flotation mechanisms in modern and fossil cephalopods, *Advance Marine Biol.,* 11, 197–268.

Erben, H. K., 1964, 1965. Die Evolution der ältesten Ammonoidea, *Neues Jahrb. Paläont., Abh.,* 120, 107–212; 122, 275–312.

Firby, J. B., and Durham, J. W., 1974. Molluscan radula from earliest Cambrian, *J. Paleontology,* 48, 1109–1119.

Fischer, A. G., 1952. Cephalopods. In R. C. Moore, C. G. Lalicker, and A. G. Fischer, *Invertebrate Fossils.* New York: McGraw-Hill, 335–397.

Flower, R. H., 1964a. The nautiloid Order Ellesmeroceratida (Cephalopoda), *New Mexico Inst. Mining Technol. Mem. 12,* 234p.

Flower, R. H., 1964b. Nautiloid shell morphology, *New Mexico Inst. Mining Technol. Mem. 13,* 79p.

Flower, R. H., and Gordon, M., Jr., 1959. More Mississippian belemnites, *J. Paleontology,* 33, 809–842.

Flower, R. H., and Teichert, C., 1957. The cephalopod Order Discosorida, *Univ. Kansas Paleont. Contrib., Mollusca 6,* 144p.

Foerste, A. F., 1930. The color patterns of fossil cephalopods and brachiopods, with notes on gasteropods and pelecypods, *Univ. Michigan, Mus. Paleont. Contrib. 3,* 109–150.

Furnish, W. M., and Glenister. B. F., 1964. Paleoecology, in Moore, 1964, K114–K124.

Haven, N., 1972. The ecology and behavior of *Nautilus pompilius* in the Philippines, *Veliger,* 15, 75–80.

Jeletzky, J. A., 1966. Comparative morphology, phylogeny, and classification of fossil Coleoidea, *Univ. Kansas Paleont. Contrib., Mollusca 7,* 42, 162p.

Kennedy, W. J., 1977. Ammonite evolution, in A. Hallam, ed., *Patterns of Evolution.* Amsterdam: Elsevier, 251–304.

Lane, F. W., 1969. *Kingdom of the Octopus.* New York: Sheridan House, 300p.

Lehmann, Ulrich, 1970. Lias-Anaptychen als Kieferelemente (Ammonoidea), *Paläont. Zeitschr.,* 44, 25–31.

Moore, R. C., ed., 1957. *Treatise on Invertebrate Paleontology,* pt. L, Mollusca 4, Cephalopoda, Ammonoidea. Lawrence, Kansas: Geol. Soc. Amer. and Univ. Kansas Press, 490p.

Moore, R. C., ed., 1960. *Treatise on Invertebrate Paleontology,* pt. I, Mollusca 1, Mollusca-General Features, Scaphopoda, Amphineura, Monoplacophora, Gastropoda-General Features, Archaeogastropoda, and some (mainly Paleozoic) Caenogastropoda and Opisthobranchia. Lawrence, Kanasa: Geol. Soc. Amer. and Univ. Kansas Press, 351p.

Moore, R. C., ed., 1964. *Treatise on Invertebrate Paleontology,* pt. K, Mollusca 3, Cephalopoda-General Features, Endoceratoidea-Actinoceratoidea-Nautiloidea-Bactritoidea. Lawrence, Kanasa: Geol. Soc. Amer. and Univ. Kansas Press, 519p.

Mutvei, Harry, 1957. On the relations of the principal muscles to the shell in *Nautilus* and some fossil nautiloids, *Arkiv Mineralogi och Geologi,* 2, 219–254.

Naef, Adolf, 1928. Die Cephalopoden: Fauna e Flora del Golfo di Napoli, *Staz. Zool. Napoli Mon.,* 35(1/2), 364p.

Reyment, R. A., 1958. Some factors in the distribution of fossil cephalopods, *Stockholm Contrib. Geol.,* 1, 97–184.

Ruzhentsev, V. E., ed., 1962. *Fundamentals of Paleon-*

tology, vol. V, Mollusca-Cephalopoda 1; Nautiloidea, Endoceratoidea, Actinoceratoidea, Bactritoidea, Ammonoidea (Agoniatitida, Goniatitida, Clymeniida). Akad. Nauk SSSR, 438p. (In Russian, English trans. 1974, Israel Prog. Sci. Trans, Jerusalem, 887p.)

Seilacher, A., ed., 1975. Paläobiologie der Cephalopoden, *Paläont. Zeitschr.* 49, 185–286.

Shrock, R. R., and Twenhofel, W. H., 1953. *Principles of Invertebrate Paleontology,* 2nd ed. New York: McGraw-Hill, 816p.

Teichert, C., 1933. Der Bau der actinoceroiden Cephalopoden, *Palaeontographica,* A, 78, 111–230.

Teichert, C., 1967. Major features of cephalopod evolution, in C. Teichert and E. L. Yochelson, ed., *Essays in Paleontology and Stratigraphy.* Lawrence, Kansas: Univ. Kansas Press, 162–210.

Voss, G. L., 1963. Cephalopods of the Philippine Islands, *U.S. Nat. Mus., Smithsonian Inst., Bull.,* 234, 180p.

Yochelson, E. L.; Flower, R. H.; and Webers, G. F., 1973. The bearing of the new Late Cambrian monoplacophoran genus *Knightoconus* upon the origin of the Cephalopoda, *Lethaia,* 6, 275–310.

Cross-references: *Ammonoidea; Aptychus; Belemnitida; Mollusca; Sexual Dimorphism.*

CHAROPHYTA

Charophyta is a phylum of fresh- or brackish-water plants, often considered as green algae. Today they are represented by only one family, the Characeae. However, because the cell walls of the living plant generally are calcified, the group is very well represented in the fossil record. The calcified female fructifications, or gyrogonites, provide abundant microfossils which are important in the analysis of the systematics and evolutionary history of the group.

Evolutionary History

The oldest undoubted charophyte is *Trochiliscus podolicus* Croft, from the Uppermost Silurian of the Ukraine. Throughout the Devonian and the Lower Carboniferous, this group shows its highest diversification in the fundamental structure of the gyrogonite (see Fig. 1). Three orders are represented:

The Trochiliscales are characterized by gyrogonites composed of dextrally coiled cells. The number of these cells is high and variable within a species. This order is composed of only one family and two genera, *Trochiliscus* and *Karpinskya.* The latter is distinguished by the presence of small apical cells very similar to the coronula cells of the Holocene charophytes.

In the Sycidiales, the gyrogonites are composed of vertical units. The order includes two families—the Sycidiaceae, in which the units are constituted of a row of small cells (genus *Sycidium*); and the Chovanellaceae, in which they are simple (genus *Chovanella*). These two orders became extinct during the Early Carboniferous.

Charales are the only order still extant. It appeared during the Middle Devonian: *Eochara* shows a structure similar to *Trochiliscus* but with a sinistral twisting of the outer cells of the gyrogonites, which is the basic feature of the order. In this group, the most important evolutionary trend is in the fixation and reduction of the number of external cells of the gyrogonite. Hence, by the Late Carboniferous, the modern structure of the gyrogonite—with five sinistrally coiled cells—is attained. This is the structure of all post-Carboniferous charophytes.

In the family Porocharaceae, the ancestral apical pore persists. The group ranges from the Upper Carboniferous to the Uppermost Cretaceous, with maximum development in the Triassic. Despite its long history, it shows little variation, but is of major phylogenetic importance as it represents the common stock from which the three post-Paleozoic families were derived.

Two of these families are characterized by the closing of the apical pore. In the Raskyellaceae, which range from the Maestrichtian to the upper Oligocene, the pore is closed by five small cells which constitute an operculum that is shed at germination. In the Characeae (Triassic-Holocene), the apical pore disappeared due to the close junction of the tips of the spiral cells at the apex. Up to the middle Cretaceous, the Characeae are very unobtrusive, but during the Late Cretaceous and Paleogene they show considerable development, giving rise to numerous genera. From the Miocene onward the group greatly decreased to reach its present much reduced state.

The Clavatoraceae deserve special attention from a phylogenetic point of view. In this family, which extends from the Upper Jurassic to the Uppermost Cretaceous, the gyrogonite remains fairly constant in form and retains the apical pore inherited from the porocharaceous stock. The group is characterized, however, by a progressive elaboration of a supplementary envelope, or utricle, surrounding the gyrogonite and composed of vegetative elements. In one of the oldest known types, the Kimmeridgian (Upper Jurassic) *Echinochara,* these elements are scarcely modified and are still recognizable as vegetative in origin. Later, the utricle becomes specialized and reaches a high structural complexity, especially during the Early Cretaceous when an important diversification produced numerous types. During the Late Cretaceous, the Clavatoraceae greatly decreased in number.

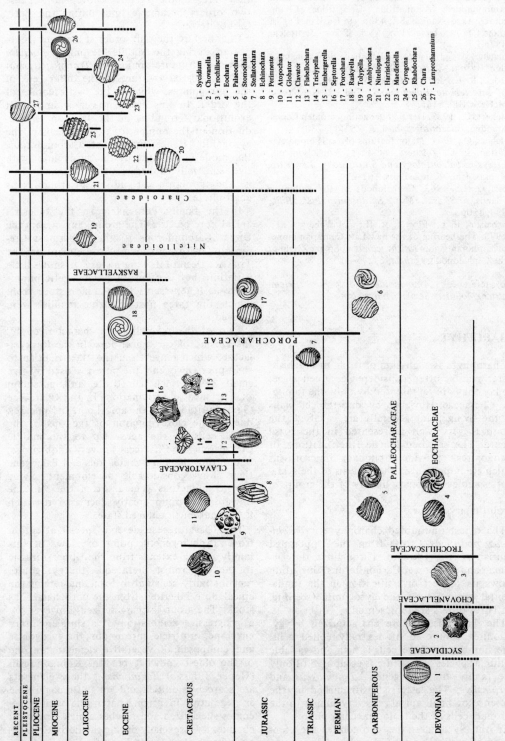

FIGURE 1. Structural evolution of the fructification in charophytes (from Grambast, 1974).

1 - Sycidium
2 - Chovanella
3 - Trochiliscus
4 - Eochara
5 - Palaeochara
6 - Stomochara
7 - Stellatochara
8 - Echinochara
9 - Perimneste
10 - Atopochara
11 - Globator
12 - Clavator
13 - Flabellochara
14 - Triclypella
15 - Embergerella
16 - Septorella
17 - Porochara
18 - Raskyella
19 - Tolypella
20 - Amblyochara
21 - Nitellopsis
22 - Harrisichara
23 - Maedleriella
24 - Gyrogona
25 - Rhabdochara
26 - Chara
27 - Lamprothamnium

They persisted until the late Maestrichtian with the genus *Septorella* but, like the dinosaurs and ammonites, became extinct before the beginning of the Tertiary. In the Clavatoraceae, it has been possible to analyze in detail three evolutionary lineages, which show a sequence of gradual changes throughout the Lower Cretaceous linking quite distinct extreme types. These lineages concern: the genera *Perimneste* and *Atopochara* (Fig. 2-1), the genus *Globator,* and the genera *Flabellochara* and *Clypeator.*

Stratigraphic Use of Charophytes

Though not recognized as such until recently, charophytes are excellent biostratigraphic index fossils. They are especially useful as they are found in nonmarine deposits. Well-preserved calcified gyrogonites often can be collected in large numbers and may, for example, be used for dating bore-hole material.

Moreover, because most charophytes are cosmopolitan in distribution, quite similar floras occur all over the world; charophytes are thus very useful for intercontinental correlation. Such correlations have already been established between western Europe and North and South America for the Cretaceous and Paleogene.

Precise charophyte zonation has now been established mainly from the Late Jurassic to the end of the Oligocene. These zonations are not based upon the evidence of morphological types or assemblages of morphological types, nor even in most cases upon the presence of characteristic species, but upon the identification of associations of natural taxa (assemblage zones) or of evolutionary stages (biochronological zones). Assemblage zones have been thus established for the uppermost Cretaceous and Paleogene on the basis of successive associations of genera or species observed in western Europe. So far, biochronological zones have been defined only for the Lower Cretaceous, based on the evolutionary stages constituting lineages analyzed in the Clavatoraceae.

It is clear that even for stratigraphic purpose the study of charophytes should be conducted from a biological point of view. The systematics of the group is rather difficult, as on one hand close convergence occurs between unrelated and chronologically distant forms and, on the other hand, a wide variability is often found within a single genus or species. It is thus essen-

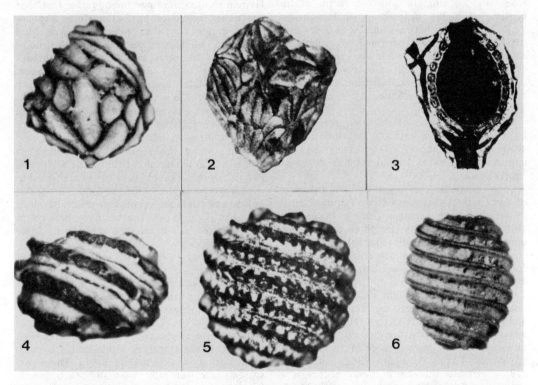

FIGURE 2. Charophyte gyrogonites. 1, *Atopochara trivolvis* Peck, X50, Barremian-Aptian; 2, *Ascidiella reticulata* Grambast and Lorch, X60, Aptian; 3, Idem., longitudinal section showing the gyrogonite cells and the outer utricle with canals. 4, *Platychara cristata* Grambast, X60, Maestrichtian; 5, *Peckichara pectinata* Grambast, X50, Maestrichtian; 6, *Rhabdochara major* Grambast and Paul, X40, middle Oligocene.

tial not to follow the typological method and to base studies on large populations whenever possible.

LOUIS T. GRAMBAST*

References

Grambast, L., 1963. Classification de l'embranchement des Charophytes, *Natur. monspel. Ser. Bot.,* **14**, 63–86.

Grambast, L., 1974. Phylogeny of the Charophyta, *Taxon,* **23**, 463–481.

Peck, R. E., 1957. North American Mesozoic Charophyta, *U.S. Geol. Surv. Prof. Paper 294A,* 44p.

Peck, R. E. and Morales, G. A., 1966. The Devonian and Lower Mississippian Charophytes of North America, *Micropaleontology,* **12**, 303–324.

Cross-reference: *Algae.*

CHITIN

Chitin, a complex polysaccharide, is one of the chief supporting tissues that keep animal cellular structures intact. It is a polymer, forming long, several-hundred-unit molecular chains (Fig. 1a), which may range in length from 0.1–1.0 μm. In purified form, chitin consists of highly ordered molecular chains, "micelles," which have a diameter of 0.01–0.03 μm. A crystal lattice is formed by precise arrangement of chitin chains in any given micelle. Chitin fibrils, which can be observed by plain light microscopy, show subdivision into smaller micelle fibrils when scanned by the electron microscope.

Three distinct types of chitin are identifiable by x-ray diffraction patterns, differing only in chain arrangement (Fig. 1e-g). They may all appear in the same animal, but in different tissues (Rudall, 1963).

Long molecular chains are grouped into sets of one (beta chitin), sets of two (alpha chitin), or sets of three (gamma chitin). In gamma chitin, a single chain yields three parallel segments by folding on itself (Fig. 1g). The "bent" chain is the only one structurally plausible following x-ray diffraction data, according to Carlström (1962).

Covalent links to protein holding the chitin chains probably determine the orientation and/or folding of such chains (Fig. 1e). There appears to be a rather exact fit of protein molecules on the pattern made by small groups of chitin chains (Rudall, 1963). Richards (1953) stressed that chitin never occurs as a separate entity in the arthropod cuticle but is

*Deceased

always only one component of the chitin-protein complex. Thus, chitin cannot be considered a "natural" compound anymore than its degradation products.

Occurrence of Chitin

The arthropod integument is composed of an alpha-chitin-protein complex in which the proportions vary rather systematically (Hackman, 1971). In arachnids, about one-half of the cuticle's dry weight is chitin; in calcified and uncalcified crustacean tissues, with rare exceptions, chitin is relatively high compared to the protein fraction; in insects, the situation is reversed—comparatively high protein and lower chitin is generally found. In dipterid insects, however, chitin is higher than the protein fraction (Richards, 1951).

Chitin is found in some coelenterates (hydrozoan polyps, milleporids, and pneumatophores of the Siphonophora), though it is absent in hydromedusae, scyphozoans, and anthozoans (Chapman, 1974).

Hyman (1958) studied chitin in the lophophorate phyla. She found chitin in secreted tubes of phoronoid worms but not in their bodies. Hyman's other findings include the fact that the entire exoskeleton of ctenostome bryozoans and noncalcareous cheilostomes consisted of chitin; whereas in calcareous cheilostomes, avicularia, opercula, and frontal membranes (plus other parts) were generally chitinous. In inarticulate brachiopods (except family Craniidae), chitin is abundantly present in shell, mantle setae, and pedicle cuticle. Valves of the Articulata are devoid of chitin.

In Mollusca, localized chitin is commonly present. Alpha chitin occurs, for example, in the radula and buccal cuticle of the gastropod *Patella.* Both alpha and beta chitin are found in cephalopods: alpha chitin is found in beaks, radula and gut linings; and beta chitin has been found in *Loligo*'s pen. Furthermore, beta chitin was reported in a decalcified skeleton of *Sepia* (Rudall, 1955).

Annelid worms have beta chitin in chaetae and gizzard lining.

Chitin is apparently absent in most sponges (Porifera) and the Echinodermata. In these organisms, collagen is an important structural element.

Collagen, a fibrous protein is of mesodermal (that is, internal) origin, while chitin is ectodermal (that is, external) in origin. These two systems are generally mutually exclusive as supporting structures. Thus, among sponges, the mesogloeal protein is typical collagen, and most of the connective tissue in echinoderms

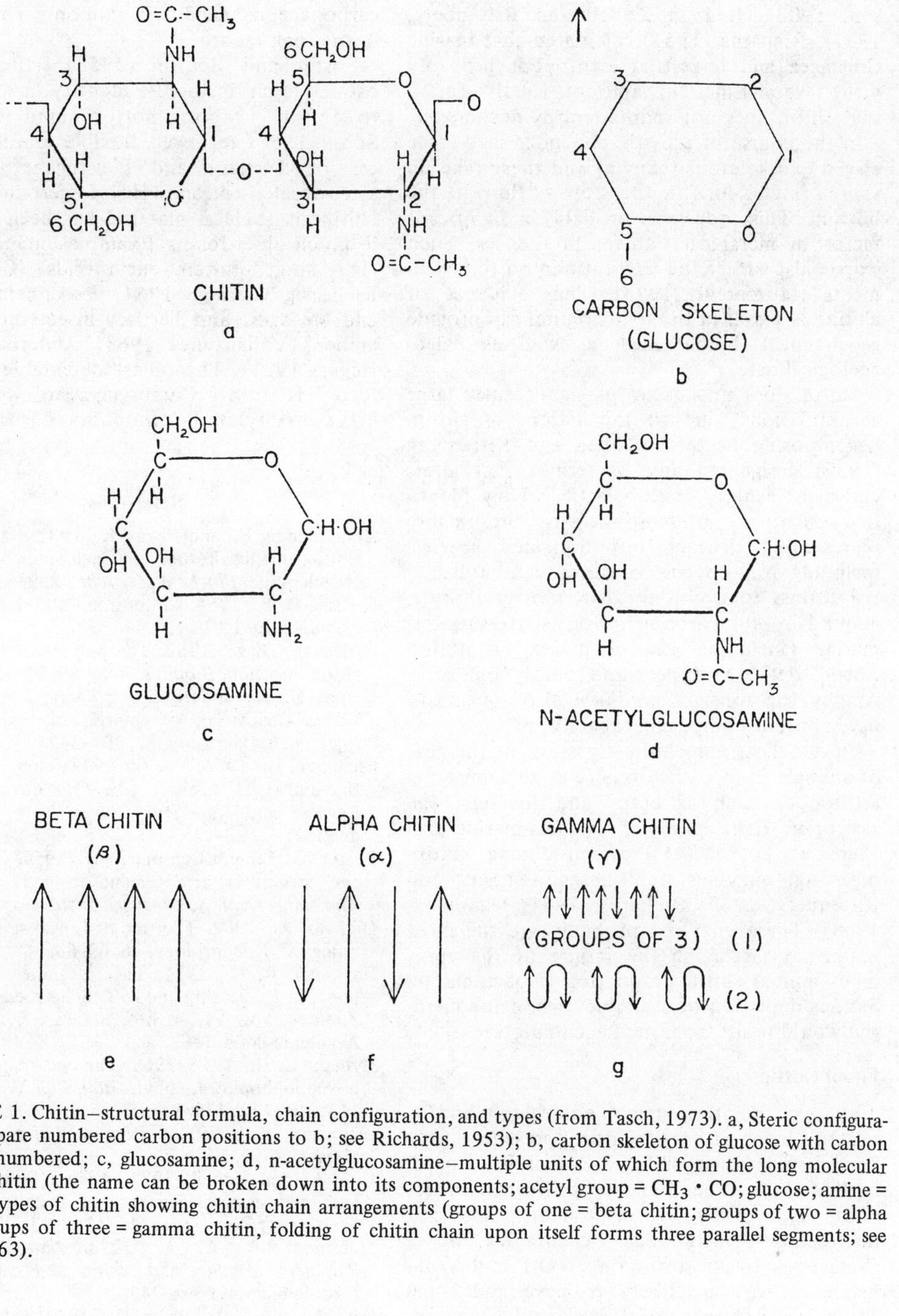

FIGURE 1. Chitin—structural formula, chain configuration, and types (from Tasch, 1973). a, Steric configuration (compare numbered carbon positions to b; see Richards, 1953); b, carbon skeleton of glucose with carbon positions numbered; c, glucosamine; d, n-acetylglucosamine—multiple units of which form the long molecular chain of chitin (the name can be broken down into its components; acetyl group = $CH_3 \cdot CO$; glucose; amine = NH_2); e, types of chitin showing chitin chain arrangements (groups of one = beta chitin; groups of two = alpha chitin; groups of three = gamma chitin, folding of chitin chain upon itself forms three parallel segments; see Rudall, 1963).

is collagen. Similarly, whereas collagen is abundant in mollusks, the chitin is localized. In arthropods, which are dominantly chitinous, the protein of the chitin-protein complex is noncollagenous. (Nevertheless, some collagen is present, e.g., in the peduncle of the barnacle *Lepas* and the subcuticular tissue of the lobster).

Chitin Accumulation and Decomposition in Nature

Of unusual interest for the study of invertebrate paleobiology is the chitin inventory in nature. Copepod crustaceans alone yield an estimated *several billion tons* annually (Johnson, 1908, cited in Zobell and Rittenberg, 1937). Richards (1951) calculated that marine Crustacea and terrestrial arthropods probably yield several times this amount. He also found that chitin does not spontaneously decompose.

In the course of growth, copepods shed some eleven exoskeletons (exuvia) and these take up various metal ions in the slow settling to the bottom. This process is probably a significant factor in metal-ion transfer in the sea. Such exuvia also affect the C/N ratio in bottom sediments (Muzzarelli, 1973). Thus evidence of chitinous fossils in the rock column can provide geochemical, biochemical, as well as paleoecological data.

Chitin does not occur in significantly large accumulations, due to the actions of chitin-decomposing bacteria. Zobell and Rittenberg (1937) designated any bacterium that alters chitin molecules "chitinoclastic." They identified 31 types of chitinoclasts, presumably representing distinct but unnamed species, including rod, coccus, vibrio, and myxobacterial forms. Such chitinoclasts, mostly aerobic, abound in all environments tested—freshwater, marine (including the deep sea [Kim and Zobell, 1972]), hypersaline muds, and soils. Among the fungi, *Penicillium* also appears to have chitin-decomposing capacity.

Chitinoclasts aid chitin digestion in the guts of animals from all habitats (e.g., decomposing arthropods such as crabs and lobsters; the gastropod *Helix pomata;* certain marine fish [Sera et al., 1974]) by producing chitin-digesting enzymes, chitinases (Zobell and Rittenberg, 1937; Richards, 1951; Jeauniaux, 1963). These enzymes probably are the most important factors in the failure of the enormous annual chitin production to accumulate. Serious depletion of available carbon and nitrogen would result from such accumulation.

Fossil Chitin

By extrapolation from equivalent and/or related living forms, chitin presence may be inferred, during life, in fossil bryozoans, coelenterates, annelid and other worms, arthropods, and other worms, arthropods, and inarticulate brachiopods. Gemmules of a Cretaceous freshwater sponge (Ott and Volkheimer, 1972) are likely to have had some chitin, as modern equivalents do; the same inference applies to demosponges (Cambrian-Holocene) (Jeauniaux, 1963).

Disappearance of chitin in almost all fossils that during life had chitinous structures may be attributed to decomposing chitinoclast bacteria, water action over extended time periods, or volatilization (gradual loss of volatiles under anaerobic conditions). This last would lead to carbonaceous residues commonly encountered in the rock record.

Foster and Benson (1958) estimated that ostracode chitin lost its identity in some 5,000 to 15,000 yr after bottom mud deposition. Sohn (1958) retrieved flexible chitinous replicas of Tertiary and Pleistocene ostracodes after decalcification in HCl. In certain instances, chitinous skeletal material has been identified in much older fossils. Examples include Ordovician and Silurian eurypterids (Caster and Kjellesvig-Waering, 1964; Rosenheim, 1905); and Mesozoic and Tertiary insects preserved in amber (Vallentyne, 1963; Abderhalden and Hegns, 1931). The oldest detectable chitin on record is from a Cambrian beard worm, *Hyolithellus* (Phylum Pogonophora; Carlisle, 1964).

PAUL TASCH

References

Abderhalden, E., and Hegns, K., 1931. Nachweis vom Chitin, in Flügelresten von Coeleopteren des oberan Mitteleocäne, *Biochem. Zeitschr.,* **259,** 320–321.

Carlisle, D. B., 1964. Chitin in a Cambrian fossil, *Biochem. J.,* **90,** 1.

Carlström, Diego, 1962. The polysaccharide chain of chitin, *Biochim. Biophys. Acta,* **59,** 361–364.

Caster, K. E., and Kjellesvig-Waering, E. N., 1964. Upper Ordovician eurypterids of Ohio, *Palaeontographica Americana,* **32,** 301–342.

Chapmen, G., 1974. The skeletal system, in L. Muscatine and H. M. Lenhoff, eds., *Coelenterate Biology: Reviews and New Perspectives.* New York: Academic Press, 93–128.

Foster, G. L., and Benson, R. H., 1958. Constituents and structural arrangement in ostracode valves, *Ann. Mtg. Geol. Soc. Amer. Abstr. Prog.,* 1958, 61.

Graham, A., 1962. The role of fungal spores in palynology, *J. Paleontology,* **36,** 60–68.

Hackman, R. H., 1971. The integument of Arthropoda, in M. Florkin and B. T. Scheer, eds., *Chemical Zoology,* vol. VI, Arthropoda Part B. New York: Academic Press, 1–62.

Hyman, L. H., 1958, 1966. The occurrence of chitin in the lophophorate phyla, *Biol. Bull.* (Woods Hole), **114,** 106–112; **130,** 94–95.

Jeauniaux, C., 1963. *Chitine et Chitinolyse.* Paris: Masson et Cie, 167p.

Kim, J., and Zobell, 1972. Agarase, amylase, cellulase, and chitinase activity at deep-sea pressures, *J. Oceanogr. Soc. Japan,* **28,** 131–137.

Muzzarelli, R. A. A., 1973. *Natural Chelating Polymers. Alginic acid, chitin and chitosan.* Oxford: Pergamon Press, 245p.

Ott, O., and Volkheimer, W., 1972. *Palaeospongilla chubutensis* n.g. et n. sp.—ein Süsswasserschwamm aus der Kreide Patagoniens, *Neues Jahrb. Geol. Paläontol., Abh.,* **140,** 49–63.

Richards, A. G., 1951. *The Integument of Arthropods.* Minneapolis: Univ. of Minnesota Press, 319p.

Richards, A. G., 1953. Chemical and physical properties of cuticle, in K. D. Roeder, ed., *Insect Physiology.* New York: Wiley, 23–41.

Rosenheim, D., 1905. Chitin in the carapace of

Pterygotus osiliensus from the Silurian rocks of Oesil, *Proc. Roy. Soc. London, Ser. B,* **76,** 398–400.

Rudall, K. M., 1955. The distribution of collagen and chitin, *in* Fibrous proteins and their biological significance, *Symp. Soc. Exper. Biol.,* **9,** 49–71.

Rudall, K. M., 1963. The chitin/protein complexes of insect cuticle, *Advances Insect Physiol.,* **1,** 257–311.

Sera, H.; Ishida, Y.; and Kadota, H., 1974. Bacterial flora in the digestive tracts of marine fish, in R. R. Colwell and R. Y. Morita, eds., *Effect of the Ocean Environment on Microbial Activities.* Baltimore: Univ. Park Press, 467–490.

Sohn, I. G., 1958. Chemical constituents of ostracods; some applications to paleontology and paleoecology, *J. Paleontology,* **32,** 730–736.

Tasch, P., 1973. *Paleobiology of the Invertebrates.* New York: Wiley, 946p.

Vallentyne, J. R., 1963. Geochemistry of carbohydrate, in I. A. Breger, ed., *Organic Geochemistry.* New York: Macmillan, 456–502.

Zobell, C. E., and Rittenberg, S. C., 1937. The occurrence and characteristics of chitinoclastic bacteria in the sea, *J. Bacteriol.,* **35,** 275–287.

Cross references: *Anthropoda.* Vol. IVA: *Chitin and Chitinous Cuticles.*

CHITINOZOA

Chitinozoa are microscopic fossils with organic-walled skeletons (tests or vesicles) measuring 0.05–2.0 mm in length and resembling minute flasks (Figs. 1–3). The major features of the test are named in Fig. 4. Most Chitinozoa occur as single tests, but frequently two or more tests are found joined together aperture-to-base in linear chains. Such chains are known in all the common genera. The Chitinozoa were named after their chitinoid appearance by their discoverer, the German micropaleontologist Alfred Eisenack (1931). Their systematic position is not known but they are treated taxonomically as animals and probably are monophyletic. They are classified according to the overall shape of the test, ornamentation, the nature of the basal margin, the tendency to occur in chains, the nature of the internal prosome-operculum structure, and size. More than 50 genera and 500 species have been described. Chitinozoa appeared during Tremadoc time, diversified rapidly in the Ordovician and Silurian Periods, and became extinct toward the end of the Devonian. Exclusively marine and distributed globally, they were an important part of microscopic marine life. They are preserved in most types of marine sedimentary rocks in concentrations of up to

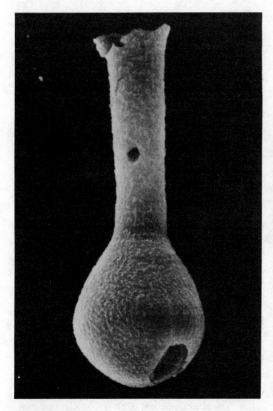

FIGURE 1. *Sphaerochitina sphaerocephala* (Eisenack) with long neck, spheroidal chamber, and textured wall surface, X275, Upper Silurian of Gotland (courtesy of S. Laufeld).

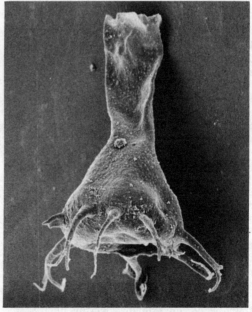

FIGURE 2. *Ancyrochitina ancyrea* (Eisenack) with long neck, conical chamber and large processes on the basal margin, X275, Middle Silurian of Gotland (courtesy of S. Laufeld).

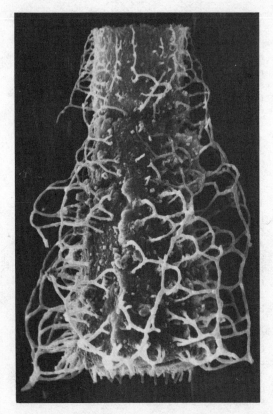

FIGURE 3. *Herchochitina downiei* Jenkins with short neck, elongate conical chamber and elaborate ornaments, X500, Upper Ordovician of Britain.

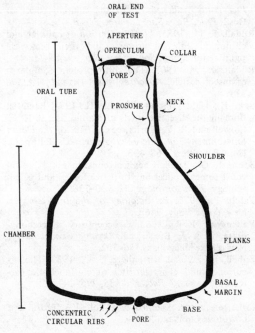

FIGURE 4. Major features of the chitinozoan test.

several hundred tests per gram, and have proved useful for dating and correlating outcrop and subsurface strata. Chitinozoa have been applied widely within the petroleum industry to the elucidation of subsurface stratigraphy. Following a period of very little publication from the 1930s through the 1950s, petroleum exploration was responsible for a rapid increase in the research and publications devoted to these microfossils in the early 1960s.

Chitinozoa are recovered from a rock sample by disaggregating and dissolving the rock's mineral constituents, generally in hydrochloric and/or hydrofluoric acids and concentrating the organic residue (containing the Chitinozoa) on sieves or by heavy liquid flotation. The residue containing the Chitinozoa is then permanently mounted on glass microscope slides.

Current research using conventional and infrared light and scanning electron microscopy is directed at the structure of Chitinozoa (Chaiffetz, 1972); the stratigraphic ranges of species at type or other reference sections (Laufeld, 1974; Legault, 1973); the global and provincial distribution of species; and the extent to which more local distribution of species is governed by facies (Mannil, 1972; Urban and Kline, 1970; Wood, 1974).

W. A. M. JENKINS

References

Chaiffetz, M. S., 1972. Functional interpretation of the sacs of *Ancyrochitina fragilis* Eisenack, and the paleobiology of the ancyrochitinids, *J. Paleontology,* **46,** 499–502.

Eisenack, A., 1931. Neue Mikrofossilien des baltischen Silurs. I, *Paläont. Zeitschr.,* **13,** 74–118.

Laufeld, S., 1974. Silurian Chitinozoa from Gotland, *Fossils and Strata,* 5, 130p.

Legault, J. A., 1973. Chitinozoa and Acritarcha of the Hamilton Formation (Middle Devonian), southwestern Ontario, *Geol. Surv. Canada, Bull.,* **221,** 1–103.

Männil, R., 1972. The zonal distribution of Ordovician chitinozoans in the eastern Baltic area, *24th Internat. Geol. Cong. Sec. 7,* 569–571.

Urban, J. B., and Kline, J. K. 1970. Chitinozoa of the Cedar City Formation, Middle Devonian of Missouri, *J. Paleontology,* **44,** 69–76.

Wood, G. D., 1974. Chitinozoa of the Silica Formation (Middle Devonian, Ohio): Vesicle ornamentation and paleoecology, *Publ. Mus. Michigan State Univ. Paleontol. Ser.,* 127–162.

Cross-references: *Micropaleontology; Plankton.*

CHORDATA

Chordata is a phylum of animals to which the vertebrates and a few related groups belong. Typically, the bilaterally symmetrical body is supported internally by a notochord or a vertebral column, running longitudinally through the segmental musculature, above the visceral cavity and below the tubular nerve cord. Specialized sensory organs for vision, olfaction, taste, hearing, and equilibrium are concentrated near the anterior end of the body adjacent to the brain and mouth. In aquatic chordates a series of gill slits are open from the pharnyx to the exterior. These admit a current of water to be strained of food particles by the gills, which also interchange oxygen for waste products in the blood stream. Gill slits also appear in the embryos of terrestrial classes. Haemoglobin is the active oxygen-carrying pigment of the blood, which is contained in a closed system of vessels, propelled by a muscular heart from which it flows anteriorly to the gill region, thence dorsally past the gills to a dorsal aorta by which it is distributed to all parts of the body; veins collect the blood from the capillary network and return it to the heart.

Protochordates display only some of the typical characters; in Cephalochordata (*Amphioxus*), brain and special sense organs are rudimentary and the notochord extends to the tip of the head. Urochordata (tunicates or sea squirts) have notochord and associated structures only in larvae and greatly hypertrophied gill baskets in the usually sessile adults. Hemichordata (*Balanoglossus, Rhabdopleura,* ?graptolites) lack a definite notochord but may possess one or more gill slits and resemble higher chordates biochemically. Aside from some problematical impressions from Oligocene shales of Switzerland which B. Peyer has interpreted as salpid tunicates, and a few other even more doubtful specimens, protochordates are unknown as fossils unless the graptolites are truly chordates (see *Vertebrata*).

Ancestry of the Chordata

No intermediate fossils connect the vertebrates with any invertebrate phylum; our ideas concerning their origin and relationships are based largely upon embryological and biochemical comparisons of various living groups. Similarities between the arrangement of internal organs of vertebrates and either arthropods or annelid worms inverted to reverse the relationship of nervous, digestive, and circulatory systems, have been noted repeatedly since Etienne Geoffroy-Saint Hilaire first compared the structural plans of vertebrate and inverted insects in 1818. Differences in the arrangement of receptor cells in the eyes; in the embryonic origin of the mouth, mesoderm, and coelom; and in enzyme systems, bar any close relationship between the chordates and the Prostomian invertebrate branch that includes the Annelida, Arthropoda, and Mollusca. Instead, the radially symmetrical and frequently sessile Echinodermata appear to be the closest invertebrate allies of the vertebrates. Significant similarities between these phyla include development of the mouth from a stomodaeum at the opposite end of the gastrula from the blastopore; mesoderm and coelom formation by out-pocketing from the archenteron or primitive gut; bilaterally symmetrical arrangement of ciliated bands on free-swimming larva; and the presence of creatine-phosphoric acid as an energy-releasing enzyme in the muscle tissues.

The earliest known vertebrates, the ostracoderms (see *Agnatha*), resembled echinoderms in being benthonic, armored, filter-feeding animals. Such resemblances as they show to a few aberrant echinoderms (the carpoids; see *Calcichordates*) are probably convergent; and the common echinoderm-vertebrate ancestor was probably a soft-bodied animal that lacked any of the characteristic features of either phylum as we know them.

JOSEPH T. GREGORY

References

Alexander, R. M., 1975. *The Chordates.* Cambridge: Cambridge Univ. Press, 480p.
Barrington, E. J. W., 1965. *The Biology of Hemichordata and Protochordata.* Edinburgh: Oliver & Boyd; and San Francisco: Freeman, 176p.
Berrill, N. J., 1955. *The Origin of Vertebrates.* Oxford: Clarendon Press, 257p.
Jefferies, R. P. S., 1968. The subphylum Calcichordata (Jefferies, 1967), primitive fossil chordates with echinoderm affinities, *Bull. British Mus. (Nat. Hist.), Geol.,* **16**, 241–339. *See also* Review, T. H. Eaton, 1970, *J. Paleontol.,* **44**, 168.
Romer, A. S., 1967. Major steps in vertebrate evolution; *Science,* **158**, 1629–1637.

Cross-references: *Calcichordates; Hemichordata; Vertebrata.*

CIRRIPEDIA

Cirripedes—more familiarly known, perhaps, by the common names of goose, acorn, and wart barnacles—form a subclass of the Crustacea. They undergo a considerable metamorphosis: on hatching from the egg many begin life as minute, free-swimming larvae and at this stage the individual is known as a "nauplius." Later, after several moults, a bivalved shell is developed which covers the body and limbs, and from its resemblance to certain ostracods the term "cypris-stage" has been applied. Finally, giving up its pelagic existence, the larva settles onto some convenient object, fixes itself head foremost by cement secreted from glands at the base of the 1st antennae, and eventually assumes the adult stage. The majority of species are hermaphrodite. Some have small complementary males, but in other cases the sexes are distinct, the females being accompanied by dwarf males.

Although Darwin (1851a) described the present day as the Age of Barnacles, Cirripedia have had a considerable geological history. Of the five orders generally recognized within the Cirripedia, the inclusion of the Apoda, known by a single specimen, has been questioned by various authors (Newman et al, 1969), and this group and the parasitic Rhizocephala (typified by *Sacculina*) are not known in the fossil record. The Ascothoracica are apparently known from cysts found on Maestrichtian octocorals (Voight, 1959) and by borings on Upper Cretaceous sea urchins (Madson and Wolff, 1965). The Acrothoracica are known as fossils by burrows in various substrates (Fig. 1); the oldest known, *Trypetesa caveata* Tomlinson, dates from the Pennsylvanian, and more than 30 fossil and Holocene species have been described. The Thoracica, on the other hand, are known by actual fossil remains and constitutes

by far the largest order, which is divided into four suborders: Lepadomorpha, Verrucomorpha, Brachylepadomorpha, and Balanomorpha. Of these, the first comprises all cirripedes supported by a more or less fleshy stalk (peduncle), while the other suborders contain the sessile forms. It is the general opinion that the sessile cirripedes originated from the stalked forms by a reduction of the peduncle and that the several groups evolved simultaneously along different lines.

In many lepadomorphs, a portion of the carapace, the capitulum, is surrounded by a number of plates, or valves, supported by the peduncle. The principle capitula valves are the paired scuta and terga, and the single carina and rostrum; sometimes a subcarina and/or subrostrum is present as well as a number of pairs of lateral valves (Fig. 2A). Complete capitula are comparatively rare in the fossil record and the majority of species are known only by isolated valves from which it is sometimes possible to make a composite reconstruction. In the balanids, the scuta and terga form opercular valves to the body cavity, which is surrounded by valves modified to form the walls, or compartments, of the "shell" (Fig. 2B,C). They are more robust and are often found with opercular valves in situ.

The earliest known lepadomorph, *Cyprilepas holmi* Wills, 1963, comes from the upper

FIGURE 1. Acrothoracica (from Tasch, 1973). Burrows of *Rogerella cragini* in host shell, *Ceritella protcori,* Cretaceous, Texas. Maximum size of burrows, 1.9 mm length X 0.7 mm width.

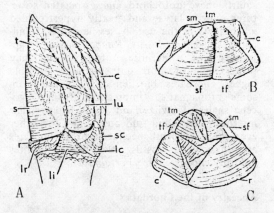

FIGURE 2. Order Thoracica (from Newman et al., 1969, courtesy of The Geological Society of America and The University of Kansas Press). A, Suborder Lepadomorpha—reconstruction of *Arcoscalpellum fossula* (Darwin), upper Senonian, England (this is an extant genus dating from the Lower Cretaceous); B,C, Suborder Verrucomorpha, *Verruca* sp., showing, wall formed, B, by the rostral and carinal latus, and, C, by the tergum and scutum. Symbols: c, carina; lc, carinal latus; li, inframedian latus; lr, rostral latus; lu, upper latus; r, rostrum; s, scutum; sc, subcarina; sf, fixed scutum; sm, moveable scutum; t, tergum; tf, fixed tergum; tm, moveable tergum.

Silurian of Estonia. It is a minute species; and, although it possesses a distinct capitulum and peduncle, the capitulum still retains a bivalved carapace with only the merest indications of the individual valves, the subsequent number and arrangement of which form an essential part of the taxonomy of the suborder. Several genera are extinct and most of the living ones are thought to have evolved from members of the Scalpellidae and Lepadidae dating from the Upper Triassic.

The Verrucomorpha (Fig. 3A), which range from the Upper Cretaceous to Holocene, include all the sessile asymmetrical barnacles. Although members of the early genus *Proverruca* possess 8 plates, these are reduced to 6 in *Verruca* through suppression of the lateral plates,

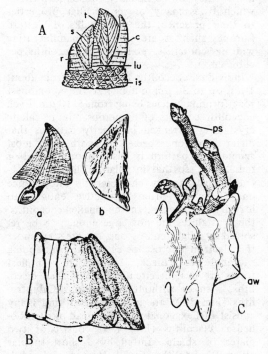

FIGURE 3. Order Thoracica. A, Suborder Brachylepadomorpha (from Newman et al., 1969, courtesy of The Geological Society of America and The University of Kansas Press)—reconstruction of *Pycnolopas rigida* (J. de C. Sowerby), Cretaceous (Albian), England (is, imbricating scales; lu, upper latus; c, carina; t, tergum; s, scutum, r, rostrum; ps, peduncle side; aw, ammonite whorl). B. Suborder Balanomorpha (from Tasch, 1973), *Balanus concavus* Bronn, Miocene, Maryland, and extant, widespread (a, exterior of left tergum; b, interior of left scutum; c, *B. concavus esceptatus* Pilsbry, Miocene, Haiti); C, Suborder Lepadomorpha (from Tasch, 1973), *Zeugmatolepas concinna* (Morris), Jurassic (Oxford Clay), England (after Darwin, 1851a)—a group of specimens attached to an ammonite.

one each of the scuta and terga leaning over to unite with the carina and rostrum, while the others become opercular valves.

The Brachylepadomorpha (Fig. 3B) is an extinct suborder ranging from the Upper Jurassic to the upper Miocene. It contains only two genera of sessile forms in one family.

There can be little doubt that the Balanomorpha was derived from pedunculate forms (Withers, 1935). It includes the families Chthalamidae, Balanidae, and Tetraclitidae, each differentiated by the number and complexity of the compartments and (in living forms) certain details of the soft parts. The chthalamid, *Cataphragmus (Pachydiadema) cretaceum* Withers from the upper Senonian of Sweden, represents the earliest known member of this suborder. As in the other suborders, there has been a tendency toward the reduction in number of the plates.

Ecologically, the Cirripedia have proved adaptable to extremes of temperature, salinity, and depth. A number have become specialized to a particular mode of life such as attachment to corals, whales, turtles, etc.; and there is much evidence pointing to early specialization among forms such as *Coronula, Tetraclita, Pyrgoma,* and other genera. Lepadomorphs have been found attached to *Eurypterus* (in the case of *C. holmi*), Jurassic and Cretaceous ammonites (Fig. 3C) and bivalves, in association with crinoids, tubes of *Rotularia,* and fossil wood.

In the United States, a considerable amount of research has been carried out on fossil thoracic cirripedes, one of the early descriptions, *Scalpellum inaequiplicatum* Schumard, being published in 1862, but unfortunately the whereabouts of the type material is not now known. Many species of balanomorphs have been described, but knowledge of the other suborders remains scanty.

Whereas Darwin's Monographs (1851–1855) of fossil and living barnacles are still the chief works of reference, both the evolution and morphology of the Cirripedia have been fully discussed by Withers (1928 1935, 1953) and Newman et al. (1969); all the works cited contain extensive bibliographies. Substrates of fossil lepadomorph barnacles have been briefly discussed by Drushchits and Zevina (1969) and Collins (1974).

J. S. H. COLLINS

References

Ahr, W. M., and Stanton, R. J., Jr., 1973. The sedimentologic and paleoecologic significance of *Lithotrya,* a rock-boring barnacle, *J. Sed. Petrology,* 43, 20–23.

Collins, J. S. H., 1974. Recent advances in the knowledge of Gault Cirripedia, *Proc. Geologists' Assoc.*, **85**, 337–386.

Darwin, C. R., 1851a. *A Monograph on the Fossil Lepadidae, or Pedunculated Cirripedes of Great Britain.* London: Palaeontogr. Soc., 88p.

Darwin, C. R., 1851b. *A Monograph on the Sub-Class Cirripedia, with Figures of all the Species. The Lepadidae; or, Pedunculated Cirripedes.* London: Ray Soc., 400p.

Darwin, C. R., 1854. *A Monograph on the Sub-Class Cirripedia, with Figures of all the Species. The Balanidae, etc.* London: Ray Soc., 684p.

Darwin, C. R., 1855. *A Monograph on the Fossil Balanidae and Verrucidae of Great Britain.* London: Palaeontogr. Soc., 44p.

Drushchits, V. V.; and Zevina, G. B., 1969. New Lower Cretaceous cirripedes from the Northern Caucasus, *Paleont. Zhur.*, 3, 214–224.

Madson, N., and Wolff, T., 1965. Evidence of the occurrence of Ascothoracica (parasitic cirripeds) in Upper Cretaceous, *Medd. Dansk Geol. Foren.*, **15**, 556–558.

Newman, W. A.; Zullo, V. A.; and Withers, T. H., 1969. Cirripedia, in R. C. Moore, ed., *Treatise on Invertebrate Paleontology*, pt. R, Arthropoda 4, Lawrence, Kansas: Geol. Soc. Amer. and Univ. Kansas Press, R206–R295.

Schumard, B. F., 1862. Descriptions of new Cretaceous fossils from Texas, *Proc. Boston Soc. Nat. Hist.*, 8, 188–205.

Seilacher, Adolf, 1969. Paleoecology of boring barnacles, *Amer. Zool.*, 9, 705–719. .

Tasch, P., 1973. *Paleobiology of the Invertebrates.* New York: Wiley, 946p. (Cirripedia, 557–563.)

Voight, E., 1959. *Endosacculus moltkiae*, n.g.n.sp., ein vermutlicher fossiler Ascothoracide (Entomostr.) als Cystenbildner bei der Oktokoralle *Moltkia minuta*, *Paläont. Zeitschr.*, 33, 211–223.

Withers, T. H., 1928–1953. *Catalogue of the Fossil Cirripedia in the Department of Geology.* 1928, *Triassic and Jurassic*, 154p.; 1935, *Cretaceous*, 434p.; 1953, *Tertiary*, 396p., London: British Mus. (Nat. Hist.).

Cross-references: *Arthropoda; Crustacea.*

CINIDARIA–*See* COELENTERATA

COCCOLITHS

The name coccolith was first used by Huxley (1858) for minute oval plates of calcite found in the globigerina ooze of the North Atlantic and now known to be produced by unicellular planktonic algae belonging to the phylum Chrysophyta class Haptophyceae. The coccoliths, eight or more in number according to the species, are attached to a membrane surrounding the living cell, thus forming a coccosphere (Fig. 1-1).

The living species are all free swimming at some stage in their life cycle, and the great majority are marine. Some species are known to have a benthic life stage that is naked of coccoliths; the benthic stage of *Cricosphaera carterae* is also the haploid chromosome phase (Rayns, 1962; Paasche, 1968). Their principal mode of nutrition is by photosynthesis; and for this reason they are confined to the upper layers of the ocean, where there is an adequate light intensity. Density of population in ocean water varies widely with the season and local conditions. In the surface layers of the North Atlantic (above 50 m), populations of a few hundred thousand individuals per liter are common in the late summer; but much greater concentrations are reached in exceptionally favorable circumstances—more than 13 million per liter have been recorded in the Oslo Fjord.

Coccoliths are abundant in modern deep-sea oozes and in many Mesozoic and Cenozoic marls. They are particularly abundant in chalk, which owes many of its peculiar properties to their presence; the soft chalks of Western Europe, such as are used for manufacturing whiting, contain several million coccoliths per cubic cm.

Individual coccoliths range in size from about 1 μm up to 30 μm in diameter, the commonest rock-forming species being from 5–10 μm. Each coccolith consists of numerous minute calcite crystallites, arranged in orderly patterns that differ from one species to another. In most genera, the pattern is based upon a spiral or radial plan, so that in polarized light, the coccolith gives a very characteristic interference image that is often distinctive enough to identify the species. In the smaller coccoliths, the crystallites are not large enough to be resolved by a light microscope, but are seen under the electron microscope as minute overlapping rhombohedra (Black, 1963). The simplest forms are oval or circular plates, called discoliths, with or without a raised rim. Calyptroliths (Fig. 1-2) are helmet-shaped, and many consist of hexagonal prisms instead of rhombohedra. Placoliths (Fig. 1-3) consist of two plates or shields united by a short tubular pillar. Rhabdoliths (Fig. 1-4) consist of a disk with a tall central spine.

Related to the coccospheres are the braarudospheres, which differ in the structure of their coccoliths. These consist of five peculiarly shaped calcite crystallites, which extinguish separately in cross-polarized light (Figs. 1-5, 1-6). The sole surviving species, *Braarudosphaera bigelowii*, first discovered living in the Bay of Fundy, is a survivor from the Upper Cretaceous. Several extinct species of *Braarudosphaera* are confined to the Mesozoic or Cenozoic, as are the numerous species of *Micrantholithus* (Fig. 1-6); they are familiar objects in early Tertiary nearshore sediment.

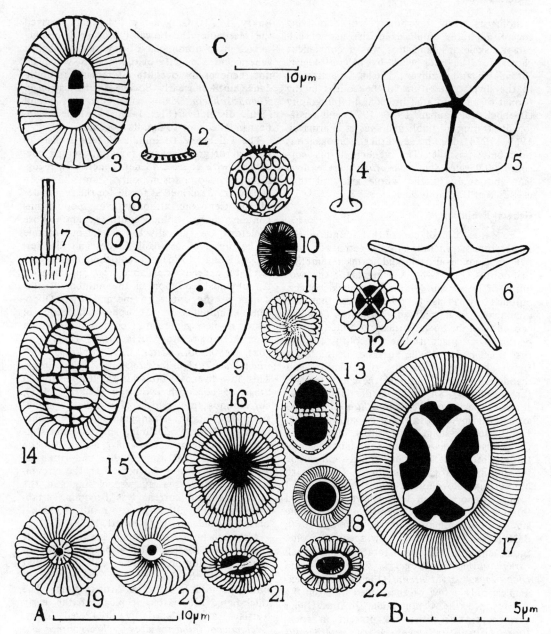

FIGURE 1. Representative coccoliths. 1, *Syracosphaera mediterranea*, a living coccosphere; 2, Calyptrolith (*Anthosphaera robusta*); 3, Placolith (*Coccolithus pelagicus*); 4, Rhabdolith (*Rhabdosphaera claviger*); 5 *Braarudosphaera bigelowii* (Upper Cretaceous–Holocene); 6, *Micrantholithus vesper* (Eocene); 7, *Parhabdolithus liassicus* (Jurassic); 8, *Stephanolithion bigoti* (Upper Jurassic); 9, *Parhabdolithus embergeri* (Lower Cretaceous); 10, *Dictyocapsa parvidentata* (Cretaceous); 11, *Watznaueria barnesae* (Cretaceous); 12, *Prediscosphaera cretacea* (Upper Cretaceous); 13, *Zygodiscus theta* (Upper Cretaceous); 14, *Arkhangelskiella* sp. (Upper Cretaceous); 15, *Zygolithus dubius* (Eocene); 16, *Cyclicargolithus floridanus* (middle Eocene–Miocene); 17, *Chiasmolithus grandis* (Eocene); 18, *Cyclococcolithina rotula* (Miocene); 19-20, *Cyclococcolithina leptopora* (Miocene-Holocene); 21, *Gephyrocapsa oceanica* (Pleistocene-Holocene); 22, *Emiliania huxleyi* (Holocene). Magnification: 3-20, scale A; 2,21,22, scale B; 1, scale C.

Biologists' early taxonomic work on living coccolithophores emphasized the use of calcareous skeletons; and this was a convenient choice for the later work by paleontologists (Bramlette and Sullivan, 1961). Organization of the diverse literature on coccoliths in the format of an annotated index and bibliography (Loeblich and Tappan, 1966, 1973) and a catalog illustrating all published species (Farinacci, 1969, 1974) has made rather homogeneous taxonomy possible. The general terms *coccoliths* and *calcareous nannofossils* are equivalent and apply to the whole array of taxa classified as Haptophyceae.

General Evolution

The earliest fossil coccoliths, found in the Lower Jurassic, include circular forms of primitive appearance and stalked forms resembling rhabdoliths (Fig. 1-7). A scheme for the evolution of Lower and Middle Jurassic coccoliths, outlined by Prins (1969), shows that the dominant, single-rimmed *Parhabdolithus*-like coccoliths of the Lower Jurassic produce *Palaeopontosphaera* placolith-like coccoliths for the first time in the Middle Jurassic. In the lower Upper Jurassic, a variety of placolith and rhabdolith genera dominate the more massive *Parhabdolithus*. In the Middle and Upper Jurassic, species of *Stephanolithion* (Fig. 1-8) are useful index fossils (Rood and Barnard, 1972). Beehive-shaped coccoliths, classified in the genus *Nannoconus,* are abundant in many Lower Cretaceous Tethyan deposits and can be the dominant rock constituent. Changes to *Nannoconus* morphology from long, appressed forms to short, open forms have been noted through the Lower Cretaceous in stratigraphic studies (Brönnimann, 1955; Trejo, 1960; Báldi-Beke, 1962). Many new species become prominent in the Lower Cretaceous, of which *Parhabdolithus embergeri* (Fig. 1-9) and *Dictyocapsa parvidententa* (Fig. 1-10) are examples; only a few of these pass up from the Albian into the Cenomanian. In the Upper Cretaceous, coccoliths are present in rock-forming quantities, and species of *Cretarhabdus* and *Prediscosphaera* (Fig. 1-12) contribute much material to the chalks of Europe, North America, and Australia; other characteristic species are shown in Figs. 1-11, 1-13, and 1-14.

Lower Paleocene coccolith assemblages show a dramatic change in generic composition from the upper Maestrichtian (Bramlette and Martini, 1964). Many characteristic Upper Cretaceous genera, most notably the stem-bearing rhabdoliths such as *Cretarhabdus, Eiffellithus, Prediscosphaera,* and *Vagalapilla,* became extinct in the Maestrichtian (Perch-Nielsen, 1969;

Bukry, 1973). Only a few (7) genera survived the Mesozoic; the Lower Paleocene is characterized by a dominance of new small placolith genera and a paucity of rhabdoliths, which do not reappear in coccolith assemblages in abundance until the early Eocene diversification in *Rhabdosphaera.* Many large (Fig. 1-17) and highly distinctive (Figs. 1-15, 1-16) species appear in Paleocene to middle Eocene.

The Paleocene to middle Eocene is also an interval of great generic radiation among coccoliths, with at least 48 new genera now recognized. By comparison, only about 23 more genera are described as fossils for the remainder of the Cenozoic, although living populations contain many holococcolith genera whose skeletons are typically too fragile or unstable to be preserved as fossils (Sachs and Skinner, 1973; Kling, 1975). In addition, extinctions tend to exceed or balance new appearances. An obvious reduction in the number of new genera in the upper Eocene and lower Oligocene culminates in the upper Oligocene, in which no new genera are recorded but in which 7 extinctions occur—*Chiasmolithus, Dictyococcites, Gemmiodiscoaster, Micrantholithus, Peritrachelina, Turbodiscoaster,* and *Zygrhablithus.* This low-diversity interval, the Clippertonian Stage of coccolith evolution (Bukry, 1973), corresponds to a period of significant alterations of climatic, oceanographic, and tectonic regimes. Widely declining marine and terrestrial paleotemperatures have a nadir for the Paleogene at that time (Wolfe and Hopkins, 1967; Douglas and Savin, 1973), the circumtropical Tethys seaway was closing and the circumantarctic current was becoming established at about 30 m yr as a result of gradual plate tectonic activity (Kennett, 1973).

Although important changes are noted at the species level for the early Miocene, the major changes of dominant species and genera that characterize the late Tertiary did not occur until the middle Miocene Aruban Stage. The thick-rayed discoasters typical of the early Tertiary are replaced by thin-rayed forms. *Cyclococcolithina* replaces *Cyclicargolithus* as a dominant cosmopolitan placolith, and evolution in the active Paleogene genera *Helicosphaera* and *Sphenolithus* is virtually suspended. Subsequent evolution in the Tertiary is highlighted mainly by the appearance of a group of star-shaped discoasters with bent rays; the appearance of horseshoe-shaped *Ceratolithus;* and finally a sequential extinction, through the upper Pliocene, of all typical late Tertiary species of *Discoaster* and *Sphenolithus.*

Quaternary assemblages closely resemble modern assemblages, being the culmination of a general trend to reduce the amount of calcite in coccoliths and to reduce the average size. Quaternary assemblages are dominated by genera bearing small placoliths such as *Crenalithus*, *Cyclococcolithina* (Figs. 1–18, 1–20), *Emiliania* (Fig. 1–22), *Gephyrocapsa* (Fig. 1–21), and *Umbilicosphaera*.

Paleoecology

As phytoplankters, the coccolithophores are directly dependent on the factors of incident solar radiation and ocean water characteristics such as turbidity, temperature, and salinity. Modern species such as cosmopolitan *Emiliania huxleyi* are tolerant to a broad range of conditions, being found in the Black Sea (salinity 17‰); the Red Sea (salinity 41‰); from the equator (temperature 30°C), with constant length of daylight, to subpolar waters (temperature 0°C), with changing length of daylight; in poorly fertile central water masses; and in highly fertile areas of coastal upwelling. Furthermore *E. huxleyi* is as numerically important in assemblages from the base of the photic zone (200 m) as at the surface (Honjo and Okada, 1974). Within these parameters of tolerance, all other living species show more restricted distributions. The general restriction of most coccolith species to tropical and subtropical oceans (Hasle, 1960; Roth and Berger, 1975) is also true for fossil distributions. For example, species of subtropical Tertiary genera *Discoaster* and *Sphenolithus* are virtually absent in the subantarctic where low-diversity placolith assemblages dominate. This same direct relation between high coccolith diversity and warm paleotemperature can be seen through time in the Mesozoic (Berger and Roth, 1975) and in the Cenozoic (Haq, 1973).

Some coccolith species such as *Braarudosphaera bigelowii* are found mainly in marginal marine deposits. A variety of fossil species in the related genus *Micrantholithus* often expand the diversity of lower Tertiary shallow-marine assemblages. The worldwide distribution of some of these species, only in shallow deposits, has been attributed to both a noncalcifying life stage in all but coastal waters (Bybell and Gartner, 1972) and differential dissolution from deep-ocean deposits (Bukry, 1978). Selective preservation by solution and overgrowth is now recognized as a fundamental factor in interpreting coccolith paleoecology and taxonomy (Winterer, et al., 1973).

Taxonomy and Preservation

Owing to their calcite composition, coccoliths, once they are separated from the coccosphere membrane, are subjected to dissolution and overgrowth that may severely alter their taxonomic characters. The structures of rim cycles of coccoliths are generally diagnostic at the generic level; and central-area structures, encircled by the rim, are widely used for distinguishing species. Because dissolution removes central-area structures first, coccolith-producing genera lose their identification keys before the more simply and massively constructed genera such as *Discoaster*. The star-shaped discoasters instead, are highly resistant to solution and can be identified to species in assemblages in which coccoliths are represented only by etched and fragmented rim cycles. However, *Discoaster* is much more susceptible to thick overgrowths of secondary calcite that obscures the form of the rays used for identifying species.

The effects of dissolution and overgrowth are not uniform within or between genera, so rankings can be determined to help interpret the history of coccolith assemblages (Roth and Thierstein, 1972; Thierstein, 1974; Roth and Berger, 1975). Part of the basis for differences in dissolution or overgrowth susceptibility seems to be organic coatings, size and thickness of crystallites, crystallographic orientation, and trace impurities in crystallite lattices (Bukry, 1971; Morse, 1974; Honjo, 1975).

Biostratigraphic Application

Owing to the astronomic abundance of coccoliths, their planktonic origin, and rapid evolution of skeletal form from Jurassic to the present, coccoliths are invaluable for long-range biostratigraphic correlation. Establishing the relative ranges of the useful stratigraphic index species and organization of this data into zonations were done first for the Cenozoic (Bramlette and Sullivan, 1961; Hay, 1964; Hay et al., 1967; Bramlette and Wilcoxon, 1967; Martini, 1971; Bukry, 1973), and rather recently for the Mesozoic (Čepek and Hay, 1969; Manivit, 1971; Thierstein, 1971, 1973, 1976; Barnard and Hay, 1975; Smith, 1975). Presently, the average duration for Cenozoic zones is 1.1 m yr, and for Mesozoic zones, 3.8 m yr. Application of techniques of probability statistics to stratigraphy can improve this resolution on a regional basis (Hay and Steinmetz, 1973).

Conclusion

Study of coccoliths was greatly accelerated by the discovery of their value as stratigraphic guide fossils (Bramlette and Riedel, 1954) and by the Deep Sea Drilling Project (1968–1975), which relied heavily on this group of oceanic nannofossils for interpretation of ages and the geologic history of ocean basins. Conversely, the wealth of comparative data about open-ocean coccolith assemblages prompted recognition of nearshore and offshore floral elements; permitted close-spaced sampling of long sections, leading to recognition of evolutionary lineages; and, probably most important, led to the recognition that differential preservation was a constant consideration in all coccolith taxonomy, paleoecology, and biostratigraphy. Because coccoliths are the most abundant calcareous microfossil on earth and represent one of the primary solar energy fixers in the oceans, they are an essential element for marine paleontology of the Mesozoic and Cenozoic.

MAURICE BLACK*
DAVID BUKRY**

References

Báldi-Beke, M., 1962. A magyarországi Nannoconuszok (Protozoa, inc. sedis) [The genus *Nannoconus* (Protozoa, inc. sedis) in Hungary]; *Geol. Hungarica, Ser. Palaeont.*, 30, 109–174.

Barnard, T., and Hay, W. W., 1975. On Jurassic coccoliths—A tentative zonation of the Jurassic of southern England and northern France, *Ecologae Geol. Helvetiae*, 67, 563–585.

Berger, W. H., and Roth, P. H., 1975. Oceanic micropaleontology: progress and prospects, *Rev. Geophys. Space Phys.*, 13, 561–585, 624–635.

Black, M., 1963. The fine structure of the mineral parts of Coccolithophoridae, *Proc. Linnean Soc. London*, 174, 41–46.

Bramlette, M. N., and Martini, E., 1964. The great change in calcareous nannoplankton fossils between the Maestrichtian and Danian, *Micropaleontology*, 10, 291–322.

Bramlette, M. N., and Riedel, W. R., 1954. Stratigraphic value of discoasters and some other microfossils related to recent coccolithophores, *J. Paleontology*, 28, 385–403.

Bramlette, M. N., and Sullivan, F. R., 1961. Coccolithophorids and related nannoplankton of the early Tertiary in California, *Micropaleontology*, 7, 129–188.

Bramlette, M. N., and Wilcoxon, J. A., 1967. Middle Tertiary calcareous nannoplankton of the Cipero Section, Trinidad, W. I.; *Tulane Studies Geol.*, 5, 93–131.

Brönnimann, P., 1955. Microfossils incertae sedis from the Upper Jurassic and Lower Cretaceous of Cuba, *Micropaleontology*, 1, 28–51.

Bukry, D., 1971. Cenozoic calcareous nannofossils from the Pacific Ocean, *San Diego Soc. Nat. History Trans.*, 16, 303–327.

Bukry, D., 1973. Coccolith stratigraphy, eastern equatorial Pacific, Leg 16 Deep Sea Drilling Project, *Initial Repts. Deep Sea Drilling Proj.*, 16, 653–711.

Bukry, D., 1978. Biostratigraphy of Cenozoic marine sediment by calcareous nanno-fossils, *Micropaleontology*, 24, 44–60.

*Deceased
**Publication authorized by the Director, U.S. Geological Survey.

Byhell, L., and Gartner, S., 1972 [1973]. Provincialism among mid-Eocene calcareous nannofossils, *Micropaleontology*, 18, 319–336.

Čepek, P., and Hay, W. W., 1969. Calcareous nannoplankton and biostratigraphic subdivision of the Upper Cretaceous, *Gulf Coast Assoc. Geol. Soc. Trans.*, 19, 323–336.

Douglas, R. G., and Savin, S. M., 1973. Oxygen and carbon isotope analyses of Cretaceous and Tertiary Foraminifera from the central North Pacific, *Initial Repts. Deep Sea Drilling Proj.*, 17, 591–605.

Farinacci, A., 1969. Catalogue of calcareous nannofossils, *Edizioni Tecnoscienza, Rome*, 1, 250p.

Farinacci, A., 1974. Catalogue of calcareous nannofossils, *Edizioni Tecnoscienza, Rome*, 7, 250p.

Gartner, S., 1977. Nannofossils and biostratigraphy: An overview, *Earth-Sci. Rev.*, 13, 227–250.

Haq, B. U., 1973. Transgressions, climatic change and the diversity of calcareous nannoplankton, *Marine Geol.*, 15, M25–M30.

Hasle, G. R., 1960. Plankton coccolithophorids from the subantarctic and equatorial Pacific, *Nytt Mag. Botanikk*, 8, 77–88.

Hay, W. W., 1964. Utilisation stratigraphique des Discoastérides pour la zonation du Paléocène et de l'Éocène inférieur, *Bur. Réch. Géol. Min. Mém.*, 28, 885–889.

Hay, W. W., 1977. Calcareous nannofossils, in A. T. S. Ramsay, ed., *Oceanic Micropaleontology, vol. 2.* London: Academic Press, 1055–1200.

Hay, W. W., and Steinmetz, J. C., 1973. Probabilistic analysis of distribution of late Paleocene–early Eocene calcareous nannofossils, *Soc. Econ. Paleontol. Mineral. Gulf Coast Sec., Calc. Nannofossils Symp. Proc.*, 58–70.

Hay, W. W.; Mohler, H.; Roth, P. H.; Schmidt, R. R.; and Boudreaux, J. E., 1967. Calcareous nannoplankton zonation of the Cenozoic of the Gulf Coast and Caribbean-Antillean area, and transoceanic correlation, *Gulf Coast Assoc. Geol. Soc. Trans.*, 17, 428–480.

Honjo, S., 1975. Dissolution of suspended coccoliths in the deep-sea water column and sedimentation of coccolith ooze, *Cushman Found. Foram. Research Spec. Pub. 13*, 114–128.

Honjo, S., and Okada, H., 1974. Community structure of coccolithophores in the photic layer of the mid-Pacific, *Micropaleontology*, 20, 209–230.

Huxley, T., 1958. Appendix A, Report on soundings, in J. Dayman, *Deep-sea soundings in the North Atlantic Ocean, between Ireland and Newfoundland, made in H.M.S. Cyclops.* British Admiralty publication, 64.

Kennett, J. P., 1973. Suggested relations between the development of the circumantarctic current and Cenozoic planktonic biogeography, *Antarctic J. U.S.*, 8, 289–290.

Kling, S. A., 1975. A lagoonal coccolithophore flora from Beliz (British Honduras), *Micropaleontology*, 21, 1–13.

Loeblich, A. R., Jr., and Tappan, H., 1966. Annotated index and bibliography of the calcareous nannoplankton, *Phycologia*, 5, 81–216.

Loeblich, A. R., Jr., and Tappan, H., 1973. Annotated index and bibliography of the calcareous nannoplankton VII, *J. Paleontology*, 47, 715–759.

Manivit, H., 1971. Nannofossiles calcaires du Crétacé

Francais (Aptien-Maestrichtien), Fac. Sci. d'Orsay, Thèse Doctorate d'Etat, 187p.

Martini, E., 1971. Standard Tertiary and Quarternary calcareous nannoplankton zonation, *2nd Planktonic Conf. Proc.,* 739–785.

Morse, J. W., 1974. Dissolution kinetics of calcium carbonate in sea water. V. Effects of natural inhibitors and the position of the chemical lysocline, *Amer. J. Sci.,* ser. 5, 274, 638–647.

Paasche, E., 1968. Biology and physiology of coccolithophorids, *Ann. Rev. Microbiol.,* 22, 71–86.

Perch-Nielsen, K., 1969. Die Coccolithen einiger Dänischer Maastrichtien- und Danienlokalitäten, *Bull. Geol. Soc. Denmark,* 19, 51–69.

Prins, B., 1969. Evolution and stratigraphy of coccolithinids from the lower and middle Lias, *1st Planktonic Conf. Proc.,* 547–558.

Rayns, D. G., 1962, Alternation of generations in a coccolithophorid, *Cricosphaera carterae* (Braarud and Fagerl.) Braarud, *J. Marine Biol. Assoc. U.K.,* 42, 481–484.

Rood, A. P., and Barnard, T., 1972. On Jurassic coccoliths: *Stephanolithion, Diadozygus* and related genera, *Ecologae Geol. Helvetiae,* 65, 327–342.

Roth, P. H., and Berger, W. H., 1975. Distribution and dissolution of coccoliths in the south and central Pacific, *Cushman Found. Foram. Research Spec. Pub. 13,* 87–113.

Roth, P. H., and Thierstein, H., 1972. Calcareous nannoplankton: Leg 14 of the Deep Sea Drilling Project, *Initial Repts. Deep Sea Drilling Proj.,* 14, 421–485.

Sachs, J. B., and Skinner, H. C., 1973. Calcareous nannofossils and late Pliocene–early Pleistocene biostratigraphy, Louisiana continental shelf, *Tulane Studies Geol. Paleontol.,* 10, 113–162.

Smith, C. C., 1975. Upper Cretaceous nannoplankton zonation and stage boundaries, *Gulf Coast Assoc. Geol. Soc. Trans.,* 25, 263–278.

Thierstein, H. R., 1971. Tentative Lower Cretaceous calcareous nannoplankton zonation, *Eclogae Geol. Helvetiae,* 64, 459–487.

Thierstein, H. R., 1973. Lower Cretaceous calcareous nannoplankton biostratigraphy, *Österreichische Geol. Bundesanst. Abh.,* 29, 1–52.

Thierstein, H. R., 1974. Calcareous nannoplankton—Leg 26, Deep Sea Drilling Project, *Initial Repts. Deep Sea Drilling Proj.,* 26, 619–667.

Thierstein, H. R., 1976. Calcareous nannoplankton biostratigraphy of Mesozoic marine sediments, *Marine Micropaleont.,* 1, 325–362.

Trejo, M., 1960. La familia Nannoconidae y su alcance estratigráfico en America (Protozoa, incertae saedis), *Assoc. Mexicana Geólogos Petroleros Bol.,* 12, 259–314.

Winterer, E. L., Ewing, J. I., and others, 1973. Leg 17 of the cruises of the drilling vessel *Glomar Challenger, Initial Repts. Deep Sea Drilling Proj.,* 17, 1–930.

Wolfe, J. A., and Hopkins, D. M., 1967. Climatic changes recorded by Tertiary land floras in northwestern North America, *in* K. Hatai, ed., *Tertiary Correlations and Climatic Changes in the Pacific.* Sendai, Japan: Sasaki Printing and Publishing Co. Ltd., 67–76.

Cross-references: *Algae; Discoasters; Plankton.* Vol. I: *Phytoplankton.* Vol. VIA: *Chalk; Pelagic Sedimentation.*

COELENTERATA

The term Coelenterata is variously used to include only the Cnidaria or these plus the Ctenophora. The two groups are considered phyla by Hyman (1940) and subphyla in the *Treatise on Invertebrate Paleontology* (Moore, 1956). Here, the former usage is adopted, and very little attention is given to the small and geologically unimportant Ctenophora.

Knowledge of the coelenterates is admirably reviewed in the two references cited above. Hyman deals with the biology of living groups; the *Treatise* deals with the biology and geologic record of fossil and living groups, with emphasis on the fossils.

Morphology

The Coelenterata (or Cnidaria) is a medium-sized phylum comprising solitary and colonial invertebrates with two body layers separated by a third layer termed mesogloea; muscle tissue is well developed, and a primitive nervous system is present in most forms. Radial symmetry (or biradial or radiobilateral) is characteristic of the phylum and is generally considered to be primary and to indicate that coelenterates are the most primitive of the tissue-level Metazoa. Hadzi, however, has theorized that coelenterates evolved from turbellarians (Phylum Platyhelminthes) and that the radial symmetry is secondary (Hadzi, 1963, and included references).

Cnidaria are characterized by the presence of stinging cells (nematocysts) and tentacles. Two body forms predominate. The polyp (Fig. 1-1A) is more or less cylindrical, consisting of a basal disc commonly attached to a substrate, a wall surrounding a single body cavity (enteron), and an oral surface on which is located the mouth and one or more circlets of tentacles. The medusa (or jelly-fish; Fig. 1-1B) is shaped like a bell or inverted bowl with tentacles on the margin; the mouth is commonly on a tubular projection descending from the center of the bell. Most cnidarians are polymorphic, and many include both polyp and medusa stages in their life cycles.

Skeletons. Cnidarians without supporting or strengthening hard parts are common, but many polyps develop chitinous, horny, or calcareous exoskeletons (Fig. 1-3). Some forms have connected or discrete spicular units apparently developed in the mosogloea.

Reproduction

Reproduction may be either sexual or asexual (Fig. 1-2). Medusae commonly produce the sex cells, and polyps commonly reproduce asex-

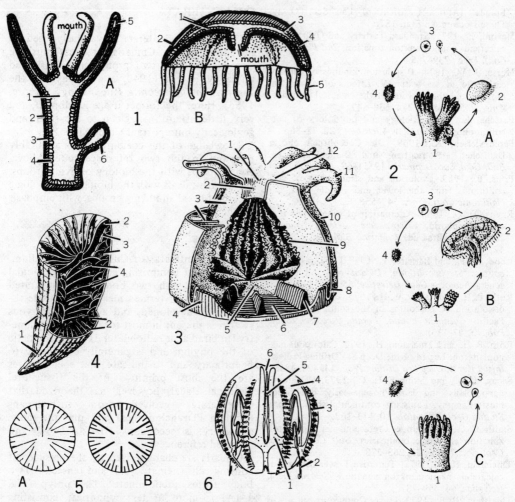

FIGURE 1. Coelenterata and Ctenophora. *1:* Polyp, A, and medusa, B (1, enteron; 2, ectoderm; 3, mesogloea; 4, endoderm; 5, tentacle; 6, bud). *2:* Life cycles of three extant classes—A, Hydrozoa; B, Scyphozoa; and C, Anthozoa (1, polyp; 2, medusa; 3, egg and sperm; 4, planula larva). *3:* Scleractinian coral polyp and skeleton, partially cut away to show morphologic features (1, mouth; 2, stomodaeum or gullet; 3, mesentery; 4, skeletal wall (theca); 5, basal disc; 6, basal plate; 7, septum; 8, edge zone; 9, mesenterial filament; 10, polyp wall; 11, tentacle; 12, oral disc). *4:* Skeleton of solitary rugose coral, partially cut away to show internal structures (1, epitheca; 2, speta; 3, dissepiments; 4, tabulae). *5:* Diagramatic transverse cuts through the skeletons of solitary corals—A, Rugosa; B, Scleractinia (Heavy lines represent 6 primary septa which divide each coral into sextants; secondary septa are inserted in only 4 sextants in the Rugosa and in all 6 in the Scleractinia). *6:* Subphylum Ctenophora—diagramatic view of representative (1, enteron; 2, tentacle [retracted]; 3, band of cilia; 4, connective tissue; 5, stomodaeum; 6, mouth). (Sources: 1, 3, 4, and 5A, from Moore et al., 1952, copyright © 1952 by McGraw-Hill Book Co., used with permission; 2 and 5B from Beerbower, 1960, copyright © 1960 by Prentice-Hall, Inc., used with permission; 6, from Moore, 1956, courtesy of The Geological Society of America and The University of Kansas Press).

ually; although, in forms lacking the medusa stage, gonads are developed by polyps. Sexually reproducing medusae or polyps may also reproduce asexually.

Coelenterates may be dioecious (male and female reproductive organs in separate individuals) or monoecious (male and female in same individual). Sexually produced individuals

pass through a planula larval stage which is characteristic of the subphylum. The planula is an elongate, ciliated, nonattached form that settles and attaches in from a few hours to several weeks, and develops into a polyp.

Asexual reproduction of polyps is by (1) complete or incomplete lateral budding below the tentacular ring, (2) the development of two or more individual polyps within the tentacular ring, (3) stolonal budding, or (4) transverse fission. Colonies are formed by methods 1, 2, or 3 if daughter polyps fail to separate from the parent. Individuals within colonies may or may not be organically connected. Colonies are of particular interest because they represent a clone of genetically similar individuals within which there is presumably less morphologic variation than within groups of related, sexually produced forms.

Solitary polyps are asexually formed if the above types of division are complete rather than incomplete.

Medusae are commonly produced (1) by budding from specialized individuals in some polyp colonies, (2) by budding from other medusae or polyps, or (3) by transverse fission of solitary polypoids.

Classification

Three living classes are universally recognized. Representatives of each are common, and each has a significant geologic record, although the most abundant fossil groups conventionally assigned to two of the classes may not be coelenterates. Extinct nominal classes are minor but of interest in broadening our knowledge of the general form and geologic history of coelenterates or coelenterate-like animals. The extant classes are discussed before the extinct ones, although the table lists the classes and their major subdivision in what is generally assumed to be the order of increasing complexity (see Table 1).

TABLE 1. Classification of the Coelenterate Phyla

Phylum CNIDARIA or COELENTERATA
*Class PROTOMEDUSAE—Precambrian(?), Cambrian–Ordovician
*Class HYDROCONOZOA—Cambrian
Class HYDROZOA
    Order Trachylinida—Cambrian(?), Jurassic–Holocene
    Order Hydroida—Paleozoic(?), Mesozoic–Holocene
    Order Spongiomorphida—Triassic–Jurassic
    Order Milleporina—Cretaceous–Holocene
    Order Stylasterina—Cretaceous–Holocene
    Order Siphonophorida—Ordovician(?)–Devonian(?), Holocene
    *Order Stromatoporoidea—Cambrian–Cretaceous
Class SCYPHOZOA
    Group Scyphomedusae—Cambrian(?), Jurassic–Holocene
    Group Conchopeltina—Ordovician
Class ANTHOZOA
    Subclass Ceriantipatharia—Tertiary–Holocene
    Subclass Octocorallia—Ordovician(?)–Triassic(?), Jurassic–Holocene
    Subclass Zoantharia
    Orders Zoanthiniaria, Corallimorpharia, and Actiniaria—Holocene
    Order Tabulata—Ordovician–Permian, Triassic(?)
    Order Heliolitoidea—Ordovician–Devonian
    Order Rugosa—Ordovician–Permian
    Order Heterocorallia—Mississippian
    Order Scleractinia—Middle Triassic–Holocene
Medusae incertae sedis—latest Precambrian(?), Cambrian–Cretaceous
Phylum CTENOPHORA—Holocene

Adapted from various sources.
*Possibly not Coelenterata.

Class Hydrozoa. The Hydrozoa are characterized by an alternation of generations in which the polyp stage predominates (Fig. 1–2A). The planula develops into a polyp which divides, or buds off other polyps, to form numerous individuals, either separate or in colonies of internally interconnected forms. Within colonies, individuals may specialize for various functions such as eating, reproduction, or defense. Chitinous exoskeletal sheaths occur in several colonial groups, and massive calcareous skeletons characterize others.

Most hydrozoans produce medusae asexually, although the medusa is reduced to an attached form in some types of colonies; sexual reproduction takes place in the medusa stage. The medusa is generally small with fourfold radial symmetry and with a characteristic peripheral shelf (vellum) projecting inward from the margin of the bell.

Siphonophorida are free-living hydrozoan colonies consisting of interconnected specialized polyps and medusae. Impressions of similar forms are known in rocks of middle Paleozoic age.

Milleporina (Fig. 2D), Stylasterina, and Spongiomorphida form massive, branching or fan-shaped calcareous skeletons (aragonite in living forms) that are not uncommon as fossils in rocks of Mesozoic and Cenozoic age.

The Paleozoic and Mesozoic Stromatoporoidea (see *Stromatoporoids;* Fig. 2E) are of uncertain affinities but may represent a hydrozoan level of development and commonly have been considered an order of the Hydrozoa. Recent work however, suggests that some or all stromatoporoids are sponges (Phylum Porifera [q.v.], Class Sclerospongiae; Hartman and Goreau, 1970). Stromatoporoids were colonial animals that secreted massive, encrusting and branching calcareous (probably calcite) skeletons with internal laminae and radial pillars.

Class Scyphozoa. The second class of the Cnidaria, the Scyphozoa, is characterized by an alternation of generations in which the medusa predominates (Fig. 1–2B). The planula develops into a small polyp that generally produces medusae by transverse fission. Full-grown medusae are relatively large and lack the peripheral shelf that characterizes the Hydrozoa. Polyps are small and serve primarily for asexual reproduction. Fourfold radial symmetry characterizes both polyps and medusae.

Scyphomedusae are rare as fossils but impressions as old as Jurassic seem to be reliably assigned to the group. Medusoids as old as Ediacaran (latest Precambrian) are recorded and some probably belong to this group.

Cambrian to Triassic Conulata (q.v.) are preserved as chitinophosphatic, elongate, four-

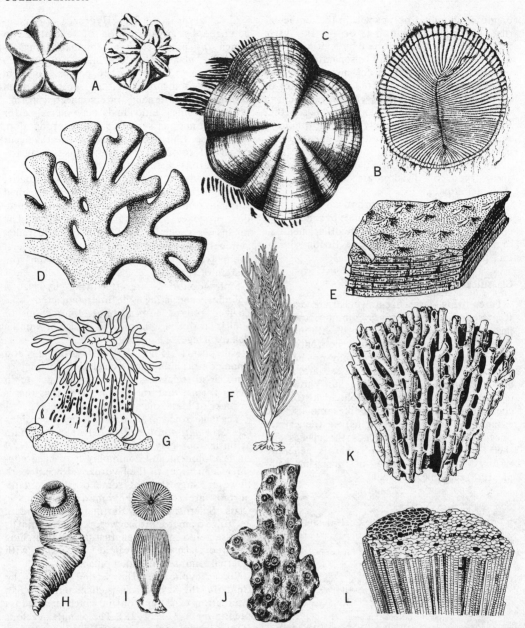

FIGURE 2. Diagrammatic illustrations of representatives of some major groups of Coelenterata and possible Coelenterata; figures vary from X0.4 to X.85. A, Class Protomedusae, upper and lower views of internal mold, early Paleozoic; B, Dipleurozoa (probably a segmented worm), impression, latest Precambrian; C, Class Scyphozoa, Conchopeltina, Ordovician; D, Class Hydrozoa, Milleporina, recent specimen; E, Hydrozoa?, Stromatoporoidea (probably a sponge), fragment of early Paleozoic specimen; F, Class Anthozoa, Octocorallia, Holocene; G, Anthozoa, Actinaria, Holocene anemone; H, Anthozoa, Rugosa, middle Paleozoic coral; I,J, Anthozoa, Scleractinia, Mesozoic and Tertiary corals; K,L, Anthozoa, Tabulata, middle Paleozoic corals. (Sources: A,B, C, H, I, and J, from Fenton and Fenton, 1958, copyright © 1958 by Carroll Lane Fenton and Mildred Adams Fenton and used by permission of Doubleday & Company, Inc. D, G, K, L, from Moore, 1956, courtesy of The Geological Society of America and The University of Kansas Press; E, from Moore, et al., 1952, copyright © 1952 by McGraw-Hill Book Co., used with permission; F, from Hyman, 1940, copyright © 1940 by McGraw-Hill Book Co., used with permission.)

sided pyramids with four internal bifurcating septa and characteristic external markings. Preserved parts are chitinophosphatic. Because of their symmetry, these fossils have been commonly referred to the Scyphozoa (e.g., Moore, 1956), although no other scyphozoans are known to have hard parts. Kozlowski (1968) showed that the similarities of this group to the Scyphozoa are superficial and suggested that they have no close living relatives and should not be assigned to any extant phylum (see *Conulata*). Ordovician Conchopeltina (Fig. 2C) have been included in the Conulata (e.g., Moore, 1956) but lack the internal septa, elongate form, external markings, and test composition of that group. Instead they are broad, low conical forms with lobate margins and chitinous(?) test; specimens with impressions apparently formed by tentacles are known. The Conchopeltina can logically be separated from the Conulata and assigned to the Scyphozoa.

Class Anthozoa. The Class Anthozoa comprises forms in which the polyp stage predominates and there is no medusa (Fig. 1–2C). A stomodaeum (gullet) extends from the mouth into the coelenteron, which is divided into vertically elongate, wedge-shaped spaces by mesenteries; these extend from oral to basal discs and are attached to the outer wall (Fig. 1–3). Anthozoans display bilateral symmetry in the development and arrangement of mesenteries, in possessing an elongate mouth, and commonly in the organization of the skeleton, but a radial arrangement is often more immediately apparent.

Corals (q.v.) are anthozoans that secrete calcareous skeletons, although the term is sometimes used to include hydrozoans (especially milleporids and stromatoporoids) with stony skeletons. The coral skeleton is deposited by the ectoderm. The skeleton of an individual (corallite) is cylindrical or cone shaped; although, when combined in colonies (coralla), the overall form may be massive, branching, or laminar. Corallites consist of a basal plate, an outer wall, septa (vertical, radiating plates), and various complete and incomplete transverse partitions such as tabulae and dissepiments, as well as innumerable minor structures (Fig. 1–4).

Subclass Octocorallia. Characterized by polyps having eight mesenteries and eight tentacles, Octocorallia (Alcyonaria) have skeletons that are horny or calcitic, rarely aragonitic, and commonly internal, although some forms have an external skeleton also. (See Fig. 2F).

Subclass Zoantharia. Tabulata (Fig. 2K,L) comprise a variety of coral groups that may be unrelated; all are essentially limited to the Paleozoic. Included forms are colonial, generally with originally calcite skeletons in which septa are reduced to spines or lacking, and have a wide variety of growth forms; colonial skeletal material (coenosteum) between corallites is lacking and interconnections (pores) between corallites are generally present.

Heliolitoidea are entirely colonial corals, known only from the Paleozoic. They are characterized by having a well-developed coenosteum, no apparent interconnections between corallites, and a set number of septa (commonly 12).

Rugosa (tetracorals) (Figs. 1–4, 2H) were solitary and colonial epithecate corals, characteristically with four positions of septal (and presumably mesenterial) insertion (Fig. 1–5A). Skeletons were solid and apparently calcitic.

Heterocorallia are a small, short-lived group of solitary corals that may be an offshoot of the Rugosa.

Scleractinia (Fig. 2I,J) are solitary and colonial corals generally with an edge zone permitting skeletal deposition outside of the theca. Skeletons are aragonitic and commonly porous; septa and mesenteries are inserted in six sectors (Fig. 1–5B).

Actiniaria (Fig. 2G), Zoanthiniaria, and Corallimorpharia are skeletonless anemones, related to the common coral groups. Widespread today, they have left no significant fossil record.

Unassigned Coelenterata. Three groups of fossils have been referred to the coelenterates but not to any of the three living classes. These are considered classes by some workers (Moore, 1956; Korde, 1963).

Protomedusae. The Protomedusae (Fig. 2A) are lobate objects (4 to 15 complete or incomplete radial lobes) that are interpreted as internal molds of medusae. They are found in rocks of Ediacaran to Ordovician age and may represent the most primitive known Coelenterates.

Hydroconozoa. A group of small, solitary, attached organisms with external skeletons are sometimes assigned as Class Hydroconozoa. All described forms have unusual features difficult to interpret, but some have septa and canals, possibly analogous to features known in scyphozoans and rugose corals (Korde, 1963). There is no convincing evidence that these are coelenterates.

Dipleurozoa. The Dipleurozoa (Fig. 2B) are a group of fossils, apparently not colonial but distinctly bilateral. Marginal scallops separate the distal ends of biradiating lines (or segments) and each bears a simple, so-called tentacle. These forms are limited to rocks of latest Precambrian (Ediacaran) age. The Dipleurozoa were considered a class of coelenterates in the *Treatise* (Moore, 1956) but are more likely a group of segmented worms.

General Ecology

Coelenterates are essentially marine, although a number of hydrozoans are widely distributed in freshwater bodies of all sizes. Many species are quite limited in their temperature, light, or salinity ranges, but different coelenterates are found at all depths and from the Arctic to the Antarctic. Hermatypic (reef-building) corals (see *Corals; Reefs and Other Carbonate Build-ups*) are restricted by light and temperature to shallow depths in tropical and subtropical oceans, but other corals are found to depths of 6000 m or more (Wells, 1957).

Geologic History

Impressions of soft-bodied coelenterates are known from latest Precambrian (Ediacaran; see *Ediacaran Fauna*) time to the Holocene (Fig. 3). Cambrian and Ediacaran medusoids have been assigned to the Hydrozoa and Scyphozoa as well as to the extinct Protomedusae, indicating a considerable diversity early in the history of the group. Polyps of Cambrian age have been referred to the Anthozoa, but polyps or polyp-like forms are extremely rare throughout the geologic record.

Fossil hydrozoans are almost limited to those forms with calcareous skeletons. Several groups are recognized but their relationships are uncertain, and in some cases their assignment to the phylum may be questioned (see Flügel, 1975). Stromatoporoids appeared in the Cambrian(?) or Ordovician and reached a maximum in the Devonian. Late Paleozoic stromatoporoid-like forms are rare and may belong to a separate group (Galloway, 1957), the Sphaeractinoidea, although other workers include these and many Mesozoic forms in the Stromatoporoidea proper (M. Lecompte, *in* Moore, 1956), and extend the range of the group through the Cretaceous. Some or all of the stromatoporoids are probably sponges rather than coelenterates. The somewhat similar Spongiomorphida are limited to the Triassic and Jurassic. Modern stony hydrozoans, the Milleporina and Stylasterina, appeared in the Late Cretaceous. Massive, laminar, and branching skeletons of all these groups were important reef elements in their own times.

Skeletal remains of questionable corals have been described from Cambrian rocks, but corals were not certainly present until the Ordovician. By Middle Ordovician time, corals were common and diverse elements of the normal marine fauna. The Ordovician witnessed the origin and separation of the important Paleozoic coral groups, the Rugosa (tetracorals), Tabulata, and Heliolitoidea. Paleozoic corals attained their maximum diversity in the Devonian but were significant indices to environment and age from the Ordovician on. All three groups participated in the building of Paleozoic reefs, but the reefs in which they seem to have played a key role are small. Rugose and tabulate corals have been reported from post-Paleozoic rocks, but none of these occurrences are certain, and both orders may have become extinct before the end of the Paleozoic.

The Scleractinia (hexacorals) are obviously related to the Rugosa, but it is not clear whether they were derived directly from late Paleozoic (or earliest[?] Triassic) rugose corals or from Paleozoic skeletonless anemones that presumably were related to somewhat earlier corals. A significant gap in the Early Triassic, separates the Rugosa from the Scleractinia. This and the lack of intermediate forms suggests that the relationship was through soft-bodied forms rather than direct.

Scleractinian corals are the "stony" or reef-building (Madreporarian) corals of modern seas. Because of their calcareous skeletons, there is a rather good record of the diversification of the group and the development of the reef-building habit. Scleractinians have been extraordinarily successful and varied, especially in the reef environment.

Octocorallia (Alcyonaria) that are obviously related to living forms are known as far back as Early Jurassic times, but the fossil record is spotty because of the spicular or noncalcareous nature of supporting structures in most of the group. Living octocorals are widespread associates of scleractinian corals in both reef and non-reef environments.

Phylum Ctenophora

The second phylum of coelenterate animals is a small group of exclusively marine, floating or swimming forms that are generally considered to be more advanced than the Cnidaria. Ctenophores (Fig. 1–6) are globular in shape, biradial, and lack nematocysts. Movement is by means of cilia arranged in bands on the body surface. They have no hard parts and no known fossil record.

Summary and Geologic Importance

Coelenterates are a phylum with a fossil record extending back to latest Precambrian (Ediacaran) time. Hydrozoans and anthozoans with calcareous skeletons are sufficiently common as fossils to furnish data for studies of their evolution and living and growing habits. Calcareous forms have been important as rock builders, especially in the reef environment.

WILLIAM A. OLIVER, JR.*

*Publication authorized by the Director, U.S. Geological Survey.

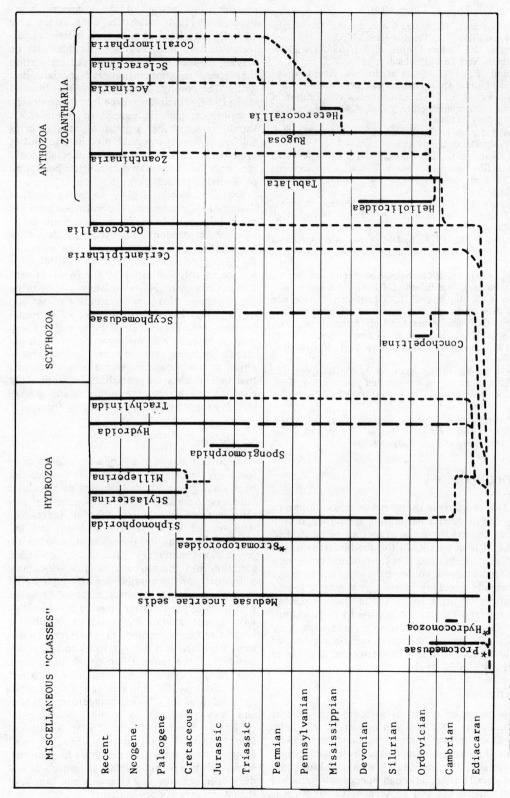

FIGURE 3. Stratigraphic ranges (solid lines) and possible relationships (broken lines) of major groups of Coelenterata. Compare with Table 1.

References

Beerbower, J. R., 1960. *Search for the Past.* Englewood Cliffs, N.J.: Prentice-Hall, 502p.

Fenton, M. A., and Fenton, C. L., 1958. *The Fossil Book.* New York: Doubleday, 482p.

Flügel, E., 1975. Fossil Hydrozoen–Kenntnisstand und Probleme, *Paläont. Zeitschr.,* 49, 369–406.

Galloway, J. J., 1957. Structure and classification of the Stromatoporoidea, *Bull. Am. Paleontol.,* 37(164), 341–480.

Hadzi, J., 1963. *The Evolution of the Metazoa.* New York: Macmillan, 499p.

Hartman, W. D., and Goreau, T. F., 1970. Jamaican corraline sponges: Their morphology, ecology and fossil relatives, *Symp. Zool. Soc. London,* 25, 205–243.

Hyman, L. H., 1940. *The Invertebrates: Protozoa Through Ctenophora.* New York: McGraw-Hill, 726p.

Korde, K. B., 1963. Hydroconozoa, a new class of coelenterates (in Russian), *Paleontol. J.,* 1963(2), 20–25.

Kozlowski, R., 1968. Nouvelles observations sur les conulaires, *Acta Paleontol. Polonica,* 13, 497–536.

Moore, R. C., ed., 1956. *Treatise on Invertebrate Paleontology,* pt. F, Coelenterata. Lawrence, Kansas: Geol. Soc. Amer. and Univ. Kansas Press, 498p.

Moore, R. C.; Lalicker, C. G.; and Fischer, A. G., 1952. *Invertebrate Fossils.* New York: McGraw-Hill, 766p.

Wells, J. W., 1957. Corals, in J. W. Hedgpeth, ed., Treatise on marine ecology and paleoecology, *Geol. Soc. Am. Mem.,* 67(1), 1087–1104; (2), 773–782.

Cross-references: *Burgess Shale; Coloniality; Conulata; Corals; Ediacara Fauna; Reefs and Other Carbonate Buildups; Zooxanthellae.*

COLONIALITY

The word *colony* means various things to various people. Even among biologists, definitions depend upon the organisms being studied and, at least partly, on the predilections of the scientist. In general, a colony of insects is very different from a colony of corals; but to many scientists a colony is simply a population of conspecific organisms living in close proximity to one another and interacting to some extent. To the specialist on marine metazoans, colony implies a more distinct interrelationship of individuals and this is the viewpoint adopted in the rest of this article.

Definition

A metazoan colony is a clone, consisting of a founding individual and one or more generations of asexually produced daughter individuals that have not separated from each other. Normally, the individuals within colonies are genetically alike, although phenotypic variation certainly exists and some genetic variation is possible. The nature of the connections between individuals varies; the connections may be only superficial so that each individual performs all normal life functions, or there may be varying degrees of tissue and organ integration that result in cooperation and specialization within the colony. The work by Boardman et al. (1973) includes excellent definitions, descriptions, and discussions of colonies. It should be noted that a group of colonies or of solitary metazoans may be a clone also, but at present the membership of disconnected clones can only be determined through observation of the living animals.

For the paleontologist, a more practical definition is necessary. The individuals within the colony must have formed preservable skeletons or impressions that held together after the removal of all "soft parts." The nature of the living connections between individuals is not necessarily indicated by the fossil, as such connections can be completely outside the skeletal mass. However, continuity of skeletal material itself and the presence of openings between individual skeletal parts may be suggestive, especially if analogous connections are known in living counterparts. Skeletal dimorphism or polymorphism where present suggests that individuals were specialized but commonly gives no insight into the nature of the specialization except by analogy with living forms.

Significance

Coloniality has been a very successful strategy for many metazoans. The reasons for this can be evaluated in both evolutionary and ecologic contexts. Asexual reproduction generally permits increase in numbers of adult individuals without the usual dangers of larval and immature life. On the other hand, asexual reproduction minimizes genetically controlled variation and thereby adaptability, since new individuals are generally genetic replicas of their single parent. However, it should be noted that phenotypic variation may be considerable. Sexual reproduction, in contrast, results in large numbers of genetically different larvae, most of which will not survive. Those that do, however, have greater potential to occupy new environmental situations. In a statistical sense, the sexually reproducing species is more flexible: it is in a better competitive position and is more likely to survive an ecological crisis (see Maynard–Smith, 1971, for discussion). Asexual reproduction then, is well adapted to a stable, predictable environment. Most of the colonial metazoan groups reproduce both sexually and asexually and thus have the advantages of both systems.

Colonial growth (1) is an efficient way to increase numbers of individuals and unit biomass; (2) provides mechanical strength through skeletal integration, larger unit size, and a platform that may lift the animal well away from the bottom; and (3) is a means toward varying degrees of cooperation and integration among individuals in a "population."

Morphology of Colonies

The general *form* of colonies is similar in quite unrelated organisms because form represents a basic response to long-term environmental pressures. Common metazoan fossil forms are illustrated in Fig. 1. Elaborate nomenclatures exist for some groups to describe form (e.g., Oliver, 1968, p. 18–19; Schopf, 1969, p. 242–244, for review of form nomenclature in corals and bryozoans respectively), but more general terms such as encrusting, massive, dome-shaped, branching, bushy, discoid, and others are useful in many cases.

Budding is the process whereby individuals multiply to form a colony. *Increase* is a skeletal term (used by coral specialists) analogous to budding in the animal. Budding can result from the parent dividing into two or more similar individuals or simply giving rise to one or more small, daughter individuals. Either process can happen in more than one way giving several modes of budding that vary in detail from group to group.

The mode of budding or increase and the nature of growth after budding determines colony *pattern*, which is a rather idealized way of looking at a colony (Fig. 2). Pattern and size, shape, and spacing of individuals determines the overall form of a colony. Although terminology differs, many of the basic modes of increase and patterns are found in several of the major groups of colonial organisms so that colony skeletons of such diverse groups as archeocyathids, corals, bryozoans, and sponges can appear to be very similar in hand specimens.

Integration

Colonies vary from aggregates of individuals that seem to be biologically independent (no connection except the nonliving skeleton) to tightly integrated groupings of highly specialized and interdependent individuals in which the nature of the individual can only be perceived with great difficulty. Both extremes occur within the Coelenterata (q.v., and below) but most colonial metazoans represent intermediate positions on the spectrum. Commonly, individuals are sufficiently interdependent that the well-being of any one can directly affect that of all others within the same colony.

Integration within colonies can be expressed in various ways:

1. *Degree of tissue or organ continuity between individuals.*—This can vary from no connection at all, to continuous body walls, to confluent body cavities, colony wide nervous system, etc.

2. *Development of common tissue.*—In many colonies, individuality within the colony is clearly recognizable. Other colonies have common or colony tissues or structures that are the result of colony-wide building and not parts of any one individual. This can be carried to the extreme where most tissue is common tissue and individual animals are indistinct, especially in the skeleton.

3. *Astogeny.*—Individual ontogeny can appear to be completely uniform or there can be successive changes in this as the colony "matures." Commonly, the development of the first (founder) member of the colony differs from all others because it developed from a sexually produced larva. In some bryozoans, successive generations are somewhat different so that individual morphology continues to change as the colony develops (Boardman and Cheetham, 1973). In graptolites, this process of change may continue throughout the life of the colony (Urbanek, 1973).

4. *Polymorphism.*—In many colonies, all individuals are morphologically alike (not identical) although such colonies can vary in degree of tissue or organ-level integration. Other colonies have two or more kinds of individuals. In living colonies, these are specialized for food ingestion, reproduction, defense, etc.; in fossil colonies, we assume that similar skeletal morphologic di- or polymorphism indicates analogous functional differentiation although this can seldom be proven. An extreme is represented by colonies composed of several kinds of individuals that are so specialized and integrated that the colony looks and behaves like a solitary animal. This is the "individual of a higher order" of Beklemishev (1969) and others.

Principal Fossil Groups of Colonial Metazoans

The following notes on various fossil groups that include colonies are intended to indicate the variety of colonial taxa and the features that all have in common. Greater detail will be found under the major articles about each group in this encyclopedia.

Archaeocyatha (q.v.) form varied massive, catenulate (chain), and branching colonies, although solitary forms are much more common (Hill, 1972, p. E6). Two or three modes of increase are known, each occurring also in other colonial groups.

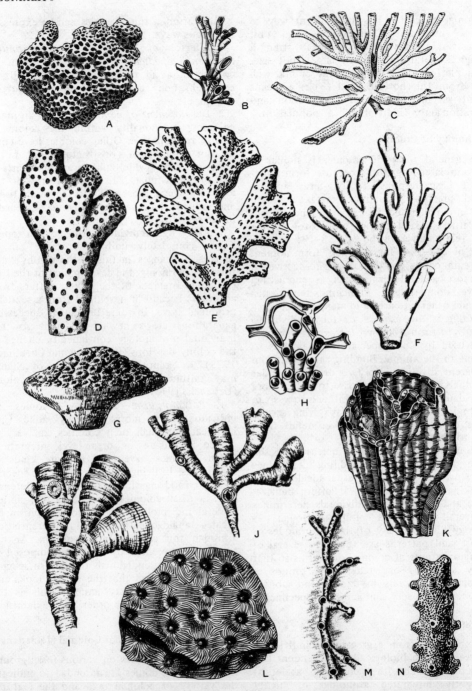

FIGURE 1. Form in fossil colonies. A,G,L, massive; D,E,F,N, massive branching; B,C,H,I,J,M, solitary branching; K, cateniform (chain); A,H,M, encrusting; C, archeocyathid; A,E,M,N, bryozoans. Corals—B,H,J,K, tabulates; I,L, rugosans; D,G, scleractinians; F, hydrozoan. Scales vary. (Sources: Illustrations A, from Moore, 1953; B,H,K, and L, from Moore, 1956; and C, from Moore, 1972; courtesy of the Geological Society of America and The University of Kansas Press. Illustrations D,F,I,J, and M, from Fenton and Fenton, 1958, copyright © 1958 by Carroll Lane Fenton and Mildred Adams Fenton and used by permission of Doubleday & Company, Inc. Illustrations E,G, and N, from Moore et al, 1952, copyright © 1952 by McGraw-Hill Book Co., used with permission).

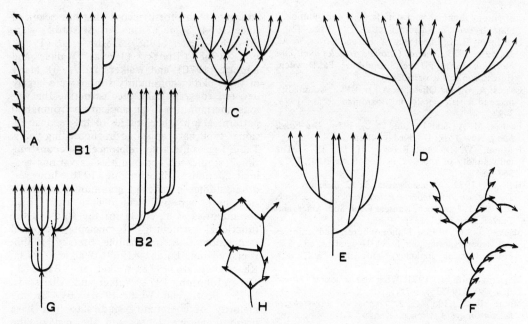

FIGURE 2. Colony patterns (after Oliver, 1968). Each line represents an individual; arrowheads indicate oral end and direction of growth; all patterns could represent branching growth with each individual forming a separate branch; most of the patterns (except H) could also represent massive growth, or branching growth with numbers of closely spaced individuals forming branches; pattern B would represent cateniform (chain) growth if maintained in one "plane;" patterns A and B2 represent dimorphism.

Corals (q.v.) and other Coelenterata (q.v.) are both solitary and colonial but colonial forms are abundant and varied. The Coelenterata display a greater range of colony morphology and integration than any other phylum. Many rugose and scleractinian corals formed colonies in which polyps were completely individualized and united only by their continuous skeleton. At the other extreme, the siphonophores and octocorals form polymorphic colonies in which highly specialized individuals are so interdependent that the colony has often been considered an "individual of a higher order." Fossil coelenterates cover most of this range, many giving clear indication of interconnecting body cavities and specialization among individuals (polymorphism). Coloniality in coelenterates has been reviewed by Coates and Oliver (1973), Bayer (1973), and Mackie (1963).

Byrozoa (q.v.) are entirely colonial animals. They display a wide spectrum of colonial attributes, and there are varied levels of integration within both hard and soft parts (Boardman and Cheetham, 1973). Polymorphism is common in byrozoans and a few may approach the extreme specialization and integration levels of the siphonophores. Numerous papers in Boardman et al. (1973) treat coloniality in this group.

Graptolithina (q.v.) are another wholly colonial group that flourished during the early Paleozoic. Urbanek (1973 and earlier) discussed the organization of graptolite colonies in great detail. He emphasized morphologic gradients along the linear colonies and interpreted these as indicating physiological gradients in the distribution of morphogenetic substances.

Porifera (q.v.), or sponges, are commonly thought of as colonies; but many contemporary workers do not agree and interpret the level of integration within sponges as reflecting a primitive individuality of the whole rather than an advanced colonial state with a high level of integration (Simpson, 1973; Hartman and Reiswig, 1973).

WILLIAM A. OLIVER, JR.*

References

Bayer, F. M., 1973. Colonial organization in octocorals. In Boardman et al., 1973, p. 69–93.
Beklemishev, V. N., 1964(1969). *Principles of Comparative Anatomy of Invertebrates,* vol. 1, Promorphology, 3rd. ed., (In Russian; English transl. 1969, Edinburgh: Oliver and Boyd, and Chicago: Univ. Chicago press, 490p.)
Boardman, R. S., and Cheetham, A. H., 1973. Degrees

*Publication authorized by the Director, U.S. Geological Survey

of colony dominance in stenolaemate and gymnolae mate Bryozoa, in Boardman, et al., 1973, 121–220.

Boardman, R. S.; Cheetham, A. H.; and Oliver, W. A., Jr., eds., 1973. *Animal Colonies, Development and Function Through Time.* Stroudsburg, Pa; Dowden, Hutchinson and Ross, 603p.

Coates, A. G., and Oliver, W. A., Jr., 1973. Coloniality in zoantharian corals, in Boardman et al., 1973, 3–27.

Fenton, M. A., and Fenton, C. L., 1958. *The Fossil Book.* New York: Doubleday, 482p.

Hartman, W. D., and Reiswig, H. M., 1973. The individuality of sponges, in Boardman et al., 1973, 567–584.

Hill, D., 1972. Archaeocyatha, in C. Teichert, ed., *Treatise on Invertebrate Paleontology*, pt. E, rev. ed., vol. 1. Lawrence, Kansas: Geol. Soc. Amer. and Univ. Kansas Press, 158p.

Mackie, G. O., 1963. Siphonophores, bud colonies and superorganisms, in E. C. Dougherty, ed., *The Lower Metazoa.* Berkeley, Calif: Univ. Calif. Press, 329–337.

Maynard-Smith, J., 1971. What use is sex? *J. Theor. Biol.,* **30,** 319–335.

Moore, R. C., 1953-1972. *Treatise on Invertebrate Paleontology.* Lawrence, Kansas: Geol. Soc. Am. and Univ. Kansas Press, pt. E, Archaeocytha, 1972; pt. F, Coelenterata, 1956; pt. G. Bryozoa, 1953.

Moore, R. C.; Lalicker, C. G.; and Fischer, A. G., 1952. *Invertebrate Fossils.* New York: McGraw-Hill, 766p.

Oliver, W. A., Jr., 1968. Some aspects of colony development in corals, J. Paleontol., **42,** pt. 2, 16–34.

Schopf, T. J. M., 1969. Paleoecology of ectoprocts (bryozoans), J. Paleontol., **43,** 234–244.

Simpson, T. L., 1973. Coloniality among the Porifera, in Boardman et al., 1973, 549–565.

Urbanek, A., 1973. Organization and evolution of graptolite colonies, in Boardman et al., 1973, 441–514.

Cross-references: *Bryozoa; Coelenterata; Corals; Graptolithina; Porifera; Reefs and Other Carbonate Buildups.*

COMMUNITIES, ANCIENT

Modern organic communities have been studied for over 100 years and much of ecological theory has as its basic framework the community concept. In spite of more recent controversy concerning the nature of communities, this concept is still the single most important unifying idea in ecology. The study of fossil assemblages within the framework of communality is relatively recent. Probably the earliest study in this vein was the now classic work by Elias (1937). Until 1960 this approach was rarely taken (see however, George, 1948; Craig, 1954; and Imbrie, 1955). The methodological works by Fagerstrom (1964) and Johnson (1960, 1964) laid the groundwork for the wide application of the community concept to the fossil record that has occurred since 1964. Some of the earliest studies were those by Ziegler (1965), Ziegler *et al.* (1968). Fox (1968), Bretsky (1969a), Walker and Laporte (1970), and Walker (1972a, b). Many of these earlier community studies were largely devoted to graphically reconstructing the various communities in an ancient environmental pattern. Many other studies of the reconstructional type have appeared in succeeding years. Table 1 gives literature references to reconstructional studies of communities as various geologic periods. Since about 1970, however, other studies of ancient community organization have appeared which have used fossil communities as a vehicle for the study of the functional structure of ancient biological associations (see for example, Bretsky, 1969b; Bretsky and Lorenz, 1969; Walker, 1972a; Rhoads et al., 1972; Walker and Bambach, 1974; Johnson, 1972; Walker and Alberstadt, 1975; Scott and West, 1976). By far the majority of community studies to date have involved marine biotas and the rest of the present discussion is restricted to the marine realm.

The definition of a community usually accepted in paleoecologic works is that suggested by Johnson (1964): "the assemblage of organisms inhabiting a specified area." The definition requires a holistic approach to fossil assemblages so that all the preserved taxa are taken into account. Many studies of fossil communities are in fact more nearly studies of taxocenes (all the members of a single taxonomic category that lived in an area, e.g., brachiopods or mollusks). Studies of the processes of burial and preservation are extremely important in reconstructing the community from the assemblage (Lawrence, 1968). These factors of definition are indicated in Fig. 1. As has been pointed out by Walker and Bambach (1974) many of the transportational and taphonomic problems involving the formation of fossil assemblages have received overemphasis. The "specified area" mentioned in Johnson's definition is often equivalent to a single sedimentary environment for, as Laporte has note, biofacies generally parallel lithofacies. This indicates that type of substratum has often been a primary controlling factor in community development. Indeed, some studies have shown that communities in similar environments of widely differing age have strikingly similar taxonomic structures (cf. Walker and Laporte, 1970).

Recently it has come to light that some fossil assemblages are actually condensations of several communities that formed an ecological successional sequence (Walker and Alberstadt,

TABLE 1. Literature references to studies that reconstruct various communities
of the Periods of the geological time scale[a]

Period	Literature References
Neogene Paleogene	Many in biological literature; Shotwell (1958)
Cretaceous	Kauffman (1974); Rhoads et al. (1972); Scott (1974)
Jurassic	Ager (1965); Hallam (1961)
Triassic	Fürsich and Wendt (1977)
Permian	Boyd and Newell (1972); Stevens (1966)
Pennsylvanian	Donahue and Rollins (1974); Rollins and Donahue (1975); West (1972)
Mississippian	Craig (1954); Ferguson (1962)
Devonian	Bowen et al. (1974); Thayer (1974); Walker and Laporte (1970)
Silurian	Ziegler (1965); Ziegler et al. (1968); Calef and Hancock (1974); Worsley and Broadhurst (1975)
Ordovician	Fox (1968); Bretsky (1969a); Walker and Laporte (1970); Walker (1972a,b); Walker and Ferrigno (1973)
Cambrian	Lochmann-Balk and Wilson (1958); Palmer (1969); McBride (1976)
Precambrian	Glaessner (1966)

[a]This table is by no means complete, but indicates only exemplary studies, and is restricted to literature in English. See list of references for complete citations; see also McKerrow (1978).

1975). The analysis of assemblages in this light promises that we may eventually get far more detailed information on ancient communities than has heretofore been expected (see Johnson, 1972). Under the proper circumstances of preservation and with careful collecting, the several stages (quasi-separate communities) of the succession may be separated as shown in the example in Fig. 2.

It has long been recognized that communities must evolve, because of the evolution within the various individual taxa that make up communities. Bretsky (1969a) has attempted to tabulate the changes that occurred in Paleozoic benthic communities on clastic substrata (Fig. 3). Olson (1966) has linked the early development of the mammals to changing community feeding organization. Shotwell (1964) has summarized the relationships between evolution of terrestrial plant communities and mammalian evolution in the late Cenozoic. Much more information is needed before community evolution can be confidently summarized in any more detail.

Many aspects of the applicability of the community concept to fossil associations have recently been summarized by Valentine (1973), Ziegler et al. (1974), and Scott and West (1976).

KENNETH R. WALKER

References

Ager, D. V., 1965. The adaptations of Mesozoic brachiopods to different environments, *Paleogeography, Palaeoclimatology, Palaeoecology*, 1, 143–172.

Bowen, Z. P.; McAlester, A. L.; and Rhoads, D. C., 1974. Marine benthic communities of the Sonyea Group (Upper Devonian) of N.Y., *Lethaia*, 7, 93–120.

Boyd, D. W., and Newell, N. D., 1972. Taphonomy and diagenesis of a Permian fossil assemblage from Wyoming, *J. Paleontology*, 46, 1–14.

Bretsky, P. W., 1969a. Ordovician benthic marine communities in the central Appalachians, *Geol. Soc. Amer. Bull.*, 80, 193–212.

Bretsky, P. W., 1969b. Evolution of Paleozoic benthic marine communities, *Palaeogeography, Palaeoclimatology, Palaeoecology*, 6, 45–59.

Bretsky, P. W., and Lorenz, D. M., 1969. Adaptive response to environmental stability: A unifying concept in paleoecology, *Proc. N. Amer. Paleontology, Conv.*, pt. E, 522–550.

```
              Living community
              on death becomes
                     |
                     |
              Fossil assemblages
      which on analysis may prove to be
           /                      \
        in situ                transported
        /                           \
  Life assemblage            Death assemblage
  (Fossil community)
              \            which may be
               \          /           \
  Indigenous        Exotic          Remanie
  derived but in    derived from    derived from
  the same environment  different but  older rocks
  as original living  contemporaneous
  community         environment
```

FIGURE 1. Classification of ecological and mechanical fossil associations (from Craig, 1966, after Craig and Hallam, 1963).

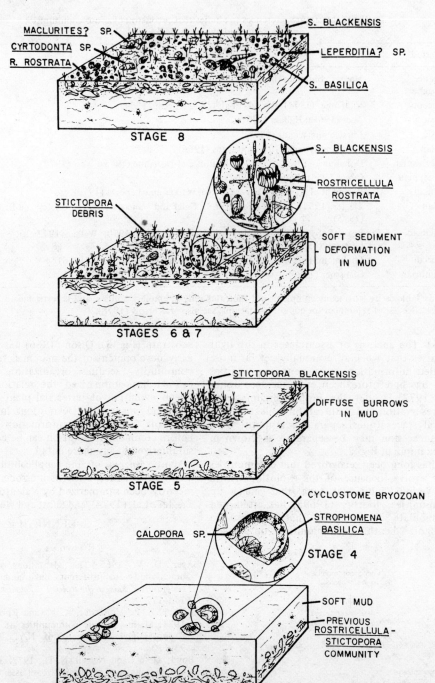

FIGURE 2. Reconstruction showing the eight stages of the *Strophomena* to *Rostricellula* succession (from Walker and Alberstadt, 1975).

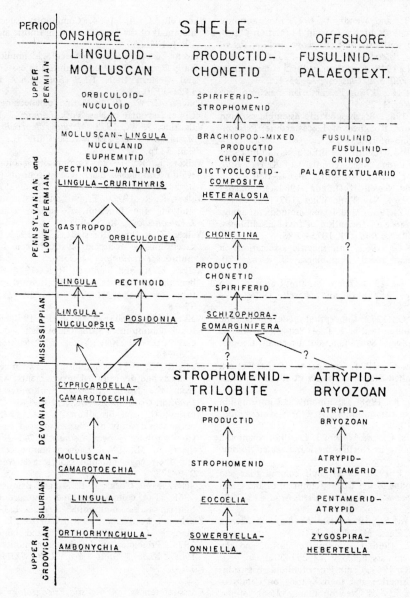

FIGURE 3. Evolution of Paleozoic marine invertebrate community structure (from Bretsky, 1969b).

Calef, C. E., and Hancock, N. J., 1974. Wenlock and Ludlow marine communities in Wales and the Welsh borderland, *Palaeontology*, 17, 779–810.

Craig, G. Y., 1954. The palaeoecology of the Top Hosie Shale (Lower Carboniferous) at a locality near Kilsyth, *Quart. J. Geol. Soc. London*, 110, 103–119.

Craig, G. Y., 1966. Concepts in paleoecology, *Earth-Sci. Rev.*, 2, 127–155.

Craig, G. Y., and Hallam, A., 1963. Size-frequency and growth-ring analyses of *Mytilus edulis* and *Cardium edule* and their palaeoecological significance, *Palaeontology*, 6, 731–750.

Donahue, J., and Rollins, H. B., 1974. Paleoecological anatomy of a Conemaugh (Pennsylvanian) marine event, *Geol. Soc. Amer. Spec. Paper 148*, 153–170.

Elias, M. K., 1937. Depth of deposition of the Big Blue (Late Paleozoic) sediments in Kansas, *Geol. Soc. Amer. Bull.*, 48, 403–432.

Fagerstrom, J. A., 1964. Fossil communities in paleoecology: their recognition and significance, *Geol. Soc. Amer. Bull.*, 75, 1197–1216.

Ferguson, L., 1962. The paleoecology of a Lower Carboniferous marine transgression, *J. Paleontology*, 36, 1090–1107.

Fox, W. T., 1968. Quantitative paleoecologic analysis of fossil communities in the Richmond Group, *J. Geol.*, 76, 613–640.

Fürisch, F. T., and Wendt, J., 1977. Biostratinomy and paleoecology of the Cassian Formation (Triassic) of the southern Alps, *Palaeogeography, Paleoclimatology, Paleoecology*, **22**, 257–323.

George, T. N., 1948. Evolution in fossil communities, *Proc. Roy. Phil. Soc. Glasgow*, **73**, 23–42.

Glaessner, M. F., 1966. Precambrian paleontology, *Earth-Sci. Rev.* **1**, 29–50.

Hallam, A., 1961. Brachiopod life assemblages from the Marlstone Rockbed of Leicestershire, *Palaeontology*, **4**, 653–659.

Imbrie, J., 1955. Quantitative lithofacies and biofacies study of Florena Shale (Permian) of Kansas, *Bull. Amer. Assoc. Petrol. Geologists*, **39**, 649–670.

Imbrie, J., and Newell, N. D., eds., 1964. *Approaches to Paleoecology*. New York: Wiley, 107–150.

Johnson, R. G., 1960. Models and methods for analysis of the mode of formation of fossil assemblages, *Geol. Soc. Amer. Bull.*, **71**, 1075–1086.

Johnson, R. G., 1962. Interspecific associations in Pennsylvanian fossil assemblages, *J. Geol.*, **70**, 32–55.

Johnson, R. G., 1964. The community approach to paleoecology, in Imbrie and Newell, 1964, 107–134.

Johnson, R. G., 1972. Conceptual models of benthic marine communities, in T. J. M. Schopf, ed. *Models in Paleobiology*. San Francisco: Freeman, Cooper, 148–159.

Kauffman, E. G., 1974. Cretaceous of the Western Interior United States, in Ziegler et al., 1974, 12.5–12.14.

Lawrence, D. R., 1968. Taphonomy and information loss in fossil communities, *Geol. Soc. Amer. Bull.*, **79**, 1315–1330.

Lochman-Balk, C., and Wilson, J. L., 1958. Cambrian biostratigraphy in North America, *J. Paleontology*, **32**, 312–350.

McBride, D. J. 1976. Outer shelf communities and trophic groups in the Upper Cambrian of the Great Basin, *Brigham Young Univ. Geol. Stud.*, **23**, 139–152.

McKerrow, W. S., 1978. *The Ecology of Fossils*. Cambridge, Mass.: MIT Press, 384p.

Olson, E. C., 1966. Community evolution and the origin of mammals, *Ecology*, **47**, 291–302.

Palmer, A. R., 1969. Cambrian trilobite distributions in North America and their bearing on Cambrian paleogeography of Newfoundland, *Mem. Amer. Assoc. Petrol. Geologists 12*, 139–148.

Rhoads, D. C.; Speden, I. G.; and Waage, K. M., 1972. Trophic group analysis of Upper Cretaceous (Maestrichtian) bivalve assemblages from South Dakota, *Bull. Amer. Assoc. Petrol. Geologists*, **56**, 1100–1113.

Rollins, H. B., and Donahue, J., 1975. Towards a theoretical basis of paleoecology: Concepts of community dynamics, *Lethaia*, **8**, 255–270.

Scott, R. W., 1974. Bay and shoreface benthic communities in the Lower Cretaceous, *Lethaia*, **7**, 315–330.

Scott, R. W., and West, R. R., eds., 1976. *Structure and Classification of Paleocommunities*. Stroudsburg, Pa. Dowden, Hutchinson & Ross, 291p.

Shotwell, J. A., 1958. Intercommunity relationships in Hemphillian (mid-Pliocene) mammals, *Ecology*, **39**, 271–282.

Shotwell, J. A., 1964. Community succession in mammals of the Late Tertiary, in Imbrie and Newell, 1964, 135–150.

Stevens, C. H., 1966. Paleoecologic implications of Early Permian fossil communities in eastern Nevada and western Utah, *Geol. Soc. Amer. Bull.*, **77**, 1121–1130.

Thayer, C. W., 1974. Marine paleoecology of the Upper Devonian of New York, *Lethaia*, **7**, 121–155.

Valentine, J. W., 1973. *Marine Paleoecology of the Marine Biosphere*. Englewood Cliffs, N.J.: Prentice Hall, 511p.

Walker, K. R., 1972a. Community ecology of the Middle Ordovician Black River Group of New York state, *Geol. Soc, Amer. Bull.*, **83**, 2499–2524.

Walker, K. R., 1972b. Trophic analysis: A method for studying the function of ancient communities, *J. Paleontology*, **46**, 82–93.

Walker, K. R., and Alberstadt, L. P., 1975. Ecological succession as an aspect of structure in fossil communities, *Paleobiology*, **1**, 238–257.

Walker, K. R., and Bambach, R. K., 1974. Feeding by benthic invertebrates: Classification and nomenclature for paleoecological analysis, *Lethaia*, **7**, 67–78.

Walker, K. R., and Laporte, L. F., 1970. Congruent fossil communities from Ordovician and Devonian carbonates of New York, *J. Paleontology*, **44**, 928–944.

West, R. R., 1972. Relationship between community analysis and depositional environments: An example from the North American Carboniferous, *24th. Internat. Geol. Cong. Proc.*, **7**, 130–146.

Worsley, D., and Broadhurst, F. M., 1975. An environmental study of Silurian atrypid communities from southern Norway, *Lethaia*, **8**, 271–286.

Zeigler, A. M., 1965. Silurian marine communities and their environmental significance, *Nature*, **207**, 270–272.

Zeigler, A. M.; Cocks, L. R. M.; and Bambach, R. K., 1968. The composition and structure of Lower Silurian marine communities, *Lethaia*, **1**, 1–27.

Ziegler, A. M.; Walker, K. R.; Anderson, E. J.; Kauffman, E. G.; Ginsburg, R. N.; and James, N. P., 1974. Principles of benthic community analysis, *Univ. Miami, Comp. Sed. Lab. Sedimenta IV*, 1.1–12.14.

Cross-references: *Actualistic Paleontology; Diversity; Dwarf Faunas; Fossil Record; Paleoecology; Population Dynamics; Substratum; Taphonomy; Trace Fossils; Trophic Groups.*

COMPUTER APPLICATIONS IN PALEONTOLOGY

The use of sophisticated equipment in paleontology has increased greatly in the past decade. It is not at all unusual to find a paleontologist using scanning electron microscopes, mass spectrometers, and computers as a part of his daily routine. Yet one finds review papers that discuss applications of the computer to various branches of science but not of other kinds of equipment that are relatively new on the scene. The reason for this is that whereas most technological improvements have led to the acquisition of more and better data, the computer has permitted more thorough analysis of data already on hand. As a result, it has provided paleontologists and other scientists as well with a means of expanding their view of problems (Olson, 1970).

Applications of the computer in paleontology as well as in other branches of science have been most successful when they have taken advantage of the ways in which the computer works best—i.e., in manipulating large masses of data, in making numerous repetitive routine calculations, and in searching for and retrieving information from files. Attempts to make computers simulate men have been less successful, largely because of the difficulty of expressing in explicit terms the decision-making processes through which the human mind goes in evaluating abstract concepts or ideas.

Data Processing

Several different ways of processing paleontological data have made use of the computer. They vary in complexity, so that the use of the computer may range from a convenience to an absolute necessity. The most simple application of the computer is in the reduction of data. For example, in studies of the ratio of calcium to magnesium in the calcite of shells (Cadot et al., 1972, 1975), data from an electron microprobe are often recorded directly onto punched cards or magnetic tape. The data are then converted from their raw form to mole-percentages of $CaCO_3$ and $MgCO_3$ by the computer. In this way, the investigator does not have to record data before making the reductions, and the chance for operator error is greatly reduced.

The increased use of descriptive statistics and statistical analysis, including univariate, bivariate, and multivariate statistics, has developed hand-in-hand with the availability of high-speed digital computers. In computing descriptive statistics such as the mean, variance, and standard error, use of the computer is a convenience. For more complicated descriptive statistics, such as measures of skewness and kurtosis, a computer is almost a necessity because computations are so tedious. Similarly, in much univariate and bivariate statistical analysis, the computer saves a great deal of time and effort; and some modern textbooks of applied statistics include a set of computer programs (Sokal and Rohlf, 1969). Most of the multivariate statistical methods, such as principal component analysis and canonical

variate analysis, simply cannot be done without a computer because of the complex manipulations of matrices involved. In any event, once one has available a working computer program, it is possible to make computations, be they of simple descriptive statistics or of the most intricate multivariate statistical analysis, free of error except for possible errors in data cards. Even errors of this kind are less likely to occur when a computer is used rather than a desk calculator or paper-and-pencil methods, since it is possible to check the cards before they are fed into the computer.

One advantage the computer gives users of statistical methods is the ability to test data to see if they meet the assumptions on which that statistical method is based. For example, most parametric statistical procedures, i.e., those that make assumptions about the nature of the population from which the data were drawn, require that data be normally distributed and that variances of all samples be equal. These assumptions, and especially the mathematically more involved assumptions of multivariate statistical methods, are too often not tested because the test computation may be more difficult than that of the statistical analysis itself. The computer makes testing the assumptions easy and obviates any attempt to avoid doing so. As a matter of fact, sets of data that meet the assumptions of most multivariate statistical methods are exceedingly rare. The more the data depart from the ideal, of course, the less reliable the results of the statistical method.

Not all computer-dependent multivariate methods are properly termed statistical. Among the more widely used of these nonstatistical methods in paleontology are cluster analysis and various ordination methods (see *Taxonomy, Numerical*). Cluster analysis has been discussed thoroughly in the literature (Kaesler, 1966; Mello and Buzas, 1968; Williams, 1971). It is a method of determining the similarity of items on the basis of their characteristics or the similarity of joint occurrences of various characteristics among the items in the study. As commonly used in paleontology, the procedure involves the computation of one of a large number of similarity coefficients for each pair of items (Q-mode) or characteristics (R-mode). These are grouped into a similarity matrix, and pairs of highly similar items or characteristics are then grouped into clusters by the computer according to one of several modes of clustering. The items that are clustered and the characteristics used for comparing them may vary widely. The items may be taxa of organisms compared on the basis of morphological characters, in which case the procedure is generally called *numerical taxonomy* (q.v.). They may be samples or outcrops compared on the basis of the organisms found in them, in which case the procedure is called *quantitative biofacies analysis*. Cluster analysis has also been used successfully in quantitative biostratigraphy. Fig. 1 shows part of Hazel's (1970) cluster analysis of beds from the Eocene and Oligocene of the Gulf Coast.

Whereas cluster analysis is most precise for pairwise comparisons, ordination methods (such as principal component analysis, principal coordinate analysis, and nonmetric multidimensional scaling) are more useful for visualizing the relationships of groups of items to each other—at the level where distortion due to averaging during cluster analysis *may* make cluster analysis somewhat unreliable. Principal component or principal coordinate ordination are the most commonly used methods in paleontology. Nonmetric multidimensional scaling has not been widely used, but it holds considerable promise for the analysis of paleontological data. This is particularly true because of the difficulty of getting meaningful quantitative data on abundances or relative abundances of fossils for the analysis of ancient communities. Nonmetric multidimensional scaling has been found to give easily interpretable results with data on presence and absence, and it is adaptable for use with rank abundances.

All of these methods involve plotting items in a reduced space that represents as well as possible the multidimensional space in which the points actually lie but which is so difficult to visualize. Fig. 2 shows a 3-dimensional ordination of 34 specimens of the fusulinid genus *Pseudoschwagerina* from 5 species or subspecies based on 33 morphological characters. The 33-dimensional space has been reduced to 3 dimensions that account for 55% of the variance in the data (Kaesler, 1970).

Simulation

Simulation offers the paleontologist a powerful tool for understanding the subject matter under consideration. This is particularly true because of the nature of computers: quick, faithful idiots that do always what we say and not necessarily what we mean. Thus it is essential that the paleontologist planning a simulation understand precisely the processes operating in the system being simulated. Often it is necessary to engage in research quite apart from the investigator's main interest in order to understand enough about the system being studied to make a meaningful simulation. As a result, the simulation is often of more benefit to the simulator than to his readers. At the same time, simulation can be misleading. As the system being simulated becomes increasingly complex or poorly understood, the

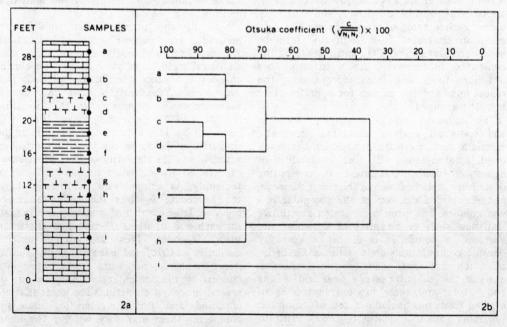

FIGURE 1. Use of cluster analysis in biostratigraphy, based on Otsuka coefficients giving overall similarity among all pairs of 9 samples, based on the occurrences of 153 species (from Hazel, 1970).

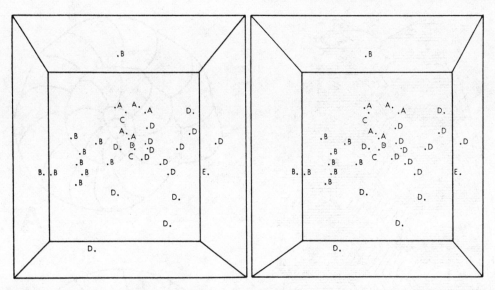

FIGURE 2. Principal component 3-dimensional ordination of 34 specimens of *Pseudoschwagerina* from 5 species or subspecies (from Kaesler, 1970).

paleontologist may be, in effect, sitting at a black box turning dials until a result is obtained that can be interpreted in a meaningful way. Other settings of the dials may produce the same results, so that the nonuniqueness of the solution can lead to a misunderstanding of the system. For example, in simulating a complex ancient ecosystem, one might be unable to separate the effects on light penetration due to suspended terrigenous detritus and microplankton. Similarly, understanding the impacts on the ecosystem of two kinds of organisms with strongly overlapping niches along one or more axes may require one to make untestable assumptions.

Nevertheless, simulation has been used to advantage in paleontology to study morphology, population dynamics, community structure, and paleoecological processes. Raup has been among the most active in simulating morphology, especially the coiling of molluscan shells (Raup, 1966, 1970). Fig. 3 shows his simulation of growth of an echinoid interambulacrum (Raup, 1968, 1970). Fig. 4 shows the morphology of and a 2-dimensional simulation of the foraminifer *Ammonia beccarii* based on only six morphological characters (Chang et al., 1974).

Craig and Oertel (1966) have used deterministic models to simulate living and fossil populations of animals. By varying recruitment, survivorship, and other parameters of the population, they were able to see variations in population structure. Fox (1970) has developed a simulation model for communities of marine organisms to test hypotheses about the distribution of the fossilized organisms through time. His model permits the paleontologist to vary several parameters of the physical environment at one time. Most recently, Kranz (1974) has simulated assemblages of bivalves under conditions of anastrophic burial. His studies make it possible to determine the effect of rapid burial of a living community of bivalves on their subsequent fossil record.

In contrast to simulation, which is a form of synthesis, harmonic or Fourier analysis of shapes of fossils seeks to determine the components of the form of the fossil. Several paleontologists have made use of this method, most recently Christopher and Waters (1974) in a study of shapes of microspores.

Data Storage and Retrieval

The means to handle large sets of data has spurred paleontologists to collect more data by studying more specimens, measuring more stratigraphic sections, and borrowing more material from museums. As a result, the use of the computer has strained to the breaking point the filing systems of organizations and institutes supporting paleontological research. Fortunately, the computer is helping to solve this problem as well. Digitizers, which are peripheral items of equipment to computers, are being used increasingly for measuring specimens automatically. With rapid measurement possible, by careful experimental design paleontologists can now measure the number of

217

FIGURE 3. Computer graphic simulation of growth of an echinoid interambulacrum (from Raup, 1968).

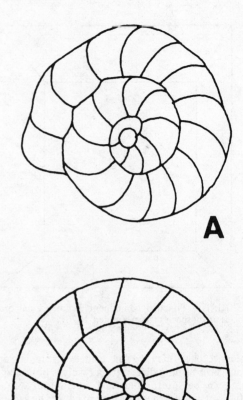

FIGURE 4. A, plan view of morphology of *Ammonia beccarii*; B, computer graphic simulation of morphology of *A. beccarii* in 2 dimensions based on 6 morphological characters (from Chang et al., 1974).

specimens needed to test a hypothesis. Previously, and still too frequently, measuring was a chore that lasted until time or patience ran out. Moreover, automated means of coping with larger and larger amounts of field data have been developed (see Cutbill, 1971). Again, these methods are based on equipment peripheral to the computer. Computer graphics are being used increasingly. These methods permit one to make complicated drawings very rapidly and inexpensively. This can be particularly useful when one is simulating morphology; steropairs can be constructed that show the organism from numerous points of view. Fig. 3 and part of Fig. 4 were drawn by plotters controlled by a computer.

Systematics collections in museums are rapidly becoming more useful as a result of information retrieval systems based on computer technology. The widely used SELGEM system developed by Mello and his associates (see Mello, 1970) permits files of data pertaining to museum collections to be sorted by the computer for numerous categories of information. As a result, questions can be asked about paleogeography, biostratigraphy, and extent of collections that might otherwise have required a drawer-by-drawer serach through the museum.

ROGER L. KAESLER

References

Cadot, H. M.; Kaesler, R. L.; and Van Schmus, W. R., 1975. Application of the electron microprobe analyzer to the study of the ostracode carapace, *Bull. Amer. Paleontology*, 65, 577-585.

Cadot, H. M.; Van Schmus, W. R.; and Kaesler, R. L., 1972. Magnesium in calcite of marine Ostracoda, *Geol. Soc. Amer. Bull.*, 83, 3519-3522.

Chang, Y. -M.; Kaesler, R. L.; and Merrill, W. M., 1974. Simulation of growth of the Foraminiferida *Ammonia beccarii* by computer, *Geol. Soc. Amer. Bull.*, 85, 745-748.

Christopher, R. A., and Waters, J. A., 1974. Fourier series as a quantitative descriptor of microspore shape, *J. Paleontology*, 48, 697-709.

Craig, G. Y., and Oertel. G., 1966. Deterministic models of living and fossil populations of animals, *Quart. J. Geol. Soc. London*, 122, 315-355.

Cutbill, J. L., ed., 1971. *Data Processing in Biology and Geology*. London and New York: Systematics Assoc. and Academic Press, 346p.

Fox, W. T., 1970. Analysis and simulation of paleoecologic communities through time, *Proc. N. Amer. Paleontological Conv.*, pt. B, 117-135.

Hazel, J. E., 1970. Binary coefficients and clustering in biostratigraphy, *Geol. Soc. Amer. Bull.*, 81, 3237-3252.

Kaesler, R. L., 1966. Quantitative re-evaluation of ecology and distribution of Recent Foraminifera and Ostracoda of Todos Santos Bay, Baja California, Mexico, *Univ, Kansas Paleontol. Contrib.*, Paper 10, 1-50.

Kaesler, R. L. 1970. Numerical taxonomy in paleontology: Classification, ordination and reconstruction of phylogenies, *Proc. N. Amer. Paleontological Conv.* pt. B, 84-100.

Kranz, P. M., 1974. Computer simulation of fossil assemblage formation under conditions of anastrophic burial, *J. Paleontology*, 48, 800-808.

Mello, J. F., 1970. Paleontologic data storage and retrieval, *Proc. N. Amer. Paleontological Conv.*, pt. B, 57-71.

Mello, J. F., and Buzas, M. A., 1968. An application of cluster analysis as a method of determining biofacies, *J. Paleontology*, 42, 747-758.

Olson, E. C., 1970. Current and projected impacts of computers upon concepts and research in paleontology, *Proc. N. Am. Paleontological Conv.*, pt. B, 135-153.

Raup, D. M., 1966. Geometric analysis of shell coiling: General problems, *J. Paleontology*, 40, 1178-1190.

Raup, D. M., 1968. Theoretical morphology of echinoid growth, *J. Paleontology*, 42, 50-63.

Raup, D. M., 1970. Modeling and simulation of morphology by computer, *Proc. N. Amer. Paleontological Conv.*, pt. B, 71-83.

Sokal, R. R., and Rohlf, F. J., 1969. *Biometry*. San Francisco: Freeman, 776p.

Williams, W. T., 1971. Principles of clustering, *Ann. Rev. Ecol. Systematics*, 2, 303-326.

Cross-references: *Biometrics in Paleontology; Diversity; Population Dynamics; Systematic Philosophies; Taxonomy, Numerical.*

CONODONTS

Conodonts, the most important and widespread, though enigmatic, Paleozoic microfossils, were first described by C. H. Pander in 1856. They range from Lower Cambrian to Upper Triassic and have a size between 0.2 and ≈6 mm. These structures are composed of an organic matrix in which calcium carbonate apatite crystallites, similar to the mineral francolithe, have been embedded. Well-preserved material is translucent and amber in color. Upon weathering, conodonts become whitish-gray and friable. Due to natural staining, or slight metamorphism of the rocks containing them, conodonts may become gray, black, or brown and opaque.

According to the outer gross morphology and notwithstanding their systematic position and their occurrence in assemblages, four main groups of conodonts can be recognized:

1. *Simple cones:* These are tooth-shaped single denticles only. Using their outer features such as their cross section, various units can be recognized. These are important index fossils from Cambrian to Devonian. (Figs. 1, 2)
2. *Bar types:* These have a thin, curved or bent shaft, which is commonly branched. They yield only few index fossils. (Figs. 3, 4)
3. *Blade types:* These are elongate, laterally compressed units formed by a row of denticles which, exclusive of their tips, are fused. Some of them are index fossils from Silurian into the Triassic. (Figs. 5, 6)
4. *Platform conodonts:* They are by far the most important units, being derived from the blade types through the development of the board flanges into a plate. Many of these are excellent index fossils from the Ordovician to the Triassic. (Figs. 7, 8)

These isolated elements are components of conodont assemblages that were formed by a variable unit of elements of 1-5 different form types. Such assemblages have been preserved under exceptional still-water conditions (Fig. 9). Some of the conodont elements occur in pairs, consisting of "right" and "left" specimens. Unpaired forms occur as well (e.g., Fig. 4) and these are believed to have been situated in the medial axis of the animal. From such evidence, it can be concluded, that the conodont-bearing animal had a distinct bilateral symmetry. Further, it is believed that is was marine

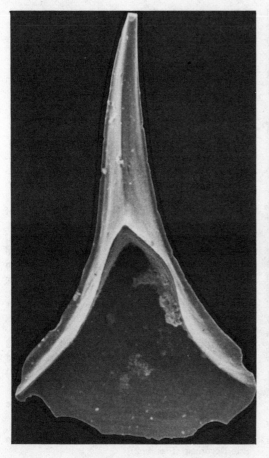

FIGURE 1. *Furnishina furnishi,* Upper Cambrian, Sweden, X76 (from Müller, 1978).

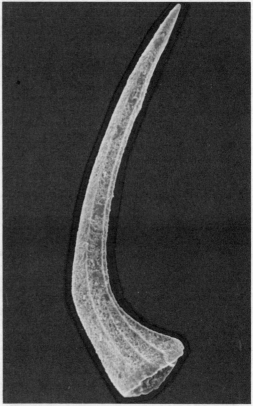

FIGURE 2. *Scolopodus rex,* Lower Ordovician, Germany, X100 (from Müller, 1978).

and free swimming. As various species have a world-wide distribution, it is concluded that some representatives of the group were pelagic, drifting in the open seas. It has been suggested that the mode of life of the chaetognaths like *Sagitta* could be taken as an ecological model for conodonts.

Systematics

The systematic position of the conodonts is still an open question. They have been assigned to various groups such as gastropods, cephalopods, annelids, nematodes, and platyhelminths, lophophorates, arthropods, vertebrates, and even algae. Comparison with any of these groups has been based on similarities in a single feature, which may be due to convergence, and because of other important dissimilarities it appears that none of these assignments is valid.

The formation of phosphatic hardparts is

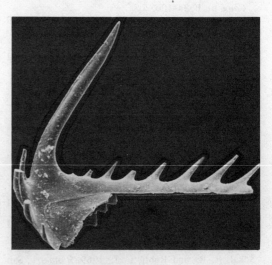

FIGURE 3. *Ligonodina,* Silurian, Germany, X70.

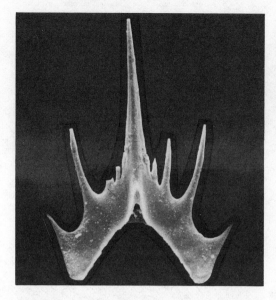

FIGURE 6. *Ozarkodina immersa,* Upper Devonian, Michigan, X73.

of little taxonomic significance. Apatitic matter is utilized in various phyla and is known from early Paleozoic brachiopods and bryozoans, conularians, and ostracods (Bradoriida), as well as in various unassigned fossils. Furthermore, the complex histology of conodonts is not related to any known living or extinct animal group.

Pander regarded them originally as fish teeth. Today we know that they were built by outer

FIGURE 4. *Hibbardella,* Upper Silurian, Germany, X100.

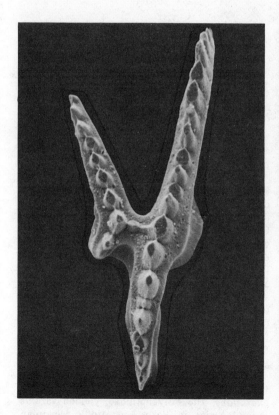

FIGURE 5. *Pterospathodus amorphognathoides,* Upper Silurian, Austria, X100 (from Müller, 1978).

FIGURE 7. *Palmatolepis hassi,* Upper Devonian, Iowa, X74 (from Müller, 1978).

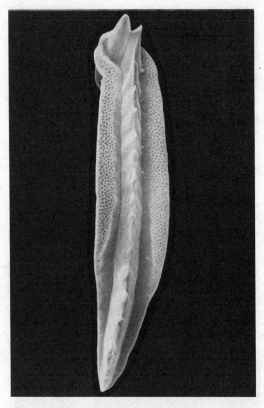

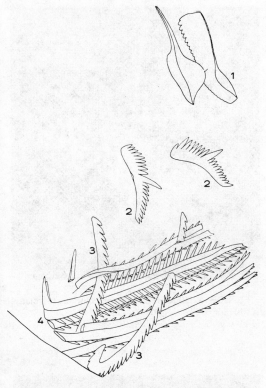

FIGURE 8. *Neogondolella prava,* Middle Triassic, Germany, X100 (from Müller, 1978).

FIGURE 9. Example of a conodont assemblage: *Gnathodus integer,* consisting of (1) one pair of platform elements *Gnathodus;* (2) one pair of blade elements *Ozarkodina;* (3) one pair of bar elements *Symprioniodina,* and (4) four pairs of bar elements *Hindeodella.* Upper Carboniferous, Germany, X16 (from Schmidt and Müller, 1964).

apposition, and thence were embedded in soft tissue during their entire growth. As conodonts grew throughout life, they must have been internal organs. Besides, they show no signs of functional wear, as in the case of chewing apparatus of vertebrates or annelids. The shape and arrangement of the elements in the apparatuses suggest that conodonts could have been supporting organs of sieving or screening devices. Histological evidence points to a function as support in tissue, and they could well have been organs for the temporary deposition of phosphate which could later be utilized either to form another conodont or for osmoregulation of the body liquid. This is suggested by recurrent resorption zones in many conodont specimens.

In spite of their doubtful systematic position, conodonts are extremely useful index fossils. They are abundant and widespread geographically and have a distinctive outer morphology and a restricted stratigraphic range. Therefore, they play an important role in oil micropaleontology, particularly in the correlation of limestone and black-shale formations. Distribution of identical or closely similar species over wide areas makes them useful tools for intercontinental correlation of rocks.

As is true of all fossil groups, abundance and time sensitivity within the conodonts varies in different epochs. The oldest described conodonts are from the Lower Cambrian. Typical Cambrian examples are *Furnishina,* a single cone (Fig. 1), and *Westergaardodina.* The first outburst of conodont evolution occurs in the Ordovician, with single cones like *Ulrichodina, Scolopodus,* (Fig. 2), *Oistodus, Drepanodus,* and the first compound conodonts like *Loxodus* (Lower Ordovician) and *Amorphognathus* (Middle-Upper Ordovician). A drop in morphologic diversity occurred in the Silurian. At this time, platform genera arose from *Spathognathodus,* a blade, e.g., *Polygnathoides* (Middle Silurian), and *Kockellela* (Late Silurian). Other genera that developed from *Spathognathodus,* the Polygnathidae, yielded various index species in Early and Middle Devonian and became of high importance for the zonation of Devonian sediments. This time may be regarded the peak of conodont evolution. *Polygnathus* gave rise to *Palmatolepis* (Fig. 7), which probably is the best single genus of all fossils for the subdivision of the Upper Devonian, having produced more than 50 species and "subspecies" in that time. *Ancyrodella* and *Ancyrognathus* are abundant in and restricted to the lower portion of the Upper Devonian. Possibly also *Ieriodus,* a useful uppermost Silurian–Devonian time maker, is related to *Spathognathodus.* The Mississippian is characterized by genera like *Siphonodella* and *Scaliognathus.* In the late Lower Carboniferious, Idiognathodontidae, e.g., *Gnathodus* and *Cavusgnathus* became prevalent, they however do not appear to match the Devonian Polygnathidae as index fossils. In the Pennsylvanian, species of *Idiognathodus* serve as index fossils. Around this time, the platform genus *Gondolella* arose from a bartype. In the Permian, conodonts were scarce again and almost died out. In the Triassic, they became widespread once more. The Triassic shows the development of the genus *Neogondolella* (Fig. 8), which is believed to have arisen from a *Spathognathodus* ancestor in earliest Triassic time. Undergoing a rapid evolutionary development, this genus then gave rise to a variety of forms which are important index fossils for this time interval. This rapid evolutionary burst seems to end abruptly before the end of the Triassic and no conodonts have yet been found in later periods. An isolated occurrence in the W African Upper Cretaceous has been proved to be a reworked Triassic fauna.

Throughout the Ordovician and Triassic periods, a variety of bar-type conodonts are present. These, however, do not seem to have such stratigraphic value as do the platform conodonts.

Conodonts occur in many types of sediments, including carbonates, shale, and sandstone. They can be isolated from limestones and dolomites by etching with 15% acetic or formic acids. They may also be obtained from slightly metamorphic carbonates, which do not yield fossils by any other means. The conodonts are commonly undectectable in the rock, and can only be observed in the etched residue. They can also be washed from shales by conventional methods. As they have a specific gravity between 2.84 and 3.10, they can be separated from common minerals with heavy liquids such as bromoform. Furthermore, they can be concentrated with a magnetic separator.

K. J. MÜLLER

References

Barnes, C. R., ed., 1976. Symposium on conodont paleoecology, *Geol. Assoc. Canada Spec. Paper 15,* 324 p.

Conway Morris, S., 1976. A new Cambrian lophophorate from the Burgess Shale of British Columbia, *Palaeontology,* **19,** 199–222.

Hass, W. H.; Rhodes, F. H. T.; Müller, K. J.; and Moore, R. C., 1962. Conodonts, in R. C. Moore, ed., *Treatise on Invertebrate Paleontology,* pt. W, Miscellanea. Lawrence, Kansas: Geol. Soc. Amer. and Univ. Kansas Press, W3–W98.

Lindström, M., 1964. *Conodonts.* Amsterdam: Elsevier, 196p.

Lindström, M., and Ziegler, W., eds., 1972. Symposium on conodont taxonomy, *Geol Palaeontol.,* SBI, 158p.

Müller, K. J., 1978. Conodonts and other phosphatic microfossils, in B. U. Haq and A. Boersma, eds., *Introduction to Marine Micropaleontology.* New York: Elsevier, 277–291.

Müller, K. J., and Nogami, Y., 1971. Über den Feinbau der Conodonten, *Mem. Fac. Sci. Kyoto Univ., Ser. Geol. Mineral.,* **38,** 1–87.

Rhodes, F. H. T., ed., 1973. Conodont paleozoology, *Geol. Soc. Amer. Spec. Paper, 141,* 296p.

Rhodes, F. H. T., and Austin, R. L., 1977. Ecologic and zoogeographic factors in the biostratigraphic utilization of conodonts, in E. G. Kauffman and J. E. Hazel, eds., *Concepts and Methods of Biostratigraphy.* Stroudsburg, Pa.: Dowden, Hutchinson & Ross, 365–396.

Schmidt, H., and Müller, K. J., 1964. Weitere Funde von Conodonten-Gruppen aus dem oberen Karbon des Sauerlandes, *Paläont. Zeitschr.,* **38,** 105–135.

Sweet, W. C., and Bergström, S. M., eds., 1971. Symposium on conodont biostratigraphy, *Geol. Soc. Amer. Mem. 127,* 499p.

Ziegler, W., and Lindström, M., 1975. Fortschrittsbericht Conodonten, *Paläont. Zeitschr.,* **49,** 565–598.

CONULATA

Conulata are steep sided, pyramidal fossils with a chitinophosphatic test or shell that is characteristically marked by fine, transverse growth ridges (Fig. 1). The cross-sectional shape is generally square or rhomboidal and internal septa may be present. Conulata are found in rocks of Cambrian to Triassic age.

Conulata have been considered to be Scyphozoans (see *Coelenterata*) by recent synthesizers (e.g., Moore and Harrington, 1956) because of the tetrameral symmetry that is characteristic of both groups and because of characters discernable in the single genus *Conchopeltis.* The latter however, lacks all of the principle characters of the Conulata proper and arguments based on this form are spurious. Kozlowski (1968) noted that none of the "evidence" of relationship between Conulata proper and the Scyphozoa were based on fundamental characters and that differences were too great to warrant assignment of the Conulata to any living group of animals. Consequently the

223

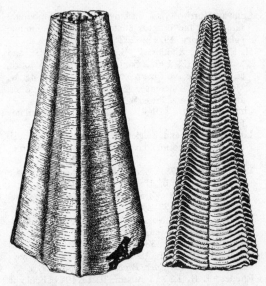

FIGURE 1. Examples of Devonian and Mississippian conularids, XI (from Fenton and Fenton, 1958; copyright © 1958 by Carroll Lane Fenton and Mildred Adams Fenton and used by permission of Doubleday & Company, Inc.

Conulata are considered to be a group of unknown affinities, but probably not Coelenterates.

WILLIAM A. OLIVER, JR.*

References

Fenton, C. L., and Fenton, M. A., 1958. *The Fossil Book*, Doubleday, 482p.

Kozlowski, R., 1968. Nouvelles observations sur les conulaires, *Acta Paleontol. Polonica*, **13**, 497–536.

Moore, R. C. and Harrington, H. J., 1956. Scyphozoa, in R. C. Moore, ed., *Treatise on Invertebrate Paleontology*, pt. F, Coelenterata, Lawrence, Kansas: Geol. Soc. Amer. and Univ. Kansas Press, F27–F66.

COPROLITE

Coprolites (Greek: *kopros*, dung; *litos*, rock) are the fossilized excrement of animals, usually in reference to vertebrates. In 1829, Rev. William Buckland coined the term to refer to the fossilized feces of terrestrial and aquatic carnivorous animals (Folk, 1965). Buckland's (1835) memoir described and discussed in detail specimens he considered to be coprolites of the *Ichthyosaurus* in the English Jurassic. However, we know today that these specimens are not coprolites in the strict sense of the word. They are actually fossilized spiral intestines, such as those possessed by sharks and some other fish (Williams, 1972).

*Publication authorized by the Director, U.S. Geological Survey

Coprolites are of two main types: petrified ones, generally phosphatic and stony; and nonpetrified ones, generally crumbly and of more recent age. Petrified coprolites range from coarse sand size up to 150 mm (6 in.) or more in length. They are usually rod shaped or cigar shaped (Fig. 1), but some specimens may be more pellet-like; and some may be twisted (Fig. 2) or convoluted, rarely spiral. They often show sphincter striations and compressions. Petrified coprolites generally are hard and compact, range from yellowish to dark brown, and take a good polish (they have occasionally been used as ornaments and cheap jewelry by persons who did not know their genesis). The composition is chiefly calcium phosphate (collophane), with impurities of calcium, magnesium, or iron carbonate, silica, iron, and aluminum oxides, and organic matter. Some have been later replaced by silica or carbonate minerals. Strontium, yttrium, arsenic, barium, vanadium, copper, and manganese are known as minor elements in various assemblages of coprolites.

In thin section, coprolites are usually dark brown, transparent to translucent, optically isotropic, with refractive index near 1.6. They may contain organic or inorganic inclusions, such as sand grains, pollen grains, and other vegetable remains; spicules; invertebrates (Fig. 1); microorganisms; and, commonly, bones, teeth, and fish scales from ingested animals.

In only a very few cases can the animal producer be positively identified. They have been ascribed to extinct marine reptiles such as ichthyosaurs, plesiosaurs, terrestrial dinosaurs, sharks and other fishes (Fig. 2), crocodiles, and carnivorous mammals.

Coprolites have been reported from rocks of Ordovician through early Holocene age, but the Mesozoic was probably their zenith. In the geologic column, nonetheless, they are rather rare curiosities, although sometimes encountered in large quantities, particularly at unconformities. Sometimes they are preserved in natural clusters in the original excreted position (Buckland, 1835). Occasionally, large accumulations can be used as a commercial source of phosphate for fertilizer.

The term "coprolite" has been extended by some authors to include fossilized fecal matter produced by any kind of animal, regardless of size or composition (Twenhofel, 1932). However, silt-to-sand feces of worms or other invertebrates, which usually consist not of phosphate but clay, carbonate, or glauconite, are more commonly known as fecal pellets rather than "coprolites." Guano is also generally excluded because of the relatively form-

FIGURE 1. Coprolite from Pierre Shale Formation of South Dakota (in collection of the University of Nebraska State Museum). Remains of crustacean invertebrates protrude from either end, scale = 1 cm.

less nature of the droppings (see Hutchinson, 1950).

Nonpetrefied coprolites, generally of Pleistocene or younger age, have been identified as being created by animals such as hyenas, other carnivores, ground sloths, and man. These are often found in cave and fissure deposits. Such coprolites can be useful in paleoecologic analysis. With analysis of the fecal material, researchers can determine the diet of the producer, and thus, obtain an excellent picture of the environment in which the producer lived. This analytical technique is not very useful with material that has undergone even a slight amount of fossilization or mineralization. It is, then, largely limited to use with post-Tertiary material from caves and fissures. The

most useful adoption of this type of paleoecologic analysis has been with coprolites of man from archaeological sites (Bryant, 1974; Bryant and Williams-Dean, 1975) and with the coprolites of extinct ground sloths from the Pleistocene of the southwestern United States (Martin, 1975). The contents of more mineralized coprolites can also be examined to determine probable dietary habits (Waldman and Hopkins, 1970), but as noted above, the process of fossilization and mineralization can alter the material in such a manner as to make an accurate analysis of the contents impossible. The data is somewhat less useful, too, when the producer is unknown, or at least doubtfully known at best.

Both types of coprolites can also be useful in taphonomic analysis of paleoecologic-paleoenvironmental problems (see *Taphonomy*). They are resistant to physical and chemical weathering; and like other fossils, they can be analyzed hydrodynamically in studies of environment of deposition and burial (Edwards and Yatkola, 1974; see also Mellet, 1974).

The most comprehensive reviews of the literature on coprolites are Häntzschel et al. (1968) and Bryant (1974). Bradley (1946) described Eocene coprolites from Wyoming, in a classic study.

PAUL EDWARDS*
R. L. FOLK

FIGURE 2. Fish coprolite from the Green River Formation (Eocene), Wyoming (in the collection of the University of Wyoming Geology Museum), scale = 1 cm.

*Deceased

225

References

Bradley, W. H., 1946. Coprolites from the Bridger Formation of Wyoming; their composition and micro-organisms, *Amer. J. Sci.,* **244,** 215-239.

Bryant, V. M., Jr., 1974. The role of coprolite analysis in archaeology, *Bull. Texas Arch. Soc.,* **45,** 1-28.

Bryant, V. M., and Williams-Dean, G., 1975. The coprolites of man, *Sci. American,* **232**(1), 100-109.

Buckland, Rev. William, 1835. On the discovery of coprolites, or fossil feces, in the Lias at Lyme Regis, and in other formations, *Geol. Soc. London, Trans.,* ser. 2, **3,** 223-236.

Edwards, P. E., and Yatkola, D. A., 1974. Coprolites of White River (Oligocene) carnivorous mammals: Origin and paleoecologic and sedimentologic significance, *Wyoming Univ. Contrib. Geol.,* **13,** 67-74.

Folk, R. L., 1965. On the earliest recognition of coprolites, *J. Sed. Petrology,* **35,** 272-273.

Häntzschel, W.; El-Baz, F.; and Amstutz, G. C., 1968. Coprolites: An annotated bibliography, *Geol. Soc. Amer., Mem. 108,* 132p.

Hutchinson, G. E., 1950. Survey of contemporary knowledge of biogeochemistry: 3. The biogeochemistry of vertebrate excretion, *Amer. Mus. Nat. Hist. Bull.,* **96,** 1-554.

Martin, Paul S., 1975. Sloth droppings, *Nat. Hist.,* **84**(7), 74-81.

Mellet, James S., 1974. Scatological origin of microvertebrate fossil accumulations, *Science,* **185,** 349-350.

Twenhofel, W. H., 1932. *Treatise on Sedimentation.* Baltimore: Williams and Wilkens, 926p.

Waldman, M., and Hopkins, W. S., Jr., 1970. Coprolites from the upper Cretaceous of Alberta, Canada, with a description of their microflora, *Canadian J. Earth Sci.,* **7,** 1295-1303.

Williams, M. E., 1972. The origin of "spiral coprolites," *Kansas Univ. Paleontol. Contrib., Paper 59,* 19p.

Cross-references: *Biogeochemistry; Paleoecology, Terrestrial; Taphonomy; Vertebrate Paleontology.*

CORALS

In paleontology, the term *coral* most commonly refers to massive calcium carbonate skeletons formed by members of the class Anthozoa and to the animals forming the skeleton. But usage is not consistent; thus, skeletonless sea anemones are not corals although belonging to the Anthozoa; whereas Octocorallia are generally termed corals, although most of them also lack the massive ectodermal skeleton. Sometimes included under the term are those hydrozoans that form massive carbonate skeletons (e.g., Milleporina, Stylasterina, and even the Stromatoporoidea; see *Coelenterata*).

The coral skeleton is external and of ectodermal origin. It is commonly made up of a wall or epitheca that encloses and unites the septa (radiating longitudinal or vertical plates) and varied transverse or horizontal elements (tabulae, dissepiments, etc.; see Fig. 1). The coral skeleton is formed by a polyp (Fig. 2) and the various structural elements reflect the morphology of the polyp in such a way that some of the fundamental characters of the polyp can be deduced from its skeleton. Thus the skeletal septa bear a clear relationship to the polyp's mesenteries both in numbers and position, and the transverse tabulae and dissepiments show the position and form of the polyp's basal disc. The contruction of a coral skeleton is an additive process so that it preserves a record of the changes that take place in the polyp during growth. Notable among these changes are modifications in the number, configuration, and arrangement of the septa, as will be noted below for some of the coral groups.

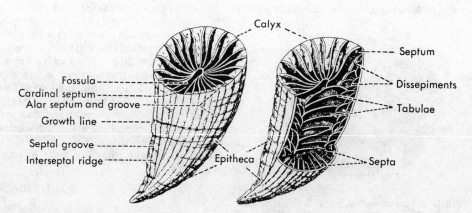

FIGURE 1. Morphology of coral skeleton; based on rugose coral (from Fenton and Fenton, 1958; copyright © 1958 by Carroll Lane Fenton and Mildred Adams Fenton and used by permission of Doubleday & Company, Inc.

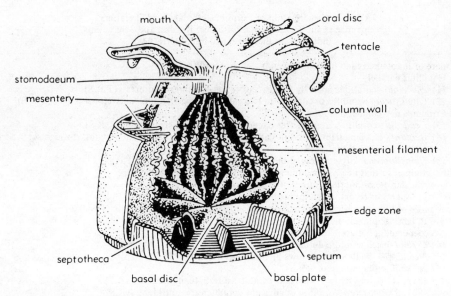

FIGURE 2. Morphology of polyp and skeleton in living scleractinian coral (from Moore, 1956, courtesy of The Geological Society of America and The University of Kansas Press).

The skeleton of a single polyp is a corallite, that of the colony is a corallum. Solitary corals have various shapes, ranging from flat (discoid) to broadly conical to cylindrical (Fig. 3). Colonial corals may be loosely or compactly arranged and a complex terminology is necessary to define various forms. These are listed in Table 1 and some are illustrated in Fig. 4.

Colonial corals increase (asexually reproduce) in several ways. In living corals, increase or budding is intratentacular (polyp division taking place inside the ring of tentacles) or extratentacular (buds forming outside the parent's ring of tentacles) with several kinds of both being recognized. In fossils, only the skeleton is available for study and different terms are used for various recognizable modes of increase (Figs. 4 and 5). *Axial increase* involves the division of the parent corallite into two or more offsets which completely occupy the cross-section area of the parent; the process may have involved the longitudinal division of the parent polyp, but in any case the parent as an individual didn't survive and the process is termed parricidal. In *peripheral increase,* one or more offsets develop near the periphery of the parent corallite but within its wall; the process was parricidal in some but apparently not in all cases. In *lateral increase,* one or a few offsets project from the side of the parent corallite; the parental diameter is not affected and the process is nonparricidal. The term *intermural increase* is commonly used in cerioid coralla where the offset appears to

form in the wall between two corallites; generally this is lateral increase within a confined space. In *coenosteal increase,* offsets arise from the coenosteum that unites corallites in coenosteoid coralla and there is no clear relationship between an offset and any one possible parent. *Syringoporoidal increase* has been described as the development of an offset from a connecting tubule between two corallites; this may not actually occur in any of the zoantharian coral groups but stolonal increase in some octocorals is analagous.

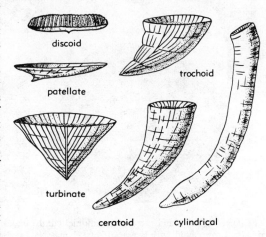

FIGURE 3. Form in solitary corals (from Moore, 1956, courtesy of The Geological Society of America and The University of Kansas Press).

TABLE 1. Classification of Form in Colonial Coral Skeletons.
(Occurrence in the Tabulata, Heliolitoidea, Rugosa, and Scleractinia
indicated by the respective initial letters.)

I. Fasciculate or branching; each branch formed by single corallite.
    A. Corallites parallel
        (1) *Phaceloid*—corallites laterally free except for supporting structures (T,R,S)
        (2) *Cateniform*—corallites in contact on two sides forming chain in transverse section (T,R,S)
    B. Corallites not parallel
        (3) *Dendroid*—corallites laterally free except for supporting structures (T,R,S)
        (4) *Reptant*—creeping habit; corallites attached to substrate, may form network through secondary
            contacts of corallites (T,R,S)
        (5) *Umbelliferous*—offsets formed in whorls at right angles to axis (T,R)
II. Corallites grouped so that neighboring ones are in contact on all sides; may be
    Massive—lenticular, hemispherical, or irregular in shape; *or*
    Laminar—thin sheets or layers, often incrusting; *or*
    Ramose—more-or-less cylindrical branches; *or*
    Foliose—flat, laminar branches.
    A. No coenosteum between corallites
        (6) *Cerioid*—each corallite defined by wall (T,R,S)
        (7) *Meandroid*—walls separate rows of confluent corallites (T,S)
        (8) No wall, septa complete (R,S)
            (8a) *Astreoid*—septa of adjacent corallites alternate in position (R,S)
            (8b) *Thamnasterioid;* septa of adjacent corallites are confluent (R,S)
        (9) *Aphroid*—no wall, septa withdrawn from periphery (R,S)
    B. Corallites separated laterally by common skeletal tissue (coenosteum)
    (10) *Coenosteoid* (H,S)

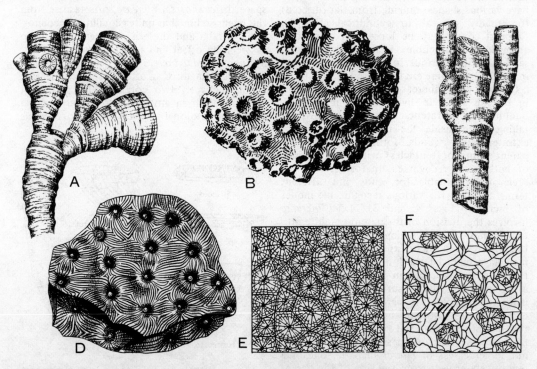

FIGURE 4. Form and increase in colonial corals; based on rugose corals; all Xl. A, dendroid with peripheral offsets; B, phaceloid with lateral offsets; C, massive, astreoid; D, massive, thamnasterioid; E, cerioid in transverse section; F, aphroid in transverse section. (Sources: A and C from Fenton and Fenton, copyright © 1958 by Carroll Lane Fenton and Mildred Adams Fenton and used by permission of Doubleday & Company, Inc. B, from Moore, et al., 1952, copyright © 1952 by McGraw-Hill Book Co., used with permission. D, E, and F, from Moore, 1956, courtesy of The Geological Society of America and The University of Kansas Press)

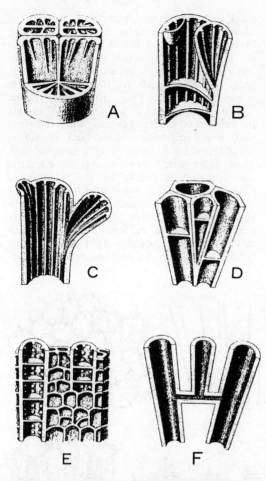

FIGURE 5. Modes of increase (from Shrock and Twenhofel, 1953, copyright © 1953 by McGraw-Hill Book Co., used with permission). A, axial; B, peripheral; C, lateral; D, intermural (commonly a form of lateral increase); E, coenosteal; F, syringoporoidal (of questionable validity). A–C based on Rugosa; D–F based on Tabulata.

Of the above types of increase, axial is intratentacular, while lateral, coenosteal, and syringoporoidal (if real) are extratentacular. Peripheral increase has been interpreted both ways and may be intratentacular in some, extratentacular in other corals.

Classification of Corals

The general relationships of the various groups that include corals are as follows:

Class ANTHOZOA
  Subclass   Octocorallia–Paleozoic(?),   Jurassic-Holocene
  Subclass Zoantharia
    Order Tabulata–Cambrian(?), Ordovician-Permian
    Order Heliolitoidea–Ordovician-Devonian
    Order Rugosa–Ordovician-Permian
    Order Heterocorallia–Mississippian
    Order Scleractinia–Middle Triassic-Holocene

The general nature of these and other anthozoan groups and the position of the Anthozoa within the phylum Coelenterata or Cnidaria are discussed in the article on Coelenterata. It should be emphasized that the Octocorallia includes many skeletonless groups and that the Scleractinia have close relatives that lack skeletons (sea anemones). It is probable that the extinct groups similarly included or were related to noncoral anthozoans.

Morphology and Distribution

The class Anthozoa comprises coelenterates in which the polyp stage predominates and there is no medusa. A stomodaeum (pharynx) extends from the mouth into the coelenteron, which is divided into vertically elongate, wedge-shaped spaces by mesenteries; these extend from oral to basal discs and are attached to the outer wall (Fig. 2). Anthozoans display bilateral symmetry in the development and arrangement of mesenteries, in possessing an elongate mouth, and commonly in the organization of the skeleton; but a radial arrangement is often more immediately apparent.

Subclass Octocorallia. Octocorallia (Alcyonaria) are exclusively colonial anthozoans characterized by polyps having eight mesenteries and eight tentacles. Skeletons are horny or calcitic, rarely aragonitic, and commonly internal, although some forms also have an external skeleton. Octocorals comprise such varied living groups as the sea pens, sea fans, organ-pipe corals, precious corals, soft corals, and horny corals. These names derive from a combination of colony form and skeletal composition. The calcareous skeleton of most octocorals consists only of isolated spicules that disassociate on the death of the colony. Octocoral spicules are common in some recent sediments and are known in rocks of Cretaceous to Holocene age. In other octocorals, the spicules are variously united to form supporting structures which, if calcified may be more easily recognized than individual spicules. One genus forms a tubular skeleton of fibrous aragonite that is quite similar in form to certain heliolitoidid corals. Fossil octocorals are not common except for local occurrences of spicules. Paleozoic and Triassic occurrences are questionable, but fossils of Jurassic to Holocene age are reliably assigned to this group.

Subclass Zoantharia. Most of the familiar fossil and living corals ("stony corals") belong to the subclass Zoantharia, characterized by the possession of paired mesenteries. Skeletons, if present, are calcareous, ectodermal in origin, and composed of elements that are united to form a characteristic unit. The Zoantharia include three orders of skeletonless anemones in addition to the following coral orders.

Tabulata (Fig. 6) comprise a variety of coral groups some of which may be unrelated and possibly not even coelenterates. Included forms are exclusively colonial with skeletons that were probably originally calcitic. Septa are incomplete or lacking and intercommunications (pores or tubes) between corallites are common. Tabulates comprise reptent and erect branching, cateniform, and massive (cerioid) forms of various sizes and shapes (Table 1). Corallites are commonly elongate and gently tapering or cylindrical with numerous transverse tabulae. Corallite diameters are generally small (0.5–5 mm) with rare larger individuals. Asexual increase in most tabulates is lateral (including "intermural") but axial and peripheral increase are known. Microstructure is fibro-lamellar or trabecular. Tabulate corals are common in rocks of Ordovician to Devonian age, less common in Mississippian to Permian rocks.

Heliolitoidea (Fig. 7) are similar to Tabulata in being exclusively colonial with skeletons that were originally calcitic and in having corallites of small diameter with many tabulae. They differ significantly in that the corallites are separated in coralla by skeletal material (coenosteum) that is apparently the deposit of colonial tissue (coenosarc) rather than of individual polyps. Twelve septa are commonly present and skeletal evidence of intercommunication between corallites (pores, connecting tubes) is lacking. Asexual increase was com-

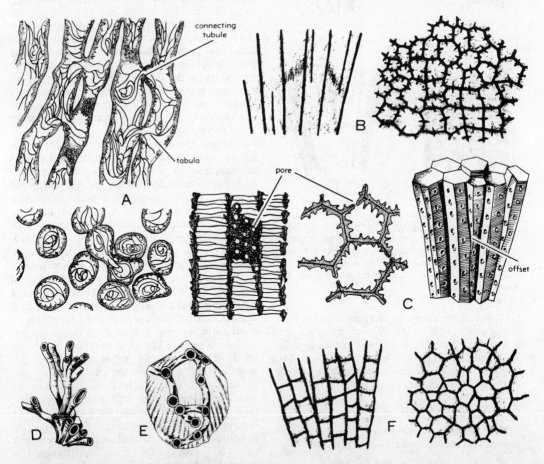

FIGURE 6. Tabulata: A, Sections of phaceloid form with connecting tubes (X5); B, Sections of cerioid form lacking tabulae and pores (X6); C, Section (X6) and fragment (X4) of cerioid form with mural pores; D, Umbelliferous form (X1); E, Reptant form encrusting brachiopod shell (X2); F, Sections of cerioid form with tabulae but lacking pores. (Sources: A, C, and D, from Moore, 1956, courtesy of The Geological Society of America and The University of Kansas Press; B, E, and F, from Fenton and Fenton, copyright © 1958 by Carroll Lane Fenton and Mildred Adams Fenton and used by permission of Doubleday & Company, Inc.)

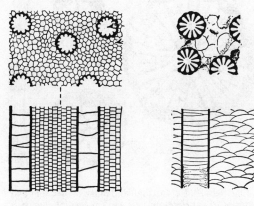

FIGURE 7. Heliolitoidea: Transverse (X4.5) and longitudinal (X2) diagrams of two Silurian examples (from Shimer and Shrock, 1944, copyright 1944, by permission of the M.I.T. Press).

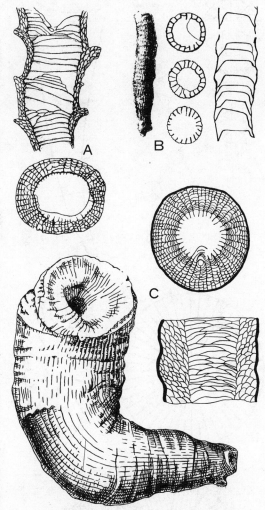

FIGURE 8. Rugosa: Transverse and logitudinal sections and exteriors of three specimens (from Moore et al., 1952; copyright © 1952 McGraw-Hill Book Co., used with permission). A, from Silurian, B and C from Pennsylvanian; sections of B are X2, are others X1.

monly coenosteal although lateral increase also occurred. Coralla are massive but may be laminar to lenticular or hemispherical or even branching in overall shape. Heliolitoidea are common in rocks of Ordovician to Middle Devonian age.

Rugosa include a more varied group of corals than do the other Paleozoic orders. Rugosa are solitary or colonial and embrace most of the shapes and colony forms mentioned in the general discussion (Table 1; Figs. 1, 3, 4, and 8). Individual rugose corals vary in size from a few mm to giants up to 15 cm in diameter and nearly a meter in length. The general shape and morphology of typical rugose corals are shown by the illustrations. Internal structures are best seen in sections cut through the coral, either transverse (at right angles) to axis or longitudinal (cut so as to include axis), and most Paleozoic corals are illustrated in this way.

The Rugosa are characterized by a high level of septal development and by the characteristic way in which septa are added during growth (Fig. 9). Major septa are inserted two or four at a time but in only four possible positions relative to an initial six protosepta and in such a way that the septa are bilaterally arranged throughout growth. With upward growth, the space between the newly added septa and the protosepta increases, leaving spaces in which the next group of septa are inserted. Minor septa are inserted after the major septa in intervening spaces, so that the number of major (including protosepta) and minor septa is commonly the same. This bilateral, four-fold plan of septal insertion is fundamental in rugose corals although large corals with many septa may appear to be radially symetrical.

Well-preserved rugose corals are composed of calcite and it is probable that this was the original mineral composition. Septa are commonly trabecular in structure although laminar septa are also known. Other skeletal elements tend to be fibro-laminar.

The Rugosa are common in rocks of Middle Ordovician to Late Permian age. For much of the Paleozoic they serve as good indices to age and environment.

Heterocorallia (Fig. 10) are a very small group of fossil corals that are known only from rocks of Mississippian age, although world wide in distribution during this time. They include only solitary corals (one or two genera)

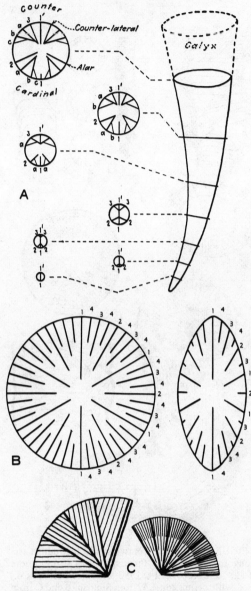

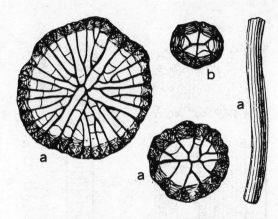

FIGURE 10. Heterocorallia: Transverse sections and exterior of (a) *Heterophyllia* and (b) *Hexaphyllia;* Mississippian; sections X5 and X7; exterior X.9 (from Moore et al., 1952, copyright 1952 McGraw-Hill Book Co., used with permission).

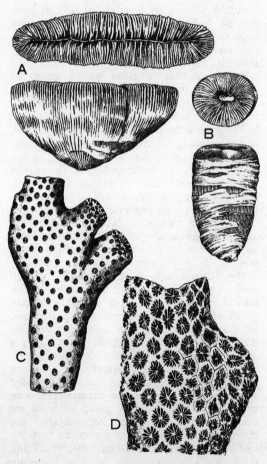

FIGURE 9. Septal insertion in Rugosa and Scleractinia: A, bilateral insertion in Rugosa; series of diagramatic transverse sections through solitary, conical coral; numbers indicate the sequence of introduction of the six primary septa (three pairs); lower case letters indicate the sequence of subsequent major septa (shown here in fours). It is important that subsequent major septa are inserted in only four positions. B, radial insertion in Scleractinia; numbers indicate successive cycles in typical form (left) and form with irregularities (right). C, results, if the outer surfaces of conical corals are rolled out on flat surface; the six primary septa are shown by heavy lines. The contrast between Rugosa septa (left) and Scleractinia (right) is pronounced. (Sources: A, from Shrock and Twenhofel, 1953, copyright 1953 McGraw-Hill Book Co., used with permission; B, from Moore, 1956, courtesy of The Geological Society of America and The University of Kansas Press; and C, from Moore et al., 1952, copyright 1952 McGraw-Hill Book Co., used with permission)

FIGURE 11. Scleractinia: Exterior views of four corals (A and B are solitary; C and D are colonial), all X.9 (from Fenton and Fenton, 1958, copyright © 1958 by Carroll Lane Fenton and Mildred Adams Fenton and used by permission of Doubleday & Company, Inc.).

and probably are an aberrent offshoot of the Rugosa, but their arrangement of septa is so peculiar that most workers (e.g., Moore, 1956) have recognized them as a separate order.

Scleractinia (or Madreporaria) include nearly all post-Paleozoic fossil and living corals. They are both solitary and colonial and are even more varied than the Rugosa in shape and colony form (Fig. 11; Table 1). Individuals within colonies may be very small; but a solitary elongate-discoidal coral with greatest diameter of 1 m is also known. General shape and morphology of selected Scleractinia are shown in the illustrations.

The Scleractinia possess well-developed septa and the sequence of their development is characteristic. There are six protosepta as in the Rugosa but additional septa are introduced in cycles of 6, 12, 24, etc., each new septum centered in one of the already existing interseptal spaces. (See Fig. 9 for diagram and comparison with Rugosa). The result is a six-fold radial symmetry, although the third and higher cycles of septa may be incomplete.

Scleractinia skeletal parts are trabecular (most septa) and fibro-laminar (transverse structures), and many groups have very porous skeletons. Skeletal composition is aragonite. Many scleractinian polyps have an edge zone (a fold of the body wall extending over the edge of the corallite) that permits the addition of skeletal materials on the outside of the corallite; this is often reflected in the presence of extensive attachment structures and in other parts not found in the Rugosa.

The Scleractinia are common as fossils in rocks of Middle Triassic to Holocene age and are abundant in modern seas, not only in coral reefs but in varied environments over large parts of oceanic space (see Ecology below).

Ecology

Living scleractinian corals are divided into two ecological groups: *hermatypic* or "reef" corals that have symbiotic algae (zooxanthellae) in their endodermal tissues, and *ahermatypic* or "nonreef" corals that lack zooxanthellae. All living corals are marine and require water movements to supply disolved oxygen; few can tolerate heavy influxes of sediment.

Hermatypic corals are restricted to shallow, tropical waters by the needs of their zooxanthellae. Corals and symbionts flourish within the temperature range 25–29°C and in depths <20 m, although they are found in depths up to 50–90 m and can survive exposure to water as cold as 16°C. Hermatypic corals are important or dominant elements in the modern reef environments.

Ahermatypic corals may occur with reef corals but are not subject to the same restrictions; they are found from the surface to depths of 6000 m and in temperatures from 1.1 to 28°C, and geographically in all oceans and seas with normal salinities.

Specific requirements of fossil corals are not known, although Mesozoic and Cenozoic Scleractinia may not have differed significantly from their living descendents. Paleozoic corals are more of a problem. They are most common in limestones and calcareous shales but are found in virtually all kinds of marine sedimentary rocks. Small reefs were formed but corals were seldom the principal builders. Indirect evidence suggests that the Paleozoic corals were much less restricted than are living hermatypic corals but there is no evidence of any being in very deep water deposits. It is probable that in their requirements the Paleozoic groups were more similar to the ahermatypic corals than to modern reef corals.

Evolutionary Trends

Certain trends are common to two or more of the coral orders and, whereas these do not indicate particular relationships, they do help explain the varying degrees of "success" implied by the fossil record. Notable trends are (1) in the increasing complexity of septal structure and in skeletal structure generally, (2) in the development of more complex or more highly integrated colonies, and (3) in finer adaptations to reef building.

A trend from simple to complex trabeculae has been noted in several families of both Rugosa and Scleractinia. Reef-building scleractinians show a trend toward more porous or fenestrate septa; a comparable development is known in only one rugosan family.

Trends from solitary to colonial habits are recognized in the Rugosa and Scleractinia and from simple to more complex colonies in these groups and the Tabulata. In general the sequence seems to have been as follows: solitary to phaceloid to cerioid, then to meandroid (in Scleractinia) or to astreoid to aphroid (in Rugosa). Different family groups went varying distances along this route. For example, half of the rugosan families that include colonial corals developed no further than the phaceloid or cerioid stage, and only a very small percentage of Rugosa genera are astreoid or aphroid. The sequences of colony form represent increasing biological integration and cooperation within the colonies.

The above-mentioned trends in septal structure and colony formation relate to a general trend toward building bigger and better reefs.

Small reefs became possible with the development of a rigid framework above the general level of sedimentation, and the colonial habit provided one such framework. In addition, branching corals formed a baffle that trapped sediments and all colonies formed elevations on which additional colonies could grow. The more integrated colonies (without internal walls between corallites) permitted the building of larger structures with less $CaCO_3$ and with less energy expended. The development of porous skeletons by the Scleractinia was even more important in increasing the building efficiency of the coral polyp. In the same group, the edge zone permitted stronger attachments to the substrate or to other corals, of further importance in developing a strong framework.

The need of all polyps to dispose of metabolic waste products must have limited the size of early reefs. The zooxanthelle (see *Zooxanthellae)* symbionts of Scleractinia helped solve this problem since they require some of the polyp discharge products for their own metabolism. The great success of the Scleractinia in building the large, modern reefs is ascribed to their development of porous skeletons, to more highly integrated colonies, and to their symbiosis (J. W. Wells, *in* Moore, 1956).

Relationships of Major Coral Groups

The groups of corals listed earlier, and especially the orders of Zoantharia, are clearly related, although the details are not all clear. The probable course of evolution is diagramed in Fig. 12. The Heliolitoidea probably evolved from a cerioid tabulate coral with similar wall structure, through the development of the coenosteoid habit. Such a tabulate genus is known and this relationship has been suggested (Flower, 1961).

The Rugosa probably evolved from the Tabulata also and may be diphyletic, as the origins of two rugosan suborders from different groups of tabulates have been logically argued (e.g., Sokolov, 1962; Flower, 1961). While Silurian and later rugose and tabulate corals are distinct, some of the Ordovician species have morphologic features of both orders. A close relationship at this time is clear but many details are still to be worked out or interpreted.

Heterocorallia are few and poorly known but their general similarity to the Rugosa suggests their origin in this order.

The origin of the Scleractinia is more controversial and two schools of thought are prominent. One notes the obvious similarities with the Rugosa and suggests direct descent,

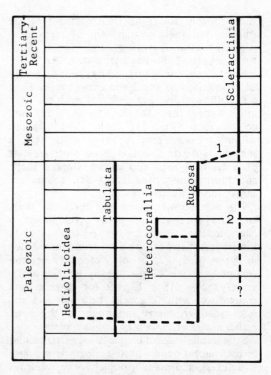

FIGURE 12. Stratigraphic ranges and probable relationships of Zoantharia coral orders.

possibly polyphyletic with different scleractinian suborders arising from different rugosan families. The second school notes the differences between Scleractinia and Rugosa, especially in the sequence and mode of septal insertion, and the time gap between the two groups (there are no known Lower Triassic corals). This school suggests that the Scleractinia evolved from a group of Paleozoic sea anemones (Actinaria possibly) so that the rugosan similarities are due to a common ancestry rather than to direct descent.

WILLIAM A. OLIVER, JR.*

References

Bayer, F. M., 1973. Colonial organization in octocorals, in R. S. Boardman, A. H. Cheetham, and W. A. Oliver, Jr., eds., *Animal Colonies.* Stroudsburg, Pa.: Dowden, Hutchinson, & Ross, 69-93.

Coates, A. G., and Oliver, W. A. Jr., 1973. Coloniality in zoantharian corals, in R. S. Boardman, A. H. Cheetham, and W. A. Oliver, Jr., eds., *Animal Colonies.* Stroudsburg, Pa.: Dowden, Hutchinson & Ross, 3-27.

Fenton, C. L., and Fenton, M. A., 1958. *The Fossil Book.* New York: Doubleday, 482p.

*Publication authorized by the Director, U.S. Geological Survey

Flower, R. H., 1961. Montoya and related colonial corals, *New Mexico Bur. Mines, Mem.* 7, 229p.

Moore, R. C., ed. 1956. *Treatise on Invertebrate Paleontology,* pt. F, Coelenterata. Lawrence, Kansas: Geol. Soc. Amer. and Univ. Kansas Press, 498p.

Moore, R. C.; Lalicker, C. G.; and Fischer, A. G., 1952. *Invertebrate Fossils.* New York: McGraw-Hill, 766p.

Shimer, H. W., and Shrock, R. R., 1944. *Index Fossils of North America.* Cambridge, Ma.: M.I.T. Press, 837p.

Shrock, R. R., and Twenhofel, W. H., 1953. *Principles of Invertebrate Paleontology,* 2nd ed., New York: McGraw-Hill, 816p.

Sokolov, B. S., 1962. Phylum Coelenterata, in *Fundamentals of Paleontology* vol. 2 (trans. from Russian, 1971, Jerusalem: Israel Prog. Sci. Trans. 222–825).

Cross-references: *Biomineralization; Coelenterata; Coloniality; Reefs and Other Carbonate Buildups; Zooxanthellae.* Vol III: *Coral Reefs.*

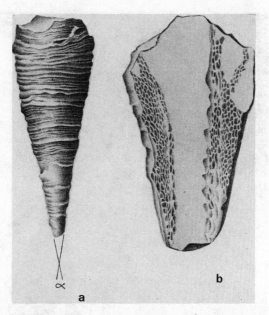

FIGURE 1. A typical cornulitid (modified from Hall, 1879). a: Surface appearance. The apical angle, α, is calculated from the formula $\alpha = 2$ arctan D/2L, where D = diameter at aperture, and L = length form apex to aperture. Surface ornament is quite variable from species to species. b: Longitudinal section, showing cellular structure usually taken as disgonostic of the cornulitidae.

CORNULITIDAE

The Cornulitidae are a group of small conical fossils that are usually found attached to the shells of other benthic invertebrates. They are found in strata ranging from Middle Ordovician to Mississippian in age, and are most abundant in Ordovician and Silurian rocks. They are found in rocks representative of a variety of marine environments, but generally are most abundant in the shelly facies of shallow oceans and epicontinental seas. Although they have been found in all sedimentary rock types, they seem to be preserved best in calcareous shales.

Morphology and Taxonomy

The form of a typical cornulitid is shown in Fig. 1. In most species, the apical angle is in the range 12 to 25°. The external form and ornament of cornulitids is quite variable, even within a species. Longitudinal striation may or may not be present. Concentric swellings or rings are regular on some species, irregular on others. In species that were attached to their substrate only along part of their length, rings tend to be more regular on the unattached portion of the shell (Fig. 2d). Most individuals of most species are curved, but some may be nearly straight, and the curvature of others appears to be dictated by the relief of the surfaces upon which they grew.

The most recent attempt at a comprehensive treatment of the taxonomy of cornulitids was by Fisher (1962). Fisher recognized 3 genera of cornulitids and reported that some 45 species had been described. According to him,

the diagnostic features of the cornultids is the structure of the walls, which are cellular. The cellular structure may be lacking in young individuals or small species, however. Cornulitids generally are attached to a host, and at least the earliest-formed part of the shell shows a flattening reflecting this habit. Tentaculitids (see *Tentaculitida*) are similar to cornulitids in appearance but are supposedly distinct from cornulitids in being free-living rather than attached, straight with a bulbous apex rather than curved, and in possessing septae which close off the apical end of the interior of the shell. Tentaculitids lack the cellular wall structure of cornulitids. The apical angles of tentaculitids vary between about 6 and 13°. Unfortunately, specimens intermediate between the two groups are not uncommon. Some free-living forms are cornulitids by morphology; some morphological tentaculitids appear to have been attached to a substrate, at least for part of their lives. Thus our understanding of the distinction between cornulitids and tentaculitids is at present incomplete.

The affinities of the cornulitids are unknown. Their general form and sessile nature suggest affinities with the tubicolar polychaetes, but no known polychaetes secrete shells with a

FIGURE 2. Representative cornulitids (from Richards, 1974). a: A solitary commensal cornulitid, *Cornulites proprius* Hall attached to the Silurian brachiopod *Stegarhynchus whitii* (Hall); the brachiopod is 1.2 cm long. b: A group of gregarious commensal cornulitids from the Upper Ordovician, showing the tendency to approach spherical form. c: A parasitic cornulitid, *Cornulites cingulatus* Hall, attached to the Devonian brachiopod *Paraspirifer bownockeri* Stewart. Note interference with shell growth of the host; cornulitid is 0.9 cm long. d: Cornulitid transitional between attached and free-living forms. The recurved portion near the apex was attached along the right side; length 1.4 cm. e: Unattached cornulitid (internal mold); length 1.6 cm.

structure comparable to that of the cornulitids. Although most authors regard cornulitids as being most likely allied with the annelids, affinities with stromatoporoids, fusulinid foraminifera, and mollusca have been suggested. It is also quite possible that they have no modern relatives.

Ecology

Examination of many well-preserved cornulitids suggests that they can be classed into four groups in terms of their interactions with their hosts: free living (no host), solitary and gregarious commensals, and ectoparasites. These groups have some implications for feeding and reproductive ecology as well. Their abundance through time is shown in Fig. 3.

Solitary Commensals. Commensalism involves the growth of individuals (Fig. 2a) or "colonies" (Fig. 2b) on hosts, apparently without interfering with the life processes of the hosts. Commensal cornulitids may occur on many host species, and in random or varied positions and orientations, including positions on the inner surfaces of brachiopod, pelecypod, and gastropod shells. In some instances, it is clear that the "host" was dead, as when a cornulitid crosses the commissure of a brachiopod, or is attached to the inside of a gastropod shell. In others, it is equally clear that the host was alive, as when a bryozoan overgrows a cornulitid attached to it. In most instances, however, it is impossible to establish whether or not the partners were alive at the same time.

The great majority of cornulitids were solitary commensals (Fig. 3). Their interaction with their hosts was never great enough to cause injury to the host. The many examples of cornulitids living in positions that would have been inaccessible when the host was living indicate that many cornulitids of this group settled and grew quite successfully on dead shells. Therefore it seems likely that the major benefit to these cornulitids was the acquisition

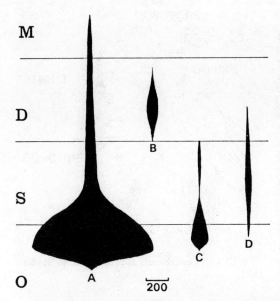

M

D

B

S

D

C

O   A   ‾200‾

FIGURE 3. Distribution through time of cornulited adaptive types (from Richards, 1974). a: Solitary commensalism, involving both living and dead skeletal substrates. b: Ectoparasitism, primarily on large spiriferid brachiopods; parasites derived their food in part from the feeding currents of the host. c: Gregarious commensalism. d: Unattached, mobile, possibly deposit-feeding habit. O, Ordovician; S, Silurian; D, Devonian; M, Mississippian.

of a stable, hard substrate on which to settle and grow. Attachment to hard substrates considerably larger than the cornulitid suggests food gathering by filter feeding.

Gregarious Commensals. Gregarious commensals are attached to each other along their length and the openings of the tests form a surface that ranges from undulatory to nearly hemispherical in ideal cases. While a group of such cornulitids may present a colonial appearance, each individual is recruited to the cluster separately, and asexual reproduction is not involved. Cornulitids that exhibit a gregarious habit are found from the Middle Ordovician to the Upper Silurian, but are uncommon after the middle Upper Ordovician. The adaptive significance of the colonial habit is not clear. One possibility is that colonialism may have enhanced reproductive success, especially in environments where scarce substrates may have forced isolation of individuals of solitary species.

Free-living Cornulitids. Free-living cornulitids (Fig. 2e) are recognized by their occurrence without any associated host. In addition, they show no smoothing of surface sculpture or flattening of the test which would reflect

attachment to a host. As a group, free-living cornulitids appear to be larger than usual. It seems probable that several different lineages of solitary commensal cornulitids independently evolved a free mode of life. Cornulitids that show no evidence of attachment are first seen in the Ordovician of Estonia. Silurian forms of quite different morphologies occur in Indiana, Maine, and Gotland. Transitional forms (Fig. 2d), which attached in early life but then grew away from their substrates, occur in the Cincinnatian (Upper Ordovician) of Indiana and Ohio, the Wenlock (Silurian) of England, and at several horizons in the Silurian of Gotland.

One benefit to be gained from a free mode of life would be greater access to food. This benefit would be maximized if the cornulitid developed the ability to utilize deposited food as well as suspended food. In environments with scarce food resources and relatively firm substrata, the benefits of a secure attachment site, and favored the evolution of free-living surface deposit-feeding species. Free-living forms, not necessarily deposit feeders, might also have evolved in environments of rapid or unpredictable sedimentation, where permanent attachement could lead to premature death by burial.

Ectoparasitic Cornulitids. At least two Devonian species of cornulitids developed parasitic relationships to their hosts, which in each case were large spiriferacean brachiopods. Both species attached themselves at the margin of the host's shell, and grew at the same rate as the host, thus keeping their shell opening projecting into the space at the edge of the host's shell, where they presumably stole food from the host's incoming feeding current. The cornulitid interacted with the host aggresively enough to force the host's mantle margin to deflect around it, producing a groove in the shell in which the cornulitid grew (Fig. 2c). The parasites were gregarious to an extent: Although most individuals of the host species escaped parasitism, those that harbored cornulitids generally had more than one; and one specimen was burdened with 24!

Parasitic cornulitids apparently relied on their host for food, for misoriented individuals are invariably small, indicating early death, stunted growth, or both. Because of the close coordination of growth rates required to keep the cornulitid in position to compete for the host's food, parasitic cornulitid species parasitized one or at most a pair of very similar host species. In each case, the extinction of the host species resulted in the extinction of the parasite as well.

Potential Use in Biostratigraphy

Cornulitids have been studied by some in the hopes that they would prove useful as guide fossils. However, what we know of their ecology makes this seem unlikely. As benthic forms, they are presumably subject to the same limitations of range as other benthic organisms, so that a given species may occur only in a small portion of the rocks deposited during the lifetime of the species. In addition, the morphological simplicity of cornulitids makes the identification of species difficult. Many currently recognized species seem to have been based only on stratigraphic position or geographic separation from other described species, not on morphological uniqueness. However, even if our understanding of cornulitid taxonomy is greatly improved, cornulitids seem destined to remain more of interest from the paleoecological perspective, and will probably not be important stratigraphically.

R. PETER RICHARDS

References

Fisher, D. W., 1962. Small conoidal shells of uncertain affinities, in R. C. Moore, ed., *Treatise on Invertebrate Paleontology,* pt. W, Miscellanea. Lawrence, Kansas: Geol. Soc. Amer. and Univ. Kansas Press, W98–W144.

Hall, J., 1879. Descriptions of the Gastropoda, Pteropoda, and Cephalopoda of the Upper Helderberg, Hamilton, Portage, and Chemung Groups, N. Y. Geol. *Surv. Paleontol.,* 5(2), 493p.

Hoare, R. D., and Steller, D. L., 1967. A Devonian brachiopod with epifauna, *Ohio J. Sci.,* 67, 291–297.

Richards, R. P., 1974. Ecology of the Cornulitidae, *J. Paleontology,* 48, 514–523.

Schumann, D., 1967. Die Lebensweise von *Mucrospirifer* Grabau, 1931 (Brachiopoda), *Palaeogeography, Palaeoclimatology, Palaeoecology,* 4, 279–285.

Cross-references: *Molusca; Tentaculita.*

CRUSTACEA

This group of arthropods is the most diverse in the animal kingdom. The crustacean body typically consists of a series of segments grouped into three regions encased in a chitinous exoskeleton. Features that are uniquely crustacean are:

1. The head bears five pairs of appendages consisting of two pairs of antennae, a pair of mandibles, and two pairs of maxillae.
2. The body consists usually of three regions; head, thorax, and abdomen.
3. The appendages are primitively polyramous and leaf-like (Fig. 1).

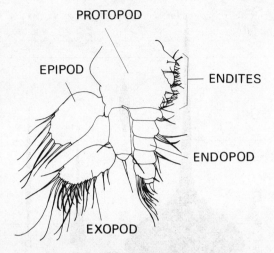

FIGURE 1. Cephalocarid appendage (from Sanders, 1957).

4. Development consists of a series of discreet larval and/or juvenile stages, initiated by a stage termed a *nauplius* (Fig. 2).

The number and combination of the body segments or somites into regions (tagmata), and the structural differentiation of the serialized paired appendages provide the basis for taxonomic subdivisions (Fig. 3). The most unifying characteristic of all is the nauplius larval stage (Fig. 2); but this and some of the subsequent stages can be suppressed in some of the more highly evolved forms, e.g., isopods, crabs, and lobsters. A cephalothoracic dorsal shield, the carapace, was believed by some to be a primitive crustacean characteristic, but this is now questionable.

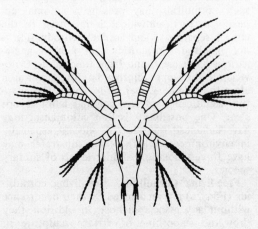

FIGURE 2. Nauplius larva of the malacostracan *Penaeus duorarum* (from Sanders, 1963).

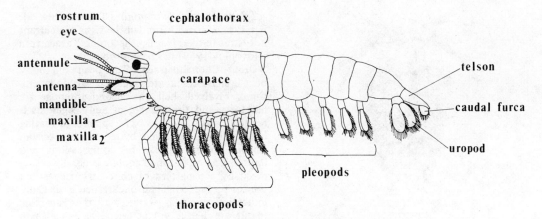

FIGURE 3. External anatomy of generalized crustacean.

Until recently (e.g., Moore, 1969), Crustacea was considered a subphylum or superclass of phylum Arthropoda. However, recent work on arthropod locomotory mechanisms (Manton, 1977) and comparative embryology (Anderson, 1973) has elucidated fundamental differences between the three living groups of arthropods–Crustacea, Chelicerata, and Uniramia, and very strongly suggests the recognition of three separate phyla. The Crustacea are the most distinctive in this scheme and have undergone the most diverse radiation in the animal kingdom. Crustacean remains occur in Cambrian strata; but, with the exception of the Malacostraca, they have left a relatively meager fossil record of their subsequent evolution.

Taxonomic subdivisions of the Crustacea were largely erected in the early 19th century, especially in the works of P. A. Latrielle. The classification now widely employed was stabilized by Calman (1909). Excluding taxa of questionable affinities or validity, there are 5 classes and 48 orders (see Schram, 1978).

Class Cephalocarida

Hutchinsoniella macracantha Sanders (1955; 1957, 1963) is the best-known species of this world-wide group. Though adapted for living in mud, this minute Holocene marine animal possesses an array of characters (Fig. 4) that indicate it may be a primitive form. *Lepidocaris rhyniensis* Scourfield, from the Devonian of Scotland, is the closest known relative, and is intermediate in form between cephalocarids and branchiopods.

Of special interest is the structural uniformity of the appendages. In the adult, the second maxillae are essentially identical to the trunk limbs; but in the larval stages, the second antennae, mandibles, and first maxillae are also similar to the adult thoracic appendages. The 8 pairs of trunk limbs have a foliaceous epipod, a foliaceous exopod, a jointed ambulatory endopod, and a protopod bearing a series of medial lobes or endites (Fig. 1). There are 11 appendageless abdominal somites and the telson bears furca. No carapace or eyes are present. There are no known fossils.

Class Branchiopoda

Though typically possessing the primitive type of appendage, the Branchiopoda (q.v.) are distinguished by having the antennules and/or the maxillae of the cephalic appendages reduced or vestigal. Foliaceous appendages are borne on the anterior trunk somites, the posterior somites are appendageless, and the telson bears furca.

The living forms are almost entirely fresh water in habit. Their general morphology, especially of tagmata and appendage anatomy, however, is suggestive of an ancient origin. The preservation of fossils alleged to be branchiopods is such that critical morphological detail is usually lacking. Orders for extinct forms have been proposed; some of which may be valid. Four extant orders are known.

Order Anostraca. The fairy or brine shrimp of Holocene lakes and transitory bodies of fresh water comprise the order Anostraca. They lack a carapace, have stalked compound eyes, and have 19 or more postcephalic somites of which only 11 typically have paired leaf-like appendages. Fossils belonging to a closely related order (Lipostraca) *Lepidocaris rhyniensis,* first occur in Devonian strata (Fig. 5) and form a nice transition from cephalocarids to Anostraca.

Order Notostraca. *Triops cancriformis* and its congeners are truly "living fossils," as the genus first occurs in Permian strata. Notostraca have a carapace forming a dorsal shield upon

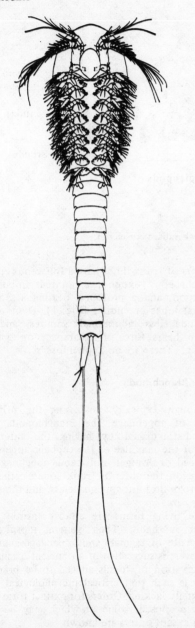

FIGURE 4. Ventral view of *Hutchinsonella macra-cantha* (from Sanders, 1963).

which sessile compound eyes are situated; the first antennae are vestigal; there are 40 to 63 pairs of foliaceous trunk appendages; and the rami of the furca are multisegmented. The large number of somites is distinctive, but it is not as great as indicated by the paired append-ages cited; as many as 6 pairs of appendages are present on the posterior appendage bearing somite. Four to 14 of the posterior trunk somites are appendageless.

Triops cancriformis minor, Triassic (Fig. 5), is a typical fossil form. On the basis of aberrant notostracan-like fossils, the order Kazacarthra (Novojilov 1959) has been proposed.

Order Chonchostraca. The head and body of Chonchostraca are enclosed within a chi-tinous bivalved shell, the second antennae are natatory, and the number of body segments is reduced with 10 to 27 appendage-bearing body somites followed by a claw-like telson. Fossils consist largely of compressions and impressions of the shell and are generally recog-nized by the pattern of concentric ridges and the form (Fig. 5). Their occurrence is allegedly indicative of fresh water, but the Paleozoic fossils are better characterized as brackish in habit. The stratigraphic range is Devonian to Holocene.

Order Cladocera. Water fleas have a carapace that does not cover the head. The eyes are sessile and coalesced. The biramous second antennae are natatory and the trunk segments are greatly reduced to from 4 to 6 appendage-bearing somites and a telson with furca. Fos-sils of these minute animals are extremely rare; their earliest occurrence is only Cenozoic.

Class Ostracoda

The minute Crustacea of the class Ostracoda have very few segments indistinctly separated and enclosed within a calcified bivalved cara-pace. There are only a few pairs of postcephalic appendages, the exact number depends on the group. Ostracods occur in both marine and fresh water and occupy a variety of ecologic niches. They have left an excellent fossil record, Ordovician to Holocene (see *Ostracoda*).

Class Maxillopoda

Subclass Mystacocarida. The microscopic copepod-like Mystacocarida (Fig. 5) live in the interstitial spaces in beach sands. Their distribu-tion is largely circum-Atlantic. The second antennae, mandible, and both pairs of maxillae are relatively large in the adults and function for locomotion as well as feeding off the organic films on sand grains. Four anterior postcephalic somites have vestigal appendages and the last five are appendageless. The telson bears furca. The structure of the mandible is among the most primitive recognized.

Subclass Copepoda. The copepods lack both a carapace and paired compound eyes. Both pairs of antennae are long and may be used for locomotion or for prehension. Typically, the trunk consists of 9 somites of which only the first 5 bear appendages. A furca is present on the telson.

There are few fossil copepods and the earliest

single order Eocaridacea. These are forms that might be related to euphausian eucarids and mysidacean peracarids as well as forms that have properties of several different advanced types. The earliest Eumalacostracan, *Eocaris oevigi* Brooks of the Middle Devonian is placed here. The Eocarida extend up into the Late Pennsylvanian.

Superorder Syncarida. Syncarids would be the most primitive of Eumalacostraca except for the total lack of a carapace. The earliest synarids occur in the Lower Carboniferous of Scotland. All the late Paleozoic syncarids are found in northern continents, while the Triassic forms are Gondwana in distribution and fresh water in habit (the essentially modern distribution of the group; Schram, 1977). One living order, however, is world wide, living interstially in ground waters.

Superorder Pancarida. Members of the single pancarid order, Thermosbanenacea, live in caves and thermal ground waters. Females brood their young up underneath the carapace, but they have a palp (lacina mobilis) on the mandible like Peracarida. Some authorities would place these as peracarids. There are no fossils known.

Superorder Peracarida. The females of the peracarid group have a seasonal brood pouch formed by medial flaps of the thoracic coxae (oöstegites). The young hatch at an advanced juvenile stage of development, by-passing the early larval stages. This is the most diverse group of malacostracans, containing the most species and occupying habitats from purely terrestrial to the abyssal depths of the sea.

Order Mysidacea. The "possum shrimp," Mysidacea, are very extensive in the late Paleozoic, and have the highly specialized suborder Pygocephalomorpha extending from the Mississippian to Permian. Living forms are largely small, natant, filter-feeding types.

Order Cumacea. The "mole shrimp," Cumacea, are largely modern, with only an Upper Permian and Jurassic fossils recognized.

Order Spelaeogriphacea. Two species of Spelaeogriphacea are known—the living *Spelaeogriphus lepidops,* found only in a pool in Bat Cave in Table Mountain outside Capetown, South Africa; and *Acadiocaris novascotia* (Copeland) from the Lower Carboniferous of Nova Scotia.

Order Tanaidacea. Members of the order Tanaidacea are a highly specialized group with a great deal of juvenile and sexual polymorphism. The earliest form, *Anthracocaris scotica* (Peach), comes from the Lower Carboniferous of Scotland.

Order Isopoda. Isopods are the most diverse and successful of the peracarids; many of them characterized by dorsal-ventral flattening of the body. The earliest form, *Hesslerella shermani* Schram, dates from the Middle Pennsylvanian.

Order Amphipoda. Denoted by the great variation of leg structure found on any one animal, Amphipoda is an important modern group, known from only a few fossil forms, the earliest being Upper Eocene.

Superorder Eucarida. The well-known group Eucarida is characterized by having the carapace fused to the thoracic segments.

Order Euphausiacea. The "krill" are important pelagic, marine, plankton feeders. No fossils are recognized though some of the Paleozoic Eocaridacea might be placed here. Eggs are shed free in the water.

Order Amphionidacea. The order Amphionidacea was recently erected on the basis of one living species in the Indian Ocean (Williamson, 1973), with eggs brooded in the space under the thorax formed by the lateral flaps of the carapace and a forward extension of a specialized first abdominal appendage.

Order Decapoda. Decapods have uniramous thoracic legs of which the anterior 3 pairs are maxillipeds and the remaining 5 pairs are "walking legs." The eggs are shed free in the more primitive forms (Dendrobranchiata), but brooded on the abdominal appendages in the advanced forms (Pleocyemata). Their earliest definite record is Late Devonian (*Paleopalaemon newberryi*). True shrimp, crabs, and lobster-like (Fig. 5) Crustacea first occur in early Mesozoic strata.

Problematic Forms

Euthycarcinus kessleri Handlirsch and its relatives, *Synaustris brookvalensis* Riek and *Kottixerxes gloriosus* Schram, have been suggested as crustaceans. The body tagmosis and appendage structure would argue against this and so these have recently been removed and tentatively placed with the merostomoidean trilibitoids.

The Burgess Shale (q.v.) of the Middle Cambrian of British Columbia contains a variety of primitive arthropod types some of which have been referred to as "pseudocrustaceans." This entire fauna is currently under intensive study at Cambridge University and final decisions as to the real affinities of these animals must await the outcome of that work.

The Cycloidea, a group of Lower Carboniferous—Upper Triassic fossils have been treated by some as possible crustaceans. But, again on the basis of body regionalization and appendage structure, affinities of this group are best sought outside the crustaceans, possibly in the merostome chelicerates.

Phylogeny

Crustacea seem to have an origin independent of any of the other arthropodous groups, based on embryology and locomotor functional morphology. Crustaceans have been frequently linked with trilobites (see *Trilobita*), with the cephalocarids as "missing links." Recent work on trilobite internal anatomy, early developmental history, locomotor functional morphology of both groups, and primitive appendage anatomy would argue against linking crustaceans and trilobites (Schram, 1978).

Morphologic evidence does indicate that cephalocarids may lie close to the ancestral crustacean stock. The Branchiopoda (q.v.) are elaborations of the primitive body plan with long-range trends to develop a carapace and reduce body segments. The main thrust of branchiopod radiation has been in fresh water. The Ostracoda (q.v.) are an extreme development of a deemphasis of segmentation. The Maxillopoda were primitively pelagic forms and still for the most part exploit maxillary filter-feeding devices. The Malacostraca have adopted rigid body regionalization and concommitantly developed great specialization of function in individual appendages.

The Malacostraca have undergone the greatest radiation. The phyllocarids are basically an early and middle Paleozoic radiation with considerable diversity of body types. The late Paleozoic saw the rise of hoplocarid and primitive caridoid-like Eumalacostra. The Mesozoic and Cenzoic has seen a great radiation of the advanced Peracarida and the Eucarida.

F. R. SCHRAM

References

Anderson, D. T., 1973. *Embryology and Phylogeny in Annelids and Arthropods.* Oxford: Pergamon Press, 495p.

Calman, W. T., 1909. Crustacea, in E. R. Lankester, ed., *A Treatise on Zoology.* London: Adam & Chas. Black, 7(3), 346p. (reprinted 1964, Amsterdam: A. Asher & Co.).

Cisne, J. L., 1974. Evolution of the world fauna of aquatic free-living arthropods, *Evolution,* 28, 337–368.

Flessa, K. W.; Powers, K. V.; and Cisne, J. L., 1975. Specialization and evolutionary longevity in the Arthropoda, *Paleobiology,* 1, 71–81.

Manton, S. M., 1977. *The Arthropoda.* Oxford University Press, 527p.

Moore, R. C., ed., 1969. *Treatise on Invertebrate Paleontology,* pt. R, Arthropoda 4, 2 Vols. Lawrence, Kansas: Geol. Soc Amer. and Univ. Kansas Press, 651p.

Sander, H. L., 1955. The Cephalocarida a new subclass of Crustacea From Long Island Sound. *Proc. Nat. Acad. Sci.,* 41, 61–68.

Sanders, H. L., 1957. The Cephalocarida and crustacean phylogeny, *Systematic Zool.,* 6, 112–129.

Sanders, H. L., 1963. Significance of the Cephalocarida, in Whittington and Rolfe, 1963, 163–175.

Schram, F. R., 1974. Convergences between Late Paleozoic and modern caridoid Malacostraca, *Systematic Zool.,* 23, 323–332.

Schram, F. R., 1977. Paleozoogeography of Late Paleozoic and Triassic Malacostraca, *Systematic Zool.,* 26, 367–379.

Schram, F. R., 1978. Arthropods: A convergent phenomenon, *Fieldiana Geol.,* 39, 61–108.

Whittington, H. B., and Rolfe, W. D. I., eds., 1963. *Phylogeny and Evolution of Crustacea.* Cambridge, Mass.: Mus. Comp. Zool. Spec. Pub., 192p.

Williamson, D. I., 1973. *Amphionides reynaudi* (H. Milne Edwards), representative of a proposed new order of eucaridan Malacostraca, *Crustaceana (Leiden),* 25, 35–50.

Cross-references: *Arthropoda; Branchiopoda; Chitin; Cirripedia; Ostracoda.* Vol I: *Zooplankton.*

D

DIAGENESIS OF FOSSILS— FOSSILDIAGENESE

The term "Fossildiagenese" was used by Müller (1963) to include those fossilization events that take place after the final burial of organic remains. As thus envisioned, the analysis of "Fossildiagenese" is an important part of work in taphonomy. Most taphonomic factors and processes operate throughout the entire history of organic remains; the use of the term "Fossildiagenese" for post-entombment studies is perhaps unfortunate, since the diagenesis of organic remains can certainly occur before, during, *and* after final entombment (Purdy, 1968).

Yet final burial *is* a distinctive event in the total environmental history of some organic remains, for there is a set of fossilization processes requiring an enclosing lithic matrix for their operation—such as the formation of molds. Perhaps the most distinctive of these processes result in the crushing, distortion, or deformation of fossil remains. The principal causes of fossil deformation are tectonic/metamorphic events, and sediment compaction due to the weight of overburden. Mathematical, crystallographic, purely graphical, and photographic techniques have been developed to analyze the amount of deformation in fossils (Ramsay, 1967; Nissen, 1964; Wellman, 1962; Sdzuy, 1966).

The literature on fossil deformation is strongly European, especially Germanic; and European workers have markedly increased their studies in this area in recent years. One major impetus for these studies was the influence of the Cloos family upon European structural geology. For example, in the oft-quoted paper of E. Cloos (1947) on deformation in the Central Appalachians regions of the United States, Cloos suggested the use of deformed crinoid stems in the analysis of rock structural patterns. This proposal did spur a number of studies of crinoid deformation in the Paleozoic rocks of the Rhine Valley region of W Germany (see Nissen, 1964). For the most recent of an ongoing European series of papers on "Fossildiagenese," see Veizer and Wendt (1976).

The uses of fossil deformation studies are manifold. The importance of these studies to descriptive paleontologic work cannot be overemphasized. A most striking example of the influence of deformation patterns upon systematics is provided by the work of A. Fanck on the bivalves of the Miocene molasse from the St. Gallen region of Switzerland (see Müller, 1963). By careful analysis of deformation patterns in these fossils, Fanck was able to reduce the 426 previously recognized species to a more realistic and more valid number—62!

Systematics also played a role in the studies of Ferguson (1962, 1963) on brachiopod deformation and sediment compaction in Carboniferous shales from Scotland. Ferguson first divided his specimens of *Crurithyris* into four distinctive groups on the basis of external form. Then, through analysis of the shell details and with the help of detailed records of the shell positions in the enclosing shale, he was able to demonstrate that the four groups represented one species and that form differences were due to the effects of compaction upon shells in different positions (Fig. 1). Deformation in these shells was marked by the development of valve corrugations, by fracturing, and by imbrication. The deformations were subtle and normally only recognizable in thin sections.

Ferguson went on to estimate shale compaction by analysis of the deformed brachiopods. Shells in presumed life positions (e.g., Fig. 1A) had their dimensions perpendicular to bedding reduced to about one-third the original; the shale laminae originally surrounding the shells were reduced to about one-half of the distance between the top and the bottom of the more resistant but crushed shells. Since these compaction factors are multiplicative, Ferguson suggested a total compaction in the shale to one-sixth of its original thickness. Crushing and compaction were less pronounced in the more calcareous layers he examined, presumably because of more rapid lithification in the calcareous shales than in the "normal" shales. Hence the details of compaction and the relationship between sediment composition and compaction factors can both be understood through studies of fossil deformation.

The use of these studies in tectonics is quite

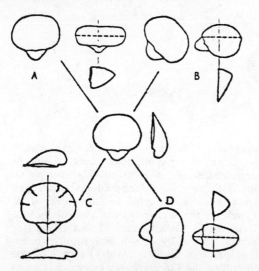

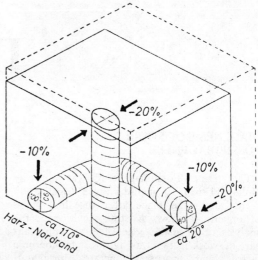

FIGURE 1. Distortion in the brachiopod *Crurithyris* and its dependence upon the position of the shell in the sediment (from Ferguson, 1962). Dorsal and lateral views of normal shell are in the center. A, brachial valve and plane of symmetry perpendicular to bedding; length greatly reduced during crushing. B, brachial valve perpendicular to bedding and plane of symmetry inclined to bedding; during crushing, ridge formed parallel to bedding. C, flat-lying shell; thickness reduced and length and width increased during crushing. D, brachial valve perpendicular to bedding and plane of symmetry parallel to bedding; width greatly reduced during crushing.

FIGURE 2. A schematic diagram of the lateral and vertical deformation of burrow systems in the flat-lying Upper Cretaceous beds of the northern flank of the Harz mountains, Germany (from Plessmann, 1966). Lateral compression was approximately 20%; vertical compression during compaction and diagenesis, an additional 10%; the total sediment was reduced by these processes to 72% of its original volume.

varied. Stress patterns and produced strains are normally far more intricate during tectonism than they are during simple deformation through compaction. Indeed, the effects of compaction and tectonism must often be evaluated simultaneously in highly deformed sedimentary rock sequences. The role of fossils in these evaluations is exemplified in the work of Plessmann (1966). By a very thorough analysis of the deformation of burrow systems, Plessmann was able to differentiate between compaction effects and the effects of lateral, compressive deformation in the Cretaceous of the Harz mountain region of Germany (Fig. 2). Both body fossils and biogenic structures can contribute to the study of deformational events.

Metamorphic events do commonly cause chemical/mineralogic or textural alteration of fossils, but these events need not cause the physical distortion of fossil remains. The review paper of Bucher (1953) is still the classic entry into the literature of fossils in metamorphic terrains. Other post-entombment processes are described in detail by Rolfe and Brett (1969) and Bathurst (1971).

DAVID R. LAWRENCE

References

Bathurst, R. G. C., 1971. *Carbonate Sediments and their Diagenesis.* Amsterdam: Elsevier, 620p.

Bucher, W. H., 1953. Fossils in metamorphic rocks: A review, *Geol. Soc. Amer. Bull.,* **64,** 275-300.

Cloos, E., 1947. Oolite deformation in the South Mountain fold, Maryland, *Geol. Soc. Amer. Bull.,* **58,** 843-918.

Ferguson, L., 1962. Distortion of *Crurithyris urei* (Fleming) from the Visean rocks of Fife, Scotland, by compaction of the containing sediment, *J. Paleontology,* **36,** 115-119.

Ferguson, L., 1963. Estimation of the compaction factor of a shale from distorted brachiopod shells, *J. Sed. Petrology,* **33,** 796-798.

Müller, A. H., 1963. *Lehrbuch der Paläozoologie Band 1. Allgemeine Grundlagen.* Jena: Gustav Fischer Verlag, 387p.

Nissen, H. U., 1964. Dynamic and kinematic analysis of deformed crinoid stems in a quartz graywacke, *J. Geol.,* **72,** 346-360.

Plessmann, W., 1966. Diagenetische und kompressive Verformung in der Oberkreide des Harz-Nordrandes sowie im Flysch von San Remo, *Neues Jahrb. Geol. Paläontol.,* **1966**(8), 480-493.

Purdy, E. G., 1968. Carbonate diagenesis: An environmental survey, *Geol. Romana,* **7,** 183-228.

Ramsay, J. G., 1967. *Folding and Fracturing of Rocks.* New York: McGraw-Hill, 568p.

Rolfe, W. D. I., and Brett, D. W., 1969. Fossilization processes, in G. Eglinton, and M. T. J. Murphy, eds, *Organic Geochemistry: Methods and Results.* Berlin & New York: Springer-Verlag, 213-244.

Sdzuy, K., 1966. An improved method of analysing distortion in fossils, *Palaeontology,* **9**, 125-134.

Veizer, J., and Wendt, J., 1976. Fossildiagenese, Nr: 23: Mineralogy and composition of Recent and fossil skeletons of calcareous sponges, *Neues Jahrb. Geol. Paläontol., Mh.,* **1976,** 558-573.

Wellman, H. W., 1962. A graphical method for analysing fossil distortion caused by tectonic deformation, *Geol. Mag.,* **99,** 348-352.

Cross-references: *Biostratinomy; Fossils and Fossilization; Taphonomy.* Vol. VI: *Diagenesis.*

DIATOMS

Diatoms, the Bacillariophyceae, are single-celled algae that secrete a shell (frustule) of opaline silica and occupy a number of marine and freshwater habitats. Microscopic in size, they range from 1 μm to more than 2000 μm in diameter and may vary in shape from extremely elongate forms to circular, triangular, and pentagonal forms. The frustule is composed of two valves or shells that fit over each other much like a pillbox and are connected by a thin silica band called the girdle band. The larger of the two valves is called the epitheca while the smaller is called the hypotheca. Normally, one sees the diatom frustule or valve in one of two possible configurations. The valve view is obtained when looking straight down on the valves; when the diatom is viewed from the side, this is referred to as the girdle view.

For purposes of discussion, we can divide diatoms into two general groups: the *Centrales* (centric forms) and *Pennales* (pennate forms). Centric forms are characterized by structures, ornamentation, etc., radiating from the central part of the valve (in valve view). Pennate forms, on the other hand, when seen in valve view, have structures that are at right angles to the longest axis. Thus, an oval form could be either centric or pennate depending upon whether the structures radiate from the central part of the valve or are at right angles to the longest axis.

When seen in valve view (Figs. 1,2), centric diatoms characteristically have a honeycombed arrangement of silica. These structures are called areolae and may range from being rounded or elliptical to having 5 or 6 sides. The arrangement and density of the areolae over the valve are conservative features and may be used to identify a species.

Other characteristic ornamentation and structures on the valve may be used to separate diatoms at the species or even the generic level. The presence of one or more pores near the central area of the valve as well as the presence

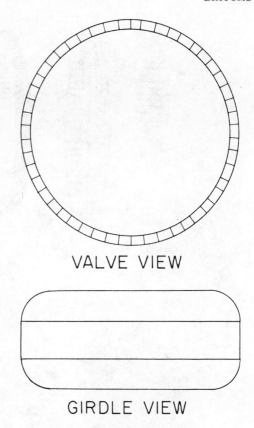

VALVE VIEW

GIRDLE VIEW

FIGURE 1. Valve view and girdle view of a centric diatom. Essentially the valve view is a "birds-eye" view, while the girdle view is a side view.

of marginal tubuli around the periphery of the valve serve to define the genus *Thalassiosira.* Further, the presence of a clear raised or depressed area (Pseudo-ocellus) serves to differentiate the genus *Actinocyclus.* Other genera of the Centrales may be separated by the shape of the valve, the presence and arrangement of spines or by the arrangement of the areolae.

The pennates also demonstrate a wide variety in structure and arrangements. One group of pennates characteristically has pores or striations at right angles to the long axis but which do not go completely across the valve. This leaves a space, running down the center, which is usually free of ornamentation. Such a clear-silica, unstructured area is called a pseudoraphe. In some groups, particularly in *Navicula,* this space has a long V-shaped slit called the raphe. Those forms which possess such a raphe are capable of some movement. Associated with the raphe may be mucous pads at the terminal ends and a nodule in the center.

Other pennates, such as in the genus *Nitzschia,*

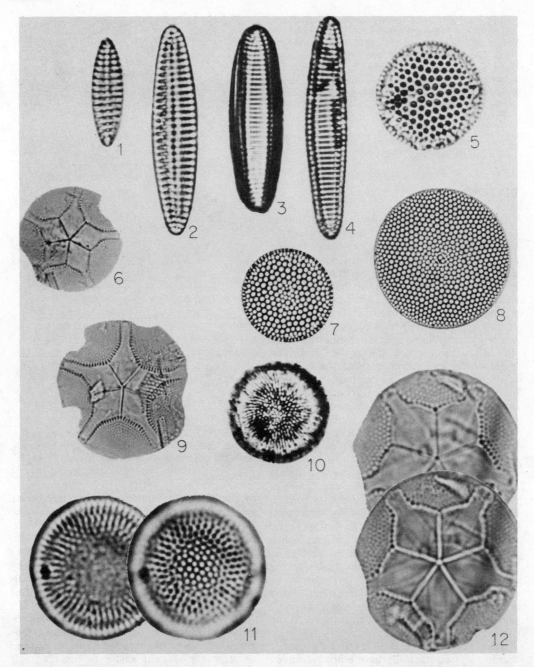

FIGURE 2. Some examples of Neogene and Pleistocene diatoms. Forms 1–4 are pennate diatoms and the remainder are centric. 1,2, *Nitzschia miocenica* Burckle; 3, *Nitzschia cylindrica* Burckle; 4, *Nitzschia pliocena* (Brun) Kanaya; 5, *Stephanopyxis dimorpha* Schrader; 6,9, 12, *Asterolampra acutiloba* Forti; 7, 8, *Coscinodiscus nodulifer* Schmidt; 10, *Thalassiosira convexa* Muchina; 11, *Thalassiosira miocenica* Schrader.

possess a raphe that is near the periphery of the valve and is raised (keeled) rather than depressed. Such groups have silica ridges (costae) that run across the valve at right angles to the long axis. Connecting these ridges are thin silica membranes (intercostal membranes), which usually contain one or more rows of pores.

Diatoms may either live singly or in colonies although they seem to derive no specific benefit from such a colonial habitat. In some genera, this association is very loose, with the valves held together by thin mucous strands. In others, the valves are side by side or attachment may be near the ends.

Since the world's ocean is always undersaturated with respect to silica, diatoms are especially susceptible to dissolution. Two mechanisms appear to offer some protection from dissolution. An organic coating around the valve seems to provide some resistance. Schrader (1971) has also pointed out that fecal pellets provide the diatoms with rapid transit to the sea floor while at the same time protecting the valves from dissolution. The fact that diatoms do dissolve, both in the water column and in surface sediments results in drastic differences between the biocoenosis in surface waters and the thanatocoenosis (or death assemblage) in the underlying sediments. Thus, genera such as *Chaetoceros,* which are very abundant in surface waters, are almost never seen in surface sediments. Diatom dissolution is also a regenerative process in that it provides the silica to surface water for continued diatom productivity.

Classification

A number of diatom classification schemes have been proposed over the years, although most diatomists tend not to be "classifiers." The scheme given here is that of Hendey (1937, 1964). It should not be taken as the last word, but rather as an important footnote to an evolving concept. It should be pointed out that most diatomists ignore the varying classification schemes and list their diatoms alphabetically.

Under Hendey's system the diatoms are listed under the Division Chrysophyta, the Class Bacillariophyceae and the Order Bacillariales.

I. Suborder Coscinodiscinae
  A. Family Coscinodiscaceae
  B. Family Hemidiscaceae
  C. Family Actinodiscaceae
II. Suborder Aulacodiscineae
  A. Family Eupodiscaceae
III. Suborder Auliscineae
  A. Family Auliscaceae
IV. Suborder Biddulphineae
  A. Family Biddulphiaceae
  B. Family Anaulaceae
  C. Family Chaetoceraceae
V. Suborder Rhizosoleniineae
  A. Bacteriastraceae
  B. Leptocylindraceae
  C. Corethronaceae
  D. Rhizosoleniaceae
VI. Suborder Fragilariineae
  A. Family Fragilariaceae
VII. Suborder Eunotiineae
  A. Family Eunotiaceae

VIII. Suborder Achnanthineae
  A. Family Acnanthaceae
IX. Suborder Naviculineae
  A. Family Naviculaceae
  B. Family Auriculaceae
  C. Family Gomphonemaceae
  D. Cymbellaceae
  E. Epithemiaceae
  F. Bacillariaceae
X. Suborder Surirellineae
  A. Family Surirellaceae

An analytical key to these families is given in Hendey (1964). If we were to apply our elementary classification to this system, Suborders I–III would belong to the Centrales and Suborders IV–X to the Pennales.

Reproduction

Most diatoms commonly reproduce asexually, with the two valves separating to form new individuals. Since the new valves are always formed within the old ones, it can readily be seen that through many generations, one of the lineages will become smaller and smaller (Fig. 3). This mode of reproduction easily explains the great range of size of many species (in some cases from 1 to 200 μm). As asexual reproduction carries the diatom to an extremely small size, one of several things may happen. The diatom may reproduce to the lowest level of vitality beyond which it dies. More generally, however, the diatom enters a sexual phase, called auxospore formation, which permits the exchange or rearrangement of genetic materials. The result of this is the formation of a new cell, usually large and close to the maximum size of the species.

The rate at which diatoms reproduce seems largely dictated by a number of external factors. Chief among these are the availability of nutrients. Similarly, auxospore formation is not entirely dictated by size alone but may also depend upon such variables as photoperiod, light intensity, and changes in the turgor pressure around the valve. Some diatom cultures, in the presence of suitable nutrients may divide two to three times a day while other, impoverished floras, may go several months without reproducing.

Given these facts it is not surprising that diatom productivity in surface waters and diatom accumulation in bottom sediments are closely related to upwelling areas and to coastal regions (Berger, 1976).

Ecology and Paleoecology

Diatoms are found in more habitats than are any other microfossil group. Besides being present in the marine environment, they

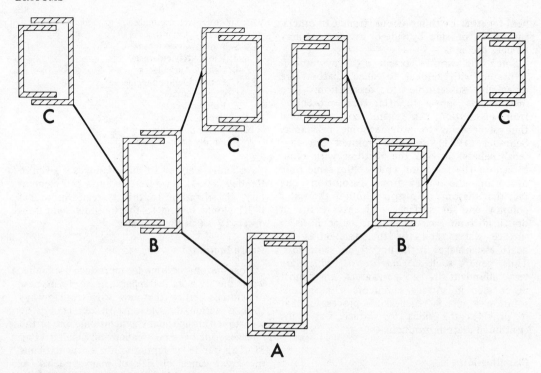

FIGURE 3. Cell division in diatoms. The two valves in specimen A separate and secrete new valves inside the old ones forming two specimens B. This process is repeated to form specimens C and so on.

occupy any number of niches in lakes, rivers, soils, periodically wetted rocks, etc. Indeed, they have even been reported in muds on the webbed feet of ducks and in the upper atmosphere as wind-blown particles.

In the marine environment, diatoms are at the base of the food chain and can readily be broken up into three groups: benthic, neritic, and oceanic. The benthic diatoms include those forms that inhabit either brackish water or normal marine environments. In this group we may also include those forms that are adapted for life attached to the underside of sea ice, as well as those forms that live on plants and animals. The neritic forms are planktic in habitat but are restricted to near-shore regions because they need a shallow-water shelf to complete their life cycle.

Oceanic forms are most abundant in the upper 40m of the water column. As mentioned previously, their distribution is dictated primarily by nutrients, including silica, in the upper water masses. If these nutrients are constantly being renewed at the surface (i.e., by upwelling) then we have a condition that allows for year-round high phytoplankton productivity.

Other factors may be responsible for the temporal and spatial distribution of individual diatom species. In freshwater environments, such factors as pH, water chemistry, and water movement (still or running water) determine species composition and abundance. In brackish-water realms, salinity seems to be an important limiting factor. In addition, we find many marine species that are able to tolerate lowered salinities as well as freshwater forms that can tolerate higher salinities. Such forms are called euryhaline and are frequently observed in the brackish-water realm (Kolbe, 1927).

Hendey (1964), along with many others, has suggested that salinity and temperature seem to be the principle agents determining the geographic range of planktic diatom species. This generalization is borne out by numerous observations from many parts of the world's ocean (i.e., strong temperatures and/or salinity gradients that limit the distribution of diatom species).

The question that we must ask about the fossil record is: Given the ubiquity of diatoms and their sensitivity to many environmental factors, how can they be used in paleoecological reconstructions? In freshwater environments the answer is obvious. Changes in water chemistry as well as pH can be quite accurately determined using changes in diatom assemblages. Further, Bradbury (1975) has used diatoms in lakes from Minnesota and South Dakota to document "limnologic changes associated with enrichment following European-American settlement." In brackish-water environments, changes in salinity can be quite accurately determined using diatoms. Miller (1964) for example, was able to demonstrate late Quaternary changes in temperature and salinity for estuarine sediments along the Swedish West coast. Such an application has considerable value in restricted environments such as the Baltic Sea or, for that matter, Long Island Sound.

In the marine environment, a number of paleoecological approaches have been utilized. Jousé et al. (1963) employed semiquantitative techniques to differentiate Antarctic and Sub-Antarctic assemblages in the Southern Ocean. Changes in these assemblages with time were then used to document changes in water mass distribution during the late Quaternary. A more rigorous treatment was proposed by Kanaya and Koizumi (1966) who used recurrent group analysis (Fager, 1957) to differentiate diatom assemblages in surface sediments of the North Pacific. Td values of diatoms (essentially an index of affinity for a cold- or warm-water assemblage) were then calculated down core. A major advance in marine diatom paleoecological studies was made by Maynard (1976), who employed the techniques of Imbrie and Kipp (1971) to define diatom assemblages in the Atlantic and to relate these assemblages to key surface-water parameters (temperature and nutrients). This approach allowed her to give an actual temperature estimate for late Pleistocene diatom assemblages.

A similar approach has been used by Cooke-Poferl et al. (1975) and Sancetta (in press). In both cases, the techniques of Imbrie and Kipp (1971) were used not only to estimate past temperature, but also to map out past changes in water mass distribution.

Fossil Record

Although individual diatom valves have been reported from the Jurassic, assemblages of diatoms do not occur in the fossil record until the middle Cretaceous. The earliest diatoms are marine and centric. By the Late Cretaceous, there was an explosion, both in abundance and taxa, which continued into the early Tertiary. The earliest known freshwater diatoms are believed to be Eocene in age (Lohman and Andrews, 1968), but the transient nature of lakes and the dearth of lacustrine sediments in the geologic record may be a factor. There has been considerable speculation on the causes of the rise of diatoms in the fossil record. The first appearance of "fully developed" assemblages in the middle Cretaceous strongly suggests that diatoms existed prior to this time as naked individuals. Harper and Knoll (1975) have speculated that the increased demand on the available silica supply caused by the rise of shell-bearing diatoms may have influenced the evolutionary strategy of the Radiolaria. Within the past ten years, and especially since the advent of the Deep Sea Drilling Program, there has been a strong effort to define diatom biostratigraphic zones and to use diatoms in paleo-oceanographic reconstructions. Parallel with this has been the increased use of paleomagnetic stratigraphy in conjunction with biostratigraphy (Burckle, 1972; see Fig. 4). More recently, the oxygen isotope method has been used to define absolute first and last appearances of species (Hays and Shackleton, 1976).

Unfortunately, since this area of study is so new to the diatoms, most of the zonal schemes that have been defined have not been adequately tested. This is particularly true of the Paleogene zonations that have been proposed (Schrader and Fenner,[*] 1976). In spite of this, it is still possible to differentiate the Paleogene using diatoms and to make rough correlations with them over a considerable distance. In the Eocene, for example, the same key species can be seen in the Southern Ocean, the Norwegian Sea, and the Kellogg and Sidney Flats Shales of California.

Indeed, many key species appear to be rather cosmopolitan during the Paleogene although we have no positive evidence that their first and last occurrences are isochronous world wide. We begin to see more provincialism in diatom floras in the Neogene. This probably resulted from the initiation of high-latitude glaciation, which resulted in the formation of new water masses and the closing off of such interocean waterways as the Pacific-Caribbean connection and the Mediterranean–Indian Ocean link.

This increasing provincialism through the Neogene has created a problem that seems rather unique to the diatoms. While low-latitude biostratigraphic zonal schemes from other microfossil groups can be applied to higher latitudes, such is not the case with the diatoms. Further, while other microfossil groups show considerable bipolarity in their species distributions, the diatoms, by and large, do not. This has necessitated the formulation of several zonal schemes to cover any interval of time. For example, in the Pliocene of the Pacific, we need a minimum of three zonal schemes: one for the North Pacific, one for the equatorial Indo-Pacific, and one for the Southern Ocean. In addition, restricted areas

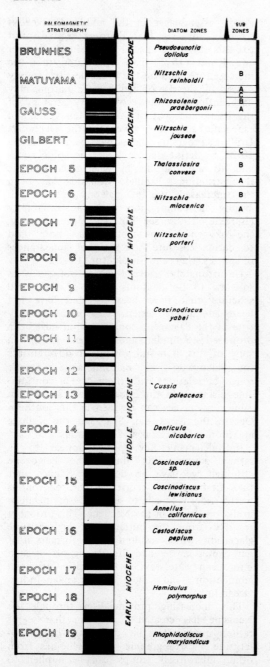

PALEOMAGNETIC STRATIGRAPHY		DIATOM ZONES	SUB ZONES
BRUNHES	PLEISTOCENE	Pseudoeunotia doliolus	
MATUYAMA		Nitzschia reinholdii	B
			A
			C
GAUSS	PLIOCENE	Rhizosolenia praebergonii	B
			A
GILBERT		Nitzschia jouseae	
			C
EPOCH 5		Thalassiosira convexa	B
			A
EPOCH 6		Nitzschia miocenica	B
			A
EPOCH 7	LATE MIOCENE	Nitzschia porteri	
EPOCH 8			
EPOCH 9			
EPOCH 10		Coscinodiscus yabei	
EPOCH 11			
EPOCH 12	MIDDLE MIOCENE	`Cussia paleaceas	
EPOCH 13			
EPOCH 14		Denticula nicobarica	
		Coscinodiscus sp.	
EPOCH 15		Coscinodiscus lewisianus	
		Annellus californicus	
EPOCH 16		Cestodiscus peplum	
EPOCH 17	EARLY MIOCENE	Hemiaulus polymorphus	
EPOCH 18			
EPOCH 19		Rhaphidodiscus marylandicus	

FIGURE 4. An equatorial Pacific diatom zonation for the early Miocene to Pleistocene. This zonal scheme is tied to the paleomagnetic reversal record.

such as the western North American coast require yet another zonal scheme. Correlation of these various zonations is a major problem in diatom studies.

LLOYD H. BURCKLE

References

Berger, W. H., 1976. Biogenous deep sea sediments: Production, preservation and interpretation, in J. P. Riley, and R. Chester, eds., *Chemical Oceanography*. London: Academic Press, 265–388.

Bradbury, J. P., 1975. Diatom stratigraphy and human settlement in Minnesota, *Geol. Soc. Amer. Spec. Paper 171*. 74p.

Burckle, L. H., 1972. Late Cenozoic planktonic diatom zones from the eastern equatorial Pacific, *Nova Hedwegia Beihefte*, **39**, 217–248.

Cooke-Poferl, K.; Burckle, L. H.; and Riley, S., 1975. Diatom evidence bearing on late Pleistocene climatic changes in the equatorial Pacific, *Geol. Soc. Amer. Abstr.*, **7**, 1038–1039.

Fager, E. W., 1957. Determination and analysis of recurrent groups, *Ecology*, **38**, 586–595.

Hays, J. D.; and Shackleton, N. J., 1976. Globally synchronous extinction of the radiolarian *Stylotractus universus*, *Geology*, **4**, 649–652.

Hendey, N. I., 1937. The plankton diatoms of the Southern Seas, *Discovery Repts.*, **16**, 151–364.

Hendey, N. I., 1964, *An introductory account of the smaller algae of British coastal waters*. London: Her Majesty's Stationery Office, 317p.

Harper, H., and Knoll, A., 1975. Silica, diatoms and Cenozoic radiolarian evolution, *Geology*, **3**, 175–177.

Imbrie, J., and Kipp, N., 1971. A new micropaleontology method for quantitative paleoclimatology: Application to a late Pleistocene Caribbean core, in K. K. Turekian, ed., *The Late Cenozoic Glacial Ages*. New Haven: Yale Univ. Press, 71–181.

Jousé, A. P.; Koroleva, G. S.; and Nagaieva, G. A., 1963. Stratigraphic and paleogeographic investigations in the Indian sector of the Southern Ocean (in Russian; English summary), *Oceanology, Russ. Acad. Sci.*, **2**, 137–160.

Kanaya, T., and Koizumi, I., 1966. Interpretation of diatom thanatocoenosis from the north Pacific applied to a study of core V20-130, *Sci. Rept., Tohoku Univ., 2nd Ser. (Geol.)*, **37**, 89–130.

Kolbe, R. W., 1927. Zur Okologie, morphologie, und systematik der brackwasser-Diatomeen, *Pflanzenforschung*, **7**, 1–146.

Lohman, K. E., 1960. The ubiquitous diatom—a brief survey of the present state of knowledge. *Amer. J. Sci.*, **258**, 180–191.

Lohman, K. E., and Andrews, G. W., 1968. Late Eocene non-marine diatoms, Beaver Divide areas, Fremont County, Wyoming, *U.S. Geol. Surv. Prof. Paper 593-E*, 26p.

Maynard, N. G., 1976. Relationship between diatoms in surface sediments of the Atlantic Ocean and the biological and physical oceanography of overlying waters, *Paleobiology*, **2**, 99–121.

Miller, U., 1964. Diatom floras in the Quaternary of the Göta River Valley, *Sveriges Geol. Undersöking*, **44**, 5–67.

Sancetta, C. Oceanography of the North Pacific during the last 18,000 years: Evidence from diatoms, *Marine Micropaleont.*, in press.

Schrader, H. J., 1971. Fecal pellets: Role in sedimentation of pelagic diatoms, *Science*, **174**, 55–57.

Schrader, H. J., and Fenner, J., 1976. Norwegian sea

diatom biostratigraphy and taxonomy, in, M. Talwani, G. Udintsev, et al., *Initial Repts. Deep Sea Drilling Proj.,* **38,** 921-1099.

Cross-references: *Algae; Micropaleontology; Plankton; Protista.*

DINOFLAGELLATES

Dinoflagellates are a group of unicellular organisms characterized by the possession of two flagella, one of which is directed transversely and the other longitudinally. Their size is typically between 20 and 350 μm, rarely more or less. The transverse flagellum arises from a pore situated in a median ventral position; it is usually ribbon-like and undulates round the cell within a groove, the transverse furrow or girdle. The longitudinal flagellum originates from a second pore situated posterior to the first. It may be ribbon-like or thread-like; its anterior portion is situated in a rather less well-defined groove, the longitudinal furrow or sulcus, and its tip trails out behind the cell. The transverse furrow most often has the form of a left-handed spiral, the two ends of which are displaced along the line of the longitudinal furrow (see Fig. 1).

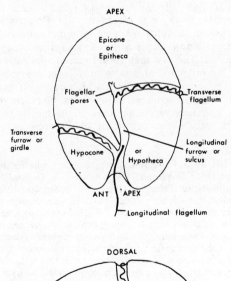

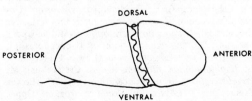

FIGURE 1. The general structure and orientation of a dinoflagellate.

Many dinoflagellates are simple cells enclosed within a soft outer pellicle (the so-called "naked" forms, such as *Gymnodinium*). A few such forms, members of the family Actiniscaceae, have an internal skeleton formed of silica or some other rigid substance: this may be in the form of a continuous layer, may contain perforations, or may consist of discrete plates or other structures.

An important group of dinoflagellates, however, have an external cover formed of cellulose or some similar material. This may be in the form of a simple bivalved shell, as in *Prorocentrum*, but more often is in the form of a series of overlapping plates constituting an "armor," the theca, very like the plate armor of medieval knights. The plates may show sculpture and often have raised margins; their arrangement (tabulation) is of prime importance in the recognition of genera (Fig. 2).

The shape of the cell is immensely variable; basically spheroidal to ellipsoidal or polygonal, it may be greatly modified by apical, antapical, or lateral outgrowths in the form of horns or wings, by the exaggerated development of the ridges (crests) delimiting plates, and by ornament in the form of granules or pits, ribbing, or simple or ramifying spines (Fig. 3). The individual cells are normally quite separate at the motile stage, but a few genera are colonial, either forming multicellular filaments (e.g., *Dinothrix*) or forming a free-swimming colony recognizable as such by the multiplication of flagella (*Polykrikos*).

Ecology

The dinoflagellates are an extremely primitive group, straddling the boundary between the animal and plant kingdoms. A minority are parasitic in varying degree, obtaining their nutrition partially or wholly from a host to which they may be attached externally or within whose body they may live. The host is sometimes another unicellular organism (e.g., the dinoflagellate *Duboscquodinium* is parasitic within Tintinnids and *Merodinium* within Radiolaria), sometimes a multicellular organism (e.g., *Blastodinium* within Copepods). Some others, the zooxanthellae, are symbionts with other organisms—protozoans (e.g., foraminifera), jellyfish, sea-anenomes, scyphozoans, gastropods, turbellarian worms, or corals. A handful of species inhabit wet sands and other sediments on the margins of the sea. Some species live attached to seaweeds or in masses (plaques) in tide ponds; and one species gives color to yellow arctic snows.

The great majority of dinoflagellates are,

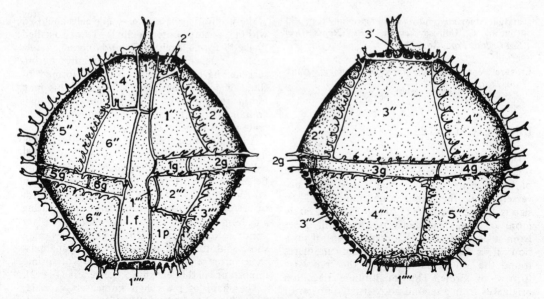

FIGURE 2. A fossil dinoflagellate cyst, *Gonyaulacysta cladophora* (Jurassic), showing a tabulation consisting of four apical paraplates (1'-4'); six pre-cingular paraplates (1"-6"); a cingulum, or transverse girdle, formed of six cingular paraplates (1g-6g); six postcingular paraplates (1"-6"); one posterior intercalary paraplate (1p); and one antapical paraplate (1" "). Paraplate 3" is normally lost in archaeopyle formation. Magnification ca. X500.

however, planktonic. They are of immense significance in the economy of present-day seas, being second only to the diatoms as a basic source of food; they are also abundant in the brackish waters of estuaries and deep lagoons and in freshwater lakes and large ponds. Most are holophytic, containing chromatophores which enable photosynthesis—the thecate dinoflagellates are dominantly holophytic. Some are holozoic or saprophytic, ingesting food into vacuoles; certain naked forms produce rhizopodia to aid in locating food particules. Others have a mixed nutrition, having at the same time both chromatophores and vacuoles containing food particles; yet others lose their chromatophores during life and become secondarily holozoic.

The flagella have a locomotory function, propelling the dinoflagellate in corkscrew fashion in a forward direction. The buoyancy of the cell is controlled by the degree of vacuolization, the number of water vacuoles reducing or increasing specific gravity relative to that of the surrounding waters and thus allowing the dinoflagellate to rise or to sink. Movement by these means is essentially in response to light conditions; many holophytic forms maintain a constant light relationship by movement downward in daytime and toward the surface at night, the diurnal migration sometimes amounting to several hundred feet. Luminiscence of seawater at night is frequently produced by dinoflagellates, such as

Noctiluca or *Pyrocystis*, that have migrated to the surface.

Reproduction has long been considered to occur dominantly by asexual means. However, sexual reproduction is now known to occur in several genera, e.g. *Noctiluca* and *Certatium*, though it has not yet been commonly reported. Binary fission occurs within a cyst formed inside the cell. The cyst may form in close application to the outer membrane or may form nearer to the center of the shell, in which case cyst and cell wall may be linked by processes. The form of the cyst may reproduce that of the motile cell or it may show significant differences. After

FIGURE 3. Fossil dinoflagellate cysts typifying the great range in morphology exhibited. A, *Deflandrea phosphoritica* (Paleocene-Oligocene); B, *Stephodinium coronatum* (Upper Cretaceous); C, *Scriniodinium dictyotum* (Upper Jurassic); D, *Heliodium patriciae* (Lower Cretaceous); E, *Dinogymnium hetercostatum* (Upper Cretaceous); F, *Spiniferites ramosus* (Upper Jurassic–Holocene); G, *Peridinium pyrophorum* (Upper Cretaceous); H, *Muderongia tetracantha* (Lower Cretaceous); I, *Pareodinia ceratophora* (Middle-Upper Jurassic); J, *Ctenidodinium ornatum* (Upper Jurassic); K, *Systematophora complicata* (Lower Cretaceous); L, *Raphidodinium fucatum* (Upper Cretaceous); M, *Pseudoceratium* (*Eopseudoceratium*) *gochti* (Lower Cretaceous); N, *Chytroeisphaeridia chytroeides* (Upper Jurassic); O, *Hystrichokolpoma sequanaportus* (Upper Cretaceous); P, *Wetzeliella symmetrica* (Oligocene). Specimens A, K, N, and O display archaeopyles. Magnification ca X300.

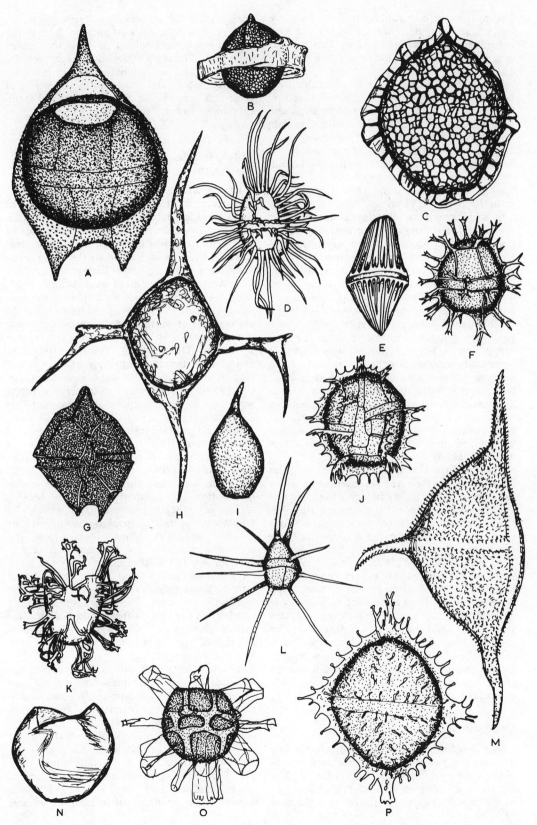

fission, the cyst may rupture to form an opening, or archaeopyle, through which the new individuals exit. Resting cysts are also formed, as a means of survival during unfavorable conditions.

Dinoflagellate distribution is controlled by a number of factors—water depth, temperature, salinity, and availability of nutrients. These factors combine to produce dinoflagellate assemblages characteristic of particular water bodies; for example, the Atlantic and North Sea assemblages are so distinct that the boundary between them can be drawn across the English Channel and can even be recognized from the air by the change in water color!

In tropical waters, dinoflagellates are to be found in fairly constant (and generally rather low) numbers throughout the year. Concentrations of dinoflagellates in temperate waters are very much greater, figures of up to 3000 individuals per liter of seawater being in no way exceptional. However, outside the tropical zone, there is considerable seasonal variation in abundance; during the periods when dropping water temperatures and/or a reduced food supply render the environment less favorable to inhabit, many species encyst until conditions improve again. Encystment may be stimulated by a drop in water temperature of as little as $2°C$: excystment is brought about when the temperature improves again. (Laboratory studies have shown, however, that excystment cannot be induced until a minimum period of a few days has passed).

Dinoflagellate blooms may occur at irregular intervals; blooms are not yet predictable, since the environmental causes inducing them are not yet fully comprehended. During such a bloom, the numerical abundance, usually of a single species, soars to totals of up to eleven millions per liter (Prakash and Viswanatha Sarma, 1964). These blooms produce "red tides" (the water being colored by the enormous dinoflagellate concentrations), which have been known to spread from the sea for a considerable distance up rivers, sheer population pressure apparently inducing the dinoflagellates to enter relatively unfavorable environments. Such blooms may cause considerable mortality among animals and fish, since certain bloom-prone dinoflagellate species contain toxic substances.

Classification

As a result of their position on the borderline between the two kingdoms, dinoflagellates are claimed by both zoologists and botanists. The former treat them as the order Dinoflagellata of the class Mastigophora, the latter as the class Dinophyceae of the division Pyrrhophyta. In general, however, present opinion is that the presence of chloroplasts is a more cogent evidence of plant affinity than is a partially holozoic habit of animal affinity; thus the dinoflagellates are customarily treated as algae and hence under the provisions of botanical nomenclature.

History of Study of Fossil Dinoflagellates

The German microscopist, Christian G. Ehrenberg, in 1836, was the first to record fossil dinoflagellates. He encountered, in thin flakes of Mesozoic flint and chert, a number of cells exhibiting clear transverse and longitudinal furrows; these he attributed to the living genus *Peridinium* (Fig. 3G). No further fossil dinoflagellates were recognized as such until 1922, when Walter Wetzel discovered them in Cretaceous flints from the shores of the Baltic Sea. Subsequent major studies of Jurassic and Cretaceous assemblages were published during the 1930s by Otto Wetzel, Georges Deflandre and Maria Lejeune-Carpentier. Otto Wetzel also recorded the first Eocene assemblages. In the 1950s, an enormous volume of new forms were described from upper Mesozoic and Tertiary horizons by Isabel Cookson, C. Downie, A. Eisenack, H. Gocht, Sarjeant and others.

Throughout this period, recognition of fossil dinoflagellates was based on their possession of a distinct tabulation or, at least, of a clear transverse furrow. Forms of similar distribution and composition, usually bearing spines or tubular processes and without a tabulation (e.g., *Hystrichosphaeridium*) or tabulate but with spines arising from the transverse girdle so as to render it clearly functionless (*Hystrichosphaera*), were termed "hystrichospheres" (q.v.) and were classed into the order Hystrichosphaeridea of uncertain systematic reference. There were, in addition, a number of genera whose morphology was neither recognizably dinoflagellate-like nor of hystrichosphere type—these were left nebulously *incertae sedis*.

A feature frequently evident in fossil tabulate dinoflagellates was the presence of some sort of opening, formed either by the loss of the group of plates comprising the apex, as in *Pseudoceratium* (Fig. 3M), of a single pre-equatorial plate, as in *Gonyaulacysta* (Fig. 2), or by schism along, or parallel to, the line of the transverse furrow, as in *Ctenidodinium* (Fig. 3J). This morphological feature was the object of a detailed study by Evitt (1961), who showed that openings of similar types and situations were present in nontabulate fossil dinoflagellates. Moreover, Evitt found identical openings in the walls of some hystrichospheres and in certain genera hitherto left "incertae sedis". He further showed (1963) that the distribution of processes in some hystrichospheres formed a regular pattern corresponding to a tabulation. The implications of this work were profound; not only were many hystrichospheres demonstrably the cysts of dinoflagellates, but also the vast majority, perhaps all, of the forms already recognized as dinoflagellates were not motile forms but cysts (Downie et al., 1963).

Recent studies indicate that superficially similar motile dinoflagellates form cysts of more than one type (see resume in Sarjeant, 1970b). This creates a number of problems in their classification. On the one hand, it means that a genus with a constant tabulation in the motile stage may nonetheless be the product of a number of lineages. If the true natural groupings are to be expressed, the genus should therefore be subdivided; but the resultant units would be recognizable only when the type of cyst formed became known, with consequent complications to research. On the other hand, a number of morphologically distinctive genera based on fossil cysts would become synonyms of modern genera based on motile dinoflagellates.

This problem is not insuperable, since the botanical code permits of the existence of form genera, morphologically defined, alongside "natural" genera. However, problems are encountered in classification about the generic level, since form genera cannot serve as nomenclatural types for families. Sarjeant and Downie (1966) proposed the classification of form genera based on cysts into a hierarchy of cyst families, independent of the arrangement into families of the motile genera. This procedure was designed to enable the erection of a firm framework for classification of fossil cysts, whose relationships to motile genera might never be accurately determined. However, it produced an adverse response from specialists and, in consequence, the cyst-family concept has since been abandoned even though, in the modified scheme more recently advanced by these authors (Sarjeant and Downie, 1974), most families still essentially or wholly comprise cysts.

A descriptive terminology used for dinoflagellate cysts evolved only slowly, but by now a number of new terms have been formulated, with the view of enabling greater precision in future taxonomic descriptions; a recently published glossary conveniently brings most of these together (Williams et al., 1978).

Geological Distribution

The first entry of dinoflagellate cysts in the stratigraphic column is difficult to establish with any certainty. Their simple morphology suggests that they must be an extremely primitive group. Motile forms are not capable of fossilization and, since it is quite possible that the earliest dinoflagellates were not capable of secreting cysts, the true antiquity of this group is unlikely ever to be known. On the basis of the character of their nucleus and the unique manner of separation of the chromatids during nuclear division (*protomitosis*), it is clear that they are of profound evolutionary importance; indeed, it has been suggested that dinoflagellates form the link between the most primitive plants—the prokaryotes (bacteria and blue-green algae [see *Protista*])—and higher organisms (the eukaryotes).

The readiest method of identifying dinoflagellate cysts is by recognizing structures indicative of a tabulation. However, many present-day dinoflagellates are without armor and, as one would expect, produce nontabulate cysts. Since armor is a complex, and probably an advanced, feature, it is likely that unarmored forms predate the armored forms and that the earliest cysts had neither traces of armor nor even a regularly formed opening.

Thus it seems that we must seek the earliest dinoflagellate cysts among the acritarchs (q.v.), an assemblage of organic-walled planktonic organisms of recognizedly problematic affinity. Certain Cambrian acritarchs, e.g., *Lophodiacrodium* (Fig. 4C), have an elongate shape, with spines at both poles and with a smooth, equatorial zone reminiscent of a transverse girdle. Openings are not yet known in these forms, however. Two Lower Ordovician genera, *Cymatiogalea* and *Priscogalea* (Fig. 4B), are characterized by large terminal openings, suggestive of an apical archaeopyle. Are these dinoflagellate cysts? In the late Silurian occurs a Tunisian genus, *Arpylorus* (Fig. 4A), which is beyond all question a dinoflagellate cyst. It has an intercalary excystment aperture, shows indication of a transverse furrow, and exhibits vestiges of a tabulation; the cyst indeed tends to break up into its constituent plates (Sarjeant, 1978b). Lister (1970) has made a detailed analysis of other Silurian genera, which he has shown to possess regularly arranged processes and regularly formed openings suggestive of a dinoflagellate tabulation pattern; thus we have two lines of evidence indicating that dinoflagellates were already present in the Paleozoic seas. The Devonian acritarchs also include forms whose morphological traits in some measure suggest a dino-

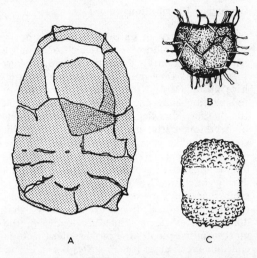

B

A   C

FIGURE 4. A *Arphylorus antiquus* (Silurian), the earliest known dinoflagellate cyst, with an intercalary archaeopyle; B–C, Paleozoic acritarchs showing features suggestive of dinoflagellate affinity. B, *Priscogalea multarea* (Ordovician), with an opening resembling an apical archaeopyle; C, *Lophodiacrodium arbustrum* (Cambrian), with an unornamented median girdle suggestive of a transverse furrow. Magnification ca X300.

flagellate affinity. Little is yet known of Carboniferous and Permian acritarchs; the species so far described afford few comparisons with the dinoflagellates and two supposed North American records of dinoflagellates from the Permian are neither impressive nor convincing.

The first post-Silurian records of undoubted dinoflagellate cysts are from the uppermost Triassic. Proximate cysts—cysts showing a close morphological correspondence to motile forms—predominate. A number of different morphological types are present, some with complex tabulation patterns, and several modes of encystment are already evident; thus it seems that these Triassic assemblages contain the product of several evolutionary lineages. Most Lower Jurassic dinoflagellate cysts so far described—and these are yet few—are of proximate character but, by the Middle Jurassic cavate cysts (forms with a distinct inner body) and chorate cysts (cysts of "hystrichosphere" type, sometimes with highly complex, often interlinked processes) are becoming frequent (see Riley and Sarjeant, 1972; Sarjeant, 1978a). By the Lower Cretaceous, chorate and proximochorate cysts, notably *Spiniferites* (Fig. 3F; ex: *Hystrichosphaera*) and *Hystrichosphaeridium,* have become the most prominant feature of the assemblages, remaining so to the present day, with the proximate cysts showing a corresponding

decline in numbers and variety. In the uppermost Cretaceous and Tertiary, cavate cysts, notably the genera *Deflandrea* (Fig. 3A) and *Wetzeliella* (Fig. 3P), are also prominent (see Harker and Sarjeant, 1975). These types disappear from marine assemblages in the later Cenozoic; chorate and proximochorate forms, with a few proximate types sometimes present, are characteristic of all Quaternary assemblages described to date (see Bradford, 1978).

From the Upper Triassic onward, dinoflagellate cysts show a diversity of type and a rapidity of evolution that render them entirely suitable for use as tools in stratigraphical correlation. They have been recorded from marine sediments of all types—limestones, sandstones, shales, and clays, from phosphate nodules, and even from salt deposits. Because of their planktonic origin, dinoflagellate cyst assemblages are less liable to rapid lateral variation than are assemblages of most other marine organisms, since the distribution of plankton is controlled by the conditions of the water body as a whole and not by bottom conditions. They are thus coming to be widely employed as stratigraphic indices by oil companies and other commercial organizations, wherever lithology or other factors reduce the utility of foraminifera.

Dinoflagellate cysts are known from nonmarine deposits, specifically from peats and lignites (see Harland and Sarjeant, 1970). However, records of nonmarine cysts are as yet infrequent. Characteristically they are marine: in horizons such as the Purbeck and Wealden of England, which show an admixture of nonmarine and marine sediments, the marine horizons are clearly defined by the incoming of dinoflagellates.

Fossil dinoflagellate cysts with calcareous shell walls, such as *Calciodinellum,* are known infrequently from deposits as early as the Upper Jurassic, but they have been relatively little studied. Forms with entirely siliceous walls (*Peridinites*) have been recorded from the Tertiary; these too are most probably cysts, though this has not yet been demonstrated beyond doubt.

WILLIAM A. S. SARJEANT

References

Bradford, M. R., 1978. An annotated bibliographic and geographic review of Pleistocene and Quaternary dinoflagellate cysts and acritarchs, *Amer. Assoc. Stratig. Palyn. Contrib.,* 6, 192p.

Chatton, E., 1952. Classe des Dinoflagellés ou Péridiniens, in P. P. Grassé, ed., *Traité de Zoologie,* vol. 1. Paris: Masson et Cie, 310-390.

Davey, R. J.; Downie, C.; Sarjeant, W. A. S.; and Williams, G. L., 1966. Studies on Mesozoic and

Cainozoic dinoflagellate cysts, *Bull. British Mus. (Nat. Hist.), Geol., Suppl.* **3**, 248p.

Deflandre, G., 1955. Classe des Dinoflagellés, in J. Piveteau, ed., *Traité de Paléontologie, vol 1.* Paris: Masson et Cie, 116-124.

Downie, C.; Evitt, W. R.; and Sarjeant, W. A. S., 1963. Dinoflagellates, hystrichospheres and the classification of acritarchs, *Stanford Univ. Publ. Geol. Sci.*, **7**, 1-16.

Downie, C., and Sarjeant, W. A. S., 1964. Bibliography and index of fossil dinoflagellates and acritarchs, *Geol. Soc. Amer. Mem. 94*, 180p.

Evitt, W. R., 1961. Observations on the morphology of fossil dinoflagellates, *Micropaleontology*, **7**, 385-420.

Evitt, W. R., 1963. A discussion and proposals concerning fossil dinoflagellates, hystrichospheres and acritarchs, *Proc. Nat. Acad. Sci.*, **49**, 158-164, 298-302.

Evitt, W. R., 1970. Dinoflagellates-A selective review, *Geoscience and Man.* **1**. 29-45.

Evitt, W. R., Lentin, J. K., Millioud, M. E., Stover, L. E., and Williams, G. L., 1977. Dinoflagellate cyst terminology, *Geol. Surv. Canada Paper 76-24*, 11p.

Harker, S. D., and Sarjeant, W. A. S., 1975. The stratigraphic distribution of organic-walled dinoflagellate cysts in the Cretaceous and Tertiary, *Rev. Palaeobot. Palynol.*, **20**, 217-315.

Harland, R., and Sarjeant, W. A. S., 1970. Fossil freshwater microplankton (dinoflagellates and acritarchs) from Flandrian (Holocene) sediments of Victoria and Western Australia, *Proc. Roy. Soc. Victoria*, **83**, 211-234.

Lentin, J. K., and Williams, G. L., 1973. Fossil dinoflagellates: index to genera and species, *Geol. Surv. Canada, Paper 73-42*, 176p.

Lister, T. R., 1970. The acritarchs and Chitinozoa from the Wenlock and Ludlow Series of the Ludlow and Millichope Areas, Shropshire, Pt. 1, *Palaeontogr. Soc. Monogr.* **124**, 1-100.

Norris, G., 1978. Phylogeny and a revised supra-generic classification for Triassic-Quarternary organic-walled dinoflagellate cysts (Pyrrhophyta). Part II. Families and sub-orders of fossil dinoflagellates, *Neues Jahrb. Geol. Paläont. Abhandl.*, **156**, 1-30.

Norris, G., and Sarjeant, W. A. S., 1964. A descriptive index of genera of fossil Dinophyceae and Acritarcha, *New Zealand Geol. Surv. Palaeontol. Bull.*, **40**, 72p.

Prakash, A., and Viswanatha Sarma, A., 1964. On the occurence of "red water" phenomenon on the west coast of India. *Curr. Sci.*, **33**, 168-170.

Riley, L. A., and Sarjeant, W. A. S., 1972. Survey of the stratigraphical distribution of dinoflagellates, acritarchs and tasminitids in the Jurassic, *Geophytology*, **2**, 1-40.

Sarjeant, W. A. S., 1964. The Xanthidia: The solving of a palaeontological problem, *Endeavour*, **24**, 33-39.

Sarjeant, W. A. S., 1967. The stratigraphical distribution of fossil dinoflagellates, *Rev. Palaeobot. Palynol.*, **1**, 323-343.

Sarjeant, W. A. S., 1970a. Xanthidiaa palinospheres and "Hystrix." A review of the study of fossil microplankton with organic cell walls, *Microscopy, J. Quekett Microsc. Club*, **31**, 221-253.

Sarjeant, W. A. S., 1970b. Recent developments in the application of fossilized planktonic organisms to problems of stratigraphy and paleoecology, *Paläobotanik*, **111**, 669-680.

Sarjeant, W. A. S., 1974. *Fossil and Living Dinoflagellates.* London and New York: Academic Press, 180p.

Sarjeant, W. A. S., 1978a. A guide to the identification of Jurassic dinoflagellate cysts, *Louisiana State Univ. School of Geosciences, Misc. Publ. 78-1*, 107p.

Sarjeant, W. A. S., 1978b. *Arpylorus antiquus* Calandra, *emend.*, a dinoflagellate cyst from the Upper Silurian, *Palynology*, **2**, 167-178.

Sarjeant, W. A. S., and Downie, C., 1966. The classification of dinoflagellate cysts above generic level, *Grana Palynologica*, **6**, 503-527.

Sarjeant, W. A. S., and Downie, C., 1974. The classification of dinoflagellate cysts above generic level: A discussion and revision, *Birbal Sahni Inst. Paleobot., Spec. Publ. 3*, 9-32.

Williams, G. L., 1977. Dinocysts: Their classification, biostratigraphy, and palaeoecology, in A. T. S. Ramsay, ed., *Oceanic Micropaleontology*, vol. 2. London: Academic Press. 1231-1325.

Williams, G. L., Sarjeant, W. A. S., and Kidson, E. J., 1973. A glossary of the terminology applied to dinoflagellate amphiesmae and cysts and acritarchs, *Amer. Assoc. Stratig. Palynol. Contrib. 2*, 222p.

Cross-references: *Acritarchs; Ebridians; Hystrichosperes; Plankton; Protista; Zoothanthellae.* Vol. I: *Phytoplankton.*

DINOSAURS

The name *Dinosauria* was proposed by Professor Richard Owen in his *Report on British Fossil Reptiles* published by the British Association in 1842. He compounded the word from the Greek *deinos* (which he translated as "fearfully great") and *sauros* (lizard). The three genera for which he created the name were then inadequately known, so that his diagnosis of the features we now know to be characteristic of the group cannot be considered valid. The name itself is not now used as a scientific term; but in its more colloquial form, "dinosaurs" provides a convenient umbrella under which different kinds of extinct reptiles, of different origin and habit, find shelter in at least the textbooks.

Dinosaurs were reptiles, with some similarities in anatomy and physiology to many reptiles known today, especially alligators and large lizards. They comprise two orders of the subclass Archosauria, as described below. More recently, some authors (e.g., Bakker, 1975; Bakker and Galton, 1974) have suggested that dinosaurs were, unlike reptiles, endothermic and might even belong, with their descendants the birds, in a separate vertebrate class (but see Bennet and Dalzell, 1973; Dodson, 1973, 1974).

Subclass Archosauria

Order Thecodontia. Suborder Pseudosuchia contains small lizard-like reptiles, 2–3 ft. (.6–.9 m) long, with small heads. The jaws had sharp teeth in sockets. The forelimbs were often shorter than the hind, so that many of these animals, such as *Scleromochlus*, were much like slightly larger forms of modern Iguanas. They are found in the Middle and Upper Triassic of Europe, North and South America and South Africa. At that geological age, they cannot be the ancestors of the almost contemporary dinosaurs, but a South African Lower Triassic form of this kind is probably the originator of the other forms of archosaurs. These are:

Order Crocodilia–Triassic to the present time (see *Reptilia*)
Order Pterosauria (or Flying reptiles)–Jurassic to Cretaceous (see *Reptilia*)
Order Saurischia–Triassic to Cretaceous
Order Ornithischia–Upper Triassic to Cretaceous

The last two orders are popularly known as the dinosaurs.

Order Saurischia. Saurischia are characterized by a tri-radiate pelvic girdle of truly reptilian (saurian) type with ilium attached to the sacral vertebrae, pubis extending forward and downward, and ischium extending downward and backward (Fig.1A); a skull with teeth in sockets; and tooth reduction, if it took place, always at the back of the jaws. Gape of jaws was wide. Forelimbs were nearly always shorter than hindlimbs; a clavicle was absent in the shoulder girdle. Skin was unarmored. There were bipedal forms and quadrupedal forms (Fig. 2).

Suborder Theropoda. These bipedal saurichia (Theropoda = beast feet) were almost invariably carnivorous–Triassic to Cretaceous in Age.

Infraorder Coelurosauria. The Coelurosauria were comparatively small and slender predators that probably roamed the mud flats. The skull was small, with large orbit; teeth sharp and recurved. Vertebrae were usually amphiplatyan (flattened on front and rear surfaces). The hind legs were usually long. These forms are almost world wide in occurrence and this, as well as some of the skeletal characters, suggests that there are probably several lines of pseudosuchian descent.

Most of the forms were small, swift-running creatures, such as *Coelophysis* from the Triassic of New Mexico, about 8 ft (2.5 m) long. *Ornitholestes* was a somewhat smaller Jurassic form known from the US. It was about 6 ft (2 m) long and was apparently a hunter of

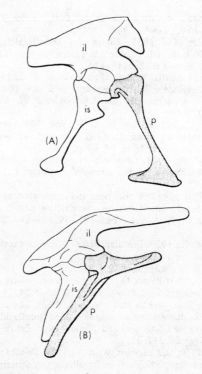

FIGURE 1. The pelvis in the two dinosaurian orders (from Colbert, 1969, used by permission). A, Saurischian pelvis, with a forwardly directed pubis. B, Ornithischian plevis, with the pubis parallel to the ischium.

small animals of the undergrowth. The evidence for this is the hand, which had only three very long fingers. This line was continued in the Jurassic by the very active predator, *Deinonychus*, from Montana. Here, the hand and forelimb elongation was such that the limb was nearly three-quarters of the length of the hindlimb. The animal, some 10 ft (3 m) long, appears to have been more ostrich-like in appearance than the usual small theropod, and shows affinities with the Cretaceous *Velociraptor* and *Dromaeosaurus* and, in another line, with *Stenonychosaurus* and *Saurornithoides*. This whole group is intermediate between Coelurosaurs and Carnosaurs and may be an off-shoot. They appear to have been good runners and raptorial, rather than carrion feeders, a conclusion that was strengthened by the discovery in the Gobi Desert of a *Velociraptor* skeleton with its teeth still embedded in that of a *Protoceratops*. The relationship with the ornithomimid (struthiomimid) group does not seem to be close.

Ornithomimus (also sometimes known as *Struthiomimus*–the "ostrich mimic," which may be a separate genus, from Alberta, Canada)

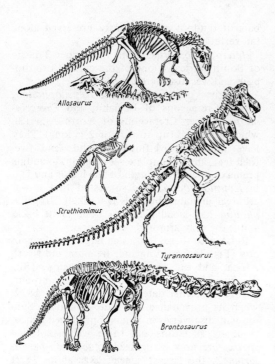

Allosaurus

Struthiomimus

Tyrannosaurus

Brontosaurus

FIGURE 2. Skeletons of representative saurischian dinosaurs (from Young, 1962, copyright © 1962 by Oxford University Press and used with permission).

was 10 ft (3 m) long, running animal, with feet like those of a bird (Fig. 2). The hand had only three fingers; and, as the skull of this dinosaur was completely toothless, the hand may have been used for plucking fruit or stealing eggs. *Ornithomimus* is often represented in an ostrich-like pose which is not sustained by reference to the type material. Skeletons are known from the closing days of the dinosaurs in Canada and from Colorado.

A line of larger flesh-eating dinosaurs—Carnosaurs—includes the fragmentary but well-known *Megalosaurus* of England and the very similar *Ceratosaurus* of the US. The latter had a horn on the nose, and although this was not characteristic of *Megalosaurus*, at least one species (*M. bradleyi*) also had one. Both of these dinosaurs were of Late Jurassic and Early Cretaceous age and are of much the same character as the better known and more formidable *Antrodemus* (*Allosaurus*) of the US (Fig. 2). This was about 30 ft (9 m) in overall length. The hands were still capable of use in feeding and the first three fingers had claws.

This line of Theropod evolution culminated in the Dinodonts (huge-teeth), the largest predators of the Mesozoic. *Gorgosaurus* and *Albertosaurus* (Upper Cretaceous of Canada), *Tarbosaurus* (Upper Cretaceous of Mongolia),

and *Tyrannosaurus* (Upper Cretaceous of US.) compose this group. *Tyrannosaurus rex* (Fig. 2) was 47 ft (14 m) long, with a skull 4 ft (1.2 m) long, in which large sharp teeth protruded 5 in. (13 cm) from the gums. Although these dinosaurs were large and may have weighed 8 tons (8100 kg), the forelimbs were very small and ended in two little fingers, an apparently useless structure.

Suborder Sauropodomorpha. Large, often very large, quadrupedal dinosaurs the Sauropodomorpha had the same pelvic characters as the Theropoda and many of the same skull characters. Their habits were mainly aquatic and herbivorous, though doubt has been thrown on a wholly aquatic role.

Infraorder Prosauropoda. The Prosauropoda are intermediate, structurally and in habit, between the Theropoda and the Sauropoda. Thus they were not aquatic, not generally quadrupedal, and are somewhat difficult to assess regarding their food.

They contain two families which, in turn, contain some familiar small and geologically early forms. While the Plateosauridae included *Plateosaurus,* from the Triassic of Germany, a clumsy and only occasional biped that lived by the margins of inlets, lakes, and watercourses and may have been a fish catcher, the Melanorosauridae includes a number of well-known American dinosaurs. The family name is, however, derived from *Melanorosaurus* from the Middle and Upper Triassic of South Africa. The American representatives are the small, active bipedal hunters long known under the name of *Anchisaurus* (6–8 ft long), *Yaleosaurus* (6 ft long) and *Ammosaurus,* all from the Connecticut Valley.

Infraorder Sauropoda. It is a far cry from the above small hunters to the ponderous Sauropods, but in Jurassic and Cretaceous times, these developed into two main groups: one large, long, and low in the shoulder, as represented by *Apatosaurus* (more often but incorrectly known as *Brontosaurus*; Fig. 2) in which the overall length of the skeleton was about 60 ft (18 m). The weight of the living animal might have been 30 tons (33,000 kg; Colbert, 1962). *Diplodocus,* with a length of 87 ft (26.5 m) and a probable weight of 40 tons (44,000 kg) is one of the greatest landliving (or freshwater) animals of all time. These named dinosaurs are from the US.

Diplodocus was exceeded in bulk, if not in length, by *Brachiosaurus,* a member of a group in which the forelegs were larger than the hind, and which were high and massive. This is probably the bulkiest dinosaur. It was 70 ft (21 m) long and the head was 40 ft (12 m) off the ground. *Brachiosaurus* has been found in the Lower Cretaceous of E Africa and the Upper Jurassic of the US.

Such ponderous animals must have been compelled to use the buoyancy of water, and their nostrils consequently were set high on the face or head. With the long neck extended, the elephantine body afloat, and the long tail floating behind, the animals pushed themselves along by thrusts of the forelegs on the muddy bottom of the water, as fossil footprints testify. Living in this way, they cropped vegetation and ingested pebbles (gastroliths) at the same time. Such a position gave a large measure of immunity from the theropods, though it is believed that the females were egg laying and thus would be compelled to lay the eggs on land.

The sauropods were at their maximum in the Northern Hemisphere in the Lower Cretaceous. Upper Cretaceous forms are known mainly from South America, Africa, and China (Charig, 1973). The pose and habitat of the sauropods have long been in question on general grounds. Were they truly aquatic or were they only occasionally so? Obviously the females, as reptiles, must have ventured some way on shore to lay their eggs, but other grounds have recently been suggested for reconsidering the problem. It is noticeable that the sauropods have a body that is higher than broad, a condition more characteristic of land-living mammals. Water-living reptiles are generally broad beamed, though the marine iguanas of the Galapagos are apparent exceptions.

The explanation for the retention of a reptilian form may be partly hereditary, since they are descended from land-living ancestors; partly because the girdle widths are reptilian and not mammalian, and, in consequence the limbs are set in columnar fashion, giving adequate columnar support. In the absence of a patella and without more adequate evidence than we have of additional muscular support, it is difficult to envisage these animals *habitually* on land. (For a recent discussion, see Coombs, 1975.)

Order Ornithischia. Either bipedal or quadrupedal, the ornithischia are herbivorous dinosaurs (Fig. 3) with a bird-like pelvis with normal ilium and ischium; but the pubis has a forward branch and also a pendent branch, the latter often paralleling the ischium (Fig. 1B). Their distribution is world wide and slightly more restricted in geological time than the Theropoda.

Suborder Ornithopoda. The herbivorous ornithopods (bird-feet) were bipedal with characteristically three-toed feet. The front of the mouth very often was without teeth and a special predentary bone developed. Ossified

bars or tendons are fequently preserved along tail vertebrae.

Heterodontosaurus, from the Upper Triassic of South Africa, is one of the earliest ornithopods and was discovered in 1962.

Primitive ornithopods are *Hypsilophodon*, Lower Cretaceous of England, and *Thescelosaurus*, Upper Cretaceous of North America, which are small (up to 6 ft or 2 m long). They have primitive five-fingered hands and five-toed feet, and a complete set of teeth, including premaxillaries. The length of the hands and feet in *Hypsilophodon* have suggested that it was arboreal in habit, but this is unlikely. *Hypsilophodon* had small plates of bone in its back, and is the only armored ornithopod.

Related to these, but much larger, are *Iguanodon* (Fig. 3) from England, Belgium and North Africa, and the American *Camptosaurus* (Fig. 3). *Iguanodon* is the oldest known dinosaur, having been first detected in England in 1822, when its provenance was quite unknown. Subsequent discoveries in Belgium produced almost complete skeletons and three well-established species are now recognized. *Iguanodon bernissartensis*, the largest species, was about 36 ft (11 m) long and, resting with its tail on the ground, the head was 16 ft (5 m) from the ground. This great herbivorous biped had efficient and rather human-like arms and

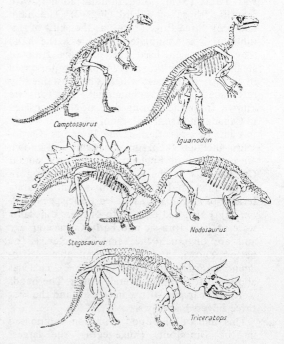

FIGURE 3. Skeletons of representative ornithischian dinosaurs (from Young, 1962, copyright © 1962 by Oxford University Press and used with permission).

hands, the thumb being represented by a single spike of bone. This was at first thought to be a horn on the nose and its purpose, in defense, offense, or in breaking down vegetation is still not clear. The three-toed feet have left footprints in Southern England, Africa, and Spitsbergen. From Bernissart came 28 skeletons of this kind and two of the smaller *Iguanodon mantelli*, which was 26 ft (8 m) long and 13 ft (4 m) high. It may be that the larger are females and the smaller "species" the male (Casier, 1960).

The American counterpart of *Iguanodon*, and the first kind of dinosaur discovered in the United States, was called *Trachodon*, a name so ill-defined that it has since been replaced with *Anatosaurus*. This is a representative of the remarkable group of bipedal, herbivorous, and semiaquatic dinosaurs known as hadrosaurs. Many forms are known from the Upper Cretaceous of the US.; Alberta, Canada; and the USSR.

They were large, like *Iguanodon*, but the skull had a duck-bill snout. The hand was four-fingered, with no thumb, and in the three-toed feet, the toes had little hoofs. There are numerous leaf-like teeth compressed on the side of the jaws, indicating a vegetable diet with a high degree of dental erosion, loss, and replacement. It has been assumed that the diet was roots of horse-tails and water-lilies but recent studies suggest that these dinosaurs were only partially aquatic and browsed on conifers and deciduous trees and shrubs (Ostrom, 1964). The hand was suited for swimming, not grasping, and the tail, flattened from side to side, would help in propulsion when the animals crossed swamps.

Many of the hadrosaurs had great outgrowths of the nasal bones forming grotesque crests or helmets or spikes. These contained extensions of the nasal passages, but their function is questioned (e.g., Hopson, 1975). They may have carried enlarged organs of smell. Hadrosaurs being 4–5 tons in weight were not too heavy for land living and it has been suggested that in the Late Cretaceous they were migratory.

Suborder Pachycephalosauria. The very successful collecting expeditions of the Polish Academy of Sciences, under Prof. Kielan-Jaworowska and her colleagues, and their brilliant analyses of the results, have added new chapters to the history of the Mongolian Cretaceous. Light has been shed on the Ornithomimid Theropods (see above) but especially on the family Pachycephalosauridae.

These animals with their remarkable boss-like skulls with other bony ornamentation, have intrigued paleontologists for a century but have so far, in *Stegoceras* (*Troodon*) and *Pachyceph-*

alosaurus from North America, and with the recently described *Yaverlandia* from England, been included in the Ornithopoda. Now, on the basis of the new Mongolian studies, Maryánska and Osmólska (1974) have elucidated the structure and habits of the group, have created three new genera (*Homalocephale*, *Prenocephale*, and *Tylocephale*), and have shown that the Pachycephalosauridae must be contained in a new Suborder of the Ornithischia.

These animals are unlike most dinosaurs in the formation of their orbits (but are like some birds and mammals) and have an unusually underdeveloped pubis. They seem to have been plant eating and insectivorous, were slow moving, and used their tails as a prop when resting, but with the back horizontal. The full study of the materials collected, with perhaps additional material from western Canada, will have important results on the classification of the Ornithischia.

Suborder Stegosauria. The Ornithischia included quadrupedal forms, all of which were armored on either body or head. Geologically oldest is *Scelidosaurus* from the base of the Jurassic of England. It was about 12 ft (3.6 m) long; the neck and tail had bony spikes; and the body had series of large bony scutes like those of a crocodile.

Stegosaurus (Fig. 3) itself carried the head low and had an arched body, the neck, body, and tail having a series of alternating plates above the backbone. The largest of these bony plates was about 3 ft (1 m) wide and was above the pelvis. Within the pelvis was a greatly enlarged "booster" of the spinal cord, to aid the movement of the animal and its tail. The tail ended in two pairs of bony spikes, probably of use as a weapon when the tail was swung sideways.

This slow-moving vegetarian had a brain of about 2.5 ounces (.07 kg) and is thus the most underbrained land animal. It is doubtful also if the armor of upstanding plates was really effective and it has been suggested that they may have helped cool the animal by radiation.

Stegosaurus is known mainly from the Jurassic of the US, but an allied form, *Omosaurus*, has been found in England, France, and Portugal.

Suborder Ankylosauria. Halfway between the stegosaurian plated forms and the bony armored bodies of the Ankylosaurs is the English *Polacanthus*. This Lower Cretaceous dinosaur, known by a headless skeleton in the British Museum and a few leg bones, had two rows of spines down the back and tail, but had a flattened mosaic of bony pieces over the hinder part of the back.

A more completely armored form, also in the

British Museum, is *Scolosaurus*, whose back and upper arms and legs were plated in bony strips and spikes. The tail was armed by bony knobs. *Scolosaurus* came from the Upper Cretaceous of Alberta.

Other American armored "tanks" are *Nodosaurus* (Fig. 3) and *Dyoplosaurus*, creatures almost completely protected against flesh-eaters like *Tyrannosaurus* and *Gorgosaurus* already mentioned. These were about 19 ft (6 m) long and were naturally low on the ground. If they were overturned, however, their soft bellies offered no protection. They were all slow-moving herbivores. For a recent review of this group, see Coombs (1978).

Suborder Ceratopsia. The most successful armored dinosaurs were the Ceratopsia or "horned-faces," with a large, rhinoceros-like body that was thick skinned but not plated. The head, however, had one, three, or more horns on the face and a long bony collar or shield extending over the neck.

Probably descended from small 5 ft (1.5 m) long dinosaurs known from the Gobi Desert, Mongolia, the Ceratopsia were almost exclusively North American (one in South America). In size, they varied from small animals (*Brachyceratops*) 5 ft (1.5 m) long, with a skull of 18 in. (.5 m) up to the giants like *Triceratops* (Fig. 3) and *Torosaurus*, with heads 6–7 ft (2 m) long and a total length of around 30 ft (about 10 m). There was usually a horn on the nose (*Monoclonius* and *Styracosaurus*) or there might be very well-developed horns above the eyes as in *Triceratops*, where the eye horns were 3 ft (1 m) long.

Sometimes the neck frill was plain, as in *Triceratops*, or it might be prolonged into a series of backward pointing spikes (as in *Styracosaurus*). There are some 13 genera of these horned dinosaurs, which appear to have been able to hold their own in the closing period of the dinosaurs. The horns could be used in defense, offense or in bringing down small trees on whose foliage the Ceratopsians fed.

Dinosaur Ecology

Dinosaurs are known by bones, complete skeletons, skin impressions, footprints, and eggs. Up to the Late Jurassic, the vegetation was mainly evergreen and the giant horsetails and the cycad trees were perhaps the main food of the herbivores. There is, however, no direct evidence for this general assumption. In the Early Cretaceous, the main elements of the modern floras were introduced, the spread of deciduous trees leading to a reduction in winter foliage (see *Paleoecology, Terrestrial*).

On the land, there were lizards and tortoises, in the water, there were many swimming reptiles, with the crocodiles emerging as perhaps the main predators on dinosaurs generally. Pterosaurs (flying reptiles) and birds were presumably of little competitive significance. The mammals were present throughout the entire span of the dinosaurs, but here the latter were probably restrictive to the former and there is no evidence that the mammals interfered in any way with dinosaurian evolution or decline.

Several dinosaur skeletons show the remains of disease as it affects the bones (see *Paleopathology*). Osteomyelitis, osteoarthritis, and the effects of injury have been diagnosed. A suspected case of osteosarcoma in a Cretaceous herbivorous dinosaur may be osteoporosis. Disease was not a factor in the decline of dinosaurs.

Evolution and Extinction

From small Pseudosuchians of limited distribution in North America, Africa, and Europe, the main groups of dinosaurs spread throughout the world, producing some of the largest land animals there have ever been (Fig. 4). The only continent from which no remains have so far

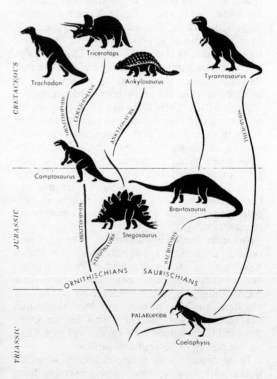

FIGURE 4. Evolution of the dinosaurs (from Colbert, 1969, used by permission).

come is Antarctica, and there is every likelihood that its Triassic deposits may yet yield evidence.

Geographically, or topographically, these reptiles inhabited low-lying regions with ample vegetation close to inland waters, lakes, and estuaries, though some of the smaller and more agile forms lived in more upland areas. None of the dinosaurs was marine and there is no evidence that they had any tolerance of salt water.

Climatically, they were suited to warm, subtropical conditions. Experiments on alligators show that they, the nearest living relatives of the dinosaurs, can stand lower temperatures than might be expected and that, like most reptiles, they do not tolerate high temperatures. It has been argued that dinosaurs may have had a heat-regulatory control and that their comparatively thick skin minimized temperature problems rather than increased them. It is, however, observable that regions inhabited by dinosaurs were warm during the time.

Throughout the groups, families and genera, there was a tendency for dinosaurs to increase in size in the course of evolution, and the largest of each group was generally the last of its line. Since increase in body size produces physical and physiological complications, this might be thought to be paralleled with decrease in the number of individuals. Yet one of the theories of dinosaur extinction is that they were victims of overcrowding. As their evolution progressed, various groups of dinosaurs were dying out, to be replaced by successors or by none at all. Thus the sauropod giants disappeared in the Northern Hemisphere while they continued in the Southern Hemisphere. Dinosaur extinction, slow but complete toward the end of the Cretaceous, 63 m yr ago, has many differing explanations.

The discovery of the physical effects of stellar explosions has suggested to some authorities that the larger-sized animals and plants at the end of the Cretaceous suffered from radiation. Other suggestions are more closely related to the animals' own physiology. Thus Bakker (1975) argues, on anatomical grounds, that dinosaurs were warm-blooded (see also Desmond, 1976). With a naked, though tough, thick skin the animals could not withstand the cold that seems to have developed on a world-wide scale about 65 m yr ago. Since the animals were too large to hibernate, or perhaps seek shelter in the forests, they were decimated.

It is likely that the disappearance of dinosaurs, which is not a unique event, was more complex. The position of the poles meant that the arctic circle was in a different position from that at present. Some cooling was inevitable in subpolar regions from seasonal lack of sunshine.

This probably had a greater effect on the vegetation than on the animals, but the latter might be compelled to migrate seasonally.

However, coincidentally, there was a considerable change in the vegetation in the later Cretaceous, the older evergreens being partially displaced by the incoming deciduous plants and trees. Furthermore, although the pouched and egg-laying mammals generally shared the fate of the dinosaurs, there was a considerable influx of placental mammals, which rapidly attained considerable size, and a period of replacement began. Undoubtedly this was aided by the effect of climate on the vegetation and its suitability for dinosaurian diet as well as the large size of the remnant dinosaur population which may have been unfitted for massive migration. Indeed there are suggestions from the later dinosaurs of hormone disturbance, producing bony exaggerations and resulting, as seen in France, in eggs with such thin shells that the survival of the young was in continual jeopardy.

While the demise of dinosaurs probably originated in the north, it soon spread southward and by about 63 m yr ago, the dinosaurs were gone, never to return.

W. E. SWINTON

References

Axelrod, D. L., and Bailey, H. P., 1968. Cretaceous dinosaur extinction, *Evolution,* **22,** 595–611.

Bakker, R. T., 1975. Dinosaur renaissance, *Sci. American,* **232**(April), 58–78.

Bakker, R. T., and Galton, P. M., 1974. Dinosaur monophyly and a new class of vertebrates, *Nature,* **248,** 168–172.

Bennet, A. F., and Dalzell, B., 1973. Dinosaur physiology, a critique, *Evolution,* **17,** 170–174.

Casier, E., 1960. *Les Iguanodons de Bernissait.* Brussels: Inst. Roy. Sci. Nat. Belgique, 134p.

Charig, A. J., 1973. Jurassic and Cretaceous dinosaurs, in A. Hallam, ed., *Atlas of Palaeobiogeography.* Amsterdam: Elsevier, 339–352.

Colbert, E. H., 1961. *Dinosaurs, Their Discovery and Their World.* New York: Dutton, 300p.

Colbert, E. H., 1962. The weights of the dinosaurs, *Amer. Mus. Novit.,* **2181,** 1–24.

Colbert, E. H., 1968. *Men and Dinosaurs.* New York: Dutton, 284p.

Colbert, E. H., 1969. *Evolution of the Vertebrates,* 2nd ed. New York: Wiley, 535p.

Coombs, W. P., Jr., 1975. Sauropod habits and habitats, *Palaeogeography, Palaeoclimatology, Palaeoecology,* **17,** 1–33.

Coombs, W. P., Jr., 1978. The families of the ornithischian order Ankylosauria, *Palaeontology,* **21,** 143–170.

Desmond, A. J., 1976. *The Hot-Blooded Dinosaurs: A Revolution in Paleontology.* New York: Dial Press, 238p.

Dodson, P., 1973. Endothermy, dinosaurs and Archaeopteryx, *Evolution*. **27**, 503–504.

Dodson, P., 1974. Dinosaurs as dinosaurs, *Evolution*, **28**, 494–497.

Galton, P. M., 1972. Classification and evolution of ornithopod dinosaurs, *Nature*, **239**, 464–466.

Galton, P. M., 1978. The Fabrosauridae, the basal family of ornithischian dinosaurs, *Paläont. Zeitschr.*, **52**, 138–159.

Hopson, J. A., 1975. The evolution of cranial display structures in hadrosaurian dinosaurs, *Paleobiology*, **1**, 21–43.

Hopson, J. A., 1977. Relative brain size and behavior in archosaurian reptiles, *Ann. Rev. Ecol. Syst.*, **8**, 429–448.

Hotton, N., III, 1963. *Dinosaurs*. New York: Pyramid Pub., 192p.

Jepsen, G. L., 1964. Riddles of the terrible lizards, *Amer. Scientist*, **52**. 227–246.

Kurten, B., 1968. *The Age of Dinosaurs*. London: World University, 225p.

Maryańska, T., and Osmólska, H., 1974. Pachycephalosauria, a new suborder of ornithischian dinosaurs, *Palaeontol. Polonica*, **30**, 45–102.

Ostrom, J. H., 1964. A reconsideration of the paleoecology of hadrosaurian dinosaurs, *Amer. J. Sci.*, ser. 5, **262**, 975–997.

Ostrom, J. H., 1966. Functional morphology and evolution of the ceratopsian dinosaurs, *Evolution*, **20**, 290–308.

Ostrom, J. H., and McIntosh, J. E., 1966. *Marsh's Dinosaurs, the Collections from Como Bluff*. New Haven: Yale Univ. Press.

Romer, A. S., 1966. *Vertebrate Paleontology*, 3rd ed. Chicago: Univ. Chicago Press, 468p.

Swinton, W. E., 1970. *The Dinosaurs*. London: Allen and Unwin, 331p.

Young, J. Z., 1962. *The Life of the Vertebrates*, 2nd ed. Oxford: Oxford Univ. Press, 820p.

Cross-references: *Aves; Extinction; Reptilia; Vertebrata; Vertebrate Paleontology.*

DISCOASTERS

Discoasters are microscopic, star-shaped, calcite nannofossils (5–40 μm) of uncertain origin that are generally classified with marine coccolith-bearing algae (see *Coccoliths*). Discoasters are known as fossils ranging in age from late Paleocene to late Pliocene. Possible living discoaster-producing organisms have been reported (Bursa, 1964, 1971), but are not yet sufficiently documented and belie the well-documented sequential extinction of various forms to the end of the Tertiary. Fossil discoasters are most abundant and diverse in ocean sediment from the tropics where they can be the dominant type of nannofossil. Classified in the organ genus *Discoaster* Tan (1927) are at least 250 species names. Of these, only about 100 appear to be distinct biologic entities. Because of diagenetic overgrowth and dissolution, diagnostic taxonomic features, such as the relative proportions, taper, and camber of rays and the form of ray tips, can be obscured and misinterpreted. Discoaster species have an average duration of 5 or 6 m yr; short-lived species of 0.5 m yr or long-lived species of 30 m yr are exceptional (Bukry, 1973).

The taxonomy and evolutionary sequence of *Discoaster*, so important to oceanic biostratigraphy, were mainly developed by Bramlette and his associates over a 15-year period at Scripps Institution of Oceanography (Bramlette and Riedel, 1954; Bramlette, 1957; Bramlette and Sullivan, 1961; Martini and Bramlette, 1963; Bramlette and Wilcoxon, 1967; and Bukry and Bramlette, 1969). Other important studies were published by Martini (1958), Stradner (1958, 1959a,b), and Stradner and Papp (1961). These results were summarized and amplified into direct evolutionary lineages by Prins (1971), who speculated that the early Paleocene coccolith genus *Fasciculithus* was the most likely ancestor for discoasters.

The earliest discoaster, *Discoaster mohleri*, and its associates are multirayed planar rosettes of up to 35 crystallites. Subsequent evolution shows a reduction in the number of crystallite rays to five or six (Bukry, 1971), which is typical for most of the Tertiary (Fig. 1).

Eocene assemblages of *Discoaster* show considerable variation in ray number and morphology following the rosette-dominated Paleocene. But Oligocene assemblages are typified by a few long-ranged conservative species. The middle Miocene to early Pliocene was again a time of experimentation in form when many new species appeared. Evolutionary changes included the development of long, thin-rayed species, many of which also had a distinct bending of the rays. These were the last significant changes in discoaster form before their disappearance in the late Pliocene.

Although most early Tertiary species where cosmopolitan, *Discoaster deflandrei* was especially so, being the dominant discoaster in the Oligocene at low latitude and virtually the only species at high latitude. Following the middle Miocene thermal high, more paleoecologic differentiation occurred between discoaster populations. Delicate species such as *Discoaster asymmetricus, D. bellus, D. brouweri, D. neohamatus, D. pentaradiatus,* and the thicker five-rayed taxa *D. berggrenii* and *D. quinqueramus* dominated warm-water assemblages, whereas, at high latitude or areas of upwelling, broader-rayed species such as *D. intercalaris, D. surculus,* or *D. variabilis* were numerically more important. Discoasters are

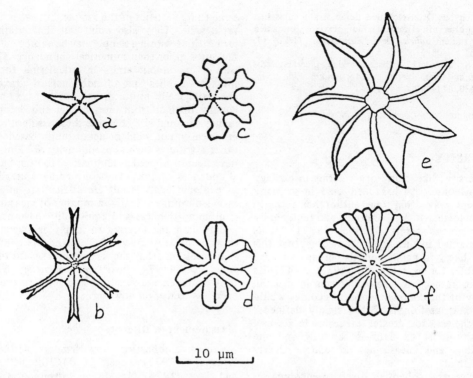

FIGURE 1. Some representative discoasters from the Tertiary. a, *Discoaster asymmetricus* Gartner; b, *D. exilis* Martini and Bramlette; c, *D. deflandrei* Bramlette and Riedel; d, *D. deflandrei* (overgrown); e, *D. lodoensis* Bramlette and Riedel; f, *D. multiradiatus* Bramlette and Riedel.

missing from subpolar sediment of post-late Miocene age.

The planktonic distribution pattern, great abundance, and rapid evolution of many species of *Discoaster* makes this group of nannofossils a fundamental means of transoceanic stratigraphic correlation for the Tertiary.

DAVID BUKRY

References

Bramlette, M. N., 1957. *Discoaster* and some related microfossils, *U.S. Geol. Surv. Prof. Paper 280-F,* F247-F255.

Bramlette, M. N., and Riedel, W. R., 1954. Stratigraphic value of discoasters and some other microfossils related to recent coccolithophores, *J. Paleontology, 28,* 385-403.

Bramlette, M. N., and Sullivan, F. R., 1961. Coccolithophorids and related nannoplankton of the early Tertiary in California, *Micropaleontology, 7,* 129-188.

Bramlette, M. N., and Wilcoxon, J. A., 1967. Middle Tertiary calcareous nannoplankton of the Cipero Section, Trinidad, W. I., *Tulane Stud. Geol., 5,* 93-131.

Bukry, D., 1971. *Discoaster* evolutionary trends, *Micropaleontology, 17,* 43-52.

Bukry, D., 1973. Coccolith stratigraphy, eastern equatorial Pacific, Leg 16 Deep Sea Drilling Project, *Initial Repts. Deep Sea Drilling Proj., 16,* 653-711.

Bukry, D., and Bramlette, M. N., 1969. Some new and stratigraphically useful calcareous nannofossils of the Cenozoic, *Tulane Stud. Geol. Paleontol., 7,* 131-142.

Bursa, A. S., 1964 [1965]. *Discoasteromonas calciferus* n. sp., an arctic relict secreting *Discoaster* Tan Sin Hok 1927, *Grana Palynol. 6,* 147-165.

Bursa, A. S., 1971. Morphogenesis and taxonomy of fossil and contemporary dinophyta secreting discoasters, *2nd Planktonic Conf. Proc.,* 129-143.

Martini, E., 1958. Discoasteriden und verwandte Formen im NW-deutschen Eozän (Coccolithophorida), *Senckenbergiana Lethaea, 39,* 353-388.

Martini, E., and Bramlette, M. N., 1963. Calcareous nannoplankton from the experimental Mohole drilling, *J. Paleontology, 37,* 845-856.

Prins, B., 1971. Speculations on relations, evolution, and stratigraphic distribution of discoasters, *2nd Planktonic Conf. Proc.,* 1017-1037.

Stradner, H., 1958. Die fossilen Discoasteriden Österreichs, *Erdoel-Z., 74,* 178-188.

Stradner, H., 1959a. First report on the discoasters of the Tertiary of Austria and their stratigraphic use, *5th World Petrol. Cong. Proc., New York, 1,* 1081-1095.

Stradner, H., 1959b. Die fossilen Discoasteriden Österreichs II. Teil, *Erdoel-Z., 75,* 472-488.

Stradner, H., and Papp, A., 1961. Tertiäre Discoasteri-

deriden aus Österreich und deren stratigraphischen Bedeutung mit Hinweisen auf Mexiko, Rumänien und Italien, *Jahrb. Geol. Bundesanst. (Wien)*, 7, 159p.

Tan, S. H., 1927. Discoasteridae incertae sedis *Koninkl Akad. Wetensch. Amsterdam Proc., Sec. Sci.*, 30, 411–419.

Cross-references: *Coccoliths; Plankton.*

DIVERSITY

Diversity, like many other terms in ecology and paleontology, has been used in so many different ways that one author has recently suggested that it be abandoned and replaced by more precise terms (Hurlbert, 1971). It is so ingrained in our science however, that this is not likely to happen and is perhaps not even desirable, for as Hedgpeth (1957, p. 47) has said in respect to ecologic terms in general, "many of the common words of ecology would lose their usefulness if too rigidly defined." Nevertheless, for greater precision in its use, diversity must be defined carefully and the sampling and measuring procedures clearly established.

The general conept of diversity encompasses both the number and the relative abundances of taxa present within a sample. The terms *diversity* and *species diversity* have been applied in ecologic literature to the number of taxa by some workers (Whittaker, 1972; Valentine, 1973), and to the combination of numbers and relative proportions by others (MacArthur, 1972; Pianka, 1974; Pielou, 1969). Because of this lack of uniformity, diversity is suitable for general useage but the specific concept in mind must be stated when more precise meaning is desired. The term *richness* is preferable to species diversity when referring to number of taxa, firstly because analysis of diversity may be based on a taxonomic level other than species or even on organisms identified to several taxonomic levels, and secondly because species diversity has been used in several other ways, as previously mentioned. The relative abundance or importance of taxa within a sample has been described by terms such as dominance, relative importance, evenness, and *equitability*. Evenness and equitability measure how uniformly abundant the taxa are. Lloyd and Ghelardi (1964) have proposed that evenness is measured in absolute numerical terms whereas equitability is relative to some hypothetical abundance distribution. This distinction has not generally been made in the paleontologic literature and equitability has been the commonly used term. Terms such as dominance and relative importance refer to the

same characteristics of the sample, but on an inverse scale. They also point out that relative proportions within a sample may be expressed in measures other than numerical abundance. The use of taxonomic units in calculating these measures implies that all individuals of a taxon are ecologically uniform, and that all taxa are equally distinct from one another and occupy niches of equal size. Because these assumptions are clearly not valid in detail, units based on other attributes such as productivity or biomass have been proposed as alternatives to numerical abundances of taxa. These are perhaps useable in ecologic analysis but are difficult to apply in paleontology. Finally, a number of measures combining richness and equitability have been widely used and referred to simply as diversity or as *dominance diversity*. To avoid the confusion resulting from the use of diversity in this way, Peet (1974) favors *heterogeneity*. The term dominance diversity would seem to cause even less confusion however.

Measurement of Diversity

Recent publications by Sanders (1968), Whittaker (1972), Fager (1972), Hill (1973), and Peet (1974) provide an introduction to the extensive literature on the measurement of diversity—on the indices that have been proposed, their attributes, and relative merits.

Diversity indices can be grouped into those that measure richness, those that measure a combination of richness and equitability (dominance diversity), and those that measure equitability. Diversity indices can also be divided into those that are or are not based upon some frequency distribution pattern assumed to be valid in general for organisms in nature. In measuring diversity, the sample size and the taxonomic level to be used (whether species, genus, etc.) depend upon the nature of the problem. Because these two factors strongly influence the calculated diversity value, they must be taken into account when comparing values from different studies or from samples of different size within a study.

To be valid, a diversity index should be ecologically sound and statistically rigorous. It is evident from the large literature dealing with diversity definition and measurement that it is difficult to simultaneously satisfy both these requirements. The application of diversity in paleontology must also take into account taphonomic processes and the biases they may produce. For example, Berger and Parker (1970) have shown that the diversity of planktonic foraminifers in deep-sea sediments is different from that of the living assemblage because of differential solution in the deep ocean; Lawrence (1968) has shown that only a small percentage (averaging about 30%, but as low as 1% in some cases) of the living assemblage is likely to be preserved in the fossil record; and Stanton (1976) has shown that this non-preservation results in a strong bias not only in diversity of different taxa but in other attributes of the community structure as well, such as the proportions of organisms with different feeding mechanisms, utilizing different food resources (see also Lasker, 1976).

Of the numerous techniques and indices that have been proposed for measuring diversity, those that have been most commonly used in paleontology are considered below.

Richness.

$$S \text{ (= number of species or taxa)}$$

the simplest measure of richness is and has been widely used in comparing lists of fossils. It suffers from being strongly dependent upon sample size.

$$(S - 1)/\log N$$

where N is the number of individuals in the sample, is one of several attempts to correct for the sample-size dependency of S. This index is based upon the commonly observed fact that if cumulative number of species is plotted against cumulative number of individuals, the equation of the curve is approximately $S = \log N$. Similarly, MacArthur and Wilson (1967) have proposed that S can be adjusted for differences in sampling area, A, by using the equation $S = CA^Z$, in which the constant C depends upon the taxa counted and the value of Z has been determined empirically to lie within the range of 0.20 to 0.35. Richness values that have been adjusted for sample size in manners such as these are more easily compared than unadjusted ones, but are imprecise then, on the other hand, because the actual diversity relations in nature do not universally correspond to the theoretical ones.

The *rarefaction method* (Sanders, 1968) compensates for the effect of sample size by graphically determining for large samples what their diversity would be at the size of the smaller samples with which they are being compared. Thus, the relative richness of different samples is determined by comparing their rarefaction curves rather than from numerical indices. The method thus is simple to interpret. It has been pointed out, however, that the method as Sanders proposed it is mathematically incorrect, so that previously published curves are erroneous, generally over-

estimating richness. The correct procedure has been described by Simberloff (1972).

Dominance Diversity.

$$\sum_{i=1}^{S} P_i^2$$

where S = number of species and P_i = proportion of ith species in the sample, measures (Simpson, 1949) the probability of two individuals randomly drawn from a sample being of the same species. Because the value of this index varies inversely with dominance diversity, variations such as its reciprocal have been proposed. Neither these variants nor the basic index have been widely used in paleontology.

$$H' = -\sum_{i=1}^{S} P_i \log P_i$$

where i = proportion of ith species in the population, is one of the measures of entropy based on information theory that are widely used in paleontology. They measure the uncertainty in predicting the identity of a randomly selected individual from a population. This, the Shannon-Weaver equation, is the most commonly used form because it is easily calculated if n_i/N, the proportion of the ith species in the sample, is substituted for P_i. The use of sample values for population values introduces a bias into the results but this is small considering the inherent inaccuracies of the paleontologic data. The relationships between the different entropy equations have been thoroughly analyzed by Pielou (1969).

Equitability.

$$J = H'/H' \text{ max}$$

is the ratio of the calculated entropy to that which the sample would have if all the species were equally abundant. Other similar ratios based on entropy indices, which have also been used, have been described and evaluated by Sheldon (1969).

Theory and Patterns of Diversity

A valuable approach to the understanding of diversity is through consideration of the environment as a multidimensional region, the hyperspace, which is partitioned into hypervolumes representing individual niches (Hutchinson, 1957). Richness, then corresponds to the number of niches, is inversely proportional to the average niche size, and is directly proportional to the magnitude of the hyperspace (determined by resource breadths, or axis lengths of the hyperspace) and to the amount of overlap between niches (MacArthur, 1972; Whittaker, 1972; Pianka, 1974). Equitability describes, within a community, relative utilization of energy flux of the organisms occupying the different niches.

Beyond this terse statement of the necessary and sufficient conditions that determine diversity, one or another of which must change if diversity is to change, is a wide range of observed diversity gradients and patterns. These occur, temporally, during succession and through the longer time spans of earth history, and spatially, along a wide range of environmental gradients of global to very localized scale.

During succession, diversity has been observed to increase in a wide variety of ecosystems. For example, Goulden (1969) reported that the diversity of lacustrine crustaceans increased in three stages: Initially, richness increased as species immigrated into the study area; then equitability increased; and, finally, richness increased by the addition of rare species. Odum (1969) tabulated many changes that take place during succession and reported that both richness and equitability increase.

During geologic time, richness of invertebrates, vertebrates, and plants have apparently increased, but with numerous fluctuations (Newell, 1967). These long-term changes have been analyzed and interpreted in detail by Valentine (1969), Tappan and Loeblich (1973), and others, and will be considered more fully later. Raup (1972) and Harper (1975), on the other hand, have pointed out that the data available about diversity through geologic time may be very difficult to measure and interpret because of sampling problems and time-dependent biases. No data are available about possible long-term changes in equitability.

Spatially, present-day total richness and richness within most higher taxonomic groups decrease globally with increasing latitude (Stehli, 1968). Regionally, both richness and equitability of marine invertebrates decrease from the stable environmental conditions of the open marine shallow shelf shoreward toward the more rigorous and variable conditions near shore and in estuaries and lagoons. Typical examples of this trend are from the Baltic Sea for both vertebrates and invertebrates (Segerstale, 1957), from the vicinity of the Mississippi Delta for benthic shelled macroinvertebrates (Stanton and Evans, 1971), from numerous estuaries and lagoons for invertebrates (Emery

et al., 1957), and from the Atlantic coast in the Cape Hatteras area for ostracodes (Hazel, 1975). Richness also increases from the shelf into deeper water of the ocean basins for benthic invertebrates (Hessler and Sanders, 1967) and foraminifers (Gibson and Buzas, 1973).

The possible causes of the observed diversity patterns have been comprehensively reviewed by Sanders (1968), MacArthur (1972), Valentine (1973), and Pianka (1974). The wide range of explanations put forth indicates that there is no simple and universal explanation but that diversity is determined by a complex interplay of historical, biological, and physical factors within the ecosystem. The factors that appear to be of prime importance are time, resources, and stability.

Time. During succession, change in diversity depends upon time for each stage of succession to develop and give way to the next. Over the much longer intervals of geologic time, diversity must be controlled by evolution as organisms accommodate themselves within the ecosystem to the environment and to other organisms. Hutchinson (1959) has proposed that the general course of this evolution is toward increased specialization and niche subdivision, resulting in increased biological interactions, thus providing opportunities for even further specialization and increase in diversity. Other time-dependent processes (in addition to decrease in niche size) by which changes in richness during geologic time can be explained include changes in hyperspace size (for example, increases during earth history as biological advances opened up first terrestrial and later aerial habitats to be colonized); changes in the number of biotic provinces, each with its own complement of approximately equivalent niches due to climatic and geographic variations; and changes in extent of niche overlap.

Resources. Any characteristic of the environment such as space or food for which organisms compete and which may be limiting, and that may be represented as an axis of the hyperspace, can be considered as a resource. Resources determine diversity because they establish the total volume of the hyperspace that potentially may be utilized. The magnitude and kinds of resources available are also important because they determine the direction of evolution leading to optimum strategies for survival of the species present. Evolutionary increase in organism interactions has had the effect of adding new dimensions of resources to the hypervolume.

Stability. Environmental stability is necessary in order for the processes of evolution and succession to be effective. For example, the global latitudinal diversity gradient may exist because there has been insufficient time for evolution to proceed in the relatively young polar environments. Regionally, the gradient of decreasing richness of marine invertebrates from open shallow marine to restricted coastal habitats has been correlated with a corresponding gradient of decreasing environmental stability and thus of decreasing time for succession to fully develop before a change in the environment returns the ecosystem to a less mature, lower-diversity stage (Gibson, 1966). Stability of trophic resources should favor the evolution of relatively specialized species with narrow niches, and consequently should lead to increased diversity. On the other hand, fluctuations in the environment might be important in increasing diversity by means of a hypothetical "diversity pump" (Valentine, 1967). This mechanism assumes that biogeographic provinces are determined by geographic and climatic factors and are characterized by a full compliment of niches and a distinctive biota. Under these conditions, fluctuations in climate or geography might have alternately created new provinces with new niches and species, and subsequently constricted or eliminated them with the increased competition leading in part to extinction and in part to niche subdivision. The postulated effect of these alternating effects through time would be a net increase in diversity.

Diversity in Paleontology

Diversity is a valuable analytical measure in paleontology because it can be interpreted in terms of evolutionary processes and the ancient biological and physical environmental conditions and because it is relatively independent of taxonomy (although influenced by taxonomic practice) and therefore of geologic time. Its application however has lagged behind the observational and theoretical understanding that has developed in ecology because of the sampling problems and biases unique to the fossil record and because of the lack of diagnostic criteria by which diversity gradients correlated with specific environmental gradients can be distinguished. Diversity has been applied in three ways in paleontologic studies: (1) Modern and ancient spatial diversity patterns have been compared in order to determine ancient environments. (2) Gross diversity changes during geologic time have been analyzed in order to understand the history of life and the processes of evolution and extinction. (3) Diversity patterns in synecology have been studied in order to perceive organism interactions within assemblages of fossils, and to understand the evolution of community struc-

ture and of evolutionary strategies of component species. The following represent only a few examples of each category:

1. Modern global diversity gradients have been applied by Stehli (1970) to the fossil record in order to estimate the positions of the earth's rotational poles during the geologic past. Examples of studies in which diversity gradients have been applied at a regional scale are these of Hallam (1972)—in a study of lower Jurassic molluscan and brachiopod faunas of the eastern North Atlantic—and Douglas (1972)—in a study of Late Cretaceous foraminifers of North America. In these studies, measure of richness were used to recognize zoogeographic patterns and broad environmental gradients. Diversity has been described and applied to more localized problems of paleoenvironmental reconstruction in numerous studies. Typical examples are supplied for Tertiary strata by Perkins (1970), Gibson and Buzas (1973), and Dodd and Stanton (1975); for Mesozoic strata by Scott (1975); for Paleozoic strata by Stevens (1971), Thayer (1974), and Shaak (1975). Emphasis in these studies has been on relating lateral or vertical diversity gradients in the fossil record with environmental factors that appear to control diversity in modern biota. However, in general, interpretations based on diversity have played a subsidiary role in paleontology to those based on taxonomic and lithologic data and on the various foraminiferal ratios, because of the difficulty of recognizing from the observed diversity the relative importance of the many significant environmental factors that may have controlled it.

More effective application of diversity in paleontology will require more precise criteria for correlating diversity to specific environmental factors and further detailed examination of the theoretical bases and models from ecology with the special conditions and problems of paleoecology in mind; such studies have been rare. Beerbower and Jordan (1969), mapped diversity of living foraminifers in Sabine Lake of Texas and Louisiana, interpreted it in relation to environmental parameters, and evaluated it as a tool in paleoecology. Stanton and Evans (1972), in a study of the benthic shelled invertebrates adjacent to the Mississippi Delta, analyzed both richness and equitability in conjunction with other components of structure such as patchiness and homogeneity in order to understand how measures of diversity reflect the physical environment and affect sampling strategy in paleoecology. Gibson and Buzas (1973) described patterns of equitability and richness of living foraminifers of the western North Atlantic and then compared them with diversity values from Miocene strata of the middle Atlantic Coastal Plain in order to test the application of diversity to paleoenvironmental analysis.

2. The capability to document changes in richness through geologic time was greatly expanded by the comprehensive synthesis of Harland (1967). Subsequent papers such as those of Banks (1970) on early Paleozoic plants, Tappan and Loeblich (1973) on oceanic plankton, and Lillegraven (1972) on Cenozoic mammals have further filled out the data base necessary for analysis of the factors causing variations in richness. Valentine (1973) has summarized the numerous mechanisms that have been proposed to account for variations in richness. In general, analysis and explanation in recent years have shifted from single, primarily physical, factors toward broader syntheses incorporating topics such as plate tectonics, atmospheric composition as determined by oceanic circulation and by the evolution of the biota, resource quantity and stability, environmental stability, and biological interactions. Any study of temporal diversity trends however must also take into account the magnitude of the sampling biases that exist in the fossil record and the relative importance of stochastic and deterministic processes in producing the evolutionary pattern we are attempting to interpret (Raup et al, 1973).

3. The interrelations of diversity, environmental stability, and taxonomic stability (taxonomic longevity) have been investigated by Bretsky and Lorenz (1970). They concluded that diversity is positively correlated with environmental stability and that taxonomic stability and environmental stability are inversely correlated because taxa in stable environments should be relatively specialized (and have low genetic variability) and thus would be relatively likely to become extinct (taxonomically unstable) in the event of environmental fluctuations. These conclusions were based on an analysis of Paleozoic faunas, but more recent studies of the genetic heterogeneity and geologic range of modern organisms indicate that the relationship of environmental stability and taxonomic stability is not so simple nor obvious (Ayala et al., 1975). Valentine (1971) suggested that organisms subjected to different regimes of environmental stability and resource level would adopt different survival strategies. These strategies would be expressed genetically within the individuals, as Bretsky and Lorenz (1970) suggested, and within the diversity of the community. Potentially then, as environmental conditions changed, strategies and diversity in terms of

both richness and equitability would also change. Whereas Valentine's analysis was largely theoretical, Bretsky (1973) has attempted to discover correlations between diversity, taxonomic stability, and relative degrees of endemism or cosmopolitanism by analyzing diversity, extinction, and geographical distribution patterns of Paleozoic bivalves. He believes that he can recognize both long-term and repetitive evolutionary and diversity trends for which unspecified environmental changes are postulated as the causes. Walker's (1972) study of an Ordovician fauna from a trophic point of view provides a further example of paleoecologic research that promises to enhance our understanding of diversity in the history of life.

ROBERT J. STANTON, JR.

References

Ayala, F. J.; Valentine, J. W.; Delaca, T. E.; and Zumwalt, G. S., 1975. Genetic variability of the Antarctic brachiopod *Liothyrella notorcadensis* and its bearing on mass extinction hypotheses, *J. Paleontology*, 49, 1–9.

Banks, H. P., et al. 1970. Symposium: Major evolutionary events and the geologic record of plants, *Biol. Rev. Cambridge Philos. Soc.*, 45, 317–454.

Beerbower, J. R., and Jordan, D., 1969. Application of information theory to paleontologic problems: Taxonomic diversity, *J. Paleontology*, 43, 1184–1198.

Berger, W. H., and Parker, F. L., 1970. Diversity of planktonic foraminifera in deep-sea sediments, *Science*, 168, 1345–1347.

Boucot, A. J., 1975. Standing diversity of fossil groups in successive intervals of geologic time viewed in the light of changing levels of provincialism, *J. Paleontology*, 49, 1105–1111.

Bretsky, P. W., 1973. Evolutionary patterns in the Paleozoic Bivalvia: Documentation and some theoretical considerations, *Geol. Soc. Amer. Bull.*, 84, 2079–2096.

Bretsky, P. W., and Lorenz, D. M., 1970. An essay on genetic-adaptive strategies and mass extinctions, *Geol. Soc. Amer. Bull.*, 81, 2449–2456.

Dodd, J. R., and Stanton, Jr., R. J., 1975. Paleosalinities within a Pliocene bay, Kettleman Hills, California: A study of the resolving power of isotopic and faunal techniques, *Geol. Soc. Amer. Bull.*, 86, 51–64.

Douglas, R. G., 1972. Paleozoogeography of Late Cretaceous planktonic foraminifera in North America. *J. Foraminiferal Research*, 2, 14–34.

Emery, K. O.; Stevenson, R. E.; and Hedgpeth, J. W., 1957. Estuaries and lagoons, in J. W. Hedgpeth, ed., Treatise on marine ecology and paleoecology, vol. 1, Ecology, *Geol. Soc. Amer. Mem. 67*, 673–749.

Fager, E. W., 1972. Diversity: A sampling study, *Amer. Naturalist*, 106, 293–310.

Gibson, L. B., 1966. Some unifying characteristics of species diversity, *Cushman Found. Foraminiferal Research, Contrib, 17*, 117–124.

Gibson, T. G., and Buzas, M. A., 1973. Species diversity: Patterns in modern and Miocene foraminifera of the eastern margin of North America, *Geol. Soc. Amer. Bull.*, 84, 217–238.

Goulden, C. E., 1969. Developmental phases of the biocoenosis, *Proc. Nat. Acad. Sci.*, 62, 1066–1073.

Hallam, A., 1972. Diversity and density characteristics of Pliensbachian–Toarcian molluscan and brachiopod faunas of the North Atlantic margins, *Lethaia*, 5, 389–412.

Harland, W. B., et al., 1967. *The Fossil Record*. London: Geol. Soc. London, 827p.

Harper, C. W., Jr., 1975. Standing diversity of fossil groups in successive intervals of geologic time: A new measure, *J. Paleontology*, 49, 752–757.

Hazel, J. E., 1975. Patterns of marine ostracode diversity in the Cape Hatteras, North Carolina, area, *J. Paleontology*, 49, 731–744.

Hedgpeth, J. W., 1957. Concepts of marine ecology, in J. W. Hedpeth, ed., Treatise on marine ecology and paleoecology, vol. 1, Ecology, *Geol. Soc. Amer. Mem. 67*, 29–52.

Hessler, R. R., and Sanders, H. L., 1967. Faunal diversity in the deep sea, *Deep-Sea Research*, 14, 65–79.

Hill, M. O., 1973. Diversity and evenness: A unifying notation and its consequences, *Ecology*, 54, 427–432.

Hurlbert, S. H., 1971. The nonconcept of species diversity: A critique and alternative parameters, *Ecology*, 52, 577–586.

Hutchinson, G. E., 1957. Concluding remarks, *Cold Spring Harbor Symp. Quantitative Biol.*, 22, 415–427.

Hutchinson, G. E., 1959. Homage to Santa Rosalia, or why are there so many kinds of animals? *Amer. Naturalist*, 93, 145–159.

Lasker, H., 1976. Effects of differential preservation on the measurement of taxonomic diversity, *Paleobiology*, 2, 84–93.

Lawrence, D. R., 1968. Taphonomy and information losses in fossil communities, *Geol. Soc. Amer. Bull.*, 79, 1315–1330.

Lillegraven, J. A., 1972. Ordinal and familial diversity of Cenozoic mammals, *Taxon*, 21, 261–274.

Lloyd, M., and Ghelardi, R. J., 1964. A table for calculating the "equitability" component of species diversity, *J. Animal Ecol.* 33, 217–225.

MacArthur, R. H., 1972. *Geographical Ecology*. New York: Harper & Row, 269p.

MacArthur, R. H., and Wilson, E. O., 1967. *The Theory of Island Biogeography*. Princeton N.J.: Princeton Univ. Press, 203p.

Newell, N. D., 1967. Revolutions in the history of life, *Geol. Soc. Amer. Spec. Paper 89*, 63–91.

Odum, E. P., 1969. The strategy of ecosystem development, *Science*, 164, 262–270.

Peet, R. K., 1974. The measurement of species diversity, *Ann. Rev. Ecol. and Systematics*, 5, 285–307.

Perkins, P. L., 1970. Equitability and trophic levels in an Eocene fish population, *Lethaia*, 3, 301–310.

Pianka, E. R., 1974. *Evolutionary Ecology*. New York: Harper & Row, 356p.

Pielou, E. C., 1969. *An Introduction to Mathematical Ecology*. New York: Wiley-Interscience, 286p.

Raup, D. M., 1972. Taxonomic diversity during the Phanerozoic, *Science*, 177, 1065–1071.

Raup, D. M.; Gould, S. J.; Schopf, T. J. M.; and

Simberloff, D. S., 1973. Stochastic models of phylogeny and the evolution of diversity, *J. Geol.*, 81, 525–542.

Sanders, H. L., 1968. Marine benthic diversity: A comparative study, *Amer. Naturalist*, 102, 243–282.

Scott, R. W., 1975. Patterns of Early Cretaceous molluscan diversity gradients in south-central United States, *Lethaia*, 8, 241–252.

Segerstråle, S. G., 1957. Baltic Sea, in J. W. Hedgpeth, ed., Treatise on marine ecology and paleoecology, vol. 1, Ecology, *Geol. Soc. Amer. Mem. 67, 751–800*.

Shaak, G. D., 1975. Diversity and community structure of the Brush Creek marine interval (Conemaugh Group, Upper Pennsylvanian), in the Appalachian Basin of Western Pennsylvania, *Bull. Florida State Mus. Biol. Sci.*, 19, 69–133.

Sheldon, A. L., 1969. Equitability indices: Dependence on the species count, *Ecology*, 50, 466–467.

Simberloff, D., 1972. Properties of the rarefaction measurement, *Amer. Naturalist*, 106, 414–418.

Simpson, E. H., 1949. Measurement of diversity, *Nature*, 163, 688.

Stanton, R. J., Jr., 1976. The relationship of fossil communities to the original communities of living organisms, in R. W. Scott and R. R. West, eds., *Structure and Classification of Paleocommunities*. Stroudsburg, Pa.: Dowden, Hutchinson, & Ross, 107–142.

Stanton, R. J., Jr., and Evans, I., 1971. Environmental controls of benthic macrofaunal patterns in the Gulf of Mexico adjacent to the Mississippi *Delta, Gulf Coast Assoc. Geol. Soc. Trans.*, 21, 371–378.

Stanton, R. J., Jr., and Evans, I., 1972. Community structure and sampling requirements in paleoecology, *J. Paleontology*, 46, 845–858.

Stehli, F. G., 1968. Taxonomic diversity gradients in pole location: The recent model, in E. T. Drake, ed., *Evolution and Environment*. New Haven: Yale Univ. Press, 163–277.

Stehli, F. G., 1970. A test of the earth's magnetic field during Permian time, *J. Geophys. Research*, 75, 3325–3342.

Stevens, C. H., 1971. Distribution and diversity of Pennsylvanian marine faunas relative to water depth and distance from shore, *Lethaia*, 4, 403–412.

Tappan, Helen, and Loeblich, A. R., Jr., 1973. Evolution of the oceanic plankton, *Earth Sci. Rev.*, 9, 207–240.

Thayer, C. W., 1974. Marine paleoecology in the Upper Devonian of New York, *Lethaia*, 7, 121–155.

Valentine, J. W., 1967. The influence of climatic fluctuations on species diversity within the Tethyan provincial system, in C. G. Adams and D. V. Ager, eds., Aspects of Tethyan Biogeography, *Systematics Assoc. Pub. 7*, 153–166.

Valentine, J. W., 1969. Patterns of taxonomic and ecological structure of the shelf benthos during Phanerozoic time, *Palaeontology*, 12, 684–709.

Valentine, J. W., 1971. Resource supply and species diversity patterns, *Lethaia*, 4, 51–61.

Valentine, J. W., 1973. *Evolutionary Paleoecology of the Marine Biosphere*. Englewood Cliffs, N.J.: Prentice-Hall, 511p.

Walker, K. R., 1972. Trophic analysis: A method for studying the function of ancient communities, *J. Paleontology*, 46, 82–93.

Whittaker, R. H., 1972. Evolution and measurement of species diversity, *Taxon,* **21,** 213-251.

Cross-references: *Communities, Ancient; Computer Applications in Paleontology; Evolution; Fossil Record; Paleobiogeography; Paleoecology; Plankton; Plate Tectonics and the Biosphere; Population Dynamics; Substratum; Trophic Groups.*

DWARF FAUNAS

Fossil animal assemblages, all components of which, regardless of phyla, are minute, have been variously called depauperate, diminutive, stunted, or dwarfed (Tasch, 1953). Ager (1963) and Hallam (1965) both preferred a term they thought to be of widest applicability—*stunting*—which denotes environmentally induced growth retardation (a correctable condition; Tasch, 1953). By contrast the term *dwarf* would be reserved for genetically controlled growth arrest (a noncorrectable condition; Tasch, 1953) of limited occurrence in the fossil record.

Growth Stages and Size

Much confusion has arisen from failure to distinguish growth stages. Arthropods that molt as they grow (ostracods and trilobites for example) yield a continuous size spectrum from small (larval stages) to large (adult stage). Similarly, each invertebrate class will display growth stages: mature and immature zones (bryozoans); closer spacing of sutures in maturity (ammonites); and so on. This must be determined initially (Cloud, 1948; Kummel, 1948; Tasch, 1953).

Size may be misleading. In protozoans (Foraminifera, among others) small size may represent optimum conditions (Lankford, 1967) and not growth-retarding conditions, but Hallam (1965) considered this last observation based on J. S. Bradshaw's experiments inconclusive. Any environmental condition such as high salinity will not affect all phyla equally; holothurians may grow to large size in high-salinity water that stunts growth of other biotic elements. There may be a low, upper limit for a given species size; or a latitudinal gradation in size (Tasch, 1953)—both natural conditions.

Modern Examples of Growth Retardation

Dwarfing of marine and nonmarine faunas, wholly or in part, in unusual or restricted environments, has been noted by several marine biologists (Hedgpeth, 1957, p. 717; Thorson, 1971, p. 63; among others). Such characteristic smallness has been attributed to life on seaweeds (Sargasso Sea); or the hypersalinity of certain lagoons; or iron enrichment of a lake and its surrounding soil (New Caledonian Lake) (see references in Tasch, 1953). Occurrence of dwarfed among normal snails along beaches of the Pacific Coast of El Salvador was attributed to sands composed of mineral magnetite, Fe_3O_4 (Schuster-Dieterichs, 1954).

Fossil Examples of "Dwarf Faunas"

There are likewise numerous reports of dwarfed fossil faunas—in part or entire—among others: Maquoketa Shale, Salem Limestone, Dry Shale, Phosphoria Formation, Shakopee Dolomite (see references in Tasch, 1953; Snyder and Bretsky, 1971). Pyritic, hematitic, limonitic and marcasitic faunas of the US, Europe, and elsewhere—most of which have stunted or dwarfed components—are widespread in the fossil record (Paleozoic, Mesozoic). Included are faunas of the Tully Limestone, Clinton Hematite Ores, and Ludlowville Formation (see references in Tasch, 1953). The most convincing documentation and analysis of a pyritic ammonite fauna was by Vogel (1959) in which 600 were found to be true dwarfs, and the remaining 100 specimens juveniles or normal individuals.

Causes of Growth Retardation

Many different explanations of causes of dwarfing or growth retardation (stunting) have been proposed for fossil forms. Among others, are the physiological effects on growth of: inadequate food supply; metallic cations (Fe, Cu); phosphorus; high or low salinity; pH; temperature; oxygen deficiency; population density; polymeric structure of water; turbidity; energy level of water.

Tasch (1953), Hallam (1965), and Mancini (1978) have evaluated the different explanations. Tasch allowed for a broad spectrum of possible causes of growth retardation, and urged a more quantitative biological approach. Hallam stressed food supply and oxygen deficiency as the two primary factors in stunting, and acknowledged the effects on growth of high and low salinity. Tasch considered many so-called "dwarf faunas" to be a consequence of hydraulic sorting—a demonstrable agent on fine to coarse sediments. Ager (1963) likewise proposed a "mechanical accumulation of small forms" as the explanation of a small-sized Devonian brachiopod-bryozoan-mollusk assemblage (cf. Bennison, 1961). Hallam (1965) found sorting a plausible agent for accumulations of small fossils in the Salem Limestone, and an explanation of a Jurassic oolite fauna (bivalves and gastropods), an accumulation of juveniles by sorting.

Many of the proposed explanations are inter-related or have multiple effects: high popula-tion density could lead to crowding (with effects on overall size of individuals—barnacles for example) but also it would enhance compe-tition for food supply. Hallam suggested that the effect of iron on stunting was indirect. That is, it would lead to an oxygen deficiency by forming sulfides. But there is also a more direct biological influence on the molecular level for ionic iron, copper, and other metallic cations not necessarily related to quantity (Tasch, 1953). It is therefore imprudent at present to pronounce definitively on the causes of stunting.

Paleoecology and Paleopathology

Any fossil fauna of diminutive size presents varied paleoecological possibilities. Bottom cur-rents may have sorted specimens yielding a uniform grade size (= *Pebble thanatoconenosis*). Water geochemistry in a restricted portion of a basin may have had growth-retarding in-fluence. Freshening of the basin by land runoff may have seasonally destroyed immature (juvenile) mollusks (or other forms) of a given species (Hedgpeth, 1957) and so on. Micro-morph assemblages or specific taxa in a portion of a formation, stunted in appearance, may actually be "substrate-related diminutives" (Mancini, 1978). Multiple hypotheses and weighting of each on the basis of all available data is the soundest approach (see Mancini, 1978; Gould, 1977).

Most growth-retarding factors have their effects ultimately on the molecular/cellular level. Was respiration, enzyme formation, circulation, general metabolism, and so on affected by low oxygen, high ionic iron, or copper for example? In these terms, all poten-tial causes are in themselves secondary causes of stunting. Primary causes must be inferred from the actual evidence preserved in given fossil faunas.

"Stunting" investigations can provide valu-able data on ancient environments; factors influencing living cell growth; marine biology; physical oceanography; water pollution; and medical pathology.

PAUL TASCH

References

Ager, D. V., 1963. *Principles of Paleoecology*. N.Y.: McGraw-Hill, 371p.

Bennison, G. M., 1961. Small *Naiadites obesus* from the Calciferous Sandstone Series (Lower Carbon-iferous) of Fife, *Palaeontology*, 4, 300-311.

Cloud, P. E., 1948. Assemblages of diminutive brach-iopods and their paleoecologic significance, *J. Sed. Petrology*, 18, 56-60.

Farrow, G. E., 1972. Periodicity structures in the bivalve shell: Analysis of stunting in *Cerastoderma edule* from the Burry Inlet (South Wales), *Palaeon-tology*, 15, 61-72.

Gould, S. J., 1977. *Ontogeny and Phylogeny*. Cam-bridge, Mass.: Harvard Univ. Press, 501p.

Hallam, A., 1965. Environmental causes of stunting in living and fossil marine benthonic invertebrates, *Palaeontology*, 8, 132-155.

Hedgpeth, J. W., 1957. Biological aspects [of estuaries and lagoons], *Geol. Soc. Amer. Mem. 67(1)*, 693-729.

Kummel, B., 1948. Environmental significance of dwarfed cephalopods, *J. Sed. Petrology*, 18, 61-64.

Lankford, R., 1967. The depauperate fauna concept relative to Foraminifera, *in Paleoecology, A.G.I. Short Course Lecture Notes*. Washington, D.C.: Am. Geol. Inst., RL-1-RL-12B.

Mancini, E. C., 1978. Origin of micromorph faunas in the geologic record, *J. Paleontology*, 52, 311-322.

Schuster-Dieterichs, O., 1954. Verzweigung durch Eisen? *Umschau*, 21, 644-646.

Snyder, J., and Bretsky, P. W., 1971. Life habits of diminutive bivalve mollusks in the Maquoketa Formation (Upper Ordovician), *Amer. J. Sci.*, 271, 227-251.

Tasch, P., 1953. Causes and paleoecological signifi-cance of dwarfed fossil marine invertebrates, *J. Paleontology*, 27, 356-444.

Thorson, G., 1971. *Life in the Sea*. N.Y.: McGraw-Hill, 241p.

Vogel, K. P., 1959. Zwergwuchs bei Polyptychiten (Ammonoidea). *Geol. Jb.*, 76, 469-540.

Cross-references: *Communities, Ancient; Paleoecol-ogy; Taphonomy.*

E

EBRIDIANS

Ebridians are unicellular marine phytoplankton that commonly range from 10 μm to 60 μm. They possess a siliceous skeleton of solid rods, hence they are found as fossils in Cenozoic sedimentary rocks, particularly those rich in siliceous remains. Their skeletons can be freed from the rock or sediment by chemical techniques. In the past, ebridians have been classified together with the silicoflagellates by botanists. Chiefly because of similarities in nuclear organization, they are now regarded as a class (Ebriophyceae) of the dinoflagellates (Divison Pyrrhophyta). Zoologists classify them among the protozoa as a separate order (Ebriida) within the class Phytomastigophora.

Ebridians are rather widespread in their occurrence but are most abundant in the nutrient-rich, cool waters of high latitudes. The living organisms have two flagella of unequal length and a single nucleus (Fig. 1). The nucleus is similar to that of dinoflagellates, which is unique among all other organisms. Within the nucleus, the organization and chemical composition are remarkably similar to prokaryotes, yet the cell organization is like that of eukaryotes. These kinds of organisms have been recently termed mesokaryotes. They have all the usual cellular structures, but the chromosomes maintain the same appearance throughout the cell cycle and the DNA of the chromosomes is not associated with protein.

Lacking chromatophores, ebridians capture their own food, reportedly diatoms, and some may possess symbiotic zooxanthellae. Reproduction is by cell division, with a daughter skeleton growing inside the parent cell until the nucleus and cytoplasm divide, freeing the complete daughter cell. Occasionally, the developing skeleton may grow together with the parent's, forming a double skeleton.

The skeletons are constructed of solid siliceous rods in a latticework that possesses tetraxial or triradial symmetry. The rods are usually angular and may be ridged or pitted. There is a wide variety of types known, of which 22 living and fossil genera are recognized. Among these are three fossil genera that are probably skeletons of radiolaria or other dinoflagellates.

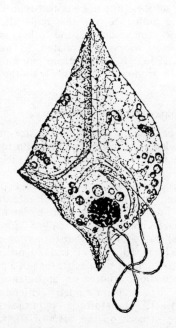

FIGURE 1. A modern ebridian, *Hermesinum adriaticum,* showing the skeleton of solid rods of silica, two unequal flagella, and a single nucleus (from Hovasse, 1934, no magnification given.)

Ebridians are first recorded in the Paleocene, during which time at least eight genera evolved. Thirteen are known from the Eocene, seven from the Oligocene, eleven from the Miocene, and two each from the Pliocene and Pleistocene. Only three genera are known to be alive today.

JERE H. LIPPS

References

Deflandre, G., 1951. Recherches sur les ébriédiens. Paléobiologie. Evolution. Systématique, *Bull. Biol. France Belgique,* **85,** 1–84.

Hovasse, R., 1934. Ebriacées, dinoflagellés et radiolaires, *Comptes Rendus Acad. Sci. Paris,* **198,** 402–404.

Loeblich, A. R., III; Loeblich, L. A.; Tappan, H.; and Loeblich, A. R., Jr., 1968. Annotated index of fossil and recent silicoflagellates and ebridians with descriptions and illustrations of validly proposed taxa, *Geol. Soc. Amer. Mem. 106,* 319p.

ECHINODERMATA

The main structural characteristics separating the phylum Echinodermata (Gr: *Echinos,* spiny; *derma,* skin) from all other animal phyla are: (1) the possession of hydraulic *tube feet,* connected by a tubular *water-vascular system;* (2) the possession of a basic *five-rayed symmetry;* and (3) the *reticular nature of the calcite* of which the skeletal elements are composed. Of these, only the first is diagnostic: all known modern echinoderms have tube feet, and it seems highly probable that all the extinct ones had them too. The nearest thing to tube feet in any other phylum is probably the lophophoral filaments of brachiopods, which also have a coelomic lumen; but their structural and functional range is far less extensive. As for the other criteria, pentamerous symmetry may well have arisen as a response to developmental requirements coupled with demands for optimal all-round command in a radial animal (Stephenson, 1974). Though there are exceptions among both extinct and modern echinoderms, only one other animal phylum, the Priapulida, shows any tendency toward five-fold symmetry.

To paleontologists, the nature of the calcite is the most useful criterion, and although it is said that some pelagic holothuroids lack a skeleton altogether, virtually all echinoderm skeletal elements show the unique reticular structure, even when fossilized (Roux, 1975). Almost every ossicle consists of a single crystal of magnesian calcite. Each crystal takes on the lattice-like structure (Fig. 1), the spaces between the struts being, in the living animal, filled with protoplasm (Fig. 2a), which maintains the skeleton in good condition throughout life. When the animal dies, the living material rots away (Fig. 2b), then sometimes the calcite is replaced or altered (Fig. 2c), and finally the intervening spaces are usually filled with secondary calcite in crystallographic continuity with the original (Fig. 2d). Sometimes the secondary calcite crystals will extend beyond the limits of the original plate. Because the organically produced calcite, or that which replaced it, is usually slightly different in appearance from the calcite filling the spaces, the pattern of the original lattice is still visible, and it is this that allows even tiny fragments of ossicles to be recognized as belonging to this phylum.

Biologically, the echinoderms are regarded as enterocoelous coelomates, that is, the body cavity develops not as a split in the mesoderm but from diverticula of the larval gut (Hyman, 1955). Phyletically, they are generally thought to be fairly close to the Minor Coelomate group of phyla, such as the Sipunculida, Phoronida, and Brachiopoda (Nichols, 1969). Members of

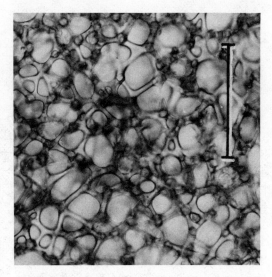

FIGURE 1. Optical micrograph of echinoderm calcite, taken from the interior of a spine of the modern regular echinoid *Heterocentrotus,* to show the reticular nature of the skeletal crystal. Scale = 100 μm.

these phyla, when adult, are bottom-dwelling detritus feeders that acquire their food by means of a system of highly ciliated arms. The arms or their filamentous offshoots are controlled by means of a special region of the coelom which surrounds the first part of the alimentary canal, sending tubular branches into the arms and filaments, so that a rise in pressure in this part of the coelom causes the soft parts to extend.

The echinoderms are similar in basic body plan to these animals. In addition, the echinoderm body itself is usually surrounded by a skeletal theca or *test,* and the arms, where they occur, are supported by skeletal elements and are therefore nonretractile. The separate coelomic cavity, including the extensions up the arms, is retained as the *water-vascular system,* and is now used to supply fluid to the *tube feet* (Nichols, 1975). Because the water-vascular fluid is separated from the surrounding sea water by only the thin tube-foot wall, the water-vascular system takes part in supplying the respiratory needs, in addition to the many other functions performed by the remarkably versatile tube feet.

In some echinoderms, the tube feet are not borne on arms but emerge from the surface of the test or theca. In all extant members of the phylum (except some that are secondarily reduced), each avenue of tube feet is underlain, not only by the water-vascular canal, but also by other coelomic canals and body spaces. The functions of these canals and spaces are by no means clear, though parts of them have been

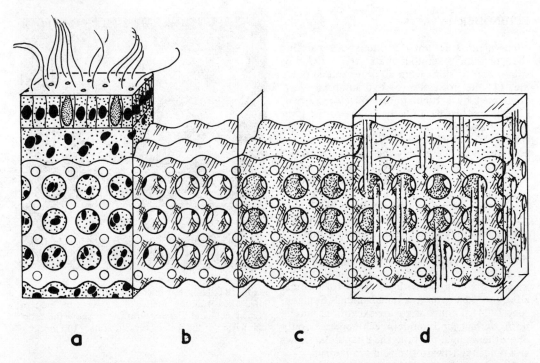

a   b   c   d

FIGURE 2. Schematic diagram of the fossilization of echinoderm calcite. a: In the living animal the calcite is interspersed with protoplasm and overlain by a ciliated and mucous epithelium. b: In the dead animal the protoplasm has rotted away, leaving only the calcite. c: The calcite may sometimes be dissolved and later completely and accurately replaced during fossilization, or it may be altered physically from the original, or, more often, it may remain unchanged. d: The fenestrae may become infilled with secondary calcite, in crystallographic continuity with the original or replaced calcite, and sometimes the secondary calcite may extend beyond the limits of the original ossicle.

shown to have an excretory function (Millott, 1967). The food grooves are also underlain by a swelling of the superficial nervous system, the *radial nerves.* Each radial unit, consisting of food groove or its derivative, coelomic canals and tube feet, spaces, and radial nerve, is termed an *ambulacrum.* Around the mouth, the radial branches of each tubular coelomic system are connected by the appropriate ring vessel, and a similar circumoral nerve ring connects the radial nerves.

Because the echinoderm body is generally almost rigid, the operation of the extensile tube feet by muscular contraction of parts of the water-vascular system would be impaired by any changes in ambient pressure, were there not provision for equalization inside the body wall. This is brought about by an opening through the wall called the *hydropore,* or, if there are many pores, the *madreporite,* which leads into an often skeletally supported *stone-canal* and thence to the circumoral water-vascular ring.

If the earliest echinoderms did indeed arise from sipunculid- or phoronid-like forebears, it seems likely that they were sedentary, possibly stalked animals with a mouth-uppermost posture. This is retained by the group that is generally regarded as the most primitive of the living classes, the Crinoidea, in which the ambulacrum lies on the upper (oral) side of the arm skeleton, not enclosed by skeletal material. This so-called "open ambulacrum" is retained in the Asteroidea and also seen in most of those extinct groups in which such interpretation of soft parts is possible. But in other living classes, such as Ophiuroidea, Echinoidea, and Holothuroidea, the ambulacral elements have to various degrees passed to the inside of the body and are protected by the overlying body wall and skeleton, when it is termed a "closed ambulacrum." When this happens, the tube feet emerge between the plates or through pores within them. The shape and orientation of these pores can often supply information about the activities of the tube feet they bore, and hence the mode of life of the animal.

As mentioned above, the name of the phylum reflects the covering of the body with various sorts of projections. These are usually spines, sometimes very simple, but sometimes modified to perform special jobs. In two classes, Asteroidea and Echinoidea, some spines become highly modified or compounded to produce small pincer-like *pedicellariae* (Campbell, 1972; Nichols, 1975), which apparently mainly help to keep the body free from settling larvae and silt, though a few are big and powerful enough to catch small organisms which may later be used as food.

Exchange of respiratory gases occurs across the walls of the tube feet; but in those groups in which these organs do not present sufficient exchange area, additional respiratory structures are found. These either circulate coelomic fluid through external thin-walled sacs, like the *papulae* of asteroids and *dipores* of some cystoids (Paul, 1972), or pump sea water through sacs or canals within the body, as in the *bursae* of ophiuroids (Hyman, 1955), *respiratory trees* of holothuroids, *dichopores* of other cystoids (Paul, 1968), and *hydrospires* of blastoids.

There are some unusual features about nutrition. Clearly, the presence of a thick endoskeleton impedes the distribution of nutrients to the external epithelium, where so many effector

organs are located. So in addition to the more usual alimentary method of obtaining food, most echinoderms have an absorptive epithelium over the whole body surface, including that of the appendages, and they absorb soluble food substances directly from the surrounding sea water (Nichols, 1975).

The first indisputable echinoderm fossils are found in the lowest Cambrian rocks. At this time, three groups appear together: Eocrinoidea, Camptostromatoidea, and Helicoplacoidea (Durham, 1963). Predating these occurrences, there is an echinoderm-like, though triradiate, animal called *Tribrachidium* in the Precambrian assemblage of Ediacara (Glaessner, 1962), but little is known about this form. Throughout geological time the phylum is of great importance, not only for zonation but also, in some deposits, as an example of the mozaic evolution of individual characters: the tests of echinoderms are sufficiently well preserved and occur in such numbers that study of evolutionary trends and principles is excellently demonstrated. The systematics and geological ranges of the echinoderms are shown in Table 1.

TABLE 1. Systematics and Geological Ranges
of the Echinodermata

Phylum ECHINODERMATA
Subphylum CRINOZOA–Ordovician–Holocene
  Class CRINOIDEA–Ordovician–Holocene
    Order Inadunata–Ordovician-Permian
    Order Flexibilia–Ordovician-Permian
    Order Camerata–Ordovician-Permian
    Order Articulata–Triassic-Holocene
Subphylum BLASTOZOA–Lower Cambrian–Permian
  Class BLASTOIDEA–Silurian-Permian
    Order Fissiculata–Silurian-Permian
    Order Spiraculata–Silurian-Permian
  Class PARABLASTOIDEA–Middle Ordovician
  Class CYSTOIDEA–Ordovician-Devonian
    Order Diploporita–Ordovician-Devonian
    Order Rhombifera–Ordovician-Devonian
  Class EOCRINOIDEA–Lower Cambrian–Upper Ordovician
Subphylum HOMALOZOA–Middle Cambrian–Upper Devonian
  Class HOMOSTELEA–Middle Cambrian
  Class HOMOIOSTELEA–Upper Cambrian–Lower Devonian
  Class STYLOPHORA–Middle Cambrian–Upper Devonian
    Order Cornuta–Middle Cambrian–Ordovician
    Order Mitrata–Ordovician-Devonian
  Class CTENOCYSTOIDEA–Middle Cambrian
Subphylum ASTEROZOA–Lower Ordovician–Holocene
  Class SOMASTEROIDEA–Lower Ordovician–Holocene
  Class OPHIUROIDEA–Lower Ordovician–Holocene
    Order Stenurida–Lower Ordovician–Upper Devonian
    Order Oegophiurida–Ordovician-Holocene
    Order Phrynophiurida–Devonian-Holocene
    Order Ophiurida–Silurian-Holocene
  Class ASTEROIDEA–Lower Ordovician–Holocene
    Order Platyasterida–Middle Ordovician–Holocene
    Order Phanerozonida–Lower Ordovician–Holocene
    Order Spinulosida–Ordovician-Holocene
    Order Euclasterida–Cenozoic-Holocene
    Order Forcipulata–Ordovician-Holocene

Subphylum ECHINOZOA–Lower Cambrian–Holocene
  Class ECHINOIDEA–Ordovician-Holocene
  Subclass Perischoechinoidea–Ordovician-Holocene
    Order Bothriocidaroida–Ordovician
    Order Echinocystitoida–Ordovician-Permian
    Order Palaechinoida–Silurian-Permian
    Order Cidaroida–Mississippian-Holocene
  Subclass Euechinoidea–Triassic-Holocene
    Superorder Diadematacea–Triassic-Holocene
    Superorder Echinacea–Triassic-Holocene
    Superorder Gnathostomata–Jurassic-Holocene
      Order Holectypoida–Jurassic-Holocene
      Order Clypeasteroida–Cretaceous-Holocene
    Superorder Atelostomata–Jurassic-Holocene
      Order Holasteroida–Jurassic-Holocene
      Order Nucleolitoida–Jurassic-Cretaceous
      Order Cassiduloida–Jurassic-Holocene
      Order Spatangoida–Upper Jurassic-Holocene
      Order Disasteroida–Jurassic-Cretaceous
  Class HOLOTHUROIDEA–Ordovician-Holocene[a]
  Subclass Dendrochirotacea
    Order Dendrochirotida
    Order Dactylochirotida
  Subclass Aspidochirotacea
    Order Aspidochirotida
    Order Elasipodida
  Subclass Apodacea
    Order Molpadiida
    Order Apodida
  Class EDRIOASTEROIDEA–Middle Cambrian-Carboniferous
  Class HELICOPLACOIDEA–Lower Cambrian
  Class OPHIOCISTIOIDEA–Ordovician-Devonian
  Class CYCLOCYSTOIDEA–Cambrian
  Class CAMPTOSTROMATOIDEA–Lower Cambrian

Of uncertain position within the phylum:
  Edrioblastoidea–Middle Ordovician
  Haplozoa (cyamoids and cycloids)–Cambrian
  Paracrinoidea–Middle Ordovician

[a]The systematics of fossil holothuroids is uncertain and does not permit meaningful infra-class geological ranges.

Subphylum Crinozoa

The Crinozoa (Ordovician-Holocene) are mainly stemmed echinoderms with cup-like theca. They bear feeding structures called *arms,* which are projections that contain an extension of the body coelom. The mouth is normally directed upwards.

Class Crinoidea. The Crinoidea (Lower Ordovician–Holocene) are mainly attached (pelmatozoic) forms, although in the Middle Triassic a group of stemless forms, the Comatulida (Fig. 3), appears for the first time (Moore, 1967; Clark, 1915 to 1967). While the stemmed forms are restricted to one spot determined by larval settlement. the comatulids are able to swim for short distances by waving their

arms; when they settle again they grip the substrate by means of cirri on the underside of the theca. The upper part of the theca is composed of a flexible membrane, the *tegmen,* at the center of which is the mouth. The tegmen is plated in many fossil crinoids. Food grooves on the upper side of the pinnules lead to similar grooves on the arms, and these in turn lead across the tegmen to the mouth. The anus also pierces the tegmen, to one side of the mouth.

There are four orders, of which the first three are confined to the Paleozoic: Inadunata, Flexibilia, Camerata, and Articulata. The oldest crinoid is the inadunate *Ramseyocrinus* from the Lower Ordovician of S Wales; this possessed four arms which branch several times; the fifth arm is possibly reduced to a short papilla carry-

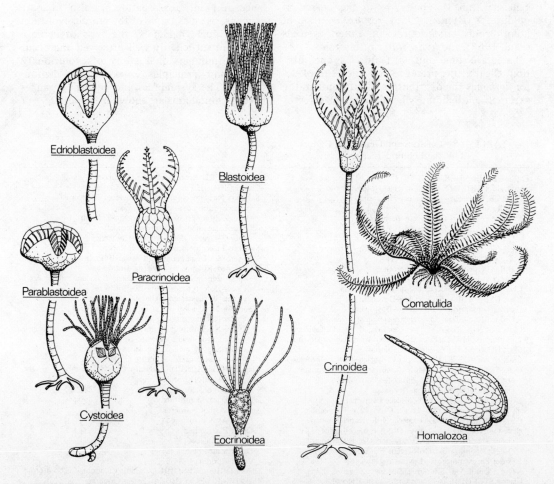

FIGURE 3. Primitive echinoderms, mainly possessing a stem, of the Subphyla Crinozoa and Homalozoa. The interclass relationships are so uncertain that no phyletic connections are implied in the diagram, except that the unstalked comatulid crinoids, here represented by *Antedon* (Holocene), are evolved from the stalked crinoids, here represented by *Ptilocrinus* (Holocene). The other drawings are based on the following genera: Eocrinoidea, *Gogia* (Middle Cambrian); Cystoidea, *Lepadocystis* (Upper Ordovician); Parablastoidea, *Blastoidocrinus* (Middle Ordovician); Paracrinoidea, *Comarocystites* (Middle Ordovician); Edrioblastoidea, *Astrocystites* (Ordovician); and Homalozoa, the homostele *Trochocystites* (Middle Cambrian).

ing the anus. The inadunate, flexible, and camerate crinoids radiate mainly in the Middle and Late Paleozoic (with Mississippian peak); the articulates first appear in the Lower Triassic. The fact that their lower arm plates articulate with the thecal plates is thought to have bestowed greater movement on the arms, and hence to have permitted the evolution of stemless (comatulid) forms.

Subphylum Blastozoa

Ranging from the Lower Cambrian to the Permian, the blastozoans were stemmed echinoderms with bud-like theca bearing feeding structures called *brachioles,* which are projections that do not carry an extension of the body coelom (Fay 1961; Moore, 1967; Sprinkle, 1973).

Class Blastoidea. The bud-shaped theca in the Blastoidea (Silurian-Permian), as in the cystoids, was borne on a stem, had five food grooves radiating downward over the theca from an upwardly directed mouth, and food-collecting brachioles arising alternately from the sides of each food groove. Between the origins of the brachioles, in most forms, are pores leading in to the *hydrospire* system, the walls of which are composed of very thin plates thrown into folds; each pair of hydrospires also has a single pore, the *spiracle,* leading from it, the five spiracles being arranged in a ring round the mouth. Thus, it appears that sea water entered the system via the pores lining the food grooves, gaseous exchange took place across the thin skeletal membrane, and the used sea water left the system via the spiracles. Because the respiratory membranes were skeletally supported, their presence is still clearly visible in sections of the blastoid body. The main evolutionary feature in the history of the blastoids was the gradual elaboration of the hydrospire system from primitive forms, such as *Codaster* (Silurian), in which the hydrospires were simply folds in the plates of the theca, to forms such as *Pentremites* (Carboniferous), in which the hydrospire folds were deeper in the body and reached by the linear series of pores mentioned above. Another trend, though a minor one, was toward bilateral symmetry, in such forms as *Eleutherocrinus* (Devonian) and *Zygocrinus* (Carboniferous).

Class Parablastoidea. The class Parablastoidea (Middle Ordovician) contains few genera, the chief one being *Blastoidocrinus* (Fig. 3), with thecal pores similar to those of some cystoids, but with blastoid-like thecal plates and brachioles (Moore, 1967). The ambulacra were covered by a roofing of plates.

Class Cystoidea. Cystoid fossils (Cambrian-Devonian) consist of a cyst-like body with

mouth at the center of the upper surface and the anus to one side of it, as in crinoids. In most forms, food grooves (five or fewer) occur on the upper part of the theca; but at intervals the food grooves lead off to the brachioles, arranged alternately down the length of the food groove. Within or between the plates are various systems of pores or canals, almost certainly respiratory, some of which are embedded in calcite while others apparently bore papula-like vesicles. During the evolution of the group, two trends are apparent: first, the food grooves increased in length, and, secondly, the pore canals became aggregated into special regions, usually lying across the interplate sutures, called *pore rhombs*. The group is generally divided taxonomically into two orders, based on the nature of the respiratory devices: Diploporita, in which the pore canals are scattered over the entire theca, and Rhombifera, in which they are aggregated into pore rhombs. Most cystoids are stalked, but some were attached directly to the sea floor.

Class Eocrinoidea. Found from the Lower Cambrian to Upper Ordovician, eocrinoids were usually stemmed forms having the typical blastozoan theca with an upwardly directed mouth (Fig. 3). Brachioles arose from the upper surface and formed a food-collected funnel. Between the thecal plates are pores, the *epispires,* which probably bore papula-like respiratory vesicles in life (Ubaghs, 1975).

Subphylum Homalozoa

The subphylum Homalozoa (Middle Cambrian–Upper Devonian) is composed of wholly extinct, asymmetrical forms, formerly called "carpoids" or "heterosteles" (Moore, 1967; Ubaghs, 1971, 1975). They all have one or more ambulacra, either borne on an arm or embedded in the body (Fig. 3). Because there are no relatives still living, their way of life is not easy to assess, but they probably lay on the sea bottom with the feeding arm, when present, raised into the current, and the stem-like stele or other projections acting to anchor the body. Other interpretations (Jefferies, 1975), which see these animals as having characteristics more appropriate to the chordates than to echinoderms, have been made, but so far these have not superceded the more usual view that they are highly specialized echinoderms (see *Calcichordata*).

Class Homostelea. The homosteles of the Middle Cambrian were spoon-shaped homalozoans (Fig. 3) with stele but no feeding arm. Two openings are present on the side of the theca opposite the stem, and the larger of these was probably the anus, which may have taken in and ejected water for respiratory purposes. Ambulacral grooves on either side of the theca lead toward the smaller of the two openings, which was therefore probably the mouth.

Class Homoiostelea. The homoiosteles (Upper Cambrian–Lower Devonian) had both stele and a feeding arm, but the structure is not well known.

Class Stylophora. The best known of the homalozoan classes is Stylophora (Middle Cambrian–Upper Devonian). Its members have irregularly shaped thecae, mostly with a single feeding arm, the *aulacophore,* which is differentiated into three regions. Notable on the theca of some are elliptical gill-slit-like *cothurnopores,* and on others grill-like *lamellipores,* the function of which is unknown, but is probably respiratory.

Class Ctenocystoidea. The ctenocystoids (Middle Cambrian) were stemless and armless homalozoans with a unique array of sieve-like plates across the mouth; these may have acted as a food filter.

Subphylum Asterozoa

Star-shaped echinoderms (Fig. 4) belong to three classes, Somasteroidea, Asteroidea, and Ophiuroidea, and range from the Lower Ordovician to Holocene (Fell, 1962; Moore, 1966; Spencer, 1914 to 1965). The fossil record permits a convincing relationship among these; and furthermore there is, in the Lower Ordovician somasteroids, a good candidate, *Chinianaster,* as common ancestor of the two major living groups. In this form, the tube feet bordering the ambulacra arise from cup-like cavities in the ambulacral plates. Such cups are retained in the ophiuroid line; but in the asteroids, the tube feet evolved internal *ampullae* to assist their protraction, and hence the position of origin of each tube foot is marked by a canal between adjacent ambulacral plates.

Class Somasteroidea. The basal stock, the somasteroids of the Lower Ordovician, from which star-shaped echinoderms probably arose had broad, flat arms with a double column of

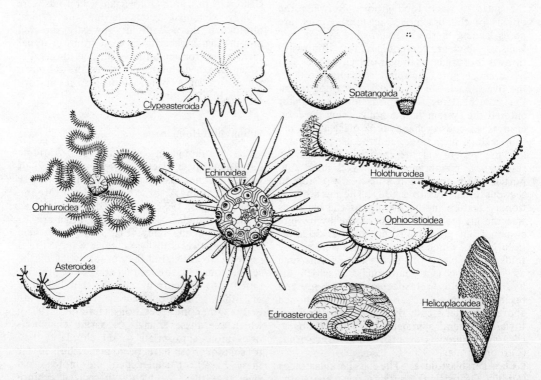

FIGURE 4. Advanced, mainly freely moving echinoderms, of the subphyla Asterozoa and Echinozoa. The drawings are based on the following genera: Edrioasteroidea, *Edrioaster* (Ordovician-Devonian); Helicoplacoidea, *Waucobella* (Lower Cambrian); Ophiocistioidea, *Sollasina* (Silurian-Devonian); Asteroidea, *Asterias* (Holocene); Ophiuroidea, *Ophiothrix* (Holocene); regular Echinoidea, *Cidaris* (Triassic-Holocene), drawn with spines on the aboral half removed; Clypeasteroida, *Clypeaster* (Pliocene-Holocene), on the left, and *Rotula* (Pliocene-Holocene), on the right; Spatangoida, *Micraster* (Cretaceous), on the left, and *Pourtalesia* (Holocene), on the right; Holothuroidea, *Holothuria* (Holocene).

ambulacral plates running down the center of each arm, and lateral ossicles called *virgals* in parallel series out to the arm's edge. There was no true ambulacral groove, though the arms may have been deformable to provide a transient burrow during burrowing and feeding. The oldest and apparently most primitive somasteroid is *Chinianaster* from the Tremadocian.

Class Ophiuroidea. The ophiuroids, or brittle-stars, diverged from the asteroids in the Lower Ordovician; they have survived to the present day (Spencer, 1914 to 1965). A series of fossils shows a gradual transition from the early asteroid condition, with ambulacral plates and external tube-foot apparatus, to a condition showing very much enlarged "vertebral ossicles" and a tube-foot apparatus still enclosed in the oral side of the arm skeleton, a condition characteristic of ophiuroids.

The first "starfish" recognizable as an ophiuroid is probably *Pradesura* (Lower Ordovician), in which the arms are thought to have been already long and sinuous and the disk relatively very small. The configuration of the ambulacrum still resembled that of an early asteroid, and not until the Silurian form *Lapworthura* do we see the typical ophiuroid ambulacral condition—in which the enlarged ambulacral plates have filled nearly the whole arm to become the characteristic "vertebrae"; and the radial water canal supplying the tube feet, until now lying "exposed" on the oral side of the ossicles, has sunk into the midline of the vertebrae.

Class Asteroidea. Members of the class Asteroidea range from the Lower Ordovician to the Holocene (Spencer, 1914 to 1965). The early asteroids did not have an ambulacral groove as such, and the arms were broad and flat. Subsequently, however, the groove deepened and the arms became narrower, finally attaining the situation seen in all asteroids of the Upper Ordovician and later, in which there is a longitudinal series of ambulacral plates defining the groove, a column of adambulacrals on each side of the groove, and two columns of marginals forming the sides of the arm, the *phanerozone* ("visible edge") condition. The upper surface is usually formed of irregular plates.

All early starfish, in common with the more primitive recent forms, apparently had pointed tube feet; and all indications are that these early forms, again like the primitive extant ones, were burrowers. Present-day phanerozones, for instance, usually live in soft substrates (Millott, 1967) with, at most, only the arm tips protruding, though sometimes the center of the disk, on which the anus opens,

is pushed above the surface for defecation. Water passes into the burrow mainly at the arm tips, and respiration occurs across the walls of the tube feet and of special thin, sac-like "blisters" of the body wall, the *papulae*. In many burrowers, special column-like ossicles called *paxillae* occur on the upper and lateral surface of the body; they bear spines that can articulate in such a way as to form a false covering to the body, beneath which currents of water can bathe the respiratory papulae. Many phanerozones feed by taking in mollusks and other benthic animals whole, digesting off the organic tissues in the stomach, then voiding the shells from the mouth (Christensen, 1970).

In the Tertiary, there appear starfish with less distinct marginal plates (the *cryptozone*, "hidden edge," condition). This apparently happened at least twice; and in each case it appears to have bestowed greater movement to the arms, and hence less ecological restriction. The Spinulosida, apparently closest to the Phanerozonida, possess special aboral spines, and mostly have suckered tube feet for more efficient locomotion, while the Forcipulata are characterized by a special type of pedicellaria, and they too mainly have suckered tube feet. In this latter group are the asteriids, which have evolved a method of feeding on bivalve mollusks, depending on their ability to exert tremendous pull on the valves with their tube feet and also their ability to intrude the stomach between the valves.

Subphylum Echinozoa

The subphylum Echinozoa (Lower Cambrian–Holocene) contains all the basically pentamerous, mostly free-living ("eleutherozoic") echinoderms that are not star shaped (generally globular) (Moore, 1966; Mortensen, 1928–1951; Durham and Melville, 1957).

Class Echinoidea. The sea urchins, sand dollars, and heart urchins (class Echinoidea, Ordovician-Holocene) are prominent members not only of the fauna of present-day seas but also of marine sedimentary rocks throughout the post-Cambrian geological column. The more primitive forms, the so-called *regular echinoids*, are globular in shape, with the mouth at the center of the under surface, and, within the mouth, a complex masticatory apparatus, *Aristotle's lantern*. The five ambulacra extend from the mouth up the sides of the animal and terminate just before the apex of the body. At the apex is the anus, and surrounding it a group of plates, the *apical disk*, consisting of two whorls of five plates each. One of these plates is the madreporite, which therefore breaks the otherwise perfect radial symmetry of the regu lar echinoids. More advanced forms, the *irregu- lar echinoids,* have acquired a secondary bilateral symmetry, and most of them were (and still are) burrowers. This trend toward irregularity is thought to have happened at least twice (Durham and Melville, 1957; see also Mintz, 1968). One of the principal features of this acquisition of bilateral symmetry was a migration of the anus, from the center of the apical disk to points successively further away down what had become the posterior inter- ambulacrum.

The first echinoid in the fossil record is the Ordovician perischoechinoid *Bothriocidaris,* which had a fairly rigid test. Slightly later, how- ever, there appears a group, the Echinocysti- toida (Kier, 1965), whose members had flexible tests made up of imbricating plates; and be- cause of the arrangement of these plates, they are generally thought to be closer to the ances- tral echinoid pattern than *Bothriocidaris.* From Silurian to Permian times there was experimen- tation in the number of columns of plates making up the test, one pattern, in the order Cidaroida, being two columns of ambulacral plates and two of interambulacral. For some reason, this pattern alone survived, and all post- Permian echinoids possess it.

In the Lower Jurassic, we see the first break in radial symmetry (Kier, 1974), the first move toward irregularity. The earliest known form to do this was *Pygomalus,* which is a member of that group of the irregular echinoids which dis- pensed with the Aristotle's lantern, the Atelo- stomata. Slightly later came *Holectypus,* a representative of the Gnathostomata which retain the lantern. The most important atelo- stomes are the Spatangoida, or heart urchins, which are remarkably well adapted for life in snug burrows deep in the substrate. Among the gnathostomes are the clypeasteroids (Durham, 1955), including the sand dollars, which also burrow, but to a lesser degree.

Class Holothuroidea. The sea cucumbers (Middle Cambrian–Holocene) are fairly com- mon representatives of modern seas, and par- ticularly so in deep water (Moore, 1966). Most of the living members are worm-like in shape (Fig. 4), the mouth and anus being at opposite ends of the cylindrical body, with the ambu- lacra, where they occur, running the length. Around the mouth is a circlet of fairly large plates, to which are attached some of the muscles of the body wall and also muscles of special large tube feet around the mouth, used for feeding and respiration. In most holothur- oids, other tube feet emerge from the ambu- lacra; but because only three ambulacra are

anywhere near the substrate (the other two lying on either side of the dorsal mid-line), often only the feet from these three are suck- ered, the others being reduced, sensory papillae. The body wall contains isolated plates, usually very small, in a variety of shapes.

Because of the nature of the skeleton, the fossil record is understandably sparse and inconsequential from a phyletic viewpoint, so very little can be deduced about the origin and evolution of the group. Fossil ossicles (Frizzell and Exline, 1955), first appear for certain in the Devonian (see *Sclerites, Holothurian*), though there are reports of holothuroid remains from the Ordovician. Fossils of complete holothuroids are also first found in Lower Devonian rocks, though some fossils from the Middle Cambrian Burgess Shale (*Eldonia*) may be holothuroids.

Class Edrioasteroidea. Although one possible member of the extinct class Edrioasteroidea (Lower Cambrian–Carboniferous) possesses a stem, the others did not (Bather, 1915; Bell, 1976). The body is flat, globular or cylindrical, typically with five straight or curved food grooves on the upper surface leading to a cen- tral mouth (Fig. 4). The food grooves are pro- tected by moveable *cover plates,* and often have two columns of pores down their sides, suggest- ing that in life they bore tube feet with ampullae, that is, actively extensile organs probably for food capture.

In spite of a fairly rich fossil record, rather little can be said about the origin and evolution of the group. The first family to appear, the Stromatocystidae, has no ambulacral pores, and so it has been suggested that their tube feet lacked ampullae. All other edrioasteroids have ambulacral pores. The edrioasteroids are among the first echinoderms to appear that show evi- dence of having borne spines.

Because of time relations, and also certain morphological features, it has been frequently suggested that the edrioasteroids may lie close to the ancestors of at least some eleutherozoan forms, i.e., free-living echinoderms such as star- fish, sea urchins, and sea cucumbers, familiar in present-day seas. The presence of ambulacral pores to the interior cannot, however, be used to indicate a relationship with the asteroids, since the earliest asteroids lacked such pores; in addition, there is evidence favoring a crinoid- asteroid relationship. But some authorities argue that the edrioasteroids may well have given rise to the echinoids, principally on grounds of similarity of plate arrangement.

Class Helicoplacoidea. Helicoplacoid fossils are known only from very early (Lower Cambrian) deposits, and probably represent a short-lived excursion into a nonbrachiate body plan (Durham, 1963). The theca is cigar shaped (Fig. 4), and plated with imbricating ossicles that could move relative to one another to allow the body to expand or contract. A single spiral food groove passed round and down the body, with a short branch joining it not far from the apex. The mouth is presumed to have been at the apex. Very little more is known about this group, so suggestions as to its mode of life are tentative; but it seems likely that its members lived in short burrows from which at least part of their thecae could be protruded. Some forms show that there was almost certainly a single column of tube feet in the ambulacrum.

Class Ophiocistioidea. The ophiocistioids (Ordovician-Devonian) had a dome-shaped body with a mouth, sometimes with teeth, in the center of the under surface (Moore, 1966). Their outstanding feature is the presence of large plated structures (Fig. 4), generally said to be tube feet but also possibly arms, which emerged from the underside.

Class Cyclocystoidea. The disk-like cyclocystoids, found only in the Cambrian, had a ring of heavy plates around the margin, an upper and lower sheet with embedded plates within, and a skirt of small plates around the periphery (Moore, 1966; Ubaghs, 1975). Not even the orientation of these little-known animals can be stated for certain.

Class Camptostromatoidea. The little-known, dome-shaped animals classed as Camptostromatoidea (Lower Cambrian) have an apron of large plates around the margin (Moore, 1967; Ubaghs, 1975). Plated arm-like structures emerge from the under-surface, where the mouth is also thought to have been situated, but how the animals lived is not known.

Of Uncertain Affinity

Edrioblastoidea. The Edrioblastoidea (Middle Ordovician) are a rare group containing a single genus, *Astrocystites* (Fig. 3), which has some features reminiscent of edrioasteroids and some of stalked echinoderms; they are not well known (Moore, 1967; Mintz, 1970).

Haplozoa. The bilaterally symmetrical cyamoids and the dome-shaped radially symmetrical cycloids (found only in the Cambrian) have echinoderm-like skeletal structure but are otherwise of uncertain relationship to the other echinoderms (Whitehouse, 1941; Moore, 1967).

Paracrinoidea. Known only in the Middle Ordovician, the Paracrinoidea are a small group with bilateral symmetry and two to four crinoid-like arms with lateral pinnules. The plates of some are thin and folded, probably to permit respiratory exchange across the actual calcite of the plates (Moore, 1967; Parsley and Mintz, 1975).

D. NICHOLS

References

Bather, F. A., 1915. *Studies in Edrioasteroidea, I-IX*. Collected papers from *Geol. Mag.*, 1898-1915, published privately.

Bell, B. M., 1976. A study of North American Edrioasteroidea, *N.Y. State Mus. Sci. Serv. Mem.* **21**, 447p.

Binyon, J., 1972. *Physiology of Echinoderms*. London: Pergamon, 264p.

Boolootian, R. A., ed., 1966. *Physiology of Echinodermata*. New York: Wiley-Interscience, 822p.

Campbell, A.C., 1972. The form and function of the skeleton in pedicellariae from *Echinus esculentus* L., *Tissue & Cell*, **4**, 647-661.

Christensen, A. M., 1970. Feeding biology of the sea-star *Astropecten irregularis* Pennant, *Ophelia*, **8**, 1-134.

Clark, A. H., 1915-1967. A monograph of the existing crinoids, *U.S. Nat. Mus. Bull. no. 82*.

Clark, A. M., 1962. *Starfishes and Their Relations*. London: British Museum (Natural History), 119p.

Clark, A. M., and Rowe, F. W. E., 1971. *Monograph of Shallow-Water Indo-West Pacific Echinoderms*. London: British Museum (Natural History), 238p.

Cobb, J. L. S., 1970. The significance of the radial nerve cords in asteroids and echinoids, *Zeitschr. Zellforsch.*, **108**, 457-474.

Durham, J. W., 1955. Classification of clypeasteroid echinoids, *Univ. Calif. Pub. Geol. Sci.*, **31**, 73-198.

Durham, J. W., 1963. Helicoplacoidea: A new class of echinoderms, *Science*, **140**, 820-822.

Durham, J. W., 1964. The Helicoplacoidea and some possible implications, *Yale Sci. Mag.*, **39**, 24-28.

Durham, J. W., and Melville, R. V., 1957. A classification of echinoids, *J. Paleontology*, **31**, 242-272.

Fay, R. O., 1961. Blastoid studies, *Univ. Kansas Paleont. Contrib.* **27**, Echinodermata, Art. 3, 1-147.

Fell, H. B., 1962. The phylogeny of sea-stars, *Phil. Trans. Roy. Soc. London Ser. B*, **246**, 381-435.

Florkin, M., and Scheer, B. T., eds., 1969. *Chemical Zoology, vol. III. Echinodermata, Nematoda, Acanthocephala*. New York and London: Academic Press, 687p.

Frizzell, D. L., and Exline, H., 1955. Monograph of fossil holothurian sclerites, *Bull. Missouri Sch. Mines Metal.* **89**, 1-204.

Glaessner, M. F., 1962. Pre-Cambrian fossils, *Biol. Rev. Cambridge Phil. Soc.*, **37**, 467-494.

Hyman, L. H., 1955. *The Invertebrates, IV. Echinodermata*. New York: McGraw-Hill, 763p.

Jefferies, R. P. S., 1975. Fossil evidence concerning the origin of chordates, *Symp. Zool. Soc. London*, **36**, 253-318.

Kier, P. M., 1965. Evolutionary trends in Paleozoic echinoids, *J. Paleontology*, **39**, 436-465.

Kier, P. M., 1974. Evolutionary trends and their functional significance in the post-Paleozoic echinoids, *J. Paleontology*, 48(3, suppl.) (*Paleontol. Soc. Mem. 5*), 1-95.

Millott, N., ed., 1967. Echinoderm biology, *Symp. Zool. Soc. London*, **20**, 240p.

Mintz, L. W., 1968. Echinoids of the Mesozoic families Collyritidae d'Orbigny, 1853 and Disasteridae Graz, 1848, *J. Paleontology*, **42**, 1272-1278.

Mintz, L. W., 1970. The Edrioblastoidea: Re-evaluation based on a new specimen of *Astrocystites* from the Middle Ordovician of Ontario, *J. Paleontology*, **44**, 872-880.

Moore, R. C., ed., 1966. *Treatise on Invertebrate Paleontology*, pt. U, 2 vols., Echinodermata 3. Lawrence, Kansas: Geol. Soc. Amer. and Univ. Kansas Press, 695p.

Moore, R. C., ed., 1968. *Treatise on Invertebrate Paleontology*, pt. S, 2 vols., Echinodermata 1. Lawrence, Kansas: Geol. Soc. Amer. and Univ. Kansas Press, 695p.

Mortensen, T., 1928-1951. *A Monograph of the Echinoidea*, 5 vols. Copenhagen. Reitzel.

Nichols, D., 1969. *Echinoderms*, 4th rev. ed. London: Hutchinson Univ. Library, 192p.

Nichols, D., 1975. *The Uniqueness of the Echinoderms*. Oxford Biology Readers, no. 53. Oxford: Oxford University Press, 16p.

Parsley, R. L., and Mintz, L. W., 1975. North American Paracrinoidea: (Ordovician: Paracrinozoa, new, Echinodermata), *Bull. Amer. Paleontology*, **68**, 115p.

Paul, C. R. C., 1968. Morphology and function of dichoporite structures in cystoids, *Palaeontology*, **11**, 697-730.

Paul, C. R. C., 1972. Morphology and function of exothecal pore structures in cystoids, *Palaeontology*, **15**, 1-28.

Paul, C. R. C., 1977. Evolution of primitive echinoderms, in A. Hallam, ed., *Patterns of Evolution*. Amsterdam: Elsevier, 123-158.

Regnéll, G., 1975. Review of recent research on "Pelmatozoans", *Paläont. Zeitschr.*, **49**, 530-564.

Roux, M., 1975. Microstructural analysis of the crinoid stem, *Univ. Kansas Paleont. Contrib.*, **75**, 1-7.

Spencer, W. K., 1914-1965. *The British Palaeozoic Asterozoa*. London: Palaeontogr. Soc. Monographs, 583p.

Sprinkle, J., 1973. *Morphology and Evolution of Blastozoan Echinoderms*. Cambridge, Mass.: Mus. Comp. Zool., Spec. Pub., 284p.

Stephenson, D. G., 1974. Pentamerism and the ancestral echinoderm, *Nature*, **250**, 82-83.

Ubaghs, G., 1971. Diversité et spécialisation des plus anciens échinodermes que l'on connaisse, *Biol. Rev. Cambridge Phil. Soc.*, **46**, 157-200.

Ubaghs, G., 1975. Early Paleozoic echinoderms, *Ann. Rev. Earth Planetary Sci.*, **3**, 79-98.

Whitehouse, F. W., 1941. Early Cambrian echinoderms similar to the larval stages of recent forms, *Queensland Mus. Mem. 12* (l), 28p.

Cross-references: *Calcichordates; Ediacara Fauna; Sclerites, Holothurian.*

ECTOPROCTA—See BRYOZOA

EDIACARA FAUNA

The Ediacara fauna, a diverse assemblage of soft-bodied invertebrates (Table 1), provides a rare glimpse of an early stage in the evolution of metazoan life. Fossiliferous beds of the late Precambrian Pound Quartzite are exposed in the Flinders Ranges of South Australia (Fig. 1). The sediments are arenaceous and vary in their degree of silicification: some places are friable due to weathering of feldspars, while others are indurated by secondary silicification. Cross-stratification indicates moving sand ripples, and ripple-marked bedding planes are common (Glaessner and Wade, 1966). The beds are flaggy, with sandstone layers separated by thin silty layers and lenses; the finer-grained strata are thought to represent sites of temporary

TABLE 1. Ediacara Faunal List

Phylum COELENTERATA
  A. Medusoid:
      Ediacaria flindersi Sprigg
      Beltanella gilesi Sprigg
      Medusinites asteroides (Sprigg)
      Cyclomedusa davidi Sprigg
      C. radiata Sprigg
      C. plana Glaessner and Wade
      Mawsonites spriggi Glaessner and Wade
      Conomedusites lobatus Glaessner and Wade
      Lorenzenites rarus Glaessner and Wade
      Pseudorhizostomites howchini Sprigg
      Rugoconites enigmaticus Glaessner and Wade
      Kimberia quadrata Glaessner and Wade
      Ovascutum concentricum Glaessner and Wade
      Chondroplon bilobatum Wade*
      "Cyclomedusa" sp.
      medusoid n.sp.
      medusa n.sp.
      Chondrophore n.gen., ln.sp.
      indeterminate medusoids
  B. Pennatulacean:
      Rangea longa Glaessner and Wade
      R. grandis Glaessner and Wade
      Pteridinium simplex (Gurich)
      Arborea arborea (Glaessner)
Phylum ANNELIDA
      Dickinsonia costata Sprigg
      D. elongata Glaessner and Wade
      D. tenuis Glaessner and Wade
      D. lissa Wade†
      D. brachina Wade†
      Spriggina floundersi Glaessner
      S.? ovata Glaessner and Wade
Phylum ARTHROPODA
      Praecambridium sigillum Glaessner and Wade
      Parvancorina minchami Glaessner
Phylum uncertain
      Tribrachidium heraldicum Glaessner
      Various trace fossils
      Coarse spicular impressions

(*Source:* After Wade, 1970, except where noted; *Wade, 1971; † Wade, 1972a)

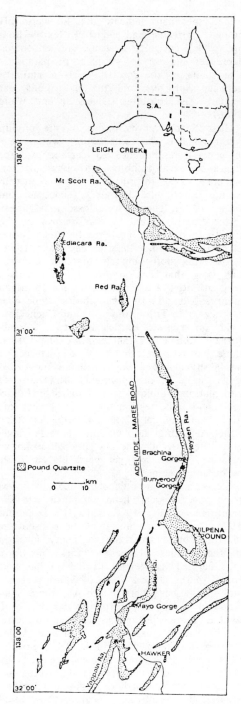

FIGURE 1. Pound Quartzite exposures near Ediacara in the Flinders Ranges (from Wade, 1972b, by permission of The Palaeontological Association).

quiescent conditions between areas of moving sand ripples. The general sedimentary environment is interpreted as littoral and shallow marine, and desiccation cracks on some of the

bedding planes indicate occasional emergence of the sandy shallows. Intermittent uplift of the area due to intrusion of contemporaneous diapirs E of Ediacara may in part explain the presence of the shoals and the deposition of fine-grained sediment without significant reworking (Wade, 1968).

The soft-bodied animals are preserved in the fine-grained sandstone as casts and molds (Wade, 1968; Glaessner and Wade, 1966), which suggests that conditions of rapid sedimentation prevailed since decomposition of such creatures occurs rapidly in oxygenated sea water (Wade, 1968). There are diverse infaunal-type trace fossils (Glaessner and Wade, 1966; Glaessner, 1971). Evidence exists of preservation in situ of some animals and of introduction of others from more or less extraneous habitats. The preferential preservation of the fauna as casts and molds is a function of the animals' resistance to decay and disintegration. Shelly fossils are absent, whereas fossils of relatively nonresistant animals (such as the medusoid coelenterates) that were partly or wholly buried in mud are preserved. When decay takes place after early diagenesis has set the sediment firm, the space once occupied by the animal is usually obliterated during late diagenesis (by compaction, for example). Only nonresistant animals are likely to decay rapidly enough to allow subsequently deposited sediment to subside, before early diagenesis, into the space they had occupied. The infilling sediment forms a cast or mold which can later be exhumed by weathering (Wade, 1968).

Trace fossils in the Pound Quartzite record the activity of various worm-like sediment and detritus feeders living in the sand or on the mud-sand interface, or grazing on the mud surface (Glaessner, 1969). Deep-water trace fossils are absent, as are suspension feeders; and none of the traces can be related with certainty to any of the known bodily preserved animals.

The phylum Coelenterata is represented by pennatulacean and medusoid (Fig. 2) members (Glaessner and Wade, 1966; Wade, 1971; Wade, 1972b). The body of the former consists of a rhachis from which spreads a frond with bilateral symmetry, hence the common name, sea pen. The rhachis may extend downward in the living position into a stalk, which in Holocene forms may end in a bulbous expansion. Fossils with a medusoid base and a pennatulacean upper part are known from Ediacara.

Both hydrozoan and scyphozoan medusae are known from Ediacara (Wade, 1972b). The floats of the circular and bilateral Precambrian hydrozoans are more strongly differentiated

FIGURE 2. *Cyclomedusa plana* Glaessner and Wade, X.85 (from Glaessner and Wade, 1966).

than those of modern forms, but the Ediacara fauna is too young to help in determining whether the circular or the bilateral float is the primitive type. Both types were solutions to the problem of increasing the volume of a single float chamber without significantly sacrificing the strength or increasing the weight.

Phylum Annelida is represented by two widely different forms: *Dickinsonia* (Fig. 3A) and *Spriggina* (Fig. 3B). *Dickinsonia* is elliptical in outline, bilaterally symmetrical, and covered with transverse ridges and grooves (Glaessner and Wade, 1966; Wade, 1972a). The animals range in length from 0.5 to 60 cm; and numerous impressions of wrinkled and

FIGURE 3. Ediacara annelids. A, *Dickinsonia costata* Sprigg, X1 (from Sprigg, 1949). B, *Spriggina floundersi* (Glaessner, X2 (from Glaessner, 1959).

folded specimens indicate that all were soft-bodied. *Spriggina* had a narrow flexible body up to 4.5 cm long; a stout, crescent-shaped carapace; and up to 40 pairs of parapodia. The shape of the head suggests a similarity between *Spriggina* and the arthropods, particularly the trilobites (Glaessner and Wade, 1966).

Most clearly representative of the phylum Arthropoda is *Praecambridium,* which bears a general resemblance to certain protaspis stages of larval trilobites, although sufficient evidence exists to indicate that the creature is not a trilobite protaspis (Glaessner and Wade, 1971). There is one fossil that appears to be closely related to *Praecambridium,* i.e. *Vendia sokolovi* from the late Precambrium of northern Russia (Glaessner and Wade, 1971). Considering both animals together confirms the view that *Praecambridium* represents a very primitive arthropod, with a soft integument showing in its segmentation some characters of both Trilobitomorpha and Chelicerata.

Of uncertain affinities are two animals: *Parvancorina* and *Tribrachidium.* The first of these has been referred to as a crustacean; it has a shield-shaped body with an anchor-like ridge. Some of the specimens show faint rib-like markings oblique to the midridge. Folded specimens indicate that their bodies were soft. The other animal, *Tribrachidium,* has three equal, radiating, hooked and tentacle-fringed arms. Because it is unlike any other known creature, it has not yet been placed in the taxonomic system.

Elements of the Ediacara fauna are known from several areas from strata of late Precambrian age (Fig. 4; Glaessner, 1971; Schopf, 1975). A similar assemblage has been discovered in the Pound Quartzite at many localities in the Flinders Ranges. A fossil provisionally defined as *Rangea* (Table 1) has been found in the Officer Basin of South Australia and in the basal part of the Arumbera Sandstone in Northern Territory, Australia. Abundant Ediacara-type fossils have been discovered in the Kuibis Quartzite of southwest Africa. *Rangea* and *Pteridinium* (Table 1), and possibly other genera, are common to both this area and the Flinders Ranges, *Rangea*-like forms also occur in England, northern Russia, and northern Siberia; and medusoids similar to those found at Ediacara have been reported from localities in England, Sweden, and the southwestern Ukraine. The form resembling *Praecambridium* is known from northern Russia, as previously mentioned.

The known age range of the Ediacara-type fauna has been determined radiometrically and lies between 600 and 700 m yr, giving a time

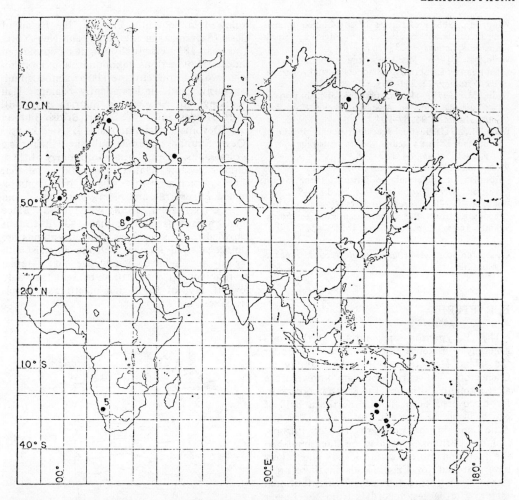

FIGURE 4. World distribution of the Ediacara-type fauna. Localities: 1, Ediacara; 2, Flinders Ranges near Ediacara; 3, Officer Basin, Punkerri Hills, South Australia; 4, Arumbera Sandstone, near Deep Well, Northern Territory; 5, Kuibis Quartzite, Southwest Africa; 6, Charnwood Forest, Leicestershire, England; 7, Tornetrask area, northern Sweden; 8, southwestern Ukraine; 9, Yarensk, northern Russia; 10, Olenek uplift, northern Siberia (from Glaessner, 1971).

span of about 100 m yr (Glaessner, 1971; Schopf, 1975). This is not deemed an anomalously long duration for Phanerozoic genera and even species, and it indicates similar rates of evolution for the late Precambrian. The development of megascopic, multicellular organisms was a relatively recent innovation, and the Ediacara fauna (and related faunas) indicate that this step took place not at the Precambrian/Cambrian boundary, but during the Proterozoic. Invertebrate biotic diversity had already reached high levels by the onset of the Cambrian. The idea of explosive evolution— the prime example of which is assumed to have taken place at the beginning of the Cambrian— is no longer an unquestioned phenomenon in light of the findings at Ediacara.

D. E. MELLOR, JR.

References

Glaessner, M. F., 1959. The oldest fossil faunas of South Australia, *Geol. Rundschau*, 47, 522-531.

Glaessner, M. F., 1969. Trace fossils from the Precambrian and basal Cambrian, *Lethaia*, 2, 369-394.

Glaessner, M. F., 1971. Geographic distribution and time-range of the Ediacara Precambrian fauna, *Geol. Soc. Amer. Bull.*, 82, 509-514.

Glaessner, M. F., and Wade, M., 1966. The late Precambrian fossils from Ediacara, South Australia, *Palaeontology*, 9, 599-628.

Glaessner, M. F., and Wade, M., 1971. *Praecambridium*—a primitive arthropod, *Lethaia*, 4, 71-78.

Schopf, J. W., 1975. Precambrian paleobiology: Problems and perspectives, *Ann. Rev. Earth Planetary Sci.*, 3, 213-249.

Sprigg, R. C., 1949. Early Cambrian(?) jellyfishes from the Flinders Ranges, South Australia, *Trans. Royal Soc. South Australia*, 71, 212-224.

Wade, M., 1968. Preservation of soft-bodied animals in Precambrian sandstones at Ediacara, South Australia, *Lethaia*, **1**, 238-267.

Wade, M., 1969. Medusae from uppermost Precambrian or Cambrian sandstones, central Australia, *Palaeontology*, **12**, 351-365.

Wade, M., 1970. The stratigraphic distribution of the Ediacara fauna in Australia, *Trans. Roy. Soc. South Australia*, **94**, 87-104.

Wade, M., 1971. Bilateral Precambrian chondrophores from the Ediacara fauna, South Australia, *Proc. Roy. Soc. Victoria*, **84**, 183-188.

Wade, M., 1972a. Dickinsonia: Polychaete worms from the late Precambrian Ediacara fauna, South Australia, *Mem. Queensland Mus.* **16**, 171-190.

Wade, M., 1972b. Hydrozoa and Scyphozoa and other medusoids from the Precambrian Ediacara fauna, South Australia, *Palaeontology*, **15**, 197-225.

Cross-references: *Annelida; Fossil Record; Precambrian Life.*

EURYPTERIDA

The Eurypterida are extinct Arthropoda generally considered an order or subclass under the class Merostomata of the subphylum Chelicerata. The eurypterids are known from all continents, although the great majority so far discovered occur in the northern half of North America extending to the Canadian Artic, and in the northern half of Europe, including Britain. This may be due simply to the fact that more paleontologists and others have searched for these fossils in these areas rather than being an actual distributional factor. The earliest record is from the Lower Ordovician of New York and the last stragglers are from the Middle Permian of Kansas. The greatest diversity in their evolution, both in regard to genera and species, and the greatest development in numbers, is found in the Upper Silurian and Lowest Devonian. However, complex, highly specialized, but still archaic forms, are known from the Upper Ordovician, which indicates that eurypterids should certainly be found in Cambrian sediments, and perhaps also in the Precambrian. After the Lower Devonian, eurypterids rapidly declined and became relatively rare.

Historical

Eurypterids were first discovered in 1818 by S. L. Mitchell in the Upper Silurian of New York, who considered the fossil to be a catfish. The prosomal appendages were mistaken for the barbels of the catfishes. DeKay in 1825 correctly recognized the arthropodal nature of the fossil and named it *Eurypterus remipes*. In the next 30 years or so, discoveries quickly

followed in Scotland, England, the Island of Oesel (Saaremaa) in the Baltic, and New York. The year 1859 essentially might be considered the start of serious taxonomic and morphologic work in the eurypterids, as almost simultaneously and independently James Hall (working on New York Silurian material), T. H. Huxley and J. W. Salter (British Silurian and Devonian material) and J. Nieszkowski (Oesel Silurian material) published their basic monographs. The external morphology of these animals became relatively well known at that time, particularly that of the two common genera *Eurypterus* and *Pterygotus* (Figs. 1 and 2; see, e.g., Holm, 1898; Wills, 1965). Since then, numerous authors have added to our knowledge. The result is that eurypterids are almost as well known as living animals with regard to detailed knowledge of external morphology. It may safely be said that only few, if any, of the higher invertebrate fossils are as well known.

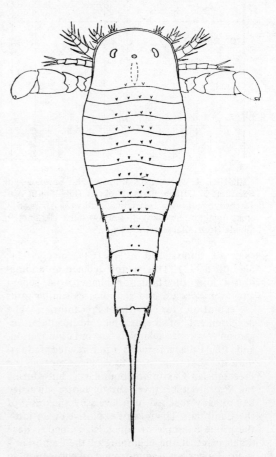

FIGURE 1. *Eurypterus pittsfordensis* Sarle (Eurypteridae) from the Upper Silurian of New York, ca. X.25 (from Kjellesvig-Waering, 1958a).

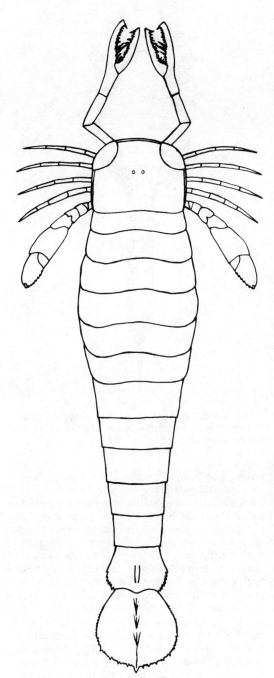

swimming, balancing, or covering themselves with the bottom sediments (Fig. 3). The other five consist of a pair of preoral chelicerae and four pairs of "walking legs." The small, simple chelicerae are composed of four joints; and, in the Pterygotidae, these structures are greatly elongated, enlarged, and armed with an inner row of formidable teeth, though still consisting of four joints. The prosoma includes a dorsal shield which contained the compound lateral eyes and nearly always had two small median ocelli developed, generally centrally located, on the highest part of the shield. The lateral or compound eyes were located anteriorly marginally or centrally, depending on the mode of living, whereas the median ocelli generally occured either between the eyes or slightly in back of them. A ventral marginal shield surrounds the prosoma which may consist of a single plate or is divided by sutures into several plates. The mouth is located centrally where the coxae of the "walking" and "swimming" legs meet. The inner edge of each coxae, in nearly all eurypterids, is armed with a gnathobase of large cutting and masticating teeth. Directly posterior to the mouth is the metastoma, a large plate of many different shapes depending on the family, and perhaps homologous to the sternum in the scorpions.

Following the prosoma is an opisthosoma comprising always twelve tergites and a telson. The anterior six tergites are considered as the mesosoma, whereas the others and the telson are included in the metasoma. On the underside of the mesosoma are five pairs of abdominal plates (miscalled sternites), which cover the rather elliptically shaped gills. The abdominal plates are joined by sutures. The genital organs occur in the first pair, which is known as the operculum. Sexual dimorphism is clear in eurypterids, but sex determination is still in disagreement. The long type of mesial appendage which may be composed of one to three joints is considered the type A appendage (Fig. 4), while the other is much shorter, commonly consisting of lesser joints, and is known as type B (Størmer, 1934, 1973, 1974). According to Størmer and others, type A is

FIGURE 2. *Pterygotus* (*Acutiramus*) *macrophthalmus cummingsi* (Grote and Pitt) (Pterygotidae) from the Upper Silurian of western New York. Individuals of this species reached a total length of 2.5 m.

Morphology

The eurypterids consist of a prosoma with six pairs of legs; in some, the posterior legs are developed into broad paddles which served in

FIGURE 3. The metastoma and complete swimming leg of *Eurypterus pittsfordensis* Sarle (Pterygotidae), X1 (from Kjellesvig-Waering, 1958a).

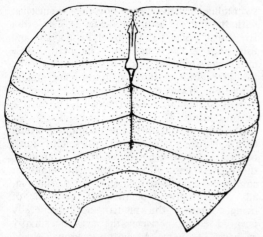

FIGURE 4. Underside of preabdomen of *Carcinosoma newlini* (Claypole) (Carcinosomatidae) showing male? (Type A) genital appendage; from the Upper Silurian of Indiana (from Kjellesvig-Waering, 1958b, pp. 295–303).

the male and type B the female, while the work of Wills (1965) indicates the opposite. The question still needs to be resolved and, in particular, we need to know more concerning these structures in many other eurypterids, particularly the stylonurids. In some eurypterids, the second and third pair of abdominal plates have median organs, although that of the third abdominal plates is very rudimentary.

Contraction of the opisthosoma, when present, occurs nearly always at the eighth tergite, from where the metasoma, or postabdomen, narrows into a rather flattened or tubular "tail." A telson is always present, which may consist of a long stiletto (e.g., *Carcinosoma*, Fig. 5), short spike or widen into various shapes such as the broad telson of the pterygotids, some of which become nearly fan shaped. In some eurypterids, such as the Megalograptidae, the telson is bounded on each side by wide pincer-like cercal blades, very much in appearance and perhaps in function (except much wider) to the terminal forceps of the Dermaptera ("earwigs"). In *Paracarcinosoma* and *Mixopterus*, the postabdomen is tubular, capable of overhead and upward thrusting movements as in the scorpions and the telson is shaped as a curved stinger, again similar to the scorpions. The presence of a pair of poison glands in this type of telson is highly questionable.

The dermal covering of the eurypterids is chitinous; and, in some, little or no ornamentation is discernible. In most, however, the exoskeleton is covered with crescentic scales, linear scales, pustules, mucrones, large wart-like

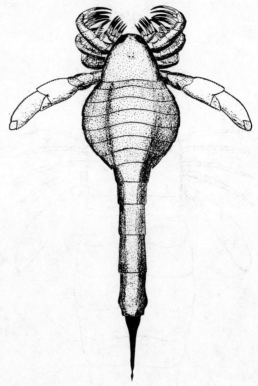

FIGURE 5. *Carcinosoma newlini* (Claypole) (Carcinosomotidae), Silurian of Indiana, ca. X0.11. (from Kjellesvig-Waering, 1958b, pp. 295–303).

bosses, and many other types of ornamentation, such as reticulated, star-like and linear patterns. Sensory hairs are common in many eurypterids; some are quite hirsute. Color pattern of the eurypterids is known in many specimens which have been preserved in an impervious matrix. Two general types are known. Most are various shades of shiny brown, ranging from a nearly buff to honey colored skin with the scales or pustules of a darker brown. The teeth of the chelicerae in the pterygotids and the teeth on the gnathobases are generally black. Occasionally some areas are concentrated into a darker brown color as in *Baltoeurypterus tetragonophthalmus* (Fischer) where the area in the front and center part of the prosoma is much darker. In this case, the dark form denotes the female (type B). Another eurypterid has been found (*Kokomopterus longicaudatus* Clarke and Ruedemann) with a median, dark brown band running throughout the length of the opisthosoma. In short, the overall color of these eurypterids is quite similar to that of the younger stages of the living *Limulus polyphemus*. The second group, which includes the superfamily Mixopteracea, is quite different. Here the color is dark brown with the scales contrastingly black. The legs, particularly at the extremities are black, and all extremities, such as the telson and cercal blades, are black. This gives a much darker overall appearance—more like the darker scorpions of today.

Little is known of the internal structure of these fossils but the intestinal tract has been preserved in at least a few specimens. This is similar to the scorpions, with the anal opening between the telson and the twelfth tergite (see Wills, 1965).

The eurypterids range in size from very small (*Hughmilleria, Parahughmilleria, Eurypterus*) forms averaging less than 6 in. (15 cm) to median-sized forms 1 m in length (e.g., *Megalograptus* [see Fig. 6], *Mixopterus, Tarsoptella*) to gigantic forms that surpassed 2 m in length (e.g., *Pterygotus, Hallipterus*). Some of the Pterygotidae, with their chelicerae outstretched would measure in excess of 2.5 m in length. Some tergites have been found with an unbroken width of over 4 cm. They are the largest Arthropoda known.

Habitats

The habitat of the eurypterids has been controversial for a considerable time. This has been due to the unfortunate assumption that all eurypterids necessarily had to inhabit

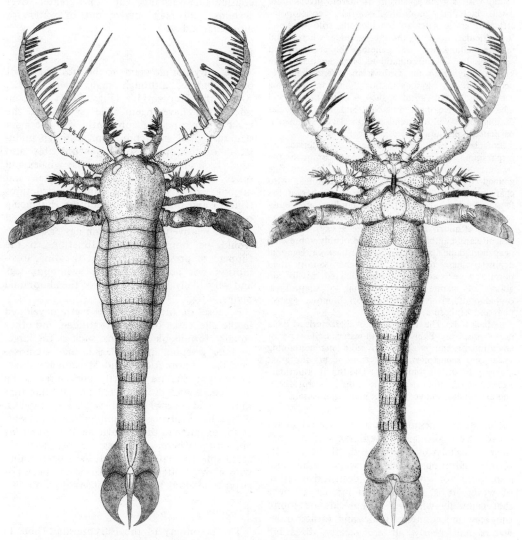

FIGURE 6. Dorsal (right) and ventral (left) views of *Megalograptus ohioensis* Caster and Kjellesvig-Waering, a female (?) (Type B), from the Upper Ordovician of Ohio, ca. X0.167 (from Caster and Kjellesvig-Waering, 1964).

a single ecological zone. They were, therefore, first considered to be marine, later (1916) through the influence of A. W. Grabau and Marjorie O'Connell they were considered to be entirely river dwellers. It appears, however, that all eurypterids were primordially and essentially marine, but that they occupied three main ecological niches or biofacies at least during their period of greatest diversity and development (Kjellesvig-Waering, 1961, p. 794; see also Andrews et al., 1974). These three zones are described below.

Biofacies 1: The first zone is characterized by the families Pterygotidae, Carcinosomatidae, and Megalograptidae. Sediments will range from clastic to dolomites. In its typical development, these eurypterids are found with a wide spectrum of marine invertebrate fossils including graptolites, starfish, cephalopods, articulate and inarticulate brachiopods, trilobites, etc. In particular, however, the Pterygotidae, when found in great abundance, will generally be accompanied only by the Carcinosomatidae. Megalograptidae are known only from the Ordovician where they also occur with the carcinosomatids. Some venturesome genera of the Pterygotidae, such as *Erettopterus,* will inhabit all three main biofacies, and certain species of this genus will be found only in clastic sediments, particularly in Biofacies 3.

Biofacies 2: The second zone is characterized by magnesium carbonate muds, which comprise the "waterlimes" of the literature, but which are fine grained dolomites, or dolomitic limestones. Also characteristic of this zone is a great number of the family Eurypteridae; but other families are represented, such as the Dolichopteridae and the Pterygotidae, although these others are always found in insignificant numbers. This zone probably represents near-shore deposits of epieric seas or backreef lagoonal deposits where tolerances for higher salinity than normal were essential. The associated fossils are always of marine origin consisting of graptolites, cephalopods, bivalves, linguloid brachiopods, gastropods and Xiphosura.

Biofacies 3: The third zone is characterized by a near-shore, very likely brackish-water, delta-type of environment and is always in sediments containing shales and sandstones. In this zone occurs the great majority of the Hughmilleriidae and the Stylonuridae. Few, if any, other fossils are found except fishes and ostracodes, and very occasionally xiphosurans.

The reason eurypterids are not found in so-called "typical marine" sediments is because their exoskeletons, composed of very soft, pliable, chitinous material, were edible and also could not withstand the destructive action of waves. It is, therefore, of interest to note that generally wherever eurypterids are found they are in sediments indicating rather quiet waters and nearly all are clearly discarded instars. However, in the rare cases where a local fauna has been overwhelmed by a catastrophe such as a volcanic ash fall as happened with the Upper Ordovician eurypterids of Ohio, it was found that the numerous specimens occurred with a typical marine fauna and apparently situated at a location several hundred miles from any shore (Caster and Kjellesvig-Waering, 1964, p. 304).

There also seems to be little doubt that some of the later eurypterids, such as those occurring in the upper Paleozoic, made temporary incursions into coal-swamp settings. These eurypterids, however, appear to belong to the third biofacies. The habits of the modern Order Xiphosurida probably best illustrate the habitat of the eurypterids, as they are marine, inhabiting relatively deep benthic depths, but also venture into estuaries and occasionally a considerable distance up the greater river systems, and may survive out of water for more than a day.

Biology

The eurypterids seem to be, for the most part, benthic, although many were nektic. All of the Pterygotidae (Figs. 2, 3) are considered to have been agile swimmers; the great telsonic paddle apparently was the main structure used in locomotion, although undulation of the opisthosoma was probably also used. It is interesting to note that whales and other mammals, such as the manatee and dugong, have somewhat similar structures (flukes) and method of locomotion. Balancing of the body apparently was accomplished by the relatively small swimming legs, which would be homologous in function to the flippers of present-day whales. In some, locomotion was achieved by the swimming legs and very likely by movement of the abdominal plates.

Because of the formidable teeth developed in the chelicerae of the Pterygotidae, and other equally formidable weapons, such as the large, spinous, grasping anterior legs in the Carcinosomatidae, Mixopteridae, and Megalopgraptidae, the eurypterids are correctly considered to be predaceous and probably fed on anything that they could overwhelm as is true of modern scorpions. Evidence of cannibalism is present, as for example in the Ordovician *Megalograptus ohioensis,* where coprolites, almost certainly ascribable to that eurypterid, have been found with chewed integument of the same species of eurypterid along with some pieces of trilobites.

Taxonomy

The taxonomy of the Eurypterida (Table 1) includes at the present time (1974) 4 suborders, 8 superfamilies, 18 families, 62 genera, and

TABLE 1. Classification of the Eurypterida–
Subphylum Chelicerata, Class Merostomata

Order Eurypterida, Burmeister, 1843
Suborder Eurypterina, Burmeister, 1843
  Superfamily Mixopteracea, Caster and Kjellesvig-Waering, 1955
    Family Megalograptidae, Caster and Kjellesvig-Waering, 1955
    Family Mixopteridae, Caster and Kjellesvig-Waering, 1955
    Family Carcinosomatidae, Størmer, 1934
  Superfamily Eurypteracea, Burmeister, 1845
    Family Hughmilleriidae, Kjellesvig-Waering, 1951
    Family Eurypteridae, Burmeister, 1845
  Superfamily Slimonoidea, Novojilov, 1962
    Family Slimoniidae, Novojilov, 1962
  Superfamily Stylonuracea, Diener, 1924
    Family Dolichopteridae, Kjellesvig-Waering and Størmer, 1952
    Family Stylonuridae, Diener, 1924
    Family Pageidae, Kjellesvig-Waering, 1966
    Family Drepanopteridae, Kjellesvig-Waering, 1966
    Family Kokomopteridae, Kjellesvig-Waering, 1966
    Family Rhenopteridae, Størmer, 1951
    Family Laurieipteridae, Kjellesvig-Waering, 1966
Suborder Woodwardopterina nov*
  Superfamily Mycteropoidea, Cope, 1886
    Family Mycteropidae, Cope, 1966
  Superfamily Woodwardopteracea, Kjellesvig-Waering, 1959
    Family Woodwardopteridae, Kjellesvig-Waering, 1959
Suborder Hibbertopterina, Kjellesvig-Waering, 1959
  Superfamily Hibbertopteracea, Kjellesvig-Waering, 1959
    Family Hibbertopteridae, Kjellesvig-Waering, 1959
Suborder Pterygotida, Caster and Kjellesvig-Waering, 1964
  Superfamily Pterygotacea, Clarke and Ruedemann, 1912
    Family Pterygotidae, Clarke and Ruedemann, 1912
    Family Jackelopteridae, Størmer, 1974

*Anterior tergites greatly expanded.

288 species and subspecies. Obviously the great diversity exhibited by the families and genera already described indicates that only a small fraction of the eurypterids have been discovered. The small number of species in relation to the genera further indicates that we may well expect considerably more species to be described. Undoubtedly, more genera will be established, particularly when the underside and prosomal appendages of some of the described species are better known. Also, there is little doubt that many of the existing genera have little or no affinities to the families to which they are at present delegated. It is to be expected that spectacular discoveries may be expected in the Eurypterida during the coming years. As evidenced by the great and unusual diversity of the order, it is highly probable that the Eurypterida is an order of greater magnitude than previously assumed by most paleontologists. Some of these discoveries have already been made and studies are underway in North America and Europe.

Future Research

Future research will likely be directed to the underside of these fossils as structures such as the metastoma, ventral prosomal shield, and the mesial appendage as well as the prosomal appendages are of primary importance to classification. Indeed, in some genera such as

Adelophthalmus, Lepidoderma, and *Anthraconectes,* it is essential to excavate, or develop by other techniques (Wills, 1965), the underside of the known specimens as it can be easily demonstrated that these eurypterids, as well as many others including the common genus *Eurypterus,* cannot always be determined from the dorsal side.

The eurypterids are not sufficiently well known or common enough to make good index fossils except in a few areas. Where they are well known, however, e.g., in the Bertie formation of New York, the individual members of the formation can be determined by the eurypterids, although the dolomite members are separated by only a few feet of intervening shale. Nevertheless, the eurypterids of each bed (Williamsville and Fiddlers Green members) are entirely different although the apparent ecology is the same. It is, therefore, suspected, at least in this section, that differentiation or evolution among eurypterids proceeded very rapidly, and it would be surprising not to find the same condition in other areas. Age determination in some unusual outpost oil wells have been accomplished only by the type of eurypterid present. The eurypterids, therefore, assume stratigraphic importance because they normally ʼoccur in the more "barren" parts of the stratigraphic column. Someday, when better known, they will be of greater "index" value.

ERIK N. KJELLESVIG-WAERING

References

Andrews, H. E.; Brower, J.; Gould, S. J.; and Reyment, R., 1974. Growth and variation in *Eurypterus remipes* Dekay, *Bull. Geol. Inst. Univ. Upsala, n.s.,* 4(6), 81–114.
Caster, K. E., and Kjellesvig-Waering, E. N., 1964. Upper Ordovician eurypterids of Ohio, *Palaeontographica Americana,* 4(32), 301–358.
Clarke, J. M., and Ruedemann, R., 1912. The eurypterids of New York, *N.Y. State Mus. Mem. 14,* 439p.
Holm, G., 1898. Über die Organisation von *Eurypterus fischeri* Eichw., *Acad. Sci. St. Petersbourg, Mem.,* ser. 8, 8(2), 1–57.
Kjellesvig-Waering, E. N., 1958a. The genera, species, and subspecies of the family Eurypteridae, Burmeister, 1845, *J. Paleontology, 32,* 1107–1148.
Kjellesvig-Waering, E. N., 1958b. Some previously unknown morphological structures of *Carcinosoma newlini* (Claypole). *J. Paleontology, 32,* 295–302.
Kjellesvig-Waering, E. N., 1961. The Silurian eurypterids of the Welsh Borderland, *J. Paleontology, 35,* 789–835.
Kjellesvig-Waering, E. N., 1966. A revision of the families and genera of the Stylonuracea (Eurypterida), *Fieldiana, Geol., 14*(9), 169–197.
Størmer, L., 1934. Merostomata from the Downtonian Sandstone of Ringerike, Norway, *Skr. Norske*

Vidensk aps-Akad., Oslo, Mat.-Nat. Kl., **1933**(10), 125p.

Størmer, L., 1955. Eurypterida, in R. C. Moore, ed., *Treatise on Invertebrate Paleontology*, pt. P, Arthropoda 2, Chelicerata. Lawrence, Kansas: Geol. Soc. Amer. and Univ. Kansas Press, P23–P41.

Størmer, L., 1973, 1974. Arthropods from the Lower Devonian (Lower Esmian of Alken an der Mosel, Germany, pts. 3, 4. *Senckenbergiana Lethaea*, **54**, 119–205, 359–451.

Wills, L., 1965. A supplement to Gerhard Holm's "Über die Organisation des *Eurypterus fischeri* Eichw." with special reference to the organs of sight, respiration, and reproduction, *Ark. Zool. ogi, Kungl. Svensk. Akad.*, ser. 2, 18(8), 93–145.

Cross-reference: *Arthropoda.*

EVOLUTION

Organic evolution is easy to understand in a general sense, but difficult to define with rigor. It is commonly defined as a change in gene frequency from generation to generation within a population of plants or animals. The difficulty here is that because of chance factors such change is almost inevitable, and if the change oscillates about a mean that undergoes no net shift in the long term, it is inconsequential. It seems more fruitful to define evolution as a change in gene frequency that persists for several generations. The change may occur in a single transition from one generation to another or may be spread over two or more generations.

Natural Selection

Natural selection, though recognized as the dominant source of evolutionary change, is not the only source and has no place in the definition of evolution. Interestingly, the word "evolution" also does not appear in Darwin's *Origin of Species,* which focuses on natural selection. In the pre-Darwinian era of biology, evolution, which literally means unfolding, referred to the emergence of the adult creature from its allegedly preformed condition in the gamete. The word "evolved" appears in its modern meaning only as the very last word of Darwin's book. With the widespread acceptance of Darwin's theory, the word "evolution" was generally adopted and applied in much the way that we use it today.

Evolution by natural selection has two basic aspects, one of which is largely random and the other of which is highly nonrandom. The largely random aspect is *mutation*, the ultimate source of variation upon which natural selection operates (see *Genetics*). Mutations are of two types, point mutations and chromosomal mutations. *Point mutations* are changes in segments of strands of DNA. They may be induced by such agents as cosmic radiation, x-radiation, ultraviolet radiation, or chemical mutagens. *Chromosomal mutations* arise from accidents in the process by which chromosomes are duplicated in the production of gametes; they are alterations in the chromosomal arrangement of genes. While point mutation produces the most basic sort of genetic change, reshuffling of chromosomes by sexual reproduction (fusion of gametes) accounts for most of the genetic variability that arises within populations. Recognition of the importance of genetic variability was fundamental to Darwin's theory; but because Mendel's seminal work of 1866 lay buried in obscurity for decades, modern genetics did not arise until after Darwin's death and its absence was a major source of disillusionment to him late in life. Without the knowledge that genetic material was transmitted in discrete units, it was difficult to understand why useful new adaptive traits should not be blended out of existence by averaging with less useful traits.

The main nonrandom aspect of evolution is natural selection, which operates upon genetic variability as expressed among individuals within species (Fig. 1). Two kinds of individuals are favored by natural selection, those that tend to survive through reproductive age and those that produce large numbers of progeny per unit time. Thus natural selection is not simply a matter of differential survival (survival of the fittest), it is a matter of differential reproduction. One of the best ways to understand how evolution occurs is to consider the conditions under which it does not occur. The absence of evolution over a series of generations represents what is called the *Hardy-Weinberg equilibrium*. The conditions for equilibrium are as follows: breeding must be random; there must be no change of gene frequencies as a result of genetic exchange with other populations; there must be no natural selection; and there must be no chance changes in gene frequencies. Given these conditions, gene frequencies will remain fixed at values that follow a binomial distribution. For example, if there are two alleles (alternative types of gene) at a given locus (location on a chromosome), then each individual in the population, inheriting an allele from each parent, may have two alleles of one kind or one of each. If the frequencies of the two alleles are p and q, the set of genotypes will follow the distribution $(p + q)^2$. The stipulation that no chance changes in gene frequency will occur is unrealistic, and yet it is generally agreed that genetic drift, as random change is commonly termed, is likely to be of importance only in small populations. The chief reason that the

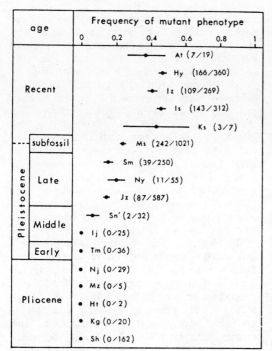

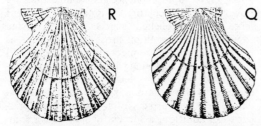

FIGURE 1. An example of the results of natural selection of the type commonly termed disruptive selection (from Hyami and Ozawa, 1975, by permission of the publisher, Universitetsforlaget). "Phenotype" refers to the morphologic character of the scallop shell. The R phenotype was the only form of the Japanese scallop *Cryptopecten vesiculosis* in the Pliocene and early Pleistocene; but in the middle Pleistocene, the Q phenotype appeared and gradually increased in relative abundance until it made up about half of typical populations that are alive today. Each bar indicates standard error of frequency. Number of individuals belonging to phenotype R and number of total individuals are in parentheses. The reason for the appearance of the dimorphism is unknown.

Hardy-Weinberg equilibrium fails to hold in nature is that natural selection is nearly omnipresent. Two factors are of primary importance in governing rates of evolution by natural selection. One is the intensity of selection pressure (the degree to which certain kinds of individuals are favored). The other is the degree of genetic variability upon which selection can operate. According to R. A. Fisher's Funda-

mental Theorem of Natural Selection, the greater the variability, the more rapid will be the change.

Of the two facets of natural selection, only differential survival among individuals tends automatically to improve adaptation within a species. Differential reproductive rate, in contrast, can produce evolution that is of no value to the species as a whole. Darwin recognized this in discussing what he termed sexual selection. As an example, female animals may favor large or especially ornate males as mates. The favored condition will then become more and more pronounced, but the evolutionary change will not necessarily offer any advantage for the species in gaining food, avoiding predators, competing with other species, or otherwise succeeding in the history of life. Nonetheless, sexual selection has led to the striking elaboration of a wide variety of morphological features in the animal world.

Units of Selection

It has been argued by some workers that selection commonly favors traits of groups of individuals rather than traits of individuals themselves. Under such circumstances, group attributes must be of overriding importance. Various objections have been directed at the concept of group selection at the population level. Among them is the claim that most traits that would seem possibly to be of group benefit can also be envisioned to be of individual benefit. The general importance of selection at the population level remains a subject of considerable debate. As will be discussed below, however, selection at the species level seems to be of great significance.

A species is generally defined as a group of interbreeding individuals that in nature is reproductively isolated from other such groups. New species arise by splitting off from preexisting species. It is commonly believed that most species arise from small isolated populations at the margins of geographic ranges of parent species. Evolution can be very rapid in such small populations, and although most undoubtedly go extinct quite rapidly in their hostile, marginal habitats, a small percentage persist by evolving into new species. Other processes of speciation can also be envisioned, but it is important to understand that speciation has a strong random aspect in that we cannot predict when, in what subpopulation, or in what subhabitat it will occur. All of these uncertain conditions have bearing on the direction taken by speciation.

The fossil record provides various kinds of evidence that most evolutionary change is con-

EVOLUTION

centrated in speciation events (Eldredge and Gould, 1972; Gould and Eldredge, 1977), Most species, once established, survive for remarkably long intervals of geologic time and change only small amounts before going extinct (Fig. 2). An average species of fossil mammals, for example, has survived for more than a million years, and perhaps closer to two million. This great duration makes it virtually impossible that gradual change within established species could have produced the great variety of mammalian orders that arose during only about 12 m yr in the early Cenozoic. It is highly unlikely that rates of evolution would somehow have been accelerated in an enormous variety of species occupying unrelated niches to the rates that would have been required. On the other hand, speciation was obviously rampant among primitive mammal groups of the early Cenozoic, as extinct reptilian groups were being replaced. Concentration of evolutionary change in speciation must account for the rapid appearance of a

variety of orders, including ones that contained such remarkable forms as bats and whales. Species of many classes of marine invertebrates last even longer than species of mammals. For many groups, average duration is in the order of 10 m yr. "Living fossil" groups offer additional evidence that evolutionary change occurs primarily in speciation. These are groups, like the coelecanths, that have survived for long intervals of geologic time with little morphological change. The important point is that during these periods of evolutionary stagnation, all groups of living fossils have been represented by very few species. They have persisted in small numbers without appreciable speciation, and it is apparently because of this lack of speciation that they have not changed.

The idea that most evolution is concentrated in speciation events would have been foreign to Darwin, who had little idea of how new species arise. Until the development of modern ideas of speciation, especially through the work of Ernst Mayr, it was generally believed that rate of evolution was unrelated to speciation. Most established species were believed to be undergoing more rapid evolutionary change than we now have evidence for. What tends to prevent large, well-established species from evolving rapdily? For one thing most species occur as arrays of loosely connected populations and these occupy differing habitats and are subjected to differing selection pressures. Moderate gene flow among populations tends to cancel out the effects of local selection pressures. In addition, a successful species is a complex homeostatic entity. Most of its adaptations are controlled by more then one gene, and most of the genes affect more than one adaptation. It is difficult to alter such an entity significantly without fatal disruption. What seems to occur in certain speciation events is what Mayr has termed a "genetic revolution".

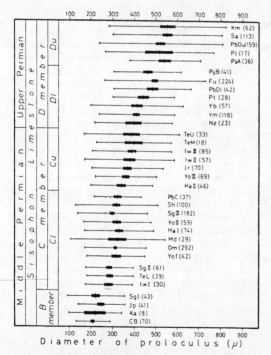

FIGURE 2. Gradual evolution within a Permian species of foraminifer (*Tepidolina multiseptata*) from eastern Asia (from Hayami and Ozawa, 1975). The proloculus, or first-formed chamber of the foraminiferal test, increased in diameter. The mean value, its range with 95% confidence, and two standard deviations are indicated. Sample size in parentheses. The slow rate of change is typical of evolution within established species. Here the species is judged not to have evolved enough to become transformed into a new species over a time span of about 15 m yr.

Evolutionary Trends

The observation that evolution tends to be concentrated in speciation has important implications for the nature of evolutionary trends. A trend is a long-term evolutionary change in a given direction. It is true that new species arise by natural selection, but we cannot predict the set of species that arise in any taxon during any interval of time. The random aspect of speciation, where most change occurs, tends to decouple large-scale evolution from small-scale evolution of the sort that occurs within established species. Large-scale trends must represent net shifts in the position of origin of new species, which themselves change little before going extinct. The process largely responsible for governing such trends is analogous to natural selection but operates at a higher level of biological organization—at the level of isolated populations and species rather than at the level of the individual (Stanley, 1975; Gould and Eldredge, 1977). By analogy with natural selection, two kinds of populations or species are favored: ones that tend to survive for long periods of time and ones that produce daughter species at high rates (Fig. 3).

To illustrate the concept of a trend, we can consider the evolution of the horse. The modern horse differs in a number of ways from *"Eohippus"* (*Hyracotherium*) of the Eocene. Among the changes that have occurred, there have been a great increase in body size, a complication of the morphology of molars, and a reduction in the number of toes, leaving a single hoof. Decades ago certain workers regarded such changes as examples of orthogenesis, or straight-line evolution. In its extreme form, the concept of orthogenesis implied that some mysterious driving force governed the direction of evolution, often leading to extinction. Even the more restrained view, that natural selection guides evolutionary trends along straight lines, has been disproven, however. Even before it was appreciated that most change is concentrated in speciation, making gradual long-term trends unlikely, G. G. Simpson showed that no single feature of horse evolution changed continuously from the Eocene to the Holocene. For example, body size, which underwent an overall, net increase, did not increase appreciably at first, and certain lines of descent exhibit reduction in size.

Evolutionary Patterns

The large-scale view of evolution provided by the fossil record offers evidence not only of crude trends in the history of many groups, but of other patterns of change as well. Rapid diversification leading to the appearance of varied new forms within any taxon is termed *adaptive radiation.* As discussed above for the Mammalia, adaptive radiation occurs by means of high rates of speciation. It may follow the appearance of an adaptive breakthrough (the appearance of a new key morphological or physiological feature) that opens the way for invasion of a set of unoccupied ecological niches. It may also follow the evacuation of a set of niches by local or worldwide extinction, as happened in the adaptive radiation of early mammals into niches vacated by Mesozoic reptiles. *Convergence* is another kind of evolutionary pattern. Here, unrelated groups evolve similar adaptations in becoming fitted to similar modes of life. *Parallelism* is the term used to

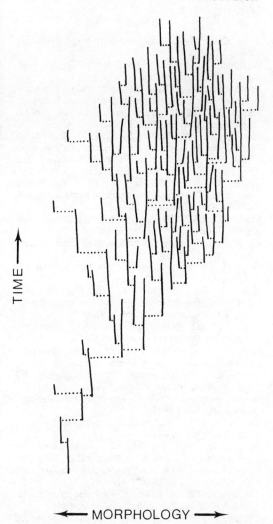

TIME →

← MORPHOLOGY →

FIGURE 3. Diagram showing a net phylogenetic trend in evolution. The nearly vertical bars represent species, which undergo minor amounts of evolutionary change. Most evolution occurs in speciation events (horizontal dotted lines), although some of these produce almost no change. Most of the overall change in average morphology is produced by species selection: species toward the right side of the diagram last longer than species toward the left side and also tend to give rise to more daughter species per unit time.

describe patterns in which groups having certain common features reflecting a common ancestry evolve independently in the same general direction. It is in elucidating patterns like long-term net trends, adaptive radiation, convergence, and parallelism that the fossil record provides documentation of the large-scale evolution of life.

STEVEN M. STANLEY

References

Avers, C. J., 1974. *Evolution,* New York: Harper & Row, 321p.

Darwin, C. A., 1859. *On the Origin of Species.* London: John Murray, 490p.

Dobzhansky, T., 1970. *Genetics of the Evolutionary Process.* New York: Columbia Univ. Press, 505p.

Ehrlich, P. R., 1974. *The Process of Evolution,* 2nd ed. New York: McGraw-Hill, 378p.

Eldredge, N., and Gould, S. J., 1972. Punctuated equilibria: An alternative to phyletic gradualism, in T. J. M. Schopf, ed., *Models in Paleobiology.* San Francisco: Freeman, Cooper, 82–115.

Gould, S. J., and Eldredge, N., 1977. Punctuated equilibria: The tempo and mode of evolution reconsidered. *Paleobiology,* **3,** 115–151.

Grant, V., 1963. *The Origin of Adaptations.* New York: Columbia Univ. Press, 606p.

Grant, V., 1971. *Plant Speciation.* New York: Columbia Univ. Press, 435p.

Hyami, I., and Ozawa, T., 1975. Evolutionary models of lineage-zones, *Lethaia,* 8, 1–14.

Jepsen, G. L.; Simpson, G. G.; and Mayr, E., eds., 1949. *Genetics, Paleontology, and Evolution.* Princeton, N.J.: Princeton Univ. Press, 474p.

Mayr, E., 1963. *Animal Species and Evolution.* Cambridge, Mass.: Harvard Univ. Press, 797p.

Mayr, E., 1970. *Populations, Species and Evolution.* Cambridge, Mass.: Harvard Univ. Press, 453p.

Moody, P. A., 1950. *Introduction to Evolution,* 3rd ed. New York: Harper & Row, 527p.

Olson, Everett C., 1971. *Vertebrate Paleozoology.* New York: Wiley, 839p.

Rensch, B., 1959. *Evolution Above the Species Level.* New York: Wiley, 419p.

Ross, H. H., 1962. *A Synthesis of Evolutionary Theory.* Englewood Cliffs, N.J.: Prentice-Hall, 387p.

Simpson, G. G., 1949. *The Meaning of Evolution.* New Haven, Conn.: Yale Univ. Press, 364p.

Simpson, G. G., 1953. *The Major Features of Evolution.* New York: Columbia Univ. Press, 434p.

Stanley, S. M., 1975. A theory of evolution above the species level. *Proc. Nat. Acad. Sci.,* **72,** 646–650.

Stebbins, G. L., 1971. *Processes of Organic Evolution.* Englewood Cliffs, N.J.: Prentice-Hall, 193p.

Tax, S., ed., 1960. *Evolution After Darwin.* Chicago, Ill.: Univ. Chicago Press, 3 vols.

Williams, G. C., 1966. *Adaptation and Natural Selection.* Princeton, N.J.: Princeton Univ. Press, 307p.

Cross-references: *Biometrics in Paleontology; Burgess Shale; Computer Applications in Paleontology; Genetics; Diversity; Ediacara Fauna; Extinction; Fossil Record; Living Fossil; Morphology, Constructional; Morphology, Functional; Plate Tectonics and the Biosphere; Species Concept; Systematic Philosophies.*

EXTINCTION

The history of life is punctuated with extinction. The fossil record, containing a 600 m yr record of diverse and abundant life, indicates that most of the species that have ever lived are now extinct. Viewed in this context, extinction is the commonplace event while survival is unusual. Extinction—the death of a species—is a phenomenon that has fascinated evolutionary biologists and paleontologists since fossils were first identified as the remains of once-living organisms. The key questions posed by fossil extinctions are these: What genetic, physiological, or ecological factors make a species susceptible to extinction? What environmental changes are responsible for the major biotic crises of the past 600 m yr?

Recognition and Measurement

It is important to distinguish between two different *types* and two different *magnitudes* of extinction. The two types of extinction are termed *phyletic* extinction and *terminal* extinction. A phyletic extinction occurs when one species evolves into another species during some amount of geologic time. In a phyletic sense, the ancestral species is extinct, but the species lineage continues and the phyletic extinction did not result in a net loss of species. Such extinctions have also been called *pseudoextinctions.* A terminal extinction occurs when the entire species population dies out without leaving any descendant species. The species lineage terminates and there is a net loss of species. Although little work has been done to assess the relative importance of the two types of extinction, most paleontologists consider terminal extinctions to be the most common. The term extinction is frequently used without qualification; so used, it usually implies an extinction of the terminal type.

The two magnitudes of extinction which can be distinguished are *background* extinctions and *major,* or *mass* extinctions. The fossil record shows that extinctions usually occur at a relatively low rate. This normal, low level of continuous extinction can be called background extinction. The fossil record also documents the fact that some intervals of time are characterized by episodes of greatly increased rates of extinction. These biotic crises are commonly termed major or mass extinctions. The term mass extinction is an unfortunate one and one that is often badly used in the popular literature. Mass extinctions are not instantaneous events but can take place during intervals of five m yr or more. This is hardly an instant when viewed from the human perspective but is a relatively brief interval in the history of life. Furthermore, as Butler et al. (1977) have shown for the major extinctions at the end of the Cretaceous, the demise of many marine groups preceded the extinction of the terrestrial dinosaurs by about one m yr.

Extinctions are usually measured by tabulating the number or percentage of the last occurrences of organisms during some time interval. The taxonomic level that is usually chosen for analyses of extinctions is the family. Species and generic stratigraphic distributions are not well enough known in all groups to permit a comprehensive analysis at these lower taxonomic levels. The time intervals that are examined may be periods (e.g., Devonian), epochs (e.g., Late Devonian), or ages (e.g., Frasnian).

Biasing Factors

Many aspects of stratigraphy, taxonomy, and the nature of the rock record can obscure the actual nature of an extinction episode. It is important to discriminate between rates and levels of extinction because the durations of the geologic time scale are not equal. The Late Cambrian, for example, is estimated to be up to 15 m yr in length, while the duration of the Late Devonian is only 6 m yr. If, for example, there were an equal number of extinctions in each of these epochs, the extinction *rate* would be much higher during the Late Devonian. The taxonomic level chosen for analysis will also influence the resulting picture of an extinction episode. Levels of species extinction are much more volatile than those of family extinctions. The higher the taxonomic level, the less is the apparent fluctuation in the extinction rate. This merely reflects the hierarchical nature of taxonomic categories.

Raup (1972) has outlined many of the biases of the rock record that can conspire to make accurate measurement of an extinction event difficult. Gaps in the stratigraphic record, poor preservation, poor rock exposure, and poor sampling may all prevent accurate estimation of the time and magnitude of the extinction (see *Fossil Record*). In addition to these factors, some species may survive longer in some regions than in others. Marsupial mammals, for example, are now largely limited to the island continent of Australia, whereas they were once more widespread. All of these biasing factors (gaps in the rock record, poor preservation, geographic distribution, etc.) will tend to present the extinction event at a time somewhat before it actually happened.

The Record of Extinction

Graphs that show the varying magnitude of extinction through geologic time are based on compilations of the stratigraphic ranges of fossils. Among the most commonly used sources of information are the *Treatise on Invertebrate Paleontology* (Moore, 1955-1975; Hill, 1972), *The Fossil Record* (Harland et al., 1967), the *Traité de Paléontologie* (Piveteau, 1952-1969), *Osnovy Paleontologii* (Orlov,

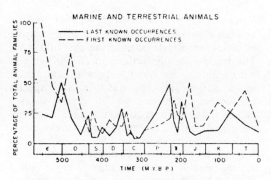

FIGURE 1. Percentage of first and last appearances of marine and terrestrial animals (from Flessa and Imbrie, 1973, after Newell, 1967; copyright © 1967 by Academic Press Inc. (London) Ltd.).

1958-1964), and *Vertebrate Paleontology* (Romer, 1966).

Fig. 1 is adapted from Newell (1967) and shows the percentage of first and last appearances of marine and terrestrial families during the Phanerozoic. Fig. 2, from the work of

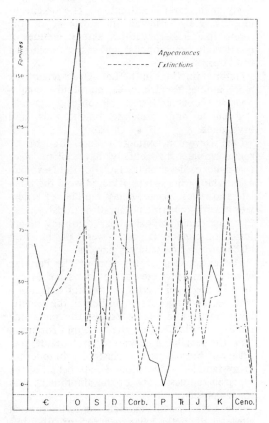

FIGURE 2. Appearances and extinctions of families of shelly, benthic marine invertebrates (from Valentine, 1969, by permission of The Palaeontological Association).

Valentine (1969), shows the number of first and last appearances of families of shelly marine benthic invertebrates during the past 600 m yr. The times of major extinction are the same in both graphs. Figs. 1 and 2 show an alternation of extinction and diversification, as if species production intensified to fill newly vacated niches following a major extinction. The coincidence of extinction episodes with the boundaries of the geologic periods is simply due to the fact that the major boundaries were placed at stratigraphic levels that showed rapid faunal change. Additional summaries of major changes in standing diversity and episodes of extinction and diversification may be found in Newell (1952, 1963), Cutbill and Funnell (1967), J. F. Simpson (1966), G. G. Simpson (1952), Flessa and Imbrie (1973), Tappan and Loeblich (1973), and Schindewolf (1955).

The most significant episodes of major extinction occurred during the Cambrian, Late Ordovician, Late Devonian, Late Permian, Late Triassic, and Late Cretaceous.

The Cambrian, the earliest of the Phanerozoic periods, is characterized by a constant, relatively high, rate of extinction. The organisms most severely affected included all of the sponge-like archaeocyathids, several families of inarticulate brachiopods, and many families of trilobites.

Major extinctions in the Late Ordovician took place among the trilobites, nautiloid cephalopods, sessile echinoderms, graptolites, ostracods, inarticulate and articulate brachiopods, and bryozoans.

Late Devonian extinctions are thought by some workers to actually have taken place somewhat earlier than the latest of the Devonian ages. Among the groups affected were the tabulate and rugose corals, many families of articulate brachiopods (including all of the atrypids), the ammonoid cephalopods, trilobites, many forms of conodonts, and several of the early placoderm and crossopterygian fish.

The extinctions at the end of the Permian were the most severe of the faunal crises evident in the fossil record. Extinction affected a wide variety of groups, including all the remaining tabulate and rugose corals, the remaining family of trilobites, all the fusulinid foraminifera, the remaining eurypterid arthropods, two orders of bryozoans, the blastoid echinoderms, many families of brachiopods, almost all of the synapsid reptiles, a few of the amphibians, and some anapsid reptiles.

The Late Triassic marked the extinction of almost half of the existing orders or suborders of reptiles, all of the labyrinthodont amphibians, a few brachiopods, and many families of ammonoid cephalopods.

The Late Cretaceous saw the demise of the dinosaur reptiles, the ammonoid cephalopods, many genera of foraminifera, many echinoids, and many genera of calcareous phytoplankton.

Within the Cenozoic, significant mammalian extinctions occurred in South America at the beginning of the Pleistocene and in all of the western hemisphere during the late Pleistocene.

Most of the extinction episodes were pervasive in their effect. Extinctions took place in all types of marine and terrestrial habitats, among sessile as well as free-living organisms, among carnivores as well as herbivores, and among single-celled as well as metazoan life. The comprehensive extent of the faunal crises implies that the causal mechanisms were not restricted to certain ecological, morphological, or taxonomic groups.

Causes of Extinction

Extinction occurs when a species fails to adapt to an environmental change. This statement will not evoke controversy, nor is it very informative. The critical questions with regard to extinction are: (1) *Why* did the organism fail to adapt to the environmental change, and (2) *What* was the nature of the environmental change?

Early workers supposed that extinction was a consequence of racial senescence, a loss of "evolutionary vigor" analogous to old age in individuals. No genetic evidence has been found that would support this idea. Orthogenesis, an irreversible evolutionary momentum, was supposed to have been responsible for maladaptive trends in the evolution of some species. The enormous horns of the Pleistocene Irish elk, *Megaloceras,* and the ever-tighter coiling among the Jurassic oyster *Gryphaea* are two of the most often cited cases of a species causing its own demise through maladaptive evolution. Recent reexaminations of these classic examples show that the horns of the elk are not out of proportion to its overall body size, and are therefore not gigantic relative to the size of the animal (Gould, 1973), and that no significant trend toward tighter coiling is present in the *Gryphaea* lineage (Gould, 1972). Early theories of racial senescence and orthogenesis essentially denied the efficacy of natural selection as a means of maintaining and improving adaptations.

More recent explanations for a species liability to extinction have focused on the degree of specialization and adaptability. In these theories, a species is supposed to become so precisely adapted to a narrow environmental range that any change in the environment will result in its extinction. This reasonable and widely held hypothesis suffers from a lack of objective tests. It is exceptionally difficult, even among living species, to quantitatively estimate the degree of specialization (or niche size). Many of a species' precise adaptations may be physiological or behavioral in nature and would not be preserved in the rock record. Elaborate morphological structures are often taken to be signs of specialization among extinct forms. Such assumptions may not be warranted until the actual function of the structure is understood. One study of morphological complexity (Flessa et al., 1975) found no relationship between the degree of limb specialization and evolutionary longevity in aquatic, free-living arthropods.

A theory of genetic specialization (Bretsky and Lorenz, 1970) holds that as organisms specialize within stable environments they tend to lose the genetic variability that allows the unspecialized to adapt to an environmental change. This theory of genetic depauperization has not found support from studies of such presumably specialized organisms as *Tridacna,* the giant clam (Ayala et al., 1973), and deep-sea organisms (Gooch and Schopf, 1973).

Some evidence exists for correlation between a species' mode of life and evolutionary longevity. Levinton (1974) has demonstrated that suspension-feeding bivalves tend to go extinct more rapidly than do deposit-feeding clams. This correlation between trophic group and evolutionary longevity may be due to the predictability of a species' food supply—deposit feeders are presumed to have a relatively constant supply of detritus from which to feed. Bretsky (1973) and Boucot (1975) have shown that bivalve and brachiopod genera that are geographically widespread tend to survive for longer intervals of time than do endemic or geographically restricted forms. Boucot attributes the evolutionary longevity of cosmopolitans to the effects of a large population size.

Van Valen (1973; see also Raup, 1975) has suggested that within an adaptive zone, extinction occurs at a stochastically constant rate—the probability of a species' extinction does not vary with the geologic age of the species. Van Valen's law suggests that the physical and biotic environment of an organism deteriorates at a constant rate and that each adaptive type has a distinct extinction rate.

In any event, all of these theories are insufficient explanation for major extinctions. The near coincidental demise of ecologically and taxonomically unrelated groups argues for a major, pervasive environmental change that occurred at a rate beyond the rate of adaptive response of many species. The causes that have been proposed are varied and in some cases, interrelated. Some of the proposed explanations are plausible but not testable, a characteristic that does not deny their possible validity but limits their usefulness. Proposed explanations can be grouped into three general categories: extraterrestrial, physical, and biological.

Extraterrestrial Changes

Among the extraterrestrial factors cited as causes of a particular extinction episode are increased levels of cosmic and x-radiation from solar and stellar sources (Terry and Tucker, 1968; Reid, et al., 1976); increased radiation during intervals of magnetic reversal (Simpson, 1966); the climatic effects of increased radiation during a nearby supernova (Russell and Tucker, 1971); and the direct effects of the magnetic field during a reversal of the earth's polarity (Hays, 1971; Crain, 1971). The direct effect of expected doses of radiation have been shown to be minimal. Marine organisms, for example, are effectively shielded by the water in which they live. Hays (1971) has shown a remarkable coincidence of Pleistocene magnetic reversals and the extinction of several species of radiolaria and has suggested that the field reversal itself had a deleterious effect on the biochemistry or behavior of the organism. Sufficiently precise and reliable paleomagnetic data are not yet available for much of the Phanerozoic to allow a test of the general applicability of this theory.

Physical Environmental Changes

Physical environmental changes have been the most popular of hypotheses to explain widespread extinctions. This popularity may be due to the fact that the physical changes may have left some record within the rocks that would permit a test of the theory. Climatic change has been suggested as being responsible for almost every extinction event. The repeated glaciations of the Pleistocene, however, did not result in many extinctions that could be attributed to the climatic change. This is especially true for the marine biota. An increase in the seasonality of the climate has been proposed by Axelrod and Bailey (1968), to explain the extinctions of the Late Cretaceous. Another popular exercise has been to relate episodes of major extinction with times of mountain building. In fact, bursts of orogenic activity do not coincide with the times of faunal crisis (Henbest, 1952). Furthermore, mountain building has only a regional environmental effect and the tectonic activity itself is spread over a broader span of time than is the extinction. A reduction in oceanic salinity was suggested by Fischer (1964) to be responsible for the Permian extinctions while Cloud (1959) has raised the possibility that many extinctions could have been caused by elevated concentrations of trace elements. Both of these suggestions await careful geochemical evaluation of the sedimentary rocks of the same age as the extinction. McAlester (1970) noted a remarkable correlation between an organism's oxygen consumption rate and the extinction rate of its fossil representatives. In this model, fluctuations in atmospheric oxygen have caused the major extinctions. Schopf et al. (1971) have questioned the reliability of some of the data used in McAlester's study. In any event, it is presently impossible to detect ancient fluctuations in atmospheric oxygen by geochemical means and the hypothesis is not directly testable. Sea-level fluctuations have often been cited as an explanation for major extinctions. Newell (1967, 1972) enumerates some of the environmental effects of reduced sea level (reduction in habitable area, increased competition, greater climatic seasonality) and has suggested sea-level changes as responsible for many Phanerozoic faunal crises.

A recently developed hypothesis is that the spectrum of environmental changes associated with continental movement may be the cause of major extinctions. For example, the slow spreading of lithospheric plates will result in reduced sea levels and a correspondingly more severe climate (Hays and Pitman, 1973). Furthermore, continental assembly will increase climatic seasonality and will bring previously isolated biotas into competition with each other (Valentine and Moores, 1972; Flessa and Imbrie, 1973; Schopf, 1974). In the oceanic realm, extinctions may be caused by oceanographic changes induced by the reconnection of isolated ocean basins (Thierstein and Berger, 1978; Gartner and Keany, 1978). Extinctions then, according to this plate tectonic hypothesis, are caused by a whole complex of environmental factors including climatic change, reduction in habitable area, competition, and altered oceanic circulation.

Biological Changes

Biological factors supposed to be responsible for widespread extinctions include predation, disease, competition, and food-web disruption. Modern man has certainly caused the extinction of many species through over hunting (predation) and habitat destruction (competition). Martin (1973) has suggested that the wave of late Pleistocene extinctions in the Americas was due to the overzealous hunting activity of the newly arrived human population. Predation, however, cannot be used as an explanation for

a major extinction that effects a wide ecological variety of both marine and terrestrial organisms. Diseases are usually limited in their effect to one or at most a few species and because of the lack of direct evidence for their presence in extinct forms, diseases are not considered serious candidates for a general explanation of major extinction. Competition is exceptionally difficult to recognize in fossil assemblages but has been indicated as the cause of the extinction of South American mammals following the late Pliocene–early Pleistocene invasion of North American mammals (Simpson, 1950). Bramlette (1965) suggested that phytoplankton extinctions during the Late Cretaceous may have caused the extinction of organisms that depended on the phytoplankton for food. Tappan (1968) and Worsley (1971) have elaborated on this hypothesis, adding that changes in phytoplankton abundance might profoundly effect atmospheric oxygen and carbon dioxide levels in the atmosphere and oceans thus causing, directly or indirectly, additional extinctions.

Conclusions

It is improbable that any single cause is responsible for all the major extinctions of the past 600 m yr. Nor is it likely that any one specific environmental change is responsible for all the extinctions that occurred during an episode of biotic crisis. Extinctions are complex, natural phenomena whose causes are undoubtedly varied. They are one of the most distinctive and enigmatic features of the fossil record.

KARL W. FLESSA

References

Axelrod, D. I., and Bailey, H. P., 1968. Cretaceous dinosaur extinction, *Evolution,* **22,** 595–611.

Ayala, F. J.; Hedgecock, D.; Zumwalt, G. S.; and Valentine, J. W., 1973. Genetic variation in *Tridacna maxima,* and ecological analog of some unsuccessful evolutionary lineages, *Evolution,* **27,** 177–191.

Boucot, A. J., 1975. *Evolution and Extinction Rate Controls.* Amsterdam: Elsevier, 427p.

Bramlette, M. N., 1965. Massive extinctions in biota at the end of Mesozoic time, *Science,* **148,** 1696–1699.

Bretsky, P. W., 1973. Evolutionary patterns in the Paleozoic bivalvia: documentation and some theoretical considerations, *Geol. Soc. Amer. Bull.,* **84,** 1–11.

Bretsky, P. W., and Lorenz, D. M., 1970. An essay on genetic-adaptive strategies and mass extinctions, *Geol. Soc. Amer. Bull.,* **81,** 2449–2456.

Butler, R. F.; Lindsay, E. H.; Jacobs, L. L.; and others, 1977. Magnetostratigraphy of the Cretaceous-

Tertiary boundary in the San Juan Basin, New Mexico, *Nature,* **267,** 318–323.

Cloud, P. E., 1959. Paleoecology–retrospect and prospect, *J. Paleontology,* **33,** 926–962.

Crain, I. K., 1971. Possible direct causal relation between geomagnetic reversals and biological extinctions, *Geol. Soc. Amer. Bull.,* **82,** 2603–2606.

Cutbill, J. L., and Funnell, B. M., 1967. Numerical analysis of the fossil record, in W. B. Harland, et al., 1967, 791–820.

Fischer, A. G., 1964. Brackish oceans as the cause of the Permo-Triassic marine faunal crisis, in A. E. M. Nairn, ed., *Problems in Paleoclimatology.* London: Interscience, 566–575.

Fischer, A. G., and Arthur, M. A., 1977. Secular variations in the pelagic realm, *Soc. Econ. Paleont. Mineral., Spec. Publ.,* **25,** 19–50.

Flessa, K. W., and Imbrie, J., 1973. Evolutionary pulsations: Evidence from Phanerozoic diversity patterns, in D. H. Tarling, and S. K. Runcorn, eds., *Implications of Continental Drift to the Earth Sciences.* London: Academic Press, 247–285.

Flessa, K. W.; Powers, K. V.; and Cisne, J., 1975. Specialization and evolutionary longevity in the Arthropoda, *Paleobiology,* **1,** 71–81.

Gartner, S., and Keany, J., 1978. The terminal Cretaceous event: A geologic problem with an oceanographic solution, *Geology,* **6,** 708–712.

Gooch, J. L., and Schopf, T. J. M., 1973. Genetic variability in the deep sea: Relation to environmental stability, *Evolution,* **26,** 545–552.

Gould, S. J., 1972. Allometric fallacies and the evolution of *Gryphaea:* A new interpretation based on White's criterion of geometric similarity, *Evol. Biol.,* **6,** 91–118.

Gould, S. J., 1973. Positive allometry of antlers in the "Irish Elk," *Megaloceras giganteus, Nature,* **244,** 375–376.

Harland, W. B., et al., eds., 1967. *The Fossil Record.* London: Geol. Soc., 828p.

Hays, J. D., 1971. Faunal extinction and reversals of the Earth's magnetic field, *Geol. Soc. Amer. Bull.,* **82,** 2433–2447.

Hays, J. D., and Pitman, W. C., 1973. Lithosphere plate motion, sea level changes and climatic and ecological consequences, *Nature,* **246,** 18–22.

Henbest, L. G., ed., 1952. Significance of evolutionary explosions in geologic time, *J. Paleontology,* **26,** 298–394.

Hill, D., 1972. *Treatise on Invertebrate Paleontology,* pt. E, rev. ed., C. Teichart, ed. Lawrence, Kansas: Geol. Soc. Amer. and Univ. Kansas Press.

Levinton, J. S., 1974. Trophic group and evolution in bivalve molluscs, *Palaeontology,* **17,** 579–585.

McAlester, A. L., 1970. Animal extinctions, oxygen consumption, and atmospheric history, *J. Paleontology,* **44,** 405–409.

McLaren, D. J., 1970. Presidential Address: Time, life and boundaries. *J. Paleontology,* **44,** 801–815.

McLean, D. M., 1978. A terminal Mesozoic "greenhouse:" Lessons from the past, *Science,* **201,** 401–406.

Martin, P. S., 1973. The discovery of America, *Science,* **179,** 969–974.

Martin, P. S.; and Wright, H. E., eds., 1967. *Pleistocene Extinctions: The Search For a Cause.* New Haven: Yale Univ. Press, 453p.

Moore, R. C., ed., 1954–1975. *Treatise on Invertebrate Paleontology,* pts. A–W. Lawrence, Kansas: Geol. Soc. Amer. and Univ. Kansas Press.

Newell, N. D., 1952. Periodicity in invertebrate evolution, *J. Paleontology,* **26,** 371–385.

Newell, N. D., 1963. Crises in the history of life, *Sci. American,* **208,** 77–92.

Newell, N. D., 1967. Revolutions in the history of life, *Geol. Soc. Amer. Spec. Paper 89,* 62–91.

Newell, N. D., 1972. The evolution of reefs, *Sci. American,* **226,** 54–65.

Orlov, Y. A., 1958–1964. *Osnovy Paleontologii.* Moscow: Akad. Nauk S.S.S.R.

Piveteau, J., ed., 1952–1969. *Traité de Paléontologie.* Paris: Masson et Cie.

Raup, D. M., 1972. Taxonomic diversity during the Phanerozoic, *Science,* **177,** 1065–1071.

Raup, D. M., 1975. Taxonomic survivorship curves and Van Valen's Law, *Paleobiology,* **1,** 82–96.

Reid, G. C.; Isaksen, I. S. A.; Hozer, T. E.; and Crutzen, P. J., 1976. Influence of ancient solar-proton events on the evolution of life, *Nature,* **259,** 177–179.

Romer, A. S., 1966. *Vertebrate Paleontology,* 3rd ed. Chicago: Univ. Chicago Press, 468p.

Russell, D., and Tucker, W., 1971. Supernovae and the extinction of the dinosaurs, *Nature,* **229,** 553–554.

Schindewolf, O. H., 1955. Über die möglichen Ursachen der grossen erdgeschichtlichen Faunenschnitte, *Neues Jahrb. Geol. Paläontol., Jahrg.,* **1954,** 457–465.

Schopf, T. J. M., 1974. Permo-Triassic extinctions: relation to sea floor spreading, *J. Geol.,* **82,** 129–143.

Schopf, T. J. M.; Farmanfarmaian, A.; and Gooch, J. L., 1971. Oxygen consumption rates and their paleontological significance, *J. Paleontology,* **45,** 247–252.

Simpson, G. G., 1950. History of the fauna of Latin America, *Amer. Scientist,* **38,** 361–389.

Simpson, G. G., 1952. Periodicity in vertebrate evolution, *J. Paleontology,* **26,** 359–370.

Simpson, J. F., 1966. Evolutionary pulsations and geomagnetic polarity, *Geol. Soc. Amer. Bull.,* **77,** 197–204.

Tappan, H., 1968. Primary production, isotopes, extinctions and the atmosphere, *Palaeogeography, Palaeoclimatology, Palaeoecology,* **4,** 187–210.

Tappan, H., and Loeblich, A. R., 1973. Evolution of the oceanic plankton, *Earth Sci. Rev.,* **9,** 207–240.

Terry, K. D., and Tucker, W. H., 1968. Biologic effects of supernovae, *Science,* **159,** 421–423.

Thierstein, H. R., and Berger, W. H., 1978. Injection events in ocean history, *Nature,* **276,** 461–466.

Valentine, J. W., 1969. Patterns of taxonomic and ecological structure of the shelf benthos during Phanerozoic time, *Palaeontology,* **12,** 684–709.

Valentine, J. W., and Moores, E. M., 1972. Global tectonics and the fossil record, *J. Geol.,* **80,** 167–184.

Van Valen, L., 1973. A new evolutionary law, *Evol. Theory,* **1,** 1–30.

Worsley, T. R., 1971. Terminal Cretaceous events, *Nature,* **230,** 318–320.

Cross-references: *Evolution; Fossil Record; Genetics; Living Fossil; Plankton; Plate Tectonics and the Biosphere.*

F

FORAMINIFERA, BENTHIC

Foraminifera constitute an Order (Foraminiferida) of the Protista (q.v.); a classification is given in Table 1. In this entry, only those forms living on the sea floor (benthic) are discussed.

Biology

Foraminifera are unicellular organisms with one or several nuclei. The protoplasm is divided into an outer, clear layer (ectoplasm) and an inner, darker region that may be pigmented and usually contains inclusions.

When foraminifera are examined under a microscope, the most obvious soft parts of the living animal are the pseudopodia. These are cytoplasmic extensions which form a reticulum (net). Granules stream along the length of the pseudopodia, often with a two-way flow along the same pseudopod. The pseudopodia extend mainly from the apertural regions. Their functions are locomotion, gathering food, building protective cysts, probably gathering detritus (in agglutinated forms), and shaping the new chamber during its formation. Recent electron microscope studies have revealed the presence of short pseudopodia emerging from wall pores. They may serve to keep the test surface clean and clear of epibionts.

The life history of very few species has been investigated. An alternation of generations has been recognized, with gamonts (sexual) and schizonts (asexual). Often many asexual phases take place between two sexual phases. Sometimes there is dimorphism of the test attributable to the alternation.

Hedley (1964), Loeblich and Tappan (1964), and Sleigh (1973) have reviewed the biology of Foraminifera.

Test

To the geologist, the most important part of the foraminiferan is the test. There is probably

TABLE 1. Classification of the Foraminifera—
Subphylum SARCODINA Schmarda, 1871, Class RHIZOPODEA von Siebold, 1845

Order FORAMINIFERIDA Eichwald, 1830
   Suborder ALLOGROMIINA Loeblich & Tappan, 1961–U. Cambrian–Holocene
      Superfamily Lagynacea Schultze, 1854–U. Cambrian–Holocene
   Suborder TEXTULARIINA Delage & Hérouard, 1896–Cambrian–Holocene
      Superfamily Ammodiscacea Reuss, 1862–Cambrian–Holocene
      Superfamily Lituolacea de Blainville, 1825–Mississippian–Holocene
   Suborder FUSULININA Wedekind, 1937–Ordovician–Triassic
      Superfamily Parathuramminacea Bykova, 1955–Ordovician–Pennsylvanian
      Superfamily Endothyracea Brady, 1844–L. Silurian–Triassic
      Superfamily Fusulinacea von Möller, 1878–U. Mississippian–U. Permian
   Suborder MILIOLINA Delage & Hérouard, 1896–Mississippian–Holocene
      Superfamily Miliolacea Ehrenberg, 1839–Mississippian–Holocene
   Suborder ROTALIINA Delage & Hérouard, 1896–Permian–Holocene
      Superfamily Nodosariacea Ehrenberg, 1838–Permian–Holocene
      Superfamily Buliminacea Jones, 1875–U. Jurassic–Holocene
      Superfamily Discorbacea Ehrenberg, 1838–Mid. Triassic–Holocene
      Superfamily Spirillinacea Reuss, 1862–Triassic–Holocene
      Superfamily Duostominacea Brotzen, 1963–L. Triassic–L. Jurassic
      Superfamily Rotaliacea Ehrenberg, 1839–U. Cretaceous–Holocene
      Superfamily Globigerinacea Carpenter, Parker & Jones, 1862–Jurassic–Holocene
      Superfamily Orbitoidacea Schwager, 1876–U. Cretaceous–Miocene
      Superfamily Cassidulinacea d'Orbigny, 1839–L. Cretaceous–Holocene
      Superfamily Nonionacea Schultze, 1854–Jurassic–Holocene
      Superfamily Robertinacea Reuss, 1850–U. Triassic–Holocene
      Superfamily Carterinacea Loeblich & Tappan, 1955–Holocene

(*Source:* After Loeblich and Tappan 1964, 1974)

no other group of organisms that offers such varied shapes or patterns of construction.

There are three main groups of wall structure: tectinous, agglutinated, and calcareous (see Lipps, 1973).

Tectinous walls (otherwise called chitinous or pseudochitinous) are composed only of organic material. Such walls are confined to the suborder Allogromiina.

Agglutinated walls (otherwise called adventitious or arenaceous) are those made of detrital particles held together by a cement. The particles may be monomineralic or heterogeneous depending partly on the animal's ability to select and partly on availability of more than one type in the sediment substrate. The secreted cement may be wholly organic (e.g., *Saccammina* (Fig. 1A, B), *Reophax*) or organic and calcareous (e.g. *Textularia, Gaudryina*).

Calcareous walls are secreted entirely by the animal. Paleozoic forms had microgranular calcitic walls which became complex and layered in the superfamily Fusulinacea. Members of the suborder Miliolina (Fig. 1C, D) have imperforate walls made of randomly oriented calcite crystallites, over which on the outer surface of the test there may be a regularly arranged layer of calcite laths. The Rotaliina (Fig. 1E, F) include a number of perforate wall types, most of which are layered. Such walls have been described as hyaline and divided into granular and radiate types, depending on whether the c-axes of the calcite crystallites are arranged randomly or radially. Recent studies of the ultrastructure using transmission and scanning electron microscopes have led to the recognition of several different patterns of structure within the broad hyaline perforate group. There is as yet no complete agreement among workers on the structure of these walls (see Hansen and Reiss, 1971; Bellemo, 1974).

The variety of morphology ranges from simple unilocular tests, through multilocular planispiral (involute and evolute), trochospiral (high and low spired), biserial, and triserial, to patterns of changing morphology with growth (e.g., planispiral to uniserial). There is also tremendous variation in aperture size, shape, position, and number, as well as in the presence of many different kinds of ornament. All these features are illustrated in a standard text such as that of Loeblich and Tappan (1964).

Ecology

Relationship with the Substrate. Modern benthic Foraminifera are known from all depths and environments from intertidal to the deepest parts of the ocean trenches (see Murray, 1973, for a review). They live on the sediment surface or within the upper few cm of soft sediment. Some live freely, clinging or even cemented to firm substrates such as shells, rocks, and various types of submarine vegetation, especially in areas subject to currents. In general, those that are attached or clinging show morphological adaptation to this mode of life, particularly flattening of one side of the test (e.g., *Cibicides*). Those that have an encrusting habit take on the form of the "host" on the side of attachment (e.g., *Acervulina*). Substrates to which Foraminifera cling or are encrusted include rocks, various kinds of submarine vegetation, and the shells and hard outer coverings of both living and dead animals, such as mollusks, bryozoans, and tunicates. Sediment-dwelling types also show morphological adaptation. In areas of wave and current disturbance, some individuals have robust tests that are well rounded (e.g., *Alveolinella*). In muddy areas, some are flattened (e.g., *Brizalina*) while others develop spines or ribs (e.g., *Bulimina* and *Uvigerina*). However, there are many representatives in all environments that do not show an obvious relationship between their morphology and the nature of the substance on which they live.

Diversity

Marginal marine environments subject to very variable conditions have low-diversity foraminiferal assemblages. Brackish and marine marshes have less than 10 species per 300 individuals. All types of lagoons (brackish, normal marine, and hypersaline), estuaries, hypersaline marshes, and nearshore shelf seas also have low diversity with generally less than 20 species per 300 individuals. Diversity increases markedly in areas of continental shelf of normal marine salinity; in these regions, there are generally more than 20 species and in some areas there are as many as 50 species per 300 individuals. Less is known of the continental-slope and ocean-basin assemblages but it seems likely that diversity remains high along the upper part of the slope although it may fall somewhat in the deeper parts of the ocean.

In addition to these trends in diversity associated with passage from shallow to deeper water, there are also latitudinal variations, which are again probably related to the variability and harshness of the environment. Thus, in general, arctic-shelf assemblages are much less diverse than those of the temperate and tropical regions.

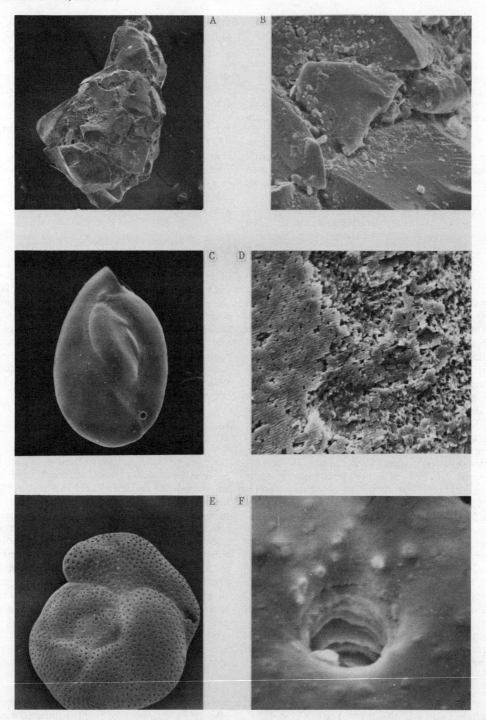

FIGURE 1. Representative benthic Foraminifera. A,B: *Saccammina,* a simple, single-chambered genus from the continental shelf off New York (A, general view, X100; B, detail of the wall showing quartz grains agglutinated together, X1250). C,D: *Miliolinella,* an example of the Miliolina having an imperforate porcellaneous wall (C, general view, X85; D, detail of a porcellaneous wall showing a superficial layer of calcite laths overlying randomly arranges calcite needles, X850). E,F: *Rosalina,* a genus of the Rotaliina, having a hyaline perforate wall and chambers arranged in a trochospiral (E, general view, X100; F, detail of a wall pore showing the layering in the wall, X4100).

Standing Crop and Biomass

The *standing crop* is the number of individuals living on a unit area of sea floor. The recorded variation is from 0 to 8000 individuals per 10 cm². Average values are 50 to 200 per 10 cm², but in regions of upwelling, where the water fertility is higher, values of 500– > 1000/10 cm² are not uncommon (e.g., mainland shelf off San Diego, California). Deltas are also regions of high standing crop because of the increased fertility due to land run-off carrying nutrients in solution and detrital organic material (e.g., Mississippi Delta).

The biomass (measured as volume of organic material) ranges from 0 to 0.9 mm³ per 10 cm³ of sea floor. The few deep-sea values are very low (< 0.1 mm³/10 cm³).

Distribution in Environments

One way of generalizing the occurence of Foraminifera in different environments is to plot the ratios of the three principal extant suborders (Textulariina, Miliolina, Rotaliina) on a triangular diagram (Fig. 2). This shows that the Miliolina are dominant only in hyper-saline and normal marine lagoons, and hypersa-line marshes. They may be present in moderate abundance in normal marine marshes but they are generally of low abundance in shelf seas. Brackish-water environments are characterized by the low abundance or total absence of this group.

The environmental distribution of some of the commoner genera is shown in a generalized manner in Fig. 3. The genera present in marginal marine environments are listed first. The same genera are commonly present in marshes and lagoons, but the different types of water masses (brackish, normal marine, hyperaline) are characterized by certain diagnostic genera: *Protelphidium* is present only in brackish and normal marine water; *Quinqueloculina* and *Triloculina* only in normal marine or hyper-saline water. Similarly, there is geographical variation of shelf genera related to the salinity and temperature of the water masses and to the type of sediment substrate. Genera re-stricted to tropical and subtropical areas are marked with a cross.

It is clear from Fig. 3 that there is a depth distribution of genera, although many of them have a great depth tolerance. Numerous ecological studies of benthic Foraminifera

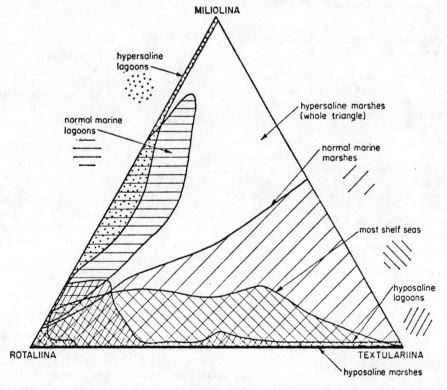

FIGURE 2. The fields occupied by modern assemblages of benthic Foraminifera based on the proportions of the three suborders (from Murray, 1973, used with permission of the publishers, Heinemann Educational Books Ltd. and Crane, Russak and Company, N.Y.).

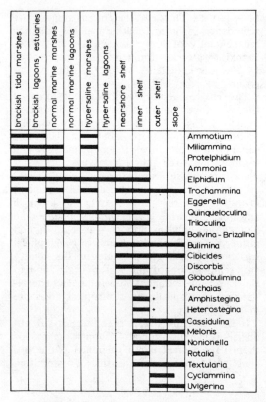

FIGURE 3. Generalized distribution of modern genera of benthic Foraminifera (+ denotes tropical or subtropical).

have been carried out to determine the depth zonation on the continental shelf and slope. In restricted geographic areas, individual species often show a marked depth restriction; but this is less obvious over a large area. Since the presence of a species on a particular area of sea floor is dependent on all the limiting environmental factors remaining within the tolerance limits, it follows that the disappearance of a species with increase in depth will be caused not necessarily by depth alone but often by changes in temperature, dissolved oxygen, substrate type, nature of the food supply, etc. In general, benthic Foraminifera in shelf seas show faunal changes at around 25 m, 80 m, 180 m, and 350 m. This defines nearshore shelf (0–25 m) inner shelf (25–80 m) outer shelf (80–180 m) and upper slope (180–350 m). However, the species that define each depth zone vary from one geographic area to another.

Relationship with Other Fauna

Benthic Foraminifera are becoming recognized as an important part of the meiofauna together with nematodes, harpacticoid copepods, and ostracods. Foraminifera are probably important consumers of organic detritus but little is known of their role in the food chain. They are known to serve as food for other elements of the meiofauna; and they must be eaten by sediment feeders; but few instances are known of Foraminifera being the principal food source of another animal.

Effects of Pollution

Benthic Foraminifera show a considerable tolerance to pollution. This is perhaps in keeping with their phylogenetic position, for pollution has more adverse effects on organisms higher up the evolutionary scale. Except in the immediate area of effluent discharge, where there may be total consumption of oxygen and consequent absence of live benthos, Foraminifera show a marked increase in abundance in the vicinity of sewer outfalls. Off the Californian coast, some species increase several hundredfold in standing crop. Probably this can be attributed to the great supply of detrital organic matter that serves as food; but it is also believed that the input of soluble nutrients leads to increased phytoplankton production, which in turn increases the food supply.

Paleoecology

The relationships between the living and dead assemblages of benthic Foraminifera in modern sediments is complex. Differences in composition between living and dead assemblages are brought about partly by normal biological processes and partly by postmortem changes (Murray, 1976).

The main biological process is reproduction. When Foraminifera reproduce, all the parent protoplasm is used to give rise to the offspring. Consequently, the parents in effect die and their empty tests are contributed to the dead assemblage. If, in any given living assemblage, some species reproduce more frequently than others or produce a greater number of offspring, it follows that the proportions of the species in the dead assemblage will differ from those of the living assemblage.

Postmortem changes further modify the dead assemblage. Processes that may operate include removal or addition of species by transport. Transport may be as bed load, in which case the individuals may become abraded, or as suspended load, in which case there is no abrasion but the forms may be size sorted and less than 0.25 mm in diameter. Also, transport may be effected by ice, by turbidity currents, by submarine sliding or slumping, by uprooting of vegetation to which Foraminifera are attached, or by wind.

Mixing of assemblages may be caused by bioturbation, especially where slow sedimentation fails to bury relict sediments on continental shelves.

However, notwithstanding all these problems, it is still possible to use benthic Foraminifera to make a paleoecological interpretation of fossil deposits. The ease of interpretation is greatest for Neogene deposits. Because of evolutionary changes, the interpretations are more subjective for Mesozoic deposits and even more so for those of the Paleozoic.

When attempting a paleoecological interpretation, it is important to see whether the assemblage shows clear evidence of transport and reworking. If it does, an interpretation will be less certain. If it does not, an interpretation should be reliable.

Evolution and Stratigraphic Value

The evolutionary history of the Foraminifera is summarized in Fig. 4. The earliest (Cambrian) representatives are simple, single-chambered forms with an agglutinated or organic wall. During the Ordovician, coiling evolved and so also did the first forms having a microgranular calcareous test. During the Silurian, the first chambered tests appeared. The Devonian was a major period of diversification. There was great expansion of the calcareous Parathuramminidae and Endothyridae.

In the Carboniferous, the Endothyridae underwent great diversification and they became sufficiently abundant to form rocks. There was great expansion too in the Palaeotextulariidae and Textulariidae. But the acme of Late Paleozoic Foraminifera came with the development of the large, complex Fusulinacea (q.v.). These forms are valuable index fossils—particularly in North America, Europe, and Asia—in rocks of Late Carboniferous (Pennsylvanian) and Permian age. However, the whole suborder Fusulinina (i.e., the Parathuramminacea, Endothyracea, and Fusulinacea) together with the Palaeotextulariidae had become extinct at the end of the Permian—with the exception of the Tetraxidae which lasted until the Triassic—(see Glaessner, 1963; Loeblich and Tappan, 1964).

In Mesozoic time, a renewal of Foraminifera took place, especially in the Jurassic, during which many of the more complex calcareous superfamilies had their origins. During this same period, a new group of larger (Lituolacean) Foraminifera emerged. In the Cretaceous, the Miliolina became important for the first time with the development of miliolid limestones and also the development of the stratigraphically important alveolinids. The

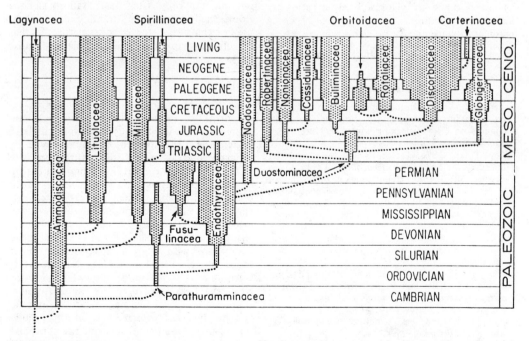

FIGURE 4. Evolutionary relationships of the foraminiferal superfamilies. Width of bars proportional to number of included families (from Tappan, 1976).

Buliminacea also underwent great expansion, producing a number of stratigraphically important genera (*Eouvigerina, Bolivinoides*).

There were further major extinctions at the end of the Cretaceous but the Tertiary faunas soon diversified and took on a modern appearance. Many of the genera known from modern environments have their earliest ancestors in the Paleogene. This period was also noted for the evolution, acme, and extinction of a group of larger Foraminifera known as "nummulites." A modern analogue of the group is *Operculina*.

It is an interesting fact that each time a group of larger benthic Foraminifera has evolved, it has had an equatorial distribution, with the exception of the Fusulinaceans (see Dilley, 1973).

Economic Significance

Much of the research into Foraminifera is promoted directly or indirectly by their value, especially in the petroleum industry, as stratigraphic and paleoecological tools. In post-Upper Cretaceous stratigraphy, benthic Foraminifera take second place to planktonic forms, but in the absence of the latter they are widely used in zonation and stratigraphic correlation. From the paleoecological point of view the benthic Foraminifera record the nature of the environment at the bottom of the sea while the planktonic forms give a guide to the nature of the surface waters. In some areas, profitable use has been made of planktic/benthic ratios to assess the proximity to and degree of communication with adjacent oceanic water.

JOHN W. MURRAY

References

Bellemo, S., 1974. The compound and intermediate wall structures in Cibicidinae (Foraminifera) with remarks on the radial and granular wall structures, *Bull. Geol. Inst. Uppsala*, n.s., **6**, 1-11.

Dilley, F. C., 1973. Larger Foraminifera and seas through time, *Spec. Papers Palaeontol.*, **12**, 155-168.

Glaessner, M. F., 1963. Major trends in the evolution of the Foraminifera, in G. H. R. Von Koenigswald; J. D. Emeis; and W. L. Buring, eds., *Evolutionary Trends in Foraminifera*. Amsterdam: Elsevier, 9-24.

Hansen, H. J., and Reiss, Z., 1971. Electron microscopy of rotaliacean wall structures, *Bull. Geol. Soc. Denmark*, **20**, 329-346.

Hedley, R. H., 1964. The biology of Foraminifera, in W. J. L. Felts and R. J. Harrison, eds., *International Review of General and Experimental Zoology*, vol. 1. New York: Academic Press, 1-45.

Lipps, J. H., 1973. Test structure in Foraminifera. *Ann. Rev. Microbiol.*, **27**, 471-488.

Loeblich, A. R., and Tappan, H., 1964. Sarcodina, chiefly "thecamoebians" and Foraminiferida, in R. C. Moore, ed., *Treatise on Invertebrate Paleontology*, pt. C, 2 vols. Lawrence, Kansas: Geol. Soc. Amer. and Univ. Kansas Press, 900p.

Loeblich, A. R., and Tappan, H., 1974. Recent advances in the classification of the Foraminiferida, in R. H. Hedley and C. G. Adams, eds., *Foraminifera*, vol. 1. New York: Academic Press, 1-53.

Murray, J. W., 1973. *Distribution and Ecology of Living Benthic Foraminiferida*. London: Heinemann, 274p.

Murray, J. W., 1976. Comparative studies of living and dead benthic foraminiferal distributions, in R. H. Hedley and C. G. Adams, eds., *Foraminifera*, vol. 2. New York: Academic Press, 45-109.

Schafer, C. T., and Pelletier, B. R., eds., 1976. First international symposium on benthonic foraminifera of continental margins. *Marit. Sediments Spec. Pub.*, **1**, 313p. + 790p.

Sleigh, M. A., 1973. *The Biology of Protozoa*. London: Edward Arnold, 315p.

Tappan, H., 1976. Systematics and the species concept in benthic foraminiferal taxonomy, in Schafer and Pelletier, 1976, **1A**, 301-313.

Cross-references: *Foraminifera, Planktic; Fusulinacea; Plankton; Protista*. Vol. I: *Zooplankton*.

FORAMINIFERA, PLANKTIC

Planktic foraminifera are one-celled animals (Protista, Order Foraminiferida Superfamily Globigerinacea) that inhabit the pelagic realm of the oceans. As zooplankton, they live suspended by passive transport in the water column and only after death fall to the seafloor. Shells of planktic foraminifera have been used extensively for reconstructing ancient oceanic environments and correlating marine rocks.

Biology

Most of the present knowledge of the biology of planktic foraminifera is ecological and distributional in nature; relatively little is known about their physiology and life cycles.

Early observations by John Murray in 1897 established that live planktic foraminifera are unicellar organisms consisting of a protoplasmic body and a shell or *test* of calcium carbonate. The protoplasm is differentiated into an outer, clear ectoplasm which generally covers the test surface and an inner, darker endoplasm with many inclusions and one or more nuclei. Pseudopodia develop from the ectoplasm as thin strands and extend outward from the test or envelop the spines found in some species. At higher magnifications, granules can be observed to stream along the pseudopodia. Communication between the ectoplasm and the endoplasm is maintained through the aperture or apertures of the test and probably through the pores which pierce the shell wall. The endoplasm contains several inclusions, often packed in the cytoplasm, the most prominent of which are the vesicular system, nuclear bodies, and the zooxanthellae. The vesicular reticulum is composed of pairs of tubules coiled about each other, resembling microvilli, and found only in planktic foraminifera. They are believed to be involved in maintaining buoyancy. A single nucleus is present in the gamont (sexual) stage whereas the agamont (asexual) stage frequently possesses several nuclei. Nuclear structures vary in their position in the shell, in size, and with the physiological activity and reproductive stage of the organism.

Epipelagic species commonly contain zooxanthellae algae, which are believed to be symbiotic with the foraminifera. However, algae in various stages of digestion in the endoplasm have been noted, indicating that they are also consumed by the foraminifera (Hansen, 1975). The zooxanthellae exhibit a circadian periodicity, entering the shell in the evening and migrating to the distal parts of the spines and pseudopodia in the daylight (Bé et al., 1977). The migration is through the aperture rather than the pores, which are closed by sieve plates or organic plugs.

The meager information available on the life cycle of planktic foraminifera suggests that they follow the typical cycle of alternation of generations, with a gamontic (sexual) and agamontic (asexual) stage. Gametogenesis has been observed (Bé et al., 1977) in a number of species.

Test

All planktic foraminifera, living and fossil, possess a secreted calcareous (calcium carbonate) shell. It is composed of nearly pure calcite although magnesium and strontium, in trace amounts, may replace the calcium.

The shell is secreted by the addition of calcite lamellae, separated by organic material, in a regular fashion. Basically it consists of two layers, an inner lamella that is confined to the individual chamber, and an outer lamella, which may also cover previously deposited parts of the test. In species inhabiting bathypelagic depths, successive growth of the outer lamella may greatly thicken the wall or form a calcite crust over much of the test. Individual calcite crystals are minute, plate-like and arranged in stacks to form larger prisms, oriented perpendicularly to the shell surface. The shell wall is pierced by numerous pores, which commonly contain organic plugs or sieve plates. However, most species have some imperforate

areas, such as the peripheral keel or apertural modifications. The distribution and size of the pores is taxonomically characteristic, but in some taxa the pores vary with water temperature.

Shell shape and size varies considerably. Most taxa have many chambers arranged in a planispiral, trochospiral, or streptospiral coil (Figs. 1 and 2). An exception is the orbuline form in which an earlier trochospiral stage may be reabsorbed leaving only a hollow, calcite sphere. Fossil genera include shells with biserially arranged chambers (Fig. 1A).

Planktic foraminifera can be divided into distinct ecological and morphological groups based on the presence and types of spines. In living and Neogene taxa, the spines are long, thin, highly flexible single crystals of calcite that develop as extensions of the shell wall (Fig. 3). Commonly, they are circular to triangular in cross section; but in *Hastigerina* they are triradiate with tiny barbs distributed near the ends of the spines. In older Tertiary and Mesozoic taxa, spines are short and heavy or the shell is covered with rugose ornamentation. Many modern and fossil species do not possess spines, at least in the adult form.

There is considerable variation in the number, size, and position of the aperture. In addition to a primary aperture, many living and fossil forms have one or more secondary or supplementary apertures. In some Cretaceous genera, a delicate grillwork of calcite (*tegilla*) is formed over the primary aperture (Fig. 2C). Bullae (blister-like covers) and/or simple, plate-like covers are present in Cenozoic taxa. Aper-

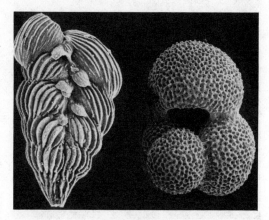

FIGURE 1. Representative-planktic Foraminifera from the Tertiary and Cretaceous. Left: A biserially arranged species, *Pseudoguembelina excolata*, Upper Cretaceous, X110. Right: A trochospiral, globigerine-shaped species, *Globigerinoides primordius*, upper Miocene, X110.

FIGURE 2. Scanning electron micrographs of Late Cretaceous planktic foraminifera. Left and center B: A planispiral species, *Globigerinelloides caseyi;* (left, general aspect, X210; center, detail of the surface showing nature of the spines, X3300). Right: Umbilical view of *Globotruncana pseudolinneiana* showing the complicated cover plate (tegilla) which covers the aperture, X30.

tural characteristics are important taxonomic features.

The perforate, hyaline lamellar calcareous walls of planktic foraminifera classify them as members of the suborder Rotaliina (Leoblich and Tappan, 1964). Recently attention has been given to wall microtexture caused by the arrangement of pores and the crystals of the outer lamellae as a basis for generic and familial classification.

Major features of the test are illustrated in Loeblich and Tappan (1964) and Boltovskoy and Wright (1976).

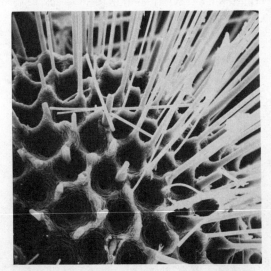

FIGURE 3. Shell surface morphology of modern planktic foraminiferan, *Globigerinoides sacculifer*, showing long, thin spines that are single calcite crystals (X640). Scanning electron micrograph courtesy of A. W. H. Bé.

Ecology and Biogeography

Planktic foraminifera are one of the common groups of plankton in the open ocean, and because of passive transport by ocean currents have global distribution. However, with rare exception, they inhabit only marine waters of normal salinity.

Small size, shell shape, thin walls, spines, and the highly porous nature of the shell all appear to be morphologic adaptations to increase buoyancy. The close similarity in shell shape between fossil and modern species suggests that the same adaptations have been repeatedly selected in their evolution. Planktic foraminifera also alter the nature of the protoplasm in order to increase flotation, by the formation of gas vacuoles, e.g., the vesicular reticulum and the bubble capsule of *Hastigerina pelagica*.

The greatest number of individuals and species live near the surface, in the upper portion of the epipelagic zone. Foraminiferal densities in the top 50 m of the water column are 10 times greater than below 200 m. Many species live in the sunlit euphotic zone in their early stages and descend to deeper depths as adults.

Most plankton undergo a daily vertical migration, rising toward the surface at night; but the diurnal migration pattern of planktic foraminifera is unclear. Several investigators have obtained different results. In an analysis of these findings, Berger (1969) showed that species in shallow, low-fertility, tropical waters exhibit little vertical variation, while species in temperate, fertile waters do exhibit diurnal migration.

Within the water column, species are depth

stratified. Spinose forms (e.g., *Globigerina* and *Globigerinoides*) are surface dwellers or prefer to live in the upper part of the euphotic zone. Most nonspinose species (e.g., *Globorotalia*) prefer to live at depths near or below the thermocline; a few inhabit depths below 500 (Bé and Toderlund, 1971). Oxygen isotope studies of Tertiary and Cretaceous planktonic foraminifera show that they have been depth stratified at least since the middle Cretaceous.

The ecology and biogeography of planktic foraminifera has recently been renewed by Bé (1977). Modern forms are distributed in rather distinct latitudinal bands and can be divided into five zones: tropical, subtropical, transitional, subpolar, and polar. These zones are closely related to global climatic belts and the distribution of major oceanic water masses.

The majority of living species inhabit tropical and subtropical waters. Species of several genera, including *Globigerinoides, Pulleniatina, Candeina, Sphaeroidinella, Hastigerina,* and *Globigerinella,* are restricted to these zones. Many of the subtropical species are found at the edge of the central water masses near the continents and are transported into transitional waters.

The composition of the assemblages in the transitional zone varies in different oceans but commonly includes *Globorotalia inflata, G. truncatulinoides,* and *Globigerina bulloides.*

Subpolar and polar waters contain 6 to 8 species, which often overlap in their biogeography. The most distinctive cold-water species is *Globigerina pachyderma,* which coils predominantly to the right in subpolar waters and to the left in polar waters. Some cold-water species undergo submergence toward the equator and can live in tropical regions at depth. These species may occur in areas of coastal upwelling together with tropical species.

Paleoecology

Globigerina ooze, a deposit composed largely of the shells of planktic foraminifera, covers 47% of the sea floor, and extensive limestone and chalk deposits represent the accumulation of fossil species. Because of the sensitivity of planktic foraminifera to environmental variation, these deposits constitute one of the best means for interpreting ancient climatic and oceanic conditions.

In deep-sea cores in many regions of the ocean, there is an alternating sequence of planktic foraminiferal faunas as a result of changing climatic conditions during the ice ages. Quantitative analysis of these faunal sequences has produced detailed climatic reconstructions (e.g., Cline and Hays, 1976)

including world maps showing the distribution of sea surface temperature and salinity during the last glacial (18,000 yr ago) and interglacial (120,000 yr ago) phases. From these data, the pattern of world climate during the ice ages can be interpreted.

Oxygen isotope studies of planktic foraminiferal shells permit the determination of the temperature of the sea water in which the shell was grown. By applying this method to fossils from Tertiary and Cretaceous deep-sea deposits, Savin et al., (1975) were able to reconstruct the temperature history of the oceans for the past 100 m yr.

Evolution and Stratigraphic use

The earliest planktic foraminifera occur in the Jurassic although possible fossils have been reported from the Triassic. It is unclear whether these early Mesozoic forms were planktic or benthic in habit (Masters, 1977).

Jurassic species are simple, small, and few in number. They are most abundant in Middle Jurassic (Oxfordian) strata and became extinct before the end of the period.

The past 120 m yr of evolutionary history are characterized by three major cycles of development. Each cycle begins with the rapid evolution of new genera and species and terminates with their decline or extinction.

The Cretaceous cycle begins with the small, *Globigerina*-like species in the Hauterivian stage and ends abruptly with the extinction of all but one or two taxa at the end of the Mesozoic. In the Cenozoic, planktic foraminifera recovered during the Paleocene, continued to diversify until the early Eocene, and then declined to a diversity low in the early Oligocene. The final cycle of evolution reached a peak in the late Miocene and planktic Foraminifera have been on the decline since the Pliocene (Berggren, 1969).

In each cycle, planktic foraminifera exhibit parallel evolutionary trends in the repeated development of similar morphological features and homeomorphic taxa. As an example, shells with marginal keels evolve independently in the Cretaceous (*Rotalipora, Globotruncana*), Paleocene (*Morozovella*), and Miocene (*Globorotalia*). Orbuline forms appear for a brief time in the middle Eocene and reevolve in the middle Miocene. Additional examples and the patterns of iterative evolution are discussed by Cifelli (1969).

The combination of rapid evolution and world-wide distribution has made the use of planktic foraminifera an important method of correlating marine deposits. The stratigraphic ranges of many genera and species have been well established (e.g., Stainforth et al., 1975), especially since the recovery of continuous sequences of Mesozoic and Cenozoic strata from the world oceans by the Deep Sea Drilling Project. Many of the volumes of the DSDP contain examples of the detailed biostratigraphic zonations based on fossil faunas that have been applied in different parts of the world (e.g., Bronnimann and Resig, 1971).

ROBERT G. DOUGLAS

References

Bé, A. W. H., 1977. An ecological, zoogeographic and taxonomic review of Recent planktonic foraminifera, in A. T. S., Ramsay, ed., *Oceanic Micropaleontology,* vol. 1. London: Academic Press, 1-100.

Bé, A. W. H., and Toderlund, D., 1971. Distribution and ecology of living planktonic foraminifera in surface waters of the Atlantic and Indian Oceans, in B. M. Funnel and W. R. Riedel, eds., *The Micropalaeontology of Oceans.* London: Cambridge Univ. Press, 105-149.

Bé, A. W. H., Hemleben, C., Anderson, O. R., Spindler, M., Hacunda, J., and Tuntivate-Choy, S., 1977. Laboratory and field observations of living planktonic foraminifera, *Micropaleontology,* 23, 155-179.

Berger, W., 1969. Ecologic patterns of living planktonic foraminifera, *Deep-Sea Research,* 16, 1-24.

Berggren, W., 1969. Rates of evolution in some Cenozoic planktonic foraminifera, *Micropaleontology,* 15, 351-365.

Boltovskoy, E., and Wright, R., 1976. *Recent Foraminifera.* The Hague: Dr. W. Junk Publ., 515p.

Bronnimann, P., and Resig, J., 1971. A Neogene Globigerinacean biochronologic time-scale of the southwestern Pacific, in E. L. Winterer et al., *Initial Reports of the Deep Sea Drilling Project, vol. VII.* Washington: U.S. Gov. Printing Office, 1235-1469.

Cline, R. M., and Hays, J., eds., 1976. Investigation of Late Quaternary paleoceanography and paleoclimatology (CLIMAP), *Geol. Soc. Amer. Mem.* 145, 464p.

Cifelli, R., 1969. Radiation of Cenozoic planktonic foraminifera, *Systematic Zool.,* 18, 154-168.

Hansen, H. J., 1975. On feeding and supposed bouyancy mechanism in four Recent globigerinid foraminifera from the Gulf of Elat, Israel, *Rev. Espan. Micropaleontol.* 7, 325-339.

Loeblich, A. R., Jr., and Tappan, H., 1964. Sarcodina, chiefly "Thecamoebians" and Foraminiferida, in R. C. Moore, ed., *Treatise on Ivertebrate Paleontology,* pt. C, 2 vols. Lawrence, Kansas: Geol. Soc. Amer. and Univ. Kansas Press, 900p.

Masters, B. A., 1977. Mesozoic planktonic foraminifera: A world-wide review and analysis, in A. T. S. Ramsay, ed., *Oceanic Micropoleontology,* vol. 1. London: Academic Press, 301-731.

Savin, S.; Douglas, R.; and Stehli, F., 1975. Tertiary Marine paleotemperatures, *Geol. Soc. Amer. Bull.,* 86, 1499-1510.

Stainforth, R. M.; Lamb, J. L.; Luterbacher, H.; Beard, J. M.; and Jeffords, R. M., 1975. Cenozoic planktonic foraminiferal zonation and characteristics of index forms, *Univ. Kansas Paleont. Contrib. Art.* 62, 425p.

Cross-references: *Foraminifera, Benthic; Micropaleontology; Plankton.*

FOSSIL RECORD

The origin and development of life through geologic time to the present has sparked the human imagination and scientific curiosity and has stimulated investigation of fossils throughout much of human history. Amateurs and trained paleontologists have searched the face of the earth and depths of the sea for all possible remains of former life, from ultramicroscopic unicellular nannofossils to huge dinosaurs and whales; and the search will continue with vigor as new discoveries continue to be made. The fossil record represents the preserved and recognized spectrum of all life during the past three billion or more years, and constitutes the evidence on which are based the reconstruction and synthesis of the phylogenetic organization and integration of past life on this planet. Rich as it is, at best, it remains a very incomplete and distorted record, with many gaps.

Investigation of the fossil record entails the study of populations, samples, and sampling methods, and thus is a statistical problem, though the average paleontologist may not approach the quest with this perspective (see *Population Dynamics*). The *total population* represents the summation of all individuals of all life that existed in the past; it is incalculable for several obvious reasons. The *sample population* represents all individuals in the total population that have been preserved throughout time and remain to the present. The *target population* represents all the individual fossils that are preserved and retrievable. These include body fossils (soft and hard parts) and trace fossils. The potentially retrievable record includes published reports and new taxa that are preserved but not yet discovered. Clearly, all fossil collections in the world and those yet to be gathered represent only a small fraction of all life that populated the globe through time. Although technological tools of resolution, e.g., the scanning electon microscope (SEM), have increased the recognition of microforms, the sample population undoubtedly contains a significant number of fossils and taxa that are impossible to recover. The basic question remains—is the "fossil record" a representative sample of all former life?

Objectives of Pursuit of the Fossil Record

What prompts such a massive attack on the world's paleontological treasure store besides the fascination of fossils and their antiquity? The utilitarian value of fossils is of course one answer (see *Biostratigraphy*). The fossil record also involves at least three scientific goals. The first is the organizational taxonomic systematics of description, identification, and classification into the traditional Linnaean hierarchy of kingdom, (super- and sub-) phylum, class, order, family, genus, species, and perhaps down to variety. With the advent of computers and large data storage and retrieval capabilities, studies are in progress to explore more satisfactory precise, modern quantitative numerical schemes to catalogue and comprehend the fossil record (see *Computer Applications in Paleontology*). From such ordered data, it is possible to establish phylogenetic patterns that outline the evolutionary trends (see *Evolution*). Such patterns might suggest a nonrandom process; yet some researchers using statistical stochastic equilibrium models think the evolutionary process may be random. They have considered patterns of phylogeny, species diversity, evolution of particular morphological forms, rates of evolution, and causes of extinction. Yet, lineages traced back in time invariably lack adequate proof of divergence at the critical junctures into common stocks. Or stating it another way, any species must have had a succession of generations of ancestors leading back to its ancestral stock even though such intermediate generations have not been found as fossils (Durham, 1971).

The second important goal concerned with the fossil record is to establish, in a sequence of sediments, the age and correlation of time-rock units (see *Biostratigraphy*). Because of evolutionary changes in the organisms, a single fossil may indicate a particular time or interval (age) in geological history or a fauna may reflect a range in time. Correlation of rock units can be made based on the contemporaneity of fossils in rock units from place to place. Zones have been established for intervals in the geological time scale on the basis of guide fossils; these zones may then be related to each other in a relative time framework, and finally may be interspersed and integrated with absolute time dates (e.g., Kauffman, 1970). In some groups, significant changes take place in shorter time so that the rate of evolution is accelerated. Zonation based on these groups involves shorter intervals, which results in more zones and closer time resolution.

The third broad objective is to establish the paleoecological significance of a single fossil or of given faunas or floras. Time-rock correlation provides a basis for the geography of ancient times, which along with environmental reconstruction is important to understanding the geological history of our earth. Natural resources found in sedimentary rock reservoirs are closely related to biogeological environments of deposition, and so the paleoenvironments represented by the strata are important explorational guides. In addition, knowledge of the geographical distribution of faunas with their latitude-longitude positions in time is critical to an evaluation of the modern seafloor spreading plate tectonics model (see *Paleobiogeography; Plate Tectonics and the Biosphere*). Interpretation of environmental reconstructions must be based on all evidence, and fossils play a vital role in terms of distribution and paleoecological significance.

In addition, hard parts of the organisms may make up a considerable volume of the physical sedimentary stratigraphic record, e.g., coquinas, biostromes, bioherms, spiculites, radiolarites, diatomaceous earth, peat, coal, and other fossiliferous sediments. Sedimentologists and petrographers are particularly interested in the details of such organic accumulations.

How Many Species?

The paleobiologist uses the fossil record and the modern living record of organisms to study rates, trends, and patterns of evolution, and relationships within and between groups. Faunal and floral diversity patterns and trends reflect evolutionary changes. Estimates of species diversity, past and present, are of considerable value to evolutionary and ecological theory as well as in tracing the history of life (Valentine, 1970). However, arrival at reliable estimates is difficult and uncertain given our present incomplete knowledge of total living biota and taxa in the fossil record.

The magnitude of the modern biosphere is reviewed below in order to put the following questions into the proper perspective. How many species are living today? How many species have lived in the past? How many marine organisms lived in the past and are living today? What proportion of organisms, past and present, are preserved and fossilizable?

The surface area of the earth is approximately 510,000,000 or 5.1×10^8 km^2 of which 70.8% or about 3.6×10^8 km^2 is sea surface. Maximum relief of the biosphere is about 34 km. The average oceanic depth is about 3795 m, so the total volume of sea water is about 1.37×10^9 km^3. Organisms that require light for photosynthesis live in only a fraction of this volume since the maximum depth in the clearest of ocean water at which plant life is expected to grow is about 200 m, which gives a photic-zone water volume of 7.2×10^7 km^3. It is estimated that the biosphere consists of 0.01×10^{20} g of carbon in the living population and 0.04×10^{20} g of carbon in dead matter, (Garrels and Mackenzie, 1972). Species diversification (see *Diversity; Evolution*) has apparently been in operation for more than 2,000,000,000 years (2 aeons). The fossil record includes marine and terrestrial species, invertebrates and vertebrates, unicells, metazoans, and metaphytes, ranging in size from a small fraction of a micron to over 30 m in length. The above factors, along with the Precambrian and Phanerozoic fossil record, circumscribe the problem of estimating total species in time.

Some estimates of the quantities of species have been made and are given as follows (for starred references, see Valentine, 1970):

1. Total number of species that have ever existed:
   a. from 50,000,000 to 4,000,000,000, probably 500,000,000 (Simpson, 1952*)
   b. at least 1,600,000 since the Cambrian (Grant, 1963)
   c. probably fewer than 100,000,000, which includes less than 6,000,000 marine forms (Valentine, 1970).
2. Total number of living species: (The following permit an average area for a single species ranging from 113 to 500 km^2.)
   a. no fewer than 1,120,000, which includes about 850,000 insect species (Mayr et al., 1953*)
   b. 1,105,000 (Easton, 1960)
   c. 1,500,000, but probably up to 4,500,000 when all are cataloged (Grant, 1963)
   d. 1,200,000, including about 1,000,000 arthropod species (Nicol, 1977)
3. Total extinct fossil species is estimated at 130,000 (Easton, 1960; Raup and Stanley, 1971). This is about 8.7% of known living species and less than 3% of probable total living organisms.
4. Potentially preservable species:
   a. 569,000 of the present living species (Teichert, 1956*)
   b. 100,000 out of 140,000 living marine invertebrate species (Valentine, 1970)
   c. 150,000 species with hard parts in the marine environment and over 4,000,000 taxa of fossilizable marine organisms since the Cambrian (Durham, 1967)
   d. 100,000 (approximately 8%) of present living species (Nicol, 1977)
5. Average rate of species turnover:
   a. 500,000 to 5,000,000 years (Simpson, 1952*)
   b. 15,000,000 years (Schilder, 1949*)
   c. 12,000,000 (Teichert, 1956*)
   d. between 5,000,000 and 10,000,000 (Valentine, 1970)

The above limited sample of census figures indicates need for resolution and the fact that we have a long way to go in documenting modern and fossil biotas.

Factors Affecting the Fossil Record

The fossil record includes body fossils (preservation of hard and rarely soft parts), chemical fossils (preservation of the original organic compounds), and trace fossils (tracks, trails, burrows, borings, and other indirect evidence of fossil organisms). The process is one of life; time; death; burial; diagenesis; preservation and/or destruction before, during, or after burial; and lithogenesis (see *Fossils and Fossilization*). A summary of some of the interrelated environmental and paleoecological factors relevant to the fossil record is given in Table 1. In the final analysis, preservation is perhaps the most important factor. Chemical stability and physical durability are differentially disposed among the different fossil groups (see *Biomineralization*). Shells and skeletal hard parts with aragonite and high magnesium calcite composition—as in the mollusks, benthic foraminifera, and algae—are usually much less durable than those with low magnesium calcite, chitin, and silica—as in pelagic foraminifera and algae, brachiopods, trilobites, ostracods, and siliceous sponges. Biological, mechanical, and chemical decomposition and disintegration take their toll and destroy a major part of the fossil record (see *Taphonomy*).

Precambrian Paleobiological Beginnings

The search for fossils in the Precambrian prior to 1947 had been virtually fruitless, with the exception of the stromatolites (q.v.) and recognition of filamentous microfossils in a thin section of chert (J. W. Gruner, 1922–1925, in Schopf, 1971). Discovery of the Ediacara fauna (q.v.; 600–700 m yr) in Australia gave renewed hope and the pace was accelerated. However, new perspective was brought into the search with the identification of a microflora found in Gunflint stromatolitic chert from Canada (S. A. Tyler and E. S. Barghoorn, 1954, 1965, in Schopf, 1971). The paleontologic strategy for field observation of Precambrian fossils focused on microforms in siliceous and carbonaceous sedimentary rocks, often stromatolitic, with the result that significant new discoveries were made on several continents. Confirmation of the paleobiologic affinity of these early fossils is based on their morphology and organic chemical composition, e.g., *n*-alkanes, isoprenoids, porphyrins, and other compounds, and carbon isotope ratio, $C^{13}:C^{12}$.

In our solar system, the earth's atmosphere with nearly 21% free (atmospheric or molecu-

TABLE 1. Some Environmental and Paleoecological Factors Relevant to the Fossil Record[a]

Chemical	Physical	Geological	Biological
Salinity	Time	Depth (aquatic)	Ecosystem
Dissolved solids:	Temperature	Water movements:	Diet and feeding habits
Inorganic	Dessication	Agitation	Food chain and web
Organic	Illumination	Waves	Trophic levels
Dissolved gases:	Pressure	Currents	Aerobic/anaerobic photosynthesis
O_2		Tides (range)	Neritic/pelagic; planktic/nektic
H_2S		Turbidity	Sessile/vagrant (mobile); benthic
CO_2		Substrates	Association:
Eh (Redox)		Lithologies	"Accidental"
pH (acid/basic)		Mineralogy (hard parts)	"Symbiotic"
		Postmortem effects	Reef community
		Continental drift	Enemies and predation
			Dispersal
			Mass extinction
			Provinciality-endemism

[a]This table is a slight modification of a tabulation by Cuffey, 1970.

lar) oxygen is unique among the planets. An important question which bears on the fossil record is the origin and history of this oxygen envelope. There is an assumption, based upon knowledge of the atmospheres of other planets, that there was no free oxygen directly after the formation of the earth. If this is so, the pattern is set for the model diagramed in Fig. 1. Incremental increases in oxygen to the atmosphere were probably largely accomplished through aerobic photosynthesis in a Precambrian world dominated by a microflora (Cloud, 1974, 1976, and references therein). When this biochemical exchange occurred and at what rate are difficult questions to answer with the present fossil record data.

A scenario for the model starts with the origin and formation of the earth with no free oxygen. Chemical evolution and possibly carbonaceous meteorite falls may have provided high local concentrations of carbon molecules, monomers and polymers, for biogenesis and eobionts. Amino acids, building blocks for proteins, were formed by natural radiation as demonstrated by the experiments of S. L. Miller and H. C. Urey, W. W. Rubey and P. H. Abelson, and E. Anders. While the origin and early development of life is a very complex physicochemical problem in a geological setting, Occam's principle of simplicity may be applied to the model. One of the simplest forms of life—that of the single, heterotrophic, anucleate cell, the prokaryote—was probably initiated during the 4th Aeon. Once formed, replication populated shallow marginal seas.

After biogenesis and prokaryote expansion, innovative developments—as deduced from the fossil record—were photosynthesis (anaerobic and aerobic), autotrophy (organisms capable of metabolizing carbon), eukaryote organization (nucleate cells), sexuality, multicellular diversification and cell differentiation, homoiothermy, coloniality, angiospermy, and others. Understanding of each of these is a major problem for which the fossil record has afforded only partial answers (see *Precambrian Life*).

Organic-looking remains, including stromatolites (q.v.), occur in carbon-rich sediments, cherts, and carbonates in the interval between $\approx 3760 \pm 70$ m yr (the age of the oldest dated minerals and rocks) and 2200 m yr (2.2 b yr; see Fig. 1). Although the earliest forms of life have been cited to occur in the Fig Tree Group and perhaps the Onverwacht Group of Africa, there is some reservation about the significance of the fossil-like forms older than 2.2 b yr. The bacterium-like rods and alga-like spheroids in the Fig Tree Group have a minimum age of about 3.1 b yr. The oldest stromatolites (q.v.) of biogenic affinity occur in the Bulwayan Group of Rhodesia.

The middle Precambrian Proterophytic (Cloud, 1974) marks the earliest preserved fossil record of diversification of cells—in stromatolites from the Transvaal Sequence of Africa dated about 2.2 b yr (Nagy, 1974). Late in this period, the classic Gunflint stromatolitic assemblage of prokaryotic microorganisms with filamentous and spheroidal forms also show cell differentiation. These blue-green algae are morphologically and perhaps biochemically similar to common modern forms.

In the late Precambrian Proterozoic, significant changes took place, as indicated in several well-preserved microfloras. The first nucleate cells occurred with asexual reproduction by mitotic cell division in the Beck Spring Dolomite

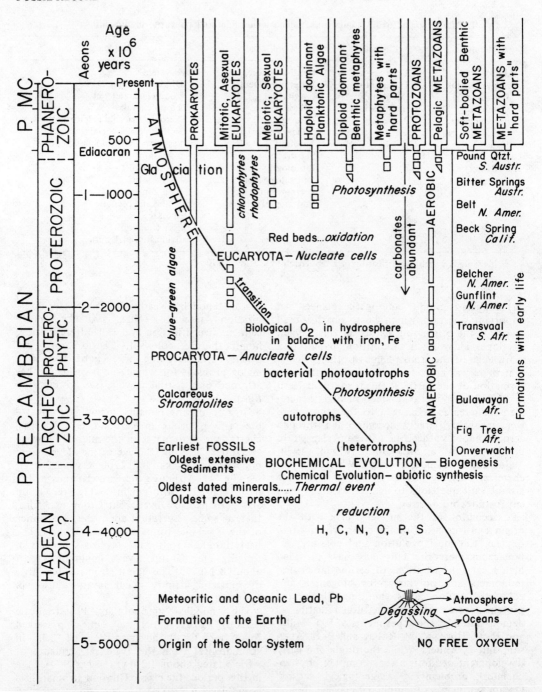

FIGURE 1. Model of physical and biological development of the primitive earth, with emphasis on the fossil record from microscopic unicellular life to megascopic metaphytes and metazoan organization in Precambrian to the Phanerozoic (after Cloud, 1974, 1976; Tappan and Loeblich, 1971; Schopf, 1971; Schopf et al., 1973, with slight modifications).

and Belt Series of North America. This is followed by development of sexes and meiotic reproduction of cells as seen in the Bitter Springs cherts of Australia. This latter event marked an increase in diversification and rate of evolution in eukaryote microorganisms which led to the emergence of pelagic, heterotrophic protists and ultimately to metazoans and metaphytes at the close of the Precambrian. Development of mobility, adaptation to a benthonic colonization, acquisition of hard parts, and cephalization set the stage for the explosive evolution of the metazoans in the Phanerozoic (Schopf et al., 1973).

Because the development of eukaryotic cells from prokaryotic ancestors was the single greatest quantum step in evolutionary history, it is important to ascertain when this transition occurred (Knoll and Barghoorn, 1975). To add new perspective and caution in evaluating this complex stage in biogenesis, Knoll and Barghoorn question the presence of unequivocal eukaryote cells in the Bitter Springs Formation and older rocks. They maintain that eukaryote-looking cells are artifacts of degraded protoplasm from undecomposed prokaryotic cyanophyte cells. This is demonstrated experimentally using evidence from unialgal cultures of modern blue-green algae. Also, the high diversity suggested by the Bitter Springs mat microflora of 45 described algal species is the result of misinterpreting degradational features as valid taxonomic characteristics for discrimination of taxa. Thus, the search is on for unequivocal late Precambrian eukaryote cells ancestral to the multicellular Ediacaran fauna (q.v.).

The Phanerozoic

The Phanerozoic (Gr. *phaneros-* visible, evident, *zoon-* animal, *zoös-* alive, living, *zoë-* life) is the interval of time that represents the growth, development, and flourish of megascopic metazoans and metaphytes (Fig. 2). Approximately 99% of all recognized fossil species are from the Phanerozoic. This period encompasses the Ediacara fauna, "ancient life" of the Paleozoic, "middle life" of the Mesozoic, and "new" or "recent life" of the Cenozoic to the present. The Phanerozoic represents the distribution and succession of animal and plant groups, after Precambrian origins, from inception to maximum development and in some cases to extinction. Because the Ediacara fauna has the first wide-spread metazoan fauna, it is here regarded as the start of the Phanerozoic even though it is dated as latest Precambrian.

Ediacara Fauna. The Ediacara Precambrian fauna is the oldest known diversified and abundant marine metazoan assemblage of apparently soft-bodied, mobile and sessile, radial and bilaterally symmetrical fossil organisms. The fauna is essentially composed of coelenterates and annelids with a few arthropods, some trace fossils, and one genus of unknown affinity (Glaessner, 1971). It includes medusoids (9 spp.), hydrozoans (Chondrophora, 3 gen., 3 spp.), Conulata (1 sp.), scyphozoans (2 gen., 2 spp.), anthozoans (Pennatulacea, 3 gen., 4 spp.), polychaetes (2 gen., 5 spp.), 1 species of Trilobitomorpha (or Chelicerata), 1 species of Crustacea (Branchiopoda). The Ediacara fauna was first found in Australia and has subsequently been recognized in SW Africa, England, Sweden, SW Ukraine (Podolia), northern Russia NE of Moscow, and in northern Siberia. They range in age in these scattered localities from about 590 to 700 m yr. Apparently this fauna has not been found in the Western Hemisphere.

The Early Cambrian and Skeletal Adaptation. A better understanding of Early Cambrian history and its fauna will provide new outlooks on the whole of earth history and biotal evolution (Zhuravleva, 1970; Sepkoski, 1978). The surge in the establishment of early metazoa can be attributed in large measure to the formation of hard parts of various compositions and types. Early trilobites had calcitic phosphatic chitinous exoskeletons as did the biconvex and conical shells of inarticulate brachiopods. Multiplated echinoderms (5 classes) had calcite skeletons and the conical archaeocyathids were also calcareous. Diversified echinoderms and pogonophorans are known from near the base of the Lower Cambrian. It is most striking that the first appearance of echinoderms includes five classes: helicoplacoids, camptostromoids, lepidocystoids, edrioasteroids, and eocrinoids (Durham, 1971). Sponges and mollusks also made their appearance in Early Cambrian.

Diversity and Trends. The origin and evolution of eukaryotic cells and sexual reproduction in the Precambrian set the stage for the first metazoans and metaphytes and their rapid increase at the start of the Phanerozoic. The increase in abundance and diversification of organisms in Cambrian time may be related to a marine transgression, which resulted in widespread shallow seas with an adequate food supply and environmental stability. This diversity is reflected in the development of provinciality, as with trilobite faunal provinces; some taxa achieve early cosmopolitan distribution, such as the trilobites and archaeocyathids.

Frequency distribution diagrams of the major groups of plants and animals (Figs. 3–6) are based on data in Harland et al. (1967), which is still the most complete compilation of di-

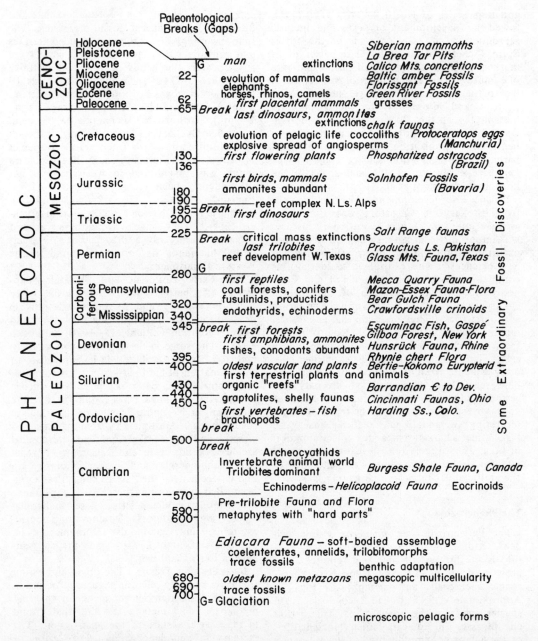

FIGURE 2. Framework of the fossil record in the Phanerozoic with some highlights of evolutionary development during the past 700 m yr. Examples of unusual preservation and importance are included. Time intervals representing paleontological breaks (major and minor extinctions) are indicated. Glacial episodes that had their effects on the fossil record are shown by G. (Data from numerous sources.)

versity data thus far published for the fossil record (Raup and Stanley, 1971). It should be kept in mind, however, that the summary is incomplete. Several important phyla are not included, such as Porifera, Bryozoa, and Annelida, and many other lower taxonomic categories are not represented. The summary

(Fig. 6) is based on 2526 taxa representing phyla, classes, orders, families, and genera of plants and animals. In many cases, the stratigraphic ranges of the taxa are not known with certainty. Interpretation of the histograms should recognize the limitations of the data.

The general trend for most groups is toward

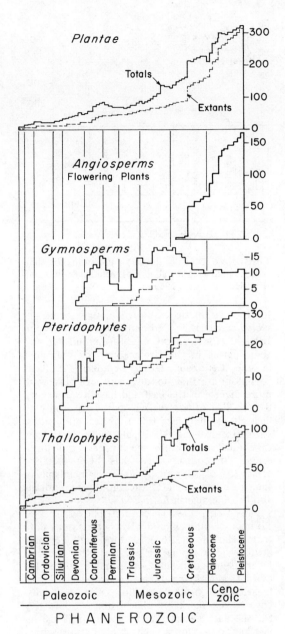

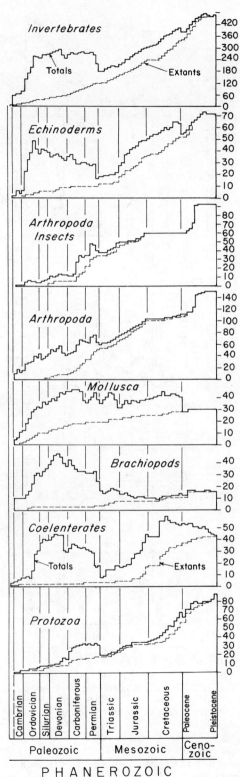

FIGURE 3. Diversity diagrams (number of taxa versus time) for major categories of plants: thallophytes, pteridosperms, gymnosperms, angiosperms, and the summation of all plant groups (after Harland et al., 1967).

FIGURE 4. Diversity diagrams for some major phyla of invertebrates: protozoans, coelenterates, brachiopods, mollusks, arthropods (including insects), echinoderms, and the summation of these invertebrates (after Harland et al., 1967).

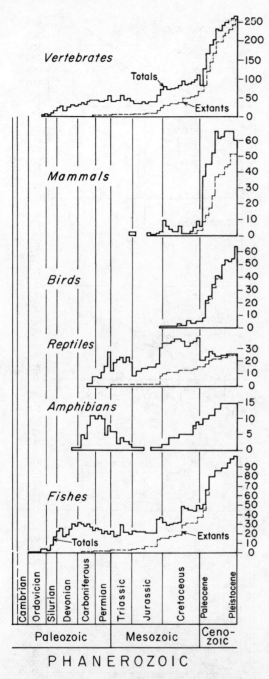

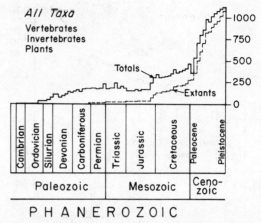

FIGURE 6. Diversity diagram representing the summation of all taxa for all groups included in Figs. 3–5 (after Harland et al., 1967).

FIGURE 5. Diversity diagrams for major vertebrate groups: fishes, amphibians, reptiles, birds, mammals, and the summation of the vertebrates (after Harland et al., 1967).

a progressive rise in total taxa during the Mesozoic and Cenozoic with increase in modern forms into the Holocene. The rate of change of this increase in the Cenozoic is much greater than in pre-Cenozoic time. In some groups,

there is dramatic explosive increase, e.g., insects since the Carboniferous, birds from Jurassic time, planktonic foraminifera and angiosperms in the Cretaceous, and mammals in the Cenozoic. Other organisms such as cephalopods, brachiopods, and crinoids were more diverse in the past and have low diversity today. Still others thrived in the past only to become extinct, e.g., the trilobites, graptolites, stromatoporoids, cystoids, blastoids, tetracorals, fusulinids, ammonites, belemnites, and dinosaurs. Of course, there has been startling discovery of supposedly extinct organisms in the deep sea such as the monoplacophorans and the coelocanth *Latimeria*.

Van Valen (1973) examined over 25,000 subtaxa of animals and plants, taking his data from Harland (1967), Moore and Teichert (1953–1975), Romer (1966), and other sources. He plotted taxonomic survivorship curves for extinct and living taxa (54 graphs) using a logarithmic ordinate-semilog plot and discovered a remarkable almost uniform linearity for the data of each and every group. This important relationship and contribution which recognizes order in the evolutionary record is known as *Van Valen's Law of constant extinction*. The law states that within an ecologically homogeneous taxonomic group, extinction occurs at a stochastically constant rate. One implication is that the sum of absolute fitness in a community is constant. If the law is true, it will give better understanding of evolutionary order in and extrapolation from the fossil record. The full impact of Van Valen's paper has not yet been realized and further testing is recommended (Raup, 1975a,b; Sepkoski, 1975; Hallam, 1976; Van Valen's comments forthcoming).

An interesting discussion of the diversity distribution quandary in the fossil record is presented by Raup (1972; Valentine, 1973a; and Raup et al., 1975). The following questions are raised: Is the evolutionary process one that leads to an equilibrium steady-state number of taxa, or should diversification be expected to continue almost indefinitely? Has equilibrium (or saturation) been attained in any habitats in the geologic past? If mass extinction has led to a significant reduction in diversity, what are the nature and rates of recovery (Raup, 1972)?

Raup presents a reasonable argument that time-dependent biases influence the data on which taxonomic diversity patterns are interpreted. These biases include direct relationship in quantitative preservation of fossils and volumes and ages of sedimentary rocks (Figs. 7,8); practical choice of biologic taxonomic hierarchy comparing diversity in family or genus versus species levels; sampling problems related to availability and accessibility; number of sites; intensity of paleontologic field pursuit and subsequent study such as monographs; and others. Biases are most significant at lower taxonomic levels and in younger rocks. Raup's conclusion is that well-skeletonized marine invertebrates had the highest taxonomic di-

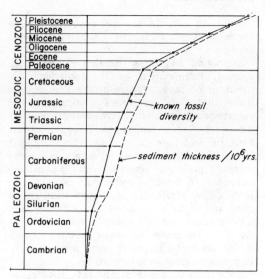

FIGURE 8. Another representation of the data given in Fig. 7 to emphasize rates of change in fossil diversity and sediment thickness.

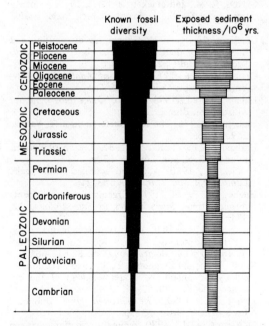

FIGURE 7. Relative numbers of taxa of fossil plants and animals. Known fossil diversity, as estimated from literature surveys, is expressed by width of blocks (data from Harland et al., 1967). Time equivalent thickness of exposed sediments per m yr is given by horizontally shaded graph (data from Holmes, 1960; after Raup and Stanley, 1971).

versity in the Paleozoic, in contrast with the traditional Cenozoic-Holocene interpretation. Valentine (1973a) compares the empirical data model with Raup's alternate bias simulation model for Phanerozoic diversity and concludes that the diversity trends suggested by the latter are historically incorrect. With present data and methods, it has not been possible to evaluate Phanerozoic taxonomic diversity with any certainty. It is suggested that future research on this problem should concentrate on quantitative assessment of the biases. One such treatment applies rarefaction techniques to successfully compensate in higher taxa for differences in sample size (Raup, 1975a). A mathematical-computer approach is taken by Flessa and Imbrie (1973) and Flessa and Levinton (1975) to determine the timing and randomness of diversification patterns in organisms (see also Sepkoski, 1978). They conclude that first appearances of taxa are episodic and are distributed nonrandom through time. The fossil record is characterized by groups of taxa whose levels of relative taxonomic diversity vary synchronously and at the same rate.

Evolutionary changes in some groups single them out for practical use as guide fossils. These are fossils that have short time span, wide geographic distribution, abundance, and generally are easily recognized. These forms include trilobites and archaeocyathids in the Cambrian; graptolites and conodonts for the early and middle Paleozoic; tournayellid, endothyrid, and fusulinid Foraminifera in the Carboniferous; ammonoids from Devonian to

Cretaceous; foraminifera and coccolith nanno-fossils in Cretaceous and Cenozoic deep-sea deposits.

Extinction and Gaps (Breaks) in the Fossil Record. Examination of the frequency distribution of taxa throughout the Phanerozoic shows that there were times of crisis and mass extinctions. Significant decrease took place in the Late Ordovician, Late Devonian, Late Permian, and Late Cretaceous. The vertebrate lineage in the Ordovician was barely under way with the primitive fishes, and land plants were unknown; however, invertebrate faunas suffered a severe setback. Late Ordovician glaciation undoubtedly affected the climates and may account for this population demise. In the Late Devonian, all major groups were affected adversely; but since there was no glaciation, other causes were responsible.

Evolutionary advancement of the major groups of animals and plants hit by mass extinctions late in Permian and Cretaceous times resulted in catastrophic breaks in the succession of life and gaps in the fossil record. The plate tectonics model and the timing of events can be integrated with Phanerozoic history to offer some reasonable explanation for these mass extinctions (Valentine and Moores, 1972). Sea level, climatic conditions, food supply, and environmental stability play an important role. Coalescence of continental plates into more and smaller units produced sea-level rise and strandline transgression and more equable stable climatic conditions, which resulted in greater diversity of organisms with endemic biotas. Major dispersal barriers leading to benthic marine provinces were changes in latitudinal thermal regimes (equatorial to polar) and deep-sea or land barriers in longitudinal directions (Valentine, 1973a). McAlester (1970) also found that fossil taxa related to Holocene representatives with high oxygen consumption had high extinction rates. Perhaps reduction in photosynthesis, hence less production of oxygen by plants, occurs when shallow-water continental margins are minimum in extent due to fewer continental plates.

RAYMOND C. GUTSCHICK

References

International scientific literature in the broad field of paleontology is vast in general treatment and especially detailed accounts of the myriads of taxa. The following list represents a sample for guidance and in-depth research.

Beerbower, J. R., 1968. *Search for the Past,* 2nd ed. Englewood Cliffs, N.J.: Prentice-Hall, 512p.

Boucot, A. J., 1975. *Evolution and Extinction Rate Controls,* vol. 1, Developments in Paleontology and Stratigraphy. Amsterdam: Elsevier, 427p.

Cloud, P., 1974. Evolution of ecosystems. *Amer. Scientist,* **62,** 54–66.

Cloud, P., 1976. Beginnings of biospheric evolution and their biogeochemical consequences, *Paleobiology,* **2,** 351–387.

Colbert, E. H., 1969. *Evolution of the Vertebrates,* 2nd ed. New York: Wiley, 535p.

Cuffey, R. J., 1970. Bryozoan-environment interrelationships—An overview of bryozoan paleoecology and ecology, *Earth Miner. Sci., Penn. State Univ.,* 39(6), 41–45.

Dott, R. H., Jr., and Batten, R. L., 1971. *Evolution of the Earth.* New York: McGraw-Hill, 649p.

Durham, J. W., 1967. The incompleteness of our knowledge of the fossil record, *J. Paleontology,* **41,** 559–565.

Durham, J. W., 1971. The fossil record and the origin of Deuterostomata, *Proc. N. Smer. Paleontological Conv.,* pt. H, 1104–1132.

Easton, W. H., 1960. *Invertebrate Paleontology.* New York: Harper & Row, 701p.

Flessa, K. W., and Imbrie, J., 1973. Evolutionary pulsations: Evidence from Phanerozoic diversity patterns, in D. H. Tarling, and S. K. Runcorn, eds., *Implications of Continental Drift to the Earth Sciences,* vol. 1. New York: Academic Press, 247–285.

Flessa, K. W., and Levinton, J. S., 1975. Phanerozoic diversity patterns: Tests for randomness, *J. Geol.,* **83,** 239–248.

Frey, R. W., ed., 1975. *The Study of Trace Fossils.* New York: Springer-Verlag, 592p.

Garrels, R. M., and Mackenzie, F. T., 1972. A quantitive model for the sedimentary rock cycle, *Marine Chem.,* **1,** 27–41.

Glaessner, M. F., 1971. Geographic distribution and time range of the Ediacara Precambrian fauna, *Geol. Soc. Amer. Bull.,* **82,** 509–513.

Grant, V., 1963. *The Origin of Adaptations.* New York: Columbia Univ. Press, 606p.

Hallam, A., ed., 1973. *Atlas of Palaeobiogeography.* Amsterdam: Elsevier, 532p.

Hallam, A., 1976. The Red Queen dethroned, *Nature,* **259,** 12–13.

Harland, W. B., et al., eds., 1967. *The Fossil Record.* London: Geol. Soc. London, 828p.

Holmes, A., 1960. A revised geological time scale, *Edinb. Geol. Soc. Trans.,* **17,** 183–216.

Hughes, N. F., ed., 1973. Organisms and continents through time, *Spec. Papers in Paleontology,* **12,** 334p.

Kauffman, E. G., 1970. Population systematics, radiometrics and zonation—A new biostratigraphy, *Proc. N. Amer. Paleontological Conv.,* pt. F., 612–666.

Knoll, A. H., and Barghoorn, E. S., 1975. Precambrian eukaryotic organisms: A reassessment of the evidence, *Science,* **190,** 52–54.

Levin, H.L., 1975. *Life Through Time.* Dubuque, Iowa: Wm. C. Brown, 217p.

McAlester, A. L., 1970. Animal extinctions, oxygen consumption and atmospheric history, *J. Paleontology,* **44,** 405–409.

Middlemiss, F. A.; Rawson, P. F.; and Newall, G., eds., 1971. Faunal provinces in space and time, *Geol. J. Spec. Issue,* **4,** 236p.

Moore, R. C., and Teichert, C., eds., 1953–1975. *Treatise on Invertebrate Paleontology,* pts. C–W.

Lawrence, Kansas: Geol. Soc. Amer. and Univ. Kansas Press.

Nagy, L. A., 1974. Transvaal stromatolite: First evidence for the diversification of cells about 2.2 X 10⁹ years ago, *Science*, 183, 514–516.

Newell, N. D., 1959. Adequacy of the fossil record, *J. Paleontology*, 33, 488–516.

Newell, N. D., 1962. Paleontological gaps and geochronology, *J. Paleontology*, 36, 592–610.

Newell, N. D., 1963. Crises in the history of life, *Sci. American*, 208(2), 76–92.

Nicol, D., 1977. The number of living animal species likely to be fossilized, *Florida Sci.*, 40, 135–139.

Olson, E. C., 1971. *Vertebrate Paleozoology*. New York: Wiley Interscience, 839p.

Orlov, Y. A., ed., 1958–1964. *Fundamentals of Paleontology*. 15 vols. Izdatel'stvo Akad. SSSR, Moskva (in Russian). Transl. 1971, Israel Prog. Sci., Smithsonian Inst. and Nat. Sci. Found., Washington, D.C.

Palmer, A. R., 1974. Search for the Cambrian world, *Amer. Scientist*, 62, 216–224.

Piveteau, J., ed., 1952– . *Traité de Paléontologie*, 7 vols. Paris: Masson.

Raup, D. M., 1972. Taxonomic diversity during the Phanerozoic, *Science*, 177, 1065–1071.

Raup, D. M., 1975a, Taxonomic diversity estimation using rarefaction, *Paleobiology*, 1, 333–342.

Raup, D. M., 1975b. Taxonomic survivorship curves and Van Valen's Law. *Paleobiology*, 1, 82–96.

Raup, D. M., and Stanley, S. M., 1971, 1978. *Principles of Paleontology*. New York: Freeman, 481p. (ch. 1. Preservation and the fossil record).

Raup, D. M., et al. 1975. Research news—paleobiology: Random events over geological time, *Science*, 189, 625–626, 660.

Romer, A. S., 1966. *Vertebrate Paleontology*, 3rd ed. Chicago: Univ. Chicago Press, 468p.

Rudwick, M. J. S., 1972. *The Meaning of Fossils*. Amsterdam: Elsevier, 287p.

Scagal, R. F., et al. 1965. *An Evolutionary Survey of the Plant Kingdom*. Belmont, Calif.: Wadsworth, 658p.

Schäfer, W., 1972. *Ecology and Paleoecology of Marine Environments*. Chicago: Univ. Chicago Press, 568p.

Schopf, J. W., 1971. Organically preserved Precambrian microorganisms, *Proc. N. Amer. Paleontological Conv., Chicago*, pt. H., 1013–1057.

Schopf, J. W.; Haugh, B. N.; Molnar, R. E.; and Satterthwait, D. F., 1973. On the development of metaphytes and metazoans, *J. Paleontology*, 47, 1–9.

Sepkoski, J. J., Jr., 1975. Stratigraphic biases in the analysis of taxonomic survivorship. *Paleobiology*, 1, 343–355.

Sepkoski, J. J., Jr., 1978. A kinetic model of Phanerozoic taxonomic diversity. I. Analysis of marine orders. *Paleobiology*, 4(3), 223–251.

Shimer, H. W., and Shrock, R. R., 1944. *Index Fossils of North America*. New York: Wiley, 837p.

Tappan, H., and Loeblich, A. R., Jr., 1971. Geobiologic implications of fossil phytoplankton evolution and time-space distribution, *Geol. Soc. Amer. Spec. Papers, no. 127*, 247–340.

Tasch, P., 1973. *Paleobiology of the Invertebrates*. New York: Wiley, 946p.

Valentine, J. W., 1970. How many marine invertebrate fossil species? A new approximation, *J. Paleontology*, 44, 410–415.

Valentine, J. W., 1973a. Phanerozoic taxonomic diversity: A test of alternate models, *Science*, 180, 1078–1079.

Valentine, J. W., 1973b. *Evolutionary Paleoecology of the Marine Biosphere*. Englewood Cliffs, N.J.: Prentice-Hall, 511p.

Valentine, J. W., and Campbell, C. A., 1975. Genetic regulation and the fossil record, *Amer. Scientist*, 63, 673–680.

Valentine, J. W., and Moores, E. M., 1972. Global tectonics and the fossil record, *J. Geol.*, 80, 167–184.

Van Valen, L., 1973. A new evolutionary law. *Evol. Theory*, 1, 1–30; see also 1975, *Nature*, 257, 515.

Zhuravleva, I. T., 1970. Marine faunas and Lower Cambrian stratigraphy, *Amer. J. Sci.*, 269, 417–445.

Cross-references: *Actualistic Paleontology; Biogeochemistry; Biostratigraphy; Burgess Shale; Communities, Ancient; Diversity; Ediacara Fauna; Evolution; Extinctions; Fossils and Fossilization; Invertebrate Paleontology; Living Fossil; Mazon Creek; Paleobiogeography; Paleoecology; Plate Tectonics and the Biosphere; Precambrian Life; Solnhofen Formation; Taphonomy; Trace Fossils; Vertebrate Paleontology.*

FOSSILS AND FOSSILIZATION

The term "fossil" derives from the Latin *fossilis*, "that which is dug up," and until the early part of the 19th century was used indiscriminately for both organic remains such as shells and organized inorganic objects such as crystals and concretions. Subsequently, its use became restricted to the identifiable remains of living organisms, or of their activities, preserved in the rocks by natural processes. As an adjective, the term has also been applied metaphorically to preserved geologic structures of purely inorganic origin, as in the expressions, "fossil dunes," "fossil moraines," or "fossil soil-profiles." These objects, however, are not truly fossils and the following discussion will be restricted to organic remains.

Kinds of Fossils

A fossil, in the strictest sense, is the body of the organism itself, or a part of it. The details of its structure are preserved to the extent that the original material has not been disorganized or removed through mechanical and chemical processes. Generally, the only parts of the body that are fossilized are the skeletal tissues, such as shells, bones, teeth, and wood, because these tissues are mechanically stronger and chemically more stable. In some instances, fossils are preserved only as impressions in the surrounding rock, the material of the organism itself having been removed subsequent to the hardening of the rock about it.

Another class of fossils consists of remains of the *activities* of organisms. These do not directly reflect the structure of the organisms that produced them. In some cases, they may be referred to particular species of organisms by comparison with the activities of living forms. In many cases, the kind of activity may be identified but not the organism that performed it. Fossils of this second class include the following, cited in order of decreasing identifiability: (1) Egg cases and similar secreted productions that are separate from the body of the organism. (2) Artifacts, including agglutinated tubes and nests (as opposed to agglutinated shells that show the form of the body), as well as tools of early man. (3) Footprints, and impressions of other parts of the body, such as tooth marks on bones and shells. In addition to recording an activity, these fossils preserve a replica of part of the animal's body and as such may be identifiable as to species (Kauffman and Kesling, 1960). (4) Coprolites and stomach contents, that is, fossil excrement or the remains of ingested food (see Zangerl and Richardson, 1963; Richardson, 1971). (5) Borings into shells, rocks, wood, and leaves. A class of organism may be identified by the nature of the boring, such as boring sponges, predaceous gastropods, rock- and wood-boring bivalves, wood- and leaf-destroying insects and fungi (see Bromley, 1970). (6) Trails and burrows made on or in soft sediment and preserved by its consolidation. Such biogenic sedimentary structures or *ichnofossils* (Seilacher, 1964) can rarely be identified as to type of animal, but provide valuable information on the environment and type of activity of the bottom fauna at the time of deposition of the sediment (see also Häntzschel, 1975). (7) Biogenic molecules, if identifiable as such, promise to yield information on organic evolution at the molecular level, even in the absence of other organized remains (Calvin, 1965). Relatively complex carbon compounds such as amino acids, purines, porphyrins, carbohydrates, and hydrocarbons (Breger, 1963; Eglinton and Murphy, 1969) have been isolated from sediments. Most are probably biogenic though they cannot be referred to a particular group of organisms. Just beginning to be explored, this field may some day yield much information of phylogenetic value as well as data on the early evolution of organic molecules. It is not yet known to what extent complex organic molecules are modified by diagenesis or whether any may have formed from inorganic processes (see Florkin, 1969; Kvenvolden, 1975).

Fossilization

As soon as an organism dies, its structure begins to disintegrate. The organic material is attacked by saprophytic plants, dominantly bacteria, and by scavenging animals, so that it is rapidly destroyed (see Schaefer, 1972). If environmental conditions about the dead organism inhibit destructive plants and animals, some of the organic material may be preserved. Such conditions include *lack of oxygen,* as in stagnant waters and bottom muds; *intense cold,* as in frozen ground; and *dryness,* as in aeolian desert deposits. Generally such preservation is incomplete, the organic material being reduced to a thin film of elemental carbon (*carbonization*). Lack of oxygen in the depositional environment most commonly brings about preservation of organic materials. Black shales frequently have carbonized films of organic structures; coal beds similarly preserve plant structures in more three-dimensional form; and at least one low-rank coal (Eocene lignite in the Giesel Valley, Germany) has yielded animal remains in which cell microstructure is preserved (Weigelt, 1935). Low temperature conditions are responsible for the preservation of Pleistocene mammoth flesh in arctic frozen ground. Dryness has been responsible for the preservation of dinosaur and other reptile skin in Mesozoic desert deposits.

Indirect preservation of soft-part structures may take place under special circumstances. Such is the mold of a rhinoceros in Oligocene lavas (Chappel et al., 1951), external molds of insects in Tertiary amber, and replacement of insect tissues by silica in Miocene lake deposits (Palmer, 1957).

Inorganic skeletal parts persist longer, but ultimately may be dissolved or mechanically disintegrated in a variety of ways. Solution by acid soil waters or disintegration by freezing and thawing on land, and the activity of shell-boring animals on the sea floor are among these. Even in the absence of mechanical disintegration, slow chemical reactions between shells and sea water may lead to changes in the chemical composition and microscopic structure of the shell. If the organism is quickly buried by sediment or volcanic debris, it may be protected from much of this destruction. The majority of fossils are preserved in this way.

Once buried, some skeletal parts may remain essentially unchanged in composition and structure for many geologic periods. This is very rare for pre-Mesozoic fossils. Generally, chemical reactions with connate waters, or with subsequently introduced meteoric waters, leads to precipitation of mineral matter in pore spaces within the fossil (*permineralization*), and/or to partial or complete replacement of the original skeletal material. Recent work has shown that changes in trace-element composition may be essentially universal, even when no other indication of alteration is apparent (Curtis and Krinsley, 1965). In some cases, the original material is completely removed without being replaced by another substance. If this takes place after the enclosing sediment has become rigid, a void will remain, whose walls will be a *mold* or *imprint* of the fossil. An imprint of the outer surface of a shell is called an *external mold,* while that of the interior is called an *internal mold.* If the surrounding sediment was still plastic at the time the shell dissolved, the void may be closed under pressure of the overburden and a *composite mold* (McAlester, 1962) may be formed, in which internal and external surface imprints are superimposed (Fig. 1). If an internal mold forms a free-standing object, as in the case of hollow shells or bones, it is sometimes loosely called a *cast,* though more properly a *steinkern.* However, the internal mold of a skull may correctly be called a cast of the brain, inasmuch as the inner surface of the skull is crudely molded to the brain surface. Footprints, trails, and burrows are entirely preserved as molds or as true casts.

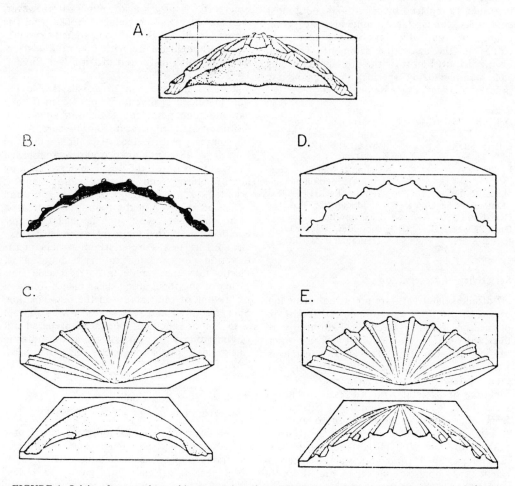

FIGURE 1. Origin of composite mold preservation (from McAlester, 1962). A: Schematic section of original shell embedded in matrix. B: Shell removed by solution leaving only impression in matrix. C: Drawing B expanded to show internal mold with muscle scars (lower block) and external mold with ribbing (lower block). D: Upper and lower blocks forced together by sedimentary compaction thus superimposing internal and external molds. E: Drawing D expanded to show composite molds; muscle scars and ribbing now appear on both upper and lower blocks; the upper block is the "negative" composite mold; the lower block is the "positive" composite mold.

Mineralogy of Fossils

The principal inorganic skeletal materials are the calcium carbonates ($CaCO_3$) aragonite and calcite (although calcite often contains a high proportion of magnesium as $MgCO_3$; Chave, 1964); apatite ($Ca_5[PO_4]_3$ [F, Cl, OH])—inarticulate brachiopods, conodonts, and vertebrates; and opal ($SiO_2 \cdot nH_2O$)—siliceous sponges, diatoms, and most radiolaria. Very much rarer are celestite ($SrSO_4$)—some radiolaria; strontianite ($SrCO_3$)—some gastropods; magnetite (Fe_3O_4)—chiton teeth; and vaterite ($CaCO_3$). In general, the carbonate minerals are most easily dissolved and are usually replaced by nonmagnesian calcite, by dolomite, or by quartz. Opaline skeletons are almost always converted to quartz and not infrequently replaced by calcite. The next most commonly precipitated minerals are compounds of iron, which are derived from iron oxides in the associated sediment. The following list gives most of the principal replacing and permineralizing species, in approximate order of abundance (see Benn, 1955).

Calcite: $CaCO_3$	Selenite: $CaSO_4$
Dolomite: $CaMg(CO_3)_2$	Glauconite: hydrous Fe and K silicate
Quartz: SiO_2	Vivianite: $Fe_3P_2O_8 \cdot 8H_2O$
Pyrite: FeS_2	Uraninite, and other complex uranium minerals
Marcasite: FeS_2	Siderite: $FeCO_3$
"Limonite": $Fe_2O_3 \cdot nH_2O$	Rhodochrosite: $MnCO_3$
Hematite: Fe_2O_3	Chalcopyrite: $CuFeS_2$
Opal: $SiO_2 \cdot nH_2O$	Diopside: $CaMg(SiO_3)_2$
Barite: $BaSO_4$	Tremolite: $CaMg(SiO_3)_4$

Selectivity of Preservation

Not all skeletal parts are preserved with equal ease. Chave (1964) has summarized recent experimental work on mechanical durability of different types of shell material and also observations on solubility of skeletal substances. Resistance to mechanical destruction by agitated waters prior to burial appears to be a function of skeletal structure rather than of mineralogy, and shows significant differences from one animal or plant group to another. Resistance of skeletal material to solution both before and after burial is related to shell mineralogy. Shells of aragonite and magnesian-calcite are more soluble than are those of calcite under the pressure and temperature conditions of depositional environments and of shallow burial such as most fossils experience. Thus, organisms with noncalcitic shells may be relatively less common as fossils than they were in the original living community.

Destruction and Distortion of Fossils

Postdepositional changes do not always lead to the preservation of buried organic remains. The recrystallization attendant upon the precipitation of calcite, silica, and especially dolomite, within the sediment may partly or wholly obliterate the original fabric of the rock including its enclosed organic remains. Bathurst (1964) has summarized recent work in this area and gives an extensive bibliography. Such diagenetic changes pass by insensible degrees into those called metamorphism. When sediments are involved in the high-temperature and high-pressure conditions associated with orogenic activity, including deep burial, igneous intrusion, and crustal distortion, fossil structures are commonly completely lost in the general reorganization of minerals. However, even in highly recrystallized rocks such as schists, fossil shells may occasionally be preserved, though usually with loss of fine structure and much distortion of form (see Bucher, 1953; Neuman, 1965).

Mechanical distortion of fossils is also the rule in folded sedimentary rocks even if they have not undergone reorganization under conditions of high temperature and pressure. A useful summary with extensive bibliography is in Rutsch (1948). Such distortion is accomplished by plastic flow of the fossil material (mainly microfracturing and recrystallization) and is oriented to the direction of compression and extension involved in the folding. It is greatest in incompetent rocks such as shales. Simple compaction of the sediment after burial may also distort fossils by shortening them in the direction perpendicular to the bedding plane. Plastic flow may grade into larger-scale fracturing of the fossil along shear planes followed by healing of the fracture in the new position. Such fractured and recemented fossils are met with among silicified specimens prepared out of limestone with acid, as well as on molds in fine-grained mudstones. (See also Bambach, 1973.)

Techniques of Collecting and Conserving Fossils

Fossils are by no means uniformly distributed in sedimentary rocks, and are commonly concentrated along certain bedding planes, or in local pockets or lenses in a given bed. Indeed the majority of sedimentary rocks are relatively barren of recognizable fossils. Weathered outcrops that have not been recently visited by paleontologists are best for collecting. For invertebrates, chunks of rock may be quarried out with hammer and chisel and then split along the bedding planes. Further preparation of fossils from solid rock involves careful removal of the matrix with needles in a pin vise, dental drills, and similar tools. Various chemical and mechanical techniques can be used to break down the matrix of the bulk rock and free the fossils in certain cases. Siliceous and organic fossils can be recovered from limestone by dissolving the rock in acid. Organic substances (e.g., chitin, cellulose) may be freed from siliceous sediments by hydrofluoric acid. Large and delicate fossils such as bones, must be removed very carefully in the field, often after hardening substances have been applied, and must be packed with great care, frequently being encased in plaster. Molds may be studied by making a cast out of latex or similar elastic material. The various techniques of preparing fossils for study are described in Kummel and Raup (1965).

A fossil specimen is worthless for scientific study without a label indicating the locality and geologic formation from which it was taken. For most studies, a large number of specimens of a given species is needed to assess the limits of morphologic variation. Fossils must also be studied in conjunction with the enclosing sediment so that the environment in which the organisms lived can be reconstructed.

Collections of fossils for reference purposes are maintained in museums; only a small portion of these collections is on public display. Reference collections may be arranged biologically or stratigraphically depending on whether their primary use is for taxonomic and evolutionary studies or for age dating and environmental reconstruction. Type specimens of named species are generally segregated in a special collection because they are the standard of reference for a particular species name, and must be preserved in perpetuity. The largest collections are to be found in the various national museums such as the U.S. National Museum and the British Museum or at the headquarters of the national geological surveys. Many of the larger universities have important collections (e.g., Peabody Museum of Natural History at Yale, Agassiz Museum of Comparative Zoology at Harvard, Sedgwick Museum at Cambridge, Ecole de Mines in Paris, etc.) as have some local museums (e.g., The American Museum of Natural History in New York, the Chicago Natural History Museum, etc.). The last two, as well as the U.S. National Museum, are particularly noted for their public displays.

Use of Fossils

Fossils provide a record of the history of life on the earth, and are the only concrete evidence for the course of organic evolution. In addition, they are useful to the geologist in two ways. They provide information on ancient environments through comparison with living relatives and through study of their functional morphology. They also date rocks in terms of the geologic periods and their subdivisions. The relative geologic time scale was originally established on the succession of fossil faunas, and fossils still provide the most useful means of correlating and dating sedimentary rocks, because absolute age determination by radiometric methods cannot be used to date most sediments directly. Furthermore, the resolution given by paleontologic dating is comparable to that obtained through radiometric means; intercontinental fossil zones are of the order of magnitude of 5 m yr long, and local zones may be still shorter.

ROBERT M. FINKS

References

Bambach, R. K., 1973. Tectonic deformation of composite-mold fossil Bivalvia (Mollusca), *Amer. J. Sci.,* **273A,** 409–430.

Bathurst, R. G. C., 1964. Diagenesis and paleoecology: A survey, in Imbrie and Newell, 1964, 319–344.

Benn, J. H., 1955. The mineralogy of fossils, *Rocks Minerals,* **30,** 3–20.

Breger, I. A., ed., 1963. *Organic Geochemistry.* New York: Macmillan, 658p.

Bromley, R. G., 1970. Borings as trace fossils, and *Entobia cretacea* Portlock as an example, in Crimes and Harper, 1970, 49–90.

Bucher, W. H., 1953. Fossils in metamorphic rocks, a review, *Geol. Soc. Amer. Bull.,* **64,** 275–300.

Calvin, M., 1965. Chemical evolution, *Proc. Roy. Soc. London, Ser.* A, **288,** 441–466.

Chappell, W. M.; Durham, J. W.; and Savage, D. E., 1951. Mold of a rhinoceros in basalt, Lower Grand Coulee, Washington, *Geol. Soc. Amer. Bull.,* **62,** 907–918.

Chave, K. E.. 1964. Skeletal durability and preservation, in Imbrie and Newell, 1964, 377–387.

Crimes, T. P., and Harper, J. C., eds., 1970. *Trace Fossils.* Liverpool: Seel House, 547p.

Curtis, C. D. and Krinsley, D., 1965. The detection of minor diagenetic alteration in shell material, *Geochim. Cosmochim. Acta,* **29,** 71–84.

Eglinton, G., and Murphy, M. T. J., eds., 1969. *Organic Geochemistry—Methods and Results.* Berlin: Springer-Verlag, 828p.

Florkin, M., 1969. Fossil shell "conchiolin" and other preserved biopolymers, in Eglinton and Murphy, 1969, 498–520.

Häntzschel, W., ed., 1975. Trace fossils and problematica, *Treatise on Invertebrate Paleontology,* pt. W, Miscellanea, suppl. 1. Lawrence, Kansas: Geol. Soc. Amer. and Univ. Kansas Press, W3–W269.

Imbrie, J., and Newell, N. D., eds., 1964. *Approaches to Paleoecology.* New York: Wiley, 432p.

Kauffman, E. G., and Kesling, R. V., 1960. An Upper

Cretaceous ammonite bitten by a mosasaur, *Contrib. Mus. Paleontol. Univ. Michigan*, **15**, 193–248.

Kummel, B., and Raup, D., eds., 1965. *Handbook of Paleontological Techniques.* San Francisco: Freeman, 852p.

Kvenvolden, K. A., 1975. Advances in the geochemistry of amino acids, *Ann. Rev. Earth Planetary Sci.*, **3**, 183–212.

McAlester, A. L., 1962. Mode of preservation in Early Paleozoic pelecypods and its morphologic and ecologic significance, *J. Paleontology*, **36**, 69–73.

Neuman, R. B., 1965. Collecting in metamorphic rocks, in Kummel and Raup, 1965, 159–163.

Palmer, A. R., 1957. Miocene arthropods from the Mojave Desert, California, *U.S. Geol. Surv. Prof. Paper 294-G*, 237–280.

Richardson, E. S., Jr., ed., 1971. Extraordinary fossils, *Proc. N. Amer. Paleontological Conv.*, pt. I, 1155–1269.

Rutsch, F. R., 1948. Die Bedeutung der Fossil-deformation, *Bull. Assoc. Suisse Géol. Petrol.*, **15.**

Schaefer, W., 1972. *Ecology and Paleoecology of Marine Environments.* Chicago: Univ. Chicago Press, 568p.

Schopf, J. M., 1975. Modes of fossil preservation, *Rev. Paleobot. Palynol.*, **20**, 27–53.

Seilacher, A., 1964. Biogenic sedimentary structures, in Imbrie and Newell, 1964, 296–316.

Weigelt, J., 1935. Some remarks on the excavations in the Giesel Valley, *Research Progress*, **1**, 155–159.

Zangerl, R., and Richardson, E. S. Jr., 1963. The paleoecological history of two Pennsylvanian black shales, with contributions by B. G. Woodland, R. L. Miller, R. C. Neavel, and H. A. Tourtelot, *Fieldiana, Geol. Mem.* 4, 352p.

Cross-references: *Biogeochemistry; Biostratinomy; Bones and Teeth; Burgess Shale; Chitin; Coprolite; Diagenesis of Fossils; Ediacara Fauna; Fossil Record; Mazon Creek; Paleobotany; Silicification of Fossils; Solnhofen Formation; Taphonomy.* Vol. VI: *Concretions; Coquina, Criquina.*

FUSULINACEA

The extinct superfamily Fusulinacea, of the protozoan order Foraminiferida, first appeared about 325 m yr ago (in Late Mississippian time) and became extinct about 225 m yr ago (near the end of Permian). During this interval, fusulinaceans evolved into 6 families, about 150 genera, and about 6000 species that have been extensively studied and used in stratigraphic correlation because of their rapid evolutionary rates (Fig. 1; Ross, 1967; Kanmera et al., 1976; Douglass, 1977). The usual method of studying fusulinaceans is by a series of thin sections pass-

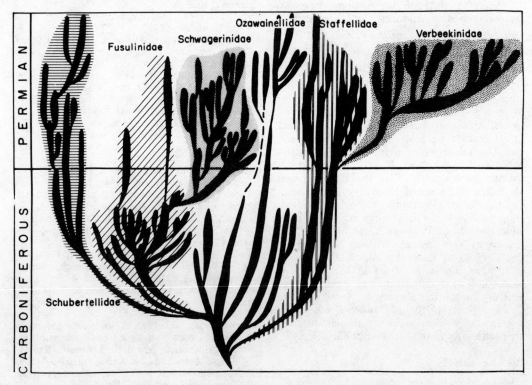

FIGURE 1. Generalized phylogeny of the superfamily Fusulinacea. Evolutionary radiation within families shows the phylogentic relationships of major genera.

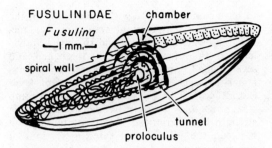

FIGURE 2. Shell of a typical genus of Fusulinidae showing construction and spiral arrangement of chambers.

ing through the proloculus and oriented in the plane of the axis of coiling and perpendicular to the axis of coiling.

Fusulinacean shells were constructed (Fig. 2) with a proloculus (the initial calcareous spherical shell) which has a small opening. Chambers were added as the volume of protoplasm increased and were added in a planispiral coil around the proloculus. In most families, successive chambers become increasingly elongate parallel to the axis of coiling and the shell resembles a grain of wheat in general shape and size.

The frontal wall of the chambers, which formed a series of septa as the shell increased in size, has a number of relatively widely spaced pores (septal pores) about 10μ in diameter. Within the shell, a tunnel is formed by resorption of the central part of the septa near the

floor of the chambers; and this feature connects all but the outer three or four chambers, presumably to permit the nucleus to move toward the outer part of the shell. A few genera in the Late Permian have multiple tunnels. In many genera, particularly during the Pennsylvanian, secondary deposits (chomata) formed rounded borders on the sides of the tunnel (Fig. 3).

Several general evolutionary trends are common in fusulinaceans (Thompson, 1964). The most noticeable evolutionary trend is a progressive increase in total size beginning from less than 1 mm in length in the Late Mississippian to a few species reaching more than 200 mm in length in the later part of the Permian. Most Pennsylvanian and Permian forms are large enough to see in rock hand specimens without the use of a microscope and are fairly easily located in the field. A second evolutionary trend is the increasing complexity of shell structures and geometry with time. The shell changes from a lenticular shape to one in which the chambers are elongated parallel to the axis of coiling. This is usually accompanied either by septa that are folded to form a corrugated frontal support for the spiral outer wall, or by septa that are reinforced by secondary organic-rich calcareous deposits. Some families and genera have both folded septa and secondary layering. Different geometries of septal folds are useful in studying evolutionary lineages; some septa have long, low, forward-projecting folds at their base, others have steep, evenly rounded septal folds, and still others

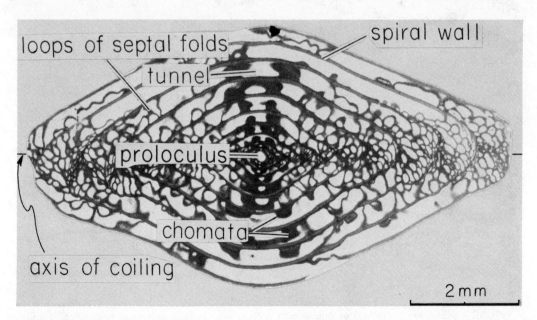

FIGURE 3. Thin section of *Triticites,* family Schwagerinidae, cut in a plane of the axis of coiling and showing a number of interior shell features.

may have square or even pointed fold crests as seen in thin section. In several lineages, opposing folds of adjacent septa touch and in some lineages resorption of these septa where they touch results in small passageways (cuniculi) between chambers.

The spiral outer wall (spirotheca) is formed by several layers (Fig. 4). The first formed layer is a thin organic-rich layer (tectum). Inside this layer there is a thicker predominately calcareous layer (diaphanotheca). These two layers constitute the primary wall. In several families, the chambers are coated by a thin veneer of secondary organic-rich calcareous material (tectorial deposits). The evolutionary trend in Early and Middle Pennsylvanian fusulinaceans results in these secondary deposits occupying first the floor of the chamber near the tunnel and chomata and then progressively coating more of the floor, then the sides, and finally the top of the chambers. This leads from an apparent three-layer (profusulinellid) wall to an apparent four-layer (fusulinellid) wall (Fig. 4a,b.).

The spirotheca has a number of closely spaced pores that pass through it and that permitted the protoplasm within the chambers to have communication with the exterior of the shell. In the schwagerinid wall (Fig. 4c), tectorial deposits are lacking and the layer equivalent to the diaphanothaca is thickened and several small pores of the tectum unite to form larger passages through the translucent layer (keriotheca). This gives the wall a honeycomb structure that apparently had considerable strength and permitted closer communication between the interior and exterior portions of the protoplasm. The neoschwagerinid wall (Fig. 4d) is also characterized by a thick keriotheca and, in addition, has pendants (septula) that strengthen the wall in both longitudinal and transverse directions. The yabeinid wall, (Fig. 4e), an evolutionary outgrowth of the neoschwagerinid wall, has primary and secondary transverse and longitudinal septula and a thin, coarsely alveolar layer. A second evolutionary development from the neoschwagerinid wall structure is the afghanellid wall (Fig. 4f) in which the spirotheca has a very thin, finely alveolar layer and primary and secondary septula are thin. In the staffellids, the original wall is recrystallized, commonly to a sparry calcite, suggesting the calcareous material may have been deposited as aragonite which has re-

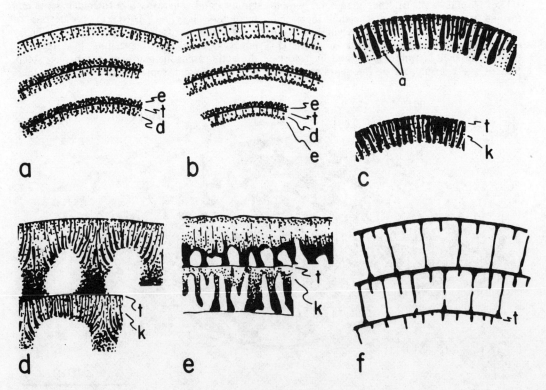

FIGURE 4. Microstructure of various fusulinacean walls as seen in axial sections. a, profusulinellid wall; b, fusulinellid wall; c, schwagerinid wall; d, neoschwagerinid wall; e, yabeinid wall; f, afghanellid wall. (t = tectum; d = diaphanotheca; e = tectorial deposits; and k = keriotheca)

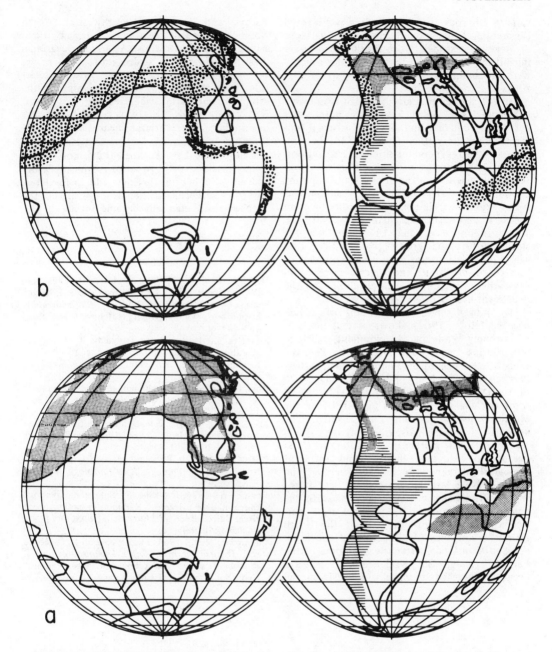

FIGURE 5. Biogeographic realms of fusulinaceans. a, near end of middle Carboniferous; b, near end of Early Permian. Light stippled pattern shows the Tethyan-Ural-Arctic realm of the Carboniferous and the Ural-Arctic realm of the Permian. Horizontal pattern is the southwestern North America-Andean realm of the Carboniferous and Permian. Coarse heavy stippled pattern is the Tethyan realm of the Permian.

crystallized during diagenesis. Some genera of the Schubertellidae also show similar recrystallized features and many of the latest Permian fusulinaceans are genera having recrystallized walls.

Fusulinaceans were marine protozoans that were commonly associated with calcareous algae and shallow-water marine invertebrates. In general, they preferred shallower water than did most of their contemporary crinoid and arenaceous foraminiferal assemblages and are commonly associated with corals, productoid brachiopods, and occasional fenestrate bryozoans. Fusulinaceans are common in a wide

335

variety of rock types including well-sorted calcarenite, light gray micrite, and quartzose or clayey limestones. Most fusulinaceans were inhabitors of the photic zone and probably were predominately in the upper part of that zone.

Growth studies of Early Permian fusulinaceans suggest they lived 5 to 6 or more years before reproducing and lived in tropical and warm temperate waters (Ross, 1972b). In addition to their general ecological associations, certain shell features in many lineages, such as the development of alveolar walls and thinning of the spiral wall, suggest that fusulinaceans had a symbiotic relationship with one or more species of photosynthetic partners (Ross, 1974). The possibility exists that the extinction of a single autotrophic group that had evolved symbiotic relationships with a large number of specialized protistans and invertebrates could cause the extinction of a large number of the dependent heterotrophic groups.

The biogeographic distribution of fusulinaceans (Ross, 1967) shows several intervals of strongly endemic development, particularly during the middle part of the Pennsylvanian, latest part of the Pennsylvanian, earliest Permian, and late Early and Late Permian. Two of these intervals are illustrated (Fig. 5) to show the changes that took place between the Middle Pennsylvanian and Late Permian. The amount of species dispersal between these realms differed markedly at different times because of climatic changes and different ecological tolerances of newly evolving fusulinacean lineages; but probably most significant was the orogenic activity associated with the continental margins in geosynclinal areas, particularly in the northern part of the North American northern Cordilleran and the Eurasian southern Urals. After the beginning of the Early Permian, the Ural–Franklin seaways became progressively more isolated from the Tethys and from the southwestern North American–Andean seaways, forming three strongly endemic realms. Although occasional invasions from one realm into another was accomplished by a few species, filtering selection was sufficiently strong to prevent the establishment of most species and only a few were more than briefly successful. By the beginning of Late Permian, nearly all the Ural–Franklin fusulinaceans became extinct; by the middle of the Late Permian time, most of the southwestern North American–Andean fusulinaceans also became extinct; and in the latest part of the Permian, the fusulinaceans of the Tethyan realm gradually became extinct.

CHARLES A. ROSS

References

Douglass, R. C., 1977. The development of fusulinid biostratigraphy, in E. G. Kauffman and J. E. Hazel, eds., *Concepts and Methods of Biostratigraphy.* Stroudsburg, Pa.: Dowden, Hutchinson and Ross, 463–481.

Kanmera, K.; Ishii, K.; and Toriyama, R., 1976. The evolution and extinction patterns of Permian fusulinaceans, *Contrib. Geol. Paleontol. Southeast Asia,* 17, 129–154.

Ross, C. A., 1967. Development of fusulinid (Foraminiferida) faunal realms, *J. Paleontology,* 41, 1341–1354.

Ross, C. A., 1972a. Paleoecology of fusulinaceans, *23rd Internat. Geol. Congr., 1968, Proc. Internat. Paleont. Union,* 301–318.

Ross, C. A., 1972b. Paleobiological analysis of fusulinacean shell morphology, *J. Paleontology,* 46, 719–728.

Ross, C. A., 1974. Evolutionary and ecological significance of large calcareous Foraminiferida (Protozoa), Great Barrier Reef, *2nd Internat. Symp. Coral Reefs, Brisbane, Australia,* 1, 327–333.

Thompson, M. L., 1964. Fusulinacea, in R. C. Moore, ed., *Treatise on Invertebrate Paleontology,* pt. C, Protista 2, vol. 1, Sarcodina. Lawrence, Kansas: Geol. Soc. Amer. and Univ. Kansas Press, 358–436.

G

GASTROPODA

The gastropods are a long lived class of mollusks dating back to the Late Cambrian. Unlike the fossil record of such groups as the Cephalopoda and Brachiopoda, their pattern of diversification at higher taxonomic levels does not show periodic restrictions or major episodic extinctions over geologic time. Rather, gastropods represent an extremely plastic or adaptive group of organisms on the rise through geologic time, exhibiting successive periods of adaptive radiation and diversification. The various estimates of living species ranging from 105,600 (Jaeckel, 1958) to 37,500 (Boss, 1971) are a reflection of an extraordinarily viable group. They represent about 80% of the diversity of the phylum Mollusca and rank second only to the insects in terms of diversity as a class.

Relative to habitats and feeding habits, gastropods must also be counted as among the most diverse and successful of animal groups. They range through almost all terrestrial, freshwater, and marine niches, and may be grazers on vegetation, carnivores, or even full parasites. They crawl, burrow, swim, float, or are sedentary.

General Characters

The generalized gastropod is characterized by having an elongate, flattened, broad foot surface; a distinguishable head bearing eyes and tentacles; and a mouth area equipped with a radular apparatus. Although bilateral symmetry may commonly be approached in external aspect, as a result of "torsion," asymmetry is seen in uneven development or twisting of most organ systems. The majority of species possess either an asymmetrical helically coiled shell or a symmetrical cap or conical-shaped shell, but the tendency toward shell reduction and loss is common throughout the class.

In most shelled forms, only the head-foot mass is protrusible and the visceral mass—which includes the mantle-covered respiratory apparatus and digestive, circulatory, reproductive, and portions of the nervous systems—remain within the shell (see Figs. 1 and 2). The head-foot mass is retracted into the shell by means of the columellar (retractor) muscles, which are attached along the columellar surface of the shell axis. During retraction, the head is first withdrawn and then the foot. When present, the operculum, which is attached at the posterior dorsal foot surface, is the last withdrawn, thus effectively sealing the aperture of the shell. A larval operculum is common to many anatomically advanced species that lack the structure when adult.

Like other mollusks, gastropods have the mantle cavity, gills, and anus lying at the posterior end of the body during early embryonic developmental stages. In early ontogeny, the drastic process known as torsion occurs. During this process, the mantle cavity and contained organs are rotated approximately 180° relative to the head-foot mass resulting in an anterior orientation of these organs above and behind the head. Internally, the long nerve connectives are twisted and crossed into a figure 8 and organs of the left assume a right-side orientation. Torsion is accomplished in a period of days and not as a single rapid event. Torsion is not to be confused with the assymetrical coiling of the shell, as it occurs equally in species having symmetrical shells. Two theories to explain the origin of torsion exist. Most commonly, torsion is viewed as a single mutation that was of prime importance to the larvae, as it allowed the head and velum to be retracted into the shell for protection. Such a mutation is likely to have occurred only once and thus leads to a monophyletic origin of the Gastropoda. The second theory is that torsion arose in a limpet-like ancestor and was selected for because it aided the coiled larvae to balance its shell after settling. Thus, torsion is seen primarily as an adaptation for gravitational stability rather than protection. In addition, this view of the gradual development of torsion would allow for a polyphyletic origin of the Gastropoda.

Classification

Higher taxonomic categories of the Gastropoda have been based most commonly upon anatomical (neontological) characteristics as: *respiratory system, heart structure, radular character,* and *arrangement of the nervous system.*

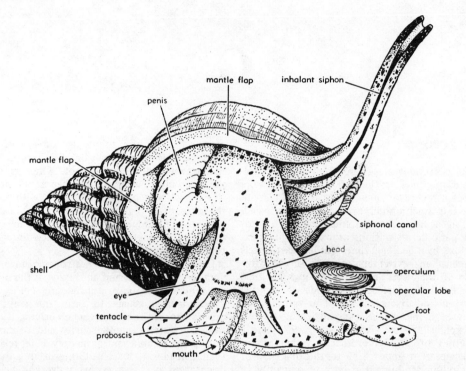

FIGURE 1. *Buccinium undatum* Linné (♂), anterior view of shell with protruded head-foot mass (from Knight et al., 1960; courtesy of The Geological Society of America and The University of Kansas Press).

Classifications based upon single-system analysis have proven generally unsatisfactory, as they are not coordinate with classifications based upon other organ structures. They all reflect phylogeny only to the extent of grouping together taxa of similar anatomical complexity relative to a single organ system. Obviously, classifications based upon organ systems have distinct disadvantages for the

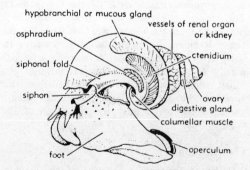

FIGURE 2. *Buccinium undatum* Linné (♀), with head-foot mass protruded but with shell treated as transparent to show positions of various organs, X0.5 (from Knight et al., 1960; courtesy of The Geological Society of America and The University of Kansas Press).

paleontologist. This is especially true in the Gastropoda because reflection of soft-part morphology is poorly represented in the shell characters (Fig. 3). Muscle-scar impressions and canal characters occassionally give some clue to soft-part anatomy and life habit, but compared to the Bivalvia this is more the exception than the rule. The lack of such all-pervasive features as the hinge line of the Bivalvia plus the rampant multiconvergence and divergence in shell form among the Gastropoda make suspect a unified classification based upon shell features. Classification of the class is still in a state of flux, as exemplified by the current shift of the Pyramidellidae from one subclass (Streptoneura) to another (Euthyneura) on the discovery that they were euthyneurous, possessed an invaginable penis, and lacked esophageal glands. With the exception of the heterostrophic larval shell, the adult shell was, however, more typical of the prosobranchs in form (Fretter and Graham, 1962).

For the paleontologist, the most acceptable arrangement remains that of Wenz (1938), Knight et al. (1960), and Taylor and Sohl (1962). Proportional size relative to numbers of included genera in major groups is shown in Fig. 4. That such arrangements will change greatly in the future is a foregone conclusion

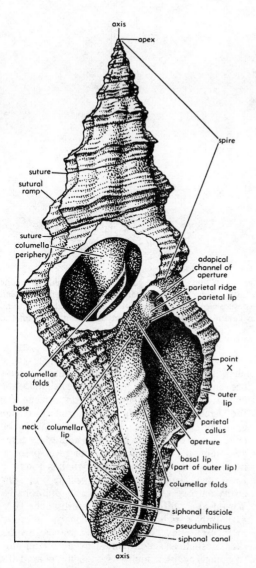

FIGURE 3. Typical gastropod shell, *Latirus lynchi* (Basterot), Miocene of France, showing terminology of its various parts (from Knight et al., 1960; courtesy of The Geological Society of America and The University of Kansas). The columella is seen through a "window" in the last whorl.

when, with the new technological tools now available, sufficient data is compiled on such features as shell structure, morphology of nuclear whorls (protoconch), chromosome counts, and chemistry of shell and soft parts. Study of these characters and others should provide a system of checks on proposed groupings relative to closeness of relationship of taxa. Computer techniques offer the advantage of interrelating the various classifications based upon unit characters.

Most of the following discussion is structured within the frame of the following subclass and order arrangement of the gastropods.

Subclass PROSOBRANCHIA (or STREPTONEURA)– visceral nerve cords cross, loop forming a figure 8; auricle anterior to ventrical
    Order Archaeogastropoda–bipectinate ctenidia, non-siphonate shell
    Order Mesogastropoda–monopectinate ctenidia where present; siphonate to nonsiphonate shells
    Order Neogastropoda–radula rachiglossate or toxiglossate, shells with well-developed siphons
Subclass OPISTHOBRANCHIA (or EUTHYNEURA)– nerve cords not crossed, auricle posterior to ventricle; contains numerous orders
Subclass PULMONATA–mantle cavity modified for air breathing

Origin and Evolution

Origin of the class Gastropoda is currently in dispute. One group proposes that the first true gastropods are represented by the Bellerophontacea which appear first in the Upper Cambrian (Knight et al., 1960). This view holds that the bellerophontids are torted and lie intermediate between planispiral monoplachophorans and helically coiled primitive gastropods. An alternative view is that the group arose from the monoplacophorans, with the helically coiled Lower Cambrian forms *Pelagiella* and *Aldanella* as the earliest representatives; and that the Bellerophontacea are untorted mollusks belonging in the Helcionellacea–Monoplacophora lineage (Runnegar and Pojeta, 1974).

Whichever of the aforementioned alternatives one accepts, most agree that gastropods arose from a bilaterally symmetrical nontorted ancestral mollusk. Whether torsion is accepted as being more advantageous to the larval animal or to the adult, the change in direction of the mantle cavity to an anterior position conveys some obvious advantages. Firstly, it allows for the head to be withdrawn into the shell before the foot mass, a protective function. In addition, it places the respiratory organs in a forward position above the head and mouth and thus aids in water circulation as the animal moves forward, prevents fouling of gills by sediment stirred up during movement, and places sense organs such as the osphradium in the direction of movement. To some degree, such advantages are countered by the problem of potential cavity and gill fouling encountered in the concomitant placement of the anus over the head. This sanitation problem has been solved in different ways in different lineages, but center on increasing the efficiency of respiratory inhalent and exhalent currents which carry off waste products (Yonge, 1947).

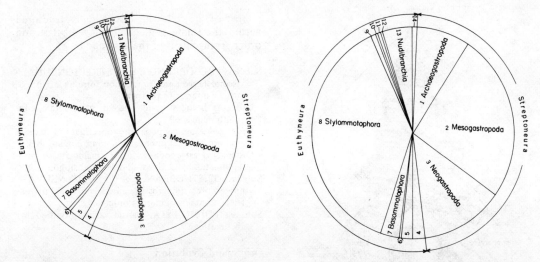

FIGURE 4. Relative size of subclasses and orders of, left, fossil and Holocene gastropods and, right, Holocene gastropods (from Taylor and Sohl, 1962). 4, Parasita and Entomotaeniata; 5, Cephalaspidea, Acochlidioidea, and Philinoglossoidea; 6. Thecosomata; 9, Sacoglossa; 10, Anaspidea; 11, Gymnosomata; Notaspidea; Soleolifera. Left, arcs are proportional to numbers of genera given in Taylor and Sohl, 1962. Right, arcs are proportional to numbers of genera in Thiele's 1929–1935 classification.

The higher streptoneurous (prosobranch) gastropods show a reduction of gills to a single monopectinate ctenidium (leaflets on only one side of main axis) to the left of the head with the anus to the right. Respiratory currents enter on the left, bathe the ctenidium, and exit on the right, voiding the waste products. In most forms of this type, as a consequence of—or concomitant with—helical coiling, the organs of the right side of the body are reduced or lost. Assumption of such characters plus modifications of the radular elements and elaboration of the reproductive system to allow for internal fertilization has led to the great radiation of the Mesogastropoda and Neogastropoda in most marine habitats as well as their invasion of both freshwater and terrestrial environments. For example, such Archeogastropoda as the emarginulids, patellids, and trochids possess a respiratory system that is poorly equipped to handle large sediment loads (see Fig. 5A), and these groups are to be found predominantly in areas of firm substrate. In contrast the neogastropod buccinids (Fig. 5B) are good examples of forms with superior respiratory apparatus which has aided in their exploitation of a wide range of substrata including tidal mud flats.

Loss of Shell. Acquaintance with the great beauty and diversity of shells leads paleontologists to equate gastropods with shells. Significant reduction of shell or total shell loss and assumption of a slug-like form (Fig. 6A) occurs in approximately one third of the gastropod superfamilies. Predictably, this multiconvergent phenomenon is most common in those groups organized into higher functional-anatomic grades. Thus, the slug form is very rare among the Archaeogastropoda and restricted to one genus of the Neritacea, a group that on several anatomic grounds has been considered by some as a separate order. At a higher level of organization, reduction and loss of shell is common to most superfamilies especially in the euthyneurous opisthobranchs and pulmonates. Within the whole group of nudibranchs (Fig. 6B), only a patelliform nuclear cap is retained in its ontogeny. Within these advanced forms, loss of shell is accompanied by detorsion and, in some, by placement of the gills behind the head and hollowing of the foot mass to house the visceral mass.

The wide variety of habits and feeding types represented among shell-less gastropods demonstrates that no one sweeping generalization as to adaptive value of shell loss pertains to all, except perhaps increased mobility. Although members of a number of groups of snails are capable of swimming, it is only among such genera as the shell-less *Pterotrachea* of the Atlantacea that efficient swimming occurs (Fig. 6C). Similarly, some authors have suggested the slug-like shape in land snails may be of advantage for squeezing through tight places in search of food, e.g., *Testacella*, which feeds on earthworms in their burrows.

Adaptive Radiation. Assessment of adaptive radiation in the Gastropoda is commonly viewed from the standpoint of modifications of

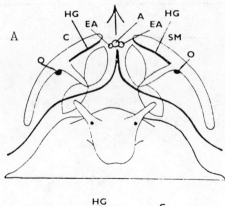

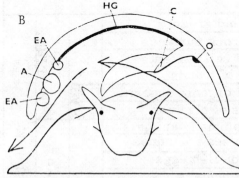

A. *Limax*

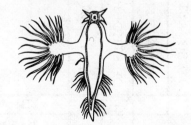

B. NUDIBRANCH *Glaucus*

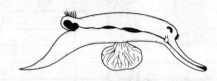

C. *Pterotrachea*

FIGURE 6. Representative shell-less gastropods.

FIGURE 5. A: Diagrammatic transverse section through the head and mantle cavity of a diotocardian prosobranch such as *Diodora*. The arrows indicate the course taken by the respiratory currents which enter symmetrically from both sides, anteriorly, and emerge dorsally through the "keyhole". On each side of the mantle cavity there is an aspidobranch ctenidium (C), an osphradium (O) and a hypobranchial gland (HG). The anus (A) and the excretory orfices (EA) open dorsally in the median line and waste products are voided into the dorsally directed exhalant water current. B: Diagrammatic transverse section through the head and mantle cavity of a monotocardian proso-branch such as *Buccinum*. The arrows indicate the course taken by the respiratory current which passes through the mantle cavity from left to right. A single osphradium (O) lies at the base of the one remaining pectinibranch ctenidium (C), in the path of the in-halant water current. The hypobranchial gland (HG) covers an extensive area of the roof of the mantle cavity above the ctenidium. The anus (A) and the ex-cretory apertures (EA) open into the mantle cavity on the right side, and discharge waste material into the exhalant water current. (From Graham, 1948.)

the buccal mass and radula. Such a perspective basically relates to solving the problem of feed-ing in various ways on a large variety of mate-rials. One should, however, also consider the role such organs as the gills play in food gather-ing and not be quite so restrictive in outlook. Recently, some advances have been made in

interpreting gastropod shell form as well (e.g., Linsley, 1977, 1978).

Most gastropods may be considered feeders on small particles. A few, such as the opistho-branch *Philine* with its gizzard modified for ingestion and crushing of whole small mollusks, feed on larger units.

Fig. 7 summarizes the presently known feed-ing habits common to the class. As indicated, the Archaeogastropoda mostly maintain a simple mode of feeding as browsers and grazers, or raspers of rock surfaces. Only one or two cases of predatory behavior are known. During their heyday of diversity, the Paleozoic, how-ever, they may have possessed a wider spectrum of habits as is indicated by the capuliform genus *Platyceras* (Ordovician–Permian), a demonstrably coprophagous symbiont on crinoids and cystoids.

The Mesogastropoda represent perhaps the greatest radiation of the Gastropoda relative to feeding types. This diversification is materially expanded by their spread into nonmarine habi-tats. The Neogastropoda, although generically diverse, have exploited few major feeding methods. Most are predaceous carnivores. Some have carried carnivorous specialization to a very

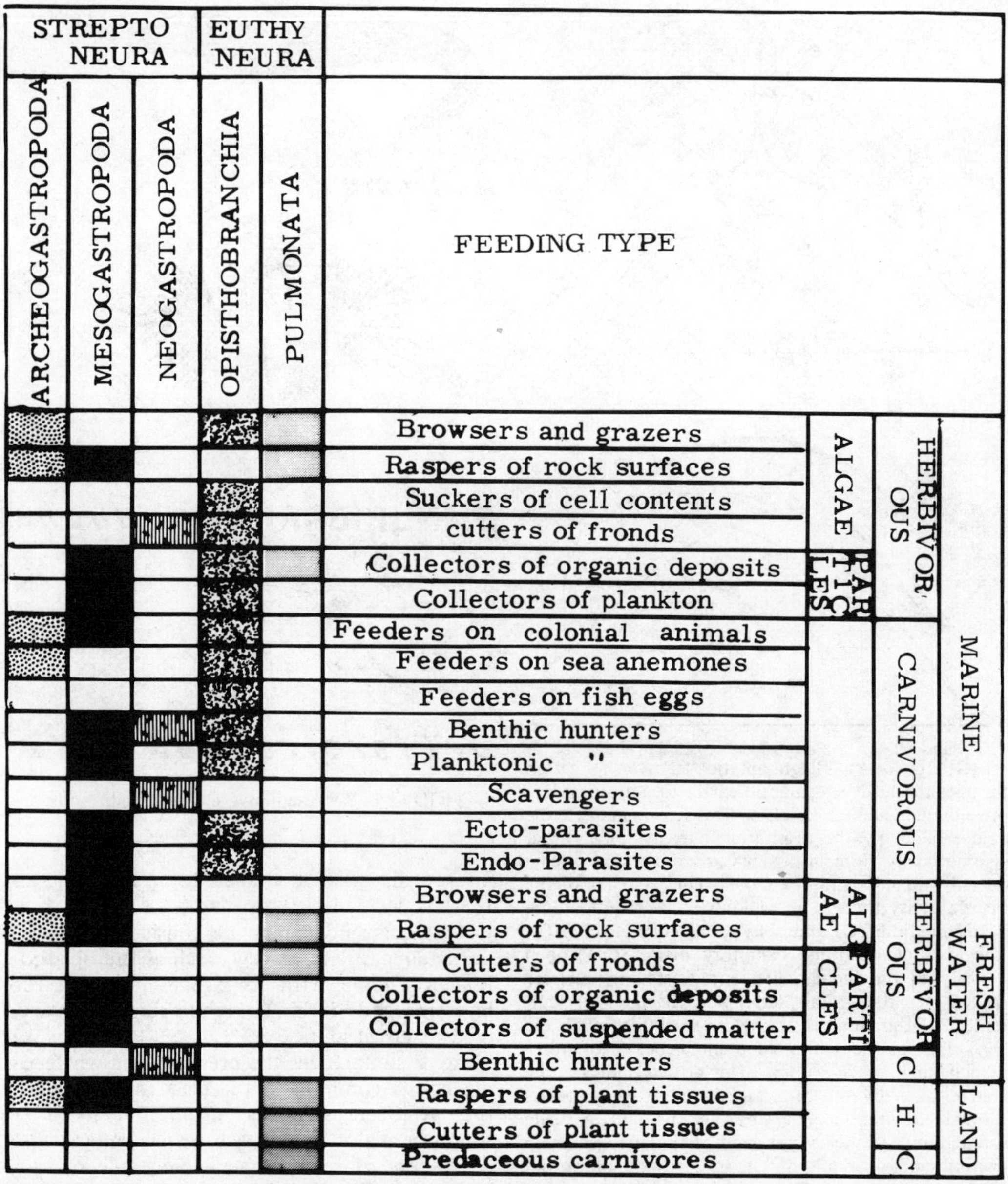

STREPTONEURA			EUTHYNEURA		FEEDING TYPE			
ARCHEOGASTROPODA	MESOGASTROPODA	NEOGASTROPODA	OPISTHOBRANCHIA	PULMONATA				
▒			▓	█	Browsers and grazers	ALGAE PARTICLES	HERBIVOROUS	MARINE
▒	█			█	Raspers of rock surfaces			
			▓		Suckers of cell contents			
		▒	▓		cutters of fronds			
	█		▓		Collectors of organic deposits			
	█		▓		Collectors of plankton			
▒	█		▓		Feeders on colonial animals		CARNIVOROUS	
▒	█		▓		Feeders on sea anemones			
			▓		Feeders on fish eggs			
	▒	█	▓		Benthic hunters			
	█		▓		Planktonic "			
	█	▒			Scavengers			
	█		▓		Ecto-parasites			
	█		▓		Endo-Parasites			
	█				Browsers and grazers	ALGAE PARTICLES	HERBIVOROUS	FRESH WATER
▒	█				Raspers of rock surfaces			
				▒	Cutters of fronds			
	█			▒	Collectors of organic deposits			
	█				Collectors of suspended matter			
	▒				Benthic hunters		C	
▒				▒	Raspers of plant tissues		H	LAND
				▒	Cutters of plant tissues		H	
				▒	Predaceous carnivores		C	

FIGURE 7. Occurrence of feeding types among the Gastropoda (compilation modified after Purchon, 1968).

advanced degree: in the Toxoglossa (*Conus* etc.), a neurotoxin is introduced into the prey when pierced by a harpoon-like everted radular tooth.

Among the Euthyneura, the opisthobranch groups show a lesser diversity of feeding types than do, for example, the Mesogastropoda. In part, this is because they have not invaded terrestrial environments and only one group, the Succineids, has managed the transition to fresh water. In the marine environment, many of the opisthobranchs show a high degree of specialization and, as in some Sacoglossa, slit and suck out the contents of algal cells. Certain sacoglossan species may feed on only one species of algae. Other opisthobranchs possess modifications of the buccal mass in the form of crushing plates, which allow for ingestion and crushing or trituration of shelled organisms, and have departed from the normal gastropod habit of small-particle feeding.

The Pulmonata are the most efficient gastropod colonizers of the terrestrial realm, even to the extent of occupying arboreal niches. Many live in fresh or brackish water and a few have presumably reverted to a marine life style (Siphonariacea). The marine forms are primarily browsers and raspers. Many terrestrial forms feed on leaf or other plant materials including fungi; others are carnivorous.

For paleontologists involved in reconstruction of ancient community structures, knowledge of gastropod feeding habits can give some indication of the presence of organisms that are seldom preserved. For example, the presence of cowries may suggest the presence of soft-bodied forms such as compound ascidians, though none may be preserved. Similarly, many pyramidellid species feed on the body fluids of worms, while epitoniids feed on anemones, and *Cerithiopsis* and *Triphorus* may suggest the presence of their prime food source the sponges. Large numbers of the Archaeogastropoda are feeders on fine algal growth (rhipidoglossans) while the docoglossan limpets are primarily algal raspers and the Columbellinidae feed on brown algae. Thus, abundance of these groups may give some indication of plant abundance and, to some degree, type of vegetation that may not be gained in some other fashion and provides an additional dimension to trophic analysis of fossil communities. Such obviously environmentally restricted gastropods as the intertidal nerites and limpets are in themselves good indicators of paleoenvironmental setting. Most gastropods, however, have sufficiently wide environmental tolerances to contribute only as part of the whole preservable spectrum of organisms available for community and paleoenvironmental interpretation. In analysis of fossil communities, normally only such herbivorous algal feeders as trochids, or suspension feeders as *Turritella* are likely to form trophic nuclei species.

Fossil Record

Acceptance of the earliest record of fossil gastropods is dependent upon one's view as to the origin of the group and would range from Lower Cambrian to Upper Cambrian. In either instance, the Gastropoda are neither common nor even moderately diverse until about the Lower Ordovician, when many of the groups of the Pleurotomariacea, Macluritacea, Euomphalacea, Platyceratacea, and Murchisoneacea become well represented. The variety of shell form present encompasses most of the major types present in the class even at this early stage and ranges from planispiral and low trochiform to the high-spired murchisoniids.

Patelliform shape first appears during the Middle Silurian as represented by the first members of the Patellacea. During the Late Paleozoic, cerithiacean forerunners, such as *Kinishbia* Winters that possess a short siphonal canal, appear as well as genera with columellar plications (*Labridens* Yochelson, *Cylindrotropis* (Gemmellaro, etc.). Thus in terms of shell form, all major types of shells are present save for the reduced or thin-shelled bullomorph typical of later opisthobranchs and the elongation of the inhalent canal typical of the higher carnivorous snails.

Several other major events in gastropod evolution transpired during the late Paleozoic. Chief among these was the invasion of both freshwater and terrestrial environments. The earliest freshwater snail appears to be the viviparid genus *Bernicia* Cox from the Lower Carboniferous of England. Invasion of the land is recorded by the pulmonate pupiform shells of the genera *Dendropupa* Owen and *Anthracopupa* Whitfield in the Upper Carboniferous (Pennsylvanian) of eastern North America. In addition, the first opisthobranch gastropods occurred by at least Early Carboniferous and perhaps as early as the Devonian. Thus, by the end of the Paleozoic, representatives of all classes of gastropods were present, and the transition from marine to freshwater to land environments was established. The stage was set for the great diversification and radiation that transpired during the later Mesozoic and early Tertiary (Fig. 8).

At the superfamily level, the period of greatest taxa introduction occurred during the Jurassic and Cretaceous: 13 superfamilies first appeared in Jurassic times and 19 during the Cretaceous. Thus, by the end of the Cretaceous, the gastropod faunas had a dominantly modern aspect (Sohl, 1964). At the family level, the greatest increase by far is during the Cretaceous (Fig. 9). Relative to extinctions, the Mesozoic was a time of changeover from the types of gastropods that dominated the Paleozoic and Triassic faunas to the modern counterparts. Thus, extinction at the family level (Fig. 9f-j) was at a proportionally high rate throughout Mesozoic time but very stable through the Tertiary.

Prosobranch history, seen at the ordinal level, shows the Archaeogastropoda diversifying rapidly at the familial level through the Ordovician and leveling off in diversity from the Devonian through most of the remainder of the Paleozoic. With the exception of the appearance of the Trochacea in Triassic times, the Archaeogastropoda have presently declined in familial and generic diversity to little more than half of their Paleozoic diversity (Figs. 8, 9). Though

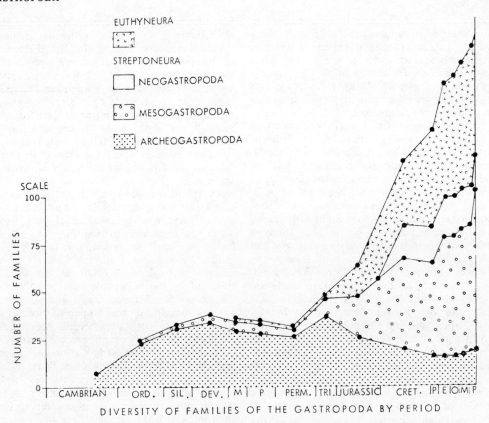

EUTHYNEURA

STREPTONEURA

NEOGASTROPODA

MESOGASTROPODA

ARCHEOGASTROPODA

SCALE

NUMBER OF FAMILIES

100—

75—

50—

25—

0—

CAMBRIAN | ORD. | SIL | DEV. | M | P | PERM. | TRI. |JURASSIC| CRET. |P| E |O|M| P

DIVERSITY OF FAMILIES OF THE GASTROPODA BY PERIOD

FIGURE 8. Diversity of families of the Gastropoda by period (from Sohl, 1977).

taxonomically less diverse, they have maintained a high individual abundance in present-day environments.

Mesogastropods first appeared in the Ordovician and remained only minor elements of all Paleozoic assemblages. During that time they did, however, solve the problem of internal fertilization, which was a necessary precursor of the late Paleozoic invasion of the freshwater and land habitats. They continued at a relatively low diversity through the Triassic; but, beginning with the Jurassic and greatly accelerated in the Cretaceous, they expanded into deposit-feeder, suspension feeder, carnivore, and other niches. They continued to diversify throughout the Cenozoic, especially at the generic taxonomic level.

The Neogastropoda, a predominantly carnivorous group, appeared early in the Cretaceous (cf. Sohl, 1969) and most major groups appear by the beginning of the Tertiary (Fig. 9C). Their Cenozoic diversification has been primarily by proliferation at the generic and specific levels.

Opisthobranchs in their earlier history mirrored the mesogastropod diversity pattern of

conservative numbers of types during the Paleozoic and showed great diversification during the Jurassic-Cretaceous times (Fig. 8).

The Pulmonata, like the opisthobranchs, first occur in rocks of the upper Paleozoic, but it is not until Jurassic times that significant diversification began. Although many new families were introduced during Jurassic and Cretaceous times, generic diversity was proportionally low. The greatest diversification of such groups as the order Stylomatophora occurred in the early Tertiary. Later proliferation was at the generic level.

In summary, the geologic record shows:

1. Gastropods are a group undergoing rather constant diversification throughout geologic time but were more conservative during the Paleozoic.
2. Archaeogastropods dominate the Paleozoic gastropod assemblages in both taxonomic diversity and in abundance of individuals. They remain numerically abundant but declined in taxonomic diversity in Late Mesozoic and Tertiary times.
3. Mesogastropods, opisthobranchs, and pulmonates occur first in Paleozoic rocks, but all remain taxonomically few in numbers until the Jurassic, then increase greatly especially in the Upper Cretaceous.

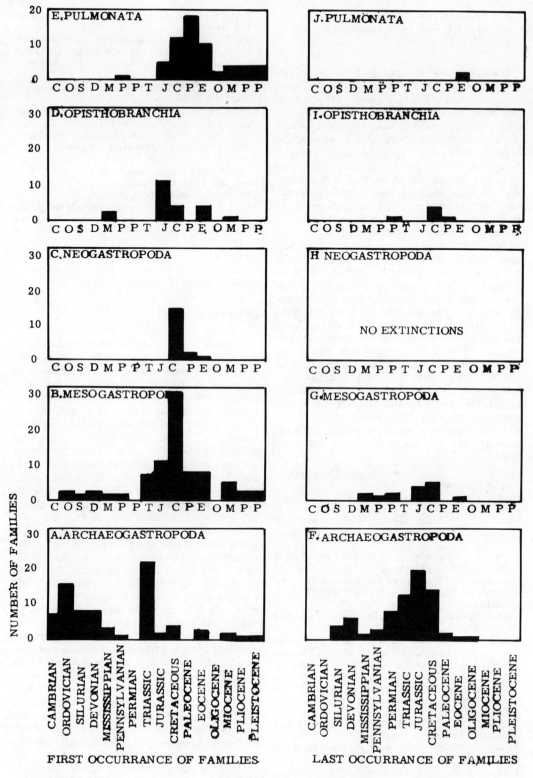

FIGURE 9. Histograms of gastropod family taxa plotted against geologic periods in which they first occur and become extinct.

3

4 The carnivorous neogastropods appear in the Cretaceous. Their great Tertiary radiation, however, is primarily at the generic level.

Biostratigraphic Utility. Because of individual abundance, taxonomic diversity, and presence in virtually all depositional environments, gastropods provide one of the largest, but as yet only partially utilized, reservoirs of biostratigraphic information (Sohl, 1977). Although in marine areas they are dominantly benthic organisms, most undergo a planktic larval stage of sufficient length to provide the potential for wide dispersal (see *Larvae of Marine Invertebrates, Paleontological Significance*). To date they have been used in zonation schemes primarily in rocks of Tertiary age. Such groups as the Turritellidae have been especially widely used both as zonal indices and as elements of assemblage zones. Generally, their utility is restricted by facies controls, but many species occur throughout biotic provinces. Seldom, however, do species seem to transgress such provincial boundaries and thus are of lesser utility for intercontinental correlation. In part because of their lesser abundance, and partly because of the presence of groups of such proven biostratrigraphic worth as ammonites, inoceramid bivalves, planktic foraminifera, conodonts, trilobites, and brachiopods, gastropods have been much less used in rocks of pre-Cenozoic age. However, their potential for use remains, and the greatest need is for more intensive study.

NORMAN F. SOHL

References

Boss, K. J., 1971. Critical estimate of the number of Recent Mollusca, *Mus. Comp. Zool. Occ. Papers,* 3(40), 81–135.

Fretter, V., and Graham, A., 1962. *British Prosobranch Molluscs.* London: Ray Society, 755p.

Graham, A., 1948. *Form and Function in the Littoral Gastropod.* Inaugural Lecture, 20.4, Birbeck College, London.

Jaeckel, S., 1958. Nachtrag. Mollusca, in W. Kukenthal and T. Krumbach, *Hanb. Zool.,* 5, 259–275.

Knight, J.; Brooks; Cox, L. R.; Keen, A. M.; Smith, A. G.; Batten, R. L.; Yochelson, E. L.; Ludbrook, N. H.; Robertson, R. R.; Yonge, C. M.; and Moore, R. C., 1960. Gastropoda, in R. C. Moore, ed., *Treatise on Invertebrate Paleontology,* pt. I, Mollusca l. Lawrence, Kansas: Geol. Soc. Amer. and Univ. Kansas Press, 351p.

Linsley, R. M., 1977. Some "laws" of gastropod shell form, *Paleobiology,* 3, 196–206.

Linsley, R. M., 1978. Shell form and the evolution of gastropods, *Amer. Sci.,* 66, 432–441.

Purchon, R. D., 1968. *The Biology of the Mollusca.* London: Pergamon Press, 560p.

Runnegar, B., and Pojeta, J., Jr., 1974. Molluscan phylogeny: The paleontological viewpoint, *Science,* 186, 311–317.

Sohl, N. F., 1964. Neogastropoda, Opisthobranchia and Basommatophora from the Ripley, Owl Creek and Prairie Bluff Formations, *U.S. Geol. Surv. Prof. Paper, 331-B,* 153–344.

Sohl, N. F., 1969. The fossil record of shell boring by snails, *Amer. Zool.,* 9, 725–734.

Sohl, N. F., 1977. Utility of gastropods in biostratigraphy, in E. G. Kauffman and J. E. Hazel, eds., *Concepts and Methods of Biostratigraphy.* Stroudsburg, Pa.: Dowden, Hutchinson and Ross, 519–539.

Taylor, D. W., and Sohl, N. F., 1962. Outline of gastropod classification, *Malacologia,* 1, 7–32.

Wenz, W., 1938. Gastropoda, in O. H. Schindewolf, *Hanb. Paläozoologie,* 6(1), 960p.

Yonge, C. M., 1947. The pallial organs in the aspidobranch Gastropoda and their evolution throughout the Mollusca, *Phil. Trans. Roy. Soc. London, Ser. B,* 232, 443–518.

Cross-references: *Larvae of Marine Invertebrates–Paleontological Significance; Mollusca.*

GENETICS

Genetics is the scientific study of biological heredity. The basic principles of genetics, known as Mendel's laws, were established by experiments carried out with garden peas (*Pisum sativum*) by the Augustinian monk Gregor Mendel, and were first published in 1866. Mendel's discoveries remained largely unknown until 1900, when similar results were obtained by Carl Correns in Germany and Hugo de Vries in Holland. Mendel's work established that hereditary information is carried in particulate factors (now called *genes*) that are passed from parents to offspring through the sex cells or gametes. There are two genes for each inherited trait, one from each parent. The genes do not fuse in the organism but rather *segregate at random* in the formation of sex cells. Thus, a pea plant with a gene for purple flower and another gene for white flower will produce gametes, half of which carry the purple gene and half the white gene. The alternative forms that a gene may have are called *alleles.* Mendel also discovered that genes controlling different traits are transmitted or *assorted independently* from each other, regardless of their mutual associations in the parental generation. Thus, if a plant carries an allele for white flower and another for purple flower, but also an allele for tall plant and a second for short plant, the four possible gametic combinations—white-tall, white-short, purple-tall, purple-short—are produced in equal proportions. Early in the 20th century, it was discovered, however, that complete independent assortment occurs only when genes are in different chromosomes.

Physical Basis of Heredity

It was first suggested in 1902, independently by W. S. Sutton and T. Boveri, that genes are carried in chromosomes (e.g., Sturtevant, 1965). When two or more genes are borne on the same chromosome they may not assort independently; they are said to be linked. In 1911, the geneticist, T. H. Morgan, working with *Drosophila* flies proposed that genes are linearly arranged in the chromosomes and that the frequency with which different genes recombine in the sex cells may be used as a measure of the distance between genes in a chromosome. This hypothesis was corroborated, and chromosomal "maps" of *Drosophila,* corn, and other organisms, including man, were gradually established. These maps gave the ordering and location of genes in the various chromosomes of an organism.

It was not until the 1950s that the molecular structure of genes was definitely ascertained. The hereditary information is encoded in the chemical substance known as deoxyribonucleic acid (DNA). A gene is simply a segment molecule of DNA. DNA consists of long chains (polymers) composed of four different units called nucleotides. Each nucleotide consists of a nitrogen-containing base linked with a five-carbon sugar (pentose) and a phosphate group. The four kinds of nucleotides are identified by the nitrogen base, which can be cytosine (C), thymine (T), adenine (A), or guanine (G). C and T are pyrimidines, A and G are purines. In 1953, J. Watson and F. Crick proposed that DNA exists as a double-helical molecule made up of two complementary polynucleotide chains with a phosphate-sugar backbone on the outside, and the nitrogen bases inside. The paired chains are held together by hydrogen bonds between bases in different chains, in such a way that A always pairs with T, and G always pairs with C. The hereditary material of all organisms (except for a few viruses) is DNA organized in a double helix. Some viruses contain ribonucleic acid (RNA) rather than DNA; other viruses have single-stranded DNA.

Genetic information is encoded in the sequence of the nitrogen bases in the nucleic acids. The nitrogen bases may be considered as letters of a genetic alphabet; ordered in different ways they may be thought of as genetic "sentences." The number of potentially different sequences of four kinds of bases in a chain with n nucleotides is 4^n. This becomes a staggeringly large number whenever the length of the chain is in the hundreds as is the case for individual genes. The basic units of information, however, are not the individual bases, but nonoverlapping groups of three consecutive bases. There are $4^3 = 64$ different triplet combinations of the 4 bases, but because of some redundancy, there are only 20 different units of information among the triplets. A polynucleotide chain with 600 nucleotides has 200 different groups of three bases. The number of potentially different messages of that length is $20^{200} = 10^{260}$. There is practically no limit to the number of different messages that can be encoded in long polynucleotide chains.

Replication of Genes

The sequence along one of the DNA strands unambiguously specifies the sequence along the complementary strand, owing to the strict determination of the base pairing between the two DNA chains (A with T, G with C). As proposed by Watson and Crick, replication of the DNA occurs through the separation of the two strands by an unwinding process. Each strand then serves as a template for a complementary strand. The result is two new double helices that, because of the rules of base pairing, are identical to each other and to the parental double helix.

The Genetic Code

There are two kinds of genes. *Structural* genes are those whose hereditary information specifies the constitution of enzymes and other proteins; all other genes are *regulatory*. Two processes, transcription and translation, mediate the transmission of information from genes to proteins. The process of *transcription* takes place in the nucleus of a cell and is similar for structural and regulatory genes. The base sequence of a DNA segment is transcribed into a complementary RNA sequence. This process is essentially similar to the replication of DNA, except that the base uracil (U) rather than thymine (T) is incorporated in the RNA.

The process of *translation* takes place in the cell cytoplasm and is mediated by the ribosomes, transfer RNA (tRNA) molecules, and several enzymes. The RNA transcribed from structural genes is called *messenger* RNA (mRNA). The mRNA's carry the information for the specific sequence of amino acids in polypeptides. One or more polypeptides make up a protein. Polypeptides are chains of many amino acids. Twenty common amino acids exist in virtually all proteins. Messenger RNA synthesized in the nucleus moves to the cytoplasm where it becomes associated with groups of ribosomes. There the base sequence in the mRNA is read with the aid of tRNA's which are specific for each kind of amino acid. The information is contained in mRNA in nonoverlapping sequences of three nucleotides, called *codons.* Each codon corresponds to a complementary sequence of three nucleotides, called an *anticodon,* in an appropriate tRNA. As successive codons of a mRNA pass through the ribosomes, tRNA's are sequentially involved in codon–anticodon pairings. The corresponding amino acids are thus brought into position, bonds form between adjacent amino acids, and a polypeptide is synthesized. Usually a single mRNA molecule is translated simultaneously into several polypeptide chains by different ribosomes. The synthesis of a polypeptide is concluded when a ribosome encounters a "terminator" codon in the mRNA. There are three terminator codons, none of which has affinity for any of the normally occurring tRNA's. The genetic code was "deciphered" in the 1960s by establishing the correspondence of the 64 possible triplets in mRNA, and the 20 amino acids or the signals for termination of protein synthesis.

Gene Regulation

Only a fraction of all genes are engaged in transcription in any one cell at any one time. In multicellular organisms, all cells carry identical sets of genes, but different subsets of genes are active in different groups of cells. Which genes become activated in which cells and at what time results from interactions among molecules in a cell, among neighboring cells, and between them and the external environment. The detailed mechanisms that regulate the activity of genes are not fully understood. In bacteria, the generally accepted model of gene regulation is the *operon* proposed by F. Jacob and J. Monod in 1961. An operon consists of an operator and several structural genes, all adjacent or in close proximity to each other. The operator, interacting with the cell environment and with the products of regulatory genes, determines when the structural genes of the operon will be transcribed. Little is known about regulation of gene activity in higher organisms. Britten and Davidson proposed an interesting but highly speculative model in 1969. A large fraction of the DNA of higher organisms consists of relatively few short sequences, each repeated many times. Many of these sequences are about 300 nucleotides in length and are interspersed with structural genes throughout the genetic complement of the organism, or genome. Britten and Davidson suggest that these highly repeated short sequences of DNA are regulatory genes which control the transcription of structural genes.

Genotype and Phenotype

The development of an organism is determined by complex networks of interactions among gene products. The manifestation of a gene varies depending on what other genes are associated with it. Genes may interact with each other and with the environment to determine a given trait in an organism. W. Johannsen introduced, in 1909, the distinction between genotype and phenotype. The *phenotype* of an organism is its appearance, its morphology, physiology, and ways of life, what we can observe. The *genotype* is the sum total of hereditary materials of the organism. The phenotype of an individual changes continuously throughout its life; the genotype, however, remains constant except for occasional mutations. Because of the complex interactions between genes with each other and with the environment, the genotype of an organism does not unambiguously specify its phenotype. Rather, the genotype determines the range of possible phenotypes which may develop; this range is called the *norm of reaction* of the genotype. What phenotypes will actually be realized depends on the sequence of environments in which development takes place.

Mutation

In the most inclusive sense, mutations are changes in the hereditary materials, not due to recombination or independent assortment of chromosomes. They can be classified into two broad categories: chromosomal mutations (also called "aberrations") and *gene mutations.* Some chromosomal mutations affect the number of genes in a chromosome: *deficiency* or *deletion* (when a segment of DNA is lost from a chromosome), and *duplication* (when a segment of DNA containing one or more genes is present more than once in the haploid [half-] complement of chromosomes). Other chromosomal mutations change the location of genes in the chromosomes: *inversion* (when a block of genes is rotated 180° within a chromosome) and *translocation* (when a block of genes changes its location in the chromosomes). Finally, some chromosomal mutations change the number of chromosomes; *fusion* (when two nonhomologous chromosomes fuse into one), *fission* (when one chromosome splits into two), *aneuploidy* (when one or more chromosomes occur a different number of times than the others), *haploidy* (when there is only one set of chromosomes), *polyploidy* (when there are more than two sets of chromosomes).

A gene or point mutation occurs when the DNA sequence of a gene is altered, and the new DNA sequence is passed to the offspring. The change may be due to the substitution of one or more nucleotides for others, or to the addition or deletion of one or more nucleotides. Nucleotide substitutions in the DNA sequence of a structural gene may result in changes in the amino acid sequence of the polypeptide encoded by the gene, although this is not always the case owing to the degeneracy of the genetic code. Additions or deletions of nucleotide pairs in the DNA sequence of a structural gene often result in a very altered sequence of amino acids in the coded polypeptide; this is because the addition or deletion of one or two nucleotide pairs shifts the "reading frame" of the DNA sequence. The effects of point mutations on the ability of organisms to survive and reproduce may range from negligible (e.g., mutations for different eye colors in man) to severe (e.g., a mutation from normal to sickle-cell hemoglobin).

Mutations are rare or ubiquitous events, depending on how we look at them. Mutation rates per gene per generation range from 10^{-9} to 10^{-6} in microorganisms such as viruses, bacteria, and unicellular organisms; they range from 10^{-6} to 10^{-4} in multicellular organisms, including man. The mutation rates of individual genes are low, but each organism has many genes, and populations consist of many individuals. Assume that a typical organism, say a *Drosophila* fly, has 30,000 pairs of genes, and that the average mutation rate per gene per generation is 10^{-5}. The average number of mutations arising for each new individual would be $2 \times 30,000 \times 10^{-5} = 0.6$. The median number of individuals in an insect species is estimated to be 1.2×10^8. If we assume that 0.6 new mutations are acquired by each individual, there would be $0.6 \times 1.2 \times 10^8 = 7.2 \times 10^7$ new mutations per generation in an insect species of medium size. Species of other groups of organisms may consist of fewer individuals than insect species, but such species will also acquire large numbers of new mutations every generation.

Even at any one given gene locus many new mutations are acquired every generation by a species. If the average mutation rate per gene per generation is 10^{-5}, about 2400 ($10^{-5} \times 2 \times 1.2 \times 10^8$) new mutations are acquired on the average per locus in an insect species of medium size. It is not surprising that different populations and different species often become independently adapted to specific environmental challenges. Many species of insects in different parts of the world have evolved resistance to DDT; industrial melanism has evolved independently in many species of moths and butterflies in industrialized regions of the world. If the appropriate genetic variants to face an environmental challenge are not already present in a population, they are likely to arise soon by mutation. The potential of the mutation process to generate new genetic variation is enormous.

Population Genetics

Individual organisms are ephemeral. In sexual outcrossing organisms, the individual is also incomplete in a sense, since it must interact with other individuals in order to reproduce. Populations are arrays of individuals, and they alone provide continuity to the genetic materials. A *Mendelian population* is an association of individuals of a sexually reproducing species within which matings take place. The most inclusive Mendelian population is the species. The sum total of the genotypes of all individuals in a Mendelian population may be conceived as the *gene pool* of the population. For diploid organisms, the gene pool of a population of N individuals consists of $2N$ genomes; i.e., $2N$ genes for each gene locus, except for sex-linked genes that exist singly in the heterogamic sex.

Genes at a given locus may exist in a variety of allelic forms. The description of a gene pool at a locus requires specification of the kinds of alleles present and the frequencies of the alleles and genotypes. In the absence of the processes by which gene frequencies change (mutation, migration, selection, genetic drift) there is a very simple relationship between allelic and genotypic frequencies in a population of random-mating individuals. Assume that at a locus there are a number of alleles: $A_1, A_2, A_3, \ldots$, whose frequencies are, respectively, $f_1, f_2, f_3, \ldots$ $(f_1 + f_2 + f_3 + \ldots = 1)$. The genotypic frequencies are given by the expansion of $(f_1 + f_2 + f_3 + \ldots)^2$; i.e., the frequency of the genotype A_1A_1 is $f_1{}^2$, the frequency of A_1A_2 is $2f_1f_2$, the frequency of A_2A_2 is $f_2{}^2$, and so on. This relationship is known as the *Hardy-Weinberg equilibrium*. Even in presence of mutation, migration, selection, and drift, the genotypic frequencies of large populations are, as a rule, very near Hardy-Weinberg equilibrium.

Evolution consists of changes in gene and genotypic frequencies through the generations. Mutation has very little effect in changing gene frequencies from one generation to the next, since mutation rates per generation are often around 10^{-5} or less. *Genetic drift* is the name used to describe changes in gene frequencies due to sampling errors from one generation to the next. Genetic drift has fairly small effects in large populations, but becomes an important factor when populations are very small. The effects of migration in changing gene frequencies become greater as the gene frequencies between neighboring populations are more different, and as the number of migrants exchanged increases. The most important evolutionary process is *natural selection*. Natural selection may be defined as differential reproduction of alternative genetic variants, and is the only process that directly promotes adaptation to the environment. Natural selection, however, can only occur if there is genetic variation. The more genetic variation in a population, the greater the opportunity for the operation of natural selection. This conclusion was formally stated by R. A. Fisher in 1930 in his *Fundamental Theorem of Natural Selection:* "The rate of increase in fitness of a population at any time is equal to its genetic variance in fitness at that time."

Genetic Structure of Populations

Two models of the genetic structure of populations have been proposed; they have been named the "classical" and the "balance" hypotheses. The *classical model* proposes that the gene pool of a population consists, at each locus, of a wild type allele with a frequency approaching one; mutant alleles exist but generally at very low frequencies. A typical individual would be homozygous for the wild type allele at most gene loci. According to the classical model, mutant alleles are continuously introduced in the population by mutation, but are generally deleterious and thus are removed from the population by natural selection. Occasionally, a beneficial mutant may arise, which will probably increase by natural selection to become the new wild type allele. The *balance model* proposes instead that there is generally no single wild type or "normal" allele. The gene pool of a population is envisioned as consisting, at most loci, of an array of alleles in moderate frequencies. A typical individual would be heterozygous at a large proportion of its gene loci. Any two individuals would differ at a great number of loci.

Evidence has gradually accumulated showing that genetic polymorphisms are widespread, and thus that the balance model is correct (see Lewontin, 1974). During the last decade, the application of techniques of gel electrophoresis to the study of protein variation has made it possible to estimate the proportion of genes that are polymorphic in natural populations. Minimum estimates indicate that between 30 and 80% of all gene loci are polymorphic in virtually all organisms studied, from bacteria, through annual plants and a variety of invertebrates, to vertebrates and man. Outcrossing sexual individuals are heterozygous at 5–20% of all their loci. In general, invertebrates have more genetic variation than vertebrates.

It has been pointed out above that a great deal of genetic variation arises every generation by mutation. Yet, this newly arisen variation represents only a small fraction of the genetic variation present in a typical population. Assume, as above, that an organism consists of 30,000 gene loci. An average individual may be heterozygous at 10% of its gene loci; i.e., it may receive from its parents two different alleles at each of 3000 gene loci. The same individual will have acquired, on the average, about 0.6 new allelic variants owing to newly induced mutations. According to these rough calculations, the amount of variation present in a population is about 5000 times greater than that acquired each generation by mutation. Although mutation is the ultimate source of genetic variation, rates of mutation are likely to have little immediate effect on rates of evolution.

Genetics and the Fossil Record

Since the science of genetics investigates the machinery of evolution (in part), the results are of great interest to paleontologists who study the record of evolution in fossil remains. Genetic questions of special paleontological relevance include modes of speciation, rates of evolution, and the relationships between genotypes and phenotypes. Two areas of presently active genetic research hold special promise for interpreting other aspects of the fossil record.

Genetic Variability. The average number of alleles per locus varies greatly among living organisms, and there have been attempts to discern a pattern to these variations. Theoretical arguments have led to the prediction that species living in environments that are highly variable spatially and/or temporally should require relatively large amounts of genetic variability in order to cope with the variety of conditions to which they are subjected. Species living in more uniform environments, on the other hand, should require less genetic variability to be well-adapted to narrow ranges of habitat conditions.

However, preliminary studies of enzymatic variability by techniques of gel electrophoresis have not supported this view. Species from uniform and variable environments alike have proven to have at least moderate amounts of genetic variability. From studies of terrestrial animals it now appears that those that range through the most habitats, and thus are subjected to broad ranges of environmental conditions, have relatively low genetic variabilities. This is generally true, for example, for mammals. Animals that tend to be more restricted in their habitat ranges, such as insects, have relatively high genetic variabilities (Selander and Kauffman, 1973). Furthermore, data from marine invertebrates suggests that animals from highly seasonal environments (the northwestern North Atlantic and Antarctica) tend to have low genetic variabilities, while those from less seasonal environments (the tropics and the deep sea) have high genetic variabilities (Valentine, 1976). Thus, it now appears that ecologically more flexible animals may have less genetic variability than ecologically more specialized ones.

The data suggest that specialized marine species such as those of the rich Permian reefs were highly variable genetically, while those in the depauperate communities of the Early Triassic had low genetic variability. Thus, extinctions would tend to carry away genetically variable lineages, and novel taxa might arise from lineages that were rather invariant genetically.

Genetic Regulation. Genes that regulate the transcription of structural genes (those that code for polypeptides) determine the timing and association of gene activities, which vary between cell types in animals, and can account for the differentiation between animal cells and for much of the morphological diversity among different animal taxa. The morphological differences between chimpanzees and man, for example, may be due chiefly to regulatory genes (King and Wilson, 1975). Britten and Davidson (1971) and Davidson and Britten (1973) have proposed a hierarchial system of genetic regulation that suggests how major morphological innovations, which could lead to novel adaptive types (including new phyla), may result from growth and repatterning of the regulatory genome. This work may greatly facilitate explanation of the relatively rapid appearance of evolutionary novelties in the fossil record, which may owe their rapid rise primarily to evolution of regulatory mechanisms (Valentine and Campbell, 1975).

FRANCISCO J. AYALA
JAMES W. VALENTINE

References

Britten, R. J., and Davidson, E. H., 1971. Repetitive and non-repetitive DNA sequences and a speculation on the origins of evolutionary novelty, *Quart. Rev. Biol.,* **46,** 111–138.
Cavalli-Sforza, L. L., and Bodmer, W. F., 1971. *The Genetics of Human Populations.* San Francisco: Freeman, 965p.
Davidson, E. H., and Britten, R. J., 1973. Organization, transcription, and regulation in the animal genome, *Quart. Rev. Biol.,* **48,** 565–613.
Dobzhansky, T., 1970. *Genetics of the Evolutionary Process.* New York: Columbia Univ. Press, 505p.
Goodenough, U., and Levine, R. P., 1974. *Genetics.* New York: Holt, Rinehart & Winston, 882p.
King, M. C., and Wilson, A. C., 1975. Evolution at two levels in humans and chimpanzees, *Science,* **188,** 107–116.
Lewontin, R. C., 1974. *The Genetic Basis of Evolu-*

tionary Change. New York: Columbia Univ. Press, 346p.
Li, C. C., 1955. *Population Genetics.* Chicago: Univ. Chicago Press, 366p.
Selander, R. K., and Kauffman, D. W., 1973. Genic variability and strategies of adaptation in animals, *Proc. Nat. Acad. Sci.,* **70,** 1875–1877.
Stent, G. S., 1971. *Molecular Genetics.* San Francisco: Freeman, 650p.
Stern, C., 1973. *Principles of Human Genetics,* 3rd ed. San Francisco: Freeman, 891p.
Strickberger, M. W., 1968. *Genetics.* New York: MacMillan, 868p.
Sturtevant, A. H., 1965. *A History of Genetics.* New York: Harper & Row, 165p.
Valentine, J. W., 1976. Genetic strategies of adaptation, in F. J. Ayala, ed., *Molecular Evolution.* Sunderland, Mass.: Sinauer, 78–94.
Valentine, J. W., and Campbell, C. A., 1975. Genetic regulation and the fossil record, *Amer. Scientist,* **63,** 673–680.
Watson, J. D., 1970. *Molecular Biology of the Gene,* 2nd ed. Menlo Park, Calif.: Benjamin, 662p.

Cross-references: *Evolution; Extinctions; Fossil Record*

GRAPTOLITHINA

Graptolithina is a now extinct Paleozoic class usually attributed to the hemichordates, living representatives of which include *Cephalodiscus* and *Rhabdopleura.* Hemichordates (see *Hemichordata*) lack the notochord, distinctive feature of the phylum Chordata (= Vertebrata), and may be broadly regarded as being intermediate between vertebrates and invertebrates. Graptolites are most abundant in Ordovician and Silurian rocks, but early, mostly benthic genera occur in Cambrian strata and persist into the Carboniferous (Fig. 1). The colonial scleroproteic skeletons are made up of fronds (stipes) that are almost always bilaterally symmetrical and composed of two peridermal layers, an inner *fusellar* layer and an outer *cortical* layer (Fig. 2; Kozłowski, 1949).

Affinities

Graptolites have been classified with most major invertebrate groups including, particularly, the Bryozoa and Coelenterata. However, a majority of present-day workers would place them with the hemichordates in view of the considerable resemblances between the graptolite skeleton on the one hand and *Rhabdopleura* and *Cephalodiscus* on the other (Kozłowski, 1966). Encrusting graptolites show a remarkable similarity to the encrusting *Rhabdopleura* skeleton, both having a structureless basal or encrusting layer and arched tube exhibiting zig-zag fusellae—both have upright,

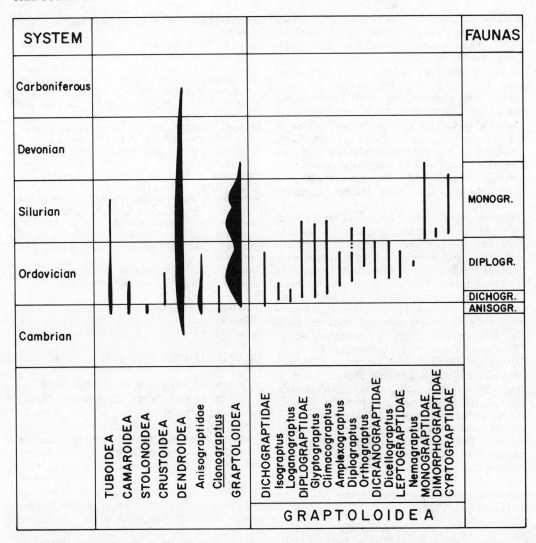

FIGURE 1. Stratigraphic range of some graptolite orders and their relative abundance; and ranges of some lesser taxa mentioned in text.

free tubes, in *Rhabdopleura* composed of complete ring fusellae, and in some tuboid graptolites of rather less complete ring fusellae. Other similarities include the presence of dorsal stolons connecting individuals of the colony, and a scleroproteic skeleton of very similar composition which, on present evidence, may be contrasted with skeletal chemistry of some other groups. In some crustoid and tuboid graptolites, the stolon may be embedded in the lower or dorsal wall as in *Rhabdopleura*. Graptolites differ from *Rhabdopleura* (but not *Cephalodiscus*) in having thicker layers of external cortical tissue, but more particularly in the mode of budding. Each graptolite zooid in turn becomes the "terminal bud." The evidence for this hinges on an understanding of the order

of fusellar secretion. *Rhabdopleura*, however, is believed to have a permanent leading or terminal bud—the relationship of the corresponding distal extremity of the scleroproteic skeleton to the terminal "zooid" is uncertain. It may be reasonably concluded that the graptolites are closer to *Rhabdopleura* than the latter is to some other hemichordates; and it is now known that rhabdopleuran hemichordates appear not long after the graptolites in the geological record, although the record of them is sparse.

Stratigraphic Distribution

Fig. 1 illustrates the longevity of the earliest known graptolites, essentially benthos, which persist into the Carboniferous. Planktic off-

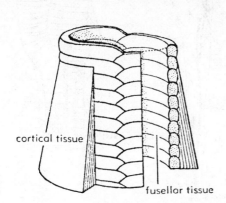 

FIGURE 2. *Left,* diagram showing fusellar tissue laid down in alternating half rings surrounded by laminated cortical tissue (from Bulman, 1970, courtesy of The Geological Society of America and the University of Kansas). *Right, Monograptus* metasicula, showing fusellae; scanning electron micrograph, X1470.

shoots appeared in the Tremadoc, became diversified and specialized in the Ordovician and Silurian, and became extinct in the Devonian. Successive evolutionary faunas have been carefully identified and defined (Bulman, 1970). These are, in ascending order: the anisograptid fauna, the dichograptid fauna, the diplograptid fauna, and the monograptid fauna. The third group is divided into four stratigraphically successive subfaunas: The *Glyptograptus-Amplexograptus* subfauna, the *Nemagraptus-Dicellograptus* subfauna, the *Dicellograptus-Orthograptus* subfauna, and the *Orthograptus-Climacograptus* subfauna. The planktic graptolites are thus extremely useful in stratigraphy, enabling subdivision of the Silurian, for example, into about one graptolite assemblage or zone for each 1 m yr (see Berry, 1977). Finer divisions are workable locally. Specific identity of planktic graptolites is maintained for approximately 0.5–2 m yr in most cases.

Abundance

Benthic graptolites—"rooted" and encrusting forms—occur in the shelly facies with the normal and usually more abundant shelly benthos. Planktic graptolites are found here also, but in relatively small numbers and variety—the more common graptolite species of the graptolitic facies described below are usually the species found in the shelly facies. These have almost always robust stipes. The graptolitic facies consists of presumed offshore, fine-grained black or grey mudstones rich in graptolites almost to the exclusion of other macrofossils; those of the latter that do occur are often themselves planktic or epi-planktic species. Some idea of the numbers of graptolites in the surface layers of the sea at any one time—accepting that graptolites in graptolitic mudstones are death assemblages not always approaching life assemblages in composition—can be obtained from a simple calculation using the Llandovery Skelgill Beds of the north of England as an example. If one takes the conservative figure of 10 graptolite rhabdosomes per square foot of bedding plane, and an areal extent of 200 square miles for the northern England surface and subsurface extent of the Skelgill Beds, then the number of rhabdosomes on one time plane would be in excess of 27 million. The Skelgill Beds average perhaps 50 ft (15 m) in thickness, and if one assumes

25 graptolite-rich bedding planes per foot thickness (again a conservative estimate to judge from x-ray studies) then the number of *preserved* graptolites in the Skelgill Beds is of the order of 40,000 million. The above figures are conservative, and the areal extent of the black graptolitic mudstone facies in the British Llandovery is certainly in excess of 200 mi^2! Comparable calculations for, say, the Utica Shale of eastern North America would be yet more spectacular, and both examples serve to illustrate the somewhat unusual nature of much of the macroplankton in the lower Paleozoic "geosynclines." Not all the rhabdosomes would be preserved, a consideration that would *increase* any such postulated figures for plankton composition, while the condensed deposition in such black-shale environments would necessitate a *reduction* of the same figures.

General Morphology

The individual zooids in a graptolite colony were connected by sclerotized stolons, partially sclerotized stolons, or unsclerotized stolons, and were contained by scleroproteic skeletal sheaths termed thecae. Thecae are of few basic types (Fig. 3): *autothecae, bithecae, stolothecae* (= immature autothecae), and less commonly *conothecae* and *microthecae*. A variable number of thecae make up individual stipes, and a variable number of stipes comprise the mature colony, termed a *rhabdosome*. Occasionally, clusters of rhabdosomes occur radiating from a common center. Such associations are termed *synrhabdosomes* and they are almost always composed of colonies of a single species; their function is uncertain.

Combinations of the above morphological elements and their detailed variations form the basis of classification of the Graptolithina:

Order Dendroidea: sessile, mostly benthic, occasionally planktonic; stipes composed of stolothecae, autothecae, and bithecae produced by regular triad budding of the stolon.
Order Tuboidea: sessile, encrusting, benthic, occasionally conical rhabdosome; stipes with less prominent stolothecae than dendroids, autothecae, bithecae, conothecae, and microthecae; stolon divisions diad, and irregular, stolons occasionally partly sclerotized.
Order Camaroidea: encrusting; indistinct stolothecae, autothecae with inflated basal part and erect, tubular distal portion; bithecae in some species.
Order Crustoidea: encrusting; autothecae, stolothecae, and bithecae produced from triad divisions of stolon; autothecae with erect distal neck and apertural modifications.
Order Stolonoidea: sessile or encrusting; stolothecae, ?autothecae; stolons exaggerated and irregular in form.

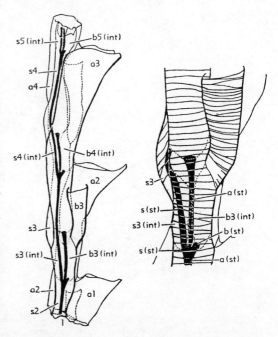

FIGURE 3. Tecal constitution of a dendroid stipe (from Bulman, 1970, courtesy of The Geological Society of America and the University of Kansas). *Left, Dendrograptus regularis* Kozlowski, X33 viewed as transparency with growth lines omitted; stolon system in solid black; stolotheca and daughter thecae in heavy outline. *Right*, portion of same with growth lines, X66. a = autotheca, b - bitheca, (int) = internal portion, s = stolotheca, (st) = stolon.

Order Graptoloidea; planktic or epiplanktic; fewer stipes than dendroids; only one type of thecae (autotheca), often with apertural modifications; no sclerotized stolons; stipes pendant to scandent, uniserial to quadriserial.
Order Dithecoidea: Middle Cambrian forms, with autothecae arranged biserially; distal portions of thecae isolate; fusellar structure proved in one genus only; other thecae and structure not observed.
Order Archaeodendrida: Middle Cambrian forms, with probable autothecae arranged singly or in groups of three possibly biserially; isolate distally and with narrow thecal bases; stolothecae possibly present; indistinct fusellar structure in one genus.

The vast majority of known graptolite species belong to the Dendroidea and Graptoloidea. Further taxonomic subdivisions are made on the form of the rhabdosome and the variation in thecae type, the latter particularly in the Graptoloidea.

Detailed Morphology

The fusellar half rings (Fig. 2) are the basic building blocks of the thecae, including almost all known complex apertural apparatuses (but

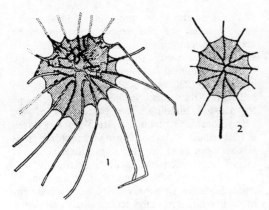

FIGURE 4. Proximal web structures in various graptolites (from Bulman, 1970, courtesy of The Geological Society of America and The University of Kansas). 1, *Loganograptus logani,* X0.65. 2, *L. Kjerulfi,* X0.45.

see Urbanek, 1970). The cortical tissue, on the other hand, serves a variety of purposes, such as: strengthening bars between stipes in the dendroid *Dictyonema;* stem-strengthening deposits in *Dictyonema* and *Dendrograptus;* the production of a thread-like attachment or tissue support in some dendroids and most graptoloids; growth of nema modifications such as vanes in graptoloids; building of web structures between stipes as in the graptoloid *Loganograptus* (Fig. 4) construction of complex

spine and rod-like networks (lacinia) outside the main rhabdosome; and in some Upper Silurian graptoloid species construction of complex thecal apparatus.

Although the graptolite thecae are intimately connected with other thecae in the colony by means of the sclerotized stolon in dendroids and the common canal in graptoloids, they are distinct structural entities: Figs. 3 and 5 illustrate the relationship of one theca to those immediately above and below in the graptoloids and dendroids, and the descriptive terminology in use.

Branching of graptolite stipes is achieved in a number of ways:

a. suppression of a bitheca at the triad in dendroids and production of two stolothecae and an autotheca. In simple dendroids, the division of stipes takes place immediately, with one stolotheca and the autotheca forming one new stipe and the other stolotheca and the *preceding* autotheca forming the other. This is dichotomous branching.

b. actual bifurcation may be delayed after (a) and compound stipes form, eventually, and usually regularly, producing new thick compound stipes. In (a), thecae open individually on the ventral side of the stipe (i.e., opposite the stolon), while in (b) thecae may open, usually ventrally, in clusters termed twigs.

c. lateral branching occurs when the new branch comes off at a high angle from the original stipe, which continues its direction of growth. Such new stipes may be thecal cladia developed from one

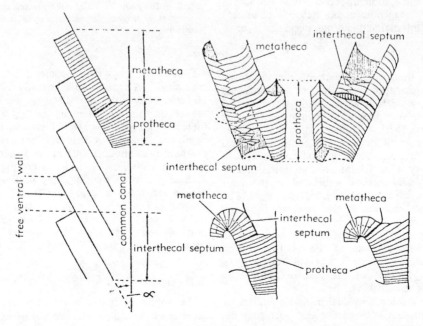

FIGURE 5. Diagrams illustrating terminology of a graptolite stipe and thecae (from Bulman, 1970, courtesy of The Geological Society of America and The University of Kansas). α = angle of inclination.

side of the thecal aperture (e.g., *Cyrtograptus*). The first individual (sicula) of the colony may also show cladial development.

The general rhabdosomal form resulting from such stipe growth varies considerably from the bushy dendroid growth of *Dendrograptus,* through the anisograptids and dichograptids with fewer but long stipes, to the diplograptids and monograptids which are two and one stiped respectively.

Ultrastructure

Recent work requires a dual terminology when discussing periderm structure: the one pertaining to structure seen under the light microscope and the other to that deduced from electron micrographs. Thus, the terms *fusellar tissue* and *cortical tissue* are applicable to those peridermal layers clearly visible under the light microscope, while the fusellar layer is composed mostly of a particular microfabric termed fusellar fabric, and the cortical layer mostly of cortical fabric. In addition, sheet fabric, virgula fabric, and others have been identified at the ultrastructural level. Further, the fibers of the cortical fabric have been shown to be closely similar to present-day collagen fibers; and yet other fibers may be of the keratin type. One of the important results of these ultrastructural studies has been to show that fusellar and cortical tissues may, in certain parts of the rhabdosome, grade into each other, thus posing considerable problems about the manner in which the graptolite zooid was able to secrete both the internal and external periderm layers.

Astogeny

The earliest formed part of the graptolite colony, produced sexually, is the prosicula (Fig. 6), a conical body usually less than 0.5 mm long. No earlier stages in the sexual cycle are known with certainty, although possible eggs have been detected in some autothecal tubes. The prosicular cone is always formed in its entirety with a faint internal spiral thread, and the only growth exhibited is the later aquisition of external longitudinal ridges. The latter may coalesce at the apex to form the nema, often a slender thread-like, tubular process probably connected in some way with the planktonic mode of life of the colony. The prosicular periderm has a different ultrastructure to that of the rest of the colony (Rickards et al., 1971).

The prosicular individual probably undergoes metamorphosis, the new zooid secreting the metasicular tube, which it constructs of fusellar rings and half rings added to the aperture at the base of the prosicular cone. These are the first secreted by the colony. The pro- and metasicula combined is termed the sicula, a strikingly uniform, elongate, conical structure known in almost all graptolite groups. It is usually from the metasicula, less commonly the prosicula, that the whole colony develops by asexual budding. The first theca of the colony develops through a lateral pore or notch in the sicula, while differing branching from this and the next few thecae produces the great variation in basic rhabdosome construction.

The branches vary not only in their number and adherence or otherwise to each other, but in their relative position to the sicula and nema, being pendent to horizontal to scandent (i.e., back to back). These variations, and changes in thecal shape (thecae are not uniform like the siculae) provide the basis for detailed classification of the graptolites. Several basic development types have been recognized by Bulman (1970):

a. Dichograptid—One theca crosses the axis of the rhabdosome immediately after production of the first theca.
b. Isograptid—Two thecae cross the axis of the rhabdosome, third theca develops from the second.
c. Diplograptid—At least three thecae cross the axis of the rhabdosome.
d. Monograptid—No thecae cross the axis, first theca grows upwards.
e. Pericalycal—Second theca originates left-handedly and grows down back of sicula.
f. Platycalycal—Third theca gives rise to two buds, with subsequent budding concentrated on the reverse side, leaving the sicula largely free on the obverse side.

Size

Conical dendroid rhabdosomes >20 cm long are unusual. A broad, conical rhabdosome some 10 cm long with 100 distal stipes and 10 thecae per cm has as many as 12,000 autothecae. Anisograptids and dichograptids are commonly larger than dendroids but have fewer thecae (up to about 3000) since the number of stipes is considerably less. Later graptolites, such as the monograptids, may have over 700 thecae and reach a length of 70 cm or more, or have as few as 5 or 6 thecae and a length of only a few mm. Most Wenlock (Silurian) horizons have at least one species with a "giant" rhabdosome of 50 cm length.

Soft Parts

The nature of the soft parts has never been ascertained by direct observation of preserved parts and must be inferred from studies of skeletal morphology and by analogy with

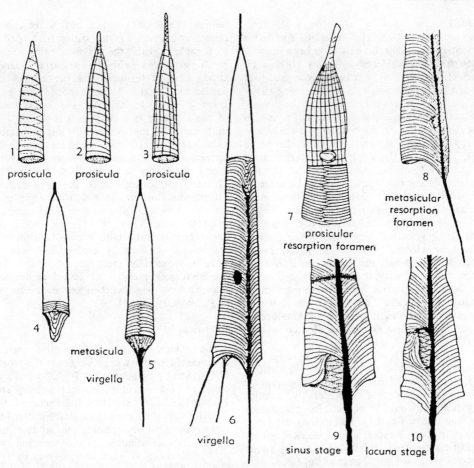

FIGURE 6. Diagrams illustrating development of sicula and initial bud (from Bulman, 1970, courtesy of The Geological Society of America and The University of Kansas). 1–3, Prosicula; 4,5, beginning of metasicula and formation of virgella; 6, completed sicula with apertural spines, virgella, and resorption foramen; 7, prosicular resorption foramen, *Didymograptus* sp., X65; 8, metasicular resorption foramen, *Orthograptus gracillis,* X45; 9, sinus stage in monograptid, *Pristiograptus bohemicus,* X65; 10, lacuna stage in same, X65.

similar recent hemichordates such as *Rhabdopleura.* Fig. 7 illustrates one of Bulman's (1970) suggestions for the nature and relationship of the zooid to the chamber.

Mode of Life

Even though benthic graptolites are rarely found in growth position, there can be little doubt that the basal discs, "roots," and encrusting bases are indeed designed for attachment to the substrate—rock, shells, and perhaps seaweeds. It is equally certain, from their cosmopolitan distribution in a variety of rock types, that the graptoloids and some dendroids were more or less planktic. A point worthy of future investigation, however, is that there is no unequivocable record of graptoloids attached to algae, while they have an un-

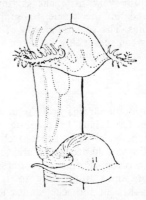

FIGURE 7. Diagrammatic restoration of thecal zooid illustrating possible relation of lophophore to apertural modification (from Bulman, 1970, courtesy of The Geological Society of America and The University of Kansas).

doubted association as plankton with dark, carbonaceous matter. The nature of this latter element of the plankton has never been proved, although it is considered by many to be algal in origin. Since it can be reasonably argued that the nema is covered in extrathecal living tissue during life, then any attachment of the colony to algae, or other colonies (as in synrhabdosomes), would be by the extrathecal tissue and not by the hard skeletal tissue of the nema itself. This might also explain the relative rarity of synrhabdosomes, particularly in monograptids, since they would usually disintegrate upon loss of the soft parts unless the nemata themselves became inextricably tangled.

It is likely that the extrathecal tissue of planktic graptolites was vacuolated tissue, at least in selected parts of the rhabdosome, and as such was responsible for the buoyancy of the colony. Mostly, the nema or vaned structure would be upward. A full interpretation of proximal-end "vane" structures has not yet been attempted and it would be premature to suggest that a few graptolites floated upside down.

A bubble of O_2 of diameter 0.20 mm is sufficient to support at the surface of distilled water a 6 mm X 0.5 mm length of *Clonograptus* stipe—in sea water a considerably smaller amount of gas would be needed. But calculating from the above figure, a 10 cm long dendroid rhabdosome of a rather broad conical nature would be supported in distilled water by about 0.8 cc of O_2, a volume that is of the same order of magnitude as the proximal "vesicles" observable upon some dendroid rhabdosomes of that size. Of course the lesser amount of gas required in sea water need not be concentrated in a proximal "vesicle" or nema vane structure, although this seems most likely. In short, it would require very little gas as tiny bubbles disseminated in the tissue to support most rhabdosomes. The specific gravity of the zooids themselves can reasonably be taken to be close to, but greater than unity.

Evolution

The benthic dendroids gave rise in the Tremadoc (Late Cambrian) to planktic dendroids. which then gave rise successively to the anisograptids and hence to the graptoloids. The changes involved a drastic reduction in the number of stipes (ultimately to 1 in *Monograptus*), a loss of bithacae, and a loss of the sclerotized stolon. Intermediates are known that exhibit partial loss of bithecae, and some tuboids have paritally sclerotized stolons embedded in the dorsal wall. The graptoloids themselves had a general tendency not only to reduce the stipe number but to change the growth direction of the stipes from pendant to scandant—again, the ultimate being achieved in *Monograptus* with its uniserial, scandant stipe. Loss of bithecae was probably accompanied by a change to hermaphroditism by the graptolites—Kozłowski's (1949) interpretation of the bithecae as male zooids is the only idea that can be applied to all known features exhibited by bithecae. Bithecal zooids that opened *inside* the autotheca can hardly have had a protective function! With the aquisition of hermaphroditism came the appearance of biform rhabdosomes in which the early thecae are of a much smaller size, or different shape, from the later "mature" thecae. Thecal form becomes very important in classification of the graptoloids, and newly evolved characters are usually proximally introduced to the colony. An increasing number of actual evolutionary lineages have been worked out in recent years (e.g., Sudbury, 1958).

Historical

The animal nature of graptolites was first appreciated by G. Wahlenberg, but the work of Hall (1865) probably set off modern research on the best possible footing. Subsequently, Lapworth (1879–1880) and Ruedemann (1904–1908) established major stratigraphic considerations. Detailed morphological studies using zoological techniques on outstanding well-preserved material were carried out by Holm (1890), Wiman (1895). Kraft (1926), and later by Bulman and Kozłowski. In more recent years, particular attention has been paid to the ultrasturcture of the periderm using electron microscopes (e.g., Berry and Takagi, 1970; Rickards et al., 1971; Urbanek and Towe, 1974).

R. B. RICKARDS

References

Berry, W. B. N., 1977. Graptolite biostratigraphy: A wedding of classical principles and current concepts, in E. G. Kauffman and J. E. Hazel, eds., *Concepts and Methods of Biostratigraphy*. Stroudsburg. Pa.: Dowden, Hutchinson & Ross, 321–338.

Berry, W. B. N., and Takagi, R. S., 1970. Electron microscope investigations of *Orthograptus Quadrimucronatus* from the Maquoketa Formation (Late Ordovician) in Iowa, *J. Paleontology*, 44, 117–124.

Boardman, R. S., Cheetham, A. H., and Oliver, W. A., Jr., eds., 1973. *Animal Colonies*, Stroudsburg, Pa.: Dowden, Hutchinson, & Ross, 603p.

Bulman, O. M. B., 1970. Graptolithina, in R. C. Moore, ed., *Treatise on Invertebrate Paleontology*, pt. V, 2nd ed. Lawrence, Kansas: Geol. Soc. Amer. and Univ. Kansas Press, 163p.

Hall, J., 1865. Graptolites of the Quebec Group, *Geol. Surv. Canada, Canadian Organic Remains*, **2**, 1–151.

Holm, G., 1890. Gotlands Graptoliter, *Svenska Vetenskaps-Akad. Handl., Bihang*, ser. 4, **16**(7), 1–34.

Kozłowski, R., 1949. Les graptolithes et quelques nouveaux groupes d'animaux du Tremadoc de la Pologne, *Palaeontol. Polonica*, **3**, 1–245.

Kraft, P., 1926. Ontogenetische Entwicklung und Biologie von *Diplograptus* und *Monograptus*, *Paläont. Zeitschr.*, **7**, 207–249.

Lapworth, C., 1879–1880. On the geological distribution of the Rhabdophora, *Ann. Mag. Nat. Hist.*, ser. 5, **3**, 245–257, 449–455, **4**, 333–341, 423–431; **5**, 45–62, 273–285, 359–369; **6**, 16–29, 185–207.

Rickards, R. B. 1975. Paleoecology of the Graptolithina, an extinct class of the Phylum Hemichordata, *Biol. Rev. Cambridge Phil. Soc.*, **50**, 397–436.

Rickards, R. B., Hyde, P. J. W., and Krinsley, D. H., 1971. Periderm ultrastructure of a species of *Monograptus* (Phylum Hemichordata), *Proc. Roy. Soc. London, Ser. B*, **178**, 347–356.

Ruedemann, R., 1904–1908. Graptolites of New York, pts. I, II, *N.Y. State Mus. Mem.*, **7**, 457–803; **11**, 457–583.

Sudbury, M., 1958. Triangulate Monograptids from the *Monograptus gregarius* zone of the Rheiddol Gorge, *Phil. Trans, Roy. Soc. London, Sec. B*, **241**, 485–555.

Urbanek, A., 1970. Neocullograptinae N. Subfam. (Graptolithina)–their evolutionary and stratigraphic bearing, *Acta Palaeontol. Polonica*, **15**, 163–388.

Urbanek, A., and Towe, K. M., 1974. Ultrastructural studies on graptolites I. The periderm and its derivatives in the Dendroidea and in *Mastigograptus*, *Smithsonian Contrib. Paleobiol.*, **20**, 18p.

Wiman, C., 1895. Uber die Graptolithen, *Bull. Geol. Inst. Univ. Uppsala*, **2**, 239–316.

Cross-references: *Chordata; Coloniality; Hemichordata.*

GROWTH LINES

Most features of living organisms reflect genetic inheritance; this fortunate state of affairs permits parents to have some expectations regarding their progeny, and gives taxonomists some claim to be living in a real world. Other features, however, are directly or indirectly a reflection of contact with the environment; these are largely ignored by the taxonomist, but can be of considerable aid to the paleoecologist. Most of these environmentally induced features, such as gastropod borings in individual shells or stunting in an assemblage, are relatively insignificant or infrequent, and thus are of limited use; but one class of such features is common and promises to be of considerable significance. These are growth lines.

Growth lines mark changes in growing shell or skeletons and these changes are usually realted to changes in the environment. Individual lines are not readily interpreted, but the pattern of growth lines formed during the lifetime of an organism may provide useful information about the organism's biology, ecology, and geologic setting. Some of the biological factors that can be determined include growth rate, age, and spawning habits; ecological factors include storm frequency, water depth, tidal effects, and the presence of seasons; and geological factors such as paleolatitude, tidal amplitudes, and even absolute age can theoretically be recorded. When growth-line patterns are examined for entire assemblages, it is even possible to determine additional factors such as average lifespan, incidence of predation, incidence of events causing catastrophic death, and questions such as whether two individuals lived or died in the same area at the same time. Growth lines, then, can be powerful tools, but like most such tools they require careful study before use.

Morphology, Formation, and Description

Growth lines are but one feature of the shells or skeletons in which they appear; to understand their formation it is first necessary to understand that of the skeleton itself.

Growth lines typically form in an accretive or *accretionary skeleton*. This is a skeleton formed by the simple (in theory) process of adding new material to old, without otherwise altering or removing any preexisting skeletal material (see Fig. 1). This type of skeleton is characteristic of a considerable number of organisms, particularily corals, brachiopods, bryozoans, barnacles, and mollusks. Even animals with other types of skeletons, such as vertebrates, commonly have specialized skeletal elements such as teeth and otoliths with accretionary structure.

The portion of an accretionary skeleton along which growth occurs can be called the *surface of deposition;* this may completely surround the skeleton, as in a pearl or a tree, or may be restricted to a part of the skeleton, as in a limpet or a horn coral. In the latter case, the area not occupied by a surface of deposition is called a *surface of accretion,* as it is a surface representation of the accretionary process that forms the skeleton. Growth lines can be thought of as forming the boundaries of growth layers, and thus can be seen in two ways—on the surface of accretion (where present) or in sections cut through growth layers within the skeleton. Growth lines visible on a surface of accretion are called *external growth lines* and

Pearl

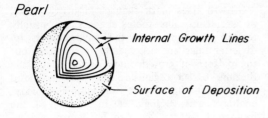

Internal Growth Lines

Surface of Deposition

Limpet

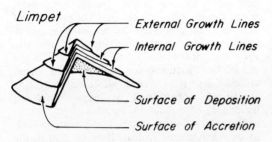

External Growth Lines

Internal Growth Lines

Surface of Deposition

Surface of Accretion

FIGURE 1. Two examples of accretionary tissues showing major features of accretionary growth. (From Clark, 1974; © 1974 by Annual Reviews, Inc. and reprinted with permission. All rights reserved.)

growth lines visible only in section are called *internal growth lines*. These basic concepts are illustrated in Fig. 1.

It is now appropriate to consider a definition for the term "growth line"; but a brief background sketch of growth line research will help illustrate the scope desireable in such a definition.

The first sort of growth lines to attract attention appear to have been the annual lines found in trees and shellfish. It is not apparent which of these was noticed first, for each predates any published scientific investigations. Leonardo da Vinci, for example, writing around the end of the 15th century, refers to growth lines in shells and plants as well-known phenomena. There is adequate precedence, then, for a definition of growth lines to include both internal (trees) and external (shells) varieties, or at least be unrestrictive in this sense.

There seems to be no restriction on the manner of expression of growth lines. Tree-rings are first noticed as differences in color or shading, but with close examination can be seen to be differences in density or spacing of cellulose walls. Upon weathering, these differences cause the rings to stand in relief as well. Growth lines in shells can also be seen as differences in color, thickness, or relief; many of these obvious differences can be traced to chemical or structural variations upon more sophisticated examination. It seems best, then, to let the term "growth

line" apply to any sort of differences between successive layers of shell material.

Often the most noticeable feature of growth lines is their periodicity; some growth lines grade so gradually from light to dark that no single point can be selected to define the line, but the number of repetitions of light and dark zones can be readily counted. In contrast to this are some bivalves which live short lives and acquire but one annual line on their shells. Here there is no repetition, but the single line has such sharp boundaries that it is readily observed; in one such species, *Argopecten irradians*, it is such a useful indicator of age that the state of Massachusetts bases its definition of legal size on the presence of a growth line. An adequate definition of a growth line must include both the abrupt and repetitive situations.

With these considerations in mind, it seems that a reasonable definition is that *growth lines* are abrupt or repetitive changes in the character of an accreting tissue (tissue in the broadest sense, including shells and noncellular skeletons). Any more restrictive definitions not only ignore aspects of the best-established growth lines, but aspects of growth lines involved in current research as well.

Some examples of growth lines might now be considered. Figs. 2-4 consist of idealized representations of actual growth-line patterns. The strips can be thought of as either cross sections or surface views, elongated in the direction of growth, and the variations in stippling can be thought of as variations in pigmentation, relief, or any other measurable character. In Fig. 2, A and B represent the two basic types of growth lines; Fig. 2A illustrates a single abrupt growth line, while Fig. 2B illustrates three growth lines that are prominent in their repetition.

It should be mentioned here that there are two approaches to working with growth lines. In the most common approach, the lines are simply noted and counted, but in the other, increasingly important, approach it is necessary also to measure the intervals between lines. This latter approach has raised the question of defining the exact position of growth lines. In Fig. 2C, for example, an investigator concerned only with counting sees three thick "growth lines"; but an investigator measuring intervals must be more detailed in his definitions, especially if the "growth lines" vary in thickness. For this reason, it is becoming accepted usage to consider growth lines to be boundaries only, lacking thickness. Any thickness or interval involved should be referred to as a *growth increment*, bounded by *growth lines*. The application of these terms to indi-

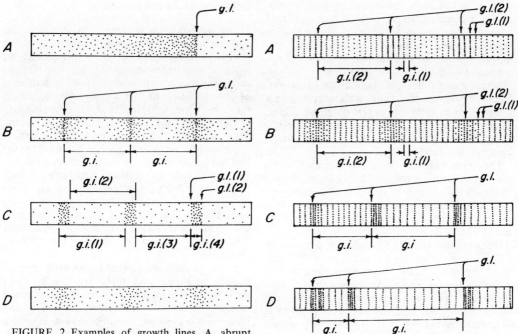

FIGURE 2. Examples of growth lines. A, abrupt growth line; B, repetitive growth lines; C, abrupt and repetitive growth lines; D, repetitive growth lines of uncertain position (zones). Position of growth lines in C may be defined as (1), resulting in increment (1); as (2), resulting in increment (2); or as both (1) and (2), resulting in increments (3) and (4) (a couplet). Increments (3) and (4) are, by one definition, bands. g.l. = growth line; g.i. = growth increment.

FIGURE 3. Examples of growth lines. A, cycles— g. l. (2)–formed by periodic darkening of g. l. (1); B, cycles (also clusters) formed by change in frequency of g. l. (1) (Evans, 1972); C, cycles (and clusters) formed by periodic variations in spacing of the fine lines; periodicity is in the number of fine lines between clusters; D, clusters—No periodicity is apparent in either number or spacing. Symbols as in Fig. 2.

vidual situations is still the choice of the investigator; thus, in Fig. 2C the growth lines might be defined at position (1), acting as boundaries of growth increment (1), at position (2), delineating increment (2), or at both positions, dividing the major increment into smaller increments (3) and (4). When this last choice is made, the investigator is viewing the increment as a *couplet* formed of two dissimilar *bands*. The growth lines in Fig. 2B may also be considered couplets, but the boundaries then become difficult to determine. Occasionally, growth-line patterns are encountered where no boundaries can be readily defined (Fig. 2D). In such cases, the term *zone* can be used in the place of lines or increments.

Not all growth lines occur as simple variations in color or relief of skeletal material; many types consist of variations in sequences of shorter-period growth lines in the same skeleton. In Fig. 3A, for example, the closely spaced growth lines vary in prominence; as this variation is also a variation in the character of the accreting tissue, it is itself a type of growth line. Other ways in which one growth

line can be formed by variations in another are shown in Fig. 3B, where the closely spaced growth lines occur twice as often in the same interval (see Evans, 1972), and in Figs. 3C and 3D, where the interval between closely spaced growth lines varies with time. Growth lines

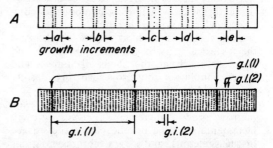

FIGURE 4. Examples of growth lines. A: Doublets (growth increments a-d); if intervals represent days, then a-d are also complex increments and e is a simple increment. B: Two orders of unrelated growth lines— g. i. (1) is, by one definition, a band. Symbols as in Fig. 2.

caused by periodic variations in the prominence, frequency, or spacing of shorter-period
growth lines are called *cycles*. Growth lines
caused by any variations, periodic or random,
in the spacing of shorter-period growth lines
are called *clusters*.

When growth lines of two or more periodicities occur on the same skeleton, they are
referred to as different *orders* of growth lines.
These can be formed by cycles or clusters,
as discussed above, or can be expressed in
completely different ways, as in Fig. 4B.

The term *band* has a second meaning when
associated with skeletons having more than a
single order of growth lines; in this case, and
particularily among workers in Paleozoic corals,
the more widely spaced growth increments are
called bands.

Occasionally two narrow growth increments
will be seen occupying approximately the same
space as the average single increment in the
vicinity; this is often interpreted as a single
increment with an intermediate growth line,
and may be called a *doublet*. This variation is
illustrated in Fig. 4A; a similar variation, where
a series of doublets forms the second type of
growth line in Fig. 3B, can be shown to occur
as postulated above (Evans, 1972). Doublets
with daily periodicity have been termed *complex increments* by Pannella and MacClintock
(1968); doublets in tree-ring research are
called *double rings,* with intermediate *false
rings* (Stokes and Smiley, 1968), although
false rings are used for other disturbance lines
as well.

Certain other terms have achieved prominence as useful descriptions of common types
of growth lines. The most famous of these is
ring; this term probably antedates "growth
line" itself and has at least three different
applications. Tree-rings, of course, are well
known as describing the internal growth lines
appearing as essentially concentric circles in
the woody tissues of trees. Rings also describe
the external growth lines, commonly indenta-
tions, which extend around cylindrical or
conical skeletons, such as rugose corals. And
the external growth lines found on discoid
skeletons, such as the shells of limpets and
bivalved mollusks, are also called rings despite
their usual lack of concentricity.

Other terms of this sort include *lamellae*
(narrow growth increments), *ridges* (growth
increments standing in relief above the surface
of the skeleton), and *ledges* (abrupt variations
in the directions of growth of a skeleton).
These terms are useful descriptions, but of
course their application depends completely
on individual workers' interpretation of
"narrow," "abrupt," and "relief."

Relationships to Environment

The use of growth lines as a source of information depends almost entirely upon
proper interpretation of their periodicity or
their cause. Unfortunately, these two factors
are commonly intermingled in discussions
of growth lines, so that a term such as "tidal"
may be used to refer to a tidal periodicity
rather than a tidal effect, such as exposure.
This can be further confused by one writer
applying the term to a 12.4-hr periodicity and
another applying it to a 14-day periodicity.
It seems best to restrict the term "tidal"
to the environmental effect, and to use terms
like "semidaily" and "fortnightly" for the
periodicity. This does not exclude combining
the terms, however; a term such as "semidaily
tidal lines" is readily understood without
ambiguity.

In the discussion of terms involving time, it
should be noted that there are two kinds of
periodicity. Periodic lines can form at fixed
intervals of time, such as every 24 hr, or can
form with a constant frequency, such as once a
year; the latter type seems more common,
because winter temperatures do not always
arrive on the same day, darkness depends on
clouds as well as sunset, and metabolic well-
being is an important and unpredictable factor.
It should also be recognized that some types of
growth lines form at random intervals of time.

One of the most difficult terms to define
precisely is the *daily growth line*. This is because there are conceivable circumstances in
which the lines could form with either fixed
intervals or fixed frequencies, and with respect
to either solar (24.00 hr) or lunar (24.84 hr)
days. To compound the confusion, there are
three terms available; daily, diurnal, and
circadian.

Sime of the difficulty can be sidestepped by
considering how such lines would be identified
and utilized; from experience, it is relatively
simple to determine that a set of growth lines
forms about once a day, but very difficult
to determine the precise periodicity. Therefore
a term which describes the former situation
will find much more application than terms
defining more precise situations. For this
reason, the term *daily growth line* is given the
braodest possible definition, and applied to
any sort of solar or lunar daily periodicity.

This leaves the two companion terms, *diurnal*
and *circadian*. Both mean "about daily," but
circadian has long been used in a specialized
sense in biological rhythm research, and because biological rhythms have some application
in growth-line research as well (see Clark,
1974, and pertinent articles in Rosenberg and

Runcorn, 1975), it seems best to maintain a consistent definition. *Circadian growth lines,* then, are defined as periodic growth lines formed at a mean interval of approximately 1 day (every 22 hr, for example). The term diurnal seems to refer to the solar day, as it is associated with the term "nocturnal" by some authorities; thus it is convenient to define *diurnal growth lines* as periodic growth lines formed with a frequency of one line per solar day.

Other periodicities are commonly encountered. *Semidaily growth lines* are defined as those formed with a frequency of two lines per solar or lunar day. *Fortnightly growth lines* are those formed at intervals of approximately two weeks. *Monthly growth* lines are formed with a frequency of one line per synodic or siderial month. *Annual growth lines* are those formed with a frequency of one line per year. Still other types of periodicities, such as weekly or semiannual, may exist but the evidence is limited.

Several different biological or environmental events are responsible for the formation of growth lines. One of the most common kinds of growth line is the *disturbance line,* caused by a cessation or severe restriction of growth. Unfortunately, there is a great deal of confusion regarding this term, because in many bivalves both annual and random growth lines form as disturbance lines. Most of the early workers recognized this, but as they could readily refer to the annual lines by their periodicity, the term "disturbance line" was more often applied to the random lines. Some recent workers have been restricting the term to random lines, further confusing the issue. It seems best to use the term as it was first intended, to refer to the manner of formation; references to the periodicity of formation should be done with terms implying time, such as annual or random. Again, there is no reason not to use combinations of terms, such as random disturbance lines.

The term *tidal growth lines* is another that has led to confusion. There are two main ways in which growth lines can result from tidal effects, and each has been called tidal. The first of these is the simple matter of exposure of an intertidal organism; this normally halts calcification, and thus produces a growth line. Depending upon the type of tides and the position in the intertidal zone, such lines can occur once or twice in a lunar day. The second effect involves the fortnightly variations in prominence of the tides; this can result in variations in the prominence or frequency of short-period growth lines (Fig. 3A,B), and produce growth lines with a period approaching 14.8 or 13.7 solar days, depending on the tidal cycles involved. With two such different tidal growth lines in existence, it seems best to define tidal growth lines simply as growth lines formed in response to tidal effects, and to refer to the different kinds by composite terms, such as "semidaily tidal lines."

Two other kinds of growth lines are strictly biological, although they may be triggered by environmental events. One, the *spawning line,* is defined as a growth line associated with the act of spawning; this is believed to be a fairly common sort of growth line in both corals and bivalves, although little evidence for the relationship seems to exist. The other is the *molting line;* this is a relatively unusual sort of growth line because organisms that molt rarely retain significant accretionary tissues, but at least one group, the barnacles, forms a permanent accretionary skeleton in addition to the temporary exoskeleton. Thus, barnacles could record the molting process as growth lines in the permanent skeleton.

In using growth lines to reconstruct environmental parameters, it is important to understand that some environmental variables, such as storms, can cause growth lines to form simultaneously in all members of a population of organisms, while other variables, such as attempts at predation, cause growth lines in single individuals only. Pannella and MacClintock (1968) have termed the former *universal events* and the latter *private events.* These events are not, however, solely responsible for differences in growth line patterns in neighboring organisms; some individuals within any population will differ from others in their response to the same stimuli.

The discussion above is intended to give a fairly general picture of growth-line formation and terminology; there are, however, numerous synonyms for many of these terms; and many other terms are (or have been) in favor with specialists.

Synopsis of Research

Growth-line research can be roughly divided into three main areas. The first, which might be called the biological approach, is mainly concerned with growth lines as useful markers in studies of the growth and life history of living organisms. The second, or ecological approach, is concerned with the causes of growth lines and the use of growth-line records to reconstruct environmental conditions in living and fossil organisms. The third, or geophysical approach, considers growth lines strictly as records of periodicities, for the common periodicities are reflections of planetary motions. When applied to fossil organisms, this latter approach should yield useful information on the history of the earth-moon system.

Comprehensive discussions of the methodology and current status of growth-line research can be found in recent review articles by Scrutton and Hipkin (1973) and Clark (1974). A broad selection of articles dealing with recent advances in the field is also available, in a single volume edited by Rosenberg and Runcorn (1975).

It would appear that growth lines, besides being common and widespread features of fossil organisms, are the basis of an exciting new field of paleontological research, with potential applications in life history, paleoecology, and even the history of the earth-moon system. However, it will probably be some time before the relationships of growth lines to the environment become well enough understood to permit other than the most cautious interpretations.

GEORGE R. CLARK, II

References

Clark, G. R., II, 1974. Growth lines in invertebrate skeletons, *Ann. Rev. Earth Planetary Sci.*, **2**, 77–99.

Evans, J. W., 1972. Tidal growth increments in the cockle *Clinocardium nuttalli*, *Science*, **176**, 416–417.

Pannella, G., and MacClintock, C., 1968. Biological and environmental rhythms reflected in molluscan shell growth, *J. Paleontology* **42**, (5, pt. 2), 64–80.

Rhoads, D. C., and Lutz, R. A., eds., 1979. *Skeletal Growth: Biological Records of Environmental Change.* New York: Plenum Press.

Rosenberg, G. D., and Runcorn, S. K., eds., 1975. *Growth Rhythms and the History of the Earth's Rotation.* New York: Wiley, 559p.

Scrutton, C. T., and Hipkin, R. G., 1973. Long-term changes in the rotation rate of the earth, *Earth-Sci. Rev.*, 9, 259–274.

Stokes, M. A., and Smiley, T. L., 1968. *An Introduction to Tree-Ring Dating.* Chicago: Univ. Chicago Press, 73p.

Cross-references: *Biomineralization; Paleoecology Population Dynamics.* Vol. II: *Tree-Ring Analysis (Dendroclimatology).*

GYMNOSPERMS

Frequently referred to by the class name "Gymnospermae," the gymnosperms are seed-bearing plants that are now assigned to several classes or divisions. These plants are distinguished from the angiosperms, or flowering plants, in possessing seeds that are not enclosed within an ovary. Some workers suggest as possible precursors of the gymnosperms an assemblage of Devonian plants referred to as the

TABLE 1. Classification of the Gymnosperms

Division	PTERIDOSPERMOPHYTA—Mississippian–Jurassic
Order	Lyginopteridales—Carboniferous–Permian
Order	Medullosales—Carboniferous–Permian
Order	Calamopityales—Carboniferous
Order	Peltaspermales—Permian–Triassic
Order	Corystospermales—Triassic
Order	Caytoniales—Jurassic
Division	CYCADEOIDOPHYTA—Triassic–Cretaceous
Order	Cycadeoidales (Bennettitales)
Division	CYCADOPHYTA—Permian–present
Order	Cycadales
Division	GINKGOPHYTA—Permian[?]–present
Order	Ginkgoales
Division	CONIFEROPHYTA—Mississippian–present
Order	Cordaitales—Mississippian–Triassic[?]
Order	Voltziales—Carboniferous–Jurassic
Order	Coniferales—Carboniferous–present
Order	Taxales—Jurassic–present
Division	GNETOPHYTA—Permian[?]–present
Order	Gnetales
Incertae Sedis	
Order	Glossopteridales—Permian
Order	Vojnovskyales—Permian
Order	Pentoxylales—Jurassic

Progymnospermophyta (see Pteridophyta); these plants had internal structure similar to that found in gymnosperms but reproduced by spores that were shed. There are about 700 living species of gymnosperms; their significance as floral components, both in numbers and in distribution, was much more considerable in the past, reaching a peak during the Mesozoic era.

It is probable that gymnospermous plants first appeared on the earth in the Devonian. The evidence is the occurrence of a seedlike body, *Archaeosperma*, from the Late Devonian (Fig. 1A). Though its higher taxonomic affinities are unclear, *Archaeosperma* is known to have had a single functional megaspore and the sporangium producing it was enveloped by an integument that was open at the tip. Somewhat more specialized seed-like structures are common in lower Carboniferous (Mississippian) deposits in Scotland. In these forms as well, sporangia, each with a single functional megaspore, are partially enveloped by a simple integumentary system in the form of a number of finger-like processes that extend around the sporangium (nucellus) from its base (Fig. 1B). In more advanced forms, these processes had fused to form an envelope extending around the sporangium to form a more nearly typical integument. Little is known of the plants that bore these seeds, but they were obviously among the simplest gymnosperms.

Division Pteridospermophyta

This group of seed plants, often called the "seed ferns," combined characters of the ferns and of more typical gymnosperms. Members

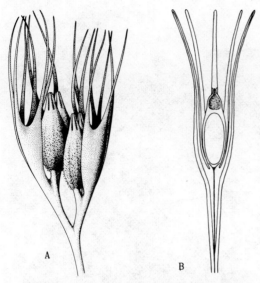

FIGURE 1. Early gymnosperms. A: *Archaeosperma arnoldii*, reconstruction of a cluster of seedlike bodies loosely surrounded by finger-like appendages; this Upper Devonian fossil is the earliest known seed (from Pettitt and Beck, 1968). B: *Genomosperma kidstoni*, longitudinal section (diagrammatic) of a Lower Carboniferous (Mississippian) seed-like body consisting of a nucellus (megasporangium) loosely invested by finger-like integumentary process (from Delevoryas, 1962; copyright © 1962 by Holt, Rinehart and Winston, Inc. and reprinted by permission).

resembled tree ferns, with branched or un-branched stems; many had frond-like leaves (Fig. 2). On the foliage were borne seeds and pollen-bearing structures. The class appeared first in the Mississippian, reached its maximum development in the late Paleozoic, and extended into the Jurassic.

Members of the order Lyginopteridales (Mississippian-Permian) had slender stems, with rather loosely constructed secondary vascular tissue, much like that in present-day cycads. On the fern-like foliage were borne small, radially symmetrical seeds. In some instances, these seeds were contained within cup-like structures. The Medullosales (Mississippian-Permian) were larger plants, also with fern-like foliage, with more than one stele in each stem. Stelar number was quite high in some forms. Seeds of this group were very large, with multiple seed coats and often with a three-angled cross-sectional configuration. Pollen-bearing organs were also large and generally had tubular sporangia fused together in various configurations. Pollen grains are among the largest known, some exceeding 500 μm in length.

A number of leaf forms that are obviously pteridospermous are known from the Paleozoic.

Some of them were probably fronds of either the Lyginopteridales or the Medullosales, but without evidence of attachment to stems their natural affinities are only conjectural. In some of these forms, seeds are borne in place of pinnules; some have seeds actually attached to the pinnules themselves.

The Calamopityales (Devonian-Mississippian) is an artificial taxon including a number of stem and petiole types that cannot be assigned with certainty to other orders. These stems may be protostelic, or a pith may be present. The wood is somewhat more compact than that of the previous two orders. Fructifications of these forms are not known, but, when they are, it may be found that some of these forms belong with the progymnosperms rather than with the gymnosperms.

A small group of Permian and Triassic seed ferns, the Peltaspermales, bore fernlike foliage. Seed-bearing structures were borne helically on an axis, and each member was peltate and stalked, with seeds borne on the underside of the shield-shaped head (Fig. 3A). Seeds of the Corystospermales (Triassic) were borne singly within helmet-shaped cupules, which, in turn, were borne on a branching axis (Fig. 3B). Micropyles of the seeds extended out beyond the opening of the cupule.

The Jurassic Caytoniales had palmately compound leaves with net venation. Seed-bearing structures were borne in two rows along an axis; they were sac-like and contained a number of seeds within (Fig. 3C). Although the seeds are almost completely covered by the fruit-like sac, pollen apparently was able to invade this sac and land directly on the ovule micropyles.

Although the Pteridospermophyta are now extinct, many botanists see in them the morphological prerequisites of an angiosperm precursor. Some of the Mesozoic forms have seed-bearing members that closely approximate angiosperm carpels.

Division Cycadeoidophyta

Plants superficially resembling those of the extant order Cycadales (Division Cycadophyta) and confined to the Mesozoic are placed in the order Cycadeoidales (often referred to as Bennettitales). The great abundance of these plants, as well as of remains of the Cycadophyta during the Triassic, Jurassic, and Cretaceous periods, is responsible for the description of the Mesozoic era as the "Age of Cycads." The Cycadeoidales, in common with the Cycadales, had leathery, generally pinnately compound leaves. Other than superficial resemblance, however, the two divisions do not

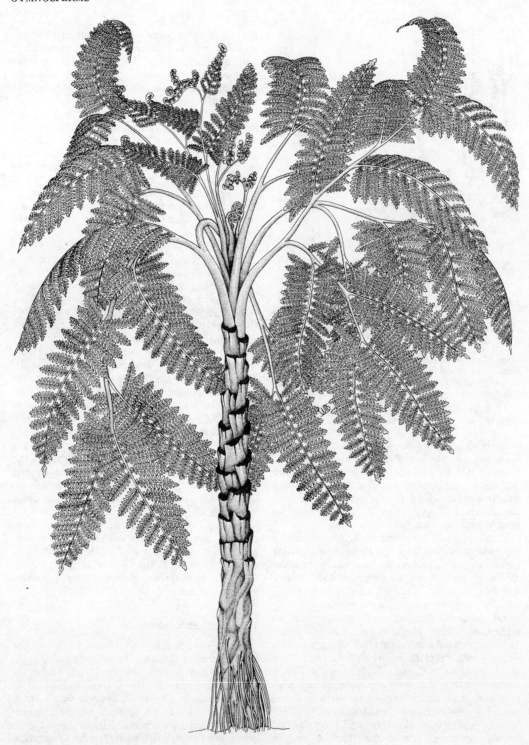

FIGURE 2. *Medullosa noei,* reconstruction of a typical Paleozoic seed fern (from Stewart and Delevoryas, 1956).

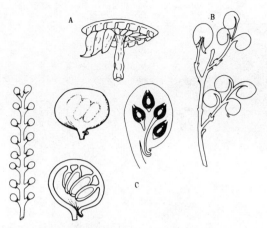

FIGURE 3. Gymnosperm seed-bearing structures (from Delevoryas, 1962; copyright © 1962 by Holt, Rinehart and Winston, Inc. and reprinted by permission). A, seed-bearing, peltate structure of a member of the Peltaspermales, *Lepidopteris ottonis;* B, seed-bearing axis of a member of the Corystospermales, *Umkomasia macleani;* C, fruit-like, seed-bearing structures of members of the Caytoniales, *Caytonia* spp. Seeds are almost completely enveloped by a saclike structure.

have much in common. One family of cycadeoidaleans, the Williamsoniaceae, typically had erect and frequently branched stems. Seeds were borne on the surface of fleshy ovulate receptacles; pollen-bearing organs were generally constructed of a number of finger-like members fused together at the base (Fig. 4). Sporangia were borne along the inner faces of the free members. In some forms, seed- and pollen-bearing structures were borne on the same cone (Fig. 5); other members had separate cones.

The second family of the Cycadeoidales, the

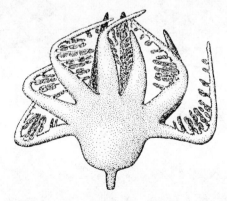

FIGURE 4. Microsporangiate fructification of *Williamsonia spectabilis* (from Delevoryas, 1962; copyright © 1962 by Holt, Rinehart and Winston, Inc. and reprinted by permission). Pollen sacs are borne on finger-like appendages.

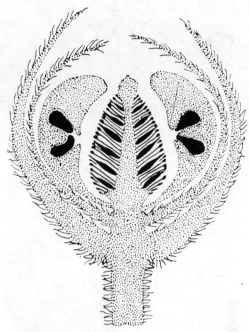

FIGURE 5. *Williamsoniella coronata,* diagrammatic longitudinal section of a cone (from Delevoryas, 1962; copyright © 1962 by Holt, Rinehart and Winston, Inc. and reprinted by permission). Seed-bearing receptacle is in the center; pollen-bearing organs surround it.

Cycadeoidaceae, had short, squat, sometimes ovoid, sometimes globose, often columnar stems with a persistent armor of leaf bases and a terminal crown of pinnately compound leaves (Fig. 6A). Among the leaf bases were borne cones, each consisting of a compact mass of pollen sacs and a fleshy receptacle covered with ovules (Fig. 6B). The cones did not expand into showy flower-like structures as is shown in many reconstructions.

The division Cycadeoidophyta, first evident in the Triassic, became extinct at the end of the Mesozoic era.

Division Cycadophyta

Members of the order Cycadales, which comprise the division Cycadophyta, were largely contemporaneous with the Cycadeoidophyta. The group appeared first in the late Paleozoic and exists at the present time in the form of 10 genera and about 100 species. As in many of the Cycadeoidales, members of the Cycadales also bear pinnately compound leaves in a terminal crown and have squat trunks, often covered with persistent bases of leaves. During the Mesozoic, however, the growth habit involved a more slender stem, with leaves not too closely arranged (Fig. 7).

367

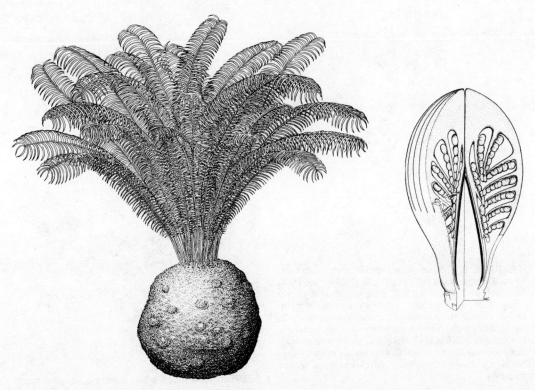

FIGURE 6. *Cycadeoidea* sp. A, reconstruction of a typical plant of the genus; B, partially sectioned, three dimensional reconstruction of a cone with inner, conical ovule–bearing receptacle surrounded by massive pollen-bearing organs. (From Crepet, 1974; reprinted by permission from E. Schweizerbart'sche Verlagsbuchhandlung).

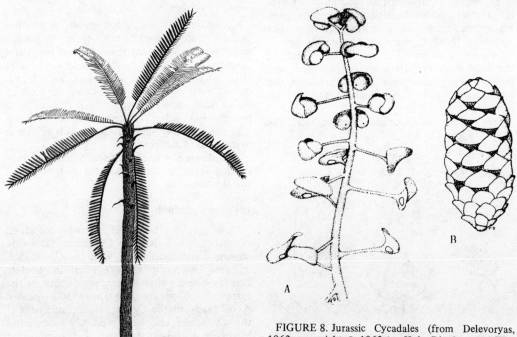

FIGURE 7. Reconstruction of an Upper Triassic cycad, *Leptocycas gracilis* (from Delevoryas and Hope, 1971).

FIGURE 8. Jurassic Cycadales (from Delevoryas, 1962; copyright © 1962 by Holt, Rinehart and Winston, Inc. and reprinted by permission). A, reconstruction of a seed-bearing cone of *Beania gracilis;* B, Pollen-bearing cone of *Androstobus manis;* pollen sacs are visible on the lower sides of reduced leaves.

Seeds are borned on modified leaves that are often clustered into cones (Fig. 8A). Pollen sacs are also borne in cones on undersides of reduced leaves (Fig. 8B).

The first recognizable remains of Cycadophyta occur in Permian rocks in the form of seed-bearing leaves. The group attained its maximum development in the Mesozoic.

Foliage of Cycadeoidophyta and Cycadophyta is common in Mesozoic rocks. Leaves of both divisions are quite similar, and can be distinguished in the fossils on the basis of epidermal characters. One order based largely on leaves, the Bennettitales, probably corresponds to foliage of members of the Cycadeoidales; the second order based on leaves, Nilssoniales, probably corresponds to the Cycadales.

Division Ginkgophyta

Another small class of gymnosperms with a past record of greater significance than present distribution would suggest are the Ginkgophyta, known at the present time by only one species, *Ginkgo biloba*. Although fossil plants showing an apparent relationship to the ginkgophytes are known from upper Paleozoic rocks, the earliest unquestioned members of the Ginkgophyta are Mesozoic. Among the late Paleozoic remains are leaves with dichotomously branched structure and with no laminar portion. Earliest Mesozoic ginkgophytes have fan-shaped and deeply incised leaves; this type of leaf grades into the more nearly typical ginkgo leaf at successively higher levels of the Mesozoic. In the upper Mesozoic and Tertiary, wood and foliage of ginkgophytes are abundant. Ovules were stalked and clustered on small branches in leaf axils; pollen sacs were borne on short appendages aggregated into cones.

Division Coniferophyta

Most of the presently existing gymnosperms are members of the division Coniferophyta. Although still an impressive group in some parts of the world even today, the coniferophytes were more important in past ages. It is suspected that the Coniferophyta may have had their origin as early as the Devonian, but the first seed-bearing members are not known until the Mississippian.

Members of the order Cordaitales were arborescent, branched, and bore extremely elongated and flattened leaves (Fig. 9). Although venation appears to be parallel, it is actually composed of distantly spaced dichotomies. Anatomically, stems of the Cordaitales were very similar to those of modern conifers, with a pith surrounded by dense wood com-

FIGURE 9. *Cordaites* sp. (from Florin, 1951). Tip of a branch showing vegetative and reproductive structures.

posed of small tracheids and narrow rays. Seed- and pollen-bearing structures were borne on slender, elongated axes. One of these fruiting axes bore reduced scale-like leaves in two rows; in the scale leaf axils were small bud-like structures. These buds had helically arranged scales; near the apex of the bud, the scales may have produced pollen sacs or ovules. Pollen and seeds were produced on separate branches. Seeds were generally flattened and bilaterally symmetrical; some were winged, and some seem to have had a softer, outer coat surrounding a harder, inner one. Pollen grains were surrounded by an equatorial bladder.

In the Voltziales (Pennsylvanian-Jurassic) are included coniferlike plants that show close affinity with modern coniferous types. These forms were arborescent, and generally had small, crowded, scale-like, helically arranged leaves (Fig. 10). In general habit they closely approximated some of the present-day araucarias. Seed cones are of particular interest because of their bearing on the interpretation of the seed cones of modern conifers. The seed-bearing structures consist of a main cone axis on which are closely spaced, helically arranged

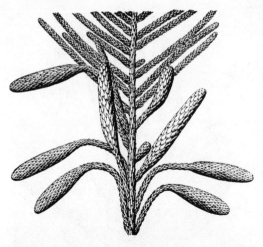

FIGURE 10. *Lebachia piniformis* (from Florin, 1944, by permission of E. Schweizerbart'sche Verlagsbuchhandlung). Portion of a branch system with vegetative leaves, pendant pollen cones, and erect seed cones.

scales or modified leaves. In the axil of each of these scales (bracts) is a small lateral shoot bearing reduced appendages, some or all of which may bear ovules. It is such a dwarf axillary shoot that has become modified by dorsiventral flattening and fusion of parts to produce the woody cone scale in cones such as those of pine. Pollen cones consisted of an axis with helically arranged microsporophylls, much as in modern conifers.

Although the oldest Coniferales extend back to the Carboniferous, modern families made their appearance during the Mesozoic. The Araucariaceae and Pinaceae are among the oldest conifer families. Remains of the Araucariaceae are recognized as far back as the Triassic. Silicified logs in the Triassic Petrified Forest National Park in Arizona include some that are presumably araucariaceous. Leafy twigs, assigned to the genus *Pagiophyllum*, appear in the Triassic and throughout the Mesozoic; these are suggestive of araucarians. Well preserved *Araucaria* cones are known from middle and upper Mesozoic and lower Tertiary strata. Cretaceous and Tertiary rocks yield remains of cone scales of *Agathis*, another araucariaceous genus.

Although the Pinaceae, a common modern-day family of conifers, probably had their origin early in the Mesozoic, the first well-documented fossils of the group occur in the Jurassic. Conceivably, the Triassic *Woodworthia*, represented by trunks associated with the Arizona *Araucarioxylon*, might have been a genus of plants with affinities with the pines. During the Cretaceous, abundant remains of pine-like wood, shoots, leaves, and cones are known. The modern genera *Pinus*, *Pseudolarix*, and *Picea* are known from the Cretaceous onward.

Another important family of the Coniferales are the Taxodiaceae, to which are assigned the present-day redwoods and big trees of the American West Coast. Several forms are known that indicate that the family originated at least as far back as the Jurassic *Elatides* is an extinct genus found in the Jurassic and Cretaceous. *Sequoia*, the genus of redwoods, is reported first in the Cretaceous, but more abundant remains appear in Tertiary deposits. *Metasequoia* (Fig. 11), found first as a fossil in Japan, has since been found living in central China. Its first appearance is in the Cretaceous. Other taxodiaceous genera also appeared in the Cretaceous, and during Tertiary times the family was a significant one among the conifers.

The family Cupressaceae is definitely recognizable from the Middle Jurassic onward.

Another relatively old family of Coniferales are the Podocarpaceae, making their appearance in the Middle Triassic, and becoming much more significant during the Jurassic. Several modern genera are found as far back as the Jurassic.

FIGURE 11. Portion of a leafy twig of a Tertiary member of the Taxodiaceae, *Metasequoia occidentalis* (from Chaney, 1951).

The Cephalotaxaceae, another family of the Coniferales now still living, began their existence at least by Jurassic times.

The Mesozoic Cheirolepidaceae are an extinct conifer family known only from seed cones. As in most typical conifer seed cones, cheirolepidaceous seeds were borne on the upper sides of woody cone scales in the axils of bracts. The difference was the presence of a flap of the cone scale that almost completely covered the ovules on the upper surface.

The Taxales, an order quite close to the Coniferales, but differing in the absence of cones, appear to have been distinct at least as far back as the Jurassic. *Taxus* and *Torreya* are both known from the Jurassic.

Division Gnetophyta

The Gnetophyta are another small group represented at the present time by three genera; many experts feel they may not be related, however, and that the division may be an artificial one. Fossil remains are skimpy, although pollen grains indicate that the group may have extended back in time into the Permian.

Incertae Sedis

There are known in the fossil record a number of obviously seed-bearing plants that do not fit comfortably into presently established systems of classification. One such group, the Glossopteridales, has been classified in the past with the ferns, seed ferns, cycadophytes, and angiosperms. Recent work suggests that *Glossopteris* and related forms were seed plants, and most probably gymnospermous. From a woody stem arose strap-shaped leaves with net venation. From the upper surfaces of some of the leaves were borne inrolled, bladelike structures on which were borne small, seed-like bodies. Glossopterids were a widespread group during the Permian, when they existed in a wide belt in the Southern Hemisphere.

An interesting order of gymnospermous plants, the Pentoxylales, has been reported from the Jurassic of India. As the name suggests, there were typically five (sometimes six) steles in the stem, alternating with an equal number of smaller one. Leaves were strap-shaped, resembling to some extent leaves of some of the fossil cycads (Fig. 12A). Seed-bearing axes were branched and terminated by clusters of fleshy seeds (Fig. 12B). As with seed cones, microsporangiate structures terminated dwarf shoots and were clustered at the ends of slender stalks.

The Permian Vojnovskyales are based on the genus *Vojnovskya* from the Soviet Union. On stout branches were borne *Ginkgo*-like,

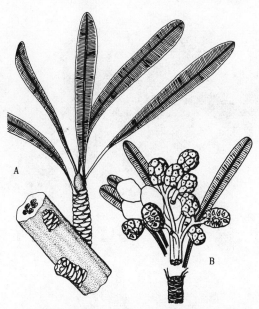

FIGURE 12. *Pentoxylon sahnii*. A: Portion of the stem with strap-shaped leaves (from Delevoryas, 1962; copyright © 1962 by Holt, Rinehart and Winston, Inc. and reprinted by permission). B: seed-bearing cones (from Sahni, 1948).

fan-shaped leaves, and on stubby fertile shoots were borne both seeds and microsporangiate appendages (Fig. 13). It is quite unrelated to anything known at present.

There are known from the fossil record, in addition, a number of gymnosperm genera of plants that are isolated by their uniqueness and presently unknown affinities. Many kinds of stems and wood fragments are also known, obviously with gymnospermous characteristics, but defying more precise classification at the present time.

Geologic Summary

Although it is generally recognized that gymnosperms probably originated in the Devonian period, the only concrete evidence in the fossil record before the Mississippian is *Archaeosperma*, a seed-like body. During Devonian times, there existed a group of fern-like plants that had secondary vascular tissues, and some having the heterosporous habit. Many workers place such plants into the division Progymnospermophyta, and emphasize the combination of fern and gymnosperm characters. By Mississippian times, true seed plants had come into existence. Some of the earliest have primitive seeds with integuments not yet fully organized into completely enveloping and protecting structures. Some of the plants that bore these simple seeds apparently lacked leaves.

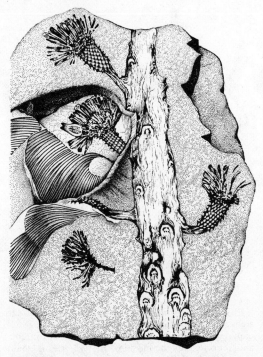

FIGURE 13. *Vojnovskya paradoxa* (from Andrews, 1961; copyright © 1961 by John Wiley & Sons, Inc. and reprinted by permission). Portion of stem with leaves and fertile branches.

Among the earliest true gymnosperms were the Pteridospermophyta, or pteridosperms (seed ferns), which first appear in the Mississippian. These seed ferns resembled, in general, arborescent ferns, but on the foliage were borne naked seeds and pollen-bearing organs. Contemporaneous with the pteridosperms in the Early Carboniferous were members of the Cordaitales, arborescent plants with strap-shaped leaves and abundant secondary wood development.

Later in the Carboniferous, seed ferns were more abundant, with two or three orders being present. The Cordaitales also were more significant floral components, and true conifers, members of the Coniferales, also began to appear. Closely related Voltziales also began to show up in the Upper Carboniferous. Thus the gymnosperms of Carboniferous forests were primarily pteridosperms, cordaiteans, coniferaleans, and voltzialeans.

These groups persisted into the Permian, and by that time it is possible that members of the small class Gnetophyta became established. Remains of the Ginkgophyta are known in the Permian as well. Also quite significant were the widespread Glossopteridales. Other gymnospermous plants of less certain position must have existed as well, as testified by the unusual Vojnovskyales.

The first members of the division Cycadophyta began to put in an appearance by Late Permian times, and in the beginning of the Mesozoic the similar appearing Cycadeoidophyta were also in existence. Thus by the early Mesozoic, in addition to new kinds of seed ferns, Ginkgoales, Coniferales, Voltziales, Cycadophyta, and Cycadeoidophyta had become well established and began the expansion that made them such important members of the floras by the Jurassic period. Toward the close of the Jurassic, seed ferns were approaching extinction, and many of the modern Coniferales had become established. Cycadlike plants, represented by the Cycadeoidales and Cycadales, were widespread.

Modern conifer families were well represented in the Cretaceous, and cycadeoids were very abundant. The latter disappeared at the close of the Cretaceous, and Lower Tertiary fossils suggest that all the modern kinds of conifers were in existence, although with a distribution quite different from that of today. Cycadophytes and Ginkgophytes assumed progressively more minor roles during the Tertiary.

During past geologic ages, from Carboniferous through the Jurassic, a number of kinds of isolated gymnosperms made brief appearances and passed off into extinction. Many of these probably had no bearing on the evolution of present-day types.

T. DELEVORYAS

References

Andrews, H. N., 1961. *Studies in Paleobotany*. New York: Wiley, 487p.

Andrews, H. N., 1963. Early seed plants, *Science*, **142**, 925-931.

Arnold, C. A., 1953. Origin and relationships of the cycads, *Phytomorphology*, **3**, 51-65.

Chaney, R. W., 1951. A revision of fossil *Sequoia* and *Taxodium* in western North America based on the recent discovery of *Metasequoia, Trans. Amer. Philos. Soc., n. s.,* **40**, 171-239.

Crepet, W. L., 1974. Investigations of North American cycadeoids: The reproductive biology of *Cycadeoidea, Palaeontographica*, **148B**, 144-169.

Delevoryas, T. 1962. *Morphology and Evolution of Fossil Plants*. New York: Holt, Rinehart & Winston, 189p.

Delevoryas, T., and Hope, R. C., 1971. A new Triassic cycad and its phyletic implications, *Postilla*, **150**, 1-20.

Florin, R., 1944. Die Koniferen des Oberkarbons und des Unteren Perms, pt. 6, *Palaeontographica*, **85B**, 365-456.

Florin, R., 1951. Evolution in cordaites and conifers, *Acta Hort. Berg.*, **15**, 285-388.

Florin, R., 1963. The distribution of conifer and taxad genera in time and space. *Acta Hort. Berg.,* **20,** 121–312.

Pettitt, J. M., and Beck, C. B., 1968. *Archaeosperma arnoldii*–a cupulate seed from the Upper Devonian of North America. *Contrib. Mus. Paleontol. Univ. Michigan,* **22,** 139–154.

Sahni, B., 1948. The Pentoxyleae: A new group of Jurassic gymnosperms from the Rajmahal Hills of India, *Bot. Gaz.,* **110,** 47–80.

Stewart, W. N., and Delevoryas, T., 1965. The medullosan pteridosperms. *Bot. Rev.,* **22,** 45–80.

Taylor, T. N., ed., 1976. Patterns in gymnosperm evolution. *Rev. Palaeobot. Palynol.,* **21,** 1–134.

Cross-references: *Paleobotany; Paleophytogeography.*

H

HEMICHORDATA

A phylum ranging from the middle Cambrian to the present day, hemichordates lack the notochord, distinctive feature of the phylum Chordata (= Vertebrata) and are broadly intermediate between vertebrates and invertebrates. The geological record, although long, is sparse. Four classes are recognized today, the pogonophorans having been excluded as a separate phylum.

Class Enteropneusta

The class Enteropneusta includes *Balanoglossus,* but all are soft bodied and are unknown as fossils, although burrow and tracks (trace fossils) have sometimes been ascribed to enteropneustons (e.g., Kaźmierczak & Pszczółkowski, 1969). The living animal has an acorn-shaped proboscis, a collar, trunk, and branchial apparatus including a double row of pores.

Class Pterobranchia

Two orders are recognized, Rhabdopleurida and Cephalodiscida, the former with one extant genus (*Rhabdopleura*) and the latter with two (*Cephalodiscus* and *Atubaria*). Three fossil rhabdopleuran genera have been described from the Ordovician: *Rhabdopleurites* Kozłowski; *Rhabdopleuroides* Kozłowski and *Eorhabdo-*

pleura Kozłowski. *Rhabdopleura* itself is known as far back as the Late Cretaceous. Of the cephaloidscids, *Cephalodiscus* may occur in the middle Eocene, while fossil representatives are *Eocephalodiscus* Kozłowski (the earliest undoubted hemichordate) from the Tremodoc and *Pterobranchites* Kozłowski from the Lower Ordovician.

Class Planctosphaeroidea

Possible larval forms of an unknown hemichordate, not known as fossils.

Class Graptolithina

A fourth, extinct class, Graptolithina (q.v.) is today included by most paleontologists in the phylum Hemichordata, largely because of the considerable similarities of encrusting graptolites to the extant hemichordate *Rhabdopleura.* The class has, however, been placed with many other phyla in times past, and may eventually be raised to the status of phylum.

Morphology and Mode of Life

Rhabdopleurans are truly colonial, with zooids attached by a contractile stalk to the stolon (Figs. 1,2); whereas cephalodiscans form free, unattached associations. The zooids of extant species differ in the two classes on the

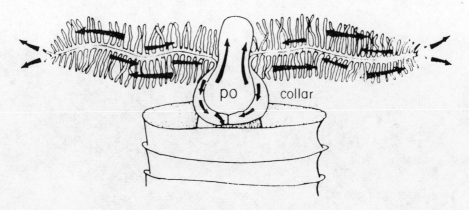

FIGURE 1. *Rhabdopleura* zooid showing direction of cilia-assisted currents. po = preoral lobe.

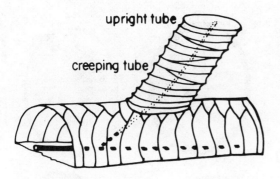

FIGURE 2. *Rhabdopleura*, portion of skeleton; stolon, black and heavy dasher; contractile stalk dotted.

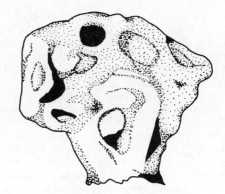

FIGURE 4. *Eocephalodiseus.*

number of arms, nature of gonads, and presence or absence of branchial pores: but the sclerotized skeletons show greater differences. *Rhabdopleua* skeletons (Fig. 2) form irregularly branching tubes attached by a basal structureless layer to the substrate or shells, and have upright zooidal tubes at intervals. The colony has a permanent, leading, terminal bud behind which daughter zooids arise and secrete their upright tubes. Cephalodiscan skeletons may be encrusting, dendroid (Fig. 3) or compact

(Fig. 4), and are constructed by the zooids, which creep out of their tubes to secrete connective and cuticular tissue.

Rhabdopleura is tolerant of a wide range of benthic marine environments and depth, as well as of latitude, mostly in the Northern Hemisphere, while *Cephalodiscus* is mostly restricted to the Southern Hemisphere (the Antarctic, and Pacific regions of the East Indies and Japan).

R. B. RICKARDS

References

Barrington, E. J. W., 1965. *The Biology of Hemichordata and Protochordata.* Edinburgh and London: Oliver and Boyd, 176p.

Bulman, O. M. B., 1970. Graptolithina, in R. C. Moore, ed., *Treatise on Invertebrate Paleontology,* pt V, 2nd ed. Lawrence, Kansas: Geol. Soc. Amer. and Univ. Kansas Press, 163p.

Kázmierczak, J., and Pszczótkowski, A., 1969. Burrows of Enteropneusta in Muschelkalk (Middle Triassic) of the Holy Cross Mountains, Poland, *Acta Palaeontol. Polonica,* 14, 299–324.

Kozʃowski, R., 1970. Nouvelles observations sur les rhabdopoeuridés (Ptérobranches) Ordoviciens. *Acta Palaeontol. Polonica,* 15, 1–14.

Cross-references: *Chordata; Graptolithina.*

HISTORY OF PALEONTOLOGY: BEFORE DARWIN

Some fossils, being objects of striking form and appearance, have been noticed and commented on in many different periods and cultures. Only within Western civilization, however, and only since the Renaissance, has this diffuse awareness of fossils crystallized into a set of coherent intellectual goals and effective technical methods, emerging in the

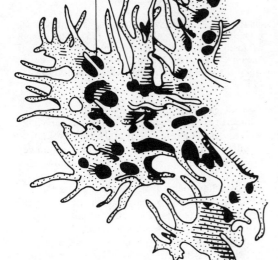

FIGURE 3. *Cephalodiscus.*

late 19th century as a distinct scientific discipline of paleontology. The discussion of the nature of fossils reached a high level of sophistication in Chinese culture at an earlier period than in the West (Needham, 1959); but unlike work in some other sciences, Chinese thought on fossils seems to have had little influence on Western thought. For reasons of space, neither early Chinese work on fossils, nor the fragmentary contributions of classical antiquity in Europe, will be considered here. From the Renaissance onward, however, there has been an unbroken tradition of discussion and research on fossils, developing within many successive and diverse intellectual cultures in the west. The opinions of those who discussed the nature of fossils in earlier periods must be understood in the light of their intellectual environment, and in terms of the empirical evidence actually available to them; their conclusions should not be judged according to their degree of compatibility with modern knowledge.

Fossils in 16th-Century "Natural Philosophy"

The term *fossil* and its cognates in other languages originally meant anything *dug up* out of the ground or found lying on the surface. It therefore included not only—or even principally—the fossils of modern science, but also distinctive rocks, gemstones, other minerals, and concretions. In the 16th century, even this wide range of objects was generally discussed in a still wider context of "stones," which included in addition such objects as meteorites, stony coral, gallstones, and prehistoric artifacts. This shows that the primary problem at this period was not to decide whether or not fossils (in the modern sense) were organic in origin, but to understand why such diverse objects were all "stony" in composition, and to form some acceptable classification for them. These objects were studied partly with utilitarian motives, in that minerals were thought to include substances of possible medicinal value, and partly in the context of a more general curiosity about natural objects and man-made artifacts and antiquities of all kinds. The careful study of fossils thus began with the assembly of collections of specimens in museums, and was greatly aided by the first attempts to publish woodcut illustrations of them in the later 16th century. This made it possible for the first time for scholars who were separated geographically to be reasonably sure that they were applying the same names to the same kinds of object.

Several scholars at this period suggested that certain fossils were the remains of organisms that had once been alive. Leonardo da Vinci's

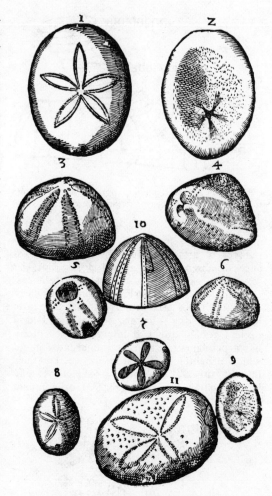

FIGURE 1. A page of illustrations of fossil sea-urchins from Ulisse Aldrovandi's (1648) large book of fossils, *Musaeum Metallicum in Libros IIII distributum.* . . . . Contrast the crudity of these 16th-century woodcuts with the more detailed 17th-century figures of Fig. 2.

conjectures of about 1508 are an early and well-known example (though they remained unpublished and had no further influence). Such suggestions were of little importance, however, because they only touched the periphery of what seemed at the time to be the real explanatory problem. Suggestions of an organic origin were confined to the small group of fossils (mostly Cenozoic marine mollusk shells) that were, in modern terms, unchanged in composition, similar in form to living organisms, and found close to present sea level.

To suggest an organic origin for such objects scarcely touched the main problem: namely, how to understand the much wider and more ambiguous range of resemblances between many other "fossils" (in the older sense) and various other natural or artificial objects. For example, concretions often had a vague resemblance to organic forms, and what we know as crinoid ossicles might look vaguely like stars or wheels. In the case of one of the greatest 16th-century naturalists, the Swiss physician Conrad Gesner (1516–1565), we can judge the extent of his biological knowledge from his published works, and therefore estimate how far his perception of resemblances between fossils (in the modern sense) and living organisms was limited by the range of organisms known to him and by the fragmentary and confusing preservation of many fossils.

It is therefore not surprising that 16th-century discussions of what were then termed "fossils" centered on the problems of explaining their "stony" substance and their resemblances in form to other classes of objects. Explanations of these problems were framed within the dominant "natural philosophies" of the time. In particular, Neoplatonic philosophy postulated a network of hidden affinities and "correspondences" ramifying through the universe, and therefore led to an expectation of resemblances between diverse natural entities. It seemed more satisfactory to explain all such resemblances in "fossils" in these terms, rather than in terms of their causal origin as living organisms, because the latter explanation could only be applied to a small minority of all the objects to be explained.

Fossils in 17th-Century "Cosmogonies"

The posing of the causal question about the origin of fossils became prominent in the 17th century within the self-consciously new "mechanical philosophy." A well-known example is found in the work of the Danish medical scholar Niels Stensen (better known as Steno, 1638–1686). Steno's essay on fossils (1667) was a detailed argument in favor of an organic origin for the objects previously known as "tongue-stones," which he compared to the teeth of living sharks. He distinguished them clearly from other "fossils" such as crystals, which are not organic in origin. This apparently "modern" argument is embedded, however, within an explanatory framework that belongs clearly to his own period. Steno assumed the validity of the traditional short time scale of the world, not from obscurantism or out of fear of persecution, but simply because it seemed obviously natural to almost everyone at that time. Thus he integrated his explanation of the organic origin of certain fossils into a traditional framework of human chronology, and attributed their emplacement to the Flood.

Steno's work was closely paralleled at the same period by the English natural philosopher Robert Hooke (1635–1703), who frequently discussed the problem at meetings of the early Royal Society of London. The Society's unofficial newsletter, the *Philosophical Transactions,* made this debate on fossils available to a wider circle of naturalists. Hooke argued that a broad range of fossils were organic in origin—for example, he believed that ammonites had been chambered mollusk shells—but he was unable to prove this to the satisfaction of his contemporaries because he could not point to similar living organisms. His explanation of the emplacement of fossils on dry land, by elevation during earthquakes, also seemed unsatisfacory. For fossils without a close resemblance to living organisms, or with problematic modes of preservation, the earlier explanations therefore continued to seem more adequate.

In effect, it was becoming clear that the central problem was that of distinguishing between fossils of organic and inorganic origin, within a wide range of natural objects. Some objects, e.g., Steno's fossil sharks' teeth and Cenozoic fossils generally, were commonly accepted as organic, and various concretions etc. as inorganic; but in between was a large class of objects whose origin was still debatable. Since the notion of extinction was generally unacceptable for basic philosophical reasons, it was difficult for fossils that were unlike living organisms (e.g., Mesozoic and earlier) to be recognized as organic. The problem was compounded, as in the previous century, by the confusing preservation of many of the commonest fossils (e.g., as molds). On the other hand, the traditional notion that every natural object must somehow display God's wisdom and design favored an organic explanation of any fossil whose structure could be analyzed in the same functional terms as living organisms.

Those fossils that were recognized as organic continued to be explained within a traditional framework of human chronology, even though this necessitated a highly philosophical (as opposed to literalistic) interpretation of what were believed to be the oldest human records, namely, the early chapters of Genesis. A long prehuman history of the world was generally unthinkable, not because it might have seemed to conflict with the Bible (much greater inconsistencies were assimilated without difficulty), but because it would have seemed to leave that history without meaning or significance. Furthermore, there was no compelling or unambiguous empirical evidence to suggest such a conclusion.

In the late 17th century, there was a spate of

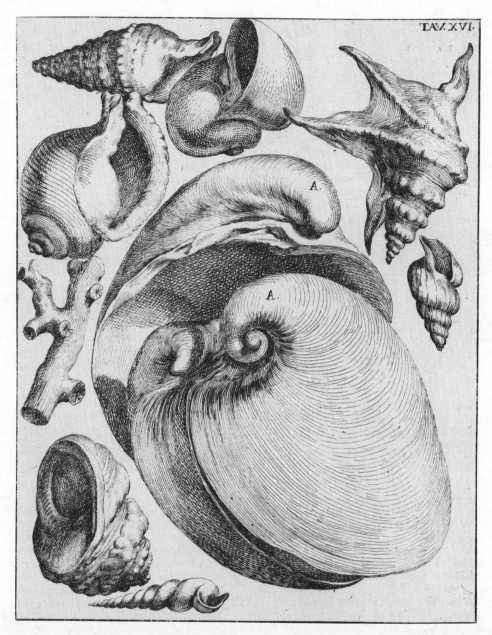

FIGURE 2. Fossil shells and corals, illustrated in one of the plates in Agostino Scilla's book (1670), *La Vana Speculazione disingannata dal Senso. Lettera responsiva circa i corpi marini, che petrificati si trovano in varii luoghi terrestri.* Being well preserved and similar to living species, their organic origin could be accepted easily.

highly speculative theorizing about the possible history of the earth (in which Isaac Newton participated). Although these theories (*cosmogonies*) were almost invariably limited within a short chronology, they did establish a conceptual framework of temporal thinking about a possible history of life. Naturalists with a wide empirical knowledge of fossils, such as the English physician John Woodward (1665–1728), were aware of stratification and its relation to the occurrence of fossils, but explained it in terms of differential settling at the time of a (far from literalistic) Flood. A few writers, such as the English naturalist John Ray (1627–1705), realized that fossils might necessitate a vast expansion of the traditional time scale, but could scarcely grasp or accept the full implications of this. Fossils, therefore, continued to be interpreted generally within a basically traditional and humanly meaningful drama extending from an original Creation and a Flood in the distant past (in which most fossils had become emplaced), through the present and toward a future End of the world.

Fossils in 18th-Century Natural History

Recent historical research has shown that, in the study of fossils, the 18th century was not a period of stagnation between the 17th and 19th centuries, although its achievements were superficially less spectacular. Large-scale theories or cosmogonies continued to be elaboraborated, for example, by the great French naturalist Georges Buffon (1707–1788); but the period was characterized more by the detailed systematic study of fossils. The problems raised by particular kinds of fossils continued to be debated, for example, in the publications of learned bodies such as the Royal Society of London and the Académie des Sciences in Paris. Detailed research demonstrated the organic origin of previously problematic fossils such as belemnites and crinoids. Often this work owed much to the discovery of better-preserved fossil specimens or new living organisms (e.g., the first living stalked crinoid). Although discoveries of the latter kind emphasized the possibility that other classes of fossils (e.g., ammonites and belemnites) might still exist alive, there was a growing suspicion that extinction would need to be reckoned with as a fact of nature. At the same time, a growing awareness of the occurrence of fossils in situ, in sedimentary formations that seemed to have been deposited slowly under tranquil conditions, increased the suspicion that their emplacement might have taken place over a time scale that would dwarf human chronology.

Such theoretical speculations are less characteristic of the century, however, than the patient accumulation of collections of fossils, primarily for their intrinsic interest as natural curiosities, and their systematic description within the broader framework of what was termed "natural history," namely, the exhaustive systematization of the whole of the natural world—animal, vegetable, and mineral.

This work was done within a social context of a much increased scale of activity in all branches of science; the mere fact that more individuals took an interest in these problems in itself helped to accelerate the pace at which new ideas and new empirical observations became available. Apart from a handful of "professionals" in state institutions such as mining academies, all this work was done by *amateurs*

(in the original French sense of the word), and was naturally dependent on the existence of a leisured class whose interests inclined them in this particular direction. For the history of what later became paleontology, therefore, the 18th century is most important for its production of work on the systematic classification of fossils, often aided by greatly improved illustrations. This established new standards of precision in the knowledge of what kinds of fossils needed explanation on a more theoretical level.

Fossils in Early 19th-Century Geology

Eighteenth-century naturalists who collected fossils were generally aware of a broad correlation between the lithological character of successive strata and the fossils they contained. It was commonplace to draw analogies between fossils as evidence for a temporal history of life (and the earth) and coins and documents as evidence for the reconstruction of human history; and it was widely recognized that these two histories, while parallel in method, were not coextensive. This rather vague insight was first given precision in the earliest years of the 19th century.

Recent historical research has shown that the well-known work of the English surveyor William Smith (1769–1839) was less isolated and original than geologists later in the century believed. Although soon superseded by better maps, Smith's geological map of 1815 was important because it quickly convinced geologists of the practical value of fossils for the correlation of strata over wide areas. Smith's social position, and his pressing need to earn a living from consulting work, discouraged him from theoretical speculations; and his understanding of fossils was highly empirical and relatively crude by the standards of the period. Indeed, fossils were only one of the criteria that he used for correlation; and they were not, until late in his life, the most important. Nevertheless, his work did encourage others to pay greater attention to the stratigraphical position of fossils, and thereby made possible a much more precise description of the temporal sequence of the forms of life.

Of much greater theoretical importance was the work of the French naturalist Georges Cuvier (1769–1832) during the same period. Cuvier was not the first to suspect that many fossil vertebrate bones had belonged to species no longer alive, but he was the first to exploit the full potentialities of the new science of comparative anatomy (see Fig. 3) to establish the fact of extinction beyond all reasonable doubt. His detailed reconstruction of extinct mammals (mainly the Pleistocene fauna)

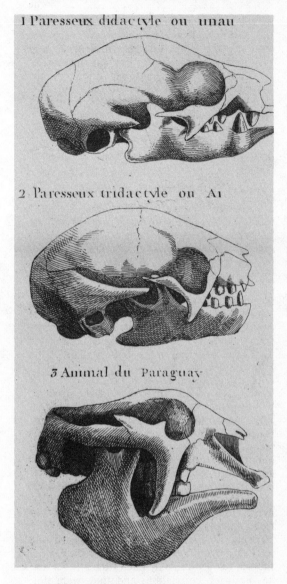

FIGURE 3. Cuvier's first use of comparative anatomy in paleontology: the skulls of two species of modern tree-sloth (1,2), compared with that of the huge fossil ground-sloth *Megatherium* from Paraguay (3), reduced to the same size. From a preliminary paper on the Paraguay fossil, published in 1796, "Notice sur le squelette d'une très-grande espèce de Quadrupède inconnue jusqu'à present, trouvé au Paraquay, et déposé au Cabinet d'Histoire naturelle de Madrid," *Magasin encyclopédique*, 2me année, **1**, 303–310.

demonstrated vividly the "revolutions" that life on earth had undergone in the past; and his *Researches on Fossil Bones* (first edition 1812) became a model for the reconstruction of other fossils. Approaching the problem from a primarily zoological viewpoint, Cuvier had a vivid

sense of fossils as the remains of animals that had once been alive. Unlike Smith, he did not regard them merely as markers of stratigraphical position, although he too realized their value for correlation. His method of anatomical reconstruction depended on a strong sense of the "designful" adaptations of all organisms. His results were therefore highly acceptable within the post-Revolutionary climate of most parts of Europe, where any new demonstration of divine design in nature was grasped eagerly. Likewise his conception of earth history as punctuated by occasional sudden "revolutions" was eagerly developed, particularly because the most recent such event seemed compatible with a moderately liberal interpretation of the Flood story in Genesis. It should be emphasized, however, that such cultural constraints did not affect the kind of time scale that Cuvier and his followers were prepared to contemplate. Although they had no firm evidence on which to base a quantitative estimate, it was now clear to all men of science that the history of life on earth had been unimaginably longer than the history of mankind.

The second quarter of the 19th century was marked by the integration of the originally biological work of Cuvier and his followers with the more pragmatic stratigraphical research that was inspired by the example of Smith's work. It was commonplace for the practical value of fossils for stratigraphy—and hence for the exploitation of natural resources—to be used as an argument to justify research on fossils. But as in some other sciences during the period of the early Industrial Revolution, it is easier to find examples of such rhetoric than to find clear evidence that these hopes were actually fulfilled. The social separation between the amateurs, who still formed the largest social group in geological research, and the civil and mining engineers, who had the strongest motives for applying the new knowledge, generally militated against practical exploitation. The first deliberately planned attempts to develop such applications arose with the establishment of the first state geological surveys (e.g., in Britain in 1935); collectors and describers of fossils were attached to some of these surveys at an early date. Such men could be termed without undue anachronism the first professional paleontologists, although the name for the science itself was not introduced until later.

The fossil vertebrates that Cuvier reconstructed included a few from the Eocene of the Paris area, which clearly differed from living animals far more than the mammoths and mastodons of the "superficial" gravels. Thus,

FIGURE 4. Lyell's illustrations (1833, *Principles of Geology* . . .) of some of the mollusk shells characteristic of his "Pliocene" period, based on the research of the French paleontologist Deshayes. With its high proportion of species still extant, the Pliocene mollusk fauna was an essential part of Lyell's endeavor to connect the present with the geological past by a continuity of slow-acting processes.

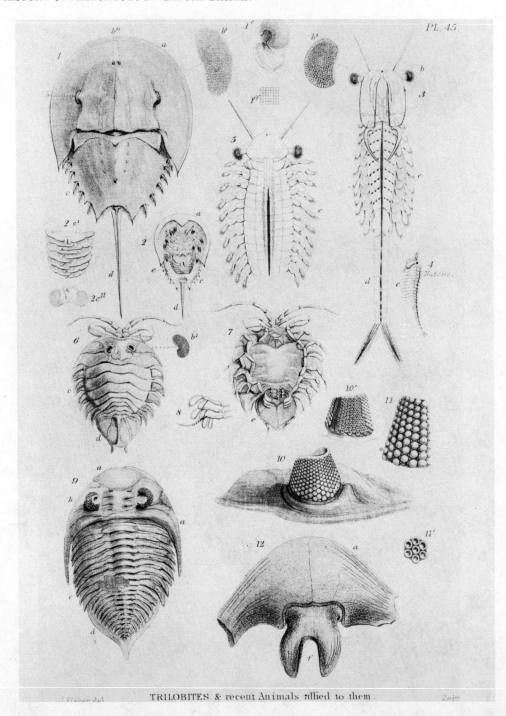

FIGURE 5. William Buckland's illustration of the "designful" adaptations of Silurian trilobites, with analogies from living arthropods (1837, *Geology and Mineralogy considered with reference to natural theology*). 9-11 show the compound eyes and all-round vision of the trilobite *Asaphus.* The xiphosuran (horseshoe "crab") *Limulus* (1,2) and the crustaceans *Branchipus* (3-5) and *Serolis* (6-7), with their compound eyes, are shown for compari-son. Buckland's analysis of trilobite adaptations was one of the most original of his examples of the designful nature of the organic world at even the earliest known period in the history of life.

even early in the century, it became clear that there had been some kind of linear change in the history of life. This became more strikingly evident as the new stratigraphy gave relative temporal precision to fossils. What were termed "Secondary" strata (roughly, Mesozoic and later Paleozoic) were found to be virtually lacking in mammals, and to be characterized by a fauna of reptiles and, among the invertebrates, groups such as ammonites that were evidently extinct. The English amateur geologist Roderick Murchison (1792–1871) established in 1839 a world-wide "Silurian System," in which vertebrates themselves were virtually absent. The later discovery of a still more ancient "Primordial" fauna in Bohemia (the Cambrian of modern geology) suggested that the origin of life itself could not be far from discovery.

All this research suggested a history of life that had been broadly "progressive" in character, though it did not seem to favor an explanation in what would now be called evolutionary terms. In an important interpretation (1837) which updated the application of "natural theology" to geology, the English geologist William Buckland (1784–1856) pointed out that even the earliest fossils then known (for example the Silurian trilobites; Fig. 5) showed as much "designful" adaptation in their structure as any living organisms. Thus it could not easily be argued that lfie had "progressed" by the gradual improvement of the forms of life from "rudimentary" beginnings. Furthermore, the fossil record did not give any empirical evidence in support of Lamarck's earlier argument that organic species were purely arbitrary divisions. More particularly, the record did not show any evidence of the temporal "transmutation" (to use the contemporary term) of one species into another.

Of course, there were also deeper motives, often unacknowledged at the time, for the reluctance to consider an evolutionary interpretation. It should be emphasized that this reluctance was by no means confined to the religious, because it touched on much broader social and metaphysical concerns that seemed equally important to those without strong religious beliefs. In effect, evolutionary theories were widely feared, even if put forward in a strictly scientific context, because they seemed to threaten the social and ethical foundations of society by implicitly eliminating the distinctive moral nature of man. Nevertheless, it should be emphasized that in the study of fossils the empirical evidence for a theory of evolution by gradual change was extremely weak at this period.

The main attraction of such a theory—which Charles Darwin (1809–1882) was privately constructing in the late 1830s—was the sheer breadth of explanatory value it could have in such diverse fields as comparative anatomy, embryology, and biogeography, much more than in paleontology. It is therefore not surprising that Darwin himself realized from the beginning of his speculations that the evidence of paleontology would need to be explained *away*. This helps to account for the less than enthusiastic reception of the *Origin of Species* (1859) among most paleontologists. Furthermore, even when they did absorb and apply the Darwinian concept to their fossil material, it made remarkably little difference to their work at the everyday level. Of course, their high-level interpretation of the meaning of faunal diversity was altered profoundly; but their practical classification of fossils and their use of them in stratigraphy was scarcely affected. For this reason, the publication of Darwin's book, although convenient as a landmark in description, is ultimately artificial as a watershed in the history of paleontology. What makes it a convenient landmark is rather the fact that by about 1860 the institutional and intellectual framework of paleontology had become well established. It was still rarely referred to as an autonomous discipline, and quantitatively the bulk of work on fossils was done within the ambit of stratigraphical geology. But it had established a set of effective technical tools for the classification and reconstruction of fossils, and a set of coherent intellectual goals in the description and interpretation of the history of life on earth.

MARTIN J. S. RUDWICK

Reference

Adams, F. D., 1938. *The Birth and Development of the Geological Sciences.* Baltimore: Williams & Wilkins, 504p.

Beringer, C. C., 1954. *Geschichte der Geologie und des Geologischen Weltbildes.* Stuttgart: Enke Verlag, 158p.

Berry, W. B. N., 1968. *Growth of a Prehistoric Time Scale Based on Organic Evolution.* San Francisco: Freeman, 158p.

Greene, J. C., 1959. *The Death of Adam. Evolution and Its Impact on Western Thought.* Ames, Iowa: State University Press, 388p.

Haber, F. C., 1959. *The Age of the World. Moses to Darwin.* Baltimore: Johns Hopkins Univ. Press, 303p.

Hooykaas, R., 1970. *Continuité et Discontinuité en Géologie et Biologie.* Paris: Editions du Seuil, 365p. (Rev. ed. of *Natural Law and Divine Miracle,* 1959, Leiden: Brill, 237p.)

Needham, J., 1959. *Science and Civilisation in China,* vol. 3, *Mathematics and the Sciences of the Heavens and the Earth.* Cambridge: Cambridge University Press, 877p.

Rudwick, M. J. S., 1972. *The Meaning of Fossils. Episodes in the History of Palaeontology*. London: MacDonald; New York: American Elsevier, 287p.

Schneer, C. J., ed., 1969. *Toward a History of Geology*. Cambridge, Mass.: M.I.T. Press, 469p.

Toulmin, S., and Goodfield, J., 1965. *The Discovery of Time*. London: Hutchinson, 280p.

Zittel, K. A. von, 1899. *Geschichte der Geologie und Paläontologie bis Ende des 19. Johrhunderts*. Munich and Leipzig: Oldenbourg, 868p. (The English translation is abridged and lacks references.)

Cross-references: *History of Paleontology: Post-Darwinian; History of Paleontology: Post-Plate Tectonics*.

HISTORY OF PALEONTOLOGY: POST-DARWINIAN

Paleontology and Evolution in the Decades Immediately Following Darwin's Origin of Species

Our understanding of the paleontological world view in the latter half of the 19th century is severely restricted because there are so few detailed modern scholarly publications to enhance the broad interpretive essay of Rudwick (1972, ch. 5 only). His essay surveys only the ten-year period following the publication of the *Origin*. Elsewhere, isolated references to late 19th-century paleontology usually form footnotes in a larger story of natural theology or developmental biology, wherein one is left with the impression that virtually all paleontologists were profoundly affected by Darwinian logic and that they had dedicated their careers to proving or disproving natural selection. The attitude that Darwin was the founder of modern paleontology appears to have remained relatively unquestioned and perhaps is accepted as common knowledge today.

Maybe the possibilities for paleontology as a final, or at least major, testing ground of Darwinian evolutionary theory did look immense in the middle parts of the 19th century. At least the works of Thomas H. Huxley leave one with the impression that the fossil record was indeed a most critical component in Darwin's synthesis. However, in the original documents of the post-Darwinian period, one notes disquiet not far beneath the surface of a few beautifully constructed examples of lineal descent that supported Darwin's thesis of a gradual morphological change through time. Some natural historians soon concluded that data from the fossil record were not going to be decisive in our understanding of an evolutionary process, Darwinian or otherwise. Paleontology, at best, could provide only confirmation of evolution. Perhaps, as has been suggested elsewhere by Peckham (1959) in more general terms, some paleontologists were converted into Darwinistic-like thinking about the fossil record, one in which the theoretical foundations were grounded in a metaphysical evolutionism, perhaps more attuned to a Spencerian doctrine than an original Darwinian one.

Herein, this author will begin to examine exactly what immediate effect, if any, Darwin did have on the "profession of paleontology." We will argue that most paleontologists became involved in the solution of detailed, but theoretically somewhat limited, biostratigraphic problems, and, on many occasions, this tended to give the appearance of a discipline that was characterized by trivial personality conflicts over the naming of species or the finer and finer subdividing of geologic time. In short, paleontology left the mainstream of biological intellectualism in the latter half of the 19th century and became, by and large, a useful subdiscipline of stratigraphic geology. Complete substantiation of this thesis awaits a more detailed reevaluation, but clearly it appears as if the general understanding of the history of paleontology over this period requires a most critical review.

The Paleontological Weltanschauung at Mid-century. By the late 1850s, most paleontologists accepted as fact that there was a progressive change throughout the history of life, conceived as a change that had taken place within major groups—the *Embranchements* of Cuvier (1817) (the Articulata, Radiata, Mollusca, and Vertebrata). The morphological unity within the groups had been optimized in the "Ideal Archetypes" of Owen (1848). Furthermore, this progressive faunal change had been accompanied by an increase in overall diversity, although there were times when the number of species seemed to have been drastically reduced. Cuvier's initial concept that the history of life had been essentially static except for interruption by only five or six major extinction events and subsequent faunal changes, hence giving the appearance of a "progressive" faunal change, had been seriously compromised by the discovery during the 30 or so years following its formulation of too many "local" extinction events.

Decidedly outside the prevailing orthodoxy were Charles Lyell's notions of a wholly nonprogressive development of life, one to which Lyell seems to have clung tenaciously. Furthermore, Lamarck's ideas about a single evolutionary lineage from Monads to Man had been so thoroughly discredited by Cuvier more than a generation earlier that its somewhat modified form proposed by Chambers (1844) was firmly discounted by virtually every scientific reviewer.

The early 19th-century school of more deistically inspired British catastrophists, who had modified in translation Cuvier's version of revolutions in the history of life, had not died away by the 1850s. Their notions of Deluge-like catastrophes, while not in conflict with their beliefs in the constancy of natural laws, were toned down by thoughts of considerably less extravagant physical perturbations throughout earth history. Natural historians thus generally accepted a few times of major extinctions; but, by and large, the fossil record appeared to show changes in life throughout geologic time that were much more gradual than catastrophic.

Few believed that earlier periods in the earth's history were unusual, environmentally or ecologically. A brief survey of "textbooks" at midcentury shows that imaginations may have been fertile but past environments were not depicted as unearthly; the lives of fossil plants and animals were not unlike those of today's organisms. There was, however, no lack of speculation on whether the physical environment had changed in a progressive linear fashion or whether it had cycled, in the Lyellian view, through time. Some favored a gradual and progressive climatic cooling throughout the geologic ages, thereby postdating the mid-eighteenth-century speculations of Georges Buffon and predating the thermodynamic calculations of Sir William Thomson (Lord Kelvin) that would soon add mathematical strength to the concept of a gradual earth cooling.

Although there did seem to be a "plan" to the history of life, no natural historian registered a clear and unambiguous proposal of a possible causal mechanism; indeed the "plan" was accepted in somewhat the same phenomenological fashion that one was asked to accept the concept of universal gravitation. Over geological time, there had occurred the *creation* of new species, sometimes presumably at the expense of older ones; but creation and extinction events were, overall, gradual.

Following the lead of Cuvier, most natural historians believed that species were indeed real entities in nature. Morphologies did vary within species, but this variability was of limited extent. Some very bitter controversies, whose stratigraphic aspects have a distinctly modern tone, ensued among naturalists, concerning the degree of specific variability, that is whether a particular suite of specimens was just a variety of a previously defined species or a new species in its own right. Species were believed to be highly adapted to their local environmental setting; and any loss in adaptive qualities would

have, as Lyell for one believed, permitted almost instantaneous faunal replacements by migration from outside the immediate area. The balance or harmony of the organic world had to be maintained, and the balancing process was continually in action.

Generalizations as to whether life was becoming more complex, morphologically or ecologically, within each of the *Embranchements,* varied considerably. There were those who believed, as had Lamarck, that there was a tendency through time toward increased morphological and anatomical complexity; or, as sometimes phrased, increased specialization followed from the more generalized forms of life. However, examples of "retrogressive" morphological changes were becoming too common to be ignored, and the ensuing semantic difficulties in these disputes of "higher" and "lower" forms of life were, as might have been anticipated, of gigantic proportions. It did appear, however, that the earliest forms of life (specimens from the "earliest Silurian" of Murchison, or the "Cambrian" of Sedgwick) were quite morphologically complex, and the faunas were already well differentiated, although this interpretation was not agreed to by all.

Perhaps the most overworked analogy in the middle portions of the 19th century was that the fossil record could be read and interpreted as "The Book of Life." Most authors alluded, in one way or another, to the pages of nature being turned gradually. Although many of the pages were presumed to be missing, the overall "plan" was evident for all to read. Gaps that existed in the fossil record were viewed by many as real breaks in the history of life, not just omissions of the record. In mid-twentieth-century terms, these natural historians believed that they had adequate and reliable samples of the fossil record; what remained were the details. Thus, again resorting to present terminology, paleontology immediately prior to the publication of *The Origin of Species* had a quite well-formed "disciplinary matrix"; and all investigators seemed to have the same general commitment, namely, the more detailed filling in of the "plan."

In chapters 9 and 10 of the *Origin,* Darwin stressed this paleontological legacy, emphasizing the need for additional stratigraphic and geographic studies in order to add more and more pages to the Book. The record was indeed incomplete, but for very good reasons, says Darwin. Most assuredly, Darwin's acceptance of a progressive and overall gradual change in the history of life was not at odds with the standard or orthodox view. What Darwin did successfully promulgate was the quite original idea that there were a whole host of reasons why the fossil record should *not* show very gradual change. Herein Darwin had employed some Lyellian logic, but did so in order to gain results that were contrary to Lyell's own conclusions regarding the progressive nature of the fossil record. Intermediate morphological links between major "classes," such as a connective link between birds and reptiles, would, Darwin believed, be found as the collecting of fossils increased. But the documentation of connective links between the *Embranchements* and those between species would be much harder to come by, and for these Darwin marshaled an array of negative evidence.

Darwin had successfully circumvented any really powerful testing of his particular brand of descent through gradual modification from the fossil record. Regardless of whether interspecific change through time was by "small quantum" jumps, or by almost imperceptible morphological change, the fossil record could be of as much help as one wanted, and Darwin clearly seems to have anticipated the fact that this type of ambiguity would continue.

Time and again throughout the 1860s and '70s Huxley tried to show that the fossil record was not at odds with the Darwinian thesis; yet at no time was there any really convincing presentation that the study of fossils indeed was the route to unequivocal proof of gradual evolutionary change. Perhaps one might wish to claim that paleontologists were seduced, principally by Huxley, into believing that their studies of the fossil record would confirm or deny some kind of evolutionary process, for indeed taxonomic and stratigraphic studies continued unabated throughout the latter 50 years of the 19th century, and monographs blossomed as if it were the springtime of the paleontological world. It seems most probable, however, that the study of the history of life in the latter half of the 19th century principally became the study of single isolated historical events, and that these events could be viewed through time as being wholly unconnected. Certainly we see lip service being given to descent with modification from ancestral forms, but the emphasis was placed on the construction of vertical range charts and the need to correlate on a finer and finer scale. It had become obvious that paleontologists were unable to reduce these rapidly accumulating detailed observational statements into any biologically meaningful axioms, Darwin notwithstanding. Instead of scientific axiomatization, one discovers the outgrowths of an evolutionary metaphysics, sometimes being copied *in toto* from developmental embryology, a disciplinary interest viewed by some of the "evolutionists" answer to Darwin's assertion on the incompleteness of the fossil record.

The Paleontological Response to Darwin. The initial paleontological response to Darwin can perhaps best be viewed in three reviews of the *Origin* by acknowledged leaders in the field. The fundamental agreement among Heinrich G. Bronn, François Jules Pictet, and Richard Owen as to the position of Darwinian logic vis-à-vis paleontology is striking, and perhaps is best summarized in Pictet's 1860 review. Each author asks the following questions of Darwin's mechanism of evolution, that is, ancestral descent by natural selection, given the presence of a fossil record:

1. Why are the earliest faunas so diverse and so morphologically complicated?

Clearly all three reviewers believed that this was tied to the *ultimate problem,* that being the origin of life. Thus, any unambiguous acceptance of Darwinian logic with regard to the fossil record seems to have found a principal stumbling block near the beginning of the fossil record. The highly diverse and morphologically complex Cambrian fauna seemed to defy an early gradualistic history for life, and was not easily explained away through appeal to the deficiencies inherent in the geological record.

2. Why do we see the sudden appearance of ordinal and familial-level taxa (e.g., birds)?

3. Where are the expected numerous interspecific intermediate types in the fossil record?

The latter should obviously exist if there are no severe limitations on the morphological variability of any given species. Herein the three reviewers believed Darwin's position to be that of a latter day apostle of Lamarck. One cannot help being struck in this instance by Owen's complete grasp of Darwin's weak appeal to negative evidence from the fossil record, and Owen concluded: "What may be, or what may not be . . . forthcoming out of the graveyards of strata" probably will be little different from what has already been discovered. Hence there shall be no great revelations from paleontology. A few interspecific and perhaps a few more numerous inter "class" connections within each of the *Embranchements* might be found, but, in turn, would not provide overwhelming support for Darwin's thesis.

4. What is the evidence regarding the "descent" or "creation" of man?

Each recognized the psychological difficulties involved in deciphering the fossil history of man, indeed one which had yet to gain full force. Data were very scarce, and although Huxley and Lyell in 1863 professed the acceptance of a long and perhaps extremely gradual evolutionary history for man, the overtones of a mechanistic or chance creation of man was a significant barrier that had to be overcome if Darwin's thesis was to receive full support from natural historians.

Paleontological Studies Deemed Critical to the Darwinian Thesis

I should now like to turn to some early studies that paleontologists believed were critical to the acceptance or rejection of the notions of Darwin. Those studies specifically concerned with the documentation of a gradual transmutation, perhaps by natural selection, fall into four categories:

1. The early history of life remained throughout the 19th century very obscure; its study was marked by a most fascinating controversy over the nature of *Eozoön canadense.*

2. The connective links between widely disparate morphological groups remained to be explored; the reported discovery of *Archaeopteryx* by Owen in 1863 provided Huxley with proof of a "missing link."

3. Perhaps the original Darwinian goal was to find a fossil record of gradually .transformed species. To Huxley, though not to all, the evolution of the Equidae was a sterling example of just the .sort of gradual transmutations that Darwin had predicted.

4. The fossil history of man was virtually unknown, but stories of extinct beasts having been found with the obvious artifacts of man were being taken more seriously in the latter half of the century.

The Earliest Record of Life. In Adam Sedgwick's 1860 review of the *Origin,* one of the major topics of controversy was the lack in the fossil record of a primordial being whose existence was believed to be essential to the Darwinian position of descent through gradual modification. "To begin then, with the Paleozoic rocks. Surely we ought on the transmutation theory, to find near their base great deposits with *none but the lowest forms of organic life.* I know of no such deposits." Louis Agassiz in his 1857 "Essay on Classification" pointed out that: "Representatives of numerous families belonging to all the four great branches of the animal kingdom, are well known to have existed simultaneously in the oldest geological formations." He derived evidence from a diversity of publications, including those of Murchison (1839, 1854), Hall (1842–1894), Barrande (1852–1911), and Sedgwick and McCoy (1855), all of which clearly demonstrated that: "Animals belonging to all the great types now represented upon the earth were simultaneously called into existence." Interestingly enough, Agassiz notes

that these discoveries of the earliest life had all come within the preceding 30 years; prior to that time, many natural historians might have been more willing to accept what was to become a more doctrinaire Darwinian position on the existence of early very primitive life.

Ten years later, Alexander Winchell (1870) again pointed out that the studies of E. Billings in Canada, J. Hall in New York, D. D. Owen in the Northwest Territories, J. Barrande in Bohemia, and G. G. Shumard in Texas had clearly demonstrated that: "One can not but be astonished that in the very onset of animalization upon our globe so high a rank and so great a variety of types [wonderfully diversified brachiopods and trilobites] should have been manifested."

This orthodox story of early life was, however, questioned by some who read the 1865 report of Sir William Logan, Director of the Geological Survey of Canada, concerning the finding of a strange new "animal" *Eozoön canadense,* preserved in rock strata that clearly predated the then oldest known fossiliferous strata of the Cambrian. Logan started a controversy that was to last until near the end of the 19th century. The implications of *Eozoön* seemed important enough to Darwin so that by the fourth edition of the *Origin* (1866) he felt that the discovery of this remote, relatively primitive, pre-Silurian life confirmed his earlier suspicions. "Thus the words, which I wrote in 1859, about the existence of living things long before the Cambrian period, and which are almost the same with those since used by Sir William Logan, have proved true." However, Darwin did recognize the difficulties posed by the discovery of *Eozoön* in such very ancient rocks and so distinctly separated from the Cambrian faunas that there were no obvious connective links. In order to overcome this deficiency Darwin again invoked the inadequacy of the fossil record but noted that: "The difficulty of assigning any good reason for the absence of vast piles of strata rich in fossils beneath the Cambrian system is very great."

Thus the blessing of *Eozoön* for Darwinian evolutionists turned out to be a mixed one, for once again Darwin had to retreat to ad hoc explanations as to why *Eozoön* was so isolated in time, and indeed the best he was able to do was appeal to the unlikely possibility that its descendants may have been very isolated in space. Of course the final retreat was to the old Lyellian position that: "The many formations long anterior to the Cambrian epoch [may perhaps be] in a completely metamorphosed and denuded condition." And Darwin concludes by saying again that the geologic record as a history of the world is imperfectly kept and once this is recognized: "The difficulties . . . are greatly diminished, or even disappear."

As it turned out Logan and J. William Dawson, the Principal of McGill University, Montreal, trusted William B. Carpenter's identification of *Eozoön* as some kind of primitive nummulitic Foraminifera. Carpenter seemed highly qualified to make such a judgment, since he had recently (1856) completed a major monograph on the fossil Foraminifera. Darwin obviously wanted to accept this identification also. Unfortunately for the evolutionists, two mineralogists at Queen's College (Galway) promptly identified *Eozoön* as inorganic, and although their questioning was not designed to provoke acrimony, they received more than their share, especially from Dawson and Carpenter. The pages of the *Quarterly Journal of the Geological Society of London, American Journal of Science,* and *Nature* carried almost bimonthly reports of the latest field and laboratory finds and then refutations from one party or the other. After almost 15 years of heated controversy Möbius (1878) dealt the organic interpretation of *Eozoön* the death blow by declaring the remains to be clearly inorganic. Some of the principals, such as Dawson, never really gave up; in his *Some Salient Points in the Science of the Earth,* published in 1893, Dawson devoted two chapters to the "dawn of life."

One of the more interesting aspects of this controversy was that the evolutionary implications of *Eozoön* seem to have been relegated to a very secondary position. O'Brien (1970) may claim that in the *Eozoön* episode: "There was something at stake for both sides in the greater [evolutionary] controversy," yet the combatants seemed to have spent more time impugning one another's scientific and professional competence than in discussing the relative evolutionary merits of *Eozoön.*

The Connective Links, Perhaps somewhat more characteristic of this period were those wonderfully logical and entertaining lectures, both "scientific" and "popular," by Thomas H. Huxley, in which he attempted to reconcile the fossil record not only to Darwin's notions of very gradual interspecific changes but also to the discovery of connective links among higher taxonomic categories. In addition to Huxley's sometimes blatant evolutionary propaganda, one could read an entire chapter in Lyell's *Antiquity of Man* (1863) wherein Lyell purports to defend the Darwinian thesis by introducing the fossil evidence of interspecific and inter "class" linkages. Unfortunately all Lyell succeeded in doing was drawing conclusions on the extreme longevity of some genera (examples are given from Davidson's 1851–1886 monographic study of British

fossil brachiopods) and the extension of the range of modern plants back into the Miocene (Oswald Heer's *Flora tertiana Helvetiae,* 1859), and giving a brief discussion of the newly reported *Archaeopteryx macrurus* Owen, which clearly demonstrated "the existence of Birds at the epoch of the Secondary rocks." Lyell adds a brief discussion of Joseph Leidy's initial studies of the fossil Equidae, but his articulations on the significance of further vertebrate finds, which perhaps could have been used to add strength to the Darwinian thesis, are always a good deal less clear than Huxley's. Lyell simply was not the believer that Huxley was.

Huxley, who was much more directly involved in demonstrating the fact of evolution, was initially concerned with the identification of morphologically intermediate fossils at higher taxonomic levels, presumably a goal more easily attained than that of demonstrating interspecific connections. He did not pretend any knowledge of the origin of the higher level interconnections or of the origin of life, and, in fact, claimed with regard to the latter that Darwin's thesis really has nothing to do with the origin of life. In his 1862 address to the Geological Society, Huxley concentrated principally on developing a foundation for an evolutionary position by attempting to demolish Owen's archetypical concept, one which included a belief that the history of life was directed toward a "progressive modification towards a less embryonic, or less generalized, type." In fact, what Huxley seems to have been attempting to do was to derive virtue from negative evidence; simply enough he says that if we accept the fact that the earliest fossiliferous rocks (i.e., Cambrian) are coeval with the beginnings of life, then only insignificant amounts of modification can be shown to have taken place within any one major group. Hence, one can discount an "inner-directed" progression, or "planned" progression of life. Herein Huxley's positive stance against an "orthogenetic" evolutionary system is in contrast to his subsequent defense of Darwin, wherein he had to assume a more circumstantial position. In fact, in order to defend the main Darwinian tenets with respect to the fossil record, Huxley was willing to accept not only descent through gradual modification but also some morphological change *per saltum* (e.g., the Ancon ram and the six-digited humans). Although "in neither case is it possible to point out any obvious reason for the appearance of the variety" (1860), it does make the Darwin-Huxley paleontological task a bit easier when reference is made to possible saltatory "jumps" in the fossil record. Huxley, in fact, felt that

Darwin had embarrassed himself, or at least placed unnecessary restrictions on his principal thesis, with the aphorism "*Natura non facit saltum,*" and Huxley would prefer to believe that: "Nature does make jumps now and then, and a recognition of the fact is of no small importance in disposing of many minor objections to the doctrine of transmutation" (1860).

By the time of his 1870 Presidential Address to the Geological Society, Huxley had marshaled what he believed to be some impressive evidence for the existence of intermediate links that occupied the intervals between families and orders of the same class. Huxley believed he did not have to defend the entire fossil record, but could emphasize some outstanding examples that are in clear agreement with his somewhat modified Darwinian position. Herein Huxley refers to the studies of E. D. Cope, which he believed established a morphological connective link between birds and reptiles. Huxley then outlines other, as he calls them, *intercalary* types, one being the genus *Anoplotherium* intermediate between pigs and ruminants. In all cases, Huxley was very careful not to leave the audience with the impression that there is any direct ancestral-descendant involvement among these connective links. Strangely enough the works of Cuvier provided many wonderful examples of such kinds of connective links for Huxley, but the highlight of the lectures must have been the recounting of the exquisite studies by Jean-Albert Gaudry on the Miocene mammalian fauna of Greece (1862–1867). Gaudry's studies provided Huxley with numerous examples on intercalary mammalian forms, and indeed Gaudry himself had even attempted to construct some of the first tentative phylogenetic diagrams from his fossil evidence. Guadry, in later publications (e.g., 1878–1890) clearly entertained evolution as a hypothesis useful to paleontologists, the working of which, however, paleontology would never be able to demonstrate.

The discovery of *Archaeopteryx,* and the subsequent report by Owen (1863) that these remains were indeed avian, prompted Huxley in 1868 to state that the real significance of *Archaeopteryx* was that it provided a perfect example of a bird exhibiting a very close approximation to a reptile, and that this intermediate link between birds and reptiles clearly was in support of the doctrine of evolution. However, a number of investigators, namely the German zoologist Rudolf Wagner, the American geologist James Dwight Dana and the expatriate Swiss naturalist Louis Agassiz, pointed out that even if *Archaeopteryx* did show characters intermediate between birds

and reptiles yet it proved nothing about evolution. Wagner believed it no more significant than the discovery of the duck-billed platypus. Dana felt that one would only have to establish a new systematic category between birds and reptiles, and that *Archaeopteryx* was related to neither. Agassiz believed that it was nothing more than a "synthetic type," many of which he had pointed out first, such as fish with reptilian characters. Huxley, however, was undaunted and continued to publish papers on the affinity between dinosaurs and birds, crocodiles and lizards, sometimes not mentioning their direct bearing on evolution, yet the implications were that all of these examples provided evidential strength for evolution.

With regard to Huxley's interpretation of *Archaeopteryx* as a link between birds and reptiles, many continued to remain skeptical; but in 1872 O. C. Marsh announced the discovery of other fossil birds with teeth, and Marsh's 1880 monograph on the "Odontornithes" has become a classic in the evolutionary paleontological literature of the 19th century.

The Case for Interspecific Transformations. Although there was opposition to Huxley's examples of *intercalary* connections in the fossil record, there were much more difficult times ahead when he tried to demonstrate interspecific, or *linear,* as he called them, transmutations. Adam Sedgwick in his 1860 review of the *Origin* berated Darwin for his attempt to link faunas between the "great stages" of geologic time, and asked, Do not each (referring to the Cambrian, Silurian, etc. periods) have a *characteristic* fauna? "Where do the *intervening* and connective types exist, which are to mark the *work of natural selection*? We do not find them." Entirely new faunas were thought by some to have characterized each major geologic period, and Darwin was not going to be allowed to casually deny Sedgwick's "conclusion grounded on positive evidence . . . [with one] derived from negative evidence." Indeed, 20 years later Dana (1880) remarked: "The transition between Species, Genera, Tribes, etc., in geological history, are, with *rare* exceptions, abrupt. . . . A survey of a history of the world's life, as all writers admit, [has] almost no cases of the gradual passage of one species to another . . . [this fact should] lead the cautious geologist to wait, before dogmatizing." But, in fact, all writers did not admit to this position for Pictet (1860) noted that Darwin's ideas in general do agree well with the known paleontological facts, especially the stratigraphic succession of species.

Whereas it may have seemed to many that

the demonstration of, as Huxley called them, "intercalary" types really provided little substantial proof of Darwinian evolution, examples of very gradual morphological change, such as that denied by Sedgwick and Dana, might prove very powerful indeed. Huxley, in his 1870 Presidential Address called these *linear* types of connections, which was meant "to express the opinion that the forms A, B and C constitute a law of descent, and that B is thus part of the lineage of C . . . [but] it is no easy matter to find clear unmistakable evidence of filiation among fossil animals . . . [whereas] it is easy to accumulate probabilities [in the intercalary sense]." However, Huxley would make the attempt, and herein he again turned to the classic studies of Gaudry and his tentativve phylogenies of the Hyaenidae, Proboscidea, and Rhinocertidae. Huxley's main emphasis, however, was placed on the lineal evolution within the Equidae; he not only detailed finds by Gaudry back into the Miocene, but also added the earlier studies of Cuvier, Leidy, and V. Kovalevsky. In the process, he arrived at a predictive hypothesis: that the further study of early Eocene deposits should uncover a most distinctive, yet clearly recognizable, ancestral "horse." Huxley's brilliant prediction was confirmed, in part, only a few years later by the superb studies of the Equidae by Marsh, and in Huxley's 1876 American Address (N.Y.C.) he integrated, for the first time, the detailed North American studies of Marsh with those earlier European studies, demonstrating in the process the much more complete evolutionary record of horses on the North American continent, which was in complete agreement with an evolutionary pattern established from more fragmentary European remains (Huxley, 1877). Alfred Russel Wallace in 1889 remarked, with what many may have considered rather excessive hyperbole: "Well may Professor Huxley say that this is demonstrative evidence of evolution; the doctrine resting upon exactly as secure a foundation as did the Copernican theory of the motion of heavenly bodies. . . ."

The presumed lineal evolution of the Equidae provided an unparalleled example, and even Huxley had to admit in 1870 that fossil evidence for the likes of the "Musk-deer" looked a good deal less complete than that of the Equidae. Huxley did present other examples of probable lines of descent in the Tertiary Ungulata (= Artiodactyla) but usually tended to twist his conclusions into a demonstration that "Every class must be vastly older than the first record of its appearance upon the surface of the globe." Hence in a way he had to explain away the general lack of common

stocks "and direct lineal ancestral forms by the manifold inadequacies of the fossil record."

Others beside Huxley attempted to support an evolutionary position and B. von Cotta (1866) while again stressing the incompleteness of the fossil record through the chance preservation of faunas, believed that instead of saltatory jumps in the history of life, migrations into and out of one particular area could easily give the appearance of disjunct faunal breaks. Von Cotta was supported by his countrymen Karl A. von Zittel and M. Neumayr, all three of whom attempted to show that lineal gradations could be found in the fossil record [see esp. Neumayr and Paul's 1875 study of the nonmarine Tertiary molluscan genus *Vivipara* (Gastropoda)]. Many continental paleontologists, however, were still impressed with Cuvier's sharp and accurate condemnation of Lamarck and Geoffroy Saint Hilaire, and many Continental Europeans held that Darwin's thesis was merely an extension of Lamarck's. Indeed, in France, with the notable exception of Gaudry, Darwinian notions were simply ignored and a brief review of the paleontological literature in the latter half of the 19th century shows that stratigraphic and taxonomic studies, many of high professional quality, continued with no noticeable change in tempo, and no reference whatsoever to Darwin or to evolution. Perhaps one of the most prophetic aspects of the Pictet (1860) review of the *Origin* was his inquiry as to whether or not Darwinian theoretics would have any effect on the practicing naturalists who "distinguish the layers of the earth and the geological strata. . . Does the increase in the limit of species variability occasioned by Mr. Darwin's research have any effect on their methods? I do not think so . . . species are fixed; they are fixed nearly as much for Mr. Darwin as they are for the most convinced partisan of absolute permanence. . . . The beautiful work of Mr. Darwin does not lead us to a new point of view."

The influence of Cuvier seems to have had an equally significant impact on some North American naturalists, perhaps because many had their intellectual roots in European soil. One might have predicted that Louis Agassiz (a former student of Cuvier) would dismiss the teachings of Darwin, and throughout the post-*Origin* years Agassiz retreated not one iota from the position he had assumed in the 1857 edition of "Essay on Classification," wherein he predicted that as additional fossil remains were studied more carefully: "The supposed identity of [the same] species in different geological formations vanishes gradually more and more." In direct contrast to the thesis of Darwin and Huxley, Agassiz believed that as paleontological studies proceeded, species would become more and more "limited in time," so much so that we would see the geologic record "circumscribed step by step within narrower, more definite . . . periods." His directive did not fall on deaf ears, for it agrees well with the increasingly more stratigraphically oriented aspects of paleontology in the latter half of the 19th century.

Of all Agassiz's encounters with Darwinian evolutionists (perhaps the best known were those with his botanist colleague at Harvard, Asa Gray), some of the least publicized were his debates with William Barton Rogers. [Rogers, who was to become the first president of M.I.T. and who had been the first Director of the Geological Survey of Virginia (1835-1841), had, along with his brother Henry D. Rogers, Geologist for the First Geological Survey of Pennsylvania, outlined in considerable detail the geological structure and stratigraphy of the Appalachian Mountains. This work was to make the Appalachians the "type" mountain belt to which all others were compared.] The Agassiz-Rogers debates in early 1860 were held before the Boston Society of Natural History; Agassiz attempted to defend his thesis of the exclusivity of fossils in a given geological formation. With both debaters using the newly acquired data on the geology and paleontology of New York State (mostly collected and compiled by James Hall)), Agassiz was unable to maintain a logically consistent position with regard to the depositional settings of the unusually complete New York Paleozoic fossiliferous section, and Rogers clearly had the better of the debates. Agassiz, however, would continue to argue against Darwin's notions of evolution; but his platform moved to other settings wherein he could continue to berate Darwin's "marvelous bear, cuckoo, and other stories" before audiences perhaps more attuned to the metaphysical and not necessarily the scientific aspects of evolution.

By the early 1870s, not only Huxley was active in the cause of evolution, but the Americans Alpheus Hyatt and Edward Drinker Cope made somewhat different kinds of contributions specifically related to North American paleontology and in general to the field of evolution. Their evolutionary interpretations were not clearly relatable to those originally expounded by Darwin, but were perhaps somewhat more Lamarckian in ancestry. Hyatt (1894), in his classic study of the Paleozoic ammonites (Cephalopoda; Mollusca), attempted to join together the then-current embryological notions of his mentor Agassiz (i.e., the geological succession of extinct forms parallels the embryological development of recent forms—

most commonly stated "ontogeny recapitulates phylogeny"). As there were no close living relatives of the cephalopods that Hyatt was studying, he decided to carry out both phases of the investigation on fossil material alone. His results showed that some of the geologically younger ammonites exhibited juvenile growth stages during the ontogenetic coiling of the shell that was reminiscent of that shown by earlier adult fossil members of the group. Hyatt wedded these facts with somewhat corrupted Lamarckian notions, in part derived from the theoretical studies of Cope, and emphasized the fact that the evolutionary trends documented from the ammonites were predictably orthogenetic. Hyatt was, of course, using a methodological approach on fossil material that had been designed principally to avoid the pitfalls of an incomplete fossil record, and indeed it seems to have had a major impact on those few North American paleontologists acutely interested in the fossil record from an evolutionary viewpoint.

In fact, the notion that the life history of present-day animals could be constructed from embryological evidence alone, as advocated by Agassiz, was not openly discounted by Darwin, although he was well aware of the alternate embryological interpretations of K. E. von Baer (see Gould, 1977a). As Darwin's ideas regarding descent through gradual modification changed throughout the six editions of the *Origin*, he assumed, as is well known, an increasingly more Lamarckian position. At the same time, Cope began to explore aspects of a "growth-force," believing that once one admitted evolution, one could not explain the obvious directional (i.e., orthogenetic) aspects of the fossil evidence without accepting some form of intrinsic directive force. "It would be incredible that a blind or undirected variation should not fail in a vast majority of these instances. . . . The amount of attempt, failure, and consequent destruction, would be preposterously large; [hence], for these and other reasons it is concluded that the useful characters . . . have been produced by the special 'location of growth-force' by use" (1872). The "influence of use" and the "location of growth-force" may have been particular undertakings of Cope, but as Osborn (1912) remarked, these studies, in which "the revolutions of the life of an individual and the life of an entire order . . . is wonderfully harmonious and precise," and the law of evolutionary acceleration and/or retardation "laid the foundations" for all future work in paleontology.

A few paleontologists may have been working on "evolutionary" problems, but their acceptance of a particular kind of directed or orthogenetic evolution produced results that may have convinced some others that evolution proceeded from youth to old age, perhaps not unlike the ideal geographic cycle of landforms as promulgated by William Morris Davis near the end of the century. As a particular species approached the "limits of phylogenetic growth," the vital life-giving force of the group was depleted; therein we have Hyatt's explanation for periods of early to late accelerated morphological change (i.e., the initial vitality of youth that gives way to an advanced or accelerated growth into evolutionary old age), and Cope's notions of early acceleration and subsequent retardations in evolutionary growth. In Cope's view, the changeover from acceleration to retardation would bring about racial death, and this could be recognized in the fossil record as a time of seemingly bizarre morphological features; that is, exaggeration of body size, and the overspecialization of certain organs, to which Hyatt's ammonite studies contributed many "type" examples. This highly modified Lamarckian thesis of the irreversibility of evolution, as propounded in slightly different forms by Cope and Hyatt, however, probably had very little influence on the "working paleontologist," who had maintained a principal interest only in the geological-stratigraphic aspects of the fossil record and not those that may have had significant biological implications.

Of some interest in support of the more geologically oriented aspects of paleontology in the latter half of the 19th century, one need only review the journals. In the pages of the *Proceedings of the Philadelphia Academy of Natural Sciences* during the late 1860s and early 1870s, there are numerous reports of fossils from the western United States by F. W. Meek, A. H. Worthen, and J. Leidy. In some cases it is obvious that the definition of stratigraphically successive species is hindered by extreme morphological variability, but few conclusions were ever drawn and those that were seemed to show little, if any, interest in any evolutionary aspects of the fossil fauna. By the next decade, syntheses such as those of Scudder (1886) and Ward (1885) included only the briefest mention of the possibility of descent by gradual modification, although Ward did remark, almost as an afterthought, that "the multiple origin of existing forms, whether plants or animals, is repugnant to modern scientific thought." In 1885, this seems not to have been a particularly daring statement, and most authors concentrated on the drafting of range charts for various taxonomic hierarchies, which became subdivided into finer and finer units, without much regard to ancestral relationships.

The Evolution of Man. Some of the most heated debates, as one might have expected, were reserved for discussion of the evolutionary history of man. Little fossil evidence was uncovered throughout the 19th century, and there was even less agreement on the interpretation of what did exist. The antiquity of man was so closely tied to biblical concepts or the physico-theological traditions of the early 19th century that many who accepted Darwin's principle of descent through gradual modification did so for all organisms but man. Wallace in 1870 doubted the adequacy of natural selection, given the information he had on the cranial capacity and other "anomalous features" seen in European and "primitive" peoples, and instead accepted the fact that some other law was fundamental in the evolution of man.

With Cuvier's pronouncement in 1813 that Scheuchzer's (1726) *Homo diluvii testis* was in reality a large amphibian, the few possible evidences of early man were discounted. Prior to the 1850s, no unequivocal human fossil was recovered, and later the age of all finds from such as the Liège caves to those of the first Neanderthal skull (1856) were the subject of much controversy. Thus in the 1840s and '50s the evidence for fossil remains greater than 6000 years old was not very impressive. However, a decade later there was a growing body of circumstantial evidence (see esp. Boucher de Perthes, 1847–1864) that man had coexisted with now-extinct mammals. Flint implements as well as mounds of ashes were found associated with the bones of those extinct animals. A paper by Joseph Prestwich (1860) prompted a most serious reconsideration of a history of man prior to the biblical 6000-year date of Creation.

Lyell (1863) clearly documented, in a popular format, the evidence for a history of man extending far into the past, although his presentation of an evolutionary doctrine never satisfied Darwin. Huxley (1863) used embryology and comparative anatomy to establish the anthropoid origin of man. In the third part of the book, on the fossil record, Huxley discussed the significance of the Neanderthal skull, and concluded that it is certainly human but more ape-like than the skull of any living human race. The evolutionary relevance of the Neanderthal skull left Huxley in somewhat of a quandary, for not only did he not know the precise age of the fossil, he could not justify its existence as an intermediate link between the anthropoids and modern man. Huxley's contribution, less than half the length of Lyell's, however was much more to the liking of Darwin (see Irvine, 1959, ch. 10). In fact, some scientists, such as

Sir John Herschel, took Lyell's book to be a refutation of Darwinian doctrine and a defense of providential direction in the process of evolution. Sir John Lubbock's *Pre-historic times . . .* (1865) was a comprehensive archaeological survey, one which prompted a vast array of books and pamphlets, many of which gave an increasingly less convincing nod to an agreement with the scriptural intrepretations of the age of mankind, Darwin's own *Descent of Man* (1871) had none of the emotional impact that his *Origin* had 12 years earlier; the "high antiquity" of man clearly had been anticipated by earlier publications.

In 1868, the first find of Cro-Magnon man, reported from southwestern France, was clearly recognized as *Homo sapiens.* Ernst Haeckel soon afterwards began to publicize the possibility of finding a "missing link" (indeed he even introduced the name *Pithecanthropus*), much as Huxley had hoped for in the Neanderthal skull, between the anthropoids and man. This is said by Coleman (1971), among others, to have inspired Eugene Dubois to undertake the discovery of such a missing link, with success in 1891 when he uncovered *Pithecanthropus* or "Java Man" (*Homo erectus*).

Summary

In his 1860 review of the *Origin,* Huxley states that prior to Darwin's explanation of the fossil record, natural history "lay between two absurdities and a middle condition of uneasy skepticism." The two absurdities Huxley referred to were: (1) A belief in a special *creative force* or (2) a Lamarckian brand of unidirectional evolution. Neither of these was acceptable to the scientific skeptics. Suppose that we were to agree with Huxley that Darwin, by introducing the concept of natural selection, had indeed found a middle ground; yet must we than admit that the concept of natural selection, or indeed evolutionary notions in general, had a major impact on the methodology of paleontology? Obviously even from this brief review of mainly secondary source materials the answer must be equivocal. Certainly Darwin had an "effect," if only in the sense that his initial evolutionary notions with regard to the fossil record could not be ignored and eventually were twisted into a Darwinistic format. Coleman (1971) noted, however, that "In no discipline did expectations appear so great but frustrations prove so common as in paleontology," and even when evidence for a gradual lineal change was found in the Equidae "it was strongly suggestive . . . not decisive."

Darwin saw in the fossil record the possibility of a confirmation of his evolutionary mechanism, natural selection, not the demonstration of it; hence, he may have been satisfied with a few scraps, such as *Eozoön, Archaeopteryx,* and the Equidae. After all, they were consistent with his major thesis. A few paleontologists in the latter portions of the 19th century, however, were not satisfied with this level of interpretation, but extended the original Darwinian concepts into an evolutionary metaphysics that relied heavily upon the documentation of orthogenesis and recapitulation in the fossil record. Studies in developmental biology began to attract more attention, and it seemed to many as if there existed a remarkable parallel between the embryology of the individual and its evolutionary history. However, in any such competition paleontology had to come in a distant second, if only because of the acknowledged incomplete nature of the fossil record. It was recognized by many in the 1880s that the process of evolution would not be discovered by looking in detail at the fossil products of change.

However, some may contend that Darwin's directive was interpreted by most paleontologists only as a call for more data. Certainly in the first half of the 19th century, chemistry and physics had been characterized by the publication of huge data monographs; and now paleontologists had the chance to fill in the stratigraphic and perhaps the "phylogenetic" gaps. Whereas Darwin's thesis with regard to the vastness of geologic time had been seriously compromised by Sir William Thomson's thermodynamic calculations, the methodology of paleontology had not been altered. Pictet's 1860 prediction that: "The beautiful work of Mr. Darwin does not lead to a new point of view," thus proved most accurate. Some may argue that the Darwinian "problem" vis-à-vis paleontology really boiled down to the natural ancestry of man; a difficulty that presumably was recognized very early by Lyell, and one which Lyell tried desperately to avoid. Hence one might be able to make the case that the original Darwinian thesis of natural selection never really critically involved the fossil record; the Darwin-Huxley synthesis needed only some general confirmation from the geologic record, an example or two, nothing more.

Geikie (1905) gives what may be a reasonable consensus of paleontology in the latter half of the 19th century: "While the whole science of geology made gigantic advances during the nineteenth century, by far the most astonishing progress sprang from the recognition of the value of fossils. To that source may be traced the prodigious development of stratigraphy over the whole world, the power of working out the geological history of a country, and of comparing it with the history of other countries, the possibility of tracing the synchronism and the sequence of the geographical changes of the earth's surface since life first appeared upon the planet." This sounds like the paleontology of a William Smith or an Alexandre Brongniart, not a Thomas Huxley or a Charles Darwin. Paleontology had found a haven in the geological sciences and the important contributions to the biological sciences began to wither away in the metaphysics of Hyatt and Cope.

In service to stratigraphic geology, however, paleontology was of the utmost importance. Goetzmann (1959) writes, "Neither Hayden nor Meek [nor Hall nor Newberry] could be said to belong to the Humboldtean tradition . . . they frankly concentrated on observation for its own sake. . . . They were satisfied, for example, if they could prove that true Permian beds, comparable to those in Europe, did, in fact, exist in Kansas, and they did not question either the meaning of this fact or its relation to every other aspect of nature." Waage (1975) disagrees with Goetzmann that Meek's paleontological studies were so mundane; yet Meek's view of paleontology was principally biostratigraphic and only, in part, ecological; the evolutionary implications of these paleontological studies were not actively pursued.

Clearly we have ended on a note of unresolved interpretations. A new and rational appraisal of these post-Darwinian years in paleontology is needed, for it seems that it was during these years that paleontologists had become more professionally restricted under the aegis of stratigraphical geology. Furthermore, such a study may put to rest some of the less sacred myths that still seem to permeate much of the peripheral historical literature.

PETER W. BRETSKY

References

Primary Sources

Agassiz, L., 1857. Essay on classification, *Contributions to the Natural History of the United States,* Vol. 1. Boston: Little, Brown & Co., 1–232.

Barrande, J., 1852–1911. *Système Silurien du Centre de la Bohème.* Prague: F. Rivnác, 8 vols.

Boucher de Perthes, J., 1847–1864. *Antiquités celtiques et antédiluviens.* Paris: Treuttel et Wurtz, 3 vols.

Carpenter, W. B., 1856. Researches on Foraminifera, *Phil. Trans. Roy. Soc. London,* **146,** 181–236.

Chambers, R., 1844. *Vestiges of the Natural History of Creation.* London: J. Churchill, 390p. (Reprinted 1969 by Leicester Univ. Press.)

Cope, E. D., 1872. The method of creation of organic forms, *Proc. Amer. Phil. Soc.,* **12,** 229–265.

Cuvier, G., 1813. *Essay on the Theory of the Earth, translated from the French . . . by Robert Kerr, with mineralogical notes by Professor Jameson.* Edinburgh: W. Blackwood, 550p.

Cuvier, G., 1817. *Le Règne animal distribué d'après son Organisation pour servir de Base à l'Histoire naturelle des Animaux et d'Introduction à l'Anatomie comparée.* Paris: Deterville, 4 vols.

Dana, J. D., 1880. *Manual of Geology,* 3rd Ed. New York: Ivison, Blakeman, Taylor, 912p.

Darwin, C., 1859. *On the Origin of Species* 1st Ed.; 6th Ed., 1872. London: J. Murray, 502p. (Facsimile ed., Harvard Univ. Press, Cambridge, 1959.)

Darwin, C., 1871. *The Descent of Man, and Selection in Relation to Sex.* London: J. Murray, 2 vols.

Davidson, T., 1851–1886. *A Monograph of the British Fossil Brachiopoda.* London: Palaeontogr. Soc. Monogr., 6 vols. (See also Cocks, 1978.)

Dawson, J. W., 1893. *Some Salient Points in the Science of the Earth.* New York: Harper, 499p.

Forster, M., and Lankester, E. R., eds., 1898–1902. *The Scientific Memoirs of Thomas Henry Huxley.* London: Macmillan, 4 vols.

Gaudry, J. A., 1862–1867. *Animaux fossiles et Géologie de l'Attique.* Paris: F. Savy, 2 vols.

Gaudry, J. A., 1878–1890. *Les Enchaînements du Monde Animal dans les Temps Géologiques.* Paris: F. Savy, 3 vols.

Gaudry, J. A., 1896. *Essai de Paléontologie Philosophique.* Paris: Masson, 250p.

Hall, J., 1842–1894. *Palaeontology of the State of New York* Albany, N.Y.: Van Benthuysen, 8 vols.

Huxley, T. H., 1860. The origin of species, *Westminster Rev.,* **17,** 541–570.

Huxley, T. H., 1862. The aniversary address to the Geological Society, *Quart. Jour. Geol. Soc. London* **18,** 40–54; *in* Foster and Lankester, 1898–1902, **2,** 512–529.

Huxley, T. H., 1863. *Evidences as to Man's Place in Nature.* London: Williams & Norgate, 159p. (Reprinted 1959, Ann Arbor, Mich.: Univ. Michigan Press)

Huxley, T. H., 1868. Remarks upon *Archaeopteryx lithographica, Proc. Roy. Soc. London,* **16,** 243–248; *in* Foster and Lankester, 1898–1902, **3,** 340–345.

Huxley, T. H., 1870. The aniversary address of the president, *Quart. Jour. Geol. Soc. London,* **26,** 29–64; *in* Foster and Lankester, 1898–1902, **3,** 510–530.

Huxley, T. H., 1877. *American Addresses.* London: Macmillan, 164p.

Hyatt, A., 1894. Phylogeny of acquired characteristics, *Proc. Amer. Phil. Soc.,* **32,** 349–647.

Lamarck, J.-B., 1809. *Philosophie Zoologique.* Paris: Chez Dentu, 2 vols. (Engl. transl. reprinted 1963, Hafner, New York.)

Logan, W. E., 1865. On the occurrence of organic remains in the Laurentian rocks of upper Canada, *Quart. Jour. Geol. Soc. London,* **21,** 45–50.

Lubbock, J., 1865. *Pre-historic Times.* London: Williams & Norgate, 619p.

Lyell, C., 1830–1833. *Principles of Geology*. London: J. Murray, 3 vols.

Lyell, C., 1863. *The Geological Evidences of the Antiquity of Man*. London: J. Murray, 518p.

Marsh, O. C., 1880. Odontornithes: A monograph of the extinct toothed birds of North America, *Mem. Peabody Mus., Yale Univ.*, **1**, 201p.

Möbius, K. A., 1878. Der Bau des *Eozoön canadense, Palaeontographica*, **25**, 175–192. (See also *Amer. Jour. Sci.*, 3, **18**, 177–185.)

Murchison, R. I., 1839. *The Silurian System*. London: J. Murray, 768p.

Murchison, R. I., 1854. *Siluria*. London: J. Murray, 523p.

Neumayr, M., and Paul, C. M., 1875. Die Congerien- und Paludinen-Schichten Slavoniens und deren Faunen, *Abh. Kaiserlich-Königlichen geol. Reichs., Wien,* **7** (3), 110p.

Osborn, H. F., 1912. Evolution as it appears to the paleontologist, *Proc. 7th Internat. Zool. Cong.*, 733–739.

Owen, R., 1848. *On the Archetype and Homologies of the Vertebrate Skeleton*. London: R. & J. E. Taylor, 203p.

Owen, R., 1860. Darwin on the origin of species, *The Edinburgh Review*, **111**, 487–532 (American edition, **225**, 251–275).

Owen, R., 1863. On the fossil remains of a long-tailed bird (*Archeopteryx macrurus* Ow.) from the lithographic slate of Solenhofen, *Proc. Roy. Soc. London,* **12**, 272–273.

Pictet, F. J., 1860. Sur *l'Origine de l'Espèce* par Charles Darwin, *Arch. Sci. Phys. Nat., Biblioteque Universelle de Genève*, 8, 233–255.

Prestwich, J., 1860. On the occurrence of flint implements associated with the remains of animals of extinct species, *Phil. Trans. Roy. Soc. London,* **150**, 277–318.

Scheuchzer, J., 1726. *Homo Diluvii Testis*. Öningen, Germany: Tiguri, 24p.

Scudder, S. H., 1886. Systematic review of our present knowledge of fossil insects including myriapods and arachnids, *U.S. Geol. Surv. Bull.* **31**, 128p.

Sedgwick, A., 1860. Objections to Mr. Darwin's theory of the origin of species, *Spectator*, March 24, 285–286; April 7, 334–335.

Sedgwick, A., and McCoy, F., 1855. *A Synopsis of the Classification of the British Palaeozoic Rocks*. London: J. W. Parker & Son, 661p.

von Cotta, B., 1866. *Die Geologie der Gegenwart*. Leipzig: J.J. Weber, 450p.

Wallace, A. R., 1889. *Darwinism*. London: Macmillan, 494p.

Ward, L. F., 1885, Evolution in the vegetable kingdom, *Amer. Naturalist,* **19**, 637–644, 745–753.

Winchell, A., 1870. *Sketches of Creation*. New York: Harper, 459p.

Selected Secondary Sources

Bartholomew, M., 1973. Lyell and evolution: An account of Lyell's response to the prospect of an evolutionary ancestry for man, *British J. Hist. Sci.*, **6**, 261–303.

Cocks, L. R. M., 1978. A review of British Lower

Palaeozoic brachiopods, including a synoptic revision of Davidson's monograph, *Palaeontogr. Soc. Monogr.*, **131**, 1–256.

Coleman, W., 1971. *Biology in the Nineteenth Century*. New York: Wiley, 187p.

Geikie, A., 1905. *The Founders of Geology*. London: Macmillan, 486p. (Reprinted 1962, Dover Pubs., New York.)

Glick, T. F., ed., 1974. *The Comparative Reception of Darwinism*. Austin, Texas: Univ. Texas Press, 505p.

Goetzmann, W. H., 1959. *Army Exploration of the American West 1803–1863*. New Haven, Conn.: Yale Univ. Press, 509p.

Gould. S. J., 1977a. *Ontogeny and Phylogeny*. Cambridge, Mass.: Harvard Univ. Press, 501p.

Gould, S. J., 1977b. Eternal metaphors of palaeontology, in A. Hallam, ed., *Patterns of Evolution*. Amsterdam: Elsevier, 1–26.

Haber, F. C., 1959. *The Age of the World, Moses to Darwin*. Baltimore: Johns Hopkins Univ. Press, 303p.

Hull, D. L., 1973. *Darwin and His Critics*. Cambridge, Mass.: Harvard Univ. Press, 473p.

Irvine, W., 1959. *Apes, Angels and Victorians*. New York; Meridian Books, 399p.

Lurie, E., 1960. *Louis Agassiz, A Life in Science:* Chicago: Univ. Chicago Press, 449p.

Lyon, J., 1970. The search for fossil man: Cinq Personnages à la Recherche du Temps Perdu, *Isis,* **66**, 68–84.

O'Brien, C. F., 1970. *Eozoön canadense,* "The Dawn Animal of Canada," *Isis,* **61**, 206–223.

Oppenheimer, J. M., 1967. *Essays in the History of Embryology and Biology*. Cambridge, Mass.: M. I. T. Press, 374p.

Peckham, M., 1959. Darwinism and Darwinisticism, *Victorian Studies*, vol. 3, 3–40.

Reingold, N., 1964. *Science in Nineteenth-Century America*. New York: Hill and Wang, 339p.

Rudwick, M. J. S., 1972. *The Meaning of Fossils*. London and New York: MacDonald and American Elsevier, 287p.

Shor, E. N., 1974. *The Fossil Feud*. Hicksville, N.Y.: Exposition Press, 340p.

Waage, K. M., 1975. Deciphering the basic sedimentary structure of the Cretaceous System in the Western Interior, *Geol. Ass. Canada Spec. Paper, No. 13*, 55–81.

Weller, J. M., 1960. Development of paleontology, *Paleontology,* **34**, 1001–1019.

Cross-references: *History of Paleontology: Before Darwin; History of Paleontology: Post-Plate Tectonics.*

HISTORY OF PALEONTOLOGY: POST-PLATE TECTONICS

To discuss paleontology since the advent of plate tectonics, which we shall take as the beginning of 1969, is really a reportorial rather than an historical task. It does seem logical to base a subdivision of the history of paleontology upon the development of plate tec-

TABLE 1. The Distribution by Paleontological Field of Publications in
Three Journals for the Years 1972 to 1974 Inclusive

Field	Journal of Paleontology	Palaeontology	Lethaia	Total
Taxonomy	177	80	12	269
Biostratigraphy	6	2	3	11
Taphonomy and environmental reconstruction	7	7	8	22
Functional morphology	22	19	20	61
Paleoecology of populations and communities	28	19	11	58
Biogeography	2	1	0	3
Evolution	2	2	3	7
Methodology and other	6	5	1	12
Total	250	135	58	443

tonics, for this event has opened the fossil record to interpretation from a powerful new basis (see *Plate Tectonics and the Biosphere*). This potential has just begun to be exploited, however, and the bulk of effort in paleontological research continues along the traditional paths. Although there is much work in progress that appears to lead toward achievement of the new potentials, we lack the perspective of time, which alone can determine how these opportunities will be met. We can, however, examine the current literature of paleontology for indications of any changes that may be occurring.

The Recent Literature of Paleontology

Periodicals and Occasionals. Some evidence of the distribution of recent effort within fields of paleontology is shown in Table 1, which samples three journals (American, British, and Scandanavian) for three years, 1972–1974. The articles published during these years are classed as to their main subject. Although they frequently cover materials in more than one field, they are tallied only according to their main thrust. The overwhelming effort has been in taxonomy—the classification and nomenclature of fossils—even though one of the journals (*Lethaia*) does not accept purely taxonomic contributions. Taxonomy has been a major field of paleontological research since well before the time of Darwin and continues so today. Historically, the other major effort in paleontology has been the study of fossils for stratigraphic purposes—to correlate and characterize rock units—but this work is not ordinarily reported in these three journals (see *Biostratigraphy*). A few papers are primarily concerned with using fossils to interpret the depositional or postdepositional environments of the rocks in which they are found.

The main purposes of papers in the first three categories of Table 1, then, are to describe fossils and apply their occurrences to understanding the geological record.

The next four categories comprise papers that employ fossils to investigate the biological processes of the past. Here the main effort is divided between functional morphology and paleoecology. The former interprets the biology of individual organisms from their fossil remains (nearly entirely skeletal); the latter deals with the relations between ancient organisms and their environments. Many of the paleoecological contributions are studies of ancient communities. Finally, research into the evolution and biogeography of fossils is reported at a low frequency in these three journals. Nevertheless, nearly 30% of the papers are devoted primarily to paleobiological topics.

The frequencies in Table 1 would obviously change if different paleontological journals were surveyed, especially those in languages other than English. In general, the proportion of papers devoted to nonpaleobiological topics, especially taxonomy, would increase to even greater predominance, unless nontaxonomic specialty journals were particularly selected. Such specialty journals began appearing principally during the closing years of the post-Darwinian period. They deal chiefly with paleoecological topics (*Lethaia; Palaeogeography, Palaeoclimatology, Palaeoecology*). In the post-plate tectonics period, still another new journal has appeared, specifically devoted to the interpretation of the biology of fossil remains (*Paleobiology*). Monographs and other occasional series remain primarily taxonomic and biostratigraphic in content, although even in these publications there appears a growing number of symposia with paleobiological themes.

Paleontological serials do not provide a full picture of the paleontological effort, however, because much research is published elsewhere. Biostratigraphic contributions appear chiefly in more geologically oriented journals, wherein they comprise the bulk of papers that are paleontological, Indeed, biostratigraphy probably outranks taxonomy in terms of total effort. Paleobiological contributions, on the other hand, frequently appear in biological or in general journals, where they will be seen by interested life scientists. The journals that are most closely identified with the paleontological profession, then, cover the middle ground and function chiefly as archives for taxonomy (an indispensible service); but there is a strong and growing component of paleobiology.

Books. It is commonly assumed that the forefront of research in any particular discipline is indicated by journal articles. Once a field is flourishing, these are summarized and organized in reviews, monographs, and specialized books; and subsequently general textbooks appear summarizing these secondary sources and introducing their topics into the regular college curriculum. This picture is often too simple. In the post-plate tectonic period of paleontology the research changes are most readily apparent in books, including textbooks at both undergraduate and graduate levels. These changes continue a trend that became particularly marked in the last two decades, and which saw the appearance of major contributions to actualistic paleontology and paleoecology.

Among the invertebrate textbooks in English that have appeared since 1968, only one includes a traditional systematic account of the morphology of fossil invertebrates (Tasch, 1973), and even this book is entitled a "paleobiology." The others are nontraditional; they include a first-course text that is heavily weighted toward functional morphology and paleoecology (Raup and Stanley, 1971), a symposium volume that is widely used for graduate courses and considers selected problems in paleoecology and biogeography in depth (Schopf, 1972), and an advanced text concerned entirely with evolutionary paleoecology in a broader format (Valentine, 1973). These works are thoroughly paleobiological. When taxonomy is treated, it is in terms of principles, but without a systematic review of the taxa themselves. In a somewhat similar vein, the latest textbook of vertebrate paleontology (Stahl, 1974) is based upon an account of the major problems facing the interpretation of vertebrate history, rather than including a systematic survey of vertebrate-fossils.

Reference works display a similar trend. The major systematic works, such as volumes of the monumental *Treatise on Invertebrate Paleontology,* continue to be produced, but at the same time such volumes as the paleobiogeographic compilations of Middlemiss et al. (1971) and Hallam (1973) and the treatment of Silurian-Devonian benthic associations by Boucot (1975) appear with increasing frequency.

The paleobiological activity displayed in book production argues for great future increases in the attention and effort expended on paleobiological topics. Many paleontology students now study the problems and methods of paleobiological interpretation from their first paleontological course, and abundant secondary materials are available for advanced study. Thus the primary paleobiological lieterature of the serials becomes accessible early in a student's career. Coupled with a growing tendency for publication by graduate students and with the increase in outlets for paleobiological research, this field can be expected to expand at an accelerating pace.

The Prospects of Paleontology

From the current activity reflected in the literature, it appears most likely that paleobiological fields will continue to strengthen. We have stressed a quantitative assessment of literature contributions; a qualitative review leads to similar conclusions. The biological topics are not only increasing in frequency, but they increasingly involve fields that had been associated little or not at all with paleontology. For example, the trophodynamics of ancient ecosystems are now being assessed, and studies are underway to determine the possibilities of estimating genetic variabilities in fossil populations with an eye toward exploring the effects on the subsequent histories of those populations. Biologists themselves have begun to contribute to this literature, enlarging the skills and techniques that are brought to bear on paleontological problems. Thus there is a diversification of paleobiological topics, which is leading to an unprecendented breadth of paleontological effort (see Gould, 1976). The traditional topics appear as vigorous as ever, and indeed they must be continued and enhanced in order to maintain an appropriate balance between the geological and biological aspects of the study of fossil material. As the scope of paleontology broadens it overlaps increasingly with biology. It is likely that the disciplinary barriers between biology and paleontology will continue to weaken and that paleobiology will join the mainstream of the life sciences.

There were several paleobiological efforts during the post-Darwinian period. Indeed, the present trend from geological to biological applications of the fossil record began then. Because it has coincided with the advent of plate tectonics and with the new approaches of population biology the present trend may prove to be permanent. Plate tectonics provides the unifying historical theme that permits paleobiological interpretation to escape the ad hoc character of previous periods and to become integrated into an expanding paradigm of the history of biological processes. Such a paradigm provides the framework and continuity required for the mature development of a scientific field.

JAMES W. VALENTINE
CATHRYN A. CAMPBELL

References

Boucot, A. J., 1975. *Evolution and Extinction Rate Controls.* Amsterdam: Elsevier, 427p.

Gould, S. J., 1976. Palaeontology plus ecology as palaeobiology, in R. M. May, ed., *Theoretical Ecology: Principles and Applications.* Philadelphia: Saunders, 218–236.

Hallam, A., ed., 1973. *Atlas of Palaeobiogeography.* Amsterdam: Elsevier, 531p.

Middlemiss, F. A.; Rawson, P. F.; and Newall, G., eds., 1971. *Faunal Provinces in Space and Time.* Liverpool: Seel House, 236p.

Moore, R. C., ed., 1953–. *Treatise on Invertebrate Paleontology.* Lawrence, Kansas: Geol. Soc. Amer. and Univ. Kansas Press.

Raup, D. M., and Stanley, S. M., 1971. *Principles of Paleontology.* San Francisco: Freeman, 388p.

Schopf, T. J. M., ed., 1972. *Models in Paleobiology.* San Francisco: Freeman and Cooper, 250p.

Stahl, B. J., 1974. *Vertebrate History: Problems in Evolution.* New York: McGraw-Hill, 594p.

Tasch, P., 1973. *Paleobiology of the Invertebrates.* New York: Wiley, 946p.

Valentine, J. W., 1973. *Evolutionary Paleoecology of the Marine Biosphere.* Englewood Cliffs, N.J.: Prentice-Hall, 511p.

Periodicals

Journal of Paleontology, 1928–. Tulsa, Oklahoma: Soc. Economic Paleontologists and Mineralogists and Paleontological Soc.
Lethaia, 1968–. Oslo, Norway: Universitetsforlaget.
Palaeogeography, Palaeoclimatology, Palaeoecology, 1965–. Amsterdam, Elsevier.
Palaeontology, 1958–. Cambridge, England: The Palaeontological Assoc.
Paleobiology, 1975–. Chicago, Ill.: Paleontological Soc.

Cross-references: *Actualistic Paleontology; Biostratigraphy; Morphology, Functional; Paleobiogeography; Paleoecology; Plate Tectonics and the Biosphere; Systematic Philosophies; Taphonomy.*

HYOLITHA

Hyolitha Henningsmoen, 1952, is a class name proposed for extinct animals thought to be mollusks; the term was defined by L. Marek in 1963. The fossils are tapering calcareous shells, closed at the narrow end, bilaterally symmetrical, and with a trigonal cross section; the lower part commonly is arched in cross section, and in profile shows logarithmic curvature at a low angle, often less than 5°. The aperture is closed by an operculum. The class is best typified by *Orthotheca* and *Hyolithes* (Fig. 1). In addition to the features listed, *Hyolithes* and allies, the most abundant members of the class, have a semicircular expansion of the lower part of the shell as a "shelf" anterior to the aperture, and paired, curved, calcareous rods extending outward between operculum and shell.

Because of its shell thickness and size, *Hyolithes* probably was benthic and capable of only limited movement, though some authors have considered it pelagic. The habit of *Orthotheca* is less certainly inferred, and while it too may have been pelagic, its size and shell thickness are comparable to that of *Hyolithes.*

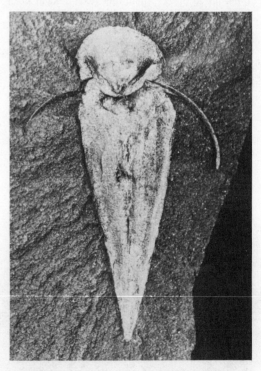

FIGURE 1. *Hyolithes carinatus* Matthew from the Middle Cambrian of British Columbia, showing shell, operculum, and rodlike outgrowths. The specimen is flattened by compaction; magnification ca. X4.

Specimens occur in all types of sedimentary rocks, though they appear to be most abundant in limestone and calcareous shale. They may be locally abundant, but are rarely major constituents of the rock.

Nearly 500 species have been described, though for many years relatively few genera were recognized. Most species have not been carefully studied and it is difficult to determine their utility as guide fossils. Both individuals and species were common in the Cambrian, were much less abundant in the Ordovician, and became increasingly rare until they became extinct in the Permian.

Hyolithids have been difficult to classify. They have traditionally been placed with the gastropods because of supposed homologies in shell shape to Holocene pelagic gastropods. Some authors have associated them with the conularids, but that group has a different shell composition. G. P. Ljaschenko in 1955 proposed the class Coniconchia, questionably Mollusca, to contain the two superorders Tentaculitoidea and Hyolithoidea. The first has been removed from Mollusca by B. Bouček and constitutes his class Tentaculitida (q.v.). Fischer, in 1962, proposed the new class Calyptoptomatida essentially for the Hyolithoidea, with certain modifications as to limits and rearrangements. Marek, in 1963, independently of Fisher's work, proposed use of Hyolitha and suggested that major subdivisions might be based on the features of the operculum rather than the tube.

Runnegar et al. (1975) have suggested the Hyolitha are a separate phylum related to the Mollusca and Sipunculoidea, while Marek and Yochelson (1976) remain convinced of the molluscan affinities of Hyolitha.

ELLIS L. YOCHELSON*

References

Fisher, D. W., 1962. Small conoidal shells of uncertain affinities, in R. C. Moore, ed., *Treatise on Invertebrate Paleontology*, pt. W., Miscellanea, Lawrence, Kansas: Geol. Soc. Amer. and Univ. of Kansas Press, W98–W143.

Ljaschenko, G. P., and Syssoiev, V. A., 1958. Class Coniconchia, in Osnovy paleontologii, spravocnik dlja paleontologov i geologov SSSR, Molljuski-golovongie II. Gos. naucno-techn. izdat, Moscow, 179–190.

Marek, L., 1963. New knowledge on the morphology of *Hyolithes, Sbornik Geol. Ved, Paleont.*, **1**, 53–73.

Marek, L., 1967. The Class Hyolitha in the Caradoc of Bohemia, *Sbornik Geol. Veds Paleont.*, **9**, 51–112.

*Publication authorized by the Director, U.S. Geological Survey.

Marek, L., and Yochelson, E. L., 1976. Aspects of the biology of *Hyolitha* (Mollusca), *Lethaia*, **9**, 65–82.

Runnegar, B., Pojeta, J. Jr.; Morris, N. J.; Taylor, J. D., Taylor, M. E.; and McClung, G., 1975. Biology of the Hyolitha, *Lethaia*, **8**, 181–191.

Cross-reference: *Tentaculita*.

HYSTRICHOSPHERES

Hystrichospheres are reproductive or resting cysts of dinoflagellates (q.v.), spherical to ellipsoidal or polygonal in shape. Characteristically they bear processes, in the form of simple or branching spines or tubes, which may be isolate or may be linked together by raised ridges on the cyst surface or by trabeculae connecting their tips. They range in size from about 30 to 250 μm. The wall is composed of a macromolecular organic substance resembling cutin or sporopollenin. The cyst usually has an opening (an archeopyle), formed either by median schism into two halves or by the loss of a part of the wall, consistently located and characteristically polygonal, which corresponds in position to a single plate or to a group of plates in the wall of the motile dinoflagellates.

History of Study

In 1836, the German microscopist Christian G. Ehrenberg discovered the presence of minute spinose spheres in flint flakes, in association with what he recognized as dinoflagellates. He mistakenly considered the spiny bodies to be the silicified zygospores of a present-day freshwater Desmid, *Xanthidium*. Despite studies in 1845 by Gideon Mantell, who demonstrated the organic nature of the shell, recognized that freshwater fossils were unlikely in marine cherts, and proposed the name *Spiniferites* for these fossils, these microfossils continued to be known as "xanthidia" for almost a century.

In 1932, however, another German, Otto Wetzel, reviewed the situation and proposed the generic name *Hystrichosphaera* (Greek: "spiny sphere") for Ehrenberg's species. The type species showed a surface pattern of crests outlining a dinoflagellate-like tabulation and even a transverse furrow, but the latter was bridged by spines and could not have contained a flagellum. Further genera were differentiated by subsequent workers, notably Deflandre, Eisenack and M. Lejeune-Carpentier, until there came to be a whole group of microfossils under the general designation "hystrichospheres," ranging in morphology from simple spheres without ornament through to forms with an elaborate meshwork of complex, linked processes. Despite early recognition of their planktonic nature, the affinities of the hystrichospheres remained a matter for dispute; they were therefore classed as an order of uncertain reference, the Order Hystrichosphaeridea.

The key to the determination of their systematic position was provided by William R. Evitt in 1961. From prolonged study of many thousands of individuals he demonstrated that the shells of undoubted fossil dinoflagellates contained openings which, in shape and position, were exactly paralleled in hystrichospheres. In a subsequent work, Evitt (1963) further showed that the pattern of process arrangement on the shells of many hystrichosphere species corresponded exactly to the pattern of tabulation of dinoflagellates (see Fig. 1). These species were reproductive cysts

formed within the outer shell of dinoflagellates; the cyst was abandoned after fission, the new individuals escaping through the archeopyle. Evitt's deductions, based on fossil hystrichospheres, have been subsequently confirmed from study of cysts in modern sediments, all stages between motile adult and abandoned cyst now being known (for references, see Sarjeant 1961, 1965).

Thus a number of genera were demonstrated to be dinoflagellates; since these included *Hystrichospaera* itself, the Order Hystrichosphaeridea was necessarily abandoned. The term *hystrichosphere* is now correctly applied to dinoflagellate cysts bearing well-developed processes; specifically, to chorate cysts—cysts in which the radius of the central cavity (the endocoel) typically forms 0.6 or less of the overall radius. The major chorate cyst genera

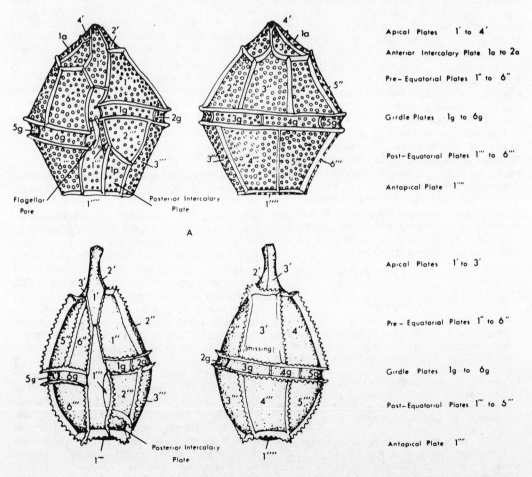

FIGURE 1. Reconstruction of the original tabulation of a dinoflagellate from the distribution of processes on a hystrichosphere. A, The probable original tabulation (which accords with that of the modern genus *Gonyaulax;* B, the cyst forming within the dinoflagellate, attached by processes; C, the abandoned cyst with an apical archeopyle, after crushing, and distortion resulting from consolidation and compaction of the sediment. The hystrichosphere shown is *Oligosphaeridium vasiformum* (Lower Cretaceous); magnification ca. X500.

were largely reexamined and redefined by Davey and Williams (1966). The residue of genera that were formerly classed as hystrichospheres but lack any clear evidence of dinoflagellate affinity, are now termed *acritarchs* (q.v.).

Geological Distribution

The earliest undoubted hystrichospheres *sensu stricto* appear in the Middle Jurassic. Cysts of this type assume great numerical importance in the early Cretaceous, after which they continue to dominate dinoflagellate cyst assemblages to the present day. Although dinoflagellate cysts of other types have been recorded from freshwater sediments, all true hystrichospheres are marine.

WILLIAM A. S. SARJEANT

References

Deflandre, G., 1947. Le problème des Hystrichospheres, *Bull. Inst. Oceanogr. Monaco.* **918**, 1–23.

Davey, R. J., and Williams, G. L., 1966. In R. J. Davey, C. Downie, W. A. S. Sarjeant, and G. L. Williams, Studies on Mesozoic and Cainozoic dinoflagellate cysts, *Bull. Brit. Mus. (Nat. Hist.) Geol.,* Suppl. **3**, 248p.

Downie, C.; Evitt, W. R.; and Sarjeant, W. A. S., 1963. Dinoflagellates, hystrichospheres, and the classification of acritarchs. *Stanford Univ. Publ., Geol. Sci.,* **7**(3), 1–16.

Eisenack, A., 1963. Hystrichosphären, *Biol. Rev. Cambridge Phil. Soc.,* **38**, 107–139.

Evitt, W. R., 1961. Observations on the morphology of fossil dinoflagellates, *Micropaleontology,* **7**, 385–420.

Evitt, W. R., 1963. A discussion and proposals concerning fossil dinoflagellates, hystrichospheres and acritarchs, *Proc. Nat. Acad. Sci.,* **49**, 158–164, 298–302.

Sarjeant, W. A. S., 1961. The Hystrichospheres: A review and discussion, *Grana Palynologica,* **2**, 102–111.

Sarjeant, W. A. S., 1965. The Xanthidia, *Endeavour,* **24**, 33–39.

Sarjeant, W. A. S., 1970. Xanthidia, palinospheres and "Hystrix." A review of the study of fossil unicellular microplankton with organic cell walls, *Microscopy* (London), **31**, 221–253.

Cross-references: *Acritarchs;Dinoflagellates;Plankton.*

I

INSECTA

Insecta, comprising the insects, is the largest class of the phylum Arthropoda, and is thought to have more species than all the other animals and plants combined. At present, more than 700,000 species of insects are known, and it seems likely that less than half the existing species have so far been named.

External Characteristics

Like other arthropods, the insects have a bilaterally symmetrical and segmented body (see Fig. 1), covered by a nonliving *exoskeleton,* which must be occasionally molted while being replaced with a larger exoskeleton as the individual grows. In molting, the insect usually steps out of the entire previous outer exoskeleton, and these highly resistant exuviae are sometimes elegantly fossilized. All insects have jointed appendages, including a single pair of antennae; most have three pairs of jointed legs, present in at least the adult stage. Legs have been lost in the juveniles of many higher insects. The exoskeleton is largely made up of protein, interlinked in most insects with chitin (q.v.), a complex polysaccharide. Calcium carbonate and other minerals, commonly prominent in the exoskeletons of marine arthropods and millipedes, are absent or limited to traces in insects. Each typical segment has a dorsal plate (*tergite*), two laterals (*pleurites*), and a ventral plate (*sternite*). The segments that make up the body of an insect are grouped in three *tagmata,* or regions—the head, thorax, and abdomen. Each tagma is a cluster of segments highly specialized for a set of functions.

The Head. The head capsule encloses six (possibly five) ancestral segments containing: a pair of usually prominent preoral appendages, the *antennae;* the *compound eyes;* usually three or two simple eyes (*ocelli*); the *mouth;* and finally three pairs of former legs modified into mouthparts—the *mandibles,* and then the first *maxillae* and the *labium* (fused second maxillae), with their own sensory appendages (*palps*). Several times during the course of insect evolution the generalized jaws and short maxillae have developed in various ways into a *beak,* usually with extreme elongation of the mandi-

bles and one or both pairs of maxillae and with some kind of sheath protecting the delicate piercing and sucking mouthparts. Many of the oldest insect fossils, belonging to the extinct paleopterous orders Palaeodictyoptera, Megasecoptera, and Diaphanopterodea, had this specialization (Carpenter, 1971, 1238–1240). Some of the diverse living groups in which sucking beaks evolved entirely independently are the leaf-hopper allies and true bugs (Hemiptera), the mammal lice (Phthiraptera), the fleas (Siphonaptera), the mosquitoes and other biting flies (Diptera), the nectar-sucking bees (Hymenoptera), and butterflies and moths (Lepidoptera). A few snouts, as in the weevils (Coleoptera) and scorpion–fly allies (Mecoptera), evolved by extreme elongation of the face, with little alteration of the actual mouthparts. Mouthparts and other head structures tend to be poorly preserved in fossils other than amber inclusions, and the true affinities of the extinct suctorial orders of Paleoptera were not convincingly worked out until fossils with visible mouthparts were finally studied.

The Thorax. The second tagma, the thorax, is set off from the head by a narrow neck. The thorax is clearly three-segmented, each segment in adults bearing a pair of legs. The legs articulate with the pleural region, from which they arose in the ancient prelegged ancestor as evaginations. The typical insect leg has the following segments, beginning nearest the body: the *coxa,* forming a socketed articulation with the thoracic wall; the *trochanter,* of one or two segments, and also having a socketlike function; the *femur,* specialized for horizontal movement and holding the foot laterad of the body; the *tibia,* specialized as a vertical element and holding the body above the substrate; and the *tarsus,* a footlike section usually divided into five or fewer subsegments, at the tip of which are one or two *ungues* (claws). Leg specializations are conspicuously related to locomotory adaptations and in some groups to predatory roles (powerful jumpers have large hind legs with huge femora (Fig. 1); water beetles and water boatmen have feathery, oar-like hind legs; preying mantids and assassin bugs have grasping forelegs, and so on). Thus, the locomotion of extinct groups whose fossils show legs

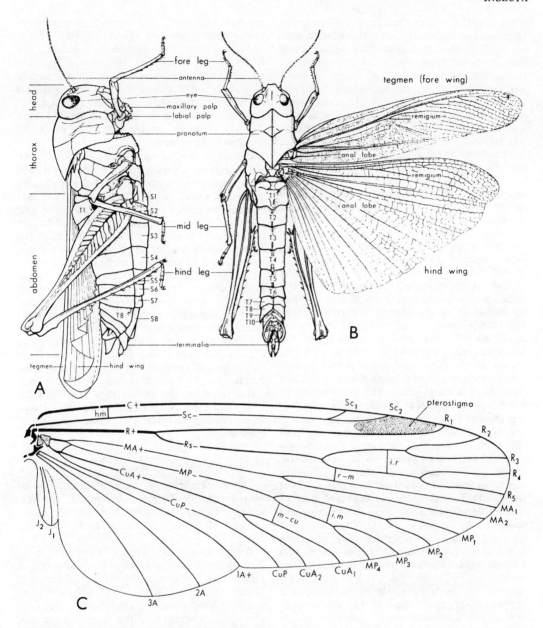

FIGURE 1. External morphology of a typical insect (from Mackerras, 1970; used by permission of Melbourne University Press). Lateral (A) and dorsal (B) views. S, *sternal* (ventral) plate, and T, *tergal* plate, of abdominal segments. In the grasshopper the thickened, leathery forewing is termed a *tegmen*, the hindwing is greatly broadened in the posterior region (anal lobe). C, diagram of wing venation; the main stems of C (costa), R (radius), MA (anterior media), and CuA (anterior cubitus) are convex in dorsal view and are marked (+); Sc (sub-costa), Rs (radial sector), MP (posterior media), and CuP (posterior cubitus) are concave and are marked (-). Outer branching varies greatly in different groups and a common pattern is shown here; the crossveins (hm, i.r., r-m, i.m., and m-cu) are often present but many alternatives are known.

well preserved can reasonably be inferred. The true legs are always limited to the three thoracic segments; but it is certain that the ancestor of insects was myriapodous, like modern centipedes, and all legs behind the first three pairs

have been lost or greatly modified into caudal structures functioning during copulatory clasping and oviposition.

The Wings. The most primitive insects (Subclass Apterygota) have never had wings. All other insects (Subclass Pterygota) are descended from winged ancestors; but many species, genera, and families, and even three whole orders (the parasitic fleas, lice, and the enigmatic Grylloblattodea) have lost all traces of wings and are therefore termed secondarily wingless. Because they are usually well preserved, wings are the principal source of information in most insect fossils; so it is fortunate that wing characters are complex, containing a large amount of information. These complexities have been so intensively studied in living as well as fossil insects that their interpretation in classification is at a high level of confidence. In some Pennsylvanian fossils there are also small but clearly wing-like structures on the prothorax (see Figs. 3B, 6D) and even on some abdominal segments, but true wings occur only on the second thoracic segment (mesothorax) and the third (metathorax). Each wing arose as an extension of the lateral region of the tergite of its segment, possibly functioning for gliding or parachuting when small. A wing is extremely thin and flat, being composed mainly of the two living layers of its flattened epithelium, and the accompanying exoskeleton which epithelia always secrete. The exoskeleton is highly resistant to decay, and the living tissue is a minor portion of a wing; hence, the preservation of wings is superior to that of other structures during fossilization. The wing struts, or "veins," are tubular enlargements extending out from the body toward the wing tips. The arrangement of these veins proves to be conservative enough so that orders, suborders, and families can be defined on venational characters. Yet it has changed sufficiently during evolution so that the details permit recognition of genera and species (e.g., Carpenter, 1954). There are eight principal veins in the primitive wing, occurring in a highly stereotyped sequence of convex and concave veins (viewed from the dorsum of the wing) (Fig. 1C). So, for example, the basic *subcosta* is concave and is preceded by the convex *costa* and followed by the convex *radius*. The convexities and concavities are as clear in rock fossils as in living insects (Martynov, 1938). Behind these eight primary (*remigial*) veins are the taxonomically less useful anal veins, which tend to have unstable numbers and no distinct convexity-concavity. The primary veins may fork one or more times and may be braced by various crossveins; some crossveins also tend to be conservative and therefore useful for higher classification. While there were originally two pairs of wings, some groups, most notably the Diptera, have lost one pair; other groups, such as Coleoptera, Dermaptera, and Protelytroptera (Fig. 6A), have had the hind pair enlarged as the sole flight mechanisms, while the forewings were converted into protective shields (*elytra*). This is unfortunate for paleoentomology, because the elytra are so thickened and altered that the ancestral forewing venation is entirely or largely obscured; and although the fine surface details are often preserved, most of the thousands of isolated beetle elytra in Mesozoic and even Tertiary deposits are not placeable to family on present knowledge.

If two pairs of wings fly independently, the first tends to create a zone of turbulence into which the second must therefore stroke with poor efficiency. In the evolution of fast and controlled flight, the reduction to a single pair of actual flying wings was one solution to this problem. With insects that use both pairs for flying, turbulence for the hindwings is avoided by a variety of coupling mechanisms that cause a forewing to operate with a hindwing as a single flight surface. For example, all Hymenoptera and many Hemiptera have on the leading edge of the hindwing a row of small hooks (*hamuli*) that couple firmly with a rigid fold on the trailing edge of the forewing. In moths with rapid flight a *frenulum,* composed of one (usually in ♂♂) or several (usually ♀♀) strong bristles at the anterior base of the hindwing, couples with a retinaculum on the underside of the forewing. These and other coupling structures can be used in interpeting relationships and the functions of fossil wings.

Metamorphosis

The postembryonic stages of insects are of two drastically different sequences, and few orders (Thysanoptera, some Hemiptera, Megaloptera) show any intermediacy between the two.

Hemimetamorphic Type. The older, more generalized pattern, as seen in a silverfish, a grasshopper, or a bug (Fig. 2A), is like that of higher vetebrates, with juveniles basically resembling adults and the adult body being only a modest elaboration of the juvenile structures. In these "hemimetabolous" insects, the most conspicuous changes from larva (traditionally called *nymph* in this group) to adult (or *imago*) are in the sudden increase in wing size and more subtly in the maturation of genitalic and ovipositional structures. However, in the three living orders with aquatic juveniles and terrestrial-aerial adults (Odonata, Ephemeroptera, and Plecoptera) the immatures have aquatic adaptations making them look very unlike the adults (Fig. 2B). In most of the hemimetamorphic insects, while the wing pads are developing they are visible as external flaps (Fig. 2A, 5C); hence the group name (Exopterygota) for all these

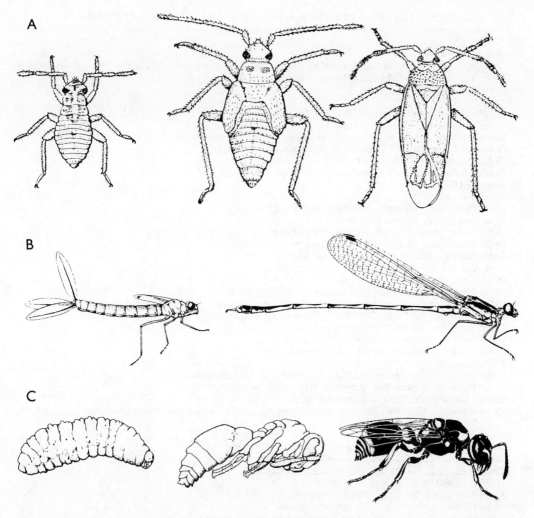

FIGURE 2. Three modes of metamorphosis. A, hemimetamorphic type (Hemiptera, a mirid bug) with juveniles and adults in same environment (from Imms, 1957; used by permission of Metheun & Co. Ltd., publishers). B, hemimetamorphic type (Odonata, damselfly) and C, holometamorphic type (Hymenoptera, a vespid wasp) with replacement of most larval tissues in formation of imaginal tissues of pupa, pupal stage nonfeeding and dormant (from Mackerras, 1970; used by permission of Melbourne University Press).

orders allocated to the Neoptera. It is remarkable that the juvenile stage of insects can never fly, possibly because a juvenile must ultimately molt in order to metamorphose into an adult, and molting over full wings is hazardous or, with some wing shapes, impossible.

Holometamorphic Type. The more specialized growth sequence characterizes the several "holometabolous" orders. These all have a feeding and growing larval stage, then a non-feeding pupal stage, finally the reproducing adult stage (Fig. 2C). The most typical ones, such as the Diptera and Lepidoptera, have the larval stage utterly unlike the adult, and the larval body functions rather like a culture medium for the small clusters of cells fated to grow into the entire adult body. These clusters (the imaginal discs) expand rapidly at the end of the growing stage by replacing the juvenile body, as it breaks down into nourishment for the imaginal growth. These orders are grouped as the Endopterygota, because the developing wings grow as internal buds in the larval thorax and do not evert until formation of the pupa. In general, the larvae share no external classificatory characters with their adults, and it is understandable that paleontologists may name larvae separately from adults unless there is strong evidence for association, such as presence in a stratum of a single, plentiful larval and a

TABLE 1. Phylum ARTHROPODA, Subphylum UNIRAMIA (= TRACHEATA)

Class INSECTA
Subclass APTERYGOTA
 Order Microcoryphia (including Monura*) (= Archaeognatha)–Pennsylvanian? Permian? Triassic–Holocene
 Order Thysanura, silverfish, etc.–Oligocene–Holocene
Subclass PTERYGOTA
Infraclass PALEOPTERA
 Order Ephemeroptera, mayflies (Fig. 3A)–Pennsylvanian–Holocene
 Order Protodonata*–Pennsylvanian–Triassic
 Order Odonata, damselflies, dragonflies (Fig. 2B)–Permian–Holocene
 Order Palaeodictyoptera* (Figs. 3B, 4)–Pennsylvanian–Permian
 Order Megasecoptera* (Fig. 5)–Pennsylvanian–Permian
 Order Diaphanopterodea*–Pennsylvanian–Permian
Infraclass NEOPTERA
 Division Exopterygota
 Order Plecoptera (= Perlaria), stoneflies–Permian–Holocene
 Order Embioptera–Eocene–Holocene
 Order Miomoptera* (Fig. 6C)–Pennsylvanian–Permian
 Order Dermaptera, earwigs–Jurassic–Holocene
 Order Protelytroptera* (Fig. 6A)–Permian only
 Order Caloneurodea* (Fig. 6B)–Pennsylvanian–Permian
 Order Grylloblattodea
 Order Phasmatodea, walking-sticks and leaves–Triassic–Holocene
 Order Orthoptera, grasshoppers, katydids, crickets, etc. (Fig. 1)–Pennsylvanian–Holocene
 Order Protorthoptera* (Fig. 6D)–Pennsylvanian–Permian, Triassic?
 Order Blattodea, roaches–Pennsylvanian–Holocene
 Order Isoptera, termites–Cretaceous–Holocene
 Order Mantodea, preying mantids–Oligocene–Holocene
 Order Zoraptera
 Order Psocoptera (= Corrodentia), bark lice, book lice–Permian–Holocene
 Order Phthiraptera (including Anoplura, Mallophaga), parasitic lice
 Order Thysanoptera, thrips–Jurassic–Holocene
 Order Hemiptera (including Homoptera), leaf-hoppers, aphids, cicadas, scale insects, bugs, etc. (Figs. 2A, 7)–Permian–Holocene
 Division Endopterygota
 Order Megaloptera, dobsonflies, alderflies–Permian–Holocene
 Order Raphidiodea, snakeflies–Pennsylvanian? Jurassic–Holocene
 Order Coleoptera, beetles (Fig. 8B)–Permian–Holocene
 Order Strepsiptera–Oligocene–Holocene
 Order Neuroptera, lacewings–Permian–Holocene
 Order Glosselytrodea* (Fig. 8A)–Permian–Jurassic
 Order Hymenoptera, sawflies, wasps, ants, bees (Fig. 2C)–Triassic–Holocene
 Order Mecoptera, scorpionflies, hangingflies–Permian–Holocene
 Order Diptera, flies, mosquitoes, midges (Fig. 8C)–Permian? Triassic–Holocene
 Order Siphonaptera, fleas–Cretaceous–Holocene
 Order Trichoptera, caddisflies–Jurassic–Holocene
 Order Lepidoptera, moths, butterflies–Eocene–Holocene

single, abundant imaginal type in the same suprageneric category.

Classification

In traditional treatments of insects, the class was considered to comprise all arthropod orders in which the adults have only three pairs of legs. Many studies in recent decades have shown that three or four of the groups of primitively wingless hexapods differ from the main array of insects in major aspects of body organization and of embryonic development and should be given separate class status; these groups are omitted in this treatment. Table 1 gives the earliest occurrence in the fossil record, but this is subject to revision whenever new deposits are studied, especially in the Mesozoic. Extinct groups are marked by an asterisk. There is no significant record for the Grylloblattodea, Zoraptera, or Phthiraptera.

Subphylum UNIRAMIA (= TRACHEATA): A single pair of antennae present (secondarily lost in Protura and some larvae); antennae and segmental appendages including mouthparts having a single main ramus (axis); first tagma enclosed in a solid head capsule, fusing the preoral region and mouthpart segments; respiration by paired tracheal trunks (openings fused middorsally in scutigeromorph Chilopoda), but tracheae absent in minute types and some endoparasites and some aquatic juveniles.

Class INSECTA (true insects): Adults hexapodous (a few have lost legs); juveniles hexapodous, except many larvae of Endopterygota; typically with three distinct tagmata—head, thorax, abdomen; abdomen with cerci, or claspers, or ovipositor, or lacking prominent appendages; second maxillae fused as labium; mouthparts never entognathous in adults but in some groups highly modified as a sheathed beak; eyes usually multifaceted, absent in a few whole groups, simple in juveniles of Endopterygota; gonopore (very rarely a pair) posterior; size, hardness, pigmentation, body form, and food diverse.

Subclass APTERYGOTA: Wings never present and presumed to have been absent in all ancestors; abdomen terminated by prominent cerci and median filament; molting continues throughout adult life; juveniles much like adults except for size and ovipositor; no copulation. (See Table 1.)

Subclass PTERYGOTA: Adults usually winged, but wings secondarily lost in many groups; molting terminates when producing the sexually mature instar, which is the only flying instar except in the Ephemeroptera.

Infraclass PALEOPTERA: Wing articulation allowing up and down strokes but not pivoting and folding of wings along abdomen at rest—except possibly as a special development in certain Paleozoic Diaphanopterodea. (See Table 1; Figs. 2B, 3, 4, 5.)

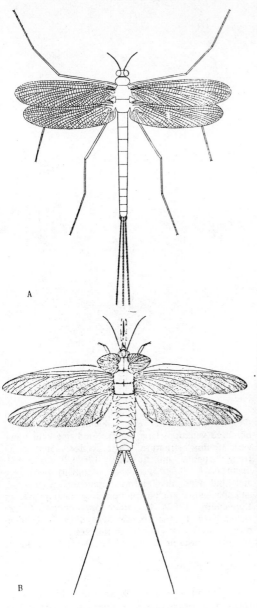

A

B

FIGURE 3. A: Order Ephemeroptera, *Protereisma* (a Permian mayfly)—differs from all living ephemerids in having hindwings similar in size to forewings (reconstruction from Carpenter, 1933). B: Order Palaeodictyoptera*, *Stenodicyta* (Upper Carboniferous)—note especially the beak, wing-like pronotal lobes, broad costal space, broad wing bases, and dense venational network between main veins (reconstruction, from Kukalová, 1970).

FIGURE 4. Order Palaeodictyoptera*, *Homaloneura* (Pennsylvanian)—photograph of wings showing remarkable preservation of color pattern (from Carpenter, 1971).

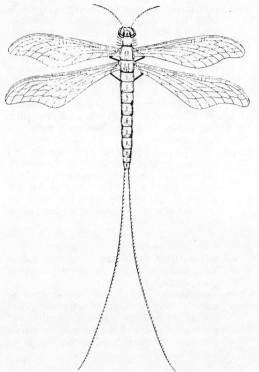

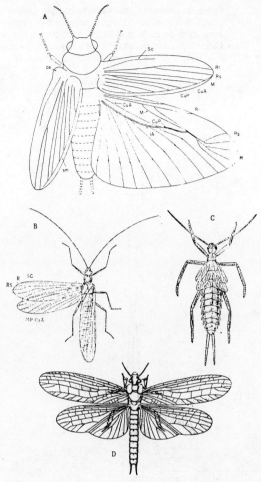

FIGURE 5. Order Megasecoptera*, *Mischoptera* (Upper Carboniferous)—note especially the narrowed wing bases and very narrow costal space (the long beak is not shown here); the wings could not be pivoted to rest along the body, unlike the related Diaphanopterodea (reconstruction, from Laurentiaux, 1953).

FIGURE 6. A: Order Protelytroptera*, *Protelytron* (Permian)—note thickened forewing with simplified obscured venation, large fanlike hindwing with transverse folding structure, short cerci (reconstruction from Carpenter and Kukalová, 1965) B: Order Caloneurodea*, *Paleuthygramma* (Permian)—note fore and hind wings alike and membranous, legs cursorial, antennae long (reconstruction from Martynova, *in* Rohdendorf 1962). C: Order Miomoptera*, *Delopterum* larva (Permian) (from Martynova, *in* Rohdendorf, 1962). D: Order Protorthoptera*, *Probnis* (Permian)—note prothoracic lobes (from Laurentiaux, 1953).

Infraclass NEOPTERA: Wing articulation allowing pivoting and folding of wings along abdomen at rest (secondarily lost in a few broad-winged forms such as butterflies).

Division EXOPTERYGOTA: Wing pads external on larvae; larval tissues mainly develop directly into imaginal tissues; no pupal stage (partial analog in Thysanoptera and certain Hemiptera); larval eyes compound. (See Table 1; Figs. 2A, 6, 7.)

Division ENDOPTERYGOTA. Wing pads internal in larvae. At metamorphosis, larval tissues mainly breaking down and being replaced by imaginal tissues. Non-feeding pupal stage preceding imaginal stage. Larval eyes not compound, except possibly in Mecoptera. (See Table 1; Figs. 2C, 8.)

Ecology

In addition to being the most numerous of all animals and the most speciated, the Insecta have produced dominant specialists in all the available nonmarine environments of the world. This has probably been true since Pennsylvanian time, possibly earlier. As animals, they depend ultimately on plants as primary food producers at the base of the food web. A very large number of insect species, concentrated in nine major living orders, feed on the foliage of green plants; these, of course, are highly deleterious to plants and have stimulated the evolution of chemical and physical mechanisms of insect resistance in virtually all green plants. A similarly large number, especially in six major ex-

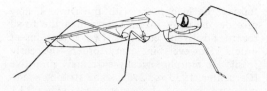

FIGURE 7. Order Hemiptera, a living reduviid bug—note sucking beak, held ventrally (from Mackerras, 1970, used by permission of Melbourne University Press).

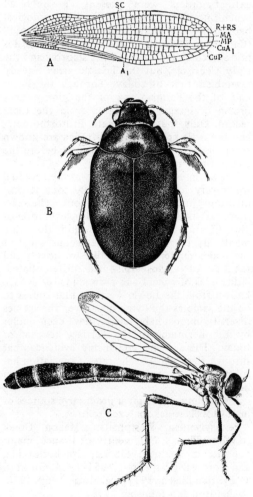

FIGURE 8. A: Order Glosselytrodea*, *Jurina* (Permian)—note greatly outbowed anterior lobe, multiple crossveins, submarginal composite vein around wing (from Martynova, *in* Rohdendorf, 1962). B: Order Coleoptera, a living hygrobiid water beetle—note swimming structure of all legs, forewings (elytra) meeting without overlap (from Mackerras, 1970). C: Order Diptera, a living robberfly—note hindwings absent except as small stalked knob, the haltere (from Mackerras, 1970). (B and C by permission of Melbourne University Press.)

tant orders, are intensely predatory; and three entire lesser orders are exclusively parasitic on vertebrates or insects, as are the larvae of several families of Hymenoptera and Diptera. With their digestive microflora, some insects feed on wood; many others ingest dead wood but actually digest the fungi and bacteria attacking the wood. And adults of thousands of species of Hymenoptera, Diptera, Lepidoptera, and Coleoptera are specialized as flower visitors feeding on nectar or pollen; most of these are highly beneficial as pollinators and have caused the evolution of production of nectar itself and of showy flowers and powerful perfumes as attractants.

Most of this range of feeding styles can be found in tropical and temperate and subpolar regions; in forests, scrub, grasslands, and deserts; in high mountains, hills, and flatlands; in continental heartlands, shores, and islands. Only in the otherwise insect-rich streams, lakes, ponds, and pitcher-plant water are pollinators, green-leaf feeders, wood ingesters, and even parasites infrequent. And green-plant feeders are of course missing in the faunas of soil and caves and understones and dead bark. Surprisingly, this ecologically most successful group of the animal kingdom is almost totally absent from the vast, rich seas of the world.

Morphological adaptations for feeding on particular foods are mainly in the mouthparts, which are usually too poorly preserved in fossils for analysis. But adaptations to broad environments tend to be more conspicuous and fossilizable, so the paleoentomologist should sometimes be able to recognize aquatic habits (gills, swimming appendages; see Figs. 2B, 8B), desert adaptations (dense hair, long legs, rough integument), cavernicoly (eyelessness, long appendages, and setae), soil specializations (such as eyelessness, short and powerful legs and segments), and origins in various other niches. The workings of some insects are also commonly fossilized and often group specific, most notably larval mines and galls and feeding damage on fossil leaves and stems, galleries in wood and bark, and nests. Ecological inferences are usually possible from a study of locomotory appendages and even the form of eyes and other large sense organs. In some insect fossils, pigmentation patterns are clearly preserved (Fig. 8C) and permit reasonable speculation on polymorphism, mimicry, sexual signals, and thermoregulation (see Carpenter, 1971).

Geologic History

Myriapods apparently near Diplopoda (Archypolypoda) are reported from the Upper Silurian and Lower Devonian of Scotland, but the oldest certain Diplopoda are from the

Pennsylvanian of Europe and North America. Surprisingly, the oldest Chilopoda fossils are from the Cretaceous of Arizona. Nearer the insects is a collembolan, *Rhyniella praecursor,* from the Devonian Rhynie Chert of Scotland. The earliest definite true Insecta are from near the base of the Pennsylvanian of central Europe, Pennsylvania, and apparently Argentina.

As usual with invertebrate taxa, the oldest insect fossils are not old enough; they represent well-differentiated members (Protodonata and Protorthoptera) of their major groups, and most of the origin and early evolution of supra-ordinal groups can only be inferred from the characters of these and living forms. There are no Apterygota from the Paleozoic other than the problematical Permian *Dasyleptus* (Order Monura). The numerous middle Pennsylvanian Pterygota include both paleopterous and exopterygote neopterous orders. Three of the ten orders represented in the Pennsylvanian are still surviving (Ephemeroptera, Blattodea, Orthoptera); they are structurally generalized. The Paleodictyoptera (Fig. 3B) show wing venation possibly of the early ancestral type, but the highly specialized sucking beak rules out this order as being near the origin of winged insects. In another very early fossil group are the giant dragonflies (Protodonata) of the late Upper Carboniferous Commentry deposit of France; but they could not have been like direct ancestors of the Neoptera, because they had already lost the MP and CuA primary veins (Fig. 1) and probably had the bizarre odonatan copulatory apparatus and larval prehensile labium. Gigantism, characteristic of a few Upper Carboniferous and Lower Permian species of Protodonata, is remarkable; *Megatypus,* from the Kansas and Oklahoma Wellington shales and from Europe, had a 26 to 30 in. wing spread (Carpenter, 1947: p.74). It is reasonably suggested that their aerial hunting niche was only available to huge insects prior to the origin of flying vertebrates, just as gigantism and perhaps surface dwelling in the earliest myriapods may have disappeared when similarly large Amphibia evolved in the late Paleozoic.

So far, paleoentomologists have not found fossils that solve these burning questions of insect phylogeny: (1) ancestry of the Thysanura; (2) ancestry of the Pterygota; (3) ancestry of the Neoptera; (4) origin of the hemipteroid line and affinities of the Zoraptera; (5) ancestry of the Coleoptera and of the Hymenoptera; and, above all, (6) which, if any, of the Exopterygota are near the ancestor of the Endopterygota. *Permotipula,* once considered "a cranefly with four wings," and *Choristotanyderus,* now so considered (Riek, 1970, p. 185), and therefore linking the Mecoptera with the Diptera, may provide the only instance in which the fossil record positively connects two extant insect orders. However, in many orders the early fossils are morphologically extremely primitive (Carpenter, 1953, p. 215); some striking examples are in the Ephemeroptera (Fig. 3A), Thysanura, and Strepsiptera.

The representation of insect orders in the late Pennsylvanian and Early Permian includes all of the major phyletic lines of the winged insects. During the Permian, 12 extant orders and all 9 extinct orders were present, possibly all synchronously. "During no other geological period has such a diverse insect fauna existed" (Carpenter, 1953). The Mesozoic fossil record is not as rich as that of the late Paleozoic and the early Cenozoic, but the Triassic array already resembled that of today. Through the latter half of the Permian, six of the nine extinct orders of insects disappeared from the fossil record. Only the Protodonata, Protorthoptera, and the little-known Glosselytrodea are known from the Mesozoic; all disappeared before the end of the Jurassic.

Known insect diversification was then gradual for a very long time and finally took its last quantum jump in the Cretaceous. The new burst was tied to the reciprocating evolution of diversification of the insects and flowering plants. By the end of the Cretaceous, most of the major modern superfamilies of insects and families of angiosperms had differentiated. Multitudes of genera and even some species are known from the Lower Tertiary that appear to be the same as the living taxa. Even the species diversity in most habitats may have been similar by the beginning of the Tertiary to that of today. There is no convincing evidence that through the Cenozoic the rate of speciation has substantially exceeded the rate of extinction in insects.

A sampling of the most productive sources of insect fossils so far analyzed follows.

Pennsylvanian. Westphalian Mazon Creek (q.v.) of Illinois; Commentry of France; major deposits in southern Siberia, Czechoslovakia, and Portugal (Carpenter, 1951–1964, Guthörl, 1934; Handlirsch, 1911; Kukalová, 1969–1970; Richardson and Johnson, 1971).

Permian. Wellington shales of Kansas and Oklahoma; New Castle Coal Measures of New South Wales and a synchronous deposit in South Africa; Czechoslovakia; many deposits in European and southern Siberian USSR (Carpenter, 1930–1950; Kukalová, 1963; Martynova, 1958; Rohdendorf et al., 1961; Tillyard, 1922–1937; Tillyard, 1926, 1935).

Triassic. Hawkesbury Sandstone of New South Wales; Ipswich Series of Queensland; several deposits in USSR in the southern Urals, central Asia, and Siberia (Martynova, 1958; Tillyard, 1917–1923).

Jurassic. Solnhofen lithographic limestone (see *Solnhofen Formation*) of Germany (Carpenter, 1932); several deposits in USSR, especially in Kazakhstan and Fergana (Rohdendorf et al., 1961.)

Cretaceous. Canadian (Carpenter et al., 1937; Wilson et al., 1967) and Lebanese amber; rock fossils very few and poor (Carpenter, 1953, p. 268).

Eocene. Green River shales of Colorado and Wyoming (Cockerell, 1920, 1921, 1925).

Oligocene. Baltic amber of northern Europe (Ander, 1942).

Miocene. Florissant shales of Colorado (Scudder, 1892); Chiapas amber of Mexico (Petrunkevitch et al., 1963, 1971).

The reference list includes some major publications on many of these deposits. Most of the principle workers in 20th century paleoentomology are also indicated by that bibliography. Essentially modern species of insects, so far mainly beetle elytra, are now being extensively used for Quaternary paleoecology by research groups in Great Britain and Canada (see Coope, 1970).

CHARLES R. REMINGTON

References

Ander, K., 1942, Die Insektenfauna des Baltischen Bernstein's nebst damit verknüpft zoogeographischen Problemen, *Kung. Fysiogr. Sällskapets Handl.*, n.f., 53, 1–82.

Carpenter, F. M., 1930–1950. The Lower Permian insects of Kansas. Parts 1–10. *Bull. Mus. Comp. Zool.*, 70(1930), 69–101; *Psyche*, 37(1930), 343–374; *Amer. J. Sci.*, ser. 5, 21(1931), 97–139, 22(1931), 113–130, 24(1932), 1–22; *Proc. Amer. Acad. Arts Sci.*, 68(1933), 411–503, 70(1935), 101–146, 73(1939), 29–70, 75(1943), 55–84, 78(1950), 185–219.

Carpenter, F. M., 1932. Jurassic insects from Solenhofen in the Carnegie Museum and the Museum of Comparative Zoology, *Ann. Carnegie Mus.*, 21, 97–129.

Carpenter, F. M., 1947. Early insect life, *Psyche*, 54, 65–85.

Carpenter, F. M., 1943–1964. Studies on Carboniferous insects from Commentry, France, *Geol. Soc. Amer. Bull.*, 4(1943), 527–554; *J. Paleontology*, 25 (1951), 336–355; *Psyche*, 68(1961), 145–153, 70(1963), 120–128 and 240–256, 71(1964), 104–116.

Carpenter, F. M., 1953. The geological history and evolution of insects, *Amer. Scientist*, 41, 256–270.

Carpenter, F. M., 1954. Extinct families of insects, in C. T. Brues, A. L. Melander, and F. M. Carpenter, *Classification of Insects*, rev. ed., *Bull. Mus. Comp. Zool.*, 108, 777–827.

Carpenter, F. M., 1971. Adaptations among Paleozoic insects, *Proc. N. Amer. Paleontological Conv.*, pt. I, 1236–1251.

Carpenter, F. M., and Kukalová, J., 1965. The structure of the Protelytroptera, with description of a new genus from Permian strata of Moravia. *Psyche*, 71, 183–197.

Carpenter, F. M., et al., 1937. Insects and arachnids from Canadian amber, *Toronto Univ. Geol. Ser.*, No. 40: 7–62.

Cockerell, T. D. A., 1920, 1921, 1925. Eocene insects from the Rocky Mountains [title varies], *Proc. U.S. Nat. Mus.*, 57, 233–260; 59, 20–39, 455–457; 66(2556), 13p.

Coope, G. R., 1970. Interpretations of Quaternary insect fossils. *Ann. Rev. Entomol.*, 15, 97–120.

Guthörl, P., 1934. Die Arthropoden aus dem Carbon und Perm des Saar-Nahe-Pfalz-Gebietes, *Abh. Preuss. Geol. Landesamtes*, 164, 1–219.

Handlirsch, A., 1906–1908. *Die fossilen Insekten und die Phylogenie der rezenten Formen*. Leipzig, 1430p.

Handlirsch, A., 1911. New Palaeozoic insects from the vicinity of Mazon Creek, Illinois, *Amer. J. Sci.*, ser. 4, 31, 297–326, 353–377.

Handlirsch, A., 1937, 1939. Neue Untersuchungen über die fossilen Insekten. I, II. *Ann. naturh. Mus. Wien*, 48, 1–140; 49, 1–240.

Imms, A. D., 1957. *A General Textbook of Entomology*, 9th ed., revised by O. W. Richards and R. G. Davies. New York: Dutton, 886p.

Kukalová, J., 1963. Permian insects of Moravia. Part 1 —Miomoptera, *Sbornik Geol. Ved. Paleontol.*, 1, 7–52.

Kukalová, J., 1969–1970. Revisional study of the order Paleodictyoptera in the Upper Carboniferous shales of Commentry, France, *Psyche*, 76(1969), 163–215 and 439–486; 77(1970), 1–44.

Laurentiaux, D., 1953. Classe des insectes, in J. Piveteau, ed., *Traité de Paléontologie*, vol. 3. Paris: Masson et Cie, 397–527.

Mackerras, I. M., ed., 1970. *Insects of Australia*. Melbourne: Melbourne Univ. Press. 1029p.

Martynov, A. V., 1938. Etudes sur l'histoire géologique et de phylogenie des ordres des insectes (Pterygota), *Trav. Inst. Paleontol. Acad. Sci. U.R.S.S.*, 7(4), 1–150.

Martynova, O., 1958. Novye nasekomye iz permskikh i mesosoiskikh otlozhenii SSSR. *Materialy k osn. paleontol.*, 2, 69–94.

Petrunkevitch, A., et al., 1963, 1971. Studies of fossiliferous amber arthropods of Chiapas, Mexico. I, II, *Univ. Calif. Publ. Entomol.*, 31,1–60; 63, 106p (and see P. D. Hurd et al., 1962, *Ciencia* (Mexico City), 21, 107–118).

Richardson, E. S., Jr., and Johnson, R. G., 1971. The Mazon Creek faunas. *Proc. N. Amer. Paleontological Conv.*, pt. I, 1222–1235.

Riek, E. F., 1970. Fossil history, in I. M. Mackerras, 1970, 168–186; *Suppl.*, 1974, 28–29.

Rohdendorf, B. B., ed., 1962. *Osnovy Paleontologii. Arthropoda, Tracheata and Chelicerata*. Moscow: Akad. Nauk SSSR, 560p.

Rohdendorf, B. B.; Becker-Migdisova, E. E.; Martynova, O. M.; and Sharov, A. G., 1961. Paleozoic insects of the Kuznetsk Basin. *Paleont. Inst. Akad. USSR Trudy*, 85, 705p.

Scudder, S. H., 1892. The Tertiary insects of North America. *U.S. Geol. Surv. Bull.*, 13, 1–734.

Tillyard, R. J., 1917–1923. Mesozoic insects of Queensland. Nos. 1–10, *Proc. Linn. Soc. N.S. Wales,* **42,** 175–200, 676–692; **43,** 417–436, 568–592; **44,** 194–212, 358–382, 857–896; **46,** 270–284; **47,** 447–470; **48,** 481–498 (summary).

Tillyard, R. J., 1922–1937. Kansas Permian insects. Parts 1–20, *Amer. J. Sci.,* ser. 5, 7–34.

Tillyard, R. J., 1926, 1935. Upper Permian insects of New South Wales. I–V, *Proc. Linn. Soc. N.S. Wales,* **51,** 1–30, 265–279; **60,** 265–279, 374–384; 385–391.

Wilson, E. O., Carpenter, F. M., and Brown, W. L., Jr., 1967. The first Mesozoic ants, with the description of a new subfamily, *Psyche,* **74,** 1–19.

Cross-references: *Arthropoda; Chitin; Mazon Creek; Paleoecology, Terrestrial; Silicification of Fossils; Solnhofn Formation.*

INVERTEBRATE PALEONTOLOGY

Invertebrate Paleontology is the study of invertebrate animals of the geologic past. Invertebrates constitute about 95% of all living animal species; and, although they share the common negative characteristic of the lack of a backbone or vertebral column (cf. *Chordata*), they show very great variation in other respects. Living invertebrates include such widely different forms as the sponges, corals, clams, snails, squids, crabs, lobsters, insects, starfish, and sea urchins. They also live in a great variety of different environments, ranging from the ocean depths to mountainous areas and from the polar regions to equatorial jungles. They show great diversity in ways of locomotion and food gathering; some are parasitic. About three-quarters of the total of about 1,120,000 species of living invertebrates are insects. Invertebrates play a major role in the economy of nature: some are prolific food sources for other animals (including man), others are pests, and many are important in plant pollination.

Invertebrates are of great importance as fossils. It was on the basis of invertebrate fossils that much of the geologic time scale was established, and they still form by far the most widely used method of rock correlation. Many microfossils are invertebrates, and these are of particular value in subsurface studies based on cores from boreholes. The chief reason for the particular value in such studies, lies in the fact that invertebrate fossils have a longer fossil history than other groups, and are especially abundant and widespread. Invertebrate fossils of various kinds are common in most rocks deposited during the last 600 m yr. Furthermore, many invertebrates were and are marine dwellers, and marine rocks are not only very common in the geologic column, but are often deposited under more favorable conditions for fossilization than strata deposited under many other conditions. Many invertebrate groups are represented by very large numbers of different species, and these in turn, are represented by enormous numbers of individuals. Invertebrate fossils therefore have a unique value in stratigraphy and historical geology. (See Fig. 1.)

Geologic History of Invertebrates

Precambrian. Fossil remains are rare in rocks of Precambrian age. Few invertebrates of this age have been described and numbers of those that have are not accepted by other paleontologists. The only generally accepted records of Precambrian invertebrates are the tracks and borings of otherwise unknown organisms from the Beltian of Montana, and a remarkable assemblage of soft-bodied invertebrates has been described by M. F. Glaessner, from the Pound Quartzite (late Precambrian) of Edicara, South Australia (see *Ediacara Fauna*). This fauna includes the remains of medusoid coelenterates, segmented worms, and forms of unknown affinities, as well as frond-like structures (*Charnia,* variously interpreted as pennatulate coelenterates and algae) that are also known from the Precambrian of England, South Africa, and other areas.

In addition to these metazoan invertebrates, Precambrian rocks have recently yielded well-preserved unicellular organisms. Bacteria and blue-green algae are found in South African rocks that are about 3.2 billion years old. Algae, fungi, and other unicellular forms are also known in Precambrian rocks of varying ages from North America, Australia, and the USSR. The oldest nucleated cells appeared about 1.5 million years ago, but there are indications of the presence of photosynthesis in bacteria some 3 billion years ago.

Cambrian. In contrast to those of the Precambrian, Cambrian faunas are represented by a large number of different species, representing a number of distinct groups. In the rocks of the Lower Cambrian, for example, there are over 500 species, representing 7 major phyla (Protista, Porifera, Coelenterata, Brachiopoda, Mollusca, Echinodermata, Arthropoda, and the problematical Archeocyatha). Later Cambrian faunas are dominated by the extinct trilobites, which achieved world-wide distribution and are of great value as index fossils (see *Biostratigraphy*). Even at this early period of history, they show differentiation into major zoogeographic provinces. Inarticulate brachiopods and archeocyathids are also abundant in some parts of the Cambrian.

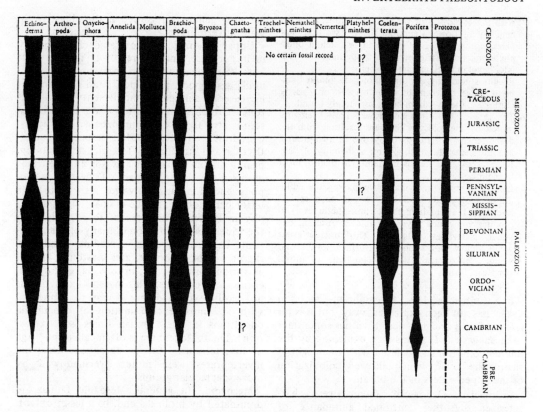

FIGURE 1. Geological range of the principal invertebrate phyla (from Shrock and Twenhofel, 1953; copyright © 1953 by McGraw-Hill Book Co., and used with permission). Width of range band roughly indicates abundance of representatives within the phylum; it is not comparable from phylum to phylum.

Comparison of Precambrian and Cambrian Faunas. The striking contrast between the general absence of Precambrian faunas and those of the Cambrian is not easily explained. It has been suggested that this change represents the initial creation of life, but the known Precambrian organisms (including a larger number of plants) and our knowledge of post-Cambrian biological processes lends no support to this. It has also been suggested that some change in the chemistry of the oceans may have resulted in the sudden acquisition of hard parts by various animal groups. There seems little evidence for this suggestion, for other groups, such as corals and bryozoans, acquired hard parts with equal "suddenness" at later periods of time and it seems extravagant to suggest successive changes of this type.

The most acceptable explanation of the "sudden" appearance of Cambrian faunas appears to be that Early Cambrian times saw the acquisition of hard parts by various groups of previously soft-bodied organisms. An indication that a large number of soft-bodied Precambrian organisms may have existed without leaving any

fossil record is provided by the Middle Cambrian Burgess Shale (q.v.), known near Field, British Columbia. In this deposit, the remains of some 130 species of animals are preserved, the great majority of them being soft-bodied forms, including jellyfish, arthropods, and worms. Not one of these has ever been found in any other locality, in spite of the fact that their abundance in the Burgess Shale clearly shows them to have been common members of Cambrian faunas. Several different lines of evidence offer support for this view.

1. The Cambrian is a very long period of time, about 90 m yr: this is half as long again as the whole Cenozoic Era, during which time enormous changes took place in the mammals. Even the Lower Cambrian represents a time span of about 30 m yr, half as long as the Cenozoic.
2. Not all the 500 species of Lower Cambrian fossils appear at the earliest moment of Early Cambrian time; they straggle in over a period of 30 m yr.
3. If this is accepted, the rate of such evolution is rapid, but not of a different order of magnitude from rates of evolution known elsewhere in the fossil record.

4. With the exception of the archeocyathids, the great majority of Lower Cambrian fossils have thin hard parts, composed of either chitin or calcium phosphate and this is true even for those organisms that later develop thick calcareous shells.
5. Recent studies suggest that among the oldest Cambrian fossils, the olenellid trilobites were only partly chitinized, the pygidium being very rarely found.
6. The rate of evolution could well have been promoted by the selective pressure resulting from the development of hard parts by one or two groups, and the generally much greater availability of "unfilled" ecological niches in Cambrian than in most later times.
7. Early Cambrian times may have marked the build-up of biologically-produced oxygen to a critical level (say 1–3% of its present atmospheric level) to produce an ozone layer capable of filtering out lethal ultraviolet radiation. This unique event would allow migration into surface waters and would "encourage" the development of novel structural forms.

In summary, the Precambrian–Cambrian "gap" has not been explained away, but it is far less profound and mystical than has sometimes been suggested. It seems best explained by the unique conditions associated with the origin of the ozone layer, which allowed colonization of the upper waters of the ocean.

Ordovician. Ordovician faunas were characterized by the continued abundance of trilobites; although most Cambrian genera were replaced by new forms, the great increase in abundance and diversity of articulate brachiopods, and by the appearance and great development of the graptolites (though some paleontologists, who regard the Tremadocian as Upper Cambrian, place their appearance in that system). All three of these fossil groups are of great value in stratigraphy. The conodonts (q.v.), of problematical affinities, first appeared in the Cambrian, but became increasingly widespread and diverse in the Ordovician. Anthozoans (see *Corals*) appeared for the first time, being represented by Rugosa and Tabulata, as did the Bryozoa, Pelecypoda, Blastoidea, Crinoidea, and Echinoidea. Few of these groups were abundant, although the ostracods and nautiloid cephalopods, both of which first appeared in the Cambrian, are locally abundant. The first vertebrates appeared in the Ordovician, and also the oldest land plants.

Silurian. The faunas of the Silurian are marked by a relative decline of the trilobites, and an increasing abundance of anthozoans, brachiopods and bryozoa. Graptolites were represented chiefly by the monograptids and crinoids; mollusks and conodonts are locally important. Primitive scorpion-like fossils, once thought to be the oldest air-breathing animals,

are now generally regarded as the oldest freshwater invertebrates.

Upper Paleozoic. Trilobites declined greatly in importance, and are relatively rare in most Late Paleozoic faunas; but anthozoans, brachiopods, gastropods, pelecypods, and cephalopods increased in abundance and diversity during this time. Crinoids and blastoids are common in some strata, as are bryozoans, ostracods, conodonts, and foraminifera (especially endothyrids and fusulinids). The oldest insects appeared in the Devonian and are an important group in the Upper Paleozoic. Terrestrial gastropods and nonmarine bivalves also occur, the latter being of great value in coal-field stratigraphy in Europe. The close of the Permian saw the widespread extinction of many of the dominant Paleozoic groups, including fusulinids, rugose corals, trilobites, and blastoids, as well as various brachiopods, bryozoans, gastropods, corals, and crinoids. This massive extinction is not easily explained, but it may be connected with an unusual combination of geographic conditions in late Permian times (see *Extinction*). The Paleozoic era is often called the "Age of Invertebrates," for in subsequent times the invertebrates were rather overshadowed by more spectacular groups.

Mesozoic. In general, Mesozoic faunas are dominated by mollusks and, to a lesser extent, by corals, foraminifera, and echinoderms. Brachiopods declined greatly in numbers and are commonly represented only by the terebratulids and rhynchonellids. Hexacorals and various new groups of foraminifera, bryozoans, crinoids, and echinoids made their appearance. The bivalves and gastropods were represented by numbers of new families, but Mesozoic marine faunas are dominated by the diversity and abundance of the ammonites and belemnites. They are found throughout the world in strata representing various environments, and they are of great value in stratigraphy. Both the ammonites and true belemnites became extinct in late Cretaceous times.

Cenozoic. An abundance of foraminifera, porifera, corals, bryozoa, crustaceans, ostracodes, insects, and especially gastropods, bivalves, and echinoids characterizes the Cenozoic fauna. Freshwater bivalves are abundant in many fluvial strata, and terrestrial gastropods are found in wind-blown loess deposits of the Pleistocene.

F. H. T. RHODES

References

Ager, D. V., 1963. *Principles of Paleoecology.* New York: McGraw-Hill, 371p.

Beerbower, J. R., 1968. *Search for the Past,* 2nd ed. Englewood Cliffs, N.J.: Prentice-Hall, 512p.

Black, R. M., 1970 *The Elements of Palaeontology.* Cambridge: Cambridge Univ. Press, 339p.

Brouwer, A., 1967. *General Palaeontology.* Chicago: Univ. Chicago Press, 216p.

Easton, W. H., 1960. *Invertebrate Paleontology.* New York: Harper & Row, 701p.

Kummell, B., 1970. *History of the Earth,* 2nd ed. San Francisco: Freeman, 707p.

McAlester, A. L., 1968. *The History of Life.* Englewood Cliffs, N.J.: Prentice-Hall, 151p.

Moore, R. C.; Lalicker, C. G.; and Fischer, A. G., 1952. *Invertebrate Fossils.* New York: McGraw-Hill, 766p.

Raup, D. M., and Stanley, S. M., 1971. *Principles of Paleontology.* San Francisco: Freeman, 388p.

Rhodes, R. H. T., 1976. *Evolution of Life,* 2nd ed. Harmondsworth, Middlesex: Penguin, 330p.

Shrock, R. R., and Twenhofel, W. H., 1953. *Principles of Invertebrate Paleontology,* 2nd ed. New York: McGraw-Hill, 816p.

Tasch, P., 1973. *Paleobiology of the Invertebrates.* New York: Wiley, 946p.

Cross-references: *Annelida; Archaeocyatha; Arthropoda; Biomineralization; Biostratigraphy; Bivalvia; Brachiopoda; Bryozoa; Burgess Shale; Cephalopoda; Coelenterata; Echinodermata; Ediacara Fauna; Extinction; Fossil Record; Graptolithina; Mazon Creek; Mollusca; Paleoecology; Precambrian Life; Solnhofen Formation.*

L

LARVAE OF MARINE INVERTEBRATES: PALEONTOLOGICAL SIGNIFICANCE

Marine benthic invertebrates vary tremendously in the morphology and duration of their larval stages. Thorson (1946, 1950) however, found that most marine bottom invertebrates can be characterized by three main types:

1. *Pelagic development.* Species with larvae that spend some time floating freely or swimming in the plankton. There are two types of pelagic larvae:
   a. *Planktotrophic larvae* feed on plankton (mainly phytoplankton) during their pelagic life and spend anywhere from a few hours to a year as part of the plankton.
   b. *Lecithotrophic larvae* feed on yolk supplies during their pelagic life with little or no dependence on the plankton for food. This type generally has a fairly short pelagic life (a few hours or days) although some may remain in the plankton for up to two months (Thorson, 1961).
2. *Direct development.* The entire development of the embryo takes place through yolk feeding in an egg capsule or egg mass deposited directly on the substrate with no pelagic stage.
3. *Viviparous development.* The developmental process occurs entirely inside the parent with no pelagic stage.

These divisions are somewhat arbitrary in that some species undergo more than one type of development, e.g., brood protection inside the parent until an advanced stage of free swimming larvae is released. Most species, however, fit fairly well into one category or another. Mileikovsky (1971), in a general review of marine invertebrate larval ecology, suggested a fourth main larval type:

4. *Demersal development.* In which the larvae undergoes a swimming and/or crawling development in a thin water layer just above the ocean bottom.

It was soon recognized by Thorson and other workers that species with a pelagic larval stage were clearly in the majority among the benthos of temperate and tropical continental shelf seas. Further work revealed the existence of a gradient in the percentage of benthic species having pelagic larval stages with latitude and water depth. In tropical areas, over 80% of benthic invertebrate species had planktotrophic larvae; while in the temperate regions, the figure was found to be closer to 60%, and in the Arctic there were very few species with pelagic larvae. Although Thorson (1950) postulated that a situation similar to the Arctic would be found in the deep sea, with most species having non-pelagic larvae, more recent work (Knudsen, 1970; Ockelmann, 1965; and Scheltema, 1972) suggest that lecithotrophic larvae may be the dominant form of development among the benthos of the deep ocean.

The dominance of species with some form of pelagic larvae can be expected to have important influences upon the rate and extent of dispersal of marine invertebrates and thus their ultimate biogeographical distribution. Thorson (1961) collected data on 195 species of marine bottom invertebrates and found that among species with pelagic larvae, most have a larval duration in the plankton of from 2 to 4 weeks (see Fig. 1). Thorson therefore discounted the view that larval dispersal serves as a major transportative mechanism across ocean basins except for minor instances. The larval durations of most of the species he studied were too short to survive the trip. Scheltema, however, (1971a, 1971b, 1977) noted that Thorson's data was taken mostly from temperate areas where the pelagic stages are generally less than two months. He was able to demonstrate that, in the tropics, there exist a considerable number of species that can delay their metamorphosis long enough to survive an Atlantic crossing in a single generation (*Teleplanic* larvae). Such larvae are capable of long-term pelagic existence of from 6 months to a year; and, with current velocities of 0.5–2.0 km/hr in the temperate and tropical North Atlantic (Scheltema, 1971b), a transatlantic crossing is easily possible. Species with pelagic larval durations of only 2 weeks can travel in an oceanic current for 168–672 km, making it possible for that species to widely establish itself around shallow shelf environments.

Long-distance larval dispersal has obvious paleobiogeographical implications. Organisms that are sessile in the adult form (e.g., corals, brachiopods, many bivalves, and other impor-

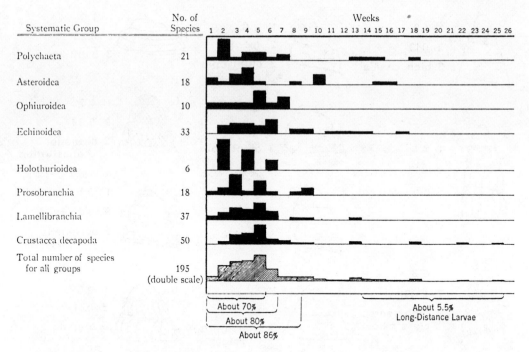

FIGURE 1. Length of pelagic larval life, in weeks, for 195 species of marine bottom invertebrates (from Thorson, 1961; copyright ©1961 by the American Association for the Advancement of Science).

tant components of the fossil record) could still attain very wide geographic ranges via pelagic larvae. Warmer temperatures, higher sea-level stands, and narrower ocean basins than at present have been common in the Phanerozoic, resulting in broader climatic belts, greater shelf areas, and smaller transoceanic distances. These conditions would favor wider dispersal in two ways: (1) benthos having short-term pelagic larvae could become geographically widespread by following continental shelves around major ocean basins; and (2) transoceanic transport could be attained by a greater percentage of pelagic larvae. This combination of warm climate, broad shelf area, and narrow distances between continents would allow geologically "instantaneous" widespread dispersal of many species, whose first appearance in the fossil record would thus be a useful biostratigraphic marker (Kauffman, 1975).

The breakup of continental masses by plate tectonics and the appearance of deep ocean basins would present an obstacle to the dispersal of species with nonplanktotrophic larvae. Species with larvae of different pelagic durations could be expected to traverse such barriers at different frequencies; and only the species with the highest dispersal potential would be able to cross major ocean barriers as land masses became increasingly separated. This would re-

sult in a gradient of increasing endemism of species between continents, with the low-dispersal forms being the first to diverge.

Scheltema (1977) has speculated on the consequences of larval dispersal in relation to biostratigraphy and evolution. He proposes that species with dispersal capacity sufficient to their becoming widely distributed, but not sufficient to maintain gene flow between scattered populations at a level that would prevent evolutionary divergence, would be widespread and rapidly evolving, and thus good potential biostratigraphic indices. Those species with wide, more or less continuous geographic ranges and a great dispersal capacity would have less chance of geographic isolation and speciation, and would be geologically longer lived, thus being less useful as index fossils.

The possibility of determining larval types in fossil mollusks by observing characteristics of their early growth stages (LaBarbera, 1974; Shuto, 1974) has great significance for the paleontologic study of relationships between larval ecology, paleobiogeography, evolution, and stratigraphy. Shuto (1974), using several lines of morphologic criteria, inferred the larval ecologies of a number of Indonesian prosobranch gastropods (see Fig. 2). He found those with planktotrophic development to have generally wider geographic ranges than

417

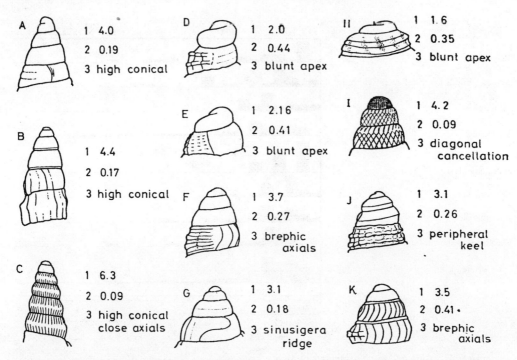

FIGURE 2. Protoconch of fossil and living prosobranch gastropods (from Shuto, 1974). 1, Number of volutions; 2, Ratio of the maximum diameter to number of volutions; 3, Remarks. A, *Tiara gerthi philippiensis* (upper Miocene), inferred planktotrophic; B, *Triplostephanus santosi* (upper Miocene), inferred planktotrophic; C, *Triphora* (*Notosinister*) *conaspera* (Pleistocene), inferred planktotrophic; D, *Ocinebrellus eurypteron* (Pleistocene), inferred lecithotrophic; E, *Merica asperella varicosa* (upper Miocene), inferred lecithotrophic; F, *Latirus* (*Dolicholatirus*) *fusiformis* (Pliocene), inferred lecithotrophic; G, *Zofra pumila* (Pleistocene), inferred planktotrophic; H, *Iwakawatrochus urbanus* (Pleistocene), inferred lecithotrophic; I, *Philbertia linearis* (Holocene), planktotrophic; J, *Nassarius* (*Nassarius*) *verbeeki* (Pliocene), inferred planktotrophic; K, *Gemmula* (*Gemmula*) *speciosa* (Pliocene), inferred planktotrophic.

those with lecithotrophic larvae, deep-water areas being the major barrier to the latter groups. He also found an apparent tendency for planktotrophic species to be geologically longer lived, probably due to greater gene flow between populations (see also Vermeij, 1978).

In an as yet unpublished study, Turner (1973) related bryozoan larval ecology to the distribution of Late Cretaceous Bryozoa in Europe and North America. Bryozoa larvae generally must settle within hours of being released; thus, wide geographic distribution of species implies the existence of a continuum of habitable regions. Similarities between New Jersey and European bryozoan faunas are due to the persistence of a continental shelf still unbroken by plate tectonics. Dissimilarities between New Jersey and Arkansas bryozoan faunas are due to a significant barrier for larval dispersal, a major delta in the upper Mississippi embayment.

Although much of the theory regarding larval ecology and its paleontological significance is

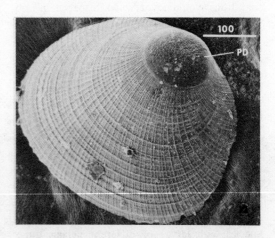

FIGURE 3. Juvenile bivalve *Nucula proxima* from the upper Miocene of Florida, showing larval shell (from LaBarbera, 1974; © 1974 by the Society of Economic Paleontologists and Mineralogists). PD = prodissoconch; scale in microns.

418

still speculative, a thorough knowledge of modern larval biology, its application to fossil counterparts, and work with well-preserved fossil material showing larval characteristics (Fig. 3) can provide many answers in what is becoming a new and exciting branch of paleontological research (e.g., Lutz and Jablonski, 1978; Hansen, 1978, Jablonski, 1978).

THOR HANSEN
DAVID JABLONSKI

References

Hansen, T. A., 1978. Larval dispersal and species longevity in Lower Tertiary gastropods, *Science,* **199,** 885-887.

Jablonski, D., 1978. Late Cretaceous gastropod protoconchs (Abstr.), *Geol Soc. Amer. Abstr.,* **10,** 49.

Kauffman, E. G., 1975. The value of benthic Bivalvia in Cretaceous biostratigraphy of the Western Interior, *Geol. Assoc. Canada, Spec. Paper 13,* 163-194.

Knudsen, J., 1970. The systematics and biology of abyssal and hadal Bivalvia, *Galathea Rept.,* **11,** 1-241.

LaBarbera, M., 1974. Larval and post-larval development of five species of Miocene bivalves (Mollusca), *J. Paleontology,* **48,** 256-277.

Lutz, R. A., and Jablonski, D., 1978. Cretaceous bivalve larvae, *Science,* **199,** 439-440.

Mileikovsky, S. A., 1971. Types of larval development in marine bottom invertebrates, their distribution and ecological significance: A reevaluation, *Marine Biol.,* **10,** 193-213.

Ockelmann, W. K., 1965. Developmental types in marine bivalves and their distribution along the Atlantic coast of Europe, *Proc. 1st European Malacological Congr.,* 25-35.

Scheltema, R. S., 1971a. The dispersal of the larvae of shoal-water benthic invertebrate species over long distances by ocean currents, in D. J. Crisp, ed., *Fourth European Marine Biology Symposium.* Cambridge: Cambridge Univ. Press, 7-28.

Scheltema, R. S., 1971b. Larval dispersal as a means of genetic exchange between geographically separated populations of shallow-water benthic marine gastropods. *Biol. Bull.,* **140,** 284-322.

Scheltema, R. S., 1972. Reproduction and dispersal of bottom dwelling deep-sea invertebrates: A speculative summary, in R. W. Brauer, ed., *Barobiology and the Experimental Biology of the Deep Sea.* Chapel Hill: N. Carolina Sea Grant Program, Univ. N. Carolina, 58-66.

Scheltema, R. S., 1977. Dispersal of marine invertebrate organisms: Paleobiogeographic and biostratigraphic implications, in E. G. Kauffman and J. E. Hazel, eds., *Concepts and Methods in Biostratigraphy.* Stroudsburg, Pa.: Dowden, Hutchinson & Ross, 73-108.

Shuto, T., 1974. Larval ecology of prosobranch gastropods and its bearing on biogeography and paleontology. *Lethaia,* **7,** 239-256.

Thorson, G., 1946. Reproduction and larval development of Danish marine bottom invertebrates. *Medd. Komm. Dan. Fisk., Havundersog. Copenhagen, Series Plankton,* **4,** 1-523.

Thorson, G., 1950. Reproduction and larval ecology of marine bottom invertebrates. *Biol. Rev. Cambridge Phil. Soc.,* **25,** 1-45.

Thorson, G., 1961. Length of pelagic life in marine invertebrates as related to larval transport by ocean currents, in M. Sears, ed., *Oceanography.* Washington, D.C.: Amer. Assoc. Adv. Sci., Pub. 67, 455-475.

Turner, R. F., 1973. The paleoecologic and paleobiogeographic implications of the Maastrictian Cheilostomata (Bryozoa) of the Navesink Formation. Ph.D. Diss., Rutgers Univ., New Brunswick, N.J., 371p.

Vermeij, G. J., 1978. *Biogeography and Adaptation: Patterns of Marine Life.* Cambridge, Mass.: Harvard Univ. Press, 332p.

Cross-references: *Bivalvia; Paleobiogeography.* Vol. I: *Zooplankton.*

LIVING FOSSIL

The term *living fossil,* first coined by Charles Darwin for the East Asian Gingko tree, refers to living species that have persisted to the present with little change through long spans of geologic time. The reasons behind this evolutionary stability (*bradytely*) are still unclear, but evolutionists generally accept Simpson's (1944, p. 41) explanation that "the final and probably the most fundamental factor . . . is that bradytelic groups are so well adapted to a particular, continuously available environment that almost any mutation occurring in them must be disadvantageous." (See also Vermeij, 1978, p. 178-186.) Living fossils are particularly interesting to the paleontologist because they provide biological data and evolutionary insights often unavailable or obscured in the fossil record.

It was originally believed that the deep sea would be the most important refuge for ancient organisms, but this has generally not been the case (Menzies and Imbrie, 1958). One notable exception is the primitive mollusk *Neopilina* (see *Monoplacophora*), which was first dredged from the Peru-Chile Trench in the Pacific in 1951 (see Filatova et al., 1974, for references on this and more recent discoveries). Other invertebrate living fossils (Fig. 1), generally confined to the shallower continental shelf, include the brachiopod *Lingula,* which dates back to the Cambrian; the crustacean *Neoglyphea,* belonging to an ancestral decapod stock thought to have gone extinct in the Eocene (Forest et al., 1976); the horseshoe crab *Limulus* (e.g., see *Solnhofen Formation,* Fig. 6); and a large number of bivalves such as *Modiolus,* which dates back to the Devonian. According to Stanley (1972, p. 206), "bivalves contain more living genera with recognized fossil records that ex-

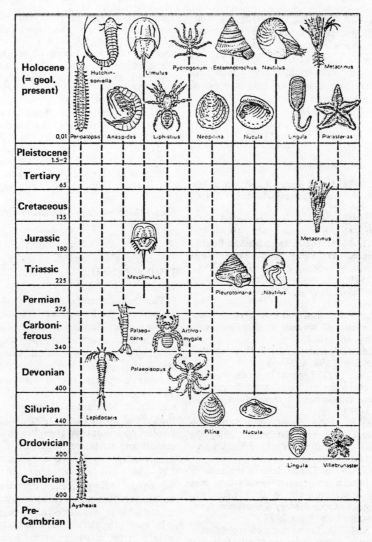

FIGURE 1. Characteristic "living fossils" among the invertebrates and their fossil relatives (from Thenius, 1973). Dotted line indicates fossil record missing.

tend back to the Jurassic and further than any other class of animals."

One of the most famous living fossils is the coelacanth fish, *Latimeria chalumnae*, discovered off the E coast of Africa in 1939. This species is the sole survivor of a crossopterygian lineage that had apparently died off in the Cretaceous (Van Wahlert, 1968). Also well known is the Australian lungfish, *Epiceratodus fosteri*, which is scarcely distinguishable from the Triassic lungfish *Ceratodus*.

On land, an important living fossil is the tuatara, *Sphenodon punctatus* of New Zealand. Though lizard-like in appearance, the tuatara is actually a rhynchocephalian reptile whose ancestors were contemporaries of the dinosaurs but survived the great extinction at the close of the Mesozoic. New Zealand also harbors the "primordial frog" *Leiopelma*, whose nearest relatives are found in Triassic deposits (Thenius, 1973). These and other vertebrate living fossils are shown in Fig. 2.

Although "living fossil" tends to have an exotic connotation, and often implies that these organisms are both rare and have an extremely limited distribution today, this need not apply. The Onychophorans, the so-called Protoarthropods (Tasch, 1973), are today widely distributed throughout the Southern hemisphere (see *Arthropoda*, Fig. 1). These bizarre animals combine features of segmented worms with those of arthropods, lacking a rigid

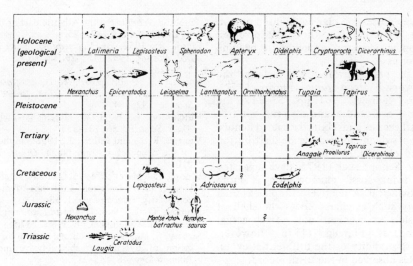

FIGURE 2. The most important "living fossils" among the vertebrates linked with fossils of the geological past (from Thenius, 1973). Dotted line indicates fossil record missing.

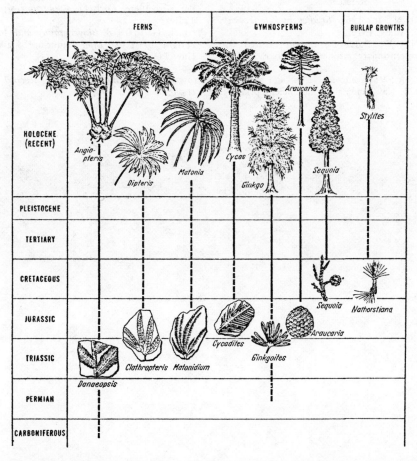

FIGURE 3. "Living fossils" among the plants and their fossil relatives (from Thenius, 1973). *Angiopteris* is a representative of the Marattiaceae and *Stylites* is a member of the Isoetaceae (quillworts).

skeleton yet possessing clawed, jointed appendages. By the Cambrian, they had diverged from creeping ancestral annelids to become walking worms.

Among the plants, living fossils include "the dawn redwood," *Metasequoia* (Florin, 1952); the Gingko (Andrews, 1961, pp. 335–347); the "palm ferns" or Cycadea; and the primitive tree ferns, Cyatheaceae and Marattiaceae (Thenius, 1973), all of which enjoyed a wide distribution in the Mesozoic (Fig. 3).

Recently, Schopf et al. (1975) have suggested that the living fossil phenomenon is more apparent than real. A relatively simple external morphology such as characterizes the shell of the horseshoe crab might not reflect actual genetic changes. Eldredge (1975), however, maintains the validity of the differential evolutionary rates observed in the fossil record.

TAMAR GORDON
DAVID JABLONSKI

References

Andrews, H. N., Jr., 1961. *Studies in Paleobotany.* New York: Wiley, 487p.

Eldredge, N., 1975. Rates of evolution revisited: Differential evolutionary rates, *Paleobiology,* 2, 174–177.

Filatova, Z. A.; Vinogradova, N. G.; and Moskalev, L. I., 1974. New finding of the ancient primitive mollusc *Neopilina* in the Atlantic part of the Antarctic Ocean, *Nature,* 249, 675.

Florin, R., 1952. On *Metasequoia,* living and fossil, *Bot. Notiser,* 1952(1), 1–29 (see also *Bot. Rev.,* 42, 215–315).

Forest, J., et al., 1976. *Neoglyphea inopinata:* A crustacean "living fossil" from the Philippines, *Science,* 192, 884.

Hutchinson, G. E., 1970. Marginalia–Living fossils. *Amer. Scientist,* 58, 531–535.

Menzies, R. J., and Imbrie, J., 1958. On the antiquity of the deep sea bottom fauna, *Oikos,* 9(2), 192–210.

Schopf, T. J. M., et al., 1975. Genomic versus morphologic rates of evolution: Influence of morphologic complexity, *Paleobiology,* 1, 63–70.

Simpson, G. G., 1944. *Tempo and Mode in Evolution.* New York: Columbia Univ. Press, 237p.

Stanley, S. M., 1972. Functional morphology and evolution of byssally attached bivalve mollusks, *J. Paleontology,* 46, 165–212.

Tasch, P., 1973. *Paleobiology of the Invertebrates.* New York: Wiley, 496p.

Thenius, E., 1973. *Fossils and the Life of the Past.* New York: Springer Verlag, 166–179.

Van Wahlert, G., 1968. *Latimeria* und die Geschichte der Wirbeltiere, *Fortschr. Evolutionsforsch.,* 4, 1–125.

Vermeij, G. J., 1978. *Biogeography and Adaptation: Patterns of Marine Life.* Cambridge, Mass: Harvard Univ. Press, 332p.

Cross-references: *Evolution; Fossil Record; Monoplacophora.*

M

MAMMALS, MESOZOIC

The term *Mesozoic Mammals* is purely informal and convenient, used to refer to mammals that lived during the Triassic, Jurassic, and Cretaceous periods, that is, during the first two-thirds of mammalian history, before the Cenozoic or so-called "Age of Mammals" began. Its use is traditional and probably derives from the necessity for a collective name to designate what long were poorly understood lineages of small, primitive mammals seemingly but distantly related to those paleontologically more familiar groups that rapidly evolved following dinosaur extinctions at the close of the Cretaceous. The term is still a useful one, but the pace of discovery of Mesozoic mammals has quickened over the past 25 years to the degree that any imperative to label a commonwealth of ignorance in respect to Mesozoic mammals is no longer valid. Simpson (1928, 1929, 1971) reviews the early history of discovery of Mesozoic mammals, and critically examines recent interpretations of Mesozoic mammalian history. (See also Lillegraven et al., 1979.)

Origin of Mammals

The origin of the Class Mammalia took place in the Mesozoic, during Triassic times, over 200 m yr ago. The oldest mammals known as fossils, from Upper Triassic rocks of South Africa, China, and western Europe, were small forms, probably predatory on insects and other small arthropods, and show that already mammals had diverged into two basic phyletic lineages, the subclasses Prototheria and Theria (see *Mammals, Placental*).

The acquisition of specifically mammal-like adaptations in the ancestral Therapsida (mammal-like reptiles; see *Reptilia*) was a gradual one, and occurred independently and at different rates in several different lineages. This remarkable parallelism in the development of "mammalness" long delayed a consensus as to how the Class Mammalia is best defined: the usual solution was that mammals were accepted as a polyphyletic taxon, that is, separate lineages of therapsids were thought to have given

rise to separate lineages of mammals. The Class Mammalia, then, was simply a grade of evolutionary advancement, whose last common ancestor was not itself a mammal.

Recent years, however, have seen the collection of new and relatively abundant Late Triassic mammals, some represented by virtually complete skulls and skeletons. Intensive study of these, of later Mesozoic mammals, and various advanced therapsids has made the case for monophyly, or at most diphyly, of the Class Mammalia relatively stronger, if not entirely acceptable without reservation (see Simpson, 1971). The critical evidence appears to come from the dentition—particularly from the nature of occlusion between upper and lower cheek teeth, and the sequence of replacement of teeth as an individual grows older. In groups of therapsids near to the ancestry of mammals (e.g., the early Triassic cynodont family Galesauridae), the upper and lower cheek teeth do not truly occlude; whereas, in even the most primitive mammals, they do: in the latter, during each chewing stroke the lowers were moved obliquely upward and transversely across the crowns of the uppers. This condition requires precise alignment of the occluding teeth, an impossible task if, as in reptiles, many generations of teeth, each containing slightly larger teeth than the last, occur at each tooth position while the animal grows. Apparently, suppression of the reptilian mode of multiple waves toward one employing only two waves (diphyodonty) allowed maintenance of the precise alignment required for true occlusion, as in mammals.

Hopson (1971 p. 19) has related the advent of diphyodonty to broader biological differences between early mammals and even the most mammal-like of therapsids:

. . . it can be reasonably argued that diphyodonty is correlated with the possession of a high degree of dependence of the young on the maternal parent for care and nourishment (milk), whereas the multiple replacement patterns of cynodonts are correlated with independence of the young from time of hatching and the presence of little or no parental care. The mammalian mode of development has probably allowed mammals to perfect their homeothermy and to evolve a much more complex nervous system than seen in reptiles.

It seems very likely, then, that the evolutionary shift from therapsids to mammals was a shift spanning a broad spectrum of anatomical, physiological, and behavioral adaptations primarily relevant to feeding and nutrition: by these criteria, both protetherian and therian mammals are "fully" mammalian.

Traditionally, Class Mammalia has included Subclass Theria (placentals, marsupials, and more primitive Mesozoic species broadly antecedent to both) and numbers of nontherian groups of high rank and of uncertain relationship to each other (Fig. 1). This classification was consistent with the prevalent opinion that mammals had had several separate phylogenetic sources from within the Therapsida. Evolution of opinion, however, toward acceptance of a monophyletic origin of the Mammalia and the recognition of an apparent phyletic unity among the non-therian groups, have necessitated revisions to the traditional approach. Foremost among these is that of

Hopson (1970), which retains the Prototheria at subclass level, but for all nontherian mammals instead of just monotremes (platypus, echidna of Australo–New Guinea) as in most previous modern classifications.

In protetherians *sensu* Hopson the lateral wall of the braincase is composed of a relatively small alisphenoid bone and an anteriorly expanded flange of the periotic bone, through which a branch of the 5th cranial nerve passed. The major cusps on the molar teeth of protetherian mammals are arranged anteroposteriorly or in patterns derived from that arrangement. Monotremes are the only living protetherians; extinct forms include triconodonts, docodonts, multituberculates, and possibly haramiyids.

In members of the Subclass Theria, the lateral braincase wall is constructed from an enlarged alisphenoid that furnished an exit for the branch of the 5th nerve; a part of the squamosal bone also contributed to the wall, but the periotic does not. Therian molars primitively bear three major cusps in a triangular pattern. Placentals, marsupials, and

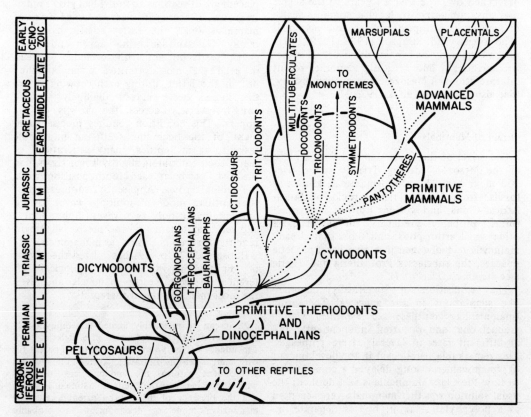

FIGURE 1. Evolutionary radiations of synapsid reptiles and early mammals (from Hopson and Crompton, 1969).

extinct groups possibly closely related to their common ancestors (Fox, 1972b, 1975), and the extinct eupantotheres and symmetrodonts are now classified as therians.

Subclass PROTOTHERIA:
Order Triconodonta

Prototherian mammals belonging to the Order Triconodonta include both the geologically oldest and the structurally most primitive mammals yet discovered: the Upper Triassic Family Morganucodontidae, and the more advanced, "typical" triconodonts, the Family Triconodontidae, known from the Upper Jurassic and Cretaceous. However, all members of the order, whether primitive or advanced, have molars that carry three major cusps in an anteroposterior line. The upper molar crowns have labial and linqual basal ridges (*cingula*) and the lowers, a lingual cingulum. The cingula are usually cuspidate, but the cingular cusps are relatively small. Triconodonts conform to the prototherian pattern of braincase construction in which the orbitotemporal wall is formed by the periotic bone rather than by the alisphenoid bone, the latter being the therian condition. Triconodonts ranged in size from shrews to domestic cats, and were probably omnivorous, insectivorous, or even piscivorous in food habits.

Family Morganucodontidae. The Family Morganucodontidae includes the Upper Triassic triconodonts *Morganucodon,* from western Europe and Yunnan, China; *Eozostrodon* from England; *Erythrotherium* and *Megazostrodon* from Lesotho; and *Sinocondon* from Yunnan. All are much alike dentally, although *Megazostrodon* and *Sinocondon* more closely resemble each other than they do the other morganucodontids, and accordingly, are sometimes classified together in a separate family, Sinocondontidae; and the separate status each of *Morganucodon, Eozostrodon,* and *Erythrotherium* has been the subject of controversy, although each is recognized as valid here. In any case, the Family Morganucodontidae includes the earliest and most primitive mammals known, and if those species that were actually phyletically basal to the evolutionary radiation of at least all prototherian and possibly all therian mammals were found, they also probably would be classified as morganucodontids. Consequently, it is fortunate indeed that the Morganucodontidae are represented by more nearly complete skeletal material than any other Mesozoic mammalian group (although much of this material is still under study and results unpublished; for a recent study, see Jenkins and Parrington,

1976), thereby giving the opportunity for extraordinary insight into a part of the origin of mammals.

As for all mammals, dental evidence is essential for determination of the phylogenetic relationships of morganucodontids. On the lower molars (Fig. 2), the central of the three major cusps is much the highest; on the uppers, the anterior and posterior cusps are closer to the height of the central cusp. Lower molars have a linqual cuspidate cingulum; upper molars have labial and lingual cingula, both of which are cuspidate. In *Morganucodon,* the best known of the morganucodontids, the dental formula appears to be (Mills, 1971):

$$\text{I}\frac{4?}{4\text{-}6}, \text{C}\frac{1}{1}, \text{P}\frac{4\text{-}5?}{4\text{-}5}, \text{M}\frac{4\text{-}5?}{4\text{-}5}$$

The "4th upper incisor" is located on the maxillary anteriorly adjacent to the canine, with consequent uncertainty about its homologies to true incisors; and shedding without replacement of some upper premolars and molars early in life makes their total number uncertain.

In nondental parts of their skeleton, the morganucodontids show an exceedingly primitive grade of structure among mammals. Both dentary-squamosal and quadrate-articular articulations were functional in the suspension of the lower jaw on the skull, and coronoid, angular, surangular, and prearticular bones are still retained in the lower jaw (K. A. Kermack et al., 1973). On its posteroventral margin, the dentary has an angular process similar to, although somewhat smaller than, that in dococontids (q.v., below); the process is likely the homologue of the mandibular angle in therians, and provided the insertion for masseter and pterygoid adductor muscles.

Among morganucodontids, the skull is known in *Morganucodon, Erythrotherium,* and *Megazostrodon,* but only notes on the braincase of *Morganucodon* have been published. A foramen pseudovale carried the mandibular branch of the trigeminal nerve (V) through the anterior lamina of the periotic bone, and the maxillary branch passed through a foramen pseudorotundum (Kermack and Kielan-Jaworowska, 1971).

Postcranial materials are known for *Morganucodon, Erythrotherium,* and *Megazostrodon,* but few results from studies on them are now available. In *Morganucodon,* the pectoral girdle has been shown to include scapula, coracoid and precoracoid ossifications in sutural contact with each other. The coracoid alone furnishes the glenoid cavity for the humerus (Kermack and Mussett, 1959).

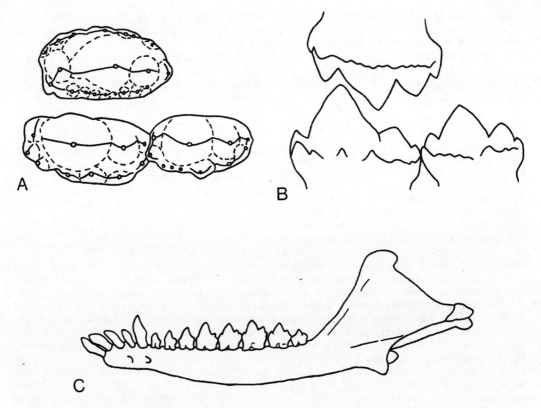

FIGURE 2. Triconodonta, Morganucodontidae: *Morganucodon* sp. A, B, crown and internal views of occluding molars, anterior to left (after Hopson and Crompton, 1969). C, external view, left mandible (after Kermack et al., 1973).

Family Triconodontidae. The Family Triconodontidae is composed of what can be called the "typical" triconodonts, from the Upper Jurassic of Europe, and from the Upper Jurassic through Upper Cretaceous of North America. Triconodontids are well understood dentally (Fig. 3): the incisors are usually small, the canines robust, and the premolars relatively simple, with a single primary cusp; the molars have three primary cusps that can differ in minor proportions but are virtually equal in height and are arranged in an anteroposterior line. At the base of the upper molar crowns is a labial and lingual cingulum; the lower molars have only a lingual basal cingulum. Successively adjacent molars interlock by a "tongue and groove" joint in which the posterior edge of the more anterior molar crown fits into a groove extending down the anterior edge of the crown of the molar following. The structure of the molars and distribution of occlusal facets worn on their crowns imply that adduction of the mandible brought the lower molars up lingually along the uppers, shearing against the uppers in vertical and lingual directions.

The Upper Jurassic and Lower Cretaceous triconodontids comprise a homogeneous group, showing minor differences in dental formulae, relative heights of primary cusps on the molars, and configurations of the basal cingula. In all, crown height below the molar cusps is relatively low, with most of the heights of the molar crowns being provided by the height of the cusps themselves; the molar crowns are transversely wide relative to their height. In the Upper Cretaceous *Alticondon,* however, the lower molars (the only teeth known) are high, transversely narrow blades, which are flat labially and are heavily ribbed down the crowns lingually (Fox, 1969). Most of the crown height in *Alticondon* is provided by the crown beneath the cusps, rather than by the cusps themselves. These features suggest that when more completely known, *Alticondon* might better be placed in its own family, separate from the Triconodontidae, although most likely descended from it.

The origin of the Triconodontidae seems securely discussed in terms of Triassic morganucodontids. Detailed comparisons between the dentitions of morganucodontids and Upper

A

B

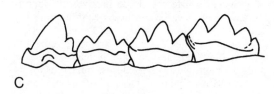

C

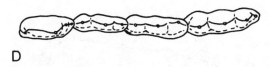

D

FIGURE 3. Triconodonta, Triconodontidae: *Priacodon* sp. (after Hopson and Crompton, 1969). A, B, internal and crown views, last premolar and first two upper molars; C, D, internal and crown views, last premolar and first three lower molars. Anterior to left.

Jurassic triconodontids reveal that "major points of resemblance are the large size of the last upper premolars, the presence of both external and internal cingula on the upper molars and only an internal cingulum on the lower molars, the presence of three main cusps in a longitudinal row, and of [the] posterior cingulum cusps" (Hopson and Crompton, 1969, p. 30). Descent of triconodontids from morganucodontids would have mostly required modification of the molars so that the three primary cusps were more nearly equal in height and cuspules on the cingula reduced. A greater difficulty for hypothesizing descent of triconodontids from morganucodontids has been the lack in triconodontids of the mandibular angle occurring in morganucodontids. But if the deep posteroventral region of the triconodontid mandible is equated with the mandibular angle in morganucodontids, and both considered to be the areas of insertion for the masseter muscle, then the tricono-

dontid angle could have been lost without degeneration of the external adductor muscles of the jaw. Fragmentary skull remains of triconodontids indicate that the lateral braincase wall was composed of an anterior lamina of the periotic, the usual prototherian condition.

Family Amphilestidae. The poorly known Family Amphilestidae, from the Middle Jurassic of Europe and the Upper Jurassic of North America, is of problematic relationships to other groups of mammals: Some specialists have thought that amphilestids were most closely allied with the Triconodonta, and hence, had prototherian affinities, while others considered amphilestids most closely related to symmetrodonts and, hence, were therians! In amphilestids the premolars are submolariform and the molars carry an enlarged central cusp and much smaller anterior and posterior cusps in an anteroposterior line. Dental formulae are not known, but there are probably more than four molars (see Mills, 1971, however). Plainly, the phylogenetic position of the amphilestids can only be clarified by the discovery of additional fossils that pertain to the group.

Order Docodonta

Family Docodontidae. The Docodontidae are a family of prototherian mammals discovered in the Middle and Upper Jurassic of Great Britain, and Upper Jurassic of Portugal and North America. Docodontids are well known dentally (Fig. 4), and seemingly comprise a clearly defined family of primitive mammals having a characteristic development of transversely enlarged molars with complex occlusal surfaces that functioned in longitudinal and transverse shear, and are suggestive of omnivorous or frugivorous habits.

In *Docodon*, the geologically youngest and best known of the docodontids, the dental formula is

$$I\frac{?}{3}, C\frac{1}{1}, P\frac{3}{3-4}, M\frac{6+}{7-8}$$

The incisors are unspecialized; the canines strong and two-rooted; and the premolars simple anteriorly and increasingly molariform posteriorly. The upper molars have a peculiar hour-glass outline in occlusal view, being expanded labially and lingually, and being wider transversely than long; the lower molars are virtually rectangular in occlusal view, and somewhat longer than wide. The dominant aspect of both upper and lower molar crowns is an enlarged labial and a lingual cusp interconnected by a high, obliquely

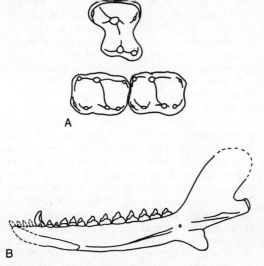

A

B

FIGURE 4. Docodonta, Docodontidae: *Docodon* sp. A, occlusal view, upper (above) and lower molars, anterior to left (after Jenkins, 1969). B, internal view, right mandible (after Simpson, 1929).

transverse crest. The sides of this crest fall abruptly away to form deep mesial and distal "half-basins" that are completed fully or in part by apposition of the half-basin of the molar at the next successive position along the tooth row (Jenkins, 1969). The two prominent cusps and transverse crest of each molar fitted within the basin formed by adjacent molars on the occluding jaw. Occlusal surfaces on the molars are coarsely grooved and ridged, an adaptation that presumably enhanced grinding and shearing of food materials (Jenkins, 1969).

The phylogenetic relationships of the docodontids were long misunderstood. Because of the transverse dimensions of the upper molars and their possession of a prominent labial and lingual cusp, docodontids were thought to be most closely related to eupantotheres (q.v., below), which also have transverse upper molars with prominent labial and lingual cusps. But detailed functional studies have shown these resemblances to be convergent and not owing to close relationship (Patterson, 1956). Structural comparisons and functional considerations indicate that docodontids were descended from a morganucodontid-like ancestor, in which the primary molar cusps were in an anteroposterior line. In comparison to this earlier stage, the expanded lingual lobe of docodontid upper molars is a new structure (although probably derived from the morganucodontid lingual cingulum) and the docodontid lower molars are transversely widened

lingually from the presumed ancestral condition (Hopson and Crompton, 1969). Docodontids are thought to have retained the reptilian (and morganucodontid) quadrate-articular articulation between the mandible and skull, as well as the dentary-squamosal articulation (Kermack and Mussett, 1958). Postcranial remains are unknown.

Order Uncertain

Family Haramiyidae. The Haramiyidae are a family known only from the uppermost Triassic of western Europe; their relationships to the Mammalia are quite uncertain. Knowledge of the family rests only on isolated teeth (Fig. 5), which, although providing some information of phylogenetic interest, do so in a wholly inconclusive way: rival hypotheses that haramiyids are best classified within the Therapsida (see *Reptilia*) or the Mammalia are both current; and it has been argued that if haramiyids are indeed mammals, their closest relationship is to multituberculates (see *Multituberculata*). But clearly the refutation of either the "reptilian" or the "mammalian" hypothesis of

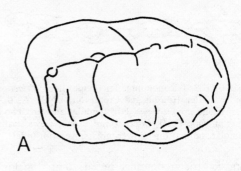

A

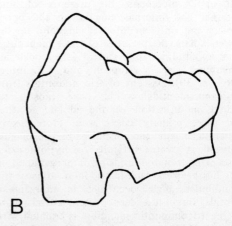

B

FIGURE 5. Haramiyidae: *Thomasia antiqua* (Plieninger), molariform tooth (after Hahn, 1973). A, side view; B, occlusal view.

relationships, and, therefore, the clarification of the phylogenetic status of the Haramiyidae, will only be effected by the field discovery and collection of additional and more complete fossils, both of haramiyids and of groups possibly related to them. Insofar as new fossil discoveries are lacking, no better conception of the haramiyids than at present will be likely, as no new empirical tests of either hypothesis will be possible.

The teeth of haramiyids have two roots that support a crown, which is longer in one dimension than the other. It is usually assumed, and probably correctly, that the long dimension is the anteroposterior one. But whatever orientation is correct for haramiyids, the occlusal surfaces of the crowns always consist of a more or less elongate central depression within a peripheral, partly cuspidate rim. However, because crowns similar to these occur both in certain therapsids and in primitive multituberculates, haramiyid coronal structure as presently understood is not decisive for determination of the relationships of the family.

Recent discovery of new haramiyid teeth and inferences from coronal striations on them thought to have been produced by occlusal contact in life have been offered as reinforcement for the view that haramiyids and multituberculates have a special relationship with each other. But the argument from this evidence, although of functional interest, has no necessary and sufficient phylogenetic implications for the relationship claimed: The fossil histories of the Therapsida and Mammalia present numerous examples of independent development of similar structural systems, some of high complexity and involving dental and nondental parts of the skeleton. With this background of parallelism as a widespread phenomenon among mammal-like reptiles and early mammals, current arguments favoring a special relationship between haramiyids and multituberculates must remain premature and will remain so until the skeletal structure of at least haramiyids is more completely known.

Subclass THERIA: Infraclass PANTOTHERIA

The Infraclass Pantotheria includes therian mammals classified in the orders Symmetrodonta and Eupantotheria, both characterized by their possession of a reversed triangle dentition (see below), but one of a primitive type: The upper molars lacked a true protocone that occluded within a basined heel on the lower molars. From detailed investigation of the structure and function of the dentition in

the two groups (see, for example, Crompton, 1971, and Fox, 1975, and references cited in each), there seems to be little doubt that the ancestry of the eupantotheres was among primitive symmetrodonts.

Order Symmetrodonta

The Order Symmetrodonta includes the geologically oldest and dentally most primitive therian mammals, and probably the ancestor of all more progressive therians was itself a symmetrodont. Paleontologists have discovered symmetrodonts in rocks as old as the Late Triassic (in England) and as young as the Late Cretaceous (in Alberta, Canada; Fox, 1972a,b), a stratigraphic range of 120 m yr, equalled only by that of the Triconodonta among mammalian orders, and making symmetrodonts and triconodonts the longest lived orders of mammals now known. Symmetrodonts were small shrew-sized mammals, probably having insectivorous food habits.

Anatomical knowledge of symmetrodonts is nearly exclusively limited to their teeth and jaws. Symmetrodont molars are highly characteristic: both upper and lower crowns formed simple occlusal triangles with a primary cusp located at each apex. The lower molar triangles are reversed in respect to the uppers and occlusion was alternate, in the sense that each molar occluded within the embrasure between two successive molars on the opposite jaw.

The Order Symmetrodonta contains three families: Kuehneotheriidae, from the Upper Triassic of England; Spalacotheriidae, from the Upper Jurassic and Lower Cretaceous of Europe, the Lower Cretaceous of Asia, and the Lower and Upper Cretaceous of North America; and Amphidontidae, from the Upper Jurassic of North America and Upper Jurassic or Cretaceous of Asia.

Family Kuehneotheriidae. The Kuehneotheriidae at present include only the type genus *Kuehneotherium*, although the Upper Jurassic *Tinodon* and *Eurylambda* could eventually be referred here, as well (see Crompton and Jenkins, 1967). *Kuehneotherium* is the geologically oldest mammal presumed to be a therian, and thus has a central position in discussions of the Mesozoic radiation of therians, the relationships of therian to prototherian mammals, and the origin of the Class Mammalia itself. Discovered fossils of *Kuehneotherium* are mostly isolated teeth (Fig. 6). Jaw fragments having articulated teeth are rare, and possible postcranial bones are known, but undescribed. Association of upper and lower teeth is inferred from detailed study of occlusal wear facets produced in life, and by

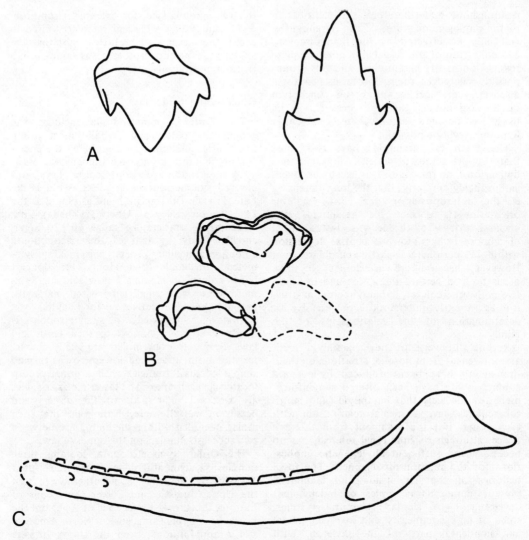

FIGURE 6. Symmetrodonta, Kuehneotheriidae: *Kuehneotherium praecursoris* Kermack, et al. (after Kermack, et al., 1968). A, external view, upper (left) and lower (right) molars, anterior to left; B, occlusal view, upper (above) and lower molars, anterior to left; C, external view, left mandible. Brackets indicate tooth positions.

comparison with articulated dentitions of other primitive therians. *Kuehneotherium* is thought to have had both quadrate-articular and dentary-squamosal articulations of the lower jaws on the skull (Kermack et al., 1968).

Unlike the lower molars of other symmetrodonts, those of *Kuehneotherium* have a small heel (talonid) placed posteriorly on the crown, and the primary cusps form a less symmetrical triangle and one more open than in spalacotheriids. The upper molars show a well-developed lingual cusp, and smaller but distinct anterior and posterior cusps, again arranged in an asymmetrical, open triangle. Minor modifications of molars of this type associated with

increasingly transverse movements of the jaws in chewing, could have led to molar structure in spalacotheriids; amphidontids; eupantotheres; and, with further changes, all "higher" therians (Crompton, 1971; Fox, 1975).

The fossils of *Kuehneotherium,* when compared with those of the Upper Triassic morganucodontids, can furnish evidence relevant to answering the questions, What are the relationships of therian to prototherian mammals; and would the last common ancestor of the two groups, if known, be itself classified as a mammal? Major differences between *Kuehneotherium* and morganucodontids are best known for the dentition and jaws: (1) The primary cusps and crests of morganocodontid molars are aligned anteroposteriorly, whereas those of *Kuehneotherium* form reversed triangles. (2) Tight occlusal fit between upper and lower molars in morganucodontids was possible only by abrasion of parts of the crowns during tooth-on-tooth contact in chewing; whereas, in *Kuehneotherium,* the molars when first erupted could be occluded tightly, although occlusal wear enhanced the fit. (3) Transverse movement of the jaws during chewing was greater in *Kuehneotherium* than in morganucodontids, since the shearing crests of *Kuehneotherium* molars had a greater transverse component to their orientation than did the shearing crests of morganocodontid molars. (4) The mandible of morganucodontids has an angular process, whereas that of *Kuehneotherium* does not. It is generally agreed that the cusp(and crest)-in-line molar construction of morganocodontids is the more primitive (similar arrangements occur on molariform teeth of some therapsids), and could have led to the condition in *Kuehneotherium* mostly by a rotation of the anterior and posterior cusps labially on the upper molars and lingually on the lowers (Crompton and Jenkins, 1968; Hopson and Crompton, 1969). The lower molar triangles then would have occluded within the embrasures between the uppers, the primitive therian condition. Possibly more formidable than dental modifications required in deriving *Kuehneotherium*-like therians from ancestral morganocudontids are aspects of the structure of the braincase: Expansion anteriorly of a lamina of the periotic bone to form the lateral braincase wall (as seen in morganucodontids) is fundamentally different from the structure of the therian braincase, in which the lateral wall is formed by the alisphenoid bone. However, the braincase in therians is not known for species older than the Late Cretaceous, and not known at all in primitive therians such as symmetrodonts and eupantotheres, so that the meaning of these differences cannot be resolved at present.

Family Spalacotheriidae. The Spalacotheriidae, or "acute-angled" symmetrodonts, are the typical symmetrodonts, on which the concept of the order was first based historically. Spalacotheriids are known from both upper and lower dentitions articulated in jaw fragments: The lower molars are tricuspid, with well-developed lingual cusps, and the trigonid angle is relatively acute; the upper molars carry a high lingual cusp and smaller anterior and posterior labial cusps (Fig. 7). The incisors are simple; the canines enlarged and double-rooted; and the premolars have a single large

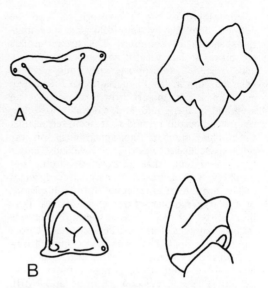

FIGURE 7. Symmetrodonta, Spalacotheriidae: *Spalacotheroides bridwelli* Patterson (after Patterson, 1955, 1956). A, occlusal and external views, right upper molar; B, occlusal and posterior views, left lower molar.

primary cusp, and much smaller anterior and posterior cusps. The mandible lacks an angular process and, in earlier species, has a reptilian-like internal mandibular groove. Spalacotheriids could have been descended from ancestors having a kuehneotheriid-like dentition, and, in turn, gave rise to no known descendants.

Family Amphidontidae. The Family Amphidontidae comprises the obtuse-angled symmetrodonts, *Amphidon* and *Manchurodon*. Both genera are based on incomplete, lower dentitions.

The lower molars of amphidontids are distinguished by being functionally unicuspid (Simpson, 1929). The accessory cusps are much reduced and, with the enlarged apical cusp, form a broadly open (obtuse) triangle in occlusal view. This pattern does not closely resemble the molar construction of other groups of mammals, but what similarities there are seem closest to symmetrodonts and provide the basis for classification of the family.

Order Eupantotheria

The Eupantotheria, an order of Mesozoic therian mammals that was probably descended from a *Kuehneotherium*-like ancestry, probably included the ancestors of "higher" Theria, marsupials and placentals. Although a moderate diversity is seen within the order (several lineages, considered families, are recognized by paleontologists), the Eupantotheria can be defined dentally by the following specializations shared in common: The upper molars are more or less greatly widened transversely and shortened anteroposteriorly, and have an enlarged cusp at the lingual apex of the crown; and the lower molar crowns consist of an asymmetrical trigonid and a small, posterior, usually unicuspid talonid that occluded against the lingual side of the apical cusp on the uppers. Occlusion was of the reversed triangle type, resembling that in symmetrodonts, but probably with a greater transverse component of jaw movement as reflected in the dimensions of the upper molars. Four eupantothere families can be defined at present—Amphitheriidae (Middle Jurassic, Europe); Peramuridae (Upper Jurassic, Europe); Paurodontidae (Upper Jurassic, Europe, North America and, possibly, East Africa); and Dryolestidae (Upper Jurassic and Lower Cretaceous, Europe and North America). Eupantotheres were probably small carnivores and insectivores, and anatomical knowledge of them is virtually restricted to jaws and teeth.

Family Amphitheriidae. The Family Amphitheriidae is based on only *Amphitherium*, from the Middle Jurassic of England, and known at present only from lower jaws with teeth (Fig. 8). The dental formula appears to be (Mills, 1964):

$$I\frac{?}{4}, C\frac{?}{1}, P\frac{?}{4}, M\frac{?}{5+}$$

The molars differ from those of other eupantotheres by virtue of their relatively elongate, narrow talonid, which is unicuspid. From technical features of the dentition, including relative sizes of the posterior premolars and anterior molars, no other known eupantothere families are likely to have been descendants of *Amphitherium*.

Family Paurodontidae. The Paurodontidae include eupantotheres having a relatively low number of molars (four). The trigonid is relatively long and narrow, the metaconid low, and the talonid robust and unicuspid. Upper molars of paurodontids have still to be discovered, and the family has left no known descendants.

Family Dryolestidae. The Family Dryolestidae comprises eupantotheres in which the numbers of cheek teeth are unreduced, there being as many as four premolars and nine molars (Fig. 9). The lower molars are anteroposteriorly compressed and the talonid short and narrow. The upper molars are transversely wide and anteroposteriorly short, and can carry an enlarged labial cusp in addition to the primary lingual cusp. Eleven genera from the Upper Jurassic and Lower Cretaceous of North America and Europe have been described; this number is probably unrealistically high, owing to separate names being given upper and lower dentitions of single species. The Dryolestidae left no known descendants.

Family Peramuridae. The eupantothere family, Peramuridae, includes only the type genus, *Peramus* (although the Tanzanian *Brancatherulum* could belong here, as well). *Peramus* (Fig. 10), from the Upper Jurassic of England, is of more than casual interest to students of early mammals because it appears to be a member of the family that likely included the ancestors of the dominant groups of mammals living today, marsupials and placentals.

The dental formula of *Peramus* is (Clemens and Mills, 1971):

$$I\frac{?}{2+}, C\frac{1}{1}, P\frac{4}{4}, M\frac{4}{4}$$

The upper molars have the major labial cusps—stylocone, paracone, and metacone—seen in higher therians, although the stylocone in

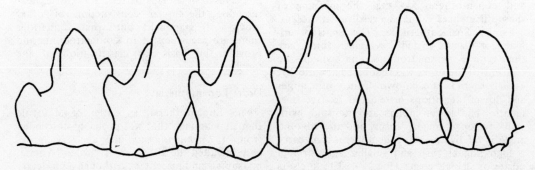

FIGURE 8. Eupantotheria, Amphitheriidae: *Amphitherium prevosti* (H.v. Meyer). External view, right lower molars. (Original, by P. L. Forey.)

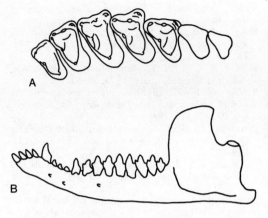

FIGURE 9. Eupantotheria, Dryolestidae, A; *Herpetairus arcuatus* (Marsh); occlusal view, right upper molars, anterior to right (after Simpson, 1929). B; *Crusafontia cueneana* Henkel and Krubs; external view, left mandible (after Krebs, 1971).

Peramus seems specialized in ways not likely in the actual ancestor of these groups. A lingual cingulum on the upper molars of *Peramus* marks the position of the presumptive protocone, and occluded against the labial side of the talonid, which was bicuspid on at least M_{2-3}. These latter structural relationships are those that would be expected in the ancestor of higher therians, which primitively have a distinct lingual protocone on the upper molars that occluded within a tricuspid basined talonid on the lowers. It is thought that *Permus* is descended from a *Kuehneotherium*-like ancestry, during which descent the upper molars increased their transverse dimensions, a true metacone appeared, and the talonid was enlarged and became bicuspid on at least some teeth.

RICHARD C. FOX

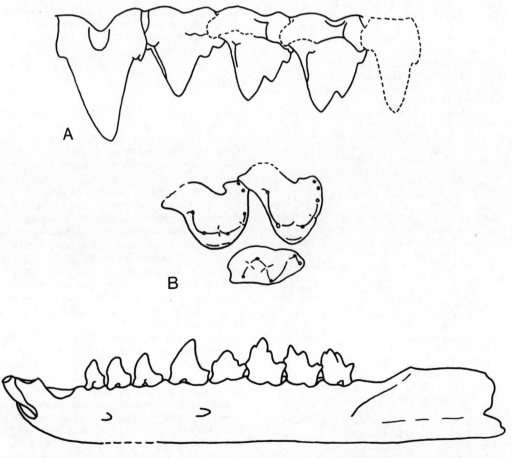

FIGURE 10. Eupantotheria, Peramuridae: *Peramus tenuirostris* Owen. (after Clemens and Mills, 1971). A, external view, left upper P^4, M^{1-4}, anterior to left; B, occlusal view, M^{2-3}, M_3, anterior to left; C, external view, left mandible.

References

Clemens, W. A., and Mills, J. R. E., 1971. Review of *Peramus tenuirostris* Owen (Eupantotheria, Mammalia), *Bull. British Mus. (Nat. Hist.), Geol.,* **20,** 89–113.

Crompton, A. W., 1971. The origin of the tribosphenic molar, in Kermack and Kermack, 65–87.

Crompton, A. W., and Jenkins, F. A., Jr., 1967. American Jurassic symmetrodonts and Rhaetic "pantotheres," *Science,* **155,** 1006–1009.

Crompton, A. W., and Jenkins, F. A., Jr., 1968. Molar occlusion in Late Triassic mammals. *Biol. Rev. Cambridge Phil. Soc.,* **43,** 427–458.

Fox, R. C., 1969. Studies of Late Cretaceous vertebrates. III. A triconodont mammal from Alberta, *Canadian J. Zool.,* **47,** 1253–1256.

Fox, R. C., 1972a. An Upper Cretaceous symmetrodont from Alberta, Canada, *Nature,* **239,** 170–171.

Fox, R. C., 1972b. A primitive therian mammal from the Upper Cretaceous of Alberta, *Canadian J. Earth Sci.,* **9,** 1479–1494.

Fox, R. C., 1975. Molar structure and function in the Early Cretaceous mammal *Pappotherium:* Evolutionary implications for Mesozoic Theria, *Canadian J. Earth Sci.,* **12,** 412–442.

Hahn, G., 1973. Neue Zähne von Haramiyiden aus Deutschen Ober-Trias Und ihre Beziehungen zu den Multituberculaten, *Palaeontographica* A, **142,** 1–15.

Hopson, J. A., 1970. The classification of nontherian mammals, *J. Mammal.,* **51,** 1–9.

Hopson, J. A., 1971. Postcanine replacement in the gomphodont cynodont *Diademodon,* in Kermack and Kermack, 1971. 1–21.

Hopson, J. A., and Crompton, A. W., 1969. Origin of mammals, *Evol. Biol.,* **3,** 15–72.

Jenkins, F. A., Jr., 1969. Occlusion in *Docodon* (Mammalia, Docodonta), *Postilla,* **139,** 24p.

Jenkins, F. A., Jr., and Parrington, F. R., 1976. The postcranial skeleton of the Triassic mammals *Eozostrodon, Megazostrodon* and *Erythrotherium, Phil. Trans. Roy. Soc. London, Ser. B* **273,** 387–431.

Kermack, D. M., and Kermack, K. A., eds., 1971. *Early Mammals,* suppl. *Zool. J. Linn. Soc.,* London: Academic Press, 203p.

Kermack, D. M.; Kermack K. A.; and Mussett, F., 1968. The Welsh pantothere *Kuehneotherium praecursoris, J. Linnean Soc., Zool.,* **47,** 407–423.

Kermack K. A., and Kielan-Jaworowska, Z., 1971. Therian and non-therian mammals, in Kermack and Kermack, 1971, 103–115.

Kermack, K. A., and Mussett, F., 1958. The jaw articulation of the Docodonta and the classification of Mesozoic mammals, *Proc. Roy. Soc. (London) B,* **149,** 204–215.

Kermack, K. A., and Mussett, F., 1959. The first mammals, *Discovery,* ser. 2, **20,** 144–151.

Kermack, K. A.; Mussett, F.; and Rigney H. W., 1973. The lower jaw of *Morganucodon, Zool. J. Linnean Soc.,* **53,** 87–175.

Krebs, B., 1971. Evolution of the mandible and lower dentition in dryolestids (Pantotheria Mammalia), in Kermack and Kermack, 1971, 89–102.

Lillegraven, J. A.; Kielan-Jaworowska, Z.; and Clemens, W. A., eds., 1979. *Mesozoic Mammals: The First Two-Thirds of Mammalian History.* Berkeley: Univ. California Press.

Mills, J. R. E., 1964. The dentitions of *Peramus* and *Amphitherium, Proc. Linnean Soc. London,* **175,** 117–133.

Mills, J. R. E., 1971. The dentition of *Morganucodon,* in Kermack and Kermack, 1971, 29–63.

Parrington, F. R., 1978. A further account of the Triassic mammals, *Phil. Trans. Roy. Soc. London, Ser. B,* **282,** 177–204.

Patterson, B., 1955. A symmetrodont from the Early Cretaceous of northern Texas, *Fieldiana, Zool.,* **37,** 689–693.

Patterson, B., 1956. Early Cretaceous mammals and the evolution of mammalian molar teeth, *Fieldiana, Geol.,* **13,** 1–105.

Simpson, G. G., 1928. A catalogue of the Mesozoic mammalia in the geological department of the British Museum. London: British Mus. (Nat. Hist.), 215p.

Simpson, G. G., 1929. American Mesozoic Mammalia, *Mem. Yale, Peabody Mus.,* **3**(1), 171p.

Simpson, G. G., 1971. Concluding remarks: Mesozoic mammals revisited, in Kermack and Kermack, 1971, 181–198.

Cross-references: *Bones and Teeth; Mammals, Placental; Marsupials; Multituberculata; Vertebrata; Vertebrate Paleontology.*

MAMMALS, PLACENTAL

The great infraclass Eutheria, commonly known as the placental mammals, counts among its members the diminutive masked shrew, the gigantic blue whale, the familiar *Homo sapiens,* and the exotic aardvark. Eutherians certainly constitute a successful and dominant group in terms of their present-day diversity and abundance. The name *placental mammals* is somewhat of a misnomer as all marsupial mammals (infraclass Metatheria) also have one or two basic kinds of placentae. However, all eutherians commonly possess a chorioallantoic placenta, whereas this type is present only among perameloid marsupials (the bandicoots) and is probably independently derived in the latter group (see Lillegraven, 1969).

Eutherians have nearly a 100 m yr history. The earliest undisputed members of the group are represented by fossils from rocks of uncertain age in Mongolia, but it is apparent that these Asian remains predate any of the Campanian (Late Cretaceous) eutherian fossils from North America and thus may exceed an age of 80 m yr. (see Kielan–Jaworowska et al., 1979). With the advent of the Cenozoic, eutherians underwent a tremendous adaptive radiation and continued to flourish, not without vicissitudes in diversity, until the present. This general pattern is illustrated in Figure 1. About 33 living and extinct orders of placental mammals are recognized, although the exact number varies with classifications of different authors.

The Eutherian Fossil Record

The world fossil record of placental mammals is strongly biased in favor of certain continents. North America shows a relatively excellent record with many faunas known from the Cretaceous through the Pleistocene. The fossil record from Europe is also quite good, although virtually nothing is known of the European Cretaceous and early Paleocene. Important faunas from crucial times are known from South America and Asia, but these continents demand much more paleontological exploration. Most disconcerting is the fossil record from Africa, where the earliest fauna of any significance, the Fayûm fauna of Egypt, dates back only to the late Eocene–early Oligocene. Because of this dearth, Africa remains a dark continent in the eyes of paleontologists interested in early eutherians; and questions concerned with the origin of several Old World groups are as yet unanswered.

Eutherian Diversity

Despite the obvious monographic problems in interpreting the world mammalian fossil record, documentation of eutherian diversity through the Cenozoic has proved instructive and a number of basic patterns have emerged. Such work was initiated primarily by Simpson (1962 and earlier papers cited therein) and the subject has been treated in detail by Lillegraven (1972), who documented rapid increases in ordinal and familial diversity during the Paleocene and Eocene which reached all-time highs of 26 orders in the late Eocene and 116 families in the early Oligocene (these figures exclude bats and mysticete and odontocete whales). This trend was followed by a decline in diversity during the later Oligocene with a secondary rejuvenation in the early Miocene (see Fig. 2).

A strong correlation apparently exists between the first appearances of many mammalian and angiosperm groups, and these factors in turn broadly fit the pattern of world wide climatic change. For example, the dramatic taxonomic turnover (i.e., the appearance of new phyla and the extinction of archaic ones) during the early Oligocene corresponds well with the global-wide decrease in

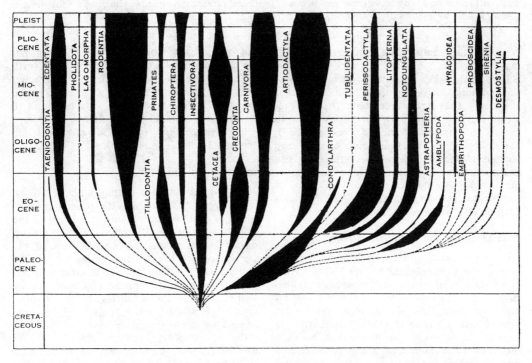

FIGURE 1. The chronologic distribution of the Eutheria (from Romer, 1966; copyright © 1966 by The University of Chicago Press).

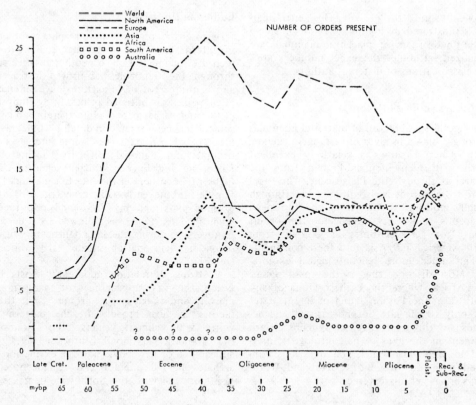

NUMBER OF ORDERS PRESENT

FIGURE 2. Graph of numbers of mammalian orders (ordinate) present through time (abscissa) for the world as a whole and for each continent (from Lillegraven, 1972).

mean annual temperature and equability at that time.

The relationship between continental size and eutherian diversity has been somewhat less clear. That larger areas tend to have predictably greater taxonomic diversities is a well-known biogeographic principle (Preston, 1962), but Lillegraven (1972) did not observe any obvious relationship between continental size and mammalian diversity. Flessa (1975), on the other hand, showed that a logarithmic plot of continental area *vs* mammalian diversity yields a strong positive correlation.

Plate Tectonics and Eutherian History

It seems that the movements as well as the sizes of the continents have played a major role in the historical course of mammalian faunal change, particularly in the late Mesozoic and early Cenozoic. As an example, the early Eocene was the time of greatest similarity, in respect to numbers of common genera, between the mammalian faunas of North America and Europe, when the two continents were still broadly connected in their northern regions. With the opening of the North Atlantic at the

end of the early Eocene (about 49 m yr ago) the mammalian faunas of Europe and North America took on radically different characteristics and underwent separate paths of faunal change. The importance of plate tectonics to mammalian faunal history (see *Vertebrate Paleobiogeography*) has been examined extensively by McKenna (1973).

Origins of the Eutheria

Perhaps no problem dealing with the evolution of placental mammals has proved as interesting as that concerned with the origin of the group itself. This multifaceted question entails, among other things, a consideration of (1) the important events that transpired in the evolution of the "key" eutherian morphological systems, such as the reproductive tract and the dentition; (2) the affinities of the earliest archaic placentals; and (3) the early biogeographic history of the group.

The morphology and physiology of the eutherian reproductive system can be regarded as highly specialized. Lillegraven (1969) has presented a detailed argument that the urogenital system and pattern of embryonic development in marsupials is, for the most part, primitive, and the highly specialized condition in eutherians can be derived from a grade of evolution of the reproductive system akin to that represented in the Metatheria. Unfortunately, the important changes that took place during the evolution of the placental reproductive tract cannot be directly documented, as the mammalian fossil record rarely provides traces of soft anatomy. When we consider fossils, we must turn to other systems for information on the origin of the Eutheria. Primary among these are teeth, which are easily preserved and, fortunately for the paleontologist, highly diagnostic. The molars of the earliest undisputed marsupials and placentals from the Late Cretaceous are clearly separable but can be traced back to a common plan represented in Early Cretaceous therians of a eutherian-metatherian grade (Fig. 3). Forms like *Aegialodon*, the Late Cretaceous relict *Potamotelses*, and deltatheroids are thought by Fox (1975) to represent a high-level taxon of therians of a primitive tribosphenic stage (condition shown in Fig. 3, in which there are three major cusps on the upper molars and a three-cusped trigonid and a basined heel, or talonid, on the lower molars) that included the ancestors of eutherians and metatherians. *Pappotherium* and *Holoclemensia* from the Early Cretaceous (Albian) of North America had already achieved an advanced stage in tribosphenic molar development characteristic

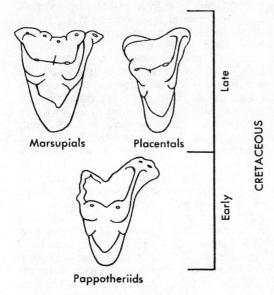

FIGURE 3. Outline drawings of upper molars showing basic similarity of primitive therian dentitions (from Lillegraven, 1969; courtesy of The University of Kansas Press).

of Late Cretaceous metatherians and eutherians. Thus it is evident that the split of marsupials and placentals from a common therian ancestor is an event that dates back at least to the earliest Cretaceous.

Most interesting is the basic difference in dental formula between metatherians and eutherians. Both groups primitively have seven or eight postcanine teeth. However, eutherians are traditionally regarded as having only three molars, whereas marsupials have four. Hence the widely accepted primitive dental formula for the Eutheria is three incisors, one canine, four premolars, and three molars above and below, or written in a shorthand commonly used by mammalogists,

$$I\frac{3}{3}, C\frac{1}{1}, P\frac{4}{4}, M\frac{3}{3}$$

while that for Metatheria is

$$I\frac{5}{4}, C\frac{1}{1}, P\frac{3}{3}, P\frac{3}{3}, M\frac{4}{4}$$

(although a greater number of molars are present in some marsupial genera, e.g., *Myrmecobius* $M\frac{5}{6}$, or *Peradercas* $M\frac{5+}{5+}$). However, the presence of five lower premolars in the Late Cretaceous eutherian *Gypsonictops* and the marsupial-like dental formula of the Cretaceous Asian genera *Deltatheridium* and *Deltatheroides* noted by Butler and Kielan-Jaworowska (1973) have led some workers to question the traditional view of the basic therian dental formula, and alternative hypotheses are now being considered by McKenna (1975).

Whatever their true dental homologies, marsupials show a distinct departure from placentals in having only the posteriormost premolar replaced by a permanent tooth during development. In placentals, all four premolars are originally deciduous and are later replaced by permanent teeth. Since it is generally assumed that the replacement of many teeth is the primitive condition in mammals, we must conclude that the replacement of only a single premolar in marsupials is an advanced specialization. Emphasizing this specialized condition, Kühne (1973) allied marsupials and monotremes as being more closely related to each other than either are to placentals because of their shared-derived dental characters. His scheme, however, largely ignores the body of evidence that both monotremes and marsupials resemble each other only in being more primitive anatomically and physiologically than their eutherian counterparts, and thus are not particularly closely related.

Aside from teeth, little is directly available in the fossil record as evidence for eutherian origins, as cranial and postcranial remains of the earliest forms are rare. Exciting new discoveries of remarkably preserved mammal skulls and skeletons from the Cretaceous of Mongolia by Polish and Russian paleontologists (Kielan-Jaworowska, 1969) are expected to shed much light on the problem of placental origins. Pending these results, it is interesting to note a few important anatomical details. The Late Cretaceous eutherian *Gypsonictops* shows a similarity with metatherians in having an inflected angular process on the lower jaw. More advanced eutherians without exception lack this feature. Placentals have lost the epipubic bones that mainly function in increasing abdominal rigidity, which in female marsupials would aid in suspension of the marsupium or pouch. The major component in the eutherian auditory bulla (the inflated area at the base of the skull that encloses the middle ear cavity) is thought to be, primitively, the entotympanic bone, whereas in marsupials the alisphenoid bone is dominant. This is interesting, as it seems that the alisphenoid was also a major bone in the braincase wall of the most primitive therians.

Taking the presently available evidence into account, it seems appropriate to conclude that eutherians arose from some Early Cretaceous (or perhaps even earlier) therian stock that had the following important characteristics: (1) a marsupial-like reproductive system, i.e., prepenile scrotum if any, gestation period shorter than estrous cycle, young carried externally, epipubic bones, paired vaginae lateral to ureters, yolk sac placenta; (2) incomplete corpus callosum of the brain; (3) "reptilian" organization of the retina; (4) alisphenoid forming a major part of the lateral wall of the braincase; (5) inflected angular process on the lower jaw; (6) replacement of all deciduous premolars with permanent premolars; and (7) primitive "tribosphenic" molars. In some of these characters (1, 2, 3, 4, and 5), this ancestral therian stock resembled present-day marsupials, but had not yet acquired specializations in the dentition characteristic of the latter group.

The early biogeographic history of the Eutheria has been the subject of much discussion and speculation. Accepting the marsupial/placental dichotomy as extremely ancient in age, it becomes apparent that this evolutionary event may have been the direct result of geographic isolation on the continental scale. Hoffstetter (1973) proposed that marsupials were isolated from eutherians in the Early Cretaceous on a chain of land masses extending from North America to Australia via South American and Antarctica and only later entered Asia by way of a Bering route in the Late Cretaceous and Europe via a Greenland route in the early Tertiary. One argument against Hoffstetter's hypothesis is the presence of possible eutherians (e.g., *Pappotherium*) in the Early Cretaceous (Albian) of North America; but in view of the meager evidence available, it is equally plausible that these forms represent offshoots of a primitive therian stock that are neither metatherians nor eutherians. Lillegraven (1969) has expressed the popular theory that eutherians arose in Asia during the Early Cretaceous and invaded North America via a Bering land route in the Late Cretaceous, contributing to the demise of many of the North American marsupial lineages at the end of the Mesozoic. In further considerations of the problem, Lillegraven (1974) noted that the Bering route was not optimal for dispersal, but acted as a "filter" route for invading eutherians from Asia during much of the Cretaceous. Despite the wide interest generated over this important biogeographic problem, there is still much to be learned; and new insight will be the direct result of further paleontological exporation rather than continued speculation.

Eutherian Systematics

It is a common view among biologists that placental mammals have enjoyed a relatively stable classification in recent decades and that, unlike the case in many other phyla, virtually all the important evolutionary problems have been solved. Such a view is far from true, and much still remains completely mysterious or highly controversial concerning the origin and relationships of the eutherian orders. Mammalian taxonomists and paleontologists have only recently applied alternative systematic methods, such as those advocated by Hennig (1966), in developing their phylogenetic classifications. Such approaches, as well as a consideration of new evidence derived from serological and protein sequence data, are expected to significantly modify certain aspects of the present eutherian classification.

The central work in mammalian systematics is Simpson's monumental "The Principles of Classification and Classification of Mammals" published in 1945. A revised classification of the Mammalia is now in preparation under the direction of M. C. McKenna at the American Museum of Natural History, in order to account for the many new paleontologic discoveries and taxonomic alterations that have been made since the time of Simpson's work.

Supra-Ordinal Categories

Much controversy exists as to how the various orders of placental mammals should be combined in a hierarchical system. Simpson (1945) recognized the following categories:

1 the cohort Unguiculata, to include the orders Insectivora (including macroscelidids), Dermoptera, Chiroptera, Primates (including tupaiids), Tillodonta*, Taeniodonta*, Edentata, and Pholidata
2. the cohort Glires, to include orders Lagomorpha and Rodentia
3. the cohort Mutica, to include the order Cetacea
4. the cohort Ferungulata, to include the orders Carnivora, Pinnipedia, Condylarthra*, Liptopterna*, Notoungulata*, Astrapotheria*, Tubilidentata, Pantodonta*, Dinocerata*, Pyrotheria*, Proboscidea, Embrithopoda*, Hyracoidea, Sirenia (including desmostylians), Perissodactyla, and Artiodactyla.

Serious objections have been raised in recent years as to the utility of the Unguiculata, as many of the orders included within it, such as edentates, clearly are no more closely related to certain unguilicates than to other groups. A close relationship between lagomorphs and rodents has not been established with any certainty, militating against the use of the cohort Glires to unite them. Finally, although carnivores and primitive ungulates may be distantly related, good structural ancestors of both groups are known from the Late Cretaceous (Lillegraven, 1969), suggesting that the time of carnivore-ungulate divergence is truly remote.

The Eutherian Orders

Brief coverage of each of the Eutherian orders, including their known geographic and geologic distribution is provided below. Abbreviations used for time units and continents are as follows: Cret., Cretaceous; Paleoc., Paleocene; Eoc., Eocene; Olig., Oligocene; Mioc., Miocene; Plioc., Pliocene, Pleist.. Pleistocene; Holo., Holocene; N. A., North America; S. A., South America; Eur., Europe; As., Asia; Afr., Africa; E., early; M., middle; L., late. Extinct orders are indicated by an asterisk. No attempt is made here to present a detailed characterization of each order, and more complete surveys may be found in Dawson (1967), Piveteau (1958, 1961), Romer, 1966), Simpson, (1945), and many other references cited by these authors.

Order Proteutheria*. (L. Cret.–M. Olig., N. A.; L. Paleoc.–L. Eoc., Eur.; M. Cret., As.) The systematic position of some of the extinct archaic eutherians has long been unclear. These forms, which include the primitive long-snouted leptictids, the dermopteran-like mixodectids, and convergently rodent-like apatemyids, were previously assigned to the catch-all order Insectivora for the lack of a better arrangement. Recently, Butler (1972) noted that the Insectivora is more easily defined with the exclusion of these archaic forms. The union of these families under the single order Proteutheria is highly tentative and future studies may lead to its dissolution. The skulls of some proteutherians are shown in Fig. 4.

Order Insectivora. (L. Cret.–Holo., N. A.; M. Paleoc.–Holo., Eur.; ?M. Cret.–Holo., As.; E. Olig.–Holo., Afr.; Holo., S. A.) No other eutherian order has caused as many systematic problems as the Insectivora. Many groups whose relationships are poorly understood have been assigned to the Insectivora, resulting in a rather chaotic arrangement of living and extinct families. Despite these problems, it now seems possible to define a nucleus of insectivore families based on their common possession of such specializations as reduction in the number of branches of the carotid arteries that supply blood to the brain, the loss of a blind pouch or caecum on the intestine, and the reduction or loss of certain bony elements in the skull. Included within the Insectivora are the living moles (Fig. 4C), shrews, hedgehogs, and tenrecs, as well as several extinct families.

Order Macroscelidea. (E. Olig.–Holo., Afr.) The curious "elephant shrews" of Africa were long regarded as relatives of the tupaiids or "tree shrews"; but most of the similarities between the two groups, notably the presence of an intestinal caecum and the retention of all major branches of the carotid artery, are primitive and not necessarily indicative of a true relationship. Macroscelidids seem best regarded as a separate order of placental mammals. Several fossil elephant shrews are known from the Miocene of Africa.

Order Anagalida*. (?Cret.–Olig., As.) Szalay and McKenna (1971) recently established an aberrant group of endemic Asian mammals, as the order Anagalida (see Fig. 4F). These forms were once considered relatives of tupaiids but are now suspected as being closely related to the Macroscelidea (Evans, 1942).

Order Scandentia. (Holo., SW As.) The tupaiids or tree shrews were long regarded as the most primitive living primates. More recently, the distinctly nonprimate features of tree shrews have prompted their allocation to the Insectivora in many classifications. However, tupaiids are excluded from true insectivores in their retention of a number of primitive mammalian features in combination with specializations of the optic and brain anatomy. A separate order for tupaiids has been proposed (Butler, 1972). Several extinct genera from the

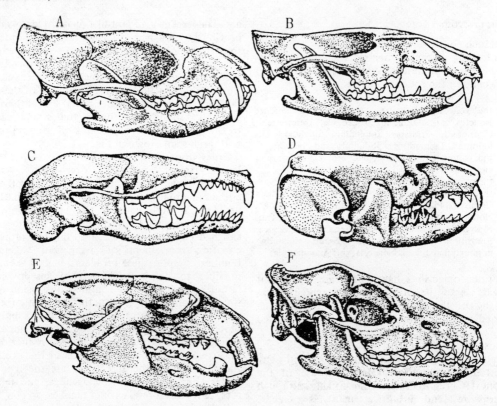

FIGURE 4. Skulls of some primitive placentals (from Romer, 1966). A, *Deltatheridium;* B, *Zalambdalestes* from the Late Cretaceous of Mongolia (skull lengths about 4.45 cm and 5.00 cm respectively); C, *Proscalops* of the Oligocene and Miocene (skull length about 2.54 cm); D, *Apternodus;* E, *Sinclairella* of Oligocene (skull lengths about 4.45 cm and 5.72 cm); F, *Anagale* of the Mongolian Oligocene (skull length about 5.72 cm). (Copyright © 1966 by The University of Chicago Press, and used by permission.)

Eocene and late Paleocene of North America and Europe have tupaiid-like features in the teeth, but the true affinities of these forms requires further study.

Order Edentata. (L. Paleoc.–Holo., S. A.; M. Plioc.–Pleist., N. A.) The edentates have remained in the shadows of paleontological consideration, yet this order apparently played a very important role in the overall phylogenetic history of eutherians. Edentates retain a number of truly primitive mammalian features—notably septomaxillary bones, ossified ribs reaching the sternum, and a low body temperature with poor thermoregulation—that have been significantly modified in all other eutherians (Hoffstetter, 1958). Further, edentates show so many truly aberrant anatomical features that the order must have diverged from the ancestral eutherian stock at some very remote time, long before their first appearance in the late Paleocene of South America.

Order Pholidota. (L. Paleoc.–E. Olig., N. A.; ?E. Olig.–Pleist., Eur.; Pleist.–Holo., As.; Afr.)

The pholidotes, whose Holocene representatives are the Manidae (pangolins), are the only known mammals with epithelial scales of a structure homologous to that of nails or claws that form a protective armor over the body. The Pholidota has a long fossil history, with a representative extinct family, Metacheiromyidae*, from the late Paleocene-early Eocene of North America. According to Emry (1970), metacheiromyids are closely related and perhaps directly ancestral to the living manids. It appears that the pholidotes originated in North America and only at some later time migrated to the Old World. A third pholidote family, the extinct Epoicotheriidae*, from the Eocene and Oligocene of North America is of uncertain affinities.

Order Chiroptera. (M. Mioc.–Holo., Old World; L. Paleoc.–Holo., world wide.) Two suborders of bats are recognized. The Megachiroptera, which includes the flying foxes (genus *Pteropus*) of Australia, are generally regarded as the more primitive suborder. The majority of bats belong to the great suborder

Microchiroptera, which is notable for many advanced specializations, such as the loss of a claw on the second digit, bizarre modifications of the shoulder girdle, and an extraordinary development of echolocating abilities. After rodents, the Chiroptera is the most diverse of the Holocene mammalian orders. Unfortunately, because of their gracile skeletons and aerial habits, bat fossils are rare; only 20 extinct genera are known compared with 175 living genera. The beautifully preserved *Icaronycteris index* from the middle Eocene of Wyoming is clearly a specialized microchiropteran suggesting that the history of the order may extend back as far as the Mesozoic.

Order Dermoptera. (L. Paleoc.-L. Eoc., N. A.; E. Eoc., Eur.; Holo., SE As.) The dermoperans or "flying lemurs" have long been considered distant relatives of primates. Although the Holocene dermopteran family Cynocephalidae does not have fossil representatives, the late Paleocene and early Eocene Plagiomenidae* might have been primitive dermopterans, as they closely resemble cynocephalids in having multicuspidate structure of the molars and bilobate incisors (Rose, 1973).

Order Primates. (?L. Cret.-Mioc., N. A.; L. Paleoc., L. Eoc-Holo., Eur.; E. Olig.-Holo., Afr.; L. Eoc.-Holo., As.; E. Olig.-Holo., S. A.) The fossil record of primates is relatively good, and excellent and abundant specimens of prosimians are known from the early Tertiary of North America and Europe. The ancestral source for the primates is still not clearly understood but evidence derived from dental and basicranial morphology points toward an affinity with primitive hedgehog-like insectivores.

Order Taeniodonta*. (E. Paleoc.-L. Eoc., N. A.; M. Eoc., As.) The early Tertiary saw the appearance and extinction of several archaic phyla. Among the most interesting of these were the taeniodonts, an aberrant group of mammals that evolved long-clawed feet; evergrowing cheek teeth; short, broad skulls; and deep jaws. The taeniodonts are subdivided into two lineages. The extinct conoryctines were primarily conservative in their evolution, while the extinct stylinodontines underwent drastic morphological change during their 33 m yr history and are often cited as one of the outstanding examples of rapid evolutionary change among the mammals. (Patterson, 1949).

Order Lagomorpha. (L. Paleoc.-Holo., As.; L. Eoc.-Holo., N. A.; L. Olig-Holo., Eur.; Mioc.-Holo., Afr.; Pleist.-Holo., S. A.; Holo., Australia.) The lagomorphs comprise an isolated mammalian order, apparently Asian in origin, that arose from some Paleocene or earlier eutherian stock. The oldest known unquestionable lagomorphs are the Eurymyli-

dae*, from the Paleocene of Asia. The eurymylid dental pattern, especially the lower molars, approaches that of later lagomorphs. Lagomorphs of the late Eocene of North America and Asia have a dental structure that could be prototypical for both the Leporidae (rabbits) and Ochotonidae (pikas). Leporids flourished in the middle and late Tertiary of North America and reached the Old World in the late Tertiary. Leporines, the Holocene leporids, are first known from the Pliocene of the Old World and probably entered the New World some time later, spreading to their present distribution.

Ochotonids are first known from the Oligocene of Eurasia and flourished during the Miocene in Asia, Africa, North America, and Europe. The Holocene genus *Ochotona* can be traced back to ancestors in the late Miocene of Asia (see Dawson, 1967).

Order Rodentia. (L. Paleoc.-Holo., N. A.; E. Eoc.-Holo., Eur.; L. Eoc.-Holo., As.; E. Olig.-Holo., S. A.; E. Olig.-Holo., Afr.) The fossil record of the Rodentia, although incomplete, suggests some basic relationships: (1) The Paramyidae* from the Paleocene and Eocene of North America and the Eocene of Eurasia are the most primitive of the rodent stock and may have been ancestral to most of the later rodent lineages. (2) Most Eocene and some Oligocene rodents and the Aplodontidae form a group having a primitive arrangement of the masseter muscle on the zygomatic arch. (3) South American caviomorphs form a group of related forms of controversial origin. (4) early geomyoids, muroids, and dipodoids were probably derived from the same ancestral stock. Outside these groupings are numerous rodent families whose affinities are not clearly known. Dawson (1967) has outlined the major events of rodent history and has cited many of the important references dealing with rodent evolution.

Order Cetacea. (M. Eoc.-Holo., all world oceans, some rivers, lakes and seas) Much remains to be learned about the evolutionary history of the Cetacea. The three suborders (sometimes recognized as orders), Archaeoceti*, Odontoceti (toothed whales), and Mysticeti (toothless, whale-bone whales), have been regarded as independent descendants from some unknown ancestral stock. However, Van Valen (1968) has argued strongly for the monophyly of the group, pointing out that the middle Eocene *Protocetus* from Africa is a clear morphological intermediate between mesonychid condylarths (see below) and modern whales.

Order Carnivora. (M. Paleoc.-Holo., N.A.; Eoc.-Holo., Eur.; L. Eoc.-Holo., As.; M. Mioc.-Holo., Afr..; M. Plioc.-Holo., S. A.) The earliest undoubted carnivores are the

Miacidae, from the middle Paleocene to late Eocene of North America, the Eocene of Europe, and the early Oligocene of Asia. Miacids differ from the more advanced carnivore families in lacking a fusion of proximal carpal (front foot) elements—the scaphoid, lunar, and centrale—and in the absence of a fully ossified auditory bulla. The origin of the Miacidae is unknown, but Lillegraven (1969) has remarked on their dental resemblance to certain Late Cretaceous palaeoryctoid insectivores. The major branches of the carnivore phylogeny are the mustelids (badgers, otters, skunks, weasels), felids (cats), hyaenids (hyaenas, aardwolfs), canids (dogs), procyonids (raccoons and allies), viverrids (civets, mongooses), and ursids (bears). Pinnipeds (seals and walruses) are recognized as a separate order of mammals in some classifications, but it seems more realistic to regard pinnipeds as carnivore families. Fossil and serological evidence suggests that the otariids (sea lions and walruses) and phocids (seals) were independently derived from separate primitive carnivore stocks.

Order Creodonta*. (M. Paleoc.–M. Olig., N. A.; M. Eoc.–M. Olig., Eur.; L. Eoc.–E. Olig., E. Plioc., As.; E. Olig.–M. Mioc., Afr.) The order Creodonta has had a tortuous taxonomic history, and many eutherian families that are now relegated to such varied orders as the Insectivora, Condylarthra*, and Carnivora were once included within the Creodonta. Recently, the Creodonta has taken on a much more restricted meaning to include early carnivorous mammals with low skulls, small braincases, well-developed sagittal and occipital crests, and $M_{\frac{1}{2}}$ or $M_{\frac{2}{3}}$ developed as carnassial teeth (strongly slicing, trenchant teeth characteristic of carnivores). Only two creodont families are now recognized, the Hyaenodontidae and the Oxyaenidae.

Order Condylarthra*. (L. Cret.–L. Eoc., N. A.; L. Paleoc.–E. Olig., Eur.; L. Paleoc.–L. Eoc., As.) The condylarths comprise a very broadly defined group of archaic mammals some of whose members stand at or near the ancestry of the perissodactyls, artiodactyls, and perhaps even cetaceans. The central condylarth stock, the Arctocyonidae*, can be traced back to the Late Cretaceous eutherian *Protungulatum*. Condylarths were diverse and abundant during the Paleocene and Eocene of North America and Eurasia and at least eight families are recognized. Most of these stocks disappeared by the end of the Eocene.

South American Ungulates—Orders Liptopterna*, Notoungulata*, Astrapotheria*, Xenungulata*, Pyrotheria*. (L. Paleoc.–L. Pleist., S. A.) During nearly the entire Cenozoic, South America was isolated from other continents,

and accordingly its mammalian fauna adopted a strikingly different character. The South American ungulates underwent a tremendous adaptive radiation during the Eocene with lineages that converged on rodents, hippopotamuses, rhinoceroses, and horses (Fig. 5). These ungulate groups reached their climax in development in the Oligocene and Miocene; but with the invasion of North American placentals in the Pliocene, the numbers of South American ungulates dwindled. By the end of the Pleistocene, these once so successful mammals had vanished entirely. For an excellent outline of South American ungulate evolutionary history see Patterson and Pasqual (1972).

Order Tubilidentata. (?L. Eoc.–Plioc., Eur.; Plioc. As.; Mioc.–Holo., Afr.) The fossil record of the tubilidentates can be traced back to the Miocene of Africa where forms closely resemble the Holocene *Orycteropus* (aardvark) in their peculiar internal columnar structure of the teeth. The origins of the Tubilidentata are obscure, but a derivation from some primitive condylarth group has been suggested (Colbert, 1941).

Order Pantodonta*. (M. Paleoc.–E. Eoc., N. A.; E. Eoc., Eur.; M.(?)L. Eoc.–M. Olig., As.) Pantodonts are an archaic order of pig-like mammals that appeared in the middle Paleocene, ranged widely in the early Eocene of North American and Europe, and virtually disappeared by middle Eocene times. However, Asian genera survived into the late Eocene and even the Oligocene.

Order Dinocerata*. (L. Paleoc.–M. Eoc., N. A.; L. Paleoc.–L. Eoc., As.) During the later Eocene in North America, a truly grotesque group of mammals flourished. These were the uintatheres, or Dinocerata, forms noted for their large size, short limbs and feet, and several pairs of hornlike bony swellings on the long, low skull. Uintatheres are unknown outside of North America except for a primitive Asiatic late Paleocene genus and *Gobiatherium* from the late Eocene of Asia.

Order Proboscidea. (L. Eoc.–Holo., Afr.; E. Mioc.–Pleist.; N. A., Eu.; E. Mioc.–Holo., As.; Pleist.; S. A.) The peculiar proboscideans (which include the mastodons, elephants, and related types) were apparently of African origin but successfully invaded other continents and by middle and later Cenozoic times were widespread in Eurasia, North America, and even reached South America. The late Eocene-early Oligocene African genus *Moeritherium*, an animal of tapir-like proportions, had a dentition that already showed some specializations found in later Proboscideans. There is some argument as to whether *Moeritherium* should

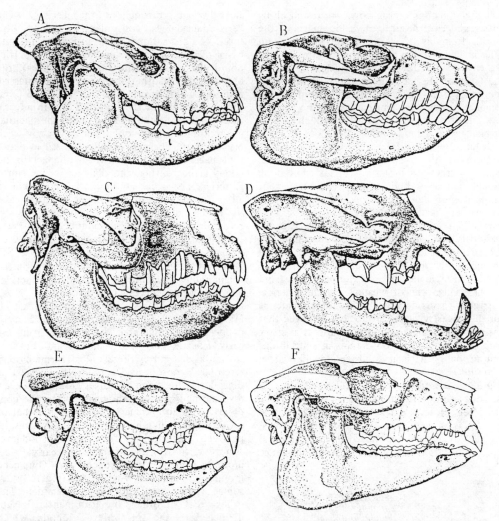

FIGURE 5. Skulls of South American ungulates (from Romer, 1966). A, *Homalodotherium,* Miocene, a toxodont (skull length about 38.1 cm); B, *Protypotherium,* Miocene, a typothere (skull length about 10.2 cm); C, *Nesodon,* Miocene, a toxodont (skull length about 40.6 cm); D, *Astrapotherium,* Miocene (skull length about 68.6 cm); E, *Notostylops,* a primitive Eocene notoungulate–Notioprogonia–(skull length about 15.2 cm); F, *Hegetotherium,* Miocene (skull length about 11.4 cm). (Copyright © 1966 by the University of Chicago Press, and used by permission.)

be regarded as a primitive proboscidean, an early sirenian, or a sole representative of a separate order intermediate between these groups.

Order Embrithopoda*. (E. Olig., Afr.) The order Embrithopoda consists of a single genus *Arsinotherium,* from the early Oligocene of Egypt. *Arsinotherium* is a large beast with massive limbs and a huge pair of horns on the nasal bones. The affinities of this curious creature are uncertain.

Order Hyracoidea. (E. Olig.–Holo., Afr.; E. Plioc.–Pleist., Eur.; ?Tertiary–Holo., As.) Although not too common today, the rabbit-sized herbivorous hyracoids, or conies, flour-

ished in Africa in the Cenozoic past, and a variety of forms are known from lower Oligocene beds of Egypt. Whitworth (1954) has suggested that the Hyracoidea evolved in isolation in Africa from some early condylarth-like stock.

Order Sirenia. (M. Eoc.–Holo., Afr.; L. Eoc.–Holo., As.; E. Mioc.–Holo., S. A.; ?M. Eoc.–Holo., N. A.; M. Eoc.–L. Plioc., Eu.) This bizarre order of marine mammals is characterized by herbivorous, wholly aquatic habits, loss of hind limbs, modification of forelimbs into paddles, horizontally expanded tail, loss of nasal and lacrimal bones in skull, and specializations in dental structure (Fig. 6). The manatees and dugongs or sea cows are found along coasts and river mouths in various parts of the world. Remains of sirenians are known from Eocene beds of Egypt. By the middle Eocene, sea cows were fully aquatic and had spread to the Gulf of Mexico, and to the East Indies by the end of the Eocene. Their distribution has remained essentially pantropical to the present day, though their greatest abundance and diversity were attained in the Miocene.

Order Desmostylia*. (L. Olig.-L. Mioc., western N. A.; E.-L. Mioc., Japan; L. Olig.-L. Mioc., eastern As.) Desmostylians were long thought to be a kind of sirenian because of similarities in dental features, but recent fossil discoveries have revealed that these strange beasts had massively developed legs (Fig. 6). Desmostylians had rather hippopotamus-like proportions and it is apparent that these animals inhabited shallow coastal waters. The earliest known members of the order are from the late Oligocene of western North America, but these fossils do not provide direct evidence for the ancestry of the group. It is believed that desmostylians share a common ancestral stock with sirenians and proboscidians in amphibious and probably African subungulates like *Moeritherium.*

Order Perissodactyla. (L. Paleoc.-Holo., N. A.; Eoc.-Holo., Eur.; As.; Pleist.-Holo., S.A.; M. Mioc.-Holo., Afr.) Today, only a few perissodactyls, or odd-toed ungulates (several species of tapirs, rhinoceroses, and horses) exist; but this group showed great abundance

and diversity throughout the Tertiary. The Perissodactyla is conveniently divided into three major subdivisions, primarily on the basis of dental characteristics. (1) The Hippomorpha includes the Equidae (horses), the old world Paleotheriidae*, and the massive rhinoceros-like Brontotheriidae*, (2) The Ceratomorpha includes the tapiroids and rhinoceratids. (3) The Ancylopoda includes the peculiar chalicotheres*, the only perissodactyls with clawed feet. There is excellent fossil evidence that perissodactyls arose from phenacodontid condylarths during the Paleocene in North America. The order subsequently underwent an adaptive radiation in the Eocene which reached its climax in the earliest Oligocene. The decrease in diversity and abundance of the Perissodactyla after the Oligocene is thought to be in part due to the rise of the artiodactyls. See Radinsky (1969) for an excellent review on the origin and early evolution of the Perissodactyla.

Order Artiodactyla. (E. Eoc.-Holo., N.A., Eur., Eoc.-Holo., As.; Pleist.-Holo., S.A.; M. Mioc.-Holo., Afr.) The earliest artiodactyls, or even-toed ungulates, are known from the early Eocene in North America and Europe. In dental structure, these oldest forms closely resemble hyopsodontid condylarths, suggesting their derivation from the latter group. However, no intermediates are known between the condylarth type of tarsus (hind foot) and that of artiodactyls, which has a characteristic "double-pulleyed" astragalus. The oldest artiodactyl suborder, Suiformes, includes the early Palaedonta*, some extinct Eocene and Oligocene groups, and the surviving Suidae (pigs), Tayassuidae (peccaries), and Hippopotamidae. The other great suborder, the Ruminantia, is subdivided into the infraorder Tylopoda, which includes the Camelidae, and the most successful of the present-day artiodactyl groups, the Pecora, which includes cervids (deer), bovids (bison), antilocaprids (antelopes), giraffids, and numerous other extinct and living families. Primitive tylopods and pecorans appeared at the end of the Eocene and seem to be phyla independently derived from earlier more primitive artiodactyls. Since Miocene times, pecorans have become the dominant artiodactyls in terms of their diversity, widespread geographic range, and abundance. Useful references on various artiodactyl groups are by Frick (1937) and Scott (1940).

MICHAEL J. NOVACEK

References

Butler, P. M., 1972. The problem of insectivore classification, in K. A. Joysey and T. S. Kemp, eds., *Studies in Vertebrate Evolution.* New York: Winchester Press, 253-265.

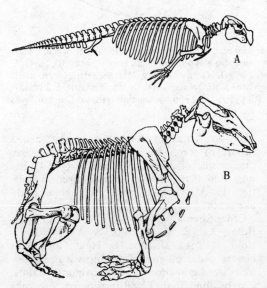

FIGURE 6. A, *Halitherium,* an Oligocene sirenian about 2.76 m in length; B, *Paleoparadoxia,* a desmostylian about 2.30 m in length (from Romer, 1966).

Butler, P. M., and Kielan-Jaworowska, Z., 1973. Is *Deltatheridium* a marsupial? *Nature,* **245,** 105–106.

Colbert, E. H., 1941. A study of *Orycteropus gaudryi* from the island of Samos, *Amer. Mus. Nat. Hist. Bull.,* 78, 305–351.

Dawson, M. R., 1967. The fossil history of the families of Recent mammals, in S. Anderson and J. Knox Jones Jr., eds., *Recent Mammals of the World, a Synopsis of Families.* New York: Ronald Press, 12–53.

Emry, R. L., 1970. A North American Oligocene pangolin and other additions to the Pholidota, *Amer. Mus. Nat. Hist. Bull.,* **142,** 455–510.

Evans, F. G., 1942. The osteology and relationships of the elephant shrews (Macroscelididae), *Amer. Mus. Nat. Hist. Bull.,* **80,** 85–125.

Flessa, K. W., 1975. Area, continental drift and mammalian diversity, *Paleobiology,* **1,** 189–194.

Fox, R. C., 1975. Molar structure and function in the early Cretaceous mammal *Pappotherium:* Evolutionary implications for Mesozoic Theria, *Canadian J. Earth Sci.,* **12,** 412–442.

Frick, C., 1937. Horned ruminants of North America, *Amer. Mus. Nat. Hist. Bull.,* **69,** 1–669.

Hennig, W., 1966. *Phylogenetic Systematics.* Urbana, Chicago, London: Univ. of Illinois Press, 263p.

Hoffstetter, R., 1958. Xenarthra, in Piveteau, ed., *Traité de Paléontologie,* vol. 6(2), 535–636.

Hoffstetter, R., 1973. Origine, compréhension et signification des taxons de rang supérieur: quelques enseignements tirés de l'histoire des mammifères, *Ann. Paléontol. (Vert.),* **59,** 137–169.

Kielen-Jaworowska, Z., 1969. Preliminary data on the Upper Cretaceous eutherian mammals from Bayn Dzak, Gobi Desert, *Palaeontol. Polonica,* **19,** 171–191.

Kielan-Jaworowska, Z.; Bown, T. M.; and Lillegraven, J. A., 1979. Eutheria, in *Mesozoic Mammals: The First Two-Thirds of Mammalian History.* Berkeley: Univ. California Press.

Kühne, W. G., 1973. The systematic position of monotremes reconsidered (Mammalia), *Zeitschr. Morphol. Tiere.,* **75,** 59–64.

Lillegraven, J. A., 1969. Latest Cretaceous, mammals of the upper part of the Edmonton Formation of Alberta, Canada, and a review of the marsupial-placental dichotomy in mammalian evolution, *Univ. Kansas Paleontol. Contrib. Article 50,* 122p.

Lillegraven, J. A., 1972. Ordinal and familial diversity of Cenozoic mammals, *Taxon,* **1,** 261–274.

Lillegraven, J. A., 1974. Biogeographical considerations in the marsupial-placental dichotomy, *Ann. Rev. Ecol. Systematics,* **5,** 263–283.

McKenna, M. C., 1973. Sweepstakes, filters, corridors, Noah's Arks, and beached Viking funeral ships in paleogeography, in D. H. Tarling and S. K. Runcorn, eds. *Implications of Continental Drift to the Earth Sciences.* London and New York: Academic Press, vol. 1, 295–308.

McKenna, M. C., 1975. Toward a phylogenetic classification of the Mammalia, in W. P. Luckett and F. S. Szalay, eds. *Phylogeny of Primates, A Multidisciplinary Approach.* New York and London: Plenum Press, 21–46.

Patterson, B., 1949. Rates of evolution in Taeniodonta, in G. L. Jepson, E. Mayr, and G. G. Simpson, eds., *Genetics, Paleontology and Evolution.* Princeton: Princeton Univ. Press, 248–278.

Patterson, B., and Pascual, R., 1972. The fossil mammal fauna of South America, in A. Keast, F. C. Erk and B. Glass, eds., *Evolution, Mammals, and Southern Continents.* Albany: State Univ. of New York Press, 247–309.

Piveteau, J., ed., 1961, 1968. *Traité de Paléontologie,* vols. 1(1, 2) and 7. Paris: Masson et Ciè, 957p. and 675p.

Preston, F. W., 1962. The canonical distribution of commoness and rarity, *Ecology,* **43,** 185–215.

Radinsky, L. B., 1969. The early evolution of the Perissodactyla, *Evolution,* **23,** 308–328.

Romer, A. S., 1966. *Vertebrate Paleontology,* 3rd ed. Chicago: Univ. Chicago Press, 468p.

Rose, K. D., 1973. The mandibular dentition of *Plagiomene* (Dermoptera, Plagiomenidae), *Breviora,* **411,** 1–17.

Scott, W. B., 1940. Pt. 4, Artiodactyla, in W. B. Scott and G. L. Jepsen, eds., The mammalian fauna of the White River Oligocene, *Trans. Amer. Phil. Soc.,* **28,** 363–746.

Simpson, G. G., 1945. The principles of classification and a classification of mammals. *Amer. Mus. Nat. Hist. Bull.,* **85,** 1–350.

Simpson, G. G. 1962. *Evolution and Geography.* Eugene, Oregon: Univ. Oregon Press, 64p.

Szalay, F. S., and McKenna, M. C., 1971. Beginning of the age of mammals in Asia: The late Paleocene Gashato fauna, Mongolia, *Amer. Mus. Nat. Hist. Bull.,* **144,** 269–318.

Van Valen, L., 1968. Monophyly or diphyly in the origin of whales, *Evolution,* **22,** 37–41.

Cross-references: *Bones and Teeth; Mammals, Mesozoic; Multituberculata; Paleobiogeography of Vertebrates; Vertebrata; Vertebrate Paleontology.*

MARSUPIALS

Marsupials (Metatheria) are the pouched mammals, that is mammals in which the females bear their young alive but in an almost foetal state and, in most species, carry them within an external pocket or pouch formed by a flap of dermal tissue on the abdomen. The most familiar marsupials are the kangaroos (*Macropus*), koalas (*Phascolarctos*), wombats (*Vombatus, Lasiorhinus*), etc., of Australia and neighboring islands, and the opossums of the Americas (*Didelphis, Marmosa,* etc.; see Fig. 1). While a living marsupial is easy to recognize in most cases because of the presence of a pouch in the female, possibly with young within, such recognition is impossible on fossil marsupials because the skin and muscles of the pouch are not preserved, and the state of the young at birth usually cannot be determined.

Fossil marsupials are recognized on anatomical peculiarities of the skeleton. Characters

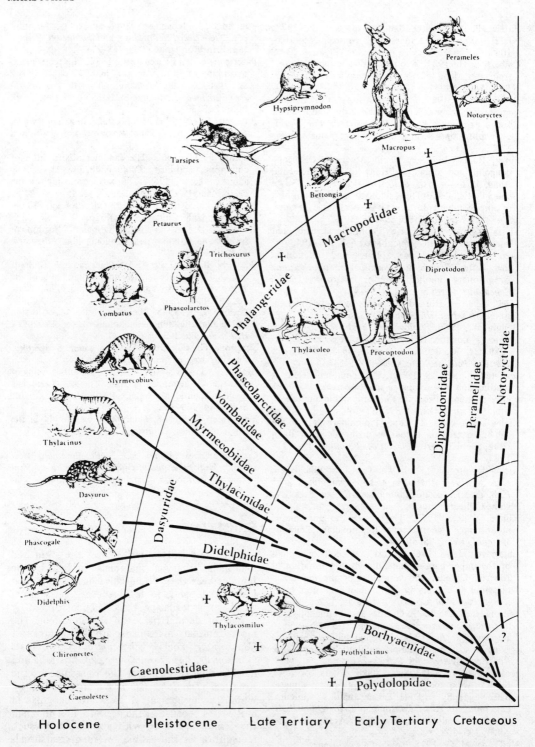

FIGURE 1. Evolution and radiation of the Marsupialia (after Thenius, 1969).

considered typical of marsupials are the presence of an inflected angle to the dentary (except in *Tarsipes*); fenestrations in the palate; marked postorbital constriction between the relatively small brain case and the region of the frontal sinuses; the lacrymal foramen opening on the facial surface; the jugal reaching the glenoid fossa; incomplete auditory bullae formed by the alisphenoid and incompletely fused to the basicranium (Fig. 2), and the presence of marsupial bones (osses marsupiales or epipubes; Fig. 3), which vary and may be as long as the ilium in *Didelphis,* or reduced to cartilaginous nubbins in the thylacine or marsupial wolf (*Thylacinus*).

Dentitions are often a basis for identification and the full dental formula may be given as

$$D.F. = I\frac{5}{4}, C\frac{1}{1}, P\frac{3}{3}, M\frac{4}{4}$$

(see Fig. 2). Reduction of the mammalian diphyodont replacement has proceeded further in marsupials than in eutherians, with only the third and possibly second premolars being

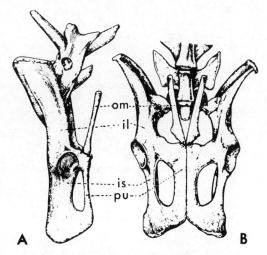

FIGURE 3. *Perameles* sp. A, lateral and B, ventral views of the pelvic girdle and posterior lumbar vertebrae to illustrate the position of the osses marsupiales (om) (from Piveteau, 1961). il, ilium; is, ischium; pu, pubis.

replaced. Thus the "milk" incisors, canines and the first premolars form part of the "permanent" dentition and are homologous with the "permanent" molars being part of the same "Zahnreihe" or toothrow. Only the replacement last premolar represents the second Zahnreihe.

Thus the true dental formula, with small letters for teeth of Zahnreihe 1 is

$$D.F. = i\frac{1-5}{1-4}, c\frac{1}{1}, p\frac{1-3}{1-3} + P\frac{3}{3}, m\frac{1-4}{1-4}$$

with p3 or p2 and p3 being replaced by P3 in adults.

Full dentitions are found only in the carnivorous or insectivorous marsupials, e.g., Didelphidae, Dasyuridae, and some Peramelidae; and reduction in numbers of both incisors and premolars, and possibly loss of canines, has taken place in the herbivores, e.g., Diprotodontidae, Phalangeridae, with possible

$$D.F. = i\frac{1-3}{1}, c\frac{0}{0}, p\frac{2}{2} + P\frac{3}{3}, m\frac{1-4}{1-4}$$

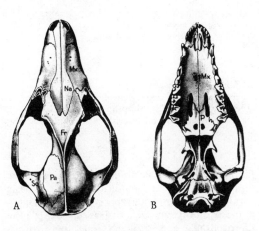

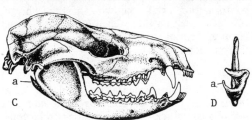

FIGURE 2. *Didelphis marsupialis.* A, dorsal and B, ventral views of skull (from Piveteau, 1961). C, lateral view of skull and mandible and D, posterior view of mandible showing inflected angle (after Romer, 1966). a, angle; bo, basioccipital; fr. frontal; mx, maxilla; na, nasal; p, palatine; pa, parietal; so, supraoccipital; sq, squamosal; t, temporal.

Within the Order Marsupialia (Table 1), Simpson (1945) lists six natural groupings, the superfamilies Didelphoidea, Borhyaenoidea (which he eliminated in 1970), Dasyuroidea, Perameloidea, Caenolestoidea, and Phalangeroidea. Attempts to unite these superfamilies at the subordinal level have proven unsatisfactory, although Polyprotodonta and Diprotodonta are

TABLE 1. Classification of Marsupials

Infraclass METATHERIA
  Order Marsupialia
    Suborder Polyprotodonta
      Superfamily Didelphoidea
        Family Didelphidae, opossums—SA, Upper Cretaceous–Holocene; Eu, Middle Eocene–lower Miocene;
            NA, Upper Cretaceous–Miocene, Pleistocene–Holocene
        Family Pediomyidae*–NA, SA, Upper Cretaceous
        Family Stagodontidae*–NA, Upper Cretaceous
        Family Borhyaenidae* South American marsupial carnivores—Upper Paleocene–Pliocene
        ?Family Necrolestidae*–SA, lower Miocene
      Superfamily Dasyuroidea
        Family Dasyuridae, Australian marsupial carnivores—Aus, NG, Pliocene–Holocene
        Family Thylacinidae, Australian wolf-like carnivores—Aus, Pliocene-Holocene
        Family Myrmecobiidae, banded anteaters or numbats—Aus, Pleistocene–Holocene
    Suborder Peramelina
      Family Peramelidae, bandicoots—Aus, NG, Pliocene–Holocene
      Family Thylacomyidae, rabbit-eared bandicoots or bilbies—Aus, Pleistocene–Holocene
    Suborder Caenolestoidea (= Paucituberculata)
      Family Caenolestidae, South American marsupial insectivores—SA, lower Eocene–Holocene
      Family Polydolopidae*–SA, upper Paleocene–lower Eocene
      Family Argyrolagidae*–SA, upper Pliocene–lower Pleistocene
      ?Family Groeberiidae*–SA, lower Oligocene
    Suborder Diprotodonta (= Phalangeroidea)
      Family Phalangeridae, phalangers—Aus, NG, Miocene–Holocene
      Family Ektopodontidae, phalanger-like marsupials—Aus, Miocene
      Family Burramyidae, pygmy possums—Aus, Miocene–Holocene
      Family Petauridae, gliders and ringtails—Aus, Pliocene–Holocene
      Family Thylacoleonidae*, marsupial lions—Aus, Miocene–Holocene
      Family Macropodidae, kangaroos and wallabies—Aus, Miocene–Holocene
      Family Vombatidae (= Phascolomidae), wombats—Aus, Miocene–Holocene
      Family Diprotodontidae*, giant quadrupedal marsupial herbivores—Aus, Miocene–Holocene
      Family Wynyardiidae*–Aus, Pliocene
      Family Phascolarctidae—Aus. Pliocene–Holocene
    Incertae Sedis
      Family Notoryctidae, Australian marsupial moles—Aus, Miocene–Holocene

Modified after Ride (1964), Clemens (1966), Romer (1966), Kirsch (1968), Simpson (1970), Calaby et al. (1974), and Archer and Kirsch (in press). Extinct families are indicated by an asterisk. SA, South America; NA, North America; Eu, Europe; Aus, Australia; NG, New Guinea.

generally used. An alternative classification is based on the union or freedom of the second and third toes of the pes as Syndactyla and Didactyla, but this classification is contradictory to the Polyprotodonta-Diprotodonta scheme. Syndactylous feet may be derived from the didelphoid condition, and appear in highly specialized form in the wallabies and kangaroos (Fig. 4).

The Caeonolestoidea and Perameloidea are anomalous within these classifications and cannot be combined into a third group with any real advantage (Fig. 5). In effect, the Didelphoidea, Borhyaenoidea, and Dasyuroidea represent a nearly intergrading morphological complex of carnivorous adaptations, with Perameloidea and Caenolestoidea representing divergent insectivorous adaptations and Phalangoroidea a spectrum of herbivorous adaptations geographically confined to Australia and showing no intergradations with the other super-

families. The Caenolestoidea and Perameloidea represent a shift in adaptation from the basic carnivorous conformation typical of Didelphoidea, Borhyaenoidea, or Dasyuroidea, to the diprotodont and syndactylous herbivorous conformation well known in the Phalangeroidea, with the Caenolestoidea having acquired diprotodonty and the Perameloidea the syndactylous pes.

History

Marsupials and placental mammals (or Metatheria and Eutheria) arose from a common mammalian stock in the middle Cretaceous, probably from the Pantotheria (Ride, 1962). Many authors feel that the different arrangements of the female urinogenital tracts in the Metatheria and Eutheria preclude any possible ancestor-descendent relationship. However, Lillegraven (1969), in his review, presents a case for Eutheria having descended from basic

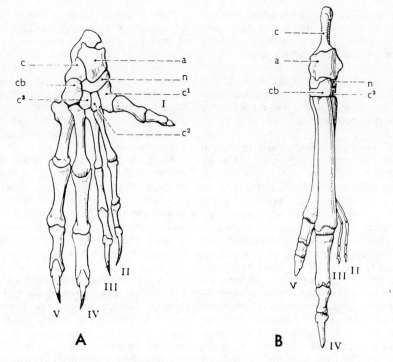

FIGURE 4. Right rear feet (pedes). A, *Phalangista vulpina,* X0.9; B, *Macropus bennetti,* to show generalized and specialized syndactyly, X0.25 (from Piveteau, 1961). Digits indicated by Roman numerals; a, astragalus; c, calcaneum; cb, cuboid; c^1, c^2, c^3, ento, meso, and ectocuneiforms, respectively; n, navicular.

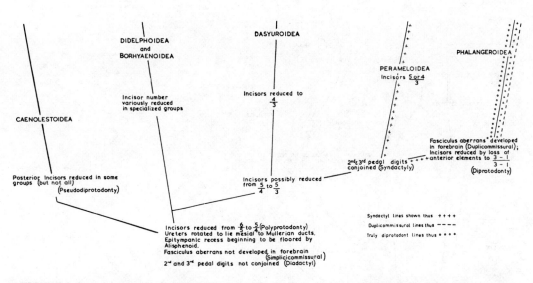

FIGURE 5. Sequence of morphological differentiation in the evolution of marsupial superfamilies (from Ride, 1962).

Metatheria, in which the reproductive innova-tions of Metatheria were further elaborated and modified. However, Lillegraven does not con-sider the nearly monophyodont condition in Metatheria, from which the basic eutherian diphyodont condition cannot be derived. The common ancestor of the two groups must thus have had basically a marsupial reproductive system and an eutherian diphyodont dental replacement.

The fossil record of Marsupialia is discon-tinuous geographically and temporally (Fig. 6). The geographic locus for the origin of the marsupials is still arguable; living or fossil forms are known from four continents (Europe, North and South America, and Australia). At present there is absolutely no evidence that any marsupials ever existed naturally during either the Mesozoic or Tertiary in Africa, Asia, or Antarctica (Fooden, 1972). Their peripheral distribution on the island continent of Australia and at the end of the Asian–North American–South American chain of continents, suggested to early workers that they represented relict populations forced to the peripheries of their range by eutherian replacement in the Eurasian heartland (Fig. 7). This, however, is unlikely because marsupials are only recorded from Europe and not from Asia.

Plate tectonic theory, with the fragmentation of Pangaea beginning in the Mesozoic (Figs. 8, 9, and 10), provides a possible explanation for the ancient and present distribution of the marsupials without resorting to land bridges or peripheral survival of relict populations. The continuing isolation of the Australian plate, and of the South American plate, from the early Paleocene to the Pliocene may account for the persistence of diverse marsupial faunas in these areas.

When Pangaea began to fragment during the Mesozoic (Fig. 8), the rifting started along what may now be called the E coast of Africa, separating to the east the land masses that would eventually be known as India, Australia, and Antarctica. The split widened during the Triassic and Jurassic to form the beginnings of the Indian Ocean (Fig. 9) and extended round southern Africa to produce the beginnings of the Atlantic Ocean. Coincident with this rifting, the Tethys sea narrowed as Africarabia, and Eurasia rotated so that their margins fused along the line of the present Elburz and Hindu Kush mountain ranges; and North America moved northwestward away from Africa, rotating about northern Greenland, to produce an early North Atlantic Basin and Caribbean Sea. There were thus two major movements at

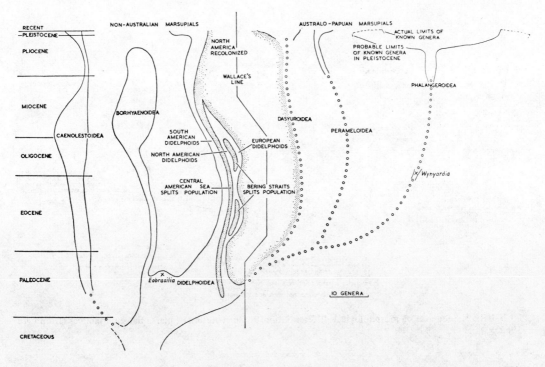

FIGURE 6. A family tree of the marsupials (from Ride, 1962; data from Simpson, 1945; de Paula Couto, 1952; and Stirton, 1955, 1957a,b). Horizontal distances between unbroken limits of each superfamily at each horizon represent known numbers of genera.

450

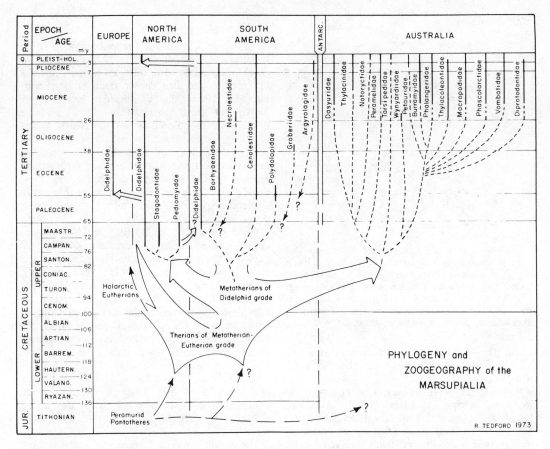

FIGURE 7. Known geological ranges and inferred phyletic and zoogeographic relationships of the families of marsupials (from Tedford, 1974). Dashed borders between continent columns show that chance for dispersal between adjacent continents decreased through the Cretaceous and early Tertiary, becoming disjunct in the late Tertiary due to interposition of oceanic crust and development of a climatic barrier of increasing severity at high latitudes in both hemispheres.

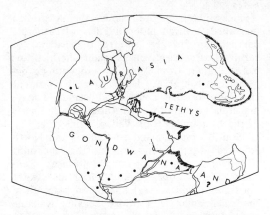

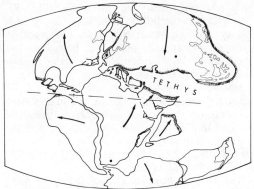

FIGURE 8. Pangaea just before its break up in the Triassic. Continental blocks are indicated by outlines of the shelves with some modern coastlines indicated for ease of identification. Dots indicate loci of mammal-like reptilian faunas. Hatchured lines indicate uncertain continental margins. Dashed line indicates position of the Triassic equator.

FIGURE 9. Pangaea in the course of its break up during the Jurassic. Outlines as in Fig. 8. Arrows indicate general directions of drifting. Dots indicate loci of finds of Jurassic mammals. Dashed line indicates location of the Late Jurassic equator. Hatchured lines indicate uncertain continental margins.

451

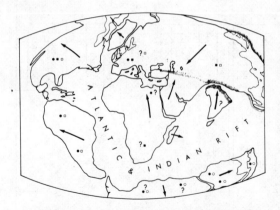

FIGURE 10. The continents in the Palaeocene. Arrows indicate general directions of Tertiary drifting by the continents. Dots indicate loci of marsupials, both fossil and recent; hollow dots of monotremes, solid squares of placentals; and question-marks, absence of Palaeocene or early Eocene fossils. Hatchured lines indicate uncertain continental margins.

work, the Atlantic and Indian Rift and the rotations of North America and Africa relative to Eurasia.

Mammal-like reptilian faunas are known from southern Africa, southern South America, India, and Antarctica in Gondwanaland, but not yet from Australasia (Fig. 8). In Laurasia, similar faunas are known from western North America, Scotland, and western China, although none of these faunas are as varied as those from the Gondwanaland. However, they seem to show that early mammals might have been originally distributed widely and this is supported by the known occurrences of the remains of early therians. The first Jurassic mammals are known from south Africa, Wales, and western China and Mongolia (Fig. 9). The lack of specimens from the New World, Antarctica, and Australasia does not say that such early mammals might not have lived on these land masses, although the known distributions may give that impression.

Marsupials thus seem to have occupied the southern and western portions of Pangaea after its break up and to have immigrated during the Eocene into Europe from North America. Such a distribution suggests an origin in the Americas (see Tedford, 1974; Clemens, 1968; Cox, 1970, 1973; but see Kirsch, 1977).

The first marsupials, known from the Late Cretaceous of North and South America (see Clemens, 1979), would thus have been able to walk from North America across what is now Greenland into Europe (as took place during the Eocene) or to Australia via Antarctica (as must have taken place during the late Mesozoic or early Cenozoic; Tedford, 1974).

All Upper Cretaceous forms belong to three families of Didelphoidea: The Didelphidae, reaching greatest diversity in North America (11 genera) and also known from South America (Sigé, 1968); the Pediomyidae, known from North and South America (Sigé, 1972); and the Stagodontidae, known only from North America. The North American record suggests that the Didelphoidea, and possibly the Didelphidae may represent the primitive stock (still represented by the existing Didelphidae and Dasyuridae) that gave rise to the other four superfamilies in Late Cretaceous–Paleocene times.

While the early marsupials seem to have occupied the southern and western portions of Pangaea, the placental mammals appear to have been predominantly Eurasian and possibly African. They colonized North America, with some herbivores reaching South America in late Mesozoic or earliest Tertiary times. Competition between the marsupials and the evolving placental carnivores resulted in the extinction of the didelphoid marsupials in Europe and North America by the end of the Miocene. The presence of a wide variety of placental herbivores in South America excluded that continent's didelphoid stock from producing a herbivore radiation, as took place in Australasia.

The South American marsupial radiation did produce many forms of carnivores, however, including the otter-like (*Lutreolina*), cat-like (*Borhyaena*), hyaena-like (*Prothylacinus, Amphiproviverra*), and sabre-tusked (*Thylacosmilus*) forms. They also gave rise to the insectivorous *Caenolestes* and mole-like *Necrolestes*.

Since the Late Cretaceous, the Australian marsupials evolved independently of eutherian competition, to give rise to the balance of carnivorous, insectivorous, and herbivorous forms known from Miocene and younger deposits (Stirton et al., 1968) of that continent and its associated continental islands. Today, marsupials are confined to Australasia and parts of the East Indies E of Wallace's line (Fig. 11), to South America, and to Central and North America where reinvasion by the opossums (*Didelphis*), mouse opossums (*Marmosa, Philander, Metachirops, Monodelphis*) and water opossums (*Chironectes*) has taken place since the Panamanian land bridge became available in late Pliocene–Pleistocene times.

Variations and Adaptations of Marsupials

The basic didelphoid marsupial is a generalized carnivore in which the skull has a well-developed rostrum, canines, and cheek teeth,

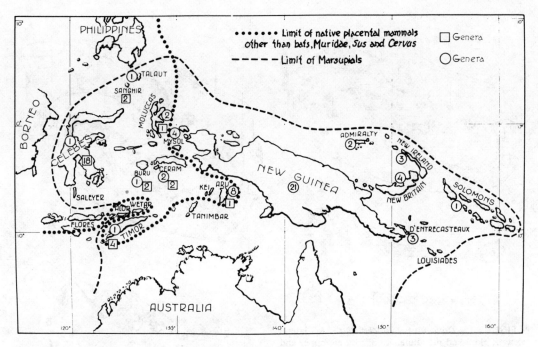

FIGURE 11. Distribution of Marsupials and Eutherians in the Indo-Australian Zone (from Simpson, 1961). Figures in boxes indicate numbers of genera present on island.

with a relatively small braincase ornamented by prominent saggital and nuchal crests (Fig. 2). The upper cheek teeth comprise a trigon with well-developed buccal styles (Fig. 12A) and the lower cheek teeth a trigonid with a large distal talonid. These teeth in occlusion comprise a functional shearing and crushing mechanism. Premolars may be laterally compressed, two-rooted, and similar to those of a dog, in the opossum, *Didelphis,* or marsupial wolf, *Thylacinus* (Figs. 12 and 13); or heavy and crushing, similar to those of hyaena-like carnivores, as in *Borhyaena* (Fig. 14) and the Tasmanian devil, *Sarcophilus* (Fig. 15). The carnivores evolved from the didelphoid pattern converged toward the placental Mustelidae or Viverridae in general conformation, with specializations in the Borhyaenidae toward large "hyaenodont-like" forms (e.g., *Borhyaena*) and mustelid-like forms (e.g., *Cladosictis,* Fig. 16) in South America and forms similar to placental Canidae (e.g., *Thylacinus*), Mustelidae (e.g., *Sarcophilus*) and Viverridae (e.g., native-cats, Dasyuridae) in Australia. In South America, there existed during the Pliocene a borhyaenid in which the canines were large and sabrelike (*Thylacosmilus*; Fig. 17). In general conformation, this animal resembled a sabre-toothed tiger in that the manible carried flanges to protect the enlarged canines, the canines were rooted high in the

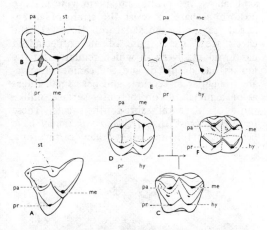

FIGURE 12. Molarization of the upper cheek teeth in marsupials (from Piveteau, 1961). Development of the oblique carnassial blade from the generalized condition (A, *Dasyurus viverrinus*) to the more specialized condition (B, *Thylacinus cynocephalus*). Specialization from the generalized bundondont stage (C, *Petauroides volans*), to a bunolophodont stage (D, *Trichosurus vulpecula*), to an eventual lophodont stage (E, *Macropus* sp.). Specialization toward a fully selendont stage is only achieved in the koala (F, *Phascolarctos cinereus*). hy, hypocone; me, metacone; pa, paracone; pr, protocone; st, style.

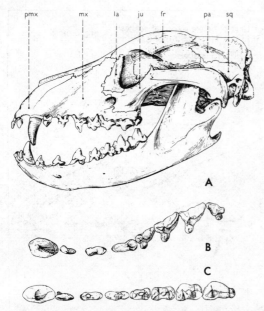

FIGURE 13. *Thylacinus cynocephalus.* A, lateral view of skull and mandible, X0.35; B, left upper and C, right lower dentitions in occlusal views, both X0.50 (from Piveteau, 1961). fr, frontal; ju, jugal; la, lacrymal; mx, maxilla; pa, parietal; pmx, premaxilla; sq, squamosal.

skull above the orbits, and there were massive postorbital bars to resist the strain of an enlarged temporal musculature. However, the cheek teeth were those of an unspecialized dasyuroid carnivore in which reduction of the molars has occurred; and the canines diverge, are triangular in section, and bear small ridges as on a metal file on the outer surface, unlike those of placental sabre-toothed cats. These features suggest that *Thylacosmilus* may have been an omnivore rather than carnivorous.

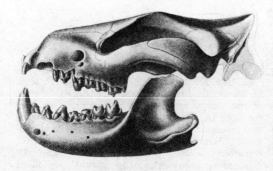

FIGURE 14. *Borhyaena excavata.* Lateral views of the skull and mandible, X0.35 (from Piveteau, 1961).

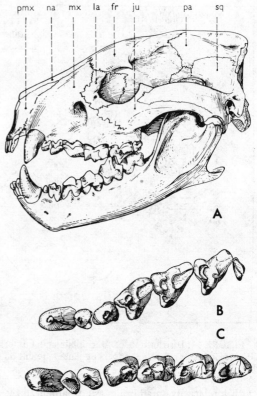

FIGURE 15. *Sarcophilus harrisi.* A, lateral view of skull and mandible, X0.4; B, left upper and C, right lower dentitions in occlusal views, both X0.6 (from Piveteau, 1961). fr frontal; ju jugal; la, lacrymal; mx maxilla; na nasal; pa parietal; pmx, premaxilla; sq, squamosal.

Canines with triangular cross sections are found in many animals that use these teeth as tools rather than as weapons, e.g., Suina; the divergence of the teeth would imply that they were used separately, perhaps on flesh, perhaps for rooting and grubbing. Finally, the generalized dentition was suited to masticating either flesh or vegetable matter as chance presented them.

The Caenolestoidea and smaller Australian Dasyuroidea include marsupial insectivores that superficially resemble the placental shrews and, in Australia, include marsupial moles (*Notoryctes*) and the numbat or banded anteaters (*Myrmecobius*). The Australian bandicoots (Peramelidae) resemble rabbits with long snouts, but are omnivorous in their diet. The molars tend to become squared and bunodont (Fig. 12D) as an adaptation to a more grinding occlusion, and the replacement premolar tends to become transversely flattened to form a shearing blade, especially in the

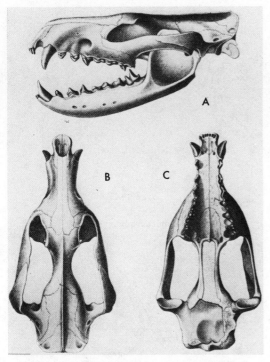

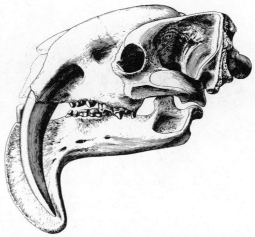

FIGURE 17. *Thylacosmilus atrox.* Lateral view of skull and mandible, X0.17 (from Riggs, 1934).

FIGURE 16. *Cladosictis lustratus.* A, lateral view of skull and mandible; B, dorsal and C, ventral views of skull (damaged in basicranial region); all X0.4 (from Piveteau, 1961).

lower jaw. Such adaptations are also seen in the South American caenolestoid Polydolopidae (e.g., *Epidolops*) and Caenolestidae (e.g., *Abderites,* Fig. 18).

The possums or phalangers (Family Phalangeridae, Suborder Diprotodonta) comprise a number of squirrel-like or lower primate-like herbivorous, insectivorous, or frugivorous forms, e.g., brushtailed possum (*Trichosurus;* Fig. 19D), glider (*Petaurus;* Fig. 18), cuscus (*Phalanger*), and a number of small forms, e.g., the pygmy gliders (*Acrobates*) and possums (*Cercartelus*).

The koala (*Phascolarctos; Fig. 19A*) is a large, arboreal vombatid that resembles a small bear. It is herbivorous, subsisting solely on a mixed diet of the leaves of a few species of *Eucalyptus* trees. The incisors are small, with three vertically oriented upper and one procumbent lower in each side, suitable for plucking off leaves. The cheek teeth are squared, four-rooted, brachyodont and selenodont in occlusal pattern (Fig. 12 F), which recalls the dentition of some of the primitive Artiodactyla. The replacement premolars are elongated mesiodistally and somewhat blade-like.

A unique evolutionary phalangerid adaptation is found in the family Thylacoleonidae, the extinct Miocene to Pleistocene carnivores exemplified by *Thylacoleo,* the marsupial "lion" (Fig. 20). In this form, the dentition consisted of large canine-like incisors, generally reduced nubbin-like premolars, large shearing carnassials derived from the upper and lower replacement premolars and first lower molar, and reduced posterior molars. The carnassials are laterally compressed, similar to those seen in the placental scimitar cat, *Homotherium,* and lack any buttressing roots as seen in modern placental carnivores' P^4's. The shearing action is enhanced by being oriented mesio-distally and not obliquely, as in primitive placental and marsupial carnivores, and the skull is massive, with a strong postorbital bar and crests, indicative of a strong musculature. There were opposable digits on the manus and pes, as in modern phalangers, and the feet were armed with large, sharp, compressed claws. As a carnivore, *Thylacoleo* could thus both grasp and claw its prey while tearing with its large incisors.

The wombats (*Vombatus,* Fig. 19B; *Lasiorhinus*) and extinct giant wombat (*Phascolonus*) are ground-dwelling, bear-like, or large marmot-like herbivores, of which the modern forms live in burrows. The incisors are rodent-like in that they are chisel-like and grow from persistent pulps, and only one is found in any one quadrant. The incisors of *Phascolonus* were larger and broad, but thin buccolingually, and appear to have been ideal as plane chisels to remove bark or cut broad-leafed vegetation. The

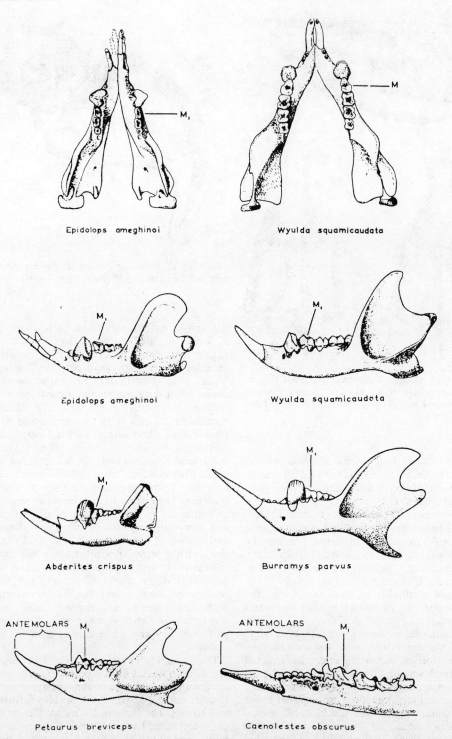

FIGURE 18. Occlusal aspects of the mandibles of *Epidolops ameghinoi* and *Wyulda squamicaudata* and lateral aspects of whole or partial mandibles and dentitions of *Epidolops ameghinoi*, *Wyulda squamicaudata*, *Abderites crispus*, *Burramys parvus*, *Petaurus breviceps*, and *Caenolestes obscurus*, to various scales to compare modification of premolars, especially last premolar (from Ride, 1962). M_1, first lower molar.

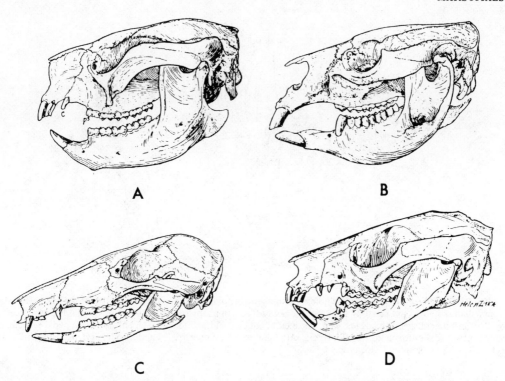

FIGURE 19. Skulls of living Australian diprotodont marsupials (from Gregory, 1951). A, *Phascolarctos cinereus,* ca. X0.3; B, *Vombatus ursinus,* ca. X0.3; C, *Potorous* sp., ca. X0.75; D, *Trichosurus vulpecula,* ca. X0.75.

cheek teeth of wombats are also persistently growing, with crowns that have waists as do those of most Artiodactyla or the koala, but the crown is bare of enamel and the exposed dentine is worn into two ridges on each tooth. The cheek teeth are curved, the uppers laterally

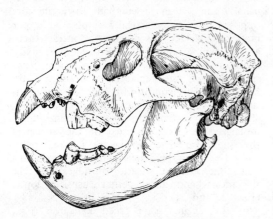

FIGURE 20. *Thylacoleo carnifex,* lateral view of skull and mandible (modified from Gregory, 1951).

and the lowers medially, so that the opposing forces in occlusion act at right angles to the axes of the pulps, similar to but reversed from the situation in the placental beaver, *Castor.* The replacement premolars are molarized with persistently growing pulps and each resembles half a molar in crown view.

The Family Diprotodontidae comprises the extinct giant quadrupedal marsupials. These range from *Diprotodon* (Fig. 21), some 7 ft (2 m) in shoulder height, to the lesser *Zygomaturus, Euryzygoma,* or *Eowenia,* of some 4 ft (1.2 m) shoulder height. These creatures had vertical chisel-like first upper incisors, with vertically implanted second and third upper incisors and single procumbant lower incisors on each side (Fig. 22). The cheek teeth, all of which were squared and transversely bilophodont (with the lophs forming shallowly crescentic ridges), were separated from the incisors by a large diastema. In *Eowenia* and *Zygomaturus,* the lower jaw moved anteroposteriorly to grind the food, as in rodents, and the upper molars' lophs were concave posteriorly to oppose the herbage moved against them and caught by the anteriorly concave lower molars'

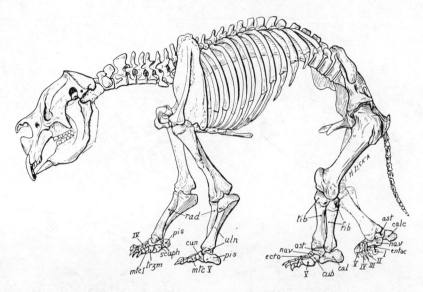

FIGURE 21. *Diprotodon australis.* Whole skeleton showing plantigrade gait, graviportal adaptations, essentially mammalian skeleton, and marsupial or epipubic bones (osses marsupiales) (from Gregory, 1951). Digits indicated by Roman numerals; ast, astragalus; calc, calcaneum; cub, cuboid; cun, manual cuneiform; ecto, pedal ectocuneiform; ento, pedal entocuneiform; fib, fibula; mtc I and mtc V, metacarpals I and V; nav, navicular; pis, pisiform; rad, radius; scaph, scaphoid; tib, tibia.

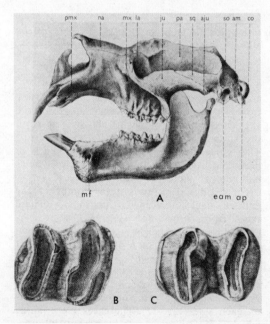

FIGURE 22. *Diprotodon australis.* A, lateral view of skull and mandible, X0.06; B, fourth upper molar, X0.50; C, lower molar, X0.50 (from Piveteau, 1961). Both teeth oriented with the distal or posterior surfaces to the left of the page, aju, jugal process; am, mastoid process; ap, paroccipital process; eam, external auditory meatus; co, occipital condyle; ju, jugal; la, lacrymal; mf, mental foramen; mx, maxilla; na, nasal; pa, parietal; pmx, premaxilla; so, supraoccipital; sq, squamosal.

lophs. In *Diprotodon,* the occlusion was more vertical with a crushing action at the end of the stroke, apparently approaching motion seen in the extinct placental proboscidean *Barytherium.* In these four marsupials, the ulnar styloid forms a ball joint with the ulnare in the wrist, about which the manus could be rotated.

The Macropodidae or kangaroos, musk kangaroos, wallabies, and extinct short-faced kangaroos comprise the marsupials that generally progress by a bounding gait. The pes is syndactylous and modified so that most of the weight falls on the fourth digit. The face is usually long, with procumbant lower incisors opposed to three vertically implanted upper incisors (Fig. 23). As in diprotodontids, there are no canines and the cheek teeth are usually four or five. In the musk kangaroos, *Hypsiprymnodon,* the replacement premolars are blade-like. This condition is reduced and eventually lost in the kangaroos and wallabies, although the premolars never become fully molarized or lose all trace of the ancestral shearing form. The molars are squared, four-cusped, with tendencies to bilophodonty (e.g., *Macropus*) or a series of ridges that link the mesial and distal paired cusps or lophs (Figs. 23 and 24).

In the extinct short-faced kangaroo (e.g., *Sthenurus, Procoptodon,* Fig. 24) the face becomes massive, the incisors short and adapted to browsing, and the diastema markedly re-

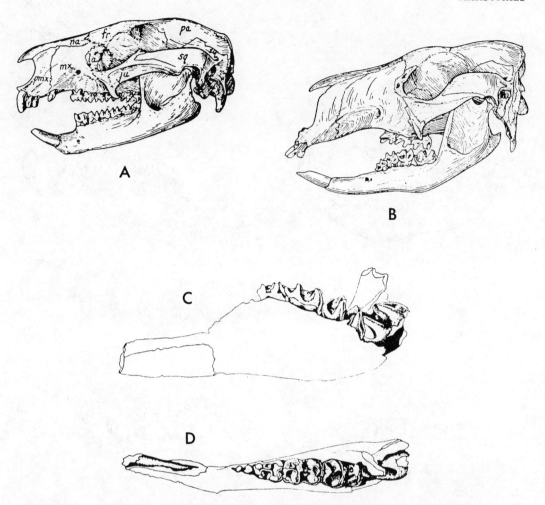

FIGURE 23. Australian Macropodinae. A, lateral view of skull and mandible of *Dendrolagus* sp., ca. X0.5; and B, lateral view of skull and mandible of *Macropus* sp., old individual ca. X0.3 (from Gregory, 1951). C, lingual and D, occlusal views of right mandibular fragment of *Macropus ferragus*, X0.75 (from Tedford, 1967; Published in 1967 by The Regents of the University of California; reprinted by permission of the University of California Press). Note that in *Macropus* the cheek teeth move anteriorly within the jaw, with the front ones being shed and the new molars erupting sequentially at the rear; see C.

duced. In the sthenurines, as in all kangaroos and diprodontids, there is a ventral process from below the orbit on the jugal, similar to the placental glyptodonts, which suggests adaptation toward a similar herbivorous diet. The cheek teeth are more complex than those in the long-faced kangaroos (compare Figs. 23 and 24), with plications and/or complex ridges forming complicated occlusal enamel patterns.

The long-faced kangaroos and wallabies live in open grassland, woodland, and rocky hillsides, with the exception of the arboreal tree kangaroos, *Dendrolagus* (Fig. 23A). In these, the diastema and facial region are elongated and the jaws less massive. The eruption of the cheek teeth is successive and, as in advanced Proboscidea, the teeth migrate anteriorly, eventually falling out in front and being replaced behind. An old individual will thus retain only the last one or two molars. The molars are bilophodont and simpler than in the short-faced kangaroos.

Both the Pleistocene short-faced and long-faced kangaroos were larger than the modern red or grey kangaroos, and may have stood as high as 10 ft (3 m) to the crown of the head.

CHARLES S. CHURCHER

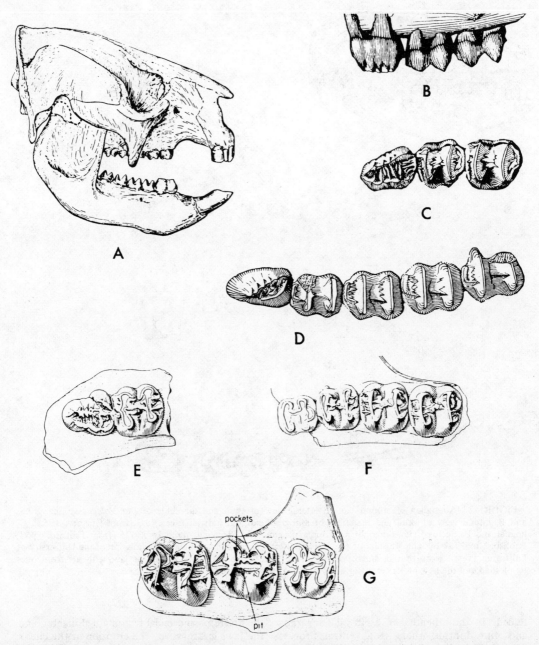

FIGURE 24. Short-faced kangaroos. *Sthenurus antiquus:* A, lateral view of skull and mandible of *Sthenurus* sp., ca. X0.3 (from Gregory, 1951). B, lateral and C, occlusal views of left maxillary P^3M^2; D, occlusal view of left mandibular cheek teeth, P_3–M_4; all X1.0 (from Bartholomai, 1963). *Procoptodon* species: E, occlusal view of left P^3–M^1, *P. rapha;* F, occlusal view of left milk P^3 and M^1–M^3, *P. pusio;* G, occlusal view of right M^1–M^3, *P. goliah;* all X1.0 (from Stirton and Marcus, 1966). Note the increased plications and ridging of the crowns of both the premolars and molars from *Sthenurus* to *Procoptodon.*

References

Bartholomai, A., 1963. Revision of the extinct macropodid Genus *Sthenurus* Owen in Queensland, *Mem. Queensland Mus.,* **14,** 51–76.

Blainville, H. M. D. de, 1816. Prodrome d'une nouvelle distribution systématique de règne animal, *Bull. Soc. Philomath. Paris,* ser. 3, **3,** 105–124.

Calaby, J. H.; Corbett, L. K.; Sharman, G. B.; and Johnston, P. G., 1974. The chromosomes and systematic position of the marsupial mole, *Notoryctes typhlops, Australian J. Biol. Sci.,* **27,** 529–532.

Clemens, W. A., 1966. Fossil mammals of the type Lance Formation, Wyoming Part III, Marsupialia, *Univ. Calif. Publ. Geol. Sci.*, 62, 122p.

Clemens, W. A., 1968. Origin and early evolution of marsupials, *Evolution*, 22, 1–18.

Clemens, W. A., 1979. Marsupialia, in Z. Kielan-Jaworowska et al., eds., *Mesozoic Mammals: The First Two-Thirds of Mammalian History*. Berkeley: Univ. California Press.

Cox, C. B., 1970. Migrating marsupials and drifting continents, *Nature*, 226, 767–770.

Cox, C. B., 1973. Systematics and plate tectonics in the spread of marsupials, *Spec. Paper Palaeontol.*, 12, 113–119.

Fooden, J., 1972. Breakup of Pangaea and isolation of Relict Mammals in Australia, South America, and Madagascar, *Science*, 175, 894–896.

Gregory, W. K., 1951. *Evolution Emerging*, vol. 1. New York: MacMillan, 1013p.

Hoffstetter, R., 1970. L'histoire biogéographique des marsupiaux et le dichotomie marsupiaux–placentaires, *Comptes Rendus Acad. Sci.*, 271, 388–391.

Hoffstetter, R., 1972. Données et hypotheses concernant l'origine et l'histoire biogéographique des marsupiaux, *Comptes Rendus Acad. Sci.*, 274, 2635–2638.

Keast, A.; Erk, F. C.; and Glass, B., eds., 1972. *Evolution, Mammals and Southern Continents*. Albany: State Univ. New York Press, 543p.

Kirsch, J. A. W., 1968. Prodromus of the comparative serology of Marsupialia, *Nature*, 217, 418–420.

Kirsch, J. A. W., 1977. The six–percent solution: Second thoughts on the adaptedness of the Marsupialia, *Amer. Sci.*, 65, 276–288.

Lillegraven, J. A., 1969. Latest Cretaceons mammals of upper part of Edmonton Formation of Alberta, Canada, and review of marsupial–placental dichotomy in mammalian evolution, *Univ. Kansas. Paleontol. Contrib.*, *Art. 50*, 1–122.

Lillegraven, J. A., 1974. Biogeographical considerations in the marsupial-placental dichotomy, *Ann. Rev. Ecol. Systematics*, 5, 263–283.

Lillegraven, J. A., 1975 (1976). Biological considerations of the marsupial–placental dichotomy, *Evolution*, 29, 707–722.

Linnaeus, C., 1758. *Systema naturae per regna tria naturae, secundum classes, ordines, genera, species cum characteribus, differentiis, synonymis, locis.* Editio decima, reformata. Stockholm: Laurentii Salvii 824p.

Müller, A. H., 1970. *Lehrbuch der Paläozoologie*, vol. 3, *Vertebraten;* pt. 3, *Mammalia.* Jena: Gustav Fischer, 855p.

Paula Couto, C. de., 1952a. Fossil Mammals from the beginning of the Cenozoic in Brazil. Marsupialia: Polydolopidae and Borhyaenidae, *Amer. Mus. Novit.*, 1559, 27p.

Paulo Couto, C. de, 1952b. Fossil mammals from the beginning of the Cenozoic in Brasil. Marsupialia: Didelphidae, *Amer. Mus. Novit.*, 1567, 26p.

Piveteau, J., 1961. Infra-classe Metatheria. Marsupialia, in J. Peveteau, ed., *Traité de Paleontologie*, vol. 6(1). Paris: Masson et Cie, 585–637.

Ride, W. D. L., 1962. On the Evolution of Australian Marsupials, in G. W. Leeper, ed., *The Evolution of Living Organisms, Symp. Roy. Soc. Victoria, Melbourne, Dec., 1959.* Parkville: Univ. Melbourne Press, 281–306.

Ride, W. D. L., 1964. A review of the Australian fossil marsupials, *J. Roy. Soc. Western Australia*, 47, 97–131.

Riggs, E. S., 1934. A new marsupial sabre-tooth from the Pliocene of Argentina and its relationships to other South American predaceous marsupials, *Trans. Amer. Phil. Soc.*, n.s., 24(1), 1–32.

Romer, A. S., 1966. *Vertebrate Paleontology*, 3rd ed. Chicago: Univ. Chicago Press, 468p.

Sigé, B., 1968. Dents de micromammifères et fragments du coquilles d'oeufs de dinosauries dans la fauna de vertébrés du Crétacé superieur de Laguna Umayo (Andes peruviennes), *Comptes Rendus Acad. Sci.*, 267, 1495–1998.

Sigé, B., 1972. La faunule de mammifères du Crétacé superieur de Laguna Umayo (Andes Peruviennes), *Bull. Mus. Natl. Hist. Natur. Sci. Terre*, 19, 375–409.

Simpson, G. G., 1945. The principles of classification and a classification of mammals, *Amer. Mus. Nat. Hist. Bull.*, 85, 350p.

Simpson, G. G., 1961. Historical zoogeography of Australian mammals, *Evolution*, 15, 431–446.

Simpson, G. G., 1970. The Argyrolagidae, extinct South American marsupials, *Bull. Mus. Comp. Zool. Harvard Univ.*, 139(1), 1–86.

Sinclair, W. J., 1906. Mammalia of the Santa Cruz beds: Marsupialia, *Rept. Princeton Univ. Exped., Patagonia*, 4, 333–482.

Stirton, R. A., 1955. Late Tertiary marsupials from South Australia, *Rec. S. Australian Mus.*, 11, 247–268.

Stirton, R. A., 1957a. A new koala from the Pliocene Palankarinna fauna of South Australia, *Rec. S. Australian Mus.*, 13, 71–81.

Stirton, R. A., 1957b. Tertiary marsupials from Victoria, Australia, *Mem. Nat. Mus. Victoria*, 21, 121–134.

Stirton, R. A., and Marcus, L. F., 1966. Generic and specific diagnoses in the gigantic macropodid Genus *Procoptodon*, *Rec. Australian Mus.*, 26, 349–359.

Stirton, R. A.; Tedford, R. H.; and Woodburne, M. O., 1968. Australian Tertiary deposits containing terrestrial mammals, *Univ. Calif. Publ. Geol. Sci.*, 77, 30p.

Tedford, R. H., 1967. The fossil Macropodidae from Lake Menindee, New South Wales, *Univ. Calif. Publ. Geol. Sci.*, 64, 156p.

Tedford, R. H., 1974. Marsupials and the new paleogeography, *Soc. Econ. Paleontol. Mineral., Spec. Publ. No. 21*, 109–126.

Thenius, E., 1969. *Phylogenie der Mammalia*. Berlin: de Gruyter, 722.

Troughton, E., 1947. *Furred Animals of Australia.* New York: Scribner, 374p.

Walker, E. P., 1964. Mammals of the World, vol. 1. Baltimore: John Hopkins Press, 644p.

Cross-references: *Bones and Teeth; Paleobiogeography of Vertebrates; Vertebrata; Vertebrate Paleontology.*

MATTHEVA

Approximately a century ago, a compressed conical fossil with two within-shell cavities

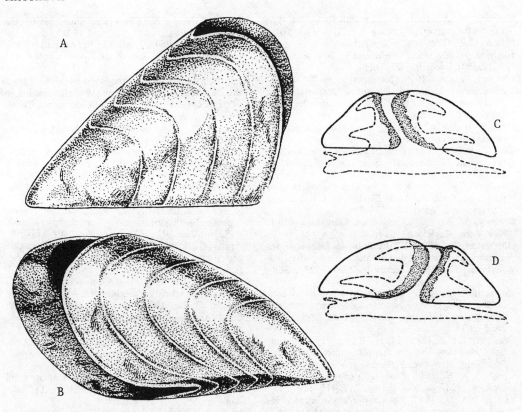

FIGURE 1. *Matthevia variabilis* Walcott (from Yochelson, 1966, courtesy of the U.S. Geological Survey). A,B, reconstruction of both hard parts, X3; C,D, alternative interpretations of the animal in life.

separated by a large septum was described from Late Cambrian limestone near Saratoga, New York. Preservation of shell was comparable to that of associated mollusks, but the genus was not easily placed in any of the conventional classes. Although the original material was redescribed, nothing more was added to the concept of *Matthevia* until the 1950s when a number of specimens were found in the western United States. These fossils all appeared to come from a relatively narrow time span, Trempealeauan stage, the youngest of the three Late Cambrian subdivisions. Further, two forms differing slightly in height and amount of compression, always seemed to be present at each locality.

In 1966, Yochelson proposed that this fossil represented an extinct class of mollusks. The two forms were interpreted as anterior and posterior hard parts of the same original with the cavities being the site of muscles (Fig. 1).

Additional field work has shown that this form occurs in the east from at least Maryland to New York (Yochelson and Taylor, 1974) and in the west from Arizona to the District of Franklin. Limited field evidence suggested that the fossils were associated with algal heads.

Presumably *Matthevia* was a herbivore like most chitons, a class first known from early Ordovician and today characteristically rocky-shore inhabitants. Rather than being able to cling with a broad foot and having a more flexible shell of eight plates, the relatively great weight of the anterior and posterior pieces helped to keep the animal in place against the force of waves and currents. If correctly interpreted, this class is an example of a widespread but short-lived experiment, which was superseded by a far more effective functional morphology evolved in a slightly younger mollusk. However, Runnegar and Projeta (1974) reinterpreted the genus as a chiton. More data is needed before either interpretation can be generally accepted.

ELLIS L. YOUCHELSON*

References

Runnegar, B., and Pojeta, J. Jr., 1974. Molluscan phylogeny: The paleontological viewpoint, *Science*, 186, 311–317.

*Publication authorized by the Director, U.S. Geological Survey

Yochelson, E. L., 1966. Mattheva, a proposed new class of mollusks, *U.S. Geol. Surv. Prof. Paper 523-B,* 1-12.

Yochelson, E. L., and Taylor, M. E., 1974. Late Cambrian *Matthevia* (Mollusca, Matthevida) in N. America, *Geol. Soc. Amer. Abstr.,* 5(1), 88.

Cross-reference: *Mollusca.*

MAZON CREEK

Jellyfish, chitons, spiders, insects, polychaete worms, larval fish, the legendary *Tullimonstrum,* and plants of extraordinary beauty are a few of the abundant and diverse organisms found in the unique fossil localities known collectively as Mazon Creek.

Mazon Creek fossils are preserved in ironstone concretions, which weather out of the middle Pennsylvanian (Westphalian D) Francis Creek Shale in northeastern Illinois (Fig. 1). Smith (1970) has described the Francis Creek Shale as a deltaic deposit that forms a clastic wedge 80 or more feet thick near the Mazon Creek localities and thins to the southwest. As the shale thins, grain size is reduced and the fauna becomes more marine. In the Mazon Creek area, the lithology is siltstone with lenses of fine-grained sandstone. Shabica (1970) has interpreted the ancient environment here as a delta front bar with distributary channels and interdistributary bays. Richardson and Johnson (1971) have reconstructed the shoreline from the locations of nonmarine and marine fossils (Fig. 1C).

Collecting in the Mazon Creek area began over a hundred years ago. The first locality was the stream bed of Mazon Creek itself. More collecting sites became available with the mining of the Colchester (No. 2) Coal, which lies just under the Francis Creek Shale. The first mines were underground, but now several square miles are undergoing strip mining. The stripped shale is deposited in spoil heaps, and erosion leaves concretions on the surface. Mazon Creek fossils have come from several mines and pits, but the principal area for collecting is now the Peabody Coal Company's Pit Eleven, which covers portions of Will and Kankakee Counties.

In Mazon Creek and in the older mines and pits north of the reconstructed shoreline, plants outnumber animal specimens by 100 to 1, and the animals are almost exclusively terrestrial or fresh-water forms. But S and W of the shoreline, plants are only about half of all specimens, and the animals are predominantly marine. The plants represent a single flora, but there are two distinct faunas: the nonmarine Braidwood fauna and the marine Essex fauna (Johnson and Richardson, 1966).

The Mazon Creek flora (Fig. 2) was described chiefly during the latter decades of the 19th and the first of the 20th century. Specimens are more diverse and abundant than in any other known North American Paleozoic flora. Peppers and Pfefferkorn (1970) list 71 genera. Specimens in the Field Museum of Natural History, Chicago, Illinois, number several tens of thousands, and there are large additional collections at other museums. The fossils are often large fern fronds up to 30 cm long and beautifully preserved. Recently, Langford (1958, 1963) and Darrah (1970) have monographed this flora.

The Braidwood fauna (Fig. 3C, D, see p. 466), with its terrestrial and freshwater animals, probably lived in the same area as the flora: the flood plains, mud flats, and swamps of a nearshore deltaic environment. Table 1 (p. 467) shows the distribution of the Braidwood fauna by species; Table 2 (p. 467) compares abundance of specimens in the major taxa for the Braidwood and Essex faunas. Arthropods are clearly the dominant animals in the Braidwood fauna. Insects account for most of the diversity (139 species) but are rare as specimens, while crustaceans are less diverse but comprise over half of all Braidwood specimens.

Arthropods are also abundant in the marine Essex fauna (Table 2 and Figs. 3A, B, 4, see p. 468). Without the Braidwood's veritable zoo of insects, diversity is spread more evenly among several groups (Table 1). The Essex fauna contains some strange animals, a few of which have not yet been described. *Tullimonstrum* is one of the strangest of all fossils (Fig. 3A), not closely related to any phylum. Its long snout terminates in a toothed "claw"; it has a segmented body, spade-like tail, and moveable anterior bar with an eye on each end (Johnson and Richardson, 1969).

Preservation of animal material in the Mazon Creek faunas is very different from preservation in most fossil localities. In most fossil beds, mineralized tissues are preserved better than chitin and soft tissues. But in Mazon Creek, chitin is preserved virtually unaltered, particularly in the crustaceans and polychaete jaws and setae (Fig. 4A, B). Soft tissues are preserved either as thin carbon films (Fig. 4B and the eyes in Fig. 4D, E), as light areas against a darker background (Figs. 3B, 4D), or as impressions (Fig. 3A) Calcium carbonate, the usual material of marine fossils, is generally absent. The shells of bivalves and gastropods are usually represented only by molds (Fig. 4C). In *Aviculopecten,* the calcite of the outer prismatic

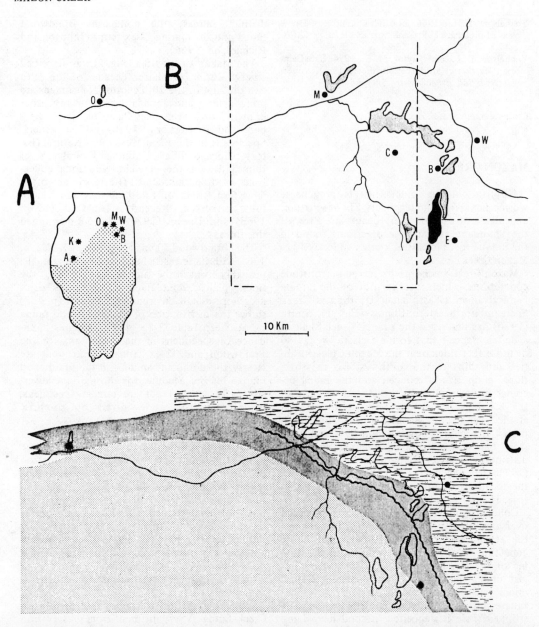

FIGURE 1. Localities of the Mazon Creek flora and faunas (from Zangrel and Richardson, 1975). A, approximate extent of the sea occupying the Illinois Basin during the deposition of the Francis Creek Shale. B, location map, northern Illinois; strip-mined areas shaded with Pit Eleven in solid black. C, approximate shoreline of the Illinois Basin in the area shown in B; light stipple, sea; dark stipple, area between sea and coal swamp (irregular broken lines); wavy line, boundary between the Braidwood fauna and the Essex fauna. *A,* Astoria; *B,* Braidwood; *C,* Coal City; *E,* Essex; *K,* Knoxville; *M.* Morris; *O,* Ottawa; *W,* Wilmington.

layer is gone, but the conchiolin sheaths that surrounded the prisms are preserved intact. However, the bone, dentine, and ganoine of vertebrates is commonly preserved (Fig. 4E), as are holothurian mouth rings and crinoid holdfasts (Richardson and Johnson 1971).

Decay and compaction usually destroy unmineralized tissue. Chitin and soft tissues decay rapidly under aerobic conditions, and mud compacts soon after deposition, squashing soft tissues. The Mazon Creek organisms must have been protected from both decay—probably by

FIGURE 2. Typical plant fossils of the Mazon Creek flora. A, *Pecopteris;* B, *Callipteridium;* C, *Neuropteris;* D, *Annularia.* All ca. X0.85.

rapid burial—and from compaction—probably by rapid concretion formation. Little is known about the chemistry of concretion formation, but it is probable that the first stage of decay, amino-acid deocmposition, releases ammonia, raising pH around the organism and causing carbonate precipitation. Zangerl (1971) buried dead fish in mud and found that iron ions migrated within a few days to form a halo around the rotting fish.

465

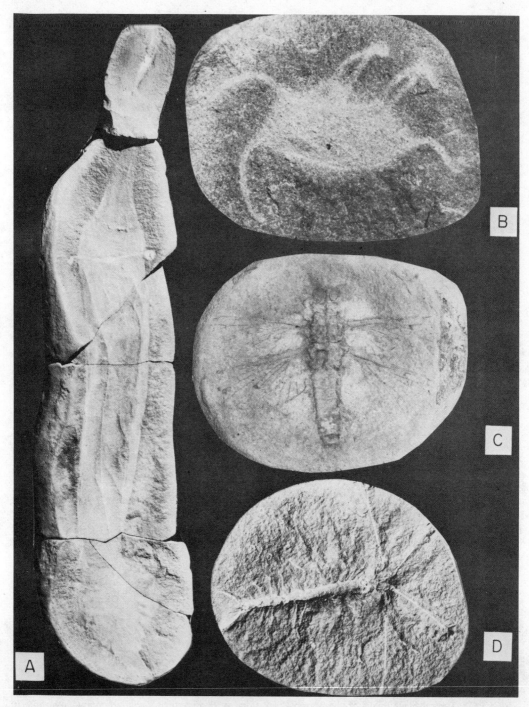

FIGURE 3. Well-preserved fossils from the marine Essex fauna (A and B) and nonmarine Braidwood fauna (C and D) (from Johnson and Richardson, 1970). A, *Tullimonstrum* (X0.6); B, *Octomedusa*, a jellyfish (X2.0); C, insect (X1.0); D, *Acanthotelson*, a crustacean (X1.0).

466

TABLE 1. Principal Members of the Mazon Creek Faunas

Taxa	No. of Species	Braidwood	Essex
COELENTERATA			
Anthracomedusa	1		X
Drevotella	1		X
Mazohydra	1		X
Octomedusa	1		X
NEMATODA	1		X
NEMERTINA	1		X
PRIAPULIDA	1		X
CHAETOGNATHA	1		X
MOLLUSCA			
Bivalvia			
Aviculopecten	3 (?)		X
Euchondria	1		X
Dunbarella	1		X
Other pectinids	2		X
Lima	1		X
Nuculana	1		X
Yoldia	1		X
"*Solenomya*"	1 (?)	X?	X?
Gastropoda			
Bellerophontids	3		X
Other gastropods	3		X
Cephalopoda			
Jeletzkya	1		X
Goniatite nautiloid	1		X
Amphineura			
Helminthochiton	1		X
ANNELIDA			
Polychaetes	15		X
TULLIMONSTRUM	1		X
BRACHIOPODA			
Lingula	1		X
Orbiculoides	1		X
ARTHROPODA			
Crustacea			
Branchiopodes	2	X	X
Ostracodes	2	X	
Acanthotelson	1	X	
Paleocaris	1	X	
Anthracaris	1		X
Belotelson	1		X
Anthracophausia	1		X
Mamayocaris	1		X
Kallidecthes	1		X
Tyrannophontes	1		X
Cryptocaris	1		X
Essoidea	1		X
Hesslerella	1		X
Kellibrooksia	1		X
Paraparchites	1		X
Merostomata			
Xiphosures	7	X	
Eurypterids	2	X	
Halicyne	1		X
Arachnida	43	X	
Mandibulata			
Myriapods	25	X	
Insects	139	X	

Modified and updated from Johnson and Richardson, 1970.

Taxa	No. of Species	Braidwood	Essex
ECHINODERMATA			
Holothurian	1		X
Crinoid	1		X
CHORDATA			
Fish			
Lamprey (*Mayomyzon*)	1	X?	X?
Chimaeroid	1		X
Cladodont	1	X	
Bandringa	1	X	
Pleurocant	1	X	
Acanthodes	1	X	X?
Gyracanthus	1		X
Conchopoma	1	X	
Ctenodus	1	X	
Esconichthys (larva)	1		X
Megapleuron	1		X
Rhizodopsis	1		X
Rhabdoderma	1		X
Amphicentrum	1	X	X?
Elonichthys	1	X	X?
Haplolepis	1	X	
Pyritocephalus	1	X	
Platysomatidae	1	X	X?
Amphibians			
Amphibamus	1	X	
Ophiderpeton	1	X	
Phlegethontia	1	X	
Cocytinus	1	X	
Branchiosaurus	1	X	
Ptyonius	1	X	
Spondylerpeton	1	X	
Reptilians			
Cephalerpeton	1	X	
Total number of species:	307	242	73

TABLE 2. Relative Abundance of Major Taxa From the Braidwood and Essex Faunas[a]

Taxa	Braidwood Fauna (percent)	Essex Fauna (percent)
Coelenterata	0.	2.3
Mollusca	18.4	15.7
Polychaetes and other vermes	1.8	20.8
Arthropoda	79.5	44.9
Echinodermata	0	5.0
Tullimonstrum	0	9.9
Chordata	0.3	1.2
	100.0	99.8

From Johnson and Richardson, 1970.
[a]The Braidwood collection consists of 1213 specimens; the Essex collection, 3214 specimens.

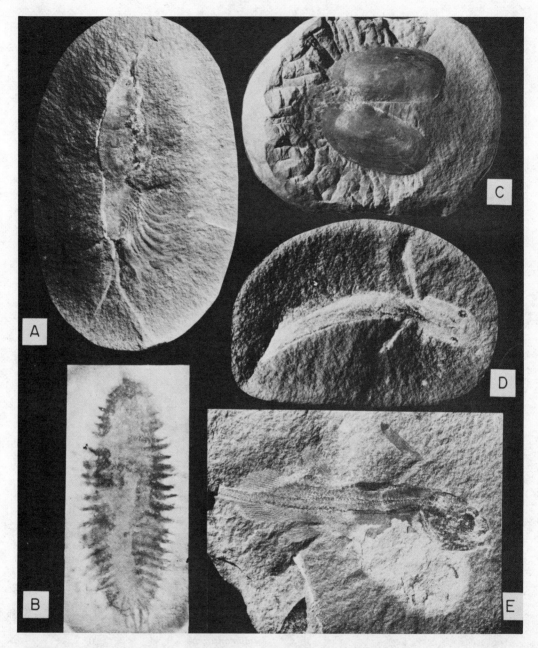

FIGURE 4. Well-preserved fossils from the marine Essex fauna (from Johnson and Richardson, 1970). A, *Kallidecthes,* a crustacean (X0.9); B, a polychaete (X1.2); C, a common bivalve (X0.9); D, *Esconichthys,* a larval fish (X0.9); E, *Rhabdodermis,* a coelacanth with attached yolk sack (X0.9).

The Mazon Creek fossils are concentrated in persistent and discontinuous lenses, mostly in the lower part of the Francis Creek Shale. Except for some plants just above the coal and for two animal specimens, fossils are found only in concretions. Does this mean that life was present on the delta only rarely in the deposition of the shale? Probably not, since plants must have been washed onto the delta front frequently. Mud conditions favorable for the formation of concretions were probably rare.

Mazon Creek provides a narrow window on the history of chitinous and soft-bodied forms.

Insects were already highly diverse. Crustaceans and polychaetes were remarkably modern in appearance. And there were strange soft-bodied beasts around, like *Tullimonstrum;* animals unique in the record of life.

IDA THOMPSON

References

Bardack, D., and Zangerl, R., 1968. First fossil lamprey: A record from the Pennsylvanian of Illinois, *Science,* 162, 1266–1267.

Carpenter, F. M., and Richardson, E. S., Jr., 1971. Additional insects in Pennsylvanian concretions from Illinois, *Psyche,* 78, 267–295.

Darrah, W. C., 1970. *A Critical Review of the Upper Pennsylvanian Floras of the Eastern United States With Notes on the Mazon Creek Flora of Illinois.* Gettysburg: Privately Printed, 220p.

Denison, R. H., 1969. New Pennsylvanian lungfishes from Illinois, *Fieldiana, Geol.,* 12, 193–211.

Johnson, R. G., and Richardson, E. S., Jr., 1966. A remarkable Pennsylvanian fauna from the Mazon Creek area, Illinois, *J. Geol.,* 74, 626–631.

Johnson, R. G., and Richardson, E. S., Jr., 1968. The Essex fauna and medusae, *Fieldiana, Geol.,* 12, 109–115.

Johnson, R. G., and Richardson, E. S., Jr., 1969. The morphology and affinities of *Tullimonstrum, Fieldiana, Geol.,* 12, 119–149.

Johnson, R. G., and Richardson, E. S., Jr., 1970. Fauna of the Francis Creek Shale in the Wilmington area, *Ill. Geol. Surv., Guidebook,* 8, 53–59.

Kjellesvig-Waering, E. N., 1963. Pennsylvanian invertebrates of the Mazon Creek area, Illinois: Eurypterida, *Fieldiana, Geol.,* 12, 85–106.

Langford, G., 1958. The Wilmington Coal Flora from a Pennsylvanian deposit in Will County, Illinois. Downers Grove, Ill.: Esconi Assoc., 360p.

Langford, G., 1963. The Wilmington Coal Fauna and additions to the Wilmington Coal Flora from a Pennsylvanian deposit in Will County, Illinois. Downers Grove, Ill.: Esconi Assoc., 280p.

Peppers, R. A., and Pfefferkorn, H. W., 1970. A comparison of the floras of the Colchester (No. 2) Coal and Francis Creek Shale, *Ill. Geol. Surv., Guidebook,* 8, 61–74.

Richardson, E. S., Jr., 1953. Pennsylvanian insects of Illinois, *Trans. Ill. Acad. Sci.,* 46, 147–153.

Richardson, E. S., Jr., and Johnson, R. G., 1971. The Mazon Creek faunas, *Proc. N. Amer. Paleontological Conv.,* pt. I, 1222–1235.

Schram, F. R., 1974. The Mazon Creek caridoid Crustacea, *Fieldiana, Geol.,* 30, 9–65.

Schultze, H-P., 1974. Osteolepididae Rhipidistia (Pisces) aus dem Pennsylvanian von Illinois USA, *Neues Jahrb. Geol. Paläont., Abh.,* 146, 29–50.

Shabica, C. L., 1970. Depositional environments in the Francis Creek Shale, *Ill. Geol. Surv., Guidebook,* 8, 43–52.

Smith, W. H., 1970. Lithology and distribution of the Francis Creek Shale in Illinois, *Ill. Geol. Surv., Guidebook,* 8, 34–42.

Thompson, I., 1976. Nine new species of whole-body fossil polychaetes from Essex, Illinois, *Fieldiana, Geol.,* 36, 46–183.

Zangerl, R., 1971. On the geologic significance of perfectly preserved fossils, *Proc. N. Amer. Paleontological Conv.,* pt. I, 1207–1222.

Zangerl, R., and Richardson, E. S., Jr., 1975. Die paläoökologische Bedeutung der Mazon-Creek- und Mecca-Faunen im zentralen Nordamerika. *C. R. 7th Congr. Internat. Strat. Geol. Carb.,* 4, 385–391.

Cross-references: *Annelida; Arthropoda; Fossil Record.*

MELANOSCLERITES

Melanosclerites are microfossils in the form of flexible rods, solid or subdivided into elements, broadening at either end into swellings of very variable shape. They are composed of a substance resembling chitin in composition and are black in color. When complete, their length may be up to 2 mm; but they are extremely brittle and most often found as fragments whose size is measurable in microns.

The swellings on the ends of the rods show a considerable variation in shape. They may be smoothly rounded or knobbly; they may simulate a tulip or the head of a classical column; or they may divide into three, four, or many divergent prongs (Fig. 1).

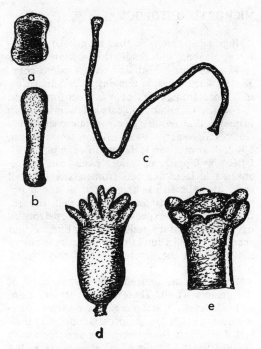

FIGURE 1. Melanosclerites (after Eisenack, 1963). a–c, *Melanosteus anceps,* Middle Devonian, Germany illustrating the range of variation (a, very short form, X180; b, normal form, X150; c, very long and thin form, X180). d, *Melanocyathus dentatus,* Ordovician, Baltic, "head" X180; e, unnamed form, Ordovician, Baltic, "head" X170.

These microfossils were first described, as "black rodlets," by Eisenack in 1932. In 1942, Eisenack coined for them the name "melanosclerites"; later (1963) he proposed the order Melanoskleritoitidea to accommodate them, recognizing this to be a grouping of uncertain systematic reference. Melanosclerites are believed by Eisenack (1963) to be supporting structures from the lower trunks of ancestral Ctenophora ("sea-gooseberries"), a group of soft-bodied animals otherwise unknown as fossils.

The known stratigraphical range of these microfossils is Ordovician to Upper Devonian. However, they have been very little studied to date and the actual range may be much more considerable.

WILLIAM A. S. SARJEANT

Reference

Eisenack, A., 1963. Melanoskleriten aus anstehenden Sedimenten aus Geschieben, *Paläont. Zeitschr.,* 37(1–2), 122–134.

Cross-reference: *Plankton.*

MICROPALEONTOLOGY

Micropaleontology is that branch of paleontology dealing with fossils that, owing to their small size, must be studied with a microscope. These ancient organic remains, called microfossils, range from giants of a few centimeters in size down to minute objects a few microns in diameter that require electron microscopy for study (Glaessner, 1945; Ericson and Wollin, 1962; Pokorny, 1963; Haq and Boersma, 1978). Often, well-preserved microfossils can be recovered in large numbers from relatively small rock or sediment samples such as those obtained from the drilling of an oil well. Indeed, the tremendous expansion of micropaleontology in this century reflects the influence that economic application in the petroleum industry has played in the development of this branch of science.

Microfossils are derived from diverse parts of the organic world. These consist of: (1) complete skeletons of microorganisms, such as foraminifera, radiolarians, and ostracodes; (2) individual elements from composite skeletons such as the plates or spines of echinoderms, the scales of fish, or coccoliths from coccolithophores; and (3) spores and pollens from plants. Also occurring in samples processed for micropaleontological study and frequently considered to be microfossils are (4) small, but identifiable, fragments of colonial organisms such as bryozoans or stromatoporoids; and (5) immature individuals or minute adults of groups usually studied as macrofossils, such as mollusks or brachiopods.

The skeletons of minute one-celled plants and animals, the algae and protozoa, comprise much of the subject matter of the micropaleontologist. Foraminifera (Order Foraminiferida), of the protozoan class Sarcodina, are the most-studied group of microfossils. Their tests, usually composed of calcium carbonate or agglutinated sand grains, are commonly less than a millimeter is size, but some have diameters of several centimeters. Known from rocks as old as Cambrian, foraminifera became abundant and diverse in the late Paleozoic and remain so to the present day. Most marine environments are inhabited by foraminifera, but individual species have limited tolerances, thus providing one of the most potent tools in paleoecology. Planktonic forms, typically characterized by wide geographic distribution and often short geologic range, are abundant in the Cretaceous and Tertiary and are valuable for the intercontinental age correlation of ancient marine sediments.

Radiolarians (Order Radiolaria), also of the class Sarcodina, have mesh or network opaline skeletons. Some are over a mm in diameter, but most are much less. All are marine and planktonic. Since larger and more compact forms float at greater depth, an abundance of these types of radiolarians in an ancient sediment indicates relatively deep-water deposition. Because of carbonate dissolution at great depths in the marine realm, radiolarian cherts are characteristic of the deeper facies of many ancient geosynclines. A few late Precambrian reports of radiolarians may be valid, but the group definitely ranges from Cambrian to the present.

The protozoan class Ciliata is represented among the microfossils only by the tintinnids (q.v.). The skeletons of these microfossils are organic cup-shaped structures of resistant organic material, which may contain agglutinated particles. Fossil tintinnids are important particularly in the Upper Jurassic and Lower Cretaceous sequence of pelagic limestones of the Mediterranean region.

Protozoans of the class Mastigophora are a heterogeneous group, often with characteristics of both plants and animals. Dinoflagellates are mastigophorans with cellulose tests that are occasionally preserved in chert or other marine deposits in which more commonly used microfossils have been destroyed by dissolution. Chrysomonadines form siliceous spherical cysts that may be fossilized in either marine or nonmarine sediments. The siliceous frame-like skeletons of silicoflagellates, 0.10 mm or a little less in size, have been recorded from Upper Cretaceous and Tertiary rocks. Coccolithophores are spherical, marine mastigophorans of planktonic habit, with calcareous plates a few microns in size, the coccoliths, embedded in the cell wall. Coccoliths are abundant in the fine matrix of globigerine ooze forming on the modern sea floor. Together with foraminifera, coccoliths compose the bulk of the porous, friable limestone known as chalk, which is so common in the Cretaceous system. There is clear evidence that coccoliths range in age from at least the Jurassic to the present. Discoasters resemble and are usually associated with coccoliths, but appear to have become extinct in the Late Tertiary.

Diatoms are algae that secrete a finely latticed siliceous skeleton, ca. 0.1–1.0 mm in size, consisting of two valves that fit together in the manner of a pillbox. Although they occur in abundance in all nontoxic, well-lighted waters, the potential use of diatoms in paleoecology and in the age correlation of sedimentary rocks is limited by the ease with which their skeletons are dissolved and mechanically destroyed. Thus diatoms were probably in existence during the deposition of rocks older than the late Cretaceous, where their first identifiable remains are found. Locally they are important rockmakers, as in California where the Monterey Formation of Miocene age contains pure diatomites as much as 200 ft thick.

Various types of lime-secreting red and green algae often occur as microfossils, especially in limestones. One group of green algae, the charophytes, are useful in the difficult problem of age correlation of some ancient freshwater deposits. They secrete a layer of calcium carbonate over certain parts of their anatomy, which are thus rendered capable of fossilization. The female spore sac, usually subspherical and ornamented with a spiral pattern, makes a distinctive microfossil with a range from Devonian to Holocene.

A number of microfossils, seemingly of protozoan or algal affinities, have defied precise systematic assignment. The vase-like chitinizoans from the lower Paleozoic may be tectinous foraminifera. Hystrichopheres are possibly the cysts of a number of different kinds of things.

Micropaleontology of the advanced plants is predominantly the study of their minute reproductive bodies, the spores and pollens. A separate subdiscipline for spore and pollen analysis (see *Palynology*) has evolved utilizing botanical knowledge and specialized techniques. Fossil seeds and wood fragments are sometimes useful in stratigraphic investigations of terrestrial sedimentary rocks.

Multicellular animals with hard parts are abundantly represented as microfossils by dispersed skeletal plates and fragments as well as by small individuals. Only two groups, however, ostracodes and conodonts, rank as microfossils of major significance in stratigraphic studies.

Ostracodes, of the arthropod class Crustacea, are second in importance in stratigraphic and paleoecological studies only to the foraminifera. They have bivalved shells of chitin impregnated with lime; most of them are 0.5–4 mm in length. Ostracodes occur in both nonmarine and marine waters; they reach their peak in both numbers and diversity in the shallow, brackish-water environment. In contrast to foraminifera, which have left an undistinguished lower Paleozoic record, ostracodes became relatively abundant and diverse soon after their first appearance in the Ordovician. They are usually subordinate in numbers to foraminifera in marine sedimentary rocks of the Mesozoic and Cenozoic.

Conodonts are microfossils of calcium phosphate averaging 1.0 mm or a little less in size. Their form may be a simple tooth or multiple cusps arrayed along a bar or blade or, again, the cusps may be suppressed while the base is broadened to form a platform. About 300 assemblages found on the bedding planes of black Carboniferous shales demonstrate that a single animal possessed an apparatus composed of many conodonts, but the nature of the animal and the function of the apparatus remain an enigma. Conodonts range from the Lower Ordovician to the Middle Triassic and many are important guide fossils, especially in the Upper Devonian and lower Mississippian.

Other assorted animal microfossils are frequently of limited, but important, stratigraphic value. The jaws of annelid worms (scolecodonts) have been used with success to zone certain lower Paleozoic strata. Colonies of stromatoporoids and bryozoans are often abundant and varied and may be identified by thin sections of relatively small fragments. These may be recovered from the drilling of oil wells and thus are of economic value in the absence of less troublesome fossils. Sponge spicules, and also plates and spines from the various groups of echinoderms, are ubiquitous but of limited value. Small adults and immature specimens of corals, brachiopods, mollusks, and shelled worms locally may be useful. Fish earbones (otoliths) have proven to be of stratigraphic value on the Gulf Coast and in the Far East. Fish scales have been used to interpret the stratigraphy and paleogeography of the Tertiary of California, and fish teeth have received attention recently in the study of the sections recovered by the Deep Sea Drilling Project.

Collecting and Processing Microfossils

Collection of microfossils from surface exposures is accomplished by first removing weathered and possibly contaminated material from the face of the outcrop. Fifty to one hundred grams of fresh rock are then selected to be taken back to the laboratory for processing. Visual inspection of the sample in the field is rarely an aid in determining the success of the collection because of the small size of most microfossils. Subsurface samples for micropaleontological study are recovered from wells drilled for oil, gas, or water. Cores may be obtained when accurate knowledge is essential, but more commonly the micropaleontologist must work with rock fragments, called ditch samples, which are pumped out of the bore hole with the drilling mud. Small sidewall cores may be obtained after the well is completed if more information is desired.

In the laboratory, the microfossils are separated from the rock matrix and concentrated. Treatment differs with the nature of the rock and composition of the microfossils (Kummel and Raup, 1965). Soft or friable rocks of any composition can usually be disaggregated by boiling water. If much clay is present, it is helpful to first thoroughly dry the sample, then boil it in water to which a spoonful of sodium bicarbonate has been added. A technique that is often successful with indurated fine-grain rocks is to soak dried, pea-sized fragments in kerosene for an hour, then drain off the kerosene and soak in water. Boiling in concentrated caustic soda solution will usually disaggregate siliceous shales but may destroy siliceous microfossils. These methods reduce the sample to a mud or sand slush which is washed over a 64 μm sieve to separate the fine-fraction material containing such fossils as coccoliths from the larger foraminifera and radiolarians. Other microfossils, such as diatoms, may span this size division and should be viewed in the total sample before determining what size separation, if any, is needed.

If much sand-sized matrix remains in the washed sample, it is desirable to concentrate the microfossils. Flotation on heavy liquids will accomplish this if the specific gravity of the fossils differs significantly from that of the matrix. Electromagnetic separation is possible for glauconite-filled fossils or for tests impregnated with iron salts. Concentration is rarely complete and it is always necessary to check the residue.

Recrystallized carbonate rocks resist the usual methods of disaggregation. Solution of carbonate with dilute hydrochloric acid will free siliceous microfossils such as radiolarians and diatoms. Conodonts can be freed from a carbonate matrix with acetic acid. Fossils also composed of carbonate such as foraminifera cannot, of course, be separated by this method and must be studied by sectioning the rock.

Radiolarians in chert are usually studied in thin translucent chips or in thin sections; but, in some cases, they can be partially freed by etching with caustic soda or hydrofloric acid. The hydrofloric acid treatment may also be used to free organic ("chitin") microfossils such as pollen, dinoflagellates, and chitinozoans.

Washed and dried residues and concentrates are scattered over a small tray and examined under 10X to 100X magnification with a stereoscopic microscope or placed on a glass slide under a cover slip (using either a temporary or permanent mounting medium) and examined under transmitted light using a compound microscope. For permanent mounting, larger microfossils may be glued (using water soluble glue) on cardboard slides, which usually have a 1 by 3-inch black-bottomed recess marked into a grid of 60 or 100 squares. Individual larger fossils can be picked up with the pointed tip of a very fine sable brush which has been slighted moistened. In quantitative studies, microfossils from a known quantity of rock (if numbers per unit weight is considered meaningful) are scattered over the bottom of a tray or slide on which a grid is marked and counted by scanning along the grid pattern. A minimum of 300 specimens are usually counted and are often merely expressed as percentages relative to the total assemblage.

Moderate magnifications of the stereoscopic microscope are adequate for the study of foraminifera, ostracodes, conodonts, and other microfossils caught on a 64 μm sieve. Higher magnifications to 2000X offered by the compound microscope are necessary for smaller microfossils and may be customary for the determination of some larger ones such as the radiolarians. The electron microscope, which can produce a useful magnification up to 200,000X, can discriminate specific differences in coccoliths that appear identical under an optical microscope. These higher magnifications have also been used to advantage in the study of the microstructure of other groups.

Some microfossils, such as larger foraminifera, stromatoporoids, or the Paleozoic classes of bryozoans, can be distinguished only by internal features of the skeleton. Very thin oriented sections of these fossils must be ground and viewed by transmitted light.

Applications of Microfossils

Microfossils are used to solve geological problems in the same manner as other fossils. The economic and scientific interest in the field of micropaleontology, however, derives from the fact that microfossils provide an abundance of useful information from a small rock or sediment sample when larger fossils are rare and usually damaged by the drilling process. In the deep ocean, macrofossils are rarely recovered, yet much of the pelagic sediments' are composed entirely of microfossils (predominantly foraminifera, radiolarians, diatoms, and coccoliths). Piston cores of the ocean floor taken from oceanographic research ships may penetrate 20 m or more and sample sediment layers that were deposited thousands, or even millions, of years ago. More recently, the Deep Sea Drilling Project has recovered thousands of meters of cored sedimet from several hundred different locations in the world ocean. These cores are only a few centimeters in diameter and they must be sampled at close intervals to reveal subtle differences in the microfauna or flora which tell a story of changing climates or of modified processes of sedimentation. Thus, microfossils play leading roles in two fields, petroleum exploration and marine geology, where small samples must be used to obtain stratigraphic and paleoecologic information.

Hundreds of students of microfossils are employed by the petroleum industry in its efforts to locate new deposits of oil and gas. Traditionally, the function of the economic micropaleontologist has been to provide datums within the stratigraphic column so that rock strata may be correlated from well to well. Ditch samples from wells are contaminated by rocks from higher in the bore hole, therefore datums are usually established on the highest occurrence within the well of certain cosmopolitan and easily recognized microfossils. The termination of the range of a microfossil within a particular stratigraphic section is caused, perhaps, as often by emigration in response to changing environmental conditions as by extinction. This means that the highest occurrence of a microfossil is not always a reliable criterion for age correlation although generally a useful one within limited areas. Microfossils of planktonic organisms are particularly valuable as age indicators because their distribution is usually widespread and their appearance and disappearance are commonly caused by evolutionary changes or major alterations in the oceanic environment, while benthic species might disappear in response to very local changes in the environment of deposition. Pollen and spores are unique in that they are disseminated by currents of wind and water and may be used to correlate ancient sediments of the terrestrial environment with those of the nearshore marine or freshwater environments.

Marine geology is another field to which micropaleontology finds major application. Beneath the floors of the ocean basins lies a blanket of sediment containing a fascinating record of earth history. The microfossils that compose much of these deep-sea sediments can be used as an aid in studying four different aspects of this history:

1. Age of the Sediment. As in oil exploration and as in the devleopment of any history, it is important to know the relative age of the sediments from different locations and to tie this history to a time scale. This is done by radiometrically dating material that is contained in or closely associated with fossiliferous sediments. By redating the stratigraphic record of evolutionary change in the microfossils to a scale of time, rates of change can be determined. Although there is much work left to be done, the biostratigraphic record of several microfossil groups has been closely related to a radiometric time scale for the Cenozoic (Berggren, 1972; Berggren and van Couvering, 1974). The Mesozoic and older microfossil records are less-well tied to time. Because of sea-floor spreading and subduction of the oceanic crust in trenches, sediments older than the Cretaceous are rarely recovered from the deep sea.

2. Oceanographic Change. The movement of planktonic species is largely controlled by near-surface currents. Thus, their distribution in the oceans closely matches that of the surface water masses (Bé and Tolderlund, 1971). Because most of the microfossils that are preserved on the sea floor have been delivered there rather rapidly in fecal pellets, the distribution of these microfossils closely matches the distribution of the living organisms in the overlying waters; thus, the distribution of individual species and groups of species from the geologic past can be used to map the location of past currents, water masses, and water-mass boundaries. By studying the changes in these oceanographic features with time, a detailed history of the near-surface waters of the oceans can be developed.

In a similar way, the distribution of deep-sea benthic organisms appears to be at least partially related to the character of the deep-water masses of the ocean. Changes in these benthic faunas, therefore, can be used to study the history of the deep ocean (Schnitker, 1974; Streeter, 1973).

Microfossil assemblages in which most of the represented species are still living may be used to make quantitative estimates of past environmental conditions (Imbrie and Kipp, 1971). This is done by first quantitatively describing the distribution of these species in the surface sediments of the ocean. Next, the mathematical description of the assemblage at each sample location is statistically related (by means of a regression equation) to a property of the overlying waters that is thought to be ecologically important (e.g., temperature). Subsurface samples are then described in terms of the assemblages found in surface sediments. Using this description and the previously established regression equation, estimates of the value of the property can be made for the time represented by the subsurface sample (CLIMAP Project Members, 1976).

3. Change in the Chemistry of Ocean Waters. The bulk of marine microfossils are formed of either calcium carbonate (e.g., foraminifera and coccoliths) or opaline silica (e.g., diatoms and radiolarians). Their distribution in the deep sea is controlled by the balance between the rate at which they are supplied to the seafloor and the rate at which they are dissolved by the bottom waters. The distribution of calcium carbonate appears to be primarily controlled by the distribution of deep, cold bottom waters that are rich in carbon dioxide and particularly corrosive with respect to aragonite and calcite. Changes in the Calcite Compensation Depth (the depth at which supply and dissolution rates balance), in the degree of preservation of the calcite microfossils, and in the rate at which carbonate dissolves as a function of depth have been used as indicators of changes in the chemical character and vertical structure of the deep ocean waters (Broecker, 1971; Berger and Winterer, 1974; van Andel et al., 1975).

The distribution of biogenic opal appears to be more closely related to the rate of supply and thus, to the patterns of primary production in the ocean (Heath, 1974). The distribution of siliceous microfossils in the past can be used, therefore, to indicate the general patterns of productivity. The preservation of the opaline microfossils, however, does vary from basin to basin at any given time in the past. In areas of comparable supply (production), biogenic opal appears to be better preserved where bottom waters are rich in dissolved SiO_2. These are the same waters that tend to be corrosive with respect to calcite. Thus, the relative degree of preservation of the carbonate and siliceous microfossils and the basin-to-basin fractionation of these two mineral forms can give information on the chemical nature of bottom waters and, thus, likely patterns of circulation (Berger, 1970).

4. Bottom-Water Movement. The deep ocean is not stagnant. The bottom waters in contact

with the sea floor do move; and in areas of restricted passage or along some continental boundaries, advective currents may reach velocities of over 1 knot. In addition, tidal currents are active throughout the deep sea and commonly reach velocities of 10 cm/sec. Such currents are capable of laterally moving the microfossils that comprise pelagic deposits. Thus, the microfossils can be viewed as sedimentary particles, each with its own hydrodynamic properties, and they can be used to study the winnowing of the sediments, changes in current velocity, and the transport and redeposition of older microfossils from outcrops on the deep-sea floor. Deep-sea density flows in pelagic sediments may result in the separation of the microfossils, with the biggest, heaviest forms on the bottom and the hydrodynamically lightest specimens on top of the sedimentary deposits resulting from such flows. These deposits are obviously of little use for either stratigraphic or paleoecological studies, but they do tell a story of sediment movement on the deep-sea floor (van Andel, 1973; Moore et al., 1973).

T. C. MOORE, JR.
R. J. ECHOLS

References

Bé, A. W. H., and Tolderlund, D. S., 1971. Distribution and ecology of living planktonic foraminifera in surface waters of the Atlantic and Indian Oceans, in B. M. Funnell and W. R. Reidel, eds., *Micropaleontology of Oceans.* Cambridge: Cambridge Univ. Press, 105–149.

Berger, W. H., 1970. Biogenous deep-sea sediments: Fractionation by deep-sea circulation, *Geol. Soc. Amer. Bull.,* **81**, 1385–1402.

Berger, W. H., and Winterer, E. L., 1974. Plate stratigraphy and the fluctuating carbonate line, *Internat. Assoc. Sedimentol. Spec. Publs.* **1**, 11–48.

Berggren, W. A., 1972. A Cenozoic time scale—Some implications for regional geology and paleobiogeography, *Lethaia,* **5**, 195–215.

Berggren, W. A., and van Couvering, J. A., 1974. The Late Neogene. *Palaeogeography, Palaeoclimatology, Palaeoecology,* **16**, 216p.

Broecker, W. S., 1971. A kinetic model for the chemical composition of sea water, *Quaternary Research,* **1**, 188–207.

CLIMAP Project Members, 1976. The surface of the ice-age Earth, *Science,* **191**, 1131–1137.

Ericson, D. B., and Wollin, G., 1962. Micropaleontology. *Sci. American,* **207**(1), 96–106.

Glaessner, M. F., 1945. *Principles of micropaleontology.* Oxford: Oxford Univ. Press and Melbourne Univ. Press, 296p.

Haq, B. U., and Boersma, A., eds., 1978. *Introduction to Marine Micropaleontology.* Amsterdam: Elsevier, 400p.

Heath, G. R., 1974. Dissolved silica and deep-sea sediments, in W. W. Hay, ed., Studies in paleo-oceanography, *Soc. Econ. Paleontol. Mineral., Spec. Publ.,* **20,** 77–93.

Imbrie, J., and Kipp, N., 1971. A new micropaleontological method for quantitative paleoclimatology: Application to a late Pleistocene Caribbean core, in K. K. Turekian, ed., *The Late Cenozoic Glacial Ages.* New Haven: Yale Univ. Press, 71–181.

Kummel, G., and Raup, D., 1965. *Handbook of Paleontological Techniques.* San Francisco: Freeman, 852p.

Moore, T. C., Jr.; Heath, G. R.; and Kowsmann, R. O., 1973. Biogenic sediments of the Panama Basin, *J. Geol.,* **81,** 458–472.

Pokorny, V., 1963. *Principles of Zoological Micropalaeontology.* New York: Pergammon Press, 651p.

Schnitker, D., 1974. West Atlantic abyssal circulation during the past 120,000 years, *Nature,* **248,** 385–387.

Streeter, S. S., 1973. Bottom water and benthonic foraminifera in the North Atlantic–Glacial-interglacial contrasts, *Quaternary Research,* **3,** 131–141.

Van Andel, T. H., 1973. Texture and dispersal of sediments in the Panama Basin, *J. Geol.,* **81;** 434–457.

Van Andel, T. H.; Heath, G. R.; and Moore, T. C., Jr., 1975. Cenozoic history and paleoceanography of the central equatorial Pacific Ocean, *Geol. Soc. Amer. Mem. 143,* 134p.

Cross-references: *Acritarchs; Algae; Biogeochemistry; Calcispheres and Nannoconids; Charophyta; Coccoliths; Conodonts; Diatoms; Dinoflagellates; Discoasters; Ebridians; Foraminifera, Benthic; Foraminifera, Planktic; Fossil Record; Fusulinacea; Palynology; Plankton; Protista; Radiolaria; Scolecodonts; Silicoflagellates; Tintinnids.* Vol. VI: *Pelagic Sedimentation.*

MOLLUSCA

Mollusca are an important and large group of invertebrate animals (some 100,000 described species are known), occupying many habitats from Greenland's icy mountains to India's coral strand. The group conforms to the definition of a phylum in that its members possess a highly characteristic structural plan but are exceptionally diverse in outward appearances and habitats. The mollusks range in size from <1 mm to over >60 ft and include such forms as snails, clams, slugs, oysters, mussels, octopuses, and tusk shells. The phylum is divisible into at least eight classes, listed below:

1. Class MONOPLACOPHORA: pratially segmented froms with multiple gills and multiple paired muscles; shell capuliform; Lower Cambrian–Holoceue; includes the living *Neopilina*
2. Class CONICONCHIA: small conoid univalved forms open at one end with or without an operculum; Cambrian through Permian; includes the hyolithids and *Tentaculites* (which may represent separate classes)
3. Class POLYPLACOPHORA: bilaterally symmetrical organisms having a shell subdivided into eight separate plates; Cambrian–Holocene; includes chitons or coal-of-mail shells
4. Class APLACOPHORA: bilaterally symmetrical worm-like animals with calcareous spicules imbedded in the epidermis; Holocene; includes sea hares
5. Class SCAPHOPODA: bilaterally symmetrical elongate forms having an external tube-shaped shell open at both ends; Devonian–Holocene; tusk shells
6. Class GASTROPODA: univalved coiled forms with or without an operculum with the gastrointestinal tract coiled in a figure 8 (torted), and with the left-hand set or organs generally missing; marine or nonmarine; Lower Cambrian–Holocene; includes snails and slugs
7. Class BIVALVIA (PELECYPODA): bivalved forms with a ventral foot and lacking a distinctive head; marine and nonmarine; Ordovician–Holocene; oysters, cockles, mussels, etc.
8. Class CEPHALOPODA: forms with paired organs and well-developed heads from which is developed a ring of prehensile tentacles; may or may not have internal or external shells; includes squid, ammonites, chambered nautilus, etc.

Definition and Morphology

The generalized Mollusca (see Fig. 1) are forms possessing an elongate body, which tends to be unsegmented and to have bilateral symmetry, and paired organs, with the exception of the gastropods. There are commonly four parts: (1) a well-developed head with tentacles and eyes—absent in the pelecypods; (2) a ventral muscular foot, which may be highly modified as in the cephalopods; (3) a well-developed visceral mass consisting of gastrointestinal tract, kidneys, a heart usually consisting of two auricles and a ventricle, and a mantle, which is a sheet of tissue completely surrounding the body; (4) finally, the calcareous shell secreted by the mantle. The mantle is domed to form the mantle cavity, which generally houses the gills and head and certain miscellaneous glands such as the shell-secreting and sensory glands.

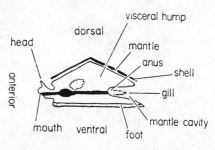

FIGURE 1. Schematic diagram of hypothetical ancestral mollusk (from Tasch, 1973). Black area = digestive tract. Compare with Figs. 2 and 3.

The head, if developed, usually possesses a mouth, eyes, and tentacles leading to a mouth cavity, which houses a chitinous lingual ribbon or some other device for the recovery of food. The foot may be highly modified, as in the case of bivalves, for digging or may be absent, as in the cephalopods where locomotion is accomplished by modifications of the mantle. The nervous system consists of centers connected by a loop, in connection with the head, and strands along the gut, foot, and mantle. In the gastropods, the nerves are coiled in a figure 8 pattern due to torsion, which has caused the mouth and anus to become juxtaposed. The molluscan heart consists of two auricles (except for the gastropods which possess one) and a single ventricle; the blood is moved through arteries into open sinuses. The reproductive system consists of paried gonads discharging their products through the kidneys. Separate sex is dominant, although hermaphroditism is also widespread. A free-swimming ciliated larva (called a trochophore) is followed by a partly protected veliger stage, which possesses primitive organs and a protoconch.

The shell, one of the most variable characters of the Mollusca (see Fig. 2) is usually composed of three layers, the first of which is an outer periostracum composed of a complex organic material allied to chitin, which serves to protect the shell from acid action. The two inner layers are composed of calcium carbonate. The outermost generally consists of prismatic columns at right angles or inclined to the surface; the inner layer tends to be thickest and is formed of overlapping laminae arranged such that the shell appears porcellaneous in structure or may be nacreous (see *Biomineralization*). There is, however, much variability in the structure of the wall and the above is a generalization. In the primitive forms, the shell is cap-shaped or conical in the gastropods, Monoplacophora (Fig. 3) and "Coniconchia." In more advanced gastropods, the shell may be coiled either symmetrically, as in the bellerophontids, or asymmetrically, as in the pleurotomarians. The pelecypods are bivalved and symmetrical perpendicular to the trace of the margin of the shell. The shells are opened by means of an elastic band composed of cutaneous material

FIGURE 2. Variations of molluscan architecture (from Weyl, 1970), showing foot (shaded), digestive tract, and shell (heavy lines). A, chiton, Polyplacophora; B, snail, Gastropoda; C, tusk shell, Scaphopoda; D, clam, Bivalvia; E, nautilus, Cephalopoda; F, squid, Cephalopoda.

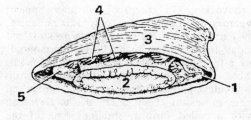

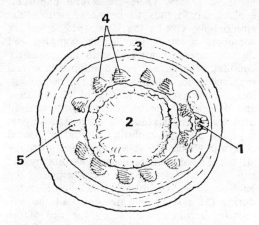

FIGURE 3. The monoplacophoran, *Neopilina,* in side and ventral views showing 1, head; 2, foot; 3, shell; 4, gills; and 5, anus. (from Solem, 1974).

(resilium) with strands of calcium carbonate that keep the shells under constant tension to remain open. Closing the shells is accomplished by either a single or multiple adductor muscles. The shell of the cephalopods if present tends to be symmetrically coiled and in two of the large orders, Nautiloidea and Ammonoidea, the shells are subdivided by means of septa into regular chambers connected by means of a perforation.

Affinities

The presence of bilateral symmetry, of a body cavity and the spiral cleavage in the fertilized egg, a trochiphore larva, and segmentation in the Monoplacophora all suggest that the mollusks were either derieved from or very closely related to several worm phyla, particularly the Annelida.

Evolution

The earliest Mollusca are representatives of the class Monoplacophora and the class Gastropoda, both of which appear in the Lower Cambrian. The Monoplacophora are cap-shaped (low, conical shelled forms) with a very wide aperture, which were probably adapted for a rock-clinging environment. Other Mollusca, such as the chitons (polyplacophora, which

appeared in the Upper Cambrian), are also rock-clinging; and it is believed this is the primitive adaptive zone. The early gastropods were probably derived from the Monoplacophora by the progressive loss of segmentation, the gradual loss of multiple gills, and lengthening of the conical shell. The monoplacophorans have a coil in their early life history, so that it seems to be a fundamental character in the mollusks. The earliest representatives of the class Gastropoda are symmetrically coiled forms, that is, the coil is in a single plane. This group, known as the bellerophontids, thrived during the early and middle Paleozoic and gradually diminished in numbers and kinds to extinction in early Triassic times. The bellerophontids are characterized by having an anal slit in the plane of symmetry, which allows the anus to extend outside and away from the mantle cavity, thus preventing fouling of the incurrent waters. This in part was necessary because of the primitive type of gill structure in the group.

From the bellerophontids an important group of gastropods arose in the Late Cambrian, the pleurotomarians. They also possess an anal slit but are differentiated on the basis of the asymmetrical shell. The asymmetrical shell was a much more successful structure and dominated the gastropod group from Paleozoic time onward. The pleurotomarians flourished in the middle and late Paleozoic, becoming almost extinct at the close of the Paleozoic but with a few representatives living today. The pleurotomarians seem to have been the stock from which all other gastropods arose. All the groups of gastropods discussed thus far had paired organs, the bellerophontids and pleurotomarians having a single pair of organs. All the groups that arose from the pleurotomarians progressively lost one set of organs, probably as a result of the asymmetrical coiling and perhaps the trend toward a higher spired shell. This gradual loss of organs can be documented in several of the Paleozoic groups, such as the murchisonids which gradually lost the medial position on the whorl of the anal slit; thus, the anus came to occupy a position at the extreme upper portion of the aperture. At this time, it is believed that a different, more complicated set of gills was developed, which allowed the sanitation problem to be solved in a different manner. The newly acquired gill permitted the development of a siphonal system allowing the organism to occupy many of the adaptive zones, such as a burrowing habitat, that had been impossible before. This occurred several times in the Paleozoic, and established groups that were to dominate the Mesozoic and Tertiary, resulting in the many diverse gastropods that make up the second largest of the invertebrate groups of animals.

The origin of the pelecypods (Class Bivalvia) is obscured. Unlike the gastropods, most of the bivalved groups usually classed as orders or suborders had appeared by Middle Ordovician time. Therefore, it is obvious that the primary radiation had occurred sometime during the Cambrian. The first pelecypod is known from the Middle Cambrian. Most of the documented evolution in the group is on the familial and generic level; and little agreement has been reached by authorities on the evolution of the higher taxa. However, the following observations can be made. The pelecypods in their early history were undoubtedly free-moving bottom dwellers. This in part is due to the primitive gill structure; however, with the development of other gill types, they were able to migrate into many different environments and dominantly today live partly or completely buried in the bottom muds or sands, or are attached by threads or cemented permanently to objects on the bottom. With the exception of a few groups such as the pectinids, they have tended to live as sessile benthic forms, locomotion not being an important avenue of exploration during the evolution of the group. There has been a gradual proliferation of genera and they appear to be at their climax today. While the gastropods show strong extinction rates at the close of the Permian and Cretaceous, the pelecypods show only a minor extinction at the end of the Cretaceous, with Tertiary replacement.

The class Cephalopoda represents mollusks that have achieved a high degree of evolution in locomotion, and includes some of the swiftest of all swimmers, such as the squid. In the fossil record, the two chief groups are the nautiloids and the ammonoids. The nautiloids, representing some 700 genera, began in the Cambrian as elongate cone-shaped shells. They dominantly have symmetrical coiled shells that have developed in several different lines early in the history of the group. The shell consists of a series of chambers separated by calcareous septa. It appears that most of the evolution of the nautiloids had to do with experimentation in size and shape of the connecting passage through the chambers. Presumably this had to do with improvement of gas exchange in the chambers in conjunction with buoyancy, but this is speculation. The nautiloids represent a type of evolutionary pattern in which there is a sudden explosion of genera immediately after they appear and move into a new adaptive zone, and then a gradual dying out for one cause or another. In the Ordovocian, 275 genera of nautiloids appear; whereas by Silurian time, only 120 remained; by Permian time, 30 genera were present; and today there are less than 10. The ammonoids are a far more important group of cephalopods; there are approximately 1800 genera. Ammonoids apparently replaced the nautiloids and were successful. They began in the Devonian with a coiled form resembling some of the nautiloid genera. There are several important evolutionary trends to be noted within the ammonoids. One is that the septa separating chambers became very highly complex and digitate on their margins with an amazingly complex evolutionary plan developed in several separate major stocks. One stock, the goniatites, became extinct by the end of the Paleozoic and only a few species of ammonites survived into the early Triassic. The early Triassic was a period of rapid evolution of the ammonoid genera in which 140 genera appeared. Virtually all the Triassic genera became extinct by the end of the Triassic and were replaced by another series of forms, which proliferated during the rest of the Mesozoic, all the ammonites becoming extinct by the end of the Cretaceous.

Other groups of cephalopods are poorly known in the fossil record in terms of abundance or distribution. The other classes likewise are known only sporadically through the fossil record and made no patterns that are distinguishable in terms of evolution. The above statement on the evolution of the mollusks is generalized; detailed hypotheses are numerous (for the latest attempt see Runnegar and Pojeta, 1974, and Pojeta and Runnegar, 1976, a rather controversial approach; see Yochelson, 1978).

Only two of the classes of mollusks, the gastropods and pelecypods, have migrated into fresh water or invaded the land. The gastropods first appear to have entered fresh waters during the late Paleozoic, but the record is spotty and no major proliferation into the adaptive zones occurred until the middle Mesozoic and Tertiary. The largest terrestrial group is represented by the Stylommatophora, with some of the prosobranch Mollusca and the Basommatophora occupying a freshwater habitat. The bivalves are represented in fresh waters by such families as the Unionidae and the Cyrenidae. It is puzzling why the fast-moving cephalopods did not manage to get into fresh waters.

Marine Mollusca dominantly occupy a shallow-water to moderately shallow-water depth zone; and, as in other groups, are principally found in warm to temperate waters. The Monoplacophora were rock clingers for the most part of their history, although the modern form *Neopilina* appears to live in soft sediments. It probably moves sluggishly along the bottom since the foot is poorly developed. Polyplacophora, like the Monoplacophora, are rock clingers and move about, if they move about at all, very slowly. The Aplacophora have been modified for swimming. The gastropods dominantly occupied a benthic habitat and are mostly slow-moving forms, creeping being the principle form of locomotion on a well-developed muscular foot. The bivalves are principally burrowers and move, if at all, very slowly through the sediment. The cephalopods have evolved from a benthic into a swimming habit and are today by far the fastest movers. In consequence, they have developed their sensory organs to a very high degree; the eye is far more efficient than in any of the other mollusks and the tentacles have been modified for grasping and tactile usage. Scaphopods are dominantly sessile benthic forms that burrow in soft sediments. Thus, we see the classes of the Mollusca have evolved into major adaptive zones based principally on locomotion.

R. L. BATTEN

References

Arkell, W. J., et al., 1957. Cephalopoda Ammonoidea, in R. C. Moore, 1957–1971, pt. L. Mollusca 4, 490p.

Cox, L. R., et al., 1969, 1971. Bivalvia, in R. C. Moore, 1957–1971, pt. N, Mollusca 6, 3 vols.

Knight, J. B.; Brooks,; Cox, L. R.; Keen, A. M.; Smith, A. G.; Batten, R. L.; Yochelson, E. L.; Ludbrook, N. H.; Robertson, R. R.; Yonge, C. M., and Moore, R. C., 1960. Mollusca general features, Scaphopoda, Amphineura, Monoplacophora, Gastropoda general features, Archaeogastropoda, mainly Paleozoic Caenogastropoda and Opisthobranchia, in R. C. Moore, 1957–1971, pt. I, Mollusca 1, 351p.

Moore, R. C., ed., 1957–1971: *Treatise on Invertebrate Paleontology:* Mollusca. Lawrence, Kansas: Geol. Soc. Amer. and Univ. Kansas Press, pts. I, K, L. and N.

Pojeta, J., Jr., and Runnegar, B., 1976. The paleontology of rostroconch mollusks and the early history of the phylum Mollusca. *U. S. Geol. Surv. Prof. Paper 968*, 88p.

Purchon, R. D., 1968. *The Biology of the Mollusca.* Oxford: Pergamon Press, 560p.

Runnegar, B., and Pojeta, J., Jr., 1974. Molluscan phylogeny: The paleontological viewpoint, *Science*, 186, 311–317.

Solem, G. A., 1974. *The Shell Makers.* New York: Wiley-Interscience, 289p.

Stasek, C. R., 1972. The molluscan framework, in M. Florkin and B. P. Scheer, ed., *Chemical Zoology. VII, Mollusca.* New York: Academic Press, 1–44.

Tasch, P., 1973. *Paleobiology of the Invertebrates.* New York: Wiley, 946p.

Teichert, C., et al., 1964. Cephalopoda general features, Endoceratoidea, Actinoceratoidea, Nautiloidea, Bactritoidea, in R. C. Moore, 1957–1971, pt. K. Mollusca 3, 519p.

Weyl, P. K., 1970. *Oceonography.* New York: Wiley, 535p.

Yochelson, E. L., 1978. An alternative approach to the interpretation of the phylogeny of ancient mollusks, *Malacologia*, 17, 165–191.

Cross-references. *Ammonoidea; Belemnitida; Biomineralization; Bivalvia; Cephalopoda; Gastropoda; Growth Lines; Hyolitha; Larvae of Marine Invertebrates, Paleontological Significance; Matthevia; Monoplacophora; Oysters; Paleoecology; Polyplacophora; Rostroconchia; Rudists; Scaphopoda; Stenothecoida; Tentaculita.*

MONOPLACOPHORA

Monoplacophora, Lemche, 1957 (Fig. 1A, B, C), constitutes a class of living Mollusca. The calcareous shell is bilaterally symmetrical, oval, and with a subcentral to anterior apex so that it is cap-shaped to spoon-shaped. Muscles or muscle scars are paired (Fig. 1E, G). The class is extinct except for one Holocene genus.

In less than two decades since description of *Neopilina galatheae* (Fig. 1), from ten specimens dredged by the Danish Deep-Sea *Galathea* Expedition in 1950–1952, the class proposal found almost universal acceptance because of the unique features of this species. The specimens were studied in detail by H. Lemche and K. G. Wingstrand in 1959, and the anatomy of preserved speciments is perhaps best known of all mollusks. *Neopilina* is generally considered to be the most primitive living mollusk.

Externally, *Neopilina* has a thin, nearly circular shell with an anterior apex; an asymmetric coiled protoconch is known from one specimen. On the ventral surface, a circular pallial cavity opens between the edge of the shell and a large central foot; in life, this pallial cavity was probably covered by the mantle. Within this cavity are five pairs of gills. The anterior mouth has lateral ciliated palps and a posterior tuft of tentacles; the anus is posterior. Within the body are six pairs of kidneys, a paired liver, two pairs of gonads, and two pairs of auricles. The viscera are attached to the shell by eight pairs of principal muscles. Two rings of nerve cords are interconnected by ten pairs of lateral-pedal commissures.

The original specimens of *Neopilina* were taken at 1963–2033 fathoms in the Guatemala Basin. Four specimens of a second species with six gills were collected from the northern part of the Peru Basin at 3072–3456 fathoms. Examples of a third species, also with five gills, were dredged in 1493–1514 fathoms in the North Pacific Basin off Baja California. A few additional specimens have been obtained

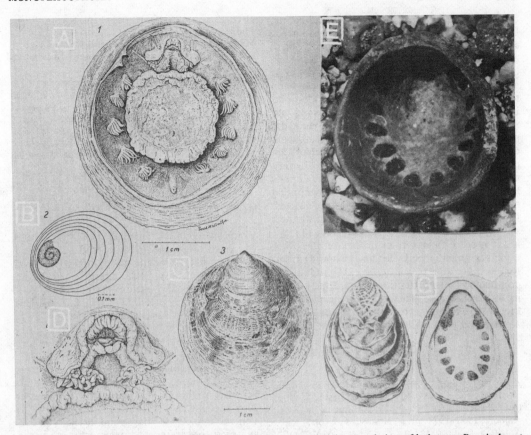

FIGURE 1. A–D: Original drawings of *Neopilina galatheae* Lemche; A, ventral view of holotype: B, apical part with larval shell; C, dorsal view of shell of holotype: D, ventral view of anterior part of body. Size given by scale (from Lemche, 1957). E, interior view of *Archaeophialia antiquissima* (Hisinger) from the Upper Ordovician of Sweden, showing the anterior tripartite paired scars and the posterior five pairs of scars, ca X2. F and G, exterior and interior of *Tryblidium reticulatus* Lindstrom from the Middle Silurian of Sweden, ca X1 (from Knight and Yochelson, 1960; courtesy of The Geological Society of America and the University of Kansas Press).

from a Sea Mount N of Hawaii and from the sourthern Atlantic and the Red Sea (for a summary, see Filatova et al., 1974). *Neopilina* remains one of the rarest of Holocene mollusks, with perhaps 150 speciments known, in spite of additional deep-sea exploration and intensive study of earlier deep-sea collections. Recently, however, monoplacophorans have been reported off San Diego, California, from only 175–225 m depth, clinging to rocks (McLean, 1976; Lowenstam, 1978).

Popular interest has been aroused by *Neopilina*, for it is truly a "living fossil." Monoplacophorans are known as fossils from rocks as young as Middle Devonian age (375 m yr ago); but until *Neopilina*, they were thought to be extinct. Again, in spite of considerable interest and intensive search by paleontologists, no fossils have been found in the gap between Devonian and Holocene.

The first fossil monoplacophoran was described by G. Lindstöm in 1880 from the Silurian of Gotland, Sweden, though placed within the gastropods. Additional species and genera have been described from all continents. The monoplacophorans from the middle Paleozoic of Czechoslovakia are well known from the description of J. Perner in the first decade of this century and revisions by R. Horný in a series of papers since 1956.

Fossil monoplacophorans are rare. Less than 50 genera are known and perhaps 100 to 150 species have been described. Most species are known from only a few specimens; this rarity makes it unlikely that these fossils will find general usage as stratigraphic guides. In terms of diversity and number of specimens, the fossil record of the monoplacophorans is comparable to that of the small molluscan class Polyplacophora.

Monoplacophorans occur in geologic settings characteristic of relatively shallow to very shallow water (e.g., Stinchcomb, 1975). Some may have been reef or near-reef dwellers; judging by associated fauna, all lived under conditions of normal salinity. Possibly the retreat from shallow-water shelf conditions to the deep ocean prevented extinction of these rare animals.

Characteristic fossil monoplacophorans have a thick shell in which the paired muscle scars are deeply impressed. The earliest undoubted genus is *Kiringella* from the Late Cambrian (525 m yr) in Siberia. Middle Cambrian beds contain some cap-shaped shells with an oval cross section. Although these are placed with the monoplacophorans, only one specimen is known that shows muscle scars, and these are not quite in bilateral arrangement. Thus, it is an open question whether the monoplacophorans constitute the most primitive molluscan stock.

Some fossil monoplacophorans show eight pairs of scars, and at one time these were thought to be homologous with the eight plates of Polyphacophora. Other genera, however, have fewer scars, and in one form the scars are reduced to a single, wide elongate pair. It is becoming increasingly apparent that the monoplacophorans underwent an evolutionary radiation from Late Cambrian onwards; by inference, Early Cambrian forms assigned to the class may not be correctly placed.

Because of their low cap-shaped shells, fossil monoplacophorans were traditionally considered to be aberrant patellid gastropods; patellid gastropods are characterized by a horseshoe-shaped muscle. In 1940, W. Wenz suggested that paired muscle scars of monoplacophorans indicated these animals had not undergone torsion of soft parts and therefore were not gastropods. This idea was amplified by J. B. Knight in 1952. His solution to these phylogenetic problems was to reduce the number of mollusk classes to four by expanding the Gastropoda into a branch that has undergone torsion and a branch that has not undergone torsion. "Amphineura" (= Aplacophora and Polyplacophora) and Monoplacophora were placed as orders in the latter branch. This scheme was abandoned when *Neopilina* was discovered. Though the emphasis is slightly different, Knight's paper is significant in demonstrating the zoological importance of a "missing link," which at that time was known only from fossils. Knight and Yochelson published a classification of Monoplacophora in 1960. Although this is still the most comprehensive source of information, rapid increase in knowledge of these curious fossils limits its usefulness as a summary of genera; the higher-level classification used is obsolete.

Interest has focussed on the anatomy of *Neopilina* because some features are reminiscent of other members of the phylum. The tentacles are similar to those in scaphopods and the labial palps to those in protobranch bivalves. Even more important is the repetition of the various soft parts and organs; this "metamerism" is commonly accepted as a prime feature of the class. Some authors contend that this is a form of segmentation, which indicates a close relationship of Mollusca to Annelida and Arthropoda. Others, while not denying that there are features in common among these phyla, doubt their close relationship and do not interpret *Neopilina* as being truly segmented.

Within the Mollusca, the Monoplacophora have assumed a key and controversial role in phylogenetic speculation. Paired muscle scars have been found in the coiled Middle Ordovician *Cyrtonella*. This in turn has lead to an increasing elaboration of the classification of Monoplacophora (see Starabogatov, 1970). The extinct Bellerophontacea, conventionally placed as bilaterally symmetrical gastropods, have been judged by some to be Monoplacophora. The degree of morphologic variation included within the class has been greatly expanded by Runnegar and Pojeta (1974). These authors apparently equate the Monoplacophora with any Paleozoic molluscan univalve assumed to have nontorted soft parts, and consider various Early Cambrian forms as intermediates that gave rise to other extant classes; they judge that all fossil mollusks, with the exception of one group, can be placed in living classes.

The limits of the Monoplacophora are not clear and it is likely that some decades of additional controversy will ensue before a concensus has been reached. Nevertheless, the finding of *Neopilina* and establishment of this class concept has resulted in a vigorous study of early Paleozoic molluscan faunas in a search for other "missing links" in molluscan phylogeny.

ELLIS L. YOCHELSON*

References

Filatova, Z. A.; Vinogradova, N. G.; and Moskalev, L. I., 1974. New finding of the ancient primitive mollusc *Neopilina* in the Atlantic part of the Antarctic Ocean, *Nature*, 249, 675.

Knight, J. B., and Yochelson, E. L., 1960. Monoplacophora, in R. C. Moore, ed., *Treatise on Invertebrate Paleontology*, pt. I, Mollusca 1. Lawrence, Kansas: Geol. Soc. Amer. and Univ. Kansas Press, 177–183.

*Publication authorized by the Director, U. S. Geological Survey

Lemche, H., 1957. A new living deep-sea mollusc of the Cambro-Ordovician class Monoplacophora. *Nature,* 179, 413–416.

Lemche, H., and Wingstrand, K. G. 1959. The anatomy of *Neopilina galatheae* Lemche, 1957 (Mollusca Tryblidiacea), *Galathea Rept.,* 3, 9–76.

Lowenstam, H. A. 1978. Recovery, behaviour and evolutionary implications of live Monoplacophora. *Nature,* 273, 231–232.

McLean J. H., 1976. A new genus and species of Monoplacophora from the continental shelf of southern California, *Bull. Amer. Malacological Union for 1975,* (abstr.), 60.

Runnegar, B., and Pojeta, J., Jr., 1974. Molluscan phylogeny: The paleontological viewpoint, *Science,* 184, 311–317.

Starobogatov, Y. I., 1970. Systematics of Early Paleozoic Monoplacophora, *Paleontol. J.,* 1970, 293–302.

Stinchcomb, B. L., 1975. Paleoecology of two new species of Late Cambrian *Hypseloconus* (Monoplacophora) from Missouri, *J. Paleontol.,* 49, 416–420.

Cross-references: *Living Fossil; Mollusca.*

MORPHOLOGY, CONSTRUCTIONAL

Definition

Constructional morphology, or Konstruktions-Morphologie, has been developed by German paleontologists as a conceptual framework for the analysis of biological form. Three fundamental factors are distinguished which interact to determine organic form in general, and the forms of skeletal hard parts that may be preserved as fossils in particular (Fig. 1). A *phylogenetic factor* depends upon the evolutionary history of the organism, transmitted as its genetic heritage by the processes of mitosis and meiosis. An *adaptive factor* recognizes that the organism is designed, by natural selection acting upon variable populations, to function efficiently and thereby survive in its ecological niche. A *morphogenetic factor* derives from the growth processes and patterns, which, under the control of biochemical programs, give rise to organic structures.

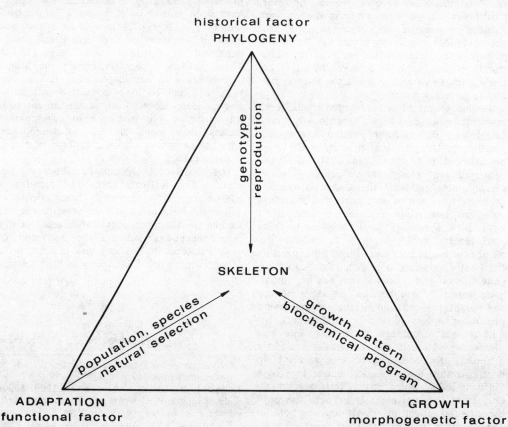

FIGURE 1. Conceptual framework for constructional morphology. Three fundamental factors interact in the genesis of organic form. Some skeletal structures are determined more immediately by one factor than another. Morphological analysis in terms of all three factors provides an integrated explanation for the origins of fossil skeletons and their constituent elements.

In terms of the classic machine analogy, these three aspects of the explanation of organic form are illustrated by three simple questions. What parts and blueprints are available? How does the machine work? How is the machine built? The interpretation of a skeletal structure in terms of all three factors provides an holistic causal explanation of form. Commonly, one factor or another predominates in the determination of a given structure. Together, these factors not only lead to the explanation of individual forms; they also prescribe boundary conditions or limits within which evolutionary modification occurs.

Origin of the Concepts

The concepts of constructional morphology have been developed and applied to research in paleontology by Seilacher (1970). No major synthesis of these ideas has yet been attempted, although they have been succinctly reviewed, with illustrative examples, by Raup (1972). Constructional morphology is known principally from lectures given by Seilacher at meetings and universities in Europe and North America, and from a series of papers written by resident and visiting scientists associated with the paleontological research group *Sonderforschungsbereich 53*, at the university in Tübingen, West Germany. Seilacher's lectures frequently generate controversy, for he has emphasized so-called nonadaptive aspects of growth patterns. Like D'Arcy Thompson, he has been accused by British and North American paleontologists, whose recent research has stressed the functional interpretation of form, with neglect, if not rejection, of the essential role of natural selection in evolution. As Gould has pointed out, Seilacher's thinking owes much to the long-standing German tradition of seeing form in terms of broad, abstract generalizations or laws which impose themselves on living organisms. However, this is largely a matter of emphasis. Seilacher's choice of the term *Konstruktions-Morphologie* to describe his conceptual model as a whole reflects his primary interest in the morphogenetic factor in the determination of skeletal form. Functional morphology and the biology of the neo-Darwinian evolutionary synthesis are nevertheless essential features of the model.

Since 1970, over eighty publications by members of the Tübingen group and its associates have appeared, under the general heading of "Konstruktions-Morphologie." Several of these are elegant and stimulating contributions to our knowledge of functional morphology and skeletal morphogenesis in diverse living and fossil organisms. For the most part, however, the roles of each of the three determinative factors in prescribing the form of a given structure have not been systematically taken up and compared. This kind of comparative analysis, presupposed in Seilacher's original formulation of constructional morphology, has considerable, still largely unrealized, potential to further our understanding of organic form.

Subsidiary Concepts and Examples

Most authors have emphasized one or another of the factors brought together in constructional morphology in their approaches to the study of biological form and its diversity. Simpson (1967), reacting strongly to various forms of vitalistic determinism including orthogenesis, which played such a large part in the paleontological thinking of the first half of this century, has insisted upon the opportunism of evolution: Historical or configurational factors predominate in the origin of species and the determination of form. Life is infinitely varied, no two species can ever evolve twice, and convergence is never complete (Jacob, 1977), even in the eyes of man and the octopus.

The role of the phylogenetic factor in evolution is illustrated by the comparison of fundamentally different structures that have evolved to serve similar purposes. The bivalve gill and the brachiopod lophophore are both ciliary pumps used in feeding and respiration, but these organs are quite dissimilar in origin, development and form. Vestigial structures, such as the splint bone in the foot of the modern horse (Fig. 2), are likewise explained in terms of their ancestry rather than their current adaptive or structural significance.

Rudwick (1961), seeking a rigorous methodology by which he could distinguish among multiple hypotheses for the adaptive significance of the skeletal structures of extinct organisms, has concentrated attention upon the functional aspects of form. For any given ethological or physiological function there is a mechanical paradigm, an optimum model of maximal efficiency. The probability that a particular structure served in one capacity rather than another is measured by the extent to which the structure approaches the paradigm for the former function and departs from that of the latter. Rudwick acknowledges the roles of phylogeny and morphogenesis, noting that the approach to a paradigm is restricted by the range of forms that the organism can generate. He has paid close attention to the inherent properties of accretionary growth. The paradigm methodology proves most fruitful where several functional adaptations appear to be equally possible, as in Rudwick's own analyses

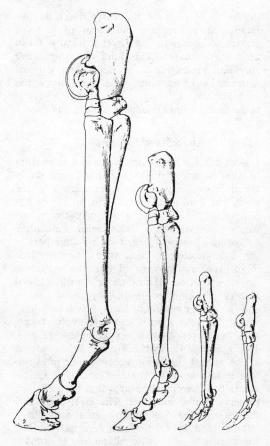

FIGURE 2. Determination of form by phylogeny. The vestigial splint bone in the foot of the modern horse reflects its derivation from ancestors which had three functional digits. Stages in the reduction of these digits are illustrated, from right to left, by *Hyracotherium* (Eocene), *Mesohippus* (Oligocene), *Merychippus* (Miocene), and *Equus* (from Matthew, 1913).

of the various functions of spines on the shell of the aberrant Permian brachiopod *Prorichthofenia*.

Adaptation is preeminent in the determination of form where a few characters of closely related and otherwise similar organisms have evolved in divergent directions, in response to the functional requirements of widely different modes of life. This pattern of evolution is evident in all the classic examples of adaptive radiation, following the achievement of a major, new, structural level or grade. Once off the ground, birds have evolved wings modified for flapping flight, hovering, soaring, and even swimming. Evolutionary convergence is also generally viewed as a consequence of functional adaptation, in this case to a common environ-

ment. For example, bryozoans and colonial corals may be said to have evolved similar massive, branching, and encrusting habits in adopting like modes of life (Fig. 3). However, while adaptation is certainly responsible for convergence of function in evolution, it is the limited number of possible solutions to common mechanical problems that facilitates convergence of form. Diverse fishes and squids are adapted to live in very similar ways, but they have for the most part solved common mechanical problems very differently, as has been elegantly shown by Packard (1972).

Raup (1968), in developing quantitative, predictive models of skeletal growth, has shown that relatively complex structures can be generated by the interaction of a few morphogenetic variables. Geometric and mechanical constraints play a significant part in computer models that simulate plate addition and growth in echinoid interambulacra. Raup concludes that similar constraints act in the morphogenesis of actual echinoid skeletons, and that the genetic programs that control their growth need not be more complex than the instructions given to his computer. Recently, Reif and Robinson (1975) have also focused their attention on the morphogenetic factor as a determinant of form. They have recognized seven major skeletal patterns, each based on a unique structural principle, which recur in disparate groups of organisms. The basic question posed by these groupings is this: does any given convergent pattern reflect an adaptive, functional convergence, or does it simply result from the limited variety of possible modes of growth? In most cases the answer to this question is not known.

The range of possible organic forms is limited by purely geometric considerations as well as the processes of growth itself. An organism may increase the size of its skeleton by addition of material to one side (external shells of molluscs and brachiopods) or on all sides (internal shells of echinoderms). Alternatively, the skeleton may grow by regeneration (vertebrates) or it may be periodically molted (arthropods). In a series of imaginative studies, Raup (1972 and references therein) has shown how variations on one pattern determine the forms of brachiopod and molluscan skeletons (Fig. 4). Skeletal materials and the processes by which they are secreted also determine elements of form. Seilacher (1972) has suggested that divaricate patterns observed in bivalve shells are based on a common principle of shell growth, although the patterns have a variety of functions and some may lack any function whatsoever.

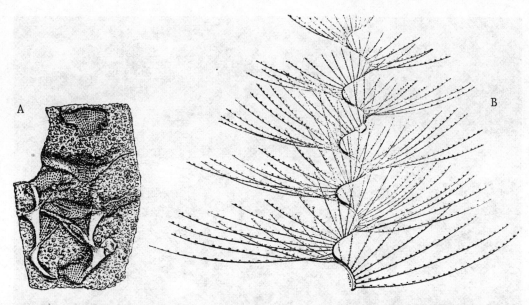

FIGURE 3. Adaptation is responsible for the convergence of function in evolution, but it is the limited number of possible solutions to common geometric and mechanical problems that gives rise to convergence of form. The well-known late paleozoic bryozoan *Archimedes* developed a delicate spiral colony (A) which is strongly convergent with that of the modern octocoral *Iridogorgia* (B), which lives attached to the bottom in deep, still water. (A from Ulrich, 1890; B from Agassiz, 1888).

Relationships Among the Factors

The object of constructional morphology is to assess the relative importance of historical, functional, and morphogenetic factors in the development of skeletal structures. This purpose may be clarified somewhat by consideration of the relationships among the three factors themselves.

Adaptation and Phylogeny. In the long run, adaptation and phylogeny are both the products of natural selection. In constructional morphology these factors operate on distinct taxonomic levels. If the object of study is a species, the ancestry of that species constitutes the historical factor, while the functional factor is reflected in the genetic and ecophenotypic variation among its constituent populations. Likewise, the phylogenetic history of a family places a common constraint on the range of functional adaptations of its genera and species. Thus, phylogeny and adaptation have independent roles in the determination of form at any given taxonomic level.

Phylogeny and Growth. Ontogeny is the product of both phylogeny and morphogenesis. There is nevertheless a fundamental difference between these two factors in their effects on development. The influence of the phylogenetic factor on form is purely configurational; later structures can only arise as modifications of

ancestral forms. There is a large element of chance in the history of any particular structure. In contrast, the morphogenetic factor depends upon the immanent properties of form and the mechanical properties of skeletal materials. The possible forms of organic structures are limited by the laws of symmetry, by the small number of distinct growth processes, and by the relationship between size, complexity, and shape. These laws, which are invariant, are responsible for most of the remarkable homologies of form observed in disparate groups of organisms.

Adaptation and Growth. Both of the dichotomies recognized above are seen in the relationship between growth and adaptation. At any given taxonomic level, the patterns of skeletal morphogenesis are prescribed by the evolutionary history of the taxon; below that level, functional adaptation produces a diversity of forms within the range allowed by the general pattern. Likewise, in terms of chance and necessity, growth patterns are conservative, limiting morphological variation to a range that does no violence to the basic geometry, while adaptation is opportunistic, tending to increase diversity up to the limits of this range. Constructional morphology provides a straightforward answer to what was once considered an important problem: what comes first in the origin of evolutionary novelties, structure or

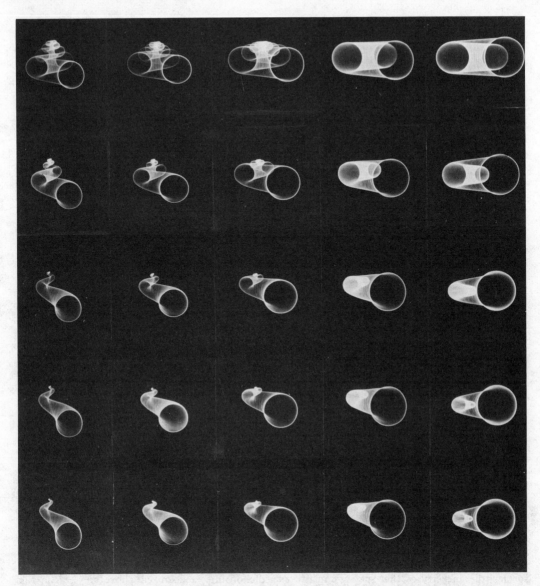

FIGURE 4. Constraints of geometry on form. An array of computer-generated shells illustrates part of the morphological ranges that are theoretically available to brachiopods, gastropods, cephalopods, and bivalves. These groups have characteristically, but not exclusively, exploited different parts of the total range. In this model, some conceivable morphologies are rarely if ever exploited by real organisms since the shells would be highly asymmetrical, structurally weak, or otherwise mechanically impractical (from Raup and Michelson, 1965; copyright © 1965 American Association for the Advancement of Science).

function? Neither. New structures and functions have invariably evolved in tandem from preexisting structures with different functions.

Other Factors. In his review, Raup (1972) has suggested that the three factors recognized by Seilacher are necessary but not sufficient to determine the forms of organic skeletons. He has introduced two additional factors—chance and ecophenotypic effects—which he thinks

should be considered in the analysis of form. We have seen above that ecophenotypic variation is a special case of adaptation—the physiological adaptation of some recent authors—where the object of study is the skeleton of a species rather than that of a genus or higher category. Chance also does not appear to be an independent factor, in that it plays a part in the expression of all three of the fundamental variables. Random factors necessarily play a large part in the course of phylogeny, which records the directional change of particular complex organisms in unique environmental circumstances. Such chance factors must often have impeded the close approach of a structure to its functional paradigm, or prevented the adoption of a possible pattern of skeletal morphogenesis, as Raup suggests. However, chance is not a factor that is complementary to the other three in determining organic form. It can either be subsumed in the historical process of phylogeny, or it can be considered as a counterfactor that works against the others. Whereas phylogeny, growth, and adaptation act together to determine organic form, chance is, of course, counterdeterminative. The object of constructional morphology is to understand the genesis of form within the framework of three fundamental limits or constraints. Chance variations constitute noise in this system, which must be filtered out if underlying patterns are to be recognized.

Conclusion

In 1970, Gould concluded that paleontologists were in the process of creating a science of form, based on rigorous studies of functional morphology, theoretical models of skeletal morphogenesis, and an understanding of the constraints exerted on form by physical laws (Gould, 1970). Since that time, a great deal more work has been done in functional morphology, while a renewed interest in the roles of geometric and mechanical factors in morphogenesis has developed outside West Germany. The concepts of constructional morphology serve to focus attention on the urgent need to integrate disparate approaches to the problem of form. Constructional morphology does not provide analytical tools for this purpose. Rather, it constitutes an intellectual framework that provokes questions about relationships among the factors responsible for the genesis of organic form, generating hypotheses that can be tested against theoretical models and the skeletons of living and extinct organisms.

R. D. K. THOMAS

References

Agassiz, A., 1888. Three cruises of the United States Coast and Geodetic Survey Steamer "Blake," *Bull. Mus. Comp. Zool. Harvard Univ.,* 15, 1–220.
Gould, S. J., 1970. Evolutionary paleontology and the science of form, *Earth Sci. Rev.,* 6, 77–119.
Jacob, F., 1977. Evolution and tinkering, *Science,* 196, 1161–1166.
McGhee, G. R., Jr., 1978. Analysis of the shell torsion phenomenon in the Bivalvia (Konstruktionsmorphologie, Nr. 91), *Lethaia,* 11, 315–329.
Matthew, W. D., 1913. Evolution of the horse in nature, in Evolution of the horse, *Amer. Mus. Nat. Hist. Guide Leaflet Ser.,* no. 36.
Packard, A., 1972. Cephalopods and fish: The limits to convergence, *Biol. Rev. Cambridge Phil. Soc.,* 47, 241–307.
Raup, D. M., 1968. Theoretical morphology of echinoid growth, *Paleontol. Soc. Mem., 2; J. Paleontology,* Suppl., 42, (5), 50–63 (see also other papers in this volume).
Raup, D. M. 1972. Approaches to morphologic analysis, in T. J. M., Schopf, ed., *Models in Paleobiology.* San Francisco: Freeman, Cooper, 28–44.
Raup, D. M., and Michelson, A., 1965. Theoretical morphology of the coiled shell, *Science,* 147, 1294–1295.
Reif, W. E., and Robinson, J. A., 1975. Geometrical relationships and the form-function complex: Animal skeletons (Konstruktionsmorphologie, Nr. 31), *Neues Jahrb. Geol. Paläontol. Monatsh,* 1975, 304–309.
Rudwick, M. J. S., 1961. The feeding mechanism of the Permian brachiopod *Prorichthofenia, Palaeontology,* 3, 450–471.
Seilacher, A., 1970. Arbeitskonzept zur Konstruktions-Morphologie, *Lethaia,* 3, 393–396.
Seilacher, A., 1972. Divaricate patterns in pelecypod shells, *Lethaia,* 5, 325–343.
Simpson, G. G., 1967. *The Meaning of Evolution,* rev. ed. New Haven, Conn.: Yale Univ. Press, 364p.
Thomas, R. D. K., 1978. Limits of opportunism in the evolution of the Arcoida (Bivalvia), *Phil. Trans. Roy. Soc. Lond.,* 284B, 335–344.
Ulrich, E. O., 1890. Paleonzoic Bryozoa, *Illinois Geol. Surv.,* 8, 283–688.

Cross-references: *Evolution; Morphology, Functional.*

MORPHOLOGY, FUNCTIONAL

Morphology is the study of the form and structure of organisms or parts of organisms, and can be divided into two subdisciplines. *Descriptive morphology* seeks to specify and portray form and structure in as precise and objective a way as possible. Its usual application is in the delineation of taxonomic categories by morphological or phenetic comparison between organisms. *Functional morphology* (and the related *constructional morphologic* approach [q.v.]) is an explicitly interpretative subdiscipline, and is the study of the form and structure of organisms or parts of organisms in terms of the biological roles that they play. Functional analysis is useful in paleoecological reconstructions of organs, organisms, and communities, and in assessing evolutionary relationships by comparing adaptations in a changing lineage of organisms.

Functional Morphology in Paleontology

The descriptive and functional morphology of extant taxa are almost always grounded on observations made on living organisms. Their biological activities can be observed directly, and experimental manipulation makes it possible to test hypotheses which seek to explain the function of particular organs. Even if living specimens are not immediately available, biologists can offer interpretative functional hypotheses knowing that in principle they can be rigorously tested on living organisms in their natural habitat.

These advantages are denied to paleontologists, who must interpret the biological functions of organs and organisms from those morphological features that happen to be preserved in the fossil record. The vagaries of fossilization force the functional morphologist to make an interpretative anatomical reconstruction of the organism under study before he can even begin to analyze its form and structure in functional terms.

Because so much is known, or can be investigated, from study of living material, functional morphology is most easily prosecuted by using direct comparison between fossil material and its most closely applicable living counterpart. For example, the early Cenozoic insects and spiders preserved in Baltic amber retain such exquisite morphological detail and are so closely similar to living species that their biological roles can be assessed without difficulty. The morphology of *Icaronycteris*, the earliest known bat, is so well preserved and so like that of living bats that a detailed and sophisticated analysis of its functional adaptations is possible (Jepsen, 1970).

The difficulty of functional analysis increases with decreasing quality of preservation, and with decreasing similarity to living organisms. Thus, for example, the biology of early primates is poorly known because they are preserved only as scattered teeth. On the other hand, the well-preserved invertebrates of the Cambrian Burgess Shale of British Columbia are difficult to interpret because they have no living counterparts, although their descriptive morphology is well advanced. There are various approaches available to aid in solutions to these problems.

Homological Comparison with Living Organisms

Taxa are grouped as units because they share morphological features. If these features have similar functional significance throughout the taxon, then the comparison of living and fossil representatives of the taxon will be highly rewarding for successful interpretation. In general, for example, one can successfully draw a functional comparison between the wings of different birds, although one might face surprises in treating hummingbirds, ostriches, penguins, and perhaps *Archaeopteryx* (Ostrom, 1974). Stanley (1975, and references therein) has done much to interpret the adaptations of extinct bivalves by direct comparison with their living relatives, and any reconstructions of the morphology and adaptations of extinct cephalopods are usually explicitly based on living genera (e.g., Denton, 1974). Geist (1966) has interpreted the function of horns in extinct bovids by direct comparison with the functions he has observed in their living descendants. Altogether, direct homological comparison is the most powerful approach used by functional morphologists.

This approach may be open to criticism where taxonomic distance widens. For example, reconstructions of dinosaurs have often been based on comparison with living lizards, and may well be erroneous (Bakker, 1971a, b); early horses and camels had adaptations not shared by their living descendants.

Analogy with Living Organisms

If direct homology with living forms is not possible, then the next best method is analogical comparison. In some cases, this makes available a rich supply of data from unrelated living organisms. Analogy between the limbs of heavy quadrupeds walking and running on firm ground reveals strong functional similarity between the reptilian dinosaurs and the mammalian elephants and rhinoceroses (Fig. 1); features of dinosaurian biology make sense in this context, whereas direct homology with lizards produces anomalies (Bakker 1971a, b). Living whales and dolphins have analogical similarity with Mesozoic reptiles such as ichthyosaurs, even to live birth. The comparison between living corals, the coral-like rudist mollusks of the Cretaceous, and the coral-like richthofeniacean brachiopods of the late Paleozoic (Fig. 2) is a more realistic functional exercise than a strictly homological comparison within the three separate phyla.

Comparison with Machines

If there is no appropriate living model, homological or analogical, for some fossil organ or organism, then one may search for a machine as an analogy. This method is often successful because the nature of the fossil record tends to preserve hard parts of organisms, and hard parts often tend to have some mechanical function.

A particularly good example is the study of the biomechanics of pterosaurs. These are the largest flying organisms ever to have evolved, and the only appropriate analog is a man-made

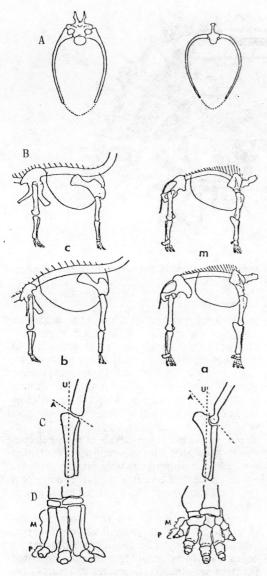

FIGURE 2. A coral, a richthofeniacean brachiopod, and a rudist bivalve mollusk (from Brouwer, 1967).

FIGURE 1. Comparison of brontosaur, on the left, and elephant, on the right (from Bakker, 1971b). A: Cross section of thorax of brontosaur (*Diplodocus*) and elephant (*Loxodonta*). B: Outline of right side of skeletons drawn to same acetabulum-to-shoulder length—c, brontosaur (*Camarosaurus*); b, brontosaur (*Brachiosaurus*); m, elephant (*Mastodon*); a, elephant (*Archidiskodon*). Full lengths of brontosaur necks not shown. C: Section through elbow to show orientation of articular surfaces of brontosaur (*Camarosaurus*) and elephant (*Elephas*)—U, axis of facets on the radius and ulna. D: Right fore feet of brontosaur (*Diplodocus*) and elephant (*Mastodon*)—M, metacarpals; P, phalanges.

glider or airplane. The vast body of information in aerodynamics and aeronautical engineering thus becomes available to help interpret pterosaurs (Bramwell and Whitfield, 1974). Other analogs that have been used in invertebrate paleontology include piping systems in heat exchangers (to interpret respiratory exchange of extinct echinoderms) and pumps and filters (in interpreting suspension feeding in several invertebrate phyla). An extreme case is the half-jocular representation of the fake genus *Mechanocythere,* an ostracod crustacean whose body parts all function in direct analogy with machines (Fig. 3).

Direct Mechanical Modelling

Successful interpretations have resulted from explicit mechanical modeling of structures found in hard parts of organisms. Thus Stanley (1975) was able to show how shape and "ornament" may affect bivalves in burrowing, by building "robot clams" to simulate burrowing activity. M. J. S. Rudwick, and R. P. S. Jefferies and P. Minton (see references in Gould, 1970) built model brachiopods and bivalves to illustrate points about feeding and floating respectively. Balsam and Vogel (1973) have built aluminum models to illustrate possible water currents flowing through sponges and the enigmatic extinct archeocyathids.

Growth and Form

As organisms grow, the mechanical and physiological systems they possess must change to operate in a changing dimensional framework. Although this consideration is intuitively obvious, it can obscure functional relationships if its implications are not fully realized. On the other hand, changing adaptations with growth can help to elucidate functional questions—a growing organism must function well at each stage of growth. Gould (1970) has written a fine review of this aspect of "the science of form."

The Paradigmatic Method

In cases where the methods already discussed are not appropriate or are insufficient to resolve the function of some structure in a

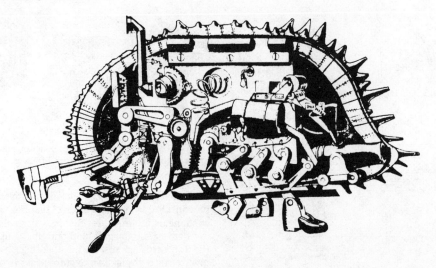

FIGURE 3. The compleat ostracode, *Mechanocythere*. Drawn for Richard H. Benson by L. B. Isham and reproduced with the kind permission of Dr. Benson (from Gould, 1970).

fossil organism, then the *paradigmatic method* may be used. This was introduced by Rudwick in 1964, and has been discussed since by a number of authors (references in Paul, 1975; DeMar, 1976). The structure is examined and the soft-part anatomy of the organism is reconstructed. Some hypothesis of the function of the structure is formed, and its validity is tested by asking, "Given the ground-plan of this particular organism, and given the mechanical and physiological properties of its tissues, what would be the optimum structure that might have evolved to perform the proposed function?" By using appropriate principles of mechanics, engineering, physiology, and so on, the investigator constructs a real or imagined model or *paradigm* as well designed as possible to perform the function. The paradigm now becomes a standard of comparison, and the degree of exactitude with which the actual structure of the fossil fulfils the "predicted" qualities of the paradigm determines the acceptability of the functional hypothesis.

The method is almost entirely interpretative, and is chiefly useful in promoting clarity of thought in functional analysis. It is subject to errors at every stage—the anatomical reconstruction, the formulation of the hypothesis, the construction of the paradigm, and the judgment of the acceptability of the hypothesis. Even given exemplary work by its users, there are inherent problems in the method.

First, evolving structures in real organisms cannot be expected to be perfectly adapted. Any organ or organ system will be favored by selection provided that it performs its function at the level required to allow its possessor to compete successfully with other organisms, irrespective of its efficiency as expressed in quantitative terms in relation to a paradigm. Thus the paradigmatic method does not permit quantification or absolute judgment, though quantitative analysis may help the investigator to make his subjective decisions about the validity of a particular hypothesis. For example, an engineering analysis of seal skeletons would show that they are very poorly adapted for land locomotion, yet they are able to move about on land to a sufficient degree to perform the vital functions of courting, mating, and giving birth.

Structures almost always have more than one function, so that an analysis of one function in isolation might give the impression that an organ was poorly adapted, or that the suggested function was incorrect. Several structures may be integrated into an adaptive complex in order to perform some function (such as feeding), and to analyze one of these structures in isolation would be futile. Both these considerations mean that it may be difficult to frame a paradigm that would be a realistic reflection of the function or functions of the structure under investigation, and yet would be simple enough to work with.

For example, the skull of a snake must be lightly constructed, so that it can be accelerated quickly for a successful strike. Yet at the same time it must be strongly built to withstand the shock of impact on the prey. Furthermore, the bones, muscles and ligaments must be so arranged that the jaws can gape widely for the strike, powerfully hinged, and yet they must distend enough to swallow a comparatively large prey. This makes the successful functional analysis of a snake skull extremely challenging (Frazetta, 1975, chs. 1 and 5). Any simplistic paradigmatic analysis would be useless in this situation.

In spite of these potential difficulties, the paradigmatic method is a very successful tool in paleontological analysis. Its great advantage is that it allows testing of a hypothesis, so that an intuitive interpretation of some feature of a fossil is no longer sufficient. Conflicting interpretations can be subjected to comparative testing. Furthermore, paradigmatic analysis is essentially an *internal* analysis of the fossil as a working organism, and is less dependent on the taxonomic preconceptions involved in choosing an appropriate homologous living species for comparison. Thus, paradigmatic analysis loses comparatively little of its utility in dealing with organisms that have no living counterparts, and some of its greatest successes have been in interpreting features like the respiratory structures of extinct echinoderms, the visual acuity of extinct trilobites, and the feeding mechanisms of brachiopods without living descendants (references in Paul, 1975).

Where one deals with individual structures, the functional hypothesis must be upheld by testing both at the level of the organ, and at the level of the whole organism. Thus the individual lenses of a trilobite eye, and the whole eye, must receive a functional explanation that will provide for favorable adaptation of the animal as an organism operating in a particular marine environment. In dealing with true colonial organisms, the functional hypothesis offered must also show how individual animals function within the colonial structure. In such cases, a functional hypothesis that survives testing can be considered robust (Cowen and Rider, 1972).

The Breadth of Functional Interpretation

Functional interpretation of fossilized structures is usually mechanical because of the nature of the fossil record. But physiological and even behavioral interpretations can sometimes be made from fossil material with a high degree of confidence. Usually this can be done when soft-part anatomy can be reconstructed with precision.

Thus, Paul has reconstructed the respiratory mechanisms of extinct echinoderms, and E. Clarkson has reconstructed the visual systems of trilobites (see references in Paul, 1975). Algal symbiosis has been proposed for some extinct brachiopods, based on anatomical reconstruction of soft tissues (Cowen, 1971).

The swimming habits of belemnite cephalopods were reconstructed by A. Seilacher from the traces left by commensal barnacles (reference in Gould, 1970). Among vertebrates, the reappraisal of dinosaur biology has resulted from reconstructions of body posture and bone structure, but this has led to the interpretation of the physiology and social behavior of dinosaurs as that of warm-blooded and sophisticated creatures (Bakker, 1971; Hopson, 1975).

Environmental Data

The functional morphologist can often draw on environmental data to aid interpretation. For benthic marine organisms in particular, the character and composition of the substrate is not only important in their lives, but retains evidence of their activity through preservation of sediment in the geological record. The lives and habits of soft-bodied organisms can often be interpreted from trails and burrows in sedimentary rocks, even though the organism itself may never have been preserved. Many trace fossils can be interpreted behaviorally, since they reflect the food-searching habits of their generators (e.g., Raup and Seilacher, 1969).

RICHARD COWEN

References

Bakker, R. T., 1971a. Dinosaur physiology and the origin of mammals. *Evolution*. **25**. 636-658.

Bakker, R. T., 1971b. The ecology of the brontosaurs, *Nature*, **229**, 172-174.

Balsam, W. L., and Vogel. S., 1973. Water movement in archeocyathids: evidence and implications of passive flow in models, *J. Paleontology*, **47**, 979-984.

Bramwell, C. D., and Whitfield, G. R., 1974. Biomechanics of *Pteranodon*, *Phil. Trans. Roy. Soc. London, Ser. B*, **267**, 503-592.

Brouwer, A., 1967. *General Palaeontology*. Edinburgh: Oliver & Boyd, 216p.

Cowen, R., 1971. Analogies between the Recent bivalve *Tridacna* and the fossil brachiopods Lyttoniacea and Richthofeniacea, *Palaeogeography, Palaeoclimatology, Palaeoecology*, **8**, 329-344.

Cowen, R., and Rider, J. R., 1972. Functional analysis of fenestellid bryozoan colonies, *Lethaia*, **5**, 147-164.

DeMar. R. E., 1976. Functional morphological models: Evolutionary and nonevolutionary, *Fieldiana, Geol.*, **33**, 339-354.

Denton, E. J., 1974. On buoyancy and the lives of modern and fossil cephalopods, *Proc. Roy. Soc. London, Ser. B*, **185**, 273-299.

Frazetta, T. H., 1975. *Complex Adaptation in Evolving Populations*, Sunderland, Mass: Sinauer Associates, 267p.

Geist, V., 1966. The evolution of horn-like organs, *Behaviour*, **27**, 175-214.

Gould, S. J., 1970. Evolutionary paleontology and the science of form, *Earth-Sci. Rev.*, **6**, 77–119.

Hopson, J. A., 1975. The evolution of cranial display structures in hadrosaurian dinosaurs, *Paleobiology*, **1**, 21–43.

Jepsen, G. L., 1970. Bat origins and evolution, in W. A. Wimsatt, ed., *Biology of Bats*, vol. 1. New York: Academic Press, 1–64.

Ostrom, J. H., 1974. *Archaeopteryx* and the origin of flight, *Quart. Rev. Biol.*, **49**, 27–47.

Paul, C. R. C., 1975. A reappraisal of the paradigm method of functional analysis in fossils, *Lethaia*, **8**, 15–21.

Raup, D. M., and Seilacher, A., 1969. Fossil foraging behavior: Computer simulation, *Science*, **166**, 994–995.

Stanley, S. M., 1975. Adaptive themes in the evolution of the Bivalvia (Mollusca), *Ann. Rev. Earth Planetary Sci.*, **3**, 361–385.

Cross-references: *Biometrics in Paleontology; Bones and Teeth; Evolution; Morphology, Constructional; Vertebrate Paleontology.*

MULTITUBERCULATA

The multituberculates are an extinct order of nontherian, mostly Mesozoic mammals, the largest order of the subclass Prototheria (which also includes triconodonts and monotremes). They were the longest lived mammalian order, originating in the Rhaetic (latest Triassic) and surviving into the early Oligocene, and were the ecological equivalents of modern small rodents, at least in the Cretaceous and Early Teritiary. For a review of the group, see Clemens and Kielan-Jaworowska (1979).

They are relatively common fossils, which is reasonable in view of their ecological role as small herbivores, and are taxonomically diverse. As a result of this diversity and their rapid evolution, they are stratigraphically useful. However, most species of multituberculates are generally recovered as loose teeth; preserved skulls and jaws are rare except in Mongolia. They are known from North America and Europe (which were a single continent until the early Eocene) in the uppermost Triassic to the lower Oligocene, and from northern Asia in the upper Cretaceous and Paleocene. Thus, their geographic distribution is exclusively holarctic so far as is known; none have been found south of the Tethys sea. In South America and Australia at least, the presence of ecological counterparts and the competitive exclusion principle suggest that none ever occurred there.

Multituberculata are divided into three suborders: the Rhaetic–Early Cretaceous Plagiaulacoidea, including three families, Haramiyidae, Paulchoffatiidae, and Plagiaulacidae; the Late Cretaceous–Oligocene Ptilodontoidea, with 3 primarily North American families, Neoplagiaulacidae, Cimolodontidae, and Ptilodontidae; and the Late Cretaceous–Eocene Taeniolabidoidea, with five included families, Taeniolabididae, Eucosmodontidae, Sloanbaataridae, Chulsanbaataridae, and Cimolomyidae. Of the last suborder, the first four families originated in Asia and the fifth may be an unrelated North American parallel. A total of 44 genera and about 128 species are known for the order.

The angiosperm radiation in the Cretaceous had a great impact on the multituberculates. Only 10 genera and about 15 species are known from the Rhaetic through the Albian, while 9 genera and 12 species are known from three Asiatic Upper Cretaceous localities and 10 genera and 27 species are known from the more thoroughly examined Upper Cretaceous of North America. With the extinction of dinosaurs, 17 new genera rapidly arose in the Paleocene, with 2 known species in Asia, 58 in North America, and 5 in Europe. With increased competition from ungulates, primates, and finally rodents, the number of species dropped to 7 in the Eocene and 1 in the Oligocene.

The Jurassic Morrison and Purbeck multituberculates represent about 14% and 10% respectively of the total mammals collected from those formations. In distinct contrast, in the Lance and Upper Edmonton faunas of the Upper Cretaceous of Wyoming and Alberta, multituberculates make up about 45% of all mammals; and at Bug Creek Anthills, in the uppermost Cretaceous of Montana, multituberculates represent 75% of all mammalian individuals. This shift in relative abundance apparently resulted from the Cretaceous development of angiosperms. By the middle Paleocene, multituberculates had declined to about 25% of all mammalian individuals, continuing to drop to 6% in the early Eocene and to 1% by the late Eocene.

Morphology

Multituberculates were small at first. *Thomasia* and most of the later genera and species would have been roughly 15 cm in head and body length and about 30 mm in skull length. The tiniest genera, *Chulsanbaatar* and *Kamptobaatar* of the Mongolian Late Cretaceous and an undescribed species of *Mesodma* from the North American Paleocene, had skulls about 15 mm long; the North American Paleocene genera *Catopsalis* and *Taeniolabis* include species with skull length of 15 cm, estimated head and body length of 1.1 m, and weight of 40 kg. Size increase in the lineage which included the latter genera was very rapid in the 2 m yr straddling the Cretaceous–Tertiary boundary, involving a three-fold increase in size.

The brain of multituberculates was smooth rather than fissured as in later Cretaceous–Eocene placental mammals of comparable size. The cochlea of multituberculates is straight with a tiny hook at the end much as in cynodonts, and is about one sixth the length of

that of Late Cretaceous–Eocene placentals of similar size. This suggests but does not prove weaker powers of pitch discrimination for multituberculates. The early Eocene *Ectypodus tardus* (Fig. 1) has a cochlea about 10% longer than that of *Mesodma formosa*, its Late Cretaceous ancestor of similar size.

The eye of all Taeniolabidoids is huge, about 30% of skull length in comparison to that of

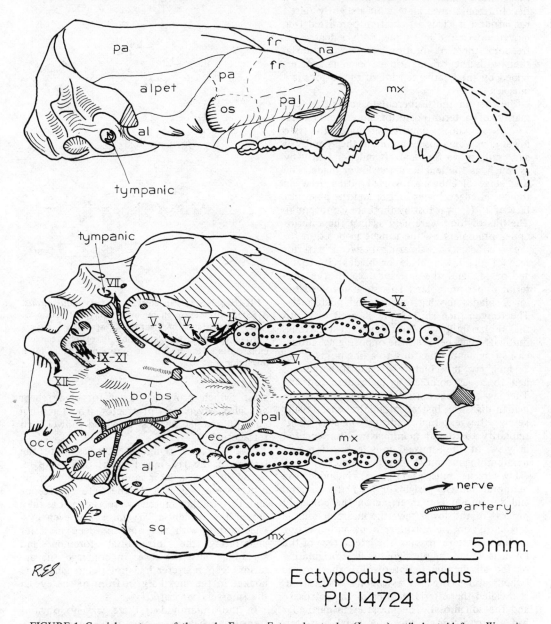

FIGURE 1. Cranial anatomy of the early Eocene *Ectypodus tardus* (Jepsen), ptilodontoid from Wyoming. *al*, alisphenoid; *alpet*, anterior lamina of the petrosal; *bo*, basioccipital; *bs*, basisphenoid; *ec*, ectopterygoid; *fr*, frontal; *mx*, maxilla; *na*, nasal; *occ*, occipital condyle; *os*, orbitosphenoid; *pa*, parietal; *pal*, palatine; *pet*, petrosal.

the Ptilodontoids, in which it is about 15% of skull length in an animal of comparable size. This suggests that Taeniolabidoids might well have been nocturnal. In both suborders the eyes were directed laterally.

The jaw articulation of Multituberculata is distinctive. The condyle of the lower jaw is a section of a sphere which occupied most of the posterior margin of the jaw. It articulated with the glenoid fossa on the squamosal, which is a flat, horizontal oval plate with a slightly raised rim around it. This articulation permitted free movement of the lower jaw with 5 degrees of freedom, three of rotation and two of translation, with the orientation of each jaw determined by the relative tension of the various jaw muscles.

Chewing in multituberculates consisted primarily of anterior-posterior movements (Fig. 2). In all known multituberculates the two lower jaws were loosely bound at the symphysis and could move fore and aft independently by as much as the length of the lower blade. The first stage of chewing involved biting between the chisel edged lower incisor and the posterior face of I^2 (I^1 was lost by the Late Cretaceous). The bits of food were held against the anterior upper premolars by the tongue in position for the second stage, shearing between exterior face of the lower blade or blades (P_1 to P_3 and/or M_b and the interior face of the posterior upper premolars. This shear involved the lower blades moving upward and posteriorly. The tongue then shifted the sheared bits into the molar mill where grinding in the longitudinal valley between the opposing pyramidal cusps fragmented the food to a fine pulp.

Thus, jaw musculature of multituberculates had to be as complex as in modern rodents. The jaw-opening muscle was a typical mammalian digastric, instead of a detrahens, passing below the external auditory meatus. This similarity to therian mammals was unexpected in view of the unusual muscle location and nonhomologous muscle in monotremes.

Jaw-closing muscles (Fig. 3) included the temporal muscles originating on the parietal and squamosal and inserting on the coronoid process of the lower jaw; the masseter muscle originating on the maxillary bone and inserting on the masseteric fossa of the lateral face of the lower jaw; and the pterygoid muscle originating on the alisphenoid (epipterygoid), the prootic (which still existed as a separate ossification, also called the anterior lamina of the petrosal), and the squamosal. The masseter in particular had three divisions, the superficial masseter, which was the main protractor; the deep masseter, which provided power to the closure of the molar mill; and the anterior deep mas-

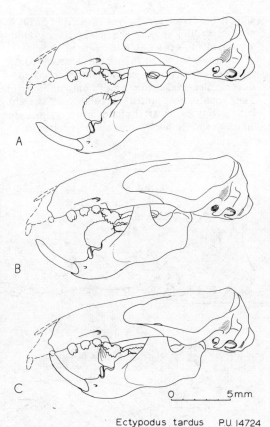

FIGURE 2. Stages in chewing in *Ectypodus tardus*. A, Start of incisor gnaw, maximum protraction of lower jaw; B, start of shear of M_b against P^4, molars starting to grind; C, completion of blade shear and molar grind, lower jaw at position of maximum retraction.

seter, which provided power for blade shear and also served as a protractor. The temporal and pterygoid muscles provided power and retraction.

Originally the temporal muscles were the largest of the jaw muscles, but they were reduced in size in later multituberculates while others were increased. In the Late Cretaceous and Tertiary, jaw muscles were as varied as they are in the Rodentia. *Stygimys* was even hystricomorph, with the anterior deep masseter penetrating the infraorbital foramen and spreading onto the face. In contrast, the anterior deep masseter had not yet reached the pocket in the maxillary in front of the eye in the Jurassic *Paulchoffatia*.

In the Haramiyidae, there was no obvious transition between premolars and molars; but in the Jurassic multituberculates, the anterior cheek teeth, particularly the lower ones were specialized for shearing. Four of these

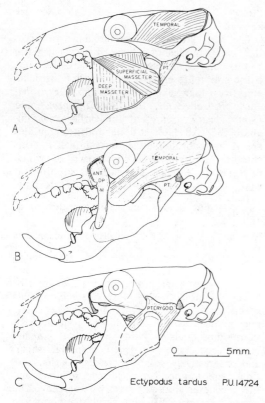

FIGURE 3. Reconstruction of the jaw musculature of *Ectypodus tardus*. A, outer two layers of the masseter muscle; B, anterior deep masseter, which originates in a pocket in front of the orbit and the temporal muscle; C, a single pterygoid muscle originates on the alisphenoid and inserts on the inside of the mandible.

lower shearing blades were present, and they have been called premolars. However, in six genera of three families of both later suborders, it can be shown that the last of these blades in fact had no deciduous precursor and erupted with the deciduous incisor. In the Jurassic species, this last tooth is the only one with an external row of cusps in addition to the principal row. Thus, by definition, this tooth is not P_4 (fourth lower premolar) as it has usually been called but rather a molar tooth. To cause minimum disturbance to the older descriptions it is now referred to as M_b or simply *the blade*. Dental formulas given here thus have one less lower premolar and one more lower molar than most of those in the literature.

Molars of multituberculates have pyramidal cusps arranged in longitudinal rows. In Jurassic and Early Cretaceous multituberculates, there were two such rows of cusps in each molar with three cusps in each row except in M^1 where there were four cusps in each row. M^2 is in-

ternal to as well as behind the M^1 so that the outer cusp row of M^2 continues the inner cusp row of M^1. In Late Cretaceous multituberculates of all families, M^1 and M^2 develop third cusp rows so that the two valleys between the cusp of the upper molars straddle the two continuous rows of cusps in M_1 and M_2. In addition, the number of cusps in a row expands to as many as 18 as the molars lengthen.

The postcranial anatomy of multituberculates is presently known from partial skeletons of *Djadochtatherium*, *Eucosmodon*, and *Ptilodus* and from many isolated bones of North American Late Cretaceous species. Although the tail is about 1.5 times the length of the body in *Ptilodus*, the number of cervical, thoracic and lumbar vertebrae is not known. The coracoid is fused to the scapula. The most distinctive feature of the combined bones is the absence of the supraspinous fossa; the blade of the scapula does not extend forward of the spine. The distal end of the humerus is typically mammalian but differs in precise mode of articulation with radius and ulna from therian mammals. The semilunar notch of the ulna is grooved to match a corresponding thin ridge on the humerus, so motion of these two bones with respect to each other is in a single plane. In contrast, the surface of articulation for the radius on the humerus is a perfect sphere and all torsion of the wrist involves only radial rotation.

The pelvis is composed of the fused ilium, pubis, and ischium, together with a separate epipubis as in cynodonts and marsupials. The acetabulum is open dorsally. In life, the typically mammalian femur was carried over much the same range of movements as a lizard's— it could be rotated to a vertical position but could also project nearly horizontally. The tibia, while mammalian, is more unusual than most of the limb bones. The proximal end of the tibia completely excluded the fibula from articulation with the femur, a prominent lateral projection well beyond the width of the shaft carried the fibular articulation on the ventral surface. The upper part of the posterior face of the shaft of the tibia is deeply excavated for the gastrocnemius muscle; for this reason, the cross section is U shaped. The ankle is fully mammalian but quite unlike that of therian mammals; the astragalus and calcaneum are very distinctive.

Evolution

The oldest species of the order, *Thomasia antigua* (Plieninger) 1847, also the first multituberculate to be described, is known from about 30 specimens from 4 Rhaetic (uppermost

Triassic) European localities no more than 1000 km apart. All four localities were coastal and on the same coast of the Tethys. Most of the specimens came from Holwell in Somerset, England; the others came from Wurtemberg, Germany and from Hallau, Switzerland. Several nominal species have been described for individual tooth types; the genera *Microcleptes* and *Haramiya* were proposed for what appear to be upper teeth, the genera *Microlestes* and *Thomasia* for what appear to be lower teeth. There is no reason for assuming the existence of more than one species at present.

This species, the only known species of the family Haramiyidae, obviously arose from the eozostrodontid (morganucodontid) triconodonts of the Late Triassic (see *Mammals, Mesozoic*) and had made considerable dental progress from that ancestor. The internal cingulum of the upper molars of eozostrodonts and a new external cingulid were enlarged so that both upper and lower molars had two rows of cusps. Wear facets indicate that anterior-posterior grinding became increasingly possible with tooth wear. Most of the known specimens of this species are described in Simpson (1928), Parrington (1946), and Peyer (1956).

The Jurassic multituberculates are known from (in chronologic order) the Kimmeridgian Guimarota coal group of Leiria, Portugal; the late Kimmeridgian Morrison Formation of (principally) Como Bluff, Wyoming; and the Portlandian Purbeck beds of Durdlestone Bay, Dorset, England.

The Portugese specimens are included in two principal genera, *Paulchoffatia* and *Kuehneodon*, with dental formulas

$$\frac{3\text{-}0\text{-}6\text{-}2}{1\text{-}0\text{-}3\text{-}3} \quad \text{and} \quad \frac{3\text{-}0\text{-}5\text{-}2}{1\text{-}0\text{-}3\text{-}3}$$

The latter genus is closest to the ancestry of the Morrison and Purbeck genera. The Morrison species. *Ctenacodon serratus* and *C. scindens* are apparently ancestral to the Purbeck species *C. minor* and *C. falconeri* respectively. The Morrison species *Psalodon fortis* is apparently ancestral to the Purbeck *Plagiaulax becklesii*. In all of these species save the last, the dental formula is the same as in *Kuehneodon*. In *Plagiaulax*, the dental formula of the lower jaw is reduced to 1-0-2-3.

The only described species from the Early Cretaceous is *Loxaulax dawsoni*, known from a handful of isolated teeth from the Wealden Sandstone of Hastings, England. At least two additional undescribed species are known from the Albian Trinity Sandstone of Texas. These species are intermediate in dental morphology between the Morrison-Purbeck species and

those of the Late Cretaceous; no jaws are known.

The next known multituberculates occur in the Late Cretaceous (?Santonian) Djadokhta Formation of Mongolia. There, *Gobibaatar* is the sole representative of the Ptilodontoidea; and five other genera, *Sloanbaatar, Kryptobaatar, Bulganbaatar, Djadochtatherium,* and *Kamptobaatar,* represent the earliest Taeniolabidoidea. All later Asiatic multituberculates are Taeniolabridoidea. Four families of this suborder are known in Asia, two of which migrated to North America. This suborder at first included forms with dental formula $\frac{2\text{-}0\text{-}4\text{-}2}{1\text{-}0\text{-}1\text{-}3}$ but later species reduced this to as low as $\frac{2\text{-}0\text{-}1\text{-}2}{1\text{-}0\text{-}0\text{-}3}$. The first upper incisor was lost in all Late Cretaceous and later species and the upper shearing premolar was reduced to one tooth on each side and the two anterior lower premolars were lost. Taeniolabidoids specialized in a distinctly rodent-like chewing method with extrudable, rooted, self-sharpening incisors with a limited enamel band; progressively reduced premolar shear; and progressively enlarged molars. At the same time, the temporal muscle was reduced and the masseter muscle greatly enlarged. The result is a very different skull shape than in ptilodontoids, particularly late forms.

In ptilodontoids, rather than the development of incisor shear as in taeniolabidoids, the emphasis was on the elaboration of blade shear with very different proportions of the various jaw muscles and skull. A comparison of the skulls of *Taeniolabis* (Taeniolabidoidea) (Fig. 4) and *Ectypodus* (Ptilodontoidea) (Fig. 1) will indicate the gross differences in most features and proportions of the skulls.

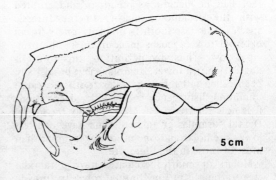

FIGURE 4. Reconstruction of the skull and lower jaw of *Taeniolabis taoensis* Cope from the early Paleocene of New Mexico, the largest known multituberculate (after Granger and Simpson, 1929).

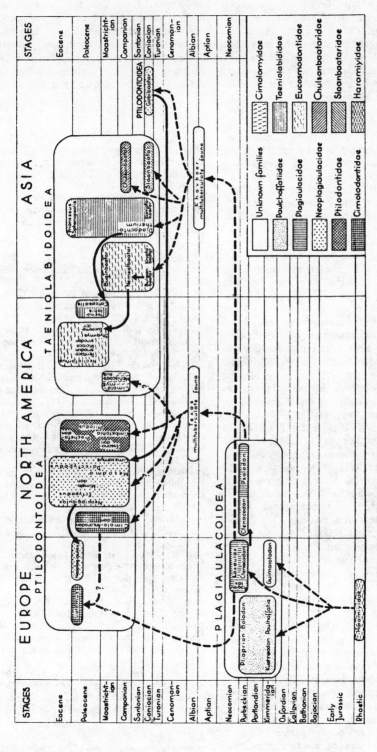

FIGURE 5. Geographical and stratigraphical distribution of the multituberculates (from Kielan-Jaworowska, 1974b). The shapes of hatched surfaces are not in proportion to the taxonomic diversity of the families. Discontinuous arrows indicate the probable phylogenetic and migration lines, solid arros the reasonably well-established migration lines. Poorly known genus *Viridomys* from Upper Milk River Formation of Canada is omitted; for *Paulhoffatia*, read *Paulchoffatia*.

Two of these genera, the large *Djadochtatherium* and the tiny *Kamptobaatar*, survived into the Campanian Barun Guyot Formation; *Bulganbaatar* gave rise to *Nemegtbaatar* and another tiny genus, *Chulsanbaatar*, appeared. In the latest Cretaceous Maestrichtian stage, at least two of these Campanian genera migrated to North America, where *Nemegtbaatar* gave rise to *Stygimys* of the latest Cretaceous and Paleocene and *Eucosmodon, Neoliotomus, Pentacosmodon*, and *Microcosmodon* of the Paleocene. *Djadochtatherium*, on arrival in North America, became the Late Cretaceous and Paleocene *Catopsalis* and its precociously gigantic early Paleocene derivative *Taeniolabis* (Fig. 4), the largest of all multituberculates. In Asia, *Djadochtatherium* and *Kamptobaatar* gave rise to the latest Paleocene giant *Sphenopsalis* and the tiny *Prionessus*, both from the Gashato Formation.

During this Cretaceous period of isolation of mammals between Asia and North America, while the radiation of the Taeniolabidoidea was taking place in Asia with only minor occurrences of Ptilodontoidea, the reciprocal events were taking place in North America (see Fig. 5). The Ptilodontoidea underwent a Late Cretaceous Paleocene radiation in North American with at most minor occurrences of taeniolabidoids. Three families of Ptilodontoidea are known from the Cretaceous of North America, while the only possible taeniolabidoids of North American origin are the Campanian-Mawstrichtian genera *Cimolomys* and *Meniscoessus*. The central and biggest family of the Ptilodontoidea is the Neoplagiaulacidae, which includes the genera *Cimexomys, Mesodma, Ectypodus, Neoplagiaulax, Parectypodus,* and two undescribed genera. Most are small with skull lengths of 15–35 mm. *Kimbetohia, Ptilodus*, and *Prochetodon* are larger genera, skull length about 60 mm, which belong to the Ptilodontidae. The Cretaceous *Cimolodon* and the Paleocene *Anconodon* and *Liotomus* are medium-sized multituberculates belonging to the Cimolomyidae. Individual species of these North American genera are short in duration and permit close biostratigraphic zonation of the North American Paleocene.

Multituberculates had apparently become extinct in Europe during the Cretaceous or early Paleocene, for a well described German middle Paleocene fauna has no multituberculates at all and the abundant multituberculates from the late Paleocene Cernay fauna of the Paris basin consists of species conspecific with or derivable directly from North American middle and early late Paleocene species.

The latest occurrences of multituberculates are in the latest Paleocene Gashato Formation in Asia, the early Eocene Sparnacian of France, and the early Oligocene of the United States. The acme of their diversity was reached in the middle Paleocene, and in North America they were reduced rapidly during the late Paleocene by competition with ungulates, primates, and ultimately rodents. In Asia, the reduction was associated with the evolution of rabbits and the subsequent introduction of rodents.

R. E. SLOAN

References

Clemens, W. A., 1964. Fossil mammals of the type Lance Formation, Wyoming. Part I. Introduction and Multituberculata, *Univ. Calif. Publ. Geol. Sci.,* 48, 106p.

Clemens, W. A., and Kielan–Jaworowska, Z., 1979. Multituberculata, in Z. Kielan–Jaworowska et al., eds., *Mesozoic Mammals: The First Two-Thirds of Mammalian History.* Berkeley: Univ. of California Press.

Granger, W., and Simpson, G. G., 1929. Revision of the Tertiary Multituberculata, *Amer. Mus. Nat. Hist. Bull.,* 56, 601–676.

Hahn, G., 1969. Beiträge zur Fauna der Grube Guimarota. Nr. 3, Die Multituberculata, *Palaeontographica, A.* 133. 1–100.

Jepsen, G. L., 1940. Paleocene faunas of the Polecat Bench Formation, Park County, Wyoming, Part I, *Proc. Amer. Phil. Soc.,* 83, 217–341.

Kielan-Jaworowska, Z., 1971. Skull structure and affinities of the Multituberculata: Results of the Polish-Mongolian Palaeontological Expeditions, Part III, *Palaeontol. Polonica,* 25, 5–41.

Kielan-Jaworowska, Z., 1974a. Multituberculate succession in the Late Cretaceous of the Gobi Desert (Mongolia): Results of the Polish-Mongolian Palaeontological Expeditions, Part V, *Palaeontol. Polonica,* 30, 23–44.

Kielan-Jaworowska, Z., 1974b. Migrations of the Multituberculata and the Late Cretaceous connections between Asia and North America, *Ann. S. African Mus.,* 64, 231–243.

Parrington, F. R., 1946. On a collection of Rhaetic mammalian teeth, *Proc. Zool. Soc. London,* 116, 707–728.

Peyer, B., 1956. Uber Zahne von Haramiyden, von Triconodonten und von wahrscheinlich synsapsiden Reptilien aus dem Rhät von Hallau, *Schweiz. Paläont. Abh.,* 72. 1–71.

Romer, A. S., 1966. *Vertebrate Paleontology,* 3rd ed. Chicago: Univ. Chicago Press, 200p.

Simpson, G. G., 1928. *A Catalogue of the Mesozoic Mammalia in the Geological Department of the British Museum.* London: British Mus. (Nat. Hist.), 215p.

Sloan, R. E., and Van Valen, L., 1965. Cretaceous mammals from Montana, *Science,* 148, 220–227.

Van Valen, L. and Sloan, R. E., 1966. The extinction of multituberculates, *Systematic Zool.,* 15, 261–278.

Cross-references: *Bones and Teeth; Mammals, Mesozoic; Vertebrata; Vertebrate Paleontology.*

O

OSTRACODA

The Ostracoda are minute aquatic crustaceans whose bivalved shells (carapaces) have been found in sediments of all ages as far back as the Upper Cambrian. Because of their great geologic range, ostracodes are studied by the biologist interested in observing how organisms change from one form to another through normal evolutionary processes, and by the geologist who finds them useful for determining the ages of ancient sediments containing their carapaces. Survival of the Ostracoda in today's oceans, lakes, and rivers permits study of their habitats for the purpose of obtaining clues to the kinds of environments inhabited by their extinct ancestors.

Many species of Ostracoda spend their lives crawling on the bottom or burrowing in sand and mud; others swim or float in the water, and a few are parasitic or commensal on crayfish and other crustacea. Ostracodes manifest a variety of food preferences: some species are scavengers, living on dead plant or animal matter; others prefer a diet of living plants; a few are more specialized in their food requirements. Although many ostracodes, because of their small size, are ingested quite accidentally by other animals browsing on plants or burrowing in bottom muds, they also form an important part of the diet of some fish, as well as of bottom-feeding invertebrate animals.

Description

The individual ostracode is composed of a soft body contained within a carapace consisting of a left and right valve, which are opened and closed by muscles that pass through the body near its center and are attached to the inner side of each valve. Valves are generally hinged at the top (dorsal margin) and are held slightly ajar during swimming, crawling, burrowing, and other life functions that require protrusion of appendages. All soft parts of the body can be completely withdrawn and the valves closed for protection.

The body is laterally compressed and during life has the head in a forward position and the dorsal side uppermost. Unlike many crustaceans, the ostracode body is unsegmented; however, a slight constriction of the body marks the boundary between the head (cephalon) and the thorax. An abdomen is lacking.

The head contains a forehead, a mouth with an upper lip and a lower lip (Hypostome) and eye organs. It also provides points of attachment for four pairs of appendages: antennules, antennae, mandibles and maxillae (see Figs. 1 and 2). Some forms are without eyes; other have a single median eye (nauplius eye) or/and two lateral eyes. A projection on the forehead of some ostracodes, called a rod-shaped organ, is thought to have a sensory function.

The thorax contains the reproductive system, the stomach, intestines and one to three pairs of appendages. In many ostracodes, the posterior part of the body terminates as a paired appendage-like structure called the *furca*. Genital organs are located on the ventral surface of the thorax in front of the furcae. Appendages on the thorax have functions that vary with different ostracodes; they are used in locomotion, in feeding, and in cleaning the inside of the carapace.

Several species of Ostracoda belonging to the subfamily Cypridininae have glands that secrete luciferin and luciferase which react in sea water to produce a blue luminescence similar to the light of a firefly.

Most ostracode carapaces measure 0.5–0.7 mm in length, which is usually their greatest dimension; however, a few attain a length of 30 mm. The carapace shows considerable variation in shape; many are bean-shaped, elliptical, or ovate; whereas a few are trapezoidal or subrectangular in lateral view. The dorsal and ventral edges are usually either convex or straight, but on some forms the ventral margin is concave. Anterior and posterior ends may be either rounded or have angular projections, and the anterior may be similar or dissimilar in outline to the posterior end.

The valves of ostracodes have hinge structures along the dorsal margin, which vary from merely a smooth contact between valves to a complex arrangement of teeth and sockets. Although, in general, an increase in complexity of the hingement seems to be an evolutionary trend, fairly complex hinges had been developed in some species prior to the end of the Paleozoic.

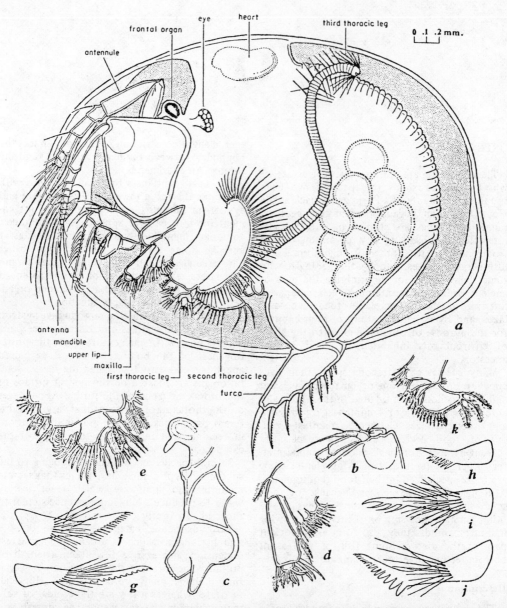

FIGURE 1. Morphology of a representative myodocopid (myodocopine) ostracode, *Cypridina norvegica* Baird (Cyprididae), Holocene (from Moore, 1961, courtesy of The Geological Society of America and University of Kansas Press). A, female with LV removed; eggs shown in one uterus at rear of body; genital lobes below base of 3rd thoracic leg. B, right antenna; inner face of endopodite and first podomere of expodite. C, frontal organ and upper lip. D, left maxilla. E, left first thoracic leg; details of protopodite and endopodite. F–J, setae from protopodite of first thoracic leg. K, left 2nd thoracic leg. (A,B,C, and K to scale in upper right corner.)

Raised bosses on the interior of the shell mark the points of muscle attachment and are called *muscle scars* (Benson, 1966, 1967). The size, shape, and arrangement of muscle scars are used extensively in classifying the Ostracoda. A major scar is left on each valve at the point of attachment of the adductor, or closing muscle. The adductor scar is composed of many smaller scars, which are designated secondary scars; these may number as many as a hundred, but fewer than 15 are normal. In some forms, additional raised bosses are located anterior to the adductor scar at points of attachment of flexor and extensor muscles of cephalic appendages. The mandibular muscle attachment leaves scars more often than do other cephalic appendages.

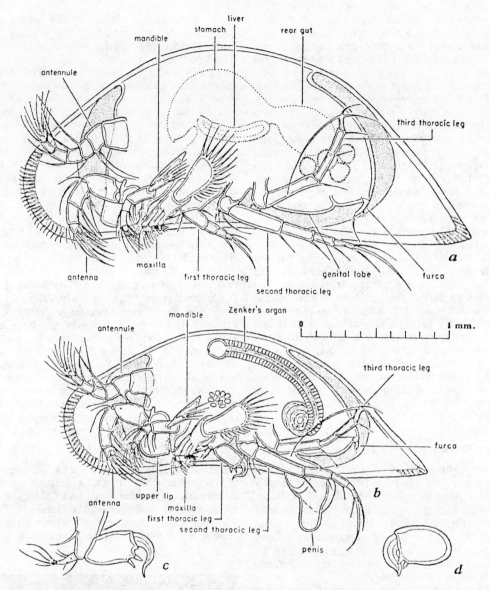

FIGURE 2. Morphology of a representative podocopid (podocopine) ostracode, *Bairdia frequens* G. W. Müller (Bairbiidae), Holocene (from Moore, 1961, courtesy of the Geological Society of America and the University of Kansas Press). A, female with LV removed; B, male with LV removed; C, male left 1st thoracic leg; D, palp of male right 1st thoracic leg.

The outer surfaces of ostracode valves may be either pitted or smooth. Many forms have raised striations and granules, or a network of intersecting ridges that form a reticulate surface. Solid or hollow spines may project from the surface and margins; and some carapaces have surface furrows, ridges, and marginal frills of varying width. Elongate depressions (sulci) and ovate protuberances (lobes) on the outer surface are reflected internally by elongate ridges and ovate depression, respectively (Sylvester-Bradley and Benson, 1971).

The ostracode carapace is composed of calcium carbonate and chiton, but the latter is seldom preserved in fossils. The chiton is present as a layer over the outer and inner surfaces of each valve and also forms a network within the valve (Kesling, 1951). The free edge of some ostracodes has a narrow inward-projecting calcareous ridge called the *infold* or *duplicature.* In the living ostracode, the *infold* is part of the *outer lamella,* and together, they form the ostracode *shell* (Kornicker, 1969).

Minute holes through which setae or hairs

having a sensory function pass through the shell when the animal is alive are termed *normal pore canals* except when they penetrate the outer edge of the infold. The latter are termed *radial pore canals*. Pore canals are absent from some species and are sparse to abundant in others.

Like other crustaceans, ostracodes grow discontinuously by a molting process. The soft-bodied animal removes itself periodically from its shell and then rapidly grows a new, and slightly larger, carapace. About seven or eight carapaces are shed before the ostracode becomes an adult and no longer molts. Each growth stage is termed an *instar* or *larval stage* with the exception of the last, wich is the *mature* or *adult stage*. Growth usually ceases and reproduction commences when the adult stage is reached.

Ostracoda are hatched from eggs, but the emerging animal only slightly resembles the adult form because many appendages and organs appear for the first time in later instars. In addition to increasing in size after each molting, the carapace may change in shape and ornamentation, so that it is difficult to determine to which species some early instars belong. Carapaces of the early larval stages are quite fragile and for this reason are preserved as fossils less often than those of adults. Some species may be parthenogenetic, as males have never been found.

Classification

The Ostracoda are a subclass of the Class Crustacea and Phylum Arthropoda. The subclass Ostracoda contains five orders: Archaeocopida, Leperditicopida, Palaeocopida, Podocopida (Fig. 2) and Myodocopida (Fig. 1); but only the last two have representatives living today (Moore, 1961; Hartmann and Puri, 1974). The remaining three orders became extinct before the end of the Paleozoic.

The extinct orders are necessarily distinguished on the basis of differences in carapace morphology because soft parts are seldom preserved as fossils. The Archaeocopida have thin shells; the Leperditicopida have thick shells with large compound muscle scars; and the Palaeocopida have small simple muscle scars and are often lobate and highly ornamented.

Orders with living representatives are classified on the basis of differences in soft parts as well as carapace morphology. Podocopida shells are strongly calcified and have a few secondary muscle scars and an inner calcareous lamellae. Carapaces of Myodocopida are often poorly calcified and contain a well to poorly developed anterior projection (*rostrum*) above a slit (*rostral incisur*) through which the antennae

of the living animal project during swimming. The antennae are modified for walking in living Podocopida, and for swimming in living Myodocopida.

The thin-shelled Archaeocopida, which lived only in the Cambrian and possibly Early Ordovician seas, and the thick-shelled Leperditicopida, which were abundant in Ordovocian, Silurian, and Devonian seas, are not subdivided into suborders. The Palaeocopida, on the other hand, contain two suborders, the Beyrichicopina and the Kloedenellocopina; both are essentially restricted to the Paleozoic and were more abundant during its earlier part. A major difference between the Beyrichicopina and the Kloedenellocopina is that representatives of the latter suborder have one valve much larger than the other, whereas valves of Beyrichicopina are almost equal in size.

The Podocopida contain three suborders: Podocopina, Metacopina, and Platycopina. The Podocopina range from the early Paleozoic to Holocene; the Metacopina range from early Paleozoic to lower Mesozoic, and may be represented in the Holocene by the genus *Saipanetta* (Maddocks, 1972). Living Platycopina are distinguished from living Podocopina by having biramous antennae; fortunately, the carapaces also have characteristic differences that make it possible to separate fossil representatives of these suborders. The suborder Metacopina have carapaces with features transitional between typical Podocopina and Platycopina (Sars, 1922).

The Myodocopida contain two suborders: Myodocopina (Fig. 3), which range from the Ordovician to the Holocene, and Cladocopina, which range from the Triassic to Holocene. The Cladocopina are much smaller than the Myodocopina, are almost circular in lateral view, and

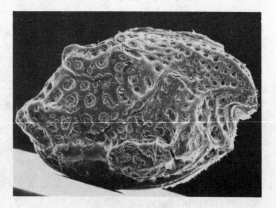

FIGURE 3. *Asteropteron* sp. from Pacific Ocean in vicinity of Los Angeles, California. Family Cylindroleberididae, Suborder Myodocopina.

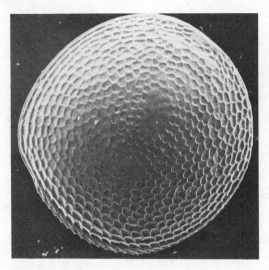

FIGURE 4. *Metapolycope* sp. Holocene ostracode from the deep sea. Family Polycopidae, Suborder Cladocopina, Order Myodocopida.

the better known ostracode genera are the marine *Bairdia* and the freshwater *Candona.*

Distribution

Ostracoda are among the more successful aquatic forms in that they inhabit almost all kinds of water bodies. They are found in shallow ponds, rivers, and swamps, as well as at great oceanic depths. Except for a few species that live in moss, the Ostracoda are not terrestrial.

In the oceans, members of the Order Podocopida dwell mostly on the bottom where they crawl and burrow in the sediment or browse on plants. The Order Myodocopida (Fig. 4), which is almost completely restricted to the marine environment, contain species capable of swimming long distances; and although most species remain at or near the bottom, some live in surface waters where they are transported by oceanic currents (Poulsen, 1962, 1965, 1969, 1973).

Freshwater Podocopida (Fig. 5) usually are capable of swimming short distances and travel from plant to plant in a manner reminiscent of a bee flitting from flower to flower. A few genera (*Notodromas, Newnhamia*) contain members capable of hanging upside-down at the water's surface. Fossil freshwater ostracodes are known from as far back as Pennsylvanian time. Carapaces of freshwater ostracodes, with a few exceptions, have a smooth outer surface and lack the ornamentation so common on marine forms. Shells of most fossil freshwater ostracodes closely resemble those of ostracodes living in today's ponds, lakes, and rivers.

do not have the deep rostral incisur developed on typical Myodocopina (Skogsberg, 1920).

Suborders of Ostracoda are subdivided into superfamilies, families subfamilies, genera, and species on the basis of differences in carapace shape, hingement, muscle scars, ornamentation, and other carapace structures, as well as appendages (Van Morkhoven, 1962, 1963; Howe, 1962). Many thousands of ostracode species have already been described, but new investigations usually discover additional species. Among

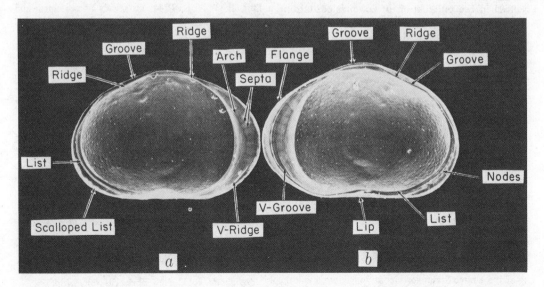

FIGURE 5. *Cypretta kawatai* Sohn and Kornicker, 1972, freshwater ostracode from Brazil; Family Cypridopsidae, suborder Podocopina, Order Podocopida (from Sohn and Kornicker, 1973). *a,* left valve; *b,* right valve (inside views).

Although ostracodes are abundant in both fresh and marine water, freshwater and marine species form two distinct assemblages. A third assemblage of ostracode species dominates estuaries and lagoons containing water with a salt content intermediate between fresh water and marine and where the salinity often fluctuates over a wide range. When one of these assemblages is recognized in ancient sediments, it is possible to estimate the salt content of the water in which the sediment was deposited.

Both living and fossil ostracodes have been described from all parts of the world. In general, more species are present in warm subtropical and tropical water than in cold waters of high latitudes. Water temperature seems to have considerable influence on distribution of many species; cold-water ostracode assemblages are composed of different species than warm-water assemblages (Hazel, 1970; Kornicker, 1975). The distribution of crawling and burrowing ostracodes is partly controlled by the physical and chemical properties of the substrate.

The abundance of living ostracodes varies from place to place. They are more abundant in the quiet waters of lakes, bays and shallow seas, where profuse plant growth assures an adequate food supply, than in the deep ocean, where food supply is low. The abundance of empty carapaces in sediment depends on the rates at which shells of other organisms and terrestrial derived sediment are deposited, as well as on the abundance of ostracodes that lived in the area.

Since 1962, students of Ostracoda have been holding international symposia that have resulted in the publication of volumes containing the presented papers (Puri, 1964; Neale, 1969; Oertli, 1971; Swain et al., 1975; Löffler and Danielopol, 1977). These contain a wealth of information, especially concerning ecology and paleoecology and are recommended reading.

LOUIS S. KORNICKER

References

Benson, R. H., 1966. Recent marine podocopid ostracodes, *Ann. Rev. Oceanog. Marine Biol.,* **4**, 213–232.

Benson, R. H., 1967. Muscle-scar patterns of Pleistocene (Kansan) ostracodes, in C. Teichert and E. L. Yochelson, eds., *Essays in Paleontology and Stratigraphy, Raymond C. Moore Commemorative Volume.* Lawrence: Univ. Kansas Press, 211–241.

Hartmann, Gerd and Puri, H. S., 1974. Summary of neontological and paleontological classification of Ostracoda, *Mitt. Hamburg. Zool. Mus. Inst.,* **70**, 7–73.

Hazel, J. E., 1970. Atlantic continental shelf and slope of the United States—ostracode zoogeography in the southern Nova Scotian and northern Virginian faunal provinces, *U.S. Geol. Surv. Prof. Paper 529-E,* 21p.

Howe, H. V., 1962. *Ostracod Taxonomy.* Baton Rouge: Louisiana State Univ. Press, 366p.

Kesling, R. V., 1951. *The Morphology of Ostracod Molt Stages.* Urbana: Univ. Illinois Press, 324p.

Kornicker, L. S., 1969. Relationship between the free and attached margins of the myodocopid ostracod shell, in J. W. Neale, ed., 1969, 109–135.

Kornicker, L. S., 1975. Antarctic Ostracoda (Myodocopina), *Smithsonian Contrib. Zool.,* **163**, 720p.

Löffler, H., and Danielopol, D., eds., 1977. *Aspects of the Ecology and Zoogeography of Recent and Fossil Ostracoda.* The Hague: Junk, 521p.

Maddocks, R. F., 1972. Two new living species of *Saipanetta* (Ostracoda, Podocopida), *Crustaceana,* **23**, 28–42.

Moore, R. C., ed., 1961. *Treatise on Invertebrate Paleontology, Arthropoda,* pt. Q. Lawrence, Kansas: Geol. Soc. Amer. and Univ. Kansas Press, 442p.

Neale, J. W., ed., 1969. *The Taxonomy, Morphology and Ecology of Recent Ostracoda.* Edinburgh: Oliver & Boyd, 553p.

Oertli, H. J., ed., 1971. Colloquium on the paleoecology of ostracodes, *Bull. Centre Rech. Pau-SNPA, Pau, France,* suppl., **5**, 953p.

Pokorný, V., 1978. Ostracodes, in B. U. Haq and A. Boersma, eds. *Introduction to Marine Micropaleontology.* Amsterdam: Elsevier, 109–149.

Poulsen, E. M., 1962. Ostracoda-Myodocopa, 1, Cypridiformes-Cypridinidae, *Dana-Report Carlsberg Found.,* **57**, 414p.

Poulsen, E. M., 1965. Ostracoda-Myodocopa, 2, Cypridiniformes-Rutidermatidae, Sarsiellidae and Asteropidae, *Dana-Report Carlsberg Found.,* **65**, 484p.

Poulsen, E. M., 1969. Ostracoda-Myodocopa, 3a, Halocypriformes-Thaumatocypridae and Halocypridae, *Dana-Report Carlsberg Found.,* **75**, 100p.

Poulsen, E. M., 1969. Ostracoda-Myodocopa, 3b, Halocypriformes-Halocypridae, Conchoecinae, *Dana-Report Carlsberg Found.,* **84**, 224p.

Puri, H. W., ed., 1964. Ostracods as ecological and palaeoecological indicators, *Zool. Station of Naples,* **33**(suppl.), 612p.

Sars. G. O., 1922. *An Account of the Crustacea of Norway. Ostracoda,* vol. 9(1–2), 1–277. Oslo; The Bergen Museum (sold by Alb. Cammermeyers Forlag).

Skogsberg, Tage, 1920. Studies on marine ostracods, Part 1 (Cypridinids, Halocyprids and Polycopids), *Zool. Beitrage Upsala (Zool. Beitrage Upsala),* 782p.

Sohn, I. G., and Kornicker, L. S., 1973. Morphology of *Cypretta kawatai* Sohn and Kornicker, 1972 (Crustacea, Ostracoda), with a discussion of the genus, *Smithsonian Contrib. Zool.,* **141**, 28p.

Swain, F. M., Kornicker, L. S., and Lundin, R. F., eds., 1975. Biology and paleobiology of Ostracoda, *Bull. Amer. Paleontology,* **65**, 687p.

Sylvester-Bradley, P. C., and Benson, R. H., 1971. Terminology for Surface Features in Ornate Ostracodes. *Lethaia,* **4**, 249–286.

Van Morkhoven, F. P. C. M., 1962. *Post-Palaeozoic Ostracoda,* vol. 1. New York: Elsevier, 204p.

Van Morkhoven, F. P. C. M., 1963. *Post-Palaeozoic Ostracoda,* vol. 2. New York: Elsevier, 478p.

Cross-references: *Arthropoda; Crustacea.*

OYSTERS

Oysters are inequivalve marine bivalve mollusks without the muscular foot found in most bivalves, and without an anterior adductor muscle, hence "monomyarian" (Fig. 1). Oysters differ from other attached, inequivalve, monomyarian bivalves by the fact that it is always the left valve that is cemented onto some solid substrate (attachment may not persist beyond youth). The left valve is larger, heavier, and move convex, sometimes even cup-like; the right valve is often nearly flat.

Zoologically, oysters are members of the class Bivalvia, subclass Pteriomorphia, order Pterioida, suborder Ostreina, superfamily Ostreacea, according to the classification used by McCormick and Moore (1969) in *Treatise on*

Invertebrate Paleontology (see also Waller, 1978). The volume on oysters in that treatise is by Stenzel (1971), who inclines toward the view that they are diphyletic (Fig. 2), the families Gryphaeidae and Ostreidae having evolved independently from two different genera of nonoyster bivalves, possibly members of the Carboniferous and Permian family Pseudomonotidae, oyster-like Pectinacea attached by the right valve.

Geologically, the earliest true oysters were several species of *Lopha* (family Ostreidae) and several species of *Gryphaea* (family Gryphaeidae) among Triassic (Carnian) marine fossils. The earliest *Gryphaea* species were arctic while the first *Lopha* species were in the Mesogean and Pacific realms. *Liostrea* appeared later in the Triassic Arctic realm, probably evolving

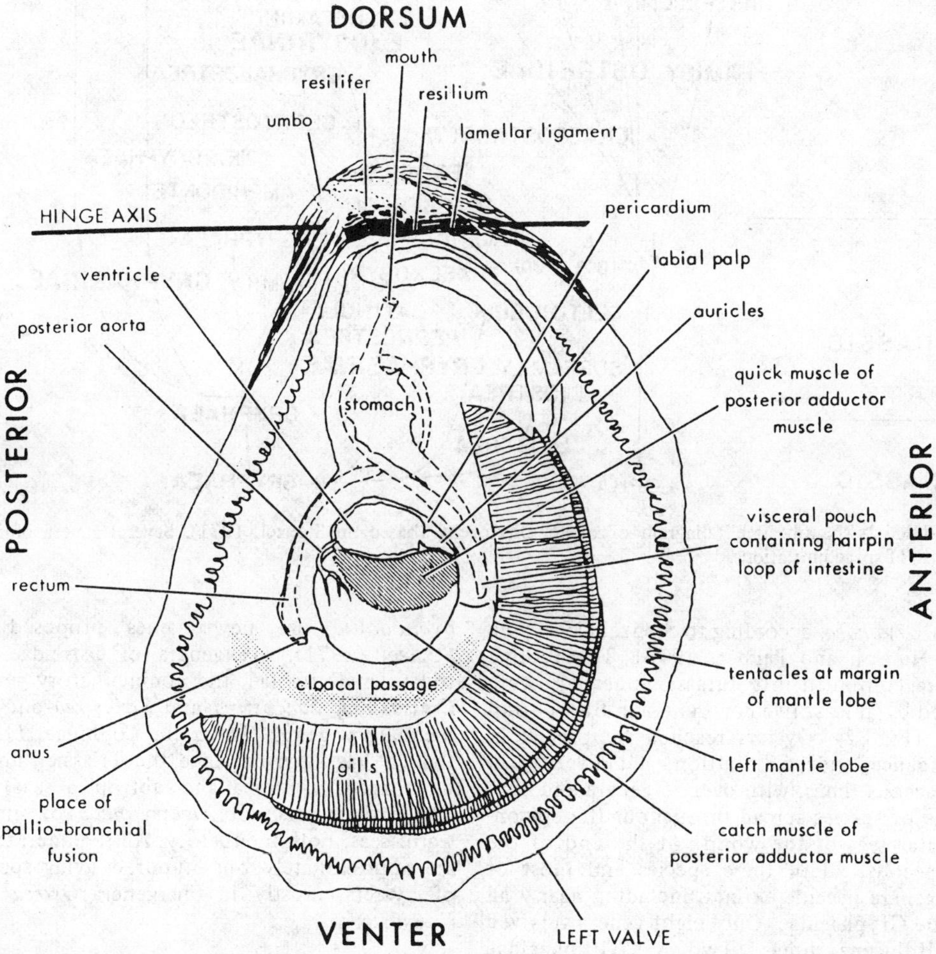

FIGURE 1. Anatomy and orientation of an oyster (*Crassostrea virginica*) (from Stenzel, 1971, courtesy of The Geological Society of America and the University of Kansas Press): The promyal passage (not shown) runs parallel to cloacal passage but dorsal to adductor muscle, between right mantle lobe and visceral mass (pericardium, stomach, intestine, etc.).

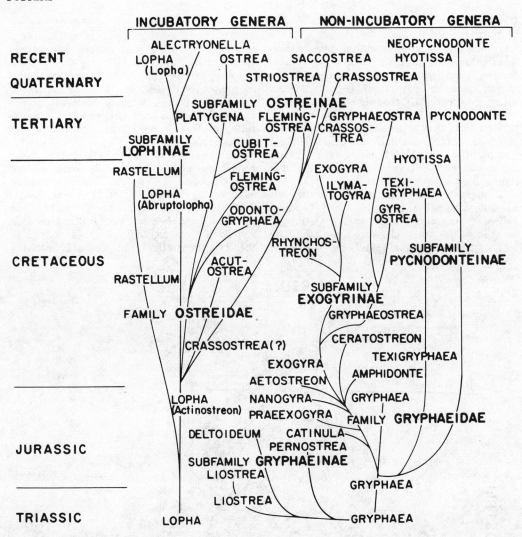

FIGURE 2. "Family tree" diagram of oyster phylogeny (based on Stenzel, 1971). Several genera omitted because of space limitations.

from *Gryphaea,* according to Stenzel (1971; see also Hudson and Palmer, 1976). These three genera continued into Jurassic times and were joined by at least five new genera in Europe and Asia (Fig. 2). Oysters reached their peak in abundance, wide distribution, and diversity in Cretaceous times, with over 25 genera and hundreds of species spread throughout the epicontinental seas of the world. At the end of the Cretaceous, all of these species and most of the genera became extinct, including nearly all of the Gryphaeidae. Only eight genera survived to Holocene times. Two are Gryphaeidae (Pycnodonteinae): *Hyotissa,* Cretaceous–Holocene, and *Neopycnodonte* which arose in the Miocene directly from *Pycnodonte* of Cretaceous origin. Both *Hyotissa* and *Neo-*

pycnodonte are new names proposed by Stenzel (1971). Six genera of Ostreidae live today: in Ostreinae, the nonincubatory genera *Crassostrea, Saccostrea* and *Striostrea* and the incubatory genus *Ostrea;* in Lophinae, *Lopha* (which has survived since the Triassic, almost unchanged, in tropical and subtropical seas) and its Tertiary offshoot *Alectryonella* of similar warm seas, both incubatory. All together, there are approximately one hundred living species of oysters, mostly in the genera *Ostrea* and *Crassostrea.*

Taxonomy

Taxonomy of oysters, especially at the species level, is very difficult because of the extreme variability of shell form due to both genetic and environmental influences. On the genetic level, it is possible by selection from one original stock to produce inbred strains, in the laboratory, which look different enough to be different species even though reared under identical conditions. Environmentally, neighboring oysters on a *Crassostrea virginica* reef may look very different, one type being elongated, narrow, and thin-shelled and another broad, rounded, and having a thick fluted or ridged shell; these differences are caused by the fact that the elongated group has grown upright in a crowded cluster and the broad group has grown separately while lying flat on a hard substrate of shells, but the individuals look different enough to be taken for different species if the shells are seen apart from their environment. Some of the differences in shell form, and their causes, were explained well by Dall (1898), in a paper that should be read by every person who intends to study either living or fossil oysters. Stenzel (1971) has also discussed oyster "ecomorphs," the apparently distinct forms due to direct effects of environment that have been called species by many authors.

Classification of oysters, at generic and higher levels, though based primarily on shells, is aided by the assumptions that can be made as to the soft parts on the basis of studies on living species. Living Gryphaeidae (*Neopycnodonte* and *Hyotissa*) have the intestine passing through the heart, possess a promyal passage, and are oviparous or nonincubatory (that is, they release eggs and sperm into the sea so that fertilization occurs outside the shell cavity instead of inside as in *Ostrea* and *Lopha*); it is assumed therefore that the extinct Gryphaeidae also had these characteristics. Oysters, living or extinct, with a cellular or vesicular shell structure (suggesting the common name "honeycomb oyster") are placed in the subfamily Pycnodonteinae to which both living genera of Gryphaeidae belong, with the extinct genera *Pycnodonte* and *Texigryphaea* of Cretaceous origin. Pycnodonteinae also have chomata: tubercles or small ridges on the inner edges of the right valve (anachomata), and pits or grooves in the inner edges of the left valve (catachomata) into which these fit when valves are closed (Figs. 3 and 4). Chomata are lacking in Gryphaeinae and in some species of Exogyrinae (the other subfamilies of Gryphaeidae). Gryphaeinae, besides lacking vesicular shell structure and chomata, differ from Pycnodonteinae by having a well-defined commissural shelf. The more typical Gryphaeinae have a beaklike umbone area of the left valve which becomes more or less strongly incurved (Fig. 5). Exogyrinae are best distinguished by

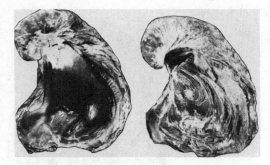

FIGURE 3. *Texigryphaea mucronata* (Gabb, 1869), a *Gryphaea*–like pycnodonteine from the Lower Cretaceous (Albian; Walnut Formation) of Texas (original, from collection of Hopkins). Note chomata along inside margin of left valve (photograph on left).

the regularly spiralled structure of the umbone region of both valves (*Exogyra;* Fig. 6), and sometimes of the entire shell (*Ilymatogyra*). The family Ostreidae is made up of oysters in which the intestine passes dorsal to the pericardium instead of through it, and the adductor muscle imprint is reniform or crescentic in contrast to its more rounded form in Gryphaeidae. (There are also other family differences.) There are two subfamilies of Ostreidae: Lophinae, with both valves bearing ribs in the form of regular, fairly sharp-crested plicae, and with numerous small tubercles scattered over the inner surface of valves, especially near margins; and Ostreinae, which lack these peculiarities. Lophinae are incubatory, the eggs being held within the mantle cavity until fertilized and the embryos being retained through early stages of larval development. Ostreinae are divided into two informal groupings of species, those that are incubatory, lack a promyal passage, and usually bear chomata (*Ostrea* and close relatives), and those that are nonincubatory, have a promyal passage, and usually lack chomata (*Crassostrea* and closely related genera). Oysters that are more closely related to *Crassostrea* than to *Ostrea* also usually have an umbonal cavity, a "cave" extending under the umbonal region inside the left valve.

Not all zoologists would accept Stenzel's taxonomic system based primarily on shell characters (usually the only features available to paleontologists). Ranson (1943, 1948, 1960) considered shell characters of secondary value and built up a system based primarily on the fine structure of the valves of the final larval stage, the prodissoconch. On this basis, Ranson would put all living oysters in three genera, *Ostrea* (including Stenzel's *Lopha* and *Alectryonella*), *Crassostrea* (including *Striostrea* and *Saccostrea*), and "*Pycnodonta*" (including the

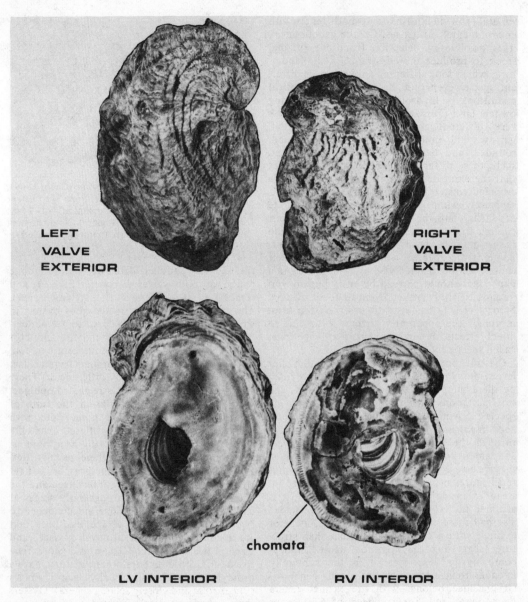

FIGURE 4. *Ceratostreon texanum* (Roemer, 1852), an exogyrine from the Lower Cretaceous (Albian: Walnut Formation) of Texas (original, from collection of Hopkins).

living species that Stenzel placed in *Hyotissa* and *Neopycnodonte,* along with the fossil species of *Pycnodonte*). However, except for differences in the names used for equivalent groups of species, a family tree drawn to show Ranson's phylogeny is much like one based on Stenzel's interpretations.

Still another approach to determining degrees of relationships is experimental hybridization. On the basis of cross-breeding experiments with oysters from various parts of the world, Menzel (1971, 1972, 1973, 1974) has found evidence that the species Stenzel (1971) put in *Crasso-strea* are more closely related to each other than any of them is to the species that Stenzel put in *Saccostrea* and *Striostrea;* he also found that *C. gigas* and *C. angulata* hybridized so readily, and their hybrids underwent mitosis and meiosis so normally, that they could be no more distant than subspecies of a single species; on the same basis, Menzel concluded that *C. rhizophorae* was probably a subspecies of *C. virginica.* It seems possible that future work of this sort may considerably reduce the number

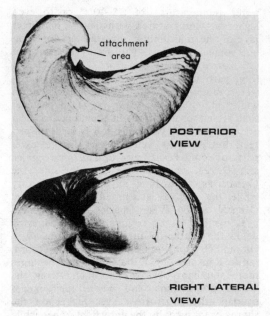

FIGURE 5. *Gryphaea (Gryphaea) arcuata* Lamarck. 1801, Jurassic, Europe (from Stenzel, 1971, courtesy of The Geological Society of America and The University of Kansas Press).

of species names in good standing, and shed further light on the grouping of species into genera.

Ecology

The extinct Gryphaeidae, like their modern descendants *Neopycnodonte* and *Hyotissa*,

were all strictly marine, often living far from shore. The only living species of *Neopycnodonte*, *N. cochlear*, is circumglobal in oceanic waters at temperatures of 12 to 14°C and depths of 27 to 1500 m, while the giant "honeycomb oyster," *Hyotissa hyotis*, lives commonly on coral reefs of the Indopacific region, and other species of *Hyotissa* live in similar situations in various tropical and subtropical regions, including the Caribbean and the Gulf of Mexico. According to Stenzel, the "hooknosed" shape of adult *Gryphaea* and similar "gryph-shaped" oysters is an adaptation to living in the soft sediments in deep waters far offshore, the larvae having attached to very small solid particles on these soft bottoms. When attached to larger solid objects, *Gryphaea* and other gryph-shaped oysters (whether related to *Gryphaea* or not) were attached by only a small part of the curved left valve and did not spread out broadly on the substrate surface like most species of *Exogyra* and modern ostreine oysters. Some *Gryphaea*-like but unrelated oysters supposedly were able to develop without the larva ever attaching to anything (in the Gulf of Mexico region, *Exogyra laeviuscula* and *Ilymatogyra arietina* of the Exogyrinae and *Odontogryphaea thirsae* of the Ostreinae).

In the Walnut Formation of the Lower Cretaceous (Middle Albian) in central Texas, millions of fossil oysters (along with frequent but less numerous ammonites, clams, snails, and sea urchins) are exposed in outcroppings over wide areas. The principal oysters are the *Gryphaea*-like *Texigryphaea mucronata* (Fig. 3), which Stenzel (1971) says is really a

FIGURE 6. *Exogyra (Exogyra) costata* Say, 1820, Upper Cretaceous, Texas (from Stenzel, 1971, courtesy of The Geological Society of America and the University of Kansas Press).

pycnodonteine, and the *Exogyra*-like *Cera-tostreon texanum* (Fig. 4); in some places they occur together in nearly equal numbers, while in other places one species predominates and the other is relatively scarce. The adult *T. mucronata* is seldom attached to anything, but even full-sized *C. texanum* specimens are sometimes still attached to shells or other hard objects. Apparently, after death both *Texigryphaea* and *Ceratostreon* shells became concentrated in shell banks, and many young oysters, either exogyrine or ostreine, attached to these shells and grew for a while before being killed by predators. Both gryphaeid oysters of the Walnut Formation often became cemented together in limy aggregations of innumerable shells of all sizes. Obviously this occurred long after death. There is no evidence that Cretaceous oysters ever formed reefs like those of the *Crassostrea* species of today. If the oysters when living were concentrated into "beds," these beds must have been like the *Ostrea edulis* "banks" of the German North Sea coast, which are deposits of mixed shells and sediments containing only about one oyster per m^2, according to Moebius (1893). Black Sea "banks" of *Ostrea taurica* are similar (Nikitin, 1934).

In the family Ostreidae, the Lophinae of today, like their ancestors ever since Triassic times, inhabit warm oceanic waters and are normally attached to such substrates as corals and gorgonians (not mangroves, with the possible exception of *Lopha cristagalli* in the Indopacific region). The Ostreinae divide into two groups which are as different ecologically as they are morphologically and physiologically. The incubatory species (*Ostrea*) live in nearshore oceanic waters or in the lower portions of estuaries where salinity, temperature, and turbidity are not too different from the conditions found in the ocean. Nonincubatory species, which according to Stenzel (1971) are divided among the genera *Crassostrea, Saccostrea,* and *Striostrea*, can live in the ocean but seldom become as abundant there as in the estuaries, to whose ever-changing salinity, wide range of temperature, and frequent high turbidity they are peculiarly adapted.

The species of *Crassostrea*, particularly, are able to withstand freshets in tidal rivers or estuaries by holding their valves closed for long periods, so that the salt water in the mantle cavity is only slowly diluted, and by being able to acclimate gradually to salinities of 10 or even 8°/$_{00}$, approximately one-fourth oceanic salinity. At the northern end of its range, *Crassostrea virginica* of eastern North America can survive freezing; and on the coast of the Gulf of Mexico, it can live in intertidal situations where it is exposed for several hours each day to water temperatures of 35–40°C or to even higher air temperatures, withstanding the direct rays of the midsummer sun (though not without some mortality). The promyal passage of *Crassostrea* and its superior ability to select food particles allows it to live in very muddy water, discarding most materials as pseudofeces while retaining and swallowing small particles that are either nutritious in themselves or bear attached bacteria (the selection is far from perfect, but it works well enough). In the lower parts of some Texas bays, *Crassostrea virginica* and *Ostrea equestris* live side by side, the *Crassostrea* flourishing in the high turbidity while the *Ostrea* is stunted and short-lived (Menzel, 1955). Where species of these genera live together, as *C. virginica* and *O. equestris* do in the southern United States and *C. angulata* and *O. edulis* do in western France (according to Ranson, 1943), the *Crassostrea* becomes more abundant upstream or inland from the *Ostrea* populations; and where their ranges overlap, the *Crassostrea* lives higher on the shore, even high in the intertidal zone, while *Ostrea* is found at deeper levels. Menzel (1955) found that in near-oceanic salinities *Crassostrea* larvae would attach in greatest numbers to suspended shells near the surface or between low- and high-water levels, while *Ostrea* larvae (and *Lopha* larvae) would attach mostly at lower levels. This explains the differences in vertical distribution of *Crassostrea virginica, Ostrea equestris,* and *Lopha frons* on offshore oilwell platforms in the Gulf, as reported by Gunter (1951).

Because of their ability to tolerate extreme temperatures, on the E coasts of continents *Crassostrea* species extend farthest toward the poles. The "*Ostrea puelchana*" mentioned by Stenzel (1971, p. N1088) as the most southern oyster in Argentina (at 42°S) is actually *Crassostrea brasiliana* according to Ranson (1948). *C. virginica* reaches 48°N in the Canadian Gulf of St. Lawrence. *C. gigas* reaches 52°N between Sakhalin and the mainland of Asia, according to Skarlato (1960) as cited by Stenzel (1971, p. N1038). On the W coasts of continents, there is a narrower range of water temperatures and the limiting factor is not extreme winter or summer temperature but the low summer temperature (too low for the reproduction of any *Crassostrea* species except in and near the tropic zone), caused by upwelling along the coasts. Species of *Ostrea* that can spawn at lower temperatures than any other oysters extend farthest poleward on W coasts. *O. lurida* reaches at least as far N as Sitka, Alaska (close to 57°N) and *O. chilensis* extends at least as far S as the island of Chiloe (43°S). All living oyster genera are represented by species in tropical and subtropical waters; in some cases, these are species that can not live in the colder waters, but *Crassostrea virginica* extends from Canada to Mexico, and farther if Menzel (1973) is right about *C. rhizophorae* being a subspecies of *C. virginica.*

Oysters feed near the bottom of the food pyramid, straining microscopic organisms (and particles rich in bacteria) from the water drawn over their gills and through the mantle chamber. This enables them to maintain dense populations and to dominate the biotic communities in which they live as trees dominate a forest. Verrill (1873) reported an extensive study of the New England oyster community, and Moebius (1877) proposed the name "biocoenosis" for such a community of plants and animals interacting with each other and with the environment, a concept that he developed further in 1893. Moebius also pointed out that offshore (North Sea) oyster communities had richer faunas, but that oysters themselves prospered better in inshore (Wattenmeer) beds where there were fewer competitors, predators, etc. Since Moebius, the oyster bed has been the classic example of a biocoenosis or biotic community, though the coral reef is a richer and more complex example. The subtidal "banks" of *Ostrea* species have been described by many Europeans and the subtidal and intertidal "reefs" of *Crassostrea* by American shellfishery biologists and ecologists.

Stenzel (1971, pp. N1043–N1048) has discussed oyster reefs in a general way, classifying them as fringe, string, and patch reefs. His "fringe reef" includes two fundamentally different types of oyster bed, the fringe of oysters alongshore, usually intertidal and attached to rocks, shells or other hard objects (including the parts of mangroves that are submerged at high tide), and the subtidal, often deep-water oyster beds of *Ostrea* and *Crassostrea* that border channels. The alongshore type, characteristic of *Saccostrea* as well as *Crassostrea,* occurs mainly in near-oceanic salinities and the channel-bordering type is more common in low-salinity estuaries. Stenzel gives European examples for what he calls the "patch reef", and indeed this is the typical form of natural oyster bed built up above the surrounding bottom but not above low tide level by *Ostrea edulis,* the commercial oyster of European waters. They are called "banks" rather than "reefs" by European authors, and are very different from the reefs of *Crassostrea* species. It was such banks that were studied as oyster biocoenoses by Moebius (1877, 1893) and by Nikitin (1934). What Stenzel calls "string reefs" are the typical *Crassostrea* reefs, crowded with dense clusters of living and dead oysters, seen

in the United States from Virginia to Texas (Figs. 7 and 8). The formation of string reefs was explained well by Grave (1901, 1905). The reefs grow rapidly in length, usually at right angles to the main currents, because oysters get more food and oxygen in the faster currents running around the ends, but grow only slowly in width because the oysters on the flanks are in slower-moving water; hence the elongated form of the reef, after it builds up to the intertidal level. Even though oysters die out on the crest when the reef builds up to high-tide level, it may still be built higher by shells and even clusters of oysters thrown up on the crest by storm waves; finally it gets high enough for seed plants to grow on the top, and becomes a "green reef" or island. Some of the Ten Thousand Islands of SW Florida were formed in this way, and many examples of the process can be seen in the bays and sounds of several Gulf and southern Atlantic states. Such oyster reefs (string reefs) occur where there are high rates of oyster reproduction and growth combined with high rates of mortality or frequent kills by storms, freshets, etc. They are most characteristic of *Crassostrea virginica,* just as the "bank" or "patch reef" is characteristic of *Ostrea edulis.* In a few places where the salinity is lowered by a large freshwater runoff, oyster reefs occur in the Gulf (Crystal River region of Florida; Marsh Island near mouth of Atchafalaya River, Louisiana), but with only these few exceptions, *Crassostrea virginica* reefs are confined to bays, sounds, and the lower stretches of tidal rivers. Farther up the estuary, in lower salinity, oysters are found only below low-tide level, and in deeper and deeper water upstream as the outflowing layer of low-salinity water dominates more over the saltier layer below it. *Crassostrea* species of North America

FIGURE 7. Live reefs ("string reefs") of *Crassostrea virginica* (Gmelin, 1791) in Apalachicola Bay, Florida, as seen from the air. (Photograph by U.S. National Marine Fisheries Service.)

FIGURE 8. Live reef of *Crassostrea virginica* in estuary near Wadmalaw Island, South Carolina, as exposed during every normal low tide period (twice daily). (Photograph by S. H. Hopkins, June 1948, showing Dr. G. Robert Lunz walking among the clusters of oysters.)

apparently grew similarly in Tertiary estuaries, from the Eocene on.

The oyster bed faunas of the Atlantic and Gulf coasts, New England to Texas, have many species in common. A compilation from many reports shows well over 600 species of oyster associates, including a number of fishes that are commonly found among oysters, in this entire stretch of coast. As Larsen (1974) showed, most species found on oyster beds are not confined to this habitat but are shared with other types of bottoms found in the estuaries. *Crassostrea virginica* flourishes in greatest abundance in two situations: in intertidal reefs in the higher salinities, where faunal diversity, competition, and predation are reduced because reefs are out of water several hours each day; and in subtidal beds in the lower salinities, where most competitors, shell borers, predators, and parasites are excluded by low salinity or frequent freshets. It was only after oysters developed enough tolerance to live and reproduce in these adverse environmental conditions

that they produced the dense populations seen today in the estuarine natural beds of *Crassostrea* and *Saccostrea*.

The enemies of oysters, from which adverse environments protect them, include several boring snails of the family Muricidae; boring sponges (family Clionidae); boring clams (Pholadidae); shell burrowing and mud-tube building annelids (Spionidae and other families); such predators as starfishes, crabs, and shell-crushing fishes (rays, skates, drums); certain predacious polyclad turbellarians (*Stylochus, Pseudostylochus*); and a number of parasites including trematodes (Bucephalidae), protozoans (Hexamita, Minchinia, and others), and fungi (*Labyrinthomyxa* and related forms). Wherever such enemies are numerous, mortality rates of oysters are high and few individuals live to reach great age or size; most may fail to live long enough to reproduce (Cheng, 1967; Mackin, 1962; Sindermann, 1970).

SEWELL H. HOPKINS

References

Adkins, W S., 1928. Handbook of Texas Cretaceous Fossils, *Univ. Texas Bull.*, **2838**, 385p.

Awati, P. R., and Rai, H. S., 1931. *Ostrea cucullata* (The Bombay Oyster). The Indian Zool. Memoirs on Indian Animal Types. III. Lucknow: Methodist Publishing House, 107p.

Baughman, J. L., 1947. *An Annotated Bibliography of Oysters with Pertinent Material on Other Shellfish and an Appendix on Pollution.* College Station, Texas: Texas A&M Research Foundation, 794p.

Cheng, T. C., 1967. Marine molluscs as hosts for symbioses, with a review of commercially important species, *Advances Marine Biol.*, **5**, 424p.

Dall, W. H., 1898. Contributions to the Tertiary fauna of Florida, with especial reference to the Silex beds of Tampa and the Pliocene beds of the Caloosahatchie River, etc. Part IV, *Trans. Wagner Free Instit. Sci., Philadelphia*, 3(4), 571–947.

Galtsoff, P. S., 1964. The American Oyster *Crassostrea virginica* Gmelin, *Fish. Bull. U.S. Fish Wildlife Serv.*, **64**, 480p.

Gardner, J. 1945. Mollusca of the Tertiary formations of Northeastern Mexico, *Geol. Soc. Amer. Mem. 11*, 332p.

Grave, C. 1901. The oyster reefs of North Carolina. A geological and economic study, *Johns Hopkins Univ. Circ. No. 151*, 9p.

Grave, C., 1905. Investigations for the promotion of the oyster industry of North Carolina. *Rept. U.S. Comm. Fish.*, No. 29, 247–341.

Gunter, G., 1950. The generic status of living oysters and the scientific name of the common American species, *Amer. Midland Naturalist*, **43**, 438–449.

Gunter, G., 1951a. The species of oysters of the Gulf, Caribbean and West Indian region, *Bull. Marine Sci. Gulf Carib.*, **1**, 40–45.

Gunter, G., 1951b. The west Indian tree oyster on the Louisiana coast, and notes on the growth of the three Gulf Coast oysters, *Science*, **113**, 516–517.

Heckel, P. H., 1974. Carbonate buildups in the geologic record: a review, *Soc. Econ. Paleontol. Mineral. Spec. Publ. 18*, 90–154.

Hill, R. T. and Vaughan, T. W., 1898. The Lower Cretaceous Gryphaeas of the Texas region, *U.S. Geol. Surv. Bull.*, **151**, 139p.

Hopkins, S. H., 1957. Annotated bibliography—oysters, *Geol. Soc. Amer. Mem. 67*, no. 1, 1129–1133.

Hudson, J. D., and Palmer, T. J., 1976. A euryhaline oyster from the Middle Jurassic and the origin of the true oysters, *Palaeontology*, **19**, 79–93.

Joyce, E. A., Jr., 1972. A partial bibliography of oysters, with annotations, *Florida Dept. Nat. Resources, Marine Research Lab., St. Petersburg, Fla., Spec. Sci. Rept. No. 34*, 846p.

Larsen, P. F., 1974. Quantitative studies of the macrofauna associated with the mesohaline oyster reefs of the James River, Virginia. Ph.D. dissertation, College of William and Mary, Williamsburg, Virginia, 182p.

Larsen, P. F., 1975. A preliminary checklist, with annotations, of potential associates of the American oyster, *Crassostrea virginica*, in New England, *Maine Dept. Marine Resources, Marine Research Lab. Publ. No. 9*, 42p.

McCormick, L. and Moore, R. C., 1969. Outline of classification, in R. C. Moore, ed. *Treatise on Invertebrate Paleontology*, pt. N, vol. 1, Mollusca 6, Bivalvia. Lawrence, Kansas: Geol. Soc. Amer. and Univ. Kansas Press, N218–N222.

Mackin, J. G., 1962. Oyster disease caused by *Dermocystidium marinum* and other micro-organisms in Louisiana, *Publ. Instit. Marine Sci. Univ. Texas*, **7**, 132–299.

Menzel, R. W., 1955. Some phases of the biology of *Ostrea equestris* Say and a comparison with *Crassostrea virginica* (Gmelin), *Publ. Instit. Marine Sci., Univ. Texas*, **4**, 69–153.

Menzel, R. W., 1971. Selective breeding in oysters, in K. S. Price, Jr. and D. L. Maurer, eds., *Proc. Conf. on Artificial Propagation of Commercially Valuable Shellfish—Oysters—October 22-23, 1969*. Newark, Delaware: Univ. Delaware, 81–92.

Menzel, R. W., 1972. The role of genetics in molluscan mariculture, *Amer. Malacological Union Inc. Bull.*, **1972**, 13–15.

Menzel, R. W., 1973. Some species affinities in the oyster genus *Crassostrea*, *Amer. Malacological Union Inc. Bull.*, (Abst.), **1973**, 38.

Menzel, R. W., 1974. Portuguese and Japanese oysters are the same species, *J. Fish. Research Board Canada*, **31**, 453–456.

Moebius, K., 1877. *Die Auster und die Austernwirtschaft.* Berlin: Verlag von Wiegand, Hempel und Parey, 126p. (Transl. by H. J. Rice in *Rept. U.S. Fish. Comm.*, No. 1880, App. H, 683–751.)

Moebius, K., 1893. Ueber die Thiere der schleswigholsteinischen Austerbänke, ihre physikalischen und biologischen Lebensverhältnisse, *Sitz. K. Preuss. Akad. Wiss., Berlin*, **8**, 67–92.

Nikitin, B. N., 1934. The Gadaut Oyster Bank: essay of ecological and fishery investigations, (in Russian with Georgian, German, and English summaries), *Trav. Station Piscicole Biol. Géorgie*, **1**, 51–179. (English section, 169–176).

Ranson, G., 1943. *La Vie des Huîtres. Histoires Naturelles*, 3rd ed. I. Paris: Gallimard, 261p.

Ranson, G., 1948. Prodissoconques et classification des ostreides vivants, *Bull. Mus. Roy. Hist. Nat. Belgique*, **24**, 1–12.

Ranson, G., 1960. Les prodissoconques (coquilles larvaires) des ostreides vivants. *Bull. Inst. Oceanogr., Monaco*, **1183**, 1–41.

Ranson, G., 1967. Les espèces d'huîtres vivant actuellement dans le monde, définies par leurs coquilles ou prodissoconques, *Rech. Trav. Instit. Pêches Marit.*, **31**, 1–146.

Sellards, E. H.; Adkins, W. S.; and Plummer, F. B., 1932. The Geology of Texas, vol. I. *Stratigraphy* (3rd printing), *Univ. Texas Bull. 3232*, 1007p.

Shier, D. E., 1969. Vermetid reefs and coastal development in the Ten Thousand Islands, Southwest Florida, *Geol. Soc. Amer. Bull.*, **80**, 485–508.

Sindermann, C. J., 1970. *Principal Diseases of Marine Fish and Shellfish.* New York: Academic Press, 369p.

Skarlato, O. A., 1960. Bivalve mollusks of the far eastern seas of the U.S.S.R. (Otryad Dysodonta) (in Russian), *Akad. Nauk SSSR, Opredel. po faune SSSR, Izdavayemye Zool. Inst. SSSR*, No. 71, 150p.

Stenzel, H. B., 1971. Oysters, in R. C. Moore, ed.,

Treatise on Invertebrate Paleontology, pt. N, vol. 3, Mollusca 6. Lawrence, Kansas: Geol. Soc. Amer. and Univ. Kansas Press, N953–N1224.

Stenzel, H. B.; Krause, E. K.; and Twining, J., 1957. Pelecypoda from the type locality of the Stone City Beds (Middle Eocene) of Texas, *Univ. Texas Pub. 5704,* 237p.

Taylor, J. W., 1927. The "mutations" of our native land and freshwater Mollusca, *J. Conchology,* London, **18,** 85–116. (P. 89, the striking differences produced in oysters by the direct effect of different environmental conditions.)

Thomson, J. M., 1954. The genera of oysters and the Australian species, *Australian J. Marine Freshwater Research,* **5,** 132–168.

Verrill, A. E., 1873. Report on the invertebrates of Vineyard Sound and adjacent waters, *Rept. U.S. Fish. Comm.,* No. 1871–1872, 295–544.

Waller, T. R., 1978. Morplology, morphoclines and a new classification of the Pteriomorphia (Mollusca: Bivalvia), *Phil. Trans. Roy. Soc. London, Ser. B,* **284,** 345–365.

Wells, H. W., 1961. The fauna of oyster beds, with special reference to the salinity factor, *Ecol. Monogr.,* **31,** 239–266.

White, C. A., 1884. A review of the fossil Ostreidae of North America, 4th *Ann. Rept. U.S. Geol. Surv. No. 1883,* 273–430.

Cross-references: *Bivalvia: Reefs and Other Carbonate Buildups.* Vol. III: *Oyster Reefs.*

P

PALEOBIOGEOGRAPHY

Paleobiogeography is the study of the geographical distribution of faunas and floras in the past. It can be divided into paleozoogeography and paleophytogeography, dealing with faunas and floras respectively.

The existence of regional patterns in the distribution of present-day animals and plants has long been recognized, but it was not until the time of Darwin and Wallace that the significance of these patterns, in terms of the history of their development, began to be realized. Darwin, in his *Origin of Species* (ch. 13) wrote: "All the grand leading facts of geographical distribution are explicable on the theory of migration, together with subsequent modification and the multiplication of new forms. We can thus understand the high importance of barriers, whether of land or water, in not only separating, but in apparently forming the several zoological and botanical provinces"— a passage that accurately anticipates modern work on the subject, especially in its emphasis on "provinces" with "barriers" between them. By the middle of the 19th century, it was being suggested by some paleontologists that comparable regional distributions of living organisms had existed in past times, and that these distributions were different from those of the present; in Darwinian terms, the geographical distributions of organisms, as well as the organisms themselves, had evolved. Nevertheless, in spite of the wealth of paleobiogeographical observations that have emerged as by-products of systematic paleontological studies during the past century, paleontologists have been slow to attempt synthesis in this field. A notable early attempt was that by Neaverson in his *Stratigraphical Palaeontology: A Study of Ancient Life-Provinces,* first published in 1928. This work, especially in its later edition (1955), divided the faunas and floras of each successive period of geological time into "faunal regions," i.e., geographical areas distinguishable by the organisms they contain. Neaverson summarized with a general review of the geographical distribution of organisms through time, which led him to a conclusion strongly hostile to continental drift. Since 1955, however, much of the increasing appreciation of the importance of paleobiogeographical studies has been generated by the wide acceptance of continental drift and the realization that paleontology has much to contribute to, as well as to learn from, Plate Tectonic theory. At a symposium on "Faunal Provinces in Space and Time" held in London in 1969 (Middlemiss et al., 1971), probably the first overall synthesis of paleobiogeography, all the members accepted, and based their thinking upon, continental drift.

Faunal Provinces

The concept of "faunal provinces," i.e., of geographical areas each with its appropriate indigenous biota, is central to paleobiogeography, as it is to neobiogeography; but it is difficult to define. Definition is easiest in the case of terrestrial faunas and floras, mainly because the land areas tend to be bounded by the sea, a largely impassable barrier to land organisms, and to contain within themselves other tracts of such difficulty as to be almost impassable, such as deserts and high mountains, which act as clearly marked barriers bounding the provinces. All the early work on neozoogeography, such as that of A. R. Wallace, was concerned with terrestrial faunas. H. R. Hewer (in Middlemiss et al., 1971) recognizes six "zoogeographical regions" in the present-day land areas, each divided into subregions (Fig. 1). The regions are large, of continental dimensions; some authors would call them provinces while others would prefer the term *realm* for units of such size, with provinces as subdivisions.

It is more difficult to distinguish such patterns of distribution among marine biota, partly because the continuity of the marine environment (in contrast to the separated land masses) may blur the boundaries of regions, and partly because marine faunal differences tend to be at a lower taxonomic level than those found on land. That geographical distribution patterns are present in modern marine faunas has been clearly shown by work in recent decades, notably by Ekman (1953). Paleontologists working on marine faunas of the past similarly observe that certain taxa are found in one place and not in another and that the

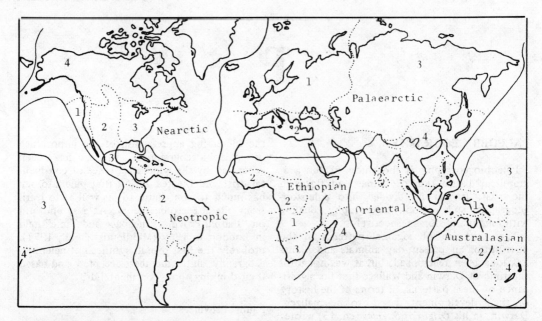

FIGURE 1. Modern zoogeographical regions (after Hewer in Middlemiss et al., 1971). Number within each region refers to provinces.

places where they are found or not found, if put together, form an area that the paleontologist may regard as a *faunal province*. In what circumstances the use of this term is justified has been much discussed. Some would reserve the term for large units, of the scale of Hewer's regions, others would happily apply it to smaller areas. Some would insist that a province must be bounded by distinct barriers but most would probably prefer to define it by its contained fauna; Valentine (1968), for example, defined a faunal province as a collection of communities associated in space and time. It is clear that for a province to be recognizable at all a substantial proportion of its fauna must be endemic to the area; pandemic taxa, invaluable as they are in stratigraphic correlation, are no help in paleobiogeography. It is often stated that the endemic faunal elements must be of high taxonomic level. It is axiomatic, as Ekman (1953) points out, that for this purpose endemic families are more diagnostic than endemic genera, and endemic genera than endemic species; but several authors have described patterns of geographical distribution that are not visible at all if higher taxa alone are considered, but are clearly marked at the level of genera or species. Where there is endemicity at a high taxonomic level it presumably means that the province has been differentiated for a longer period of time than in cases where the endemicity is at a lower

taxonomic level only, but the latter are no less valid as distinguishable biological provinces at a given moment of time.

Topographic and Ecologic Barriers

A more important question is to what extent a faunal province, to be recognized as such, must be independent of environmental control. Ideally it ought to be possible to find two kinds of biogeographical distributions, one strictly geographical, consisting of biota bounded by topographical barriers in the broad sense, the other ecological, consisting of biota favored by the conditions of life in that particular area. The latter is the *biome* of ecologists and the *magnafacies* of paleontologists. Two geographical provinces might each contain a similar range of broad environments (magnafacies) that would have broadly similar faunas adapted to them (faunal facies) but differing in the two provinces at the specific or generic level, or at higher levels according to how long the geographical separation had been established. A. Brouwer has presented this concept in the form of a matrix (see Fig. 2).

Faunal provinces and ecologically controlled geographic groupings are thus conceptually independent. Some authors, failing to find geographical groupings of biota that are clearly independent of ecological control, have declared that faunal provinces do not exist.

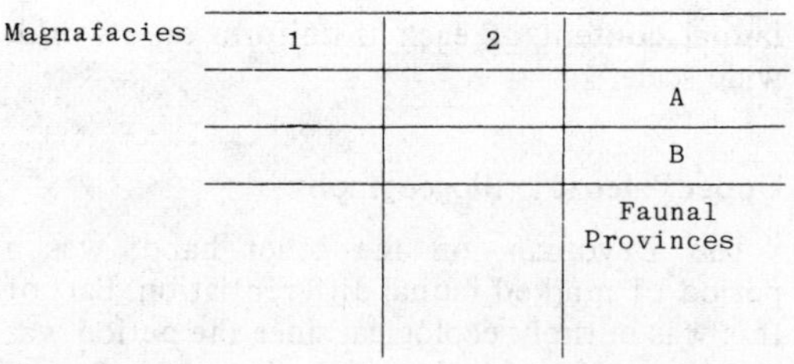

Magnafacies	1	2	
			A
			B
			Faunal Provinces

FIGURE 2. The relationship between magnafacies and faunal provinces (after Middlemiss and Rawson in Middlemiss et al., 1971).

Nevertheless, maintenance of this independence seems to be impossible in the face of the facts of nature, which are vastly more complex than the simple conceptual model. Taxonomic differentiation must require barriers between parts of a population, but such barriers are not necessarily simple topographic ones—they may be ecological or physiological. The evolution of a fauna is channeled by the environment and not solely by geographical isolation, thus the taxonomic content of the fauna and the facies of the fauna (reflecting the environment) cannot be separated; the isolation factor and the environmental factor cannot be separately analyzed. Also, since different taxa react differently to the environment, nearly all barriers are pervious to some taxa; only the most absolute of geographical barriers (a rarity in marine conditions) are impervious. It follows that faunal provinces may be clearly recognizable in certain phyla or classes in which endemicity is high, but not in others whose taxa may be more or less pandemic at the particular time under consideration. In practice, most of the faunal provinces of the past that appear to have been well established, according to the literature, have been based upon the particular group of organisms studied by the author concerned.

Statistical Methods of Study

Recognition of faunal differentiation between two areas can often be safely based upon the simple plotting of taxa on a map, but when large and complex faunas are involved statistical methods are usually necessary. Williams (1962), for example, in demonstrating the similarities between the Scottish and E North American Caradocian brachiopod faunas, was faced with the problem of estimating the degree of morphological likeness between two faunas, one of about 90 species, the other of about 750 species, collected from rock successions over 4000 miles apart, which seemingly duplicated one another for a sufficiently long period of time to allow identical changes to take place in each of them.

The most generally used technique has been that of the coefficient of association, which in this context means the ratio number of taxa common to the two areas: total number of taxa in both areas (see Ross, 1974; Campbell and Valentine, 1977). The distribution of the coefficients can then be conveniently analyzed by the method of cluster analysis (e.g., Rowell et al., 1973). Another statistical concept that has been used in paleobiogeographical studies is that of diversity (q.v.), which is usually taken to mean the number of separate taxa present irrespective of the number of individuals. Thus an abundant population consisting of only one or two species would show low diversity. High diversity implies an unspecialized environment, with optimum conditions for a large number of different forms of life. The contrast between contemporaneous faunas of high and of low diversity could be brought about by many different ecological factors including temperature, salinity, nature of sea bottom, or topographic isolation. On a large scale (world-wide or continent-wide) at the present day, however, the most obvious cause of diversity differences is latitude, the tropics having more diverse faunas than cooler latitudes, as authors from Wallace onward have shown. Stehli (1968) had demonstrated diversity contours, approximately parallel to latitudes. Several authors have attempted to establish or confirm palaeolatitudes by plotting such diversity gradients for fossil biota.

Lower Paleozoic Biogeography

Too little is known about Precambrian life for any discussion of its paleobiogeography to be profitable. The same could almost be said of the Cambrian; only the trilobites were then sufficiently abundant and differentiated to form the basis for such a study. J. W. Cowie (in Middlemiss et al., 1971) recognizes two major "realms" characterized respectively by the redlichiine trilobites (Mediterranean, Middle East, Asia, Australia) and the olenelline trilobites (Europe, North and South America), with the latter realm subdivided into the Acado-Baltic province (Europe apart from extreme NW Scotland but including the extreme eastern fringe of North America) and the Pacific province (remainder of North and South America but including extreme NW Scotland). A. R. Palmer (in Hallam, 1973) finds a generally similar distribution but points to unexpected differences between the faunas of Antarctica and South America.

In the Ordovician, brachiopods, trilobites, and graptolites have been utilized in discussions of paleobiogeography, and all indicate that provincial differentiation of faunas was well developed at this time. Williams (1969) has shown that, at least in the European–North American area, Ordovician brachiopods fall into three faunal provinces that remained fairly distinct throughout the period: (1) an American province–North America, NW Ireland, Scotland; (2) a Baltic province–SE Ireland, N Wales, Baltic Europe; (3) an Anglo-Welsh province, with affinities to central Europe. By Late Ordovician (Ashgillian) times the boundaries between the provinces had changed in that the "American" fauna was restricted to mid and W North America while in central Europe the isolated Bohemian province had developed, leaving eastern North America, western and Baltic Europe as a large, uniform "North European" province. In 1939, Lamont had already recognized this trend toward breakdown of provincial boundaries near the end of the Ordovician. Stubblefield (1939), considering trilobites, also noted that the "Scots-Irish" province had no species in common with the "Anglo-Welsh" province up to the Ashgillian but that then the faunas became mixed "as if some previous barrier had disappeared." More recently H. B. Whittington (in Hallam, 1973; and H. B. Whittington and N. F. Hughes, in Ross, 1974) has regarded the northern hemisphere trilobite faunas as peripheral to the continental masses of the Canadian and Baltic Shields and as warm-water faunas living close to the Ordovician equator. He points out the great differences between these faunas and those peripheral to the southern continents (Gondwanaland), which must have been cool-water dwellers, separated from the northern faunas by a wide and deep ocean. O. M. B. Bulman (in Middlemiss et al., 1971) emphasizes the existence of "Pacific" and "European" graptolite provinces during the Arenigian and Llanvirnian, the distinction fading after the Llanvirnian and yielding place to a cosmopolitan fauna.

The changes in the boundaries of provinces toward the end of the Ordovician foreshadowed the lack of provincial differentiation that is characteristic of the Silurian. C. H. Holland (in Middlemiss et al., 1971) has reviewed the main elements in the Silurian fauna and confirms earlier conclusions by Boucot (see Boucot in Ross, 1974) and others that "the predominant impression is of cosmopolitan faunas." There are two main faunal facies, the "shelly" (brachiopods, trilobites, corals, crinoids) and the "graptolitic" (graptolites, orthoconic nautiloids and certain bivalved mollusks), but the faunal content of each is uniform on a worldwide scale.

Upper Paleozoic Biogeography

The Devonian, on the other hand, was a period of marked faunal differentiation. Part of this was entirely ecological since the period was also one of marked environmental contrasts, including the wide-spread development of the nonmarine "Old Red Sandstone" facies. In the marine facies M. R. House (in Middlemiss et al., 1971) recognizes (1) an Austral province (South Africa and southern South America); (2) an Appalachian province (eastern North America and northern South America); and (3) an Old World Province (the rest of the world but subdivided into Rhenish-Bohemian, Uralian, Tasman, New Zealand, and Cordilleran subprovinces). The distinctiveness of the Appalachian province may be largely due to its restricted nearshore lithofacies, while the Austral province cannot be recognized after the Lower Devonian, so that the subdivisions of the Old World province assume greatest importance (Fig. 3). These provinces are recognized essentially on the benthic faunas, especially brachiopods; ammonoid genera tended to have a world-wide distribution, although House (in Hughes, 1973) has pointed out the difficulty in analyzing ammonoid distribution that is caused by the effects of postmortem drifting of shells.

The Carboniferous was, like the Devonian, a time of strongly differentiated depositional facies, but provincial faunal differentiation had again become faint. Generally speaking, each depositional facies (marine limestones, marine shales, coal measures, etc.) had its distinctive faunal facies; but these were rather uniform throughout the world, so that the succession of changes with time observable within these facies had world-wide validity. Hill (1948) was able to recognize three main provinces among coral faunas: (1) China and the East Indies; (2) the Mississippi area; (3) the European area, including the Urals and Siberia. These, however, contained many pandemic genera.

With the Permian, provincial differentiation again became prominent. This is especially well seen in the floras (see W. G. Chaloner and S. V. Meyen, and E. P. Plumstead, in Hallam, 1973). The earlier Carboniferous was apparently a time of floral uniformity, but toward the end of the Carboniferous, and particularly in the Permian, at least four geographic floras are distinct: (1) the Arctocarboniferous flora of North America, Europe, and central Asia; (2) the Cathaysian flora of China and SE Asia, perhaps

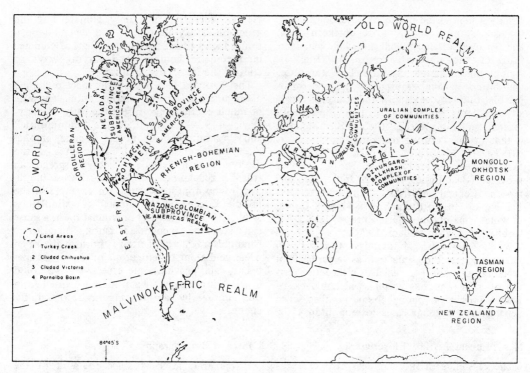

FIGURE 3. Devonian faunal provinces recognized on the basis of brachiopods (from Boucot, 1975).

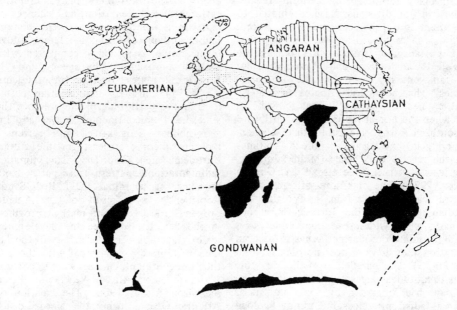

FIGURE 4. Geographical distribution of the major floras in Pennsylvanian and Permian rocks (from Ager, 1963, copyright © 1963 by McGraw-Hill Book Co., and used with permission).

strictly a subdivision of the Arctocarboniferous; (3) the Angaraland flora of northern Asia; and (4) most distinct of all, the Gondwanaland flora of the southern continents (Fig. 4). Among marine faunas, Stehli, J. B. Waterhouse, and others have attempted to locate Permian paleolatitudes by means of the study of diversity gradients, especially of brachiopods. As regards faunal provinces, the main impression is that in the Permian the situation characteristic of the Mesozoic was first established in that the faunas are divisible into a Tethyan province (East Indies, northern India, Mediterranean, Central America, with branches extending up to northern Siberia) and a Boreal province (mainly northern Europe) (see D. J. Gobbert, Fig, 2, in Hallam, 1973).

A. J. Charig (in Middlemiss et al., 1971) has reviewed the situation as regards land vertebrates in the Permian to Cretaceous interval of time and has emphasized the lack of faunal provinces among them, families, even genera, being in many cases cosmopolitan.

Mesozoic and Tertiary Biogeography

The existence of more or less distinct Tethyan and Boreal faunal provinces or realms is the dominant feature of paleobiogeography during the Triassic, Jurassic, and Cretaceous, although many authors, using the term "realm" for these major divisions, have attempted to divide them into provinces and W. J. Arkell, in 1956, recognized a third—the Pacific realm. The importance of this is that during the Mesozoic the major geographic faunal distributions ran generally E-W, parallel to paleolatitudes. The boundaries fluctuated widely. Arkell applied the term "Boreal Spread" to times when the boreal fauna spread southward into southern Europe and California (Callovian-Kimmeridgian) and "Tethyan Spread" to times when the reverse took place (Middle Jurassic, Hauterivian, Aptian). (See also W. A. Gordon in Ross, 1974.) The maximum differentiation of these realms is found in the Volgian and Berriasian, straddling the Jurassic–Cretaceous boundary, when widespread nonmarine facies tended to isolate the marine faunas to such an extent that they have almost nothing in common (Fig. 5). The parallelism of the Mesozoic realms to latitudes suggests climatic control and they have been regarded by some authors as not true faunal provinces but purely ecologically controlled distributions. Detailed distribution patterns, however, show many features not easily explained by any one controlling factor (such as Tethyan species appearing in Greenland during the Early Cretaceous) and it

seems likely that not only climate but the geographical lay-out of land and sea and depositional facies (itself partly a product of climate) interacted in complex and varying ways to control the faunal distribution.

The Tethyan and Boreal provinces remained distinguishable in the marine fauna, in spite of much overlapping, during the early Tertiary. Davies (1934) called them the "nummulitic" and "nonnummulitic" in the Eocene, emphasizing the importance of nummulites in the Tethyan region. Periods of Tethyan spread can still be recognized by the northward migration of nummulite species, for example in the middle Eocene. B. M. Funnell (in Middlemiss et al., 1971) shows that provinciation in a sense can even be recognized among planktonic foraminifera in the Tertiary; both in the open ocean and along coastlines major faunal changes occur, and provinces are effectively delimited at the boundaries of water masses divided from one another by abrupt temperature or salinity changes.

Causes of Biogeographic Changes

In general, at least since the end of the Cambrian, there seem to have been alternating phases of provincialism and cosmopolitanism. The causes appear to be linked with the movements of the lithospheric plates, alterations in patterns of sedimentation, and changes of climate in a complex manner. The merging of faunal provinces during the later Ordovician, for example, is probably connected with the mutual approach of "European" and "North American" plates. Holland (in Middlemiss et al., 1971) considers that Silurian cosmopolitanism was partly due to the presence then of widespread epicontinental shelves and limited geosynclines. This agrees with current plate tectonic theory, at least for the American–European area. The Devonian climatic and sedimentational pattern arose partly from the presence of a united "Old Red Sandstone Continent" as a result of the Caledonian orogeny. Resurgence of marked provincialism in the Permian followed the development of major glaciation. The uniformity of the Triassic land vertebrate fauna agrees with the concept that the continental masses were in conjunction at that time (Alfred Wegener's "Pangaea"; see also Sutton, 1968). The creation of the Atlantic Ocean during the Mesozoic by the separation of the American plates from those of the Old World is emphasized by the distribution of two groups of large benthic foraminifera (F. C. Dilley, in Middlemiss et al., 1971): the early Cretaceous *Orbitolina* occurs in a belt

FIGURE 5. Ammonite provinces at the end of the Jurassic (after Casey in Middlemiss et al., 1971), T, Tethyan; V, Volgian; P, Portlandian.

corresponding to Mesozoic Tethys, with its Old World and American portions torn apart; to *Alveolina* of the Upper Cretaceous, occupying the same belt in the Old World, the ocean now presented an impassable barrier (Fig. 6).

In the Tertiary, South America, with its indigenous mammal faunas, presents a good example of a faunal province bounded by absolute geographical limits. At the end of the Pliocene, its isolation was ended by the establishment of the isthmus of Panama; cross migration of mammals with North America immediately began and at the same time the Carribean and Pacific marine faunas began to differentiate. Similarly the early Tertiary

Tethyan marine fauna differentiated into Indian Ocean and Mediterranean provinces from the early Miocene with the impingement of the African–Arabian block upon Asia Minor during the closing of Tethys by continental drift.

The biogeographical provinces recognized at the present day on both land and sea seem to be due, at least in part, to the aftermath of the rapid climatic changes of the Pleistocene glaciations, which caused isolation and changes in annual temperature ranges and hence long-term changes in the distributions that were present before and even during the Pleistocene.

FRANK A. MIDDLEMISS

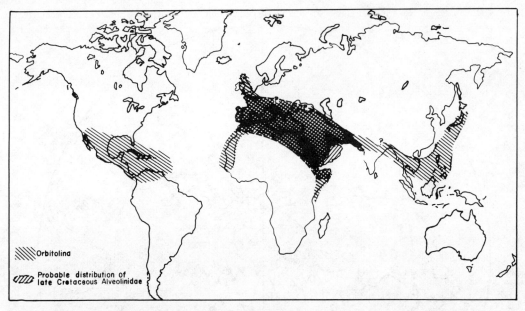

Orbitolina

Probable distribution of late Cretaceous Alveolinidae

FIGURE 6. Distribution of *Orbitolina* in Early Cretaceous and the Alveolinidae in the Late Cretaceous (after Dilley in Middlemiss et al., 1971).

References

Ager, D. V., 1963. *Principles of Paleoecology.* New York: McGraw-Hill, 371p.

Boucot, A. J., 1975. *Evolution and Extinction Rate Controls.* Amsterdam: Elsevier, 427p.

Campbell, C. A., and Valentine, J. W., 1977. Comparability of modern and ancient faunal provinces, *Paleobiology, 3,* 49–57.

Cox, C. B.; Healey, I. N.; and Moore, P. D., 1976. *Biogeography: An Ecological and Evolutionary Approach,* 2nd ed. New York; Wiley, 194p.

Davies, A. M., 1934. *Tertiary Faunas.* 2 vols. London: Thomas Murby, 406p. and 252p.

Ekman, S., 1953. *Zoogeography of the Sea.* London: Sidgwick & Jackson, 418p.

Gray, J., and Boucot, A. J., eds., 1979. *Historical Biogeography, Plate Tectonics, and the Changing Environment.* Corvallis: Oregon State Univ. Press.

*Hallam, A., ed., 1973. *Atlas of Palaeobiogeography.* Amsterdam: Elsevier, 531p.

Hill, 1948. Distribution and sequence of Carboniferous coral faunas, *Geol. Mag., 85,* 121–148.

*Hughes, N. F., ed., 1973. Organisms and continents through time, *Spec. Paper, Palaeontology, 12,* 334p.

Lamont, A., 1939. Distribution and migration of brachiopods in the British and Irish lower Palaeozoic faunas, *Irish Naturalists' J.,* 7(6), 172–178.

MacArthur, R. H., 1972. *Geographical Ecology: Patterns in the Distribution of Species.* New York: Harper & Row, 269p.

*Middlemiss, F. A.; Rawson, P. F.; and Newall, G., eds., 1971. *Faunal Provinces in Space and Time.* Liverpool: Seel House Press, 236p.

Neaverson, E., 1955. *Stratigraphical Palaeontology: A Study of Ancient Life-Provinces,* 2nd ed. Oxford University Press, 806p.

Ross, C. A., ed., 1974. Paleogeographic provinces and provinciality, *Soc. Econ. Paleontol. Mineral. Spec. Pub.,* 21, 233p.

Ross, C. A., ed., 1976. *Paleobiogeography.* Benchmark Papers in Geology, Vol. 31. Stroudsburg, Pa.: Dowden, Hutchinson & Ross, 427p.

Rowell, A. J.; McBride, D. J.; and Palmer, A. R., 1973. Quantitative study of Trempealeauian (latest Cambrian) trilobite distribution in North America, *Geol. Soc. Amer. Bull., 84,* 3429–3442.

Simberloff, D., 1972. Models in biogeography, in T. J. M. Schopf, ed., *Models in Paleobiology.* San Francisco: Freeman, Cooper, 160–191.

Stehli, F. G., 1968. Taxonomic diversity gradients in pole location: The Recent model, in E. T. Drake, ed., *Evolution and Environment.* New Haven Conn.: Yale Univ. Press, 163–227.

Stubblefield, C. J., 1939. Some aspects of the distribution and migration of trilobites in the British lower Palaeozoic faunas, *Geol. Mag., 76,* 49–72.

Sutton, J., 1968. Development of the continental framework of the Atlantic, *Proc. Geologists' Assoc., London,* 79, 275–303.

Valentine, J. W., 1968. The evolution of ecological units above the population level, *Palaeontology, 42,* 253–267.

Williams, A., 1962. The Barr and Lower Ardmillan Series (Caradoc) of the Girvan District, south-west Ayrshire, with descriptions of the Brachiopoda, *Geol. Soc. London Mem. No. 3,* 1–265.

Williams, A., 1969. Ordovician Faunal Provinces with reference to brachiopod distributions, in A. Wood, ed., *The Pre-Cambrian and Lower Palaeozoic Rocks of Wales.* Cardic: Univ. Wales Press, 117–154.

*Major recent symposia are marked with an asterisk.

Cross-references: *Diversity; Larvae of Marine Inverte-brates–Paleontological Significance; Paleobiogeography of Vertebrates; Paleogeographic Maps; Paleophytogeography; Plate Tectonics and the Biosphere.* Vol. II: *Zone–Climatic.*

PALEOBIOGEOGRAPHY OF VERTEBRATES

Biogeography has always been one of the cornerstones of evolutionary biology. Charles Darwin devoted two chapters in *The Origin of Species* to the subject of distribution, and his approach strongly influenced biogeographic analysis for the next hundred years. Although Darwin was expressing many ideas that were in the writings of numerous contemporaneous biologists, coming from Darwin they have had added impact: "The view of each species having been produced in one area alone, and having subsequently migrated from that area as far as its powers of migration and subsistence under past and present conditions permitted, is the most probable." This concept of distributional history was clearly linked with his strong opinion that it was highly improbable that "our continents which now stand quite separate have been continuously, or almost continuously united with each other, and with the many existing oceanic islands."

These two ideas—that continents have not moved relative to one another and that organisms, therefore, must originate in one area and disperse to another—have dominated the thinking of most workers since the time of Darwin, and later they were adopted and promoted by three of the most influential biogeographers of the 20th century, Matthew (1915), Simpson (1947, 1965), and Darlington (1957, 1965). The biogeographic hypotheses postulated by these workers to explain vertebrate distribution patterns ruled out interchange of faunas among the southern continents, except perhaps by chance dispersal, and instead called for the origin and differentiation of major groups in the large land masses of North America and Eurasia followed by dispersal to South America, Africa, and Australasia. The influence of continental drift and prior continental ligations, with the notable exception of narrow land bridges, was either denied or assumed to have occured so long ago that they could be discounted.

The publication of Wegener's *The Origin of the Continents and Oceans* (1915) and DuToit's *Our Wandering Continents* (1915) presented an alternative viewpoint: the earth's crust is not stable but dynamic, and through time the continents have shifted their positions and interconnections. Both Wegener and DuToit had short discussions on biotic distribution in which they suggested that continental drift offered an explanation for the disjunct distributions of many plants and animals.

From 1920 until the 1960s, biologists debated whether or not it was necessary to invoke continental drift to explain biogeographic patterns. A critical history of this debate has not yet been written, but it is fair to say that the viewpoints of the stabilists prevailed, at least in the scientific circles of North America. Nevertheless, some biologists continued to insist that the distributions of plants and animals were consistent with continental drift and that additional, unnecessary ad hoc hypotheses had to be proposed if biogeographic explanations were to rely solely on dispersal from one region to another, all upon a stable geography (notes on the history of biogeographic thinking can be found in Nelson, 1973, and Croizat et al., 1974).

An advance in biogeographic theory accompanied a new conceptual and methodological approach to the analysis of phylogenetic relationships introduced by Willi Hennig in the 1950s and 1960s. This method of analysis—termed phylogenetic systematics (see *Systematic Philosophies*)—emphasizes a precise definition of phylogenetic relationship and a rigorous approach to the construction of hypotheses about the phyletic branching sequences of the organisms being studied (Hennig, 1966a,b, Brundin, 1966). From these branching sequences, then, it is possible to reconstruct the distributions of the unknown common ancestors of each lineage and thus postulate a biogeographic history for the group (Nelson, 1969).

In recent years, another theoretical approach has been used in biogeographic analysis that does not place emphasis on reconstructing centers of origin and pathways of dispersal. This approach, termed the *panbiogeographic* or *vicariance model* of biogeography, was first formulated most comprehensively by the botanist Croizat (see particularly 1958, 1962). The principles are very simply stated (Nelson, 1974; Croizat et al., 1974; Cracraft, 1975):

1. The components of a biota evolve together and significant changes in geography or climate must effect parallel influences on the components of that biota.
2. Biogeographic analysis consists of attempting to reconstruct the history of biotas through time; thus, we seek generalizations about the similarities in distributional history of floras and faunas rather than the particulars of individual taxonomic components.
3. Biotic distribution is to be looked upon as the result of subdivision or vicariance of ancestral biotas

rather than as origin in one region and dispersal to another.

4. Evidence for dispersal is seen in the sympatry of individual or group distributions, and thus the problem of dispersal can be considered only after the generalized vicariance patterns have been established.

See Ball (1976) for a further treatment of biogeographic theory).

If an attempt is made to examine the inter-relationships of the higher taxa of vertebrates in terms of their phyletic (cladistic) affinities, repeated patterns are discovered when the distributions of close relatives are plotted on a world map. In many cases, these joint patterns seem to reflect ancestral vertebrate faunas that were fragmented during the breakup and drift of the continents during the Mesozoic and Cenozoic.

Biogeographic Patterns of Fossil Vertebrates

To a great extent the distributions of fossil and living vertebrates can be explained by or are consistent with our current knowledge of paleogeography and paleoclimatology. Exceptions to this statement are primarily the result of (1) inadequate understanding of phyletic relationships, (2) lack of distributional data on fossil groups because of the absence of particular strata at a given locality or the absence of sufficient paleontological sampling, or (3) gaps in our basic knowledge about the configurations of the continents or the chronology of their movements. Nevertheless, we recognize a number of major patterns that provide evidence for ancestral biotas that were once common to two or more previously united landmasses (reviews can be found in Keast, 1972, 1973; Hallam, 1973; Cox, 1974; Cracraft, 1973, 1974, 1975).

The relative positions of the continents prior to the late Paleozoic formation of the supercontinent Pangaea are still uncertain in a number of cases. Likewise, the vertebrate faunas are not sufficiently known to demonstrate convincingly intercontinental biogeographic patterns; and, prior to the upper Permian and Triassic, terrestrial vertebrate faunas for any reasonably restricted time interval are lacking on a world-wide scale (see Cox, 1974, for a review of those patterns that might be present). This situation improves for paleontological samples taken from Mesozoic and Cenozoic sediments, but here too the data are still incomplete. Mesozoic fossils provide several important lines of evidence that support biogeographic patterns recognized for the modern vertebrate faunas (Colbert, 1971, 1973; Cox, 1974). The Triassic tetrapod fauna is cosmopolitan and strongly indicates that Pangaea had not yet dispersed sufficiently to subdivide and isolate this fauna. This finding is compatible with the geological data in that Pangaea apparently did not commence significant breakup until the end of the Triassic (see *Paleogeographic Maps*). Jurassic faunas are very poorly known, but the few that have been studied suggest that faunas in Laurasia and Gondwanaland, the two main components of Pangaea, were still similar. Seemingly, this further reflects the fact that interconnections among the continents were still present. While vertebrate faunas of the Cretaceous continue to show some similarities, further breakup of Laurasia and Gondwanaland and the transgressions of seas upon the continents progressively divided and isolated these faunas. By the end of the Cretaceous, South America was isolated from Africa, India from Antarctica and Africa, and New Zealand from Antarctica. Although North America and Eurasia were still united across the North Atlantic, both were divided by extensive epicontinental seas. Thus, provinciality of the continental vertebrate faunas increased through the Cretaceous.

Modern Vertebrate Distribution and Continental Drift

An analysis of the distributions of living vertebrates reveals a number of well-defined biogeographic patterns that reflect the imprint of Mesozoic and Cenozoic continental breakup. Each intercontinental pattern is composed of groups that are found on each of the continents in question and of groups that are restricted to one of the continents but whose closest relatives are located on the other continent comprising the pattern. Five major biogeographic patterns can be recognized among the continents, representing ancestral faunas that were divided by the breakup of Laurasia or Gondwanaland.

1. **Africa–South America.** The vertebrate links between these two continents are particularly strong. They share at least six families of freshwater fishes in addition to the important order Siluriformes (catfishes), and at least five families of amphibians. Moreover, four families of reptiles in South America have their closest relatives in Africa; within the ratite birds, the South American rheas are closely related to the African ostriches; and finally, monkeys and caviomorph rodents are also shared. These similarities, along with those of the plants and invertebrates, suggest an ancestral biota that was subdivided and later differentiated following continental separation. Present geological evidence indicates an early

Late Cretaceous separation of Africa and South America.

2. **South America–Australia.** Links between these two continents are not as numerous as those between Africa and South America, but they do include a representative sample of vertebrates, thus indicating substantial land connections at one time: lungfishes, osteoglossid fishes, several families of frogs, two families of turtles, ratite ·and galliform birds, and marsupials and monotremes. Again, additional similarities in various plant and invertebrate groups also support the vicariance (subdivision) of a biota that was once continuous from southern South America, across E Antarctica, to Australia. Separation of these continents, and thus isolation of the faunas, seems to have taken place early in the Cenozoic.

3. **South America–New Zealand.** Although the links between these two continents are not strong within vertebrates, several different taxa show relationships indicating a trans-Antarctic pattern: leiopelmatid frogs, ratite birds, penguins, and a group of gruiform birds including the Eurypygidae of South America and the Rhynochetidae of New Caledonia and Aptornithidae of New Zealand. Relationships between South America and New Zealand are strong in some invertebrate and plant groups. New Zealand drifted northward from W Antarctica begining in the Late Cretaceous, thus the ancestral biota was vicariated, split, long ago.

4. **North America–Eurasia.** Many groups of vertebrates are shared and constitute what might be termed a Laurasian faunal element. Final separation of North America and Europe took place in the early Eocene; NW North America and Siberia have been joined repeatedly as sea levels fluctuated during the Cenozoic. Many of the vertebrates probably were once widespread across Laurasia and were isolated after final separation; others appear to have dispersed across the Bering Land Bridge during the Cenozoic.

5. **Tropical Old World.** Many similarities are shown in the tropical vertebrates of Africa and southern Asia (Cracraft, 1973). These faunas were fragmented by the early Cenozoic collison of India with Asia but particularly by the Cenozoic climatic changes that took place in Africa and southwestern Asia. Components of this vicariated fauna include numerous freshwater fishes (e.g., catfishes, perciforms), frogs (bufonids, ranids, microhylids), reptiles (agamids, elapids, viperids), many families of birds and mammals, to name just a few.

Following the breakup of Gondwanaland, continental movements brought faunas of some southern continents in contact with those to the north in Laurasia. Two examples of faunal merging are particularly important. First, South American and North America were connected via southern Central America in the Pliocene after a long period of being broadly separated by ocean (geology and biogeography reviewed in Rosen, 1976; see also Webb, 1976). Northern faunal elements that penetrated into South America included salamanders, possibly scincid and colubrid reptiles, passerine birds, and placental mammals. Southern faunal elements extending northward were fishes of the families Cichlidae and Characidae, catfishes, numerous frogs, iguanid and teiid lizards, suboscine birds, and marsupials. The second example of faunal merging is that provided by the Cenozoic collision of Australia with the southeastern portion of Asia. Numerous asian groups appear to have invaded Australia: ranid frogs; agamid, scincid, and varanid lizards; pythons, colubrid and elapid snakes; probably all oscine birds (songbirds) and many nonpasseriform faimilies, and murid rodents. Australian faunal elements seem not to have penetrated very far north into Asia.

While our understanding of vertebrate biogeography is steadily increasing, some cases remain enigmatic: a prime example is the fauna of Madagascar. Recent investigators have concluded that Madagascar has remained in the same position relative to Africa since the beginning of the Mesozoic. This finding is in contrast to previous opinions of many geologists and biologists who have argued that Madagascar drifted to its present position. Indeed, the biological evidence would seem to suggest a Cretaceous or early Cenozoic connection to Africa. Large dinosaurs are known from the Late Cretaceous; aepyornithid (elephant-birds) ratites were once present on both Africa and Madagascar; there are some similarities in the frog faunas; and the unique mammal fauna is said to show links to Africa (also possibly India). On the other hand, Madagascar lacks many of the groups that are characteristic of the African fauna, particularly in freshwater fishes, amphibians, and reptiles. If Madagascar had been united to Africa at one time, we might expect greater faunal similarities; and yet, if Madagascar has always remained isolated, it is difficult to believe that each of the faunal elements arrived by rafting or some other mode of chance dispersal. Hopefully, new geological data will be found to help solve this problem.

JOEL CRACRAFT

References

Ball, I. R., 1976. Nature and formulation of biogeographic hypotheses, *Systematic Zool.* 24, 407–430.
Brundin, L., 1966. Transantarctic relationships and

their significance, as evidenced by chironomid midges, *Kungl. Svenska Vetan Handl.*, **11**, 1–472.

Colbert, E. H., 1971. Tetrapods and continents, *Quart. Rev. Biol.*, **46**, 250–269.

Colbert, E. H., 1973. Continental drift and the distributions of fossil reptiles, in Tarling and Runcorn, 1973, 395–412.

Cox, C. B., 1974. Vertebrate palaeodistributional patterns and continental drift, *J. Biogeography*, **1**, 75–94.

Cracraft, J., 1973. Vertebrate evolution and biogeography in the Old World tropics: Implications of continental drift and palaeoclimatology, in Tarling and Runcorn, 1973, 373–393.

Cracraft, J., 1974. Continental drift and vertebrate distribution, *Ann. Rev. Ecol. Systematics*, **5**, 215–261.

Cracraft, J., 1975. Historical biogeography and earth history: Perspectives for a future synthesis, *Ann. Missouri Bot. Gard.*, **62**, 227–250.

Croizat, L., 1958. *Panbiogeography*. Caracas: Published by author, 3 vols.

Croizat, L., 1962. *Space, Time, Form: The Biological Synthesis*. Caracas: Published by author.

Croizat, L.; Nelson, G. J.; and Rosen, D. E., 1974. Centers of origin and related concepts, *Systematic Zool.* **23**, 265–287.

Darlington, P. J., Jr. 1957. *Zoogeography: The Geographical Distribution of Animals*. New York: Wiley, 675p.

Darlington, P. J., Jr., 1965. *Biogeography of the Southern End of the World*. Cambridge, Mass.: Harvard Univ. Press, 236p.

DuToit, A. L. 1937. *Our Wandering Continents*. London: Oliver and Boyd, 366p.

Hallam, A., ed., 1973. *Atlas of Palaeobiogeography*. Amsterdam: Elsevier, 531p.

Hennig, W. 1966a. The Diptera fauna of New Zealand as a problem in systematics and zoogeography, *Pacific Insects Monogr.*, **9**, 1–81.

Hennig, W., 1966b. *Phylogenetic Systematics*. Urbana: Univ. Illinois Press, 263p.

Keast, A., 1972. Continental drift and the evolution of the biota on southern continents, in A. Keast; F. C. Erk; and B. Glass, eds. *Evolution, Mammals, and Southern Continents*. Albany: State Univ. New York Press, 23–87.

Keast, A., 1973. Contemporary biotas and the separation sequence of the southern continents, in Tarling and Runcorn, 1973, 309–343.

Matthew, W. D., 1915. Climate and evolution, *Ann. New York Acad. Sci.*, **24**, 171–318.

Nelson, G. J., 1969. The problem of historical biogeography, *Systematic Zool.*, **18**, 243–246.

Nelson, G. J., 1973. Comments on Leon Croizat's biogeography, *Systematic Zool.*, **22**, 312–320.

Nelson, G. J., 1974. Historical biogeography: An alternative formalization, *Systematic Zool.*, **23**, 555–558.

Rosen, D. E., 1976. A vicariance model of Caribbean biogeography, *Systematic Zool.*, **24**, 431–464.

Simpson, G. G., 1947. Holarctic mammalian faunas and continental relationships during the Cenozoic, *Geol. Soc. Amer. Bull.*, **58**, 613–688.

Simpson, G. G., 1965. *The Geography of Evolution.. Collected Essays*. Philadelphia: Chilton Books, 249p.

Tarling, D. H., and Runcorn, S. K., eds., 1973. *Implications of Continental Drift to the Earth Sciences*, vol. 1. New York: Academic Press, 309–343, 373–412.

Webb, S. C., 1976. Mammalian faunal dynamics of the great American interchange, *Paleobiology*, **2**, 220–234.

Wegener, A. 1915 (1966). *The Origin of Continents and Oceans*. New York: Dover, 246p.

Cross-references: *Paleogeographic Maps; Systematic Philosophies.*

PALEOBOTANY

Paleobotany is the study of plant remains preserved in the geologic record. Traditionally the study involves plant megafossils (leaves, cuticles, fruits, seeds, stems, roots) in contrast to pollen and spores (see *Palynology*). These megafossils are usually encountered as four types of preservation (see Schopf, 1975). The *impression* is an imprint of the organism or its parts showing size, shape, and general features of external morphology. The *compression* consists of organic remains flattened between strata. By suitable techniques it is possible to remove and clear compressed material such as leaves and fronds, revealing cellular details of the epidermis, stomatal apparatus, venation, trichomes, glands, indusia, sporangia, and other features. The *mold and cast* is formed when plant parts—petioles, stems, roots, cones, and seeds—are covered by sediments, and subsequently removed by microbial action creating a cavity or mold. The mold becomes filled with sediment and, if indurated, a cast is formed showing three-dimensional size and shape and reverse images of surface detail. The *petrifaction* forms when plant structures such as stems, roots, cones, seeds, petioles, and foliage become impregnated with carbonate or silicate minerals. If crystal size remains small and infiltration is slow and complete, considerable cellular detail can be preserved (Taylor and Millay, 1977). Petrified material may be mounted and ground into thin sections or prepared by the peel technique. The latter consists of immersing a section of petrified plant remains in acid capable of dissolving the matrix (HF for silicates, HCL for carbonates). The cellulose is unaffected and will emerge in microrelief. The surface is covered with acetone and a layer of acetate sheet. The acetone softens the acetate and slowly plant material is embedded in the partially dissolved sheet. After drying, the acetate may be peeled away providing a section revealing histological detail.

In recent years, the traditional study of plant megafossils has been supplemented by new techniques and emergence of new philosophies concerning the role of paleobotany in understanding the origin and development of the earth's vegetation. In particular, the availability of absolute dating techniques has refined the stratigraphy of fossil-bearing deposits, allowing for more critical study of evolutionary trends, development of major vegetation types, climatic changes, migration of floras, and estimates of absolute rates of evolution.

Techniques of organic geochemistry have recently been applied to problems of paleobotany, particularly those concerned with the Precambrian (J. W. Schopf et al., 1968). Certain organic compounds, e.g., amino acids, porphyrins, and sugars, may be preserved in sediments. Techniques are available for recovery and these substances may be identified by usual methods of organic chemistry. This affords a procedure not only for verifying the biological origin of fossils of questionable affinity, but for detecting the existence of living organisms in the absence of structurally preserved fossils. The technique is being applied to both terrestrial and extraterrestrial materials; for application to later periods, see Niklas and Gensel (1977), and Niklas and Giannasi (1977).

Paleopalynology is considered a discipline separate from paleobotany but there is considerable interaction between the two fields. Precambrian paleobotany involves palynological techniques; and study of the comparatively sparse megafossil floras of the Mississippian, Permian, Triassic, and Jurassic are supplemented by microfossil assemblages. The principle contributions of palynology to traditional paleobotanical investigations are in age assignment and correlation of megafossil-bearing strata; confirmation of identifications (especially in Cenozoic deposits) by recovery of microfossils belonging to the same taxon; providing suggestions for identification of unknown megafossils; providing information on past vegetation in regions, portion of a section, or rock types where megafossils are absent; and in amplifying the species list on which most interpretations are based. Evidence from plate tectonics, morphology of modern plants, and application and development of new techniques (e.g., SEM; permineralization studies, Oehler and Schopf, 1971) are commonly integrated into paleobotanical investigations. A trend in modern paleobotanical research is to focus attention on particular problems (e.g., origin of gymnosperms) and use a multiplicity of approaches to generate relevant data (e.g., Beck, 1960, 1970; Namboodiri and Beck, 1968). In these studies, distinctions between traditionally separate disciplines are not especially significant.

Stratigraphic Distribution of Fossil Plants

Precambrian. The oldest known structurally preserved fossil plants are from the Precambrian Fig Tree formation of South Africa (ca. 3.2 b yr, Engel et al., 1968; 3.1 b yr, J. W. Schopf and Barghoorn, 1967), the Gunflint Chert exposed along the N shore of Lake Superior (ca. 2 b yr, Tyler and Barghoorn, 1954; Barghoorn and Tyler, 1965), and the Bitter Springs formation of central Australia (ca. 1 b yr, J. W. Schopf, 1968, 1972). These fossils resemble fungal spores and hyphae, unicellular, colonial and filamentous blue-green algae, and numerous forms of unknown biological affinity. Among the principal genera are *Eobacterium, Archaeosphaeroides, Huronispora, Entosphaeroides, Eoastrion, Kakabekia, Eosphaera, Biocatenoides,* and *Eomycetopsis.* The collective assemblages reveal that self-replicating biological systems originated between 4 and 3 b yr ago; transition from anoxic to oxygen-rich environments sufficient for algal photosynthesis occurred about 3 b yr ago; and development of the eucaryotic organization estimated at about 1.3 b yr. By the close of the Precambrian, chlorophycean (green algal) forms developed that probably served as progenitors to land vascular plants present by the close of the Silurian (J. W. Schopf, 1970).

Paleozoic. The oldest known vascular plant is *Cooksonia* from the upper Silurian Downtonian of Wales. The plant consists of a slender dichotomously branched stem, without lateral appendages, and bearing terminal sporangia containing trilete cutinized spores. Other presumed vascular plants of the Silurian and Lower Devonian are *Zosterophyllum, Taeniocrada, Hostimella,* and *Steganotheca* (Banks, 1972; Andrews, 1974). Associated nonvascular, algal-like forms are represented by *Prototaxites, Nematothallus, Pachytheca,* and *Parka.* The genus *Eohostimella* is interpreted by J. M. Schopf et al. (1966) as an erect, nonvascular, land plant. Clearly the impression from these floras is an assemblage of marsh-inhabiting organisms with structural features transitional between land vascular and aquatic nonvascular plants.

Above the basal Devonian, during Siegenian times, early forms of lycopods appear (*Protolepidodendron, Drepanophycus, Baragwanathia*). The psilopsids include *Dawsonites, Psilophyton* (Hueber, 1967, 1971; Hueber and Banks, 1967; Gensel, 1977), *Hedeia,* and *Yarravia.* Later, during the Emsian, the sphenopsids evolved, along with *Pseudosporochnus, Protopteridium, Tomiphyton, Rhynia* (Satterthwait and Schopf, 1972), *Horneophyton, Asteroxylon, Kaulangiophyton,* and plants of uncertain affinities including *Platyphyllum* and *Enigmophyton*.

In the upper Devonian, *Callixylon* appears, often associated with *Archaeopteris*. The diagnostic feature of *Callixylon* is the radial clusters of pits on the tracheid walls, and on the basis of anatomical characters it has been placed in the gymnosperms. *Archaeopteris* was regarded as a fern because of its large frond-like leaves and cryptogamic (nonseed) reproduction. Some species were heterosporous (Arnold, 1939). In 1960, Beck reported the connection of *Callixylon* and *Archaeopteris* and proposed a new class, the Progymnospermopsida, to include these and other ancient gymnospermous plants. According to Beck (1960), these plants likely "comprise the ancestral complex from which the major groups of gymnosperms evolved. Certain primitive features, especially of the Aneurophytales, suggest that the Progymnospermopsida are descended directly from some psilophyte-like ancestors."

Thus among major evolutionary events of the Devonian were development of a diverse and extensive land vegetation; establishment of early lines of psilopsid, lycopsid, sphenopsid, filicinean, and gymnosperm evolution; development of the megaphyllous leaf; and appearance of heterospory, an early stage in development of the seed.

Among plant fossils characteristic of late Paleozoic times (Mississippian and Pennsylvanian) are structures resembling fronds and pinnules of ferns. As early as 1883, D. R. J. Stur noted the sterile condition of most specimens and suggested they many not be ferns. In 1903, F. Oliver and D. H. Scott "announced their epoch-making discovery of the identity of the seed *Lagenostoma lomaxi* and the stem *Lyginopteris oldhamia* and proposed the name Pteridospermae for those Cycadofilices that bore seeds" (Arnold, 1947). The seed ferns, now included in the gymnosperms, persisted into Triassic times as a highly evolved group called the Caytoniales. Other important members of the Carboniferous coal-swamp vegatation were *Lepidodendron, Sigillaria, Calamites*, the Sphenophyllales, Coenopteridales, and Cordaitales. Miller and Brown (1973) have recently reported embryos, previously unknown in Paleozoic seeds, from conifer ovules in the Lower Permian of Texas.

A significant change in vegetation took place during the transition between the Permian and Triassic. Major groups of plants, such as the Lepidodendrales, Calamitales, Sphenophyllales, Coenopteridales, and Cordaitales became extinct; and new forms appeared (Lycopodiales,

Equisetales, Filicales, Cycadales, Cycadeoidales, Ginkgoales, and Coniferales). Climatic and orogenic events were important in the replacement of Carboniferous coal-swamp vegetation by more xeric-adapted gymnosperms of the Triassic and Jurassic, as evidenced by numerous glacial and evaporitic deposits of Permian age. The Gondwana flora, characterized by *Glossopteris*, was widespread in the southern hemisphere during Permian and early Mesozoic times (J. M. Schopf, 1970; Plumstead, 1958).

Mesozoic. The dominant elements in Triassic and Jurassic vegetation were the gymnosperms, which were capable of surviving under the arid conditions of lower Mesozoic times. The flora of the upper Triassic Santa Clara formation of Sonora, Mexico includes *Ctenophyllum, Taeniopteris, Pterophyllum, Zamites, Asterocarpus, Thaumatopteris, Mertensides*, and *Cladophlebis*. Among genera of the Triassic flora of Arizona are *Lonchopteris, Neocalamites, Dadoxylon, Baiera, Araucarioxylon*, and *Woodworthia*. The cycadeoides, earlier considered progenitors of the angiosperms, have been studied by Delevoryas (1963, 1968) and are regarded as a separate line of evolution. Triassic megafossil floras are not common, and recent studies, by Delevoryas (1970; Delevoryas and Hope, 1971) on deposits in North Carolina are important contributions to knowledge of early Mesozoic vegetation.

The Jurassic floras of Yorkshire (Harris, 1961–1969) are among the most extensive known, and generally represent a continuation of Upper Triassic vegetation. Among the common plants are *Neocalamites, Todites, Phlebopteris, Coniopteris, Sagenopteris, Caytonanthus, Caytonia, Nilssonia, Pseudoctenis, Ctenzamites, Ctenis, Stenopteris, Androstrobus*, and *Beania*. Delevoryas (1969; Delevoryas and Gould, 1971) has recently initiated studies on a Jurassic flora from Oaxaca, Mexico. The Jurassic is of special paleobotanical interest because of numerous reports of pre-Cretaceous angiosperms in that Period (see *Angiospermae*). Although it may seem logical to expect these plants in the Jurassic, there is no convincing evidence for angiosperms older than Cretaceous (Scott et al., 1960, 1972). By the end of the Cretaceous, however, they had displaced much of the lower and middle Mesozoic gymnosperm vegetation and had become the dominant plant group. Recently, studies have been made on Cretaceous floras from Coahuila, Mexico (Weber, 1972, 1973), further documenting the widespread distribution of aniosperms by that time.

Cenozoic. Paleobotanical investigations of the Cenozoic are generally concerned with modernization of vegetation at the generic level, migration of floras, pathways of distribution, climatic change, and the impact of fluctuating physical conditions on vegetation (e.g., emergence of the Isthmian region between North and South America; plate tectonics). Dilcher (1973) and Wolfe (1971, 1972) have discussed new methods for more critical identification of leaf remains and environmental reconstruction; Hickey (1973) has recently provided a morphological classification of leaf forms; and Walker (1971) and others have clarified taxonomic relationships between primitive aniosperm taxa critical in the early evolution of aniosperms. Consequently, considerable new information is being generated on the fossil record of various angiosperm groups and environmental conditions under which the floras developed. From these studies, it appears that during the Paleocene and Eocene, climatic conditions presently typical of equatorial latitudes extended further N and S, as evidenced by tropical elements in the Wilcox and associated floras of the Mississippi Embayment region (Dilcher, 1969; Dilcher and Dolph, 1970) and in the Eocene London Clay flora (Chandler, 1961–1964). Tropical to warm-temperate conditions continued to prevail until about the middle Oligocene, when a trend toward cool-temperate conditions becomes evident. By the middle Miocene, a broad-leaved deciduous forest occupied much of the middle-altitude, mid-latitude northern hemisphere (Chaney and Axelrod, 1959; Ferguson, 1971). Toward the end of the Miocene, rise of the Cascades, Coast Ranges, northern Andes and other mountain systems created arid conditions to the lee and stimulated development of modern desert vegetation (Axelrod, 1958). The trend toward cool-temperate conditions eventually culminated in the Pleistocene glaciations, and during this period final modernization of floras occurred. The latter record is most extensively preserved as fossil pollen and spores and is treated in the literature on palynology (q.v.).

ALAN GRAHAM

References

Andrews, H. N., 1974. Paleobotany, 1947-1972, *Ann. Missouri Bot. Gard., 61,* 179-202.

Arnold, C. A., 1939. Observations of fossil plants from the Devonian of eastern North America. IV. Plant remains from the Catskill Delta deposits of northern Pennsylvania and southern New York, *Contrib. Mus. Paleont. Univ. Mich., 5,* 271-314.

Arnold, C. A., 1947. *An Introduction to Paleobotany.* New York: McGraw-Hill, 433p.

Axelrod, D. I., 1958. Evolution of the Madro-Tertiary Geoflora, *Bot. Rev., 24,* 433-509.

Banks, H. P., 1972. The stratigraphic occurrence of early land plants, *Palaeontology,* 15, 365-377.

Barghoorn, E. S., and Tyler, S. A., 1965. Microorganisms from the Gunflint chert, *Science,* 147, 563-577.

Beck, C. B., 1960. Connection between *Archaeopteris* and *Callixylon, Science,* 131, 1524-1525.

Beck, C. B., 1970. The appearance of gymnospermous structure, *Biol. Rev. Cambridge Phil. Soc.* 45, 379-400.

Chandler, M. E. J., 1961-1964. *The Lower Tertiary Floras of Southern England,* 4 Vols. London: British Mus. (Nat. Hist.).

Chaney, R. W., and Axelrod, D. I., 1959. Miocene floras of the Columbia Plateau *Carnegie Inst. Wash. Publ. 617,* 237p.

Delevoryas, T., 1963. Investigations of North American Cycadeoids: Cones of Cycadeoidea, *Amer. J. Bot.,* 50, 45-52.

Delevoryas, T., 1968. Some aspects of cycadeoid evolution, *J. Linnean Soc. (Bot.),* 61, 137-146.

Delevoryas, T., 1969. Glossopterid leaves from the middle Jurassic of Oaxaca, Mexico, *Science,* 165, 895-896.

Delevoryas, T., 1970. Plant life in the Triassic of North Carolina, *Discovery,* 6, 15-22.

Delevoryas, T., and Gould, R. E., 1971. An unusual fossil fructification from the Jurassic of Oaxaca, Mexico, *Amer. J. Bot.,* 58, 616-620.

Delevoryas, T., and Hope, R. C., 1971. A new Triassic cycad and its phyletic implications, *Postilla,* 150, 1-21.

Dilcher, D. L., 1969. *Podocarpus* from the Eocene of North America, *Science,* 164, 299-301.

Dilcher, D. L., 1973. A paleoclimatic interpretation of the Eocene floras of southeastern North America, in A. Graham, ed., *Vegetation and Vegetational History of Northern Latin America.* Amsterdam: Elsevier, 39-59.

Dilcher, D. L., and Dolph, G. E., 1970. Fossil leaves of *Dendropanax* from Eocene sediments of southeastern North America, *Amer. J. Bot.,* 57, 153-160.

Engel, A. E. J.; Engel, C. C.; Kremp, G. O. W.; and Drew, C. M., 1968. Alga-like forms in Onverwacht Series, South Africa: Oldest recognized lifelike forms on Earth, *Science,* 161, 1005-1008.

Ferguson, D. K., 1971. *The Miocene Flora of Kreuzau, Western Germany.* 1. The leaf-remains. Amsterdam: North-Holland, 297p.

Gensel, P. G., 1977. Morphologic and taxonomic relationships of the Psilotaceae relative to evolutionary lines in early land vascular plants, *Brittonia,* 29, 14-29.

Harris, T. M., 1961-1969. *The Yorkshire Jurassic Flora.* 3 Vols. London: British Mus. (Nat. Hist.).

Hickey, L. J., 1973. Classification of the architecture of dicotyledonous leaves, *Amer. J. Bot.,* 60, 17-33.

Hueber, F. M., 1967. *Psilophyton:* The genus and the Concept, in D. H. Oswald, ed., *International Symposium on the Devonian System, Vol. 2.* Calgary: Alberta Soc. Petrol. Geol., 815-822.

Hueber, F. M., 1971. *Sawdonia ornata:* A new name for *Psilophyton princeps* var. *ornatum, Taxon,* 20, 641-652.

Hueber, F. M., and Banks, H. P., 1967. *Psilophyton*

princeps: The search for organic connection, *Taxon,* **16**, 81-160.

Miller, C. N., and Brown, J. T., 1973. Paleozoic seeds with embryos, *Science,* **179**, 184-185.

Namboodiri, K. K., and Beck, C. B., 1968. A comparative study of the primary vascular system of conifers, *Amer. J. Bot.,* **55**, 447-472.

Niklas, K. J., and Gensel, P. G., 1977. Chemotaxonomy of some Paleozoic vascular plants. II. Chemical characterization of major plant groups, *Brittonia,* **29**, 100-111.

Niklas, K. J., and Giannasi, D. E., 1977. Flavonoids and other chemical constituents of fossil Miocene *Zelkova* (Ulmaceae), *Science,* **196**, 877-878.

Oehler, J. H., and Schopf, J. W., 1971. Artificial microfossils: Experimental studies of permineralization of blue-green algae in silica, *Science,* **174**, 1229-1231.

Plumstead, E. P., 1958. Further fructifications of the Glossopteridae and a provisional classification based on them, *Trans. Geol. Soc. S. Africa,* **61**, 51-76.

Satterthwait, D. F., and Schopf, J. W., 1972. Structurally preserved phloem zone tissue in *Rhynia, Amer. J. Bot.,* **59**, 373-376.

Schopf, J. M., 1970. Petrified peat from a Permian coal bed in Antarctica, *Science,* **169**, 274-277.

Schopf, J. M., 1975. Modes of fossil preservation, *Rev. Palaeobot. Palynol.,* **20**, 27-53.

Schopf, J. M.; Mencher, E.; Boucot, A. J.; and Andrews, H. N., 1966. Erect plants in the early Silurian of Maine, *U.S. Geol. Surv. Prof. Paper 550D,* 69-75.

Schopf, J. W., 1968. Microflora of the Bitter Springs formation, late Precambrian, central Australia, *J. Paleontology,* **42**, 651-688.

Schopf, J. W., 1970. Precambrian micro-organisms and evolutionary events prior to the origin of vascular plants, *Biol. Rev. Cambridge Phil. Soc.,* **45**, 319-352.

Schopf, J. W., 1972. Evolutionary significance of the Bitter Springs (late Precambrian) microflora, *24th Internat. Geol. Cong. Proc.,* sec. 1, 68-77.

Schopf, J. W., and Barghoorn, E. S., 1967. Alga-like fossils from the early Precambrian of South Africa, *Science,* **156**, 508-512.

Schopf, J. W.; Kvenvolden, K. A.; and Barghoorn, E. S., 1968. Amino acids in Precambrian sediments: An assay, *Proc. Nat. Acad. Sci.,* **59**, 639-646.

Scott, R. A.; Barghoorn, E. S., and Leopold, E. B., 1960. How old are the angiosperms? *Amer. J. Sci.,* ser. 5, **258**, 284-299.

Scott, R. A.; Williams, P. L.; Craig, L. C.; Barghoorn, E. S.; Hickey, L. J.; and MacGintie, H. D., 1972. "Pre-Cretaceous" angiosperms from Utah: Evidence for Tertiary age of the palm woods and roots, *Amer. J. Bot.,* **59**, 886-896.

Taylor, T. N., and Millay, M. A., 1977. Structurally preserved fossil cell contents, *Trans. Amer. Micr. Soc.,* **96**, 390-393.

Tyler, S. A., and Barghoorn, E. S., 1954. Occurrence of structurally preserved plants in Pre-Cambrian rocks of the Canadian shield, *Science,* **119**, 606-608.

Walker, J., 1971. Pollen morphology, phytogeography, and phylogeny of the Annonaceae, *Contrib. Gray Herb. No. 202,* 130p.

Weber, R., 1972. La vegetacion Maestrichtiana de la formacion Olmos de Coahuila, Mexico, *Bol. Soc. Geol. Mexicana,* **33**, 5-19.

Weber, R., 1973. *"Salvinia coahuilensis"* Nov. Sp. del Cretacico superior de Mexico, *Ameghiniana,* **10**, 173-190.

Wolfe, J. A., 1971. Tertiary climatic fluctuations and methods of analysis of Teritary floras, *Palaeogeography, Palaeoclimatology, Palaeoecology,* **9**, 27-57.

Wolfe, J. A., 1972. An interpretation of Alaskan Tertiary floras, in A. Graham, ed. *Floristics and Paleofloristics of Asia and Eastern North America.* Amsterdam: Elsevier, 201-233.

Cross-references: *Algae; Angiospermae; Bryophyta; Charophyta; Gymnosperms; Paleoecology, Terrestrial; Paleophytogeography; Palynology; Palynology, Paleozoic and Mesozoic; Palynology, Tertiary; Plankton; Pteridophyta.* Vol. VI: *Coal.*

PALEOECOLOGY

Paleoecology is the study of the relationships between fossils and the ancient environments in which the fossilized organisms lived. In view of the inadequacies of the fossil record and the difficulties of distinguishing between the environment in which a fossil lived and that in which it was buried, it is not, however, desirable to impose too rigid a definition on the subject. The present author has expressed it more simply as "the study of how and where animals and plants lived in the past." Attention must also be given to related topics, such as the way in which the remains of animals and plants are disturbed and distributed after death before being incorporated in the sediment and tranformed into fossils (see *Taphonomy*). What is more, in this field of research, the paleontologist is closely in contact with the sedimentologist, who is equally concerned with the problems of environment. In certain circumstances, for example, in the development of reefs that may form petroleum traps (Fig. 1.), paleoecological studies may be of great economic value.

A further complication is that the study of paleoecology may operate in two opposite directions: The presumed sedimentary environment may be used to make deductions about the habits and habitat of a particular fossil, or the presumed habits and habitat of the fossil may be used to make deductions about the nature of the environment. The dangers of cyclic reasoning are obvious.

The principles of paleoecology are essentially the same as those of ecology, but compared with the ecologist the paleoecologist studies his subject with his eyes blindfolded and both hands tied behind his back. In that the paleo-

FIGURE 1. Silurian coral *Favosites* with corallites filled with tar-like substance, significant because of the common association of oil accumulations and reefs, Thornton Quarry, near Chicago, Illinois, X1.

ecologist is trying to do the same thing as the ecologist, great use is made of ecological studies of the present world in interpreting that of the past. This is dangerous however, since a narrow uniformitarian approach may have many pitfalls. Quite obviously, genera may have changed their habits (e.g., the bivalve mollusk *Pholadomya*, which is a deep-sea form at the present day was certainly an inhabitant of shallow water during the Mesozoic Era). What is more, many of the forms the paleoecologist must handle have no direct relations of any kind living today. Even the physicochemical conditions of various environments may have changed; we cannot be sure, for instance, of the composition of the atmosphere or the salinity of the sea in the distant past. In practice, unless he is dealing with Cenozoic sediments, the paleoecologist can rarely make direct comparisons with the present day.

Paleoecology may be subdivided into two fields: *paleoautecology*, which is concerned with the ecology of individual fossil species or groups; and *paleosynecology*, which is concerned with the ecology of fossil communities. These fields may be investigated in a number of different ways, either directly, by first-hand observation of fossils and the way they occur in the rocks, or indirectly, by analogy with related living forms and living communities.

It would be impossible to consider systematically all the varied environments that have ever existed through earth history, even if they had been adequately studied; and the ecology of individual fossil groups is best considered under the taxa concerned. Here, the writer will

follow the methodological approach used in his book on the subject (Ager, 1963).

Morphology

The morphology of a fossil may tell us something of its ecology in three ways: (1) from the known physicochemical processes involved in its formation, (2) by comparison with living forms; and (3) by deductions about its form and function.

Physicochemical Processes. On the broadest scale, the physicochemical processes involved in the formation of a particular fossil may be ecologically significant by reason of its gross mineralogy. Thus an aragonitic shell would dissolve in sea water (under present-day atmospheric pressures) at temperatures lower than $5°C$. At an elemental level, the ratio of strontium to calcium in a fossil shell may be a clear indication of the original salinity in which it lived. But it is in the proportion of different isotopes of the same element in a fossil that we have the most promising source of ecological information in recent years. Thus from the ratio between the isotopes O^{16} and O^{18} in a fossil it is theoretically possible to deduce the temperature at which that organism lived, though the method cannot yet be regarded as completely reliable since the differences in ratio are so small compared with the range of analytical error.

Comparative Morphology. The comparative morphology of fossil and living forms is the most obvious method of deducing the habits and habitat of the former. This may either be by using morphology to deduce relationships and thence to deduce ecology, or by the recognition of analogous morphological features. Thus a fossil bird bearing a close resemblance to modern ducks was probably related to them and probably lived in the same way; on the other hand the duck-billed dinosaurs were certainly not related to the ducks but may have fed in the same way. The limitations of this approach are obvious.

Deductions About Form and Function. The use of morphological features to make deductions about form and function is even more fraught with difficulties and dangers. This applies particularly to completely extinct stocks where no modern comparisons are possible. Nevertheless, the logical, and (if possible) experimental, examination of the possible function of particular features of a fossil may, in certain circumstances, throw strong weight behind one supposed function rather than another.

Thus two entirely different interpretations have been placed on the lateral processes at the apertural end of fossil hyolithids. The matter is not yet settled, although attempted reconstructions have suggested that one relationship of form and function is more likely than the other. Much use has recently been made of the principle of the paradigm, whereby one argues that the probable function of a given structure was that which it could have performed most efficiently. Thus the spines on an extinct brachiopod may have been for protection, for sensory perception, for hydrodynamic stability, or for sieving the food supply. It would seem to be a fairly simple matter to test the efficiency of the spines on a particular brachiopod for each of these functions, though surprisingly enough, different investigators still come to different conclusions. (See Morphology, Constructional; Morphology, Functional).

Orientation

There are three possible types of orientation that may be observed in fossils: (1) life orientations, (2) death orientations, (3) postmortem orientations.

Life Orientations. Life orientations are those in which the fossil maintains the attitude it held in life. A preponderance of a certain orientation in a particular fossil may tell us something of the life habits of that organism and may also tell us that the fossil assemblage as a whole has been comparatively little disturbed after death. Common examples of such life orientations are hemispherical compound corals or calcareous algae still resting on their flat sides, and burrowing bivalves with their hinge lines slightly inclined to the vertical and their posterior gape uppermost (Fig. 2). A clear

FIGURE 2. Burrowing bivalve *Pleuromya* fossilized in life position in a Jurassic limestone. Dunraven, Wales, X0.55.

dominance of such attitudes must be observed before it can be assumed that the organism is really unmoved, and even then there are possible errors due to the probability that life orientations (as in the case of the hemispherical coral) may also be the position of greatest mechanical stability and that, therefore, the specimen may have been transported for some distance after death.

Death Orientations. Death orientations are those in which the fossil is found in the same position as that in which it fell to the ground or the sea floor on death. Such orientations apply particularly to pelagic organisms, which could not possibly become fossilized in life position, but which must be regarded as equally in situ if a death orientation can be distinguished from a mechanical one. An obvious example of such an orientation is the ammonite in the Solenhofen Limestone (Upper Jurassic) of southern Germany which shows the impression made by the venter when the shell landed on the sea floor and before it toppled on to one side.

Postmortem Orientations. Postmortem orientations, which include the vast majority of fossil occurrences, show merely the orientation in which the organisms happened to come to rest after their dead bodies were moved by currents or other mechanical (or biological) means. Almost all fossils, with the exception of burrowing forms and those which in life were firmly cemented in one place, are likely to have been moved after death to a greater or lesser degree. The paleoecologist must look for evidence of such movement before deciding whether a given fossil assemblage was more or less indigenous.

Elongated fossils—such as belemnites and many turreted gastropods—very commonly show a parallel orientation due to current action, and planoconvex ones—such as certain bivalves—are often found in their position of greatest mechanical stability.

The disarticulation of bivalve shells may also be considered here (though not strictly an aspect of orientation). This provides an easily observed measure of the degree and severity of transport. The degree of wear or comminution may also be a measure of distance of transportation, though it is by no means an infallible guide. This may have more ecological significance than is immediately obvious. Thus one is sometimes able to trace a progression in the field from finely comminuted shell material to whole but separated valves and thence to whole and articulated valves. This clearly indicates the direction toward the environment where the bivalve mollusk or brachiopod concerned actually lived.

Activity

One of the branches of paleoecology that has developed most rapidly in recent years is the study of the activity of past animals by means of *trace fossils* (q.v.). These are the marks such as tracks, trails, burrows, and borings made by animals in sediment. A whole new subject called *paleoichnology* has grown up around trace fossils, with a wealth of unnecessary jargon. These traces appear to have little or no stratigraphical value (though claims have been made for them); their great value is their ecological and environmental significance. Here, at least, we have something that tells us about living creatures rather than dead bodies; and we have fossils that cannot possibly have been derived or transported from a different environment unless in obviously derived blocks. Distinctive groups of trace fossils can be recognized in different sedimentary facies and appear to have persisted in the same environment for the greater part of Phanerozoic time.

The actual traces themselves may reveal a great deal about the mode of life of the animals concerned (e.g., Fig. 3). In a sense this is ethology (the study of habits) rather than ecology; but, clearly, how a trilobite progressed and fed is essential evidence if we are to understand its relationship to its environment.

Unfortunately the vast majority of trace fossils give us no evidence of the identity of the creature that made them. This is especially true of the great diversity of burrowing structures characteristic of shallow-water (especially littoral) sediments. Even at the present day, it is often very difficult to find out which animal made a particular trail on the beach or sea floor. In the rocks, it is far more difficult and it is only the rare miracle of preservation that shows us the limulid dead at the end of a particular trail or a trilobite still resting in the hollow it scooped out for itself millions of years ago.

A second great disadvantage of trace fossils is that we know that a single animal may leave different kinds of traces. Thus a modern *Littorina* leaves a straight trail when it is heading for the sea, but a rambling one when feeding. Similarly, a resting trilobite leaves a trace fossil called *Rusophycus*, while a moving trilobite leaves a quite different one called *Cruziana*. A trace fossil may also differ according to the nature of the sediment being traversed. Thus callianassid crabs in fine-grained sediment produce the smooth Y-shaped burrows called *Thalassinoides*, while the same creature in coarse-grained sediment tends to produce the tuberculate burrow *Ophiomorpha*.

Conversely, different genera and even different phyla, may produce remarkably similar trace fossils.

Seilacher (1964) recognized three main groups of trace fossils related to bathymetric facies. These were (1) the *Nereites* facies, as in the flysch, with a dominance of browsing trails made by deposit feeders; (2) an intermediate *Zoophycos* facies; and (3) a shallow-water *Cruziana* facies characterized by a great variety of burrowing and other forms. He also recognized a group of what he called "facies breaking types," of which the best-known are the radiating burrows of *Chondrites*, which turn up in a great variety of sedimentary facies.

FIGURE 3. Track left by a straight-shelled nautilioid touching down on the sea floor while feeding. This is an example of a trace fossil providing evidence of the mode of life of a long extinct group of fossils, Carboniferous, Ireland.

This bathymetric distribution of "ichno-facies" seems to work pretty well in sediments of all ages except that the intermediate form *Zoophycos* certainly ranges from shallow-water deltaic sands to deep-water geosynclinal deposits, at least in Paleozoic sediments.

More recently it has proved possible to subdivide further the shallow-water assemblage. Thus within the shallowest-water assemblages of the Jurassic in N France, Ager and Wallace (1970) recognized a succession from deeper, calmer water with the horizontal U-shaped burrow known as *Rhizocorallium* through shallower, higher energy conditions with *Thalassinoides* into intertidal conditions with the vertical U-burrow *Diplocraterion*. But it is important to emphasize that depth is not the only controlling factor, since the deeper-water forms are also found in what seems to have been quiet, protected conditions in shallower water (Fig. 4). Food supply is also a very important consideration. Browsing trails are dominant in deep-water conditions simply because in that environment it is no use waiting for the food to come to you, as do the suspension feeders, it is necessary to go and look for it, as do the detritus feeders.

The actual nature of the sea floor does not seem to be very important, except in special circumstances (e.g., Fig. 5). Similar trace fossils occur in all kinds of sediments, though they are more obvious in some (such as sand/shale alternations) than others (such as limestones). It is important to realize that the nature of the sediment that has been burrowed

does not necessarily tell one anything about the environment in which the animal that made the burrow lived. In other words, the environment of the sea floor is not necessarily the same as that of the sediments below the sea floor. This is particularly obvious when one is dealing with borings into what was already hard rock.

Trace fossils may seem to have received an undue amount of attention here, but the author believes that they are one of the most promising fields in paleoecology, not only in the marine sphere, but more and more in the brackish freshwater and terrestrial fields as well.

Relationship to Sediments

The relationship between fossils and the sediments in which they occur is the most obvious line of investigation in paleoecology; but, although general observations are common, detailed studies are unfortunately still rare. An important conclusion reached recently in marine ecology that has a considerable significance for paleoecologists is that during the pelagic larval stage of most marine benthos, there is a high degree of selectivity of the type of sea floor on which to settle. What is more, the physical environment is more important than the biological.

Two possibilities must be considered carefully before any deductions are attempted about the relationship between organisms and sediment: firstly the possibility that the fossils may have been transported after death, and secondly the possibility that the fossils had a

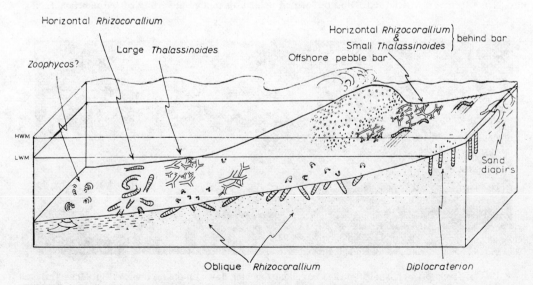

FIGURE 4. Block diagram showing the postulated environments of the more important trace fossils in the uppermost Jurassic of northern France (from Ager and Wallace, 1970, by kind permission of the *Geological Journal* and The Seel House Press).

FIGURE 5. Worm burrows filled with glauconitic sand avoiding the centers of incipient chert nodules, indicating that the nodules were already forming when the sediment was fresh on the sea floor. Lower Cretaceous, Isle of Wight, England, X0.66.

mode of life that made them independent of the conditions of sedimentation (e.g., a planktonic mode of life in the sea). However, the fossils may be related in some way to (1) the composition of the sediments surrounding them, (2) the grain size of the sediment, and (3) structures within the sediments.

Sediment Composition. The composition of the sediments may be considered in a very broad way (e.g., limestone/shale/sandstone) or in a very detailed way (e.g., the proportion of the various clay minerals present). Significant relationships have been observed at all levels. A very promising line of research at the moment is the relationship between the distribution of certain fossils and the abundance of certain trace elements in the sediments. Thus boron, which may be a useful guide to ancient salinities, varies inversely in certan Pennsylvanian sediments with the abundance of the marine fauna. The main problems here are firstly, that this is only a relative measure within a particular sequence (there are no absolute values) and secondly that there may well be earlier boron within the sediment concerned.

Sediment Grain Size. The grain size of the sediment is a notoriously misleading character in paleoecological analysis, since the fossils (especially microfossils) are commonly sorted mechanically in the same way as the sediment, and any apparent relationship is purely a postmortem effect. Nevertheless, true life relationships of this sort probably do exist. The writer has found, for example, that the Cretaceous brachiopod *Cyclothyris* always occurs in coarse sandy sediments and was probably adapted to life on such a sea floor.

Sediment Structure. Sedimentation structures such as ripple marks, slumped beds, and cross-bedding are commonly associated with particular fossil assemblages. Seilacher (1964), for instance, contrasted the types of trace fossils found associated with oscillation ripplemarks and those found associated with turbidite structures. Ager and Wallace (1970) noted the association of certain types of burrow with quicksand diapirs.

A fact of considerable importance to paleoecologists that has emerged from modern marine studies is that fine-grained bottom sediments normally support a much lower abundance of organisms than do coarse-grained deposits. This is particularly true of fossilizable (as distinct from soft-bodied) animals.

Living Together

An essential part of the environment of a living organism is the living community in which it finds itself. In the marine invertebrate communities with which most paleoecologists are chiefly concerned, the biological environment does not now seem to be as important as was formerly thought. Nevertheless, within any community there are multiple and complex interrelationships that concern the individual. These are usually difficult to recognize in the fossil record, but glimpses can be had from time to time. There are three basic relationships between organisms living together: *antagonism,* in which one or both suffers; *symbiosis,* in

535

which neither suffers but one or both benefits; and *toleration* in which neither is affected. These relationships may be conveniently (albeit somewhat artificially) subdivided as follows:

Antagonistic relationships
1. Antibiosis: in which one suffers and the other is not affected
2. Exploitation: in which one suffers and the other gains
3. Competition: in which both suffer

Symbiotic relationships
4. Commensalism: in which one gains and the other is not affected
5. Mutualism: in which both gain.

Tolerant relationships
6. Between species
7. Within species

Antibiosis. The relationship in which one organism suffers incidently as the result of the actions or merely the existence of another organism is known as antibiosis. The best-known examples at the present day are perhaps the mass mortalities of fish or other marine organisms brought about by the proliferation of microorganisms such as diatoms. Occurrences of one fossil species in vast numbers have been attributed to such a cause.

Exploitation. A far more common relationship, especially in the forms of predation and parasitism, is that of exploitation (see, e.g., Figs. 6 and 7). Attempts are being made to work out the predation patterns within fossil communities. Very often, vital links are missing due to nonfossilization. Direct evidence of predation comes from scarred fossils, from the contents of stomachs, and from coprolites. Evidence of supposed parasitism is known in fossils of all ages but is most convincing when the host organism has developed secondary structures in response to the irritation.

Competition. Competition is the natural state of affairs in the organic world and is usually detrimental to the organisms immediately involved, though through natural selection it may in the long run be beneficial to the stock. It is most intense between organisms that have similar ecological requirements; as a result, two members of the same genus rarely live in the same habitat. This has an important effect on the fossil record in that, except in special circumstances, more than one species of a particular genus is not to be expected in a single bed.

Looked at from this viewpoint, many instances in the fossil record of one taxonomic group replacing another in the same environment may be more readily understood. Thus

FIGURE 6. Association of small bivalves and brittle-stars on an Upper Jurassic bedding-plane. The fragile nature of the brittle-stars makes it unlikely that these fossils were transported after death. This could be a natural association of the two forms, and possibly an example of predation.

the nonmarine bivalve *Anthraconauta* appears to have suddenly replaced *Naiadites* in the Pennsylvanian coal-swamp environment, and the brachipods *Septaliphoria* and *Lacunosella* appear to have replaced one another in particular Jurassic reefs.

FIGURE 7. Parasitism in the fossil record. Above, Cretaceous bivalve *Exogyra* with borings by another bivalve (*Lithophaga*) along growth lines; Isle of Wight, England, X0.45. Below, Eocene bivalve *Cardita planicosta* with borings in the posterior end taken as evidence of its orientation and partial exposure on the sea floor in life; Isle of Wight, England, X0.3.

Commensalism. Although well known at the present day, commensalism is difficult to prove in the past. The most likely evidence is provided by one fossil attached to another (e.g. the corals in Fig. 8), though this may also be a record of parasitism or (more likely) of the settling of one organism on the dead shell of another. Only when there seems to have been a definitely localized attachment without apparent harm to the host can commensalism be regarded as probable.

Mutualism. A comparatively rare relationship at the present day, mutualism is almost impossible to prove in fossil associations, though cases have been claimed. Best-known are the intergrowths of two organisms to their presumed mutual advantage as in the bryozoan/algal association postulated by Condra and Elias (1944) as forming the distinctive Mississipian fossil *Archimedes*.

Toleration. Between species, toleration is an essential part of the ecological framework of any community. There are great paleoecological possibilities in the study of fossil associations, even if they were not involved in any direct relationships of the kinds discussed above. Johnson (1962) has demonstrated simple statistical tests of the significance or otherwise of fossil A being found with fossil B, and these are more likely to be valuable than any strained comparisons with present-day communities.

Toleration *within* a species bears considerably

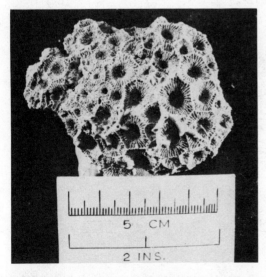

FIGURE 8. Devonian coral *Hexagonaria* with smaller tabulate coral *Aulopora* growing along the walls between corallites. From the position of the smaller corals it would appear that this was an association during life. USA.

on its mode of life and thence on its probable mode of occurrence as fossils. Thus gregarious organisms are likely to be found in concentrations. The absolute or relative density of particular fossils or fossil communities can be measured under certain circumstances. Sexual dimorphism has often been postulated for certain groups, in explaining pairs of "species" that are habitually found together in reasonable proportions.

There are many other curious ways in which different species may be found together in the same association, but the only one known commonly among fossils is that in which one species makes use of the dead remains of another.

The Problem of the Fossil Assemblage

Following naturally from the consideration of how fossils may have lived together is that of fossil assemblages as possible living communities. Probably no association of fossils has ever been found that really approaches the *biocoenose* or living community of the ecologist. Every fossil association is a *thanatocoenose* or death assemblage to a greater or lesser degree. The great problem confronting all paleoecologists lies in distinguishing the extent to which the living community has been obscured in the fossil assemblage. All the methods so far mentioned may be employed in this connection. In addition, great use is now being made or quantitiative methods of testing the degree to which a fossil assemblage may have been transported and sorted after death. Within an assemblage, the relative abundance of different age groups in a particular species is very significant, but consideration must also be given to the amount of wear, the degree of disarticulation of the skeleton, or the parts of a shell. Paleontologists pay much lip service to the need for comparisons with living communities, but very little has in fact been done in this respect. In practice, living communities can only be a real guide to the very top of the fossil record, where relationships are close enough to be meaningful.

More fruitful, from the paleoecologists' point of view, have been the recent efforts, especially in Germany (e.g., Schäfer, 1962; Reineck et al., 1967) to study what has been called "the paleontology of the present" for the express purpose of elucidating the records of the past (see *Actualistic Paleontology*). This goes beyond the studies of the ecologist, who is naturally only concerned with living material, and deals also with the fate of the dead body, which is all the paleontologist has to study. Schäfer (1962) defined a number of "biofacies" on the basis of the work of the Senckenberg-am-Meer Institute on the North Sea coast. Such "biofacies" take into consideration not only the fate of the dead organisms but also the trace fossils, the sediments and the sedimentary structures. Ager and Wallace (1970) showed how they could be applied to the interpretation of Jurassic sections in northern France.

One point of particular interest that has emerged from modern studies of this kind has been the demonstration, in many different places, that the shells of marine invertebrates are not normally transported far on the sea floor after death. This contradicts the pessimistic presumptions of many paleontologists in the past, but is in line with the conclusions commonly reached after careful study of the distinctiveness of most faunal assemblages. In other words, faunal and floral associations recognized by eye and by statistical analysis, usually prove to be separate and unmixed. One does not usually find more than the rare specimen in the "wrong" assemblage. It must be pointed out, however, that at the present day, nowhere near all the possible combina-

tions of species in a particular environment actually occur.

But on the whole, and especially if one is dealing with older assemblages, it usually proves more rewarding to analyze the significant associations statistically without preconceived ideas. Wallace (1969) showed how "constellation diagrams" could be produced for fossil assemblages (Fig. 9). These are diagrams that express the relative degrees of association of the different species in a fossil assemblage.

Thus, by utilizing every possible quantitative method and significance test communities can be reconstructed even where no direct deductions are possible about the nature of the original community from any modern counterpart. A great deal may be learned by analyzing the general features of an assemblage. Thus a molluscan fauna may show a predominance of large thick-shelled forms, or of undersized forms which from their growth lines can be seen to be adults. Such features are clearly of ecological significance.

Another particularly interesting characteristic of an assemblage is its specific diversity, which in a marine environment, for example, may be

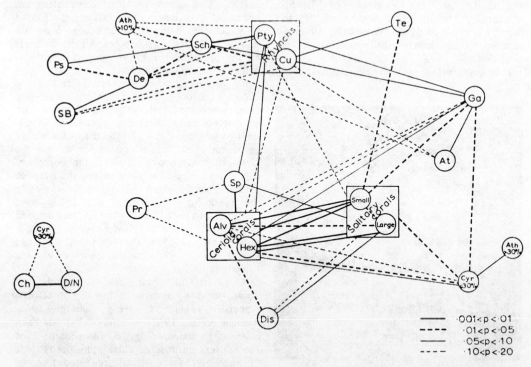

FIGURE 9. Constellation diagram showing the relative degrees of association of the species in a single bed of Devonian limestone in north France (from Wallace, 1969, by kind permission of the author and the Palaeontological Association). Alv, *Alveolites;* At, *Atrypa;* Ath, *Athyris;* Ch, *Chonetes;* Cu, *Cupularostrum;* Cyr, *Cyrtospirifer;* De, *Delthyris;* Dis, *Disphyllum;* D/N, *Douvillina/Nervostrophia;* Ga, Gastropod; Hex, *Hexagonaria;* Pr, *Productella;* Ps, *Pseudaviculopecten;* Pty, *Ptychomaletoechia;* Sch, *Schizophoria;* Sp, *Spinatrypa;* SB, Stick Bryozoans; Te, *Tenticospirifer.*

related to depth, salinity, or latitude. More sophisticated statistical methods have now been developed for expressing diversity as a measurable character of fossil populations, though the elucidation of the causes still remains very much an observational and deductive matter. Above all else, there is the climatic control that gives the low diversity of polar regimes (e.g., the shoals of cod, the pine forests, and the herds of reindeer) contrasting with the high diversity of the tropical coral reef or rain forest. Climatic belts in the past have been deduced on this sort of evidence, notably by Stehli (e.g., Stehli, 1968), but the data provided so far has not been very convincing. More important from the paleoecologists' point of view are the differences in diversity relating to local differences in environment. Thus, shallow marine environments of normal salinities have a very high level of diversity (especially when the bottom allows the growth of vegetation); whereas extremes of salinity, including fresh water, have low diversities. It must be remembered that diversity indices should be something more than a mere counting of heads and should reflect the different relative abundance of species. Thus, shallow-marine faunas, though high in number of species present, are usually dominated by one or only a few species (see *Diversity*).

Population Dynamics

More and more paleoecological studies are concentrating on the population rather than the individual. Once we have overcome the problem of relating the fossil assemblage to the once-living community, we can then begin to see these communities as living, self-contained, migrating, and evolving organizations. As we consider these dynamic properties for whole groups of organisms, the many parallel lines of evidence may lead to the same conclusion, and the hypotheses that bedevil paleoecology become several degrees less hypothetical. Thus, if a single species is small in size, it may just be a naturally small species. But if all the species in a fossil population are small, and assuming that there is evidence from their growth lines, etc. that they are normal adults and not just mechanically concentrated juveniles, then we must conclude that the whole population has been affected by its common environment. This would be "stunting" in the true sense of the word and of great paleoecological interest (see *Dwarf Faunas*).

One of the many snags about fossil assemblages that has only recently become apparent is that a population may change because of environmental changes outside the area studied. This situation arises when a benthic species (for example) depends for its continuance in a particular area on a continuous supply of pelagic larvae from elsewhere. This might explain some of the anomalies of the fossil record when a species disappears without any evident change in the sedimentary environment.

An even more important factor may be changes within the nonfossilizable elements in a community. There is the well-known case of the killing off by disease of the shallow-water eel grass of Atlantic coasts in the 1930s. This had immediate effects on some elements of the animal community, especially on the mollusks that normally fed on the grass, though the population as a whole was curiously little affected.

We may consider fossil populations as a whole in the same way that we consider individual species. In particular we may consider their geographical distribution and lateral variation. In the last few years, a number of works have been published in this field that are already paleoecological classics. One may mention in particular the work of Ziegler and his collaborators on the Silurian faunas (mainly brachiopods) of Wales and the Welsh borderlands (e.g., Ziegler et al., 1968). They found it possible to distinguish distinct communities, which they related to the depth of water in which they lived. Similarly, but dealing with a greater variety of fossils, Bretsky (1970a, 1970b) recognized several distinct communities in the Ordovician of the Applachians.

Such studies, though forward looking in outlook, are still very much in the classic tradition that goes back to J. Barrande and his "communities" in the Paleozoic rocks of Bohemia. For our fossil populations to become dynamic, it is necessary to consider them evolving and migrating through time; and it is a major undertaking to get to know properly a whole population at a single level or a single group through time. A major contribution has been made by McAlester, who demonstrated (1963) by means of pie diagrams how bivalve faunas changed their internal proportions through Devonian times, probably in relation to contemporary environments, and changed within themselves as one species evolved into another. One of the most exciting areas of research in dynamic paleoecology is the discovery of how one species or genus took over the ecological niche of another. Usually this happened between closely related forms (as in McAlester's bivalves), but sometimes it is possible to demonstrate, or at least claim, that one group of organisms was replaced by a totally different group as one slowly climbs the stratigraphical column. The large recumbent productid brachiopods of the late Paleozoic, for example, may have been replaced ecologically by the large recumbent molluscan oysters of the early Mesozoic.

We have yet to find really convincing "seres" and "climax" populations in the fossil record similar to those we know in modern ecology. Probably the whole ecological succession or "sere" is too fleeting a thing to make such showing in the rocks, though we see hints of it in certain reef developments.

It is always very difficult to prove that the fauna or flora was changing to fit a changed environment rather than that it was changing in tune with a changing environment.

General Conclusions

Paleoecology can never hope to emulate ecology in the understanding of fossil assemblages as dynamic and intimately interwoven living communities. Most of the new techniques now being developed (especially in the field of geochemistry) will provide much more information about ancient environments, but this is only half the battle, since paleoecology is the study of relationships, not of environments. The fundamental advances in paleoecology will be, on the theoretical side, the elucidation of evolution in terms of adaptation to environment and on the practical side, the disentagling of lateral variation in response to environment from vertical (and therefore stratigraphically significant) evolutionary changes. The application of modern statistical techniques and the development of computer programs to handle a great diversity of information will probably lead to great advances in the study of the relationship between past life and its past environments. But the first essential continues to be accuracy in the observation and recording of data before plunging into a sea of hypotheses.

DEREK V. AGER

References

Ager, D. V., 1963. *Principles of Paleoecology*. New York: McGraw-Hill, 371p.

Ager, D. V., and Wallace, P., 1970. The distribution and significance of trace fossils in the uppermost Jurassic rocks of the Boulonnais, northern France, in T. P. Crimes and J. C. Harper, eds., *Trace Fossils*. Liverpool: Seel House, 1–18.

Bretsky, P., 1970a. Late Ordovician benthic marine communities in North Central America, *N.Y. State Mus. Scientific Bull.*, Albany, **414**, 34p.

Bretsky, P., 1970b. Upper Ordovician ecology of the Central Appalachians, *Bull. Peabody Mus. Yale Univ.*, **34**, 150p.

Condra, G. E., and Elias, M. K., 1944. Study and revision of *Archimedes* (Hall), *Geol. Soc. Amer. Spec. Paper 53*, 243p.

Craig, G. Y., 1966. Concepts in palaeoecology, *Earth-Sci. Rev.*, **2**, 127–155.

Imbrie, J., and Newell, N., 1964. *Approaches to Paleoecology*. New York: Wiley, 432p.

Johnson, R. G., 1962. Interspecific associations in Pennsylvanian fossil assemblages, *J. Geol.*, **70**, 32–55.

Ladd, H. S., ed., 1957. Treatise on marine ecology and paleoecology, vol. 2, Paleoecology, *Geol. Soc. Amer. Mem. 67*, 1077p.

Lawrence, D. R., 1971. The nature and structure of paleoecology, *J. Paleontology*, **45**, 593–607.

McAlester, A. L., 1963. Pelecypods as stratigraphic guides in the Appalachian Upper Devonian, *Geol. Soc. Amer. Bull.*, **74**, 1209–1224.

Reineck, H. E.; Gutmann, W. F.; and Herweck, G., 1967. Das Schlickgebiet sudlich Helgoland als Beispiel rezenter Schelfablagerungen, *Senckenbergiana Lethaea*, **48**, 219–275.

Reyment, R. A., 1971. *Introduction to Quantitative Paleoecology*. Amsterdam: Elsevier, 226p.

Schäfer, W., 1962. *Actuo-Paläontologie nach Studien in der Nordsee*. Frankfurt-am-Main: Waldemar Kramer, 666p. (English transl., 1972. *Ecology and Paleoecology of Marine Environments*. Chicago: Univ. Chicago Press, 568p.

Seilacher, A., 1964. Biogenic sedimentary structures, in Imbrie and Newell, 1964, 296–316.

Speden, I. G., 1966. Paleoecology and the study of fossil benthonic assemblages and communities, *New Zealand J. Geol. Geophys.*, **9**, 408–423.

Stehli, F. G., 1968. Taxonomic diversity: Gradients in pole location: The recent model, in E. T. Drake, ed., *Evolution and Environment*. New Haven: Yale Univ. Press, 163–227.

Valentine, J. W., 1973. *Evolutionary Paleocology of the Marine Biosphere*. Englewood Cliffs, N.J.: Prentice-Hall, 511p.

Wallace, P., 1969. Specific frequency and environmental indicators in two horizons of the Calcaire de Ferques (Upper Devonian), northern France, *Palaeontology*, **12**, 366–381.

Warme, J. E., 1971. Paleoecological aspects of a modern coastal lagoon, *Univ. Calif. Pubs. Geol. Sci.*, **87**, 110p.

Zangerl, R., and Richardson, E. S., Jr., 1963. The paleoecological history of two Pennsylvanian black shales, *Fieldiana, Geol. Mem. 4*, 352p.

Zeigler, A. M.; Cocks, L. R. M.; and Bambach, R. K., 1968. The composition and structure of Lower Silurian marine communities. *Lethaia*, **1**, 1–27.

Cross-references: *Actualistic Paleontology; Biogeochemistry; Communities, Ancient; Diagenesis of Fossils; Diversity; Dwarf Faunas; Fossils and Fossilization; Growth Lines; Larvae of Marine Invertebrates—Paleontological Significance; Morphology, Constructional; Morphology, Functional; Paleobiogeography; Paleoecology, Inland Aquatic Environments; Paleoecology, Terrestrial; Paleotemperature and Depth Indicators; Plate Tectonics and the Biosphere; Population Dynamics; Substratum; Taphonomy; Trace Fossils; Trophic Groups.* Vol I: *Benthonic Zonation; Benthos, Marine; Marine Ecology; Mass*

Mortality in the Sea. Neritic Zone (and Sedimentation Facies). vol IVA: *Ecology; Paleosalinity.* Vol. VI: *Paleobathymetric Analysis; Sedimentation: Paleoecologic Aspects.*

PALEOECOLOGY, INLAND AQUATIC ENVIRONMENTS

Paleoecology is the study of the interaction of ancient plants and animals with each other and with their physicochemical environment. The stratigraphic and paleontologic records of lakes, rivers, ponds, swamps, and similar aquatic environments permit the paleoecologist to study the ecological and environmental interactions and the evolution of inland aquatic ecosystems through long intervals of time, which are unavailable in modern ecological studies. Nevertheless, this important potential of paleoecology is limited by the numerous problems that obscure the fossil record. For example, communities of organisms are seldom, if ever, preserved intact, and the composition of a fossil assemblage is biased toward those species with durable skeletons that live in habitats where burial and preservation is likely. In addition, an assemblage of fossils is time-averaged, because it may contain a mixture of individuals from different communities that lived at different places and times. Time-averaging can obscure seasonal variations in community structure and biogeography. The aquatic paleoecologist must, therefore, determine the extent to which his samples represent the original community or ecosystem by analyzing the pre- and postburial history (taphonomy) of a fossil asemblage. (See Fagerstrom, 1964; Johnson, 1960; Lawrence, 1968, 1971; Seilacher, 1973.)

The stratigraphic record of inland aquatic environments has been studied extensively by stratigraphers, sedimentologists, and geochemists to document or reconstruct the history of ancient alluvial, paludal (swamp), and lacustrine systems. However, such studies have been primarily concerned with the physical and chemical processes that produce distinct associations of sedimentary rocks or economically important minerals. Paleobiologists, on the other hand, have used the stratigraphic record almost exclusively for taxonomic analyses, with fewer studies of the ecology, biostratigraphy, and evolution of terrestrial and aquatic organisms. However, the stratigraphic record of inland aquatic environments is the product of biological, physical, and chemical processes that should be integrated in a temporal framework for complete paleoecological analyses.

The purpose here is to provide an overview of the principles and concepts of inland aquatic paleoecology through examples of innovative, integrated paleoecological analyses.

Methods and Materials.

The methods of freshwater paleoecology are almost entirely borrowed from other disciplines—principally stratigraphy, sedimentology, ecology, and limnology. Organic and inorganic geochemistry and geophysics also play important analytical roles in interdisciplinary paleoecological studies. Traditionally, the materials of freshwater paleoecology are lake sediments and their associated fossils. The basin morphometry, drainage area, sedimentology, water chemistry, and biology of lakes are the bases for the interpretation of their deposits (Frey, 1969, 1974).

In the stratigraphic record, this information must be inferred by stratigraphic, sedimentologic, and geochemical methods. Facies relationships, sedimentary sequence, rock type and texture, sedimentary structures, paleocurrent patterns, and ion and isotope ratios (Picard and High, 1972) assist in inferring the basin morphometry, drainage-area characteristics, water depth and chemistry, and distribution of paleoenvironments. The diversity, abundance, functional morphology, ecology, and distribution of fossil organisms (Table 1; Frey, 1964) provide the primary biological data for paleoecology. Biogenic products such as fossil pigments (Sanger and Gorham, 1972); distinctive types of organic matter such as tannins and lignins (Anderson and Kirkland, 1969); and some basic elements of living matter such as carbon, hydrogen, nitrogen, and phosphorus (Hutchinson and Wollack, 1940) also provide indirect information about the ancient lacustrine and adjacent terrestrial paleoenvironments.

Lacustrine sediments are more frequently studied than fluvial deposits in inland aquatic paleoecology. Fluvial deposits typically lack the preserved biological diversity necessary for paleoecological interpretation of ancient river systems because of dissolution and destruction of organisms within the permeable fluvial sands and silts. Further, hiatuses and redeposition of previously deposited material complicate the depositional history of these deposits. Ironically, the development and character of riverine and riparian environments are often better recorded in sediments from the lakes into which the river flowed than they are from the fluvial deposits themselves.

Even though lake sediments are partly products of limnological and ecological processes,

TABLE 1. Major Groups of Organisms and Their Utility in Freshwater Paleoecology.[a]

Fossil Groups	Utility in Paleoecology		
	Habitat	Water Chemistry	Seasonality
Blue-green algae (pigments)			
Dinoglagellates			
Chrysophyceae			Tippett, 1964
Diatoms*	Bradbury, 1971	Bradbury, 1971	Tippett, 1964
			Evans, 1972
			McLeroy & Anderson, 1966
Vascular plants*	Watts & Winter, 1966	Watts & Bright, 1968	Anderson & Kirkland, 1969
(pollen, seeds, leaves)			
Rhizpoda	Harnisch, 1927		
	Steinecke, 1929		
Porifera	Hutchinson, et al., 1970	Hutchinson, et al., 1970	
(sponge spicules)			
Mollusca	Altena, 1957	Hanley, 1974	
Bivalves*	Altena & Kuiper, 1945	LaRocque, 1960	
Gastropods*	Hanley, 1976	Gibson, 1967	
	Herrington & Taylor, 1958		
	Hibbard & Taylor, 1960		
	Kozlovskaya, 1961		
	LaRocque, 1960, 1966		
	Taylor, 1954, 1960b, 1966		
	Taylor & Hibbard, 1955		
	Sparks, 1961		
	Leonard, 1950		
	Ryder, Fouch & Elison, 1976		
	Eagar, 1948, 1973		
	Miller, 1966		
	Berry & Miller, 1966		
	Gibson, 1967		
	Frye, Leonard, Willman & Glass, 1972		
Ostracoda*	Delorme, 1970	Lüttig, 1959	Anderson & Kirkland, 1969
		Anderson & Kirkland, 1969	
		Delorme, 1970	
Cladocera*	Whiteside, 1970		
Conchostracans	Tasch and Zimmerman. 1961	Tasch, 1963	Tasch, 1958
Midges			Anderson & Kirkland, 1969
Chironomids*			
Chaoborus*			
Bryozoa	Deevey, 1942		
Fish			

[a]The most commonly studied groups are marked with (*). For more information see summaries by Frey, 1964, 1969, 1974.

their study may contribute substantial knowledge about many past events and processes that may lie somewhat outside the realm of strictly defined aquatic paleoecology. For example, pollen analysis is frequently an integral part of lake-sediment studies. The pollen contained in lake sediments is largely derived from terrestrial plants, and its study can be the basis for stratigraphic correlations and for histories of past floras and vegetation. Because the terrestrial vegetation is a product of the regional climate, just as are lake chemistry and biology, pollen studies are quite relevant to interpretations of ancient aquatic ecosystems. The same reasoning could apply to the terrestrial leaf, seed, and insect fossils that generally occur in lacustrine sediments.

Pleistocene, Tertiary, and pre-Tertiary lacustrine deposits are well represented in the stratigraphic record in the western United States (Feth, 1961, 1963, 1964). Allen (1965) provided a useful list of fluvial deposits in the stratigraphic record.

The Time Framework for Paleoecology

An accurate, absolute time scale is required to measure the rates of biological and environmental change in ancient aquatic ecosystems. A time scale may be established by radiometric dating of lake deposits or associated material in several ways. Radiocarbon (^{14}C) dates the organic material of lake sediments between 350 and 50,000 years old. Lava flows and ash deposits associated with lake sediments can be reliably dated by the isotope ^{40}A if they are greater than 500,000 years old. To some extent, the daughter isotopes of the uranium decay series could fill the time gap between 50,000 and 500,000 years, but these techniques need refinement before this important part of the Pleistocene can be studied in a reliable time framework.

| | Utility in Paleoecology | | | |
Temperature	Productivity	Oxygen	Community Structure and Ecologic Interaction	Paleobiogeography
Tippett, 1964	Gorham & Sanger, 1975 Norris & McAndrews, 1970 Hutchinson, *et al.*, 1970 Bradbury & Waddington, 1973	Abbott & VanLandingham, 1972		
Anderson & Kirkland, 1969 MacGinitie, 1953, 1969				Bertsch, 1951 Deevey, 1955
Hibbard & Taylor, 1960 Johansen, 1904 Leonard, 1950 Sparks, 1957 Taylor, 1954, 1960b, 1965 Taylor & Hibbard, 1955 Gibson, 1967			Hanley, 1976	Herrington & Taylor, 1958 Taylor, 1960a, 1960b, 1965, 1966 van der Schalie, 1944, 1963
Lüttig, 1955 Delorme, 1970	Delorme, 1970	Stark, 1970 Delorme, 1970		McKenzie, 1970
Megard, 1964 Goulden, 1964		Goulden, 1964	Goulden, 1966 Tsukada, 1967	Deevey, 1955
				Tasch & Volkheimer, 1970
Anderson & Kirkland, 1969	Deevey, 1942 Frey, 1955			
			Perkins, 1970	Hubbs, Miller, & Hubbs, 1974

Unfortunately, the range of error in a stratigraphic series of radiometric dates is often too great to trace ecological interactions over decades or centuries—time intervals of importance to theoretical ecologists investigating the development, diversification, and competitive interactions of aquatic communities. Fortunately, some lakes contain annually deposited layers of sediment (varves) that provide very accurate time control of a lake's history, much as growth rings establish the age and developmental history of an individual tree. Paleoecological studies of varved lake deposits can be precisely quantitative, because absolute rates of deposition can be calculated. In addition, events that produce stratigraphic markers such as ash falls (Wilcox, 1965), distinctive airborne pollen assemblages (Swain, 1973), and paleomagnetic variations in minerogenic lake sediments (Mackereth, 1971) could be accurately time-calibrated in varved sediments.

The absolute chronology of varved lacustrine sediments has even greater application outside the sphere of paleoecology. In addition to precisely timing paleoecological processes, such deposits could fix volcanic eruptions of tephra and paleomagnetic fluctuations in time and space. Varves have even been used to study the cyclic nature of solar events (Anderson, 1961), the variation of the ^{14}C content of the atmosphere in the Holocene (Stuiver, 1970), and the retreat of glacial environments in northern Europe (De Geer, 1940).

Tasch (1958, 1963) has used thin layers of conchostracan fossils in Permian pond deposits to approximate the sedimentation rate and temporal duration of these shallow, saline, aquatic environments. Although such "biological clocks" are based on many assumptions about fossil conchostracan behavior, the approach holds promise in some types of unvarved sediments.

Occasionally, a time scale can be established by the correlation of distinctive sedimentary units produced by known historical events, such as the introduction of mine wastes in lake basins, the fallout of radioactive material from atomic testing (Pennington et al., 1973), or the addition of other unique pollutants into lakes. Time scales based on these events could be very useful in studying lacustrine paleoecology during the last few hundred years, when radiocarbon dating lacks precision in determining age differences of small magnitude.

Applications of Inland Aquatic Paleoecology

Integrated Studies. Paleoecology is a highly synthetic discipline that requires interdisciplinary integration to produce a complete biological and environmental picture of an inland aquatic ecosystem through time. For practical reasons, this is rarely attempted; although the combined talents of teams of investigators concentrating on one lake basin and its sedimentary record have approached this theoretical goal. The underlying rationale for such comprehensive studies stems from man's intellectual curiosity about the origin, development, and fate of freshwater communities and environments and his interest in the economic potential of ancient lake and river deposits. This knowledge of history greatly aids our understanding of modern aquatic ecosystems and allows predictions about their future. In addition, the individual specialized studies that make up the integrated, comprehensive works are frequently useful in solving more restricted paleoecological problems. Outstanding works of this nature have been done under the guidance of G. E. Hutchinson (Yale University) and his associates. The 24,000-year history of Lago di Monterossi in central Italy (Hutchinson et al., 1970), for example, was interpreted from the stratigraphic analysis of over 40 elements and numerous other mineralogical and compositional parameters of the sediments. The organic history of the lake was traced by the study of green and blue-green algae, diatoms, chrysophycean cysts, sponge spicules, turbellarians, bryozoans, cladocerans, insect fossils, and the pollen of both aquatic and terrestrial plants. Further, the chemistry and mineralogy of the drainage area and the modern limnology of the lake were investigated. The entire environmental and biological history of Lago di Monterossi was then integrated with the Etruscan, Roman, and later history of the site to determine the effect of man on this ecosystem. As complete as this study appears, many other biological, chemical, and sedimentological parameters can be incorporated in lake-sediment studies, depending upon the suitability of the deposits and the inclination of the investigators. The varved deposits of an early Pleistocene lake in the High Plains of Texas were analyzed in detail by Anderson, Kirkland, and others (1969). Studies of ostracods, leaves, terrestrial and aquatic insects, fish, and pollen were combined with a detailed petrologic and sedimentologic analysis of the laminations that included chemical and oxygen- and carbon-isotope studies. The evidence was imaginatively integrated to show that the laminations were seasonal deposits (varves). Once the origin of these distinctive sediments was understood, larger cyclic variations in the same paleontologic and chemical components of the deposit were utilized in a paleoclimatic analysis that was related to possible solar fluctuations by a statistical analysis of cyclic patterns.

These two examples are chosen as highly integrated paleoecologic studies of inland aquatic environments. Not only were many parameters used in the reconstruction of the past environment, but also they were well synthesized and integrated to produce a picture of a dynamic ecosystem through time. This important requirement is too often omitted from otherwise complete and detailed paleoecological studies.

Lake Typology. Specific applications of aquatic paleoecology are exceedingly diverse, and only a few illustrations can be discussed here. Paleolimnology has contributed significantly to our understanding of lake evolution, classification, and typology. At one time it was thought that as lakes aged they underwent a change from oligotrophy (poor in nutrients) to eutrophy (rich in nutrients), and this developmental sequence was utilized in a graduated classification system for lakes. Obviously, knowledge of a lake's trophic history can be used to test this hypothesis, and paleolimnologic studies based on the faunal succession and chemical composition of lake sediments of existing lakes has given limnologists many insights into the natural development of lakes with time. Hutchinson and Wollack (1940), Deevey (1942), and Patrick (1943) applied this type of study to the sediments of Linsley Pond, Connecticut. Their studies showed that the trophic aging of Linsley Pond was not uniform but changed early and rapidly from an oligotrophic state to a eutrophic state. The lake then remained in equilibrium for several thousand years until human disturbance of its watershed further increased its productivity and sedimentation rate in the last 200 yr. This developmental sequence is not universal, however. Lago di Monterossi (Hutchinson et al., 1970) began as a productive lake and became less productive with time until disturbance of its drainage basin by Roman road construction flushed more nutrients into the water and reversed its trend toward oligotrophy. Paleoecological studies have thus helped limnologists understand the causes of productivity changes and thereby evaluate productivity in lake-classification schemes.

Influence of Man. The recognition that man's impact on lake ecosystems is preserved in the sedimentary record has generated a series of specialized applications of aquatic paleoecology. In the last decade, state and federal governments have become concerned about the quality of aquatic environments and their susceptibility to degradation by waste disposal or by clearing, cultivating, or settling in their drainage areas. Because it became essential to know what the characteristics and quality of these environments were before man settled in the area under question, the paleolimnologic record was utilized to establish precultural limnological and ecological baselines for planning and lake-rehabilitation purposes. This type of study in Shagawa Lake, a highly polluted lake in northeastern Minnesota (Bradbury and Waddington, 1973), involved an interdisciplinary investigation of microfossils—pollen, cladocerans, and diatoms—and sediment chemistry—organic matter, biogenic silica, iron, phosphorus, and recently fossil algal pigments (Gorham and Sanger, 1976). This region of Minnesota was settled in 1886 when iron ore deposits were discovered along the shores of Shagawa Lake. The mine wastes that were pumped into the lake form a stratigraphic marker that can be used to date the subsequent limnological changes. As the drainage area of the lake was cleared for timber and settlement sites, nutrient-rich forest soils were washed into the lake, and its productivity began to increase. Different species of diatoms came to dominate the algal flora. When phosphate detergents were introduced in the late 1940s, the summer diatom plankton was replaced by blue-green algae and the lake was no longer useful as a water supply or for recreation. The paleoecologic record that documented these changes provides a means of evaluation of the success of a pollution-control facility recently established on Shagawa Lake.

Many other lakes have been studied in similar ways (Bradbury, 1975), and it is hoped that increased use of the most recent freshwater paleoecologic record will allow man to assess fully his environmental impact and to correct undesirable trends before irreparable damage is done to lakes, rivers, and wetlands.

Paleoclimate. For many years, one of the principal goals of paleoecology has been that of paleoclimatic analysis. Lakes clearly form only under appropriate geomorphic, hydrologic, and climatic conditions, and the influence of climate on lake chemistry and biology is well documented in some areas. Moyle (1945) and Birks (1973) illustrated the changes in aquatic macrophytes along a transect of decreasing precipitation across northern Minnesota and into the North Dakota prairie. Under favorable circumstances, rather detailed paleoclimatic interpretations may be made from lake sediments and their contained fossils. Watts and Bright (1968) interpreted the fossil seeds of aquatic plants, pollen, and mollusk remains as characterizing the paleoclimate and surrounding vegetation of Pickerel Lake in South Dakota during postglacial time. This study has been complemented by an additional investigation of diatoms by Haworth (1972), who suggested that the productivity of the lake increased in response to nutrient input from soil erosion during dry periods.

In E Africa, D. A. Livingstone (Duke University) and his students (Kendall, 1969; Richardson and Richardson, 1972) have documented paleoclimatic changes for the terminal Pleistocene and Holocene in considerable detail. Analyses of diatoms, pollen, and geochemistry of cores from the Rift Valley lakes and other smaller basins nearby reveal climatic changes that coincide in time with glacial and interglacial environments of northern Europe, but that differ in climatic character. When glaciers advanced in north-temperate regions, the climate of equatorial Africa was cool and arid. During interglacial times, a warm, moist climate prevailed in these regions (Hecky and Degens, 1973; Butzer et al., 1972).

Longer paleoclimatic records based on freshwater paleoecology are also noteworthy. Horie (1974) and many associates have undertaken a remarkably complete study of a 200-m core from Lake Biwa in Japan that possibly represents continuous deposition for the last 700,000 yr. Although dating of the core and more refined analyses are still in progress and the diverse studies have yet to be synthesized, pollen, diatom, and isotope evidence indicate that several fluctuations between warm and cool climates occurred. Such information will be helpful in reconstructing terrestrial climatic changes during the Pleistocene.

Species Interaction. The stratigraphic record allows the paleoecologist to study species interactions within a community through long periods of time not available to ecologists working in the modern environment. Although aspects of community structure are often obscured by the time-averaged and biased nature of the paleontologic record, the structure of fossil and living cladoceran associations appears very similar (Goulden, 1966). Goulden demonstrated that diversity and equitability of cladocerans in a small pond changed in response to Mayan agriculture in Guatemala 200 yr ago. Initially, cladoceran diversity was low; but it gradually increased through time until agricultural disturbance caused species replacements that again resulted in an inequitable distribution and lowered diversity. Tsukada (1967) showed similar instabilities and disequilibria caused by volcanic ash falls in Holocene cladoceran assemblages of Lake Nojiri, Japan. Analysis of community structure and its evolution, particularly in Mesozoic and Tertiary deposits, is a significant but poorly studied aspect of theoretical paleoecology. Future studies must utilize the structure of modern communities as analogs for the ancient counterparts. Perkins (1970), for example, found that equitability of fossil predatory and planktivorous fish in the Green River Formation (Eocene) in the fossil basin, Wyoming, was similar to modern populations of lakes and ponds.

Pre-Quaternary Inland Aquatic Paleoecology

Paleoecological analyses in Paleozoic, Mesozoic, and Tertiary alluvial and lacustrine strata typically lack the interdisciplinary integration and analytical refinement apparent in Quaternary studies. This is primarily because of the information loss in older paleontologic and stratigraphic records, and because of the independent research of stratigraphers, sedimentologists, geochemists, and paleontologists. Interdisciplinary analyses of such deposits must be based on integration of biological, physical, and chemical data derived from the study of modern aquatic ecosystems.

Mollusks and ostracods, the dominant preserved freshwater invertebrates in pre-Quaternary deposits, are poorly studied. Mollusks have frequently been objects of taxonomic analysis, but their paleoecologic utility has not been fully realized. The application of mollusks in Pleistocene paleoenvironmental interpretations is well known. (See summary in Frey, 1964; Hibbard and Taylor, 1960; LaRocque, 1966; Taylor, 1965.) Taylor 1960b, 1965) provided an excellent discussion on the principles of paleoecologic interpretation of late Pliocene and Pleistocene nonmarine mollusks. Although the analysis of evolution, functional morphology, ecophenotypic variation relative to environmental parameters, and factors affecting the structure of molluscan associations have great potential in paleoecologic studies, these topics are seldom studied in fossil nonmarine mollusks.

Pre-Quaternary lake deposits that are almost entirely composed of the siliceous frustules of diatoms (diatomites) are relatively common in the western United States and in many other parts of the world. Their taxonomic and paleoecologic study can provide many insights into Tertiary paleoclimates and aquatic environments. Diatomites associated with the basalt flows (Miocene Yakima Basalt) of southern Washington have been the object of a comprehensive taxonomic study (Van Landingham, 1964). Andrews (1970, 1971) determined the water depth and alkalinity of Miocene lakes in Nebraska by qualitatively comparing their diatom floras to modern analogs. A more quantitative approach to paleoecologic interpretation of freshwater diatomites, utilized by Abbott and Van Landingham (1972), provided such paleolimnologic information as water depth, habitat, current strength, pH, salinity, and nutrient and oxygen content of a Miocene lake in Oregon.

Pollen and leaf fossils are also abundant in pre-Quaternary lake deposits and are commonly used in paleoecologic reconstructions (MacGinitie, 1953, 1969).

Paleoecological studies of pre-Quaternary alluvial and lacustrine deposits are numerous, but only a few innovative, interdisciplinary examples are considered here.

Mazon Creek Faunas (Pennsylvanian). The Mazon Creek faunas (q.v.), composed of nonmarine, marginal-marine, and open-sea invertebrates and vertebrates occur in the Francis Creek Shale Member (Wanless, 1969) of the Carbondale Formation (middle Pennsylvanian) in Illinois and Indiana (Zangerl and Richardson, 1963). In this deposit, animals with thin chitinous skeletons and animals without skeletons are preserved along with carbonate-shelled mollusks and bony and cartilaginous vertebrates (Richardson and Johnson, 1971, p. 1224). Zangerl and Richardson provided a highly innovative, interdisciplinary analysis of this deposit. The composition, biogeographic and stratigraphic distribution, and taphonomy of invertebrates, vertebrates, and plants, together with a stratigraphic, chemical, and mineralogical analysis of the enclosing shale, were utilized in a paleoenvironmental reconstruction. The deposit was inferred to be a large clastic wedge deposited in several environments: (1) near shore and slightly above sea level on an aggrading deltaic plain with ponds; (2) below sea level on the lower deltaic plain or deltaic slope; and (3) in estuaries and bays of varying salinities (Richardson and Johnson, 1971, p. 1233).

Fort Union Formation (Paleocene). Studies of alluvial deposits typically lack integration of biological and sedimentological data. Royse (1970), however, integrated sedimentological and faunal data in a detailed analysis of the Tongue River and Sentinel Butte Members of the Fort Union Formation (Paleocene) in the Williston Basin, North Dakota. Channel, floodplain, and backswamp facies were differentiated by granulometric analyses of 500 sediment samples. Primary sedimentary structures, rock types, and paleocurrent data permitted inferences to be made regarding paleostream velocities, paleochannel forms, mode of sediment transport, sediment-dispersal patterns, paleoslope, and sedimentary provenances. Further, Royse (1970, p. 60) concluded that the flora and molluscan fauna of the Tongue River and Sentinel Butte Members support an alluvial origin for the strata that enclose them.

Green River Formation (Eocene). Perhaps the best-studied lacustrine system in the stratigraphic record is the Green River Formation (Eocene) in Wyoming, Colorado, and Utah. The paleobiology, stratigraphy, sedimentation, mineralogy, and geochemistry of the Green River have been studied for over 50 years. (See bibliography of Mullens, 1973, listing 597 papers.) In spite of this extensive research, many aspects of the lacustrine system remain unclear: origin of oil shale, the character and distribution of paleoenvironments within the lacustrine system, and paleolimnology. On the basis of a lifetime of interdisciplinary research, Bradley (1963) summarized the stratigraphy, basin genesis, hydrography, morphometry, climate, water chemistry, paleontology, paleobotany, productivity, and energy relations of this ancient lacustrine system. Bradley and Eugster (1969) further discussed lacustrine geochemistry and paleolimnology of the Wilkins Peak Member of the Green River Formation. The Green River was characterized as a large, meromictic lake having freshwater and saline stages.

MacGinitie (1969) extensively revised the taxonomy and biostratigraphy of the flora of a part of the Green River Formation to interpret its ecological and paleoclimatic significance. On the basis of a comparison with modern related plants, he described four terrestrial floral habitats. MacGinitie further integrated paleoclimatic evidence from the fossil flora, varves, and the distribution of evaporites and red beds in the underlying and laterally intertonguing Wasatch Formation. The paleoclimate was inferred to have been subtropical, bordering on subhumid, with an average annual temperature of 67°F (19°C). Winters were relatively dry, late spring and early summer were the rainy periods, and late summer and fall were hot and dry. Average annual precipitation at lake level was approximately 38 in. (71 mm).

Hanley (1974) delineated the morphologic bases for species recognition in mollusks from the complexly intertonguing Green River and Wasatch Formations in southwestern Wyoming by comparison with concepts used to discriminate biological species in related, living mollusks. The ecology, structure, functional morphology, paleobiogeography, and biostratigraphy of molluscan associations indicate fluvial, alluvial-plain, pond, and littoral and sublittoral lacustrine environments (Hanley, 1976).

On the basis of an integration of stratigraphic, sedimentologic, geochemical, mineralogic, and paleontologic data, Baer (1969) inferred deltaic and lacustrine paleoenvironments for interbedded sediments in the lower part of the Green River Formation of central Utah.

Further consideration of stratigraphy, sedimentary structures, and geochemistry led Eugster and Surdam (1973), Surdam and Wolfbauer (1975), and Eugster and Hardie (1975) to propose that the Green River Formation of Wyoming was deposited in and around a playa-lake complex with broad mud flats, and that the lacustrine environment was characterized by significant changes in size, salinity, and alkalinity.

Additional innovative research integrating biological, physical, and chemical aspects of all stages in the evolution of this lacustrine system will be required to resolve this controversial paleoecological problem.

Florissant Lake Beds (Oligocene). The Florissant Lake Beds (Oligocene) were deposited in central Colorado. Although the lake beds occupy only 15 square miles (39 km^2), they have been discussed in 226 papers (McLeroy and Anderson, 1966), most of which concern the flora and insect fauna of the lake beds. McLeroy and Anderson provided an excellent integrated analysis of Florissant paleolimnology, paleophysiography, and paleoclimate by using plant megafossils (summarized from MacGinitie, 1953), pollen, and diatoms. The petrography and stratigraphic relations of diatomite laminae and the ecology of living diatoms suggest that these laminae formed rapidly in the spring at relatively low temperatures in neutral to slightly alkaline lake water. The stratigraphy, inferred geochemistry, and paleobotany of sapropel laminae indicate that they formed by autumnal accumulation of organic debris. These integrated analyses establish the diatomite-sapropel laminae couplets as varves. Paleoclimate, preservation of varves, and the absence of benthonic organisms suggest that the lake was meromictic.

Summary

Inland aquatic paleoecology is a synthetic discipline that requires an integration of physical, chemical, and biological data to fully interpret ancient lacustrine, alluvial, and paludal systems. Observation and interpretation of existing lacustrine and fluvial environments provide valuable analogs for interpretation of ancient deposits. Studies of Quaternary lake sediments clearly document the utility of freshwater paleoecology in interpretation of ancient lacustrine systems and their terrestrial climatic regimes. Interpretation of Paleozoic, Mesozoic, and Tertiary inland aquatic ecosystems requires innovative interdisciplinary research and cautious application of data derived from modern analogs. Analyses of species interactions and functional morphology and of the structure and evolution of ancient freshwater communities are challenging but inadequately studied aspects of freshwater paleoecology.

J. PLATT BRADBURY*
JOHN H. HANLEY*

References

Abbott, W. H., and Van Landingham, S. L., 1972. Micropaleontology and paleoecology of Miocene non-marine diatoms from the Harper District, Malheur County, Oregon, *Nova Hedwigia,* **23,** 847–906.

Allen, J. R. L., 1965. A review of the origin and characteristics of Recent alluvial sediments, *Sedimentology,* **5,** 89–191.

Altena, C. O. van R., 1957. Pleistocene Mollusca, *Verh. Geol. Mijnb. Genoot. Ned. Kolon., Geol. Ser.,* **17,** 121–138.

Altena, C. O. van R., and Kuiper, J. G. J., 1945. Plistocene land–en zoetwatermollusken uit den ondergrond van Velzen, *Zool. Meded.,* **25,** 155–199.

Anderson, R. Y., 1961. Solar-terrestrial climatic patterns in varved sediments, *Ann. New York Acad. Sci.,* **95,** 424–439.

Anderson, R. Y., and Kirkland, D. W., eds., 1969. Paleoecology of an early Pleistocene lake on the High Plains of Texas, *Geol. Soc. Amer. Mem. 113,* 215p.

Andrews, G. W., 1970. Late Miocene non-marine diatoms from the Kilgore area, Cherry County, Nebraska, *U.S. Geol. Surv. Prof. Paper 683-A,* 24p.

Andrews, G. W., 1971. Early Miocene non-marine diatoms from the Pine Ridge area, Sioux County, Nebraska, *U.S. Geol. Surv. Prof. Paper 683-E,* 17p.

Ash, S. R., ed., 1978. Geology, paleontology, and paleoecology of a Late Triassic lake, western New

Mexico, *Brigham Young Univ. Geol. Studies,* **25**(2), 100p.

Baer, J. L., 1969. Paleoecology of cyclic sediments of the lower Green River Formation, central Utah, *Brigham Young Univ. Geol. Studies,* **16,** 3–95.

Berry, E. G., and Miller, B. B., 1966. A new Pleistocene fauna and a new species of *Biomphalaria* (Basommatophora: Planorbidae) from southwestern Kansas, U.S.A., *Malacologia,* **4,** 261–267.

Bertsch, K., 1951. *Geschichte des deutschen Waldes,* 3rd ed. Jena: Fischer, 118p.

Birks, H. H., 1973. Modern macrofossil assemblages in lake sediments in Minnesota, in Birks and West, 1973, 173–189.

Birks, H. J. B., and West, R. G., eds., 1973. *Quaternary Plant Ecology.* Oxford: Blackwells, 173–189, 289–307.

Bradbury, J. P., 1971. Paleolimnology of Lake Texcoco, Mexico. Evidence from diatoms, *Limnol. Oceanogr.,* **16,** 180–200.

Bradbury, J. P., 1975. Diatom stratigraphy and human settlement in Minnesota: *Geol. Soc. Amer. Spec. Paper 171,* 74p.

Bradbury, J. P., and Waddington, J. C. B., 1973. The impact of European settlement on Shagawa Lake, northeastern Minnesota, U.S.A., in Birks and West, 1973, 289–307.

Bradley, W. H., 1963. Paleolimnology, in D. G. Frey, ed., *Limnology in North America.* Madison: Univ. Wisconsin Press, 621–652.

Bradley, W. H., and Eugster, H. P., 1969. Geochemistry and paleolimnology of the trona deposits and associated authigenic minerals of the Green River Formation of Wyoming, *U.S. Geol. Surv. Prof. Paper 496-B,* B1–B71.

Butzer, K. W.; Isaac, G. L.; Richardson, J. L.; and Washbourne-kamau, 1972, Radiocarbon dating of East African lake levels, *Science,* **175,** 1069–1076.

Deevey, E. S., 1942. Studies on Connecticut lake sediments III, the biostratonomy of Linsley Pond, *Amer. J. Sci.,* **240,** 233–264, 313–334.

Deevey, E. S., 1955. Some biogeographic implications of paleolimnology, *Verh. Internat. Ver. Limnol.,* **12,** 278–283.

De Geer, G., 1940. Geochronologia Suecica Principales, *K. Svenska Vetensk. Handl.,* ser. 3, **18**(6), Stockholm: Almqvist & Wiksells, Test and Atlas, 367p.

Delorme, L. D., 1970. Paleoecological determinations using Pleistocene freshwater ostracodes, in Oertli, 1970, 341–347.

Eagar, R. M. C., 1948. Variation in shape of shell with respect to ecological station. A review dealing with Recent Unionidae and certain species of the Anthracosiidae in Upper Carboniferous times, *Trans. Roy. Soc. Edinburgh,* **63,** 130–147.

Eagar, R. M. C., 1973. Variation in shape of shell in relation to palaeoecological station in some non-marine Bivalvia of the coal measures of south-east Kentucky and of Britain, *Septième Congrès Internationale de Stratigraphie et de Géologie du Carbonifère,* vol. 1, 386–416.

Eugster, H. P., and Hardie, L. A., 1975. Sedimentation in an ancient playa-lake complex: The Wilkins Peak Member of the Green River Formation of Wyoming, *Geol. Soc. Amer. Bull.,* **86,** 319–334.

Eugster, H. P., and Surdam, R. C., 1973. Depositional environment of the Green River Formation of Wy-

*Publication authorized by the Director, U.S. Geological Survey.

oming: A preliminary report, *Geol. Soc. Amer. Bull.*, **84**, 1115–1120.

Evans, G. H., 1972. The diatom flora of the Hoxnian deposits at Marks Tey, Essex, *New Phytologist*, **71**, 379–386.

Fagerstrom, J. A., 1964. Fossil communities in paleoecology: Their recognition and significance, *Geol. Soc. Amer. Bull.*, **75**, 1197–1216.

Feth, J. H., 1961. A new map of western Conterminous United States showing the maximum known or inferred extent of Pleistocene lakes, *U.S. Geol. Surv. Prof. Paper 424-B*, B110–B111.

Feth, J. H., 1963. Tertiary lake deposits in western conterminous United States, *Science*, **139**, 107–110.

Feth, J. H., 1964. Review and annotated bibliography of ancient lake deposits (Precambrian to Pleistocene) in the Western States, *U.S. Geol. Surv. Bull. 1080*, 119p.

Frey, D. G., 1955. Langsee: A history of meromixis, *Mem. Ist. Ital. Idrobiol. Dott Marco de Marchi*, suppl. 8, 141–164.

Frey, D. G., 1964. Remains of animals in Quaternary lake and bog sediments and their interpretation, *Ergebnisse Limnol.*, **2**, 1–114.

Frey, D. G., 1969. The rationale of paleolimnology, *Mitt. Internat. Verein. Limnol.*, **17**, 7–18.

Frey, D. G., 1974. Paleolimnology, *Mitt. Internat. Verein. Limnol.*, **20**, 95–123.

Frye, J. C.; Leonard, A. B.; Willman, H. B.; and Glass, H. D., 1972. Geology and paleontology of late Pleistocene Lake Saline, southeastern Illinois, *Illinois State Geol. Surv. Circ. 471*, 44p.

Gibson, G. G., 1967. Pleistocene non-marine Mollusca of the Richardson Lake deposit, Clarendon Township, Pontiac County, Quebec, Canada, *Sterkiana*, **25**, 1–36.

Gorham, E., and Sanger, J. E., 1976. Fossilized pigments as stratigraphic indicators of cultural eutrophication in Shagawa Lake, northeastern Minnesota, *Geol. Soc. Amer. Bull.*, **87**, 1638–1642.

Goulden, C. E., 1964. The history of the cladoceran fauna of Esthwaite Water (England) and its limnological significance, *Archiv Hydrobiologie*, **60**, 1–52.

Goulden, C. E., 1966. La aguada de Santa Ana Vieja: An interpretive study of the cladoceran microfossils, *Archiv Hydrobiologie*, **62**, 373–404.

Hanley, J. H., 1974. Systematics, paleoecology, and biostratigraphy of nonmarine Mollusca from the Green River and Wasatch Formations (Eocene), southwestern Wyoming and northwestern Colorado. Ph.D. dissertation, Univ. Wyoming, Laramie, Wyoming, 285p.

Hanley, J. H., 1976. Paleosynecology of nonmarine Mollusca from the Green River and Wasatch Formations (Eocene), southwestern Wyoming and northwestern Colorado, in R. W. Scott and R. R. West, eds., *Classification and Structure of Paleocommunities*. Stroudsburg, Pa.: Dowden, Hutchinson, & Ross, 235–261.

Harnisch, Otto, 1927. Einige Daten zur rezenten und fossilen testaceen Rhizopodenfauna der Sphagnen, *Archiv Hydrobiologie, (Plankt.)*, **18**, 345–360.

Haworth, E. Y., 1972. Diatom succession in a core from Pickerel Lake, northeastern South Dakota, *Geol. Soc. Amer. Bull.*, **83**, 157–172.

Hecky, R. E., and Degens, E. T., 1973. Late Pleistocene–Holocene chemical stratigraphy and paleolimnology of the Rift Valley Lakes of Central Africa, *Techn. Rept. Woods Hole Oceanogr. Inst. (WHOI-73-28)*, 114p.

Herrington, H. B., and Taylor, D. W., 1958. Pliocene and Pleistocene Sphaeriidae (Pelecypoda) from the central United States, *Univ. Mich. Mus. Zool. Occ. Paper 596*, 1–28.

Hibbard, C. W., and Taylor, D. W., 1960. Two late Pleistocene faunas from southwestern Kansas, *Contrib. Mus. Paleontol. Univ. Michigan*, **26**, 1–223.

Horie, S., ed., 1974. *Paleolimnology of Lake Biwa and the Japanese Pleistocene*. Published privately, Otsu Hydrobiological Station, Kyoto University, 288p.

Hubbs, C. L.; Miller, R. R.; and Hubbs, L. C., 1974. Hydrographic history and relict fishes of the North-Central Great Basin, *Calif. Acad. Sci. Mem. 8*, 259p.

Hutchinson, G. E., and Wollack, A., 1940. Studies on Connecticut lake sediments II, chemical analyses of a core from Linsley Pond, North Branford, *Amer. J. Sci.*, **238**, 493–517.

Hutchinson, G. E., et al., 1970. Ianula: An account of the history and development of Lago di Monterossi, Latium, Italy, *Trans. Amer. Phil. Soc.*, n.s., **60**(4), 1–178.

Johansen, A. C., 1904. Om den fossile kvartære Molluskfauna i Danmark og dens Relationer til Forandringer i Klimaet. Land- og Ferskvandsmolluskfaunaen, *Disputats., Kobenhavn, Gyldendalske Boghandel*, 1–136.

Johnson, R. G., 1960. Models and methods for analysis of the mode of formation of fossil assemblages, *Geol. Soc. Amer. Bull.*, **71**, 1075–1086.

Kendall, R. L., 1969. An ecological history of the Lake Victoria basin, *Ecol. Monogr.*, **39**, 121–176.

Kozlovskaya, L. S., 1961. Znacheniye presnovodnykh molluskovov izuchenii Golotsena SSSR, in M. I. Neistadt and V. Gudelis, *Voprosy Golotsena*. Vilnius, Lith. SSR, 157–175.

LaRocque, A., 1960. Molluscan faunas of the Flagstaff Formation of central Utah, *Geol. Soc. Amer. Mem. 78*, 100p.

LaRocque, A., 1966. Pleistocene Mollusca of Ohio; *Ohio Geol. Surv. Bull.*, **62**(1), 1–111.

Lawrence, D. R., 1968. Taphonomy and information losses in fossil communities, *Geol. Soc. Amer. Bull.*, **79**, 1315–1330.

Lawrence, D. R., 1971. The nature and structure of paleoecology, *J. Paleontology*, **45**, 593–607.

Leonard, A. B., 1950. A Yarmouthian molluscan fauna in the midcontinent region of the United States, *Univ. Kansas Paleontol. Contrib., Mollusca 3, Article*, **8**, 1–48.

Lüttig, G., 1955. Die Ostrakoden des Interglazials von Elze, *Paläont. Zeitschr.*, **29**, 146–169.

Lüttig, G., 1959. Die Ostrakoden des Spätglazials von Tatzmannsdorf (Burgenland), *Paläont. Zeitschr.*, **33**, 185–197.

MacGinitie, H. D., 1953. Fossil plants of the Florissant beds, Colorado, *Carnegie Inst. Washington Pub. 599*, 198p.

MacGinitie, H. D., 1969. The Eocene flora of northwestern Colorado and northeastern Utah, *Calif. Univ. Pub. Geol. Sci., No. 83*, 140p.

McKenzie, K. G., 1970. Paleozoogeography of freshwater Ostracoda, in Oertli, 1970, 207–237.

Mackereth, F. J. H., 1971. On the variation of the direction of the horizontal component of remanent magnetisation in lake sediments, *Earth Planetary Sci. Lett.,* **12**, 332–338.

McLeroy, C. A., and Anderson, R. Y., 1966. Laminations of the Oligocene Florissant lake deposits, Colorado, *Geol. Soc. Amer. Bull.,* **77**, 605–618.

Megard, R. O., 1964. Biostratigraphic history of Dead Man Lake, Chuska Mountains, New Mexico, *Ecology,* **45**, 529–546.

Miller, B. B., 1966. Five Illinoian molluscan faunas from the southern Great Plains, *Malacologia,* **4**, 173–260.

Moyle, J. B., 1945. Some chemical factors influencing the distribution of aquatic plants in Minnesota, *Amer. Midland Naturalist,* **34**, 402–420.

Mullens, M. C., 1973. Bibliography of the geology of the Green River Formation, Colorado, Utah, and Wyoming, to March 1, 1973, *U.S. Geol. Surv. Circ. 675,* 20p.

Norris, G., and McAndrews, J. H., 1970. Dinoflagellate cysts from post-glacial lake muds, Minnesota, U.S.A., *Rev. Paleobot. Palynol.,* **10**, 131–156.

Oertli, H. J., ed., 1970. Colloqium on the Paleoecology of Ostracodes, *Bull. Centre Rech. Pau,* suppl. **5**, 207–237, 341–347.

Olsen, P. E.; Remington, C. L.; Cornet, B.; and Thomson, K. S., 1978. Cyclic change in Late Triassic lacustrine communities, *Science,* **201**, 729–733.

Richardson, E. S., Jr., and Johson, R. G., 1971. The Mazon Creek faunas, Proc. N. Amer. Paleontological Conv., pt. I, 1222–1235.

Richardson, J. L., and Richardson, A. E., 1972. History of an African Rift Lake and its climatic implications, *Ecol. Monogr.,* **42**, 499–534.

Royse, C. F. Jr., 1970. A sedimentologic analysis of the Tongue River–Sentinel Butte interval (Paleocene) of the Williston Basin, western North Dakota, *Sed. Geol.,* **4**, 19–80.

Ryder, R. T.; Fouch, T. D.; and Elison, J. H., 1976. Early Tertiary sedimentation in the Western Uinta Basin, Utah, *Geol. Soc. Amer. Bull.,* **87**, 496–512.

Sanger, J. E., and Gorham, E., 1972. Stratigraphy of fossil pigments as a guide to the post-glacial history of Kirchner Marsh, Minnesota, *Limnol. Oceanogr.,* **17**, 840–854.

Seilacher, A., 1973. Biostratinomy: The sedimentology of biologically standardized particles, in R. N. Ginsburg, ed., *Evolving Concepts in Sedimentology.* Baltimore: Johns Hopkins Univ. Press, 159–177.

Sparks, B. W., 1957. The non-marine Mollusca of the interglacial deposits at Bobbitshole, Ipswich, *Phil. Trans. Roy. Soc. London, Ser. B,* **241**, 33–44.

Sparks, B. W., 1961. The ecological interpretation of Quaternary non-marine Mollusca, *Proc. Linnean Soc. London,* **172**, 71–80.

Stark, D. M., 1971. A paleolimnological study of Elk Lake in Itasca State Park, Clearwater County, Minnesota. Ph.D. Thesis, Univ. of Minnesota, 178p.

Steinecke, F., 1929. Die Nekrozönosen des Zehlaubruches. Studie uber die Entwicklung des Hochmoors an Hand der fossilen Mikroorganismen, *Schr. Phys.-okon. Ges. Königsb.,* **66**, 193–214.

Stuiver, M., 1970. Long-term C14 variations, in I. U. Olsson, ed., *Radiocarbon Variations and Absolute Chronology, Proc. 12th Nobel Symposium.* New York: Wiley, 197–213.

Surdam, R. C., and Wolfbauer, C. A., 1975. Green River Formation, Wyoming: A playa lake complex, *Geol. Soc. Amer. Bull.,* **86**, 335–345.

Swain, A. M., 1973. A history of fire and vegetation in northeastern Minnesota as recorded in lake sediments, *Quaternary Research,* **3**, 383–396.

Tasch, P., 1958. Permian conchostracan-bearing beds of Kansas *J. Paleontology,* **32**, 525–540.

Tasch, P., 1963. Paleolimnology: Part 3–Marion and Dickinson Counties, Kansas, with additional sections in Harvey and Sedgwick Counties: Stratigraphy and biota, *J. Paleontology,* **37**, 1233–1251.

Tasch, P., and Volkheimer, W., 1970. Jurassic conchostracans from Patagonia, *Univ. Kansas Paleont. Contrib. Paper 50,* 23p.

Tasch, P., and Zimmerman, J. K., 1961. Comparative ecology of living and fossil conchostracans in a seven county area of Kansas and Oklahoma *Univ. Wichita Bull.,* **36** (Univ. Studies no. 47), 14p.

Taylor, D. W., 1954. A new Pleistocene fauna and new species of fossil snails from the High Plains, *Univ. Michigan Mus. Zool., Occ. Paper 557,* 1–16.

Taylor, D. W., 1960a. Distribution of the freshwater clam *Pisidium ultramontanum:* A zoogeographic inquiry, *Am. J. Sci.,* ser. 5, **258-A**, 325–334.

Taylor, D. W., 1960b. Late Cenozoic molluscan faunas from the High Plains, *U.S. Geol. Surv. Prof. Paper 337,* 94p.

Taylor, D. W., 1965. The study of Pleistocene nonmarine mollusks in North America, *in* Wright and Frey, 1965, 597–611.

Taylor, D. W., 1966. Summary of North American Blancan nonmarine mollusks, *Malacologia,* **4**, 1–172.

Taylor, D. W., and Hibbard, C. W., 1955. A new Pleistocene fauna from Hooper County, Oklahoma, *Oklahoma Geol. Surv. Circ.,* **37**, 1–23.

Tippett, Roger, 1964. An investigation into the nature of the layering of deep-water sediments in two eastern Ontario lakes, *Canadian J. Bot.,* **42**, 1693–1709.

Tsukada, M., 1967. Fossil Cladocera in Lake Nojiri and ecological order, *Quaternary Research* (Tokyo), **6**, 101–110.

Van der Schalie, H., 1944 (1945). The value of mussel distribution in tracing stream confluence, *Michigan Acad. Sci., Arts, and Letters, Papers,* **30**, 355–373.

Van der Schalie, H., 1963. Mussel distribution in relation to former stream confluence in northern Michigan, U.S.A., *Malacologia,* **1**, 227–236.

Van Landingham, S. L., 1964. Miocene non-marine diatoms from the Yakima region on south-central Washington, *Nova Hedwigia,* **14**, 78p.

Wanless, H. R., 1969. Marine and non-marine facies of the Upper Carboniferous of North America, *C. R. 6e Congr. Internat. Strat. Geol. Carbonif.,* pt. I, 293–336.

Watts, W. A., and Bright, R. C., 1968. Pollen, seed and mollusk analysis of a sediment core from Pickerel Lake, northeastern South Dakota, *Geol. Soc. Amer. Bull.,* **79**, 855–876.

Watts, W. A. and Winter, T. C., 1966. Plant macrofossils from Kirchner Marsh, Minnesota–A paleoecological study, *Geol. Soc. Amer. Bull.,* **77**, 1339–1360.

Whiteside, M. C., 1970. Danish Chydorid Cladocera: Modern ecology and core studies, *Ecol. Monogr.,* **40**, 79–118.

Wilcox, R. E., 1964. Volcanic-ash chronology, in Wright, and Frey, 1965, 807–816.

Wright, H. E., Jr., and Frey, D. G., eds., 1965. *The Quaternary of the United States.* Princeton: Princeton Univ. Press, 597–611, 807–816.

Zangerl, R., and Richardson, E. S. Jr., 1963. The paleoecological history of two Pennsylvanian black shales, *Fieldiana, Geol. Mem. 4,* 352p.

Cross-references: *Algae; Branchiopoda; Charophyta; Diatoms; Mazon Creek; Ostracoda; Paleoecology.* Vol. IVA: *Limnology; Lake Geochemistry; Paleohydrology; Paleolimnology.*

PALEOECOLOGY, TERRESTRIAL

Ecology is that field of biology concerned with the interactions among individuals of the same and different species and between organisms and their abiotic environment. Traditional paleoecology, because of its origin as a subdiscipline of geology, has been concerned primarily with attempts to reconstruct ancient depositional environments. The organisms themselves have been of interest insofar as their study furthered this goal.

Paleobotanists and palynologists have thus generally been more interested in using plant fossils to interpret ancient environments and climate than in considering the natural histories of the prehistoric plants themselves (e.g., Axelrod, 1965; Srivastava, 1970; Habib and Groth, 1967; Habib, 1968; Axelrod and Bailey, 1969; Wolfe 1971; Daley, 1972; Krassilov, 1973a,b; Gibbard and Stuart, 1975). The same is true for paleontologists working with terrestrial invertebrates (exceptions include the extremely interesting speculations about plant-invertebrate coevolution; e.g., Hughes and Smart, 1967; Kervan et al., 1975). Most vertebrate paleontologists have also approached paleoecology from this point of view.

In recent years, however, many workers have begun to approach paleoecological studies in a manner that facilitates ecological consideration of ancient organisms themselves. This review will concentrate on these kinds of studies. Because vertebrate paleontologists have been most active in this field of paleoecology, discussion will be limited to papers dealing with fossil vertebrates.

Paleoecological interpretation is more difficult for terrestrial as opposed to marine or even freshwater environments for the simple reason that dry-land situations serve as sites of sediment accumulation (and thus fossil preservation) only under rather special circumstances. Most fossils of terrestrial animals are found in water-deposited sedimentary situations and thus represent a small, probably biased, sample of the living dry-land fauna. The primary job of the terrestrial paleoecologist is elucidating and attempting to correct for the effects of such biased sampling. Such problems constitute the subject matter of *taphonomy* (q.v.), the study of the fate of organisms from the time they die until their bodies are either destroyed by biological or abiotic agencies or become fossilized (Efremov, 1940, 1953).

Catastrophic Versus Attritional Mortality

There are several potential sources of mortality in terrestrial vertebrates—predation, starvation, disease, intraspecific combat, physical accident—some of which are more likely than others. From the viewpoint of the paleoecologist, these agents of mortality can be categorized as attritional or catastrophic. Attritional mortality represents normal, day-to-day deaths of animals dying over a long period of time. Catastrophes, on the other hand, sample the fauna over a very short period of time, often in the form of mass kills of animals that become fossilized in a high degree of completeness. Catastrophic mortality agents include floods; drought; wind, sand or snow storms; volcanic eruptions, and mass panic in herding animals (as when living African ungulates attempt to cross watercourses along which predators wait to ambush them, resulting in stampedes and the drowning of many herd members).

Mass Assemblages as Sources of Paleoecological Information

In late Tertiary continental formations of western North America, large-mammal faunas often occur as quarriable concentrations of bone in coarse-grained sediments. The bones are hydraulically equivalent to the enclosing sediments and show signs of water transport, such as breakage, abrasion, and size sorting. These mass occurrences are usually dominated by grazing animals, such as horses and antilocaprids. Herds of such animals probably made frequent trips to watercourses for drinking or bathing, and may occasionally have been stampeded into the water by lurking predators. Herding animals also probably sought shelter in flood-plain forests during violent storms and were sometimes killed by flash floods. Water-dependent herbivores also tend to concentrate in lowland areas during drought, and many may die when food proximal to water sources is exhausted (Shipman, 1975). Catastrophic deaths of this nature were probably infrequent compared to attritional mortality in the preferred upland habitat, but proximity to the depositional environment would increase chances of fossilization for the victims of catastrophe.

This suggests that sampling error is a serious problem in attempts to reconstruct terrestrial faunas from mass assemblages. Voorhies (1970) noted that at Verdigre, Nebraska, the most abundant mammal represented in the Pliocene Valentine Formation is the antilocaprid *Merycodus*, with about 500 individuals preserved. In the same formation at the Devil's Gulch Horse Quarry, 100 miles W of Verdigre, there is a large number of mammalian fossils—but no *Merycodus*. The Verdigre assemblage had a catastrophic origin and represents animals that died over an estimated period of about a week. The numerical dominance of *Merycodus* in the quarry may not reflect its dominance in the living community of the region, but only that a herd of these highly mobile animals happened to be around when calamity struck. Similarly, if the Devil's Gulch quarry is also catastrophic in origin, the absence of *Merycodus* may only mean that no herds were in the region at the time of the kill. Differences between the two fossil faunas do not necessarily reflect ecological differences between the living communities.

Another potential source of error associated with mass assemblages is sample size. At the Verdigre quarry, bones of the browsing horse *Hypohippus* were almost as common as those of the grazing horse *Protohippus* after the first season of excavation, but, after three seasons of work, *Protohippus* bones outnumbered *Hypohippus* elements by twenty to one. Most fossils of the carnivore *Leptocyon* were collected during the second season of work, and nearly all the oreodonts during the third. Mass assemblages are thus particularly susceptible to "pocketing" effects.

Even where mass assemblages do not necessarily represent mass deaths there may be serious sampling errors. Examples include the famous Rancho La Brea tar pits of the California Pleistocene and the Cleveland–Lloyd quarry of the Utah Jurassic where carnivorous mammals and flesh-eating dinosaurs respectively are preserved in unusually high numbers relative to their prey species. Such "predator trap" situations are unlikely to reflect the predator/prey ratio of the living community.

In short, a death assemblage composed of large numbers of complete skeletons is likely to represent highly localized conditions of death and/or transport, and extrapolating from these assemblages to the living community can be risky. The catch is that the high degree of preservation of such assemblages makes them particularly attractive to collectors.

Attritional Mortality Situations as Sources of Paleoecological Information

Attritional mortality, producing more or less disarticulated and fragmented specimens, is more likely to reflect actual community composition, at least for large animals that lived near the depositional environment. Small vertebrates, with high population turnover rates, actually produce more carcasses per unit time than larger species, but these more delicate carcasses are much more readily destroyed by carnivorous animals or removed/destroyed by physical agents. Consequently, small vertebrates are grossly under-represented in most fossil assemblages, while bones of larger species have a greater chance of preservation.

Animals that die in the water have a greater probability of preservation than those dying in upland situations. In addition to the catastrophic situations described earlier, animals can accidentally drown or otherwise die in or near water because they habitually frequent watercourses and their margins.

Complete or nearly complete carcasses can be floated from the place of death by current action as long as they remain lighter than water, as when decompositional gases remain trapped inside the body. For example, terrestrial dinosaurs are sometimes found preserved in marine sediments, perhaps reflecting a post-mortem downstream cruise to a final estuarine resting place. If a carcass is dehydrated by exposure to sun and air (e.g., dinosaur "mummies") it becomes unattractive to scavenging carnivores and can be carried by flood waters for some distance without disarticulation.

Ultimately, a floating carcass sinks and begins to disarticulate as decomposition continues. Mammalian carcasses submerged in water may disarticulate in a couple months, but can last longer. The process is naturally accelerated if aquatic predators not particularly offended by fragrant carcasses, such as crocodiles and sharks, are around. As disarticulation proceeds, the skeleton becomes increasingly susceptible to the winnowing effects of current action and bones are removed as a function of their size, shape, and density (Table 1). Dinosaurs from the Late Cretaceous Oldman Formation of western Canada are preserved in just such a progressive sequence of decompositional classes (Fig. 1).

Once the bones of animals are buried they are

still susceptible to various agents of destruction. Bioturbation by plant roots and burrowing animals can readily destroy bones that are wet and friable due to subsurface conditions. On flood plains, bones have the best chance of survival if the sediment increment covering them after floods exceeds the average depth of bioturbation (10–50 cm). Sediment compaction can distort or crush buried bone, particularly in clay-rich sediments that lose large amounts of water during compaction. After lithification, clay units often undergo a good bit of expansion and contraction during weathering, with destructive effects on enclosed bones. Sandy units are less destructive to enclosed bones in this regard. Bones dissolve in acidic soils or sediments and have optimal chances of preservation in alkaline, calcium carbonate-rich sediments. Too much carbonate, however, can also be destructive.

Behrensmeyer (1975) concluded that the

TABLE 1. Elements of the Skeleton of Coyote- and Sheep-sized Mammals Grouped According to their Characteristic Susceptibility to Transport (from Voorhies, 1969.)[a]

GROUP I Immediately removed, transported by saltation or flotation)	GROUP II (Removed gradually, transported by traction)	GROUP III (Lag deposit)
RIBS	FEMUR	SKULL
VERTEBRAE	TIBIA	MANDIBLE
SACRUM	HUMERUS	
STERNUM	METAPODIA	ramus
	PELVIS	
scapula[b]	RADIUS	
phalanges		
ulna	scapula[b]	
	ramus	
	phalanges	
	ulna	

[a]This classification holds only for species of the size of coyotes or sheep. For smaller (see Dodson, 1973) or larger (see Behrensmeyer, 1975) species, the order of transport will differ, depending on the sizes, densities, and shapes of the bones involved.

[b]Elements in lower-case type are intermediate between the two groups in which they appear.

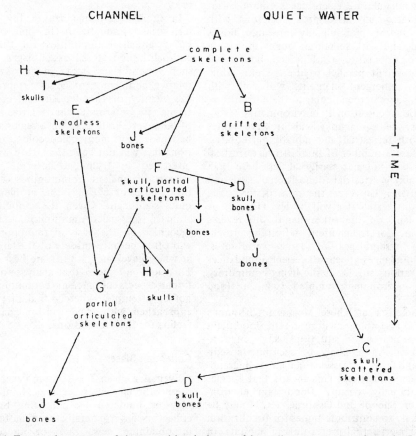

FIGURE 1. Temporal sequence of decompositional classes of large-dinosaur carcasses from the Late Cretaceous Oldman Formation of Canada (from Dodson, 1971). Note the differences in sequence depending on whether the carcass decomposes in quite or running water. All the decompositional classes are represented in the Oldman fauna. While predator-scavenger animals will accelerate skeletal disarticulation and produce specimens of a given decompositional class sooner than rotting and current action alone, this does not appear to have happened to the Oldman dinosaurs.

optimal assemblage of disarticulated bones for reconstructing community composition is where: (1) the bones represent a wide range of dispersal potentials—as determined by bone size, density, and shape; (2) the numbers of different skeletal elements are not significantly altered from the numbers of these bones in an average skeleton; (3) the bones are hydraulically dissimilar to the associated sedimentary particles; and (4) the bones retain fresh, unweathered or unabraded surfaces. The bones of a given carcass need not be associated, though this is obviously desirable. It is also useful, as Voorhies (1969) has noted, if the bones show no preferred orientation with respect to paleocurrent. Most animals represented in such an assemblage were preserved in more or less the same habitat in which they lived. Many North American early Teritiary flood-plain faunas appear to meet these criteria.

Similarly, primarily autochthonous (rather than carried from elsewhere by stream action), but less reliable, are assemblages composed of disarticulated but associated skeletons with smaller and/or hydraulically unstable bones removed, and with remaining bones showing a preferred orientation. Submerged bones will probably orient parallel to the current, while partically emergent elements will roll with their long axes perpendicular to the current.

Aquatic depositional environments (e.g., channels, delta margins) will preserve bones from both terrestrial and aquatic vertebrates. The absolute number of bones from terrestrial animals will not necessarily be less here than in a more terrestrial depositional situation (flood plain). However, the ratio of terrestrial to aquatic vertebrates will be less (Fig. 2). A flood plain, on the other hand, will reverse this ratio. For fragmented, definitely autochthonous assemblages of large vertebrates, fossil abundances appear to reflect abundances of the various species in the living community of the environment sampled (e.g., a flood plain).

But there is a catch here. Most autochthonous terrestrial assemblages will represent flood plain or other near-water situations, such environments being more likely to accumulate sediments and bury carcasses than are topographically higher habitats. This means that species that habitually occupy flood-plain environments (cf. Sheppe and Osborne, 1971) may be more likely to contribute bones for fossilization than are species that prefer upland habitats. In that case, upland species will be under-represented if one attempts to reconstruct the relative abundance of the various fossil vertebrates of a large region (consisting of many kinds of lowland and upland habitats) on

the basis of fossils preserved in flood-plain situations.

Several factors may diminish this problem. Because large vertebrates are fairly mobile animals, many species will regularly pass between upland and lowland habitats. If large herds of upland herbivores visit flood plains for drinking, feeding, bathing, or shelter from storms, individuals may frequently die there due to attritional ("natural") or catastrophic mortality agents. This will increase their relative abundance in future fossil assemblages, *perhaps* to the point where their numbers in the flood-plain assemblages, relative to those of lowland species, *may* approximate those of the regional fauna. One might expect this sort of thing in assemblages of animals that died due to the effects of drought (cf. Shipman, 1975). However, the chances of sampling error (especially if the upland forms occur as mass assemblages within the overall flood-plain setting) are such as to make such an assumption risky.

In short, while one can readily distinguish terrestrial vs. aquatic or (in some cases) forest vs. savannah communities on the basis of authochthonous fossil assemblages, and while such assemblages may broadly reflect the relative abundance of large vertebrate taxa in the vicinity of the depositional environment, trying to determine the relative abundance of large vertebrate taxa on a regional basis will be trickier. For many paleoecological considerations this will not be critical. However, if one is interested in trying to interpret functional aspects of ancient communities (e.g., energy flow, nutrient cycling) the problem becomes quite significant. Given the mobility of large animals, the paleocommunities that one can recognize on the basis of taphonomic criteria will often be only subsets of the larger systems at whose level such studies are best conducted. This does not mean that analyses of the functional aspects of ancient communities should not be attempted—merely that they should be approached with great caution and a realistic feeling for their limitations.

Collecting Biases

Even if a given fossil assemblage lacks great preservational biases, ecological interpretation may be hindered by collecting biases. Most collectors have generally been more interested in obtaining new, rare, or unusually well-preserved specimens than in conducting a census for paleoecological purposes. Russell (1967) believed that field workers in the Upper Cretaceous formations of western Canada often left poor specimens of the com-

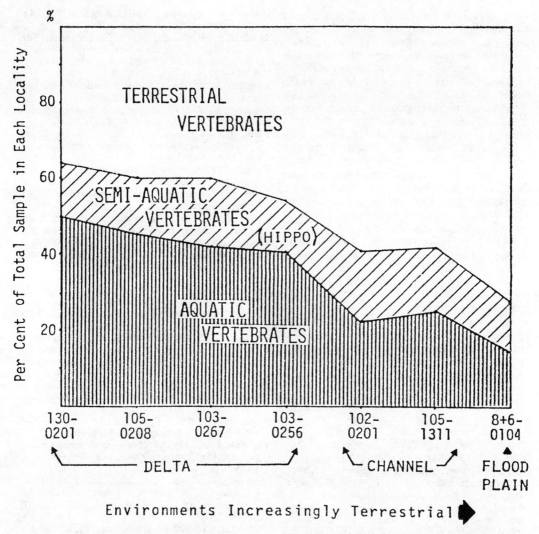

FIGURE 2. Ratios of aquatic, semiaquatic, and terrestrial vertebrate groups represented in assemblages from the Pliocene–Pleistocene Koobi Fora Formation of Kenya (from Behrensmeyer, 1975). The abundance of a given taxon is based on the number of 10-square-meter sample grids in which surface-collected fossils of that taxon are found at each locality, and the relative abundance as given in this figure is the percentage of the cumulative grid frequencies (number of grids) for each locality. The sample localities are arranged from the more aquatic (130–0201, 105–0208, etc.) to the least aquatic (8+6–0104) depositional environments on the basis of geological criteria alone. Aquatic animals include crocodiles and turtles (except *Geochelone*) semiaquatic includes only hippopotamus; all other groups are considered to be terrestrial.

mon hadrosaurs and ceratopsians behind, but would grab any and all specimens of the much less abundant tyrannosaurs they encountered. Although Béland and Russell (1978) now doubt the existence of such biases in the Canadian collections, this does not rule out their existence in other areas. As previously noted, collectors are also attracted to mass assemblages. For these reasons, simply counting the number of museum specimens in the course of a paleoecological study is risky. It is to be hoped that future collecting can be done so that potentially important paleoecologic data is not lost.

Status of Vertebrate Paleoecology

In spite of the difficulties involved in the study of vertebrate paleoecology, much progress has been made. Appreciation of the necessity for taphonomic studies for meaningful paleoecology is spreading rapidly. A number of interesting studies of the ecology of ancient vertebrates have resulted. By developing criteria for determining relative or even absolute onto-genetic ages of specimens in large fossil samples, and taking taphonomic factors into account, it has been possible to study the population dynamics of many kinds of fossil mammals. Olson's (1952, 1962, 1966) concept of the chronofauna provides valuable insight into the structural (and to a certain extent the functional) changes that occurred during the establishment of terrestrially based vertebrate communities and the accompanying evolution of the therapsid reptiles. His work also provides examples of the long-term persistence of given community types that may be of great interest to theoretical ecologists interested in questions of ecosystem stability. Bakker's (1972, 1975) innovative considerations of dinosaur and therapsid bioenergetics and paleoecology, although controversial, may well point the way to at least a partial understanding of the ecological energetics of ancient vertebrate communities (e.g., Farlow, 1976a). A number of workers have even speculated about the behavior of fossil vertebrates (e.g., Galton, 1971; Ostrom, 1972; Barghusen, 1975; Farlow and Dodson, 1975; Hopson, 1975; Farlow, 1976b). Terrestrial paleoecology is a wide-open field for research.

JAMES O. FARLOW

References

Axelrod, D. I., 1965. A method for determining the altitudes of Tertiary floras, *Palaeobotanist,* 14, 1–4.

Axelrod, D. I., and Bailey, H. P., 1968. Cretaceous dinosaur extinction, *Evolution,* 22, 595–611.

Axelrod, D. I., and Bailey, H. P., 1969. Paleotemperature analysis of Tertiary floras, *Palaeogeography, Palaeoclimatology, Palaeoecology,* 6, 163–195.

Bakker, R. T., 1972. Anatomical and ecological evidence of endothermy in dinosaurs, *Nature,* 238, 81–85.

Bakker, R. T., 1975a. Dinosaur renaissance, *Sci. American,* 232(4), 58–78.

Bakker, R. T., 1975b. Experimental and fossil evidence of the evolution of tetrapod bioenergetics, in D. Gates and R. Schmerl, eds., *Perspectives in Biophysical Ecology.* Berlin: Springer-Verlag, 365–399.

Barghusen H. R., 1975. A review of fighting adaptations in dinocephalians (Reptilia, Therapsida), *Paleobiology,* 1, 295–311.

Behrensmeyer, A. K., 1975. The taphonomy and paleoecology of Plio-Pleistocene vertebrate assemblages east of Lake Rudolf, Kenya, *Bull. Mus. Comp. Zool. Harvard Univ.,* 146, 473–578.

Behrensmeyer, A. K., 1978. Taphonomic and ecologic information from bone weathering, *Paleobiology,* 4, 150–162.

Béland, P., and Russell, D. A., 1978. Paleoecology of Dinosaur Provincial Park (Cretaceous), Alberta, interpreted from the distribution of articulated vertebrate remains, *Canadian J. Earth Sci.,* 15, 1012–1024.

Bond, G., and Bromley, K., 1970. Sediments with the remains of dinosaurs near Gokwe, Rhodesia, *Palaeogeography, Palaeoclimatology, Palaeoecology,* 8, 313–327.

Chaloner, W. G., 1967. Spores and land-plant evolution, *Rev. Palaeobot. Palynol.,* 1, 83–93.

Chaloner, W. G., 1970. The rise of the first land plants, *Biol. Rev. Cambridge Phil. Soc.,* 45, 353–377.

Clark, J., and Guensberg, T. E., 1970. Population dynamics of *Leptomeryx, Fieldiana, Geol.,* 16, 411–451.

Clark, J.; Beerbower, J. R.; and Kietzke, K. K., 1967. Oligocene sedimentation, stratigraphy, paleoecology, and paleoclimatology in the Big Badlands of South Dakota, *Fieldiana, Geol.,* 5, 1–158.

Coope, G. R., 1967. The value of Quaternary insect faunas in the interpretation of ancient ecology and climate, in Cushing and Wright, 1967, 359–380.

Cushing, E. J., and Wright, H. E., Jr., eds. 1967. *Quaternary Paleoecology.* New Haven: Yale Univ. Press, 433p.

Daley, B., 1972. Some problems concerning the Early Tertiary climate in southern Britain, *Palaeogeography, Palaeoclimatology, Palaeoecology,* 11, 177–190.

Dodson, P., 1971. Sedimentology and taphonomy of the Oldman Formation (Campanian), Dinosaur Provincial Park, Alberta (Canada), *Palaeogeography, Palaeoclimatology, Palaeoecology,* 10, 21–74.

Dodson, P., 1973. The significance of small bones in paleoecological interpretation, *Univ. Wyoming Contrib. Geol.,* 12, 15–19.

Efremov, J. A., 1940. Taphonomy: A new branch of paleontology, *Pan-Amer. Geol.,* 74, 81–93.

Efremov, J. A., 1953. Taphonomie et annales géologiques, *Ann. Centre d'Etude Doc. Paléontol.,* 4, 164p.

Farlow, J. O., 1976a. A consideration of the trophic dynamics of a Late Cretaceous large-dinosaur community (Oldman Formation), *Ecology,* 57, 841–857.

Farlow, J. O., 1976b. Speculations about the diet and foraging behavior of large carnivorous dinosaurs, *Amer. Midland Naturalist,* 95, 186–191.

Farlow, J. O., and Dodson, P., 1975. The behavioral significance of frill and horn morphology in ceratopsian dinosaurs, *Evolution,* 29, 353–361.

Galton, P, M., 1971. A primitive dome-headed dinosaur (Ornithischia: Pachycephalosauridae) from the Lower Cretaceous of England and the function of the dome of pachycephalosaurids, *J. Paleontology,* 45, 40–47.

Geist, V., 1972. An ecological and behavioral explanation of mammalian characteristics, and their implication to therapsid evolution, *Zeitschr. Säugetierkunde,* 37, 1–15.

Gibbard, P. L., and Stuart, A. J., 1975. Flora and vertebrate fauna of the Barrington Beds, *Geol. Mag.,* 112, 493–501.

Gill, E. D., 1968. Palaeoecology of fossil human skeletons, *Palaeogeography, Palaeoclimatology, Palaeoecology,* 4, 211–217.

Gould, S. J., 1970. Land snail communities and Pleistocene climates in Bermuda: A multivariate analysis of microgastropod diversity, *Proc. N. Amer. Paleontological Conv.*, pt. E, 486-521.

Gould, S. J., 1974. The origin and function of "bizarre" structures: Antler size and skull size in the "Irish Elk," *Evolution*, 28, 191-220.

Gradzinski, R., 1970. Sedimentation of dinosaur-bearing Upper Cretaceous deposits of the Nemegt Basin, Gobi Desert, *Paleontol. Polonica*, 21, 147-229.

Guthrie, R. D., 1967. Differential preservation and recovery of large mammal remains in Alaska, *J. Paleontology*, 41, 243-246.

Guthrie, R. D., 1968. Paleoecology of the large-mammal community in interior Alaska during the Late Pleistocene, *Amer. Midland Naturalist*, 79, 346-363.

Habib, D., 1968. Spore and pollen paleoecology of the Redstone Seam (Upper Pennsylvanian) of West Virginia, *Micropaleontology*, 14, 199-220.

Habib, D., and Groth, P. K. H., 1967. Paleoecology of migrating Carboniferous peat environments, *Palaeogeography, Palaeoclimatology, Palaeoecology*, 3, 185-195.

Hopson, J. A., 1975. The evolution of cranial display structures in hadrosaurian dinosaurs, *Paleobiology*, 1, 21-43.

Hughes, N. F., and Smart, J., 1967. Plant-insect relationships in Palaeozoic and later time, In W. B. Harland et al., eds., *The Fossil Record*. London: Geol. Soc. London, 107-117.

Hunt, R. M., Jr., 1978. Depositional setting of a Miocene mammal assemblage, Sioux County, Nebraska (U.S.A.), *Palaeogeography, Palaeoclimatology, Palaeoecology*, 24, 1-52.

Kervan, P. G.; Chaloner, W. G.; and Saville, D. B. O., 1975. Interrelationships of early terrestrial arthropods and plants, *Palaeontology*, 18, 391-417.

Krassilov, V. A., 1969a. On the reconstruction of extinct plants, *Paleontol. J.*, 3, 1-8.

Krassilov, V. A., 1969b. Types of palaeofloristic successions and their causes, *Paleontol. J.*, 3, 296-308.

Krassilov, V. A., 1973a. Climatic changes in eastern Asia as indicated by fossil floras. I. Early Cretaceous, *Palaeogeography, Palaeoclimatology, Palaeoecology*, 13, 261-273.

Krassilov, V. A., 1973b. Climatic changes in eastern Asia as indicated by fossil floras. II. Late Cretaceous and Danian, *Palaeogeography, Palaeoclimatology, Palaeoecology*, 13, 157-172.

Krassilov, V. A., 1975. *Paleoecology of Terrestrial Plants: Basic Principles and Techniques*. New York and London: Wiley, 283p.

Kurtén, B., 1953. On the variation of population dynamics of fossil and Recent mammal populations, *Acta Zool. Fennica*, 76, 5-118.

Kurtén, B., 1958. Life and death of the Pleistocene cave bear, a study in paleoecology, *Acta Zool. Fennica*, 95, 1-59.

Mahabalé, T. S., 1968. Spores and pollen grains of water plants and their dispersal, *Rev. Palaeobot. Palynol.*, 7, 285-296.

Martin, P. S., and Wright, H. E., Jr., eds., 1967. *Pleistocene Extinctions: The Search for a Cause*. New Haven: Yale Univ. Press, 453p.

Mellet, J. S., 1974. Scatological origin of micro-vertebrate fossil accumulations, *Science*, 185, 349-350.

Müller, A. H., 1957. *Lehrbuch der Paläozoologie. Vol. I: Allgemeine Grundlagen*. Jena: Gustav Fischer Verlag, 322p.

Munthe, K., and Mcleod, S. A., 1975. Collection of taphonomic information from fossil and recent vertebrate specimens with a selected bibliography, *Paleobios, Contrib. Univ. Calif. Mus. Paleont., Berkeley*, 19, 12p.

Olson, E. C., 1952. The evolution of a Permian vertebrate chrono-fauna, *Evolution*, 6, 181-196.

Olson, E. C., 1962. Late Permian terrestrial vertebrates, U.S.A. and U.S.S.R., *Trans. Amer. Phil. Soc.*, 52, 1-224.

Olson, E. C., 1966. Community evolution and the origin of mammals, *Ecology*, 47, 291-302.

Oshurkova, M. V., 1974. A facies-paleoecological approach to the study of fossilized plant remains, *Paleontol. J.*, 8, 363-370.

Ostrom, J. H., 1970. Terrestrial vertebrates as indicators of Mesozoic climates, *Proc. N. Amer. Paleontological Conv.*, pt. D, 347-376.

Ostrom, J. H., 1972. Were some dinosaurs gregarious? *Palaeogeography, Palaeoclimatology, Palaeoecology*, 11, 287-301.

Parrish, W. C., 1978. Paleoenvironmental analysis of a Lower Permian bonebed and adjacent sediments, Wichita County, Texas, *Palaeogeography, Palaeoclimatology, Palaeoecology*, 24, 209-237.

Rasnitsyn, A. P., and Ponomarenko, A. G., 1967. Procedure for approximately estimating the variety of local faunas of the past, *Paleontol. J.*, 1, 95-101.

Russell, D. A., 1967. A census of dinosaur specimens collected in western Canada, *Nat. Mus. Canada Nat. Hist. Papers*, 36, 13p.

Rybníčková, E., and Rybníček, K., 1971. The determination and elimination of local elements in pollen spectra from different sediments, *Rev. Palaeobot. Palynol.*, 11, 165-176.

Samoilovich, S. R.; Mtchedlischivili, N. D.; Griazeva, A. S., Evseeva, G. V., and Lubomirova, K. A., 1971. Method and principles of compiliation of paleovegetation maps from palynological data, *Paleontol. J.*, 5, 236-242.

Saunders, J. J., 1977. Late Pleistocene vertebrates of the western Ozark Highland, Missouri, *Illinois State Mus. Rept. Invest.*, 33, 118p.

Schopf, J. M., 1975. Modes of fossil preservation, *Rev. Palaeobot. Palynol.*, 20, 27-53.

Sheppe, W., and Osborne, T., 1971. Patterns of use of a flood plain by Zambian mammals, *Ecol. Monogr.*, 41, 179-205.

Shipman, P., 1975. Implications of drought for vertebrate fossil assemblages, *Nature*, 257, 667-668.

Shotwell, J. A., 1955. An approach to the paleoecology of mammals, *Ecology*, 36, 327-337.

Shotwell, J. A., 1958. Intercommunity relationships in Hemphillian (Mid-Pliocene) mammals, *Ecology*, 39, 271-282.

Shotwell, J. A., 1963. The Juntura Basin: Studies in earth history and paleoecology, *Trans. Amer. Phil. Soc.*, 53, 3-77.

Sloan, R. E., 1969. Cretaceous and Paleocene terrestrial communities of western North America,

Proc. N. Amer. Paleontological Conv., pt. E, 427–453.

Smart, J., and Jughes, N. F., 1973. The insect and the plant: Progressive palaeoecological integration, *Symp. Roy. Entomol. Soc. London,* **6**, 143–155.

Spotila, J. R.; Lommen, P. W.; Bakken, G. S., and Gates, D. M., 1973. A mathematical model for body temperatures of large reptiles: Implications for dinosaur ecology, *Amer. Naturalist,* **107**, 391–404.

Srivastava, S. K., 1970. Pollen biostratigraphy and paleoecology of the Edmonton Formation (Maestrichtian), Alberta, Canada, *Palaeogeography, Palaeoclimatology, Palaeoecology,* **7**, 221–276.

Stanley, S. M., 1974. Relative growth of the titanothere horn: A new approach to an old problem, *Evolution,* **28**, 447–457.

Toots, H., 1965. Reconstruction of continental environments: The Oligocene of Wyoming. Ph. D. Thesis, Univ. of Wyoming, 176p.

Vaughn, P. P., 1970. Lower Permian vertebrates of the Four Corners and the midcontinent as indices of climatic differences, *Proc. N. Amer. Paleontological Conv.,* pt. D, 388–408.

Voorhies, M. R., 1969. Taphonomy and population dynamics of an Early Pliocene vertebrate fauna, Knox County, Nebraska, *Contrib. Geol. Univ. Wyoming Spec. Paper, no. 1,* 69p.

Voorhies, M. R., 1970. Sampling difficulties in reconstructing Late Tertiary mammalian communities, *Proc. N. Amer. Paleontological Conv.,* pt. E, 454–468.

Weighelt, J., 1927. *Rezente Wirbeltierleichen und ihre paläobiologische Bedeutung.* Leipzig: Max Weg, 227p.

Wolfe, J. A., 1971. Tertiary climatic fluctuations and methods of analysis of Tertiary floras, *Palaeogeography, Palaeoclimatology, Palaeoecology,* **9**, 27–57.

Wolff, R. G., 1973. Hydrodynamic sorting and ecology of a Pleistocene mammalian assemblage from California, *Palaeogeography, Palaeoclimatology, Palaeoecology,* **13**, 91–101.

Wolff, R. G., 1975. Sampling and sample size in ecological analyses of fossil mammals, *Paleobiology,* **1**, 195–204.

Woodward, G. D., and Marcus, L. F., 1973. Rancho La Brea fossil deposits: A re-evaluation from stratigraphic and geological evidence, *J. Paleontology,* **47**, 54–69.

Cross-references: *Amphibia; Aves; Bones and Teeth; Communities, Ancient; Coprolite; Dinosaurs; Fossil Record; Gymnosperms; Insecta; Mammals, Mesozoic; Mammals, Placental; Marsupials; Morphology, Functional; Multituberculates; Paleobiogeography of Vertebrates; Paleobotany; Paleoecology, Inland Aquatic Environments; Paleopathology; Palynology; Palynology, Paleozoic and Mesozoic; Palynology, Tertiary; Reptilia; Sexual Dimorphism; Taphonomy; Trace Fossils; Vertebrata; Vertebrate Paleontology.*

PALEOGEOGRAPHIC MAPS

The first attempt to draw maps showing past continental distributions was made by Wegener (1924), who, of course, proposed that continents had drifted relative to one another through geological time. For some unknown reason his maps were not well drawn. Several attempts were made to improve maps of the world as it may have been in past periods (e.g., Du Toit, 1937; King, 1962; Carey, 1958), though most of the effort was devoted to improving maps of the southern continents (Gondwanaland) before and after its breakup.

In 1965, Bullard, Everett, and Smith presented the first maps based on the computer-fitting of the edges of all the continents around the Atlantic Ocean. The maps were drawn by hand from numerical data. In 1970, a possible reassembly of Gondwanaland was drawn by computer using a modified version of R. L. Parker's map-making program SUPERMAP (Smith and Hallam, 1970). The Atlantic Ocean and southern continent reconstructions were joined together to form Wegener's Pangaea (Briden et al., 1971; Smith, 1971), thus for the first time reproducing by precise methods one of his world maps.

Until paleomagnetic and ocean-floor magnetic anomaly data were available, continental distributions intermediate in time between the existence of Pangaea (Permian–Triassic time) and the present-day were based mostly on inspiration (e.g., Argand, 1924). Currently the spreading histories of the Atlantic Ocean (Le Pichon and Hayes, 1971; Pitman and Talwani, 1972) and Indian Ocean (McKenzie and Sclater, 1971; Sclater and Fisher, 1974) are sufficiently well known for world maps of intermediate periods to be drawn. There are uncertainties, some not negligible, discussed below.

All the maps in this article have been traced directly from computer-drawn output (Briden et al., 1974). The data on which they are based are either geophysical (paleomagnetic pole positions; ocean-floor spreading anomalies) or topographic (the 500-fathom or 1000-m submarine contours), published up to early 1972.

Lambert Equal-Area Projections of the Entire Sphere

The advantages of the Lambert projection used for Figs. 1 to 9 are that it enables the entire sphere to be projected economically on one map and that it is an equal-area projection. The equal-area properties may be particularly useful for many global density distributions. The maps all have their centers of projection on the equator or paleoequator. The "front" hemisphere on the maps is familiar to paleontologists: it is bounded by

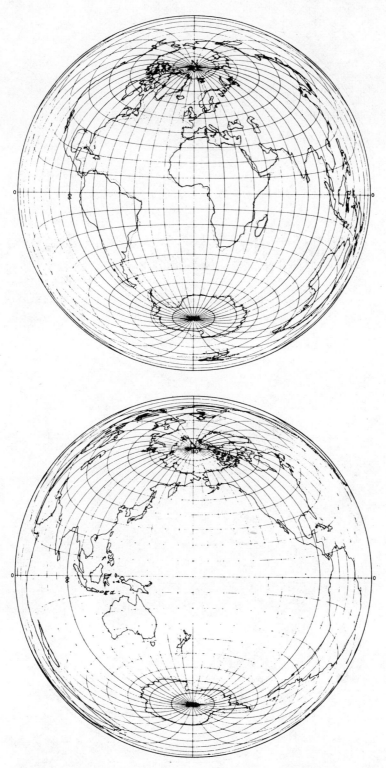

FIGURE 1. Present-day world. Lambert equal-area projection; latitude-longitude grid at 10° intervals. (from Briden et al., 1974, copyright 1974 by The University of Chicago Press). Above, front view; below, back view.

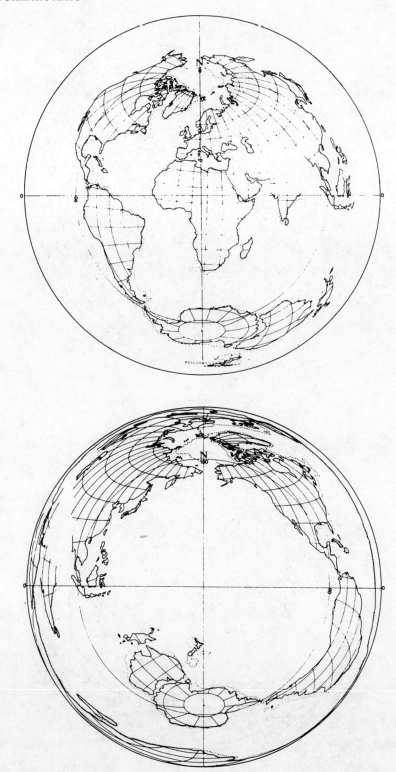

FIGURE 2. "Tertiary" (Eocene) world, about 50 ± 5 m yr B. P. (from Briden et al., 1974, copyright 1974 by The University of Chicago Press).

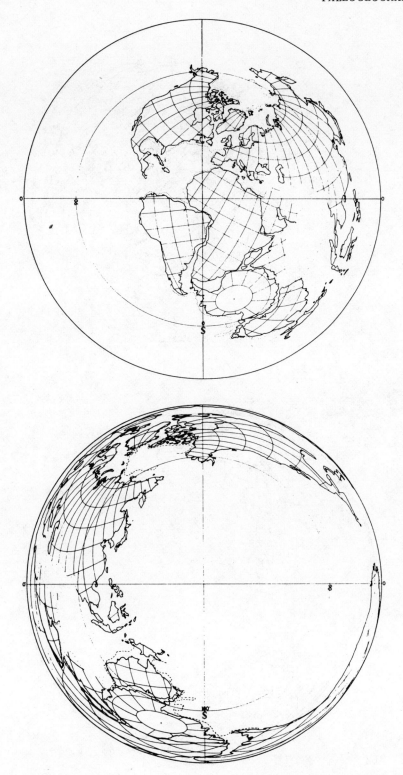

FIGURE 3. "Cretaceous" world, about 100 ± 10 m yr B.P. (from Briden et al., 1974, copyright 1974 by The University of Chicago Press). Recent sea-floor data suggest that Gondwanaland had broken up by this time.

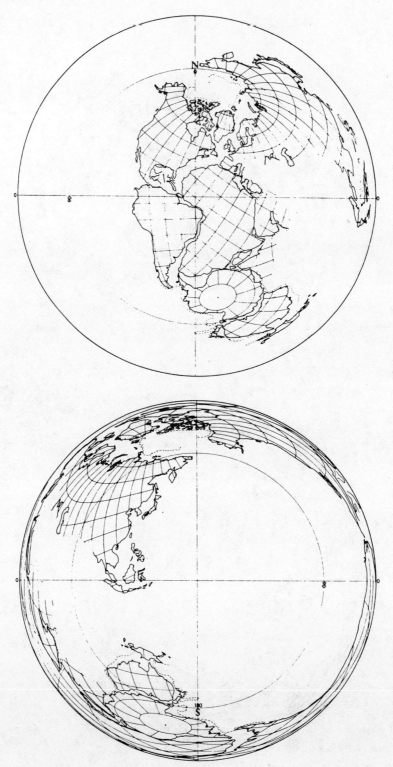

FIGURE 4. "Jurassic" world, about 170 ± 15 m yr B.P. (from Briden et al., 1974, copyright 1974 by The University of Chicago Press).

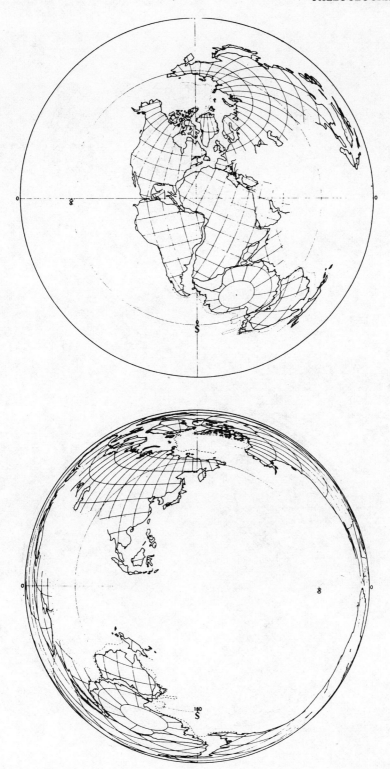

FIGURE 5. "Triassic" world, about 220 ± 20 m yr B.P. (from Briden et al., 1974, copyright 1974 by The University of Chicago Press).

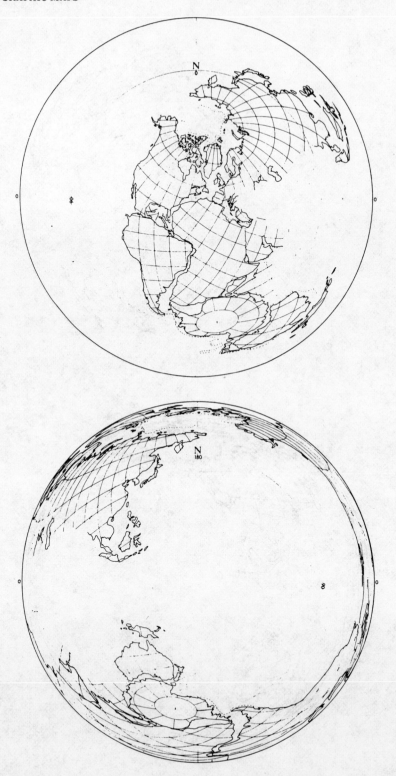

FIGURE 6. "Permian" world, about 250 ± 25 m yr B.P. (from Briden et al., 1974, copyright 1974 by The University of Chicago Press).

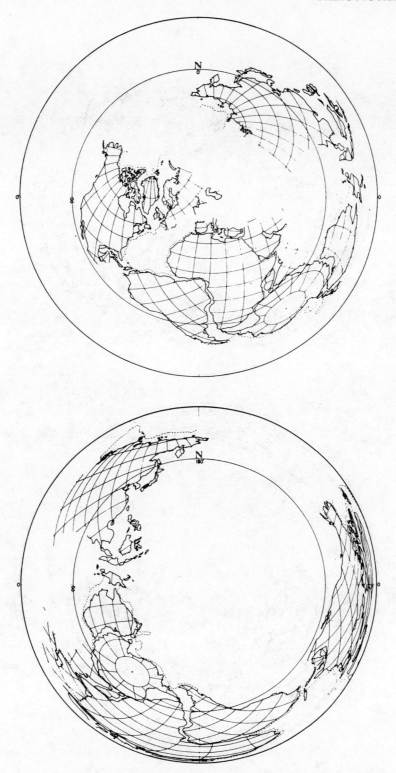

FIGURE 7. "Lower Carboniferous" (Mississippian) world, about 340 ± 30 m yr B.P. (from Briden et al., 1974, copyright 1974 by The University of Chicago Press). Asia east of the Urals has been positioned by using Permian paleomagnetic data. Longitude separation of continental fragments is arbitrary.

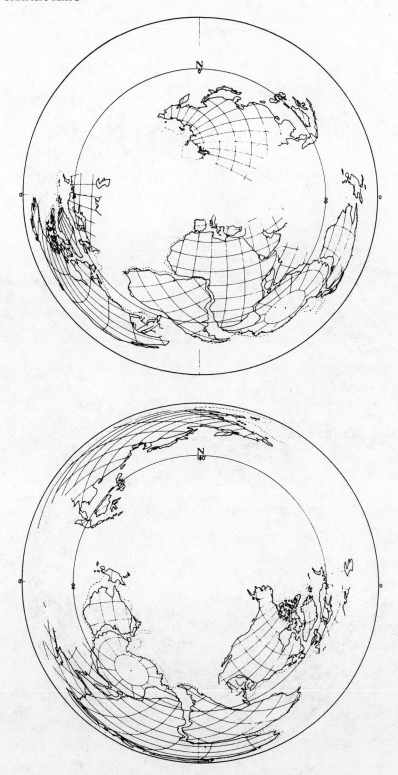

FIGURE 8. "Lower Devonian" world, about 380 ± 35 m yr B.P. (from Biden et al., 1974, copyright 1974 by The University of Chicago Press). Longitude separation of continental fragments is arbitrary.

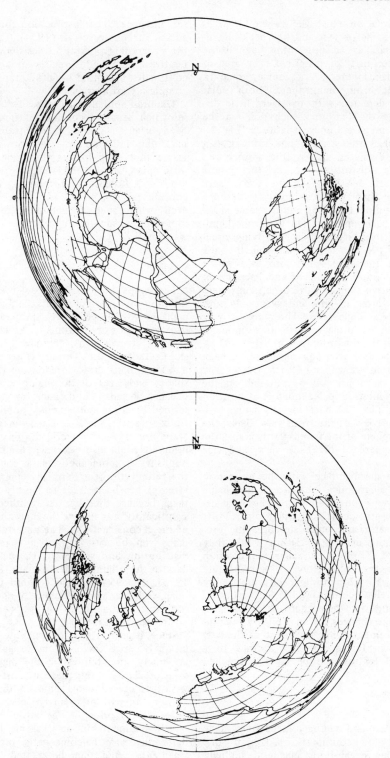

FIGURE 9. "Cambrian/Lower Ordovician" world, about 510 ± 40 m yr B.P. (from Briden et al., 1974, copyright 1974 by The University of Chicago Press). Longitude separation of continental fragments is arbitrary. Recent paleomagnetic data suggests that the time interval to which this composite applies is Upper Cambrian/Lower Ordovician.

the inner circle on which the geographic poles lie. On the same projection, the "back" hemisphere appears as an outer ring beyond the familiar inner ring, or primitive. The area of the back hemisphere is, of course, equal to that of the front hemisphere. Both polar regions are depicted with relatively little distortion. The ovals around each pole on the present-day maps are lines of latitude (Fig. 1). The equator is the straight line running as a horizontal diameter across the projection. Considerable distortion occurs for points with low latitudes lying on the back hemisphere. The distortion is so great at the edge of the projection that, for clarity, some small areas have been omitted. On the front hemisphere, the tropics plot as a strip lying equatorwards of the two larger ovals. On the back hemisphere, the strip splits into two where it joins the outer margin of the back hemisphere. The two strips on the back hemisphere follow the outer margin and eventually link up again as a single strip. The simplest way to visualize the projection is to examine a projection of the present-day world (Fig. 1).

The members of each pair of maps have been labelled "front view" and "back view." The front view is a projection centered on the equator, or inferred paleoequator, at an arbitrary longtitude. The back view is a diametrically oposite view of the front view. Thus, the front hemisphere of the front view becomes the back hemisphere of the back view, and vice versa.

The main disadvantage with the projection is that the geographical outlines change shape when moved longitudinally. These shape changes do not occur for longitudinal shifts in the Mercator and polar stereographic projections (Smith et al., 1973) nor would they occur for a Lambert equal-area projection centered on the poles. However, approximate shape changes for longitudinal shifts are readily found by reference to Fig. 1. The latitude/longitude grids on these maps show the progressive shape changes for an arbitrary longitudinal displacement. Once the shape of the new grid has been found, the new outline of a geographical feature can be rapidly sketched in by eye.

The Maps

The Permian and younger maps were made by reassembling continents relative to one another using ocean-floor magnetic anomaly data or the fit of the appropriate continental edges using data published to early 1972. The mean paleomagnetic pole of the reassembled continents was used as the geographic pole

for the map. The justification for this is discussed in Drewry et al. (1974). The fixing of the geographic pole also fixes the equator and relative longitude differences among the various continents. Absolute longitudes are, of course, completely arbitrary.

The pre-Permian maps are really "composites" in which separate continental fragments have been oriented geographically from paleomagnetic data and then projected onto a coordinate frame that encompasses the globe. Longitude differences among the various fragments are arbitrary. Longitude differences within each continental fragment are correctly shown, even when such fragments include two or more pieces, such as the components of Gondwanaland.

Limitations of the Maps

The maps purport to show the latitudinal distributions of continents and oceans for eight geological periods: Cambrian/Lower Ordovician, Lower Devonian, Lower Carboniferous (Mississippian), Permian, Triassic, Jurassic, Cretaceous, and Tertiary (Eocene). Permian and later maps show, in addition, the probable shapes and sizes of the major continents and oceans. Permian and later maps could be refined by using subsequently published ocean-floor spreading data and paleomagnetic data. Maps portraying the world during shorter time intervals could also be drawn. Similarly, newly published paleomagnetic data could be used to refine the pre-Permian composites. No attempt has been made to modify the original maps because the new data neither alter them significantly nor resolve the major uncertainties of the reconstructions. The uncertainties in the maps can be best gauged by examining the distribution and scatter of the paleomagnetic data in the original series of maps (Smith et al., 1973). The 95% confidence limit for the Permian and later maps is typically about 5°, and for large fragments in the pre-Permian composites the limit is about 10°.

Since it is not yet known how to position uniquely areas involved in orogenesis, there are many uncertainties in the maps. For the sake of clarity only, the continents are shown with their present geographical outlines whether or not any part of them has been involved in orogenesis. In principle, the latitudinal positions of all fragments in an orogenic belt can be determined by paleomagnetic methods. The fragments could then be treated in the same way as the stable continental areas have been treated in the pre-Permian composites. At the present time, the paleomagnetic data are mostly inadequate for this method to be used. The extent of the uncertainties in any pair of maps can be assessed by comparing the maps with the Mercator maps published previously, which show the approximate outcrop areas of Phanerozoic orogenic belts (Smith et al., 1973).

Some Unsolved Problems

The pre-Permian composites present some unsolved problems. One of these is that of geographical sequence. Because the continental fragments have been individually projected onto a global map frame without reference to other continents, their relative geographical sequence from west to east may not be correct. Thus, although the sequence of continents in the Cambrian/Lower Ordovician map is geometrically capable of evolving into the Permian map via the Lower Devonian and Lower Carboniferous (Mississippian) configurations, this does not prove the correctness of the sequence.

The basic reason for this uncertainty is that the longitude separation of continental fragments now joined together by younger orogenic belts is at present undeterminate. Until some method is discovered whereby the characters of orogenic belts can be quantitatively related to the presumed former subduction zones within them, there is no way of estimating how much oceanic crust has been consumed in such zones, and no way of determining the separation of the continental fragments prior to collision. In exceptional circumstances, it may be possible to place relatively narrow limits on the separation of continental fragments. For example, a large proportion of the latitude band at about 20°S on the Cambrian/Lower Ordovician maps is almost entirely occupied by continental material (Fig. 9). If the sequence is correct, then the longitude separations of the fragments are restricted by the fact that all fragments must be accommodated without overlap in the 20°S latitude band.

A particularly controversial part of the lower Paleozoic and Devonian maps concerns the shape and width of any oceans that formerly lay within the Caledonian/Appalachian orogenic belt. A paleomagnetic assessment, discussed in part by Briden et al. (1973), is that, by Early Ordovician time, the belt was not more than 1000 km wider than its width on a North Atlantic reconstruction prior to the opening of the Atlantic Ocean in early Mesozoic time. Alternatively, any motions within the belt were largely along transform faults during Early Ordovician time. The pattern of closure and the plate boundaries of that age are obscure. There is no evidence that Greenland, northern Canada, and the Baltic regions moved relative

to each other between Ordovician and late Paleozoic time. Yet there is a period during Silurian-Devonian time when the whole of the British Isles, including areas on both sides of the presently favored Caledonian suture, moved relative to the Greenland-Canada-Baltic region. The British paleomagnetic data were used to orient the whole of Europe, of Greenland, and of North America in the Lower Devonian map, and it may have been incorrect to do so. The proper Lower Devonian positions of most of North America, of Greenland, and of the Baltic regions may be closer to those on the Lower Carboniferous map than on the Lower Devonian map presented here.

The main uncertainties in the Permian and later maps are the positions of some of the southern continents relative to one another in Triassic, Jurassic, and earlier Cretaceous time. The relative positions of Australia-Antarctica to India, and of both relative to Madagascar and Africa-South America are poorly known in the Triassic to earlier Cretaceous interval. Indeed, there is still some controversy about the precise position of Madagascar and India-Australia-Antarctica relative to Africa-South America when they were all joined together to form part of Gondwanaland. Some of these problems will probably be solved within the next few years.

A. G. SMITH

References

Argand, E., 1924. La tectonique de l'Asie, *13th Internat. Geol. Cong., Brussels,* 171–372.
Briden, J. C.; Drewry, G. E.; and Smith, A. G., 1974. Phanerozoic equal-area world maps, *J. Geol.,* **82,** 555–574.
Briden, J. C.; Morris, W. A.; and Piper, J. D. A., 1973. Palaeomagnetic studies in the British Caledonides—VI. Regional and global implications, *Geophys. J. Roy. Astron. Soc.,* **34,** 107–134.
Briden, J. C.; Smith, A. G.; and Sallomy, J. T., 1971. The geomagnetic field in Permo-Triassic time, *Geophys. J. Roy. Astron. Soc.,* **23,** 101–117.
Bullard, E. C.; Everett, J. E.; and Smith, A. G., 1965. The fit of the continents around the Atlantic, *Phil. Trans. Roy. Soc. London, Ser. A.,* **258,** 41–51.
Carey, S. W., 1958. A tectonic approach to continental drift, in *Continental Drift, a Symposium.* Hobart: Tasmania Univ., 177–355.
Drewry, G. E.; Ramsay, A. T. S.; and Smith, A. G., 1974. Climatically controlled sediments, the geomagnetic field and trade wind belts in Phanerozoic time, *J. Geol.,* **82,** 531–553.
Du Toit, A. L., 1937. *Our Wandering Continents.* Edinburgh: Oliver & Boyd, 366p.
Irving, E., 1977. Drift of the major continental blocks since the Devonian, *Nature,* **270,** 304–309.
Kanasewich, E. R.; Havskov, J.; and Evans, M. E., 1978. Plate tectonics in the Phanerozoic, *Canadian J. Earth Sci.,* **15,** 919–955.

King, L. C., 1962. *The Morphology of the Earth.* Edinburgh: Oliver & Boyd, 699p.

Le Pichon, X., and Hayes, D., 1971. Marginal offsets, fracture zones and the early opening of the South Atlantic, *J. Geophys. Research,* 76, 6283-6293.

McKenzie, D. P., and Sclater, J. G., 1971. The evolution of the Indian Ocean since the Late Cretaceous, *Geophys. J. Roy. Astron. Soc.,* 24, 437-528.

Morel, P., and Irving, E., 1978. Tentative paleocontinental maps for the Early Phanerozoic and Proterozoic, *J. Geol.,* 86, 535-561.

Pitman, W. C., and Talwani, M., 1972. Sea-floor spreading in the North Atlantic, *Geol. Soc. Amer. Bull.,* 83, 619-646.

Sclater, J. G., and Fisher, R. L., 1974. Evolution of the East Central Indian Ocean, with emphasis on the tectonic setting of the Ninetyeast Ridge, *Geol. Soc. Amer. Bull.,* 85, 683-702.

Smith, A. G., 1971. Continental drift, in I. G. Gass, P. J. Smith, and R. C. L. Wilson, eds. *Understanding the Earth,* Open University Set Book. Sussex: Artemis Press, 213-232.

Smith, A. G., and Hallam, A., 1970. The fit of the southern continents, *Nature,* 225, 139-144.

Smith, A. G., Briden, J. C., and Drewry, G. E., 1973. Phanerozoic world maps, *Spec. Paper Palaeont., No. 12,* 1-42.

Wegener, A. L., 1924. *The Origin of Continents and Oceans.* New York: Dutton (English translation of 1922 German edition; reprinted by Dover Books), 212p.

Cross-references: *Paleobiogeography; Paleobiogeography of Vertebrates.*

PALEONTOLOGY

Paleontology is the study of prehistoric life. It includes the study of both fossil animals (paleozoology) and fossil plants (paleobotany). The study of microscopic fossils of both groups (micropaleontology) involves different techniques from the study of larger specimens (macropaleontology). Invertebrate paleontology is frequently regarded as a separate subdiscipline from vertebrate paleontology.

Data of Paleontology

Fossils, and their enclosing rocks, provide the chief data of paleontology. Fossils are the remains of, or direct indication of, life of the geologic past, and are preserved by natural means in the crust of the earth. The study of their enclosing rocks provides important information about the environment in which fossil organisms lived.

Development of Paleontology

Fossils have been known since the time of primitive man, and a number of Greek scholars (Herodotus and Xenophanes, for example) recognized their true nature. The earliest descriptions of fossils are those of the middle ages, including works by Conrad Gesner [in 1565] and Edward Lhuyd [in 1699]. Numbers of these early writers failed to recognize fossils as the remains of once-living organisms, but Robert Hooke [1635-1703] accurately interpreted fossils, and suggested their use in reconstructing past climates. The scientific study of fossils dates back chiefly to the early 19th century; and the works of (J. B. P. A. M. de) Lamarck, G. Cuvier, A. d'Orbigny, and William Smith are among the most notable early contributions. The study of fossils received particular impetus from the publication of the Darwin-Wallace theory of evolution by natural selection [1858 and 1859] and the demonstration by William Smith in 1815 of the value of fossils in the study and correlation of strata (see *History of Paleontology—Before Darwin; History of Paleontology: Post-Darwinian).*

Fossils

Fossils include both the remains and direct indications of life of the geologic past. Remains of organisms only rarely include the complete preservation of both the hard and soft parts of the body, as in the case of Pleistocene mammoths preserved by deep freeze in frozen ground in Siberia and Alaska. Insects and leaves preserved in amber (fossil resin) are more common and are geologically older than frozen specimens, but they are preserved only as a thin film of carbon, coating the outer surface of a cavity that is a replica of the original form. More commonly, the impression of soft tissues may be preserved as a thin film of carbon, especially in fine-grained shales, the other volatile components having been lost by distillation. The remains of leaves in many continental rocks are preserved in this way. In the great majority of fossils, the soft parts of the organisms are lost in the process of fossilization, and only the hard parts (bones and shells) are preserved. Rarely, these may be preserved almost unchanged, examples being many fossil teeth and some chitinous and phosphatic invertebrate fossils; but usually the fossil remains are leached as a result of the removal of their more soluble components by water. Leached but otherwise unaltered fossil remains tend to be confined to younger rocks (Mesozoic and Tertiary); those in older strata have generally been altered in composition, sometimes by the deposition of new material in the pore spaces and sometimes by the entire removal of the original substance and its replacement by some new compound.

Most fossils belong to one or other of these categories. The chief replacing compounds are calcium carbonate, iron salts, and silica. In some cases the replacement leaves the original microstructure intact (the petrified logs of Arizona are an example) but usually this structure is lost, although the details of the surfaces of the organism are often beautifully preserved.

In some cases, especially in permeable rocks such as sandstone, the original organism is completely dissolved and is represented only by a natural cast, surrounding a cavity in the rock. This may later be filled to give a natural mold. Other types of direct evidence for the former existence of prehistoric life include tracks, trails, burrows, and borings of many different animal groups; coprolites and gastroliths (stomach stones); and artifacts (the stone implements of prehistoric man). Coal and petroleum owe their origin to the action and preservation of once-living organisms and are often described as fossil fuels.

The fossil record is a very incomplete representation of the great number of prehistoric organisms. About 91,000 species of fossils are known at present, but it is estimated that the total number of organisms that have lived in the geologic past may be of the order of 500 million (other estimates range from 50 million to 4000 million). More than half the classes of living organisms are unrepresented as fossils. The record is not only incomplete, but also biased. Factors favoring fossil preservation include the possession of hard parts, deposition in an environment of rapid burial by sediments, and occurrence in well-surveyed areas. The recognition of this structural, environmental, and geographic bias is important in paleontologic interpretation.

Paleontologic Methods

Fossil collecting must be done with great care if the results are to be of maximum value. Specimens must be very accurately located, both geographically and stratigraphically. Most fossil collections are made by rigorous sampling of rock sequences, either as continuous channel samples, or at measured intervals. Recent studies have shown that significant changes may occur within a few inches of strata. Studies of the lateral variation in fossiliferous strata are also essential. Microfossils are particularly useful because they can be extracted from subsurface cores and drilling fragments. The collection of large vertebrates often requires minor quarrying and wrapping of specimens, which are not removed from the enclosing rock until they arrive at the laboratory.

The extraction of fossils from the rock matrix may be easy or difficult. In rare cases, specimens may be readily removed with a rock splitter, or even a hammer and chisel, but the removal of most specimens requires many hours (or sometimes weeks) of careful preparation. Power tools are useful for larger specimens; and, in softer strata, fossils may sometimes be removed by boiling under pressure or with suitable reagents. In calcareous rocks, phosphatic and chitinous fossils may be extracted by dissolving the matrix in acetic, monochloracetic, or formic acids. Chitinous fossils and spores may be removed from other rocks by the use of stronger acids, including hydrofluoric acid. Alternate freezing and thawing and ultra sonic treatment are also useful. Microfossils are extracted from rock residues by washing, sieving, centrifuging, separation in heavy liquids, and other methods. Stains of various kinds are used for some fossils, and many are coated with a thin film of ammonium chloride to facilitate photography. Electron microscopy has been used to study fossil coccoliths and shell structure.

Modern paleontology generally involves several or all of the following methods of analysis: biometric analysis; analysis of position, preservation and function; chemical analysis; pathologic and microbiologic analysis; distributional analysis; comparison with living organisms; taxonomic analysis. Each of these is described below.

Biometric Analysis. Biometric studies which must be based upon adequate samples, have been recently used in a study of faunal association, in which the combined presence of certain faunal elements in different samples was analyzed statistically to discriminate between contemporary but distinct faunal assemblages. By applying statistical studies of relative mortality (based on age and aspects of morphology), it has become possible to recognize the effects of natural selection in certain fossil groups. Biometric studies are important in the evaluation of morphological difference, including those arising from relative growth and sexual dimorphism. The estimation of the homogeneity of samples or their degree of difference is of great value in taxonomy (see *Biometrics in Paleontology; Taxonomy, Numerical*).

Preservation, Position, and Function. Analysis of preservation, position, and function contributes information to the study of the habits of fossil organisms. It is always important to attempt to distinguish the original members of a life assemblage (a biocoenose) from those which have been added to or subtracted from it by sedimentation and other physical processes (see *Taphonomy*). The detailed study of gastric residues, for example, may provide information about prey-predator relationships, as may the study of mutilation of speciments. The orientation and degree of disarticulation of speciments and their relative density of occurrence, and the wear of certain functional parts such as teeth, are also useful supplementary studies. The analysis of living positions of aquatic organisms, especially burrowers, in the rock matrix, and their association with tracks or trails can be used to reconstruct the assemblage association. Recent studies of functions have included experiments with models of extinct brachiopods, in which ciné photography has revealed the pattern of circulating currents in the shell, and other experiments on transportation of empty shells and on the effects of relative streamlining and buoyancy.

Chemical Analysis. Chemical analysis of fossils may include organic, isotopic, and inorganic studies. It has recently been shown that numbers of amino acids are preserved in fossil shells and skeletons. Of the seventeen amino acids that make up the protein of a living pelecypod, *Mercenaria,* seven have been detected in a middle Tertiary fossil representative. Isotopic studies of oxygen isotopes (O^{16}/O^{18}) have been used on fossil mollusks to determine ancient temperatures. Consistent latitudinal temperature variation has been demonstrated in rocks of Cretaceous age; changing climatic patterns in time have also been detected. Carbon isotype ratios of Pleistocene fossil wood have been used as a measure of geologic age. Studies of calcite/aragonite ratios and calcium, strontium, and magnesium ratios may provide indications of temperature and salinity relationships, but present work is inconclusive (see *Biogeochemistry*).

Pathologic and Microbiologic Analysis. Pathologic studies have recently shown that the stoop generally attributed to Neanderthal man, may be pathologic, resulting from acute spinal arthritis (see Paleopathology). Microscopic examination has revealed the effects of parasitic infection by bacteria and fungi on Carboniferous spores and a scorpion and a Devonian fish.

Distributional Analysis. Studies of ecological distribution of fossils may be carried out by the sedimentary and chemical study of the enclosing rocks (see *Paleoecology*). Distributional and morphologic differences in Carboniferous nonmarine pelecypods have been shown to correlate with changes in quartz, sulphur, and carbon content of the rock in which they occur. Studies of boron, uranium, thorium and strontium contents of the enclosing rocks also provide potential ecologic indicators, and the abundance of certain fossils may often be related to the lithology of the rock matrix. Temporal distribution of fossils may be analyzed and may sometimes reveal correlations that arise from evolutionary associations or ecological replacement of migratory routes. Studies of small-scale geographic distribution of Jurassic brachiopods have provided evidence of subspecific patterns, while larger-scale studies of many other groups, some on a global basis, have been used with varying success in paleoclimatic reconstructions and as evidence both for and against continental drift. Such studies are often limited in value by lack of outcrops, extinction of many previously large and widespread fossil groups, and possible changes in tolerance and habits of surviving groups.

Comparison with Living Organisms. Wherever possible, the study of fossil material should be supplemented by a study of closely related living groups. In extinct groups of high taxonomic grade (e.g., placoderms, trilobites, graptolites), this may be of limited value, although the existing "laws" of genetics, biochemistry, physiology, and ecology will still contribute to an understanding of these groups. One of the major problems in paleontology is the degree to which it is proper to extrapolate the habits and characteristics of living organisms backward in geologic time. Generally, this must be done with great caution.

Taxonomic Analysis. Taxonomic analysis combines all the foregoing data into an interpretation of the affinities and evolutionary relationships of fossils. Through the use of a vast literature and collections and type specimens from most parts of the world, taxonomic analysis is concerned with the identification of fossilized organisms. In some cases, the fossil studies may represent new and undescribed genera and species, the description, illustration, and nomenclature of which is subject to International Zoological and Botanical Codes.

Importance of Paleontology

Paleontology has an intrinsic value of its own, as has any subject. This value is heightened by its historical foundation, by the philosophical implications of the time span of prehistoric life (at least 2700 m yr), and by the fact that man himself forms a part of the material of paleontologic study. The evolution and relationships of man have a particular interest and significance in many other disciplines.

One of the most important practical aspects of paleontology is its value in rock correlation (the demonstration of contemporaneity in time). Rocks of the same lithology, deposited under the same conditions, at the same period of time, generally contain the same fossils. Geology has therefore been able to use the sequence of fossils to provide a scale of geologic time. Although the ages provided by fossils are only relative, the scale so constructed has been qualified by the use of radio-active methods of time measurement. By means of fossils, any rock of unknown age may be assigned to its position in the geologic time scale, and the ages of two formations may thus be compared. Recent studies, especially in micropaleontology, have demonstrated that such correlation may often be done with great precision. Correlation is important in geologic mapping of all kinds, and also has considerable economic value in petroleum exploration, water supply and engineering geology.

Paleontology is also of value in the construction of maps showing the geography of earlier periods (paleogeographic maps). Fossils often provide the only reliable indication of the conditions of deposition of the rocks in which they occur. On a much smaller scale, fossils may provide a detailed ecologic guide to the character of an ancient sedimentary environment. The reefs in the Permian Guadeloupe rocks of Texas have been intensively studied in this way, and the information has been used in petroleum exploration and reservoir studies.

The fossil record provides the only evidence of the course of evolution, and, although lacking many of the advantages available in the study of living organisms (*neontology*), paleontology is able to provide an account of the evolutionary histories of most of the major groups of organisms. Sometimes, as in the case of horses and elephants, this may be a very detailed reconstruction, but often the inadequacy of the fossil record leads to considerable gaps in our knowledge. In spite of this, the fossil record provides a unique record of the evolution of living things.

F. H. T. RHODES

References

Ager, D. V., 1963. *Principles of Paleoecology.* New York: McGraw-Hill, 371p.

Andrews, R. N., 1964. *Studies in Paleobotany.* New York: Wiley, 487p.

Beerbower, J. R., 1968. *Search for the Past,* 2nd ed. Englewood Cliffs, N.J.: Prentice-Hall, 512p.

Black, R. M., 1970. *The Elements of Palaeontology.* Cambridge: Cambridge Univ. Press, 339p.

Brouwer, A., 1967. *General Palaeontology.* Chicago: Univ. Chicago Press, 216p.

Colbert, E. H., 1969. *Evolution of the Vertebrates,* 2nd ed. New York: Wiley 535p.

Dott, R. H., Jr., and Batten, R. L., 1976. *Evolution*

of the Earth, 2nd. ed. New York: McGraw-Hill, 504p.

Kummel, B., 1970. *History of the Earth,* 2nd ed. San Francisco: Freeman, 707p.

McAlester, A. L., 1968. *The History of Life.* Englewood Cliffs, N.J.: Prentice-Hall, 151p.

Moore, R. C.; Lalicker, C. G.; and Fischer, A. G., 1952. *Invertebrate Fossils.* New York: McGraw-Hill, 766p.

Raup, D. M., and Stanley, S. M., 1971. *Principles of Paleontology.* San Francisco: Freeman, 388p.

Rhodes, F. H. T., 1976. *Evolution of Life.* Harmondsworth, Penguin, 330p.

Romer, A. S., 1966. *Vertebrate Paleontology,* 3rd ed. Chicago: Univ. Chicago Press, 687p.

Shrock, R. R., and Twenhofel, W. H., 1953. *Principles of Invertebrate Paleontology.* New York: McGraw-Hill, 816p.

Tasch, P., 1973. *Paleobiology of the Invertebrates.* New York: Wiley, 946p.

Cross-references: *Evolution; Fossil Record; Fossils and Fossilization; Invertebrate Paleontology; Paleobotany; Paleoecology; Vertebrate Paleontology.*

PALEOPATHOLOGY

Paleopathology has a variety of definitions. One of the most satisfactory is offered in *Gould's Medical Dictionary* (5th ed.): "The science of the diseases which can be demonstrated in human and animal remains of ancient times." Such a definition is sufficiently broad to cover the field and does not place restrictions resulting from the use of the term "prehistoric" which has different temporal terminations and definitions within each continent or major geographic region.

Moodie's *Paleopathology, an Introduction to the Study of Ancient Evidences of Disease,* published in 1923, is still the standard reference although it is out of date and contains a great deal of irrelevant material. Each year sees the publication of papers dealing with individual specimens or small suites of specimens showing some pathological condition. There has been no modern review of the subject in the strictly geological sense (thus excluding nonfossil man), dealing in terms of modern medical knowledge and techniques. Bibliographic coverage of the subject may be found in Sigerist (1951), Pales (1930), and in the Subject Indexes of the G.S.A. bibliographies of vertebrate paleontology (*Geol. Soc. Am. Spec. Paper,* nos. 27 and 42; *Mem.,* nos. 37, 57, 84, 92, 117, 134, and 141).

Pathological conditions have been reported in fossil plants; invertebrates (see Tasnádi-Kubacska, 1962), and vertebrates, including specimens of prehistoric man (e.g., Jarcho, 1966). It is not surprising that the vast majority of papers and reports dealing with paleopathology refer to vertebrates. Pathological conditions are more easily recognized in the higher animals, and there is greater general interest in those animals more closely related to man. In fact, the earliest reference in the scientific literature to a paleopathological condition is to a fractured femur of the cave bear, *Ursus spelaeus* Blumenbach, by E. J. C. Esper in 1774. Esper described the condition as an *osteosarcoma* or bone tumor, but it is now considered to be a healed fracture showing *callus* and some *necrosis*.

The most commonly recognized conditions include mechanical injuries and attendant infections or healings, if the animal survived the injury (Fig. 1); the marks of combat or the attacks of predators (Fig. 2); "arthritic" conditions, primarily *spondylitis deformans, osteoarthritis,* and *synostoses* (Figs. 3 and 4); *dental caries* or *anomalies;* and parasitic or bacterial lesions. All of these may be found in ancient man, plus: diseases of the soft parts, which may be preserved in either artificial or natural mummies; abnormalities that may be depicted by drawings, carvings, and ceramics; and the results of primitive bone surgery, either successful or unsuccessful.

Confining paleopathology to terms of geologic time, the following pathologic conditions are to be found throughout the literature.

Auxesis. Hypertrophy and *hyperostosis* may be found as improper but common synonyms for *auxesis,* which is an increase in bulk or size of the bone or portions of a bone. Moody reports such occurrences in the dinosaur *Triceratops* and the nothosaur *Proneusticosaurus.*

Dental Anomalies. Misplaced, rotated, malformed (including *dental exostosis*) or extra teeth are common among fossil vertebrates.

Dental Caries. Cavities in the teeth of vertebrates are occasionally reported, but it is quite probable that the majority of these actually refer to postmortem damage to the tooth.

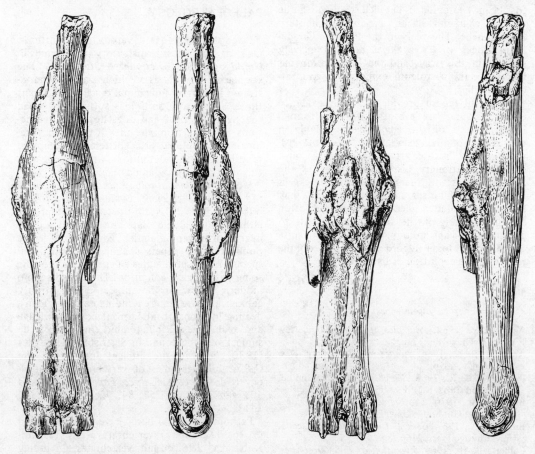

FIGURE 1. Fractured metacarpals of a middle Miocene cervid, *Blastomeryx* Cope from Wounded Knee, South Dakota, showing *callus.*

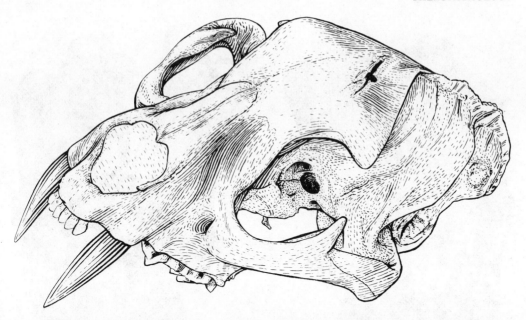

FIGURE 2. Cranium of the Oligocene cat *Nimravus bumpensis* Scott and Jepsen, from the White River Badlands, South Dakota. The left frontal has been pierced by the upper canine of a sabre tooth, probably *Eusmilus* Gervais. This wound did not result in death as extensive *callus* has formed.

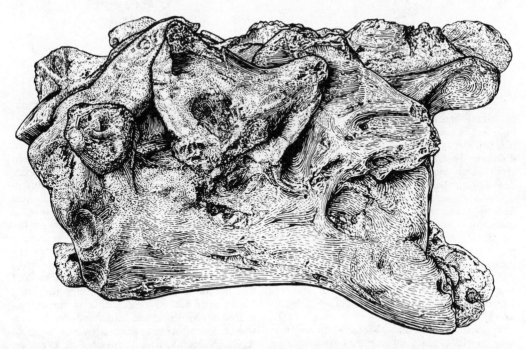

FIGURE 3. Three lumbar vertebrae of the Pleistocene sabre tooth *Smilodon californicus* Bovard from the asphalt pits of Rancho La Brea, fused together with *spondylitis deformans*.

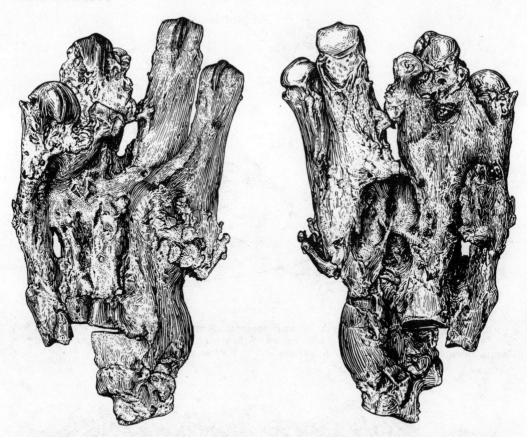

FIGURE 4. The metacarpals of the Pleistocene wolf, *Canis dirus* Leidy from Rancho La Brea, with fracture, *callus,* and *synostoses.*

Exostosis. Swellings on the surface of the bone may generally be classed under this heading. References to *Osteitis deformans* in the literature may be the result of overclassifying diseases. *Dental exostosis* produces malformed teeth which generally may be considered under dental anomalies.

Fracture and Callus. Broken and naturally healing bones are found in most vertebrate collections. Specimens in which there has been an overlap of the bone ends and subsequent healing with a shortening of the limb are not uncommon. Fig. 1 shows the metacarpus of *Blastomeryx,* a primitive cervid, from the early Miocene of Wounded Knee, South Dakota, which has been broken and healed with an excessive growth of *callus.*

Necrosis. Pitting and disintegration of bone near injuries is generally referred to necrosis.

Nearthrosis. The development of a new joint at the point of a fracture of dislocation, i.e., nearthrosis, has been reported in dinosaur ribs by von Huene.

Odontoma. Tumors associated with the teeth, containing dentine, or exhibiting tooth-like structures may be referred to odontoma.

Opisthotonos, Emprosthotonos, Pleurothotonos, Scoliosis. Many articulated skeletons are found in attitudes that have been interpreted as evidence of these conditions. Natural poisoning or diseases producing muscular spasms have been proposed as possible reasons for these conditions. It would appear to be far simpler to explain these attitudes on the basis of postmortem positioning of the carcass by water currents, by the contraction of the powerful muscles of the back and limbs through dessication under arid conditions, and the natural contraction of muscles following death. Long-necked forms including plesiosaurs, pterodactyls, dinosaurs, birds, and the diminutive early Miocene camel *Stenomylus* from Agate, Nebraska, have been cited as examples of these conditions.

Osteoma. Commonly, any bony tumor is designated as osteoma.

Osteomalacia. Possibly osteomalacia is incorrectly used to classify long-bone distortions presumably resulting from nutritional deficiencies. This diagnosis should be used with caution; postmortem deformation of fossil remains is extremely common, and, in the limb bones, could easily resemble this malady.

Osteomyelitis. Inflammation of the marrow and its effects on the bone have been reported. This condition, osteomyelitis, has been observed particularly in conjunction with fractures.

Osteoperiostitis and Periostitis. Inflammation of the bones and periostium produces bony lesions, which have been found in conjunction with arthritic conditions.

Osteosarcoma. Presumed bony tumors (osteosarcoma) may generally be the result of *callus* or *exostosis*.

Osteoarthritis. Large-joint arthritis (osteoarthritis) is generally located in the pelvic region.

Pachygnathy and Pachyostosis. Conditions of massive bone growth have been referred to both pachygnathy and pachyostosis. Marine mammals normally show the latter as an adaptive feature and not as a diseased condition.

Parasitism. The results of parasitic infestation have been noted primarily in marine invertebrates.

Regeneration. Although not technically a pathologic condition, regeneration has been noted among fossil invertebrates. It is seen in certain lower vertebrates in neozoology, and so its report among the fossil forms of these animals may be anticipated.

Rheumatoid Arthritis. Small-joint (rheumatoid) arthritis is generally located in the wrists, ankles, fingers, and toes.

Spondylitis Deformans. The ankylosing of vertebrae is found throughout the vertebrate classes. Fused vertebrae is one of the most common pathologic conditions in the fossil record. A mosasaur and a bronthothere skeleton at the Museum of Geology, South Dakota School of Mines and Technology, are excellent examples of this condition. Many specimens of the Sabre-tooth cat from the asphalt pits at Rancho la Brea, in the collections of the Los Angeles County Museum, illustrate this malady (Fig. 3).

Synostoses. The co-ossification of normally separate bones is found in many animals. This condition may result from *callus* (Fig. 4) or from an arthritic condition.

Other Pathologies. Many anomalies, including spasmatic attitudes in articulated skeletons, nonadaptive polydactyly, and adaptive pachyostosis are attributed to pathologic conditions. Great care should be taken in diagnosing the ailments of fossil patients. The postmortem alteration of bones is likely to produce damage that will resemble the ravages of disease.

Although there is as yet no evidence of fly-borne diseases in the fossil record, it should be noted that the flea *Palaeopsylla klebsiana* has been found in the Baltic amber of Eocene age; the tse-tse fly, *Glossina oligocena*, is recorded from the Eocene Brown Coal of Germany and the Oligocene of Colorado (Tasnáda-Kubacska, 1962). Other flies and blood-sucking midges have been recorded as far back as the Miocene. These were potential bearers of disease and may have been the cause of extinction in some mammals that have disappeared for no apparent reason, such as the Oligocene brontotheres (Osborn, 1929, pp. 869–874; Tasnáda-Kubacska, 1962) and the Pleistocene horses and camels of North America.

J. R. MACDONALD

References

Jarcho, S., ed., 1966. *Human Palaeopathology.* New Haven, Conn.: Yale Univ. Press, 182p.

Kerley, E. R., and Bass, W. M., 1967. Paleopathology: Meeting ground for many disciplines, *Science,* 157, 638–644.

Moodie, R. L., 1923. *Paleopathology, an Introduction to the Study of Ancient Evidences of Diseases.* Urbana, Ill.: Univ. Illinois Press, 567p.

Moodie, R. L., 1967. General considerations of the evidences of pathological conditions found among fossil animals, in D. Brothwell and A. T. Sandison, eds., *Diseases in Antiquity.* Springfield, Ill.: Thomas, 31–46.

Osborne, H. F., 1929. The titanotheres of ancient Wyoming, Dakota and Nebraska. *U.S. Geol. Surv. Monogr. 55,* 953p.

Pales, L., 1930. *Paléopathologie et Pathologie Comparative.* Paris: Masson et Cie, 352p.

Sigerist, H. E., 1951. *A History of Medicine,* Vol. 1, *Primitive and Archaic Medicine.* New York: Oxford Univ. Press, 564p.

Tasnádi-Kubacska, A., 1962. *Paläopathologie.* Jena: Fischer, 240p.

Cross-references: *Bones and Teeth; Morphology, Functional; Vertebrata; Vertebrate Paleontology.*

PALEOPHYTOGEOGRAPHY

Phytogeography is the study of plant distribution, and paleophytogeography is the study of plant distribution through time. As early as 1750, Linnaeus noted similarity in vegetation between the eastern United States and eastern Asia (Graham, 1966, 1972a). Further interest in the arrangement of plants over the surface of the earth resulted from voyages of exploration made during the 18th and 19th centuries. Cook sailed the ship *Endeavour* around the world on three voyages between 1768 and 1780 to study the natural history of distant lands (Merrill, 1954), and Darwin served as naturalist on the *Beagle* [1831-1836]. As plant collections from these and other voyages were studied, certain patterns in distribution emerged (Fig. 1). There are floristic relationships between the dry regions of western North America and South America (symposium proceedings, *Quart. Rev. Bio.*, **38**, 1963), the deciduous forests of eastern United States and eastern Mexico (Graham, 1972b, 1973), and between Africa and South America (Thorne, 1973; Raven and Axelrod, 1974). As patterns became evident, the need for explanations arose, and the science of plant geography had its beginnings.

Sources of Pytogeographic Data

In the older literature, distribution data was obtained primarily from herbarium labels. These early collections often reflected only a portion of the range of a taxon, and many identifications have subsequently been revised. Consequently, the tendency to map broad, world-wide occurrences of major groups (families, genera) produced patterns illustrating the course of collecting expeditions or centers of botanical research more accurately than the distribution of the taxon. Refinements in geology have frequently shown that areas assumed to be ice free during the Pleistocene were glaciated, land bridges and continental masses to be geophysically impossible as proposed, and that the fixed position of continents no longer provides a sound physiographic context for theories of phytogeography. The numerous "theories" so evident in early plant-geography literature suffered from these deficiencies and from attempts to widely apply them beyond a few specific examples. Thus Hulten's (1937) Equiformal Progressive Areas, Willis' (1922) Age and Area, Fernald's (1925) Nunatak hypothesis, and Simroth's Schwingpolen hypothesis (see Wulff, 1950) are largely of historical interest. A current trend is to rely more selectively on data from recent monographs. Individually, these studies result in new collections, revised taxonomy, and annotations of older material. Collectively, they provide a more reliable base for determining broad pat-

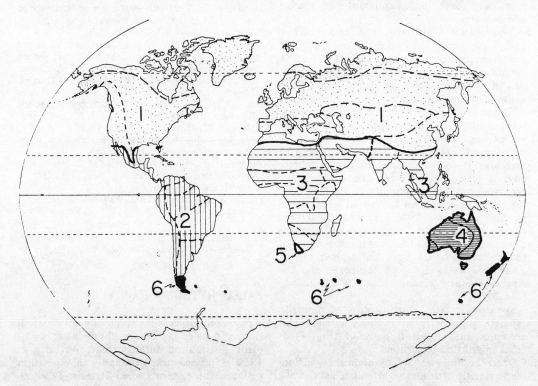

FIGURE 1. Present phytogeographic subdivisions (from C. A. Ross, Paleogeography and provinciality, *S.E.P.M. Spec. Pub. 21*, 1-17; copyright 1974 by The Society of Economic Paleontologists and Mineralogists, and used with permission). 1, Boreal Kingdom; 2, Neotropical Kingdom; 3, Palaeotropical Kingdom; 4, Australian Kingdom; 5, Cape Kingdom; 6, Antarctic Kingdom; dashed lines are province boundaries.

terns of distribution and for estimating causal factors for the arrangement of organisms over the landscape.

Another trend is to utilize sources of information outside phytogeography per se. Most modern studies consider data from plate tectonics (e.g., Hughes, 1973; Raven and Axelrod, 1972) and patterns evident in other groups of organisms (Fittkau et al., 1968–1969; Lent, 1967; Meggers et al., 1973). Statistical tests are commonly applied to biogeographic data (MacArthur and Wilson, 1967); and computer technology is being used increasingly to generate distribution, select-purpose, and multicorrelation maps of great accuracy and detail (Tralau, 1969; Lieth, 1972). In turn, results of plant distribution studies are used in tracing human migrations and pre-Columbian contacts (Riley et al., 1971) and to detect centers of origin for agriculturally important crop plants and domesticated animals (Sauer, 1969). There is close interaction between taxonomy and plant geography (Valentine, 1972; Walker, 1971, 1972); and, as noted by MacArthur and Wilson (1967, p. v), there is little fundamental distinction between the fields of biogeography and ecology.

Factors of Plant Distribution

The principal controlling factor in plant distribution is climate, and particularly the interaction between climate and the ecological amplitude of individual species. The important aspects of climate, with reference to plant geography, are minimum winter temperatures, amount and distribution of rainfall, and day length. On a more local level, the range of plants is influenced by edaphic factors. The color, texture, pH, depth, water-holding capacity, organic content, and mineral composition of soils vary depending on the composition of bedrock and length of time the parent material has been exposed to weathering. The importance of soil in plant distribution is most apparent in regions where glaciation has not deposited a veneer of soil over bedrocks of varying composition. On the Texas Gulf Coastal Plain, for example, the relationship between vegetation and soil types is particularly evident.

Among biological factors, the presence of pathogens, pollinators, mycorhiza-forming fungi, and human activities are important. One significant human influence has been the parceling of vegetation into small isolated units ("woodlots") separated by areas of intense cultivation. This not only limits the effective size of breeding populations, with its consequent evolutionary implications, but limits the migration of plants lacking means of long-distance dispersal. Fence-row communities and gallery forests are frequently the only avenues of dispersal for these plants in the intensely cultivated regions of the midwest and central states.

Barriers are an important factor in plant distribution. Most important are oceanic barriers; valleys, highlands, rivers, brackish-water inlets, outcrops of different soil types and closed communities should also be mentioned. The effectiveness of these barriers, however, differs for each plant species depending on the dispersal potential of the propagules, and has also varied through geologic time. The importance of historical factors is particularly evident in the isthmian region of Panama and in the Australasian area. Recent studies in the Canal Zone have shown that interchange between certain groups of North and South American terrestrial vertebrates (Whitmore and Stewart, 1965) and plants (Graham, 1972b, 1973) correlate with provincialization patterns of marine invertebrates (Woodring, 1966) and the proposed uplift of the Isthmian region during late Pliocene times. The important biogeographic boundary in Australasia designated as Wallace's Line marks generally the contact between the Australian and Asian plates (Raven and Axelrod, 1972).

Undoubtedly, the most profound impact on biogeographic thought has been the confirmation of plate tectonics, and especially the detailing of relative continental positions at specific intervals of geologic time. When the paleobotanical record for a plant group is accurately known (i.e., the time and place of its earliest occurrence, if not its origin), these data may be integrated into a general reconstruction of paleophysiography and paleoenvironmental conditions during initial stages in the evolution and radiation of the group, and allow interpretation of modern distribution patterns in light of subsequent continental fragmentation.

For example, SE Asia and Australasia has frequently been suggested as the site of origin for the angiosperms because primitive types are concentrated in that region. This view is no longer tenable, however, because as noted by Raven and Axelrod (1974, p. 632), "This region is a composite one geologically that came into existence only with the Miocene arrival of the Australian plate in the vicinity of Asia. The persistence in Australasia of equable climates during the Tertiary, coupled with the increasing isolation of some of its components (i.e., New Caledonia, Fiji), has afforded some primitive angiosperms unusual opportunities for survival." Raven and Axelrod suggest W Gondwanaland as a likely site of origin for angiosperms because of varied environmental conditions favoring the genesis of new phenotypes, continental configurations in the Early to middle Cretaceous that allowed effective radiation, and the ancient occurrence there of fossil tricolpate pollen with progressively later appearances in temperate Laurasia and Arctic regions. These concepts could only be developed as the time of angiosperm origin became accurately fixed, continental positions known for this early period of angiosperm history, and the post-middle Cretaceous movements of continents established.

Another subject of biogeographic interest is the floristic and faunal relationships between Africa and South America, and between North and South America. There has been a tendency to attribute the common occurrence of taxa between Africa and South America to chance dispersal, parallel or convergent evolution, or poor taxonomy. Similarity between the biotas of North and South America has been more readily accepted because of the greater land continuity between the continents. However, the positions and configuration of the continents as late as the early Eocene placed South America about equal distance from Africa and North America. Depending on the time of appearance of a particular phyletic line, the occurrence of families and genera in Africa and South America does not necessarily require any unique explanation not equally applicable to disjunctions between North and South America.

The wide acceptance of plate tectonic theory, however, cannot be applied uncritically to problems of biogeography. Pollen of the family Gramineae (grasses), for example, is first reported from the Maastrichtian (uppermost Cretaceous) and is common by the lower Eocene. Thus the past position of continents becomes an important factor in unraveling the distributional history and geographic relationships for members of the family. In contrast, pollen of the Compositae appears in the Oligocene, and pollen of the Tribe Cichorieae is not encountered until the upper Miocene. Thus, views on the origin and early radiation of the Compositae must deal with the present relative position of continents.

The complex interactions between fluctuating historical events, current influence of prevailing climate, soil, physical barriers, and biological factors, and the myriad of ecological amplitudes among plant species has produced the array of patterns and relationships evident in modern vegetation (cf. discussion of distribution types in Good, 1964; Cain, 1944; Polunin), 1960; Gleason and Cronquist, 1964). The study of plant geography involves identification of these dynamic patterns, establishing methods and pathways of dispersal, and determining causal factors for the present mosaic of vegetation.

ALAN GRAHAM

References

Cain, S. A., 1944. *Foundations of Plant Geography*. New York: Harper and Row, 566p.
Fernald, M. L., 1925. Persistence of plants in unglaciated areas of boreal America, *Mem. Amer. Acad. Arts Letters*, **15**, 239–342.
Fittkau, E. J.; Illies, J.; Klinge, H.; Schwabe, G. H.; and Sioli, H., eds., 1968–1969. *Biogeography and Ecology in South America*. 2 vols. The Hague: Junk.
Gleason, H. A., and Cronquist, A., 1964. *The Natural Geography of Plants*. New York: Columbia Univ. Press, 420p.
Good, R., 1964. *The Geography of Flowering Plants*. New York: Wiley, 518p.
Graham, A., 1966. Plantae rariores camschatcenses: A translation of the dissertation of Jonas P. Halenius, 1750, *Brittonia*, **18**, 131–139.
Graham, A., ed., 1972a. *Floristics and Paleofloristics of Asia and eastern North America*. Amsterdam: Elsevier, 278p.
Graham, A., 1972b. Some aspects of Tertiary vegetational history about the Caribbean Basin, in *Mem. Symp. I Congreso Latinoamericano y V Mex. Bot.*, Mexico, D. F., 97–117.
Graham, A., ed., 1973. *Vegetation and Vegetational History of Northern Latin America*. Amsterdam: Elsevier, 393p.
Hughes, N. F., 1973. Organisms and continents through time, *Spec. Papers Palaeontol.*, no. 12, 334p.
Hulten, E., 1937. *Outline of the History of Arctic and Boreal Biota During the Quaternary Period*. Stockholm: Bokforlags Aktiebolaget Thule, 168p.
Humboldt, A. von, 1805. *Essai sur la Géographie des Plantes*. Paris: Levrault, Shoell et Cie, 155p.
Kremp, G. O. W., 1974. A re-evaluation of global plantgeographic provinces of the late Paleozoic, *Rev. Palaeobot. Palynol.*, **17**, 113–132.
Leith, H., 1972. A computer model of the world vegetation, in *Mem. Symp. I Congreso Latinoamericano y V Mex. Bot.*, Mexico, D. F., 451–458.
Lent, H., ed., 1967. *Atas do Simposio sobre a Biota Amazonica*. 7 vols. Rio de Janeiro: Pub. pelo Conselho Nac. Pesquisas.
MacArthur, R. H., and Wilson, E. O., 1967. *The Theory of Island Biogeography*. Princeton, N.J.: Princeton Univ. Press, 203p.
Meggers, B. J.; Ayensu, E.; and Duckworth, W. D., 1973. *Tropical Forest Ecosystems in Africa and South America: A Comparative Review*. Washington, D.C.: Smithsonian Inst. Press, 350p.
Merrill, E. D., 1954. *The Botany of Cook's Voyages*. Waltham, Mass.: Chronica Botanica, 383p.
Polunin, N., 1960. *Introduction to Plant Geography*. London: Longmans, Green, 640p.
Raven, P., and Axelrod, D. I., 1972. Plate tectonics and Australasian Paleobiogeography, *Science*, **176**, 1379–1386.
Raven, P., and Axelrod, D. I., 1974. Angiosperm

biogeography and past continental movements, *Ann. Missouri Bot. Gard.*, **61**, 539–673.

Riley, C. L.; Kelley, J. C.; Pennington, C. W.; and Rands, R. L., eds., 1971. *Man Across the Sea—Problems of Pre-Columbian Contacts.* Austin: Univ. Texas Press, 552p.

Sauer, C. O., 1969. *Agricultural Origins and Dispersals —The Domestication of Animals and Foodstuffs.* Cambridge: M.I.T. Press, 175p.

Thorne, R. F., 1973. Floristic relationships between tropical Africa and tropical America, in Meggers et al., 1973, 27–47.

Tralau, H., ed., 1969–. *Index Holmensis—A World Index of Plant Distribution Maps.* 12 vols. Stockholm: Scientific Publ.

Valentine, D. H., ed., 1972. *Taxonomy, Phytogeography and Evolution.* New York: Academic Press, 446p.

Walker, J., 1971. Pollen morphology, phytogeography, and phylogeny of the Annonaceae, *Contrib. Gray Herb., no. 202,* 130p.

Walker, J., 1972. Chromosome numbers, phylogeny, phytogeography of the Annonaceae and their bearing on the (original) basic chromosome number of angiosperms, *Taxon,* **21,** 57–65.

Whitmore, F. C., and Stewart, R. H., 1965. Miocene mammals and Central American seaways, *Science,* **148,** 180–185.

Willis, J. C., 1922. *Age and Area—A Study in Geographical Distribution and Origin of Species.* Cambridge: Cambridge Univ. Press, 259p.

Woodring, W. P., 1966. The Panama Land Bridge as a sea barrier. *Trans. Amer. Phil. Soc.,* **110,** 425–433.

Wulff, E. V., 1950. *An Introduction to Historical Plant Geography.* Waltham, Ma.: Chronica Botanica, 223p.

Cross-references: *Paleobiogeography; Paleobotany; Palynology, Paleozoic and Mesozoic; Palynology, Tertiary.* Vol. II: *Vegetation Classification and Description; Zone–Climatic.*

PALEOTEMPERATURE AND DEPTH INDICATORS

Determination of temperature at the time marine sediments were deposited poses problems for geochemists, paleoecologists, and biologists; determination of depth of deposition, on the other hand, is purely a problem of paleoecology or of sedimentation physics coupled with a knowledge of the prior and post tectonics of the region. Relatively exact values for sea temperatures that prevailed during specific periods of geologic time can be obtained through analyses of carbonate skeletons of organisms that lived during that period. Composition of carbonate compounds and various bound metals from sea water relating to such variables as O^{18}/O^{16} ratios (Urey, 1948), magnesium concentrations (Chave, 1954), strontium values (Lowenstam, 1961), and aragonite/calcite ratios (Lowenstam, 1954) appear sensi-

tive to environmental temperatures at time of their deposition. For instance, mass spectrographic analyses of fossil shells can be converted to temperatures (in °C) by the following formula: $T = 16.5 - 4.3(\delta - A) + 0.14(\delta - A)^2$, where δ is the instrumentally determined difference (in per mil) of the O^{18} to O^{16} ratio between the sample and a standard reference gas (Epstein et al., 1953). Considerable literature, some of it contradictory, exists as to both concentrations of these factors during the life cycle of organisms and also as to changes occurring through diagenesis (Urey et al., 1951; Odum, 1957; Lowenstam, 1954, 1961; Lowenstam and Epstein, 1954; and Epstein and Lowenstam, 1953). A summary of these data can be found in Emiliani (1972) and Van Donk (1977).

Although all relationships between carbonate secretion and uptake of various isotopes and elements as related to stages of the life cycle of organisms are not fully understood, a close approximation of average water temperatures extant during the lives of carbonate skeletal organisms can be obtained. It is misleading, however, to use analyses of this kind without knowledge of the Holocene autecology of the animals that deposited the shells. For instance, it might be erroneous to base average marine-water temperatures on forms that might have been fresh water, brackish, or more intertidal than subtidal in habitat. It would be even more dangerous to use shells of animals that live only below the thermocline for determination of average surface sea temperatures.

The most accurate and promising method of determining paleotemperature in geologic eras lacking closely related representatives of modern organisms is by geochemical means. Fossils from pre-Tertiary formations are difficult to use as temperature indicators on the basis of the somewhat maligned term "uniformitarianism," because their ecology can only be approximated. For instance, it has been possible to reconstruct major climatic trends in the Mesozoic by using geochemical analyses of such completely extinct forms as ammonites and belemnites, whose ecology is more or less unknown (Urey et al., 1951; Lowenstam, 1954; Bowen, 1961, 1966). Oxygen isotope paleotemperature analyses are useful in determining climatic changes over both long and short periods, particularly during the Pleistocene (Emiliani, 1955, 1971, 1972; Valentine and Meade, 1961).

Within a late Tertiary to Holocene frame reference, it is often practical to use the principle that "the present is the key to the past" in order to interpret either temperature or climatic changes. The majority of modern shallow-water species of mollusks, for instance, have remained relatively unchanged in external morphology and by inference, therefore, in their life processes since middle Miocene times. Many have close relatives from Eocene times. One can hypothesize average temperatures where fossil assemblages lived if the present temperature limitations are known for Holocene species of skeletal organisms that have the same or closely related species as fossil ancestors. Most molluscan species have restricted temperature or latitudinal ranges, and in fact are so restricted that they have been used a number of times to delimit temperature-dependent zoogeographic provinces along almost every N- and S-aligned coast. On this basis, Valentine and Meade (1961) using geographic ranges of Holocene mollusk species represented in Pleistocene deposits were able to determine shifts in average water temperatures during various glacial and interglacial stages. The sections studied lie at the boundary of two present-day climatic zones (warm-temperate and subtropical). Since a fluctuation of a few degrees in yearly mean water temperature over a period of years in this locality can cause a shift in faunal composition to either Panamanian or pure Californian, Valentine and Meade presumed that population changes of this nature must have occurred as a result of periodic glaciations during the Pleistocene. In truth, they did find that there were sections of the Pleistocene that contained a dominance of cold-water or northward-ranging species and other sections that contained mostly tropical or southward-ranging species. These findings were then substantiated by geochemical methods, and the components of these two kinds of assemblages were analyzed for O^{18}/O^{16} ratios. Those presumed to be warm-water species proved to have lived in water temperatures of $17°$–$20°C$, much warmer than present-day waters; while the cold-water species apparently lived in temperatures ranging between $5°$ and $12°C$, much colder than today and more typical of waters far to the north.

Boundaries between temperature-dependent zoogeographic provinces are the easiest places to detect climatic changes, even in colder regions (see Bloom, 1963). On the other hand, in the tropics temperature or climatic changes are almost impossible to detect on the basis of paleoecology alone, since water temperatures apparently have always been above $20°C$ and the faunas have always been tropical in aspect.

Temperature changes and climatic shifts may occur in time spans of less than a decade up to 50 years; although short-term changes would be difficult to detect in older sediments unless the rate of deposition was extremely rapid. In a study of benthic faunas in the northern Gulf of Mexico, Parker (1959, 1960) found that when a prolonged drought accompanied by higher mean temperatures occurred at the boundary between two zoogeographic provinces the species composition changed also. At first, only larvae of tropical species appeared seasonally in the former colder waters, and eventually adults were found throughout the year. Finally, when temperatures (and associated higher rainfall) returned to "normal," tropical species disappeared, leaving empty shells to provide a puzzle for future paleoecologists.

A similar phenomenon on a much greater time scale was hypothesized by Parker (1960). Subfossil shells were found on a series of submerged strand lines on nondepositional portions of the Texas continental shelf. Each strand line seemed to be associated with an interrupted rise of sea level during the last 17,500 years (Curray, 1960, 1961). Curray (1960), using microbathymetry to map ancient stream patterns and plotting sources and later distribution of certain heavy mineral groups, demonstrated that one set of strand lines was formed by predominately eastward-flowing longshore currents and the others by westward-flowing currents. Parker (1960) identified shells from these deposits and found that when the currents flowed from E to W, the species were predominately cold-water forms from the NE (Fig. 1). On the other hand, deposits associated with currents flowing SW to E contained shells of tropical species now known only from the Campeche Gulf region (Fig. 2).

An approximate index of paleotemperature can be based upon the relative diversity of species within a typical marine assemblage. Parker (1964) made a comparison between the number of species per m^2 of level bottom in the tropics and in boreal and cold-temperate regions. From five to ten times the number of species were found on tropical level bottoms than were found on cold-climate bottoms. Thorson (1957), on the other hand, stated this was true only for epifaunal or reef assemblages, but indicated that the number of species on level bottoms (infaunal) were the same from poles to tropics. His later unpublished studies in the Gulf of Thailand refuted his older hypothesis. In comparisons of this nature, one must make sure that other environmental factors are equal. Adverse conditions of any sort will tend to reduce the number of species in any one spot.

No reliable means of determining the depth of water at which sediments were deposited has yet been discovered. Depth of deposition can be hypothesized, however, on the basis of certain physical principles associated with dispersal and deposition of sedimentary particles. Such

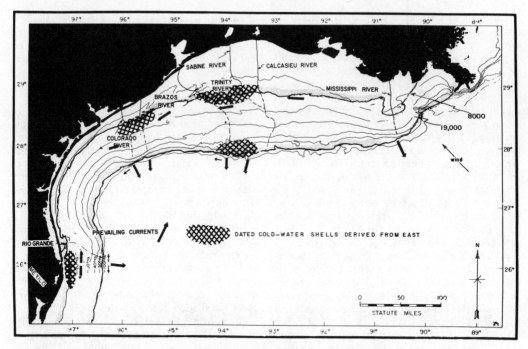

FIGURE 1. Paleogeography of shoreline deposits, NW Gulf of Mexico, at approximately 19,000 yr B.P. for outer shoreline and 8,000 yr B.P. for inner shoreline. Climate determined by cold-water species remains.

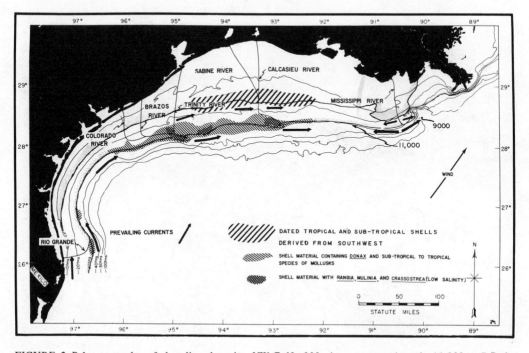

FIGURE 2. Paleogeography of shoreline deposits, NW Gulf of Mexico, at approximately 11,000 yr B.P. for outer shoreline and 9,000 yr B.P. for inner shoreline. Tropical climate determined from Caribbean-type mollusk remains.

phenomena as graded-bedding, certain kinds of ripple marks, and cross-bedding have erroneously been used to characterize shallow-water deposits. The presence of evaporites can usually be considered a reliable indication of whether sedimentation occurred at the shoreline or in deep water. However, evaporitic compounds may form through processes that do not involve evaporation, such as, the Dead Sea deep brines described by Degens and Ross (1969). Evaporites also may form behind barriers as in the enclosed Delaware Basin of Permian times (Shaw, 1964), and more specifically in the Mediterranean Sea when the Straits of Gibraltar was closed. More precise determinations of water depth can be based on the kinds of organisms imbedded in sediments to be analyzed for depth. Sufficient autecological knowledge concerning Holocene organisms is now available so that many species and genera can be used as precise depth or shoreline indicators. Even shape and certain shell morphological features can be used to give approximate depths where certain organisms have lived. The pelecypod genus *Donax,* for instance, is almost entirely confined to surf zones throughout the world. Crustaceans such as *Callianassa, Ocypode,* and *Uca* are also con-

fined to shorelines and extremely shallow water. Certain representatives of two molluscan families (Mactridae and Corbiculidae) are remarkably similar in external morphology. These resemblances seem to be associated with the ability to live in shallow, very low-salinity water, thus morphology alone can be used to indicate shoreline (Parker, 1956, 1959, 1960). Likewise, molluscan morphology is a reliable indicator for very deep waters. Many molluscan genera and families have representatives in all depths of water; but while those that live in shallow water are robust, highly colored, and heavily muscled, deep-water members of these same families are usually colorless or white, extremely thin shelled or covered with thick periostracum, and are weakly muscled. These specializations appear to be adaptations to fewer predators, a very uniform environment, and no turbulence in the case of deep-sea forms, versus many predators, a variable environment, and considerable turbulence acting upon the shallow-water species.

The continental shelf in most warm-temperate to tropical portions of the world can be divided into parallel depth zones, marked by distinct assemblages of marine invertebrates. Charac-

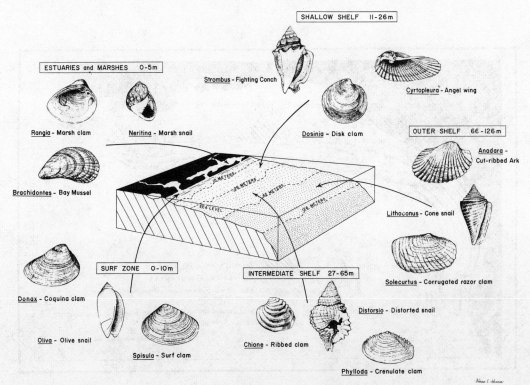

FIGURE 3. Characteristic or index mollusk species for various depth facies on a typical continental shelf and coastal zone (from Parker, 1976; copyright ©1976 by Dowden, Hutchinson & Ross, Inc.).

teristic shells for each depth zone are shown on Fig. 3. Depth separation of assemblages is more or less distinct in warmer waters, because there is a considerable difference between surface and bottom temperatures, associated with differing degrees of mixing between surface and bottom waters (Parker, 1960). Parallel depth communities are not so evident in cold-temperate to arctic regions, because there is little difference between surface and bottom temperatures even at slope depths. Parallel assemblages in increasing depths also can be associated with changes in sediment size on depositional coasts. Since grain size is a function of settling velocity, coarse-grain sediments usually fall out first, and thus into shallow water. Therefore, there is a decrease in grain size from shore to shelf-break.

Many invertebrates are restricted to distinct sediment types, thus depth assemblages may exist as a function of sedimentation physics. The following depth zones on the continental shelf are characterized by distinct faunal assemblages: (1) 0–10 m, (2) 11–26 m, (3) 27–65 m, and (4) 66–126 m (Fig. 3). The depth limits of each assemblage may shift slightly in either direction, depending upon latitude or sediment type.

The continental slope, abyssal, and hadal sea floor also can be divided into distinct depth zones, each with distinctive faunas. These zones seem to be more absolute than the shelf zones, as bottom water temperatures are more stable, and the habitats are extremely uniform over large distances. Assemblages have been de-

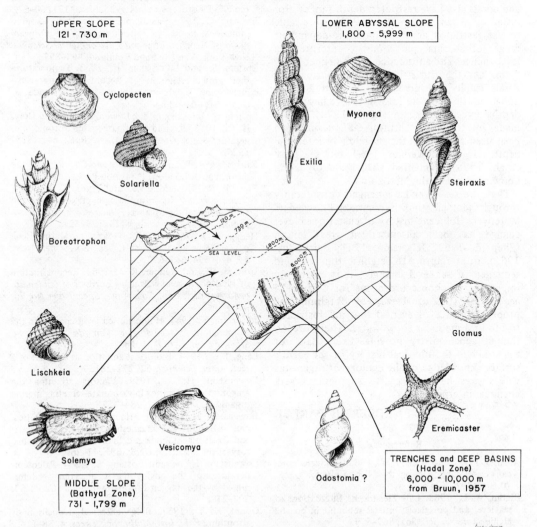

FIGURE 4. Characteristic or index invertebrate species for various depth facies on a typical continental slope, trench, and abyssal sea floor (from Parker, 1976; copyright © 1976 by Dowden, Hutchinson & Ross, Inc.).

scribed for the following depths along the Pacific coast of Middle America (Parker, 1963, 1964) and for other deep-sea deposits throughout the world: (1) the upper slope, 121–730 m; (2) middle slope, 731–1,799 m; (3) outer slope and abyssal plains, 1,800–5,999 m; and (4) trenches and deep abyssal basins–the hadal zone–6,000 m to >10,000 m (Bruun, 1957). Index species for the various deep-sea zones are shown on Fig. 4. Although fossiliferous uplifted deep-sea sediments are rare, an excellent example of bathyal and abyssal fauna can be found on the Burica Peninsula of Panama, where sediments of Pliocene-Miocene age contain a full range of bathyal-abyssal fauna (Olsson, 1942; Parker, 1976, Fig. 7).

Although mollusks are excellent depth indicators, certain bryozoans, echinoids, crustaceans, and corals also have restricted depth ranges. For instance, Allan and Wells (1962) were able to use restricted depth-temperature-dependent ranges of ahermatypic corals to interpret sea level changes and subsidence off the Niger River delta. On the other hand, some hermatypic corals usually considered to be restricted to the upper photic zone are known to live happily at depths exceeding 90 m (Lang, 1974). Tubes made by the ghost shrimp, *Callianassa*, have been used to indicate the shoreline or intertidal depth (Hoyt and Weimer, 1963; but see Frey et al., 1978). Cemented *Callianassa* tubes are known at least to the Mesozoic.

That the sea level has changed considerably through geologic time is common knowledge. Moreover, Holocene history of eustatic sea-level change has been discussed at great length (Shepard, 1963; Jelgersma, 1961; Fairbridge, 1961). In attempting to establish the time and sequence of sea-level changes, it is extremely important to choose organisms for Carbon-14 age dating (the usual way of obtaining the proper sequence of events) that are restricted to life in a meter or so of water. Therefore, although geochemistry provides exact data for the study of the physical history of the earth's surface, knowledge of the ecology of organisms will always be necessary to supplement geochemical interpretations.

ROBERT H. PARKER

References

Allan, J. R. L., and Wells, J. W., 1962. Holocene coral banks and subsidence in the Niger Delta, *J. Geol.,* 70, 381–397.

Bloom, A. L., 1963. Late Pleistocene fluctuations of sea level and postglacial crustal rebound in coastal Maine, *Amer. J. Sci.,* 261, 862–879.

Bowen, R., 1961. Paleotemperature analyses of Mesozoic Belemnoidea from Germany and Poland, *J. Geol.,* 69, 75–83.

Bowen, R., 1966. *Paleotemperature Analysis.* Amsterdam: Elsevier, 265p.

Bruun, A., 1957. Deep sea and abyssal depths, in Hedgpeth, 1957, 641–672.

Chave, K. E., 1954. Aspects of the biogeochemistry of magnesium. I. Calcareous marine organisms, *J. Geol.,* 62, 266–283.

Curray, J. R., 1960. Sediments and history of Holocene transgression continental shelf, northwest Gulf of Mexico, in Shepard et al., 1960, 221–266.

Curray, J. R., 1961. Late Quaternary sea levels: A discussion, *Geol. Soc. Amer. Bull.,* 72, 1707–1712.

Degens, E. T., and Ross, D. A., eds., 1969. *Hot Brines and Recent Heavy Metal Deposits in the Red Sea.* New York: Springer-Verlag, 600p.

Emiliani, C., 1955. Pleistocene temperatures, *J. Geol.,* 63, 538–578.

Emiliani, C., 1966. Isotopic paleotemperatures, *Science,* 154, 851–857.

Emiliani, C., 1971. Paleotemperature variations across the Plio-Pleistocene boundary, *Science,* 171, 60–62.

Emiliani, C., 1972. Paleotemperatures–Isotopic determinations, in R. W. Fairbridge, ed., *The Encyclopedia of Geochemistry and Environmental Sciences.* New York: Van Nostrand Reinhold, 891–897.

Epstein, S., and Lowenstam, H., 1953. Temperature-shell growth relations of Recent and interglacial Pleistocene shoal water biota from Bermuda, *J. Geol.,* 61, 404–438.

Epstein, S.; Buchsbaum, R.; Lowenstam, H.; and Urey, H. C., 1953. Revised carbonate-water isotopic temperature scale, *Geol. Soc. Amer. Bull.,* 64, 1315–1325.

Fairbridge, R. W., 1961. *Physics and Chemistry of the Earth,* vol. 4. London: Pergamon Press, 99–185.

Frey, R. W.; Howard, J. D.; and Pryor, W. A., 1978. *Ophiomorpha:* Its morphologic, taxonomic, and environmental significance, *Palaeogeography, Palaeoclimatology, Palaeoecology,* 23, 199–229.

Hedgpeth, J. W., ed., 1957. Treatise on Marine Ecology and Paleoecology, vol. I, *Geol. Soc. Amer. Mem.* 67, 461–534, 641–672.

Hoyt, J. H., and Weimer, R. J., 1963. Reconstruction of Pleistocene sea levels using burrows of *Callianassa major* (abstr.), *Geol. Soc. Amer. Spec. Paper No. 76,* p. 84.

Jelgersma, S., 1961. Holocene sea level changes in the Netherlands, *Mededelingen Van de Geologische Stichting,* ser. C, 6(7), 101p.

Lang, J. C., 1974. Biological zonation at the base of a reef, *Amer. Scientist,* 62, 272–281.

Lowenstam, H. A., 1954. Factors affecting the aragonite:calcite ratios in carbonate-secreting marine organisms, *J. Geol.,* 62, 284–322.

Lowenstam, H. A., 1961. Mineralogy, O^{18}/O^{16} ratios, and strontium and magnesium contents of recent and fossil brachiopods and their bearing on the history of the oceans, *J. Geol.,* 69, 241–260.

Lowenstam, H. A., and Epstein, S., 1954. Paleotemperatures of the post-Aptian Cretaceous as determined by the oxygen isotope method, *J. Geol.,* 62, 207–248.

Odum, H. T., 1957. Biogeochemical deposition of strontium, *Publ. Inst. Marine Sci. Texas,* 4, 38–114.

Olsson, A. A., 1942. Tertiary and Quaternary fossils from the Burica Peninsula of Panama and Costa Rica, *Bull. Amer. Paleontol.,* 27(106), 82p.

Parker, R. H., 1956. Macro-invertebrate assemblages as indicators of sedimentary environments in east Mississippi Delta region, *Bull. Amer. Assoc. Petrol. Geol.*, **49**, 295–376.

Parker, R. H., 1959. Macro-invertebrate assemblages of central Texas coastal bays and Laguna Madre, *Bull. Amer. Assoc. Petrol. Geologists*, **43**, 2100–2166.

Parker, R. H., 1960. Ecology and distributional patterns of marine macro-invertebrates, northern Gulf of Mexico, in Shepard et al., 1960, 302–380.

Parker, R. H., 1963. Deep sea benthic invertebrate communities *Proc. 16th Internatl. Cong. Zool.,* vol. 4, 309.

Parker, R. H., 1964. Zoogeography and ecology of some macro-invertebrates, particularly mollusks, in the Gulf of California and the continental slope off Mexico, *Viden, Medel. fra Dansk Naturhist. Foren.,* **126,** 178p.

Parker, R. H., 1976. Classification of communities based on geomorphology and energy levels in the ecosystem, in R. W. Scott, and R. R. West, eds., *Structure and Classification of Paleocommunities.* Stroudsburg, Pa.: Dowden, Hutchinson & Ross, 67–86.

Shaw, A. B., 1964. *Time in Stratigraphy.* New York: McGraw-Hill, 365p.

Shepard, F. P., 1963. *Essays in Marine Geology in Honor of K. O. Emery.* Los Angeles: Univ. of S. Calif. Press, 1–10.

Shepard, F. P.; Phleger, F. B.; and van Andel, Tj. H., eds., 1960. *Recent Sediments Northwest Gulf of Mexico, 1951–1958.* Tulsa: Amer. Assoc. Petrol. Geol., 221–266, 302–380.

Thorson, G., 1957. Bottom communities (sublittoral or shallow shelf), in Hedgpeth, 1957, 461–534.

Urey, H. C., 1948. Oxygen isotopes in nature and in the laboratory, *Science,* **108,** 489–496.

Urey, H. C.; Lowenstam, H. A.; Epstein, S.; and McKinney, C. R., 1951. Measurement of paleotemperature and temperatures of the Upper Cretaceous of England, Denmark, and the southeastern United States, *Geol. Soc. Amer. Bull.,* **62,** 399–416.

Valentine, J. W., and Meade, R. F., 1961. Californian Pleistocene paleotemperatures, *Univ. Calif. Publ. Geol. Sci.,* **40**(1), 1–46.

Van Donk, J., 1977. O[18] as a tool for micropaleontologists, in A. T. S. Ramsay, ed., *Oceanic Micropaleontology,* Vol. 2. London: Academic Press, 1345–1370.

Cross-references: *Actualistic Paleontology; Biogeochemistry; Diversity; Paleobiogeography; Dwarf Faunas; Paleoecology.* Vol. IVA: *Paleotemperatures —Isotopic Determinations.*

PALYNOLOGY

Palynology, as introduced and originally defined by H. A. Hyde and D. A. Williams in 1944, is taken from the Greek verb *paluno,* meaning to strew or sprinkle, and the noun *palē* or fine dust, and is intended to apply to "the study of pollen and other spores and their dispersal, and application thereof." The term, first published in the *Pollen Analysis Circular,* an informal mimeographed publication issued in the United States during the 1940s and early 1950s, came about as a response to a query made in an early issue by Ernst Antevs, the well-known varve chronologist. Antevs questioned whether the term "pollen analysis," the forerunner of palynology, was indeed a satisfactory one. Pollen analysis, then understood to apply largely to studies dealing with bog and lake sediments, was not appropriate in designating, for example, the morphological and taxonomic aspects of spores and pollen grains. Because of its limited coverage, therefore, and because of ambiguity between the term and what it was intended to cover, pollen analysis had become antiquated and a better term was needed that could be applied to all pollen and spore work. Other terms such as "pollen science" were suggested but never achieved the general acceptance of the term "palynology" which is now deeply established in the literature. The original definition of palynology has, however, undergone modification in certain quarters, particularly with regard to petroleum exploration work, to include not only fossil pollen and spores but other fossil groups such as hystrichospherids, coccoliths, microforaminifers, dinoflagellates, and chitinozoans. But in concerning itself with such a broad spectrum of microfossils, palynology becomes indistinguishable from micropaleontology.

Covering all studies dealing with pollen and spores, paynology finds application in many fields. It is invaluable in geology in the stratigraphic correlation of freshwater deposits where foraminifers, for example, of use in the case of marine sediments, are absent. Since spores and pollen chronologically range from the Paleozoic to the present day, they enter into a wide variety of problems covering many time units. The value of palynology in paleoclimatology lies in the reconstruction along uniformitarian lines of past vegetation and, in turn, climate from fossil assemblages contained in sediments of different age. Palynology serves the interests of archaeology in tracing the foods and environments of prehistoric man. Medical science investigates pollen and spores from the standpoint of hay fever (pollinosis) and allergies. In fact, during the flowering period of ragweed, many newspapers publish daily pollen counts for the benefit of pollinosis sufferers. In plant systematics, palynology often enters into the determination of taxonomic relationships held likely on gross morphological grounds. Even in the field of criminology, the identification of certain pollen grains found in mud on a suspect's shoes has led to his prosecution and conviction.

It was the invention of the compound microscope in the latter half of the 17th century that enabled Marcello Malpighi and Nehemiah Grew to observe and describe pollen grains and spores in some detail for the first time. Not until the 19th century, however, did the works of J. E. Purkinje, H. von Mohl, C. J. Fritzsche, H. Fischer, and a number of other Europeans contribute more significantly to pollen morphology. It was also during the 19th century that microscopists began to take notice of pollen and spores in peats, lake sediments, lignites, coals, shales, and other sedimentary material. At first these microfossils were looked upon only as curiosities. Later, however, at the beginning of the 20th century, G. Lagerheim and his students in Sweden calculated the quantitative changes in pollen types at different depths in bog peat and foresaw their meaning in terms of vegetational history. Somewhat later, one of Lagerheim's students, E. J. Lennart von Post, building on his teacher's work, not only diagrammed the percentage variations of pollen in a number of bog sections but also correlated the diagrams, successfully showing the pattern of postglacial succession and the climatic sequence that followed the reduction of the Scandinavian ice sheet. Von Post's studies, made public in an Oslo lecture in 1916, earned for him the reputation of father of modern pollen analysis.

Since the publication of von Post's findings in 1918, pollen analysis and subsequently palynology have captured the minds of hundreds of investigators (see Manten, 1966; Traverse, 1974). Interest first spread in Europe largely as a result of Erdtman's efforts to bring the potentialities of the field to English-speaking people; and, in 1927, V. Auer introduced the science to North America in his investigation of the bogs of southeastern Canada. Subsequent noteworthy contributions to our knowledge of Pleistocene vegetation and climate are exemplified by the further studies of Auer in Tierra del Fuego, by L. Cranwell and E. J. L. von Post [1936] in New Zealand, and by H. P. Hansen [1947] in the Pacific Northwest of North America. In the case of pre-Pleistocene sediments, results have been slow to appear, but the pioneer efforts of R. Potonié, L. R. Wilson, R. P. Wodehouse, and a few others have been examplary. Morphological studies, at the same time, have been carried forward, particularly by Erdtman, Wodehouse, Iversen, and Faegri. The leading contributor is undoubtedly Erdtman, who has written a number of reference texts and has edited the serial publications *Literature on Pollen Statistics, Grana Palynologica,* and *Grana.* Other important journals are *Pollen et*

Spores, issued since 1959 by the Muséum National d'Histoire Naturelle in Paris, and *Review of Palaeobotany and Palynology* published since 1967 by Elsevier in Amsterdam.

Perhaps the greatest surge of interest in palynology has taken place since World War II. Largely responsible are the petroleum companies, which have installed laboratories employing full-time palynologists to serve in connection with their exploration programs. Indicative of the interest generated by the field was the first International Conference on Palynology held in Tucson, Arizona in 1962. It was attended by 245 palynologists from 22 countries, and its program listed 137 papers for presentation. Later conferences have been held at Utrecht (The Netherlands) in 1966 and at Novosibirsk (USSR) in 1971.

Pollen Grains and Spores

Pollen grains are, of course, microscopic, but anyone whose home is located near forests of pine, spruce, hemlock, or birch, for example, is likely to have observed in springtime the yellow dust of pollen covering the ground or floating on the surface of a pond. Many plants like those mentioned produce pollen in vast quantities. A mature pine will shed billions of grains in a single season, and a figure of 75,000 tons is given for pollen of the spruce forests of coastal and southern Sweden during peak production. Plants achieve a maximum annual yield after which, for a week or so, the pollen is in the air in large numbers. When production subsides, small amounts continue to settle out only for a time, although it is possible to pick up pollen grains from the atmosphere even in winter, the grains having been reworked by wind long after they were shed. Pollen normally is not carried in any quantity far from its source by wind, but grains have been "caught" some distance from their points of origin by aircraft flying polar and transoceanic trajectories. These are invariably the so-called wind-pollinated types, which depend on air currents to ensure transfer to the female flower; insect-pollinated types are found less frequently.

Pollen grains are the male reproductive bodies in the seed plants or spermatophytes (known also as phanerogams). They originate in the anthers of the flowering plants or angiosperms (grasses, daisies, oaks, for example) and in the microsporangia of the gymnosperms (exemplified by the cycads, ginkgo, and pines). The function of pollen is to serve as a means for the transfer of the contained sperms to the female reproductive part (i.e., pistil of the flower in angiosperms or the carpellate cone in gymnosperms). By differentiation of the sporogenous tissue in the anther of the stamen, microspore mother cells form; these, by a series of cell divisions, give rise to microspores. The nucleus in each microspore next divides and a binucleated pollen grain is formed. Subsequent divisions of one of these nuclei produces a pair of sperms. After pollination, when pollen reaches the female part, tube formation occurs, the sperms moving along as the tube digests its way through tissue of the pistil. Inside an ovule within the ovary, one of the sperms fuses with the enclosed egg and fertilization is accomplished. Of course, of the countless number of pollen grains shed by the seed plants only a fraction enter into fertilization or even pollination. Countless more disintegrate on the ground surface or are preserved in the accumulating sediments of bogs, lakes, and other wet places. The value of pollen and spores to the earth sciences rests in the preserving qualities of water bodies.

A pollen grain is basically characterized as to its type, shape, size, surface sculpture, and the structure of its wall. Although several type classifications exist, at least 26 classes are recognized. These take into account grains usually occurring freely or in groups; those equipped with wings or bladders (sacci); those with or without apertures by which germ tube formation occurs; and, in the case of those with apertures, the number and combinations of germ pores and furrows (colpi; see Figs. 1 and 2). A wide variety of shapes exists. Variations of the basic spheroidal shape are most common; less common are such examples as kidney and crescent shapes. Grains also appear circular, triangular, and polygonal. Their size is found most frequently to lie between 20 and 80 μm but it ranges from <10 μm in such angiospermous families as the Cunoniaceae, Eucryphiaceae, and Piperaceae to 200 μm or more in the Onagraceae, Nyctaginaceae, and Asclepiadaceae. Perhaps of greatest diagnostic value in the identification of species are surface sculpture and wall structure. Some 12 kinds of sculpture or combinations of these are recognizable, representing grains that are smooth, pitted, and grooved or exhibit more or less isodiametric (club-shaped, pointed, etc.) or elongated (striate, reticulate, etc.) elements. Structurally, the wall is made up of three enveloping layers; the innermost consisting of living cellular material, the middle or intine, and the outermost or exine. On fossilization, both the innermost and middle layers decompose; what remains is only the exine, which is usually extremely resistant to decay. The exine, in turn, is frequently divisible into an inner homogeneous layer called the endexine and an outer layer, the ektexine, forming the sculpture elements. On occasion, a middle layer or mesine is distinguishable. Further detail of the ektexine may be discernible in the nature of a tectum or additional membrane outside of the endexine. Grains possessing a tectum are said to be tectate; those simply exhibiting ektexine elements are referred to as intectate.

Spores by comparison are produced by the so-called lower plants or cryptogams comprising such groups as the algae, fungi, mosses, liverworts, horsetails, clubmosses, quillworts, and ferns. Spores differ from pollen grains in that when germination occurs their walls break and the growing prothallus pushes out of the spore. By contrast, germination in pollen does not cause the grain to rupture its wall and the prothallus remains enclosed. In a strictly botanical sense, however, pollen grains are spores and are analogous to the microspores of the vascular cryptogams (ferns and fern allies). They may be thought of among the seed plants as germinated microspores.

Spores, in addition, do not possess apertures (pores and furrows), another characteristic that distinguishes them from certain pollen grains. These differences underlie the distinction between spores and pollen in living plants; but in certain microfossil material, particularly of early and pre-Mesozoic age, classification of extinct entities is often not possible. For this reason, such terms as *polospore* have been introduced to apply to microfossils of indeterminate affinity; another term, *sporomorph,* applies collectively to pollen and spores.

Certain cryptogams are homosporous (yield only one kind of spore) while others are heterosporous (yield both microspores and megaspores). Because of their relatively large size (for example, ca. 250–900 μm in the case of the quillwort *Isoetes*), megaspores often do not come under consideration in palynological work. Moreover, the vascular cryptogams receive greatest attention, palynologically speaking; the fungi, mosses, and other nonvascular groups have been studied much less up to the present time.

Spores are spheroidal, tetrahedral, and elongate (generally biconvex or planoconvex) and when mature occur singly or in tetrads. Their size range is comparable to that of pollen. Surface markings termed *laesurae* represent scars from the time the spores were in contact in the form of a tetrad. In tetrahedral or trilete spores, the laesurae are triradiate while in elongate or monolete spores, they form a single straight line. Some spores lacking visible markings of this kind are termed *alete*. Each laesura is in a sense an aperture because when the

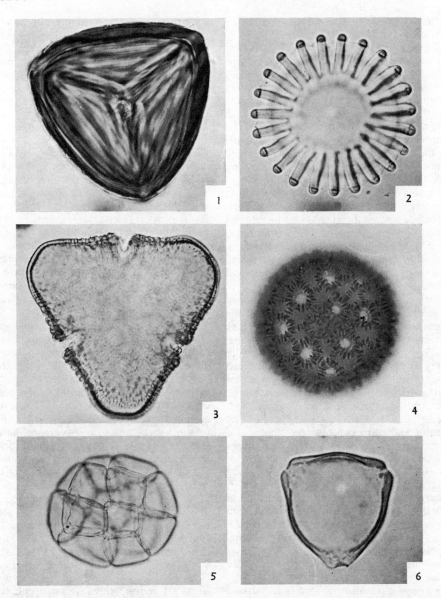

FIGURE 1. Modern pollen and spore types (from Graham, 1963). 1: *Anemia rutifolia,* a fern spore; the triradiate scar is characteristic of the spores of many ferns. 2: *Polygala fruticosa;* this polar view shows the numerous colpi, one of which serves as the place of exit for the pollen tube. 3: *Bombax ellipticum,* a tropical plant. Note the short colpi and their position between the apices of the triangular grain. Usually the apertures are situated at the apices. 4: *Pachysandra procumbens,* a plant of eastern Asia. A pollen type of this kind has recently been found in the Miocene of western United States, approximately 18 m yr old. This grain is *periporate,* that is, the pores are several and scattered over the surface. 5: *Albizia julibrissin,* a subtropical plant often cultivated in warmer parts of the country. It belongs to the legume family, one tribe of which is characterized by having the pollen grains united into *polyads.* 6: *Myrica pensylvanica,* a plant of northeastern North America, related to the wax myrtles. This grain is typical of pollen that is distributed by the wind. Note the small size and smooth surface. This grain is *triporate.* Mostly ca X100.

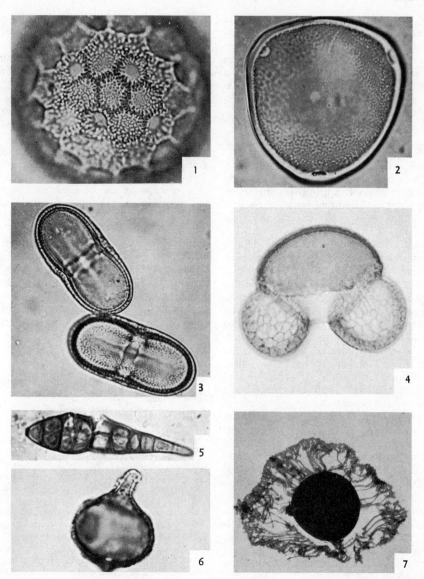

FIGURE 2. Modern and fossil pollen and spore types (from Graham, 1963). 1: *Polygonum longistylum,* plant belonging to the buckwheat family; the coarse reticulum is composed of ridges (muri) and low areas (lumina); within the lumina are pores, and since here are several of them, this grain, like *Pachysandra,* is periporate. 2: *Carya illinoiensis;* the pecan and the hickory, both members of the genus *Carya,* have pollen of this type. The grain is *triporate,* but the pores are not situated exactly on the equator, being slightly above the equator and in one hemisphere. In *Juglans* (the walnut), a member of the same family, there are several pores, but these are scattered over the surface of one hemisphere only, the other half of the grain containing no pores. Grains of this type are said to be *heteropolar.* 3: *Bifora americana,* a member of the Umbelliferae, the carrot family; the dumb-bell shape of the grain is characteristic of the pollen of the Umbelliferae. In this case, a pore is situated in the center of the long colpus, and the grain is *tricolporate.* 4: *Pinus contorta.* The pollen of pine is adapted to wind pollination as evidenced by the air bladders. The furrow is located on the lower side of the body, between the bladders. 5: A fossil fungal spore from the Miocene of Oregon. 6: *Sequoia gigantea,* the giant sequoia; the pollen tube germinates from the tip of the *papillus.* 7: *Triletes rotatus;* from the Carboniferous coal pebbles around Ann Arbor. Mostly ca X100.

developing prothallus emerges, splitting takes place along it. The spore wall, like the wall of a pollen grain, is usually seen to consist of an intine, exine, and, in addition, a layer outside the exine called the perine, observed in certain spores. Many spores appear strikingly ornamented by sculpture elements much in the same manner as pollen grains.

Application to the Earth Sciences

Most pollen and spores are remarkably resistant to decay and are found preserved in sedimentary basins where conditions are unsuitable for decomposition. The number and variety of types encountered in sediments can be large because of the enormous quantities picked up by the wind and carried to the site of deposition. The amount of a particular type, however, is usually not closely proportional to the abundance of mature plants comprising the regional vegetation. Overrepresentation and underrepresentation are recognized by palynologists; in the case of plants now in existence, the degree of proportionality can be ascertained in a general way for the purpose of interpreting the vegetation from a pollen record. Some pollen grains transported from afar or others eroded out of older sediments and redeposited can constitute contamination.

Sedimentary basins have existed throughout geologic time, and where plants were present and reproducing sexually, pollen and spores became incorporated with the sediments under conditions of deposition. Spores of vascular plants occur in rocks as old as the Silurian; pollen of gymnospermous affinity is recorded since the Devonian and of angiosperms since at least the late Early Cretaceous. Samples of sediments collected stratigraphically from sections of former or existing sedimentary basins, are processed in the laboratory with techniques that bring about dissolution of the preserving matrix and release of the pollen and spores. Sonification or boiling in potassium hydroxide solution are simple methods for deflocculation of relatively unconsolidated material. Acetolysis mixture consisting of acetic anhydride and sulfuric acid is an effective material for removing quantities of cellulose and hemicellulose in peats. Siliceous sediments require hydrofluoric acid treatment or differential flotation using bromoform or a similar reagent possessing a relatively high specific gravity. Coals are usually processed by first macerating them in hydrofluoric acid and then treating with a strong oxidizing mixture such as Schulze solution consisting of potassium chlorate and nitric acid. Repeated washing and centrifuging concentrates the microfossils, which are then mounted on microscope slides. Identifications are made with the aid of reference collections and available literature, and the proportion of each type of grain or spore is calculated and plotted in the form of a diagram that becomes the basis for interpretation and correlation work.

Pollen and spores can be identified in the case of extant species because of the constancy of morphological characters. Certain species are known to produce more than one kind of pollen, but they are rare. In the case of microfossils of extinct plants, artificial names need to be assigned. Although constancy of characters prevails at the species level, differences among most species of a genus are not sufficiently distinctive to enable identifications to be made using light microscopy. As a consequence, pollen grains can be identified only at the generic level and in more limiting cases only to family. Electron microscopy over the past decade has proven exceedingly valuable in revealing differences in the structure and sculpture of pollen and spore walls of closely related species. And future work in palynology will increasingly rely on this means for the identification of pollen and spores that are at present problematical.

CALVIN J. HEUSSER

References

Erdtman, G., 1952. *Pollen Morphology and Plant Taxonomy: Angiosperms.* Stockholm: Almqvist & Wiskell, 539p.

Erdtman, G., 1957. *Pollen and Spore Morphology and Plant Taxonomy: Gymnospermae, Pteridophyta, Bryophyta. Illustrations.* Stockholm: Almqvist & Wiskell, 151p.

Erdtman, G., 1965. *Pollen and Spore Morphology and Plant Taxonomy: Gymnospermae, Bryophyta. Text.* Stockholm: Almqvist & Wiskell, 191p.

Faegri, K., and Iversen, J., 1975. *Textbook of Pollen Analysis,* 3rd ed. Copenhagen: Munksgaard, 295p.

Graham, A., 1963. Palynology, with special reference to palynological studies in Michigan, *Michigan Bot.,* 2, 35–44.

Kremp, G. O. W., 1968. *Morphologic Encyclopedia of Palynology.* Tuscon: Univ. Arizona Press, 263p.

Manten, A. A., 1966. Half a century of modern palynology, *Earth-Sci. Rev.,* 2, 277–316.

Traverse, A., 1974. Paleopalynology, 1947–1972, *Ann. Mo. Bot. Gard.,* 61, 203–236.

Tschudy, R. H., and Scott, R. A., eds., 1969. *Aspects of Palynology.* New York: Wiley-Interscience, 510p.

Wodehouse, R. P., 1935. *Pollen Grains: Their Structure and Identification in Science and Medicine.* New York: McGraw-Hill, 574p.

Cross-references: *Angiospermae; Bryophyta; Gymnospermae; Paleobotany; Paleoecology, Terrestrial; Paleophytogeography; Palynology, Paleozoic and Mesozoic; Palynology, Tertiary; Plankton; Pteridophyta.*

PALYNOLOGY, PALEOZOIC AND MESOZOIC

Palynology is the study, for any scientific purpose, of those entities, living or fossil, that are assumed to be spores or pollen grains of vascular plants. Fossil spores and pollen from Paleozoic and Mesozoic sediments can seldom be confidently referred to low-level natural taxa. For this reason, botanists have usually found palynological study of older geological systems less satisfying than that of the Upper Cretaceous and Cenozoic. Most major publications dealing with Paleozoic and Mesozoic palynology are geologically inspired. Many arise from studies whose original aims were primarily utilitarian and local. As data accumulate, however, it becomes increasingly obvious that palynology will make an important and unique contribution to our understanding of the phytogeography of the past (see, e.g., Hart, 1974).

The resistant outer coats (exines) of spores and pollen grains are by far the most common identifiable fragments of fossil plants. They are especially abundant in coals, dark-colored shales and siltstone. Their minute size protects them from abrasion during sedimentary transport. A few grams of lithologically suitable material may yield thousands of delicately preserved plant microfossils. Pollen and spore exines are believed to consist of oxidative polymers of carotenoids and carotenoid esters (Shaw, 1971). They are highly resistant to biological decay and chemical attack. Because of this, they are easily and rapidly removed from the sedimentary matrix by chemical techniques. Kummel and Raup (1965) have described some of the laboratory techniques that are most commonly used. The chemical stability of plant microfossils is not entirely a benefit. It enables them to survive weathering processes and redeposition, and this explains their high incidence as reworked fossils.

Most pollen grains and spores that occur commonly in sedimentary rocks were wind-distributed during the initial stages of their transport. Prior to deposition, however, they were almost all also carried by water for distances that vary between a few meters and hundreds of kilometers. They are thus subject to the same sedimentary processes that affect other particles in the fine-silt range (Brush and Brush, 1972). Variations in size, density, and buoyancy place constraints on the random distribution of plant microfossils and preclude the independence of facies that is sometimes claimed for them. Spores and pollen are, nevertheless, the only terrestrial "organisms" that are widely distributed in marine sediments. Because of this, they provide the main practicable paleontological method for the direct correlation of marine and nonmarine successions.

Classification of Spores and Pollen

Illustrations and descriptions in many early papers dealing with systematic palynology are inadequate as a basis for stable nomenclature. Many attempts have been made to devise comprehensive systems of suprageneric categories, but only one, that originally proposed by R. Potonié and Kremp (1954), has been widely adopted. Potonié and Kremp's scheme provides a hierarchical key based almost entirely on arbitrary morphographical features.

Four morphographic characters of spore and pollen exines provide the principal bases for classifying them. In descending order of importance, they are:

1. Form of the germinal aperture of apertures, which in spores are usually controlled by the trace of the tetrad scar and in pollen by localised areas of weakness in the exine;
2. Internal structural details of the exine.
3. Shape, especially in polar view.
4. Form and distribution of surficial sculptural elements.

Final authority for the legitimacy and validity of specific and generic names rests with the International Code of Botanical Nomenclature, which incorporates the rules of typification and publication. For nomenclatural purposes, genera of spores and pollen grains are normally treated like other fossil remains of identifiable fossil plant organs, and generic names are formed by adding suffixes such as *-spora, -sporites, -pollis,* and *-pollenite.* Some genera of Mesozoic dispersed spores have been given so-called "half-natural names" (e.g., *Matonisporites* and *Microcachryidites*) when their authors wished to draw attention to their resemblances to spores of modern genera. A more extreme practice, especially of Russian authors, is to assign fossil spores of Mesozoic age to either genera of extant plants or to fossil genera, such as *Caytonia,* that were established on the basis of other plant organs.

Paleozoic Assemblages

Lower Paleozoic. Spore-like spheroidal bodies, some bearing what may be faint triradiate markings, occur in Proterozoic and Cambrian sediments. They derive from thallophytes, almost certainly. The first record of trilete spores with heavily thickened, somewhat differentiated, exines is from the Late Ordovician and Early Silurian in the United States and

North Africa (Gray, 1971; Gray and Boucot, 1972). There is no direct evidence that these were produced by vascular plants, but their facies distribution suggests strongly that their parent plants were nonmarine, or grew in shallow coastal waters. Little published information exists on plant microfossil assemblages from the Silurian. A variety of unsculptured and sculptured forms is known to occur in Wenlockian and Ludlovian strata in the Welsh Borderland (Richardson and Lister, 1969).

Devonian. Rapid colonization of terrestrial habitats by vascular plants is reflected by explosive diversification of spore assemblages during the Early and early Middle Devonian. All these older forms are trilete and radially symmetrical. Some possess complexly structured exines and strongly developed sculpture. The striking genus *Emphanisporites* (Fig. 1–2) occurs first in Lower Devonian strata and has a world-wide distribution in the Middle and lower Upper Devonian. Some of the most remarkable spore assemblages in the whole of the geological column occur in Middle Devonian sediments. Apart from their extraordinary diversity, they are notable for the large size of many of their component spores. Heterospory had undoubtedly developed in certain groups as early as the Middle Devonian. However, it seems probable that the microspores and megaspores of Devonian plants did not exhibit the extreme size differences that characterized male and female spores of many Carboniferous heterosporous lycopods. Large, thick-walled, trilete spores bearing anchor-shaped appendages (Fig. 1–16) are common in Middle Devonian assemblages in many parts of the world. They are associated with genera such as *Calyptospora*, with a spinose exine, differentiated into clearly delimited inner and outer layers. Most basic patterns of exine sculpture are found among the trilete spores of the diverse Middle Devonian assemblages.

Important phytogeographic changes occurred at the close of the Middle Devonian and also during the Late Devonian. *Archaeoperisaccus*, the oldest bilaterally symmetrical spore genus, characterizes lower Upper Devonian strata in high present-day northern latitudes and *Geminospora* is found in Frasnian sediments throughout the world. Assemblages from the uppermost Devonian are essentially similar to those from the basal Carboniferous. A characteristic element of most is the foveolate, cavate, species *Spelaeotriletes lepidophytus* which ranges from Famennian to Tournaisian in the USSR, western Europe, North America, and Western Australia.

Carboniferous. Palynological investigations of Carboniferous strata have a long history. Vir-

tually all the pioneer studies in stratigraphic palynology were carried out on assemblages from coals of that age in Russia, Europe, and the United States. As a result, the Carboniferous, at least in the coal measures developed in the northern hemisphere, is among the best-documented systems from the palynological viewpoint (cf. Smith and Butterworth, 1967; Potonié and Kremp, 1955, 1956).

Early Carboniferous assemblages from Tournaisian and Viséan strata are diverse and consist almost entirely of radially or trigonally symmetrical trilete spores, many with structurally differentiated exines and intricate sculpture (Playford, 1962–1963). Heavy equatorial thickenings are a common modification of the exine (e.g., *Triquitrites*, Fig. 1–3); and coarse verrucae, muri, and bacula are characteristic sculptural elements. Although fronds attributed to seed plants are common in the Viséan, undoubted pollen grains are extremely rare prior to the Namurian and not common before the Westphalian (or Desmoinesian). The earliest identifiable gymnosperm pollen grains are radially or bilaterally symmetrical monosaccate forms resembling *Cordaitina* (Fig. 1–17) and *Florinites* (Fig. 1–10). Structurally, they are analogous to the pollen grains of several genera of extant conifers, but were almost certainly produced by pteridosperms. *Schopfipollenites*, the huge sulcate pollen grain of certain medullosean pteridosperms, first appears in the Namurian; but monosulcate pollen is quantitatively negligible in pre-Permian assemblages.

Most plant microfossil assemblages that have been described from Upper Carboniferous-Pennsylvanian strata have been recovered from coals or sediments closely associated with coals. They therefore represent specialized swamp floras and are usually dominated by lycopod spores among which *Densosporites* (Fig. 1–8) and *Lycospora* (Fig. 1–4) are the most common genera. Monolete spores, with a single linear tetrad scar, became important components of plant microfossil assemblages for the first time in the Late Carboniferous (cf. *Torispora*, Fig. 1–5, and *Laevigatosporites*, Fig. 1–6). Their appearance and diversification coincides with the rise of the true ferns. Uppermost Carboniferous strata in Russia and the United States contain a diversity of monosaccate and disaccate pollen grains. Among the latter group is a striking array of forms with parallel bands of thickened exine arranged transversely on the proximal face (cf. *Protohaploxypinus*, Fig. 1–14, and *Lunatisporites*, Fig. 2–13). Pollen grains of this type are referred to a suprageneric taxon, the *Striatiti*, and are most common and diverse in Upper Permian assemblages, although they range into the Triassic.

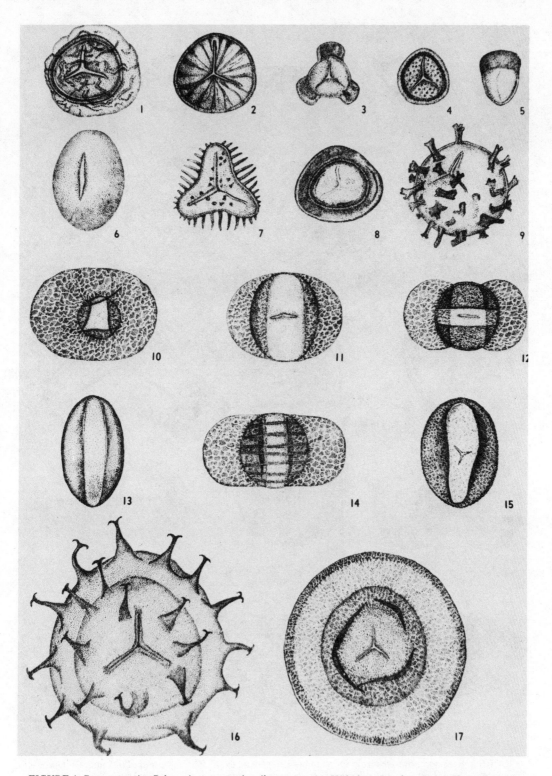

FIGURE 1. Representative Paleozoic spore and pollen genera, ca. X1000. 1, *Diaphanospora*, Upper Devonian-Lower Carboniferous; 2, *Emphanisporites*, upper Lower-lower Upper Denovian; 3, *Triquitrites*. Catboniferous; 4, *Lycospora*, Carboniferous; 5, *Torispora*, Upper Carboniferous; 6, *Laevigatosporites*, Carboniferous-Permian; 7, *Diatomozonotriletes*, mainly Lower Carboniferous; 8, *Densosporites*, Carboniferous; 9, *Raistrickia*, Carboniferous; 10, *Florinites*, Carboniferous-Permian; 11, *Limitisporites*, Permian-Triassic; 12,*Lueckisporites*, Permian-Triassic; 13, *Cycadopites*, Permian-Mesozoic; 14, *Protohaploxypinus*, Permian-Triassic; 15, *Marsupipollenites*, Permian; 16, *Ancyrospora*, Middle to Upper Devonian; 17, *Cordaitina*, Permian-Triassic.

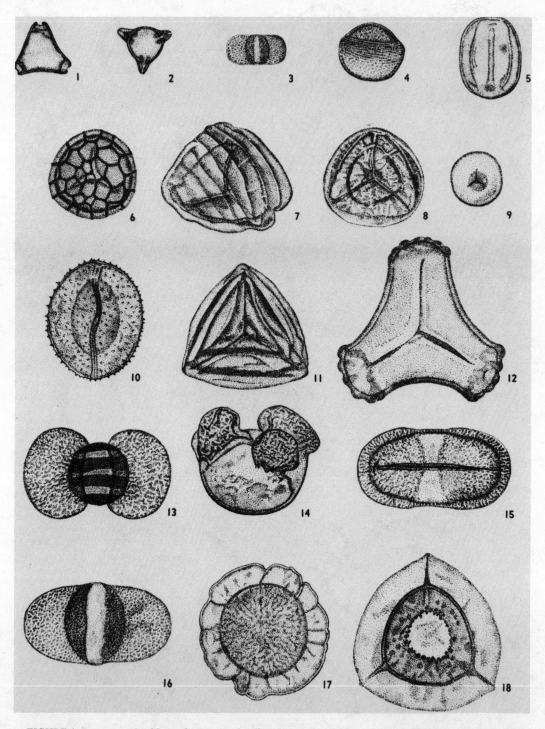

FIGURE 2. Representative Mesozoic spore and pollen genera, ca. X1000. 1, *Vacuopollis*, Upper Cretaceous; 2, *Turonipollis*, Upper Cretaceous; 3, *Vitreisporites*, Permian–Cretaceous; 4, *Classopollis*, Upper Triassic-Cretaceous; 5, *Eucommiidites*, Jurassic–Cretaceous; 6, *Lycopodiumsporites*, Jurassic-Holocene; 7, *Contignisporites*, Jurassic–Lower Cretaceous; 8, *Coronatispora*, Jurassic–Lower Cretaceous; 9, *Stereisporites*, Jurassic–Holocene; 10, *Aratrisporites*, Triassic; 11, *Cicatricosisporites*, uppermost Jurassic-Tertiary; 12, *Trilobosporites*, Cretaceous; 13, *Lunatisporites*, Upper Permian–Triassic; 14, *Microcachryidites*, Jurassic–Holocene; 15, *Ovalipollis*, Triassic; 16, *Falcisporites*, Permian–Jurassic; 17, *Callialasporites*, Jurassic–Lower Cretaceous; 18, *Aequitriradites*, Cretaceous.

Permian. Spectacular morphological diversification of gymnosperm components is the most obvious characteristic of Permian spore and pollen suites. Pollen grains of seed plants became, for the first time, dominant elements of almost all palynological assemblages. Saccate pollen is ubiquitous, varied, and abundant. Sulcate genera, such as *Marsupipollenites* (Fig. 1-15) and *Cycadopites* (Fig. 1-13) are more sporadically distributed, but locally common.

Although Permian assemblages throughout the world display essential similarities, floral provincialism on a gross scale is clearly demonstrated by palynological data. Plant microfossil suites from the Gondwanaland countries show a fundamental unity, as do those from the red-bed successions of western Europe and the American mid-continent (e.g., Hart, 1974). Clearly defined floral provinces are also recognizable in the USSR, one W of the Urals and the other in the Tunguss Basin, Siberia.

Triassic. In detail, Early Triassic spore and pollen associations differ notably from those of the Late Permian. Few well-defined Permian species range into the Triassic, and there is clear evidence that the massive Permian-Triassic invertebrate extinctions were accompanied by widespread modifications of terrestrial floras. In their broad characters, however, Triassic spores and pollen associations, prior to the Rhaetian, are basically similar to those of the later Permian. The oldest Triassic assemblages, particularly those from marine strata, often have atavistic features and consist mainly of lycopod spores. One of the lycopod genera, *Aratrisporites* (Fig. 2-10), ranges from Lower Triassic to basal Jurassic in widely separated parts of the world. Gymnosperm pollen grains dominate Triassic assemblages from late Early to near the close of Triassic time. Saccate forms, probably derived from corystosperms or other "pteridosperm" groups, are the most typical spermatophyte elements (cf. *Falcisporites*, Fig. 2-16; *Ovalipollis*, Fig. 2-15), but the monosulcate genera *Cycadopites* and *Ephedripites* are locally abundant. The latter suggests great antiquity for the extant family Ephedraceae, but it should be emphasized that the pollen evidence is so far unsupported by other ephedralean remains older than Oligocene.

Like the Permian, the Triassic appears to have been a period characterized by sharply delineated phytogeographic regions. Gondwanaland assemblages are, in most of their details, distinct from those of the northern hemisphere, although typical Keuper species occur in northern Malagasy and northwestern Australia. The Rhaetic marked the decline of Triassic "pteridosperm" formations and the establishment of stable cosmopolitan floras which characterized the Jurassic and Early Cretaceous.

Jurassic. Conifer pollen predominates in Jurassic assemblages. *Classopollis* (Fig. 2-4) and *Callialasporites* (Fig. 2-17) are probably araucariacean and have a mondial distribution. Trisaccate pollen of podocarpacean origin (Fig. 2-14) is, on the other hand, confined to the Gondwanaland countries and possibly asiatic Russia.

Fern spores are important components of Jurassic assemblages. Extant fern families, the Osmundaceae, Gleicheniaceae, Matoniaceae, and Schizaeaceae, among others, are represented by abundant dispersed spores. Spores closely similar to those of certain modern species of *Lycopodium* (Fig. 2-6) are common in Middle and Upper Jurassic strata and bryophytes may be represented by *Stereisporites*. Monosulcate pollen is surprisingly rare in view of the abundance of bennettitalean fronds in Jurassic plant fossil collections. Perhaps the rather fragile pollen of the Bennettitales was selectively destroyed, although if the group was insect- or self-pollinated, little pollen may have been released to the atmosphere.

The Jurassic is thought to be a crucial period in the evolutionary history of the angiosperms and there were early hopes that palynology would further illuminate this controversial subject. So far, its contributions have been almost wholly negative. Pollen with exine structures and apertures of the angiosperm type occurs in the Lower Cretaceous. No pollen grain from older strata can be regarded with any conviction as angiospermous.

Cretaceous. The continental and paralic facies that represent the earliest Cretaceous in many parts of the world are ideal for palynological investigation. For this reason, some of the best-illustrated and comprehensive palynological studies deal with plant microfossil assemblages of the Early Cretaceous (cf. Dettmann, 1963, Couper, 1958). As in the Jurassic, these appear to have been produced by floras that were essentially cosmopolitan at a low taxonomic level. Although the Jurassic/Cretaceous boundary is not marked by major floral modifications, several distinctive taxa appeared toward the close of the Jurassic and early in the Cretaceous. The schizaeaceous form *Cicatricosisporites* (Fig. 2-11) is a notable example. In Western Europe, the USSR, United States, and Australia, it first occurs at about the same stratigraphic level, which correlates with a horizon high in the Upper Jurassic. It ranges throughout the Cretaceous and, in some places, also the Tertiary. Other distinctive typical Early Cretaceous genera that have a wide distribution are *Aequitriradites* (Fig. 2-18), *Trilobosporites*

(Fig. 2–12), *Pilosisporites, Rouseisporites,* and the megaspore *Balmeisporites. Gleicheniidites* is a fern spore that is abundant in the upper part of the Lower Cretaceous in widely separated parts of the world.

Botanically, the Mesozoic ended at the close of the Early Cretaceous. From Cenomanian times onward, pollen grains of the dicotyledonous angiosperms are the principal components of palynological assemblages. Pollen grains that can be confidently referred to extant angiosperm families are prominent in Late Cretaceous assemblages and it appears that the broad patterns of modern phytogeography were established by the Early Tertiary. Thus typically southern families such as the Proteaceae and Casuarinaceae and the southern beeches *Nothofagus* are represented in Australian sediments from the Upper Cretaceous onwards but are absent from northern hemisphere assemblages. Similarly, typically northern taxa, such as *Vacuopollis* (Fig. 2–1), *Turonipollis* (Fig. 2–2), and *Aquilapollenites,* are not found in the countries of the southern hemisphere.

BASIL E. BALME

References

Brush, G. S., and Brush, L. M., 1972. Transport of pollen in a sediment-laden channel: A laboratory study, *Amer. J. Sci.,* **272,** 359–381.

Couper, R. A., 1958. British Mesozoic microspores and pollen grains, *Palaeontographica,* **103B,** 75–179.

Cramer, F. H., and Diez, M.D.C.R., 1974. Early Paleozoic palynomorph provinces and paleoclimate, *Soc. Econ. Paleontol. Mineral. Spec. Publ.,* no. 21, 177–188.

Dettmann, M. E., 1963. Upper Mesozoic microfloras from south-eastern Australia, *Proc. Roy. Soc. Victoria,* **77,** 1–148.

Gray, J., 1971. Earliest Silurian spore tetrads from New York: Earliest New World evidence for vascular plants, *Science,* **173,** 918–921.

Gray, J., and Boucot, A. J., 1972. Palynological evidence bearing on the Ordovician-Silurian paraconformity in Ohio, *Geol. Soc. Amer. Bull.,* **83,** 1299–1314.

Hart, G. F., 1974. Permian palynofloras and their bearing on Continental drift, *Soc. Econ. Paleontol. Mineral. Spec. Publ.,* no. 21, 148–164.

Kummel, B., and Raup, D., eds. 1965. *Handbook of Palaeontological Technique.* San Francisco: Freeman, 852p.

Playford, G., 1962–1963. The Lower Carboniferous microfloras of Spitzbergen, *Palaeontology,* **5,** 550–678.

Potonié, R., and Kremp, G. O. W., 1954. Die Gattungen der paläozoischen Sporae dispersae und ihre Stratigraphie, *Geol. Jahrb.,* **69,** 111–194.

Potonié, R., and Kremp, G. O. W., 1955–1956. Die sporae dispersae der Ruhrkarbons, ihre Morphographie und Stratigraphie mit Ausblicken auf Arten anderer Gebiete und Zeitabschnitte," *Palaeon-*tographica, **98B,** 1–136, **99B,** 85–191, **100B,** 65–121.

Richardson, J. B., and Lister, T. R., 1969. Upper Silurian and Lower Devonian spore assemblages from the Welsh Borderland and South Wales, *Palaeontology,* **12,** 201–252.

Shaw, G., 1971. The chemistry of sporopollenin in J. Brooks et al., eds., *Sporopollenin.* London: Academic Press, 305–348.

Smith, A. H. V., and Butterworth, M. A., 1967. Miospores in the coal seams of the Carboniferous of Great Britain *Palaeont. Assoc. London, Spec. Paper Palaeont.,* **1,** 324p.

Tschudy, R. H., and Scott, R. A., eds., 1969. *Aspects of Palynology.* New York: Wiley-Interscience, 510p.

Cross-references: *Angiospermae; Bryophyta; Gymnospermae; Paleobotany; Paleoecology, Terrestrial; Paleophytogeography; Palynology; Palynology, Tertiary; Plankton; Pteridophyta.*

PALYNOLOGY, TERTIARY

Early Tertiary Palynology

The Paleocene, Eocene, and Oligocene, collectively designated as the Paleogene or early Tertiary, were times of considerable change in both the living and the physical world. The task of reconstructing the evolving floras of this period is one of enormous complexity. Until the 1930s, knowledge of plants of the past rested mainly on interpretations of leaf fossils. Pioneering studies of Tertiary spores and pollen by German brown-coal paleobotanists in the 1930s and by American and Dutch oil-geology paleobotanists in the 1940s and 1950s, however, did much to expand the older "pollen analysis," concerned largely with Quaternary pollen, into the multidisciplinary science of Palynology. In the thirty-five year period 1930–1965, over 20,000 publications in the new discipline contributed fresh data on fossil floras, paleoecology, and biostratigraphy. The continuing annual output of publications now taxes the capacities of palynologists to keep abreast of this information explosion (Manten, 1968).

The Plant World of the Paleogene. Before the close of the Cretaceous, angiosperms had become the dominant element in the floras of the northern and southern hemispheres. Their rise to dominance began sometime in the Early Cretaceous over 120 m yr ago; but antecedants, place of origin, and early dispersal routes remain conjectural (Stebbins, 1974). In bold outlines, the vegetation of most of the early Tertiary world consisted of extensive warm-temperate to tropical forests, thickets, and swamps supporting woody dicots, conifers, palms, and ferns in rich variety. Herbaceous genera, especially of the great herbaceous families of grasses and composites, were absent or but sporadically represented until post-Oligocene time. While ferns and gymnosperms continued to be well represented, their record relative to that of the angiosperms is one of gradual contraction and decline.

Legacy from the Cretaceous. *The Foundational Stock of Angiosperm Families.* There are about 300 families of flowering plants recognized in the living flora. The fossil record cannot give complete enough evidence for accurate estimates of the number of families that had evolved by the end of the Cretaceous, but the leaf fossil evidence indicates that 50 or more angiosperm families may have been in existence by the middle Cretaceous and that perhaps another 25 had emerged by the close of the period. The record of Cretaceous pollen, on the contrary, is much less informative because of the high percentage of fossil genera of unknown botanical affinities. Probably fewer than 20 Cretaceous families have been recognized on the basis of pollen alone, and fewer still on the basis of both leaf and pollen. This latter group includes the families:

Aquifoliaceae	Magnoliaceae	Platanaceae
Fagaceae	Menispermaceae	Proteaceae
Hamamelidaceae	Myrtaceae	Liliaceae
Juglandaceae	Nyssaceae	Palmae

Thus estimates must be approached with at least two cautions. First, it cannot be assumed that vegetation and reproductive structures of plants evolved simultaneously. While, for example, the leaves of a Late Cretaceous angiosperm might have acquired distinctive and specific morphological features, its pollen might have retained a more primitive morphology with resemblances to more than one group of plants. Secondly, fossil floras are never complete records of the total vegetation of a particular time or place. Because of restricted habitat distributions, many plants, such as entomophilous herbs of xeric localities, or epiphytes of mountain forests, have low probabilities of contributing leaves or pollen to basins of deposition and thus to their preservation as fossils. There is, however, strong inferential evidence that perhaps one-quarter of the angiosperm families were in existence at the beginning of the Tertiary period. From this stock, numerous phylogenetic lines emerged throughout the early Tertiary, marking that time as a phase of major angiosperm diversification. By the end of the Oligocene, most of the major evolutionary lines within the angiosperms appear to have been established. A list of early Tertiary angiosperm families recognized on the basis of fossil pollen is given in Table 1. This is

TABLE 1. Early Tertiary Angiosperm Families Recognized on the Basis of Fossil Pollen

Aceraceae	Hamamelidaceae	Rhamnaceae
Alangiaceae	Juglandaceae	Rhizophoraceae
Amaranthaceae	Lauraceae	Rutaceae
Anacardiaceae	Leguminosae	Salicaceae
Apocynaceae	Linaceae	Santalaceae
Aquifoliaceae	Loranthaceae	Sapindaceae
Araliaceae	Lythraceae	Sapotaceae
Bombacaceae	Magnoliaceae	Simaroubaceae
Buxaceae	Malpighiaceae	Sonneratiaceae
Caprifoliaceae	Malvaceae	Symplocaceae
Casuarinaceae	Meliaceae	Tamaricaceae
Chenopodiacea	Monimiaceae	Tiliaceae
Combretaceae	Moraceae	Ulmaceae
Cornaceae	Myricaceae	Vitaceae
Corylaceae	Myristicaceae	Winteraceae
Ctenolophonaceae	Myrtaceae	
Cunoniaceae	Nyssaceae	Araceae
Cyrillaceae	Nymphaeaceae	Cyperaceae
Elaeagnaceae	Olacaceae	Gramineae
Ericaceae	Oleaceae	Liliaceae
Euphorbiaceae	Onagraceae	Palmae
Fagaceae	Polygalaceae	Sparganiaceae
Guttiferae	Polygonaceae	Typhaceae
Haloragidaceae	Proteaceae	Zosteraceae

not to suggest, however, that all early Tertiary spores and pollen can be identified with living families and genera. Indeed, for most Paleogene floras, botanical affinities can be ascertained for only a fraction of the contained angiosperm pollen (Leopold, 1974). Problems here may be due to extinction, to difficulty in comparing fossil with living species because of the large number of living species, and to difficulties in identifying intermediate stages in the lineages of rapidly evolving genera. One group in particular of Late Cretaceous–early Tertiary angiosperms that has given palynologists the greatest taxonomic and phylogenetic difficulties is the Normapolles group of dicots designated by Pflug (1953) on the basis of their characteristic and complex pollen. This group reached its maximum development in latest Cretaceous time in Euamerica. Its Paleogene history is one either of extinction or of evolution into the lineages of several modern genera of wind-pollinated angiosperms. For example, Tschudy (1973) has worked out the lineage of one Normapolles genus of unknown affinities, from its middle Cretaceous appearance until its extinction in the Eocene (Fig. 1). Nichols (1973) has shown possible lines leading to the pollen types of several wind-pollinated genera of the walnut family, such as *Engelhardtia, Platycarya,* and *Carya,* presumably arising out of Cretaceous Normapolles ancestry (Fig. 2).

Pollen Wall Evolution. Spores of pteridophytic plants, and pollen of seed plants

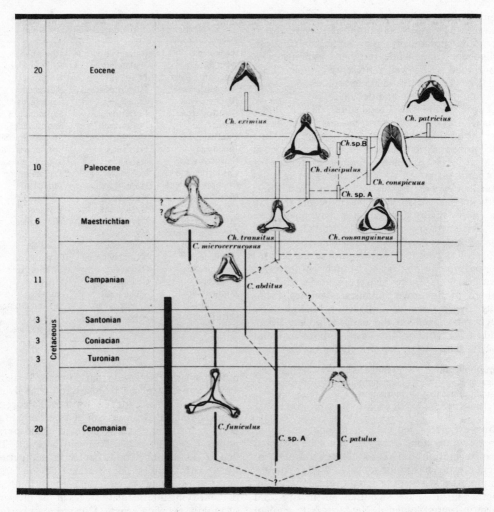

FIGURE 1. Suggested Complexiopollis lineage in Mississippi Embayment (after Tschudy, 1973).

differ from all other fossil microscopic structures in the unique chemistry and architecture of their walls (exines). Resistance to degradation, at least in anaerobic environments, is due primarily to polymeric exine substances known collectively as sporopollenin. Less resistant inner-wall cellulosic substances are early degraded in the fossilization process. Electron microscopic studies of thinly sectioned fossil pollen grains have shown clearly that by early Tertiary time angiosperm pollen was similar in basic exinal structure to that of modern pollen (Ehrlich and Hall, 1959). In all liklihood, the essential chemical and structural features of angiosperm pollen had evolved much earlier in Cretaceous time. Clues to possible phylogenetic sequence in the evolving exines of angiosperms were discovered by Walker and Skvarla (1975) from comparative studies of pollen of living members of the ranalean complex, generally considered to include the most primitive families of extant flowering plants. The Walker-Skvarla interpretation and attending nomenclature are illustrated in Fig. 3.

Germinal Aperture Evolution. While the combination of relative abundance, small size, and highly resistant walls has accounted for the impressive fossil record of spores and pollen, exploitation for paleobotanical and biostratigraphic purposes can begin only after accurate identifications have been made. Competent identification requires skilled observation not only of the more obvious criteria of size and shape, but of the finer details of exine ornamentation and apertural characteristics. This latter feature particularly is of major taxonomic and phylogenetic importance. In the germination of spores and pollen, breaching of exine walls by emerging protoplasmic

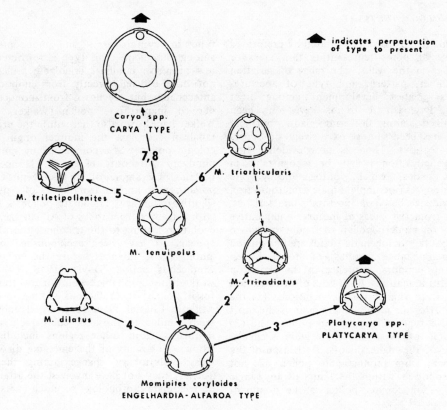

FIGURE 2. Morphological trends among Late Cretaceous–early Tertiary Juglandaceous pollen (after Nichols, 1973).

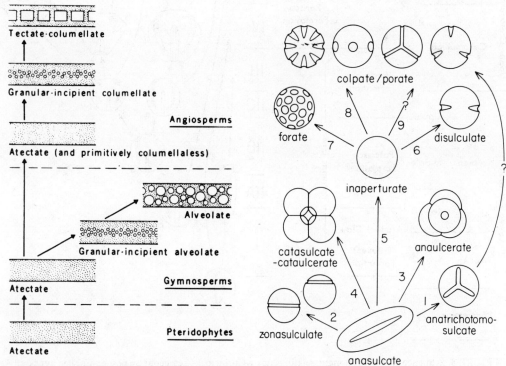

FIGURE 3. Generalized diagram of suggested major evolutionary trends in spore-pollen exine structures (after Walker and Skvarla, 1975).

FIGURE 4. Evolutionary relationships of Ranalean pollen aperture-types (after Walker, 1974).

contents is most often facilitated by preformed furrows or pores representing thin areas or openings in the walls. The range of variation in number, pattern, and type of apertures reaches greatest development among dicot pollen. Phylogenetic trends have long been recognized among flowering plants, but emphasis has been almost exclusively on sporophyte features. There is now evidence that phylogenetic trends may be recognizable in pollen grains, which, ontogenetically, are gametophytes (incomplete male gametophytes). This evidence, based on apertural characteristics, comes from two areas of inquiry: comparative studies on modern pollen of families in which sporophytic phylogenetic trends are known and stratigraphic-range studies of fossil pollen revealing temporal sequence in the evolution of apertural features. Both lines of investigation have led to similar basic conclusions: (1) the single-furrow or monocolpate grain is the most primitive type of angiosperm pollen, and (2) the three-furrow or tricolpate grain is characteristic of the higher dicots. Derivation of the tricolpate from monocolpate pollen has not been followed within the limits of any single family. Monocolpate pollen is the dominant type in gymnosperms and monocots, but among the dicots, this type is restricted to a few primitive families. Tricolpate pollen may not have evolved directly from monocolpate antecedants but, de novo, from monocolpate-derived inaperturate pollen (Walker, 1974). Walker has shown also that within the primitive ranalean complex, the indirect origin of tricolpate pollen was accompanied by numerous blind-end offshoots (Fig. 4). Monocolpate pollen—for example, of the fossil genus *Clavatipollenites*—is known from rocks of Aptian or slightly older age. Tricolpate pollen is known with certainty from rocks of Albian age. Thus, evolution leading to the tricolpate morphology appears to have been accomplished broadly among dicot alliances before the end of the Cretaceous period. Doyle (1973, 1977) has worked out probable evolutionary trends in fossil pollen of Cretaceous sediments of the Atlantic Coastal Plain (Fig. 5). Trends here seem similar to those of comparable age assemblages from other regions including the tropics. For many of the emerging dicot families, evolutionary change during the early Tertiary may not have involved the attainment of tricolpate morphology so much as its modi-

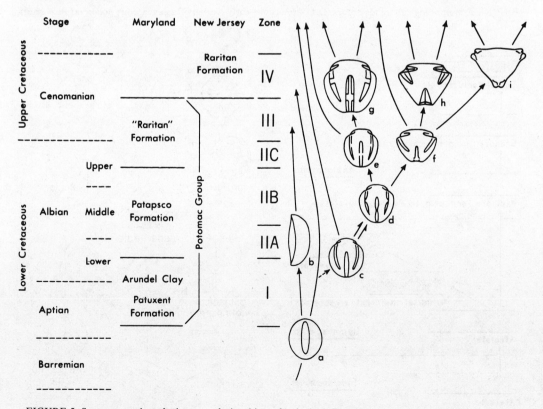

FIGURE 5. Sequence and evolutionary relationships of principal Cretaceous angiosperm pollen types of Atlantic Coastal Plain (from Doyle, 1977).

fication into several derivative types as well as considerable variation in exine surface sculpturing. Tschudy (1969) has given clear definitions and illustrations of the major categories of exinal wall sculpturing.

Summary. The palynological record of the early Tertiary epochs greatly extends data on evolution of the flowering plants as well as on biostratigraphy and geographic distribution. Knowledge of the Paleogene plant world has been expanded by palynological studies, many of which were made in connection with oil and coal exploration and stratigraphic correlation studies. In broad survey, early Tertiary pollen floras may be defined by three characteristics: (1) rich diversity of species, (2) absence of well-marked provincialism, and (3) low percentages of genera still living in the vicinity of the fossil sites.

Penny (1969) has illustrated and summarized worldwide palynological characterization of the Early Tertiary, and Germeraad et al. (1968) have summarized results of twenty years of applied palynological investigations of Tertiary sediments in South America, Africa, and Asia.

J. S. PENNY

Late Tertiary Palynology

The Miocene and Pliocene epochs, recording the terminal part of the Tertiary (termed Neogene or late Tertiary), represent times when the earth's climate was cooling, but nevertheless times of rather wqrm conditions in the mid and high latitudes. The study of fossil pollen and spores as well as evidence from fossil leaves indicate that, during the Miocene, temperate conditions existed at least as far N as the Arctic Circle. The Neogene climate, being warmer than today's and generally warmer than that of the Quaternary but cooler than any part of the early Tertiary, was therefore a transitional one.

Compared to floras of early Tertiary age, pollen and megafossils of the Neogene appear to be assignable to living families or genera. Furthermore, Neogene assemblages to a greater extent than earlier ones seem to suggest modern vegetation associations.

As in the early studies on Paleogene pollen, pioneer work in Neogene palynology began in the 1930s in Germany on the deltaic deposits of the Rhine River and in the 1940s by workers in Venezuela and Colombia. The literature regarding Neogene palynology has grown vastly since then, especially in the mid-north latitudes, and some work is completed in low latitudes and in the southern hemisphere.

As in other parts of the Cenozoic record, evidence from palynology has added greatly to the understanding of Neogene floras and vegetation. This is so for three reasons: (1) Pollen can be found in many sediments that do not contain fossil leaves, and this means that pollen can extend the general understanding of fossil floras geographically into areas where leaves have not yet been found or studied. (2) Because of the abundance of pollen grains in sediments (up to several thousand per gram of mud), pollen lends itself to statistical analysis of closely sampled horizons through a geologic section, so that trends in relative frequency of taxa can be documented and interpreted ecologically. (3) Fossil pollen and spore types generally include at least some of the taxa represented from the same beds as megafossils but usually also include many forms not shown in the leaf record; hence the pollen record supplements the floristic record from megafossils.

Methods. For work in the Neogene, a palynologist needs to have a modern pollen reference collection that includes a broad spectrum of woody types from a large region. He can judge generally how to build his reference collection by consulting fossil leaf records from the sedimentary beds or region in question. For example, in preparation for analysis of the Miocene and Pliocene of Alaska, an examination of Neogene leaf records of S Alaska (e.g., Wolfe et al., 1966) suggests that one should have collections of modern pollen from at least all the major woody plant genera and main herbaceous families from mid and north latitudes N of 33°–34° in the Pacific Basin.

Preparation methods discussed in detail by Gray (1965) are somewhat more complex than those used in Quaternary pollen studies because of the greater degree of induration of Neogene sedimentary rocks. Field collections should as much as possible be accompanied by detailed geologic stratigraphy, geologic mapping, etc.

Interpretation of Neogene pollen assemblages taken as individual floras from mappable units hinges on the analyst's ability to identify the nearest equivalent figures that may be modern analogues of the fossil suites.

Interpretation of the vertical changes in relative and actual composition of the pollen succession depends on the possibility of inferring general or specific climatic or environmental meaning for each of the phases or zones of the sequence. A useful tool in this regard is a transect of modern pollen rain samples (i.e., contemporary soil or mud samples) taken across various vegetation zones of a region; the transect serves to characterize contemporary pollen rain of different environments and altitudinal zones (see McAndrews and Wright, 1969).

Broad Ecological and Floristic Trends. Certain trends or characteristics can be seen in floras that span the late Tertiary in a single area (see Leopold, 1969b). Most of these features stem from the fact that, in a given region, Miocene floras generally indicate warmer conditions than those of the Pliocene age.

Older Neogene pollen floras tend to be larger and more diverse (i.e., have more taxa represented) than younger ones. This may be a reflection of the fact that floras as well as modern pollen surface samples (i.e., pollen rain) from warmer regions are richer in taxa than those from cooler areas. In comparison with Miocene floras, younger Neogene floras generally have a higher proportion of plant genera that now grow near the site or in the immediate area, i.e., they appear more like the modern regional flora than do older Neogene floras. Compared to older Tertiary assemblages, Neogene floras show more regional variation of differentiation (geographic provincialism). Finally, certain very young highly evolved families such as the Compositae do not appear in the geologic record until the upper Oligocene or lower Miocene. In many mid-continent sites, Compositae pollen becomes a common component in the Miocene, while in coastal areas, the group begins to be common in the Pliocene.

In the discussion below are some examples of Neogene floral successions from middle and high north latitudes where such sequences are best known from pollen evidence. A few examples of work from the Southern Hemisphere and from low latitudes are mentioned.

North Pacific Basin. The Neogene floristic sequence of the North Pacific Basin provided by leaf data and supplemented by pollen records indicates a great similarity of floras over a very broad area, namely along the coasts of China (Hu and Chaney, 1940), Japan (Tanai, 1967), Alaska (Wolfe et al., 1966; Wolfe and Tanai, in press), and the W coast of North America including western Washington and Oregon. Though the cooling trends are similar during the Neogene in Japan and Alaska, for example, still the effects of latitude modify the patterns. The warmest conditions of the Neogene were clearly reached in the middle Miocene in Japan (Fuji, 1969-1972), and the same is generally clear in southern Alaska. The Japanese forests of the middle Miocene show an altitudinal zonation that included evergreen oaks and laurels in the lowlands, deciduous broadleaved forests on slopes, and mixed conifer forest in the mountains; the leaf floras show strong affinity with con-

temporary forests of mainland China, but also are reminiscent of middle Miocene floras of Oregon (Tanai, 1972). The Alaskan middle Miocene forests of Cook Inlet suggest recent mixed hardwood forest of northern China (Wolfe and Tanai, in press).

The Neogene pollen floras of the two areas include cool-loving conifers (*Picea, Abies*), cosmopolitan (*Pinus*), and warm-loving conifers (*Keeteleria* in Japan and Taxodiaceae); the floras include many warm to subtropical broadleaved deciduous trees of the Fagaceae, Ulmaceae, Tiliaceae, Juglandaceae, and Aceraceae, plus *Nyssa, Liquidambar,* and *Ilex.*

Miocene pollen floras from Shantung Province of Northern China (Sung et al., 1964), from near Vladivostok (Baikovskaya, 1974), from elsewhere in Japan (Sato, 1963), and from Oregon (Gray, 1964) supplement this picture. According to a well-dated Pliocene site of Seward Peninsula in Alaska (Hopkins, et al., 1971) and other sites, a mixed conifer forest persisted as far north as the Arctic Circle through at least part of the Pliocene. These data suggest that the areas of high and Arctic latitudes were much warmer than now during much of the Neogene.

Eastern and Western Europe. Recognizable provinces of Neogene floras in eastern and western Europe broadly defined begin as very large areas of homogenous floras, and grow more and more diverse, i.e., show more localization or provincialism by the upper Pliocene with latitudinal differentiation becoming more delineated (see Andreeva et al., 1970; Pokrovskaya, 1956; and summary by Leopold, 1969).

West of the Ural Mountains in North Poland and White Russia the early half of the Miocene (Tortonian-middle Sarmatian) seems to represent a northern province with a rich warmtemperate flora containing a few subtropical elements; these floras are dominated by *Pinus* or Taxodiaceae. A southern province S of about 55°N latitude but N of the Caucasas records a diverse broadleaved deciduous forest with evergreen subtropical plants. This flora is quite similar to those of the early Miocene of northern Germany and southern Poland. A third province is seen S of the Urals, where a simpler pollen flora of warm-temperature character is dominated by Juglandaceae.

The late Miocene (late Sarmation and Meotian) is known from the S European part of the USSR, the Crimea and Caucasas area; the floras show a shift to impoverished broadleaved and pine forest of warm-temperate character with rich herbaceous flora. In contrast, the late Miocene of Europe (Brussumian) is still a rich, warm-temperate broadleaved forest with subtropical elements and some

Taxodium swamp. Zagwijn (1967) had compared it to the mixed mesophytic forest of S China.

The Pliocene brought Arctic-Alpine communities to the northwesternmost USSR (near Murmansk at latitude 68°N), but European-type conifer forests occur elsewhere N of 60° latitude.

The early and middle Pliocene of North Germany and the Netherlands at 53°N latitude (Reuverian) still record a rich warm-temperate and largely woody flora with thermophilous forms representing about 10–20% of the pollen counts. The *Taxodium* swamp environment in the coastal plain was making its last appearance. The following interval (the Praetiglian), which is now considered to be of late Pliocene age by Berggren and Van Couvering (1974), shows a change to a shrubby cold-tolerant vegetation of glacial aspect.

Mid-Continent Asia and Some Comparisons with Cordilleran USA. In central Asia, the early Miocene began, as in the examples discussed above of Neogene succession, with the existance of very large provinces of vegetation types, and the Pliocene ended with more sharp latitudinal zonation and more provincial vegetation patterns. This sequence particularly involved the shrinking of the mixed hardwood forests southward, associated with the Neogene cooling. The central parts of a large continent such as Asia, however, developed Neogene features not seen along coastal areas except in the rain shadows of coastal mountains—namely the development of steppe.

The following discussion is based only on pollen analysis and hence the conclusions can only deal with the floras in the broadest terms; but pollen evidence tells us much ecologically about vegetation types and general climate. Specific pollen records and summaries are provided by Andreeva (1970) and Pokrovskaya (1956).

Pokrovskaya (1956) recognizes 5 broad provinces of Miocene vegetation based on pollen analysis in central Asia. In northern Siberia, N of about 60° latitude and ranging to the Arctic Ocean are two forest provinces of undifferentiated Miocene age both of temperate character. In northwestern Siberia is a moderately diverse broadleaved hardwood and *Pinus-Picea* forest zone with rich herbaceous flora ranging N to about latitude 73°. In northeastern Siberia, the forests are also highly deciduous but more diverse, with large participation of broadleaved tree genera. *Pinus, Picea,* and *Tsuga.* Betulaceae pollen is prominent.

South of about 60°N latitude, in southwestern Siberia, forest zones of undifferentiated Miocene age are of warm-temperate character, and record diverse broadleaved and conifer forests in which *Pinus* is the dominant pollen type, with locally prominent *Alnus, Betula,* and *Pterocarya.* In southeastern Siberia near Lake Baikal, early and middle Miocene records show broadleaved and *Picea-Tsuga* forest also warm-temperate in nature with local importance of *Juglans, Ulmus,* and Betulaceae pollen.

In the far south in Kazakstan, in the Turgai subprovince just NE of the Sea of Aral, the early Miocene assemblages were of warm-temperate character indicating diverse broadleaved forest with Juglandaceae as the dominant pollen type. Many semishrub forms, Gramineae, and *Artemisia* are present. Widespread districts of steppe are thought to have occurred on highlands.

The above means that, as far as is known, the Neogene in central Asia began with extensive mixed temperate and warm-temperate forests ranging over a tremendous sweep of latitudes, from 43° to 73°N. Of course, this is in contrast with the spectra from Europe and much of the North Pacific Basin, which housed from few to many subtropical groups.

The late Miocene records from eastern Siberia S to Lake Baikal indicate coniferous forests of cool-temperate aspect with large areas of *Picea* and *Tsuga* but with insignificant amounts of only a few hardwoods (i.e., Betulaceae, *Fagus, Ulmus, Fraxinus,* and *Juglans*).

In the far south, the late Miocene of the Turgai subprovince shows extensive warm-temperate broadleaved deciduous with large districts of *Alnus, Pinus,* and *Betula,* and probably with extensive grassy steppe developed on highlands.

Pliocene records from the USSR are hard to interpret because of the difficulties of correlation and defining stratigraphic age. Tikhomirov (1963) indicated that during the preglacial "Eopleistocene" (late Pliocene and early Pleistocene according to Berggren and Van Couvering, 1974) the forests of Siberia had become a depleted version of the earlier communities in that they were now dominantly coniferous with some relictual areas in the mountains of northeastern Siberia where late Tertiary types of broadleaved trees remained. Some districts of Arctic-Alpine tundra are known (as in the Byrranga Mountains, Tamyr Peninsula) but the evidence suggests that mixed conifer forests (*Picea, Larix, Pinus, Tsuga*) grew on the continental shelf to the ancient coastline at about 80°N latitude. South of this northerly zone (S of about 72°N latitude) was a highly complex pattern of vegetation provinces largely relating to topography. Vast forests of *Picea-Larix, Pinus-Larix,* and *Larix-Betula* covered much of Siberia S to at least 60°N. In the south, steppe had become more widespread. Pollen records from the "Pliocene" (upper Pavlodar zone) of the Turgai subprovince indicate for example that steppe elements occupy 80% of the counts and include grasses and shrubs with a prominence of *Artemisia* and Chenopodiaceae. Steppe was more widespread also in the area immediately N of the Caspian Sea, now a desert area.

As in the mid-continent of Asia, the Rocky Mountain Cordilleran area of the USA saw a development of steppe in the lowlands beginning in Miocene and particularly in Pliocene time. Contrary to examples from central Asia discussed above, Rocky Mountain floras lost most of their Tertiary-type tree genera by Miocene time. A summary of this aspect of Rocky Mountain floristic development is provided by Leopold (1969a, 1967).

The Neogene development of mid-continental desert (Axelrod, 1950) and dry steppe areas undoubtedly was related to the disjunction of hardwood forest types that in the early Miocene had been contiguous and so very widespread from E to W across Asia and the USA (Leopold and MacGinitie, 1972).

Southern Hemisphere and Low Latitude Floras. Relatively recent work on Neogene pollen floras of the Southern Hemisphere comes from scattered localities and does not yet provide a picture of regional floral sequences; few publications assign fossil pollen to living plant genera, and therefore few provide botanical analysis of the material.

A well-illustrated survey and record of stratigraphic ranges for various tropical pollen and spores of Tertiary age is found in Germeraad et al. (1968). This large work covers parts of tropical South America, Africa, and Asia; but all taxa are assigned to form or organ genera. The publication documents the late Tertiary evolution and expansion of Gramineae, Compositae, and various mangrove types. This work and that by Muller (1964, 1969) record the Tertiary evolution of the Sonneratiaceae. Muller (1966) has also described the distribution of montane conifer and *Alnus* pollen in the late Tertiary of Borneo.

Miocene pollen and algal sequences from deep cores on Eniwetok, Marshall Islands, provide a basis for inferring strand and raised limestone island environments from sediments now 2000 ft (610 m) below sea level; hence a history of gradual subsidence of the area is implied. The bulk of the Eniwetok flora is quite similar to forms occuring in the Western Caroline Islands some 15° longitude farther W (Leopold, 1969). The Cenozoic paleobotanical literature for

South America has been reviewed by Graham (1973), who has also completed some important studies of Miocene and Oligocene pollen floras of the Caribbean (Graham and Jarzen, 1969). He postulates a continuous and gradual southward migration of temperate woody genera from the N into Central America and northern South America by means of a landbridge linking the islands of the Caribbean during the late Cenozoic.

Results so far suggest that many tropical genera now in low-latitude floras existed in their respective areas as early as the Miocene or Oligocene.

E. B. LEOPOLD

References

Andreeva, E. M., et al., 1970. *Paleopalynology,* vol. 2. *Assemblages of spores, pollen and other microfossils characteristic of various stratigraphical subdivisions from the Upper Precambrian to the Holocene in the USSR.* Trans. by Kenneth Syers. National Lending Library for Service and Technology, Boston Spa, Yorkshire England, 498-796.

Axelrod, D. I., 1950. Studies in late Tertiary paleobotany, *Carnegie Inst. Wash. Pub.,* **590,** 323p.

Baikovskaya, T. H., 1974. *Late Miocene floras from South Premorya.* Leningrad: Akad. Nauk SSSR, Botanical Inst. vm B. L. Komarova Uzdatelstvo, 136p.

Berggren, W. A., and Van Couvering, J. A., 1974. The Late Neogene, *Palaeogeography, Palaeoclimatology, Palaeoecology,* **16,** 1-216.

Cranwell, L. M., ed., 1964. *Ancient Pacific Floras.* Honolulu: Univ. Hawaii Press, 114p.

Doyle, J. A., 1973. The monocotyledons: Their evolution and comparative biology. V, Fossil evidence on early evolution of the monocotyledons, *Quart. Rev. Biol.,* **48,** 399-413.

Doyle, J. A., 1977. Spores and pollen: The Potomac Group (Cretaceous) angiosperm sequence, in E. G. Kauffman and J. E., Hazel, eds., *Concepts and Methods of Biostratigraphy.* Stroudsburg, Pa.: Dowden, Hutchinson & Ross, 339-363.

Ehrlich, H. G., and Hall, J. W., 1959. The ultrastructure of Eocene pollen, *Grana Palynologica,* **2,** 32-35.

Fuji, N., 1969-1972. Fossil spores and pollen grains from the Neogene deposits in Noto Peninsula, central Japan, I-IV, *Trans. Proc. Palaeontol. Soc. Japan,* **73** (1969), 1-25; **74** (1969), 51-80; **76** (1969), 185-204; **86** (1972), 295-318.

Germeraad, J. H.; Hopping, C. A.; and Muller, J., 1968. Palynology of Tertiary sediments from tropical areas, *Rev. Palaeobot. Palynol.,* **6,** 189-348.

Graham, A., 1972. *Floristics and Paleofloristics of Asia and Eastern North America.* Amsterdam: Elsevier, 272p.

Graham, A., ed., 1973. *Vegetation and Vegetational History of Northern Latin America.* New York and Amsterdam: Elsevier, 393p.

Graham, A., and Jarzen, D. M., 1969. Studies in

Neotropical paleobotany, I. The Oligocene communities of Puerto Rico, *Ann. Missouri Bot. Gard.,* **56,** 308-357.

Gray, J., 1964. Northwest American Tertiary palynology: The emerging picture, in Cranwell, 1964, 21-30.

Gray, J., coordinator, 1965. Techniques in palynology, in B. Kummel and D. Raup, eds., *Handbook of Paleontological Techniques.* San Francisco: Freeman, 471-706.

Hopkins, D. M., et al., 1971. A Pliocene flora and insect fauna from the Bering Strait region, *Palaeogeography, Palaeoclimatology, Palaeoecology,* **9,** 211-231.

Hu, H. H., and Chaney, R. W., 1940. A Miocene flora from Shantung Province, China, *Carnegie Inst., Washington, Publ. 587,* 147p.

Leopold, E. B., 1967. Late Cenozoic patterns of plant extinction, in P. S. Martin and H. E. Wright, Jr., eds., *Pleistocene Extinctions.* New Haven: Yale Univ. Press, 204-246.

Leopold, E. B., 1969a. Late Cenozoic palynology, in Tschudy and Scott, 1969, 377-438.

Leopold, E. B., 1969b. Miocene pollen and spore flora of Eniwetok Atoll, Marshall Islands, *U.S. Geol. Surv. Prof. Paper 260-II,* 1133-1185.

Leopold, E. B., 1974. Pollen and spores of the Kisinger Lakes fossil locality, *in* H. D. MacGintie, An Early Eocene flora from the Yellowstone-Absaroka Volcanic Province, Northwest Wind River Basin, Wyoming, *Univ. Calif. Publ. Geol. Sci.,* **108,** 49-66.

Leopold, E. B., and MacGintie, H. D., 1972. Development and affinites of Tertiary floras in the Rocky Mountains, in Graham, 1972, 147-200.

McAndrews, J. E., and Wright, H. E., Jr., 1969. Modern pollen rain across the Wyoming Basins and the northern Great Plains (U.S.A.). *Rev. Palaeobot. Palynol.,* **9,** 17-43.

Manten, A. A., 1968. A short history of palynology in diagrams, *Rev. Palaeobot. Palynol.,* **6,** 117-188.

Muller, J., 1964. A palynological contribution to the history of the mangrove vegetation in Borneo, in Cranwell, 1964, 33-42.

Muller, J., 1966. Montane pollen from the Tertiary of NW Borneo. *Blumea,* **14**(1), 231-235.

Muller, J., 1969. A palynological study of the genus *Sonneratia* (Sonneratiaceae). *Pollen Spores,* **11,** 223-298.

Nichols, D. J., 1973. North American and European species of *Momipites* ("*Engelhardtia*") and related genera, *Geosci. Man,* **7,** 103-117.

Penny, J. S., 1969. Late Cretaceous and early Tertiary Palynology, in Tschudy and Scott, 1969, 331-376.

Pflug, H. D., 1953. Zur Enstehung und Entwicklung des angiospermiden Pollens in der Erdgeschicte. *Palaeontographica,* **B95,** 60-171.

Pokrovskaya, I. M., ed., 1956. Atlas of Miocene spore-pollen complexes of various areas, USSR, *Vsesoiuznyi geol. inst. Mater., Moscow,* n.s., **13,** 460p.

Sato, S., 1963. Palynological study on Miocene sediments of Hokkaido, Japan, *Hokkaido Univ., J. Fac. Sci.,* ser 4, **12,** 1-110.

Stebbins, G. L., 1974. *Flowering Plants: Evolution Above the Species Level.* Cambridge, Mass.: Harvard Univ. Press, 399p.

Sung, T. -C., et al., 1964. Tertiary sporo-pollen complexes of Shantung, *Mem. Inst. Geol. Paleontol., Acad. Sinica,* **3,** 286-290.

Tanai, T., 1967. Miocene floras and climate in East Asia, in *Klimaänderungen im Tertiär aus paläobotanischer Sicht,* vol. 10. Berlin: Zentralen Geol. Inst., 195-305.

Tanai, T., 1972. Tertiary history of vegetation in Japan, in Graham, 1972, p. 235-255.

Tikhomirov, B. A., 1963. Principal stages of vegetation development in northern USSR as related to climatic fluctuations and the activity of man, *Canadian Geographer,* **7,** 55-71.

Tschudy, R. H., 1969. The plant kingdom and its palynological representation, in Tschudy and Scott, 1969, 5-34.

Tschudy, R. H., 1973. Complexipollis pollen lineage in Mississippi embayment rocks, *U.S. Geol. Surv. Prof. Paper 743-C,* 15p.

Tschudy, R. H., and Scott, R. A., eds., 1969. *Aspects of Palynology,* New York: Wiley-Interscience, 5-34, 331-438.

Walker, J. W., 1974. Aperture evolution in the pollen of primitive angiosperms, *Amer. J. Bot.,* **61,** 1112-1137.

Walker, J. W., and Skvarla, J. J., 1975. Primitively columellaless pollen: A new concept in the evolutionary morphology of angiosperms, *Science,* **187,** 445-447.

Wolfe, J. A., et al., 1966. Tertiary stratigraphy and paleobotany of the Cook Inlet region, Alaska, *U.S. Geol. Surv. Prof. Paper 398-A,* 29p.

Wolfe, J. A., and Tanai, T., 1978. The Middle Miocene Seldovia flora from the Kenai Group, Alaska. *U.S. Geol. Surv. Prof. Paper.*

Zagwijn, W. H., 1967. Ecologic interpretation of a pollen diagram from Neogene beds in the Netherlands, *Rev. Palaeobot. Palynol.,* **2,** 173-181.

Cross-references. *Angiospermae; Gymnosperms; Paleobotany; Paleophytogeography; Palynology; Palynology, Paleozoic and Mesozoic.*

PELECYPODA—See BIVALVIA

PISCES

Pisces, a class of vertebrates comprising the true fishes, have the jaws supported by a skeleton derived from primitive gill arches. Typically, they have two sets of paired fins, pectoral and pelvic, as well as dorsal, caudal, and anal fins in the midline. They breathe by means of gills, which are carried on the outer side of the gill arches; and they generally have the skin protected by scales. There are four distinct subclasses whose interrelationships are still controversial. The geological history of Pisces has been reviewed by Romer (1966), Moy-Thomas and Miles (1971), and Thomson (1977).

Subclass Placodermi

Members of the subclass Placodermi (Devonian to lowest Mississippian) are characterized by possession of bones and a neck joint, and by the position of the gills far under the head. Commonly, there are bony head and thoracic shields that articulate in the neck region. But in primitive forms, the shields are little developed and an articulation between the dermal bones is lacking; instead, there is an endoskeletal neck joint between the neurocranium and fused anterior vertebrae. The surface of dermal bones lacks dentine, and its ornamentation characteristically is formed of semidentine or bone. Placodermi have been considered as relatives of Chondrichthyes, but are more probably a distinct group of Pisces according to Denison (1978); they are grouped in the following orders.

Order Arthrodira. (Lower to Upper Devonian). The Arthrodira were primitively bottom-dwelling fishes with a strong armor that commonly had a tuberculate surface; they had a pair of well-developed lateral spines on the thoracic shield, behind which were attached narrow-based pectoral fins (*Arctolepis, Phlyctaenius*). From these evolved in the Middle and Upper Devonian variously adapted genera, which, in most cases, showed reduction or loss of the pectoral spines and enlargement of the pectoral fins; *Coccosteus* (Fig. 1) is the best known genus at an intermediate evolutionary stage. The Upper Devonian *Dunkleosteus* was a specialized arthrodire; it reached a length of perhaps 15 ft (4.6 m) and had scissors-like, shearing jaws. Upper Devonian shales near Cleveland, Ohio and Wildungen, Germany contain famous faunas of Arthrodira.

Order Petalichthyida (Lower to Upper Devonian). Petalichthyida had the posterior part of the cranial shield elongate as compared to Arthrodira, and the orbits were dorsal.

Lunaspis and *Macropetalichthys* are typical genera.

Order Phyllolepida (Upper Devonian). A broad, flat shield with an ornamentation of concentric ridges defined the order Phyllolepida, to which *Phyllolepis* is the only genus certainly referable.

Order Ptyctodontida (Lower Devonian to lowermost Mississippian). Members of the order Ptyctodontida had a short skull and shoulder girdle. The upper jaws were firmly attached to the cranium and there were crushing or shearing tooth plates. They are considered by some to be ancestral to Holocephali (Westoll, 1962). Best known are *Rhamphodopsis* and *Ctenurella*.

Order Antiarcha (Lower to Upper Devonian). Antiarcha were aberrant, bottom-dwelling Placodermi with a strongly developed shield, flattened body, dorsally placed eyes and nostrils, and oar-like pectoral flippers instead of fins. They were commonly, but not exclusively, freshwater stream dwellers, and were almost worldwide in distribution. *Asterolepis* and *Bothriolepis* are well known.

Order Rhenanida (Lower to Upper Devonian). The flat-bodied ray-like Rhenanida had enlarged pectoral fins and dorsal eyes and nostrils; the skull was covered with a few dermal bones and numerous small tesserae. *Gemuendina* is a well-known Lower Devonian genus.

Other. In addition, Placodermi include some poorly known Lower Devonian forms mostly from Europe and Asia. *Palaeacanthaspis* and *Radotina* have a short shoulder girdle and a head variably covered with bones and small tesserae; they are referred to the Order Acanthothoraci. From the German Hunsrückschiefer comes the small *Pseudopetalichthys* with large, dorsal eyes and peculiar plate-like jaws. From the same beds comes *Stensioella* which has few dermal bones and may be the most primitive of known placoderms.

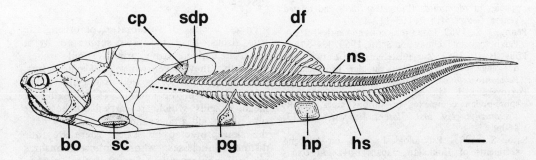

FIGURE 1. Placodermi, Arthrodira, *Coccosteus cuspidatus*, Devonian (from Miles and Westoll, 1968). bo, branchial opening; cp, carinal process; df, dorsal fin; hp, postanal plate; hs, haemal spine; pg, pelvic girlde; sc, scapulacorocoid; sdp, submedian dorsal plate. Scale bar is 10 mm.

Subclass Chondrichthyes

Subclass Chondrichthyes (Middle Devonian to Holocene) differs from other major groups of Pisces in that their skeletons are formed of cartilage and bone is typically absent. The cartilage is sometimes calcified, usually as a coating of prisms of calcium phosphate, and then may be preserved as a fossil. Their skin is set with scattered placoid scales formed of dentine with an enamel-like surface, or with denticles that may be simple or compound structures with a bony base. Chondrichthyes are predominantly, but not exclusively, marine. Their evolutionary history is poorly known and many genera are based on fin spines, scales or teeth. They are here divided into three major groups, Elasmobranchii, Holocephali, and Bradyodonti. Their classification is discussed by Saint-Seine et al. (1969), Schaeffer (1967), and Moy-Thomas and Miles (1971).

Elasmobranchii. Sharks and rays lack opercula and have the gill slits opening directly to the outside, have the palatoquadrates loosely attached to the neurocranium, and typically have a successional dentition formed of labio-lingual tooth rows. They are mostly predaceous forms with relatively large olfactory organs and small eyes. The classification of early forms is difficult and disputed. Paleozoic genera, appearing in the Middle Devonian, lack vertebral centra, have a terminal mouth and weakly developed rostrum, and have cartilaginous fin radials extending to the fin margins. In many, the teeth have a large central cusp flanked by small side cusps; such teeth have been referred to numerous species of *"Cladodus"* ranging from Middle Devonian to Permian. *Cladose-*

lache (Fig. 2), with two dorsal fins and broad-based pectoral fins, is well known from numerous entire specimens from the Upper Devonian Cleveland shale of Ohio. The predominantly Mississippian genus *Ctenacanthus* is known commonly from isolated, ornamented fin spines, though nearly complete skeletons have been found. The Xenacanthidae (Middle Devonian to Triassic) had paired fins with a long, central axial skeleton, a long, tapering, symmetrical tail, and an elongated dorsal fin; they occur commonly in freshwater deposits. Complete specimens of *Pleuracanthus* are known from the Permian of Germany, while isolated teeth with two large cusps separated by a smaller cusp are referred to *Diplodus* and *Xenacanthus*. The Edestidae (Mississippian-Permian) are specialized elasmobranchs, sometimes referred to Bradyodonti, that show a tendency to develop large symphysial tooth whorls; *Edestus* and *Helicoprion* are well known. The Hybodontidae (Permian-Paleocene) are intermediate to modern sharks; they retain a primitive jaw suspension and unreduced notochord, but are advanced in their narrow-based, flexible paired fins with a concentrated skeleton and shortened fin radials; *Hybodus* and *Asteracanthus* are well known Mesozoic genera. The Cretaceous Ptychodontidae, known from their button-like teeth, apparently were derived from hybodontids. Modern sharks, Order Selachii, have subterminal mouths with protrusible jaws and hyostylic suspension, vertebral centra, narrow-based fins with well-developed ceratotrichia and reduced radials, and simple placoid scales. The oldest known is *Palaeospinax* from the Lower Jurassic of England. *Heterodontus* and *Hexanchus,* Holo-

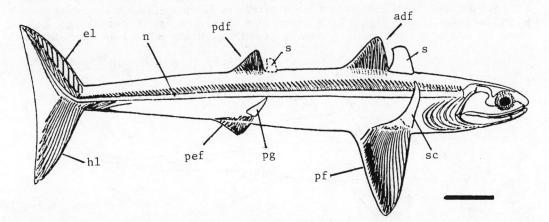

FIGURE 2. Chondrichthyes, Elasmobranchii, *Cladoselache fyleri,* Devonian (from Morris, 1938). adf, anterior dorsal fin; el, epichordal lobe; hl, hypochordal lobe; n, notochord; pdf, posterior dorsal fin; pef, pelvic fin; pf, pectoral fin; pg, pelvic girdle; s, spine; sc, scapulacorocoid. Scale bar is 50 mm.

cene genera first appearing in the Jurassic, retain many primitive features. Most modern sharks are referred to Suborders Galeoidea and Squaloidea which first appear in the Upper Jurassic. Entire fish are rare as fossils, but teeth are common and may be of stratigraphic value. Well-known genera going back to the Early Cretaceous are the sand sharks, *Carcharias*, and the mackerel sharks, *Isurus* and *Lamna*; appearing in the early Cenozoic are the white sharks, *Carcharodon* and the tiger sharks, *Galeocerdo*. Order Batoidea (Jurassic–Holocene) are flat-bodied, mostly bottom-living rays and skates. They swim by an up-and-down movement of their greatly expanded pectoral fins, and their tail is reduced. Their teeth are flattened and form a pavement often used for crushing mollusks. Modern Batoidea dating from the Cretaceous are *Raja*, the rays, *Myliobatis*, the eagle rays, and Pristidae, the sawfishes.

Holocephali (Mississippian to Holocene). Holocephali differ from Elasmobranchii in having short jaws with an autostylic suspension (that is with the palatoquadrates fused to the neurocranium), opercular flaps covering the gill openings, and moderately large eyes. The Order Chimaerida first appeared in the Mississippian with the genus *Helodus* (Fig. 3), which had an elasmobranch-like, successional dentition. Typical members of the order, the chimaeras or ratfishes, appeared in the Jurassic, and have large, crushing dental plates, two dorsal fins, the first with a spine, and the tail tapering to a point. Early forms were the Jurassic *Myriacanthus* and *Squaloraja*; while *Edaphodon*, more closely related to modern genera, appeared in the Early Cretaceous. The recently discovered Order Iniopterygia (Pennsylvanian) primitively had a shark-like dentition, but developed tooth whorls and complex dental plates (Zangerl and Case, 1973). It has a single dorsal fin without a spine, enlarged, dorsally

attached pectoral fins, and a tail that is circular in side view. *Iniopteryx* and four other genera are known from the central United States. All Holocephali are marine and modern forms mostly live in the deep sea.

Bradyodonti (Upper Devonian to Permian). An artificial assemblage of pavement-toothed sharks of uncertain relationships have been assigned to the Bradyodonti. These are known mostly from teeth found in marine rocks of Mississippian and Pennsylvanian age. The teeth are flat-crowned, composed of tubular dentine, and were probably used for crushing hard food such as mollusks (Fig. 4). Cochliodontidae evolved large tooth plates with a twisted surface. Petalodontidae had teeth with long roots and a lozenge-shaped crown. Psammodontidae had four-sided tooth plates. According to one theory, some Bradyodonti were related to Holocephali (Patterson, 1965).

Subclass Acanthodii

Subclass Acanthodii (Lower Silurian–Lower Permian) includes small fishes with large eyes. Typically, the head was covered with small plates and the fusiform body was covered with minute, closely fitting scales. The scales lack a pulp cavity and grew by the addition of concentric layers of bone and dentine. The tail, as in most primitive Pisces, is heterocercal, with an upturned muscular and skeletal lobe above a distinct hypochordal lobe. The paired and median fins have anterior spines. Many Acanthodii are known only from isolated scales and fin spines, and some of these are of stratigraphic value. Silurian and some later acanthodians lived in the seas, but by the Early Devonian many also lived in fresh waters. Primitive Acanthodii, such as the Lower Devonian *Climatius* (Fig. 5), had strongly sculptured, superficially inserted fin spines, numer-

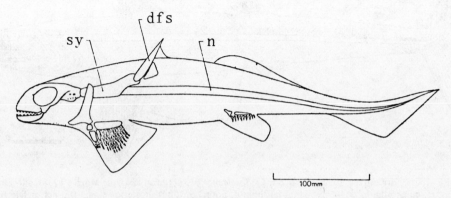

FIGURE 3. Chondrichthyes, Holocephali, *Helodus simplex*, Pennsylvanian (from Moy-Thomas and Miles, 1971, by permission of Chapman & Hall Ltd). dfs, dorsal fin-spine; n, notochord; sy, synarcual.

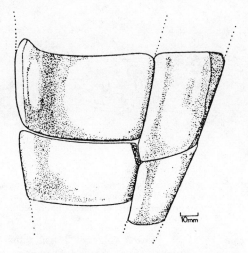

FIGURE 4. Chondrichythyes, Bradyodonti, *Psammodus rugosus,* teeth in coronal view (from Moy-Thomas and Miles, 1971, by permission of Chapman & Hall Ltd).

ous intermediate spines between the pectoral and pelvic fins, and two dorsal fins. A group of predators, typified by the Lower Devonian *Ischnacanthus*, had large, anchylosed jaw teeth, two dorsal fins, and only two sets of paired fins. The latest group of Acanthodii, best known from the Pennsylvanian and Lower Permian *Acanthodes*, had an elongated body, lacked teeth, had a single dorsal fin and deeply inserted fin spines. *Gyracanthus*, a common Mississippian and Pennsylvanian genus, had fin spines covered with ridges meeting in a chevron pattern.

Subclass Osteichthyes

The higher bony fishes, Subclass Osteichthyes, are divided into three goups, Crossopterygii, Dipnoi, and Actinopterygii. The fossil representatives are described in more detail by Lehman (1966) and Moy-Thomas and Miles (1971).

Crossopterygii. The lobe-finned fishes, Crossopterygii, were primitively a group of fast-swimming predators with a fusiform body, heterocercal or upturned tail, paired fins with fleshy lobes and central axial skeleton, two dorsal fins, rather small eyes, and sharp-pointed teeth with infolded structure.

Order Rhipidistia (Lower Devonian–Permian) are important because they were ancestral to land vertebrates. They had internal and external nostrils; presumably they had lungs as well as gills for breathing; and some had homologues of the humerus, radius, and ulna of land vertebrates in their pectoral fin skeleton. Their scales were primitively rhombic, thick, and of the cosmoid type, which means that the dentine layer overlying the bone contained a canal network that opened to the surface by pores. In later forms, the scales became thinner due to the reduction or loss of the dentine layer, and the primitively heterocercal tail evolved into a symmetrical or diphycercal fin. This order is typically, but not exclusively, a freshwater group. Characteristic genera are *Osteolepis* (Middle–Upper Devonian Fig. 6), *Eusthenopteron* (Upper Devonian), and *Rhizodopsis* (Mississippian–Pennsylvanian). *Porolepis* (Lower–Middle Devonian), *Glyptolepis* and *Holoptychius* (Middle–Upper Devonian) represent a family distinguished by a deeper body, relatively longer and more slender paired fins, and a more complicated infolding of the teeth.

Order Struniiformes (Middle and Upper Devonian), including only *Strunius* and *Onychodus*, was made a distinct order by Jessen (1966), but its rank and relationship are controversial. Its members are characterized by a truncate snout, large orbits, reduced operculars, and specialized lower jaws with a median, symphysial tooth whorl. The body is short and deep, the median fins lack fleshy lobes, and the tail is symmetrical with a long axial lobe.

Order Coelacanthini, or Actinistia (Middle

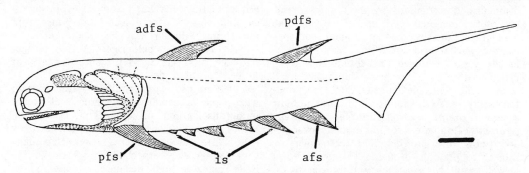

FIGURE 5. Acanthodii, *Climatius reticulatus,* Devonian (from Watson, 1937). adfs, anterior dorsal fin-spine; afs, anal fin-spine; is, intermediate spines; pfs, pectoral fin-spine; pdfs, posterior dorsal fine-spine. Scale bar is 10 mm.

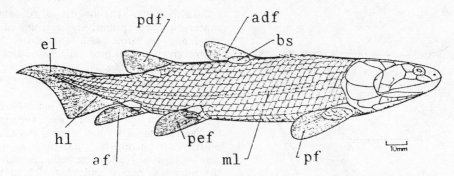

FIGURE 6. Osteichthyes, Crossopterygii, Rhipidistia, *Osteolepis macrolepidotus*, Devonian (from Moy-Thomas and Miles, 1971, by permission of Chapman & Hall Ltd). adf, anterior dorsal fin; af, anal fin; bs, basal scute; el, epichordal lobe; hl, hypochordal lobe; ml, main lateral-line; pdf, posterior dorsal fin; pef, plevic fin; pf, pectoral fin.

Devonian–Holocene), a slowly evolving side branch of Crossopterygii is characterized by fan-shaped fins with narrow bases and a symmetrical tail with a projecting axial lobe. The head was short and deep, the eyes relatively large, the jaws short, and the lateral teeth reduced or lost. There were no internal nostrils as in Rhipidistia. Scales were thin and there was a tendency to reduction of ossification. Paleozoic genera lived mostly in fresh waters, but the Mesozoic forms were mostly marine, as is the only living genus, *Latimeria*. Well known fossil genera are *Rhabdoderma* (Mississippian–Pennsylvanian), *Diplurus* (Triassic), and *Macropoma* (Upper Cretaceous).

Dipnoi (Lower Devonian to Holocene). Lungfishes or Dipnoi primitively resembled early Crossopterygii in their body shape, fins, and scales, and presumably had a common ancestry (Westoll, 1949). The tail, heterocercal in the earliest genera, soon evolved into a symmetrical, tapering fin, confluent with the elongated dorsal and anal fins. The paired fins were leaf-shaped with a long fleshy lobe and a central axial skeleton. The scales were primitively rhombic and cosmoid as in Crossopterygii, but soon became thin and round. Dipnoi are distinguished particularly by the reduction and loss of the marginal teeth and the bones that bore them, and by the development of specialized palatal and lower jaw teeth, usually in the form of ridged or denticulated tooth plates. Although a few early lungfishes have been found in marine deposits, they are characteristically a freshwater group. The modern genera (*Protopterus*, *Lepidosiren*, and *Neoceratodus*) are adapted for life in rivers, swamps, and pools that are deficient in oxygen or subject to periodic drought. They have lungs as well as gills for breathing, and some aestivate in burrows during the dry season. The habits of many fossil lungfishes were probably similar as is suggested by burrows found in Paleo-

zoic rocks in North America. Fossil genera include: *Dipterus* (Fig. 7b) and *Scaumenacia* (Devonian); *Sagenodus* and *Ctenodus* (Mississippian and Pennsylvanian): *Ceratodus* (Triassic-Cretaceous).

Actinopterygii (?Upper Silurian; Lower Devonian–Holocene). Actinopterygii, the ray-finned fishes, is by far the largest and most varied group of Pisces, and includes the great majority of living fishes. It is distinguished from other Osteichthyes especially by (a) the structure of the paired fins, which are supported mostly by dermal fin rays and have only small fleshy lobes with a short internal skeleton of separate radials; (b) the structure of the scales, which primitively were of the ganoid type with a layered bony base, overlain by thin dentine and topped by layers of an enamel-like substance known as ganoine; and (c) the presence typically of only one dorsal fin. Relationship to the Crossopterygii and Dipnoi may not be close, and some have claimed that there are closer ties to the Acanthodii. At this time, they are conveniently separated into four major groups, Chondrostei, Subholostei, Holostei, and Teleostei, although it is recognized that these are for the most part not phyletic units but grades of evolutionary development. Scattered scales and bones from the Upper Silurian have been considered by Gross (1968) to belong to Actinopterygii, but the first certain representatives are of Lower Devonian age.

1. Chondrostei are primitive Actinopterygii with a heterocercal tail, ganoid scales, and numerous dermal fin rays. The mouth had a long gape with the jaw articulation far back (Fig. 7a). Paleoniscoidei (Lower Devonian-Cretaceous) were especially important in the late Paleozoic and declined rapidly in the Mesozoic. *Cheirolepis* is a well-known Devonian genus; *Rhadinichthys* and *Elonichthys* are late Paleozoic genera; *Paleoniscus* is restricted

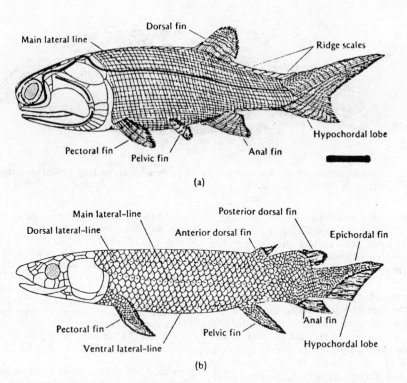

FIGURE 7. a, Actinopterygii, Chondrostei, *Moythomasia nitidia,* Devonian (from Stahl, 1974). Scale bar is 10 mm. b. Osteichthyes, Dipnoi, *Dipterus valenciennesi,* Devonian (from Stahl, 1974). Scale bar is 10 mm.

to the Permian. Platysomoidei (Mississippian-Lower Triassic) are relatives of the Paleoniscoidei, with deep bodies and elongate dorsal and anal fins. *Platysomus* and *Amphicentrum* are late Paleozoic representatives. Acipenseroidei (Jurassic-Holocene) includes the modern sturgeons, *Acipenser,* and paddlefishes, *Polyodon,* both of which date back to the Cretaceous. They are generally considered as degenerate survivors of Paleoniscoidei, with greatly reduced ossification. The African group, Polypterini (Eocene-Holocene), including the living *Polypterus,* is considered by some to include greatly modified descendants of Paleoniscoidei; but others make it a separate group, Brachiopterygii, equivalent in rank to Actinopterygii.

2. Subholostei (Triassic and Jurassic) are intermediate in age and structure between Chondrostei and Holostei. The tail, while still heterocercal, has the scale-covered lobe reduced; the number of fin rays was reduced and the mouth was shortened somewhat. *Redfieldia* and *Perleidus* are typical Triassic genera. *Cleithrolepis* (Triassic) was a deep-bodied form comparable to Platysomoidei. *Saurichthys* (Triassic and Lower Jurassic) had a long pike-like body and an elongated beak.

3. Holostei is an important Mesozoic group, though it appeared first in the Upper Permian, and two genera, *Amia,* the bowfin, and *Lepisosteus,* the garpike, still survive. The tail fin is described as abbreviated heterocercal, that is, it was nearly symmetrical (Fig. 8). The dermal fin rays were reduced so that they corresponded in number to the endoskeletal rays. The gape of the jaw was short. The ganoine was lost on the scales of some. Typical Holostei are *Semionotus* (Triassic) and *Lepidotes* (Triassic to Cretaceous). A specialized family, Pycnodontidae (Triassic to Eocene), had deep bodies and knobby crushing teeth on the palate and lower jaws. The Jurassic genus, *Aspidorhynchus,* had a long, slender body and an elongate rostrum.

4. Teleostei are the successful fishes of the present day. They may be distinguished by their homocercal tail which is superficially symmetrical, though internally there is still a trace of the upturned axial skeleton. The scales have lost all ganoine and dentine, are formed of thin bone, and are usually rounded and deeply overlapping. Primitive Teleostei (Leptolepidae) appeared in the Jurassic with the genus *Leptolepis.* Other primitive teleosts appeared in the Cretaceous, including the Elopiformes (tarpons), Clupeiformes (herrings), and Salmoniformes (salmon and trout). Teleostei have undergone a tremendous adaptive

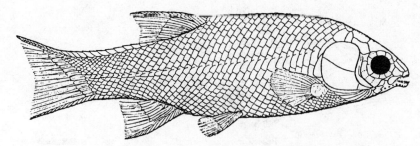

FIGURE 8. Actinopterygii, Holostei, *Acentrophorus varians,* Permian (from Moy-Thomas and Miles, 1971, by permission of Chapman & Hall Ltd), ca. X0.75.

radiation and are commonly divided into about 30 orders, hundreds of families, and thousands of genera. Some well-known groups are the Ostariophysi (catfishes and suckers), Anguilliformes (eels), and Esocoidei (pikes). The more advanced Teleostei, or Acanthopterygii, are distinguished by having stiff spines developed on the fins and by having the pelvic (or ventral) fins moved forward under the pectoral fins. The Beryciformes are the earliest spiny-finned Teleostei, dating from the Cretaceous. Other well-known groups are the Percoidei (perches and bass), Pleuronectiformes (flat-fishes), and Scombroidei (mackerel and tuna). Teleostei are predominantly marine, some even abyssal, but a number are adapted to brackish and fresh waters. Others are eury-haline and can tolerate a wide range of salinity. Scales of teleosts are identifiable, though some-times only to family or genus, and are of use in stratigraphic and ecological studies. Otoliths or "ear bones," because of their density and composition, are sometimes the only parts of a fish that is preserved; they also have potential value as stratigraphic and ecologic indicators.

ROBERT H. DENISON

References

Denison, R. H., 1978. Placodermi, in H. -P. Schultze, ed., *Handbook of Paleoichthyology, Vol. 2.* Stutt-gart: Gustav Fischer, 128p.

Gross, W., 1968. Fragliche Actinopterygier-Schuppen aus dem Silur Gotlands, *Lethaia,* **1,** 184–218.

Heyler, D., 1969. Acanthodii, in Piveteau, 1969, 21–70.

Jessen, H. L. 1966. Struniiformes, in Piveteau, 1966, 387–398.

Lehman, J. P., 1966. Actinopterygii, Dipnoi et Cros-sopterygii, Brachiopterygii, in Piveteau, 1966, 1–420.

Miles, R. S., 1973. Relationships of acanthodians, in P. H. Greenwood, R. S. Miles, and C. Patterson, eds., Interrelationships of fishes, *Zool. J. Linnean Soc.,* **53**(Suppl. 1), 63–103.

Miles, R. S., and Westoll, T. S., 1968. The placoderm fish *Coccosteus cuspidatus* Miller ex Agassiz from the Middle Old Red Sandstone of Scotland, *Trans. Roy. Soc. Edinburgh,* **67,** 373–476.

Morris, J. E., 1938. The dorsal spine of *Cladoseloche, Cleveland Mus. Nat. Hist., Sci. Pub.,* 8(1), 1–6.

Moy-Thomas, J. A., and Miles, R. S., 1971. *Palaeozoic Fishes,* 2nd ed. Philadelphia: Saunders, 259p.

Obruchev, D. V., ed., 1967. *Fundamentals of Paleon-tology,* vol. 11, *Agnatha, Pisces.* Translated from Russian. Jerusalem: Israel. Program for Scientific Translations, 168–825.

Patterson, C., 1965. The phylogeny of chimaeroids, *Phil. Trans. Roy. Soc. London, Ser. B,* **249,** 101–219.

Piveteau, J., ed., 1966, 1969. *Traité de Paléontologie.* Paris: Masson et Cie, vol. 4, no. 3, 1–420; no. 2, 21–776.

Romer, A. S., 1966. *Vertebrate Paleontology,* 3rd ed. Chicago: Univ. Chicago Press, 24–77.

Saint-Seine, P.; Devillers, C.; and Blot, J., 1969. Holocéphales et élasmobranches, in Piveteau, 1969, 693–776.

Schaeffer, B., 1967. Comments on elasmobranch evolution, in P. W. Gilbert, ed., *Sharks, Skates and Rays.* Baltimore: Johns Hopkins Press, 3–35.

Stahl, B. J., 1974. *Vertebrate History: Problems in Evolution.* New York: McGraw-Hill, 594p.

Stensiö, E. A., 1969. Placodermata, in Piveteau, 1969, 71–692.

Thomson, K. S. 1977. The pattern of diversification among fishes, in A. Hallam, ed., *Patterns of Evolu-tion.* Amsterdam: Elsevier, 377–404.

Watson, D. M. S., 1937. The acanthodian fishes, *Phil. Trans. Roy. Soc. London, Ser. B,* **228,** 49–146.

Westoll, T. S., 1949. On the evolution of the Dipnoi, in G. L. Jepsen, E. Mayr, and G. G. Simpson, eds., *Genetics, Paleontology and Evolution.* Princeton: Princeton Univ. Press, 121–284.

Westoll, T. S., 1962. Ptyctodontid fishes and the ancestry of Holocephali, *Nature,* **194,** 949–952.

Zangerl, R., 1973. Interrelationships of early chon-drichthyans, in P. H. Greenwood, R. S. Miles, and C. Patterson, eds., Interrelationships of Fishes, *Zool. J. Linn. Soc.,* **53**(Suppl. 1), 1–14.

Zangerl, R., and Case, G. R., 1973. Iniopterygia, a new order of chondrichthyan fishes from the Penn-sylvanian of North America, *Fieldiana, Geol. Mem.,* **6,** 1–67.

Cross-references: *Agnatha; Bones and Teeth; Chor-data; Vertebrata; Vertebrate Paleontology.*

PLANKTON

The Plankton (Greek: wanderer) consists of the passively floating or drifting plants and animals inhabiting the water column, or those whose movement is negligible as compared to the effect of the ocean currents, as distinguished from the nekton or swimming animals, and the benthos or bottom-living organisms.

Composition of the Plankton

Living phytoplankton consists of the nannoplankton and microplankton dominated by diatoms, dinoflagellates, coccolithophorids, and silicoflagellates, and lesser numbers of Chrysophyta, Haptophyta, blue-green and green algae (especially the Class Prasinophyceae), with a few red algae (Tappan and Loeblich, 1973a). The zooplankton shows a greater size range— from the nannoplankton <60 mμ in diameter and microplankton up to some millimeters, to the macroplankton, such as the jellyfish, that may attain a diameter of 1 m. The modern zooplankton is highly diverse taxonomically, including protozoans (foraminifers, radiolarians, and tintinnids), invertebrate metazoans (jellyfish, hydrozoans, nemertineans, annelids, rotifers and chaetognaths, pteropods, cephalopods, some crinoids, arthropods—particularly the ostracodes, copepods, euphausids, and decapods), and even vertebrates (pelagic fish). In addition to this permanent component of the plankton, various invertebrates in the low-latitude warm waters have pelagic larval stages. Pelagic larvae characterize the sponges, coelenterates (planula larva and medusa), annelids (trochophore larva), crustaceans (nauplius larva), mollusks (modified trochophore and veliger larval stages), and echinoderms. These benthic organisms are directly affected by the nature of the plankton habitat during their life cycle, as well as directly or indirectly through the food chain. In contrast, the high-latitude benthos generally lacks a pelagic larval stage, and interacts with the plankton only through the food chain.

Fossil plankton assemblages differ somewhat from the modern ones, being strongly biased toward the components that possess mineralized loricas, tests, or other skeletal structures. Thus, the copepods, euphausids, and chaetognaths that dominate the modern zooplankton are unlikely to be preserved; the fossil zooplankton is dominated by the shelled Protozoa (foraminifers, radiolarians, tintinnids), pelagic crinoids, pteropods and cephalopods, and the extinct chitinozoans, graptolites, and conodonts. The fossil phytoplankton (e.g., diatoms, coccolithophorids, and silicoflagellates), like the zooplankton, is commonly modified by selective preservation or dissolution of the silica or lime. The dinoflagellates, chrysomonads, and prasinophycean green algae are represented solely by their resting cysts rather than the motile cells; whereas planktic blue-green, red, and green algae are rarely preserved other than under exceptional conditions.

The Plankton Habitat

The total volume of space occupied by the oceanic plankton exceeds that of any other habitat on the earth. It extends areally over some 70% of the earth's surface, and ranges vertically through the entire water column of up to 10,500 m over the oceanic trenches (thus some 18% greater vertical distance than from sea level to the top of Mount Everest). Vertically, the plankton habitat may be subdivided: the *neuston* inhabits the surface water layer and includes dinoflagellates, colorless flagellates, diatoms, and ciliates, associated with bacteria; fungi; and fragments of various other phytoplankton and zooplankton. Cells may be very abundant in this surface micro layer. Beneath the surface few centimeters, to a maximum depth of 200 m is the photic zone occupied by most of the remaining phytoplankton. Various species may be restricted to particular depths within this zone, and some phytoplankton may have temporary resting or encysted stages during which depth restrictions are less important. At the lower depth limit for phytoplankton growth, the 24-hour photosynthetic uptake of CO_2 equals that produced by respiration. Some phytoplankton may subsist temporarily by uptake of dissolved organic matter where photosynthesis is not possible, as perhaps in higher latitudes during the winter season, but this is probably of little importance in the general energy budget of the sea, and the phytoplankton is generally present only in a dormant state when not actively photosynthesizing.

The actively growing zooplankton also concentrates in the higher water layers near its food supply, but may migrate vertically to feed in the upper waters, and then sink to cooler and deeper waters to rest and thereby conserve energy. The deeper-dwelling plankton species subsist on the limited amount of organic matter that sinks downward from the photic zone of production.

Although the vertically separated habitats are subject to great differences in pressure and available light, and may differ as greatly in temperature, salinity, dissolved nutrients, oxygenation, and pH as do the latitudinal zones of the surface plankton, the bottom sediments

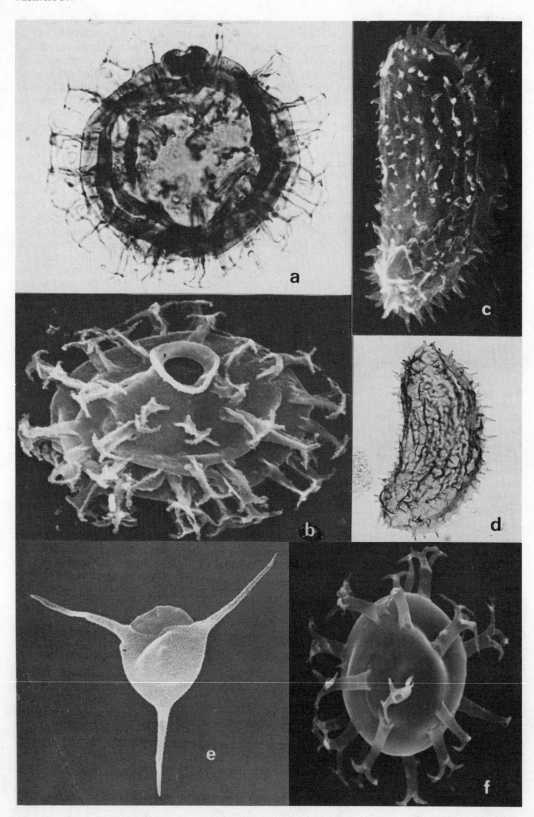

(and thus the fossil record) generally form a composite of all the habitats represented in the overlying water column. Some fossil plankton has been depth-correlated on the basis of· the oxygen isotope ratios of the skeletal calcium carbonate (Emiliani, 1954); morphologic features such as wall thickening (Be and Ericson, 1963); general test shape (Douglas and Savin, 1978); or correlation with associated benthic invertebrate faunal assemblages (Gray and Boucot, 1972).

Geographic distribution of the plankton is broadly latitudinal, being temperature related, but is modified by oceanic circulation patterns. Salinity and density may vary latitudinally, depending on precipitation and evaporation rates and proximity to land masses and river discharge systems. Nutrient supply and regeneration reflect influx from land, rate of utilization, and effectiveness of recycling, as well as the degree of upwelling, stratification, or stagnation of the water column. High-latitude species must tolerate high and constant salinity, extremely low temperature, and a period of minimal light; whereas low-latitude species are subjected to high temperatures and high salinity. Cosmopolitan phytoplankton species must be eurythermal in order to remain in the photic zone; whereas zooplankton may live near the surface at high latitudes, and in deeper, cooler water at low latitudes.

Origin and Geologic Occurence of the Plankton

Prior to the attainment of an oxygenic atmosphere, the absence of an ozone layer that shielded the earth's surface from ultraviolet radiation probably limited early life to water greater than 40 m in depth but within the photic zone above 100 m (Tappan, 1974). As turbulence and current movement would subject free-floating organisms to the dangers of excess radiation at the surface, a planktic habitat probably was not feasible. The appearance of red beds coincident with the cessation of deposition of the banded iron formations of the Precambrian suggests that an oxygenic atmosphere had been attained by the middle Precambrian.

Phytoplankton. Apparently planktic unicells referred to the chroococcacean blue-green algae occur in the 1.9 b yr old Gunflint Formation, together with various filamentous benthic species referrable to the Oscillatoriaceae and Nostocaceae. Restudy of other Gunflint taxa (*Eosphaera* and *Huroniospora*) has resulted in their tentative reassignment to the exclusively aerobic eukaryotic red algae, the Porphyridiaceae (Tappan, 1976), indicating the attainment of eukaryotic cell organization and mitotic cell division and providing additional evidence of an oxygenic atmosphere.

Chlorophycean and prasinophycean green algae are also present in the upper Precambrian, but the dominant microplankton of the Precambrian and Paleozoic is known solely by resistant organic-walled resting cysts (Fig. 1) variously referred to as Acritarcha, Hystrichophyta, or Undifferentiated Phytoplankton, as their exact affinity is uncertain and some may represent now extinct higher taxa. They flourished and diversified in the early Paleozoic, but decreased in numbers and diversity within the late Devonian, being represented solely by a few relicts in the upper Paleozoic and Triassic (Tappan and Loeblich, 1972, 1973a,b).

As the acritarchs declined, their place was filled by other groups. Dinoflagellates (Fig. 2) have a sparse record in the Paleozoic and early Mesozoic (Silurian, Permian, and Triassic), extending some 200 m. yr. before their major diversification. This group expanded rapidly in the Jurassic and Cretaceous, which saw their acme. Still diverse and important components of the plankton, the dinoflagellates have declined somewhat through the Cenozoic.

Important changes in the nature of the phytoplankton also characterized the Mesozoic, with the evolution of the silica and calcium carbonate depositing diatoms and coccolithophorids, respectively. These groups became progressively more important, and caused a major change in the nature of the sediments being deposited. The calcareous nannoplankton (Fig. 3a–g) rose to major importance in the Jurassic; although there are a few isolated reports of older taxa, most of these appear questionable. Maximum diversity of this group is recorded in the Upper Cretaceous, and the abundance was reflected in the extent of the Cretaceous coccolith chalks. It was followed by

FIGURE 1. Acritarchs, resting cysts of phytoplankton *incertae sedis*. a,b: *Polyancistrodorus columbariferus* Loeblich and Tappan (a, X900; b, scanning electron micrograph, showing excystment opening or pylome, X1250), Middle Ordovician Bromide Formation, Oklahoma (from Loeblich and Tappan, 1969). c,d: *Holothuriadeigma heterakainum* Loeblich (c, scanning electron micrograph, X1300; d, X880), Middle Silurian Neahga Shale, New York (from Loeblich, 1970). e: *Villosacapsula setosapellicula* (Loeblich) Loeblich and Tappan (scanning electron micrograph, showing flaplike excystment opening or epityche, X830), Middle Ordovician Bromide Formation, Oklahoma. f: *Ordovicidium elegantulum* Tappan and Loeblich (scanning electron micrograph, X780). Middle Ordovician Bromide Formation, Oklahoma.

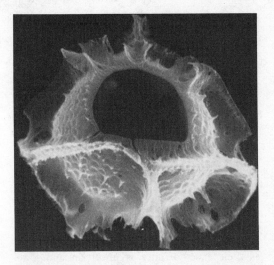

FIGURE 2. Dinoflagellate cyst (from Drugg, 1970). *Muratodinium fimbriatum* (Cookson and Eisenack) Drugg (scanning electronic micrograph of dorsal side, showing excystment opening or archeopyle, X750), lower Eocene Wilcox Group Alabama.

a major wave of extinction (Bramlette and Martini, 1964), during which two-thirds of the genera and a higher percentage of the species disappeared. As a result, deposition of the coccolith oozes was interrupted, and disconformities at the Cretaceous-Tertiary contact are commonly present in oceanic sequences as well as in other nonclastic marine deposits (Tappan, 1968).

The geologically oldest record of the diatoms (Fig. 3h–j) is mid-Cretaceous, but diatoms remained rare and primitive (only the centric diatoms were present) through the latest Cretaceous. Diversification and increase in abundance has occurred through the Cenozoic, resulting in the deposition of diatom oozes, and the formation of diatomites in many of the marginal seas of the Pacific borders, the Mediterranean, and in the oceanic areas of upwelling. Pennate diatoms appeared in the Paleocene, and thereafter formed a steadily increasing percentage of the microfloras.

Silicoflagellates, Order Dictyochales of the Chrysophyta, have an internal siliceous skeletal framework. Their requirements were similar to those of the marine diatoms, and their fossil remains also occur in the diatomaceous deposits. The oldest members of this group are of middle Cretaceous age, but like the diatoms, they become more important in the Tertiary, reaching their maximum diversity in the Miocene (Loeblich et al., 1968).

Zooplankton. No undoubted zooplankton record is known for the Precambrian or Cambrian, the most ancient being the Ordovician polycystine radiolarians. The earliest radiolarians were Spumellarians, but by the Carboniferous the Nasellaria also were present. A late Paleozoic decrease and Mesozoic and Cenozoic increases in the radiolarians paralleled the phytoplankton decrease and increases.

Planktonic foraminifera (Fig. 4a) probably first appeared in the early Mesozoic, but became widespread during the Cretaceous. The nearly complete extinction at the close of the Cretaceous left a sparse Paleocene assemblage, but this group again diversified in the Tertiary. Tintinnids (Fig. 4b,c) have a scattered record in the Paleozoic; they reached their maximum in the Late Jurassic and Early Cretaceous. Records in the Upper Cretaceous and Tertiary are sparse, and as yet none are known in the Neogene.

In general, zooplankton diversity closely paralleled that of the phytoplankton throughout their geologic history, as does their geographic distribution at present.

Interaction with the Physical Environment

Adaptations for Suspension. The pelagic habitat allows for maximum occupation of the lighted areas of the ocean by the phytoplankton, and for proximity to this rapidly regenerated food source for the grazing zooplankton and higher trophic levels. It also provides the possibility of current-borne dispersal into all available areas, and local interchange with the atmosphere of gases such as oxygen, carbon dioxide, and nitrogen (for blue-green algae), to supplement the nutrients recycled in the sea.

Invasion of the plankton habitat required rapid adaptation to its distinctive features, and perhaps this rapidity explains why the immediate precursors have not yet been determined for any of the major groups of the plankton. Some normally benthic protozoa and algae respond to unfavorable bottom conditions such as oxygen deficiency or to increased temperature by forming intracellular gas *vacuoles*, thereby reducing the total density of the cell and allowing it to float temporarily. Perhaps such an adaptation in the geologic past, during periods of relatively warm equable climates, stratified seas, and poorly oxygenated bottom waters, led to the origin of some of the major groups of plankton (Tappan and Loeblich, 1973a). Major invasions of this habitat occurred in the early Paleozoic (acritarchs, radiolaria), within the Jurassic (dinoflagellates, coccolithophorids, planktic foraminifera, tintinnids), and Paleocene (expansion of previous groups following their nearly complete extinction, and the addition of ebridians, discoasters,

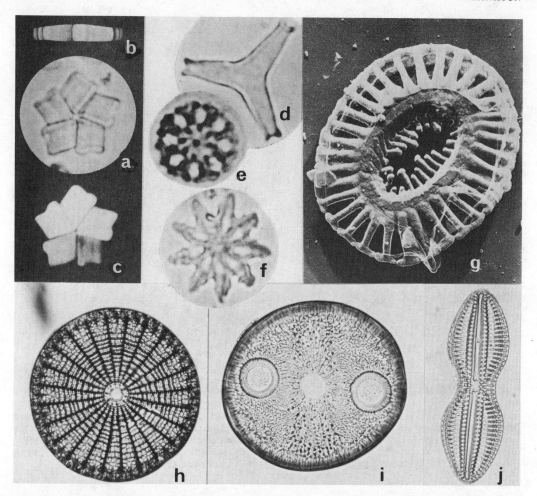

FIGURE 3. Calcareous nannoplankton (a–g) and diatoms (h–j). a–c, *Micrantholithus concinnus* Bramlette and Sullivan (a, plan view; b, sketch of side view; c, in polarized light, crossed nicols), Paleocene Lodo Formation, California; d, *Discoaster tribrachiatus* Bramlette and Riedel, lower Eocene Lodo Formation, California; e, *Discoaster nonaradiatus* Klumpp, middle Eocene Canoas Formation, California; f, *Discoaster limbatus* Bramlette and Sullivan, Paleocene Lodo Formation, California (a–f, from Bramlette and Sullivan, 1961; X2025). g, *Emiliania huxleyi* (Lohmann) Hay and Mohler (transmission electron micrograph of carbon replica, X1475), Holocene, mid-oceanic ridge, South Atlantic Ocean (from Black and Barnes, 1961). h, *Arachnoidiscus ornatus* Ehrenberg (centric diatom, X335), upper Miocene Monterey Formation, Santa Monica Mountains, California. i, *Auliscus pruinosus* Bailey (centric diatom, X430), upper Miocene Monterey Formation, Santa Monica Mountains, California. j, *Diploneis crabro* (Ehrenberg) Ehrenberg (pennate diatom, showing central groove or raphe, X395), Pliocene Sisquoc Formation, near Lompoc, California.

and the pennate diatoms). Extensive reductions in plankton diversity and sweeping extinctions took place in the Late Devonian, Permian, and latest Cretaceous.

Various convergent adaptations in size, shape, density, and motility have evolved to aid in flotation, retard sinking, or otherwise maintain a favored position in the water. The adaptations variously facilitate nutrient collection, waste disposal, or protection against grazers or predators, as well as aiding suspension, hence their primary adaptive value is not always readily determined. A high surface-to-volume ratio is particularly effective in aiding suspension, and so many components of the plankton are small, ranging down to only a few mμ in diameter, whereas others have greatly elongated projections that increase the surface area. For organisms with a density near that of sea water, energy expenditure that is required to maintain flotation is minimal, hence adaptations may involve cell inclusions of fat globules or gas bubbles; low-density cell sap; and very light skeletal elements that are thin, hollow, or

619

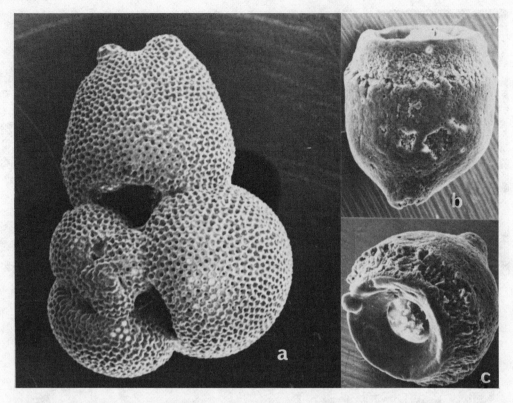

FIGURE 4. Zooplankton. a, Foraminiferida, *Globigerinoides sacculifer* (Brady) (scanning electron micrograph of spiral side, showing secondary apertures and pitted perforate surface, X90), Holocene, bottom sediments, Sahul Shelf of Timor Sea, Northwest Australia. b,c, Tintinnida, *Remanellina eocenica* Tappan and Loeblich (scanning electron micrographs, X220; b, side view; c, oblique apertural view showing circumoral shelf), upper Eocene Moodys Branch Formation, Mississippi (from Tappan and Loeblich, 1968).

perforated. For such organisms, the movement due to water currents, turbulence, wind-driven convection cells in the surface waters, and variations of specific gravity may be of greater importance than the motility of the organism itself. However, the movement resulting from morphologic modifications that produce spinning or spiralling, or movement due to flagella, cilia, etc., may also aid in nutrient renewal in the immediate vicinity of the plankton. Both density and viscosity of the water are greater in cold than in warm water, allowing dominance by the generally larger and passively floating diatoms in the higher latitudes; whereas the tropical phytoplankton has a high proportion of the smaller motile flagellates and ciliates, such as coccolithophorids, dinoflagellates, and tintinnids. Similarly, the pseudopodial length of surface-living planktic foraminifera is considerably greater than of those species living deeper in the water; and the flanges, keels, setae, spines, and processes that produce spinning or retard sinking are most pronounced in low-latitude

taxa and greater in individuals during the summer season than in winter individuals.

Geochemical effects of the plankton. The nature and abundance of the plankton at various geologic times has had a profound effect on the nature of the sediments formed, as well as the character of the sea water and atmosphere. The balance between phytoplankton photosynthesis and carbon fixation and the carbon recycling by the community trophic structure has fluctuated through geologic time. These variations have affected the pCO_2, pH, and pO_2 of oceans and atmosphere, as well as the depth of the oxygen minimum zone and carbonate compensation level in past geologic periods. Petroleum deposits, as well as much of the dispersed organic carbon in sediments consist of the carbon reduced by phytoplankton photosynthesis that escaped recycling, its burial thereby allowing the accumulation of oxygen in the atmosphere. Changes through geologic time in the rate of carbon fixation, resulting from fluctuations in the abundance of the phytoplankton, may have caused fluctuations in atmospheric oxygen levels in the past as well.

Phytoplankton photosynthesis, like that of other plants, selectively utilizes the lighter isotope of carbon (C_{12}), leaving the heavier C_{13} to accumulate. As a result, geologic periods of maximum phytoplankton abundance are characterized by isotopically light organic matter, and isotopically heavy limestones and other calcium carbonate material in the sediments.

Most of the Phanerozoic cherts apparently have been the result of opaline silica deposition by various microplankton organisms, radiolarians, and diatoms particularly. Certainly a large amount of silica was removed from solution in the formation of the approximately 5000 m of Upper Cretaceous and Cenozoic diatomite in California, and similar deposits of this age are known to occur over large areas of the ocean floor.

The rapid expansion of the coccolithophorids in the Jurassic and of the planktonic foraminifera in the Cretaceous led to the accumulation of extensive oceanic calcareous oozes as well, although the calcareous deposits are selectively removed by solution in the deeper regions of the ocean basins. Stable oxygen isotopic ratios for the calcium carbonate deposited by planktic foraminifera have been utilized to determine paleotemperatures of the surface waters, and of varying depths in the water column.

Biotic Interactions—The Plankton Ecosystem

Plankton communities are perhaps more physically controlled than benthic ones, because of the stresses associated with limited light; nutrients; the necessity of flotation or otherwise maintaining a favored position in the water column; lateral displacement by currents; and fluctuations in temperature, salinity, and turbulence. Regions of highest stress—in areas of upwelling or in high latitudes with temperature, nutrient, and light extremes—have relatively low diversity, but those species present are generally cosmopolitan types that may become abundant. Similar patterns may be observed worldwide, species being small, highly variable in morphology, with rapid reproduction, numerous offspring, short life cycles, and short food chains in which large diatoms are consumed directly by grazing fish, without intervening microzooplankton trophic levels. Examples of such plankton systems in the fossil record are in the Lower Triassic (low diversity but local abundance of organic microplankton); in the diatomites of the Upper Cretaceous and Tertiary; and

in the plankton assemblages of the lower Paleocene, where very few planktic taxa are present, though these may be abundant and have sufficiently plastic genotypes to hinder species allocation (Tappan and Loeblich, 1973a,b).

In contrast, communities of low stress—as in lower latitudes with vertically stratified waters characterized by little mixing and lesser productivity—contain many endemic species; but diversity increases almost indefinitely with the area sampled. Total biomass is greater, food chains are more complex (for example, coccolithophorid-tintinnid-copepod-fish associations), and the biota shows a greater size range and greater total diversity. Individuals per species are fewer, and intraspecific variation is restricted; taxa commonly are morphologically complex, with a larger degree of inert matter in walls, ornate shells, and skeletal structures. Planktonic foraminifera in such a community have large apertures, thin shells, marginal flanges, elongate spines, and similar "ornamental" structures, many of adaptive value as an aid to flotation in the less dense warmer water. Upper Cretaceous, Eocene, and Miocene plankton assemblages supply geologic examples of such ecosystems. These may contain calcareous nannoplankton (coccolithophorid), ebriid-silicoflagellate, and chrysomonad assemblages of tiny size, accompanied by larger dinoflagellates, prasinophycean green algae, radiolaria, and planktonic foraminifera, as well as planktonic invertebrates (e.g., ammonites, belemnites, pteropods, crinoids, ostracodes) and vertebrates (fish, marine reptiles, and mammals).

HELEN TAPPAN*

References

Barron, J. A., 1973. Late Miocene-Early Pliocene paleotemperatures for California from marine diatom evidence, *Palaeogeography, Palaeoclimatology, Palaeoecology*, 14, 277-291.

Bé, A. W. H., and Ericson, D. B., 1963. Aspects of calcification in planktonic Foraminifera (Sarcodina), *Trans. N.Y. Acad. Sci.*, 109, 65-81.

Black, M., and Barnes, B., 1961. Coccoliths and discoasters from the floor of the South Atlantic Ocean, *J. Roy. Microsc. Soc.*, 80, 137-147.

Bolli, H. M.; Loeblich, A. R., Jr.; and Tappan, H., 1957. Planktonic foraminiferal families Hantkeninidae, Orbulinidae, Globorotaliidae, and Globotruncanidae, *Bull. U.S. Nat. Museum*, 215, 1-50.

Bramlette, M. N., and Martini, E., 1964. The great change in calcareous nannoplankton fossils between the Maestrichtian and Danian, *Micropaleontology*, 10, 291-322.

*Support is acknowledged from the Earth Science Section, National Science Foundation, N.S.F. grant GA-40419.

Bramlette, M. N., and Sullivan, F. R., 1961. Coccolithophorids and related nannoplankton of the Early Tertiary in California, *Micropaleontology*, 7, 129–188.

Brönnimann, P., and Renz, H. H., eds., 1969. *Proc. 1st. International Conference Planktonic Microfossils, Geneva, 1967.*, Leiden. Brill, vol. 1, 422p.; vol. 2, 745p.

Douglas, R. G., and Savin, S. M., 1978. Oxygen isotopic evidence for the depth stratification in Tertiary and Cretaceous planktic Foraminifera, *Mar. Micropaleontol.*, 3, 175–196.

Drugg, W. S., 1970. Some new genera, species, and combinations of phytoplankton from the Lower Tertiary of the Gulf Coast, USA, *N. Amer. Paleontological Conv. Proc.*, pt. G, 809–843.

Emiliani, C., 1954. Depth habitats of some species of pelagic Foraminifera as indicated by oxygen isotope ratios, *Amer. J. Sci.*, 252, 149–158.

Farinacci, A., ed., 1971. *Proc. II Planktonic Conf. Roma 1970.* Rome: Edizioni Tecnoscienza, vol. 1, 676p.; vol. 2, 677–1369.

Funnell, B. M., and Riedel, W. R., eds., 1971. *The Micropalaeontology of Oceans, Proc. Symp. Micropalaeontology of Marine Bottom Sediments, Cambridge, 1967.* London: Cambridge Univ. Press, 828p.

Gray, J., and Boucot, A. J., 1972. Palynological evidence bearing on the Ordovician-Silurian paraconformity in Ohio, *Geol. Soc. Amer. Bull.*, 83, 1299–1314.

Lipps, J. H., 1970. Plankton evolution, *Evolution*, 24, 1–22.

Loeblich, A. R., Jr., 1970. Morphology, ultrastructure and distribution of Paleozoic acritarchs, *N. Amer. Paleontological Conv., Proc.*, pt. G, 705–788.

Loeblich, A. R., Jr., and Tappan, H., 1969. Acritarch excystment and surface ultrastructure, with descriptions of some Ordovician taxa, *Rev. Española Micropal.*, 1, 45–57.

Loeblich, A. R., III; Loeblich, L. A.; Tappan, H; and Loeblich, A. R., Jr., 1968. Annotated index of fossil and Recent silicoflagellates and ebridians with descriptions and illustrations of validly proposed taxa, *Geol. Soc. Amer. Mem. no. 106*, 319p.

Sarjeant, W. A. S., 1974. *Fossil and Living Dinoflagellates.* London, New York, San Francisco: Academic Press, 182p.

Tappan, H., 1968. Primary production, isotopes, extinctions and the atmosphere, *Palaeogeography, Palaeoclimatology, Palaeoecology*, 4, 187–210.

Tappan, H., 1971. Microplankton, ecological succession, and evolution, *N. Amer. Paleontological Conv. Proc.*, pt. H, 1058–1103.

Tappan, H., 1974. Molecular oxygen and evolution, in O. Hayaishi, ed., *Molecular Oxygen in Biology: Topics in Molecular Oxygen Research*. Amsterdam: North-Holland Publ., 81–135.

Tappan, H., 1976. Possible eucaryotic algae (Bangiophycidae) among early Proterozoic microfossils, *Geol. Soc. Amer. Bull.*, 87, 633–639.

Tappan, H., and Loeblich, A. R., Jr., 1968. Lorica composition of modern and fossil Tintinnida (ciliate Protozoa), systematics, geologic distribution, and some new Tertiary taxa, *J. Paleontol.* 42, 1378–1394.

Tappan, H., and Loeblich, A. R., Jr., 1971a. Geobiologic implications of fossil phytoplankton evolution and time-space distribution, *Geol. Soc. Amer. Spec. Paper no. 127*, 247–340.

Tappan, H., and Loeblich, A. R., Jr., 1971b. Surface sculpture of the wall in Lower Paleozoic acritarchs, *Micropaleontology*, 17, 385–410.

Tappan, H., and Loeblich, A. R., Jr., 1972. Fluctuating rates of protistan evolution, diversification, and extinction, *24th Internat. Geol. Congr.*, sec. 7, 205–213.

Tappan, H., and Loeblich, A. R., Jr., 1973a. Evolution of the oceanic plankton, *Earth-Sci. Rev.*, 9, 207–240.

Tappan, H., and Loeblich, A. R., Jr., 1973b. Smaller Protistan evidence and explanation of the Permian-Triassic crisis, in A. Logan, and L. V. Hills, eds., The Permian and Triassic Systems and their mutual boundary, *Canadian Soc. Petrol. Geol., Mem. no. 2*, 465–480.

Wicander, E. R., 1975. Fluctuations in a Late Devonian-Early Mississippian phytoplankton flora of Ohio, U.S.A. *Palaeogeography, Palaeoclimatology, Palaeoecology*, 17, 89–108.

Cross-references: *Acritarchs; Algae; Anellotubulates; Biogeochemistry; Chitinozoa; Coccoliths; Diatoms; Dinoflagellates; Discoasters; Diversity; Ebridians; Extinction; Foraminifera, Benthic; Foraminifera, Planktic; Fusulinacea; Hystrichospheres; Larvae of Marine Invertebrates—Paleontological Significance; Melanosclerites; Palynology; Precambrian Life; Protista; Pyritospheres; Radiolaria; Silicoflagellates; Tasmanitids; Tintinnids.* Vol. I: *Fertility of the Oceans; Marine Microbiology; Nutrients of the Sea; Pelagic Biogeochemistry; Pelagic Distribution; Pelagic Life; Phytoplankton; Planktonic Photosynthesis; Zooplankton.*

PLATE TECTONICS AND THE BIOSPHERE

Natural selection acts chiefly to maintain or enhance the adaptations of populations of organisms to their environments. When the environment changes significantly, organisms either adapt to the new conditions or become extinct. As a result, the patterns of the ecological relations of organisms should conform closely with the environmental patterns at any time, barring lag effects. Since the processes of plate tectonics create major changes in the earth's environment, they should affect the biotic patterns and the course of evolution and extinction.

Plate Tectonics and Environmental Change

Geographic changes that result from plate tectonic processes must cause great environmental changes. For example, when continents drift into new latitudes, climatic patterns in both marine and terrestrial environments will be affected. As climatic zones sweep across a moving continent, the effects must vary among local environments, where they are modified by local conditions. The sequence of detailed climatic changes will thus vary from locality to locality within the environmental mosaic. As complex as such changes must be, they are easier to infer than are the effects of the continental motions on the climatic zones themselves. A glance at modern climatic patterns demonstrates that they are much influenced by continental geography; continents perturb the flow of winds and ocean currents and thus the transport of heat and other factors. Their low heat capacity is in marked contrast with the oceans, and strong temperature and pressure gradients (seasonal gradients into the continental interiors and daily gradients in coastal regions) arise across continental margins. Thus, the changing positions of continents within a climatic zone will affect the nature of the climate itself. For that matter, latitudinal biogeographic zones may be created or destroyed by continental patterns; if a continental configuration greatly restricts the poleward flow of heat, polar cooling will ensue and a high latitudinal thermal and climatic gradient, with a relatively high biogeographic zonation, will result. Conversely, if heat transport is facilitated by continental patterns, then the poles will be warmer, the latitudinal gradient will be gentler, and fewer latitudinal biogeographical zones may be present. However, there are certainly limitations to the climatic changes we may postulate. For example, there are limits to the temperature of the poles no matter what the continental pattern, and the necessity to conserve momentum places such constraints on the zonal wind patterns (Lamb, 1972) that ancient trade wind belts can be inferred with some confidence from stratigraphic data (Drewry et al., 1974).

Another example of the environmental effects of drift involves the sizes of continental masses. At times, continents have been large and few in number, or even assembled into a single supercontinent, creating a very large superocean. At other times, continental masses have been fragmented into numbers of widely dispersed smaller continents and the oceans have therefore been divided into smaller, often partially isolated basins. The resulting variation in the size and distribution patterns of continents and oceans must cause considerable variations in the degree of continentality versus the extent of maritime conditions, both locally and on hemispheric or planetary scales.

Two other plate-tectonic sources of major environmental change should be noted. Eustatic changes in sea level may alternately flood and expose vast regions of the continental platforms, as has been well-established empirically. Obvious environmental consequences of eustatic variations include the addition of wide tracts of benthic habitats to the marine environment and subtraction of those tracts from the terrestrial realm, during transgressions, and opposite trends during regressions. This should result in changes not only in habitable area but in habitat heterogeneity. A further effect of changing sea levels is to enhance or reduce the continentality (versus the maritime nature) of the climatic regime. Widespread epicontinental seas will greatly moderate the climate, reducing the climatically effective size of the continents. Thus, transgressions should increase marine species diversity within communities. At the same time, transgressions may reduce spatial heterogeneity on the larger scales by connecting oceanic arms or shelf regions that had formerly been separated by barriers to gene flow. The resulting reduction in endemism would lower species diversity. Sea-level changes must often be due to changes in ocean basin volume caused by plate-tectonic processes, but perhaps owing to the manifold causes of sea-level variation, convincing general correlations of eustatic and plate-tectonic histories have not yet been developed.

One last major effect of geographic change is to alter the nature of ocean currents. Raising or lowering the planetary thermal gradients will speed or slow the average surface circulation; while increasing the supply of cool dense surface water will lower bottom and mid-water temperatures. The paths of surface currents will obviously depend greatly upon land–sea geography, and may be altered slowly with gradual continental motions or suddenly as continents assume key configurations that switch the current patterns.

Other lesser sources of environmental change are related to plate-tectonic events. Topographic relief, which can enhance spatial heterogeneity of the environment, is commonly achieved through collissions or interactions of plate elements with continental margins. The extent and positions of island-arc and intraplate volcanic chains change to create new patterns of oceanic habitats and new dispersal routes or barriers for shallow-water or terrestrial organisms. In many such ways, the environment is constantly altered, in global patterns and in local regimes, and occasionally the change may be unusually great, as when some threshold state or condition is achieved or exceeded.

Biosphere Structure and Environmental Patterns

Many different features of the biosphere should be sensitive to the environmental changes brought about through plate-tectonic events, owing to processes of ecology and of evolution. The fossil record should reflect the historical sequence of these changes. By far the richest and most extensive fossil assemblages consist of marine invertebrates from epicontinental seas and other shallow habitats. It is therefore appropriate to examine the ecological structure of the present continental shelves and shallow island habitats, for purposes of comparison with conditions in the past.

The major subdivisions of the shallow-marine biosphere are biotic provinces, regions of oceanic or suboceanic scale within which are found biotic associations of characteristic compositions. They are separated by barriers to dispersal that restrict many species within provincial areas. On continental shelves today, provincial borders occur at regions where marine temperature regimes change markedly, as at climatic boundaries, or where provinces are separated by topographic barriers—land or deep-sea—that shallow-water organisms cannot ordinarily cross. There are over 30 shelf provinces at present. Many of them form latitudinal chains of climatically restricted biotas along the N-S trending shelves that border the world's major coastlines. It is likely that there are more provinces today then ever before; this high provinciality arises from the number, shape, and dispersion pattern of the present continents and oceanic islands. It results in a high species diversity, since each province possesses a large native element. Whenever in the past continents were (a) sutured together to reduce deep-sea barriers, or (b) were crossed by E-W trending (climatically monotonous) epicontinental seaways to reduce land barriers, or (c) were situated so that E-W coastlines were more predominant, or (d) permitted a lower latitudinal temperature gradient, then provinciality would have been reduced and species diversity lowered.

The next smaller ecological units of the shallow shelf subdivide the provinces; these are communities, characteristic associations of species that are found together where similar environmental conditions occur. Exposed rocky shore and protected muddy shore associations are examples. The ecological structure of shallow-marine communities, as indicated for example by the number, sizes, and trophic organizations of their component populations, varies in fairly regular pattern. The causes of this variation are in dispute. This is partly because the structure is presumably created by natural selection, which acts to promote the fitness of individual genotypes. The special ecological properties of communities involve relations among populations, and thus emerge at a higher level of integration than that at which they are selected. The processes that forge the links between these levels are not yet understood for many ecological properties. Thus, it is not yet possible to predict from first principles the community structure that will result from a given environmental change. Instead, reliance must be placed upon empirical correlations of community structure with the environment.

These correlations suggest that spatial and temporal patterns of environmental variability are closely associated with community structure, and indeed there are general theoretical reasons to expect such an association (May, 1974). Presumably, spatial patchiness of environments provides more habitat types for exploitation; it has been suggested that larger numbers of communities may occur, and larger number of species are associated within communities, in the more patchy environments, other things being equal. There is evidence that temporal environmental variability influences the numbers and kinds of species associated in communities. In the relatively stable low latitudes where seasonality is low and many factors vary but little, there are high proportions of specialized species, high proportions of consumers (species that feed upon living organisms), and high species diversity (total number of species) compared with the relatively unstable high latitudes. In unstable regions, there seems to be a higher proportion of species that are opportunistic, flexibly adapted generalists, a higher proportion that are recuperators (eat detritus, bacteria, or dead organisms), and low species diversity. As a result, communities in these very different regimes have strikingly different structures.

There is little agreement on which environmental factors might underly these structural differences. In the sea, the factors most frequently considered are temperature, salinity, storminess and allied climatic features, and food (see for example Sanders, 1968, and Valentine, 1973). On land, the complex patterns of rainfall, frost, and humidity would also have to be considered. All of these factors will clearly be affected by the environmental changes arising from plate tectonic events, and community structures should be altered accordingly.

Communities are composed of species populations. In theory, constant evolutionary activity within species lineages would be expected to accompany the constant environmental alterations. Evolution adapts species to the ambient environmental regime, not to some possible future conditions, and thus environmental changes would tend to be generally deleterious. If environmental changes were slow, slight, or resulted in an increased capacity to support species (that is, in a rise in potential diversity), then the chance that the average lineage would be able to maintain its adaptation should have been relatively high, and extinction low. However, if changes were rapid, severe, or resulted in a lowered capacity of the environment to support species, then the average lineage would have had much smaller chance to adapt successfully and numerous extinctions are likely to have ensued. In either event, lineages that happened to possess adaptations that were of special utility in the novel conditions might do particularly well, perhaps diversifying under appropriate circumstances to increase their relative significance in the biosphere (Simpson, 1953).

Plate Tectonics and the History of the Shallow-Marine Biota

Some idea of the sequence of shallow-marine conditions that may have affected biosphere structure and composition can be inferred from gross continental geographies. Phanerozoic continental reconstructions have been presented by Smith et al. (1973) and Briden et al. (1974) (see *Paleogeographic Maps*). Modifications of these geographies are required as new evidence is developed, but they are the most useful general paleogeographic models at present. They tend to support simplified reconstructions that have been employed to assess the relations of plate tectonics to the fossil record (Valentine and Moores, 1970, 1972). There appear to have been at least four major continents during the Cambrian. One continent, Gondwana, persisted throughout the Paleozoic and usually included the South Pole, but was so large that it commonly extended across the equator or at least into very low latitudes. The north pole appears to have had little important continental influence. The continents gradually coalesced around Gondwana and finally formed a single supercontinent, Pangaea, about 225 m yr ago. These Paleozoic geographies appear to indicate relatively low-latitudinal thermal gradients in the sea, since: (1) there was no circumpolar current in the south, and therefore transport of warm water from low latitudes was probably much greater than at present; and since (2) the North Pole seems to have been free of barriers to heat flow. On land, the episodic appearance of glacial deposits testifies to a rather wide climatic variety, at least at times. During the Mesozoic and Cenozoic, Pangaea

fragmented to produce the present dispersed condition and large number of continents. The continents have tended to drift northward, leaving Antarctica-Australia, and finally Antarctica, alone on the South Pole surrounded by a circumantarctic current and insulated from warm-water influences. At the same time, the Arctic Ocean has come into being, surrounded and nearly isolated by large land masses. These trends may well be responsible for polar cooling leading to the high-latitudinal thermal gradient of the present.

Thus, there are two major paleogeographic systems during the Phanerozoic. Each of them is characterized by a different type of biosphere structure, and (as has been well-known for over a century) by strikingly different faunal compositions. The first system appeared in the late Precambrian and lasted throughout the Paleozoic. It is characterized by a very large southern continent, to which smaller continents joined during the era, and probably by a low marine climatic gradient. Thus, marine biogeographical barriers, either topographic or climatic, were few and this system is marked by relatively low shallow-marine endemism and provinciality (usually no more than four or five provinces are present at any time; see papers in Hughes, 1973 and Hallam, 1973). Shallow-marine diversity levels should have been low compared with today; appropriately low levels are recorded (Fig. 1).

During the late Paleozoic, as continents were sutured, species diversity may have declined below mid-Paleozoic levels, although it is difficult to determine this from the spotty fossil

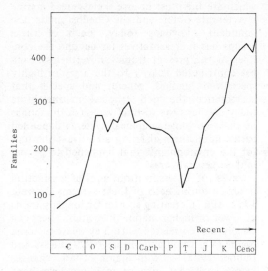

FIGURE 1. Diversity level of families of well-skeletonized shelf invertebrates during the Phanerozoic, plotted by epoch (after Valentine, 1973).

record, in which species numbers are impossible to determine directly. At any rate species diversity can be estimated to have fallen drastically in a series of extinctions near the close of the Paleozoic (Fig. 1). These extinctions correlate both with the final assembly of Pangaea and with a major regression. Presumably these two events would cause a lowering in provinciality, a lowering in shallow-water habitat area, and a considerable increase in continentality on the remaining shallow epicontinental sea floor. These are all conditions that should lead to a lowering of diversity. It can therefore be postulated that the Permian-Triassic extinctions represented a biotic adjustment to a lowered species capacity from several causes (Valentine and Moores, 1970; Valentine, 1973; Schopf, 1974; Simberloff, 1974; Valentine et al., 1978).

The second paleogeographic system grew out of the fragmentation of the supercontinent Pangaea. It developed into a geography of numerous well-dispersed continents with long N-S trending coastlines and zonal climates; therefore, there is high provinciality and high diversity (Fig. 1). The early Mesozoic fauna is of low diversity and can be interpreted as consisting by and large of ecologically flexible species that were best adapted to survive the conditions across the Permian/Triassic boundary. Radiations from among these survivors have produced hosts of new species; in addition, some of the more characteristic animal groups of the modern seas originate and diversify. The latter include the Scleractinia, which includes all living stony corals; the Neogastropoda, chiefly predatory gastropods, which are the most diverse skeletonized marine invertebrate order; and the Cheilostomata, the dominant byozoans today. Each of these groups has representatives far beyond the confines of the present tropics; nevertheless, each has contributed heavily to the array of highly specialized families, genera, and species that characterize the rich tropical marine biota today. Perhaps the distinctiveness of the faunas of the two biogeographical systems depended partly upon their differing adaptive potentials for the environments that happened to prevail in these separate eras.

Waves of diversification and of extinction occurred during each of these systems (Newell, 1952, 1967), creating shorter or longer periods of lower or higher marine diversities. These can usually be correlated with key plate-tectonic events (Valentine and Moore, 1972; Flessa and Imbrie, 1973) or with major eustatic changes in sea level that may or may not have been owing to plate tectonics. Most of these faunal events are now under study by specialists, and

it is still too early to attempt a recital of their individual characters. Although important syntheses remain to be made from data already in hand, continued major progress in understanding the relations between plate tectonics and the history of life depends upon refinement of biogeographies and of paleoenvironmental interpretations on one hand, and upon solutions to some of the evolutionary problems associated with population and community biology on the other.

JAMES W. VALENTINE

References

Boucot, A. J., and Gray, J. S., eds., 1979. *Historical Biogeography and Plate Tectonics.* Corvallis: Oregon State Univ. Press.
Briden, J. C.; Drewry, G. E.; and Smith, A. G., 1974. Phanerozoic equal-area world maps, *J. Geol., 82,* 555–574.
Dobzhansky, T., 1950. Evolution in the tropics, *Amer. Scientist, 38,* 209–221.
Drewry, G. E.; Ramsay, A. T. S.; and Smith, A. G., 1974. Climatically controlled sediments, the geomagnetic field, and trade wind belts in Phanerozoic time, *J. Geol., 82,* 531–553.
Flessa, K. W., and Imbrie, J., 1973. Evolutionary pulsations: Evidence from Phanerozoic diversity patterns, in D. H. Tarling, and S. K. Runcorn, eds., *Continental Drift, Sea Floor Spreading, and Plate Tectonics: Implications for the Earth Sciences.* London: Academic Press, 245–285.
Hallam, A., ed., 1973. *Atlas of Paleobiogeography.* Amsterdam: Elsevier, 531p.
Hughes, N. F., ed., 1973. Organisms and continents through time, *Spec. Paper Palaeontol., no. 12,* 334p.
Lamb, H. H., 1972. *Climate Present, Past and Future,* vol. 1. London: Methuen, 613p.
May, R. M., 1974. *Stability and Complexity in Model Ecosystems,* 2nd ed. Princeton: Princeton Univ. Press, Monogr. Pop Biol., no. 6, 265p.
Newell, N. D., 1952. Periodicity in invertebrate evolution, *J. Paleontology, 26,* 371–385.
Newell, N. D., 1967. Revolutions in the history of life, *Geol. Soc. Amer. Spec. Paper no. 89,* 63–91.
Sanders, H. W., 1968. Marine benthic diversity: A comparative study, *Amer. Naturalist, 102,* 243–282.
Schopf, T. J. M., 1974. Permo-Triassic extinctions: Relation to sea-floor spreading, *J. Geol., 82,* 129–143.
Simberloff, D., 1974. Permo-Triassic extinctions: Effects of area on biotic equilibrium, *J. Geol., 82,* 267–274.
Simpson, G. G., 1953. *The Major Features of Evolution.* New York: Columbia Univ. Press, 434p.
Smith, A. G.; Briden, J. C.; and Drewry, G. E., 1973. Phanerozoic world maps, *Spec. Paper Palaeontol., no., 12,* 1–42.
Valentine, J. W., 1973. *Evolutionary Paleoecology of the Marine Biosphere* Englewood Cliffs, N.J.: Prentice-Hall, 511p.
Valentine, J. W., and Moores, E. M., 1970. Plate-tectonic regulation of faunal diversity and sea level: A model, *Nature, 228,* 657–659.

Valentine, J. W., and Moores, E. M., 1972. Global tectonics and the fossil record, *J. Geol.,* 80, 167–184.

Valentine, J. W.; Foin, T. C.; and Peart, D., 1978. A provincial model of Phanerozoic marine diversity, *Paleobiology,* 4, 55–66.

Cross-references: *Diverstiy; Extinction; Fossil Record; History of Paleontology, Post-Plate Tectonics; Larvae of Marine Invertebrates—Paleontological Significance; Paleobiogeography; Paleoecology.*

POLYPLACOPHORA

The Polyplacophora, or chitons, differ from all other mollusks in several anatomical particulars. The most striking is that the shell consists of eight somewhat overlapping calcareous plates or valves (Fig. 1) instead of a single shell or pair of shells, as in the gastropods and the bivalves. These valves are embedded in a tough, muscular girdle, which holds them in place while allowing considerable freedom of movement. The shells of some chitons are covered by the girdle more extensively than those in others and in one species—the Giant Chiton of the North American West Coast—the shells are completely covered and internal. The ability to roll up in a ball, like a pill bug, when disturbed or removed from the substratum, has given rise to the name "sea cradle" for them.

Chitons occur at all depths in the sea, and they live from the coldest polar waters to tropical areas. They are most abundant, however, on temperate rocky shores. Here they find a hard substratum for attachment, protection from possible enemies, and usually food in abundance. Some species live on the outer coast and are able to withstand the buffeting of heavy surf; others seek protected habitats in areas where the water is quieter behind offshore reefs or rocks, in embayments, and in tidepools. Other species live mainly or wholly subtidally and can be found only when the tides are lowest, or by SCUBA diving or dredging; a few live higher up in the middle and low intertidal zones.

Sexes are separate in the chitons. Each species has its particular breeding season when the sex products are released into the open water (in most instances), the young hatching into minute, ciliated veliger larvae that swim for a short period before settling down to the bottom to develop their eight articulating valves

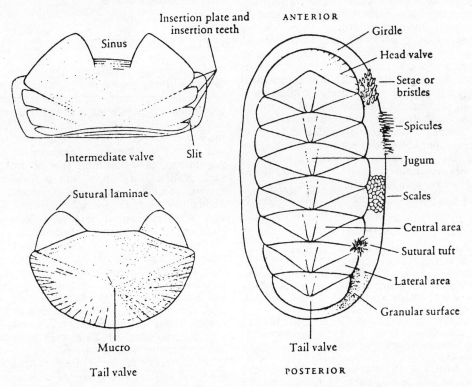

FIGURE 1. Hard-part morphology of the Polyplacophora (from A. M. Keen and E. Coan, 1974, with the permission of the publishers, Stanford University Press; Copyright © 1963, 1974 by the Board of Trustees of the Leland Stanford Junior University.)

and take on the characters of adults. They are provided with a lingual ribbon, or radula, with which they scrape or tear food from the substratum. The teeth are capped with a hard material containing so much iron that the radula can be picked up with a magnet. Some chitons are strictly vegetarians; other are omnivorous, eating diatoms, very young barnacles, or other marine animals as well as marine algae.

Today chitons have two major centers of distribution in which they reach unusually high species diversities. One includes Australia, New Zeland, and Tasmania; the other is the west coast of North America, which in the area north of the Mexican border alone supports a chiton fauna of close to 100 species. Thirty species have been reported from a single favorable spot in the Californian intertidal zone, and in Monterey Bay as many as 15 species have been brought up in a single dredge haul.

The chitons have had a long geologic history, with scattered plates dating back at least as far as the Late Cambrian (Bergenhayn, 1960). Little is known of chiton evolution or paleoecology, however (although comparative morphologic work with living species has been productive, e.g., Beedham and Trueman, 1967, 1969). This is both because the plates disarticulate on the death of the animal, and because the preferred environment (rocky shore) is not conducive to the preservation of fossil material. Thus, the paleontologic study of the Polyplacophora is still primarily in the descriptive stage. From the sparse data available, it appears that the chitons were a relatively homogeneous group through the Paleozoic (about 75 species known) and Mesozoic (about 30 species known; see Smith, 1973b), and diversified in Cenozoic times (about 250 species known). However, this is as likely a reflection of preservational effects as of evolutionary trends. Over 500 living species have been described.

A. G. SMITH*

References

Abbott, R. T., 1974. *American Sea Shells,* 2nd ed. New York: Van Nostrand, 392–407.

Beedham, G. E., and Trueman, E. R., 1967. The relationship of the mantle and shell of the Polyplacophora in comparison with that of other Mollusca, *J. Zool. London,* **151,** 215–231.

Beedham, G. E., and Trueman, E. R., 1969. The shell of the chitons and its evolutionary significance in the Mollusca (abstr.), *Proc. Malac. Soc. London,* **38,** 550–551.

Bergenhayn, J. R. M., 1960. Cambrian and Ordovician loricates from North America, *J. Paleontology,* **34,** 168–178.

*Deceased.

Boyle, P. R., 1977. The physiology and behavior of chitons (Mollusca: Polyplacophora), *Ann. Rev. Oceanogr. Marine Biol.,* **15,** 461–509.

Burnett, R., et al., 1975. The biology of chitons, *Veliger,* **18**(suppl.), 128p.

Glynn, P. W., 1970. On the ecology of the Caribbean chitons *Acanthopleura granulata* Gmelin and *Chiton tuberculatus* Linné. Density, mortality, feeding, reproduction and growth, *Smithsonian Contrib. Zool.,* **66,** 21p.

Haas, W., 1972. Untersuchungen über die Mikro- und Ultrastruktur der Polyplacophorenschale, *Biomin. Research Rept.,* **5,** 52p.

Hyman, L. H., 1967. *The Invertebrates,* vol. 6, *Mollusca 1.* New York: McGraw Hill, 70–142.

Keen, A. M., and Coan, E., 1974. *Marine Molluscan Genera of Western North America: An Illustrated Key.* Stanford: Stanford Univ. Press, 208p.

Sirenko, V. I., and Starobogatov, Ya. I., 1977. On the systematics of Paleozoic and Mesozoic chitons, *Paleont. J.,* **11,** 285–294.

Smith, A. G., 1960. Amphineura, in R. C. Moore, ed., *Treatise on Invertebrate Paleontology,* pt. I, Mollusca 1. Lawrence, Kansas: Geol. Soc. Amer. and Univ. Kansas Press, 41–76.

Smith, A. G., 1973a. Polyplacophora—A selected bibliography, *Of Sea and Shore,* **4**(4), 201–206, 208.

Smith, A. G., 1973b. Fossil chitons from the Mesozoic—A checklist and bibliography, *Occ. Papers Calif. Acad. Sci., no. 103,* 30p.

Cross-reference: *Mollusca.*

POPULATION DYNAMICS

The study of population dynamics is concerned with the growth and regulation of populations of organisms, regulation referring to adjustment of population size and structure to ambient environmental conditions. It may include problems of great scope, such as the stability of communities composed of many populations, as well as more specific problems, such as determination of the probability of survival as a function of age in a population. Population dynamics are of interest in connection with population ecology (survivorship, recruitment, life-history phenomena), community ecology (population interactions, theory of community structure), biogeography (dynamics of dispersal and population growth as related to genesis of distribution patterns), and general evolutionary biology (natural selection, which is after all a phenomenon of differential survival and reproduction in populations, and the evolution of life-history phenomena and community structures).

Development of generalized mathematical models describing the dynamics of populations and communities has led to numerous conclusions and theories on the design and evolution of life-history phenomena and the structure of communities. Population parameters in these models, notably the maximum per capita growth rate, are related to the vital statistics of mortality and fecundity by a few basic mathematical expressions.

Apart from theory, study of the population dynamics of extinct organisms is primarily limited to study of survivorship patterns from mortality statistics on fossil assemblages. Vital statistics on reproduction, and on survivorship of very young stages, are very difficult if not impossible to obtain from paleontological samples. Nevertheless, mathematical relationships among vital statistics and the population's growth rate sometimes make it possible to roughly bracket population parameters that are not themselves directly measurable. Studies of population dynamics can (1) provide important insights on the life history of extinct organisms and on selective pressures behind evolutionary change, (2) clarify the evolution of life-history phenomena in a taxonomic group, (3) elucidate the dynamics behind the structure of a community type, and (4) help explain the dynamics behind the distribution of an extinct species. Though opportunities for them are plentiful enough, studies of population dynamics for modern species are few, and for extinct species they are rarities in an extreme.

Population Growth and Regulation

The growth and regulation of a population is an extremely complex process. A population's growth rate, the prime focus for study of population dynamics, is the sum of the birth rate, death rate, and immigration or emigration rate. The latter, though of much interest in connection with dispersion patterns and biogeography, is very difficult to treat in a general case and has generally been neglected in demography, as it is here (but see MacArthur and Wilson, 1967; Levin, 1974; and Goel and Richter-Dyn, 1974).

Interpretations of population dynamics are couched in terms of mathematical models that necessarily present a highly simplified representation of the processes involved. Two sorts of models are used: deterministic models, in which parameters of the population and its environment are assumed constant, and stochastic models, in which at least one parameter is taken as a random variable. While deterministic models help in seeing the basic patterns, appropriately designed stochastic models, though generally less tractable, should give a more realistic representation of natural populations. For a review of current work, see Goel et al. (1971).

Deterministic Models. The simplest, most generalized deterministic equation describing

population growth is:

$$\frac{dN}{dt} = rN \qquad (1)$$

where N is the size or density of the population and the "actual rate of increase" r is the per capita birth rate minus the per capita death rate. The dependence of the growth rate on N expresses the idea that population growth is intrinsically exponential.

Assuming for simplicity that the population's growth rate is determined for all practical purposes only by population density, let $r = F(N)$. Regardless of its particular form, this function can be expressed as the sum of a Taylor series expansion, $F(N) = a_0 + a_1 N + a_2 N^2 + \ldots$. Making the simplest possible assumptions, suppose that terms of higher than first order are negligible, so that $F(N)$ varies linearly with N (making this a *linear model*), and that $a_0 > 0 > a_1$ so that $F(N)$ approaches zero as some limiting density is approached. Equation (1) becomes the famous Verhulst-Pearl logistic equation:

$$\frac{dN}{dt} = \frac{r_m N(K - N)}{K} \qquad (2)$$

where the "intrinsic rate of increase" $r_m \equiv a_0$ and $r = r_m(K - N)/K$, and where the "carrying capacity of the environment" K is $-a_0/a_1 > 0$. In integrated form, (2) becomes

$$N_t = \frac{K}{1 + Ce^{-r_m t}} \qquad (3)$$

This describes a sigmoid curve for N, population density at time $t = t$, which (starting from an initial density at time $t = 0$ that is reflected in the integration constant C) approaches the limit K.

The logistic equation (2,3) is very useful as a general model in ecological and evolutionary theory. However, it rests on assumptions that are often unrealistic—such as that the population maintains a stable age distribution throughout the course of growth (i.e., age-specific survival rate does not fluctuate from one generation to the next). Without introduction of a time delay between generations or variation in r_m or K, the logistic cannot express the fluctuations in density that are often observed in natural populations. In pursuit of more realistic representations, more recent models (e.g., May, 1974) have stressed nonlinear equations for $F(N)$, that is, equations including higher-order terms in the Taylor series.

A modified version of the logistic equation is

used to describe the growth and regulation of populations in communities:

$$\frac{dN_i}{dt} = \frac{r_m N_i (K_i - N_i - \sum \alpha_{ij} N_j)}{K_i} \quad (4)$$

where N_i and N_j are respectively the population density of the particular ith species and the jth species among s species in the community, and $\alpha_{ij} \geq 0$ is the "interaction coefficient" describing the effect of the jth species on the growth of the population of the ith species. For simplicity, the alphas are assumed constant with changing densities.

Analysis of models like (4) has led to the suggestion that, contrary to widespread belief (MacArthur, 1955; Hutchinson, 1959), increase in species diversity does not contribute to community stability, in the sense of tending to decrease fluctuations in population densities; instead, it may lead to decreased stability (May, 1973). Actually, for communities as for single populations, the response of population growth parameters to changing population densities may be highly nonlinear and hence very difficult to model in a general case. Though it appears that stability permits diversity (May, 1973), the problem of the diversity–stability relationship is far from resolved.

In ecological theory, species are frequently categorized with reference to the logistic model as being r-selectionists—"opportunistic" or "colonizing" species that are characterized by high r values in the early successional stages and disturbed situations in which they typically occur—and K-selectionists—"equilibrium" or "climax" species that typically occur late in succession and persist for long periods of time— (MacArthur and Wilson, 1967). These two categories represent ends of a spectrum.

Stochastic Models. Every deterministic model has as an analog at least one stochastic model in which one or more of its parameters are made random variables. Perhaps the most general stochastic model is that in which r in equation (1) is determined by a large number of independent factors so as to be normally distributed with a mean near zero for a well-established population. The logarithm of N_t/N_0 then becomes normally distributed. This model leads to one explanation for log-normal distribution of species abundance (Preston, 1948, 1962), a widespread pattern in communities. At least for communities of opportunistic species, the pattern may result from randomness in population growth (MacArthur, 1960). The importance of this species abundance pattern is that a "canonical" log-normal distribution (a one-parameter distribution in which the logarithmic variance is a particular function of the logarithmic mean) predicts the approximate fourth root dependence of species diversity on area that is generally encountered in species–area curves for geographical islands (Preston, 1962; MacArthur and Wilson, 1967), a pattern of much interest in biogeography and paleobiogeography.

The Life Table

A life table is a tabulation of age-specific vital statistics on mortality and fecundity for a cohort (group of individuals born at the same time) that applies under a particular set of environmental conditions. While these vital statistics may be expressed as continuous functions of age, it is generally more convenient in making and interpreting measurements to take them as discrete functions of age intervals. The statistics can then be presented in a table, hence the name "life table." In practice, vital statistics determined for a population represent sample values, not probabilities.

While numerous other vital statistics are used in demography the most basic ones are:

l_x – "age-specific survival rate," the number or proportion of individuals surviving to age x out of some initial number of individuals in a cohort. The initial number or proportion at birth or hatching, l_0, is commonly taken as 1, 100, or 1000. "Life table statistic" is another name for l_x.

m_x – "age-specific fecundity rate," the average number of eggs or young produced per unit time by an individual of age x.

In human demography, these parameters are strictly applied only to the female component of the population. Their meanings can easily be modified to apply to an entire population. In paleontology, it is often convenient to express survivorship in terms of the "age-specific mortality rate" $d_x = -dl_x/dx$ in the continuous case, $d_x = l_x = l_x - l_{x+1}$ in the discrete case. For further review, see Deevey (1947), Andrewartha and Birch (1954), Pielou (1969), or Poole (1974).

Survivorship curves (plots of l_x against x) can be loosely grouped into three types, as shown in Fig. 1 (Deevey, 1947):

Type I. Probability of survival decreases with age, as in man and certain insects.

Type II. Probability of survival is constant with age, as in certain birds. (This condition may apply to particular stages in a life cycle.)

Type III. Probability of survival increases with age, as in certain barnacles and probably as in many other marine invertebrates in which females or hermaphrodites produce thousands to millions of eggs.

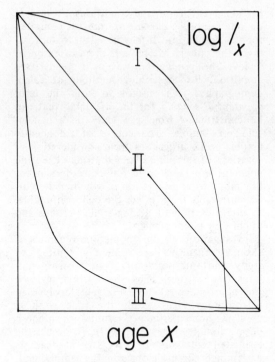

FIGURE 1. The three basic types of survivorship curve. Semilogarithmic plots of survival rate (l_x) as a function of age x. Probability of survival may decrease (Type I), remain constant (Type II), or increase (Type III) with age.

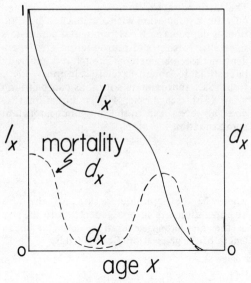

FIGURE 2. A plot illustrating the relationship between the survivorship curve (l_x) and the mortality rate curve (d_x), which is the slope of the survivorship curve times minus one.

Though it is difficult to generalize from the relatively very few cases studied, vertebrates and insects, which generally produce relatively few eggs or young at a time, would seem to commonly fall in Types I and II. Marine invertebrates, organisms that generally produce relatively many eggs or young at a time, would seem to commonly fall in Types II and III. The relationship between a Type I survivorship curve and the derived mortality rate curve (d_x) is illustrated in Fig. 2. An age-specific fecundity rate (m_x) curve of the form typically associated with Type I survivorship curves is shown in Fig. 3.

The fundamental relationship between the intrinsic rate of increase r_m and survival and fecundity rates is

$$\int_0^\infty e^{-r_m x} l_x m_x \, dx = 1 \qquad (5a)$$

$$\sum_{x=0}^\infty e^{-r_m x} l_x m_x \approx 1 \qquad (5b)$$

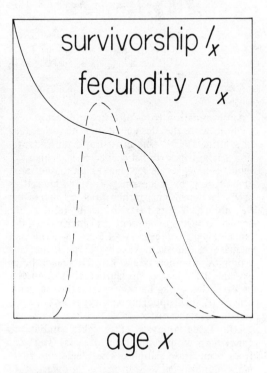

FIGURE 3. A plot illustrating the type of fecundity rate curve (m_x) typically associated with Type III survivorship curves (l_x).

The intrinsic rate of increase, properly defined only for a population with a stable age distribution, is dependent on environmental parameters since survivorship and reproduction are dependent on those parameters. Though it is generally impossible to obtain explicit solutions for r_m from (5a), approximate solutions can be found using (5b) (see Leslie, 1945). Equation (5a) does, however, lead to the rough but important approximation

$$r_m \approx \frac{\ln R_0}{T} \qquad (6)$$

where the "net reproductive rate" R_0, the ratio of population size in one generation to the size in the preceeding generation, and the mean length of the generation T are given by

$$R_0 = \int_0^\infty l_x m_x \, dx \qquad (7a)$$

$$\approx \sum_{x=0}^\infty l_x m_x \qquad (7b)$$

$$T \approx \frac{1}{R_0} \int_0^\infty x l_x m_x \, dx \qquad (8a)$$

$$\approx \frac{1}{R_0} \sum_{x=0}^\infty x l_x m_x \qquad (8b)$$

Approximation (6) is of much interest in connection with the design of life histories (see Lewontin, 1965). One most important adaptation for an opportunistic species is a high r_m value. For $R_0 > 1$, r_m can be increased most dramatically by decreasing T, on which it is inversely dependent, rather than by increasing R_0, on which it is logarithmically (and hence not so strongly) dependent. In other words, if a female lays a few eggs at an early age, this can increase r_m more drastically than if she lays prodigious numbers late in life. This result helps explain the selective pressures behind neoteny in evolution, and helps explain the neotenic origins of such opportunistic types as copepod and cladoceran crustaceans.

Life Table Analysis. Life table studies are concerned with estimation of vital statistics from population samples. The ideal *cohort life table* applies to a *stationary population* (a population that has both a stable age distribution and a constant size through time) under particular environmental conditions. From data on one sample, a *current life table* is compiled for the population. It may be taken to approximate the cohort life table if there is reason to believe the population was in a steady-state condition. For describing methods of paleodemography, it is useful to coin the term *apparent life table* for the life table that can be constructed from raw data on a paleontological sample. The course of paleodemographic work progresses from consideration of a variety of sampling biases that may keep the apparent life table from reflecting the current life table, to consideration of whether the current life table may reflect the cohort life table, to interpretation of demographic patterns and life history phenomena.

Investigations of the population dynamics of extinct organisms are generally limited to study and interpretation of survivorship patterns (see Kurtén, 1964, for further review). The approaches taken to life table analysis in paleontology are basically the same as those used in nonexperimental studies of modern populations (see Deevey, 1947). To obtain the estimated survivorship curve, age at death is determined for individuals in the sample, and, based on the taphonomy of the assemblage, those data are treated as appropriate to the nature of the sample in calculating apparent l_x values.

The two fundamental problems that must be solved before undertaking a paleontological life table study are (1) the nature of the sample as determined from study of the taphonomy of the fossil assemblage, which is the key to analysis of the raw data on age at death, and (2) a means of determining the age of individuals. Two types of fossil assemblage lend themselves to life table studies: *census assemblages,* representing a natural census of organisms that were killed and buried at one time, and *time-averaged assemblages,* representing organisms that died over some period of time, preferably much longer than the mean length of the generation. Unfortunately, most fossil assemblages seem to represent variously biased mixtures of these two types (Johnson, 1960, 1964; Fagerstrom, 1964). The problem is to determine the manner of accumulation, preservation, and perhaps collection of the specimens to be studied, in order to identify the biases on this sample and ascertain whether a life table study is feasible. In mixed assemblages, it may be possible to sort out remains or organisms killed in the burial event from remains of organisms that died in prior time. Arthropods present a special problem in that they may be represented by remains of both actual organisms and molted exoskeletal parts. It is an exceptional fossil assemblage that readily lends itself to life table studies.

Age at death can be determined in several ways. Periodic growth rings in mollusk shells, echinoderm plates, fish scales and otoliths, and mammal teeth and horns can be used to determine age directly (Deevey, 1947; Kurtén, 1964). For vertebrates, tooth replacement and tooth wear can also be used in age determination (Kurtén, 1953). For seasonally reproducing organisms in census assemblages, the length of the period between successive generations can sometimes be distinguished as time units in size-frequency distributions (see Craig and Hallam, 1963; Hallam, 1967). If the reproductive period is short enough, and if the growth rate is high enough and not too variable, age classes may appear as distinct peaks in the size-frequency distribution. However, the method may break down for older individuals as a result of decelerated growth rate. For many seasonally reproducing organisms, reproduction takes place once a year, or alternatively twice a year, and not necessarily at even intervals. To precisely determine age, it is also necessary to determine at what phase of the reproductive cycle the census event occurred. For arthropods, particularly ostracod crustaceans (e.g., Kurtén, 1953), age is sometimes measured in terms of molt stages (instars). However, this scale is, in general, markedly nonlinear over the life cycle as the rate of molting tends to decrease with age, among other complications (Passano, 1961).

Once problems of sample type, age determination, and—for census assemblages—census time are resolved, construction of an apparent survivorship curve becomes a simple matter. For a time-averaged assemblage, a frequency plot of age at death represents a mortality rate (d_x) curve. The apparent survivorship (l_x) curve is obtained by summation according to the relationships:

$$l_x = \int_0^x d_x \, dx \qquad (9a)$$

$$\approx \sum_{x=0}^x d_x \qquad (9b)$$

For a census assemblage of a population that reproduced continually at a constant rate, a frequency plot of age at death represents an apparent survivorship curve. Data from a census assemblage of a population with markedly seasonal reproduction must be treated in a slightly different manner. If the laying of eggs or the birth of young were restricted to a relatively very short period, an apparent survivorship curve can be constructed by fitting a curve to a plot of number of individuals versus age for each age class. A frequency plot of age at death for a population that reproduces continually but at a seasonally fluctuating rate should show less-marked peaks and troughs. An apparent survivorship curve can be obtained by taking a moving average of a number of individuals over a moving age interval of a length appropriate for making the apparent survival rate a monotonically decreasing function of age.

A survivorship curve that does not decrease monotonically with age is an anomaly. It may result from one or another bias on the collection, or it may result from a high or markedly fluctuating immigration rate of individuals from other populations or a fluctuating emigration rate. In the last case, the apparent survivorship curve can be discarded as representing a current survivorship curve. Construction of current and cohort life tables rests on the assumption that the population studied exists in a habitat that forms a closed system, or at least that immigration and emigration rates are constant and more or less equal. This assumption may be violated in all sorts of natural migration, dispersal, and colonization phenomena. For a time-averaged assemblage, a high and constant immigration rate for older individuals could lead to a fluctuating apparent survivorship curve, while fluctuating immigration or emigration rates could lead to a fluctuating apparent survivorship curve for a census assemblage. For the many marine invertebrates that have pelagic larval stages and benthic adult stages, or have juvenile and adult stages living in somewhat different habitats, there is a very good reason why fossil assemblages and the survivorship curves surmised from them may represent only a part of the life history. Quite apart from differential preservation, only certain stages may have been represented in one type of depositional environment.

Aside from these problems, age-specific bias relating to size in the accumulation and preservation of the assemblage and in collection of the sample is the main problem in obtaining a current life table from statistics in the apparent life table. Paleontological collections often show a size-dependent bias. The hard parts of young individuals much smaller than adults may be intrinsically less preservable, and may be less readily collected and identified, than corresponding hard parts of adults. Arthropods present a special case in that the molted shells (exuvia) left behind in the growth of an individual are not only smaller than the shell occupied by the animal at a given time, but also may

633

be in general less preservable owing to resorption of mineral material from the shell prior to molting. It is necessary to distinguish remains of actual animals from molted parts in getting an approximation to the current life table. In a perfectly preserved, time-averaged assemblage, data on actual arthropod animals would give a mortality rate curve, and data on exuvia would give a survivorship curve.

Owing to a differential nonrepresentation of very young stages, only an incomplete life table commencing at some later age could be constructed in many instances from paleontological data, though early survivorship might be guessed at by extrapolation.

For a time-averaged assemblage that accumulated over a relatively very long period of time without other biases, it is probably reasonable to assume that the apparent life table approximates a cohort life table, though it may represent various cohort life tables as averaged over changing environmental conditions. For a census assemblage, deciding whether the current life table represents a cohort life table may ultimately come down to making an educated guess.

Populations that had a Type II survivorship curve lend themselves to analysis from all types of fossil assemblage, provided there is no age-dependent bias. The Type II survivorship curve represents an exponential distribution of the form $l_x = e^{-kx}$, where the constant k is the slope of the line in a semilogarthmic plot like that in Fig. 1. Owing to the relationship $d_x = -dl_x/dx = ke^{-kx}$, the same sort of plot for d_x curve (the age-frequency distribution in a time-averaged assemblage) would show exactly the same slope as the l_x curve (the age-frequency distribution in a census assemblage). Only the intercepts for $l_0 = 1$ and $d_0 = k$ are different. Moreover, for a mixed assemblage with a "census" component $l_x = a \cdot e^{-kx}$ and a "time-averaged" component $d_x = b \cdot ke^{-kx}$, the sum $l_x + d_x$ obtained through totally ignoring taphonomy should show the same slope k in a semilogarithmic plot provided that a and b are not functions of age x. This is a most convenient result. Under somewhat idealized circumstances, the parameter k determining survival rate l_x can be determined from any type of assemblage independent of taphonomy so long as it has no age-dependent bias over the age range of interest. Kurtén (1953) noted this effect in analyzing a sample of ostracods with a more or less exponential age distribution.

Once what is taken to be a cohort life table for survival rate has been obtained, other vital statistics can sometimes be roughly estimated. By definition, a cohort life table applies to a population for which $R_0 = 1$. Equation (7) is satisfied, and the fecundity rate (m_x) distributions satisfying it remain to be determined. Anatomical clues as to which individuals in the population were reproductively mature may assist in narrowing down the possible distributions to reasonable approximations. In the record of shell growth, some mollusks give a record of the timing of reproductive periods (Rhoads and Pannella, 1970). Analogies with living relatives of the fossil organisms might also be used in estimating the fecundity rate distribution. Taken as a working hypothesis, the assumption that the apparent survivorship curve represents something approaching a cohort survivorship curve could be used in making rough, order-of-magnitude estimates of fecundity rates necessary for the population's maintenance of a steady-state condition.

Studies of the population dynamics of fossil organisms have pursued only some of the potential lines of work here sketched out. A few examples suffice to illustrate the information obtainable on ecology and evolution. Kurtén (1957, 1958) analyzed selection and selective pressures on bear populations through life-table studies, demonstrating instances of Darwinian selection and a biogeographically interesting case of selection for maintenance of an abnormal trait in a peripheral population.

Levinton and Bambach (1970) found that in soft-bottom communities, similar survivorship patterns in modern and fossil deposit-feeding bivalves can be correlated with similar modes of life. They found that a modern suspension-feeding bivalve living in the same type of environment had a higher juvenile mortality rate. This species thus might appear to be a numerical dominant in a time-averaged death assemblage even though these bivalves formed a minor component of the living community at most times. Snyder and Bretsky (1971) explained bivalve distribution patterns in terms of population dynamics by relating different survivorship patterns to differing depositional environments (see also Richards and Bambach, 1975, for a study of Paleozoic brachiopod population dynamics and depositional environments). From study of an exceptional census assemblage in which the trilobite's larvae were represented, Cisne (1973) found that very young *Triarthrus eatoni* had a relatively long pelagic life before becoming nektobenthic. This helps explain the cosmopolitan distribution of the particular genus (see Thorson, 1961) (see *Larvae of Marine Invertebrates—Paleontological Significance*).

J. L. CISNE

References

Andrewartha, H. G., and Birch, L. C., 1954. *The Distribution and Abundance of Animals.* Chicago: Univ. Chicago Press, 782p.

Cisne, J. L., 1973. Life history of the Ordovician trilobite *Triarthrus eatoni, Ecology,* 54, 135–142.

Craig, G. Y., and Hallam, A., 1963. Size-frequency and growth-ring analyses of *Mytilus edulis* and *Cardium edule,* and their palaeoecological significance, *Palaeontology,* 6, 731–750.

Deevey, E. S., 1947. Life tables of natural populations of animals, *Quart. Rev. Biol.,* 22, 283–314.

Fagerstrom, J. A., 1964. Fossil communities in paleoecology: Their recognition and significance, *Geol. Soc. Amer. Bull.,* 75, 1197–1216.

Goel, N. S., and Richter-Dyn, N., 1974. *Stochastic Models in Biology.* New York: Academic Press, 269p.

Goel, N. S.; Maitra, S. C.; and Montroll, E. W., 1971. *On the Volterra and Other Nonlinear Models of Interacting Populations.* New York: Academic Press, 164p.

Hallam, A., 1967. Interpretation of size-frequency distributions in molluscan death assemblages, *Palaeontology,* 10, 25–42.

Hutchinson, G. E., 1959. Homage to Santa Rosalia or Why are there so many kinds of animals? *Amer. Naturalist,* 93, 145–159.

Johnson, R. G., 1960. Models and methods for the analysis of the mode of formation of fossil assemblages, *Geol. Soc. Amer. Bull.,* 71, 1075–1086.

Johnson, R. G., 1964. The community approach to paleoecology, in J. Imbrie and N. D. Newell, eds., *Approaches to Paleoecology.* New York: Wiley, 107–134.

Kurtén, B., 1953. On the variation and population dynamics of fossil and recent mammal populations, *Acta Zool. Fennica,* 76, 1–122.

Kurtén, B., 1957. A case of Darwinian selection in bears, *Evolution,* 11, 412–416.

Kurtén, B., 1958. Life and death of the Pleistocene cave bear, a study in paleoecology, *Acta Zool. Fennica,* 95, 1–59.

Kurtén, B., 1964. Population structure in paleoecology, in J. Imbrie and N. D. Newell, eds., *Approaches to Paleoecology.* New York: Wiley, 91–106.

Leslie, P. H., 1945. The use of matrices in certain population mathematics, *Biometrika,* 35, 213–245.

Levin, S. A., 1974. Dispersion and population interactions, *Amer. Naturalist,* 108, 207–228.

Levinton, J. S., and Bambach, R. K., 1970. Some ecological aspects of bivalve mortality patterns, *Amer. J. Sci.,* 268, 97–112.

Lewontin, R. C., 1965. Selection for colonizing ability, in H. G. Baker and G. L. Stebbins, *The Genetics of Colonizing Species.* New York: Academic Press, 77–91.

MacArthur, R. H., 1955. Fluctuations of animal populations, and a measure of community stability, *Ecology,* 36, 533–536.

MacArthur, R. H., 1960. On the relative abundance of species, *Amer. Naturalist,* 94, 25–36.

MacArthur, R. H., and Wilson, E. O., 1967. *Theory of Island Biogeography.* Princeton: Princeton Univ. Press, 203p.

May, R. M., 1973. *Stability and Complexity in Model Ecosystems.* Princeton: Princeton Univ. Press, 265p.

May, R. M., 1974. Biological populations with non-overlapping generations: Stable points, stable cycles, and chaos, *Science,* 186, 645–647.

May, R. M., ed., 1976. *Theoretical Ecology: Principles and Applications.* Philadelphia: Saunders, 317p.

Passano, L. M., 1960. Molting and its control, in T. H. Waterman, ed. *The Physiology of Crustacea,* vol. 1. New York: Academic Press, 473–536.

Pielou, E. C., 1969. *An Introduction to Mathematical Ecology.* New York: Wiley-Interscience, 268p.

Poole, R. W., 1974. *An Introduction to Quantitative Ecology.* New York: McGraw-Hill, 532p.

Preston, F. W., 1948. The commonness, and rarity, of species, *Ecology,* 29, 254–283.

Preston, F. W., 1962. The canonical distribution of commonness and rarity, *Ecology,* 43, 185–215, 410–432.

Rhoads, D. C., and Lutz, R. A., eds, 1979. *Skeletal Growth: Biological Records of Environmental Change.* New York: Plenum.

Rhoads, D. C., and Pannella, G., 1970. The use of molluscan shell growth patterns in ecology and paleoecology, *Lethaia,* 3, 143–161.

Richards, R. P., and Bambach, R. K., 1975. Population dynamics of some Paleozoic brachiopods and their paleoecological significance, *J. Paleontology,* 49, 775–798.

Snyder, J., and Bretsky, P. W., 1971. Life habits of diminutive bivalve molluscs in the Maquoketa Formation (Upper Ordovician), *Amer. J. Sci.,* 271, 227–251.

Thorson, G., 1961. Length of pelagic life in marine bottom invertebrates as related to larval transport by ocean currents, in M. Sears, ed., *Oceanography.* Washington D. C.: Amer. Assoc. Adv. Sci. Publ. 67, 455–474.

Cross-references: *Actualistic Paleontology; Biometrics in Paleontology; Communities, Ancient; Diversity; Paleoecology.*

PORIFERA

Porifera (the sponges) is a phylum of simple multicellular animals, sometimes placed in the separate subkingdom Parazoa. They are aquatic organisms, drawing water into the body for feeding and respiration; water circulation is maintained by flagellated cells (choanocytes, or collar cells) which resemble individual choanoflagellate Protozoa. Most sponges have an internal skeleton, which may consist of (a) calcareous spicules; (b) siliceous spicules; (c) siliceous spicules and organic matter (spongin) cementing spicules together, or forming fibers in which spicules are imbedded; or (d) spongin fibers only. A few have calcareous nonspicular structures in addition to spicules, and a few have no skeleton. Sponges are found in deposits of all ages throughout the Phanerozoic, and may occur in Precambrian rocks as well.

Structure

Water Circulation. The simplest larval sponges are saccular, with a fixed base and an open top (Fig. 1-1). Water is drawn in through pores in the external surface, enters a central cavity (paragaster or spongocoel), and leaves through its terminal opening (osculum). The choanocytes line the central cavity. Some adult sponges, termed ascons, are similar except for forming branching or branching and anastomosing tubes. In sycons (Fig. 1-2), choanocytes are restricted to hollow radial flagellated chambers, which may be separate, or fused together with tubular inhalant canals (epirhyses, prosochetes) left between groups of chambers, for the ingress of water. Leucons (Fig. 1-3), which are commonest, have further exhalant canals (aporhyses, apochetes) through which numbers of chambers discharge. The name "rhagon," sometimes misused as a substitute for leucon, applies properly to larvae with flagellated chambers but no canals, from which some leucons develop. Adult leucons may have no central cavity, and openings of exhalant canals may then form multiple oscules.

Spicular Skeleton. Calcareous spicules are formed from crystalline calcite, and siliceous types from opaline silica. Calcareous spicules are needle-like, or stellate with three or four radiating needle-like rays. Siliceous spicules are needle-like, stellate with three to many rays, or of special symmetrical or irregular shapes (cf. Figs. 2-9). Names given to spicules may be

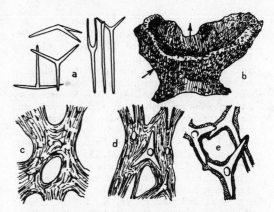

FIGURE 2. Pharetronid Calcarea (from Moret, 1952). a, isolated spicules; b, general appearance of *Pachytilodia* (Cretaceous); c-e, skeletal fibers magnified, showing imbedded spicules and cementing matter.

based on the number of rays (e.g., tetractin, with four rays), on symmetry (e.g., monaxon, for needle-like spicules; tetraxon, for tetractins with rays arranged like tetrahedral axes), or other characters (e.g., calthrops, for tetractins shaped like a calthrops; desmas, arbitrary name for articulated spicules). Some names refer to function, e.g., basalia for spicules protruded for basal attachment.

Many sponges with siliceous spicules show differentiation of larger and smaller types, classed as megascleres and microscleres respectively. These are often also of different types morphologically, and some morphological types are only known as microscleres. In monaxial and stellate types of megascleres, silica is secreted concentrically around organic axial filaments, which grow terminally until growth of the spicule is almost complete. After death, axial filaments are replaced by hollow axial canals. In other types, notably in desmas, axial filaments cease to grow during early ontogeny of spicules, and most of the silica secreted forms anaxial outgrowths of ray-like or other shapes. Some microscleres show axial structures, but many microscleres and some megascleres appear to be entirely anaxial.

Sponges with calcareous spicules (Class Calcarea; see below) may have some of the spicules united by a secondary calcareous cement, to form rigid skeletal frameworks. Some with siliceous spicules form similar rigid frameworks, in which the main internal megascleres are united by articulation (desmas of lithistid Demospongia) or fusion (in dictyonine Hexactinellida, some lyssacines). Most fossils showing the form of the body have rigid skeletal frameworks; but forms with loose spicules, which are normally dispersed after death, have

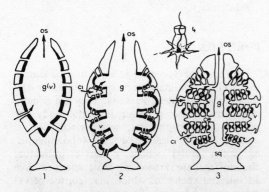

FIGURE 1. Sponge types in diagrammatic longitudinal sections (from Moret, 1952). 1, Ascon; 2, sycon; and 3, leucon. Choanocytes, 4, cover the internal surfaces shown in thick black. Arrows show direction of flow; g, paragaster (cloaca, spongocoel); v, flagellated chambers; ci, inhalant canals; ce, exhalant canals; os, osculum.

sometimes been preserved in sediments deposited under quiet conditions.

Sizes of spicules are difficult to summarize; but most stellate types found internally have rays between about 2 μm to 2 cm long, and the size range of other types is similar. Needle-like megascleres and rays of stellate forms are mostly more than about 0.1 mm long. Megascleres protruded as basalia may, however, reach abnormal sizes through continuing growth of one ray, which is imbedded in the body. In Class Hexactinellida, such spicules may reach a length of 20–50 cm; and in one case (living *Monoraphis*) a single giant basal needle attains a length of over 2m and is 1 cm thick.

Microscleres are rare in fossil sponges, but are sometimes preserved loose in sediments (e.g., in the upper Cretaceous of Germany and the upper Eocene of New Zealand).

Nonspicular Mineral Structures. Living *Murrayona* and *Petrobiona* of Class Calcarea have rigid calcitic skeletal frameworks, formed by union of small spherulitic bodies (sclerodermites), in addition to loose calcareous spicules. Several genera of class Demospongia, with siliceous spicules, form similar aragonitic structures, which may more or less resemble the skeletons of fossil Stromatoporoidea, Chaetetida or Tabulata.

Classification

The Porifera comprises four classes, three of which are extant.

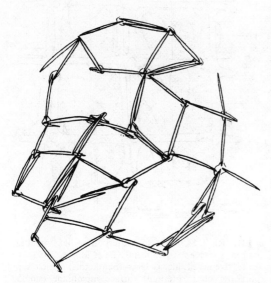

FIGURE 3. Skeleton of a monaxonid demosponge, in which a net-like structure consists of spicules (oxeas) bound together at their tips by spongin (from Hyman, 1940, copyright © 1940 by the McGraw-Hill Book Company).

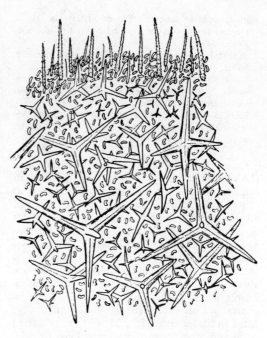

FIGURE 4. Skeleton of a simple choristid demosponge (from Hyman, 1940, courtesy Dr. M. W. de Laubenfels). The internal megascleres are tetraxons (calthrops), but spiny monaxons (acanthostyles) are present at the surface. Note also the microscleres (microstrongyles).

Class Calcarea. The Calcarea (Mississippian–Holocene) have calcareous spicules; some have rigid spicular or nonspicular frameworks; none have spongin. Living forms have gelatinous internal mesenchyme, which is covered externally by epithelioid pinacocytes except where choanocytes are present. Nearly all fossil forms have rigid skeletal frameworks [Spinctozoica (= Thalamida), Pharetronida; Fig. 2].

Class Demospongia. Members of the class Demospongia (Cambrian–Holocene) have soft parts like those of Calcarea, but have siliceous spicules, alone or accompanied by spongin, or a spongin skeleton only, or no skeleton. The spicules are needle-like to many-rayed (polyactinal) or of special or irregular shapes, and are usually divided into megascleres and microscleres or consist of megascleres only; tetractin spicules are typically tetraxons; most megascleres are monaxons, tetractins, or of types that are partly or wholly anaxial. Some megascleres have three rays or more than four, but not normally more than six. A few genera have aragonitic structures in addition to spicules, or spicules of types that are normally microscleres only. Three main types of megascleric skeletons occur in modern and fossil genera:

1. monaxonid (= monactinellid of Zittel; Fig. 3); all megascleres monaxons, not articulated (Cambrian–Holocene).
2. choristid (= tetractinellid of Zittel, *not* Marshall, Sollas; Figs. 4,5): with tetractin megascleres, or rarely three-rayed triactins, and usually also monaxons, not articulated (Upper Ordovician; Lower Carboniferous–Holocene).
3. lithistid (Fig. 6): main internal megascleres are articulated desmas, formed mainly or wholly by anaxial secretion (Ordovician–Holocene).

Class Hexactinellida. The Hexactinellida (Lower Cambrian–Holocene) have siliceous spicules with six or fewer rays, arranged as though following the axes of a cube; tetractins are always of this type, never tetraxons. There is no spongin. Modern adults have chambers suspended in three-dimensional networks of syncytial filaments (trabeculae), water-filled interspaces, and no mesenchyme. Spicules are always divided into megascleres and microscleres in modern forms. The internal megascleres are six-rayed (hexactinal) and always fused in dictyonines (Fig. 7), and are of varied types and often loose in lyssacines (Fig. 8).

Class Heteractinellida. Spicules of unknown composition, which are mainly polyactinal, and of megascleric sizes, are the only known characters of the Heteractinellida (Lower Cambrian–Middle Triassic; Fig. 9). Spicules may be (a) "umbrella-spiclules" or polyaenes, with two to nine rays arranged like spokes at the end of a

FIGURE 6. Skeleton of a lithistid demosponge, *Iouea* de Laubenfels–Cretaceous (from Moret, 1952). a, nonarticulated megascleres (dichotriaenes), which support the external (dermal) tissues; b, the rigid internal skeleton, consisting of articulated desmas (dicranoclones); c, small accessory desmas, which supplement the loose dermal megascleres in this genus.

shaft formed by a further ray; (b) mesopolyaenes, with three- to six-spoke rays emitted from the center of a two-rayed shaft; (c) star or flower-like polyactins, or neoasters; (d) forms intermediate between polyaenes and neoasters; or (e) like polyaenes or mesopolyaenes but lacking the shaft rays. This class was united with the Hexactinellida by de Laubenfels (1955), as Hyalospongea; but the spicules are unlike those of Hexactinellida, resemble various demosponge spicules, and are calcareous according to Finks (1960).

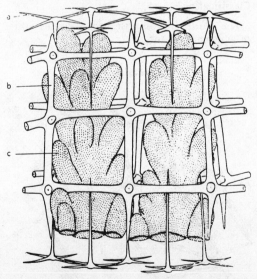

FIGURE 7. Skeleton of a dictyonine hexactinellid, *Farrea*–Cretaceous–Holocene (from Moret, 1952). a, loose five-rayed megascleres (pentactins) which support the outer (dermal) surface membrane (similar spicules support the inner, gastral, surface at the bottom); b, lattice-like dictyonal framework, consisting of fused six-rayed spicules (dictyonalia); c, the flagellated chambers that lie within the meshes of the framework.

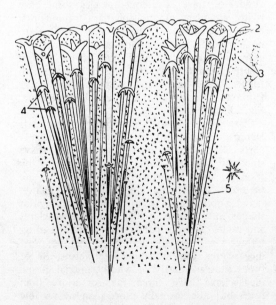

FIGURE 5. Skeleton of a more advanced choristid, in which the megascleres are tetraxons developed as two types of triaenes (2, 4) with a radial arrangement; also showing two types of microscleres (3, 5) (from Hyman, 1940, courtesy Dr. M. W. de Laubenfels).

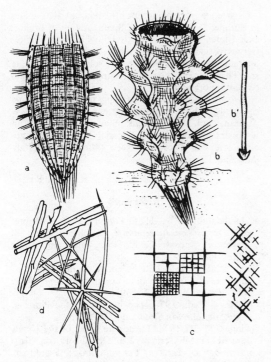

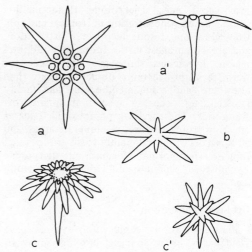

FIGURE 9. Spicules of Heteractinids. a,a', umbrella-spicule (polyaene) of *Chancelloria*—Cambrian (a, from above; a', vertical section); the sutured rays and pits on ray bases are features of the genus. b, eight-rayed meso-polyaene of *Astraeospongia* (Devonian). c,c', polyacins (neoasters) of *Asteractinella* (Carboniferous).

FIGURE 8. Lyssacine Hexactinellida (from Moret, 1952). a, *Dictyospongia* (Devonian); *Hydnoceras* (Devonian-Carboniferous); b', a grapnel-like spicule (anchorate pentactin) from the root-tuft (b, bottom); c, arrangements of cross-shaped four-rayed spicules (stauractins) in the skeletons of *Protospongia* (Cambrian–Ordovician; left) and *Diagonella* (Ordovician; right); d, six-rayed and needle-like spicules (hexactins, rhabdodiactins) of *Proeuplectella* (Cretaceous), with some of the spicules connected by secondary synapticula.

A fifth class, Sclerospongiae was proposed by W. D. Hartman and T. F. Goreau (*in* Fry, 1970) for living forms with aragonitic structures (here regarded as Demospongia) and the fossil Chaetetida and Stromatoporoidea, supposed to be still represented by living *Ceratoporella* and *Astrosclera*; but the modern forms do not resemble the fossils completely, and are normal monaxonid Demospongia apart from their secretion of aragonite. Subclass status within Class Desmospongia hence seems more appropriate (e.g., Vacelet, Vasseur, and Lévi, 1976).

Ecology and Paleoecology

Modern sponges are mainly marine, but one family (Spongillidae; monaxonid Demospongia) occurs in fresh water. Calcarea range from shorelines to deep water (deepest 2195 m), but are most common at depths of <100 m. Marine Demospongia range from shorelines

to abyssal. Dictyonine Hexactinellida live mainly on the continental slopes, and lyssacines mainly on the continental slopes and deep ocean floors. Few of either are known from a depth of <200 m, and almost none from <90 m, except for lyssacines living under ice off Antarctica.

Facies distributions of fossil Calcarea and Hexactinellida, consistent with modern depth habits, have been noted by various writers (e.g., Finks, 1960); but one Paleozoic lyssacine is known from a shale above a coal seam. Many fossil Calcarea are forms with rigid frameworks, of a group not known later than Upper Cretaceous (elasmostomatid Pharetronida); but many occur as associates with calcareous algae or reef-corals, implying shallow-water habits, and a few were frame builders in (Permian) algal or (Triassic) coral reefs (Rigby, 1971). Examples from reef crests of Permian reefs of Texas may have lived at barely subtidal depths. Much in contrast, several genera of the extant minchinellid Pharetronida occur with Hexactinellida in the Chalk (Upper Cretaceous) of Europe; but related modern forms are photophobic, and today *Petrostoma* is known only between 195–392 m.

Modern sponges occur at all latitudes, but show varying distributions. Demospongia with spongin skeletons (Keratosa) occur mainly in warm seas; but monaxonids are as common off Antarctica as in warm regions. Lithistid Demospongia are most numerous in warm seas, but reach cold ones. Dictyonine Hexactinellida show marked concentration in tropical and subtropical regions, despite deep-water distribution, and are commonest in the IndonesiaPhillipines region. Lyssacines show less preference for warm seas, though more numerous there than elsewhere, and some members of one family (Rossellidae) occur in special abundance off Antarctica. The surviving pharetronid Calcarea are from tropical to warm-temperate areas, and all survivors of several mainly fossil groups (e.g., lithistid Helomorina, Megamorina; dictyonine Lychniscosa; Sphinctozoid Calcarea) are tropical.

Geologic Distribution

Many formations, some whole sytems, have yielded few or no sponges, but individuals or species are sporadically numerous in others. Hexactinellida and lithistid Demospongia predominate numerically, but pharetronid and sphinctozoid Calcarea are conspicuous in some (Pennsylvanian–Cretaceous) faunas. Paleozoic Hexactinellida are all lyssacines (mainly Dictyospongiidae), but nearly all Mesozoic examples are dictyonines. Mesozoic rarity of lyssacines, which now outnumber dictyonines, may mark restriction of most to deep oceans, since a few of diverse modern types occur in Chalk faunas. Most North American fossil sponges are Paleozoic, those from Europe mainly Upper Jurassic or Cretaceous. Cenozoic occurrences are rare. Main occurences are outlined below.

Cambrian: Early monaxonids, lyssacines, and heteractinellids in Burgess Shale, British Columbia.
Ordovician, Silurian: early lithistids (Anthaspidellidae, Astylospongiidae, Hindiidae) numerous in some North American formations (especially Chazyan, Niagaran).
Devonian: large lyssacine (dictyospongiid) fauna in Chemung Formation, New York).
Carboniferous, Permian: many drifted siliceous spicules in some sediments in Britain, Ireland (Mississippian), Indiana, and Illinois (Pennsylvanian), representing mainly choristids and lyssacines; various lithistid, lyassacine, and heteractinellid genera in Carboniferous of Ireland, Britain, Belgium, and USSR; many Calcarea, lithistids, and lyssacines in the Texas region (Pennsylvanian-Permian); mainly lithistid fauna in Timor (Permian). Last anthaspidellids, hindiids, and dictyspongiids.
Triassic: Calcarea in Alpine Trias and Nevada; few early dictyonines (Germany, Hungary); last heteractinellids.
Jurassic: many lithistids and dictyonines in Upper Jurassic of SE France, Germany, Switzerland, and Poland; pharetronid Calcarea numerous in some shelf and reef limestones of Middle and Upper Jurassic of western Europe (Gwinner, 1971).

Cretaceous: very numerous lithistids and dictyonines in Europe, mainly Aptian of Spain; Albian of France; and Upper Cretaceous of Ireland, Britain, France, Germany, Czechoslovakia, Poland, Sweden, and USSR. Sporadic shallow-water Calcarea, with Sphinctozoida and Pharetronida. Few dictyonines in Texas Albian (Washita, Fort Worth).
Cenozoic: loose spicules of many monaxonids, some choristids, lithistids, and lyssacines in Eocene of W Australia and New Zealand; small lithistid faunas in W Australia (Eocene) and Algeria (Miocene); dictyonines rare, but a few in Algeria (Miocene) and Italy (Miocene, Pliocene).

R. E. H. REID

References

de Laubenfels, M. W., 1955. Porifera, in R. C. Moore, ed., *Treatise on Invertebrate Paleontology*, pt. E. Lawrence, Kansas: Geol. Soc. Amer. and Univ. Kansas Press, 21-122.

Finks, R. M., 1960. Late Paleozoic sponge faunas of the Texas region. The siliceous sponges, *Amer. Mus. Natl. Hist. Bull., 120*, 1-160.

Fry, W. G., ed., 1970. The biology of the Porifera. *Zool. Soc. London Symp.*, no. 25, 512p.

Gwinner, M. P., 1971. Carbonate rocks of the Upper Jurassic in SW-Germany, in G. Müller, ed., *Sedimentology of Parts of Central Europe*. Frankfurt: Verlag Waldemar Kramer, 193-207.

Hartman, W. D., 1977. Sponges as reef builders and shapers, *Amer. Assoc. Petrol. Geol. Stud. Geol., 4*, 127-134.

Hyman, L., 1940. *The Invertebrates: Protozoa Through Ctenophora*. New York: McGraw-Hill, 726p.

Lagneau-Hérenger, L., 1962. Contribution à l'étude des Spongaires siliceux du Crétacé inferieur, *Soc. géol. France Mém.*, n.s., *95*, 252p.

Moret, L., 1952. Spongiaires, in J. Piveteau, ed., *Traité de Paléontologie*, vol. I. Paris: Masson et Cie, 333-374.

O'Connel, M., 1919. The Schrammen collection of Cretaceous Silicispongiae in American Museum of Natural History, *Amer. Mus. Natl. Hist. Bull., 41*, 1-261.

Reid, R. E. H., 1968. Bathymetric distributions of Calcarea and Hexactinellida in the present and the past, *Geol. Mag., 105*, 546-559.

Rezvoi, P. D.; Zhuravleva, I. T.; and Koltun, V. M., 1962 (transl. 1971). Phylum Proifera in Yu. A. Orlov, ed., *Osnovy Paleontologii (Fundamentals of Paleontology): Porifera, Archaeocyatha, Coelenterata, Vermes*, Jerusalem: Israel Program for Scientific Translations, 15-83.

Rigby, J. K., 1971. Sponges and reef-related facies through time, *Proc. N. Amer. Paleontological Conv.*, pt. J, 1374-1388.

Seilacher, A., 1961. Die Sphinctozoa, eine Gruppe fossiler Kalkschwamme, *Akad. Wiss. Lit. Mainz, Abh. Math.-Nat. Kl., Jg., 1961*, 723-790.

Simpson, T. L., 1973. Coloniality among the Porifera, in R. S. Boardman, A. H. Cheetham, and W. A. Oliver, Jr., eds., *Animal Colonies*. Stroudsburg, Pa.: Dowden, Hutchinson & Ross, 549-565.

Vacelet, J.; Vasseur, P.; and Lévi, C., 1976. Spongiares

de la pente externe des récifs coralliens de Tulear (sud-ouest de Madagascar), *Mém. Mus. Nat. d'Hist. Nat., N.S., A, Zool.,* 49, 1-116.

Wiedenmayer, F., 1978. Modern sponge bioherms of the Great Bahama Bank, *Eclogae Geol. Helv.,* 71, 699-744.

Cross-references. *Coloniality; Reefs and Other Carbonate Buildups; Stromatoporoids.* Vol. IVA: *Silica-Biogeochemical Cycle.* Vol. VI: *Sponges in Sediments; Spongolite.*

PRECAMBRIAN LIFE

Perhaps foremost among recent advances in paleontology has been the discovery that the fossil record of the earliest, most primitive organisms to have inhabited the earth—organisms that lived during the Precambrian eon, the earliest seven-eighths of geologic time—is far longer and far more impressive than had previously been imagined. Studies of these Precambrian fossils have begun to fill a major gap in understanding of biologic history. For the first time, there is reason to believe that the tree of life can be traced to its roots, that evidence of the evolutionary progression can be traced back through the geologic past to a time approaching that of the beginnings of life itself.

In retrospect, it may seem somewhat surprising that prior to the past decade the Precambrian fossil record was all but unknown. Certainly, this lack of knowledge was not due to a lack of interest—geologists and biologists alike have long speculated on such matters as, When did life begin? By what course did early life evolve? In what ways did its evolution affect and interact with the developing primitive environment? Prior to about 1965, however, speculations on these problems were little more than just that—speculative, and for the most part untested, guesses. Since that time, the situation has begun to change, and to change markedly. An impressive, if still rather limited array of evidence has been uncovered; a generalized picture of the early evolutionary progression is coming into focus (Cloud, 1968, 1974; Schopf, 1970, 1975, 1978; Margulis, 1970; Barghoorn, 1971). Virtually all of this new evidence has come from discoveries of cellular microfossils preserved in ancient sediments; earlier workers were unsuccessful largely because they sought evidence of megascopic organisms where only microscopic fossils would ultimately be found. In marked contrast with the more recent, Phanerozoic eon, it has become apparent that the Precambrian was "The Age of Microscopic Life."

Stromatolitic Microbiotas

The major breakthrough in the field has come from application of a newly developed set of techniques—the specialized laboratory procedures of micropaleontology first developed during the 1950s for studies of Phanerozoic deposits. Unlike their Phanerozoic analogues, however, Precambrian microfossils are apparently of relatively rare occurrence; the fossils themselves can be detected only after thorough, time-consuming, laboratory investigation. Thus, even with the application of these new techniques, the search for Precambrian life seemed destined to be a protracted, and at best highly inefficient, venture. In the early to mid-1960s, however, a key discovery was made—it was demonstrated that distinctive, laminated, centimeter- to meter-sized mounds, structures termed "stromatolites" and long known to be abundant in Precambrian terrains, were in fact reef-like remnants formed as a result of the precipitation (due to photosynthesis) and binding of mineral matter on the growth surfaces of microbial communities (e.g., Logan, 1961). Moreover, it was discovered that at least some Precambrian stromatolites contained the petrified, cellular remains of the diverse community of fossil microorganisms—the Precambrian microbiota—responsible for stromatolite formation (e.g., Barghoorn and Tyler, 1965). Thus, over the past decade a successful strategy has emerged, a strategy based on a combination of field work and laboratory study and concentrating chiefly on potentially fossiliferous, stromatolitic rock units, a strategy that for the first time has provided effective means for investigating the early history of life.

The recent upsurge of interest in the paleontology of the Precambrian is well illustrated by the data summarized in Fig. 1. Of the 18 ancient microbiotas known as of 1974 (listed in Fig. 6), all but one were first reported during the past decade, more than 50% have been discovered since 1973, and fully one-third were first reported in 1974. Although discoveries are thus being made at an increasingly rapid pace, it is important to note that much of the information currently available is still of a preliminary kind, not yet followed up by detailed analyses nor confirmed by multiple, independent investigations. For example, as is illustrated in Fig. 2, only two of the 18 Precambrian microbiotas known in 1974 had been studied in detail (those of the Bitter Springs and Gunflint Iron Formations); extended monographic descriptions of three of the other assemblages (those of the Hector Formation and of the Skillogalee and Beck

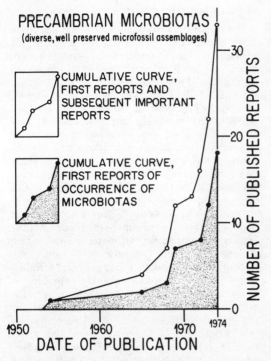

FIGURE 1. Cumulative curves showing the number of published reports describing Precambrian microbiotas and their dates of publication (the 18 microbiotas here considered are listed in Fig. 6).

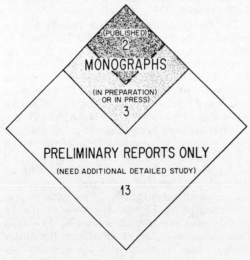

FIGURE 2. Diagram summarizing the status of studies (in 1974) of the 18 microbiotas known from Precambrian sediments (microbiotas are listed in Fig. 6).

Spring Dolomites) are currently in the press or are in preparation [recently published: Licari, 1978]. For the remaining 13 microbiotas—70% of those now known—published data are limited to reports a few paragraphs in length, illustrated with a few photomicrographs. From these observations it follows that at present only the most tentative of conclusions can be drawn regarding the early history of life; understanding of early evolution will doubtless be subject to change as new data accumulate and as new discoveries are made.

The temporal distribution of microbiotas and of stromatolites reported from the Precambrian is summarized in Fig. 3. As now known, the early fossil record can be divided conveniently

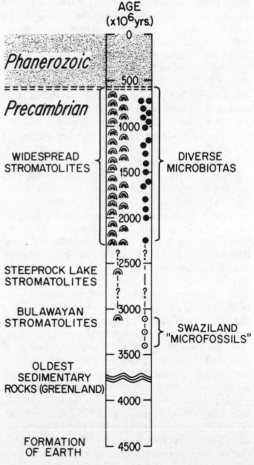

FIGURE 3. Geologic column summarizing the temporal distribution of stromatolites, microbiotas, and fossil-like microstructures known from sediments of Archean (3750 to <2500 m yr) and Proterozoic (2500 to <600 m yr) age (from Schopf, 1975; copyright © 1975 by Annual Reviews, Inc. All rights reserved).

into two phases of roughly equal duration: an earlier, "Archean" phase, spanning the time between the age of the oldest known sedimentary rocks (ca. 3750 m yr) and the first widespread appearance in the rock record of cratonal, platform-type deposits (ca. 2500 m yr ago); and a subsequent, "Proterozoic" phase extending from the close of the Archean and the approximately synchronous appearance in the rock record of widespread stromatolites (ca. 2300 m yr ago) to the beginning of the Phanerozoic (ca. 600 m yr ago). Although both of these phases of the fossil record have been studied rather intensively by paleobiologists during the past decade, they differ significantly in the quality and quantity of evidence of past life that they are now known to contain.

The Archean Fossil Record

Despite the studies of recent years, the history of Archean life remains very much a mystery. Nevertheless, several lines of reasoning seem to indicate that living systems were extant during the Archean, and that life probably originated more than 3000 m yr ago.

First, it is now known that microfossils—or at least microscopic objects that closely resemble microfossils—occur in several Archean deposits (for detailed reviews, see Schopf, 1975, 1976). The oldest of these occur in carbonaceous sediments of the Swaziland Sequence of the eastern Transvaal, South Africa, sediments ranging in age from about 3100 to 3400 m yr. These microscopic bodies, however, are of very simple morphology (Fig. 4). Although

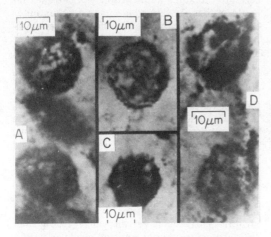

FIGURE 4. Photomicrographs showing microfossils or microfossil-like spheroids in petrographic thin sections of carbonaceous chert from the Swaziland Sequence (ca. 3200 m yr old) of South Africa (from Schopf and Barghoorn, 1967; copyright © 1967 by the American Association for the Advancement of Science).

they apparently occur in considerable abundance, they do not exhibit a degree of structural complexity (such as being demonstrably multicelled or having well-defined, regular surface ornamentation) that would make their biological interpretation wholly convincing. Moreover, and perhaps most perplexing, they bear at least superficial resemblance to organic spheroids that are known to be entirely of nonbiologic origin, such as the spheroidal bodies produced abiotically in laboratory experiments designed to simulate events that may have occurred on the primitive earth (e.g., Fox, 1974) and the spheroidal "organized elements" known to occur in some carbonaceous meteorites (Rossignol-Strick and Barghoorn, 1971). Thus, while it is possible, and perhaps likely, that at least some of these fossil-like objects are actually microfossils (and while it is true that they have been so regarded by many workers), the evidence is as yet equivocal (Schopf, 1976). Further studies are needed to define their true nature.

Second, laminated stromatolites are known to occur in at least two Archean units (Fig. 3). The oldest of these structures, occurring in limestones of the Bulawayan Group of Rhodesia, are apparently between about 2900 and 3200 m yr in age. As is shown in Fig. 5, the Bulawayan stromatolites are composed of closely spaced, stacked layers of calcareous and organic material (Schopf, et al., 1971), an organization essentially indistinguishable from that exhibited by younger stromatolites, both fossil and modern, that are known to have been formed by communities of photosynthetic microorganisms. Moreover, this type of organization seems quite difficult to explain as having resulted solely from inorganic (e.g., accretionary) processes. Thus, although neither the Bulawayan stromatolites nor the younger, Steeprock Lake stromatolites of Canada (Hofmann, 1971) are known to contain structurally preserved microfossils, both deposits appear to be of biologic origin. Relatively little, however, can be said regarding the organisms—presumed to have been primitive, microscopic, and photosynthetic—that were responsible for their formation.

Third, the carbon isotypic composition of carbonaceous material occurring in Archean sediments at least as old as 3300 m yr is consistent with, and perhaps indicative of, a biological origin (Oehler et al., 1972). The ^{13}C and ^{12}C ratio exhibited by this particulate organic matter is comparable to that of younger carbonaceous materials of demonstrable biogenicity—specifically, organic materials formed as a result of the biological fixation of carbon dioxide—and differs distinctly from that of

FIGURE 5. Limestone stromatolites from the Bulawayan Group (ca. 3100 m yr old) of Rhodesia (from Schopf et al., 1971). A, weathered surface showing well-defined laminations; B, photomicrograph of petrographic thin section showing alternating organic (dark) and calcareous (light) laminations.

carbon-bearing substances known to be of nonbiological origin. Although relatively little is known regarding the geochemical cycle of Archean carbon, it seems a highly difficult matter to account for the isotopic composition of this carbonaceous material without the intervention of biological activity.

And fourth, the occurrence of relatively abundant, diverse, and morphologically complex microorganisms, and of widespread stromatolites, in deposits of early Proterozoic age (Fig. 3) evidences an episode of prior, Archean, evolution. Although there are no obvious means of estimating accurately the time required for this early evolutionary development, the Bulawayan stromatolites and the carbon isotopic data (and, possibly, the Swaziland "microfossils") suggest that living systems originated earlier than 3000 or perhaps 3300 m yr ago. Moreover, the evidence strongly suggests that some organisms of this time had the physiological capability of fixing carbon dioxide—that is, that the Archean biosphere included autotrophs—and that these CO_2-fixers were probably photosynthetic. Geologic data seem to indicate, however, that the Archean atmosphere was probably oxygen deficient, and that appreciable quantities of atmospheric oxygen did not accumulate until the early to middle Proterozoic (e.g., Cloud, 1974). It therefore seems reasonable to surmise that photoautotrophs of the Archean probably did not produce oxygen as a by-product of photosynthesis; in terms of their physiological capabilities; they may thus have been comparable to modern, obligately anaerobic, photosynthetic bacteria. And finally, the absence of unusually thick deposits of organic matter in Archean terrains, and the similarity in total organic content exhibited by sediments of the Archean and those of younger geologic age, indicate that the Archean biosphere almost certainly included heterotrophs—organisms that could utilize and recycle organic compounds that had been synthesized in the anoxic environment by either abiotic or biologic processes. Thus, unlike the more complex ecosystems of later geologic time, the role of "primary producer" in the Archean was presumably played by both nonbiologic (abiotic syntheses) and biologic (autotrophic) processes.

In sum, it seems likely that the Archean biota was comprised entirely of primitive, anaerobic, microscopic organisms; that the biota included both heterotrophs and autotrophs; and that at least some of the autotrophs were photosynthetic stromatolite formers. Although it can be presumed that the majority, and perhaps all, of these early organisms were of bacterial affinities, little is as yet known regarding their antiquity, morphology, physiological variability, ecology, or evolutionary history.

The Proterozoic Fossil Record

In rather striking contrast with comparable studies of the Archean, the search for evidence of Proterozoic life has met with substantial

PRECAMBRIAN MICROBIOTAS
(DIVERSE, WELL-PRESERVED MICROFOSSIL ASSEMBLAGES)

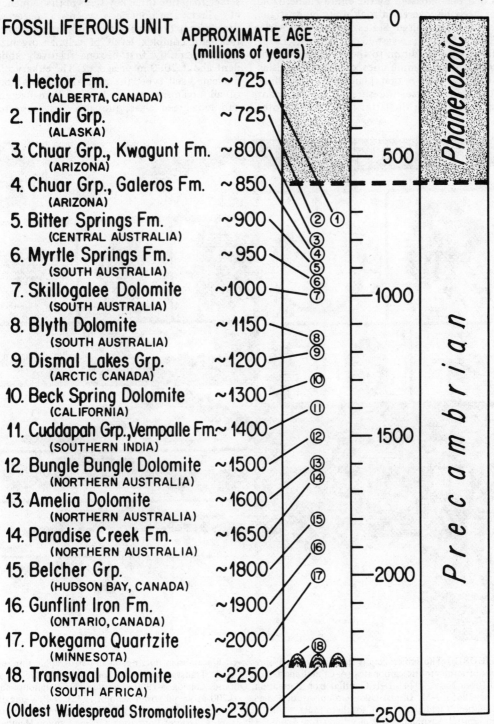

FIGURE 6. Geologic column summarizing the temporal distribution of the 18 microbiotas known from Proterozoic sediments (see Schopf, 1975, for references to original reports).

success. In recent years, more than a score of microbiotas, *in toto* spanning a time range (2250–725 m yr) about three times as long as that encompassed by the entire Phanerozoic, has been discovered in rock units of Australia, Canada, the United States, India, and Africa (Figs. 6 and 7; reviewed in Schopf, 1975). Moreover, in addition to these benthic, mostly stromatolitic communities of microorganisms, it is now known that planktonic, unicellular algae were diverse, widespread, and abundant, especially during the latter third of the Proter-

ozoic (Schopf, 1975). Although the course of Proterozoic evolution is not as yet understood in detail, these recent discoveries have served to define three key steps in the sequence: (1) Free atmospheric oxygen, a necessary precursor to the development of aerobic respiration and complex levels of cellular organization, apparently first became relatively abundant about 2000 m yr ago. (2) The eukaryotic, nucleated-cell type, the cellular building block of all organisms more advanced than bacteria and blue-green algae, probably first appeared

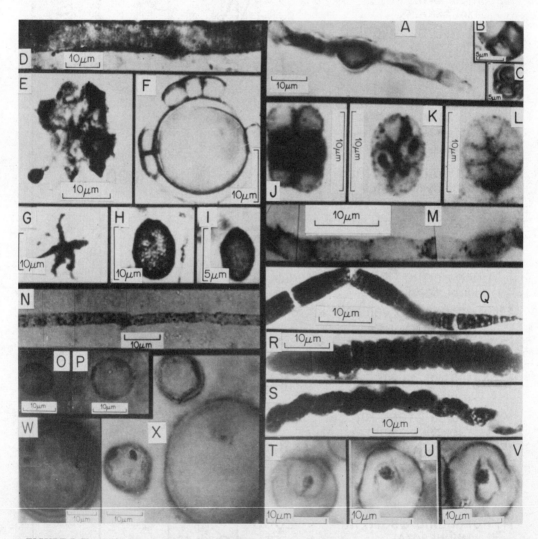

FIGURE 7. Photomicrographs showing cellularly preserved Precambrian microfossils in petrographic thin sections of stromatolitic carbonates (A–C) and cherts (D–X). A–C: Transvaal Dolomite, South Africa, ≈2250 m yr old (from Nagy, 1974). D–I: Gunflint Iron Formation, Ontario, Canada, ≈1900 m yr old (from Barghoorn and Tyler, 1965). J–M: Dismal Lakes Group, Arctic Canada, ≈1200 m yr old (from Schopf, 1975). N–P: Vempalle Formation, Cuddapah Group, southern India, ≈1400 m yr old (from Schopf, 1975). Q–V: Bitter Springs Formation, central Australia, ≈900 m yr old (from Schopf, 1972). W–X: Galeros Formation, Carbon Canyon Member, Chuar Group, eastern Grand Canyon, Arizona, ≈850 m yr old (from Schopf, 1975; copyright © 1975 by Annual Reviews, Inc. All rights reserved).

about 1500 m yr ago. And (3) as a result of the development of eukaryotic sexuality and of ecologic interactions among evolving stocks of plants and animals, the megascopic, multicellular level of organization first appeared toward the close of the Proterozoic, probably about 700 m yr ago.

Atmospheric Evolution. About one-fifth of the present atmosphere is composed of oxygen. This highly reactive gas, a by-product of the "green-plant-type" of photosynthesis, is used to biologically oxidize (i.e., to "burn" intracellularly) organic compounds—and to thus produce biochemically stored energy—in the process of aerobic respiration. Aerobic respiration, in turn, serves as the major energy-yielding biochemical process occurring in extant living systems. Thus, the majority of modern microorganisms and virtually all higher forms of life are dependent on atmospheric oxygen. As is noted above, however, the Archean atmosphere appears to have been oxygen deficient. The Archean biota would be expected to have been entirely anaerobic; oxygen-utilizing, aerobic organisms could not have evolved until sufficient quantities of oxygen became available in the environment. When might this have occurred? And from what source was the oxygen derived?

Geologic evidence indicates that although oxygen deficient, the Archean atmosphere was not entirely devoid of free oxygen; the occurrence of oxidized iron minerals, such as hematite, in early sedimentary rocks—including those of the Isua Iron Formation of West Greenland, the oldest sediments now known (Moorbath et al., 1973)—indicates that at least trace amounts of oxygen were locally present. This oxygen, however, was presumably produced nonbiologically, through the photodissociation of water vapor induced by ultraviolet light. And, the small amounts of oxygen thus produced would have been consumed rather rapidly by reaction with previously unoxidized volcanic gases and minerals available in the environment. Some other source of oxygen, a quantitatively important source that could produce oxygen at a rapid rate and in large quantity, seems needed to account for the onset of widespread oxygenic conditions. The obvious choice is green-plant photosynthesis.

Amoung oxygen-producing photosynthesizers of the modern biota, the most primitive are blue-green algae, a group composed of unicellular and filamentous microorganisms that, like bacteria, divide by fission, do not have their genetic material organized into a discreet nucleus, and lack the intracellular organized bodies (e.g., plastids, pyrenoids, etc.) characteristic of higher organisms. Because of their evident similarities, blue-green algae and bacteria are taxonomically grouped together as "prokaryotic" (prenucleated) organisms; they differ significantly in intracellular organization as well as in aspects of biochemistry from the protists, fungi, plants, and animals, the nucleated "eukaryotes," that comprise the majority of living species. In essence, blue-green algae can be regarded as bacteria that "discovered" the process of green-plant photosynthesis. Thus, it seems reasonable to suspect that the onset of oxygenic conditions could have been a result of the first widespread appearance of these primitive microorganisms. When might this have occurred?

Of the various generalizations that have emerged from Precambrian studies of recent years, one of the most interesting is that although stromatolites are all but absent from the Archean, they first become abundant and widespread in the early Proterozoic, in deposits about 2300 m yr in age (Figs. 3 and 6). This date now seems reasonably well established in both Africa and Australia, the two areas of the world in which the geochronology of early stromatolitic deposits is best defined; and it may also be true of Asia and North America (Schopf, 1975). Modern stromatolites are known to be formed by blue-green algal-dominated microbiological communities (e.g., Logan, 1961); thus, it is tempting to interpret these early Proterozoic structures as evidencing the occurrence of blue-green algae. And, although few microbiotas are known from stromatolites of this age, those that have been detected—as in the Transvaal (Fig. 7A–C; Nagy, 1974) and the Gunflint (Fig. 7D–I; Barghoorn and Tyler, 1965; Awramik and Barghoorn, 1977) assemblages—seem to support this supposition; without exception, they are dominated by filamentous microorganisms that are similar in morphology to modern blue-greens. Thus, the paleobiologic data now available suggest that communities of microscopic, oxygen-producing blue-green algae (or their more primitive evolutionary precursors) may have first become widespread about 2300 m yr ago. With this development, increasing quantities of oxygen would have been added to the environment and the transition from an oxygen-deficient to an oxygenic atmosphere would have commenced.

The transition in atmospheric oxygen content, however, would not have been instantaneous and probably, even by geologic standards, it would not have been rapid. Prior to the accumulation of an oxygenic atmosphere, oxygen would first have been consumed in the oxidation of previously unoxidized materials present in the environment. Geologic data suggest that the onset of oxygenic conditions may have thus been delayed for a period 200 to 400 m yr. The occurrence in early Proterozoic stream deposits of unoxidized pyrite and uraninite, minerals readily oxidized in an oxygen-rich environment, suggests that the atmosphere was still oxygen-deficient as late as 2200 m yr ago (Cloud, 1974, 1976). The major iron ores of the Precambrian, composed predominantly of oxidized hematite, were deposited between about 2200 and 1900 m yr ago (Goldich, 1973); massive quantities of oxygen were consumed in this major episode of ore deposition. Nevertheless, as early as 1800 or 1900 m yr ago, atmospheric oxygen concentrations may have been relatively high (i.e., perhaps 1% of the present level); oxidized terrigenous deposits, red beds formed on the land surface as a result of subaerial oxidation, are first known from rocks of this age (Cloud, 1974). Thus, geological, geochemical, and paleobiological evidence can be interpreted as suggesting that a transition in atmospheric oxygen content probably occurred about 2000 m yr ago. This transition, a development triggered by the widespread appearance of primitive, oxygen-producing photosynthesizers and an event that well illustrates the interplay between biological and geological processes on the primitive earth, was to have a profound and lasting impact on the subsequent course of biological evolution.

Development of the Nucleated Cell. As is mentioned above, most living species—indeed, well over 99%—are comprised of nucleated, eukaryotic, cells. The development of this cell type, derived from prokaryotic, presumably blue-green algal ancestors—or, conceivably, from a symbiotic combination of bacterial and blue-green algal unicells (Margulis, 1970; McQuade, 1977)—was thus among the most important of evolutionary events to have occurred during the history of life. Two types of questions come to mind regarding this signal evolutionary development. First, how—by what process or series of processes—did it occur? And second, when did the eukaryotic lineage first appear? Unfortunately, it appears unlikely that the fossil record will ever yield direct evidence in answer to the first of these questions. Fossils provide a record only of the products of the evolutionary process, not of the dynamic process itself. Moreover, in this case the evolutionary transition involved a sequence of changes in biochemistry and intracellular organization. It is thus unlikely that intermediate forms will be discovered or, if detected, that they would be interpreted as such. On the other hand, the endpoints of the evolutionary continuum, the prokaryotic pre-cursor(s) and the eukaryotic descendant(s), are at least potentially preservable and detectable in ancient sediments. In principle, therefore, paleontology might provide evidence of the *time,* if not of the *mode,* of origin of the nucleated cell type.

Unlike many prokaryotes, virtually all eukaryotic cells are oxygen requiring. Moreover, this aerobic mode of existence appears to be a fundamental aspect of eukaryotic organization. It follows, therefore, that nucleated cells could have originated only after free oxygen had become relatively abundant in the environment; judging from geologic evidence, it thus seems unlikely that eukaryotes could have existed prior to perhaps 2000 m yr ago. At the more recent end of the time spectrum, it is well known that multicellular, eukaryotic algae and invertebrate animals had become abundant, diverse, and morphologically quite complex by the close of the Precambrian, some 600 m yr ago (Glaessner, 1971). Thus, it may be surmised that some time during the Proterozoic—probably between about 1800 and perhaps 700 m yr ago—the nucleated cell type first made its appearance. In recent years, evidence provided by newly discovered Precambrian microbiotas has helped to further define the timing of this event.

At present, among the earliest assured eukaryotes are those occurring in bedded black cherts of the approximately 900 m yr old Bitter Springs Formation of central Australia; nine taxa of eukaryotic microorganisms, including possible fungi and several types of unicellular algae, have been described from this extraordinarily well preserved assemblage (Schopf, 1968, 1972; Schopf and Blacic, 1971). Most importantly, a number of these taxa (and literally thousands of individuals) have been shown to contain dense, intracellular, organic bodies (Figs. 7T–V, 8), structures of regular size, shape, and distribution that may represent remnants of intracellular organelles (e.g., nuclei, pyrenoids, etc.). The eukaryotic affinity of these organisms thus seems reasonably well established, though this has recently been disputed (Knoll and Barghoorn, 1976; for a reply see Schopf and Oehler, 1976). Intracellular structures of rather similar form have been recently reported as well from at least nine other Precambrian assemblages (Schopf, 1975). Apparently the oldest, reasonably convincing such structures now known are those occurring in the microbiota of the Bungle Bungle Dolomite of northern Australia (Diver, 1974)—an assemblage thought to be about 1500 yr in age. Thus, it now seems evident that although megascopic, multicellular eukaryotes did not appear until quite late in the

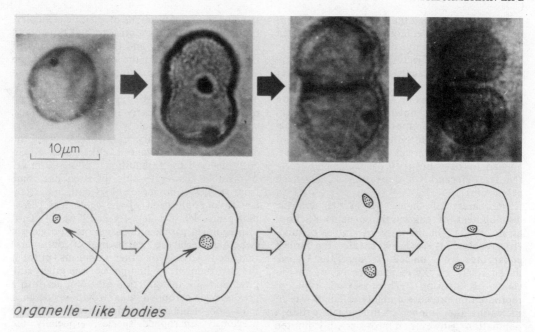

organelle-like bodies

FIGURE 8. Photomicrographs showing algal unicells in petrographic thin sections of carbonaceous chert from the ≈900 m yr old Bitter Springs Formation of central Australia (from Schopf, 1968). The cells have been preserved in what may be various stages of mitotic (asexual) division.

Precambrian, the eukaryotic lineage extends far into the geologic past, probably back to as early as 1500 m yr ago. This realization, however, one of the more important conclusions to be drawn, albeit tentatively, from recent studies, raises yet another difficult question. Namely, why was there such an extensive time lag, a gap apparently on the order of nearly a thousand million years in duration, between the development of the eukaryotic cell and the emergence of megascopic, multicellular eukaryotes? Although relevant data remain few, this question, a question that bears directly on one of the most classic of paleontologic problems—the origin of the Metazoa—may also be nearing solution.

Development of Advanced Eukaryotes. The beginning of the Phanerozoic eon, about 600 m yr ago, is marked in the fossil record by the first widespread appearance of numerous types of invertebrate animals and megascopic algae. This "explosion" of advanced life forms, an event that apparently occurred within a period less than 50 m yr in duration, has puzzled paleontologists for more than a century. In Charles Darwin's day, for example, when the Precambrian was thought to be entirely devoid of fossils, this abrupt beginning of the fossil record presented a major dilemma, a problem that Darwin thought would require convincing explanation if his theory of evolution were to be ultimately accepted. In subsequent years,

and especially in the early decades of the 1900s, it thus became fashionable to construct hypothetical postulates to explain the apparent absence of Precambrian fossils. To cite just a few: (1) Precambrian sedimentary rocks were claimed to be entirely of oceanic origin; life, however, was postulated to have first evolved in terrestrial lakes, invading the marine environment only at the beginning of the Phanerozoic. (Interestingly, the converse of this hypothesis—that Precambrian rocks were entirely of freshwater origin but that early life forms were entirely marine—also gained acceptance in some quarters.) (2) It was suggested that the earliest rocks of the Phanerozoic might represent the oldest unmetamorphosed sediments of the geologic record. Thus, although evidence of life could have once existed in earlier deposits, such evidence was assumed to have been obliterated by normal geologic processes. And (3) it was postulated that all Precambrian animals lacked fossilizable hard parts. The "explosion" at the beginning of the Phanerozoic was thus viewed as having resulted from the near simultaneous acquisition of hard exoskeletons by various animals groups, rather than as reflecting the (presumably much earlier) origin of the animals themselves.

A number of other postulates, some quite bizarre, could be added to this short list. The important point, however, is that none of these suggestions provides the complete answer. All, for one reason or another, fall short. In recent years, as new paleobiologic data have been amassed, a decidedly different strategy has emerged, one of rejecting the ad hoc assumptions and special causes called upon by earlier workers and of relying instead on concepts derived from studies of living organisms and on acceptance of the known fossil record "as is." According to this approach, two sequential evolutionary events—the development of sexuality and the origin of cropping protists—appear to have been of particular import in the late Precambrian derivation of advanced eukaryotes.

The earlier of these two events was the development of eukaryotic sexuality (Schopf et al., 1973). On the basis of studies of modern organisms, it seems evident that the earliest eukaryotes were unicellular algae that reproduced solely via the asexual process of mitosis (body-cell division). This asexual type of reproduction provides a highly efficient way of increasing the number of individuals within a population. However, it provides only limited means (the occurrence of rare mutations) for the introduction of genetic variability; each mitotically produced cell is an exact copy of the parent. Variability, however, is the keystone of evolutionary advance—the concept of the "survival of the fittest" presupposes that important differences, differences that determine relative fitness, exist among members of a population; this variability is minimized in asexual organisms. In sexually reproducing populations, however, the situation markedly differs. Sexually produced offspring exhibit traits of both parents but differ from either; sexually reproducing species are thus far more varied than their asexual analogues. Because of this difference in variability, evolution (resulting from natural selection of relatively better adapted individuals) can proceed at a decidedly more rapid pace in sexually reproducing populations. Thus, with the origin of eukaryotic sexuality, both the diversity of living species and their rate of evolutionary change can be expected to have increased.

When did eukaryotic sexuality first appear? Sexual reproduction of this type requires the occurrence of meiosis (germ-cell division), a division pattern more complex than that of mitosis, one that seems certain to have been derived from the previously established asexual, mitotic system. And, although the data are as yet quite fragmentary, the earliest evidence suggestive of meiosis—sporelike cells that may be of meiotic origin—dates back only to about 900–1000 m yr ago (Schopf and Blacic, 1971). Moreover, it seems possible that these simple sporelike cells may represent a relatively early stage in the establishment of meiotic sexuality. If so, rates of evolution prior to that time could have been quite slow and the major portion of the time lag between the development of the nucleated cell and the emergence of advanced eukaryotes will have been explained. But, what of the remainder of the time gap? If the origin of sex resulted in vastly increased variability, and if increased variability resulted in markedly increased evolutionary rates, why was the development of advanced eukaryotes apparently delayed for an additional 200–300 m yr? A possible answer to this question has come from studies of modern ecology and, in particular, from application of the ecological "cropping principle" (Stanley, 1973, 1976).

In ecologic parlance, a *cropper* is an organism (whether herbivore, carnivore, or omnivore) that feeds actively on other organisms (prey) in the environment. At first glance, one might suppose that the introduction of a cropper into a previously uncropped area would result in a reduction of the number of species present—the previously uncropped species, especially the rarer varieties, might be cropped to extinction. In fact, however, in modern ecosystems just the opposite occurs: in communities of primary producers, the one or few species of photosynthesizers that are best adapted to the local environment tend to monopolize available space; since a cropper tends to feed on the most abundant, dominant species of a community, the introduction of a cropper decreases the abundance (and dominance) of the principal prey; this, in turn, frees space that can be used for the expansion of other, previously subordinate varieties of organisms and for the introduction of entirely new species. In short, the introduction of a cropper into a previously uncropped ecosystem results in an increase, rather than in a decrease of diversity among the "croppees"; this increased diversity, in turn, permits the establishment of increasingly specialized types of croppers. Thus, a sort of ecological feedback system develops that results ultimately in an increase of biological diversity at all levels in the food web. Among modern ecosystems, those that are well cropped are maximally diverse, composed of many species of which each is represented by relatively few individuals. How can this "cropping principle" be applied to late Precambrian evolution?

All microbiotas known from the late Precambrian are dominated by photosynthetic, primary producers. And, like their modern counterparts, they tend to be dominated by one or only a few species of microorganisms, principally by various types of filamentous blue-green algae. During the late Precambrian,

TABLE 1. Benchmarks in Precambrian Evolution

	Approximate Age $(\times 10^6 \text{ yr ago})$
Precambrian-Cambrian "boundary"	600
Origin of megascopic life	700
Development of cropping protists	800
Development of meiotic sexuality	1,000
Origin of mitotic, nucleated cells	1,500
Development of an oxygenic atmosphere	2,000
Widespread O_2-producing microbiotas	2,300
Oldest known stromatolites	3,100
Origin of microscopic life	3,300
Oldest known sedimentary rocks	3,750
Formation of earth	4,500

as sexual eukaryotes became increasingly abundant and diverse, many new types of microorganisms must have first appeared. Although most of these, like their evolutionary precursors, were primary producers, the ability to photosynthesize was lost in several evolving stocks. The resulting single-celled protists, perhaps first appearing about 800 m yr ago, were the earliest croppers. The most primitive were probably omnivores, feeding on organic detritus and small cells. As their cropping resulted in an increase of the diversity at lower levels of the food web, however, other, specialized and morphologically more complex croppers—herbivores and ultimately carnivores—made their appearance. Thus, this rather simple ecologic model appears to provide a possible explanation for the explosive evolution of the latest Precambrian: the development of the cropping habit in protozoans led, within a relatively short segment of geologic time, to a marked increase of biologic diversity and ecologic complexity at all levels of the food web, culminating in the appearance and rapid diversification of megascopic croppers (invertebrate animals) and coevolving, megascopic, "croppees" (advanced, multicellular algae).

Summary

In Table 1 are summarized the major benchmarks in Precambrian evolution that seem suggested by the evidence now at hand. In viewing this list, however, it should be recognized that the nature, timing, and relative importance of events there noted are as yet far from firmly established—Table 1 summarizes the state-of-the-art (as of 1974) rather than a set of immutable, clearly defined, truths. Still, important progress has been made in recent years. Questions regarding the existence of Precambrian life, questions that only a decade ago were popular and pressing, are no longer raised. The question now is not whether evidence of early life exists, but what does it say? What does it tell us of the course of early evolutionary advance? And at present, one principal observation stands out: after more than a century of search, the evolutionary continuum that ties together the modern and the ancient is beginning to be deciphered, a continuum that is now known to extend back through the geologic past to a time equal to, and very probably greater than, half the currently accepted age of this planet.

J. WILLIAM SCHOPF

References

Awramik, S. M., and Barghoorn, E. S., 1977. The Gunflint microbiota, *Precambrian Res., 5*, 121–142.

Barghoorn, E. S., 1971.The oldest fossils, *Sci. American, 224*(5), 30–42.

Barghoorn, E. S., and Tyler, S. A., 1965. Microorganisms from the Gunflint chert, *Science, 147*, 563–577.

Cloud, P. E., Jr., 1968. Pre-Metazoan evolution and the origins of the Metazoa, in E. T. Drake, ed., *Evolution and Environment*. New Haven: Yale Univ. Press, 1–72.

Cloud, P. E., Jr., 1974. Evolution of ecosystems, *Amer. Scientist, 62*, 54–66.

Cloud, P. E., Jr., 1976. Beginnings of biospheric evolution and their biochemical consequences, *Paleobiology, 2*, 351–387.

Diver, W. L., 1974. Precambrian microfossils of Carpentarian age from Bungle Bungle Dolomite of Western Australia, *Nature, 247*, 361–362.

Fox, S. W., 1974. From proteinoid microsphere to contemporary cell: Formation of internucleotide and peptide bonds by proteinoid particles, *Origins of Life, 1*, 227–238.

Glaessner, M. F., 1971. Geographic distribution and time range of the Ediacara Precambrian fauna, *Geol. Soc. Amer. Bull., 82*, 509–514.

Goldich, S. S., 1973. Ages of Precambrian iron-formations, *Econ. Geol., 68*, 1126–1134.

Hofmann, H. J., 1971. Precambrian fossils, pseudofossils and problematica in Canada, *Geol. Surv. Canada Bull., 189*, 146p.

Knoll, A. H., and Barghoorn, E. S., 1975. Precambrian eukaryotic organisms: A reassessment, *Science, 190*, 152–154.

Licari, G. R., 1978. Biogeology of the Late pre-Phanerozoic Beck Spring Dolomite of eastern California, *J. Paleontology, 52*, 767–792.

Logan, B. W., 1961. Cryptozoon and associated stromatolites from the Recent, Shark Bay, western Australia, *J. Geol., 69*, 517–533.

Margulis, L., 1970. *Origin of Eukaryotic Cells*. New Haven: Yale Univ. Press, 349p.

McQuade, A. D., 1977. Origins of the nucleate organisms, *Quart. Rev. Biol., 52*, 249–262.

Moorbath, S.; O'Nions, R. K.; and Pankhurst, R. J., 1973. Early Archaean age for the Isua Iron Formation, West Greenland, *Nature, 245*, 138–139.

Nagy, L., 1974. Transvaal stromatolite: First evidence

for the diversification of cells about 2.2×10^9 years ago, *Science,* 183, 514–516.

Oehler, D. Z.; Schopf, J. W.; and Kvenvolden, K. A., 1972. Carbon isotopic studies of organic matter in Precambrian rocks, *Science,* 175, 1246–1248.

Rossignol-Strick, M., and Barghoorn, E. S., 1971. Extraterrestrial abiogenic organization of organic matter: The hollow spheres of the Orgueil meteorite, *Space Life Sci.,* 3, 89–107.

Schopf, J. W., 1968. Microflora of the Bitter Springs Formation, late Precambrian, central Australia, *J. Paleontology,* 42, 651–688.

Schopf, J. W., 1970. Precambrian micro-organisms and evolutionary events prior to the origin of vascular plants, *Biol. Rev. Cambridge Phil. Soc.,* 45, 319–352.

Schopf, J. W., 1972. Evolutionary significance of the Bitter Springs (late Precambrian) microflora, *24th Internat. Geol. Cong., Proc.,* sec. 1, 68–77.

Schopf, J. W., 1975. Precambrian paleobiology: Problems and perspectives, *Ann. Rev. Earth Planetary Sci.,* 3, 213–249.

Schopf, J. W., 1976. Are the oldest "fossils," fossils? *Origins of Life,* 7, 19–36.

Schopf, J. W., 1978. The evolution of the earliest cells, *Sci. American,* 239 (Sept.), 110–138.

Schopf, J. W., and Barghoorn, E. S., 1967. Alga-like fossils from the early Precambrian of South Africa, *Science,* 156, 508–512.

Schopf, J. W., and Blacic, J. M., 1971. New microorganisms from the Bitter Springs Formation (late Precambrian) of the north-central Amadeus Basin, Australia, *J. Paleontology,* 45, 925–960.

Schopf, J. W.; Haugh, B. N.; Molnar, R. E.; and Satterthwait, D. F., 1973. On the development of metaphytes and metazoans, *J. Paleontology,* 47, 1–9.

Schopf, J. W., and Oehler, D. Z., 1976. How old are the eukaryotes? *Science,* 193, 47–49.

Schopf, J. W.; Oehler, D. Z.; Horodyski, R. J.; and Kvenvolden, K. A., 1971. Biogenicity and significance of the oldest known stromatolites, *J. Paleontology,* 45, 477–485.

Stanley, S. M., 1973. An ecological theory for the sudden origin of multicellular life in the late Precambrian, *Proc. Nat. Acad. Sci.,* 70, 1486–1489.

Stanley, S. M., 1976. Ideas on the timing of metazoan diversification, *Paleobiology,* 2, 209–219.

Cross-references: *Ediacara Fauna; Fossil Record; Protista; Pseudofossils; Stromatolites.* Vol. IVA: *Oxygen: Evolution in the Earth's Atmosphere; Precambrian Atmosphere–Geochemical History; Precambrian Hydrocarbons.*

PROTISTA

The Kingdom(s) of the Protista

The Protista include single-celled organisms that have been regarded as animals (protozoans), as plants (algae), or as a distinct kingdom of organisms, or considered to include representatives of 2, 3, or as many as 16 separate kingdoms (see references in Whittaker, 1969, and Tappan, 1974a). As the classification and phylogenetic interpretation of these organisms by various workers have been based on different aspects of morphology, cytology, ultrastructure, biochemistry, physiology, life cycles, and the fossil record, the concept of the Protista has varied widely. Some definitions exclude the Prokaryota or Monera (bacteria and blue-green algae), regarding as the greatest discontinuity in the living world the distinction between these organisms that lack a membrane-bound nucleus and other cell organelles, and all other Eukaryota (Protozoa, Algae, Metazoa, and Metaphyta). Others regard the Prokaryota as the Lower Protista. Most biologists regard all protozoans and eukaryotic algae as Protista; but some exclude the red, green, and brown multicellular algae; some include the fungi and others do not.

Four kingdoms of organisms were recognized by H. F. Copeland—the Mychota (prokaryotes or bacteria and blue-green algae, and hence equivalent to the Monera), the Protoctista (including the Rhizopoda and flagellate protozoa, Sporozoa and Ciliophora, Fungi, Mycetozoa, the Pyrrhophyta, Rhodophyta, and Phaeophyta *sensu lato,* in which the classes and orders of the Chrysophyceae, Bacillariophyceae, and the Oomycetes were combined with the Phaeophyta *sensu stricto*), the Plantae (including both the algal Chlorophyta and the higher plants), and the Animalia.

Whittaker recognized the Monera and Eunucleata (phytoflagellates and protozoans) as subkingdoms of the Protista in 1959, whereas the multicellular algae were considered as subkingdoms (Rhodophyta, Phaeophyta, and Euchlorophyta) within the kingdom Plantae. Higher plants were placed with the green algae in the Euchlorophyta. The kingdoms Animalia, Plantae, and Fungi were defined by their respective modes of nutrition, i.e., ingestion, photosynthesis, and absorption; but members of the kingdom Protista may utilize any of these. In 1969, Whittaker separated the Monera from the eukaryotic Protista, and recognized five kingdoms. A different organization was proposed in 1960 by E. C. Dougherty and M. B. Allen, who included within the Protista the Monera (bacteria and blue-green algae), Mesoprotista (for the red algae), and the Metaprotista (including the remaining algae and protozoa and excluding only the Metaphyta and Metazoa).

At another extreme, L. S. Dillon divided all organisms into 14 kingdoms on a cytological basis in 1963; 13 of these included, in part, members of Dougherty and Allen's Protista, Leedale's (1974) multikingdom scheme contains 20 kingdoms (16 of which are or include members of the Protista *sensu lato*), Monera (the prokaryote bacteria and blue-green algae), Rhodophyta, Heterokontae (including the Xanthophyta, Chrysophyta, Bacillariophyta, Phaeophyta, and Oomycota), Eustigmatophyta, Haptophyta, Cryptophyta (Cryptomonads), Dinophyta (dinoflagellates), Chytridiomycota, Euglenophyta, Zoomastigina, Myxomycota, Sarcodina, Ciliophora, Sporozoa, and Plantae (which includes the algal Chlorophyta and Charophyta together with the higher plants). Nonprotist kingdoms are Porifera, Mesozoa, Animalia, and Fungi (exclusive of Chytrids and Myxomycetes).

The distinctive dinoflagellate nucleus is intermediate in mode of division between that of the prokaryotes and the eukaryotes; has continuously condensed chromosomes and a nuclear membrane; and undergoes mitotic division, although without the typical eukaryotic spindle fibers and centromeres. The DNA is associated with no (or very little) RNA and combined protein, and the DNA fibrils are only one-fourth to one-tenth the diameter of the typical eukaryote DNA fibrils. Based on these distinctive features, Dodge (1966), p. 113) proposed a separate kingdom Mesocaryota, to separate the dinoflagellates from the Prokaryota and more typical Eukaryota.

The bases for classification at levels below the kingdoms also have varied considerably. Bacteria have generally been left to the microbiologists; phycologists have studied both prokaryotic and eukaryotic algae; while protozoologists also have claimed as protozoans the flagellated eukaryotic algae and the myxomycetes. Phycologists generally have emphasized the nature of the photosynthetic pigments for classification and evolutionary reconstruction (see references in Tappan 1974a), considering the algae as a classic example of parallel evolution in gross morphology. Thus, the single class Chlorophyceae of the Division Chlorophyta includes the flagellate Volvocales, the coccoid Chlorococcales, the siphonate Siphonocladales and the Zygnematales. Other flagellated algae are variously placed in the Cryptophyceae, Dinophyceae, Chrysophyceae, and Haptophyceae of Christensen's Chromophyta and the Euglenophyceae and Prasinophyceae of the Chlorophyta. In contrast, some phycologists regard an algal "class" as necessarily encompassing all these morphological variations (Fott, 1971) leaving unassigned as "residual flagellates" those not similarly represented by filamentous and thallose forms. Protozoologists (Honigberg et al., 1964), on the other hand, place all flagellate taxa in a single class, Phytamastigophorea (orders Chrysomonadida, Silicoflagellida, Cocco-

lithophorida, Heterochlorida, Cryptomonadida, Dinoflagellida, Ebriida, Euglenida, Chloromonadida, and Volvocida), and disclaim completely the coccoid, filamentous, and siphonous taxa that are aligned with these groups by the phycologists.

Protistan Phylogeny

Classically, bacteria have been regarded as the most primitive living organisms, having given rise early to the blue-green algae, these in turn being ancestral to the red, green, and other eukaryotic algae. Various lineages secondarily lost their plastids (as have some living euglenids, dinoflagellates, diatoms, coccolithophorids, and other apochromatic flagellates) and thereby "became" protozoans and fungi. Metazoans arose from the protozoa, and metaphytes from the green algae.

In a radically different interpretation, Lynn Margulis (references in Tappan 1974a; Taylor, 1974; McQuade, 1977) regards all eukaryotic cells as the end result of a series of symbioses, one symbiont producing the (9 + 2) homologue as flagella, flagellar basal body, and intranuclear and extranuclear division centers, and other endosymbionts becoming the mitochondria of the protozoa. Eventually various prokaryotic algae, as obligate endosymbionts, became the plastids of the eukaryotic algae and higher plants, a suggestion that was first proposed by A. F. W. Schimper, C. Mereschkowsky, and others and repeatedly reintroduced on various bases over the last 90 years (see references and discussion in Taylor, 1974). The unrelated monad, filamentous, and siphonate hosts that acquired similar green symbionts, for example, resulted in the present morphologic variety of the Chlorophyta; this division thus consists of unrelated hosts, with closely related plastids. Following Copeland's 1956 classification in part, but with five kingdoms, as in Whittaker (1969), Margulis bases a suggested phylogeny of the lower eukaryotic algae on the homologies of the "host" and the presumed stage of protozoan development at which the symbiosis was believed to have arisen. Various groups of "green algae" were regarded as of wholly independent origin, the desmids arising early with the amoeboid protozoans, the Sphaeropleaceae and Oedogoniales somewhat later with the higher plant lineage, and the Euglenids and Siphonaleans arising at a more advanced "protozoan" stage of development; still later, the Chlorellaceae, Volvocales and Vaucheriales branched off as they acquired endosymbionts. The Chryosphytes had a similarly polyphyletic origin. The end result was not far different from Dillon's 14-kingdom proposal.

Instead of regarding the bacteria as primitive, Bisset (1973) suggested that the inner layer of the bacterial cell envelope is equivalent to the nuclear membrane of the eukaryotes. Rather than primitive, the bacteria are "highly specialized organisms, making use of their small size and consequent low bulk-surface ratio and rapid metabolism. They are almost certainly derived from other, more primitive, flagellate protista, and the origin of their organs must be sought in the general structural pool." The systematic association of bacteria and blue-green algae as prokaryotes Bisset regarded as based on an invalid assumption.

Although some consider protistans to be too simple and primitive to have shown any evolutionary tendencies, or even to have undergone any evolution at all, paleontologists have utilized many fossil protistans as index species for stratigraphic zonation, age dating, and correlation. For these workers, protistan evolution is an obvious and well-documented fact, with rates of evolution comparable to those in Metazoans.

The fossil record indicates that the protistans have been present for some 3 b yr, and the eukaryotic protists for at least one-third and possibly two-thirds that time, hence their evolutionary history spans a period some five times that of the Metazoa, and 7.5 times as great as for any known Metaphyta. As often observed (see references in Tappan, 1974a), protistan specialization has been biochemical, cytological, morphological, and physiological, without involving compartmentalism. Protists differ from other organisms mainly in being smaller; but these size limitations affect their chemistry, metabolism, movement and dispersal, prey–predator relationships, and even reproduction, virtually every cell being a potential reproductive cell. The direction of evolution thus contrasts sharply with that of the higher organisms, in which major trends involve a progressive increase in size, higher levels of structural organization and differentiation, greater efficiency of particular functions, and more elaborate nervous systems. Detailed study of the protistan fossils also shows that some supposedly geologically long-lived taxa represent a variety of misidentified forms, which in reality are highly distinctive for the various geologic periods (e.g., Foraminiferida, Radiolaria, Pyrrhophyta or Dinophyceae, Dictyochales, Bacillariophyceae, Coccolithophoraceae, etc.). Throughout their evolution, times of major diversification have alternated with periods of severe reduction in diversity and abundance, even to the elimination and extinction of certain groups (Tappan, 1968; Tappan and Loeblich, 1972).

Nature of the Protistan Fossil Record

Electron microscopy has emphasized the great diversity represented in the living protists, in the complexity and variety shown by nuclei, flagella, plastids, pyrenoids, eyespots, trichocysts, and other organelles. Although these structures are not recognizable in the fossils, a similar variety of organic or mineralized tests, loricae, frustules, thecae, spicules, skeletons, and resting cysts does record the diversification of many of the major groups of the protists through time (see Fig. 1; Loeblich, 1974; Loeblich and Tappan, 1964; Tappan and Loeblich, 1968, 1971, 1972, 1973).

Cellulose, of which many dinoflagellate thecae and green and brown algal walls are constructed, is readily hydrolyzed by bacterial action, and hence is rarely directly preserved, although the walls and even mucilage sheaths may be exceptionally preserved in cherts. "Sporopollenin," a class of biopolymers derived by oxidative polymerization of carotenoids and/or carotenoid esters, occurs in the highly resistant higher plant spore and pollen walls, and also is characteristic of the cyst wall of dinoflagellates, prasinophycean algae, and acritarchs (an informal term to include fossil cysts of phytoplankton of unknown relationship). The highly stable and resistant compounds forming these cysts are preserved in rocks from the early Proterozoic to the present, a span of some 2.5 b yr. Although they may be destroyed by severe oxidation (weathering) of the enclosing rocks, or by excessive heat or pressure, the organic-walled microfossils withstand rock temperatures up to about 300°C and remain recognizable.

Calcareous sediment-trapping mats of blue-green algae and green algae form stromatolites (q.v.) over a similar range in time, and may occur both in the marine environment, and in freshwater streams and lakes where they form travertine and lake balls. Extracellular and intracellular calcite may be deposited by Paleozoic dasyclad green algae, Cryptonemiales (red algae), charophytes, the coccoliths and discoasters of the Haptophycean Coccolithophoraceae, the lorica of some Tintinnida, tests of Foraminiferida, and cysts of a few dinoflagellates. The less stable form of calcium carbonate, aragonite, occurs in the foraminiferal superfamilies Duostominacea and Robertinacea (Loeblich and Tappan, 1964, 1974) the red algal Nemalionales, the brown alga *Padina,* and the Caulerpalean and Cenozoic dasycladalean green algae (Tappan and Loeblich, 1971). Silica, in the unstable hydrated opaline form, is biologically deposited by the Chrysophyta (resting cysts, plates, and scales); diatom (Bacillariophyceae) frustules; the intracellular skele-

		Monera			Rhodophyta			Chromophyta							Chlorophyta				Protozoa		
		Schizophyta	Cyanophyta (Cyanophyceae)	Rhodophyceae (Bangiophycidae / Florideophycidae)			Dinophyceae	Chrysophyceae	Haptophyceae	Bacillariophyceae	Xanthophyceae	Phaeophyceae		Euglenophyceae / Prasinophyceae	Chlorophyceae				Rhizopodea	Actinopodea	Ciliatea

(Ordinal-level column headers, left to right): Bacteria · Chroococcales · Stigonematales · Nostocales · Porphyridiales · Bangiales · Nemaliales · Cryptonemiales · Gigartinales · Rhodymeniales · Ceramiales · Dinophysiales · Peridiniales · Ebriales · Ochromonadales · Dictyochales · Cysts, incertae sedis · Prymnesiales · Biddulphiales · Bacillariales · Vaucheriales · Desmarestiales · Dictyotales · Laminariales · Fucales · Euglenales · Halosphaerales · Volvocales · Chlorococcales · Ulotrichales · Zygnematales · Oedogoniales · Siphonocladales · Caulerpales · Receptaculitales · Dasycladales · Charales · Arcellinida · Gromiida · Foraminiferida · Radiolaria · Acantharia · Heliozoid · Heterotrichida · Tintinnida · Hypotrichida

(Stratigraphic rows): Cenozoic — Quaternary; Tertiary (Pliocene, Miocene, Oligocene, Eocene, Paleocene); Mesozoic — Cretaceous, Jurassic, Triassic; Paleozoic — Permian, Pennsylvanian, Mississippian, Devonian, Silurian, Ordovician, Cambrian; Precambrian (PreЄ) — Late, Middle, Early.

FIGURE 1. Geologic range to the ordinal level for Protista (modified from Loeblich, 1974). X denotes known occurrences, ● represents interpolated range.

tons of the Silicoflagellates; Ebriales; *Actiniscus,* and related genera of the dinoflagellate Gymnodiniales; the green alga *Pediastrum;* the skeleton of the protozoan Radiolaria; and the lorica plates and scales of the thecamoebian Gromiida and Arcellinida of the Rhizopoda. In spite of their small size, fossil representatives of a wide variety of protists are known from all aquatic habitats and in rocks of all ages from the very oldest sediments yet recognized.

Evolutionary History

Bacteria may represent the nearest living analogue to the primitive moneran, which may have been heterotrophic, deriving its energy from consuming abiotically formed organic compounds. As the supply of the latter became limited, the appearance of chemosynthetic and photosynthetic bacteria that produced new organic compounds allowed both their survival and that of some associated heterotrophs. In spite of their small size and nondescript morphology, fossil bacteria have been reported from the early Precambrian Fig Tree Series of South Africa, and in sedimentary iron ores, bauxite and pyrite, coal, oil, and salt, as well as sandstones, shales, cherts, and limestones from the Precambrian to the present. Parasitic and pathogenic bacteria have been reported in fossilized plant tissues, fossil bone, and fossilized coprolites.

Living photosynthetic bacteria have relatively inefficient pigments, bacteriochlorophyll, and chlorobium chlorophylls; the development of chlorophyll *a* by the blue-green algae was a major evolutionary advance in the early Proterozoic. The blue-green algae are structurally little advanced over certain bacteria, but are much superior biochemically and physiologically, as the possession of chlorophyll *a* allowed splitting water molecules as a hydrogen source for carbon dioxide fixation, releasing oxygen as a photosynthetic by-product. This evolutionary advance allowed the gradual accumulation of oxygen in the atmosphere, with all the attendant evolutionary changes in the biota for adaptation to an aerobic existance (Tappan, 1968, 1974b). Stromatolites as well as cellularly preserved blue-green algae also occur widely in the geologic record. Possible coccoid blue-green algae occur in the Fig Tree Group (3.1 b yr) and filamentous forms occurred by 2.2 b yr.

The earliest appearance of eukaryotes is more difficult to date accurately, as their recognition as such is based on varying types of evidence.

Possibly the earliest and most primitive euka-ryotes are red algae (Rhodophyta). They resemble the blue-green algae in having chloro-phyll *a* as the main photosynthetic pigment and in having accessory phycobilin pigments, of which phycocyanin is dominant in blue-green algae and phycoerythrin in the reds. Both store starch as a reserve, and neither has any flagel-late stage in the life cycle. Spore tetrads that are probably referrable to the Florideophycidae occur in the late Precambrian Bitter Springs Formation, Australia (Schopf and Blacic, 1971); and other red algal remains, including the more highly specialized lime-depositing groups, occur throughout the Phanerozoic. Cer-tain unicellular fossils of the middle and late Precambrian (*Eosphaera, Huroniospora* and *Palaeocryptidium*) also appear referrable to the red algae (Tappan, 1976), as they show indica-tions of both binary fission and budding, as well as morphological similarity to the unicellu-lar Bangiophycidae such as the living *Porphy-ridium*. Hence, the eukaryotic cell evolved prior to 1.9 b yr ago (Gunflint Formation). As the living unicellular Bangiophycidae apparently do not have sexual reproduction, whereas it is well developed in the Florideophycidae, sexual re-combination and meiosis probably first ap-peared in the Protista 1.9–0.9 b yr ago.

Whether red algae were ancestral to other eukaryotes, or represent a parallel but unre-lated evolutionary sequence, is unknown. Green algae also are recognizable in the Precambrian, including probable Prasinophyceae (tasmanitids, leiospheres) and Chlorophyceae, represented by the Chlorococcales (*Botryococcus*), and Dasycladales. As was true of the red algae, the most complete record of the green algae is that of the calcareous taxa, or those with resistant cell or cyst walls. Brown algae (Phaeophyceae) have a spotty record, as they are represented only by carbonaceous films on bedding planes. The Euglenophyceae are still rarer, only iso-lated occurrences in Eocene and Pliocene lake deposits are known.

In contrast to the limited record of the multi-cellular red, green, and brown algae, many phytoplankton groups have an excellent fossil record. Acritarchs (Fig. 2), or algal cysts whose relationship is uncertain, are particularly abun-dant and valuable as index fossils in the upper Precambrian and lower Paleozoic marine strata, becoming rare after the Devonian (Loeblich, 1970; Tappan, 1968; Tappan and Loeblich, 1972). Dinophyceae (Fig. 3), known only from the encysted state, have isolated occurrences in the Paleozoic, but are abundant and distinctive through the Mesozoic and Cenozoic in marine

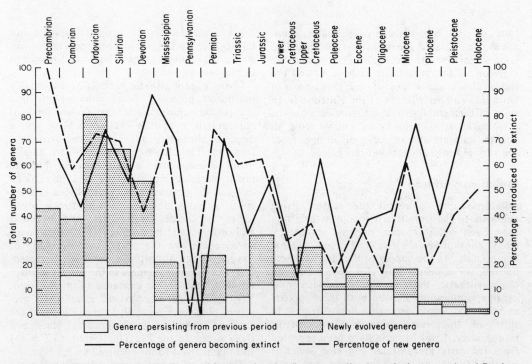

FIGURE 2. Generic diversity of the combined Acritarcha (incertae sedis phytoplankton cysts) and Prasino-phyceae (from Tappan and Loeblich, 1972). Duration of geologic periods not to scale.

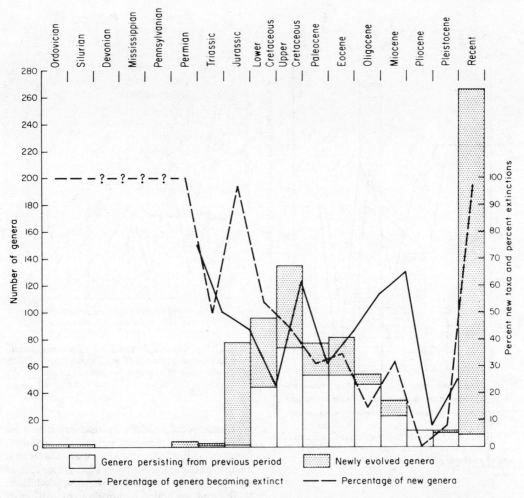

FIGURE 3. Generic diversity of the dinoflagellates (Dinophyceae) (from Tappan and Loeblich, 1972). Duration of geologic periods not to scale.

sediments. A few freshwater fossils are known, and living dinoflagellates also occur in this habitat. Although a few dinoflagellates have calcified cysts, the great majority consist of resistant "sporopollenins." A few dinoflagellates have internal siliceous skeletons, such as *Actiniscus* and related genera of the Miocene to present; and the Ebriales, also variously regarded as a class or order of the dinoflagellates (Loeblich, 1974), similarly are well represented from Paleocene to present by their internal siliceous skeleton.

Other siliceous phytoplankton include representatives of the Chrysophyceae, both the silicoflagellates (see Fig. 4; Loeblich III et al., 1968) and the tiny flask-like cysts of other groups of chrysophytes occurring from the Cretaceous to the present, and of the Bacillario-

phyceae or diatoms, which first appeared in the Cretaceous and expanded rapidly in the Cenozoic, invading a wide variety of aquatic habitats.

The Haptophyceae (Fig. 5), including the coccolithophorids and probably related calcareous nannoplankton, are abundant and serve as excellent biostratigraphic markers in the Mesozoic and Cenozoic, with scattered reports of occurrences in the upper Paleozoic.

Animal protistans, or Protozoa, undoubtedly were present in the Precambrian as a first step prior to the evolution of the Metazoans (Tappan, 1974b); but no recognizable protozoan fossils are known from Precambrian strata. The earliest representatives probably were unpreservable, as Protozoa lack the resistant cell wall of certain plant protists and the highly specialized forms with mineralized test, lorica, or

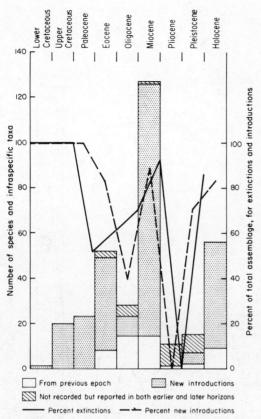

FIGURE 4. Species diversity of silicoflagellates (Dictyochales) (from Tappan and Loeblich, 1972). Duration of geologic periods not to scale.

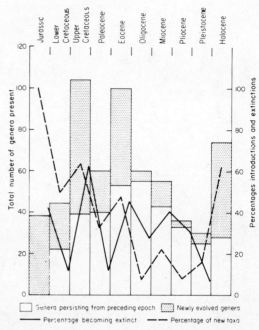

FIGURE 5. Generic diversity of calcareous nannoplankton (Haptophyceae) (from Tappan and Loeblich, 1972). Duration of geologic periods not to scale.

skeleton did not appear until the Paleozoic. Benthic Foraminiferida (Fig. 6) are known from the Late Cambrian, and became much more diverse in the late Paleozoic and post-Paleozoic; planktonic taxa appeared in the Mesozoic (Loeblich and Tappan, 1964, 1974). The exclusively marine Radiolaria occur from the Ordovician to the present, being progressively more useful stratigraphically and highly distinctive in appearance in the Cretaceous and Tertiary. The Tintinnida are known from scattered lower Paleozoic occurrences, and temporarily became important components of the plankton in the upper Jurassic and Cretaceous, only to subside into near obscurity in the Cenozoic, with only a few taxa known from Eocene and Oligocene strata (Tappan and Loeblich, 1968). The only freshwater fossil occurrence is Pleistocene in age. Other Protozoa have very minor records, sufficient only to suggest their antiquity.

The classification and suggested evolutionary relationships of modern Protista are based on data concerning pigments and storage products, flagellar, ciliar, or pseudopodial characters, nuclear phenomena, variations in cytokinesis, as well as life cycles, cell coverings, skeletons, shells, and general cell morphology. Although most of these characters cannot be documented among the fossils, sufficient evidence is present in the fossil record (e.g., cysts, spicules, tests, or skeletons) to indicate the great antiquity and early diversification of the major groups, as well as periods of major biotic turnover and extinction of some protistans, and their eventual replacement by other expanding groups. Many problems concerning the method of origin, and the nature and even timing of protistan evolution and diversification, phylogeny, and interrelationships, are being intensively studied from many viewpoints; and, while some questions probably will remain unanswered, the state of our knowledge of the Protista has increased exponentially during the last decade or two.

HELEN TAPPAN*

*Support is acknowledged from the Earth Sciences Section, National Science Foundation, NSF Grant GA-40419.

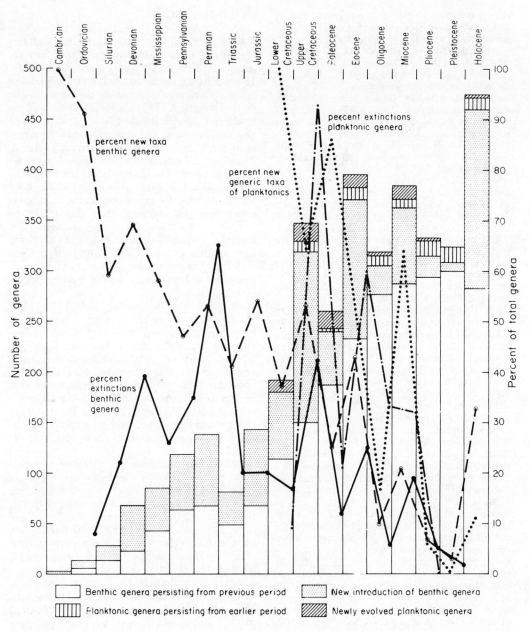

FIGURE 6. Generic diversity of Foraminiferida (from Tappan and Loeblich, 1972). All taxa are benthic through Jurassic, planktic and benthic taxa indicated separately for Cretaceous to Holocene or present. Duration of geologic periods not to scale.

References

Bisset, K. A., 1973. Do bacteria have a nuclear membrane? *Nature,* **241,** 45.

Dodge, J. D., 1966. The Dinophyceae, in M. B. E. Godward, ed., *The Chromosomes of the Algae.* London: Edward Arnold, 96–115.

Fott, B., 1971. *Algenkunde,* 2nd ed. Jena: Gustav Fischer, 581p.

Honigberg, B. M.; Balamuth, W.; Bovee, E. C.; Corliss, J. O.; Gojdics, M.; Hall, R. P.; Kudo, R. R.; Levine, N. D.; Loeblich, A. R., Jr.; Weiser, J.; and Wenrich, D. H., 1964. A revised classification of the Phylum Protozoa, *J. Protozool.,* **11,** 7-20.

Leedale, G. F., 1974. How many are the kingdoms of organisms? *Taxon,* **23,** 261-270.

Loeblich, A. R., Jr., 1970. Morphology, ultrastructure

and distribution of Paleozoic acritarchs, *Proc. N. Amer. Paleontological Conv. 1969,* pt. G, 705–788.

Loeblich, A. R., Jr., 1974. Protistan phylogeny as indicated by the fossil record, *Taxon,* **23,** 277–290.

Loeblich, A. R., Jr., and Tappan, H., 1964. *Treatise on Invertebrate Paleontology,* pt. C, Protista 2, Sarcodina, chiefly "Thecamoebians" and Foraminiferida, 2 vols. Lawrence: Geol. Soc. Amer. and Univ. Kansas Press, 900p.

Loeblich, A. R., Jr., and Tappan, H., 1974. Recent advances in the classification of the Foraminiferida, in R. H. Hedley, and C. G. Adams, eds., *Foraminifera,* vol. 1. London, New York, and San Francisco: Academic Press, 1–53.

Loeblich, A. R., III; Loeblich, L. A.; Tappan, H.; and Loeblich, A. R., Jr., 1968. Annotated index of fossil and Recent silicoflagellates and ebridians with descriptions and illustrations of validly proposed taxa, *Geol. Soc. Amer. Mem., no. 106,* 319p.

McQuade, A. D., 1977. Origins of the nucleate organisms, *Quart. Rev. Biol.,* **52,** 249–262.

Schopf, J. W., and Blacic, J., 1971. New microorganisms from the Bitter Springs Formation (Late Precambrian) of the North-Central Amadeus Basin, Australia, *J. Paleontology,* **45,** 925–960.

Tappan, H., 1968. Primary production, isotopes, extinctions and the atmosphere, *Palaeogeography, Palaeoclimatology, Palaeoecology,* **4,** 187–210.

Tappan, H., 1974a. Protistan phylogeny: Multiple working hypotheses, *Taxon,* **23,** 271–276.

Tappan, H., 1974b. Molecular oxygen and evolution, in O. Hayaishi, ed., *Molecular Oxygen in Biology, Topics in Molecular Oxygen Research.* Amsterdam and Oxford: North-Holland, 81–135.

Tappan, H., 1976. Possible eucaryotic algae (Bangiophycidae) among early Proterozoic microfossils, *Geol. Soc. Amer. Bull.,* **87,** 633–639.

Tappan, H., and Loeblich, A. R., Jr., 1968. Lorica composition of modern and fossil Tintinnida (ciliate Protozoa), systematics, geologic distribution, and some new Tertiary taxa, *J. Paleontology,* **42,** 1378–1394.

Tappan, H., and Loeblich, A. R., Jr., 1971. Geobiologic implications of fossil phytoplankton evolution and time-space distribution, *Geol. Soc. Amer. Spec. Paper, no. 127,* 247–340.

Tappan, H., and Loeblich, A. R., Jr., 1972. Fluctuating rates of protistan evolution, diversification and extinction, *24th Internat. Geol. Cong.,* sec. 7, 205–213.

Tappan, H., and Loeblich, A. R., Jr., 1973. Evolution of the oceanic plankton, *Earth-Sci. Rev.,* **9,** 207–240.

Taylor, F. J. R., 1974. II. Implications and extensions of the serial endosymbiosis theory of the origin of eukaryotes, *Taxon,* **23,** 229–258.

Whittaker, R. H., 1969. New concepts of kingdoms of organisms, *Science,* **163,** 150–160.

Cross-references: *Acritarchs; Algae; Calcispheres and Nannoconids; Charophyta; Chitinozoa; Coccoliths; Diatoms; Dinoflagellates; Discoasters; Ebridians; Foraminifera, Benthic; Foraminifera. Planktic; Hystrichospheres; Paleoecology; Palynology; Plankton; Precambrian Life; Silicoflagellates; Stromatolites; Tasmanitids; Tintinnids.*

PSEUDOFOSSIL

A pseudofossil is a natural object, structure, or mineral of inorganic origin which may resemble or be mistaken for a fossil (Gary et al., 1972). Where question exists as to the organic or inorganic origin of a specimen, the terms *dubiofossil* (Hofmann, 1971), *problematicum(a),* or *problematic fossil* are used. Some true fossils of one taxonomic group may appear to be representative of another only very distantly related. These are also considered pseudofossils by some paleontologists. For example, Andrews (1966) has described a fossil reptile coprolite (q.v.) which superficially resembles a cone of an evergreen tree. He suggests that this might be termed a "pseudo-plant fossil." A celebrated instance of misidentification of this kind was the labeling in the 18th century of a fossil skeleton of an Amphibian as *Homo diluvii testis,* presumed to have been a human drowned in the Noachian flood.

Although the misleading nature of most pseudofossils is readily apparent, some others have been the subject of debate amont paleontologists for many years. This has been particularly true of certain questionable objects found in rocks of Precambrian age. Precambrian rocks are relatively barren of organic remains, so the search for evidence of former life has been intense. The result has been the discovery and naming of a large number of unusual structures that subsequently have proved to be pseudofossils.

Pseudofossils are found in all kinds of rock—sedimentary, igneous, and metamorphic. In sedimentary rocks, many pseudofossils are actually sedimentary structures formed during the transportation and deposition of mud and sand; ripple marks preserved in sandstone have been mistaken for fossil tadpole nests (Hantzschel, 1975), and tool marks (drag marks) formed by pebbles or other debris scraped across soft sediments by currents have been confused with marks made by grazing animals. (The preceding are examples of pseudo-trace fossils.) Rill marks formed by small branching rivulets of water on a beach or in a stream bed may bear some resemblance to plant fossils. Structures interpreted as fragments of arthropods have later been identified as clay galls or desiccation curls (Fig. 1) or mud curls (Cloud, 1973; Häntzschel, 1975). These are created when a surface layer of mud dries and forms flakes that have a tendency to curl; they then may be incorporated in other sediments. Small mounds or pits in shale or fine sandstone appear to be organic in origin but are often caused by escape of water, air, or gas from soft sediments or by the impact of raindrops.

FIGURE 1. Casts of Precambrian pseudofossils formed by curling and desiccation of a thin film of sediment, ca. X0.1 (cover of *Geotimes,* Dec. 1972, photo courtesy of U.S. Geological Survey, from Häntzschel, 1975).

Pseudofossils also form during or after diagenesis (conversion of sediments to rock). Concretions or nodules (Figs. 2, 3) assume subspherical or discoidal shapes which, to the untrained eye, are very much like fossils. They are, in fact, segregations of the minor mineral constituents of some sedimentary rocks, usually being composed of quartz (SiO_2), calcite ($CaCO_3$), hematite (Fe_2O_3), goethite ($HFeO_2$), or gypsum ($CaSO_4 \cdot 2H_2O$). Some minerals such as quartz, calcite, gypsum, barite ($BaSO_4$), and pyrite (FeS_2) form radiating crystals or rosettes that have been mistaken for fossils (Fig. 4). Radiating forms thought by some to be fossil medusae of Phylum Coelenterata have been explained by others as pyrite rosettes formed inorganically (Cloud, 1973). Dendrites (Fig. 5) are dark, finely branching deposits found on bedding surfaces or fracture planes of rocks and are often mistaken for ferns, mosses, or other plant fossils. They are deposits of manganese or iron oxide, perhaps precipitated through the action of bacteria (Billy and Cailleux, 1968), and close examination indicates that they branch in a very irregular fashion, quite unlike plants. Their presence in fractures also suggests that they are not fossils. Stylolites (Fig. 6) are irregular seams commonly

FIGURE 2. Calcareous concretion, ca. X0.5 from Pettijohn, 1957; used with permission of Harper & Row, Inc.

FIGURE 3. "Box work" consisting of vein fillings which have weathered out of a septarian nodule (from Pettijohn, 1957; used with permission of Harper & Row, Inc.). Length of specimen about 12.5 cm.

661

FIGURE 4. Molds of radiating crystal aggregated (?gypsum, ?ice) originally interpreted as algae; ca. X12.5 (from Häntzschel, 1975, courtesy of The Geological Society of America and The University of Kansas).

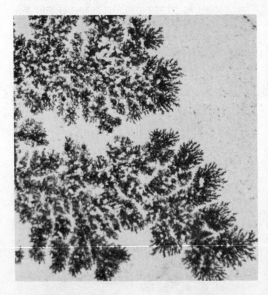

FIGURE 5. Dendrites, ca. X0.5 (from Müller, 1963).

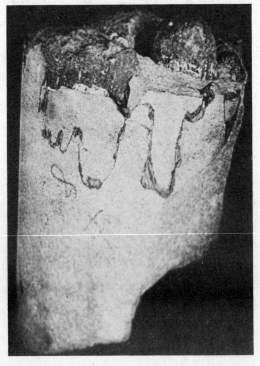

FIGURE 6. Stylolites in sandstone; diameter of core about 5 cm (from Pettijohn, 1957; used with permission of Harper & Row, Inc.).

observed in limestone, dolomite, marble, and sometimes sandstone, which may be confused with fossil corals. They are created by irregular solution under pressure. Also caused by pressure are organic-looking cone-in-cone structures (Fig. 7), nested inverted cones of calcite with ribbed or grooved sides, usually found in shale.

Some pseudofossils are found in igneous and metamorphic rocks. Pseudo-algae (Fig. 8) found in geodes from a lava flow, now a rhyolitic welded tuff, were precipitated within cavities formed as the lava solidified (Brown, 1957). Banded structures (Fig. 9) of calcite or dolomite $[CaMg(CO_3)_2]$ and serpentine $[Mg_3Si_2O_5(OH)_4]$ formed by metamorphism have been called "Eozoon canadense" and erroneously interpreted as stromatoporoids, stromatolites, sponges, and giant Foraminifera (Hofmann, 1971).

A slickenside is a polished and smoothly striated surface that results from friction along a fault plane (Gary et al., 1972). These surfaces

bear a superficial resemblance to fossil coal-forming plants such as cordaites. Whereas plant fossils are always deposited parallel to bedding, slickensides nearly always cut across bedding.

Weathering of rocks at the surface of the earth produces some forms that have been thought to be fossils, particularly where rocks are irregular in composition or contain silica nodules. Solution and replacement of limestone along fractures (Fig. 10) has produced connected solid rods and other structures originally interpreted as algal (Gutstadt, 1975).

Man-made pseudofossils are properly called *artifacts*. These may be created accidentally, as in the preparation of replicas for examination by electron microscopy. A famous 18th-century hoax involved the manufacture by envious colleagues of exotic "fossils" which were foisted upon a fellow faculty member who, unfor-

FIGURE 7. Cone-in-cone structure showing transverse ridges and longitudinal striations; length of cone about 7.5 cm (from Pettijohn, 1957; used with permission of Harper & Row, Inc.).

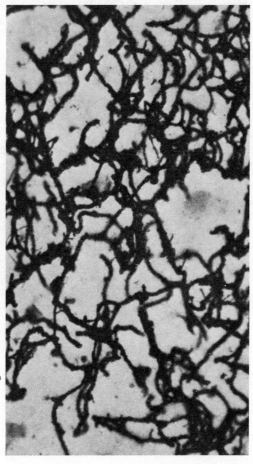

FIGURE 8. Pseudo-algae from geode, showing irregularity of branching and thickness; ca. X20 (from Brown, 1957).

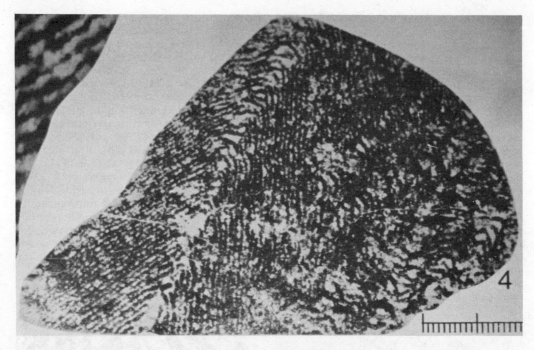

FIGURE 9. "Eozoon canadense," polished surface composed of dark green bands of serpentine and spinel and light colored bands of dolomite; length of specimen ca. 8 cm (from Hofmann, 1971).

FIGURE 10. Pseudofossils resulting from solution and replacement along fractures, ca. X0.5.

tunately, published a learned treatise based on the mischievously deceptive artifacts (Jahn and Woolf, 1963).

The most comprehensive catalogue of pseudofossils is that of Häntzschel (1975), which includes an extensive bibliography. Other useful surveys, mainly of pseudofossils of Precambrian age, are those by Cloud (1968, 1973), Glaessner (1962), and Hofmann (1971). These works deal with pseudofossils that have been seriously considered to be biological remains and have been assigned Latin names. The more common pseudofossils, easily identifiable by anyone acquainted with the fossil phyla and modes of preservation of fossils, are discussed in most textbooks of paleontology or paleobotany (e.g., Andrews, 1966; Matthews, 1962; Müller, 1963).

ALLAN M. GUTSTADT

References

Andrews, H. N., Jr., 1966. *Studies in Paleobotany.* New York: Wiley, 487p.

Billy, C., and Cailleux, A., 1968. Dépôts dendritiques d'oxydes de fer et manganese par action bactérienne, *Compte Rendus Acad. Sci.,* Paris, D, **266,** 1643–1645.

Brown, R. W., 1957. Plantlike features in thunder-eggs and geodes, *Ann. Rept. Smithsonian Inst.,* **1956,** 329–339.

Cloud, P., 1968. Pre-Metazoan evolution and the origins of the Metazoa, in E. T. Drake, ed., *Evolution and Environment.* New Haven, Conn.: Yale Univ. Press, 1–72.

Cloud, P., 1973. Pseudofossils: A plea for caution, *Geology,* **1,** 123–127.

Gary, M.; McAfee, R., Jr.; and Wolf, C. L., eds., 1972. *Glossary of Geology.* Washington, D.C.: Amer. Geol. Inst., 857p.

Glaessner, M. F., 1962. Pre-Cambrian fossils. *Biol. Rev.,* **37,** 467–494.

Gutstadt, A. M., 1975. Pseudo- and dubiofossils from the Newland Limestone (Belt Supergroup, late Precambrian), Montana, *J. Sed. Petrology,* **45,** 405–414.

Häntzschel, W., 1975. Trace fossils and problematica, in C. Teichert, ed., *Treatise on Invertebrate Paleontology,* pt. W, Miscellanea, suppl. 1. Lawrence, Kansas: Geol. Soc. Amer. and Univ. Kansas Press, 269p.

Hofmann, H. J., 1971. Precambrian fossils, pseudofossils, and problematica in Canada, *Geol. Surv. Canada Bull.,* **189,** 146p.

Jahn, M. E., and Woolf, D. J. (translators), 1963. *The Lying Stones of Dr. Johann Bartholemew Adam Beringer being his Lithographiae Wirceburgensis.* Berkeley and Los Angeles: Univ. California Press, 221p.

Karcz, I.; Enos, P.; and Langille, G., 1974. Structures generated in fluid stressing of freshly deposited clays resemble ichnofossils, *Geology,* **2,** 289–290.

Matthews, W. H., III, 1962. *Fossils.* New York: Barnes and Noble, 337p.

Müller, A. H., 1963. *Lehrbuch der Paläozoologie,* vol. 2 Invertebraten, pt. 3, Arthropoda 2–Stomochorda. Jena: Gustav Fischer, 698p.

Pettijohn, F. J., 1957. *Sedimentary Rocks,* 2nd ed. New York: Harper & Row, 718p.

Cross-references; *Fossils and Fossilization; Precambrian Life; Pyritospheres.*

PTERIDOPHYTA

Pteridophyta is an older name for those land plants that possess a vascular conducting system and reproduce by spores rather than by seeds. Ferns, club mosses, horsetails, and whisk ferns (Psilotaceae) are the extant members, some of which have a long geologic record. The catchall group Psilophytales (redefined below) included the earliest fossil precursors. For convenience, the fossil record of pteridophytes is discussed in terms of a classification modified slightly from Foster and Gifford (1974) and Bierhorst (1971). All Classes discussed below belong to the Division Tracheophyta; the first three formerly comprised Psilophytales (Banks, 1968).

Class Rhyniopsida

The first megafossils proven to have a vascular system and resistant, trilete spores isolated from attached sporangia (spore-containing organs; see Fig. 1) lived in the Upper Silurian (Pridolian Stage) strata in Wales approximately 405–395 m yr ago. The occurrence of resistant, trilete spores as microfossils in earlier stages of Silurian to 435 m yr ago leads to speculation on an earlier origin of vascular plants. The question remains unsettled.

The earliest megafossils were simple anatomically, histologically, and morphologically, lacking both leaves and roots, branching only dichotomously, bearing terminal sporangia, and with a slender, solid, centrarch (first cells to differentiate located in center of strand) primary xylem strand. Rhyniopsida became extinct no later than the Late Devonian.

Class Zosterophyllopsida

By the Early Devonian (Gedinnian Stage), approximately 395–390 m yr ago, a new type of leafless vascular plant had evolved with lateral sporangia and exarch (first cells to differentiate located at periphery of strand) primary xylem. Zosterophylls became worldwide in their distribution but persisted only until Late Devonian.

Class Trimerophytopsida

In the Siegenian Stage of the Early Devonian (390–370 m yr ago), rhyniophytes had evolved into a more advanced group with complex

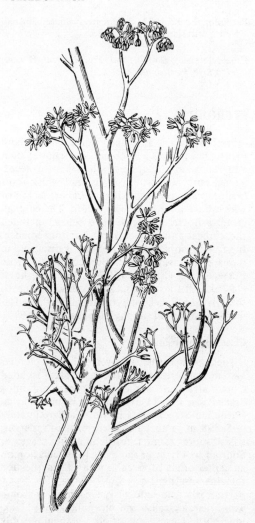

FIGURE 1. *Psilophyton dawsonii.* Reconstruction of an Early Devonian trimerophyte showing the complexity of its branching, its large clusters of sporangia terminal on some branches and its lack of leaves, (from Banks et al., 1975). Its recurved ultimate branchlets may be an early stage in the evolution of large leaves.

lateral branching, possible precursors of large leaves, and large clusters of sporangia terminating branch systems (Fig. 1). These characteristics suggest precursors of higher pteridophytes such as ferns and progymnosperms. Trimerophytes disappeared in the Middle Devonian.

Class Lycopsida

Also by Siegenian time, possible descendants of zosterophylls had appeared. The first lycopods bore small leaves with adaxial sporangia (located on upper side of leaf) and exarch

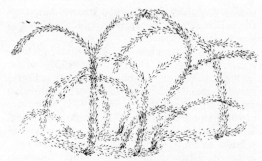

FIGURE 2. *Colpodexylon.* Reconstruction of Late Devonian herbaceous lycopod that showed many features of modern descendants; X1/8 (from Banks, 1970).

primary xylem, all characters so similar to modern genera as to demonstrate the lycopods have had the longest continuous fossil record of any group of vascular plants (Figs. 2 and 3). Zosterophylls and lycopods seem independent of all other groups of pteridophytes. During the Middle and Late Devonian, arborescent lycopods began to appear; and these gave rise to the well-known lepidodendrids, a dominant element in the Pennsylvanian Coal Swamps. Secondary xylem, ligules (tongue-like membranous structure located on leaf near its base), leaf cushions (large pad of parenchymatous tissue on stem surface; it bears a leaf), heterospory, and a close approach to the seed habit all evolved in these giant club mosses, which then disappeared in the Permian. Herbaceous genera have persisted continuously since the Devonian and ligulate, eligulate, homosporous (one kind of spore that produces a bisexual gametophyte), and heterosporous (two kinds of spore, one producing male, one producing female gametophyte) genera exist today.

Class Cladoxylopsida

The class Cladoxylopsida was erected for fossils found in Mississippian strata that were preserved only as petrifactions showing a polystelic (numerous separate xylem strands) stem. From Middle and Upper Devonian strata compression specimens of complexly branched plants with polystelic stems and with various patterns of fertile branches (e.g., *Pseudosporochnus, Calamophyton*) have been added to this group. Some think these aberrant plants were a line parallel to ferns, some relate them to medullosan seed ferns, some speculate that evolution of their xylary tissues could have resulted in the primary xylem configuration characteristic of Carboniferous and younger sphenopsids. They disappeared by the end of the Pennsylvanian.

FIGURE 3. Stem of Devonian lycopod *Colpodexylon* showing places of attachment of pseudo-whorled leaves and some attached three-forked leaves; X2.

Class Sphenopsida

Middle Devonian strata include the Hyeniales, possible ancestors of the class Sphenopsida. Upper Devonian strata include more obvious sphenopsids (e.g., *Pseudobornia, Sphenophyllum, Eviostachya*). By Carboniferous time, arborescent calamites (Fig. 4) had evolved secondary xylem, advanced heterospory, and a wide range of cones of considerable complexity. The whorls of leaves (Fig. 5) and vertically ridged pith casts are conspicuous fossils in Pennsylvanian strata. Like lepidodendrids, these arborescent swamp plants declined by the end of the Paleozoic. Herbaceous representatives persisted in marked abundance through the Mesozoic, but today there remains only the genus *Equisetum* (Fig. 6).

Class Coenopteridopsida

Coenopterids appeared during the Late Devonian and became extinct during the Permian. They are known primarily from petrifactions. Most characteristic are their complex fronds (large, often much-divided leaves) whose anatomy is the chief means of identification of genera. Pinnules (one of the smallest subdivisions of a frond; see Fig. 7, which is just a portion of one frond) have been found on some fronds, other fronds even produced branches. Some pinnules bore nomosporous

FIGURE 4. *Calamites.* Cast of pith from stem of Pennsylvanian arborescent sphenopsid showing a node with a whorl of small scars, one large circular branch scar and internodal longitudinal ridges that alternate at the nodes; X2/3.

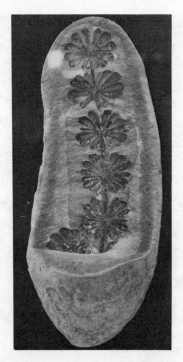

FIGURE 5. *Annularia*. Nodule containing twig of a sphenopsid showing whorls of leaves; X1.

FIGURE 7. *Cladophlebis*. Widespread Mesozoic representative of the fern family Osmundaceae; X1.

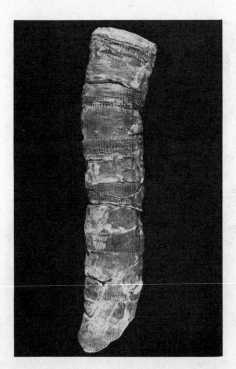

FIGURE 6. *Equisetum*. Cast of Jurassic specimen from Yorkshire. Several nodes and internodes present; X1/3.

sporangia, others were heterosporous. Coenopterids were significant small plants on the floor of Carboniferous Coal Swamps and were epiphytic as well. They were more closely allied to ferns than to any other group and some ferns, e.g., Osmundaceae, may have evolved from them. Trimerophytopsida may have been ancestral to this group.

Class Ophiglossopsida

The adder's tongue ferns are primarily an extant group, comprising three genera of relatively primitive construction, possibly derived from coenopterid ferns, possibly from early progymnosperms or their precursors, the trimerophytes. Probable representatives of the Ophiglossopsida are found in early Tertiary deposits.

Class Marattiopsida

The Marattiopsida were represented by Upper Carboniferous tree ferns with a slender trunk (*Psaronius*) covered by a mantle of adventitious roots and terminated by a crown of large, compound fern-like fronds (*Pecopteris*). Abaxial (on lower surface of leaf) sporangia were arranged in clusters (synangia) as in living representatives. Stem anatomy was monocyclic to a complex polycyclic stele. These ferns were a dominant element in Pennsylvanian–Permian forests. The class continued through the Mesozoic and Cenozoic. There are seven extant genera. Some early Jurassic representatives were essentially identical to modern genera and they lived as far north as Greenland, in sharp contrast to their present tropical distribution.

Class Filicopsida

The phylogenetic relationships and systematic treatment of Filicopsida are dynamic subjects on which authors vary widely. The following are frequently used familial categories among the still-evolving true ferns.

Schizaeaceae *sensu lato*—lived in Pennsylvania forests; bore pecopterid foliage; were widespread in Mesozoic, modern-appearing by the Cretaceous, and are represented today by five genera. Their apically located annulus has remained unchanged. Possible precursors were trimerophytes, possible derivative groups are mentioned below.

Gleicheniaceae—may have been present in the Pennsylvanian, bearing foliage of the sphenopterid type; ranged widely in the Mesozoic; and are currently restricted to one (or more) genus of tropical and south-temperate distribution.

Osmundaceae—possibly derived from coenopterids; typically produced an erect stem; appeared in the Permian, were widespread in Mesozoic, and now are restricted to three genera (Fig. 7). Fossils illustrate a splendid anatomical series from protosteles (a solid strand of xylem) to dictyosteles (primary xylem arranged as a circle of strands around a central pith). The group seems to be an evolutionary dead end.

Matoniaceae—appeared in the Triassic, were world wide by Mesozoic, and essentially modern by the Cretaceous. The two extant genera are now restricted to the tropics. They are possibly related to Gleicheniaceae and Dipteridaceae.

Dipteridaceae—appeared in the Triassic, and the one extant genus, like the Matoniaceae is now reduced from a worldwide to a small tropical distribution. They are apparently related to Gleicheniaceae and Matoniaceae.

Cyatheaceae—(including Dicksoniaceae) appeared in the Triassic and are possibly derived from the Schizaeaceae. They include the tree ferns and are relatively numerous today.

Hymenophyllaceae—may date back to the Jurassic; now comprise two genera with numerous tropical species; may be derivatives of Schizaeaceae; and may be precursors to the water ferns.

Salviniaceae and Marsileaceae—the first with two, the latter with three widely distributed extant genera constitute the heterosporous water ferns that arose in the Cretaceous (possibly earlier). They may be derivatives of the Hymenophyllaceae.

Polypodiaceae, *sensu lato*—possibly originating in the Jurassic; are the most highly evolved, systematically complex, numerous ferns; perhaps derived in part from Matoniaceae.

Class Progymnospermopsida

Progymnosperms reproduced by spores and had a mesarch (first cells to differentiate located in middle of primary xylem) protostele like pteridophytes, but their secondary tissues were gymnospermous. They appeared early in the Middle Devonian with three-dimensional, nonlaminated, lateral appendages. By the Late Devonian, planation (flattening of a three-dimensional system) and webbing (production of laminae on cylindrical axis to produce a leaflike organ) had produced leaves, and stelar evolution had produced a eustele. Homospory gave way to heterospory. Fertile lateral branch systems became complex. These seem to be the pteridophytes whence both cycad- and conifer-type of gymnosperms evolved. They may have arisen by modification of trimerophytes and they persisted perhaps into Mississippian.

HARLAN P. BANKS

References

Andrews, H. N., Jr., 1961. *Studies in Paleobotany.* New York and London: Wiley, 487p.

Banks, H. P., 1968. The early history of land plants, in E. T. Drake, ed., *Evolution and Environment.* New Haven, Conn. and London: Yale Univ. Press, 73–107.

Banks, H. P., 1970. *Evolution and Plants of the Past.* Belmont, Calif.: Wadsworth, 170p.

Banks, H. P.; Leclerq, S.; and Hueber, F. H., 1975. Anatomy and morphology of *Psilophyton dawsonii*, sp. n. from the late Lower Devonian of Quebec (Gaspé) and Ontario, Canada. *Palaeontogr. Amer.,* 8(48), 127p.

Bierhorst, D. W., 1971. *Morphology of Vascular Plants.* New York: Macmillan, 560p.

Boureau, E., ed., 1964. *Traité de Paléobotanique,* vol. III, *Sphenophyta, Noeggerathiophyta.* Paris: Masson et Cie, 544p.

Boureau, E., ed., 1967. *Traité de Paléobotanique,* vol. II, *Bryophyta, Psilophyta, Lycophyta.* Paris: Masson et Cie, 845p.

Boureau, E., ed., 1970. *Traité de Paléobotanique,* vol. IV, *Filicophyta.* Paris: Masson et Cie, 519p.

Edwards, D., and Davies, E. C. W., 1976. Oldest recorded *in situ* tracheids, *Nature, 263,* 494–495.

Foster, A. S., and Gifford, E. M., Jr., 1974. *Comparative Morphology of Vascular Plants.* San Francisco: Freeman, 751p.

Harland, W. B., et al., eds. 1967. *The Fossil Record.* London: Geological Society of London, 827p.

Cross-references: *Gymnospermae; Paleobotany; Palynology, Paleozoic and Mesozoic.*

PYRITOSPHERES

Pyritospheres are organic structures, mainly spheroidal, associated with pyrite framboids and grains.

Spheroidal bodies of organic composition, isolated from framboidal pyrite spherules from rocks by solution of the pyrite in nitric acid, were named *Pyritosphaera barbaria* Love 1957 (gen. and sp. nov.; Figs. 1–1 to 1–6). The size range is that of pyrite framboids, 2–50 μm commonly; and the structure is vesicular, of cavities .25–2 μm, with the whole body being described as having a smooth to jagged or nearly spinose outer surface. The cavities originally contain pyrite grains, often of euhedral form, especially cubes and pyritohedra, and it is these shapes, with some collapse, that give the form of the outer surface. The organic material is therefore a continuous matrix to the individual pyrite grains of the pyrite framboid.

Originally found in the course of palynological investigation, the status of fossil microorganism was given on the combined grounds of regularity of morphology and demonstrable organic composition (translucent pale brown; nonbirefringent; ultimately destroyed by nitric acid; insoluble in hydrofluoric acid; C–H chemical analysis). On the basis of the exclusive association with framboidal pyrite, itself almost a characteristic component of sapropelic and carbonaceous sediment, it was tentatively suggested that some microorganism was represented, of a hydrogen-sulfide producing activity, such as a sulfate reducer. No present-day organism, however, could be matched in morphology.

From the extent of the geological ages—at least Cambrian, if not lower Proterozoic, to Tertiary—of samples from whose framboidal pyrite these organic spherules could be detected, but most particularly from study of pyrite in recent sediments, this hypothesis has been laid aside. In a wide range of Holocene sediments of marine and lacustrine origin, it has not been found possible to demonstrate corresponding organic material in abundant framboids. The bodies are therefore now regarded as fortuitous matrix material to the pyrite grains of the framboid. Possibly they originated by migration of organic material and its deposition, maybe polymerization, about the pyrite. Rickard (1970) has postulated that oil globules are the locus of formation and control the sphericity of framboidal pyrite spheroids; if the tangible organic matrices of ancient framboids were to represent this material, not so far physically revealed in Holocene framboids, then a considerable degree of metamorphism must be involved.

Pyrite spherules and single-crystal grains, Holocene and ancient are frequently found inside tests of foraminifera, ostracods, and cells of plant tissues, and inside a wide variety of simple and multiple organic sacs and cells. The organic material is left when the pyrite framboids are dissolved in the laboratory, when the cells and sacs frequently show an internal ornament that may often be recognized as molded by the outside of a framboid, or a polygonal form reflecting the euhedral pyrite grain. In the latter case the forms were assigned to *Pyritella polygonalis* Love 1967 (gen. sp. nov.), of unknown affinity (Figs. 1–7, 1–8). They are now regarded as representing a systematic modification of organic material at the site of very early diagnetic pyrite formation.

Another form of organic material associated with the pyrite is that produced when the matrix form ("*P. barbaria*") is evicted from its original site on recrystallisation of the pyrite framboid to a simple euhedral crystal grain. Around the latter it then forms a closely fitting and correspondingly shaped envelope, revealed by dissolving the pyrite. When such a pyrite framboid was originally within a sac or cell then this results in the presence of a double sac (Love, 1965). Such double sacs have been recognized from the Mt. Isa Shale (lower Proterozoic, Queensland, Australia). Matrix spheroids ("*P. barbaria*") and associated sacs have been isolated from pyrite framboids and other grains from Cambrian to Tertiary.

A definitive description of framboidal and other very early diagenetic pyrite in rock has been given by Love and Amstutz (1965).

The origin of framboidal pyrite has been discussed by several authors but to no agreed conclusion; and, while laboratory synthesis has been achieved, this points at present to the existence of alternative routes none of which appears to correspond with all the conditions of *in-situ* formation in sediments. It is however, agreed that here their origin results from the reaction of H_2S or sulfide products of anaerobic bacterial activity with Fe of the sediment, first giving mackinawite (FeS), which is converted to pyrite (FeS_2), and the whole process occurs very early in diagenesis in the upper 2 m of sediment or less when sufficient deoxygenation occurs. Frequently nanno- and microorganism tests or cells form a suitable microenvironment in advance of reduction of the whole sediment or in the absence of it. As the deoxygenation requires an excess of early decomposable organic matter over available diffusing oxygen, pyrite within cells and sacs is an indicator that these were added to or were within the sediment immediately after death. Instances have been observed of the distinction

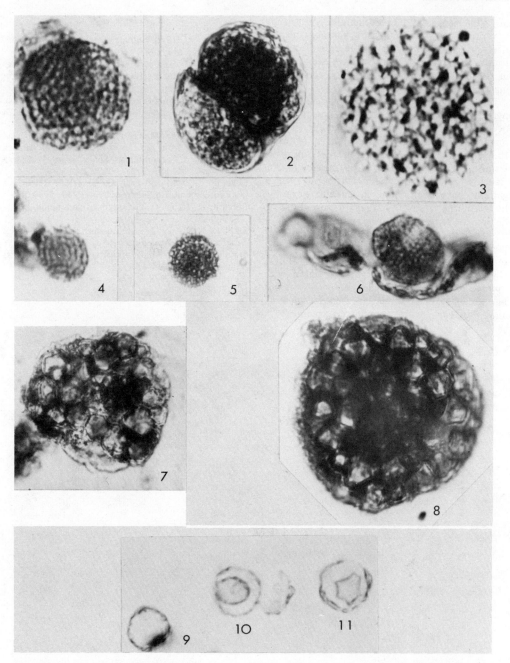

FIGURE 1. 1–6 *Pyritosphaera barbaria* Love 1957, various forms (1, X1860; 2, X2800; 3, X2100; 4,5, X465; 6, X2475; 1,3,4,5 Carboniferous; 2,6, Liassic). 7–8, *Pyritella polygonalis* Love 1957. (Carboniferous; 7, X1325; 8, X1860). 9–11, organic forms after pyrite spherules (9, empty sac; 10, sac with internal *P. barbaria;* 11, sac showing polygonal form with internal strongly polygonal sac, from recrystallisation of framboidal pyrite; all X2325.

of indigenous and derived nannoplankton by the presence and consistent absence, respectively, of pyrite infilling. In the case of derived forms, any pyrite that might have been present in the first deposit would be decomposed during erosion and transport.

The organic material so often associated with pyrite spheroids is a distinctive aspect of the occurrence of very early diagenetic pyrite. Spheroidal, granular forms of pyrite, of much the same size range, and convergent upon the framboidal texture possibly because of a relationship in origin as colloidal spherular segregations, have been found in igneous rocks and hydrothermal deposits. They are usually of a distinctive range of textures apart from the lack of associated organic material. From the Orgveil meteorite, however, Staplin (1962, 1963) has reported forms in isolated organic material, *Clausisphaera* and *Caelestites,* associated with "pyrito-organic?" material. These show some resemblance, respectively, to *"Pyritosphaera"* and *"Pyritella,"* which remains unexplained.

L. G. LOVE

References

Berner, R. A., 1970. Sedimentary Pyrite Formation, *Am. J. Sci.,* 268, 1–23.

Kobluk, D. R., and Risk, M. J., 1977. Algal borings and Framboidal pyrite in Upper Ordovician brochiopods, *Lethaig,* 10, 135–143.

Love, L. G., 1956. Micro-organisms and the presence of syngenetic pyrite, *Quart. Geol. Soc. London,* 113, 429–440.

Love, L. G., 1965. Micro-organic material with diagenetic pyrite from the Lower Proterozoic Mount Isa Shale and a Carboniferous Shale, *Proc. Yorkshire Geol. Soc.,* 35, 187–202.

Love, L. G., and Amstutz, G. C., 1965. Review of microscopic pyrite, *Fortschr. Miner.,* 43, 273–309.

Rickard, D. T., 1970. The origin of framboids, *Lithos,* 3, 269–293.

Staplin, F. L., 1962. Microfossils from the Orgueil meteorite, *Micropaleontology,* 8, 343–347.

Staplin, F. L., 1963. Comments on extra-terrestrial taxa, *Taxon,* 12, 14–15.

Cross-reference: *Pseudofossil.*

R

RADIOLARIA*

Radiolaria are free-living marine pseudopod-bearing protozoans characterized by (1) a radiating network of slender pseudopodia, (2) an organic "central capsule membrane" within the cytoplasm, and (3) a skeleton composed of transparent glass (opal) with or without organic inclusions or of crystalline strontium sulfate (celestite). Only the skeleton is preservable under some conditions, and Radiolaria are best known to paleontologists for the variety of minute, exquisitely structured skeletons of remarkable delicacy, and often great complexity, that are characteristic of most, but not all, modern species. The term *Radiolaria* has had various usages in the classifications proposed over the years, and care must be taken to ascertain which organisms are included within this taxon by different taxonomists. In fact, modern cytological studies have demonstrated that the groups traditionally included in the Radiolaria differ markedly in cell morphology. It is becoming increasingly clear that Radiolaria may be no longer justifiable as the name of a formal taxon, and in the succeeding discussion we will use "radiolarian" (uncapitalized) as an informal name in collective reference to the traditional "radiolarian" groups (Acantharia, Polycystina, and Phaeodaria).

Radiolarians are exclusively holoplanktonic with the exception of one benthic acantharian. Most are solitary and approximately spherical in shape, but members of two families (Collosphaeridae and Sphaerozoidae) form colonies in which hundreds of individuals are enclosed in gelatinous masses. The preponderance of radiolarians fall within the size range of microplankton (20-200 μm), although many are larger and members of one polycystine family (Orosphaeridae; see Friend and Riedel, 1967) produce skeletons as large as 5 mm in diameter. By far the largest known solitary species is a phaeodarian described by Haeckel (1887), *Coelothamnus maximus,* with a skeletal diameter of 32 mm. Radiolarians are present in all of the major oceans and mediterranean seas;

they appear to be well adapted to pelagic waters ranging from polar to equatorial and to depths ranging from the surface to the abyss; although their maximum abundance occurs at the base of the epipelagic zone (Petrushevskaya, 1971b). However, with only a few notable exceptions, they are infrequent in shallow coastal waters.

The ectoplasm of radiolarians emits many radiating, slender pseudopodia. Unsupported pseudopodia may be straight (filopodia), or they may form an anastomosing reticulation (rhizopodia or reticulopodia). Some pseudopodia (axopodia) are stiffened by an internal organic axis (axoneme) composed of a complex bundle of microtubules. Seston in the water column, which serves as food, becomes trapped in the pseudopodial network, and is transported into the main body by cytoplasmic streaming along the pseudopodia. Most radiolarians contain a perforated central capsule membrane, which separates the cytoplasm into two distinct regions. The extracapsular ectoplasmic layer is highly vacuolated and of varying thickness and structure. The endocapsular cytoplasm (endoplasm) is densely packed with most of the cell organelles (one or many nuclei, mitochondria, endoplasmic reticulum, Golgi complex, oil globules, as well as small crystals and concretions of diverse composition and uncertain functions). Although absent in phaeodarians, many epipelagic Acantharia and Polycystina contain symbiotic algae (e.g., dinoflagellates; Taylor, 1974) usually called *zooxanthellae.* Little is known about the relationship of these algae and their radiolarian hosts, but it probably involves the usual benefits of shelter for the algae and energy production for the hosts. Gas-exchange reactions, which could modify the specific gravity of vacuolar liquids, are believed to play a role in the diel vertical migration observed for some radiolarians. Reproductive phenomena are imperfectly known. Much of what is known about the biology and ecology of living radiolarians is based on studies of Phaeodaria and some larger skeletonless Spumellaria (Polycystina). Riedel and Holm (1957) cautioned against the generalization of information concerning specific radiolarians to other members of this heterogeneous group. English summaries of this diverse literature are provided

*Preparation of this entry was supported by National Science Foundation Grant OCE 75-21374.

673

TABLE 1. Variations in Radiolarian Taxonomy

Haeckel (1883, 1887)	RADIOLARIA			
Campbell (1954)	Porulosida		Osculosida	
	Acantharia	Spumellaria	Nassellaria	Phaeodaria
Dreyer (1913)	RADIOLARIA			
Enriques (1932, simplified)	Acantharia	Polycystina		Phaeodaria
		Spumellaria	Nassellaria	
Trégouboff (1953)	ACANTHARIA	RADIOLARIA		
Deflandre (1952, 1953)		Spumellaria	Nassellaria	Phaeodaria
Cachon-Enjumet (1961)	ACANTHARIA	RADIOLARIA		PHAEODARIA
		Spumellaria	Nassellaria	
Riedel (1967)	ACANTHARIA	RADIOLARIA		
	(implied)	Polycystina		Phaeodaria
		Spumellaria	Nassellaria	

by: Campbell (1954), Riedel and Holm (1957), Orlov (1962), Pokorny (1963), Grell (1973), Sleigh (1973), and Casey (1977). Reviews of general radiolarian literature are presented by Berger (1976) and Kling (1978).

Taxonomy

In modern classifications, radiolarians are placed in the Sarcodina (pseudopod-bearing protozoans) as a subdivision of the Actinopoda, which also includes Heliozoa as well as some minor groups (e.g., Honigberg et al., 1964). The name Radiolaria was first proposed by Müller (1859) for marine Sarcodina with radial symmetry, which was used to distinguish them from foraminifera. However, many organisms included in this group do not possess radial symmetry of any type, and Haeckel (1862) emended its definition to include organisms containing a central capsule membrane.

The most important comprehensive taxonomic monograph on radiolarians is the famous *Challenger* Report, in which Haeckel (1887) presented a complete classification for the Radiolaria based on his studies of plankton and sediment samples collected on the *Challenger* Expedition (1873–1876). Haeckel (1883, 1887) subdivided the Radiolaria into four groups: Acantharia, Spumellaria, Nassellaria, and Phaeodaria, all of which are still accepted today and have been the basis of all subsequent classifications. His use of Porulosida and Osculosida is highly artificial and should be discarded; modern cytological data supports the arrangement of Dreyer 1913. Campbell (1954) reexamined the entire Haeckelian classification in order to accommodate later work and to conform to the Code of Zoological Nomenclature, but his volume is flawed by innumerable errors. The main variations in radiolarian taxonomy are summarized in Table 1.

Levine (1963) calculated the number of described species of living and fossil radiolarians at 4791 and 2389 respectively. These numbers are inflated by unwarranted taxonomic splitting, and more conservative species counts would be: about 200 for the Acantharia; more than 600 for the Phaeodaria (Reshetnayak, 1966); and a far higher but uncertain number, perhaps more than 1500, for present and extinct Polycystina.

The Three Main Groups of Radiolarians. The taxonomic rank of the major radiolarian groups is a matter of continuing debate, and it will suffice to treat them informally here. Although a few species from each group have been described as skeletonless, most taxonomists have utilized skeletal morphology to a variable degree for the definition of taxa below ordinal rank.

1. Acantharia. A central capsule membrane is present in most species and is composed of light fibrillar material uniformly perforated by many small pores. Symbiotic algae of uncertain identification are usually endocapsular. The exclusively celestite composition of the skeleton has been well confirmed (e.g., Odum, 1951), although the older literature includes many erroneous statements. Primarily, the skeleton consists of diametral (10 or 16) or radial (20 or 32) spines arranged in fixed geometric configurations (Fig. 1, 1). Members of some families have lattice shells joined to these

FIGURE 1. Line drawings of modern radiolarian skeletons; X200 (from Haeckel, 1862, 1887). 1, Acantharian, *Hexaconus serratus*. 2, Phaeodoarian, *Polypetta tabulata*. 3–6, Spumellarians (3, *Euchitonia mulleri;* 4, *Actinomma asteracanthion;* 5, *Collosphaera huxleyi;* 6, *Sphaerozoum ovodimare*). 7–9, Nassellarians (7, *Archicircus rhombus;* 8, *Lamprocyclas maritalis;* 9, *Callimitra agnesae*).

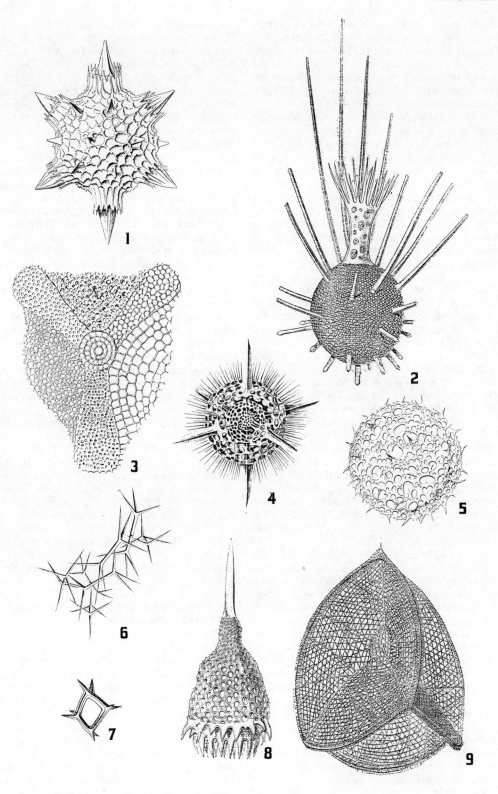

centrogenous spines. Such skeletons are superficially similar to those of Polycystina and Phaeodaria. Haeckel's (1887) *Challenger* monograph is of little value for the taxonomy of Acantharia, which was completely revised by the basic work of Schewiakoff (1926) and summarized by Tregouboff (1953).

2. Phaeodaria (Tripylea). The central capsule membrane appears to be composed of two layers when viewed by transmitted light, but EM studies reveal it to be a single thick layer of dense organic material. The central capsule membrane contains only three perforations: the principal aperture (astropyle) and two accessory apertures (parapylae). There are no algal symbiotes, although concentrations of a brown granular pigment ("phaeodium") are found outside of the astropyle. Some species construct agglutinated skeletons in which foreign particles are embedded in the ectoplasm. However, most Phaeodaria produce skeletons composed of opal with admixed organic matter, which range from dispersed spines to lattice shells of diverse architecture (Fig. 1, 2). Lattice bars and spines commonly take the form of thin-walled hollow tubes filled with fluid. Haeckel's (1887) taxonomic treatment of this group has received little revision. However, later more detailed studies have been presented by Immermann et al. (1904–1926), Haecker (1908), and Reshetnyak (1966).

3. Polycystina. The central capsule membrane is formed by the juxtaposition of dense organic plates, and it is penetrated by many small pores. Symbiotic dinoflagellates are extracapsular. The name Polycystina was proposed by Ehrenberg (1838) for fossil radiolarian skeletons consisting of solid opaline processes. This group includes the two taxa Spumellaria and Nassellaria.

A. Spumellaria: Small pores are uniformly distributed over the surface of the central capsule membrane. The Spumellaria display great variation in skeletal development and include the largest Polycystina, which may be solitary or colonial, and naked or having isolated spicules dispersed throughout the ectoplasm. Most Spumellaria are smaller, solitary, and have skeletons with varying degrees of organization (Fig. 1, 3–6; Fig. 2, 1). A few species have two to five concentric lattice shells interjoined by lattice spines. These specimens are classical examples of spherical symmetry. More commonly, however, symmetry is radial with unequal development in one axis. On other specimens, uniformly trelliced lattice shells are replaced by dense developments of chaotically arranged lattice bars in a "spongy" configuration.

B. Nassellaria: Perforations of the central capsule membrane are concentrated at one pole as a pore plate, which forms the base of the endocapsular axonematic cone. The nassellarian skeleton is characterized by bilateral symmetry and monaxial development. Consequently, the lattice shell commonly assumes a conical form, which may be divided by transverse constrictions into a uniserial arrangement of incomplete chambers (Fig. 1, 7–9; Fig. 2, 2–5). However, this basic configuration may be highly modified, and some Nassellaria have skeletons that are reduced to simple rings and branched spines.

Whereas the two main polycystine groups, Spumellaria and Nassellaria, are morphologically distinct and undeniably closely related, taxonomic subdivision within each group is presently in a state of confusion. Taxonomists since Haeckel's time have attempted to erect classifications for the Polycystina based on evolutionary interpretations. Haeckel was one of the earliest defenders of Darwinism, and his classification (Haeckel, 1887) was based on a hypothetical phylogeny. Although his classification of the Polycystina has come into serious question only in recent years, it is now apparent that Haeckel's dendrograms are of little value. The first major reassessment of polycystine taxonomy is that of Riedel (1967). Petrushevskaya (1971a) and Petrushevaskaya and Kozlova (1972) have proposed emended classifications for the Nassellaria, in which Riedel's nine families are recognized, but further subdivided into 15 and 20 taxa, respectively, of varying rank. Yet additional revision is necessary, especially for the Spumellaria, and the proper assignment of genera and species within these higher categories will require years of effort. The first two volumes of a catalogue of the Polycystina have been published by Foreman and Riedel (1972). Mesozoic Polycystina have been the subject of taxonomic treatments by Foreman (see Foreman, 1977, and references therein), Pessagno (see Pessagno, 1977, and references therein), and Dumitrica (1970). Paleozoic Polycystina are reviewed by Holdsworth (1977).

New insights into the taxonomy of the Polycystina have been provided by cytologists, who insist that classifications based exclusively on skeletal morphology are artificial and strongly at variance with comparative cytoplasmic studies (Hollande and Enjumet, 1960). Cachon and Cachon, (1974) have proposed "cytological" subdivisions of the Spumellaria and Nassellaria, but these schemes are difficult to apply routinely because they require cytological examination of thin sections. Although

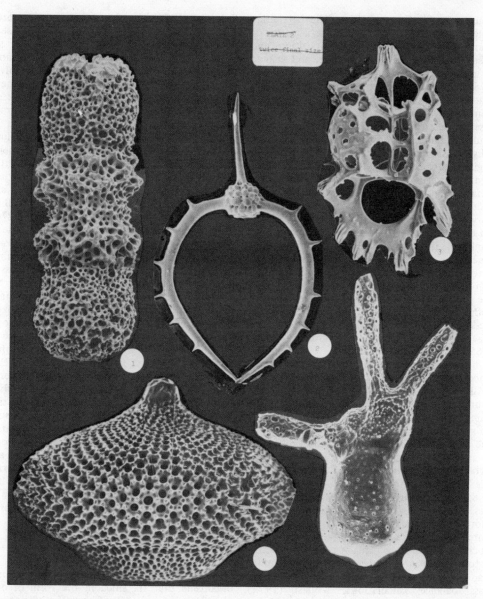

FIGURE 2. Scanning electron photomicrographs of Miocene Polycystina (from Goll, 1972b). 1, Spumellaria: *Cannartus* (?) *petterssoni*, X290: 2-5, Nassellaria; 2, *Dorcadospyris dentata*, X125; 3, *Tholospyris infericosta*, X125; X300; 4, *Cyclampterium* (?) *brachythorax*, X290; 5, *Acrobotrys tritubus*, X290.

their work is unquestionably of great importance and may foreshadow major revisions in polycystine taxonomy, the number of species that have been analyzed at the present time is too small to verify the validity of such classifications. Most taxonomists favor the relative ease of recognition afforded by skeletal morphology. The Polycystina have recieved greater attention in the past from paleontologists than from biologists, and the traditional classifica-

tions of this group reflect the concentration on study of preservable structures.

Skeletons and Preservation

Skeletal structures produced by Acantharia and Phaeodaria are rarely preserved over the course of geologic time. Sea water is strongly undersaturated in strontium, and Acantharia must maintain their skeletons against an enormous concentration gradient. The crystalline skeleton dissolves very rapidly (in a matter of hours) after protoplasmic decomposition, and fossil Acantharia are extremely rare and suspect of misidentification (see Orlov, 1962). Berger (1968) concluded that whole skeletons of Phaeodaria dissolve somewhat more rapidly than polycystine skeletons, but the rarity of observations of fossil Phaeodaria may be a result of the relatively low numbers of living individuals and the susceptibility of their fragile, and often weakly articulated, skeletons to disintegration. Scarce occurrences of Phaeodaria are reported from recent sediments (Ling, 1966; Reshetnyak, 1971), especially in the Norwegian Sea (Stadum and Ling, 1969); but most Phaeodaria are known only from the plankton. Isolated Cretaceous occurrences of Phaeodaria have been reported from central Europe and Puerto Rico; however, reexamination of each occurrence has led to their respective recognition as hystrichosphaerids (see Deflandre, 1960) or Spumellaria and Nassellaria (Pessagno, 1977). Descriptions of Miocene Phaeodaria from Romania (Dumitrica, 1964, 1965) and Kamchatka (Runeva, 1974) appear to be valid, and are the oldest known occurrences of these organisms.

All the modern oceans are also highly undersaturated with respect to dissolved silica (silicic acid), and exposed polycystine skeletons are subjected to extensive dissolution pressures. The degree of silicic acid undersaturation of the ocean ranges from essentially 100% in warm surface water, where phytoplankton deplete this nutrient to unmeasurable concentrations, to 85% for Antarctic bottom water. The total mass of polycystine opal suspended in the photic zone is small compared to that represented by diatoms. However, because of their much larger surface area to volume ratios, diatom skeletons dissolve much more rapidly than those of the Polycystina. On the basis of mass balance calculations, Wollast (1974) and Heath (1973) concluded that 97% of the opal taken up biologically dissolves before reaching the sea floor and only 0.3% is preserved in the sediments. The rate of opal dissolution varies directly with water temperature and pH, and inversely with pressure and dissolved silica concentration (Hurd, 1972, 1973). Hurd (1972) concluded that opaline particles descending through the water column dissolve 16 times more slowly in cold, silica-rich Antarctic bottom water than in equatorial surface waters of the Pacific. Nevertheless, large volumes of polycystine skeletal material reach the sea floor in equatorial regions encapsulated in fecal pellets (Adelseck and Berger, 1975).

Johnson (1976) hypothesized that the preservation of siliceous skeletal debris in marine sediments may be adversely effected by the cooccurrence of some unstable terrigenous silicates that react with interstitial dissolved silica. Lewin (1961) found that certain metal coatings on diatom frustules were quite effective in retarding dissolution. Electron probe analyses of fossil and recent Polycystina by Stanley (1973) indicate magnesium, calcium, sodium, and aluminum as minor elements present in concentrations up to 4% by weight.

After burial, biogenic opal must come into equilibrium with sediment pore waters and other silicate phases, which may take up silicic acid under certain conditions (Wollast, 1974). Moreover, sediments are believed to release silicic acid to the water column by diffusion across the ocean-sea floor interface (Schink, 1968). These factors lead to the progressive destruction of polycystine skeletons over the course of geologic time. Yet there are many fossil occurrences of well-preserved Cenozoic Polycystina. Hurd (1973) calculated that opal particles dissolve in the sediments as much as eight orders of magnitude more slowly than particles suspended in the water column, thus implying that other silica-containing minerals must be active in regulating interstitial silicic acid concentrations. This retarded dissolution may result from close packing of particles such that silicic acid diffusion halos overlap. Alternatively, the strong negative charge on opal surfaces may attract iron, magnesium, and calcium alumino-silicate authigenic minerals. Coatings of organic carbon may produce the same result. The preservation or dissolution of biogenic opal is the subject of growing interest for chemists and geologists in recent years.

Fossil Record and Biostratigraphy

In addition to radiolarians, other marine organisms that fix silica in the form of opal include: diatoms, silicoflagellates, ebridians, the Hyaliospongia, the coralline sponges, many Demospongia, some dinoflagellates, and a few gastropods (see *Biomineralization*). However, Polycystina have a common sedimentary co-occurrence only with diatoms, silicoflagellates, and sponge spicules. Most Polycystina are easily recognizable by their solid skeletal bars and spines, whereas hollow central canals are present in sponge spicules as well as in the delicate rings and stellate lattice plates of silicoflagellates (Riedel, 1963). The larger, more massive rings, spines, and open-lattice shells produced by Polycystina differ markedly from diatom frustules, which commonly take the form of discoid, elliptical, triangular, or rod-like sheets of opal perforated by minute pores arranged in concentric or radiating rows.

Much of the sea floor of the open ocean is covered by sediments composed largely of the skeletal remains of pelagic plankton. Deep-sea sediments containing more than 30% by weight biogenic opal are collectively referred to as *siliceous oozes*. Oozes in which polycystine debris are the dominant constituent are called *radiolarian oozes*. In modern oceans, siliceous oozes accumulate under portions of surface-current divergence zones where calcareous skeletal debris undergo dissolution (Berger, 1974). Riedel (1959) and others found diatom oozes in the Pacific S of the southern Subtropical Convergence and radiolarian oozes in the equatorial Pacific. However, Polycystina are almost invariably present in significant numbers in diatom ooze, and such sediments characterize the sea floor underlying much of the Antarctic circumpolar belt (Hays, 1965). Radiolarian oozes are widely distributed in the Indian Ocean (Nigrini, 1967). True radiolarian ooze has not been observed in the Atlantic, although significant concentrations of polycystine skeletons have been mapped in sediments E of the mid-Atlantic Ridge (Goll and Bjørklund, 1971, 1974).

In general, well-preserved Polycystina become progressively less abundant in older sediments, and Paleocene occurrences are quite rare. However, there are notable exceptions, such as during the Jurassic–Cretaceous and Eocene, when extensive accumulations of Polycystina were preserved. Many Paleozoic and Mesozoic occurrences have undergone mineralization during which structural details are lost. In the course of this transformation from opal to cristobalite, pores in the lattice shells become filled and only the gross structure of these specimens is discernible. However, isolated occurrences of well-preserved ancient Polycystina have been found in coprolites and concretions. For a fuller treatment of the polycystine fossil record, the reader is referred to Kobayashi and Kimura (1944), Campbell and Holm (1957), Grunau (1965), Ramsay (1973), and Reidel and SanFillipo (1977).

The process of transformation of Polycystina and other forms of skeletal opal to chert has been investigated by Heath and Moberly (1971), Heath (1973), Wise and Weaver (1974), and Hurd and Theyer (1975). Methods for the separation of Polycystina from matrix material and preparation techniques are described by Cambell (1954), Burma (1965), and Pessagno and Newport (1972).

The relative paucity of fossil Polycystina in sediments of shallow-water marine origin has had two major consequences. Not only have Polycystina had limited application to problems of local stratigraphic correlation, but efforts to use Polycystina for global biostratigraphy have been hampered as well. Fossil Polycystina occur in very few of the European stratotype sections of the Tertiary stages. Hence, determination of series boundaries has been difficult in sedimentary sections where Polycystina are the only well preserved fossils. This problem has been eliminated to a large degree in recent years by the collection of long sequences of cores drilled in deep-sea sediments containing well-preserved calcareous and siliceous microfossils. Analysis of these drilled cores, and the geomagnetic reversal histories of long piston cores, has led to the establishment of radiolarian zonations that can be correlated to zonations based on other fossil groups as well as stage/series boundaries and radiometric ages with increasing accuracy (Moore, 1972; Goll, 1972a; Theyer and Hammond, 1974a,b; for additional references see Hays and Donahue, 1972; Riedel and Sanfillipo, 1977; Foreman, 1977).

Evolution

Polycystina of possible Precambrian age have been reported in deposits from France, Australia, and India (Kobayashi and Kimura, 1944; Rao and Mohan, 1954) but the age assignment of some of these rocks has been questioned and their identity is subject to doubt (Deflandre, 1952). As with many Paleozoic and even younger occurrences, these oldest specimens are very poorly preserved cristobalitic recrystallizations in silicified rocks. In the past, this type of preservation required examination by thin section, and it is very difficult to distinguish fossils from inorganic cavity fillings by this technique. Like many other fossil-forming organisms, the gross identification of Polycystina is based on the presence of certain anatomical properties (central capsule, pseudopodia, etc.) that are never preserved, and the recognition of fossils as Polycystina is largely a subjective determination of the degree to which they resemble modern polycystine skeletons. Our understanding of Paleozoic Polycystina rests largely on very rare occurrences where specimens can be partially or completely separated from their carbonate or phosphate matrix, and detailed structures have been accurately preserved in the recrystallization process. The oldest such occurrence presently known is the Valhallfonna formation of Spitzbergen, in which latticed skeletons of unquestioned polycystine origin are preserved in Lower

Ordovician calcite micrites (Fortney and Holdsworth, 1971). Interestingly, abundant benthic invertebrate fossils not only attest to a shallow-water depositional environment of these rocks, but also enable confident age assignment for the polycystine horizons.

A few Paleozoic Polycystina have vague nassellarian affinities (such as *Cyrtentactinia* Foreman, 1963 and the *Pylentonematidae* Deflandre, 1973) and have been regarded as intermediate between Nassellaria and Spumellaria. However, skeletal types obviously homologous with modern Nassellaria do not occur in sediments older than Mesozoic (Fortney and Holdsworth, 1971). Assuming the validity of some reported Paleozoic Nassellaria, Tappan and Loeblich (1973) tabulated the number of spumellairan and nassellarian species in Phanerozoic eras and illustrated the gradual increase in relative frequency of Nassellaria during the Mesozoic and Cenozoic such that they represent over 70% of the modern faunas. However, the modern dominance of Nassellaria may be partially the result of selective preservation (Caulet, 1974).

Misconceptions about the rate of radiolarian evolution has resulted in reports that they are a conservative group that has changed little over the course of their long history (Campbell, 1954). These erroneous conclusions gave rise to a long-pervasive opinion that Polycystina have minimal applicability as biostratigraphic indices. The notion of conservative or slow polysystine evolution has its roots in studies of poorly preserved fossil material by early paleontologists. Examinations of specimens preserved in cherts and jasper were followed by extensive speculation about their original morphology (e.g., Ruedemann and Wilson, 1936). Rigid adherence to Haeckel's classification led to the assignment of these "reconstructed" Paleozoic and Mesozoic Polycystina to genera defined to accommodate modern faunas. Hence, it is not uncommon to find stratigraphic ranges for spumellarian taxa given as Cambrian to Holocene (Campbell, 1954). In fact, examination of well-preserved Paleozoic Spumellaria reveals that they are structurally very distinctive from younger representatives, and new families are being erected for them. Fortney and Holdsworth (1971) concluded that the conservative nature of radiolarian evolution, "is essentially mythical." Studies of well-preserved and diverse Cretaceous and Cenozoic faunas by many workers reveals that Polycystina are evolving at least as rapidly as other groups of marine plankton.

Several observers have remarked on the general time-progressive trend among Cenozoic tropical Polycystina toward reduced skeletal weight resulting from larger lattice pores, more slender lattice bars, and simplified structure. After reviewing this evidence, Harper and Knoll (1975) postulated that this evolution is an adaptation to the reduced availability of dissolved silica in sea-water resulting from long-term concomitant increases in diatom biomass. This interesting model is difficult to test, but may deserve further consideration.

Polycystina have played a central role in the controversey concerning the synchroneity of evolutionary events and reversals in the earth's magnetic field throughout geologic history. The evolutionary extinctions of some Polycystina in deep-sea cores from both high and low latitudes closely approximate horizons of magnetic field reversals (for references see Hays and Donahue, 1972). a casual relationship between these two phenomena cannot be demonstrated, but is has been suggested that these extinctions may have resulted from any of three possible causes associated with magnetic reversals: (1) increased cosmic radiation; (2) climatic changes; (3) physiological sensitivity of organisms to the earth's magnetic field. The first of these explanations invokes a reduction or absence of the magnetic field during the transition phase of dipole reversal. Exposure to much higher radiation dosages during these relatively unshielded intervals is said to result in increased mutation frequencies and, in some case, extinction. This model has been challenged on theoretical grounds by Black (1967), Waddington (1967), and Harrison (1968). There is little direct evidence of climatic shifts severe enough to produce extinctions at magnetic boundaries, but the direct influence of magnetism on animals and plants is thought to be a fruitful area for further study. Correlation between magnetic reversals and diatom extinctions (Hays and Donahue, 1972) is much weaker than for Polycystina, and similar evidence in other groups of protist plankton is lacking.

The present stage of polycystine studies is largely descriptive, and comparatively little effort has been devoted to deciphering evolutionary processes from their fossil record. Polycystina provide an excellent history of organic evolution in the marine pelagic ecosystem because: (1) they produce internal skeletons of complex morphology; (2) each skeleton is the sole inorganic product of an entire single individual; and (3) they are abundant in Cenozoic deep-sea sediments, and thousands of specimens of each species are available for examination. With the acquisition of many drill cores (Deep Sea Drilling Project) and long piston cores, earlier evolutionary speculations are being replaced by studies in which age relationships of fossil taxa are understood with relative accuracy (Sanfilippo and Riedel, 1970; Moore, 1972; Goll, 1972). It has long been recognized that Polycystina display long-term phyletic trends, and the evolutionary rates and coefficients of variation for such gradational changes have been calculated for a few taxa (see Kellogg, 1976, for references). Yet, the polycystine fossil record has been the subject of widely differing interpretations. Riedel and Sanfilippo (1974b) conclude that Polycystina have evolved orthogenetically, while Kellogg (1976) found evidence of character displacement. From these divergent observations it appears that much can be read into the preserved history of polycystine skeletal development. However, it seems reasonable to conclude that the latitude of these opinions results largely from the lack of attention this field has received until recent years, and that Polycystina have great potential for additional studies.

Paleoecology

Before 1965, much of the limited paleo-ecological analysis of Polycystina was directed toward attempts to ascertain whether sediments containing their remains were of deep-water or shallow-water origin. This literature has been synthesized and annotated by Campbell and Holm (1957), but the question has never been answered satisfactorily.

Despite the paucity of biogeographical data on living Polycystina, numerous investigations demonstrate that their skeletal remains in surface sediments of the sea floor accurately reflect the major zooplanktonic distribution provinces. These studies include Hays (1965) and Petrushevskaya (1967) for the Antarctic; Nigrini (1967, 1968, 1970) for the Indo-Pacific; and Petrushevskaya and Bjørklund (1974), and Goll and Bjørklund (1971, 1974) for the Atlantic. Epipelagic plankton, in turn, are controlled in their distribution by chemical and physical properties and biological associations of surface waters. However, confidence concerning the reliability of fossil Polycystina as indicators of past oceanic environments must be tempered by the observations of anomalies between surface-water populations of Polycystina and their skeletal remains in sea-floor sediments (Casey, 1971; Renz, 1976).

Radiolarian paleoecological research has been conducted principally in polar-subpolar regions. Hays (1965) initiated the study of Polycystina as climatic indices by delineating alternating stratigraphic intervals characterized by cold and warm assemblages in piston cores from the Antarctic and Subantarctic. Using these same

assemblages or more simplified schemes based on the relative frequencies of two or three particularly sensitive species, Huddlestun (1971), Keany (1973), and Fillon (1973) have found evidence of four to six intervals of warmer temperatures in the Subantarctic during the past 700,000 years that are in good agreement with interpretations based on planktonic foraminifera. However, the exact chronology of these climatic oscillations is problematical, and there are major discrepancies in the inferred temperatures of the Subantarctic Pliocene and early Pleistocene (Bandy et al., 1971). Radiolaria in surface sediments of the Atlantic and west Indian sectors of the Antarctic Ocean have been investigated by Q-mode factor analysis in order to determine assemblages that reflect the physical setting of the overlying water column (Lozano and Hays, 1976; Hays et al., 1976). By the application of this procedure to subsurface sediments, these authors have compiled an isotherm map of the most recent glacial maximum (ca. 18,000 yr B.P.) as well as an inferred paleotemperature curve for the last 230,000 years.

The first paleoecological study of North Pacific Radiolaria is that of Nigrini (1970), who used recurrent group analysis to recognize six modern assemblages. These assemblages were examined in a long piston core, and a "temperature" curve was computed which correlated well with earlier diatom results. This study was soon followed by others (Moore, 1973, 1978; Sachs, 1973a,b; Robertson, 1975) from which it has been possible to deduce much about North Pacific surface waters and location of the Alaskan Gyre for the late Pleistocene back to 450,000 yr B.P.

There have been comparatively fewer paleo-ecological studies of tropical Radiolaria, because climatic fluctuations are regarded to have been much less severe than in higher latitudes. Nevertheless, Johnson and Knoll (1974) have demonstrated a good correlation between curves of weight percent carbonate and a "climatic index" based on recurrent group analysis of radiolarian assemblages in two piston cores from the eastern equatorial Pacific. Applying a similar statistical technique, Casey (1972) has compiled paleotemperature curves for the past 5 m yr for submarine drilled cores (Experimental Mohole) and sections cropping out in Southern California.

Inherent in all of the studies reviewed above is the underlying assumption that surface-water temperatures control the distributions of modern Polycystina. There is no literature to either confirm or negate this basic premise. However, these studies provide ample evidence that the sediments contain the record of measurable shifts in radiolarian faunas, and these fluctuations probably represent responses to Neogene climatic changes.

ROBERT M. GOLL
E. GEORGES MERINFELD

References

Adelseck, C. G., and Berger, W. H., 1975. On the dissolution of planktonic foraminifera and associated microfossils during settling on the sea floor, in Dissolution of deep-sea carbonates, *Cushman Found. Foram. Research, Spec. Publ. No. 13,* 70–81.

Bandy, O. L.; Casey, R. E.; and Wright, R. C., 1971. Late Neogene planktonic zonation, magnetic reversals, and radiometric dates, Antarctic to the Tropics, in *Antarctic Oceanology 1,* Antarctic Research Ser., vol. 15, Washington: Amer. Geophysical Union, 1–26.

Berger, W. H., 1968. Radiolarian skeletons: Solution at depth *Science,* 159, 1237–1239.

Berger, W. H., 1974. Deep-sea sedimentation, in C. A. Burk and C. L. Drake, eds., *The Geology of Continental Margins.* New York: Springer-Verlag, 213–241.

Berger, W. H., 1976. Biogenous deep-sea sediments: Production, preservation and interpretation, in J. P. Riley and R. Chester, eds., *Chemical Oceanography,* 2nd ed. London: Academic Press, vol. 5, 266–388.

Black, D. I., 1967. Cosmic ray effects and faunal extinctions at geomagnetic field reversals, *Earth Planetary Sci. Letters,* 3, 134–143.

Burma, B. H., 1965. Radiolaria, in B. Kummel and D. M. Raup, eds., *Handbook of Paleontological Techniques.* San Francisco: Freeman, 7–14.

Cachon, J., and Cachon, M., 1974. Les systèmes axopodiaux. *Ann. Biol.,* 13, 523–560.

Cachon-Enjumet, M., 1961. Contribution à l'étude des Radiolaires Phaeodariés, *Arch. Zool. exp. gén.,* 100, 151–237.

Campbell, A. S., 1954. Radiolaria, in R. C. Moore, ed., *Treatise on Invertebrate Paleontology,* Part D, Protista 3. New York: Geol. Soc. Amer., 11–163.

Campbell, A. S., and Holm, E. A., 1957. Radiolaria, *Geol. Soc. Amer., Mem. No. 67(2),* 737–743.

Casey, R. E., 1971. Distribution of polycystine Radiolaria in the oceans in relation to physical and chemical conditions, in Funnell and Riedel, 1971, 151–159.

Casey, R. E., 1972. Neogene radiolarian biostratigraphy and paleotemperatures; southern California, the experimental Mohole, Antarctic core E 14–8, *Palaeogeography, Palaeoclimatology, Palaeoecology,* 12, 115–130.

Casey, R. E., 1977. The ecology and distribution of Recent Radiolaria, in Ramsay, 1977, 809–845.

Caulet, M. J., 1974. Les Radiolaires des boues superficielles de la Méditerranée, *Bull. Mus. Natl. Hist. Nat.,* ser 3e, 249, 217–288.

Deflandre, G.. 1952. Classe des Radiolaires, in J. Piveteau, ed., *Traité de Paléontologie,* vol. 1. Paris: Masson, 303–313.

Deflandre, G., 1953. Radiolaires fossiles in Grassé, 1953, vol. 1, pt. 2., 389–436.

Deflandre, G., 1960. A propos du développement des recherches sur les Radiolaires fossiles, *Rev. Micropaleontologie,* 2, 212–218.

Deflandre, G., 1973. Sur quelques nouvelles espèces d'*Archocyrtium,* Radiolaires Pylentonemidae du Viséen de Cabrières, *Comptes Rendus Acad. Sci. Paris,* 277, 149–152.

Dreyer, F., 1913. Die Polycystinen der Plankton-Expedition, *Ergebn. Plankton-Exped. Humboldt-Stiftung,* 3, (L d and e), 1–104.

Dumitrica, P., 1964. Asupra prezentei unor radiolari din familia Challengeridae (Ord. Phaeodaria) in Tortonianul din Subcarpati, *Studii Cerc. Geogr. Geof. Geol., Bucarest,* 9, 217–222.

Dumitrica, P., 1965. Sur la présence de Pheodaires dans le Tortonien des Subcarpathes roumaines, *Comptes Rendus Acad. Sci., Paris,* 260, 250–253.

Dumitrica, P., 1970. Cryptocephalic and cryptothoracic Nassellaria in some Mesozoic deposits of Romania, *Rev. Roum. Géol. Géolphys., Géogr.-Géol. Ser.,* 14, 45–124.

Ehrenberg, C. G., 1838. Über die Bildung der Kreidefelsen und des Kreidemergels durch unsichtbare Organismen, *Abh. Kgl. Akad. Wiss. Berlin, Jahrg.,* 1838, 59–147.

Enriques, P., 1932. Saggio di una classificazione dei Radiolari, *Arch. Zool. Torino,* 16, 978–994.

Fillon, R. H., 1973. Radiolarian evidence of late Cenozoic oceanic paleotemperatures, Ross Sea, Antarctica, *Palaeogeography, Palaeoclimatology, Palaeoecology,* 14, 171–185.

Foreman, H. P., 1963. Upper Devonian Radiolaria from the Huron member of the Ohio shale, *Micropaleontology,* 9, 267–304.

Foreman, H. P., 1977. Mesozoic Radiolaria from the Atlantic Basin and its borderlands, in Swain, 1977, 305–320.

Foreman, H. P., and Riedel, W. R., 1972. *Catalogue of Polycystine Radiolaria,* vols. 1–2. New York: Amer. Mus. Natl. Hist.

Fortney, R. A., and Holdsworth, B. K., 1971. The oldest known well-preserved Radiolaria, *Boll. Soc. Paleontol. Ital.,* 10, 35–41.

Friend, J. K., and Riedel, W. R., 1967. Cenozoic orosphaerid radiolarians from tropical Pacific sediments, *Micropaleontology,* 13, 217–232.

Funnell, B. M., and Riedel, W. R., eds., 1971. *The Micropalaeontology of Oceans.* Cambridge: Cambridge Univ. Press, 151–159, 309–317, 343–349.

Goll, R. M., 1972a. Systematics of eight *Tholospyris* taxa (Trissocyclidae: Radiolaria), *Micropaleontology,* 18, 443–475.

Goll, R. M., 1972b. Leg 9 synthesis, Radiolaria. *Initial Rept., Deep Sea Drilling Proj.,* 9, 947–1058.

Goll, R. M., and Bjørklund, K. R., 1971. Radiolaria in surface sediments of the North Atlantic Ocean, *Micropaleontology,* 17, 434–454.

Goll, R. M., and Bjørklund, K. R., 1974. Radiolaria in surface sediments of the South Atlantic, *Micropaleontology,* 20, 38–75.

Grassé, P. P., ed., 1953. *Traité de Zoologie.* Paris: Masson, 271–436.

Grell, K. G., 1973. *Protozoology,* Berlin: Springer-Verlag, 554p.

Grunau, H. R., 1965. Radiolarian cherts and associated rocks in space and time, *Eclogae Geol. Helvetiae,* 58, 158–208.

Haeckel, E., 1862. *Die Radiolarien (Rhizopoda Radiolaria). Eine Monographie.* Berlin: Reimer, 572p.

Haeckel, E., 1883. Über die Ordnugen der Radiolarien, *Sitzungsber. Jenaischen Gesellsch. Med. Naturwiss., Jahrg 1883*, 1–19.

Haeckel, E., 1887. Report on the Radiolaria collected by H. M. S. Challenger during the years 1873–1876, *Reports Voyage Challenger, Zool*, 18, 1803p.

Haecker, V., 1908. Tiefsee Radiolarien, *Wiss. Ergebn. der Deutschen Tiefsee-Expedition 1898-1899*, 14, 1–706.

Harper, H. E., and Knoll, A. H., 1975. Silica, diatoms, and Cenozoic radiolarian evolution, *Geology*, 3, 175–177.

Harrison, C. G. A., 1968. Evolutionary processes and reversals of the Earth's magnetic field, *Nature*, 217, 46–47.

Hays, J. D., 1965. Radiolaria and late Tertiary and Quaternary history of Antarctic Seas, in *Biology of the Antarctic Seas 2*, Antarctic Research Ser., vol. 5, Washington: Amer. Geophysical Union, 125–184.

Hays, J. D., and Donahue, J. G., 1972. Antarctic Quaternary climatic record and radiolarian and diatom extinctions, in *Antarctic Geology and Geophysics*, Internat. Union of Geological Sci., ser. B., no. 1, 733–738.

Hays, J. D.; Lozano, J. A.; Shackleton, N.; and Irving, G., 1976. Reconstruction of the Atlantic and western Indian Ocean sectors of the 18,000 BP Antarctic Ocean, *Geol. Soc. Amer. Mem. No. 145*, 337–372.

Heath, G. R., 1973. Dissolved silica and deep-sea sediments, *Soc. Econ. Paleontol. Mineral., Spec. Publ. 20*, 77–93.

Heath, G. R., and Moberly, R., 1971. Cherts from the Western Pacific, Leg 7, Deep-Sea Drilling Project, *Initial Reports of the Deep-Sea Drilling Proj.*, vol. 7. Washington: U.S. Gov. Printing Office, 991–1007.

Hill, M. N., ed., 1963-1974. *The Sea*. New York: Wiley, vols. 3 and 5.

Holdsworth, B. K., 1977. Paleozoic Radiolaria: Stratigraphic distribution in Atlantic borderlands, in Swain, 1977, 167–184.

Hollande, A., and Enjumet, M., 1960. Cytologie, évolution et systématique des Spaeroidés (Radiolaires), *Arch. Mus. Natl. Hist. Nat.*, ser 7, 7, 1–134.

Honigberg, B. M., Balamuth, W.; Bovee, E. C.; Corliss, J. O.; Gojdics, M.; Hall, R. P.; Kudo, R. R.; Levine, N. D.; Loeblich, A. R., Jr.; Weiser, J.; and Wenrich, D. H., 1964. A revised classification of the phylum Protozoa, *J. Protozool.* 11, 7–20.

Huddlestun, P., 1971. Pleistocene paleoclimates based on Radiolaria from subantarctic deep-sea cores, *Deep-Sea Research*, 18, 1141–1144.

Hurd, D. C., 1972. Factors affecting solution rate of biogenic opal in seawater, *Earth Planetary Sci. Letters*, 15, 411–417.

Hurd, D. C., 1973. Interactions of biogenic opal, sediment and seawater in the central equatorial Pacific, *Geochim. Cosmochim. Acta*, 37, 2257–2282.

Hurd, D. C., and Theyer, F., 1975. Changes in the physical and chemial properties of biogenic silica from the central equatorial Pacific, in *Analytical Methods in Oceanography*. Washington: Amer. Chem. Soc., 211–230.

Immermann, F.; Borgert, A.; Schmidt, W. J.; and Popofsky, A., 1904-1926. Die Tripyleen Radiolarien der Plankton-Expedition, in *Ergebn. Plankton-Expedition Humboldt-Stiftung*, vol. 3, pts, Lh-1 to Lh-13, 1–716 (each family by one author).

Johnson, D. A., and Knoll, A. H., 1974. Radiolaria as paleoclimatic indicators: Pleistocene climatic fluctuations in the equatorial Pacific Ocean, *Quaternary Research*, 4, 206–216.

Johnson, T. C., 1976. Controls on the preservation of biogenic opal in sediments of the eastern tropical Pacific, *Science*, 192, 887–890.

Keany, J., 1973. New radiolarian palaeoclimatic index in the Plio-Pleistocene of the Southern Ocean, *Nature*, 246, 139–141.

Kellogg, D. E., 1976. Character displacement in the radiolarian genus *Eucyrtidium, Evolution*, 29, 736–749.

Kling, S. A., 1978. Radiolaria, in B. U. Haq and A. Boersma, eds., *Introduction to Marine Micropaleontology*. New York: Elsevier, 203–244.

Kobayashi, T., and Kimura, T., 1944. A study of the radiolairan rocks, *J. Fac. Sci. Imp. Univ. Tokyo*, sec. 2, 7, 75–178.

Levine, N. D., 1963. Protozoology today, in *Progress in Protozoology*. New York: Academic Press, 39–40.

Lewin, J. C., 1961. The dissolution of silica from diatom walls, *Geochim. Cosmochimi. Acta*, 21, 182–198.

Ling, H. -Y., 1966. The radiolarian *Protocystis thomsoni* (Murray) in the northeast Pacific Ocean, *Micropaleontology*, 12, 203–214.

Lozano, J. A., and Hays, J. D., 1976. Relationship of radiolarian assemblages to sediment types and physical oceanography in the Atlantic and western Indian Ocean sectors of the Atlantic Ocean, *Geol. Soc. Amer. Mem. No. 145*, 303–336.

Moore, T. C., Jr., 1972. Mid-Tertiary evolution of the radiolarian genus *Calocycletta, Micropaleontology*, 18, 144–152.

Moore, T. C., 1973. Late Pleistocene-Holocene oceanographic changes in the northeastern Pacific, *Quaternary Research*, 3, 99–109.

Moore, T. C., 1978. The distribution of radiolarian assemblages in the modern and ice-age Pacific, *Marine Micropaleontology*, 3, 229–266.

Müller, J., 1859. Über die Thalassicollen, Polycystinen und Acanthometren des Mittelmeeres, *Abh. Kgl. Akad. Wiss. Berlin, Jahrg.*, 1858, 1–62.

Nigrini, C., 1967. Radiolaria in pelagic sediments from the Indian and Atlantic Oceans, *Bull. Scripps Inst. Oceanog.*, 11, 1–106.

Nigrini, C., 1968. Radiolaria from eastern tropical Pacific sediments, *Micropaleontology*, 14, 51–63.

Nigrini, C., 1970. Radiolarian assemblages in the North Pacific and their application to a study of Quaternary sediments in core V20-130, *Geol. Soc. Amer. Mem. No. 126*, 139–183.

Odum, H. T., 1951. Notes on the strontium content of sea water, celestite Radiolaria, and strontianite snail shells, *Science*, 114, 211–213.

Orlov, Yu. A. (Ed.), 1962, Radiolaria, in *Fundamentals of Paleontology*, vol. 1. Jerusalem: Israel Program Sci. Transl., 552–712.

Pessagno, E. A., 1977. Radiolaria in Mesozoic stratigraphy, in Ramsay, 1977, 913–950.

Pessagno, E. A., Jr., and Newport, R. L., 1972. A technique for extracting Radiolaria from radiolarian cherts, *Micropaleontology*, 18, 213–234.

Petrushevskaya, M. G., 1967. Radiolaria of orders Spumellaria and Nassellaria of the Antarctic region, in *Studies of marine fauna*. Biological reports of the Soviet Antarctic Expedition, (1955–1958), vol. 3. U.S.S.R.: Akad. Nauk (in Russian), 2–186.

Petrushevskaya, M. G., 1971a. On the natural system of polycystine Radiolaria (class Sarcodina), in *Proc. 2nd Planktonic Conference, Roma 1970*, vol. 2. Rome: Edizioni Technoscienza, 987–991.

Petrusevskaya, M. G., 1971b. Supmmellarian and nasselarian Radiolaria in the plankton and bottom sediments of the central Pacific, in Funnell and Riedel, 1971, 309–317.

Petrushevskaya, M. G., and Bjørklund, K. R., 1974. Radiolarians in Holocene sediments of the Norwegian-Greenland Seas, *Sarsia*, 57, 33–46.

Petrushevskaya, M. G., and Kozlova, G. E., 1972. Radiolaria: Leg 14, Deep-Sea Drilling Project, in *Initial Reports of the deep-Sea Drilling Pro*, vol. 14, Washington: U.S. Gov. Printing Office, 495–647.

Pokorný, V., 1963. *Principles of Zoological Micro-palaeontology*, vol. 1. Oxford: Pergamon, 652p.

Ramsay, A. T. S., 1973. A history of organic siliceous sediments in oceans, *Palaeont. Assoc., Spec. Paper 12*, 199–234.

Ramsay, A. T. S. ed., 1977. *Oceanic Micropaleontology*. London: Academic Press, vol. 1, 1–808; vol. 2, 809–1453.

Rao, S. R. N., and Mohan, K., 1954. Microfossils from the Dogra slates (Pre-Cambrian) of Kashmir, *Curr. Sci. (India)*, 23, 11–12.

Renz, G. W., 1976. The distribution and ecology of Radiolaria in the central Pacific plankton and surface sediments, *Bull. Scripps Inst. Oceanog.*, 22, 1–267.

Reshetnyak, V. V., 1966. Deep-sea Radiolaria Phaeodaria of the north-west part of the Pacific Ocean, in *Fauna SSSR* (in Russian), n. s., Moscow-Leningrad, Nauka, no. 94, 208p.

Reshetnyak, V. V., 1971. Occurrence of phaeodarina Radiolaria in recent sediments and Tertiary deposits, in Funnell and Riedel, 1971, 343–349.

Riedel, W. R., 1959. Siliceous organic remains in pelagic sediments, *Soc. Econ. Paleontol. Mineral., Spec. Pub. 7*, 80–91.

Riedel, W. R., 1963. Paleontology of pelagic sediments, in Hill, 1963, vol. 3, 866–887.

Riedel, W. R., 1967. Actinopoda, in W. B. Harland et al., in *The Fossil Record*. London: Geol. Soc., London, 291–298.

Riedel, W. R., and Holm, E. A., 1957. Radiolaria, *Geol. Soc. Amer., Mem. No. 67*, 1069–1072.

Riedel, W. R., and Sanfilippo, A., 1974. Radiolaria from the western Indian Ocean, DSDP Leg 26, in *Initial Reports of the Deep-Sea Drilling Proj.*, vol. 26. Washington: U.S. Gov. Printing Office, 771–814.

Riedel, W. R., and Sanfillipo, A., 1977. Cainozoic Radiolaria, in Ramsay, 1977, 847–912.

Robertson, J. H., 1975. Glacial to interglacial oceanographic changes in the northwest Pacific, including a continuous record of the last 400,000 year. Ph.D. dissertation, Columbia Univ., 355p.

Ruedemann, R., and Wilson, T. Y., 1936. Eastern New York Ordovician cherts, *Gol. Soc. Amer. Bull.*, 47, 1535–1586.

Runeva, N. P., 1974. Iskopayemyye Phaeodaria mlotsena Kamchatki, *Doklady Akademii Nauk*, 215, 969–971.

Sachs, H. M., 1973a. North Pacific radiolarian assemblages and their relationship to oceanographic parameters, *Quaternary Research*, 3, 73–78.

Sachs, H. M., 1973b. Late Pleistocene history of the North Pacific: Evidence from a quantitative study of Radiolaria in core V21-173, *Quaternary Research*, 3, 89–98.

Sanfilippo, A., and Riedel, W. R., 1970. Post-Eocene "closed" theoperid radiolarians, *Micropaleontology*, 16, 446–462.

Schewiakoff, W., 1926. Die Acantharia des Golfes von Neapol, *Flora e Fauna del Golfo di Napoli, Mon. 37*, 755p.

Schink, D. R., 1968. Observations relating to the flux of silica across the sea floor interface, *Amer. Geophys. Union Trans.*, 49, 335.

Sleigh, M. A., 1973. *The Biology of Protozoa*. New York: Am. Elsevier, 315p.

Stadum, C. J., and Ling, H. Y., 1969. Tripylean Radiolaria in deep-sea sediments of the Norwegian Sea, *Micropaleontology*, 15, 481–489.

Stanley, E. M., 1973. Minor element abundance in radiolarian tests, *Geol. Soc. Am. Abstr.*, 5(1), 110.

Swain, F. M., ed., 1977. *Stratigraphic Micropaleontology of Atlantic Basin and Borderlands*. Amsterdam: Elsevier Press, 603p.

Tappan, H., and Loeblich, A. R., 1973. Evolution of the oceanic plankton, *Earth-Sci. Rev.*, 9, 207–240.

Taylor, D. L., 1974. Symbiotic marine algae: Taxonomy and biological fitness, in *Symbiosis in the Sea*. Columbia: Univ. South Carolina Press, 245–262.

Theyer, F., and Hammond, S. R., 1974a. Paleomagnetic polarity sequence and radiolarian zones, Bruhnes to polarity epoch 20, *Earth Planetary Sci. Letters*, 22, 307–319.

Theyer, D., and Hammond, S. R., 1974b. Cenozoic magnetic time scale in deep-sea cores: Completion of the Neogene, *Geology*, 2, 487–492.

Trégouboff, G., 1953. Classe des Acanthaires and Classe des Radiolaires, in Grasse, 1953, vol. 1, 271–320 and 321–388.

Waddington, C. J., 1967. Paleomagnetic field reversals and cosmic radiation, *Science*, 158, 913–915.

Wise, S. W., and Weaver, F. M., 1974. Chertification of oceanic sediments, *Internat. Assoc. Sedimentologists, Spec. Pub. No. 1*, 301–326.

Wollast, R., 1974. The silica problem, in Hill, 1963–1974, vol. 5, 359–392.

Cross-references: *Micropaleontology; Plankton; Protista.*

RECEPTACULITOIDS

The receptaculitoids are a poorly understood group of calcareous fossils ranging from Ordovician through Devonian. The skeleton comprises a multitude of radiaxial elongate elements, *meroms* (from German, *Merom;* Greek *meros*), whose extremities are usually expanded to form an outer wall and less frequently an inner wall

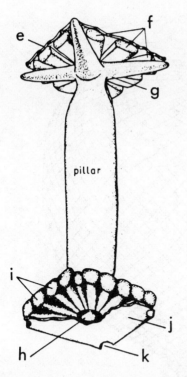

FIGURE 1. Common merom components. 3, facet; f, distal secondary branches; g, cruciform ray; h, proximal termination; i, proximal secondary branches; semicontinuous innermost wall layer; k, inner wall perforation. There is sometimes a perforation (see Fig. 5) at point h.

(Figs. 1 and 7). One end of the skeleton, usually the more tapered one in conical or pyriform examples, is open (Figs. 2 and 3). The interpretation of the group as photosynthesizing organisms is presently favored, and the entire living body is referred to herein as a *thallus.*

Description

Receptaculitoids were probably spherical, pyriform, or conical during life. The closed end is acutely conical (Figs. 2 and 3) or rounded with a small conical protrusion or a broad depression at the closed pole. The plant body, or thallus, largely comprised a central soft mass from which a stalk led to a holdfast. Whorled calcified meroms surrounded the central mass and formed an exoskeleton. New meroms arose at the base of the thallus.

There is significant variation in meroms along the length of the thallus. In the basal zone (Fig. 2) the meroms are thin and cylindrical; nearer the aperture, they are long and thin in many species. Toward the base of the

thallus, *facets* (see below; Fig. 1e) are poorly developed and the proximal (inner) expansions of meroms are not in contact. This portion of the larger skeletons does not hold together after death.

Higher in the thallus, the meroms proximal expansions come into mutual contact and form an inner wall. Before attaining their full size, facets often display marginal crenulations. Incomplete coalescence of inner and outer wall units within the basal zone allows dispersal of these meroms after death. The apical zone (Fig. 2) consists of mature meroms with inflated pillars almost in contact; occasionally even with crudely hexagonal cross section. In many species, this is the only portion of the skeleton that remains intact after death of the plant. The typical bowl-shaped *Receptaculites* fossil comprises only the meroms coherent around the closed pole, which is sometimes too acutely conical to be considered the base of an organism lacking external soft tissue. The more conical forms are preserved fairly frequently in growth position with the closed pole upwards, as in the type strata of *Hexabactron.* The fully calcified forms, preservable as cones, frequently grew in reef, mound, or hard-ground carbonate environments and were sometimes colonial. Paleogeographic evidence suggests that these conical forms may represent warmer water species than those that produced bowl-shaped fossils.

Arrangement of Meroms and Facets. During growth, existing spirals could not cover the rapidly expanding area above the maximum diameter of the thallus. Thus new spirals were intercalated. Each new spiral intercalation is marked by a pair of unusually shaped facets: usually a small triangular one adjacent to a larger pentagonal one on its abapertural side. The larger pentagonal facet is termed an interpositum (Fig. 4), but the smaller triangular facet more conveniently marks the origin of a new spiral, especially as its center is usually at a half whorl position.

The pattern of intersecting sets of dextral and sinistral spirals formed by the facet sides is the best known characteristic of receptaculitoids (Fig. 4). Similar patterns higher in the plant kingdom, as in pine cones, certain cacti, and the sunflower, usually have the ratio of dextral and sinistral spirals formed by adjacent members of the Fibonacci series: 1, 1, 2, 3, 5, 8, 13, 21, etc. The Fibonacci patterns of spirals are known to result where new apical growth components are added one at a time at the largest gap available. In receptaculitoids, the opposing spirals are equal or subequal in frequency, suggesting that the meroms were added in circlets rather than one after the other

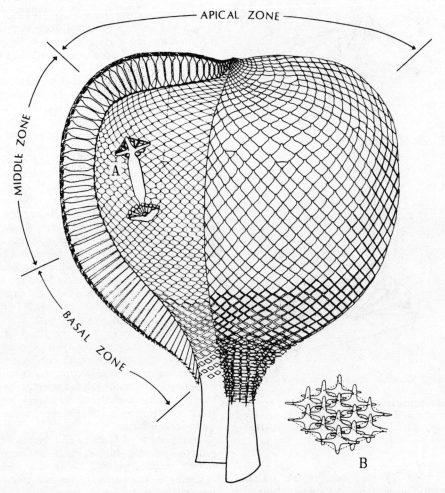

FIGURE 2. Schematic reconstruction to illustrate features common in the Receptaculitaceae. The right half of the diagram presents the external view showing merom arrangement into spirals. The sectioned half shows the meroms in lateral view and the expanded meromal bases upon the surface of the central mass. A, enlarged merom (see Fig. 1). B, shows the interlocking of cruciform rays; the latitudinal rays (horizontal) overlie the adapertural rays (pointing downwards), while the abapertural rays (pointing upwards) are the outermost.

(Gould and Katz, 1975). Occasionally, however, meroms are aligned along a tightly compressed spiral growth curve.

The closed polar whorl often contains eight (rarely four) meroms. The spiral rows of facets extending outward from this central whorl must bifurcate to fill the surface area of the thallus. This is because surface area expands away from the pole until after the maximum diameter is reached, while the growth size of the facets is limited. Most intercalation of new spirals occurs within the first 20 to 25 whorls out from the apex.

Beneath the maximum diameter of the thallus, there was an ever-decreasing area to be filled by further growth of existing spirals and spiral intercalations, which were no longer needed, are few or absent. But, although the area to be covered decreased away from the apex, at least the same number of facets are usually to the found in each lower whorl. Thus, compressive tendencies may be observed. The most common compressional feature is the replacement of rhomboidal wall units by irregular hexagonal ones. This is common within the inner wall (see Fig. 5) and sometimes present in the outer wall. In such irregular hexagons, the shorter sides are meridional.

A wide range of suprafacetal structures characterize the lower, apertural, portions of the skeletons. They include patterns of knobs in many species and goblet-like structures in *Ischa-*

FIGURE 3. Steikern of *Ischadites murchisoni* showing the basal foramen, tapered apex, circular pillar tops, and criciform rays; X1.4 (from Rauff, 1892a).

FIGURE 4. Apical polar view of facet arrangement based on *Ischadites koenigi* (from Rauff, 1892a). Shading accentuates spiral slignments, with bifurcations at interposita (i). The polar whorl comprises eight facets; X2.3.

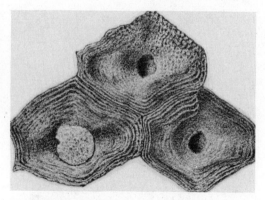

FIGURE 5. Proximal (inner) side of inner wall in *Receptaculites neptuni,* showing central perforations entering pillars; X8.5 (from Rauff, 1892a).

dites iowensis. An outer covering of soft tissue may have been necessary for photosynthesis.

Outer Wall Structures. *Distal secondary structures* (Fig. 1) are situated beneath the facets, each set usually comprising about 12 ill-defined lobes. Facets may be absent where distal secondary branches are well developed.

Stellate structures are rays tangential to the surface of the skeleton, situated below the facets and distal secondary branches. In a few forms, each merom has six (or three) stellate rays at 60° (or 120°) separation. In most receptaculitoids, the stellate structures are quadrate and cruciform, with approximately longitudinal and latitudinal ray orientation (Figs. 2 and 3).

Each cruciform ray set comprises an abapertural longitudinal ray pointing toward the closed pole, an adapertural longitudinal ray, and two latitudinal rays. Normally, the abapertural ray is longest, has an oval cross section, and may extend over the surface of a neighboring facet (Fig. 6). The latitudinal rays are of equal length, have circular cross section, and are slightly proximal to the pole-pointing ray. The adapertural ray is slightly more proximal (Fig. 7).

The outer wall is often thickened by deposition around and between the cruciform rays.

Hollow spheres, situated between the cruciform rays and frequently paired, may be *reproductive cysts.* They are known, as spheres or hollows in cruciform-ray-embedding calcite, in at least five species. Large funnel-shaped pores in the outer wall of *Hexabactron,* six surrounding each pillar, may be comparable.

Pillars. The inner and outer walls are connected by pillars (Fig. 1). Average pillar separation varies from several diameters to zero, when pillars may assume hexagonal cross

FIGURE 6. Near polar portion of *Ischadites koenigi* as viewed in Fig. 4 (from Rauff, 1892a). The prominent cruciform rays are abapertural. The central facet is an interpositum; X8.

section. Abnormally large pillar diameters may represent fusion of two or more pillars.

Inner Wall. An inner wall, comprising proximal merom expansions, is present almost invariably in *Receptaculites*. In some *Ischadites*, it occurs occasionally in the closed polar region. It is more common in *Sphaerospongia* and *Hexabactron* but in the former is usually confined to the closed polar region. In *Receptaculites*, the inner wall is often rendered bipartite through the presence of 4–12 proximal secondary branches on each merom (Fig. 1i). Inner-wall perforations enter the merom, either centrally (Fig. 5) or from rhomb corners. In the latter case, paired perforations are sometimes present. Some species may also carry pores that communicate with the intermeromal space. The thin innermost wall layer (Fig. 1j) obscures intermeromal sutures.

Central Axis, Stalk, and Holdfast. Evidence for a central axis derives mostly from taphonomy. In thalli buried with the closed pole downward, the central cavity usually filled with sediment, but in some cases a basal void resulted. Presumably, by the time the central axis decayed completely, the overlying sediment had in these cases become too firm to flow into the resultant space. In other examples, only late phases of the polyphase intermeromal fillings are noted in the central cavity. There appear to be few specimens known that may preserve structure within the central axis (e.g., Rauff, 1892a, Pl. 6, Figs. 10, 10a, 12).

Ontogeny

Intercalated new spirals are common around the closed pole in many species. In the lower portion of the thallus below the maximum diameter, intercalations appear to be rare and the number of meroms per whorl may increase downward. This implies that the lower whorls are the youngest and that whorls were generated near the base of the central axis at the "aperture" (Gould and Katz, 1975). Campbell et al. (1974) postulated that the generative area in *Hexabactron* might even have been below the sediment surface during life. As a thallus grew, the increasing size of the central axis caused the whorl circumference to increase. The interposita represent first attempts to fill the increased area by latitudinal expansion. They are coupled to slightly younger small triangular new facets which raise the number in the next whorl. Progressively larger numbers of facets thus occupied the whorls generated around the thickening stalk during life. Since the surface area decreases below the maximum diameter of the plant, there is no necessity for intercalation of new spirals in this lower portion.

FIGURE 7. Diametrical meridional section through *Receptaculites neptuni*. (from Rauff, 1892a). High-level abapertural cruciform rays (d) point (left) toward the closed pole while the adapertural rays (pr) slope downward. Latitudinal rays (l) are at an intermediate level, X3.4.

Calcification, especially of pillars, would appear to present a barrier to continuing growth. If calcification were delayed or incomplete (e.g., granular) until stable adult form was attained, then one would not expect to find young forms preserved. This may be the case for most of the abundant and well-known species, although specimens claimed to be young forms of *Ischadites iowensis* have been illustrated.

Taxonomic Position

The two most popular interpretations of the receptaculitoids, as sponges or as algae, are quite dissimilar. If they are viewed as sponges, the meroms are spicules, the central cavity was empty during life, and the open end was uppermost. If interpreted as algae, the central cavity contained the major body bulk, the meroms are branches, and the open end was downmost in life. The sponge interpretation (e.g., Foster, 1973) has generally been prevalent; but in recent years the algae interpretation has been elaborated, principally by Nitecki, and an algal calssification is given here. Algal affinity, suggested as long ago as 1877 (Munier-Chalmas), enables many features to be explained through analogies with higher calcareous algae such as dasycladaceans (see *Algae*), though there are differences in scale. Zhuravleva and Miagkova (1972) have raised a new kingdom, Archaeata, to accomodate receptaculitoids and the Archaeocyatha.

Classification (Algal)

CLASS SQUAMULIFERI Suschkin 1958 (nom. transl. herein, ex Squamuliferida)
Order Receptaculitales Suschkin 1962 (nom. transl., Rietschel, 1969, ex Receptaculitida)
Family Receptaculitaceae Eichwald 1860
(non. transl. Byrnes, 1968 ex Recaptaculidae)
Recaptaculites, Ischadites, Ehlersopongia, Acanthochonia, Dictyocrimus (= *Dictyocrinites*), *Lepidolites,* (*Palaeospongia?*)
Family Tettragonaceae Rietschel 1969
Tettragonis (more often spelled *Tetragonis*)
Family Sphaerospongaceae Rietschel 1969
Spaerospongia (= *Polygonosphaerites*), *Hexabactron*
Family Soanitaceae Miagkova, 1965 (nom. transl. herein, ex Soanitidae)
Soanties, Calathium (?including *Calathella*)
Family Amphispongaceae Rauff, 1894 (nom. transl. herein, ex Amphispongiidae)
Amphispongia (?*Pirania*)
Order Dasycladales
Family Cyclocrinitaceae
Cyclocrinus (=*Cyclocrinites*), *Mastropora, Coelosphaeridium, Pasceolus, Apidium, Cerionites, Nidulites, Anomaloides* (= *Anomalospongia*), *Lunulites*

Receptaculitaceae. Members of the family Receptaculitaceae are typical receptaculitoids as described herein. Most facets are rhomboidal. At the present time there is little known to separate the genera *Ehlersospongia, Selenoides, Acanthochonia,* and *Dictyocrinus* from *Isochadites,* only the last name being in general use. This family is generally equivalent to Nitecki's (1969a) Tribe Receptaculiteae. The family ranges from Middle Ordovician to Upper Devonian.

Tettragonaceae. The genus *Tettragonis* (Tetragonis) is very close to the Receptaculitaceae, but in shape and completeness the skeletons also resemble those of the Sphaerospongaceae. These are mainly known from the Silurian but possibly range from the Middle Ordovician to the Upper Devonian.

Sphaerospongaceae. Dominance of hexagonal facet shape is characteristic of the Sphaerospongaceae. The merom calcification tends to proceed a shorter distance below the facet in *Sphaerospongia* than in other receptaculitoids. This genus may have lived on into the Permian, whereas other receptaculitoids were extinct before the Carboniferous. The genus first appears in the Upper Ordovician and is common in the Middle Devonian. *Hexabactron* is confined to the Silurian.

Soanitaceae. Equivalent to Tribe Calathieae of Nitecki (1969b, 1970, 1971), the family Soanitaceae exhibits strong calcification of the inner wall and the proximal secondary branches. The cruciform rays are often fused to form a continuous mesh. Facets are usually absent but strong irregular calcification may be present at a suprafacetal level. These are the earliest receptaculitoids, appearing in the Early Ordovician and becoming extinct before the close of that period.

Amphispongaceae. In the Amphispongaceae equivalent to Tribe Amphispongieae of Nitecki (1971), there are few or no facets, no inner wall and the meroms differ greatly along the length of the thallus. *Amphispongia* is confined to the Silurian.

Cyclocrinitaceae. Following Nitecki (1970), the Cyclocrinitaceae might be considered as receptaculitoid rather than dasycladacean. The Aphrosalpingidea, a Lower Cambrian to Upper Silurian problematical group usually compared with the Archaeocyatha, also resemble the Cyclocrinitaceae.

Occurence

Receptaculitoids are known from about 600 localities. Of Gondwana continents, the major center was Australia, where the range is Ordovician to Devonian with a Devonian maximum.

A similar range exists in South America (Argentina and Bolivia), but there is little documentation. African occurences are Ordovician (Mauritania) and Devonian (Morocco). In southwestern Asia (Turkey, Iran, Afghanistan, Pakistan) occurences are mostly or solely Upper Devonian. Indian examples, Ordovician–Silurian, are little known. Ordovician examples from elsewhere in Asia (Siberia, Mongolia, Manchuria, Vietnam) are very poorly known. In Europe, Ordovician occurences are of British-Baltic extent; Silurian records are widespread but few in number; Devonian occurences are both numerous and widespread (Great Britain, Belgium, Germany, Bohemia, Silesia, Spain, Russia). An anomalous record from the Permian of Sicily requires confirmation. Receptaculitoids are best known from North America. There they are common and widespread in the Ordovician (30 states) compared to the Silurian (10 states) and Devonian (7 states).

JOHN G. BYRNES

References

Alberstadt, L. P., and Walker, K. R., 1976. A receptaculitid–echinoderm pioneer community in a Middle Ordovician reef, *Lethaia*, **9**, 261–272.

Byrnes, J. G., 1968. Notes on the nature and environmental significance of the Receptaculitaceae, *Lethaia*, **1**, 368–381.

Campbell, K. S. W.; Holloway, D. J.; and Smith, W. D., 1974. A new receptaculitid genus, *Hexabactron*, and the relationships of the Receptaculitaceae, *Palaeontographica*, A, **146**, 52–77.

Etheridge, R., and Dun, W. S., 1898. On the structure and method of preservation of *Receptaculites australis* Salter, *Rec. Geol. Surv. N.S.W.*, **6**, 62–75.

Fisher, D. C., and Nitecki, M. N., 1978. Morphology and arrangement of meromes in *Ischadites dixonensis*, an Ordovician receptaculitoid, *Fieldiana, Geol.*, **39**, 17–31.

Foster, M., 1973. Ordovician receptaculitids from California and their significance, *Lethaia*, **6**, 35–65.

Geinitz, E., 1888. Receptaculitidae und andere Spongien der mecklenbergischen Silurgeschiebe, *Zeitschr. Deutsch. Geol. Gesell.*, **40**, 17–23.

Gould, S. J., and Katz, M., 1975. Disruption of ideal geometry in the growth of receptaculitids: A natural experiment in theoretical morphology, *Paleobiology*, **1**, 1–20.

Gübel, C. W., 1875. Beitrage zur Kenntniss der Organisation und systematischen Stellung von *Receptaculites*, *Abh. math.-phys. Classe König. bayer. Acad. Wiss.*, **12**, 167–215.

Kesling, R. V., and Graham, A., 1962. *Ischadites* is a dasycladaceous alga. *J. Paleontology*, **36**, 943–952.

Nitecki, M. N., 1969a. Redescription of *Ischadites koenigii* Murchison, 1839. *Fieldiana, Geol.*, **16**, 943–952.

Nitecki, M. N., 1969b. Surficial pattern of receptaculitids, *Fieldiana, Geol.*, **16**, 361–376.

Nitecki, M. N., 1970. North American cyclocrinitid algae, *Fieldiana, Geol.*, **21**, 182p.

Nitecki, M. N., 1971. Amphispongieae, a new tribe of Paleozoic dasycladaceous algae, *Fieldiana, Geol.*, **23**, 11–21.

Nitecki, M. N., 1972. The paleogeographic significance of receptaculitids, *24th Internat. Geol. Cong. Proc.*, **7**, 303–309.

Rauff, C. F. H., 1892a. Untersuchugen über die Organisation und systematische Stellung der Receptaculitiden. *Abh. math.-phys. Classe bayer. Akad. Wiss.*, **17**, 645–722.

Rauff, C. P. H., 1892b. Kalkangen und Receptaculiten. *Sitz. -Ber. niederrhein. Ges. Naturu. Keikkde.*, **1892**, 74–90.

Rietschel, S., 1969. Die Receptaculiten. *Senckenbergiana Lethaea*, **50**, 465–517.

Rietschel, S., 1970. Rekonstruktionen als Hilfsmittel bei der Untersachung von Receptaculiten (Receptaculitales, Thallophyta), *Senckenbergiana Lethaea*, **51**, 429–447.

Zhuravleva, I. T., and Miagkova, Ye. I., 1972. Archaeata: A new group of Paleozoic organisms (in Russian), *Paleontologiya (Dokl. Sovets. Geol.)*, 7–14.

Cross-references: *Algae; Reefs and Other Carbonate Buildups.*

REEFS AND OTHER CARBONATE BUILDUPS

Much has been written about carbonate buildups ever since geologists have noted that traceable bedding planes within carbonate units, or the boundaries of entire carbonate units, commonly circumscribe positive topographic features that have a nature and appearance differing to some degree from that of the surrounding rock. Most literature involves specific features or suites of specific features that have been variously termed reefs, banks, mounds, bioherms, knolls, or complexes modified by any of the preceeding terms.

Heckel (1974; see also Wilson, 1975, pp. 20–24) defines a carbonate *buildup* as a carbonate mass or a local portion of a carbonate unit that (1) differs in nature to some degree from equivalent deposits and surrounding and overlying rocks (this involves both constituent composition and internal fabric as is most obvious in the case of skeletal buildups); (2) is typically thicker than equivalent carbonates; and (3) probably stood topographically higher than the surrounding sediment during some time in its depositional history. Although involving greater interpretation than the first two criteria, the third criterion is necessary to make the term "buildup" useful by ruling out channels and other depression fillings.

Carbonate buildups may be small enough to be viewed in at least two dimensions on a small outcrop, or large enough to be viewed in a nearby mountain range or to be detected only by regional mapping. Large, broad carbonate buildups have been termed *platforms* or *shelves*. In composition, carbonate buildups may be organically controlled accumulations of various sorts of skeletal secretions; hydrodynamically controlled piles of skeletal debris and oolite; accumulations of lime mud of obscure origin; or, commonly, combinations of two or more of these groups of constituents and having a mixed origin. This usage of the term "buildup" remains relatively descriptive in that it does not imply any particular origin; thus it is more encompassing than the usage of Stanton (1967) and others.

Bioherm and Biostrome

The term *bioherm* was applied by Cumings (1932) to mound-like structures that are of *strictly organic origin*. This term is thus strongly genetic as well as descriptive. Whereas it may be relatively easy to determine if a rock body is mounded (i.e., a buildup), it may be less easy to determine if it was of strictly organic origin. Although the examples given by Cumings seem straightforward because they are accumulations of organic remains of corals, algae, bivalves and crinoids, such a feature might also be formed hydrodynamically by waves and currents sweeping the remains into piles. Nelson et al. (1962) specifically restrict *bioherm* to in-place accumulations of organisms and give criteria for their recognition. "Bioherm" has been applied to accumulations primarily of lime mud that are inferred to have been caused by growth of organisms no longer preserved (Troell, 1962; Pray, 1969). Many of these features, however, contain fewer recognized skeletal remains than do the surrounding sediments and thus do not resemble the features originally listed as bioherms by Cumings (1932).

The term *biostrome* was defined more descriptively as applying to "distinctly bedded structures that do not swell into lens-like or reef-like form but . . . consist mainly or exclusively of the remains of organisms" (Cumings, 1932). There were no restrictions as to origin, and its use has encountered less of the previously mentioned difficulties.

All bioherms probably are buildups, but only buildups of strictly organic origin are bioherms. Although one might reasonably conclude from definition and general usage that biostromes are not buildups, the suggestion by Nelson et al. (1962) that a breadth-to-height ratio of 30:1 might be a useful limit between biostromes and bioherms would thus allow biostromes to have positive topographic relief and

include those that do as buildups. On the other hand, the breadth-to-height ration of 30:1 also makes the Great Barrier Reef a biostrome (from data in Nelson et al., 1962, pp. 239–240). It would seem more reasonable to restrict *biostrome* to a bed of skeletal remains that exhibits no topographic relief, because this criterion is more likely to separate features of significantly different origin than is a rather arbitrary limit of the breadth-to-height ratio. Such a restriction would definitely exclude biostromes from buildups.

Reef and Bank

The word *reef* was originally applied by navigators to those narrow ridges (ribs) of rock, shingle (coarse detritus), or sand lying near enough to the water surface such that a vessel might strike or ground upon them (Cumings, 1932). Variations in its usage by sedimentary geologists is traced by Nelson et al. (1962) and is further extended in Fig. 1 (see also Heckel, 1974).

For contrast with "reef," the term *bank* was used by Lowenstam (1950) for structures formed by organisms that are incapable of raising their substrate; thus banks may be topographically well defined, but are unconsolidated and have low-angle slopes inasmuch as they lack debris-binding elements. In his view, reef organisms actively raise their foundation and thus control their environment, whereas bank organisms play only a passive role of producing sediment and are totally controlled by their environment.

Strict adherence to Lowenstam's (1950) recognition of hexacorals (in conjunction with red algae) as the only reef builders in modern seas and to the resulting limits of framework rigidity generally attributed to his reef definition would probably restrict existence of reefs only to open oceanic regions. Furthermore, it would exclude not only most organic buildups in the geologic record (Wilson, 1975) but also large portions of modern viable hexacoral buildups such as Alacrán reef (Kornicker and Boyd, 1962) from the *reef* category. Strict adherence to his definition and comments also excludes from the function of active reef building many organisms significant in buildup formation throughout the geologic record, even though they apparently built their substrates into the face of less powerful waves in shallower, more enclosed bodies of water. Yet this is precisely the regime in which many carbonate buildups were formed in the geologic past.

One could draw the line here at the limits of rigid organic framework, and many have. But

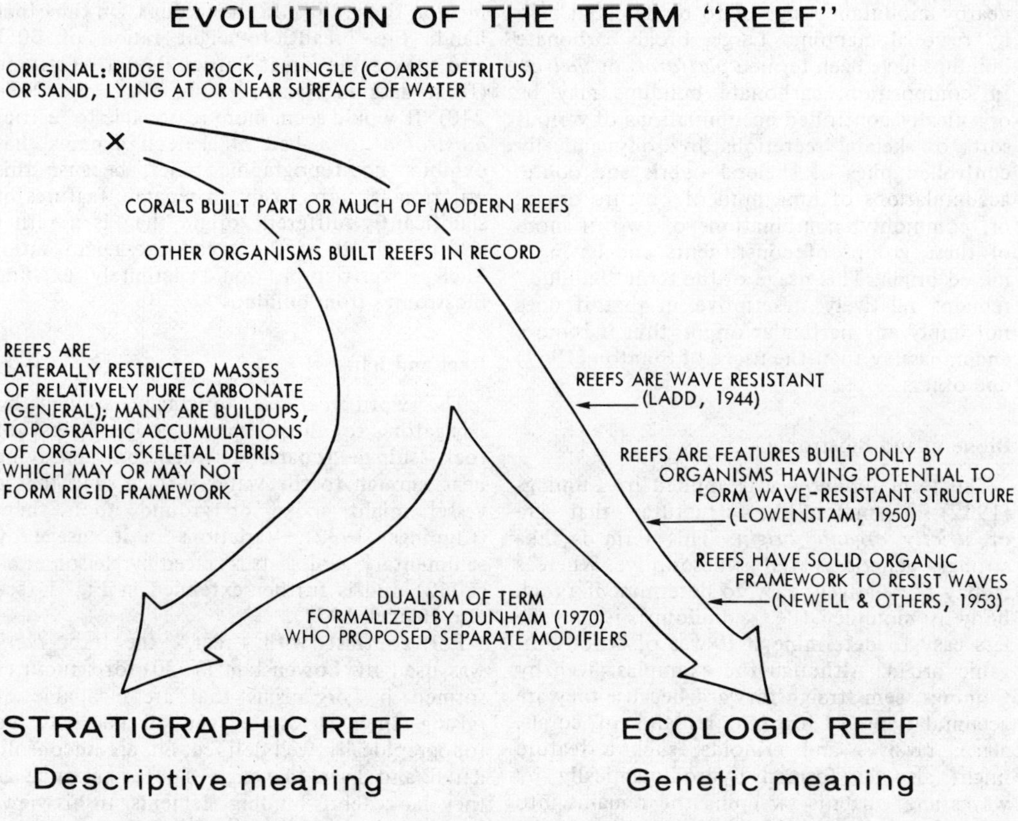

EVOLUTION OF THE TERM "REEF"

ORIGINAL: RIDGE OF ROCK, SHINGLE (COARSE DETRITUS)
OR SAND, LYING AT OR NEAR SURFACE OF WATER

CORALS BUILT PART OR MUCH OF MODERN REEFS

OTHER ORGANISMS BUILT REEFS IN RECORD

REEFS ARE
LATERALLY RESTRICTED MASSES
OF RELATIVELY PURE CARBONATE
(GENERAL); MANY ARE BUILDUPS,
TOPOGRAPHIC ACCUMULATIONS
OF ORGANIC SKELETAL DEBRIS
WHICH MAY OR MAY NOT
FORM RIGID FRAMEWORK

REEFS ARE WAVE RESISTANT
(LADD, 1944)

REEFS ARE FEATURES BUILT ONLY BY
ORGANISMS HAVING POTENTIAL TO
FORM WAVE-RESISTANT STRUCTURE
(LOWENSTAM, 1950)

REEFS HAVE SOLID ORGANIC
FRAMEWORK TO RESIST WAVES
(NEWELL & OTHERS, 1953)

DUALISM OF TERM
FORMALIZED BY DUNHAM (1970)
WHO PROPOSED SEPARATE MODIFIERS

STRATIGRAPHIC REEF
Descriptive meaning

ECOLOGIC REEF
Genetic meaning

FIGURE 1. Evolution of usage of the term *reef* among sedimentary geologists into separate descriptive and genetic meanings (from Heckel, 1974; copyright 1974 by Society of Economic Paleontologists and Mineralogists; used with permission). Restricted meaning of "ecologic reef" and broad meaning of "stratigraphic reef" illustrate shortcomings of strongly genetic or strongly descriptive definitions.

the aforementioned exclusions imply that many carbonate buildups that had no rigid framework, but did resist breaking waves in the surf zone of epicontinental seas (and thereby exerted substantial control over the surrounding environment), are basically different from other buildups in the same regime, but have massive frame builders and pervasive binders (and thus can be called reefs in the strict sense), when in fact there may be little or no significant difference between them at all. This is particularly well illustrated when comparing upper Paleozoic buildups, in which large frame-building and pervasive binding organisms were uncommon, to middle Paleozoic buildups in which the dominant skeletal constituents have been more readily analogized to modern reef builders.

Some more recent discussions of what constitutes a reef have recognized these problems. Harrington and Hazlewood (1962) believe that too much emphasis is placed on *rigidity* of the structure to be called a reef. Kornicker and Boyd (1962) would agree and have defined a reef as "a concentration of carbonate skeletons, in growth position, which significantly influences adjacent sedimentation because of its relief above the surrounding sea floor."

Recognizing the dual usage into which the term "reef" had evolved (Fig. 1), Dunham (1970) proposed separation of the two usages by adding modifiers to the basic term. The compound term *stratigraphic reef* applies to "thick laterally restricted masses of pure or largely pure carbonate rock." By contrast, the term *ecologic reef* applies to "rigid, wave-resistant topographic structures" that are produced by "actively building and sediment-binding organisms." Dunham's modifiers separate the more objective, descriptive usage (stratigraphic) from the more subjective, genetic usage (ecologic) of the term "reef," and renders the term more useful for communication between different groups of geologists.

Dunham (1970) also suggested that examination of reef talus will yield relatively objective criteria by which one may distinguish between the two types of reefs in the fossil record.

Contemporaneous talus derived from the buildup gives clues as to the nature of binding of loose material in the buildup. If the blocks of talus are pervasively bound by sparry cement, then the binding, and implicitly the wave resistance of the buildup, resulted from inorganic processes, and the feature is a *stratigraphic* reef. If, on the other hand, the blocks of talus are organism-encrusted grains of sediment and large pieces of colonial framework and/or encrusting organisms, then the binding is organic and the buildup is an *ecologic* reef. If both types of talus are present, then the buildup is a stratigraphic reef with ecologic reef stages locally or temporally significant.

However, Dunham's definition of "stratigraphic reef" seems to encompass too many features of dissimilar nature to be a useful category for many purposes. His definition does not exclude a sucession of localized biostromes, because no criterion of topographic relief is stated or implied. Thus not all stratigraphic reefs are buildups. This greater inclusiveness of "stratigraphic reef" is useful for some purposes, however, because it readily covers all thick, laterally restricted carbonate masses, whether or not they can be shown to have had topographic relief. Restriction of "stratigraphic reef" to those features having topographic relief might be suggested, but the term "buildup" should suffice for positive topographic features. Nevertheless, with the term "reef," we seem to be left with either an extremely restricted definition in "ecologic reef" or an extremely inclusive definition in "stratigraphic reef."

In practical terms in the geologic record, a *reef* was defined by Heckel (1974) as a buildup that displays (1) evidence of potential wave resistance or growth in turbulent water, which implies wave resistance, and (2) evidence of control over the surrounding environment. Influence of the reef upon the surrounding environment is typically reflected in development of flank beds (Kornicker and Boyd, 1962) and often in differentiation of the growing buildup into distinct facies that reflect different effects of different parts of the buildup on waves and currents. These facies often are termed "back-reef," "reef-margin," and "fore-reef" (or "flank") facies from their physiographic position relative to the margin of the buildup. ("Reef margin" is preferable to "reef," which is too general, and to "reef core" or "reef wall," which imply too stongly a rigid organic framework.) The binding of loose sediment that allows a reef to resist waves and influence its environment could be organic, inorganic or both.

To identify a reef in this proposed sense, one needs to identify some evidence that water was turbulent during growth of the buildup. Such evidence may range from contemporaneous, buildup-derived, wave-washed talus to abraded-grain calcarenites along the buildup margin. This seems to be the most practical way to define and identify ancient buildups that grew in the face of potentially destructive water turbulence. If no evidence of turbulent water is apparent, then the positive topographic feature can be called a bank or, if uncertain, merely a buildup, which is the more inclusive term.

Braithwaite (1973) notes that the high rates of change of environments and the juxtaposition of contrasted habitats are responsible for the commonly abrupt variations in reef biota that result in a prominant zonation (see Stoddart, 1969). Therefore this faunal zonation, which is general, rather than precise in its boundaries, might be an additional criterion for recognizing reefs. By analogy with modern deep-water coral structures, Braithwaite suggests that fossil structures lacking zonation probably were below wave base.

Classification

Reefs as redefined can be further classified in a relatively objective fashion by focusing on the evidence for wave resistance. (Fig. 2). If contemporaneous talus reflecting a solid lithified framework that resisted waves is present, the buildup can be called a "framework reef." If the talus consists of organically encrusted sediment and large pieces of colonial organisms, it is an "organic-framework reef"; this is the practically identified, approximate counterpart of Lowenstam's "reef" and Dunham's "ecologic reef." If the talus consists of fragments of spar-cemented sediment, then the buildup is an "inorganic-framework reef" or "spar-cemented-framework reef."

In the absence of talus, evidence of wave resistance is much less obvious. Preserved rigid organic frameworks surrounded by evidence of water agitation would be ideal but are quite rare in the record (thereby aggravating much of the previous reef–bank controversy). Without contemporaneous talus it is difficult to determine time of inorganic cementation. Rims of drusy spar around grains and in cavities can form very soon after deposition, as seen in beachrock and in cavities remaining within organic encrusted frameworks. Because such rims could also form much later, they are not definite proof but can only be suggestive of early lithification. If marine sediment filters into interstices after an early rim of drusy spar, however, there is stronger evidence of suf-

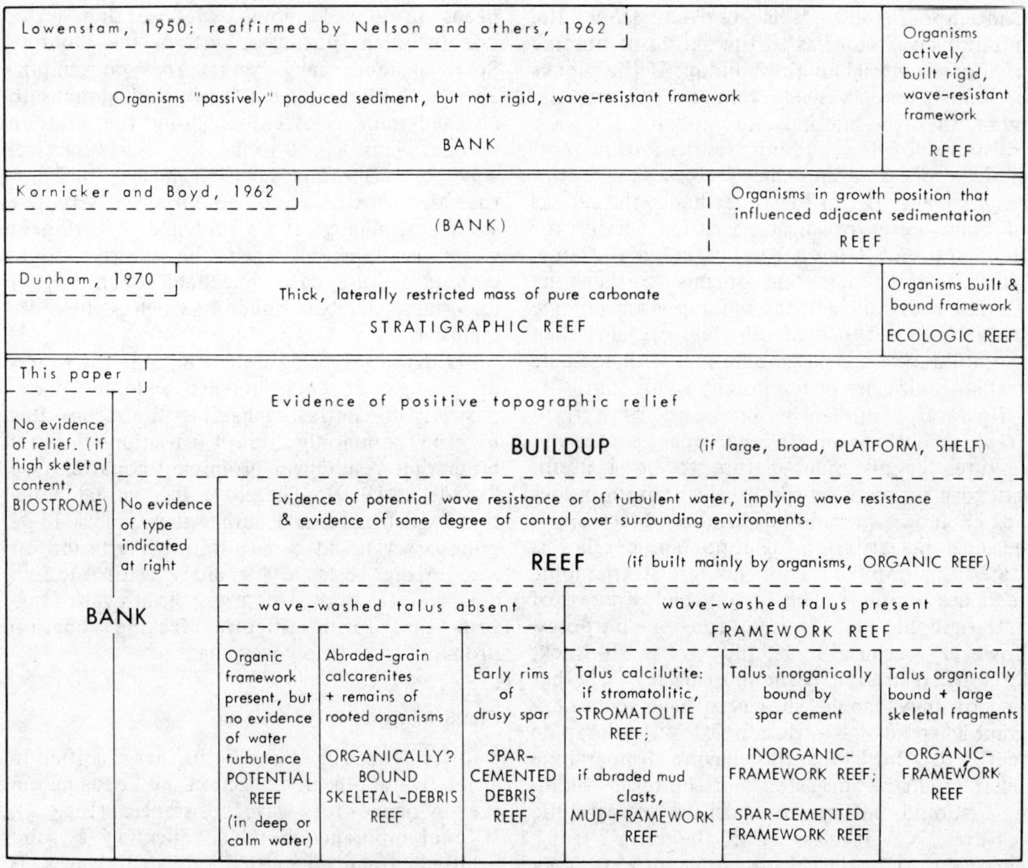

Lowenstam, 1950; reaffirmed by Nelson and others, 1962	Organisms "passively" produced sediment, but not rigid, wave-resistant framework BANK	Organisms actively built rigid, wave-resistant framework REEF
Kornicker and Boyd, 1962	(BANK)	Organisms in growth position that influenced adjacent sedimentation REEF
Dunham, 1970	Thick, laterally restricted mass of pure carbonate STRATIGRAPHIC REEF	Organisms built & bound framework ECOLOGIC REEF

This paper

No evidence of relief. (if high skeletal content, BIOSTROME)

Evidence of positive topographic relief — **BUILDUP** (if large, broad, PLATFORM, SHELF)

No evidence of type indicated at right

Evidence of potential wave resistance or of turbulent water, implying wave resistance & evidence of some degree of control over surrounding environments.

REEF (if built mainly by organisms, ORGANIC REEF)

BANK

wave-washed talus absent			wave-washed talus present FRAMEWORK REEF		
Organic framework present, but no evidence of water turbulence POTENTIAL REEF (in deep or calm water)	Abraded-grain calcarenites + remains of rooted organisms ORGANICALLY? BOUND SKELETAL-DEBRIS REEF	Early rims of drusy spar SPAR-CEMENTED DEBRIS REEF	Talus calcilutite: if stromatolitic, STROMATOLITE REEF; if abraded mud clasts, MUD-FRAMEWORK REEF	Talus inorganically bound by spar cement INORGANIC-FRAMEWORK REEF; SPAR-CEMENTED FRAMEWORK REEF	Talus organically bound + large skeletal fragments ORGANIC-FRAMEWORK REEF

FIGURE 2. Usage of terms *reef, bank, buildup* (and modifiers) in previously proposed and here proposed schemes of definition (from Heckel, 1974. Copyright 1974 by Society of Economic Paleontologists and Mineralogists; used with permission). Terms are in capital letters; criteria are in small letters. Usage proposed in Heckel (1974) and here is largely hierarchical in that more general terms (above dashed lines) include more specific terms (below dashed lines), which allows refinement of terminology as progressively more evidence becomes available.

ficiently early lithification of the mass for it to have provided wave resistance to the buildup. A buildup displaying such evidence could then be called a "spar-cemented-debris reef."

Little, if any, direct evidence of binding and baffling by rooted organisms is preserved in the record. Presence of much pelmatozoan material, particularly on high angles of slope (Harbaugh, 1957) is suggestive. Presence of encrusting, epibiotic, organisms with no preserved object of attachment is suggestive of the former presence of decayed nonskeletal organisms (Davies, 1970) but these may or may not have been rooted. These organic means of binding sediment are thus cryptic and can only be inferred in the record.

Abraded-grain calcarenites give evidence of consistent, or at least periodic, water agitation. Abraded-grain calcarenites composed of pelmatozoan debris, epibiotic material (now free),

sponge spicules, etc., where occurring around the margin of a buildup, provide evidence that the buildup was populated by attached organisms, experienced water turbulence, and thus had a certain degree of wave resistance. Abraded-grain calcarenites may result from much movement of sediment on or around a buildup having minimal resistance to waves, but they provide the only evidence available that a buildup may have been growing in the face of water turbulence. Such a buildup could be called an "organically bound, skeletal-debris reef." In such cases, we are dealing with lesser degrees of wave resistance, so less-adequate evidence is available. These buildups are gradational into those that need no wave resistance (and thus are not reefs). Transitional features are difficult to classify, but all can be called buildups whether or not evidence of wave resistance is found.

DESCRIPTIVE TERMINOLOGY FOR CARBONATE BUILDUPS

PREDOMINANT CONSTITUENT	SKELETAL GRAINS	LIME MUD	NONSKEL. GRAINS
DOMINANT ROCK TYPES	PACKSTONE, GRAINSTONE BOUNDSTONE	WACKESTONE, MUDSTONE	(OVER 70% TOTALLY NONSKEL. GRAINS)
GENERAL TERM	SKELETAL BUILDUP	LIME-MUD BUILDUP	OOLITE (etc) BUILDUP
DISTINCTION AS TO SHAPE	SKEL. MOUND, KNOLL, BAR, BARRIER REEF, ATOLL, etc.	LIME-MUD MOUND, LIME-MUD BAR	OOLITE MOUND, OOLITE BAR
DISTINCTION AS TO TYPE OF SKEL. MATERIAL	e.g. SPONGE MOUND, CORAL-STROMATOPOROID PATCH REEF, BRACHIOPOD KNOLL, DIVERSE SKELETAL ATOLL		
DISTINCTION AS TO DOMINANT HABIT OF SKELETAL MATERIAL	Use ENCRUSTED for encrusting or otherwise permanently attached skeletal material e.g. ENCRUSTED BRYOZOAN MOUND, ENCRUSTED OYSTER REEF Use LOOSE for solitary colonies, unattached, whole or disarticulated skeletal material e.g. LOOSE FORAM MOUND, LOOSE GREEN ALGAL-PELMATOZOAN REEF Use ABRADED for material exhibiting abrasion e.g. ABRADED DIVERSE SKELETAL BAR Use MIXED for buildups in which no one form or component is dominant e.g. MIXED DIVERSE SKELETAL-LIME MUD-PISOLITE BARRIER REEF		

FIGURE 3. Suggested descriptive terminology for all carbonate buildups, based on dominant constituent composition, general shape, and certain characteristics of skeletal constituents (from Heckel, 1974; copyright 1974 by Society of Economic Paleontologists and Mineralogists; used with permission). Distinction between modifiers "encrusted" and "loose" are based upon preserved appearance of skeletal material in record.

One may wish to recognize the potential of a buildup to be a reef when direct evidence of water agitation is lacking. One could recognize a "potential reef" if obvious indication of wave resistance such as welded or encrusted organic framework is present. This would cover deep-water coral buildups discussed by Teichert (1958) and shown to form rigid masses by Stetson et al. (1962).

Descriptive Classification for All Carbonate Buildups

In addition to the proposed redefinition and classification of reefs, a special type of buildup, a more descriptive classification for the general category of carbonate buildups, has been given by Heckel (1974). This classification is based on the objective characters of constituent composition of the buildup and on type and habit of the coarser material present (Fig. 3). This terminology provides immediate general designations for all buildups, utilizing descriptive aspects that are relatively easy to determine. Some of the terms infer origin with a fair degree of certainty, thus providing a useful descriptive terminology. More genetic modifiers can be added to the basically descriptive name as soon as sufficient work justifies them.

Possible Modes of Origin for Carbonate Buildups

Carbonate buildups may be either constructed in place or derived from irregular erosion of a tabular layer or rock. Among the constructional modes of origin (which account for most buildups), we may distinguish that which involves various types of organic activity, that which involves water movement or hydro-dynamic activity, and that which involves movement on a slope or gravitational activity. A clear distinction may not always be drawn among the various modes of origin. Only the organic mode of origin will be discussed in any detail. For a more complete discussion, see Heckel (1974).

The term "organic" as used here refers to the subjective concept of origin in contrast to the term "skeletal," which refers to composition. Thus an organic buildup forms as a direct or indirect result of growth of organisms. Organic buildups include accumulations of skeletal material and organically induced accumulations of any type of carbonate material.

Many organisms grow in clumps and therby cause a local buildup of organically produced carbonate sediment. Localization of organic growth involves the limited areal extent of favorable conditions for larval settling and luxuriant adult growth. Advantages from building up the substrate locally include: (a) providing suitable habitats for organisms unable to live elsewhere, (b) inducing better water circulation patterns for nutrient and oxygen replenishment and waste removal, and (c) improving or maintaining (during subsidence) position in the euphotic zone for organisms that require sunlight.

Certain organisms encrust and attach to one another and bind loose sediment into a coherent, rigid mass, which is readily preserved as such in the rock (see Fig. 4). Other organisms that do not encrust or attach to one another also flourish in local accumulations in reaction to similar sorts of favorable environmental conditions. These buildups involve production of enough sediment for the local raised substrate to grow into, or remain in, an environment favorable to organic proliferation as outlined above.

The size range of loose sediment produced by organisms extends from large unattached colonies and coarse gravel-size shells down to fine-grained mud resulting from disintegration of algal secretions. Many lime-mud buildups may be in-place accumulations of organic

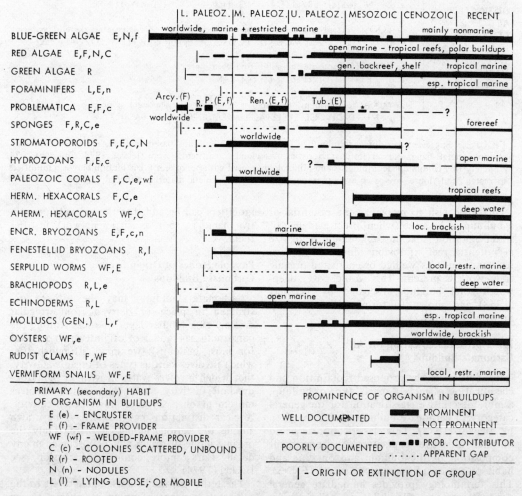

FIGURE 4. Geologic history of major skeletal contributors to carbonate buildups. (from Heckel, 1974; copyright 1974 by Society of Economic Paleontologists and Mineralogists; used with permission). Abbreviations not explained on figure: Arcy., Archaeocyathids; P., *Pulchrilamina*; R., Ren., *Renalcis* and related forms; Tub., *Tubiphytes*; *, other references to specific problematica.

debris resulting from localized flourishing of green algae or of other organisms whose secretions disintegrate readily.

Organisms that secrete no preservable hard parts may induce local accumulation of carbonate sediment of any origin by baffling, binding, burrowing, and providing substrate for epibiotic carbonate producers.

In certain environments, marine grasses grow thickly in local patches and act as baffles causing currents to slow down and sediment to settle out in local buildups (Ginsburg and Lowenstam, 1958; Scoffin, 1970; Davies, 1970). Before marine grasses evolved in the Late Cretaceous (Davies, 1970), large non-skeletal algae probably caused similar effects. A community of unpreservable organisms also provides habitats for many other, particularly epibiotic, organisms unsuited to live elsewhere, many of which contribute more sand-size and mud-size carbonate to the buildup (Davies, 1970; Land, 1970).

Above the level of low tide, nonskeletal blue-green algal mats bind sediment into stromatolites (Logan et al., 1964). In environments of high tidal range and/or strong wave action, turbulent water persistently scours channels between patches of mats which accrete sediment to form local buildups termed heads. In the subtidal environment, mats formed by various nonskeletal algae and other small organisms similarly trap and bind sediment (Neumann et al., 1970), but these tend to be ephemeral and do not appear responsible for thick accumulations of modern sediment (Scoffin, 1970).

Modern Distribution of Carbonate Buildups

The modern distribution of carbonate buildup-forming assemblages is controlled by environments for which physical factors can be measured and evaluated. Although a great number of ecologic factors influence occurrence of organic assemblages, we can focus upon three major controlling biogeographic factors. Figure 5 shows the general relationship of distribution of modern skeletal buildup types to water depth, temperature, and salinity (assuming that other factors such as nutrient and oxygen replenishment are favorable). Seven modern buildup assemblages correspond to four major environmental regimes (Roman numerals on Fig. 5) and to subdivisions of two of them.

The shallow marine environment (1 in Fig. 5) exhibits two temperature-based subdivisions characterized by significant differences in major skeletal contributors. Buildups in warm shallow-marine water of the tropics and subtropics (approximately between $30°N$ and S latitudes) are characterized by the classic coral-reef assemblage consisting chiefly of hermatypic hexacorals, coralline red algae, calcareous green algae, foraminifers, and mollusks (Stoddart, 1969). In many buildups, the corals and red algae produce a bound (or partially bound) framework, with milleporid hydrozoans and massive alcyonarian corals contributing substantially to it in places. Other invertebrate groups are typically present, and most buildups exhibit an extremely rich biota with an enormous number of species. Hermatypic corals and calcareous green algae are essentially restricted to this buildup assemblage because they require both sunlight and fairly high temperatures. Today hermatypic coral reefs cover a total area of some $619,000$ km^2 (Smith, 1978), and this figure has probably been considerably higher at other times in geologic history.

Many species of other groups also are restricted to the shallow-marine buildup assemblage in warm water where secretion of calcium carbonate is enhanced by higher organic metabolism and reduced solubility of carbon dioxide. The latter factor promotes inorganic precipitation of calcium carbonate as well as accounts for concentration of oöids in the same general regime. Carbonate mud also is concentrated here because of both this factor and the restriction of calcareous green algae to warm water, as well as because of the abundance of other skeletal material that provides disintegration products. Thus lime-mud buildups also occur toady only in warm shallow water.

In cold, shallow-marine water (Ib in Fig. 5) of polar and temperate seas, buildups are dominated by red algae. The associated biota is characterized by bryozoans, serpulids, and other marine invertebrate groups, with corals present only in the temperate portion. Green algae that live in cold water do not calcify. Hermatypic corals are absent, and the rest of the invertebrate fauna exhibits a reduction in number of species.

In the deep marine environment (II in Fig. 5), which is generally cold at all latitudes, buildups are dominated by ahermatypic hexacorals, which form a welded framework. Associated fauna includes stylasterid hydrocorallines and most other invertebrate groups comprising a relatively large number of species. These are the only buildups that modern brachiopods or crinoids inhabit to any significant degree. No algae of any type or hermatypic corals occur on deep-water buildups because they require sunlight to live. The effective depth limit for

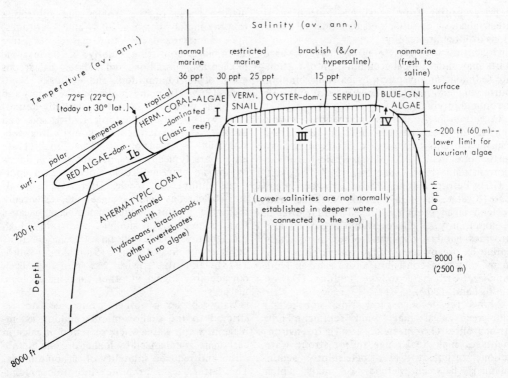

FIGURE 5. Relation of major modern skeletal buildup assemblages to major biogeographic factors of environment (from Heckel, 1974; copyright 1974 by Society of Economic Paleontologists and Mineralogists; used with permission). Roman numerals identify major environmental regimes described in text.

growth of algae ranges from about 100 ft (30 m) in polar regions to 300 ft (90 m) at the equator, and the extreme depth limit ranges from about 250 to 600 ft (75 to 180 m) respectively; tropical red algal-coral reef building takes place today mainly above 300 feet (Teichert, 1958). (In deeper portions of shallow-water tropical reefs, sclerosponges, encrusting foraminifers, bryozoans, brachiopods, and ahermatypic corals become conspicuous as algae, and hermatypic corals become rare.) Modern, ahermatypic coral-dominated buildups that are independent of shallow tropical reefs are known from about 200 ft (60 m) in depth (in cold temperate zones) down to at least 8500 ft (2600 m).

In nearshore environments of restricted salinity (III in Fig. 5) in which water and biota are essentially derived from the normal marine environment, buildups are dominated by euryhaline organisms that can form a welded framework. The number of species is greatly reduced from that of normal marine environments, because most marine organisms are stenohaline and cannot live where salinity varies much beyond the normal marine range of about 30 to 40‰ (Heckel, 1972a). Steno-

haline marine organisms that do not inhabit restricted marine, brackish, or hypersaline environments include corals, calcareous hydrozoans, echinoderms, calcisponges, and most calcareous red and green algae. Of the more tolerant groups, chiefly mollusks, bryozoans, and serpulid worms, only a few members are euryhaline and live outside the normal marine environment. Those species that do tolerate greater salinity variation populate the restricted environment in extremely large numbers owing to lack of predation by stenohaline organisms as well as to reduced competition for space and nutrients. Because of their position along coastlines, all restricted marine and marine-derived environments have very shallow water. (Even where immense amounts of fresh water flood onto steep clastlines, it spreads out above the marine water because of its lesser density.)

Within the shallow, restricted marine (and marine-derived) environments of modern coastlines, three types of organisms are, or have recently been, forming buildups. Although many other ecologic factors besides salinity are critical to their distribution, the three types of organisms—vermetid gastropods, oysters, and serpulid worms—apparently distribute their reef building across the entire regime, with vermetid gastropods occupying the least restricted and serpulid worms inhabiting the most restricted environments. The dominant reef builders in these environments are usually accompanied by a small number of accessory organisms typically representing a few species of encrusting bryozoans, serpulid worms, oysters, barnacles, and mussels.

In nonmarine bodies of either fresh or saline water (IV in Fig. 5), buildups are dominated by blue-green algae. These can live in environments of any salinity but are intensively grazed by certain fish and snails and thus are eliminated from most restricted marine environments as well as from normal marine environments. They do form stromatolite heads today in a hypersaline lagoon on the western Australian seacoast, where grazers and burrowers are absent, and thus enter the most restricted end of regime III (Fig. 5). Many blue-green algal buildups have no skeletal associates, but those forming in fresh water may be accompanied by charophytic green algae, freshwater snails, clams, diatoms, worm tubes, ostracodes, and demosponges. Some nonmarine lakes reach depths below which blue-green algae or charophytes cannot live. For example, Lake Superior is 1290 ft (393 m) deep, and Lake Baikal in Siberia is nearly 5000 ft (1500 m) deep. It is not known if any other organisms fill the buildup-forming niche in this rare environment.

Recognition of Environment of Ancient Buildup Assemblages

By analogy to skeletal composition of modern carbonate buildups, we can trace the environmentally restricted assemblages back to various times in the geologic past.

Shallow, warm, marine water is the environment having the greatest amount of buildup formation that is likely to be preserved in the readily accessible geologic record. The shallow, warm, marine (or classical tropical reef) assemblage is characterized today by great abundance and diversity of biota and also often by association with large amounts of nonskeletal carbonate, including mud, as well as by its particular biotic assemblage. Thus, even without considering taxonomic composition of the assemblage, we might expect that ancient buildups having an extremely rich biota and associated with dominantly carbonate sequences reflect shallow, warm, marine environments and represent the tropical reef assemblage of that time in the past. Furthermore, because life itself probably arose in the warm, favorable surface waters of the sea, it is also likely that this teeming milieu is where the first buildup-

forming assemblages evolved and derived advantage from their newly found capacity for building up the substrate.

There are also a number of critical life requirements and distributional patterns for certain organisms that most likely have remained stable throughout geological time and thus constitute criteria that can be used to interpret environments of ancient buildup formation:

1. All calcareous algae live only in shallow water. So do all hermatypic hexacorals, but it is impossible to determine when a particular coral stock became symbiotic with the unpreserved Zooxanthellae (q.v.). Thus, designation of ancient corals as hermatypic is done mainly by associating them with other indicators such as calcareous algae.
2. All corals, echinoderms, calcisponges, calcareous hydrozoans, and most all calcareous red and green algae are stenohaline and live only in the normal marine environment.
3. All other carbonate-secreting organic groups are most diverse in the marine environment.
4. Within shallow marine water the biotic assemblage is more diverse in warm than in cold water.
5. Deep-marine assemblages may be more diverse overall than certain shallow-marine assemblages but are not as diverse as the total shallow-reef assemblage. For example, Teichert (1958) reports a quite low diversity of frame-building organisms in modern deep-marine buildups in comparison with tropical shallow-marine buildups.
6. Buildup assemblages in both restricted marine and nonmarine environments have a very low diversity of organisms. Diversity decreases further with greater short-term variability and instability of the environment.

Much of the rationale behind the above criteria has been summarized by Heckel (1972a).

Ancient Buildup Assemblages in the Major Environmental Regimes

The shallow-marine assemblage is best identified by presence of algae along with stenohaline marine invertebrates. Using these criteria along with evidence of water agitation, we can detect a succession of shallow-marine buildup assemblages going back, not only to the time of rise of skeleton-secreting organisms, but also into the Precambrian (Fig. 6). Because of a generally rich biota and common association with thick, extensive limestone sequences, we can assume that these assemblages flourished mostly in warm water.

A cold-water, shallow-marine assemblage is suggested by isolated, strongly red algal-dominated buildups in California and parts of Europe throughout most of the Tertiary (Howe, 1934). The shallow cold-water assemblage is especially distinct today because of the temperature limit for hermatypic corals and

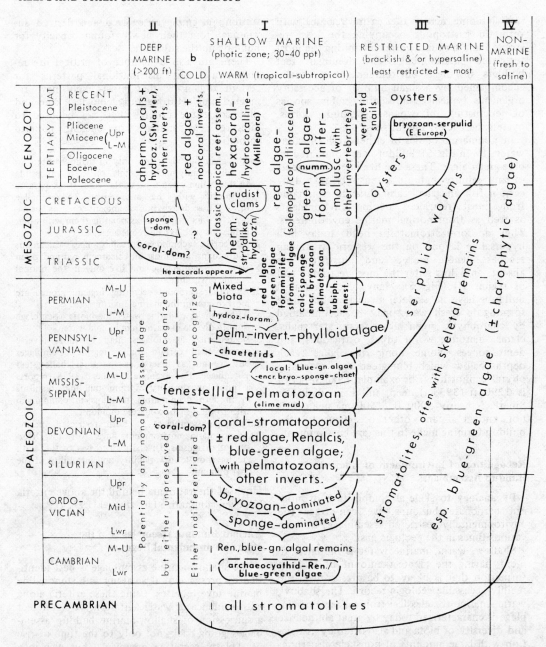

FIGURE 6. Geological history of major skeletal buildup assemblages in major environmental regimes (from Heckel, 1974; copyright 1974 by Society of Economic Paleontologists and Mineralogists; used with permission). Because this scheme is based mainly on information from North America and Europe, modifications may be expected as more information from other areas becomes available.

because of the appearance of ahermatypic corals only in deeper water. But it is uncertain that corals have been consistently absent from cold shallow water, particularly for the Paleozoic. Also, it is probable that, considering climatic variation, different land configurations, and less restricted oceanic circulation,

warm water having tropical temperatures like those of the present may have extended much farther poleward in the distant past. If so, then no truly cold shallow-marine buildup assemblage may have been differentiated in much of the past. Nevertheless, detailed studies of buildup assemblages of different ages might result in identification of shallow cold-water buildups. It is interesting that supposedly cold-water marine biotas of middle Paleozoic age in southern Africa and South America occur almost entirely in terrigenous noncarbonate rocks.

The deep-marine buildup assemblage is best identified by lack of algae. It inhabits an environment that is established only in deeper basins on cratonic areas, on continental slopes, and in ocean basins. In light of probable large-scale sea-floor spreading and subduction of ocean bottom, it is likely that little, if any, really deep-oceanic Paleozoic buildups remain in the record. Buildups from continental slopes are most likely to be preserved in orogenic regions where deep-water sediments are brought up in mountain building, such as in the Cordilleran region of western North America and the Tethyan region of southern Europe and south-central Asia. Perhaps the Triassic coral masses in western North America (Smith, 1912, 1927; Muller, 1936) represent this assemblage, assuming that no associated calcareous algae are found. The Jurassic sponge-foraminifer-alga assemblage in southern Germany (Aldinger, 1968; Gwinner, 1971) may have deeper attributes than does the classical shallow-reef assemblage, but it was still within the photic zone (200 to 600? ft) if the so-called "algal" crusts are indeed algal. Lack of both algae and evidence of wave action on at least some Mississippian Waulsortian buildups suggests that some of them may have formed in deeper water (Pray, 1969). Further detailed study of buildup assemblages in needed, particularly in orogenic belts and deep cratonic basins, to assess adequately the history of deep-water buildup assemblages.

Restricted marine and marine-derived, brackish to hypersaline environments are best identified by lack of stenohaline marine groups and by greatly reduced biotic diversity, even though having lateral stratigraphic proximity to normal marine environments. Exact composition of the assemblages forming buildups in this regime has varied markedly in the recent past as illustrated by the upper Miocene–Pliocene bryozoan-dominated mounds of eastern Europe. This variation may be related to other factors that are more significant in controlling individual buildup assemblages in this suite of highly variable environments than in the more stable normal marine regime.

Nonmarine buildups are probably best identified by their stratigraphic isolation from marine environments in conjunction with their blue-green algal composition, because blue-green algae live as well in some restricted marine environments today and lived in more normal marine

environments in the past. Presence of strictly nonmarine organisms such as charophytes, however, would be diagnostic. One should expect eventually the discovery of more blue-green algal and charophyte-dominated buildups in nonmarine sequences older than those already known.

Synopsis of Geologic History of Buildup Assemblages

With the background given above, we can briefly review the succession of carbonate buildup assemblages through geologic time, and place emphasis on assemblages in the widely identified shallow-marine and locally identified restricted-marine environments (Fig. 6; see Heckel, 1974; Wilson, 1975). The waxing and waning of tropical reef building throughout the past several hundred million years and possible causes are outlined by Newell (1971).

In the Precambrian, blue-green algal stromatolites produced buildups throughout the world in the shallow-marine environment (Cloud and Semikhatov, 1969) and in any other shallow-water regime that was available (e.g., lacustrine in Morocco according to Choubert, *fide* Stubblefield, 1960). In Early Cambrian time, the archaeocyathid-*Renalcis*-blue-green algae assemblage appared, and formed marine buildups on a worldwide scale (Hill, 1965; McKee and Gangloff, 1969; Brasier, 1976; James and Kobluck, 1978). After extinction of the archaeocyathids early in the Middle Cambrian, blue-green algae again dominated buildup formation. At this time, they left a large number of skeletal remains locally in stromatolitic buildups as well as produced strictly organo-sedimentary stromatolites (Ahr, 1971). In the Early Ordovician, a sponge-dominated assemblage began forming buildups, while blue-green algae continued to produce stromatolites (Toomey and Ham, 1967; Toomey, 1970). With the appearance of many new groups in the Middle Ordovician, the sponge-dominated assemblage became overshadowed by an encrusting bryozoan-dominated assemblage (Pitcher, 1971; Walker and Ferrigno, 1973). Blue-green algae participated in these assemblages, but purely stromatolitic buildups apparently were relegated to less-common, more restricted environments at this time.

Evolving from the welter of other groups first appearing or becoming well established in the Middle Ordovician [and forming buildups of various compositions in the Lake Champlain region (Pitcher, 1964; Kapp, 1975) and Tennessee (Alberstadt et al., 1974)], a stromatoporoid- and coral-dominated assemblage established itslef by Late Ordovician time

(Twenhofel, 1950; Copper, 1974). It gained broad distribution in the Silurian (Manten, 1971; Lowenstam, 1957; Heckel and O'Brien, 1975; Scoffin, 1971; Shaver et al., 1978) and attained nearly worldwide distribution (North America, Eurasia, North Africa, and Australia) by Devonian time (Jamieson, 1971; Klovan, 1974; Krebs, 1974; Dumestre and Illing, 1967; Playford and Lowry, 1966). It typically includes pelmatozoans and other, less common invertebrates. Red algae, *Renaleis* (which may be a benthic foraminifer; Riding and Brasier, 1975), and blue-green algae are common associates in the area from northwestern North America through Eurasia to western Australia, particularly in Devonian rocks where green algae also are found locally in very shallow facies. Different growth forms of stromatoporoids characterized different portions of reef complexes. Within the assemblage, coral-versus-stromatoporoid dominance also apparently reflects local environmental differences. Local assemblages ought to be investigated in more detail in order to detect large-scale trends that possibly may be related to climatic zonation. Local assemblages that appear to lack algae, such as the coral-dominated buildups in the Onondaga Limestone of eastern New York (Oliver, 1956), should be searched further for algae to determine if they might not represent deeper-water environments.

After the abrupt extinction of reef-dwelling corals and stromatoporoids late in the Devonian, a variety of assemblages participated in buildup formation at different times and places for the remainder of the Paleozoic era. During the Mississippian, the fenestellid bryozoan-pelmatozoan assemblage became conspicuous in Waulsortian mud-rich buildups in North America, Europe, and North Africa (Lees, 1964; Troell, 1962; Pareyn, 1960; Dupont, 1969). Those buildups having little evidence of algae, water turbulence, or emergence, may represent a deep marine environment, perhaps even one below the photic zone (Pray, 1969). Further detailed comparison of these buildups is needed, however, for if all were deep-water buildups, then shallow-marine Mississippian environments are vastly underrepresented with respect to buildups. A composite assemblage of pre–coral-stromatoporoid assemblage buildup formers appeared, however, during late Mississippian time in Great Britain (Wolfenden, 1958), Arkansas (Jackson, 1972), and in Algeria (Pareyn, 1960), where blue-green algae, encrusting bryozoans, and sponges participated in buildup formation along with chaetetid and other corals locally. Green algae continued their history in the back-reef facies.

During early to middle Pennsylvanian time,

chaetetid corals dominated scattered buildups. Phylloid algae (which include both ancestral coralline red algae and codiacean green algae) appeared and dominated buildup formation on a large scale from at least middle Pennsylvanian time to Early Permian (Wray, 1968, 1971; see also Ball et al., 1977). Margins that are well defined on these buildups were often dominated by pelmatozoans and other invertebrates as well as by algae (Heckel, 1972b; Toomey and Winland 1973; Frost, 1975). The resulting back-reef position of phylloid algae in conjunction with their common occurrence locally to the exclusion of most invertebrates (particularly stenohaline echinoderms) suggests that these algae may have extended into the restricted marine regime of lower salinities. Nevertheless, serpulid worms apparently established themselves as the dominant reef builder in certain restricted shallow environments at about this time (Haack, 1921). From Pennsylvanian time into Permian, a hydrozoan-encrusting foraminifer assemblage dominated buildups locally from NW North America to Siberia (Davies, 1971). A mixed and variable assemblage, including calcisponges, pelmatozoans, nonphylloid red and green algae, stromatolitic algae, foraminifers, fenestellid and other bryozoans, *Tubiphytes,* and many unattached invertebrates, constituted most Permian buildups (Newell et al., 1953; Newell at al., 1976; Malek-Aslani, 1970). Increasing restriction of the Late Permian Zechstein Sea of Germany resulted in fenestellid-dominated mounds becoming pervasively bound by stromatolitic algae (Füchtbauer, 1968). Except for fenestellids and *Tubiphytes,* the major Permian buildup constituents persisted into the Late Triassic buildups of the Alps (Bosellini and Rossi, 1974; Zankl, 1968; Fürsich and Wendt, 1977). Appearance of the newly evolved hexacorals in these, however, was a major milestone in development of the modern shallow-marine buildup assemblage.

The hexacoral–calcisponge–red-algae–hydrozoan–bryozoan–foraminifer–stromatolitic-algae-pelmatozoan–green-algae biota in Alpine Triassic reef complexes is a transitional assemblage. Calcisponges, bryozoans, and pelmatozoans soon dropped out of the buildup assemblage that remained in the shallow marine regime. The Jurassic sponge-foraminifer-algal assemblage formed buildups under conditions less than favorable for the typically normal shallow-marine assemblage, apparently in slightly deeper water (Aldinger, 1968; Gwinner, 1971). The Triassic coral masses of western North America (Muller, 1936; Squires, 1956) and the Jurassic and Cretaceous reeflike masses of northern Europe (Hadding, 1941, *fide* Teichert,

1958) may have formed in deeper and/or colder water. Oysters (q.v.) arose in the early Mesozoic, evolved into brackish-water environments during the Cretaceous (e.g., Laughbaum, 1960), and have eclipsed the serpulid worms in dominating a variety of buildup assemblages ever since. Rudists (q.v.) encroached upon hexacoral domination of framework on shallow-marine buildups during the Cretaceous (e.g., Perkins, 1974; Coates, 1977; and many papers in Debout and Loucks, 1977) but became extinct by the end of the period. Nummulitic foraminifers (Arni, 1965) flourished in reef-shoal areas for a time during the Eocene. The replacement of corallinacean for solenoporid red algae (Wray, 1971) and of hydrocorallines for stromatoporoid-like hydrozoans throughout Cretaceous time and into early Tertiary time essentially completed the transformation of the ancient shallow-marine buildup assemblage to its modern form.

Appearance of red algae-dominated buildups in western North America and southern Europe (Howe, 1934) suggests establishment of the modern, cold-water, shallow-marine assemblage during the Tertiary period. This, in conjunction with appearance of modern hermatypic coral genera through the Tertiary, finally differentiated more definitely the dominant shallow-marine assemblage as being tropical to subtropical. Nonalgal, ahermatypic, coral-dominated buildups might be predicted in Tertiary sediments at or below the edge of continental shelves and in young orogenic belts. Vermiform gastropods (vermetids and vermiculariids) arose in the Eocene and by at least Miocene time established themselves as sporadic buildup formers in the near-normal end of the restricted marine regime (see Shier, 1969, and see Burchette and Riding, 1977, for a possible Carboniferous record). Apparently, local variations in conditions attending the late Miocene restriction and subsequent freshening of the Sarmatian Sea allowed the bryozoan-serpulid assemblage, rather than oysters, to build reefs in eastern Europe during late Tertiary time. In the meantime, blue-green algae no longer even particiuplated in marine buildups during the Cenozoic, but they have been involved in forming extensive nonmarine buildups at least since the Eocene (Bradley, 1929) and probably since the Cambrian.

PHILIP H. HECKEL
DAVID JABLONSKI

References

Ahr, W. M., 1971. Paleoenvironment, algal structures, and fossil algae in the Upper Cambrian of central Texas, *J. Sed. Petrology*, **41**, 205–216.

Alberstadt, L. P.; Walker, K. R.; and R. P., Zurawski, 1974. Patch reefs in the Carters Limestone (Middle Ordovician) in Tennessee, and vertical zonation in Ordovician reefs, *Geol. Soc. Amer. Bull.*, **85**, 1171–1182.

Aldinger, H., 1968. Ecology of algal-sponge reefs in the Upper Jurassic of the Schwäbische Alb, Germany, in Müller and Friedman, 1968, 250–253.

Arni, P., 1965. L'évolution des Nummulitinae en tant que facteur de modification des dépots littoraux, *Mém. Bur. Rech. Géol. Min.*, **32**, 7–20.

Ball, S. M.; Pollard, W. D.; and Roberts, J. W., 1977. Importance of phylloid algae in development of depositional topography—reality or myth? *Stud. Geol., Amer. Assoc. Petrol. Geologists*, **4**, 239–259.

Bosellini, A., and Rossi, D., 1974. Triassic carbonate buildups of the Dolomites, northern Italy, *Soc. Econ. Paleontol. Mineral. Spec. Pub. 18*, 209–233.

Bradley, W. H., 1929. Alga reefs and oolites of the Green River Formation, *U.S. Geol. Surv. Prof. Paper 154*, 203–223.

Braithwaite, C. J. R., 1973. Reefs: Just a problem of semantics? *Bull. Amer. Assoc. Petrol. Geologists*, **57**, 1100–1116.

Brasier, M. D., 1976. Early Cambrian intergrowths of archaeocyathids, *Renalcis* and pseudo stromatolites from South Australia, *Palaeontology*, **19**, 223–245.

Burchette, T. P., and Riding, R., 1977. Attached vermiform gastropods in Carboniferous marginal marine stromatolites and biostromes, *Lethaia*, **10**, 17–28.

Cloud, P. C., and Semikhatov, M. A., 1969. Proterozoic stromatolite zonation, *Amer. J. Sci.*, **267**, 1017–1061.

Coates, A. G., 1977. Jamaican Cretaceous coral assemblages and their relationships to rudist frameworks, *Mém. Bur. Rech. Géol. Min.*, **89**, 336–341.

Copper, P., 1974. Structure and development of early Palaeozoic reefs, *2nd Internat. Symp. on Coral Reefs Proc.*, vol. 1. Brisbane: Great Barrier Reef Committee, 365–386.

Cumings, E. R., 1932. Reefs or bioherms? *Geol. Soc. Amer. Bull.*, **43**, 331–352.

Davies, G. R., 1970. Carbonate bank sedimentation, eastern Shark Bay, Western Australia, *Mem., Amer. Assoc. Petrol. Geologists*, **13**, 85–168.

Davies, G. R., 1971. A Permian hydrozoan mound, Yukon Territory, *Canadian J. Earth Sci.*, **8**, 973–988.

Debout, D. G., and Loucks, R. G., eds., 1977. Cretaceous carbonates of Texas and Mexico, *Rep. Invest. Bur. Econ. Geol. Univ. Texas*, **89**, 332p.

Dumestre, A., and Illing, L. V., 1967. Middle Devonian reefs in Spanish Sahara, in D. H. Oswald, ed., *International Symposium on the Devonian System* vol. 2. Calgary: Alberta Soc. Petrol. Geol., 333–350.

Dunham, R. J., 1970. Stratigraphic reefs versus ecologic reefs, *Bull. Amer. Assoc. Petrol. Geologists*, **54**, 1931–1932.

Dupont, H., 1969. Contribution à l'étude des Faciès du Waulsortien de Waulsort, *Mem. Inst. Géol. Louvain*, **24**, 94–164.

Elloy, R., 1972. Reflexions sur quelques environments recifaux du Paleozoique, *Bull. Centre Rech. Pau*, **6**, 1–105.

Frost, J. G., 1975. Winterset algal-bank complex, Pennsylvanian, eastern Kansas, *Bull. Amer. Assoc. Petrol. Geologists*, **59**, 265–291.

Füchtbauer, H., 1968. Carbonate sedimentation and subsidence in the Zechstein Basin (northern Germany), in Müller and Friedman, 1968, 196–204.

Fürsich, F. T., and Wendt, J., 1977. Biostratinomy and palaeoecology of the Cassian Formation (Triassic) of the Southern Alps, *Palaeogeography, Palaeoclimotology, Palaeoecology*, **22**, 257–323.

Ginsburg, R. N., and Lowenstam, H. A., 1958. The influence of marine bottom communities on the depositional environment of sediments, *J. Geol.*, **60**, 310–318.

Gwinner, M. P., 1971. Carbonate rocks of the Upper Jurassic in SW Germany, in *Sedimentology of Parts of Central Europe*. 8th Internat. Sedimentol. Cong. Heidelberg, Guidebook, 193–207.

Haack, W., 1921. Zur Stratigraphie und Fossilführung des mittleren Bundsandsteins in Norddeutschland, *Preuss. Geol. Landesanst. Jahrb.*, **42**, 560–594.

Harbaugh, J. W., 1957. Mississippian bioherms of northeast Oklahoma, *Bull Amer. Assoc. Petrol. Geologists*, **41**, 2530–2544.

Harrington, J. W., and Hazlewood, E. L., 1962. Comparison of Bahamian land forms with depositional topography of Nena Lucia dune-reef-knoll, Nolan County, Texas: Study in Uniformitarianism, *Bull. Amer. Assoc. Petrol. Geologists*, **46**, 354–373.

Heckel, P. H., 1972a. Recognition of ancient shallow marine environments, *Soc. Econ. Paleontol. Mineral. Spec. Pub. 16*, 226–286.

Heckel, P. H., 1972b. Pennsylvanian stratigraphic reefs in Kansas, some modern comparisons and implications, *Geol. Rundschau*, **61**, 584–598.

Heckel, P. H., 1974. Carbonate buildups in the geologic record: A review, *Soc. Econ. Paleontol. Mineral. Spec. Pub. 18*, 90–154.

Heckel, P. H., and O'Brien, G. D., eds., 1975. Silurian reefs of Great Lakes Region of North America, *Amer. Assoc. Petrol. Geologists, Reprint Ser.*, vol. 14, 243p.

Henson, F. R. S., 1950. Cretaceous and Tertiary reef formations and associated sediments in Middle East, *Bull. Amer. Assoc. Petrol. Geologists*, **34**, 215–238.

Hill, Dorothy, 1965. Archaeocyatha from Antarctica and a review of the phylum, *Trans-Antarctic Expedition 1955–1958, Sci. Repts.*, **10**, (Geol. 3), 151p.

Hoffman, A., and Narkiewicz, M., 1977. Developmental pattern of Lower to Middle Paleozoic banks and reefs, *Neues Jahrb. Geol. Paläont. Mh.*, **1977**, 272–283.

Howe, M. A., 1934. Eocene marine algae (Lithothamnieae) from the Sierra Blanca Limestone, *Geol. Soc. Amer. Bull.*, **45**, 507–518.

Jackson, K. C., 1972. Ecology of Pitkin algal mounds and associated sediments (abstr.), *Geol. Soc. Amer. Abstr.*, **4**, 281–282.

James, N. P., and Kobluk, D. R., 1978. Lower Cambrian patch reefs and associated sediments: Southern Labrador, Canada, *Sedimentology*, **25**, 1–35.

Jamieson, E. R., 1971. Paleoecology of Devonian reefs in western Canada, *Proc. N. Amer. Paleontological Conv.*, pt. J, 1300–1340.

Kapp, U. S., 1975. Paleoecology of Middle Ordovician stromatoporoid mounds in Vermont, *Lethaia*, **8**, 195–207.

Klovan, J. E., 1974. Development of western Canadian Devonian reefs and comparison with Holocene analogues, *Bull. Amer. Assoc. Petrol. Geologists*, **58**, 787–799.

Kornicker, L. S., and D. W. Boyd, 1962. Shallow-water geology and environments of Alacrán reef complex, Campeche Bank, Mexico, *Bull. Amer. Assoc. Petrol. Geologists*, **46**, 640–673.

Krebs, W., 1974. Devonian carbonate complexes of central Europe, *Soc. Econ. Paleontol. Mineral. Spec. Pub. 18*, 155–208.

Land, L. S., 1970. Carbonate mud: Production by epibiont growth on *Thalassia testudinum, J. Sed. Petrology*, **40**, 1361–1363.

Laughbaum, L. R., 1960. A paleoecologic study of the upper Denton Formation, Tarrant, Denton, and Cooke counties, Texas, *J. Paleontology*, **34**, 1183–1197.

Lees, A., 1964. The structure and origin of the Waulsortian (Lower Carboniferous) "reefs" of west-central Eire, *Phil. Trans. Roy. Soc. London, Ser. B*, **247**, 483–531.

Logan, B. W.; Rezak, R.; and Ginsburg, R. N., 1964. Classification and environmental significance of algal stromatolites, *J. Geol.*, **72**, 68–83.

Lowenstam, H. A., 1950. Niagaran reefs of the Great Lakes area, *J. Geol.*, **58**, 430–487.

Lowenstam, H. A., 1957. Niagaran reefs in the Great Lakes area, *Geol. Soc. Amer. Mem. 67* (2), 215–248.

McKee, E. H., and Gangloff, R. A., 1969. Stratigraphic distribution of archaeocyathids in the Silver Peak Range and the White and Inyo Mountains, western Nevada and eastern California, *J. Paleontology*, **43**, 716–726.

Malek-Aslani, Morad, 1970. Lower Wolfcampian reef in Kemnitz Field, Lea County, New Mexico, *Bull. Amer. Assoc. Petrol. Geologists*, **54**, 2317–2335.

Manten, A. A., 1971. *Silurian Reefs of Gotland*. Amsterdam: Elsevier, 539p.

Müller, G., and Friedman, G. M., eds., 1968. *Recent Developments in Carbonate Sedimentology in Central Europe*. Berlin: Springer-Verlag, 196–204, 215–218, 250–253.

Muller, S. W., 1936. Triassic coral reefs in Nevada, *Amer. J. Sci.*, **231**, 202–208.

Nelson, H. F.; Brown, C. W.; and Brineman, J. H., 1962. Skeletal limestone classification, *Mem., Amer. Assoc. Petrol. Geologists*, **1**, 224–252.

Neumann, A. C.; Gebelein, C. D.; and Scoffin, T. P., 1970. The composition, structure and erodability of subtidal mats, Abaco, Bahamas, *J. Sed. Petrology*, **40**, 274–297.

Newell, N. D., 1971. An outline history of tropical organic reefs. *Amer. Mus. Novit.*, **2465**, 37p.

Newell, N. D.; Rigby, J. K.; Fischer, A. G.; Whiteman, A. J.; Hickox, J. E.; and Bradley, J. S., 1953. *The Permian Reef Complex of the Guadalupe Mountains Regions, Texas and New Mexico*. San Francisco: Freeman, 236p.

Newell, N. D.; Rigby, J. K.; Driggs, A.; Boyd, D. W.; and Stehli, F. G., 1976. Permian reef complex, Tunisia, *Brigham Young Univ. Geol. Stud.*, **23**, 75–112.

Oliver, W. A., 1956. Biostromes and bioherms of the

Onondaga Limestone in eastern New York, *N.Y. State Mus. Sci. Service Circ. 45*, 23p.

Pareyn, C., 1960. Les récifs carbonifères du Grand Erg occidental, *Bull. Soc. Geol. France*, ser. 7, **1**, 347–364.

Perkins, B. F., 1974. Paleoecology of a rudist reef complex in the Comanche Cretaceous Glen Rose Limestone of central Texas, *Geoscience and Man*, **8**, 131–173.

Pitcher, M., 1964. Evolution of Chazyan (Ordovician) reefs of eastern United States and Canada, *Bull. Canadian Petrol. Geol.*, **12**, 623–691.

Pitcher, M., 1971. Middle Ordovician reef assemblages, *Proc. N. Amer. Paleontological Conv.*, pt. J, 1341–1357.

Playford, P. E., and Lowry, D. C., 1966. Devonian reef complexes of the Canning Basin, Western Australia, *Geol. Surv. Western Australia Bull.*, **118**, 150p.

Pray, L. C., 1969. Micrite and carbonate cement: genetic factors in Mississippian bioherms (abstr.), *J. Paleontology*, **43**, 895.

Riding, R., and Braiser, M., 1975. Earliest calcareous Foraminifera, *Nature*, **257**, 208–210.

Scoffin, T. P., 1970. The trapping and binding of subtidal carbonate sediments by marine vegetation in Bimini Lagoon, Bahamas, *J. Sed. Petrology*, **40**, 249–273.

Scoffin, T. P., 1971. The conditions of growth of the Wenlock reefs of Shropshire (England), *Sedimentology*, **17**, 173–219.

Shaver, R. H., et al., 1978. The search for a Silurian reef model, Great Lakes area, *Indiana Dept. Nat. Res., Geol. Surv. Spec. Rept.*, **15**, 36p.

Shier, D. E., 1969. Vermetid reefs and coastal development in the Ten Thousand Islands, southwest Florida, *Geol. Soc. Amer. Bull.*, **80**, 485–508.

Smith, J. P., 1912. The occurrence of coral reefs in the Triassic of North America, *Amer. J. Sci.*, ser. 4, **33**, 92–96.

Smith, J. P., 1927. Upper Triassic marine invertebrate faunas of North America, *U.S. Geol. Surv. Prof. Paper 141*, 135p.

Smith, S. V., 1978. Coral-reef area and the contributions of reefs to processes and resources of the world's ocean, *Nature*, **273**, 225–226.

Squires, D. F., 1956. A new Triassic coral fauna from Idaho, *Amer. Mus. Novit.*, **1797**, 27p.

Stanton, R. J., 1967. Factors controlling shape and internal facies distribution of organic carbonate buildups, *Bull. Amer. Assoc. Petrol. Geologists*, **51**, 2462–2467.

Stetson, T. R.; Squires, D. F.; and Pratt, R. M., 1962. Coral banks occurring in deep water on the Blake Plateau, *Amer. Mus. Novit.*, **2114**, 39p.

Stoddart, D. R., 1969. Ecology and morphology of Recent coral reefs, *Biol. Rev. Cambridge Phil. Soc.*, **44**, 433–498.

Stubblefield, C. J., 1960. Sessile marine organisms and their significance in pre-Mesozoic strata, *Quart. J. Geol. Soc. London*, **116**, 219–238.

Teichert, C., 1958. Cold- and deep-water coral banks, Bull. Amer. Assoc. Petrol. Geologists, **42**, 1064–1082.

Toomey, D. F., 1970. An unhurried look at a Lower Ordovician mound horizon, southern Franklin Mountains, west Texas, *J. Sed. Petrology*, **40**, 1318–1334.

Toomey, D. F., and Ham, W. E., 1967. *Pulchrilamina*, a new mound-building organism from Lower Ordovician rocks of west Texas and southern Oklahoma, *J. Paleontology*, **41**, 981–987.

Toomey, D. F., and Winland, H. D., 1973. Rock and biotic facies associated with middle Pennsylvanian (Desmoinesian) algal buildup, Nena Lucia Field, Nolan County, Texas, *Bull. Amer. Assoc. Petrol. Geologists*, **57**, 1053–1075.

Troell, A. R., 1962. Lower Mississippian bioherms of southwestern Missouri and northwestern Arkansas, *J. Sed. Petrology*, **32**, 629–664.

Twenhofel, W. H., 1950. Coral and other organic reefs in geologic column, *Bull. Amer. Assoc. Petrol. Geologists*, **34**, 182–202.

Walker, K. R., and Ferrigno, K. F., 1973. Major middle Ordovician reef tract in east Tennessee, *Amer. J. Sci., Cooper Vol.*, **273-A**, 294–325.

Wilson J. L., 1975. *Carbonate Facies in Geologic History*. Berlin: Springer-Verlag, 471p.

Wolfenden, E. B., 1958. Paleoecology of the Carboniferous reef complex and shelf limestones in northwest Derbyshire, England, *Geol. Soc. Amer. Bull.*, **69**, 871–898.

Wray, J. L., 1968. Late Paleozoic phylloid algal limestones in the United States, *23rd Internat. Geol. Cong. Proc.*, **8**, 113–119.

Wray, J. L., 1971. Algae in reefs through time, *Proc. N. Amer. Paleontological Conv.*, pt. J., 1358–1373.

Zankl, H., 1968. Sedimentological and biological characteristics of a Dachsteinkalk reef complex in the Upper Triassic of the northern Calcareous Alps, in Müller and Friedman, 1968, 215–218.

Cross-references: *Algae; Annelida; Brachiopoda; Bryozoa; Coelenterata; Coloniality; Corals; Echinodermata; Oysters; Porifera; Rudists; Stromatolites; Stromatoporoids; Zooxanthellae.* Vol. III: *Algal Reefs; Atolls; Biological Erosion of Carbonate Coasts; Coral Reefs; Faro; Fringing Reef; Great Barrier Reefs; Microatoll; Oyster Reefs.* Vol. VI: *Algal Reef Sedimentology; Coral Reef Sedimentology; Reef Complex.*

REPTILIA

Reptilia is the class of backboned animals comprising living crocodiles, alligators, gavials, lizards, snakes, turtles, tortoises, and the nearly extinct *Sphenodon* of New Zealand, plus many extinct kinds including dinosaurs, therapsids, pterosaurs, ichthyosaurs, mosasaurs, plesiosaurs and many other less well-known animals (Fig. 1). Living reptiles constitute only a very small fraction of the total class, ancient reptiles having been much more diverse. Of the 191 reptilian families (placed in 17 orders) recognized by Romer (1966) and Colbert (1969), only 38 (assigned to 4 orders) are still represented by extant species. These figures indicate some measure of the former diversity and significance of the class, and the present impoverished state of the existing reptilian fauna.

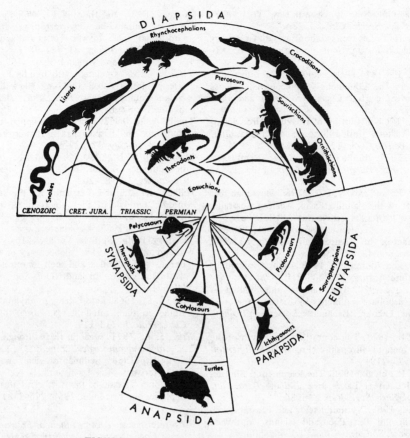

FIGURE 1. Reptilian radiation (from Colbert, 1969).

Descended from anthracosaurian amphibian stock, and having given rise to both birds and mammals, reptiles occupy an intermediate systematic position between Class Amphibia and the classes Aves and Mammalia. Living and extinct reptiles are distinct from one or the other of these other vertebrates in the following features:

1. A relatively deep skull with reduced number of skull bones (over the amphibian condition), reduced otic notch, and a single occipital condyle (except in some advanced therapsids)
2. A solitary ossicle (stapes) in the middle ear
3. Articulation of the lower jaw with the skull by means of the articular and quadrate bones
4. Lower jaws constructed of several bones in addition to the tooth-bearing dentary bone
5. Vertebrae composed largely or entirely of pleurocentra, the intercentra being reduced or lost, plus a neural arch
6. Two or more sacral vertebrae articulated with enlarged ilia
7. Dominance of scapula and coracoid in the shoulder region and reduction of the cleithrum, clavicles, and interclavicles
8. Skin covered with horny scales

Living reptiles (and presumably extinct reptiles as well) are further characterized by:

a. A three- (four- in crocodiles, and probably in dinosaurs and pterosaurs) chambered heart
b. Fully developed lungs and an absence of gills
c. Cold blooded (poikilothermy) or variable body temperature
d. Internal fertilization and the absence of a true larval stage
e. Egg characterized by amniotic and allantoic membranes and a large yolk

Except in certain lizards and snakes (and extinct ichthyosaurs), where eggs are retained in the body cavity until development is complete, all living reptiles deposit shelled eggs, generally burying them in shallow nests in soil or sand. The incubation period varies, ranging up to 15 months in *Sphenodon*. The embryo is protected from shock and injury by the outer leathery shell, and from dessication by a sac of fluid—the amnion. A second membraneous sac, the allantois, receives embryo waste products and in some forms also serves as an embryonic respiratory structure. Young reptiles hatch as essentially fully developed miniature replicas of the parents. There is no larval stage or metamorphosis as in amphibians. Gills are never present and respiration is largely or entirely by means of well-developed lungs and costal or diaphragmatic pumping.

Reptiles are distinct from birds, which possess a four-chambered heart, constant and high body temperature, and feathers. All but a few extinct birds are also edentulous. Reptiles are distinguished from mammals, which also have a four-chambered heart and uniform high body temperature, by hair and an absence of scales, viviparity, single-bone construction of the lower jaws, three ossicles in the middle ear, and a jaw articulation with the skull formed by the squamosal and dentary bones. Most mammals also possess highly differentiated teeth, whereas only extinct mammal-like reptiles (therapsids and pelycosaurs) among reptiles have evolved truely differentiated dentitions.

The oldest known reptiles are of Early Pennsylvanian age. By Permian times, reptiles had achieved significant diversification; and by Jurassic times, every reptilian order had made its appearance. All but one (Pelycosauria) of the 17 reptilian orders were present during part or all of the Mesozoic, but only 4 orders survived that era as the close of the Mesozoic marked an interval of crisis for reptiles: group after group declined and disappeared before Cenozoic times.

Classification

The class Reptilia has gone through extensive systematic evolution as various workers have proposed a number of different classifications. Because the vast majority are known only from fossil evidences, most classifications are subdivided into major categories almost entirely on the basis of skeletal anatomy. The classification currently favored by most neo- and paleoherpetologists appears in Table 1.

The six subclasses cited in Table 1 are distinguished chiefly by the construction of the skull, particularly in the temporal region (Fig. 2). Primitively, the temporal region was completely roofed over by dermal bones, as it was in labyrinthodont amphibians and primitive stem reptiles (Cotylosauria), and still is in most turtles. This imperforate condition is referred to as the anapsid condition (lacking "arches" or bridges between fenestrae). As more advanced forms evolved, the temporal region was modified in a variety of ways, probably in conjunction with improvements in the feeding apparatus and jaw musculature and mechanics. The degree of ossification of the several "temporal" bones was reduced at several points, resulting in small to large fenestrations that "exposed" the underlying jaw muscles, and thus either permitted greater space for these muscles to bulge during contraction, or provided better areas of muscle attachment or better muscle leverage. Whatever their functional significance (still under debate), the various resultant patterns of temporal fenestration (or nonfenestration) have been adopted as major anatomic criteria for subdividing the class.

Anapsid reptiles have a nonfenestrated skull roof, although some turtle skulls are deeply emarginated (from behind) in the upper temporal area. Archosaurs possess two (diapsid condition) temporal fenestrae (an upper and a lower opening) on each side, and their more primitive(?) relatives the lepidosaurs (snakes and lizards) display a modified diapsid condition where one or both bony arches delimiting the lower fenestra borders are lost. *Sphenodon,* the only surviving truely diapsid lepidosaurian is characterized by double temporal arches which define two distinct fenestrae. Crocodilians are the only surviving archosaurian diapsids.

The remaining reptilian skull types are marked by a single temporal fenestra on each side, the position of which varies. In Subclass Synapsida, this opening is low on the side of the skull and is bounded by the jugal (sometimes the quadratojugal), the squamosal, and postorbital bones. A condition termed "euryapsid" refers to the presence of a dorsal opening surrounded by the squamosal, postorbital, and parietal bones, but without a typical lower temporal opening. Some genera (such as *Araeoscelis*) appear to show this as a primitive condition; but other groups, such as the nothosaurs and plesiosaurs, seem to have developed it secondarily by loss of the lower bar from the diapsid condition. There is no evidence that the ichthyosaurs, which also show this condition, are closely related to nothosaurs or plesiosaurs.

The classification in Table 1 is the result of work by many scholars and thus represents both a synthesis and a series of compromises of many opinions. Not expressed formally here are some recent and somewhat controversial suggestions by several leading students of Reptilia. Olson (1947) suggested, for example, division of the class into two major categories: Subclass Parareptilia, including diadectomorphs and their "probable" descendants (turtles), and Subclass Eureptilia, which included the captorhinomorph cotylosaurs and all other reptiles as their probable descendants. More recent studies by Olson (1965, 1966) and Romer

TABLE 1. Class Reptilia

	Order Cotylosauria*	Suborder Captorhinomorpha* Suborder Procolophonia*
Subclass Anapsida	Order Chelonia	Suborder Proganochelydia* Suborder Amphichelydia* Suborder Cryptodira Suborder Pleurodira
	Order Mesosauria*	
Subclass Lepidosauria	Order Eosuchia*	Suborder Younginiformes* Suborder Choristodera* Suborder Thalattosauria* Suborder Prolacertiformes*
	Order Squamata	Suborder Lacertilia Suborder Ophidia
	Order Rhynchocephalia	
	Order Thecodontia*	Suborder Proterosuchia* Suborder Pseudosuchia* Suborder Aetosauria* Suborder Phytosauria*
	Order Crocodilia	Suborder Protosuchia* Suborder Archaeosuchia* Suborder Mesosuchia* Suborder Sebecosuchia* Suborder Eusuchia
Subclass Archosauria	Order Pterosauria*	Suborder Rhamphorhynchoidea* Suborder Pterodactyloidea*
	Order Saurischia*	Suborder Theropoda* Suborder Sauropodomorpha*
	Order Ornithischia*	Suborder Ornithopoda* Suborder Stegosauria* Suborder Ankylosauria* Suborder Ceratopsia*
Subclass Synaptosauria*	Order Sauropterygia*	Suborder Nothosauria* Suborder Plesiosauria*
	Order Placodontia*	
Subclass Ichthyopterygia*	Order Ichthyosauria*	
Subclass Synapsida*	Order Pelycosauria*	Suborder Ophiacodontia* Suborder Sphenacodontia* Suborder Edaphosauria*
	Order Therapsida*	Suborder Phthinosuchia* Suborder Theriodontia* Suborder Anomodontia*

*Extinct group.

(1964) concluded that diadectids are not reptiles at all, but seymouriamorph amphibians. Along quite different lines, Goodrich (1916) noted that living reptiles have an aortic arch that is quite distinct from that of living mammals and neither can be derived from the other. This has led to the opinion by some that Reptilia should be subdivided into two major groups: a "reptile segment" (termed the *sauropsids*), comprising all living reptiles and their extinct relatives, plus the Eosuchia, Thecodontia, Saurischia, Ornithischia, and Pterosauria, and a "mammal-like segment" (termed *theropsids*) that includes synapsids, euryapsids, plus the problematical Mesosauria. The significance of this twofold classification is a presumed early split in the primitive reptilian stock that resulted in the evolution of (a) many reptile-like groups and (b) the very different mammal-like pelycosaurs and therapsids and ultimately their mammalian descendants. Birds are descendant from sauropsids, either directly from some early thecodont, or more probably from a theropod dinosaur. Few students today accept either Olson's or Goodrich's dual lineage classification, although Huene (1956) advocated a slightly modified version of Goodrich's scheme. Romer (1967) has presented the most

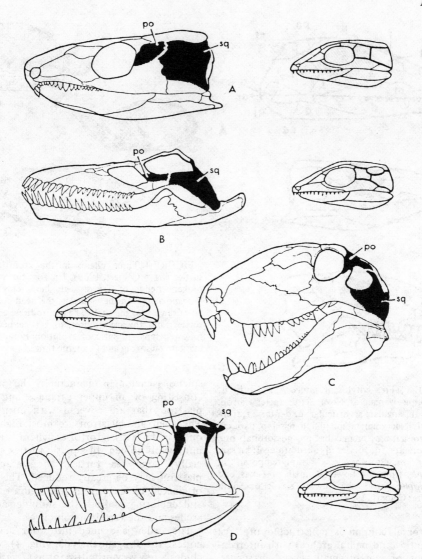

FIGURE 2. Reptilian skull types, illustrated by the skulls of characteristic genera (from Colbert, 1969). A, Anapsid skull (*Captorhinus,* a cotylosaur) with no temporal opening. B, Euryapsid skull (*Muraenosaurus,* a plesiosaur) with a superior temporal fenestra bounded below by the postorbital (po) and squamosal (sq) bones. C, Synapsid skull (*Dimetrodon,* a pelycosaur) with a lateral temporal fenestra bounded above by the postorbital and squamosal bones. D, Diapsid skull (*Euparkeria,* a thecodont) with superior and lateral temporal fenestra separated by the postorbital and squamosal bones. Not to scale.

persuasive arguments for the classification used here.

Subclass Anapsida. All reptiles with non-fenestrated temporal regions of the skull are placed in Subclass Anapsida.

Order Cotylosauria. The so called *stem reptiles,* the Cotylosauria (Fig. 3), are the earliest and most primitive representatives of Reptilia. The order is divided into two major categories— the Captorhinomorpha and the Procolophonia. Until recently, most students included two additional groups, the seymouriamorphs (in-

cluding *Seymouria,* a structural intermediate between amphibians and reptiles) and diadectomorphs. However, recent studies by Spinar (1952), Romer (1964), and Olson (1965) have shown seymouriamorphs to be anthracosaurian labyrinthodonts, and the position of diadectomorphs remains subject to dispute.

Captorhinomorphs were small to medium-sized, carnivorous animals ranging from 1 to 2 ft (30–60 cm) in length (*Captorhinus*) up to nearly 6 ft (2 m) (*Labidosaurus*). In contrast to more advanced reptiles, there is no otic notch.

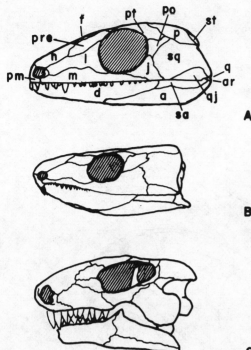

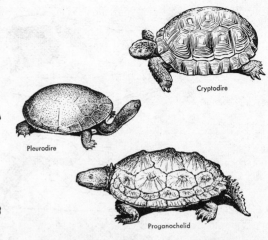

FIGURE 4. Three phases in the evolution of the turtles (from Colbert, 1969). The first turtles were the proganochelids of Triassic age, here represented by *Proganochelys*. In these turtles, the head could not be withdrawn into the shell. The pleurodires, or side-neck turtles, and the cryptodires, or vertical-neck turtles, followed separate paths of evolutionary radiation during Late Mesozoic and Cenozoic times.

FIGURE 3. Three cotylosaur skulls. A, *Paleothyris,* Middle Pennsylvanian of Nova Scotia (after Carroll, 1969). a, an, angular; ar, articular; d, dentary; f, frontal; j, jugal; l, lacrymal; m, maxilla; n, nasal; p, parietal; pf, postfrontal; pm, premaxilla; po, postorbital; pre, prf, prefrontal; q, quadrate; qj, quadratojugal; sa, san, surangular; sq, squamosal; st, supratemporal. B, *Captorhinus,* from the Lower Permian of Texas; and C, *Procolophon* from the Lower Triassic of South Africa (after Colbert, 1969).

The vertebral column is constructed of large pleurocentra (and neural arches) with intercentra absent or merely small ventral crescents. The limbs and pectoral and pelvic arches are robust, indicative of a high degree of terrestrial activity. The oldest known captorhinomorphs are *Hylonomus* and *Archerpeton* from the Lower Pennsylvanian of North America, and the last survivors come from Middle and Upper Permian strata of North America, Europe, North Africa, and India.

The Procolophonia includes a variety of small, lizard-like animals (the procolophonids) and the much larger and ponderous pareiasaurs. All were herbivores and show a high degree of dental and jaw specialization for plant eating. The pareiasaurs are exclusively Permian and are known only from Europe, Asia, and Africa. Their specific ancestry is in question.

Order Chelonia. Several hundred living turtle species and a much larger number of extinct species are among the Chelonia. The most conspicuous chelonian character is the bony shell consisting of an upper carapace and a lower plastron that are covered with horny scutes. The shell and plastron are most highly developed in terrestrial tortoises where the two are immovably joined. In aquatic turtles and, especially, large sea turtles, the carapace and plastron may be greatly reduced. The carapace originated by expansion and fusion of the ribs and enclosing of both the shoulder and pelvic arches and the vertebral column.

The order is divided into four major categories: the Proganochelydia (Fig. 4), which includes a few very primitive forms (*Triassochelys*) from the Triassic of Europe; the Amphichelydia, a large group of chiefly Mesozoic forms of primitive construction; and modern turtles, the Cryptodira and Pleurodira. The cryptodires are the most diverse and abundant of the two extant groups. Both are believed to have evolved from an amphichelydian ancestry. The cryptodires are distinct from pleurodires in the construction of the neck vertebrae, which restricts neck flexure and withdrawal of the head to a vertical plane. Pleurodires withdraw the head into the shell by a lateral flexion of the neck and hence are known as "side-neck" turtles. Turtles have diversified into many marine, fresh-water, and terrestrial niches and today are nearly universal in distribution, except for very high latitudes. Fossil turtles are most abundant from Mesozoic and Tertiary sediments of Europe, Asia, and North America, but are also known from South America and Africa. The specific ancestry of chelonians is not known.

Order Mesosauria. The Mesosauria are a minor group of small, elongate, aquatic reptiles known only from the Lower Permian of South Africa and South America. The nature of the temporal fenestration is subject to dispute. Vertebral construction indicates a possible relationship with cotylosaurs, but the precise systematic position of mesosaurs cannot be established at present because of the rarity of specimens and the poor quality of their preservation. Their peculiar distribution, in western South Africa and southern Brazil in freshwater deposits, has been cited as evidence of Permian juxtiposition of those continents.

Subclass Lepidosauria. Members of Subclass Lepidosauria are primitive diapsid (or modified diapsid) reptiles distinct from the advanced archosaurian diapsids. The subclass consists of three orders including lizards and snakes; rhynchocephalians; and a number of primitive, extinct forms from the Late Paleozoic and Mesozoic.

Order Eosuchia. The eosuchians (Fig. 5) include some of the earliest and most primitive, but not necessarily closely related, diapsids. Some are presumed ancestral to later advanced diapsids, but these relationships are not certain. Best known perhaps is *Youngina* (Suborder Younginiformes), a small, primitive, lizard-like diapsid from the Late Permian of South Africa. Because of its place in time and its primitive anatomy, *Youngina* (Fig. 5A) has been considered the possible ancestor of lizards and

rhynchocephalians. *Prolacerta* (Suborder Prolacertiformes), of the Early Triassic of South Africa is a possible lizard ancestor with its incomplete lower temporal arch (that is entirely missing in all lizards). *Prolacerta* (Fig. 5B) is very probably descended from Younginiformes. Champsosaurs (Suborder Choristodera), a third kind of eosuchian, were amphibious, crocodile-like reptiles of the Late Cretaceous and Early Tertiary. Although similar to modern crocodilians in superficial appearance, champsosaurs were not closely related to ancient or living crocodilians. Another eosuchian group is the poorly known thalattosaurs, small to large aquatic reptiles of Triassic age with a slit-like upper temporal fenestra and an open or emarginated lower fenestra. Both the thalattosaurs and the champsosaurs became extinct without descendants.

Order Rhynchocephalia. Members of the order Rhynchocephalia consist of several dozen extinct species of Mesozoic age and a solitary living species—*Sphenodon punctatum*—a lizard-like animal that is close to extinction on a few islands off the coast of New Zealand (Fig. 6). A principal distinction between rhynochocephalians and lizards is the existence of two complete temporal arches instead of just one as occurs in most lizards. Other differences are an overhanging beak of the upper jaw, non-socketed teeth fused to the jaw margins (occurs in some lizards) and a large parietal foramen. Fossil rhynchocephalians are known from nearly all continents of the world, but they seem never to have been abundant.

Order Squamata. The most important group

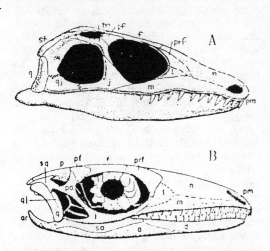

FIGURE 5. Order Eosuchia (From Romer, 1966). A, *Youngina,* Upper Permian of South Africa; length of skull about 6 cm. B, *Prolacerta,* Early Triassic of South Africa; length of skull about 6 cm. sp = splenial; see Fig. 3 for other abbreviations.

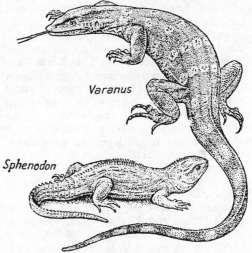

FIGURE 6. Living representatives of subclass Lepidosauria (from Young, 1962). *Sphenodon,* of order Rhynchocephalia; *Varanus,* of order Squamata.

of living reptiles, the Squamata includes all living lizards (Suborder Lacertilia) and snakes (Suborder Ophidia) and their fossil relatives. Earliest known members are Late Permian in age; and even at that time they are anatomically distinct from their more primitive eosuchian ancestors (Carroll, 1977). Lizards are distinct from other diapsids in the loss of the lower temporal arch. This resulted in a freely movable (streptostylic) quadrate, which is correlated with high mandibular flexibility and gape characteristic of lizards. Lizards are also distinct from most other reptiles in having determinent growth, controlled (as in mammals) by ossification of the epiphyses. Snakes, as highly specialized descendants of lizards, have carried this skull and jaw flexibility even farther. Here the upper temporal arch has also been lost, producing additional articular freedom in the mandibles and skull. Other cranial regions have been loosened and together these are responsible for the excessive jaw gape of nearly all snakes. Other principal features of snakes are the loss of limbs and a great multiplication of body segments increasing the number of vertebrae.

Lizards and snakes are represented by more than 5000 living species and as a group they range over most of the world. Of the major kinds of lizards, iguanids and teiids are widespread in the New World, agamids, lacertids and varanids (Fig. 6) are widespread in the Old World, and skinks and geckos are nearly universal in distribution. Of the principal kinds of snakes, colubrids are the most diverse and widespread (almost worldwide); the boids (constrictors), next most important in diversity, are found in North and South America, Africa, and Asia. Poisonous snakes (Viperidae and Elapidae) are found on all continents (except Antarctica).

An important group of extinct squamatans are the Mosasauridae of the Late Cretaceous. They were large, marine lizards, probably of varanid affinities. These fish- or mollusk-eaters retained almost no vestige of their terrestrial ancestry, except in the construction of the skull and vertebrae, having evolved a large sculling tail and large flipper-like appendages. They are known chiefly from the Niobrara chalk of central North America and similar-aged Late Cretaceous marine sediments of Europe, Africa, Asia, and New Zealand.

Another notable group of lizards are some Triassic forms recently discovered that paralleled the modern gliding lizard *Draco*. These ancient fliers, grouped in the family Kuehneosauridae, are known by several specimens from both North America and Europe. As in *Draco*, these Triassic lizards had large membranes along each side of the body that were supported by greatly elongated ribs.

Subclass Archosauria. The advanced diapsids, such as the dinosaurs, pterosaurs, and crocodilians and their ancestral and related forms comprise the subclass Archosauria. Although archosaurian and lepidosaurian diapsids are grouped separately, some students believe the two are related by a common ancestry. Thus some classifications place archosaurians and lepidosaurians together in the Subclass Diapsida. Others are of the opinion that there is no evidence that links the two beyond the fact that both are diapsid in skull construction.

Reig (1967) has suggested that archosaurs arose from Permian sphenacodontid pelycosaurs, but the evidence he presents is not conclusive. A possible archosaur ancestor, *Heleosaurus*, has recently been described from the Upper Permian of South Africa (Carroll, 1976). The general anatomy of the genus resembles that of the eosuchian *Youngina*, but the small number of blade-like teeth and the presence of dermal armor indicate relationship to the primitive thecodont *Euparkeria*.

Order Thecodontia. The thecodonts are a primitive diapsid group probably evolved from eosuchians. The order includes ancestors of the dinosaurs, crocodilians, probably also of pterosaurs and possibly that of birds, plus a variety of sidelines. Present classifications subdivide the thecodonts into four suborders: Proterosuchia, Pseudosuchia, Aetosauria and Phytosauria. The Proterosuchia include a variety of primitive Triassic forms, exemplified by *Proterosuchus* and *Erythrosuchus* from South Africa. These were quadrupedal, unarmored, and in all probability carnivorous. The Pseudosuchia are more advanced forms, trending toward bipedal posture and locomotion (Fig. 7), that included the ancestors of the Saurischia and most probably also of the Ornithischia. *Euparkeria*, of the Early Triassic of South Africa, is generally recognized as an appropriate structural ancestor of the carnivorous saurischians. The Aetosauria include a variety of moderately armored, apparently secondarily quadrupedal forms. All are from Middle or Upper Triassic strata, from Europe, North America, or South America and presumably became extinct without descendants. The phytosaurs (Fig. 7) were a widespread aquatic group that in many ways paralleled the later crocodiles. In fact, it is probable that these animals filled a crocodile–alligator ecologic niche during Triassic times only to become extinct at the end of that period after the appearance of the first primitive crocodilians. Phytosaurs were not ancestral to crocodiles, but both groups probably originated from primitive thecodonts.

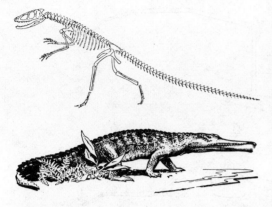

FIGURE 7. Thecodontia (from Colbert, 1969). Above, skeleton of a generalized thecodont, *Hesperosuchus* about 1.2 m in length, from the Upper Triassic of Arizona. Below, *Rutiodon,* a phytosaur from the Upper Triassic of North America, about 2 m in length.

Order Crocodilia. Third most important of living reptiles are the crocodilians. Twenty-three living species, belonging to three families (Crocodylidae, Alligatoridae, and Gavialidae), and several hundred fossil species are included in this order. The geographic distribution of fossil remains indicate a far greater former distribution for Crocodilia than is known today. A nearly worldwide range during Late Mesozoic and Tertiary times has dwindled to a limited Holocene distribution (Africa, southern and eastern Asia, southeastern North and Central America, and northern South America).

Adult crocodilians are highly aquatic, occupying marsh and stream environs of major river systems of the world (Mississippi, Amazon, Congo, Nile, Yangtze). Most living and fossil species are long and low-bodied, with short legs, large and long-snouted heads, and powerful sculling tails. All are active predators, preying on terrestrial animals, which they drag into or catch in the water. Gavials (*Gavialis*) and the false gavial (*Tomistoma*) are fish eaters, their very long and extremely narrow snouts being specializations for fish predation. Crocodilians show some advances over other living reptiles, particularly in the development of an almost completely divided four-chambered heart.

The earliest known true crocodilians are *Protosuchus,* a small, two-foot long crocodile-like reptile from the Late Triassic of southwestern United States and *Erythrochampsa* and *Notochampsa* from South Africa. Presumably, from protosuchian ancestors evolved a variety of advanced types, generally grouped together as the Mesosuchia, which were abundant and widespread during the Mesozoic. The sebecosuchians were a specialized dead-end lineage of deep-skulled, Late Mesozoic–Early Tertiary crocodilians from South America. Modern crocodilians and their immediate ancestors are placed in the Suborder Eusuchia. A recent revision by Walker (1970) subdivides known crocodilians into two suborders: the Paracrocodylia, which include various ancient kinds with crocodilian tendencies, but are apparently separated from the main crocodilian lineage, and the Crocodylia which include all proper crocodilians and their primitive ancestral stock (Protosuchia).

Order Pterosauria. Most distinctive and bizarre of all reptiles were pterosaurs or flying reptiles of the Jurassic and Cretaceous. True diapsids, they only superficially resembled birds or bats. Pterosaurs were highly modified in their skeletal structure to fit them for flight. Most bones were hollow or lightened by extra fenestrations. The most conspicuous pterosaurian feature is the excessive length of the fourth finger, which in some specimens is more than four or five times as long as the arm. A large wing membrane extended from this finger and the arm bones to the side of the body and, perhaps in some forms, back to the relatively short and weakly developed hind limbs. Very rarely, impressions of this wing membrane are preserved in pterosaur specimens from the Solnhofen limestones of Bavaria (q.v.). The underdeveloped hind limb of pterosaurs inducates that terrestrial locomotion was not of prime importance and it has been suggested that they hung bat-like from precipices, or swam duck or gull-fashion on the water.

The order is divided into two major groups: the chiefly primitive, long-tailed, toothed forms (rhamphorhynchoids) of Late Triassic and Jurassic age (Fig. 8A), and the advanced pterodactyloids with reduced tails, reduced or absent dentition and bird-like beaks. The latter apparently replaced the more primitive forms in Late Jurassic times and dominated the Cretaceous skies. Pterosaurs came in all sizes, species of *Pterodactylus* are known not much larger than a sparrow, and *Pteranodon* (Fig. 8B) had a wing span in excess of 25 ft (7.6 m). Isolated fragments indicate other kinds may have been even larger. Pterosaurs probably evolved from some Triassic thecodont. They became extinct at the end of the Cretaceous without leaving any descendants. Extinction probably resulted from an inability to compete with more efficient birds that were abundant and diverse by Late Cretaceous times.

Order Saurischia. The Saurischia are one of two orders of dinosaurs (q.v.), and include the great *Brontosaurus*-like sauropods and their ancestral stock, and all carnivorous dinosaurs (theropods). Derived from a thecodont ancestry, saurischians were important elements in

713

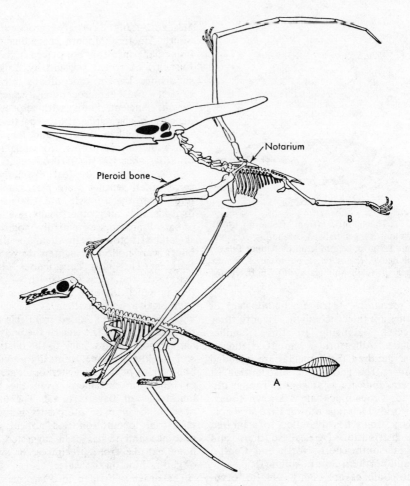

Notarium

Pteroid bone

B

A

FIGURE 8. Pterosaurs or flying reptiles (from Colbert, 1969). A, *Rhamphorhynchus* of Jurassic age, with a wing spread of 60 cm. B, *Pteranodon* of Cretaceous age, with a wing spread of about 6 m.

reptile faunas from Late Triassic until the end of the Cretaceous. The principal distinction between saurischian and ornithischian dinosaurs is in the structure of the pelvis, the three pelvic bones of the former being arranged in a triangular pattern, as viewed in lateral aspect, whereas it is quadrangular in ornithischians (see Fig. 9).

Suborder Theropoda include a variety of lightly constructed, hollow-boned, bipedal predators (Infraorder Coelurosauria) such as *Coelophysis*, *Ornitholestes*, and *Ornithomimus*, and the larger more advanced bipedal carnivores (Infraorder Carnosauria), including such well-known forms as *Allosaurus* and *Tyrannosaurus*. Coelurosaurs ranged from Late Triassic to Late Cretaceous and were nearly worldwide in distribution. Carnosaurs had a similar duration and distribution. Theropods, as the probable ancestral group of *Archaeopteryx* and later

birds (Ostrom, 1975), are the only dinosaurian group with living descendents.

Suborder Sauropodomorpha includes the all-time giants among terrestrial animals and only certain whales exceeded them in bulk. Included here are *Apatosaurus* (= *Brontosaurus*), *Diplodocus*, and *Brachiosaurus* among others. *Diplodocus* is credited with being the longest of all land animals [known specimens approximate 90 ft (27 m) in length] and *Brachiosaurus* is the heaviest (a recent estimate indicates a possible live weight in excess of 80 tons). The great size, plus the location of the nostrils high on the head have led to the belief that sauropods were probably amphibious, spending much of their time in water. Some students have challenged this, though. Sauropods also were of nearly worldwide occurrence, having been found on all continents except Antarctica. Their time span was from Early Jurassic to Late Cretaceous.

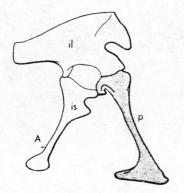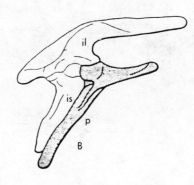

FIGURE 9. The pelvis in the two dinosaurian orders (from Colbert, 1969). A, Saurischian pelvis, with a forwardly directed pubis. B, Ornithischian pelvis, with the pubis parallel to the ischium. il, ilium; is, ischium; p, pubis.

Also included among the sauropodomorphs are the sauropod ancestors, the Prosauropoda of Triassic age. Best known of these is *Plateosaurus* of Europe. Although both theropods and sauropodomorphs have saurischian pelves, it has long been recognized they are only distantly related and it is not at all improbable that the two groups arose from different thecodont ancestors.

Order Ornithischia. The second order of dinosaurs, Ornithischia, was independently derived from a thecodont ancestry (unknown at present) and is only distantly related to saurischians. Pelvic construction is tetraradiate, rather than triangular, with a long posterior process of the pubis paralleling the ischium. Ornithischians were herbivorous and either bipedal or secondarily quadrupedal. The order is subdivided into four suborders: Ornithopoda, including numerous bipedal types (*Camptosaurus, Iguanodon, Corythosaurus*), some of which may have been amphibious (hadrosaurs); Stegosauria or plated dinosaurs (*Stegosaurus*); Ankylosauria or armored dinosaurs (*Euoplocephalus = Ankylosaurus, Palaeoscincus*); and Ceratopsia or horned dinosaurs (*Triceratops, Protoceratops*). The last three groups are quadrupedal, a condition that has long been presumed to be a secondary adaptation from three different bipedal, ornithopod ancestors, probably during the Triassic. Ornithopods ranged from Late Triassic to Late Cretaceous, the oldest known forms being *Pisanosaurus* from South America and *Heterodontosaurus* and *Fabrosaurus* from South Africa. The suborder was worldwide in distribution, but the most diverse collections are from North America and Eurasia. Stegosaurs were largely North American, European, and African in distribution and seem to have been restricted to

the Jurassic and earliest Cretaceous. Ankylosaurs were of Cretaceous age and chiefly North American and Eurasian in distribution. Ceratopsians were of Late Cretaceous age and, with the exception of *Protoceratops* from Mongolia, were restricted to western North America. Like most saurischians, the last ornithischians became extinct at the end of the Cretaceous without any evolutionary issue.

Subclass Synaptosauria. Two major groups of Mesozoic marine reptiles, the ichthyosaurs and plesiosaurs, exhibit a large upper temporal opening, but apparently not a lower opening, and have been grouped together as "Euryapsida." This common character does not necessarily indicate close relationship, however, and these groups are now placed in separate subclasses. Associated with the plesiosaurs (common in the Jurassic and Cretaceous) are the Triassic nothosaurs and placodonts.

It has been suggested that a heterogeneous assemblage of Permian and Triassic tetrapods, alternatively referred to as the Araeoscelidia or the Proterosauria, might be related to the ancestry of nothosaurs and plesiosaurs. The included genera, *Araeoscelis, Trilophosaurus* (Fig. 10A), *Proterosaurus,* and the long-necked *Tanystropheus* (Fig. 10B; Wild, 1974) were apparently terrestrial forms, with the common presence of an upper, but not an orthodox, lower temporal opening. None of these forms show specific affinities with the subsequent aquatic groups and as the anatomy of each genus becomes better known, it is increasingly apparent that they constitute an artificial assemblage.

Order Sauropterygia. The sauropterygians were moderate to large-sized, marine reptiles termed plesiosaurs and nothosaurs. The nothosaurs were primitive fish-eating forms of

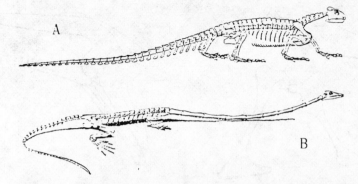

FIGURE 10. Subclass Synaptosauria, Order Araeoscelidia (from Romer, 1966). A, *Trilophosaurus,* Triassic of North America, length about 3 m. B, *Tanystropheus,* European Triassic; average specimens about 75 cm long.

Triassic age. Nothosaurs were closely related predecessors of the Jurassic and Cretaceous plesiosaurs. The latter were broad-bodied reptiles with large paddle-like flippers, short nonsculling tails and short (pliosaurs) or extremely long (plesiosaurs; Fig. 11A, B) flexible necks. Although the head was rather small, the mouth was large and armed with many long, sharp teeth, indicating a fish diet. Plesiosaurs are known from marine sediments of all continents except Antarctica and South America, whereas nothosaurs are known only from Triassic sediments of North America, and Eurasia.

Piveteau (1955) referred to specimens from the Upper Permian of Madagascar as possible

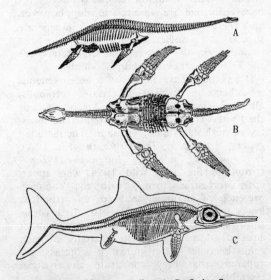

FIGURE 11. Marine reptiles. A, B, Order Sauropterygia (from Romer, 1966): A, *Muraenosaurus,* a long-necked Jurassic plesiosaur, about 6.3 m in length. B, *Thaumatosaurus,* a short-necked Jurassic plesiosaur, ventral view, about 3.3 m long. C, Order Ichthysauria: *Ichthyosaurus,* European Jurassic.

plesiosaur antecedents. Additional, excellently preserved material demonstrates clear relationships between these forms and nothosaurs and plesiosaurs on one hand, and eosuchians on the other. Specimens from the Upper Permian and possibly lowest Triassic of Madagascar show an almost complete transition from probably terrestrial eosuchians, through genera showing modest aquatic adaptation, to fragmentary remains of presumably totally aquatic nothosaurs.

Order Placodontia. The highly specialized aquatic, mollusk-eating reptiles grouped here in the order Placodontia are often classified as a suborder of sauropterygian. But their very different feeding adaptations and appendicular structures warrant separation from that order. Most placodonts have procumbent anterior teeth and greatly expanded flat tooth crowns that seem explicable only as crushing plates—presumably for crushing clams or other shellfish. Placodonts are exclusively Triassic in age and so far are known only from Eurasia and North Africa.

Subclass Ichthyopterygia. *Order Ichthyosauria.* The various reptilian forms classified as Ichthyosauria were among the reptiles most highly specialized for an aquatic life. Ichthyosaurs were remarkably fish-like in body shape, fin design, and form of the appendages (see Fig. 11C). The body was streamlined and fusiform with a large fish-like vertical tail fin, a large dorsal fin, and large pectoral and pelvic fins. The head was large, often with huge eyes, and long snouted—with large jaws armed with numerous needle-like teeth indicating a predaceous habit. They may have fed on fish, but at least some specimens contain remains of ancient squids. The skull construction, long thought to be unique (the so-called parapsid condition) has recently been shown to have been of euryapsid design, although this does not necessarily mean a close relationship to other euryapsids.

Ichthyosaurs frequently are compared with sharks and mammalian porpoises as examples of evolutionary convergence of three superficially structurally similar, but remotely related, animals. Ichthyosaurs appeared fully evolved in Middle Triassic times and endured with little change throughout the Mesozoic. There is no fossil evidence of a pre-ichthyosaur stage and thus no concrete indication of ichthyosaurian ancestry. It is most likely, however, that they had an independent derivation from cotylosaurian stock. Marine in habit, they are known to have reproduced by internal fertilization and oviviparity, by several remarkable specimens from the Holzmaden limestones of Germany that show unborn juveniles preserved within the adult body cavity. Ichthyosaur remains have been found in marine deposits of every continent of the world except Antarctica and Africa.

Subclass Synapsida. At present, only two orders of Synapsida are recognized, the pelycosaurs or primitive mammal-like reptiles and the extremely diverse therapsids or advanced mammal-like reptiles. They are distinguished from all other reptiles by the "synapsid" skull condition, where a single temporal fenestra occurs low on each side in the temporal region.

Order Pelycosauria. An important group of primitive synapsids abundant during the Permian, Order Pelycosauria, included ancestral forms of the higher therapsids and their descendants, the mammals. The order is divided into three subgroups—sphenacodonts (typified by *Dimetrodon*), edaphosaurs (including the bizarre, yard-armed-neural-spined *Edaphosaurus*), and their primitive ancestors, the ophiacodonts (Fig. 12). Pelycosaurs first appeared during the early Pennsylvanian, probably as an early offshoot of captorhinomorphs, but they did not reach their peak until Permian times. The earliest and most primitive pelycosaurs are classified as Ophiacodontia. They were initially small lizard-like forms distinguished from captorhinomorphs by little more than the presence of a lateral temporal opening. More advanced forms (Suborder Sphenacodontia) are believed descendant from ophiacodonts and were the dominant terrestrial predators during the Late Pennsylvanian and Early Permian, with their large jaws and spike-like teeth. A second lineage derived from ophiacodonts, the Edaphosauria, was a group of medium-sized (5–8 ft; 1.5–2.5 m) herbivores. Pelycosaurs are known primarily from the Upper Carboniferous and Permian of North America, to a lesser extent from Eurasia and Africa. They became extinct at the end of the Permian. Most students agree that the ancestry of the therapsids (and ultimately the mammalian class) lies within the

sphenacodontid pelycosaurs, close to *Dimetrodon*.

Order Therapsida. The Therapsida was a large diverse group of advanced synapsid or mammal-like reptiles that was abundant during Permian and Triassic times (Fig. 13). The order is subdivided into three suborders: Phthinosuchia, Anomodontia, and Theriodontia. The phthinosuchids were the most primitive therapsids and in many respects were little advanced over sphenacodontid pelycosaurs. They are known chiefly from Permian deposits of Russia. Like the sphenacodontids, they all seem to have been carnivorous. Phthinosuchids are believed to have given rise to two distinct radiations, the herbivorous anomodonts and the largely carnivorous theriodonts. Important anomodont trends were the primitive and often ponderous dinocephalians such as *Jonkeria* and *Moschops*—large, heavy-bodied and spraddle-legged animals that were abundant in South Africa and Europe during Permian times. Another major anomodont lineage was the dicynodont radiation, a diverse group in which teeth became reduced and lost, except for two large tusks in the upper jaws. The rest of the upper jaw and the lower jaw were covered with modified beaks, rather like those of turtles (Fig. 13D). Anomodonts are known from Permian and Triassic sediments, chiefly from South Africa, but also from Europe, Asia, and North and South America. A very important recent discovery was that of dicynodont (and theriodont) remains in Antarctica, which is generally considered as significant evidence that Antarctica was much closer to (if not actually continuous with) Africa and South America during Permian-Triassic times.

Theriodonts included a variety of carnivorous mammal-like reptiles, the most advanced of which (Infraorder Cynodontia; Fig. 13B) gave rise to mammals. Some were very mammal-like in skeletal construction and were characterized by such mammalian features as tooth differentiation, enlargement of the brain case, ventral (rather than lateral) position of the limbs, reduction of accessory mandible bones and reorganization of the dentary bone. Some were marked by prominent "zygomatic" arches as in mammals, a high saggital crest, and high coronoid processes on the lower jaws, all of which relate to powerful jaw musculature. Theriodonts are best known from Permian, Triassic, and Lower Jurassic rocks of the Old World, particularly of South Africa; but important new finds have recently been made in South America and also in Antarctica. Therapsids became extinct in the Jurassic, the majority dying out without issue by the end of the Triassic.

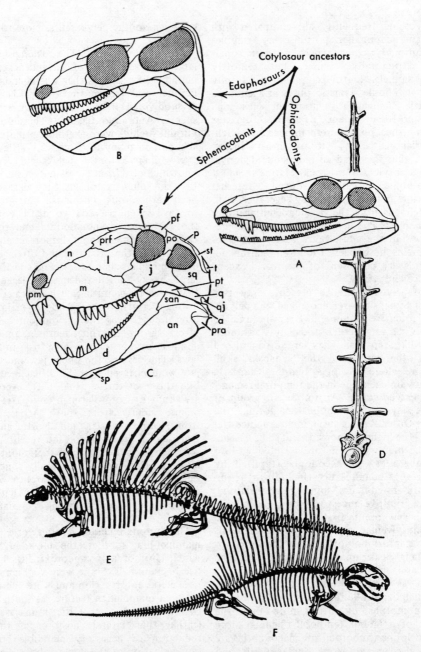

FIGURE 12. Pelycosaurs (from Colbert, 1969). A, skull of *Varanosaurus;* B, skull of *Edaphosaurus;* C, skull of *Dimetrodon;* D, dorsal vertebra of *Edaphosaurus* in posterior view; E, skeleton of *Edaphosaurus;* F, skeleton of *Dimetrodon.* These reptiles were about 2–3 m in length. pra, prearticular; pt, pterygoid; sp, splenial; t, temporal; see Fig. 3 for other abbreviations.

Evolution

The transition from amphibian to reptile probably took place during Early Carboniferous times, since true reptiles are known from Lower Pennsylvanian rocks. Early reptiles can be differentiated from contemporary amphibians by the structure of the palate and back of the skull roof, indicating a reorganization of the jaw musculature and an increase in its mass. Early reptiles were of small body size (approximately 100 mm from the snout to the base of the tail), probably reflecting an insectiverous diet and change from amphibian reproductive habits. It is reasonable to suppose that Pennsylvanian reptiles had already achieved the definitive rep-

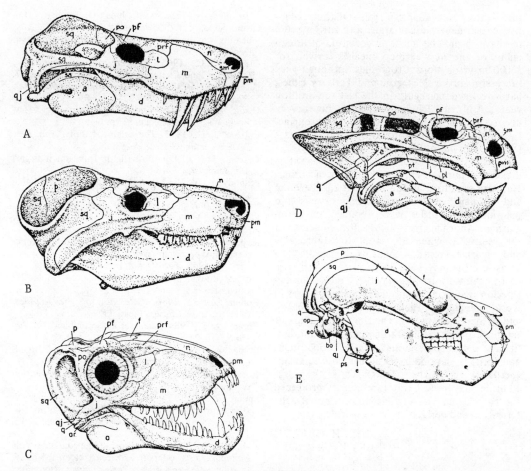

FIGURE 13. Therapsida, mammal-like reptiles, (from Romer, 1966). A, skull of the therocephalian *Lyco-suchus*, length about 28 cm; B, skull of the cynodont *Cynognathus*, length about 46 cm; C, skull of the primitive therapsid *Phthinosuchus*, length about 20 cm; D, skull of *Dicynodon*; E, skull of the tritylodont *Bienotherium*, length about 13 cm. bo, basioccipital; e, epipterygoid; eo, exoccipital; op, opisthotic; pl, palantine; ps, parasphenoid; sm, septomaxillary; pt, pterygoid; see Fig. 3 for other abbreviations.

tilian amniote egg and developmental sequence. Modification of reproductive processes to one of internal fertilization and the development of the large-yolked amniote egg, with its protective shell and its respiratory and antidessication membranes, were key steps in the rise of higher tetrapods, making it possible for reptilian eggs to be deposited and to develop outside of the aquatic environment. These changes presumably occurred via intermediate forms, analogous to plethodontid salamanders, that laid nonamniotic eggs on land, with the young developing without an aquatic larval stage. Once released of this link to the aquatic heritage, reptiles were free to diversify into many new terrestrial and aerial niches. Some however, returned to a highly specialized aquatic existence.

The oldest known reptilian remains (*Hylonomus*) are of Early Pennsylvanian age. No appropriate amphibian ancestor is known prior to that time, but the Middle-Pennsylvanian anthracosaur *Gephyrostegus* is judged by Carroll (1969) as close anatomically to that expected in reptilian ancestry. Captorhinomorph cotylosaurs represent the initial "conservative" reptilian rediation, which probably gave rise to Eosuchians (and their derivatives the squamatans and rhynchocephalians) and thecodonts (and their descendants the crocodiles, dinosaurs, pterosaurs, and birds). Turtle origins are not known. Pelycosaurs and their descendants (therapsids and mammals) also evolved from a captorhinomorph ancestry. The large marine groups are also ultimately derived from the captorhinomorphs—synaptosaurians via eosuchians and the ichthyosaurians via still-unknown intermediates. Each reptilian order appears to represent a distinctive evolutionary trend into a new adaptive type. Of these adaptive experiments, squamatans and turtles seem to have been the most successful, although many of the now-extinct lineages thrived for a 100 m yr or more. Crocodiles, although very successful during Mesozoic and Tertiary times, have declined markedly in the last few million years. And the Rhynchocephalia, never very diverse or abundant, are now close to extinction with a few small populations of one species (*Sphenodon punctatum*).

Reptilian evolutionary radiation was at a maximum during Permian and Mesozoic times. Major crises arose at the close of the Permian and Triassic when many archaic types and mammal-like reptiles were reduced or extinguished. Another crisis occurred at the close of the Mesozoic when dinosaurs and pterosaurs and marine reptiles became extinct. The Triassic crisis may have been related to the rise of the "novel" archosaurs; the Cretaceous extinction may have resulted from the changes in world geography, degeneration of climates, and/or major changes in the flora. Of the major reptilian trends, lepidosaurs survive in modern lizards and snakes, archosaurs in crocodilians and birds, synapsids in mammals, and anapsids (perhaps) in turtles. Thus, while the class Reptilia may now be at low ebb, it produced the two higher classes that now dominate the terrestrial and aerial vertebrate world.

JOHN H. OSTROM
ROBERT L. CARROLL

References

Bellairs, A. d'A., 1957. *Reptiles: Life History, Evolution and Structure*. London: Hutchinson, 159p. (New York: Harper paperback, 1960; 2nd ed., 1968.)

Bellairs, A. d'A. and Fox, C. B., 1976. *Morphology and Biology of Reptiles*. London: Academic Press, 290p.

Carroll, R. L., 1969. Problems of the origin of reptiles, *Biol. Ref.*, 44, 393–432.

Carroll, R. L., 1976. Eosuchians and the origin of archosaurs, in C. S. Churcher, ed., *Athlon. Essays on Palaeontology in Honour of Loris Shano Russell*. Toronto: Roy. Ont. Mus. Life Sci. Misc. Pub., 58–79.

Carroll, R. L., 1977. The origin of lizards, in S. M. Andrews, *et al.*, eds., *Problems in Vertebrate Evolution*. London: Academic Press, 359–396.

Colbert, E. H., 1961. *Dinosaurs, Their Discovery and Their World*. New York: Dutton, 300p.

Colbert, E. H., 1969. *Evolution of the Vertebrates*, 2nd ed. New York: Wiley, 535p.

Goodrich, E. S., 1916. On the classification of the Reptilia, *Proc. Roy. Soc. London*, 89, 261–276.

Gow, C. E., 1975. The morphology and relationships of *Youngina capensis* Broom and *Prolacerta broomi* Parrington, *Palaeontol. Africana*, 18, 89–131.

Huene, F. von, 1956. *Paläontologie und Phylogenie*

der niederen Tetrapoden. Jena: G. Fischer, 716p.

Olson, E. C., 1947. The Family Diadectidae and its bearing on the classification of reptiles, *Fieldiana, Geol.*, 11, 2–53.

Olson, E. C., 1965. Relationships of *Seymouria, Diadectes*, and Chelonia, *Amer. Zool.*, 5, 295–307.

Olson, E. C., 1966. Relationships of *Diadectes, Fieldiana, Geol.*, 14, 199–227.

Olson, E. C., 1971. *Vertebrate Paleozoology*. New York: Wiley-Interscience, 839p.

Ostrom, J. H., 1975. The origin of birds. *Ann. Rev. Earth Planetary Sci.*, 3, 55–77.

Ostrom, J. H., 1979. Bird flight: How did it begin? *Am. Scientist*, 67, 46–56.

Peabody, F. E., 1952. *Petrolacosaurus kansensis* Lane, a Pennsylvanian reptile from Kansas, *Paleontol. Contrib. Univ. Kansas, Vertebrata*, 1, 41p.

Piveteau, J., 1955. L'origine des Plésiosaures, *Comptes Rendus Acad. Sci.*, 241, 1486–1488.

Reig, O. A., 1967. Archosaurian reptiles: A new hypothesis on their origins, *Science*, 157, 565–568.

Reisz, R., 1975. *Petrolacosaurus kansensis* Lane, the oldest known diapsid reptile. Ph.D. thesis, McGill University.

Romer, A. S., 1956. *Osteology of the Reptiles*. Chicago: Univ. Chicago Press, 772p.

Romer, A. S., 1964 *Diadectes* an amphibian? *Copeia*, 1964, 718–719.

Romer, A. S., 1966. *Vertebrate Paleontology*, 3rd ed. Chicago: Univ. Chicago Press, 468p.

Romer, A. S., 1967. Early reptilian evolution re-viewed, *Evolution*, 21, 821–833.

Spinar, Z. V., 1952. Revise nekterých morauských Diskosauriscido (Revision of Some Moravian Discosauriscidae), *Roz. Ustred. Ustav. Geol.*, 15, 1–159.

Walker, A. D., 1970. A revision of the Jurassic reptile *Hallopus victor* (Marsh) with remarks on the classification of crocodiles, *Phil. Trans. Roy. Soc. London*, Ser. B, 257, 323–372.

Wild, R., 1974. Die Triasfauna der Tessiner Kalkalpen. XXIII: *Tanystropheus longobardicus* (Bassoni), *Schwiez. Paläontol. Abhand.*, 95, 162p.

Young, J. Z., 1962. *The Life of the Vertebrates*, 2nd ed. Oxford: Clarendon Press, 820p.

Cross-references: *Amphibia; Aves; Bones and Teeth; Dinosaurs; Paleoecology, Terrestrial; Vertebrata; Vertebrate Paleontology*.

ROSTROCONCHIA

Rostroconchs are an extinct class of mollusks characterized by a bilaterally symmetrical, univalved, untorted, and uncoiled larval shell (protoconch, Fig. 1-10, p. 722) and a bivalved adult shell. They lack a ligament, the structure that holds together the two valves of such other bivalved mollusks as pelecypods (= Class Bivalvia) and some gastropods. In rostroconchs, one or more of the calcareous shell layers are continuous across the dorsum so that a dorsal commissure is lacking (Figs. 1-1, -14). Rostroconchs have a single center of calcification (protoconch) and become bivalved by accentuated growth of the lateral lobes of the mantle and shell; pelecypods and bivalved gastropods become bivalved by secretion of the shell from two centers of calification. Comarginal growth lines of rostroconchs cross the dorsum at nearly right angles to the hinge axis (Figs. 1-14, -15, -17), rather than being reflected toward the beak, because of the single center of calcification.

Distribution

Rostroconchs are known only from Paleozoic rocks and range in age from the Early Cambrian through the Late Permian. They are known from all continents except Antarctica.

Classification

The order Ribeirioida (Figs. 1-1-7; Fig. 2) has previously generally been allied to the bivalved arthropods. The members of the order show molluscan affinities in the presence of a protoconch (Fig. 1-1), comarginal growth increments (Fig. 1-2), growth increments on the muscle scars (Fig. 1-4), and a pallial line in some forms. Rostroconchs that have all shell layers continuous across the dorsal margin, and that have an anterior pegma (Figs. 1-1-7), a dominant posterior growth component (Figs. 1-1-7), and musculature consisting of anterior and posterior median pedal retractor muscles (Figs. 1-4, -5) connected by linear right and left side muscles (Figs. 1-5; 2) are placed in the Ribeirioida. The stratigraphic range of the order is Lower Cambrian through Upper Ordovician.

The order Conocardioida (Figs. 1-8-17) has traditionally been placed in the class Bivalvia. Its relationship to the ribeirioids is shown by the presence of a pegma (Figs. 1-12, -13) in primitive members of the group. All rostroconchs that have external (Figs. 1-8, -15, -17) and internal (Figs. 1-13) ribs, marginal denticles (Figs. 1-11, -13, -16), an anterior gape (Figs. 1, -11, -16), dorsal clefts (Figs. 1-14, -17), and a musculature of right and left pedal retractor muscles and a pallial line are placed in the Conocardioida. The stratigraphic range of the order is Lower Ordovician through Upper Permian.

Diversity

The approximately 375 known species of rostroconchs are placed in 34 genera; the 15 known Cambrian species are placed in 9 genera; about 93 Ordovician species are known, which are placed in 20 genera; and about 265 species of the post-Ordovician Paleozoic are placed in 7 genera.

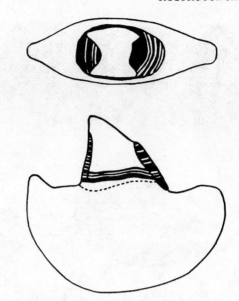

FIGURE 2. *Pauropegma jelli* (Pojeta and Runnegar). Dorsal and left lateral views of an internal mold, showing anterior and posterior median pedal retractor muscle scars, right and left side muscle scars, and prominent anterior and posterior pegmas which produced the large conical umbonal filling. Anterior of shell is to left.

Rostroconchs underwent their greatest radiation in the Early Ordovician; 5 families, 16 genera, and about 50 species are known. By the Late Ordovician, 5 families were still represented, but only 6 genera and 19 species are known. By contrast, the other major group of bivalved mollusks, the pelecypods, was represented in the Cambrian by one family, genus, and species (*Fordilla troyensis*). By the Early Ordovician, there were 6 families, 16 genera, and about 45 species. In the Middle and Late Ordovician, pelecypods radiated explosively and were represented by about 16 families, 140 genera, and 1400 species. The decline in diversity of rostroconchs after the Early Ordovician may be due to competition with pelecypods for living space and foodstuffs.

JOHN POJETA, JR.
BRUCE RUNNEGAR

References

Pojeta, J., Jr., 1971. Review of Ordovician pelecypods, *U.S. Geol. Surv. Prof. Paper 695,* 46p.

Pojeta, J., Jr., and Runnegar, B., 1976. The paleontology of rostroconch mollusks and the early history of the Phylum Mollusca, *U.S. Geol. Survey Prof. Paper 968,* 88p.

Pojeta, J., Jr., Runnegar, B., Morris, N. J., and Newell, N. D., 1972. Rostroconchia: A new class of bivalved mollusks, *Science,* 177, 264–267.

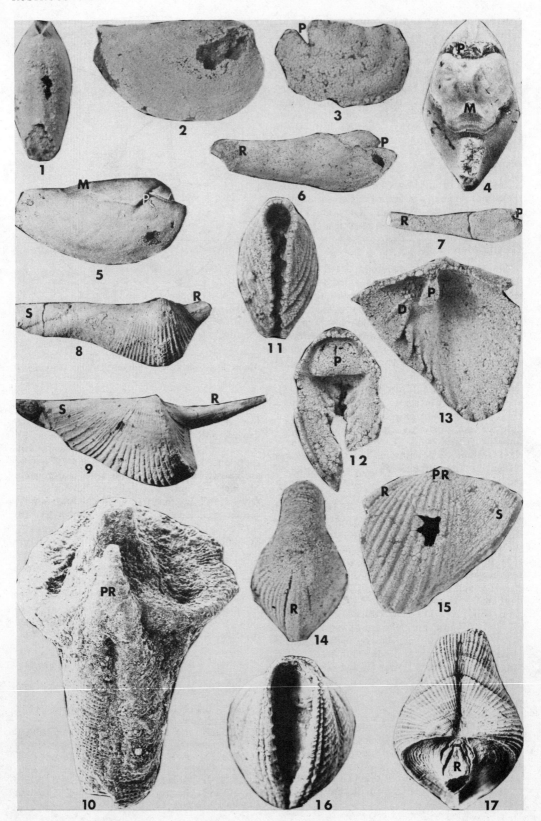

FIGURE 1. Various rostroconch mollusks illustrating diversity of form and basic morphology of the class. 1–7, Ribeirioida; 8–17, Conocardioida. 1–2, Dorsal and right side views of *Ribeiria australiensis,* showing protoconch, lack of dorsal commissure, and growth increments. 3, Internal mold of *Cymatopegma semiplicatum;* left-side view showing impression of pegma (P). 4, Internal mold of *Technophorus marija;* dorsal view, showing growth increments on the posterior median pedal retractor muscle scar (M), and pegma (P), and lack of a dorsal commissure. 5, Internal mold of *Ribeiria apusoides;* right-side view, showing posterior median pedal retractor muscle scar (M), impression of pegma (P), and side muscle scar between M and P. 6, Internal mold of *Pinnocaris robusta;* right-side-view, showing impression of pegma (P), and rostrum (R). 7, Internal mold of *Pinnocaris wellsi;* right-side view, showing impression of pegma (P), and rostrum (R). 8, Left side view of *Conocardium elongatum* showing rostrum (R) and snout (S). 9, Left-side view of *Conocardium pseudobellum,* showing rostrum (R) and snout (S). 10, Dorsal view of *Hippocardia*?, showing protoconch (PR). 11–15 Anterior, anterior interior, right-side interior, dorsal, and right-side exterior views of *Eopteria struszi,* showing anterior gape, pegma (P), marginal denticles elongated as internal ribs (D), rostrum (R), dorsal rostral clefts on either side of rostrum, lack of dorsal commissure, protoconch (PR), and snout (S). 16–17, Anterior and dorsal views of *Pseudoconocardium lanterna* showing anterior gape lined by marginal denticles, rostrum (R), and dorsal rostral clefts on either side of rostrum.

Pojeta, J., Jr., Gilbert-Tomlinson, J., and Shergold, J. H., 1977. Cambrian and Ordovician rostroconch molluscs from northern Australia, *Bur. Miner. Resour. Australia Bull.,* **171,** 54p.

Runnegar, B., 1978. Origin and evolution of the Class Rostroconchia, *Phil. Trans. Roy. Soc. London, Ser. B,* **284,** 319–333.

Runnegar, B., and Pojeta, J., Jr., 1974. Molluscan phylogeny: The paleontological viewpoint, *Science,* **186,** 311–317.

Cross-references: *Bivalvia; Mollusca; Monoplacophora.*

RUDISTS

The evolution of modern types of coral reefs started early in the Mesozoic Era, when the rootstocks of living hermatypic corals, hydrozoans, crustose algae, sponges, and other reef-forming organisms first appeared. During the Triassic and Jurassic Periods, 136–225 m yr ago, these reef-associated organisms evolved rapidly; radiated into most of the major ecological niches that they occupy today; and, by the end of the Triassic, combined to form extensive "barrier" reefs (aligned bioherms), isolated bioherms, and patch reefs of modern proportions. These were reefs of great ecological complexity, composed of diverse organisms and zoned into distinct communities. In all probability, the complex symbiotic relationships between hermatypic corals (especially Thamnasteroids, Microsalenids, Latomeandroids, Synastreids, and Amphastreids) and intracellular zooxanthellae had formed by Jurassic time. All evidence indicates that reef communities dominated by hermatypic corals, hydrozoans, and crustose algae, were already well on their way toward the evolution of modern reefs by the end of the Jurassic Period.

The Cretaceous Period, however, brought bizarre and unexpected changes in the composition of reef-framework communities, which interrupted the evolutionary trend toward modern reefs for 70 m yr. Once-dominant hermatypic corals retreated abruptly from the main reefal environments, for the most part, to continue their evolution on hardgrounds and skeletal sands of protected lagoons and shallow wave-swept flats. In their place, a remarkably similar group of bivalves known as the rudists evolved rapidly to dominate the main reef environments.

Functional Morphology

The rudists (Superfamily Hippuritacea) were a highly aberrant group of small to very large (up to 2 m in height), massive-shelled, rapidly growing, gregarious epifaunal bivalves that dominantly lived cemented to bioclastic substrates, or to each other, in shallow-marine (photic, inner sublittoral), tropical to subtropical, carbonate-platform environments. Perkins (1969, pp. N751–N764) Dechaseaux (1952), Douville (1935), Klinghardt (1931), Milanović (1933), Toucas (1903–1904), and Skelton (1976, 1978) give details of rudist morphology and its functional significance, and Yonge (1967) interprets rudist form and habit using living Chamidae as a model. The shells (Fig. 1) were slightly to grossly inequivalved, the lower (attached) valve normally being larger and more massive than the upper (free, cap) valve. The attached valve was either left or right, depending upon the group, and coiled to straight (cylindrical or conical). The free valve was small and flat, conical or rounded, or was large, highly convex, and coiled. Both valves, though exceptionally thick among bivalves, were of low density in most species owing to shell structure in both the inner (aragonitic) and outer (calcitic) shell layers. These layers consisted largely of an open meshwork of thin-

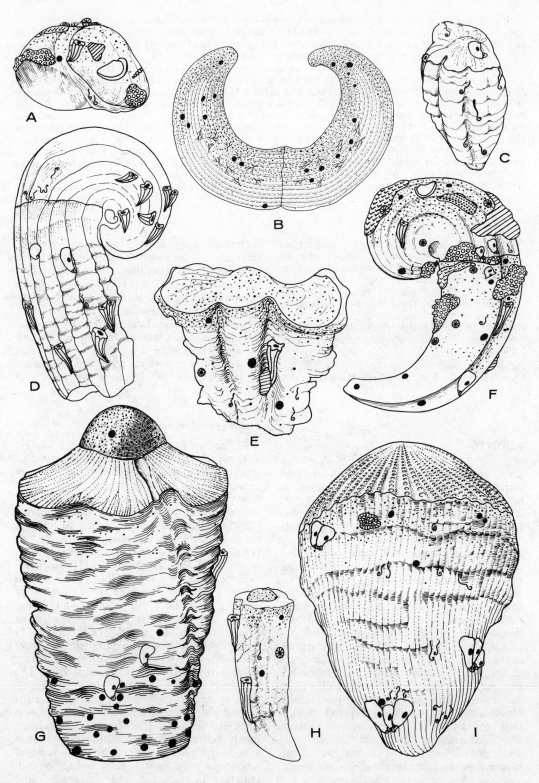

walled cells, which presumably contained fluid (sea water and/or pallial fluid) during life. This open shell structure was, in some taxa, accompanied by large elongated tubules and pallial canals through the shell.

The rudist animal occupied a shallow to deep, cylindrical to cone-shaped body cavity largely confined to the lower valve. In some taxa, the main part of this body cavity, containing the viscero-pedal mass, was closed off at the base repeatedly by concave plates (tabulae or dissepiments) as the shell grew upward, and the animal retained its dorsal living position just below the aperture. Complex infoldings of shell material, roughly similar to coral septae, deep grooves for the inhalant and exhalant siphons, and two elongated sockets for teeth of the upper valve, subdivided the body cavity in the lower valve of various groups. Two teeth and a deep socket in the free valve, and a single tooth bounded by two sockets in the attached valve, characterize most rudists.

The free valves of most rudists were normally smaller and morphologically distinct from the attached valves, forming a cap or operculum-like structure over the body cavity of the attached valve. Most free valves contain a shallow internal depression continuous with the body cavity of attached valves, mainly to accommodate mantle material and muscles used in opening and closing the shell and projecting or retracting the mantle. Three basic types of free valves evolved in the rudists (Fig. 1), reflecting distinct feeding methods and mantle characteristics. The first, or normal bivalve type, is thick, usually convex and cap shaped, and lacks major perforations (Fig. 1D,F). It covers all of the dorsal surface of the attached valve, and is hinged along one side so that it flips open through expansion of the ligamental pad within the shell. The second, found primarily in the Radiolitidae, is a thin shield over the dorsal surface of the attached

valve (Fig. 1C,E), partially or wholly covering it with a translucent screen, or else simply a small plug for the living cavity (Fig. 1G), leaving the mantle exposed to light over the upper surface of the attached valve, suggesting that the mantle may have been utilized for the nurturing of intracellular zooxanthellae. This type of valve had long hinge teeth and sockets, and had to be raised vertically. The third type of cap (Hippuritidae; Fig. 1I) is a thin perforated sediment and/or light screen having very long teeth, and it too could only be raised vertically above the attached valve, possibly in a pumping fashion to clean the mantle cavity and to circulate fresh water within it.

The open cellular structure of the rudist shell permitted rapid growth without great expenditure of calcium carbonate, and this resulted in the construction of very large massive shells in short periods of time. Filling of these cells with fluid would have provided the necessary density to make the rudist shells stable on the sea floor as exposed epifaunal organisms. Rapid growth was probably aided by a very large mantle, which had an exceptionally active secretory function compared to other bivalves, and would have been further enhanced by the presence of intracellular symbiotic zooxanthellae in the mantle tissue.

Most rudistids had small attachment bases cemented to the substrate during part or all of life, and grew erect, perpendicular to the bottom. This growth habit had the advantage of elevating the feeding, respiratory, sensory, and excretory areas of the shell well above the substrate into actively circulating waters overlying the main zone of sediment transport. This prevented fouling and enhanced feeding and respiration. It was this ability for rapid upward growth, and their strong tendency toward a gregarious life habit, that allowed rudists to outcompete Mesozoic corals in their

FIGURE 1. Characteristic Late Cretaceous rudistid bivalve morphotypes from the Tropical Caribbean Province (from Kauffman and Sohl, 1974). All were reefoid framework builders. Epibionts and boring organisms on the bivalves are shown in typical density and distribution; note their low level of development compared with that on modern reef-dwelling shell surfaces, and thus lack of potential binding of rudist frameworks. Common borers include lithophagid bivalves (large black spots), clionid sponges (small black dots), worms, and bryozoans (dendritic pattern, linear dots). Common epizoans include other rudists (as in D), oysters (with subcentral black spot, = muscle scar), serpulid worms (coiled and sinuous lines), hermatypic corals (open circles), solitary corals (circle with radiating lines), "stromatoporoids" (box-like pattern), bryozoans (crowded dots within colony outline), hydrozoans (inclined-line pattern), and crustose algae (check pattern). Note zonation of borers and epibionts relative to light (D, F, G) and, in prone forms, relative to current (B; current from bottom of figure). Adaptive types and representative taxa are as follows: broadly cemented oyster-like morphotype, *Plagioptychus* (A); large, massive free-living recumbent taxa, *Titanosarcolites* (B); small to moderate size, erect, highly gregarious forms, *Thyrastylon* (C), *Biradiolites* (H), and *Bournonia* (E); moderate to large-size, suberect to semirecumbent taxa with curved valves, offset center of gravity, *Chiapasella* (D) and *Antillocaprina* (F); very large, barrel-shaped taxa with highly modified upper valves as small plugs, *Durania* (G)—with mantle blood vessel markings on upper surface of lower valve indicating continuous mantle exposure, probable symbiotic zooxanthellae—or thin sieve-like covers, *Barrettia* (I).

prime habitats. In order to maintain stability in this erect posture, in the face of current and wave action, many rudists grew clustered together and possessed external flanges or axial ribs that interlocked between adjacent individuals and provided mutual support. Some forms had varying amounts of mutual cementation for additional stability.

Other rudists lay recumbent on the substrate during life, on the flattened side of the shell. In some of these, broad cementation surfaces provided stability. Free-living forms were normally large and massive and commonly possessed large flanges and external ribs on the down-facing side, which stabilized them on the substrate in the face of currents.

Paleoecology

Throughout their history, the rudistids preferentially lived in tropical to subtropical marine environments (Tethys, including the Caribbean, and the tropical Indo-Pacific during the Mesozoic). A very few more generalized Cretaceous taxa, mainly of the family Radiolitidae, spread widely into the warm to mid-termperate climatic zones that existed during the Cretaceous as far north as southern Canada and the British Isles. Rudistids showed a strong preference for the shallow-water, photic habitats of tropical-subtropical carbonate platforms in water depths primarily less than 50 m, and within reach of normal wave base (15–20 m). A few occurrences, however, indicate that some single individuals and clusters may have been able to survive depths as great as 200 m. Some Cretaceous taxa flourished in hypersaline lagoons (Perkins, 1974), and others were relatively common in sand-mud, clastic sedimentary environments that had at least moderate turbidity. In general, however, the rudists preferred clear, nonturbid, normal marine waters in areas having a low sedimentation rate. This environmental preference is compatible with the suggestion that, like hermatypic corals, some rudists (e.g., radiolitids) had symbiotic zooxanthellae in their mantle tissue, large areas of exposed mantle, and thus low tolerance for turbid conditions, poor light penetration, and sediment fouling. Also, like hermatypic corals, rudists may have been able to flourish in relatively nutrient-poor marine waters.

The rudistids clearly lived in a variety of ways and subenvironments within their preferred habitat (Douvillé, 1935; Klinghardt, 1931; Kauffman and Sohl, 1974; Philip, 1972; Yonge, 1967). Their evolution was directed toward increasing diversification and niche partitioning of the shallow Mesozoic carbonate platforms and of reefoid framework environments (see subsequent disscussion and Figs. 3–6). Mainly they lived in three ways: (a) prone or semi-prone on the substrate, commonly in dense groupings; (b) erect to suberect, in gregarious to semigregarious association, having the bases of the lower valve lightly attached to and/or partially buried in the substrate; and (c) firmly cemented by a broad base to hard substrates in growth forms grossly resembling those of oysters. Whereas form (c) rudists (e.g., *Plagioptychus*) were common early colonizers of sedimentary hardgrounds and/or hard shell surfaces (including other rudists), most rudistids (forms a,b) preferred relatively firm bioclastic carbonate sand and coarser-grained substrates, or existing rudist-dominated frameworks that originated on such substrates, as a growing surface. This environmental preference exercised strong control on the distribution of rudists, the formation of the reef-like frameworks (Kauffman and Sohl, 1974; Philip, 1972; Bein, 1976), and the structure of rudistid-dominated paleocommunities and their successional relationships.

Because Cretaceous rudistids lived predominantly erect to suberect in gregarious association to gain mutual support, and thus to offset the small size of their individual attachment bases and high centers of gravity, they became principal builders of reefoid frameworks. Only those rudists that had broad attachment bases, or the most massive barrel-shaped taxa that had heavy bases, commonly lived isolated in the erect-growth posture, and then mainly in the protected areas of lagoons, etc., that had low current or wave velocities. The inherent instability of the preferred erect-growth orientation in rudists, coupled with strong environmental preference for current-swept, shallow-water environments, created problems, especially for the initial colonizers of rudist framworks that grew unprotected on mobile clastic substrates. Thus, early in the history of rudist frameworks, complex community successions evolved to offset these problems by utilizing varied adaptive morphotypes at different stages of framework development. These have been described in detail by Kauffman and Sohl (1974) and Kauffman (1974) for the Caribbean Cretaceous.

The pattern of succession, especially in the Late Cretaceous, was commonly as follows: Rudist barrier zones probably formed first among different types of frameworks, commonly on elevated, current-swept bioclastic sand and shell-debris surfaces such as bar and bank crests already occupied by diverse mollusks and small massive heads of corals, "stromatoporoids," and hydrozoans. In most places, the colonizing rudists were large species having a recumbent to semiprone life habit, and thus were mechanically stable on current-swept substrates. Small erect forms (usually Radiolitidae), growing individually or in small clusters, formed part of the colonizing community only in places where large shell debris, limestone cobbles, and coral or stromatoporoid heads were dense enough to create small eddy pockets protected from the main thrust of current and wave action.

The second phase of the barrier framework succession involved dense clustering and intergrowth of large recumbent colonizer rudists; smaller recumbent forms; and, in certain places, small erect rudist clusters to form a biostrome having many protected depressions between large rudists and associated organisms. This settling base was ideal for the spat of erect-growing rudists; widespread settlement of these dominantly radiolitid and hippuritid taxa in Upper Cretaceous frameworks subsequently took place over the entire biostrome; small shells were allowed to grow upward, shielded from strong current action which might have uprooted them, until they were large enough to have established relatively broad firm attachment bases. These upright rudists anastomozed as clusters, providing mutual support as they continued to grow upward beyond their protected niches, forming a dense mat; this completed a third phase of barrier framework formation. Finally, this mat provided excellent settlement surfaces for additional generations of the same kinds of erect rudistids. The collective generation-upon-generation upward growth of the structure elevated it well above the sea floor, and continued until storms or rapid sedimentation killed most of the framework builders.

Emplacement of a rudist barrier framework seems to have, in turn, created relatively quiet-water backreef and lagoonal environments in which an entirely different suite of rudists could form dense biostromes and elevated frameworks without going through a colonizing phase by recumbent forms. These environments were the sites of direct colonization by diverse upward-growing rudists of several types; in the Late Cretaceous, these were straight, small, slender, thin-shelled gregarious forms in soft lagoonal muds; very large barrel-shaped forms isolated or clustered in outer lagoon slope sands; and smaller, thick-shelled clustered forms having interlocking ribs, mutual cementation surfaces, and/or firm attachment bases in shallower lagoonal environments (data based on Caribbean studies; Kauffman and Sohl, 1974). In the Early Cretaceous, these forms included various erect taxa that had slightly to moderately and openly coiled

or twisted shells with offset centers of gravity and that were highly unstable in the erect growth position (e.g., *Monopleura, Toucasia*; Perkins, 1974). Topographic buildup of these frameworks depended on continued colonization of old surfaces by young of the same taxa, so that ultimately these "patch reefs" were monospecific to paucispecific in terms of rudists. Kauffman and Sohl (1974) described these structures from the Caribbean Upper Cretaceous in considerable detail.

Detailed mapping of rudist frameworks suggests that most were relatively short lived in geological time, lasting only for a few generations of adults (probably <100 yr) and were easily destroyed by storm waves or intense currents. On the other hand, true coral reefs today are normally long lived and highly storm resistant; these facts suggest that binding and mutual cementation between individual rudists in Cretaceous frameworks was minimal. Studies of rudist framework paleocommunities by Kauffman and Sohl (1974), Kauffman (1974), Perkins (1974), and Coates (1977) for the Caribbean Province, and Philip (1972) and Bein (1976) for the Tethys Province, clearly demonstrate this situation. Early Cretaceous rudist frameworks were very simply composed, dominated by one to two rudist species and rarely associated corals, algae, boring sponges, and ostreiform bivalves. They were basically unbound by corals, hydrozoans, crustose algae, bryozoans, etc., which bind reefs today. In the Late Cretaceous, at the time of extensive radiation of rudists into diverse living habits and niche partitioning of frameworks, reefal paleocommunities were more complex. Many frameworks were still monospecific to paucispecific in rudist composition, whereas others became more diverse, composed of several rudist species occupying different microniches and stages in the framework succession. Diverse epifaunal bivalves, some gastropods, and echinoids similar to those of modern coral reefs, were commonly associated with these structures. A detailed study of rudist epibionts and borers shows that binding organisms were still sparsely distributed as small clusters and colonies on and around dominant rudists and rarely, if ever, cemented individual rudists together into solid frameworks (Kauffman and Sohl, 1974). Moderately common but spatially restricted epibionts on rudists in framework communities included both solitary and colonial corals, hydrozoans, bryozoans, crustose algae, "stromatoporoids," oysters (*Ostrea, Pycnodonte*) and other bivalves (*Spondylus, Plicatula,* other rudists), and serpulid worms. Common boring organisms on rudists were the sponge *Cliona,* bivalves like *Litho-*

phaga, polydorid worms, and bryozoans. Distinct light zonation of borers and epibionts has been found on many types of rudists, suggesting that most of the rudist lower valve was exposed above the substrate during life (Kauffman and Sohl, 1974).

Evolutionary History

The rudistids had a short but dramatic evolutionary history (Fig. 2) restricted to an 80-m-yr-long period of the latest Jurassic–Cretaceous (Dechaseaux, 1952; Douvillé, 1935; Kauffman, 1974; Toucas, 1903–1904; Yonge, 1967). Rare specimens have been reported from the Paleogene, but their age is in doubt.

Most workers believe that the origin of the rudists was among the aberrant Paleozoic and early Mesozoic bivalve family Megalodontidae— the so-called pachydonts and megalodonts— which were characterized by medium to large sized, inflated, thick, normally equivalved shells having massive umbonal areas and hinge plates bearing one to a few heavy teeth. Some taxa had internal plates partitioning the body cavity into visceral and mantle-gill areas. These bivalves shared several characteristics with all early and many more advanced rudists: massive shells, large size, thickened hinge plates having a reduced heterodont dentition, external ligament, internal plates, similarly positioned posterior adductor muscles, a preference for Tropical-Subtropical carbonate shelf environments, and, in later megalodonts, a dominantly epifaunal to semi-infaunal life habit. Dechaseaux (1969) noted that one of the earliest rudistid bivalves, *Diceras,* is especially close to the megalodontids in regard to shell and hinge structures and musculature.

The evolution of the rudistids was preceded by a long Paleozoic and early Mesozoic evolution in the ancestral Family Megalodontidae (Fig. 3) in which the major trends were toward progressively larger, thicker shells, more massive hinge areas (especially anteriorly), and a mode of life change (based on functional morphology of the shell) from a dominantly infaunal to a dominantly epifaunal habitat. Silurian megalodontids were mainly shallow infaunal to semi-infaunal in habitat, and most possessed small elongated shells of only moderate thickness; some large forms arose by Late Silurian time. Probable epifaunal taxa first appeared in the Devonian but were relatively small and of secondary importance throughout the middle and late Paleozoic in comparison with semi-infaunal adaptive forms. These latter forms became progressively larger and thicker-shelled through the Paleozoic.

The large massive forms derived stability from their heavy shells, which permitted a more exposed semi-infaunal mode of life by the late Paleozoic, with at least half the shell extending above the sediment. Continuation of these evolutionary trends across the Permian/Triassic boundary into the Jurassic produced giants among the megalodontids, which had shells so massive and large that an infaunal habitat was no longer necessary for protection against scouring currents and predators. They bcame dominantly epifaunal and shallow semi-infaunal by the Triassic. Further, evolution of a massive hinge plate and greatly thickened umbonal and postumbonal shell regions produced a strong dorsal center of gravity and led to a reorientation of many shells in life position with the umbonal area down and the ventral (feeding) margins facing upward into the water column, above the zone of current scour. In this orientation, the hinge line was parallel to moderately inclined to the sediment-water interface. Thus, many Mesozoic Megalodontidae probably lived like the modern Tridacnidae (especially *Hippopus,* a common inhabitant of mobile sands), and shared many of their structural adaptations. Fischer (1964) has demonstrated such an orientation in the Lofer cyclothems of the Alpine Triassic and further noted that most specimens gape dorsally, as do living *Tridacna* and *Hippopus.* From this he infers the possible development of symbiotic zooxanthellae in megalodontid mantle tissue that could have been permanently exposed (like *Tridacna,* etc.) in the space between these gapes.

These observations on the evolution of Megalodontidae are relevant to understanding evolu-

FIGURE 2. Generalized diagram depicting the evolution of adaptive morphological types and principal families among the rudist bivalves (from Kauffman, 1974). Morphotypes and their living orientation are generalized in small sketches, with downward-pointing arrows indicating probable erect to semi-erect juvenile growth followed by recumbent to semiprone adult life orientation among aberrantly coiled shells. Patterns represent different rudist families, keyed at bottom. The evolutionary record of each family can be determined by tracing its key pattern vertically through time (left column). Vertical multipatterned bars indicate evolutionary record and degree of homeomorphy among diverse families within each adaptive morphotype; the width of each column reflects the number of genera whose species dominantly (but not always) fall within that morphotype (smallest width equals one genus: rest of columns scaled accordingly). Note great radiation of form first in the earliest Cretaceous (Berriasian, Valanginian Stages), and especially in the Barremian–Cenomanian Stages of the late Lower and early Upper Cretaceous sequence.

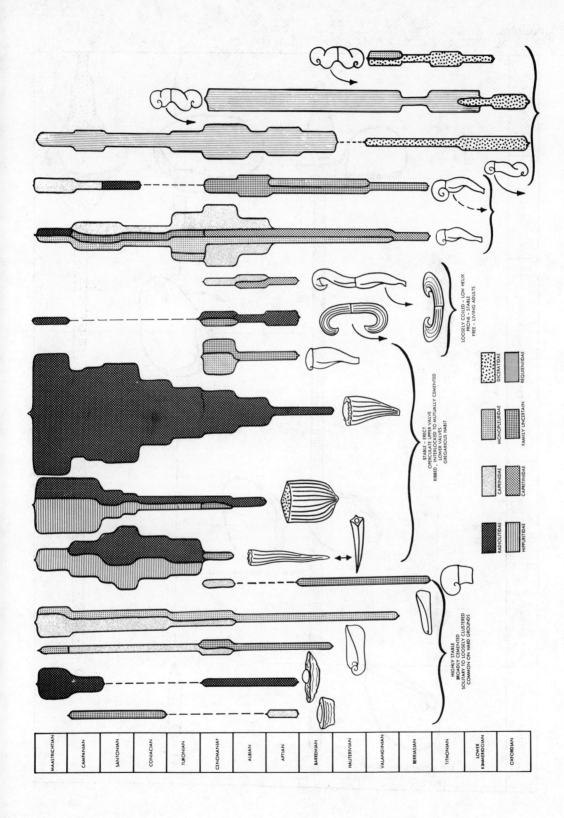

729

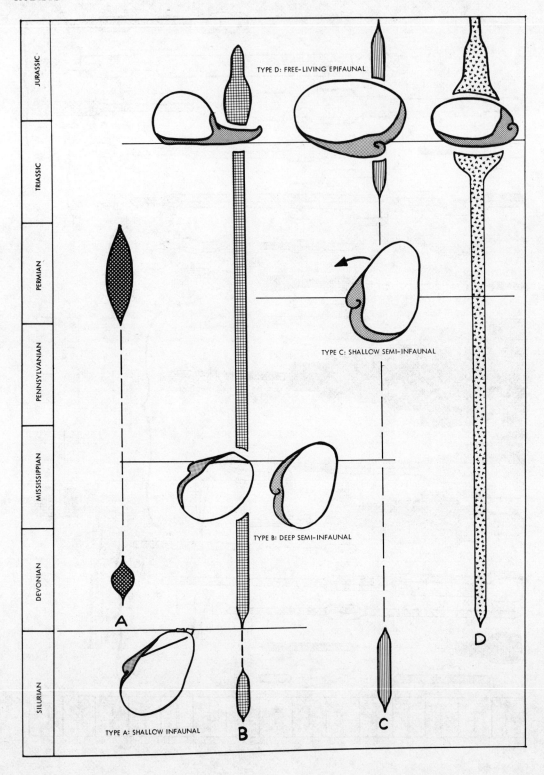

TYPE D: FREE-LIVING EPIFAUNAL

TYPE C: SHALLOW SEMI-INFAUNAL

TYPE B: DEEP SEMI-INFAUNAL

TYPE A: SHALLOW INFAUNAL

A

B

C

D

JURASSIC

TRIASSIC

PERMIAN

PENNSYLVANIAN

MISSISSIPPIAN

DEVONIAN

SILURIAN

tion in the rudistids because many characters thought to be typically rudistid were present in the megalodontids by Triassic time, including the massive shell and reduced hinge, the apparent ability to produce large amounts of shell material in short periods of time (which would be aided by the presence of zooxanthellae in the mantle), reduced but massive heterodont hinge structures, musculature, accessory internal plates, and an epifaunal life habit with the "dorsal" (umbonal) part of the shell in contact with the substrate (Fig. 3). The stage was thus set for the Late Jurassic evolution of the Diceratidae, the first rudist family.

The rudistids differ from the megalodontids primarily in possessing unequal valves, the lower (left or right) valve commonly being attached to the substrate, the upper one slightly to greatly reduced in size and highly modified. The rudists also had a much greater tendency toward gregariousness. The upward-growing cemented valves of most taxa, the gregarious life habit, and the rapid rate of shell calcification and growth in rudistids were all important features that allowed them to diversify quickly into many ecological niches of the reef environment, to outcompete corals and other reef-forming organisms for space, light, and resources, and by the end of the Early Cretaceous to dominate the carbonate-platform reef biotopes ecologically and numerically.

The earliest rudists (Late Oxfordian Stage, Late Jurassic), the diceratids (Fig. 4B) were not reef-formers but did make up dense biostromes much like those of the ancestral megalodontids (Fischer, 1964, figs. 30–33). Their doubly coiled and only moderately inequivalved shells were attached early in life. As these shells continued to grow along an irregular helical plane, however, the center of gravity moved upward and outward, became offset from the base, and the adult shells reclined into a prone or semiprone living position close to the substrate (Figs. 2 and 4B). An adaptive advantage over other reef-forming organisms had not yet been gained, and corals continued to dominate Late Jurassic reefs. In the latest Jurassic (Tithonian Stage), the diceratids gave rise to a more advanced rudist family, the Requieniidae, which had a similar morphological plan, but which had shells more inequivalved (attached one largest, upper one commonly reduced to a cap) and less coiled; the lower valve remained attached for most of the life history of some requieniids, although these too lived semiprone as adults on the substrate, forming mainly biostromal mats and clusters. These new adaptations involving greater uncoiling, cementation and increase in relative size of the attached valve, heralded a major Early Cretaceous radiation of reef-forming rudistids having an erect growth posture—and ecological displacement of the corals. Both the Diceratidae and Requieniidae continued into the Cretaceous, but neither was an important component of large reef-like rudistid frameworks. Some requieniid rudistids (e.g., *Toucasia*) were involved in the construction of small low lagoonal bioherms on the Texas Lower Cretaceous platform (Perkins, 1974). Thus, by the Early Cretaceous, the stage was set for the evolution of framework-building rudistids.

Two major bursts of evolution characterized the Early Cretaceous of the tropical-subtropical Tethyan and related seaways in many parts of the world. The first was in the lower part of the Early Cretaceous ("Early Neocomian," Urgonian, Valanginian Stages) and involved several new kinds of adaptations, in at least two new families (Caprotinidae, Monopleuridae) and the genus *Bicornucopia* (family uncertain; Neocomian), which led to the formation of the earliest rudist reefoid frameworks. These groups evolved into highly inequivalved forms which had largely uncoiled, twisted, firmly cemented lower valves and coiled but greatly reduced caplike upper valves. These rudists grew erect and were highly gregarious. The center of gravity remained roughly over the attachment base, and mutual support was gained from the gregarious life habit, so that for the first time the rudists had achieved an evolutionary grade that allowed them to compete with other reef-forming organisms— through rapid upward growth and crowding—and to form elevated reeflike frameworks. In addi-

FIGURE 3. Major life-habit adaptations and their evolutionary history among the Megalodontidae, the ancestors to the rudists. The four adaptive types are represented by sketches (A–D) that tie to vertical range bars; the width of each bar represents the number of contemporaneous genera whose species primarily fall into each morphotype and life-habit group in each geological period (left column); the smallest width represents one genus and the rest are scaled accordingly. Heavy shading in each sketch represents approximate position and extent of thickest calcium deposits (hinge plate, umbones, etc.) in shell; and horizontal line represents the sediment-water interface. Note changes in dominance from infaunal and deep semi-infaunal habitats in Paleozoic megalodonts, to epifaunal and shallower semi-infaunal habitats in the early Mesozoic Era. Morphotypes: A, shallow infaunal, as in *Eomegalodus* (Devonian); B, deep semi-infaunal, as in *Pinzonella* (Permian); C, shallow semi-infaunal, as in *Neomegalodon* (Triassic) and *Pachyrismella* (Jurassic); D, free living epifaunal, as in *Paramegalodus* (Triassic) and *Durga* (Jurassic).

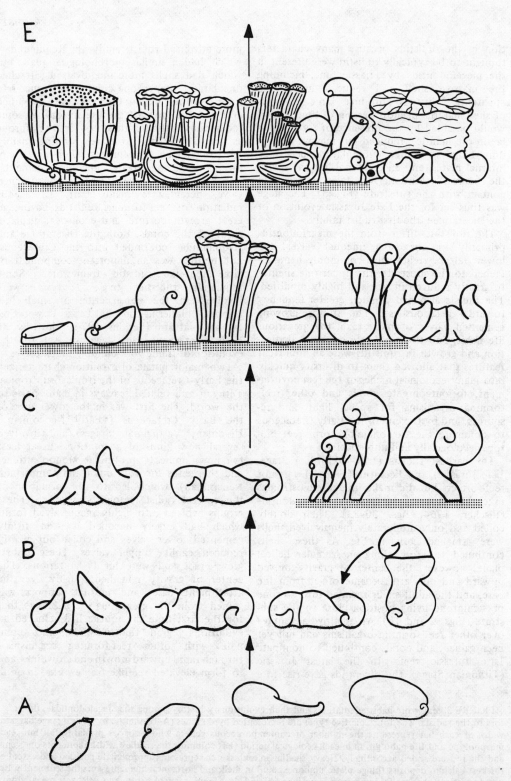

FIGURE 4. Major steps in the history of rudist evolution, as portrayed diagramatically by principal morphotypes and life orientations. A, the principal evolutionary trends (left to right, oldest to youngest) among the ancestral megalodontid bivalves. B, the latest Jurassic adaptive trends among the earliest rudists (Diceratidae, Requieniidae). C, earliest Cretaceous (Berriasian, Valanginian) adaptations and the origin of erect growth habits. D, the principal range of adaptations among Barremian rudists. E, the scope of adaptive forms that arose in the Aptian through Cenomanian stages and were already present from previous radiations and that continued to exist through the Late Cretaceous in a diversity of tropical carbonate-platform environments.

FIGURE 5. Diversity and classification of rudist-dominated reefoid frameworks, schematically showing the life habit, relative abundance, and distribution of major adaptive morphotypes in the Late Cretaceous (from Kauffman and Sohl, 1974). The barrier and patch reefs involving numerous corals, "stromatoporids," hydrozoans, and other binding organisms rarely if ever formed. This vision of "rudist reefs" (top) widely suggested in the literature, has not yet been documented by careful field mapping. Corals and other reef "binders" normally play a minor role in rudist framework communities.

tion, low, erect, broadly cemented oyster-like rudists, which were dominantly encrusters of hard substrates (shells, other rudists, hardgrounds), evolved at this time among the Monopleuridae and the possibly related genus *Anodontopleura*. Recumbent Diceratidae and Requieniidae continued to form low biostromes and clusters, mainly in more protected (lagoonal, etc.) shallow-water environments. Figure 4C diagramatically represents the stage of rudist evolution at this time. The Diceratidae became extinct in the Valanginian, but their adaptive forms continued to the end of the Cretaceous among the Requieniidae.

The Late Early Cretaceous brought a second burst of rudist evolution, possibly tied to worldwide transgressions of tropical seaways onto shallow continental platforms dominated by carbonate sedimentation. This was the greatest evolutionary development at higher taxonomic levels among rudistids and gave rise to most of the major families and morphological adaptations that dominated the Late Cretaceous (Fig. 2). New Barremian forms included the dominant Late Cretaceous Family Radiolitidae, which had largely to totally uncoiled, cemented, erect-growing lower valves and greatly reduced, fragile, caplike upper valves. In many radiolitid taxa, strong longitudinal ribs, flanges, or costae evolved which interlocked in gregarious framework-building taxa to give additional support in an upright growth habit. The small radiolitid cap and characteristics of the upper surface of lower valves in radiolitids suggest that the mantle remained partially exposed all the time, and thus may have contained symbiotic zooxanthellae. Low, cemented, oyster-like taxa evolved among the Caprotinidae at this time. Figure 4D represents the scope of evolutionary adaptations during the Barremian.

The Aptian Stage brought considerable diversification among rudists (Fig. 2), including forms that allowed widespread niche partitioning of reef environments and complex community successions. Also, new carbonate-platform environments were exploited by the rudistids in the latest Early Cretaceous and, increasingly, through the Late Cretaceous. Very large, mainly free-living recumbent rudists having subequal massive coiled valves evolved in both the Caprotinidae and Radiolitidae (Fig. 2) during the Aptian. These were to become the early colonizers of rudist-framework community successions during much of the later Cretaceous. Large, platterlike, broadly cemented, low-growing radiolitids also evolved, possibly as an adaptation to spreading out mantle tissue containing symbiotic zooxanthellae over a broad surface. The new Family

Caprinidae first evolved as an additional line of largely erect-growing, strongly cemented forms having reduced caps—the principal reefoid framework adaptation—and quickly diversified into many forms in the Albian and Cenomanian. These were largely homeomorphic after earlier osyter-like and asymmetrically erect-growing monopleurids and caprotinids, and after recumbent, largely free-living caprotinids.

The Albian and Cenomanian stages were again times of widespread transgression of shallow warm seas over continental platform areas and were characterized by a major diversification of taxa within adaptive morphological plans that had already arisen earlier among rudists (Figs. 2, 4E). This period of rudist evolution produced a greater diversity of higher taxa than any other time during their evolutionary history and produced continued radiation of new adaptive forms among erect-growing, framework-building rudists. Very large barrel-like forms with weak basal attachment, reaching nearly 2 ft (60 cm) in diameter later in their evolution, arose among the Radiolitidae. Very tall, slender, straight, highly gregarious rudists with small attachment bases evolved in the genus *Palus* (family uncertain) and later (Turonian) both in the radiolitids and hippuritids. Most of these grew erect to semi-erect in dense association to form prominent frameworks that had individuals a meter or more in length but only a few centimeters in width. Some were dominantly recumbent as adults, growing one on top of the other in low frameworks. Most Albian–Cenomanian groups persisted to the end of the Cretaceous; only the Caprotinidae became extinct toward the end of the Cenomanian.

This tremendous diversification of forms (Figs. 2, 4E) allowed development of many types of reef-like frameworks (Fig. 5) in the Albian and Cenomanian. These included barrier structures, bioherms, banks, and patch reef-like clusters, as well as extensive biostromal units. Virtually all major carbonate-platform environments within the shallow photic zone

FIGURE 6. Schematic diagram showing cross section through a Late Cretaceous (Maestrichtian) Caribbean shallow carbonate-platform environment with diverse rudist-dominated frameworks, and typical morphotypes, shown in preferred spatial distribution—based on field mapping and/or sedimentary rock analysis (from Kauffman and Sohl, 1974). "Barrier" frameworks were typically discontinuous and short lived, being easily destroyed or buried periodically by high-energy, shallow-water current and wave activity; lagoons and other backreef areas were thus, for the most part, relatively open, well circulated, and of normal salinity.

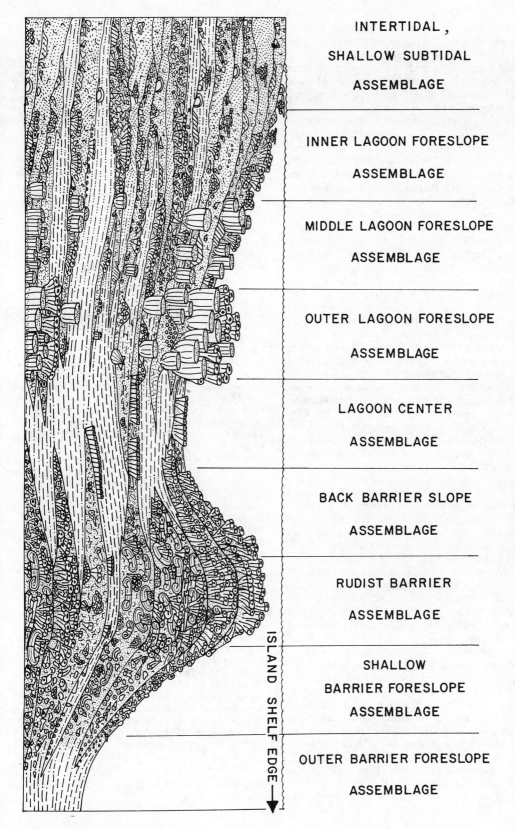

INTERTIDAL,
SHALLOW SUBTIDAL
ASSEMBLAGE

INNER LAGOON FORESLOPE
ASSEMBLAGE

MIDDLE LAGOON FORESLOPE
ASSEMBLAGE

OUTER LAGOON FORESLOPE
ASSEMBLAGE

LAGOON CENTER
ASSEMBLAGE

BACK BARRIER SLOPE
ASSEMBLAGE

RUDIST BARRIER
ASSEMBLAGE

SHALLOW
BARRIER FORESLOPE
ASSEMBLAGE

OUTER BARRIER FORESLOPE
ASSEMBLAGE

ISLAND SHELF EDGE

MODEL OF CARIBBEAN RUDIST FRAMEWORK DISTRIBUTION

were occupied by abundant rudistids. Corals, "stromatoporoids," sponges, hydrozoans, crustose algae, bryozoans, and other cemented mollusks—the principal reef-formers of the earlier Mesozoic, and of the Cenozoic including the Holocene had a very secondary ecological role during this time.

The post-Cenomanian Late Cretaceous evolution of the rudists was dramatic. It mainly involved tremendous radiation of new lower-level taxa, morphological diversification among already established families and adaptive types (Figs. 2, 4E), widespread niche partitioning and diversity increase within major framework types, and the formation of new kinds of frameworks in more marginal environments. The Hippuritidae, a dominant Late Cretaceous family, arose in the Turonian and diversified rapidly into erect-growing, largely gregarious, framework-building forms with a large lower valve and a small perforated thin cap. This cap represented a new adaptation for screening larger sediment particles from water taken in for feeding and respiration, for sensory perception, and/or providing a protective cover for the upper mantle surface while still allowing light through for the growth of intracellular zooxanthellae. All Late Cretaceous rudist families except the Caprotinidae (Cenomanian and older) persisted until the end of the Cretaceous (Fig. 6), but the great radiation of the Radiolitidae and Hippuritidae, culminating in the latest Cretaceous (Maestrichtian), overshadowed all other families.

The rudistids became extinct abruptly at the end of the Maestrichtian (Tertiary reports are questionable), at the peak of their evolutionary history when they were more abundant, diverse, and occupied a greater variety of habitats than at any other time in their history. This is one of the most dramatic events of the great Late Cretaceous extinction, and its cause remains somewhat of a mystery, as favored tropical carbonate-platform environments persisted into the Paleocene. The great global marine regression of the latest Cretaceous may have destroyed most of the prime rudist habitats, considerably weakened existing populations, and greatly restricted their range. This, coupled with a major cooling trend at the end of the Cretaceous, may have created an environmental crisis for rudists and triggered their extinction. Also, as all other *equally dramatic* marine extinctions at the end of the Cretaceous were among the pelagic microbiota (foraminifera, coccoliths) or other tropical benthos, the rudist larvae—probably long-lived pelagic larvae like so many of the living tropical benthos—may have been greatly affected by the same environmental factors (temperature?, salinity?, radiation?) that brought about massive and abrupt extinction of other elements in the upper part of the oceanic water column. This may have been the *primary* cause of the great extinction of the rudistids.

ERLE G. KAUFFMAN
NORMAN F. SOHL*

References

Bein, A., 1976. Rudistid fringing reefs of Cretaceous shallow carbonate platform of Israel, *Bull. Amer. Assoc. Petrol. Geologists,* **60,** 258–272.

Coates, A. G., 1977. Jamaican coral-rudist frameworks and their geologic setting, *Amer. Assoc. Petrol. Geologists Stud. Geol.,* **4,** 83–91.

Coogan, A. H., 1977. Early and middle Cretaceous Hippuritacea (rudists) of the Gulf Coast, *Bur. Econ. Geol., Univ. Texas, Rept. Inv.,* **89,** 32–70.

Dechaseaux, C., 1952. Classe de Lamellibranches, in J. Piveteau, ed., *Traité de Paleontologie,* T. II. Paris: Masson et Cie, 220–364. (Rudistae, 323–364)

Dechaseaux, C., 1969. Origin and extinction, in Moore, 1969, N765.

Douvillé, H., 1935. Les Rudistes et leur évolution, *Bull. Soc. Géol. France,* Sér. 5, **5,** 319–358.

Fischer, A. G., 1964. The Lofer cyclothems of the Alpine Triassic, *Geol. Surv. Kansas Bull.,* **169**(1), 107–149.

Kauffman, E. G., 1974. Cretaceous assemblages, communities, and associations: Western Interior United States and Caribbean Islands, in A. M. Ziegler et al., *Principles of Benthic Community Analysis (notes for a short course).* Sedimenta IV, Comp. Sed. Lab., Div. Mar. Geol. Geophys., Rosensteil Sch. Mar. Atmos. Sci., Univ. Miami, 12.1–12.27.

Kauffman, E. G., and Sohl. N. F., 1974. Structure and evolution of Antillean Cretaceous rudist frameworks, *Verhandl. Naturf. Ges. Basel,* **84,** 399–467.

Klinghardt, F., 1931. *Die Rudisten, Teil III. Biologie und Beobachtungen an anderen Muschelen.* Berlin: privately printed, 60p.

Milanović, B., 1933. Les problèmes paléobiologiques et bio-stratigraphiques des Rudistes, *Mém. Serv. Géol. Yugoslavie,* 194p.

Moore, R. C., ed., 1969. *Treatise on Invertebrate Paleontology,* pt. N. Mollusca 6, Bivalvia vol. 2. Lawrence, Kansas: Geol. Soc. Amer. and Univ. Kansas Press, N751–N765.

Perkins, B. F., 1969. Rudist morphology, in Moore, 1969, N751–N764.

Perkins, B. F., 1974. Paleoecology of a rudist reef complex in the Comanche Cretaceous, Glenrose Limestone of central Texas, *Geoscience and Man,* **8,** 131–173.

Philip, J., 1972. Paléoécologie des formations à rudistes du Crétacé supérieur: L'exemple du sud-

*Publication authorized by the Director, U.S. Geological Survey

est de la France, *Palaeogeography, Palaeoclimatology, Palaeoecology, 12,* 205–222.

Skelton, P. W., 1976. Functional morphology of the Hippuritidae. *Lethaia, 9,* 83–100.

Skelton, P. W., 1978. The evolution of functional design in rudists (Hippuritacea) and its taxonomic implications, *Phil. Trans. Roy. Soc. London, Ser. B, 284,* 305–318.

Toucas, A., 1903–1904. Etudes sur la classification et l'évolution des Hippurites, *Soc. Géol. France, Paléont. Mém., 30,* 128p.

Yonge, C. M., 1967. Form, habit, and evolution in the Chamidae (Bivalvia) with reference to conditions in the rudists (Hippuritacea), *Phil. Trans. Roy. Soc. London, Ser. B, 252,* 49–105.

Cross-references: *Bivalvia; Reefs and Other Carbonate Buildups; Zooxanthellae.*

S

SCAPHOPODA

The Scaphopoda, a small but distinct class of the Mollusca, range with certainty back in time to the Early Devonian; but their origins in the early Paleozoic still remain an unsolved problem. Slender annulated tubes are known from the Ordovician that may be either scaphopods or pteropods.

No agreement exists on the number of species there are in the Scaphopoda. About 750 are usually acknowledged by most modern workers (400 fossil and 350 living); but the present writer has more than 1200 described and named forms in his card index—626 fossil and 643 living. Allowing that 20% of these may be synonyms, the truth probably lies somewhere between the two extremes, perhaps in the region of 1000. The class name derives from the Greek and roughly means "shovel-foot" (see Fig. 1) while the vernacular "tusk-shell" requires no explanation (see Fig. 2).

Scaphopods are fully marine animals—avoiding estuaries and any fresh water—and the only class of the Mollusca that is exclusively infaunal, using the "shovel-foot," like the bivalves, as a means of locomotion. The slender ciliated captaculae are used for catching a variety of marine animals, mainly foraminifera.

Systematically, they seem to occupy a position between the Bivalvia and the Gastropoda. The digging foot, absence of eyes, the fused mantle open at both ends, and the bilateral symmetry aligns them with the bivalves while the univalve shell, presence of rudimentary head with a buccal mass and radula, the predatory habit, and most aspects of the nervous system relate them to the gastropods. Pojeta and Runnegar (1976, p. 43-44) have suggested that the Scaphopoda are derived from the extinct molluscan class Rostroconchia (q.v.).

The Animal

Anatomically the scaphopods are distinct from most other mollusks in that they are bilaterally symmetrical elongated animals in which mantle fusion occurs along the mid-ventral line but remains open at both ends (Fig. 2). The anterior opening is for the foot and captaculae and the posterior is for circulation of water and elimination of feces. In these arrangements one sees similarities with the infaunal bivalve razor shells, but the mode of feeding is totally different.

As a result of their infaunal habit, scaphopods have no eyes, but statocysts are present in the foot. The proboscis, with two bundles of laterally placed prehensile ciliated filaments called captaculae, lies above the foot which, together with the captaculae, can be thrust out into the sediment where the foot serves for locomotion while the captaculae catch infaunal organisms, bringing them to the frilled and probably selective lips of the proboscis. The mantle cavity, being continuous from end to end, is open at the posterior for the discharge of feces and admission of water for respiration which, in the absence of a gill, takes place at the mantle surface. Behind the radula a straight oesophagus passes food posteriorly into the stomach, which receives ducts from a bilobed and much divided liver (Fig. 3d). The liver, having a lung-like appearance, deceived early 19th century workers into believing that it was a respiratory organ. The correct digestive function was demonstrated by Lacaze-Duthiers in the late 19th century. The convoluted intestine continues in an anterior direction and terminates ventrally at the anus opening posteriorly.

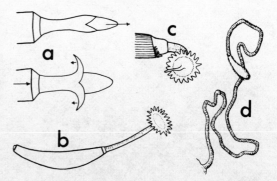

FIGURE 1. a, the dentaloid foot showing function of the epipodial collar during burrowing. b,c, the siphonodentaloid foot with pedal disc expanded (b, *Cadulus*; c, *Entalina*, with median filament). d, enlarged view of captaculum. (Modified after Trueman, 1968, and Pilsbry and Sharp, 1897-1898.)

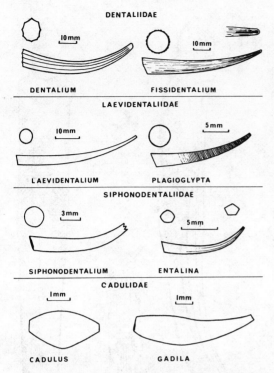

FIGURE 2. Diagrammatic sketches to illustrate characters of leading genera in each of the four scaphopod families.

The sexes are separate and a single ovary or testis opens into the right kidney.

From singly laid eggs, floating trochophore larvae develop and are succeeded by a veliger stage. After 5–6 days, the veliger loses its ciliated velum and settles on the sea floor to begin its benthonic life within the sediment.

The character of the foot divides the class into two orders. In the Dentalioida, the foot is short and conical in shape with an epipodial collar, interrupted dorsally, which expands during locomotion and gives the foot a three-lobed appearance (Fig. 1a). In the Siphonodentaloida, the epipodial collar forms a round crenulated pedal disc, which is terminal (Fig. 1b) except in the two genera *Pulsellum* and *Entalina* (Fig. 1c). In these, a distal median filament on the pedal disc is present. This is probably homologous with the conical foot of the dentaloids since both are anterior to the epipodial collar of the one and pedal disc of the other.

The Shell

Structure. The scaphopod shell is tubular, tapering, more or less curved, open at both ends, and composed of four layers (Fig. 4). The outermost layer is a chitinous periostracum while the three inner layers are calcareous and composed entirely of aragonite. The outer calcareous layer is prismatic with the long axes of the prisms normal to the tube axis. The middle layer is of crossed-lamellar structure very similar to that of the Bivalvia. The innermost calcareous layer is of fine crossed-lamellar structure deposited in concentric conical layers. The chitinous and prismatic layers, being primary deposits, thicken with growth toward the anterior end of the shell. Crossed-lamellar and concentric shell layers strengthen (by secondary deposition) posteriorly and consequently thicken in that direction. Resorption of the apex with growth results in the inner concentric layer lying discordantly on the other layers at the apex.

The earliest known scaphopods have resorbed apices, that of *Prodentalium* being slightly notched while *Plagioglypta* has a simple apex (see Figs 2 snd 4). Long slits and fissures accompanying resorption of the apex are apparently developments of Late Cretaceous—early Paleogene times.

Biometrics. In describing and comparing the gross morphological shell characteristics of the dentaliid scaphopods, an increase in precision and some brevity can be achieved if diagnostic characters are given a simple numerical snd symbolic expression. A more sophisticated mathematical expression, though impressive, would be too powerful and unwieldy a tool for this simple function.

Thus, if d^1 = apertural diameter, d^2 = apical diameter, l = length of median axial line joining d^1 and d^2; then

$$\frac{d^1 - d^2 \times 100}{l} = E$$

expansion rate of the tube (see Fig. 4).

Similarly, if ch = length of chord joining d^1 and d^2, h = the greatest distance between the chord and the dorsal surface of the shell; then

$$\frac{h \times 100}{ch} = A$$

the arcuation of the shell (note, this is not the curvature of mathematicians) (see Fig. 4).

Finally, the notation $\hat{6}, \hat{12}, \overline{24}$, represents use of a ribbing formula that reads "six sharp apical ribs, 12 rounded medial, and 24 obsolescent ribs at the aperture." Apical characters may be symbolized by (O) = simple, (V) = notched, (T) = slit or fissured, (P) = pipe (see Fig. 4).

Thus the principle diagnostic shell features of a dentaliid scaphopod, say *Fissidentalium vernedei,* may be summarized by —A4.3, $E6.3(T)35 - 38 - 39$.

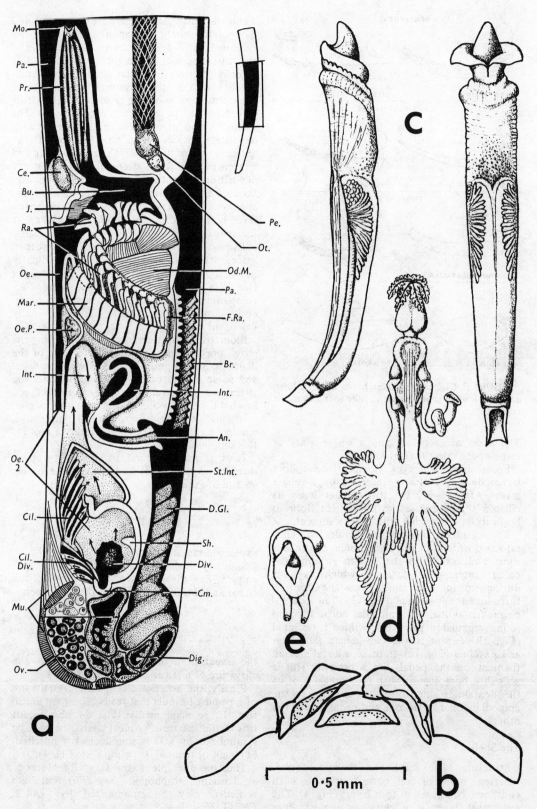

Mo.
Pa.
Pr.

Ce.
Bu.
J.
Ra.

Oe.

Mar.

Oe.P.

Int.

Oe.
2

Cil.

Cil.
Div.

Mu.

Ov.

Pe.

Ot.

Od.M.
Pa.

F.Ra.

Br.

Int.

An.

St.Int.

D.Gl.

Sh.

Div.

Cm.

Dig.

a

c

d

e

b

0·5 mm

FIGURE 3. Anatomy of *Antalis:* a,b, detailed anatomy of *Antalis entalis* (after Morton, 1959). c,d,e, gross anatomy and details of *Antalis vulgaris* (after Lacaze-Duthiers, 1856–1857). a: Median section of *A. entalis*—An., anus; Br., folds of the mantle wall serving as the respiratory organ; Bu., buccal cavity; Ce., cerebral ganglion; Cil., ciliated fold of the roof of the stomach leading to the intestine; Cil.Div., ciliated folds associated with the caecum and digestive diverticula; Cm., caecum of the stomach; D.Gl., digestive gland tubules spreading round the side of the mantle cavity; Dig., digestive gland in section; Div., digestive diverticulum leading from the stomach; F.Ra., formative cells of the radula at the bottom of the radular sac; Int., intestine; J., jaw; Mar., marginal teeth of left side within the radular caecum; Mo., mouth; Mu., retractor muscles attached posteriorly to the shell; Od.M., muscles of the odontophore; Oe., oesophagus; Oe.P., oesophageal pouch; Oe.2, extent of interrupted section of the oesophagus; Ot., otocyst; Ov., ovary; Pa., pallial cavity; Pe., pedal ganglion; Pr., proboscis; Ra., radula (lateral teeth of left side and median teeth); Sh., vestige of gastric shield; St.Int., anterior part of stomach, tapering forward to intestine. b: Single row of teeth from the radula. c: Lateral view on the left, ventral view on the right of *A. vulgaris*. d: Dorsal view of alimentary tract showing the much divided "liver" or digestive gland of *A. vulgaris*. e: Ventral view of gut of *A. vulgaris*.

Feeding

The feeding habits of scaphopods have been well reported by W. Clark, Dinamani, and Morton (see Palmer, 1975) and more recently by Bilyard (1974). All agree that these mollusks have a varied diet, but one consisting mainly of Foraminifera and, occasionally, small infaunal

bivalves. These they catch by means of the captaculae (see Fig. 1d), which penetrate the sediment in a broad cone around the foot. Captured prey is brought to the frilled lips of the proboscis, where a certain amount of selection takes place, before passing into the buccal mass to be ground up by the radula. About 14 genera of Foraminifera have been reported from the buccal pouches of dentaliid scaphopods and, in addition, Bilyard (1974) found, in the buccal pouches of *Antalis entalis stimpsoni,* eggs, ostracods, bivalve spat, marine mites, copepod eggs, nematodes, and small worms. Clarke, in 1849, found two bivalves, *Kellia suborbicularis* (Montagu) and juvenile *Goodalia triangularis* (Montagu), in addition to six kinds of Foraminifera in the buccal pouch of *Antalis entalis* (da Costa). Dinamani (1954) reported "large diatoms, single algal cells, and unidentified particles of detritus" in the stomach and intestine of *Dentalium conspicuum* Melville.

Finally, Palmer (1975) suggested that the fossil Foraminifera *Epistomina,* associated with the Upper Jurassic scaphopod *Prodentalium calvertensis* Palmer, formed the principal part of the diet of the scaphopod.

Evidently scaphopods have a varied diet, taking anything edible, and may be described as being as much omnivorous as carnivorous.

Habitat

Scaphopods occupy mud, silt, or sand with the concave (dorsal) side of the tube uppermost and the posterior tip protruding about 10 mm

$$E, \text{expansion rate of shell} = \frac{d - d \times 100}{l} \%$$

$$A, \text{arcuation of shell} = \frac{h \times 100}{ch} \%$$

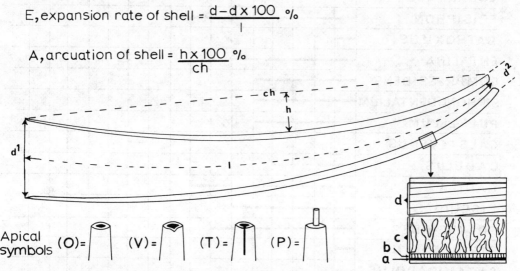

FIGURE 4. Measurable parameters of generalized scaphopod shell shown in median section, together with formulae for arcuation and expansion and symbols for expressing character of apex. a–d: Section of shell showing the four layers—a, chitinous periostracum; b, prismatic layer; c, crossed-lamellar; d, fine crossed-lamellar in concentric conical layers. Layers b,c,d are of aragonite.

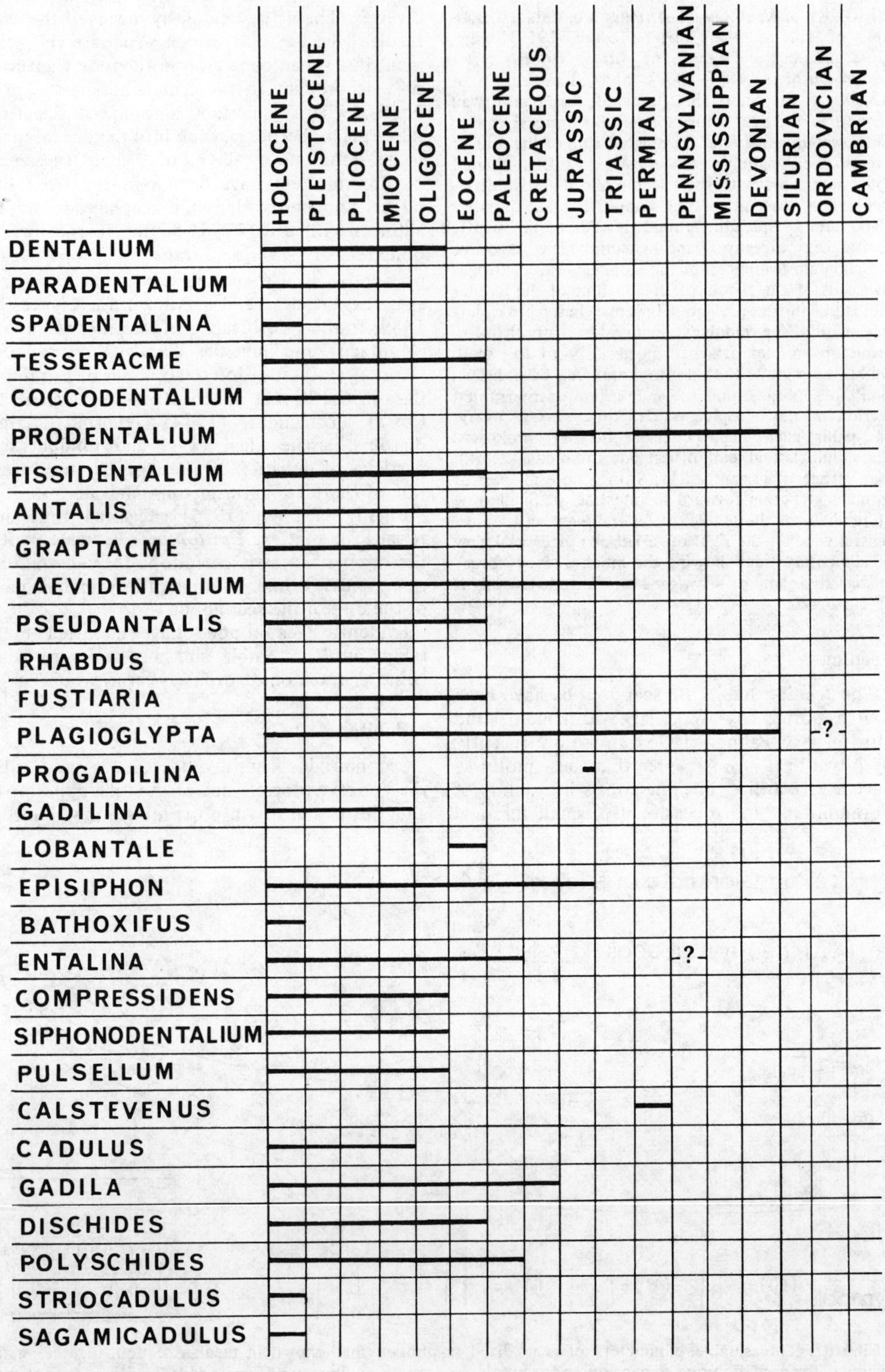

FIGURE 5. Table indicating probable ranges of genera through geological time. Thick lines indicate certain ranges, thin lines uncertain.

above the sediment-water interface. Their depth range is fairly wide, being between a few meters just off shore in *Antalis* down to more than 6000 m in the case of *Fissidentalium candidum*. Most are sublittoral, although a few species can tolerate exposure during spring tides, and occasionally a living specimen may be cast up on a beach during a violent storm.

Their tolerence of reduced salinities, being considerably less than their tolerance of depth, results in their virtual absence from estuaries and river outflows. So their presence in sedimentary rocks may be taken, in the absence of cephalopods, brachiopods, or echinoderms, as strong evidence of euryhaline conditions of deposition.

Classification

Scaphopods are currently divided into the two orders Dentalioida and Siphonodentalioida. Each has two families: Dentaliidae and Laevidentaliidae in the first; and Siphonodentaliidae and Cadulidae in the second (Palmer 1974).

CLASS SCAPHOPODA
    Order Dentalioida
        Family Dentaliidae
            (9 genera)—typically ribbed scaphopods that seem to divide naturally into two groups (Fig. 2): the symmetrical, paucicostate forms which include *Dentalium, Paradentalium, Tesseracme,* and *Spadentalina* and the unsymmetrical multicostate forms which are typified by *Fissidentalium* and *Prodentalium,* the oldest group (Fig. 5).
        Family Laevidentaliidae
            (10 genera)—dentaloid scaphopods lacking longitudinal sculpture being either smooth or annulated. These again divide into two groups (Fig. 2): smooth forms with or without apical slits and annulated forms with or without apical slits, e.g., *Plagioglypta,* the oldest group (Fig. 5).
    Order Siphonodentalioda
        Family Siphonodentaliidae
            (5 genera)-scaphopods with vermiform foot lacking apertural constrictions. These divide into two groups: *Entalina* and *Pulsellum* with a medial filament on the pedal disc (Figs. 1c and 2) and *Siphonodentalium* which lacks the filament but has apical lobes or notches. *Calstevenus* is a *Pulsellum*-like Permian fossil.
        Family Cadulidae
            (6 genera)—scaphopods with vermiform foot and a constricted aperture in the adult, a compact group that does not divide easily. *Cadulus* (Figs. 1 and 2) has the greatest diameter near the center. *Gadila* and the rest have it near the anterior aperture (Fig. 2).

Range and Evolution

The earliest undoubted siphonodentaloid is a *Gadila* (see Fig. 2) from the Lower Cretaceous but Emerson records *Entalina*, with a query,

from the Triassic; *Calstevenus* from the Permian, though similar to *Pulsellum,* is not certainly a siphonodentaloid.

In the Laevidentaliidae, the annulated *Plagioglypta* (see Fig. 2) ranges from the Devonian to the present day and gave rise to the trigonal *Progadilina* in the Lower Jurassic and the apically slit *Fustiaria* in the Eocene.

Laevidentalium (Fig. 2), smooth and with either simple or notched apex, ranges from the Triassic to the present day. It gave rise to the apically slit *Pseudantalis* in the Eocene; the acicular *Rhabdus* and the trigonal *Gadilina* in the Miocene; the laterally compressed *Bathoxifus* in the Holocene; and to two cylindrical, nearly straight, forms with an apical pipe—*Episiphon*—in the Jurassic and *Lobantale,* with two lateral internal ridges, in the Eocene.

The greatest uncertainty lies in the ranges of genera in the Dentaliidae, due to the indiscriminate use of the name *Dentalium* in describing fossils. The picture is far from clear but it appears that the unsymmetrical multicostate *Prodentalium/Fissidentalium* lineage was the central stock from which the symmetrical paucicostate group, typified by *Dentalium elephantinum* Linné 1758, were derived during the Late Cretaceous or early Paleogene.

It is unlikely that the partially costate *Antalis* ranges further back than the Late Cretaceous, although this depends upon the interpretation of *Antalis,* type species *Dentalium entalis* Linné 1758. In the strict sense, the genus can hardly be applied beyond the Paleocene.

The problems facing paleontologists when studying fossil scaphopods do not just involve the rarity of complete individuals, or the tendency in dentaloids for the character of ribbing to change quite radically with development. It is these, quite normal, difficulties combined with the practice of mature scaphopods to grow by absorption the juvenile, and often quite differently ornamented, part of the shell. Matching the juvenile shells with mature specimens can sometimes prove perplexing.

C. P. PALMER

References

Bilyard, G. R., 1974. The feeding habits and ecology of *Dentalium stimpsoni* Henderson (Mollusca: Scaphopoda), *Veliger,* 17, 127-138.

Bretsky, P. W., and Bermingham, J. J., 1970. Ecology of the Paleozoic scaphopod genus *Plagioglypta* with special reference to the Ordovician of eastern Iowa, *J. Paleontology,* 44, 908-923.

Dinamani, P., 1954. Feeding in *Dentalium conspicuum, Proc. Malac. Soc. London,* 36, 1-5.

Emerson, W. K., 1961. A classification of the scaphopod mollusks, *J. Paleontology,* 36, 461-482.

Emerson, W. K.; 1978. Two new Eastern Pacific species

of *Cadulus*, with remarks on the classification of the scaphopod mollusks, *Nautilus*, **92**, 117–123.

Haas. W., 1972. Micro- and ultrastructure of Recent and Fossil Scaphopoda, *Proc. 24th Internat. Geol. Cong.* **7**, 15–19.

Lacaze-Duthiers, H., 1856–1857, Histoire de l'organisation et du development du dentale, *Ann. Sci. Nat. France, ser. 4, Zoologie*, **6**, 320–385; **7**, 171–225; **8**, 18–44.

Morton, J. E., 1959. The habits and feeding of *Dentalium entalis*, *J. Marine Biol. Assoc. U.K.*, **38**, 225–238.

Palmer, C. P., 1974. A supraspecific classification of the scaphopod Mollusca, *Veliger*, **17**, 115–123.

Palmer, C. P., 1975. A new Jurassic scaphopod from the Oxford Clay of Buckinghamshire, *Palaeontology*, **18**, 377–383.

Pilsbry, H. A., and Sharp, B., 1897–1898. Scaphopoda, in G. W. Tryon, ed., *Manual of Conchology*, **17**, 1–348.

Pojeta, J., Jr., and Runnegar, B., 1976. The paleontology of rostroconch mullusks and the early history of the Phylum Mollusca, *U.S. Geol. Survey Prof. Paper 968*, 88 p.

Trueman, E. R., 1968. The burrowing process of *Dentalium* (Scaphopoda), *J. Zool.* **154**, 19–27.

Yancey, T. E., 1973. A new genus of Permian siphonodentalid scaphopods, and its bearing on the origin of the Siphonodentaliidae, *J. Paleontology*, **47**, 1062–1064.

Cross-reference: *Mollusca*.

SCLERITES, HOLOTHURIAN

The Modern Animal

Holothurians are "polyp-like" soft-bodied animals commonly called sea cucumbers because of their similarities in size, elongate shape, and tough skin over a soft interior. They are exclusively marine, invertebrate echinozoans (see *Echinodermata*) that inhabit all water depths in modern oceans from tropical to subarctic latitudes. The organisms are predominantly free-living, benthic, sediment-feeding forms, although a few are pelagic, filter-feeding types. Some are sessile and live in burrows; others are, rarely, pelmatozoans attached to a firm substrate.

The Holothurian Skeleton

The echinoderm skeleton is always internal (endoskeleton) and is of mesodermal origin (Raup, *in* Boolootian, 1966). However, Pawson (1966) refers to the calcareous exoskeleton of most fossil echinoderms. Among extant holothurian genera, only two orders have animals encased in more or less rigid, external, placoid skeletons of overlapping or contiguous calcareous plates. In most other holothurians, the plates are reduced in size and increased to large numbers of microscopic calcareous elements lodged within connective and dense layer tissue inside the animal.

The calcareous spiculate endoskeleton in the individual holothurian animal may consist of four distinct elements: (1) circumesophageal ring (10 or more plates), (2) innumerable microscopic body sclerites of extreme diversity in shape and size, (3) five tooth-like anal plates, and (4) a madrepore plate. The body sclerites make up the major part of the skeletal system.

Sclerites

Holothurian sclerites (Greek: *skleros* = hard), sometimes called spicules or ossicles, are the solid calcareous pieces or elements embedded in the dermal layer of the animal. Sclerites range in size from less than 0.1 mm (silt size) to more than 1 mm (coarse-sand size) in maximum diameter. They are known to approximate 10% of the body weight of an individual equivalent to more than 20,000,000 sclerites (Hampton, 1959).

Common descriptive names are applied to the great variety of sclerite shapes, such as sieve plates, wheels, hooks, anchors, rods, tables buttons, spectacles, racquets, rosettes, crosses, discs, and other types (Fig. 1). They may be fenestrate as in the sieve plates, or have cross-bars, eyes, flukes, hubs, rims, shanks, sockets, spears, spokes, stirrups, and other descriptive morphology.

This variation in type, shape, and size in individual or characteristic association of sclerites is useful taxonomically to recognize and identify species and higher categories in a holothurian classification. While it is easy to determine with certainty the sclerite assemblage of a living animal, it is much more difficult to establish the zoologic affinity of a fossil assortment of sclerites. As a result, problems arise in applying uniformity to discrete fossil sclerites and collections of assorted plates.

It is perplexing to imagine the biologic function of the sclerites in view of their small size, odd shapes, and scattered distribution inside the body of the animal. Apparently they reinforce the skin and internal organs into a flexible skeletal network much like the spicules in a sponge. Sclerites are often fenestrate for lightness in weight, effective pervasion of the tissue, and stress resistance to force. In some holothurians, they are imbricate in the wall to form a protective armor. In others, they penetrate the skin externally to aid in locomotion and for grasping or anchoring to seaweed.

The modern holothurian, *Molpadia intermedia*, has brown to red, ovoid to ellipsoid dermal granules (maximum diameter 10–

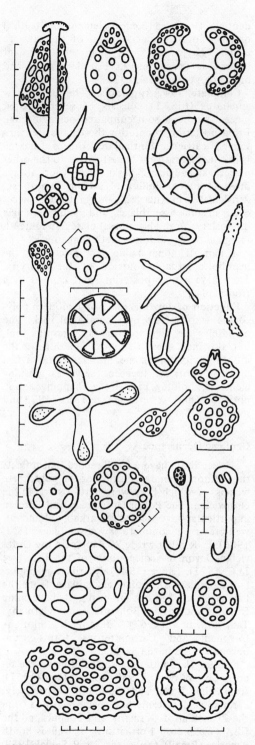

FIGURE 1. Outlines of select holothurian sclerites ranging in age from Ordovician at the bottom to Holocene (Recent) at the top (after Deflandre-Rigaud; Frizzell and Exline; Gutschick, Canis, and Brill; Pawson; and others). Size scales vary but each unit interval represents 0.1 mm.

350 μm) with concentrically laminated structure of iron, phosphate-rich, and silica composition (Lowenstam and Rossman, 1975). Members of the family Molpadiidae are unique in that calcareous deposits, anchors, anchor plates, tables, and racquet-shaped plates of the juvenile may be transformed into this phosphate material with the passage of time (Pawson, 1963). The Molpadiidae have very high potential for sizable reservoirs of iron and phosphate in the oceans.

The Fossil Record

Holothurian remains are known from the Ordovician period (Gutschick, 1954) to the present. Most surprising is the fact that in the entire fossil record, only three occurrences of entire individual holothurians have been reported to date. These have been found in the Lower Devonian and Upper Jurassic rocks of Germany and middle Pennsylvanian strata of Illinois, U.S.A. The Pennsylvanian soft-bodied holothurian fossils are found in siderite concretions. They are common, about the size of a fountain pen, and their body is lined with two distinct sizes of fishhook-type sclerites. No other sclerite types are present. In contrast, sclerite assemblages are common in Phanerozoic rocks and must be relied upon for evolutionary trends, biostratigraphy, and paleoecology of the holothurians.

Sclerites are $CaCO_3$ secretions, which become enlarged into single calcite crystals, as indicated by their ubiquitous optical homogeneity. These stable mineralized plates are often perserved in clays, shales, and marls from which they can be carefully water-washed, perhaps boiled or disaggregated with the use of a dispersing agent, and recovered. They are also common in some limestones that can be dissolved in dilute acetic or formic acid and the sclerites recovered even though they are of delicate calcite composition like the matrix rock.

Future studies will continue on the biology including skeletal morphology, ecology, burrowing, and epizoic characteristics of modern animals. Search will continue in the fossil record for whole animals and much more extensive discovery and documentation of sclerite faunas. A prime objective will be to reconcile the paleobiology of fossil holothurians with living forms.

RAYMOND C. GUTSCHICK

References

Boolootian, R. A., ed., 1966. *Physiology of Echinodermata.* New York: Wiley-Interscience, 822p.

Deflandre-Rigaud, M., 1963. Contribution à la connaissance des sclerites d'holothurides fossiles, *Mus. Hist. Nat. Paris, Mém.,* n.s., ser. C, **11**(1), 123p.

Frizzell, D. L., and Exline, H., 1955. Monograph of fossil holothurian sclerites, *Missouri Univ., Sch. Mines and Met., Bull. Tech. Ser.,* 89, 204p.

Frizzell, D. L.; Exline, H.; and Pawson, D. L., 1966. Holothurians, in R. C. Moore, ed., *Treatise on Invertebrate Paleontology,* Pt. U, Echinodermata 3. Lawrence, Kansas: Geol. Soc. Amer. and Univ. Kansas Press, U641–U672.

Gutschick, R. C., 1954. Holothurian sclerites from the Middle Ordovician of northern Illinois, *J. Paleontol.,* 28, 827–829.

Gutschick, R. C., 1959. Lower Mississippian sclerites from the Rockford Limestone of northern Indiana, *J. Paleontology,* 33, 130–137.

Gutschick, R. C., and Canis, W. F., 1971. The holothurian sclerite genera Cucumarites, Eocaudina, and *Thuroholia*—re-study of *Eocaudina* and *Protocaudina* from the Devonian of Iowa, *J. Paleontology,* 45, 327–337.

Gutschick, R. C.; Canis, W. E.; and Brill, K. G., Jr., 1967. Kinderhook (Mississippian) holothurian sclerites from Montana and Missouri, *J. Paleontology,* 41, 1461–1480.

Hampton, J. S., 1959. Statistical analysis of holothurian sclerites, *Micropaleontology,* 5, 335–349.

Lowenstam, H. A., and Rossman, G. R., 1975. Amorphous, hydrous, ferric phosphatic dermal granules in *Molpadia* (Holothuroidea): Physical and chemical characterization and ecologic implications of the bioinorganic fraction, *Chem. Geol.,* 15, 15–51.

Pawson, D. L., 1963. The holothurian fauna of Cook Strait, New Zealand, *Zool. Publ. Victoria Univ. of Wellington, N.Z.,* 36, 38p.

Pawson, D. L., 1966. Phylogeny and evolution of holothuroids, in R. C. Moore, ed., *Treatise on Invertebrate Paleontology,* pt. U, Echinodermata 3(2). Lawrence, Kansas: Geol. Soc. Amer. and Univ. Kansas Press, U641–U646.

Cross-reference: *Echinodermata.*

SCOLECODONTS

Chitinous, marine, annelid worm jaws, $50\mu m$ to a few millimeters in size, when found as fossils are collectively referred to as "scolecodonts." They constitute the armour of the pharynx that can protrude (eversion) for food capture or retract for digestion of food. Acid-resistent (acetic, hydrochloric, and hydrofluoric acids) scolecodonts are released from limestones as either disjunct jaw pieces or as entire jaw apparatus (Kielan-Jaworowska 1966, 1968, and earlier papers [hereafter K-J], Lange, 1949; Kozlowski, 1956; and others, see Tasch, 1973; Jansonius and Craig, 1971, 1974, [hereafter J&C]. This has resulted in a dual nomenclature.

Structure

Disjunct Jaw Pieces. Individual jaw elements have a basal cavity (fossa or myocoele) that generally narrows as it extends to tip of the denticle. Denticles (conical elements or "teeth") on the dorsal (inner) margin of the jaw are usually larger anteriorly and tend to slant backwards (posteriorly). (For definitions and morphology, see J&C, 1971).

Complete Jaw Apparatus. The entire jaw apparatus (Fig. 1) consists of several paired pieces (although some unpaired peices occur). From posterior to anterior these pairs are: vari-shaped carriers; denticulate forceps (maxilla I, M_1) comparatively large relative to the other pieces; an adjacent basal plate (not always present) and excluding M_1, from two to six denticulate maxilla (M_2–M_6). Other pieces occur in some assemblages, and include: a single intercalary tooth above basal plate; lateral teeth marginal to M_1; anterior teeth (unpaired).

A pair of noneversible jaw pieces are the mandibles situated outside of the pharynx. They have a long posterior shaft capped by a frontal plate.

Function of Jaw Pieces. Some modern polychaete worms bore into coral-algal rock. The mandibles are used for this—as a rasping tool, but also may serve a support function for the mouth cavity and some muscles. The several maxillae working together serve as grasping elements, holding prey by a snapping action until the pharynx retracts (see Tasch, 1973 and references therein).

Geologic Occurrence

Scolecodonts have a fossil record that reflects the cosmopolitan distribution of the polychaete worms, of which they were part; it starts in the Ordovician. The Polish Ordovician and Silurian in particular have yielded remarkably abundant, well-preserved complete jaw assemblages (K-J, 1966). E. R. Eller wrote 20 papers on the Ordovician–Devonian scolecodonts of the U.S. (see J&C, 1971). Lange (1949) described an important complete Devonian assemblage from the Parana Basin of Brazil. Other reports have come from Scandinavia, Bohemia, and Germany. Taugourdeau (1967) described Silurian–Devonian and Carboniferous forms from borehole cores in the Sahara.

Further Carboniferous scolecodonts were reported by Sylvester (1959) and by Eller (1967), from the USA and Scotland respectively. Permian scolecodonts have been recorded from the Zechstein of Germany and Poland, and the USA (Wellington Formation), as well as South America (Brazil). (Sylvester, 1959; Siedel, 1959; Tasch and Stude; 1966; Szaniawski, 1968).

Mesozoic forms are known from the Triassic (Muschelkalk) and Jurassic Solenhofen limestone of Germany (Kozur, 1963) and the Upper Cretaceous of New Jersey (Charletta and

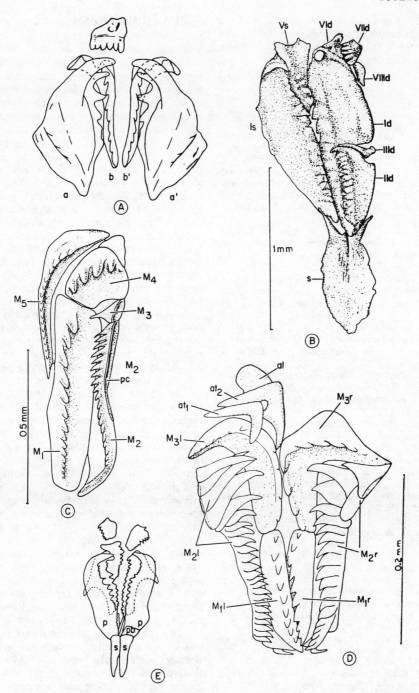

FIGURE 1. Some natural fossil scolecodont assemblages (from Tasch, 1973; used with permission). A: maxilla I, *Eunicites* sp., Permian, Kansas—a,a[1], *Arabellites comis;* b,b[1], *A. Falciformis;* unpaired piece, *Eunicites* sp. B: *Polychaetaspis wyszogrodensis* Kozlowski—s, supports; Is, Id, left and right forceps; IId, basal piece; IIId, intercalary tooth; Vs, unpaired piece; VId, right posterior maxilla; VIId, right anterior maxilla; VIIId, right lateral tooth (= paragnath). C: *Vistulella kozlowski* Kielan-Jaworowska, Ordovician, Poland, dorsal view—M_1, M_5, maxillae; pc, pulp cavity (= fossa). D: *Xanioprion borealis* K = J, Ordovician, Poland, schematic—M_1l, M_1r, left (l) and right (r) maxilla; at_1, at_2, anterior teeth; al, attachment lamella. E: *Paulinites paranaensis*, Lange, Devonian, Brazil, dorsal view—s, carrier; pb, basal plate of right forceps; p,p, left and right forceps; other components (2 asymmetrical subtriangular dental plates; an unpaired piece, and 2 oblong asymmetrical paragnaths).

747

Boyer, 1974) and of Lebanon, as well as of the North Apennine flysch (Roger, 1946, see Tasch, 1973; Corradini and Serpagli, 1968). Tertiary scolecodonts are rare.

PAUL TASCH

References

Charletta, A. C., and Boyer, P. S., 1974. Scolecodonts from Cretaceous greensand of the New Jersey coastal plain, *Micropaleontology,* 20, 354-368.

Corradini, D. and Serpagli, E., 1968. Preliminary report on the discovery and initial study of large amounts of "scolecodonts" and polychaete jaw apparatuses from Mezozoic formations, *Soc. Paleont. Ital., Boll.,* 7, 3-5.

Eller, E.-R., 1967. Scolecodonts from the Charlestown Main Limestone, Lower Carboniferous at Cults, Fifeshire, Scotland, *Ann. Carnegie Mus.,* 39, 69-73.

Jansonius, J., and Craig, J. H., 1971. Scolecodonts: 1. Descriptive terminology and revision of systematic nomenclature, etc, *Bull Canadian Pet. Geol.* 19, 251-302.

Jansonius, J., and Craig, J. H., 1974. Some scolecodonts in organic association from Devonian strata of western Canada, *Geoscience and Man,* 9, 15-26; 11, 151.

Kielan-Jaworowska, Z., 1966. Ploychaete jaw apparatuses from the Ordovician and Silurian of Poland and a comparison of modern forms, *Palaeontol. Polonica,* 16, 152p.

Kielan-Jaworowska, Z., 1968. Scolecodonts versus jaw apparatuses, *Lethaia,* 1, 39-49.

Kozlowski, R., 1956. Sur quelques appareils masticateurs des annélides polychétes ordoviciens, *Acta Palaeontol. Polonica,* 1, 165-204.

Kozur, H., 1967. Scolecodonten aus dem Muschelkalk des germanischen Binnenbecken, *Deutsche Akad. Will. Berlin, Mh.,* 9, 842-886.

Lange, F. W., 1949. Polychaete annelids from the Devonian of Parana, Brazil, *Bull. Amer. Paleontol.,* 33, 1-102.

Roger, J., 1946. Les Invértebrés des Couches a Poissons du Crétacé Supérieur du Liban, *Mém. Soc. Géol. France,* No. 51, 92p.

Seidel, S., 1959, Scolecodonten aus den Zechstein Thuringens, *Freiberger Forscheingshefte,* 76, 1-32.

Sylvester, R. K., 1959. Scolecodonts from central Missouri, *J. Paleontology,* 33, 33-49.

Szaniawski, J., 1968. Three new polychaete jaw apparatuses from the Upper Permian of Poland, *Acta Palaeontol. Polonica,* 13, 255-281.

Tasch, P., 1973. *Paleobiology of the Invertebrates.* New York: Wiley, 973p. (Polychaetes, 456-470.)

Tasch, P., and Stude, J. R., 1966. Permian scolecodonts from the Ft. Riley Limestone of southeastern Kansas, *Wichita State Univ. Bull.,* 42,(3); *Univ. Studies* No. 68, 3-35.

Taugourdeau, P., 1967. Scolécodontes du Siluro-Dévonien du Cotentin, *Bull. Soc. Géol. France,* Sér. 7, 9, 457-475.

Cross-reference: *Annelida.*

SEXUAL DIMORPHISM

Sexual dimorphism is the morphologic (phenotypic) expression of sexuality in the form of male and female. Recognition of this phenomenon in the fossil record depends entirely on the preserved skeleton and is thus restricted to tertiary sex characters or attributes (the primary and secondary attributes being respectively the gametes and sex organs), which are fully developed only in the mature individual. Hermaphroditic species are usually excluded since "maleness" and "femaleness" occur in the same individual (e.g., some Bryozoa and Porifera), mostly in that sequence (e.g., many gastropods). Although sexuality depends on various, and in each species different, determining, influencing, and differentiating processes, the male to female ratio in most populations is approximately one to one. However, intersexes and asexual casts (e.g., in insects) may also occur. Dimorphs are often distinguished in size; if the size ratio is higher than about two as is often the case in the invertebrates, the larger dimorph is consistently the female. In the vertebrates, the male is commonly slightly to moderately larger.

The criteria for recognition of sexual dimorphism in fossils are that the morphotypes (1) occur together in the same geological horizon (bed) and in similar quantities; (2) have identical immature growth stages, while the mature and adult stages differ significantly (statistically); and (3) show identical stratigraphic and geographic distributions and evolutionary trends.

However, for a variety of reasons, these criteria are not always realized, strongly limiting recognition of dimorphism. (1) Temporary segregation of the sexes is common among living nektonic invertebrates and vertebrates, so that their skeletons may be preserved in different localities and rock types (lithofacies); adult sex ratios may deviate markedly from unity; and preservation and recovery of fossils are strongly biased in favor of larger size. (2) The required ontogenetic record in fossils is usually known only in skeletons with accretionary (marginal) growth. However, in skeletons of crustaceans, arthropods, and vertebrates where molting and remodeling occur, analogy with tertiary sex attributes of living relatives usually has to suffice. Colonial organisms (with interconnected tissue and often skeletons), of course, pose no problem with regard to correspondence, but may display additional polymorphs that, in the fossil, are difficult to distinguish from the "sex" morphs (e.g., some coelenterates); they are not considered here. Finally, (3) evidence for vertical and lateral distribution and for evolutionary trends is often scarce and controversial in the fossil record.

In the fossil record, good evidence for sexual dimorphs of noncolonial metazoans exists in the phyla Brachiopoda(?), Mollusca, Echinodermata, Crustacea, Arthropoda, and Vertebrata (Westermann, 1969). While the character and magnitude of differentiation is usually within the range of phenotypic variation commonly attributed to (biological) species, this does not appear to be the case in the fossil cephalopod mollusks with dimorphism frequently as great as the differences between conventional genera or subgenera. Classification and nomenclature at the species and generic levels will then depend on the correct interpretation of the dimorphs, un'ss the "morphospecies" (based on morphology alone) rather than "biospecies" (the species of neontologists) is used as the basic unit of classification. This problem has been particularly evident and common in the extinct ammonites.

Brachiopoda

Although most are dioecious (sex separate), the shells of living brachiopods (lamp shells) are generally indistinguishable according to sex. However, bimodal distribution in shell thickness has been recorded from several late Paleozoic species and interpreted as sexual dimorphism (e.g., *Pugnax*, *Pleuropugnoides*, *Leiorhynchus*, *Cyrtospirifer*). Brood pouches (marsupia) indicate female shells in several living and fossil species (e.g., *Megousia*).

Mollusca

Shells of chitons, scaphopods, gastropods, and bivalves are only rarely markedly dimorphic, although many more species are dioecious than hermaphroditic. It is in fact in the hermaphroditic living, and apparently also fossil, gastropods that marked dimorphism, mainly in size and shape, has been shown to exist (e.g., the Cretaceous *Euspira* and *Calliomphalus*). Only a few dimorphic genera of fossil bivalves are known (e.g., the Paleogene *Astarte* and *Venericardia*); these are distinguished by type of ornamentation, shell thickness, or shape.

In the living cephalopods, male and female shells where present are similar as in *Nautilus*, the only extant ectocochliate (Saunders and Spinosa, 1978), and the same is true for *Sepia* and *Spirula* with internal shells. (The female "shell" of *Argonauta* is only a specialized brood chamber without equivalent in the male, which,

however, is much smaller.) In some Paleozoic and Mesozoic nautiloids, however, sexual dimorphism is indicated, either in size or in shape of the body chamber (e.g., Oncocerida, Ascocerida, Nautilida).

Very marked and therefore most significant dimorphism is well established mainly in the Jurassic and Cretaceous Ammonitina (e.g., Makowski, 1963; Westermann, 1964; Moore, 1964). Almost throughout their history, hundreds of genera (and some species) are known to occur side by side in pairs of large macroconchs and small microconchs, the former also frequently characterized by adult reduction of ribbing, the latter by apertural projections (rostra or lappets). The size ratio ranges from about two to five or six, and the macroconchs are considered the female shells (Fig. 1).

Arthropoda

In the Trilobites of the Paleozoic, the cephalon (head shield) of the dorsal carapace may display traits similar to those of the dimorphic living amphipods; e.g., consistent differences in the shape of the compound eyes and in ornamentation (relief) (e.g., *Weberides*, *Phacops*). In the eurypterids, also Paleozoic, sexual dimorphism may be expressed in the operculum, walking legs, and ornamentation of the integument. Sexual dimorphism is also particularly apparent in the last, adult instar (molting stage) of living and fossil ostracodes in which the female develops brood pouches (e.g., Guber, 1971).

Echinodermata

In the echinoids (sea urchins), the different size of the genital pores (for release of sperm and eggs) and genital plates is clear evidence of the sex, the larger size belonging to the female. Brood pouches may also be present in the females.

Vertebrata

As in the skeleton of living vertebrates, sexual dimorphism in fossil vertebrates is mostly recognized by moderate difference in size (ratio ≤1.5; see Fig. 2), the male being the larger one in the quadrupeds (but the smaller one in many fishes and birds). In addition, the males are usually more robust with thicker bones. Classification is sometimes a problem for the paleontologist when the sexual differences exceptionally reach the magnitude of variation normally attributed to species.

In some late Paleozoic amphibians (e.g., *Seymouria;* Vaughn, 1966), single fossil popula-

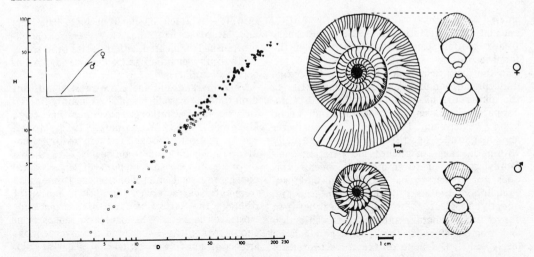

FIGURE 1. The dimorphic Jurassic ammonite *Stephanoceras intinsae* ♀, ♂, from Queen Charlotte Islands, Canada (from Hall, 1975). Plot of whorl height (H) against diameter (D) for 18 female and 8 male shells from a single locality.

tions include two sets of skulls, one with rounded contours, oval orbits, and small teeth, the other more robust with more angular contours and orbital outlines as well as larger teeth. By close analogy to the living frog *Rana*, the former are considered to belong to the females, the latter to the males. Among fossil reptiles of the same period, several pelycosaurians (e.g., *Dimetrodon, Cotylorhynchus, Cicynodon*) existed in two size groups in otherwise very similar samples; the size differences are only about 10–20%. Sexual dimorphism has been reported for some dinosaurs as well (e.g., Dodson, 1975, 1976; Kurzenov, 1972).

In the mammals, horns, antlers, and tusks are often additional male attributes. However, these features may also be present in both sexes of closely related species, as is well known from living antelopes. More reliable evidence is the quantitative difference between the sexes in the size of antlers, horns, and tusks since they tend to be relatively larger in the males even considering their overall larger body size, i.e., growth of these attributes is positively allometric.

G. E. G. WESTERMANN

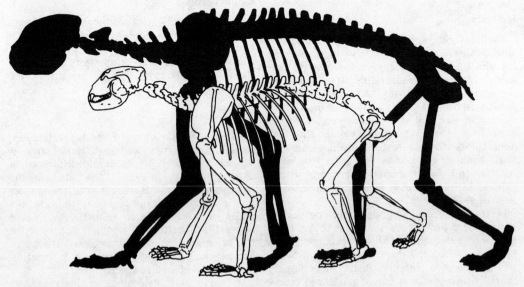

FIGURE 2. Sexual dimorphism in the North American Pleistocene bear *Tremarctos floridanus*, ♀, ♂ (from Kurtén in Westermann, 1969). Male skeleton in silhouette compared with female skeleton in outline.

References

Coombs, H. C., 1975. Sexual dimorphism in chalicotheres (Mammalia, Perissodactyla), *Syst. Zool.,* **24**, 55–62.

Dodson, P., 1975. Taxonomic implications of relative growth in lambeosaurine dinosaurs, *Syst. Zool.,* **24**, 37–54.

Dodson, P. 1976. Quantitative aspects of relative growth and sexual dimorphism in *Protoceratops, J. Paleontology,* **50**, 929–940.

Guber, A. L., 1971. Problems of sexual dimorphism, population structure and taxonomy of the Ordovician genus *Tetradella* (Ostradoca), *J. Paleontology,* **45**, 6–22.

Hall, R., 1975. Sexual dimorphism in Jurassic ammonites from Queen Charlotte Islands, *Geoscience Canada,* **2**(1), 21–25.

Kurzanov, S. M., 1972. Sexual dimorphism in protoceratopsians, *Paleontol. J.,* **1972**, 91–97.

Linsley, R. M., 1974. Sexual dimorphism in Paleozoic Gastropoda (abstr.), *Geol. Soc. Amer. Abst.,* **6**, 867.

Makowski, H., 1963. Problem of sexual dimorphism in ammonites. *Paleontol. Polonica,* **12**, 1–92.

Moore, R. C., ed., 1964. *Treatise on Invertebrate Paleontology,* pt. K, Mollusca 3. Lawrence, Kansas: Geol. Soc. Amer. and Univ. Kansas Press, 519p.

Saunders, W. B., and Spinosa, C., 1978. Sexual dimorphism in *Nautilus* from Palau, *Paleobiology,* **4**, 349–358.

Vaughn, P. P., 1966. *Seymouria* from the Lower Permian of southeastern Utah and possible sexual dimorphism in that genus, *J. Paleontology,* **40**, 603–612.

Westermann, G. E. G., 1964. Sexual-Dimorphismus bei Ammoniten und seine Bedeutung für die Taxionomie der Otoitidae (Ammonitina; M. Jura), *Palaeontographica A,* **124**, 33–73.

Westermann, G. E. G., ed., 1969. *Sexual Dimorphism in Fossil Metazoa and Taxonomic Implications.* Stuttgart: E. Schweizerbert'sche Verlagsbuchhandlung (Nägele u. Obermiller), Internat. Union Geol. Sci., ser. A, no. 1, 251p.

Cross-references: *Ammonoidea; Biometrics in Paleontology; Bones and Teeth; Population Dynamics; Vertebrate Paleontology.*

SILICIFICATION OF FOSSILS

Silicified fossils are fairly widespread in various types of sediments. They are of particular interest to paleontologists when they occur in limestones, because they then can be isolated by etching the rock with 5% hydrochloric acid, this technique being simpler than the conventional hammer and needle. Furthermore, here the sample may be studied quantitatively; specimens can be observed three-dimensionally; and large quantities can be obtained with little expense. In order to obtain sufficient silicified fossils one has to pay more than the usual attention to them during geological field work.

Solutions causing silicification may originate in different ways: (1) *By disintegration of silicates* either by hydrolysis or exchange of ions. Quartz, although common in occurrence, is not easily soluble in water. Thus, it seems to play only a minor role in silicification. (2) *Organic silica solutions* are formed, for example, by diatoms, higher plants, Radiolaria, and sponges. As organic opal skeletons are about 15 times more soluble than quartz, they are commonly dissolved shortly after their formation and redeposited in or on the fossil. (3) *Volcanism*—since silicic acids, H_4SiO_4 and H_2SiO_3, are more stable in low concentrations than in higher, they may be transported as such over long distances in volcanic waters. In addition, volcanic glass and ash are unstable and break down rapidly, thus freeing silicic solutions.

Silicate deposition is a result of a change in pH of the depositional environment. Laboratory analysis of the process is not possible as the concentrations involved are extremely low, always being well below the saturation point. Evaporation may be an important factor in silicate precipitation. Silicification is rare in humid climates.

The silicification process is selective. In most rocks, only certain groups are silicified while others remain unaltered. For some unknown reason, there is a variation in groups silicified at different localities. In general, however, it can be stated that the Brachiopoda, Ostracoda, Trilobita, and Bryozoa are most commonly silicified, while Mollusca and Echinodermata are only rarely silicified.

Kinds of Silification

Incrustation. The fossil may be *coated* with silica; in some instances only one side may be covered. This coating is usually very thin and requires a specially gentle and careful treatment.

Pseudomorphosis. Hard parts and in exceptional cases even soft parts of an organism may be *replaced* by silica. Here in some cases typical concentric structures of deposition, "silicification rings," may be observed.

Impregnation. Silica may interpenetrate the original substance. This process is commonly observed in the silicification of woods. In rare cases, not only hard, but also the soft parts are permeated with silica and thus become beautifully preserved (Fig. 1).

The Time of Silification

The process of silicification can be classified according to the point at which it occurred in the geologic history of the fossil.

FIGURE 1. *Dasyhelea australis antiqua* Palmer, silicified Diptera from the Miocene of California (from Palmer, 1957).

Syngenetic Silification. Silification may occur during the formation of the sediments. Silicification of soft parts can only occur in this way. An example of this type of preservation process is the fauna of Arachnida, small Crustacea, Copepoda, and Insecta (Fig. 1) in

the Barstow Formation (Miocene) of California (Palmer, 1957). Here the fossils are enclosed in carbonate concretions and are thus excellently preserved. As such fossils cannot be detected through mere surface observation, all occurrences of such concretions should be investigated by routine trial dissolution in acids.

One classical locality, the Devonian Old Red of Rhynie, Scotland, has produced some of the oldest known insects and various important Psilophyta. These fossils are preserved in a peat bog that silicified shortly after deposition.

In both these localities, it is probable that volcanism has contributed to the silicification process, providing the silica source.

Early Diagenesis. Silification may occur during the period of lithification of the sediments, e.g., in reefs, near-reef facies and in the chert concretions widespread in the Upper Cretaceous chalk of northern Europe. The source of silica in this case is probably the opal skeletons of various organisms. The chemistry of its migration is little understood.

Epigenetic silicification. The most widespread and most common type of silification seems to occur long after the diagenetic induration of limestone (Müller, 1964). For example, the silicification of Devonian fossils may have occurred beneath a Tertiary or Holocene land surface. Here there is erratic deposition of

FIGURE 2. Partly dissolved-block of limestone with cluster of partly exposed specimens of *Echinauris opuntia* (Waagen), ca X1.5 (from Grant, 1968).

silica restricted to the areas (commonly to a depth of only a few cm) in contact with the topsoil or neighboring joints or in caves. Such material can be successfully obtained, for example, in freshly ploughed soils. Impure (carbonaceous) porous limestones seem best suited for this type of silicification. In this case, the silica generally originates through the dissolution of silicates in various soil types; it could be influenced by organic products formed by plant growth.

Field Observation and Techniques

It is not difficult to spot silicification in the field in most cases, as silicified fossils may be partly or entirely freed by weathering. Dolomitized fossils, which may also form a relief on a weathered surface, can be distinguished from silicified fossils by testing with a drop of HCl.

It must be noted, however, that the finest examples of silicified fossils cannot be detected on the rock surface; their presence is revealed only through trial etching. Many localities containing silicified fossils have been discovered accidently through routine etching procedure, e.g., for separation of conodonts.

For a trial etching, a small sample is placed on a plastic etching screen, which is then placed in a plastic bucket containing 5% hydrochloric or 15% acetic acid. After the reaction is complete, the dissolved material is carefully washed away by immersing the screen in water. What remains is then dried and screened for microfossils under the microscope.

Larger pieces of rock must be partly covered with acid-resistant paint, so that the acid can attack only the upper surface. Otherwise, the fragile isolated fossils will be destroyed by their own weight. If carefully treated, it is possible in some cases to etch blocks up to several hundred kilograms, thus obtaining large and completely silicified specimens (Fig. 2).

K. J. MÜLLER

References

Boyd, D. W., and Newell, N. D., 1972. Taphonomy and diagenesis of a Permian fossil assemblage, *J. Paleontology*, **46**, 1-14.

Cooper, G. A., and Grant, R. E., 1972. Permian brachiopods of West Texas, I, *Smithsonian Contrib. Paleobiol.*, **14**, 231p.

Cooper, G. A., and Whittington, H. B., 1965. Use of acids in preparation of fossils, in B. Kummel and D. Raup, eds., *Handbook of Paleontological Techniques*. San Francisco and London: Freeman, 294-300.

Grant, R. E., 1968. Structural adaptation in two Permian brachiopod genera, Salt Range, West Pakistan, *J. Paleontology*, **42**, 1-32.

Hayami, I.; Moeda, S.; and Fuller, C. R., 1977. Some Late Triassic Bivalvia and Gastropoda from the Domeyko Range of North Chile, *Trans. Proc. Palaeontol. Soc. Japan*, **108**, 202-221.

Müller, K. J., 1964. Uber die Verkieselung von Fossilien, *Zeitschr. Deutsch. Geol. Gesell.*, **114**, 647-656.

Palmer, A. R., 1957. Miocene arthropods from the Mojave desert, *U.S. Geol. Surv. Prof. Paper 294-G*, 237-280.

Whittington, H. B., and Evitt, W. R., 1954. Silicified Middle Ordovician trilobites, *Geol. Soc. Amer. Mem., no. 59*, 137p.

Cross-references: *Diagenesis of Fossils—Fossildiagenese; Fossils and Fossilization; Taphonomy.*

SILICOFLAGELLATES

Silicoflagellates are small (usually less than 100 μm), unicellular marine phytoplankton particularly abundant in nutrient-rich, upwelling areas. They move like some animals, hence zoologists and many paleontologists have considered them to be protozoa; yet they possess photosynthetic chloroplasts, hence botanists have claimed them as algae. They possess a siliceous skeleton that surrounds most of the cell material except for fine pseudopodia and a single anterior flagellum. The skeletons are relatively resistant to solution and so are found commonly as fossils. They can be retrieved from marine sedimentary rock by chemical extraction techniques (Lipps, 1973; Mandra et al., 1973).

The Living Organism

Living silicoflagellates (Marshall, 1934; Van Valkenberg, 1971) usually possess a single nucleus with nucleolus, numerous yellow-brown chloroplasts, various other cytoplasmic granules and vacuoles contained mostly within the framework of the skeleton (Fig. 1). In individuals from rapidly reproducing culture stocks, the cytoplasm is differentiated into an area surrounding the nucleus and containing numerous organelles. Extending from this to or just beyond the skeletal rods are cytoplasmic processes embedded in a structureless, viscous wall material. Exterior to the skeleton are small thin pseudopodia and an anterior flagellum. The flagellum seems to lack typical flagellar structures internally, but nevertheless it beats near its tip. Pseudopodia are relatively long and thin and become more abundant as the organism settles to the bottom of culture dishes. Some cells in culture have been observed to be multinucleate and possess no skeleton.

The skeleton is made of hollow rods arranged in a lattice of various types (see Figs. 2-5). It

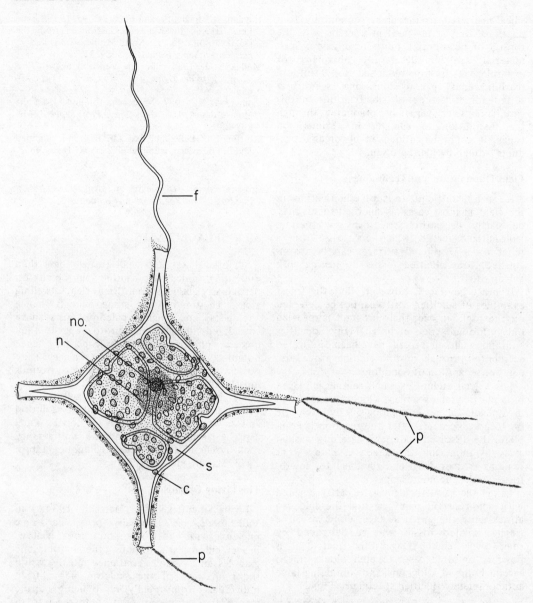

FIGURE 1. A living silicoflagellate, *Dictyocha fibula* (composite drawing based on VanValkenberg, 1971, and Marshall, 1934). s, hollow skeleton; c, chloroplasts; f, flagellum; p, pseudopodia; n, nucleus; no, nucleolus.

may range from a simple ring to a relatively complex dome-like structure. Basically, each skeleton has a "basal ring" from which rods arise on one side to form an arch or dome resulting in an overall hemispherical shape. Each part of the skeleton has been given a special name (Fig. 6). The skeletons are predominantly silica, present probably as the monomeric acid $Si(OH)_4$. The rods of the skeleton may be smooth or possess various small spines, ridges, or nodes (Figs. 3,4).

Fossil Silicoflagellates

Silicoflagellates have been described ever since C. G. Ehrenberg first noticed them as fossils in 1837. Many genera, species, and sub-specific taxa have been described, but many of these are probably variants or misidentifications of others. At the generic level, bits and pieces of radiolarians, diatoms, and other microfossils have been described previously as silicoflagellates. All but eight (Fig. 5) of these can either

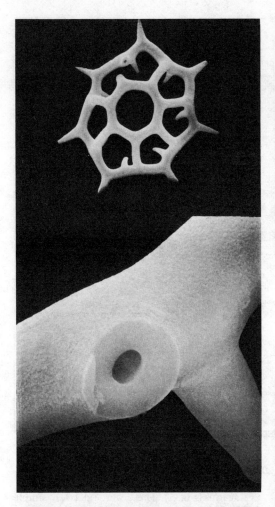

FIGURE 2. Scanning electron micrographs of the skeleton of *Distephanus speculum* (Ehrenberg) showing the general morphology of this common species (above, X900) and the hollow skeletal rods (below, X8600); from late Miocene rocks near San Felipe, Baja California, Mexico. Courtesy of Dr. Y. Mandra, San Francisco State University.

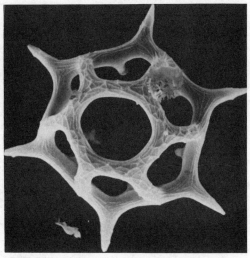

FIGURE 3. Scanning electron micrograph of *Distephanus speculum* (Ehrenberg) with sculptured skeleton (X2035); from the late Eocene, New Zealand. Courtesy of Dr. Y. Mandra, San Francisco State University).

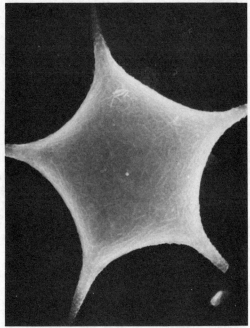

FIGURE 4. Scanning electron micrograph of *Vallacerta siderea* (Schulz) from the Late Cretaceous of the Arctic Ocean (X630). Illustration courtesy of Dr. H. Y. Ling, University of Washington, and reproduced with permission from *Science,* 180, cover, 29 June 1973. Copyright 1973 by the American Association for the Advancement of Science.

be excluded from the silicoflagellates or placed in synonomy with other valid genera. At the specific and subspecific level, many varieties have been described that probably have no biological or practical significance. Caution must be exercised in the evaluation of species limits, for great variation has been demonstrated in skeletons obtained from clonal cultures (VanValkenberg and Norris, 1970), as well as those found in nature.

Biostratigraphically, silicoflagellates are not generally useful because there are not many species and those few have long geologic ranges. The first silicoflagellates appeared in the

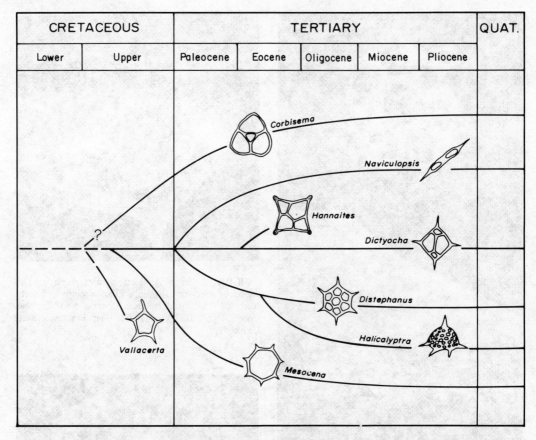

FIGURE 5. Geologic range and inferred phylogeny of silicoflagellate genera (modified after Lipps, 1970b).

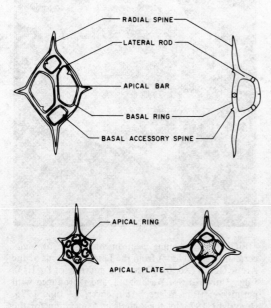

FIGURE 6. Terminology applied to silicoflagellate skeletons.

Cretaceous (Fig. 5). These were relatively simple forms from which others evolved. Four genera are known from the Upper Cretaceous and all but one of these are alive today. In the Tertiary, at least four other genera evolved, and again all but one are alive today. Species, although not so long-ranging, still do not provide as sound a basis for detailed zonation as do other planktic microfossils. For example, a recent zonation of the Upper Cretaceous to the lower Miocene used only 8 silicoflagellate zones (two of which in fact were based on microfossils that are probably not silicoflagellates) compared with 21 coccolith zones (Bukry and Foster, 1974). Nevertheless, silicoflagellates occasionally provide useful supplementary data, especially in sediments deposited beneath upwelling regions, or in areas where calcareous microfossils are not present.

Ecology, Paleoecology, and Evolution

Silicoflagellates appear to be distributed according to water masses and nutrient supply in the modern oceans (Lipps, 1970b; Martini,

1977). For example, the genus *Dictyocha* is found in low and middle latitudes in water warmer than about 10°C, while *Distephanus* predominates in water at higher latitudes generally cooler than about 20°C. Such biogeographic distinctions seem to have existed for a long geologic time, and are in accord with species diversity patterns known for other plankton.

Abundance, however, is dependent on the supply of required nutrients in the surficial waters. These nutrients, especially silica, are most abundant in the euphotic zone where upwelling brings them from deeper depths, and where the organisms can also photosynthesize. Silicoflagellates flourish in such areas. Indeed they have probably always been most abundant in upwelling regions because they are very commonly found in Cretaceous through Holocene sedimentary rocks that were deposited below the nutrient-rich regions.

The evolutionary record of silicoflagellates shows decreased species diversity and abundance in the Paleocene and Oligocene; these times of low diversity can be attributed to unique ecologic events. At these times, it is likely that upwelling intensity declined, perhaps because of altered thermal gradients or current patterns in the sea that produced more homogeneous environments (Lipps, 1970a). Decreased nutrient supply and homogeneous seas would eliminate ecologic barriers that keep species separate, and this probably resulted in their extinction.

JERE H. LIPPS

References

Bukry, D., and Foster, J. H., 1974. Silicoflagellate zonation of Upper Cretaceous to Lower Miocene deep-sea sediment, *J. Research U.S. Geol. Surv.,* **2,** 303–310.

Cornell, W. C., 1974. Silicoflagellates as paleoenvironmental indicators in the Modelo Formation (Miocene), California, *J. Paleontology,* **48,** 1018–1029.

Deflandre, G., 1950. Contribution a l'étude des silicoflagellides actuels et fossiles (Suite et fin), *Microscopie, U.Z.,* December, 191–210.

Glezer, Z. I., 1966. Kremnevye zhgutikovye vodorosli (Silikoflagellaty). Silicoflagellatophyceae, *Flora Sporovykh Rasteniy SSSR, Flora Plantarum Cryptogammarum URSS,* vol. 7. Moscow, Leningrad: Akad. Nauk SSSR, Botanicheskiy Institut V. L. Komarova, 330p.

Lipps, J. H., 1970a. Plankton evolution, *Evolution,* **24,** 1–22.

Lipps, J. H., 1970b. Ecology and evolution of silicoflagellates, *Proc. N. Amer. Paleontological Conv.,* pt. G, 965–993.

Lipps, J. H., 1973. Microfossils, in P. Gray, ed., *Encyclopedia of Microscopy and Microtechnique.* New York: Van Nostrand Reinhold, 308–312.

Loeblich, A. R., III; Loeblich, L. A.; Tappan, H.; and Loeblich, A. R., Jr., 1968. Annotated index of fossil and Recent silicoflagellates and ebridians with descriptions and illustrations of validly proposed taxa, *Geol. Soc. Amer. Mem. no. 106,* 319p.

Mandra, Y. T., 1968. Silicoflagellates from the Cretaceous, Eocene, and Miocene of California, U.S.A., *Proc. California Acad. Sci.,* ser. 4, **36,** 231–277.

Mandra, Y. T.; Brigger, A. L.; and Mandra, V. T., 1973. Chemical extraction techniques to free fossil silicoflagellates from marine sedimentary rocks, *Proc. California Acad. Sci.,* ser. 4, **39,** 273–284.

Marshall, S. M., 1934. The Silicoflagellata and Tintinnoinea, *British Mus. (Nat. Hist.) Great Barrier Reef Expedition, 1928-1929, Sci. Repts.,* **4,** 623–664.

Martini, E., 1977. Systematics, distribution and stratigraphical application of silicoflagellates, in A. T. S. Ramsay, ed., *Oceanic Micropaleontology, Vol. 2.* London: Academic Press, 1327–1343.

VanValkenburg, S. D., 1971. Observations on the fine structure of *Dictyocha fibula.* II. The Protoplast, *J. Phycology,* **7,** 118–132.

VanValkenberg, S. D., and Norris, R. E., 1970. The growth and morphology of the silicoflagellate *Dictyocha fibula* Ehrenberg in culture, *J. Phycology,* **6,** 48–54.

Cross-references: *Algae; Coccoliths; Plankton; Protista;* Vol. I: *Phytoplankton.* Vol. IVA: *Silica-Biogeochemical Cycle.* Vol. VI: *Pelagic Sedimentation.*

SOLNHOFEN FORMATION

Although fossils are generally rare in the Solnhofen Formation (Upper Jurassic, lower Tithonian), the exquisite preservation of the Solnhofen fossils make this one of the most famous localities in paleontology. The fossils are enclosed in a matrix of even-bedded, very fine-grained limestone (Fig. 1), which has been quarried for construction purposes since Roman times and for lithography since the end of the 18th century. The formation is named for the town of Solnhofen, a village on the Altmühl River, Bavaria, and outcrops in the area between the Danube and Altmühl Rivers, extending from Donauwörth in the west to Regensburg in the east (Fig. 2). The quarry districts include, from W to E, Daiting (abandoned), Solnhofen-Langenaltheim (important), Eichstätt (important), Gungolding-Pfalzpaint (minor), Zandt (abandoned), and Kelheim-Painten (minor).

The Solnhofen Formation comprises four members, from bottom to top: Rögling Member, Lower Solnhofen Lithographic Limestone Member, Upper Solnhofen Lithographic Limestone Member, and Mörnsheim Member (Fesefeldt, 1962; Barthel, 1970). Quarries operate

FIGURE 1. Typical quarry, Eichstätt district.

primarily in the Upper Solnhofen Lithographic Limestone Member.

The Solnhofen Formation is, when unweathered, light gray to dark bluish gray micrite (particle size under 5 μm). Bed thicknesses range from <1 mm to 300 mm, and individual beds may be traced considerable distances (up to 8 km). Bedding planes are perfectly smooth except when disrupted by subaqueous slumping, load casts, and other penecontemporaneous structures. Layers of 95–98% pure $CaCO_3$ ("Flinze") alternate with less resistant impure layers ("Fäulen") consisting of 85% $CaCO_3$ plus clay minerals and wind-blown quartzose silt and in some cases up to 5% bituminous matter (Barthel, 1978). Graded

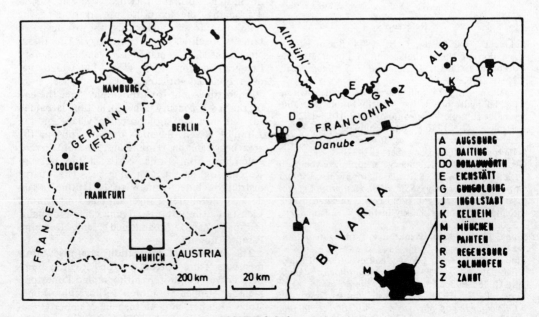

A	AUGSBURG
D	DAITING
DO	DONAUWÖRTH
E	EICHSTÄTT
G	GUNGOLDING
J	INGOLSTADT
K	KELHEIM
M	MÜNCHEN
P	PAINTEN
R	REGENSBURG
S	SOLNHOFEN
Z	ZANDT

FIGURE 2. Index map.

bedding indicative of turbiditic deposition is present in lagoonal basins near reefs (see below).

The fine carbonate matrix has preserved imprints of the finest details, though preservation varies with the original composition of the organism. Chitinous and apatitic structures are well preserved throughout, while proteinaceous material, such as horn and feathers, has largely decomposed. However, in some instances even muscular tissue has been phosphatized (as Frankolite), and muscular striation has been observed in fishes and squids. Carbonaceous material has, in most cases, been destroyed. In some cases, body and shell cavities are filled by calcite; pyrite is almost absent.

The state of preburial decay of the individual fossils varies with the distance the carcass has been transported and the force of episodic current. Compaction of the sediment has flattened the fossils, most of which are found attached to the lower face of a bed.

Flora and Fauna

Almost 700 species of plants and animals have been cited from the Solnhofen Formation. Modern taxonomic studies suggest that about 400 of these would be considered valid today. Most of the fauna is nektonic marine in origin, with some allochthonous benthic forms also present. Terrestrial animals, dominantly insects, are much scarcer. The few tetrapods, pterosaurs, and birds occur in areas closer to shore (Eichstätt, Kelheim quarries).

Some of the more common or important fossil genera of the Solnhofen Formation are listed below. (u = unique; er = extremely rare; r = rare; mf = moderately frequent; f = frequent; a = abundant; l = locally.)

Plants

Coccolithophores: *Watznaueria* (Fig. 3: a, in certain layers), *Cyclagelosphaera* (a, in certain layers)
Rhodophyceae(?): *Phyllothallus* (mf)
Pteridophyta (absent)
Gymnosperms
  Pteridospermae: *Cycadopteris* (lmf)
  Bennettitales: *Zamites* (r), *Sphenozamites* (er)
  Ginkgoaceae: *Ginkgoites* (r)
  Conifers: *Brachyphyllum* (lf), *Paleocyparis* (lf)
  Angiosperms (finds reported but still unconfirmed by specialists)

Invertebrates

PHYLUM PROTOZOA
  Foraminifera (f, in certain seams)
PHYLUM PORIFERA: *Ammonella* (r), *Tremadictyon* (r)
PHYLUM COELENTERATA: *Rhizostomites* (Fig. 4; lf), other genera rare, corals (f in reef tract, absent in lagoon)

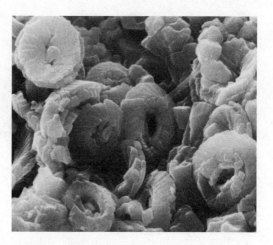

FIGURE 3. Solnhofen cocoliths, *Watznaueria*, X3500.

PHYLUM MOLLUSCA
  Classes Gastropoda and Bivalvia (mf; all allochthonous: only oysters are attached to ammonite shells)
  Class Cephalopoda
    Nautiloidea: *Pseudaganides*
    Ammonoidea: *Glochiceras* (lf), *Neochetoceras* (lf), *Taramelliceras* (lf), *Lithacoceras* (lf), *Subplanites* (lf)
    Belmnoidea: *Hibolithes* (r)
    Teuthoidea: *Plesioteuthis* (mf), *Acanthoteuthis* (r), *Celaeno* (r), *Leptoteuthis* (r), *Trachyteuthis* (r)
PHYLUM BRACHIOPODA: *Septaliphoria* (r), *Loboidothyris* (r)
PHYLUM ANNELIDA: *Eunicites* (r), *Ctenoscolex* (Fig. 5; r)

FIGURE 4. *Rhizostomites admirandus* Haeckel, maximum diameter 320 mm. Specimen in Bayer Staatslg. Paläont., Munich.

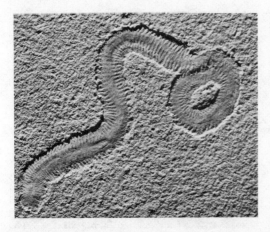

FIGURE 5. *Ctenoscolex procerus* Ehlers, maximum distance 71 mm. Specimen in Bayer Staatslg. Paläont., Munich.

PHYLUM ARTHROPODA, Superclass Chelicerata
  Class Merostomata: *Limulus (Mesolimulus)* (Fig. 6; mf); ichnogenus *Kouphichnium* named prior to *Mesolimulus*
  Class Arachnida: *Sternarthron* (er)
Superclass Crustacea
  Class Ostracoda: unnamed genus, possibly of freshwater origin (r)
  Class Cirripedia: *Archaeolepas* (er)
  Class Malacostraca, Subclass Eumalocastraca
    Isopoda: *Urda* (er), *Naranda* (er)
    Stomatopoda: *Sculda* (er)
    Decapoda: *Cycleryon* (r), *Eryma* (mf), *Glyphaea* (r), *Mecochirus* (f), *Aeger* (Fig. 7; f), *Antrimpos, Palpipes* (larva; f)

FIGURE 7. *Aeger tipularius* Schlotheim, rostral length 31 mm.

Superclass Hexapoda, Class Insecta (generally r, but many genera): *Pycnophlebia, Lithoblatta, Chresmoda, Procalosoma, Cerambycinus, Pseudosirex, Isophlebia, Libellulium, Stenophlebia, Hexagenites, Kalligramma* (Fig. 8; u), *Eocicada, Mesobelostomum*
PHYLUM ECHINODERMATA, Subphylum Crinozoa
  Class Crinoidea: *Millericrinus* (r), *Pterocoma* (lmf), *Saccocoma* (a), *Solanocrinus* (r)
Subphylum Asterozoa
  Class Ophiuroidea: *Geocoma* (lf), *Ophiocten* (lf),
  Class Asteroidea: *Lithasie* (r)
Subphylum Echinozoa
  Class Echinoidea: *Pedina* (r), *Tetragramma* (r), cidaroids near reef (r)

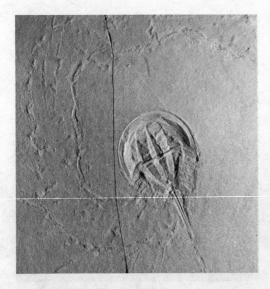

FIGURE 6. *Limulus (Mesolimulus) walchi* Desmarest and tracks; prosoma width 92 mm. Specimen in Solenh. Akt.-Ver. Maxberg.

FIGURE 8. *Kalligramma haeckeli* Walther, maximum wingspread 252 mm. Specimen in Bayer Staatslg. Paläont., Munich.

FIGURE 9. *Caturus bellicianus* Thiollière, length 354 mm. Specimen in Bayer Staatslg. Paläont., Munich.

Vertebrates

PHYLUM CHORDATA, Superclass Pisces

Class Chondrichthyes (all r): *Hybodus, Hexanchus, Paleoscyllium, Protospinax, Squatina, Aellopos, Ischyodus*

Class Actinopterygii

Holostei (generally r, but many genera): *Lepidotes, Gyrodus, Gyronchus, Mesturus, Proscinetes, Caturus* (Fig. 9; mf), *Furo, Ionoscopus, Urocles, Histionotus, Macrosemius, Propterus, Astheno- cormus, Hypsocormus, Orthocormus, Aspido- rhynchus, Belonostomus, Pholidophorus, Pleuro- pholis*

Teleostei: *Anaethalion* (mf-la), *Leptolepis* (mf-la), *Thrissops* (mf-la), *Pachythrissops* (r)

Class Crossopterygii: *Coccoderma* (r-er), *Libys* (r-er), *Holphagus* (r-er)

Class Amphibia: *Cyrtura* (Stegocephalian?; u)

Class Reptilia (all r-u)

Subclass Anapsida

Chelonia: *Platychelys, Plesiochelys, Eurysternum, Idiochelys*

Subclass Lepidosauria.

Squamata: *Bavarisaurus, Eichstättisaurus, Palaeo- lacerta, Proaigialosaurus, Ardeosaurus*

Rhynchocephalia: *Homoeosaurus, Kallimodon, Acrosaurus, Pleurosaurus*

Subclass Archosauria

Crocodilia: *Steneosaurus, Alligatorellus, Atopo- saurus, Dakosaurus, Geosaurus*

Theropoda: *Compsognathus*, the smallest known dinosaur (u)

Pterosauria: *Scaphognathus, Rhamphorhynchus, Anurognathus, Ctenochasma, Gnathosaurus, Pterodacylus* (Fig. 10)

Subclass Sauropterygia: *Stretosaurus*, single tooth, 230 mm! (u)

Subclass Icthyopterygia: *Macropterygius* (er)

Class Aves, Subclass Archaeornithes: *Archaeo- pteryx* (er)

Class Mammalia (absent)

Traces. Short tracks on bedding planes have been attributed to four preservable genera: the horseshoe crab *Limulus* (Fig. 6), the decapod crustacean *Mecochirus*, the gastropod *Rissoa*, and the bivalve *Solemya*. Episodic bottom currents have also produced "traces;" Seilacher (1963) has described the marks produced by ammonite shells rolling over the lagoon floor. The intricately coiled "*Lumbricaria*," pre- viously thought to be benthic worms, are now considered to be the coprolites of nektonic cephalopods.

Paleogeography

Like many other remarkable fossil deposits, the Solnhofen fauna owes its fine preservation

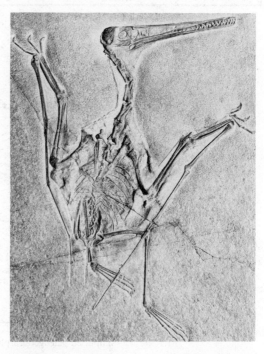

FIGURE 10. *Pterodactylus kochi* (Wagner), skull length 83 mm. Specimen in Bayer Staatslg. Paläont., Munich.

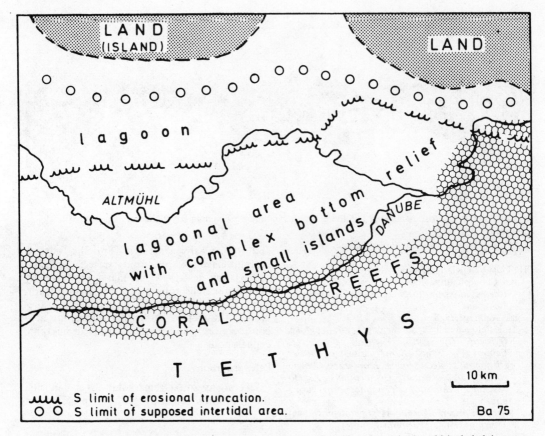

FIGURE 11. Outline of paleogeography; present positions of rivers Danube and Altmühl included (compare with Fig. 1).

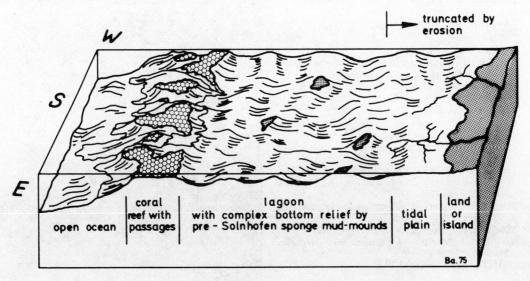

FIGURE 12. Simplified paleogeomorphologic diagram (after Barthel, 1970).

to local paleogeographic conditions. The most recent model for the Solnhofen paleoenvironment is a synthesis of current knowledge of regional paleogeography (Freyberg, 1968), sedimentology (Straaten, 1971), and biostratinomy (Barthel, 1964, 1970, 1972).

Figure 11 is a restoration of the paleogeography, showing from S to N: open ocean (the Tethys seaway), coral reefs, lagoon, and coast. A major point of debate has been whether the sea episodically withdrew from the area (Abel, 1927; Mayr, 1967), or was present throughout the time of deposition (Gümbel, 1891; Fesefeldt, 1962; Barthel, 1964, 1970, 1972; Keupp, 1977). The absence of land animal or insect tracks, and of algal crusts, birds-eye limestones, and tidal channels all point to a purely subtidal origin for the deposits. Also significant is the fact that autochthonous benthic organisms, including scavengers, are absent except in some marginal areas (Janicke, 1969; Barthel, 1964, 1970, 1972).

The following model (see Figs. 11 and 12) is discussed in greater detail by Barthel (loc. cit.):

1. Pre-existing relief of sponge-algal mud buildups.
2. Uplift to the N; cessation of sponge growth along axis of uplift.
3. Development of a coral reef barrier on top of dead sponge mounds at optimal depths. Sediment supply to lagoon is maintained via passages through reefs, primarily as suspended material driven into lagoon by episodic storms, forming the "Flinze." The "Fäulen" represent times of slow deposition between storms and contain some terriginous material (clays and wind-blown dust) and coccoliths.
4. Lagoon bottom relief of mounds, ridges, and basins ("Wannen" of Fesefeldt, 1962) prevents a continuous water exchange between basins and open sea.
5. Hot climate (as suggested by flora, fauna, and sediments; Straaten, 1971; Barthel, loc. cit.) initiates evaporation of lagoon waters, causing a dense brine to settle into the basins. Anoxic to temporarily euxinic conditions exclude benthic organisms and aid in preservation of fossils.
6. Because water exchange can still take place between lagoonal basins and open sea during storms, evaporite precipitation does not occur. Between storms, the dense, stagnant, high-salinity water stays in the basins, protecting the bottom sediments from disturbance by surface currents, giving rise to laminated bedding. Sponge mound relief suggests depths of up to 60 m, more than sufficient to permit this salinity stratification.
7. Seasonal runoff of rain water into the lagoon transports insects and other terrestrial organisms into the lagoonal basins.

Paleontology

The Solnhofen fauna is of great interest to the paleontologist because of the preservation of soft tissues and delicate structures. This adds to our very sketchy knowledge of the soft-bodied members of ancient faunas and of the anatomy of extinct organisms. The unusual preservation also gives insight into the evolutionary history of rarely preserved organisms. *Archaeopteryx*, the outstanding Solnhofen fossil (see *Aves*, Fig. 1), reveals its intermediate position between archosaurs and birds by preservation of its plumage. It was the first "missing link" between major taxa to be discovered, and is still under study, giving rise to new ideas on the origin of flight (e.g., Ostrom, 1975).

K. WERNER BARTHEL
DAVID JABLONSKI

References

For an extensive bibliography, see Kuhn (1961, 1973); a classical review is given in Walther, 1904.

Abel, O., 1927. *Lebensbilder aus der Tierwelt der Vorzeit*, 2nd ed. Jena: Fischer, 714p.

Barthel, K. W., 1964. Zur Enstehung der Solnhofener Plattenkalke (unteres Untertithon), *Mitt. Bayer. Staatssamml. Paläontol. hist. Geol.*, 4, 37–69.

Barthel, K. W., 1970. On the deposition of the Solnhofen lithographic limestone, *Neues Jahrb. Geol. Paläontol. Abh.*, 135, 1–18.

Barthel, K. W., 1972. The genesis of the Solnhofen lithographic limestone (Low. Tithonian): Further data and comments, *Neues Jahrb. Geol. Paläontol., Mh.*, 1972, 133–145.

Barthel, K. W., 1978. *Solnhofen Ein Blick in die Erdgeschichte*. Thun: Ott, 393p.

Fesefeldt, K., 1962. Schichtenfolge und Lagerung des oberen Weissjura zwischen Solnhofen und der Donau (Südliche Frankenalb), *Erlanger Geol. Abh.*, 46, 80p.

Freyberg, B. von, 1968. Übersicht über den Malm der Altmühl-Alb, *Erlanger Geol.*, 70, 40p.

Gümbel, C. W. von, 1891. *Geognostische Beschreibung des Konigreiches Bayern. IV. Abth. Geonostische Beschreibung der Fränkischen Alb.* Kassel: T. Fischer, 763p.

Hemleben, C., 1977. Autochthone und allochthone Sedimentanteile in den Solnhofener Plattenkalken, *Neues Jahrb. Geol. Paläont., Mh.*, 1977, 257–271.

Janicke, V., 1969. Untersuchungen über den Biotop der Solnhofener Plattenkalke, *Mitt. Bayer. Staatssamml. Paläontol. Hist. Geol.*, 9, 117–181.

Keupp, H., 1977. Die Solnhofener Plattenkalke—ein Blaugrünalgen–Laminit, *Paläont. Zeitschr.*, 51, 102–106 (Engl. summary, *Proc. 3rd Internat. Coral Reef Symp.*, 2, 61–64).

Kuhn, O., 1961. Die Tier- und Pflanzenwelt des Solnhofener Schiefers, *Geologica Bavarica*, 48, 68p.

Kuhn, O., 1973. Die Tierwelt des Solnhofener Schiefers, 4th ed. Zimsen-Wittenberg: N. Brehm-Bücherei, pt. 318, 40p.

Mayr, F. X., 1967. Paläobiologie und Stratinomie der Plattenkalke der Altmühl-Alb., *Erlanger Geol. Abh.*, 67, 40p.

Ostrom, J. H., 1975. The origin of birds, *Ann. Rev. Earth Planetary Sci.,* **3**, 55-77.

Seilacher, A., 1963. Umlagerung und Rolltransport von Cephalopoden-Gehäusen, *Neues Jahrb. Geol. Paläontol., Mh.,* **1963**, 593-615.

Straaten, L. M. J. U. van, 1971. Origin of Solnhofen Limestone, *Geol. Mijn.,* **50**, 3-8.

Walther, J., 1904. Die Fauna der Solnhofener Plattenkalke, *Jenaische Denkschr.,* **11** (Haeckel anniv. vol.), 135-214.

Cross-reference: *Fossil Record.*

SPECIES CONCEPT

A species is a group of animals or plants all of which are similar enough in form to be considered as minor variations of the same organism. Members of the group normally interbreed and reproduce their own kind over considerable periods of time.

A species is the basic unit in modern biological classification. Each species has a double name in the usually accepted system of biological nomenclature, e.g., *Homo sapiens,* man. The first word designates the next higher group above the species, the genus, to which the species belongs; the second word is peculiar to the species. Few biologists believe that the categories above species are objective, but rather are set up at convenient levels of classification. If the classification is natural, each of the supraspecific categories should include a distinct part of the phylogenetic tree and the higher the category the further it should extend back into evolutionary history.

History

Preevolutionary Views. The word *species* is now usually understood to have the biological significance of its definition above, although in the broader sense a species is any group whose members have some characteristics in common. All groups of biological classification could be called a species in this sense and the adoption of the term in the modern sense was a slow process. John Ray, a 17th-century naturalist, did much to establish its modern use. Ray's successors, e.g., Linnaeus, who was responsible for the modern biological classificatory system, equated species with distinct kinds of organisms as created by God. Both subspecies and genera (and larger groups) were considered human abstractions.

Postevolutionary Views. Belief in evolution as opposed to special creation changed this concept of the species. The acceptance of evolution meant that biologists considered that one species could give rise to another. That this was a crucial point in the arguments for evolution is implicit in the title of Charles Darwin's work, *The Origin of Species.* As soon as a species is not considered as a fixed unit, one must decide the degree of discontinuity in form between two subgroups that will allow the establishment of separate species. This may be determined by the point of view and field in which the investigator works.

The Species Problem

Nonequivalence of Species. Species groups into which animals are organized are not necessarily equivalent throughout the animal kingdom. Reproductive processes and behavior vary greatly between major groups, and species organization will differ among animals that reproduce asexually or sexually, scatter their gametes or copulate, or exhibit social behavior. Species organization is most obviously different between the Protozoa and the Metazoa and, accordingly, definition of a species is usually restricted to bisexual metazoans.

Varied Approaches Toward Defining Species

A major difficulty in defining the species is the outlook of the investigator. The systematist may think of a species as a category in his classification, the biologist or naturalist as a natural group of animals observed in the field, and the paleontologist as a step in the changes that occur in his evolutionary series.

The Species of Systematists. The systematist often works on preserved material and, although he always tries to make the classification natural, he has not necessarily agreed with the biologist in making the species a natural group. The systematist, of course, often must classify large quantities of material with limited biological data. It is always easier to collect material than to study its ecology and behavior. In some ways, the approach of the systematist may be likened to that of the paleontologist.

The ideal of a systematist is that the species should be easily recognized and accepted as such by others. The description is mainly morphological, although it may involve the determination of protein variability between animals (electrophoresis; see *Genetics*), and the distinction between related species must be large enough to be accepted as of specific rank. The amount of difference varies considerably in the different animal phyla. Morphological differentiation should be made on an extensive range of specimens although single specimens may be taken as types representing the typical form of the species. The stability of species and the absence of intergrading between them is important to the systematist. Evolutionary change is relatively slow in relation to the experience of an individual, so that the species remains stable.

Finally, the systematist often finds evidence of specific difference from a study of geographical distribution. This evidence should be such as to indicate common descent from a common ancestral group. A species must be a single group with a single origin.

The Species of the Biologist. A definition that would be acceptable to most naturalists is that an animal species is a group of animals organized into interbreeding populations or demes, which are often differentiated into groups by morphological distinction within the species but which will all live together and naturally interbreed. This definition meets with difficulties when the geographical range is large and the species contains considerably differentiated groups. It is usually impossible to apply the test, "Do the two forms interbreed freely?" and the decision is often a matter of individual judgement, although it implies that the two forms will remain distinct if the opportunity for interbreeding should occur. The herring gull (*Larus argentatus*) and the lesser black-backed gull (*L. fuscus*), which both occur in England and behave there as distinct species, are illuminating exceptions (Mayr, 1970): It has been shown that there is a circumpolar series of 10 recognizable forms or subspecies differentiated by leg and mantle color, forming a "ring species." The two forms in Britain are terminal links of the chain, and because their breeding behavior places them in different species, the boundary between the two species must be placed rather arbitrarily in the chain (Fig. 1). Where subspecies have been established in isolated groups within species, it is frequently difficult to decide on their status without the test of interbreeding by the remixing of the isolated populations, as has occurred in the kingfisher, *Tanysiptera hyrocharis* of New Guinea.

The systematist's species often coincide with the naturalist's because the former's first criterion is morphological distinctness, which is generally maintained by noninterbreeding, or the difference would disappear in a hybrid population. Thus, morphological distinctness is usually a good criterion for the establishment of separate species as long as it is realized that two morphologically similar

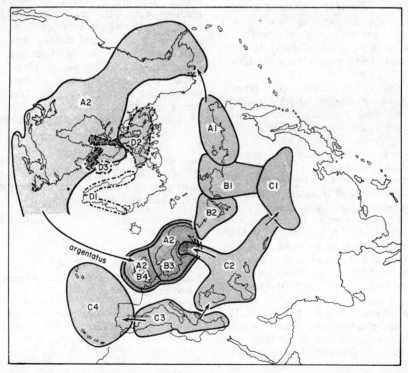

FIGURE 1. "Ring species" in gulls of the *Larus argentatus* group (from Mayr, 1970; copyright © 1963, 1970 by the President and Fellows of Harvard College. Reprinted by permission of the publishers). The subspecies of A, B, and C evolved in Pleistocene refuges. When A expanded, post-Pleistocene, probably from a north Pacific refuge, it spread across all of North America and into western Europe (*L. argentatus*). Here it became sympatric with *L. fuscus* (B3, B4), the westernmost of a chain of Eurasian populations.

animals may be separate biological species, e.g., *Drosophila pseudo-obscura* and the similar *D. persimilis*. Biological species may include forms that are not good systematic species.

The Species of the Paleontologist. The outlook of the paleontologist differs from that of the modern systematist because he is presented with a series of forms in stratigraphic sequence (Imbrie, 1957; Einor, 1972). Thus, the paleontologist can study the course of evolutionary divergence much more directly than can the biologist. Unfortunately he cannot test for interbreeding between groups and may be limited to only a few specimens or even fragments. Morphological distinctness accordingly plays a large part in the paleontologist's assessment of species. Where relatively large collections are available, the paleontologist usually resorts to a statistical analysis of the proportions of the fossils (see *Biometrics in Paleontology; Computer Applications in Paleontology*).

Because biological criteria are limited almost exclusively to hard parts, the paleontologist must define a species morphologically and tends to use the category of the species in the same way higher systematic divisions are used. The species difference is that grade of morphological change in an evolving series that is acceptable to the majority of competent paleontologists as of specific status. It is not surprising that this may vary in different groups.

This comparison of morphological features involves the assumption that nothing interferes with random cross breeding in the species. When populations are isolated from each other in any way, differences in small characters will arise. This has been demonstrated, for example, in bivalves of the genus *Carbonicola* when specimens from two localities have been compared.

Genetic Differences of Species. A genotype may change either by rearrangement of the material of a chromosome or by changes in the gene (see *Genetics*). A comparison of the structure of the genotype between *Drosophila pseudo-obscura* and *D. miranda* shows that the rearrangements of the chromosomes are numerous and complex. Other pairs of species of *Drosophila*, e.g., *D. melanogaster* and *D. simulans*, show much less rearrangement. It is clear that chromosomal mutations have been numerous in the evolution of at least some of the species of *Drosophila*, but it is not certain that they are as important in other organisms. In *Drosophila*, the rearrangements that differentiate races or subspecies are of the same kind as those that differentiate species but smaller and less numerous. Genetic differentiation is of the same kind at all stages of specific divergence. One exception to this is the development of hybrid sterility, which may be independent of other characters and isolates a section of a previously interbreeding population, so allowing divergence to take place.

Gene mutation also plays a large part in specific differentiation but the genetic control of differences is generally polygenic.

Origin of Species

The division of a deme into two parts allowing either to evolve into a distinct species is possible only if there is some barrier to free interbreeding. Thus, the exchange of genetic material throughout the deme is prevented or greatly reduced, and divergence may take place in either section of the population.

Isolation barriers may be maintained (1) geographically or spatially, (2) ecologically, or (3) genetically. Geographical barriers physically maintain sections of the population away from each other, e.g., Darwin's Finches on the Galapagos Islands where migration of the finches between the islands is a relatively rare event. By contrast, in the oceans or over great areas of land, gradual morphological changes may occur, forming a cline, or morphological gradient—e.g., with respect to size in Indian birds, the largest occur in the Himalayas, the smallest in Ceylon and Malaya. Any geographical deterrent to free interbreeding produces some degree of isolation. Ecological barriers involve isolation by habitat, season, or mating preferences. Genetic barriers may involve either the inviability of hybrid or their lack of ability to produce functional sex cells.

Any two species are likely to be separated by more than one of these processes (for reviews, see Mayr, 1963, 1970; Bush, 1975; White, 1978). The geographic and ecological barrier may often occur first allowing genetic or sexual compatibility to develop.

E. R. TRUEMAN

References

Bush, G. L., 1975. Modes of animal speciation, *Ann. Rev. Ecol. Syst.,* **6,** 339–364.

Dobzhansky, T., 1970. *Genetics of the Evolutionary Process.* New York: Columbia Univ. Press, 505p.

Einor, O. L., 1972. The problem of species in paleontology, *Proc. 23rd Internat. Geol. Congr., Internat. Paleont. Union,* 53–68.

Ghiselin, M. T., 1974. A radical solution to the species problem, *Syst. Zool.,* **23,** 536–544.

Imbrie, J., 1957. The species problem with fossil animals, in E. Mayr, ed., *The Species Problem.* Washington D.C.: A.A.A.S., 125–153.

Levin, D. A., 1979. The nature of plant species. *Science,* **204,** 381–384.

Mayr, E., 1963. *Animal Species and Evolution.* Cambridge: Harvard Univ. Press, 797p.

Mayr, E., 1969. The biological meaning of species, *Biol. J. Linnean Soc.,* **1,** 311–320.

Mayr, E., 1970. *Population, Species and Evolution.* Cambridge: Harvard Univ. Press, 453p.

Scudder, G. G. E., 1974. Species concepts and speciation, *Canadian J. Zool.,* **52,** 1121–1134.

Sokal, R. R., 1973. The species problem reconsidered, *Syst. Zool.,* **22,** 360–374.

Sylvester-Bradley, P. C., ed., 1956. The species concept in paleontology, *Syst. Assoc. Pub.,* **2,** 145p.

Weller, J. M., 1961. The species problem, *J. Paleontology,* **35,** 1181–1192.

White, M. J. D., 1978. *Modes of Speciation.* San Francisco: Freeman, 456p.

Wiley, E. O., 1978. The evolutionary species concept reconsidered, *Systematic Zool.,* **27,** 17–26.

Cross-references: *Biometrics in Paleontology; Evolution; Systematic Philosophies; Taxonomy, Numerical.*

STENOTHECOIDA

Stenothecoid fossils constitute a small group of peculiar Early Cambrian shells. Except at one or two localities, specimens are rare, but they are known from North America, Greenland, southern Europe, and Siberia and may be worldwide in occurance. Specimens are composed of calcium carbonate and show prominent growth lines; and, when preserved in the same rocks as more conventional mollusks, stenothecoids are similar in preservation;

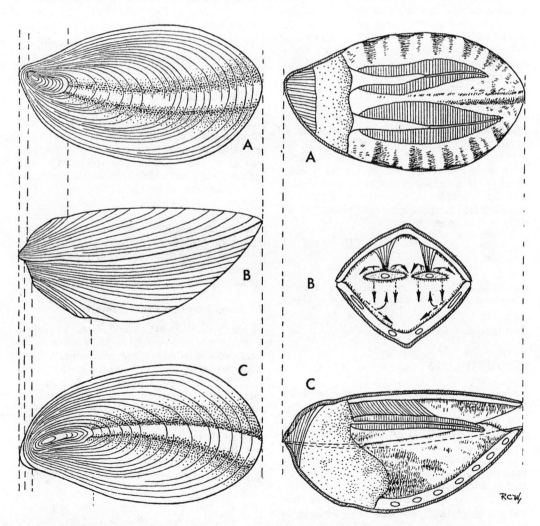

FIGURE 1. *Stenothecoides knighti* Yochelson (from Yochelson, 1969). Left, reconstruction in dorsal, side, and ventral views X9; right, interpretation of soft parts, dorsal, with valve removed, cross-section and transverse section.

however, their morphology is unlike that of other mollusks.

Specimens are up to 2 cm in length, but narrow; the median line is curved, each side sloping downward from it. Some appear to curve toward the right and others toward the left. A few internal molds show a major groove below the external mid-line with shorter and shallower grooves extending laterally from it, more or less paired. Because of this apparent pairing, the grooves were considered as allied to muscle scars and the stenothecoids were placed with the Monoplacophora, bilaterally symmetrical cap-shaped univalves.

Several specimens that are bivalved are now known. There is no hinge line. A single tooth in one valve and a socket in the other may be present. The valves are unequal in size. The specimens thus are asymmetrical across the valves and between the valves. This morphology combined with the absence of a hinge line indicates that these cannot be Bivalvia (Pelecypoda, or clams in a loose sense).

Yochelson (1969) assigned the fossils to a new class called Stenothecoida. In an interesting example of multiple discovery in science, N. A. Akserina independently named a class Probivalvia for essentially the same material. Stenothecoids are interpreted as living with the deeper valve downward and opening the upper valve by transferring fluid within the mantle (Fig. 1). If correctly interpreted, they constitute an example of a relatively short-lived unsuccessful morphologic experiment of an early molluscan radiation.

ELLIS L. YOCHELSON*

Reference

Yochelson, E. L. 1969. Stenothecoida, a proposed new class of Cambrian Mollusca, *Lethaia*, **2**, 49–62.

Cross-reference: *Mollusca.*

STROMATOLITES

Algal stromatolites are among the most peculiar fossils of the geologic record. Firstly, they do not represent the skeletal remains of a particular organism, as do shells or corals, but are calcareous, dolomitic, or siliceous biosedimentary structures resulting from the life processes and the rock-building activity of primitive algal and/or bacterial communities. Thus they are really trace fossils rather than fossil organisms. They are also remarkable

because they record the first significant and preserved manifestations of life on earth, and yet are still forming today, spanning more than 3 billion years of geological history.

They are widely studied by scientists in such fields as paleontology, stratigraphy, paleogeography, sedimentology, biology, and even astronomy.

The term "stromatolith" (in German, derived from the Greek "stroma": blanket, mat and "lithos": rock) was first applied by Ernst Kalkowsky in 1908 to stratiform, columnar, branched, etc. calcareous masses showing a fine and prominent lamination in the Triassic of Northern Germany.[1] He further proposed the word "stromatoid" for sets of planar laminae that build up the stromatolite, but this term was not retained in the geological literature.

Morphology

The morphology of stromatolites is extremely variable: they may form flat stratiform deposits or domal, bulbous, or columnar bodies (Fig. 1). The columns may be strong or slender, simple or variously branched. The size of stromatolites ranges from a few tens of microns to tens of meters. The basic common feature of all these structures is their internal lamination, which is generally prominent on weathered surfaces as well as in thin section. The thickness of the laminae, each of which records the rhythmic growth of the structure, vary from a few microns to a few millimeters, and sometimes more. Their configuration (flat, concave, convex, smooth, crinkled, etc.) and the way they pile up on top of each other as the stromatolite grows, as well as their microstructure as shown in thin section, are fundamental characters in the description of stromatolitic structures.

Historical Development of Concepts

Reference to what would later be called stromatolites dates back to 1825 when J. H. Steel reported curious limestone objects from the Upper Cambrian of Saratoga Springs, New York. The Petrified Gardens in Richtie Park, 3 miles W of Saratoga Springs, nowadays constitutes a major tourist attraction where masses of stromatolites outcrop beautifully on a glaciated ledge (Fig. 2). It was not until 1883, however, that these structures were reconsidered, illustrated, and described by James Hall who called them *Cryptozoon proliferum* (the suffix "zoon"

---

* Publication authorized by the Director, U.S. Geological Survey.

[1] The term "stromatolith," independently of Kalkowsky's definition, was applied in 1916 to complex, sill-like igneous intrusions that are interfingered with sedimentary strata (Amer. Geol. Inst., 1973, *Glossary of Geology*).

FIGURE 1. Cross section of a columnar stromatolite, showing prominent lamination (Ordovician, Oklahoma, USA). Height of specimen about 25 cm.

suggests that Hall considered them as animal organisms, although he clearly wrote they were "sea plants"). From then on, stromatolites became true paleontological objects while their affinities remained much debated: they were indeed said to be fossil sponges, protozoans, stromatoporoids, red algae, etc. It was finally C. D. Walcott who, in 1914 proved that stromatolites resulted from the activity of blue-green algal communities. He reached this conclusion after comparing the structure of Precambrian stromatolites with that of modern algal concretions growing in freshwater lakes and streams. Like stromatolites, these algal tufa form various calcareous masses characterized by a conspicuous concentric lamination. Inasmuch as they result from *in situ* precipitation of calcium carbonate on mats of blue-green algae, a similar

origin was proposed for stromatolites. Walcott's theory gained acceptance and stromatolites were considered as ordinary fossils, receiving a binomial Linnean nomenclature. Scores of genera and species were soon named and described from the various geological horizons.

One main problem remained however: whereas the bulk of the modern analogues were growing in freshwater settings, many fossil stromatolites were typically associated with marine faunas. The lack of modern marine stromatolites presented an embarrassing situation.

In 1933, an expedition to the Bahamas led Maurice Black to Andros Island, where he discovered extensive coastal areas colonized by a variety of algal laminated structures, many of which resembled fossil stromatolites. Black's

FIGURE 2. Stromatolites outcropping on a glaciated ledge (Petrified Garden, New York), the *Cryptozoon proliferum* of James Hall.

discovery was fundamental and had a tremendous impact on later developments—first because he had found the first extensive laminated algal structures associated with marine sediments; and second, because these structures were composed of sedimentary particles trapped and bound by the sticky blue-green algal filaments, rather than by *in situ* precipitated carbonate as found in freshwater tufas. Finally, Black found that the Bahamian structures were not built by one single alga but by a complex community of unicellular (coccoid) and filamentous algae.

From then on, views concerning the genesis and habitat of fossil stromatolites changed radically and were based on Black's observations. Stromatolites were said to be organosedimentary structures resulting from the sediment-trapping and binding activity of blue-green algal mats; the intertidal flats were said to be their typical environment of formation. After World War II, the pioneering work of R. N. Ginsburg (1955) and B. W. Logan (1961), on the intertidal flats in Florida and Western Australia respectively, reenforced these two assumptions. This current of ideas was tempered by Monty in 1967 (see references in Hofmann, 1973), who showed that the bulk of the stromatolitic structures growing in the freshwater environments of Andros Islands, Bahamas, resulted from *in situ* precipitation, whereas the trapping and binding mechanisms of mineralization became increasingly dominant in mats bathed by brackish and marine waters. It was further claimed that the intertidal flats were but one of the various possible environments of stromatolitic growth.

Modern Concepts

Nowadays, stromatolites (see review in Monty, 1977) can be defined as millimeter- to decameter-sized biosedimentary structures showing, in vertical section, a prominent concentric lamination reflecting their rhythmic growth. The basic unit of a living stromatolite is the algal film or mat composed of monospecific or polyspecific communities of blue-green algae and/or bacteria. Environmental parameters as well as the biological activity of the mat algae will determine the type of mineralization: trapping and binding of ready-made detrital particles, *in situ* precipitation of calcite, of silica, etc. The algal mat may undergo periodical differentiations that will give rise to a laminated fabric. This may involve, for example, the seasonal calcification of the constitutive algal filaments yielding a succession of calcified and uncalcified layers, or the seasonal alternation of two communities with different mineralogic properties. In other cases, the lamination will result from periodic input of detrital particles onto the mat; these grains are soon trapped by the algae, which permeate and glide through the deposit to form a new algal mat on top of it. This will yield a succession of organic-rich and sediment-rich layers in the stromatolite.

Analysis of modern stromatolites has revealed that the time-significance of the lamination could be daily, synodic, or yearly.

The organic layers are very rarely preserved in the fossil record and the lamination will principally result from the superposition of the alternation of distinct mineral layers. For a stromatolite to be preserved and fossilized, it has to be cemented, either as it grows or soon after; if it is not cemented, it will rot or collapse and rapidly disappear.

It is now evident from both modern and fossil occurrences that stromatolites are not restricted to the intertidal zone, but may have formed in a wide range of environments, from lacustrine fresh water to rather deep marine settings. It has also been shown in the last few years that bacteria could play a significant role in the building of certain stromatolites, such as the silicous ones forming around hot springs or the ferromanganese ones growing on the deep-sea floor.

Classification of Stromatolites

The fact that modern stromatolitic structures have been shown to be built by complex and versatile communities of algae rather than by one organism, along with the increasing amount of data revealing the great variability and intricate morphological intergradations shown by fossil stromatolites, forced some paleontologists to drop the rigid Linnean bionomial nomenclature. A new system of classification, based on the general spatial characteristics of the lamination, was proposed in the 1960s to distinguish and name stromatolites by alphabetical formulae. However, this very simplified scheme, if handy for gross characterization of stromatolites in the field, could not account for their enormous morphologic and structural diversity.

While western geologists considered stromatolites as sedimentary structures shaped and controlled by environmental parameters (water agitation, dessication, etc.), Russian geologists considered detailed observations on overall morphology and microstructure, studied by serial sectioning and graphic reconstruction techniques, to be of taxonomic value. They developed a binomial nomenclature, not based on genera and species, but on the concepts of "group" and "form." These were not taxonomic categories referring to unique biological entities, but plain nomenclatural categories based on such criteria as growth morphology, characteristics of the branching, shape of the laminae, microstructure, etc. (Fig. 3). Such a classification is in fact similar to the one used for other trace fossils (or ichnofossils), namely burrows, tracks, and trails. Such classifications are useful in reference and information retrieval. However, they become dangerous or equivocal when they are not internally consistent, which unfortunately has too often been the case. For instance, the criteria for defining the groups, or forms within given groups, vary widely not only between different authors but also with one author according to the group or form considered. Other problems arise as to the relative weighting of the selected criteria. For example, though the microstructure of algal mats is in many cases specific to a given group, it may be severely altered by sedimentological and diagenetic events—and in fact many of the stromatolitic microstructures described are actually diagenetic fabrics. These problems are particularly acute with Precambrian stromatolites, which, in the absence of any other fossils, are the only available means of dating and correlating sedimentary rocks. Correct dating and correlation requires correct identification and utilization of the same terminology in the definition of groups and forms.

Paleontological History

The oldest known stromatolites are about 3 b yr old. The group as a whole reached its maximum development and greatest diversity during Precambrian times, which could be called the "Age of Blue-green Algae" or the "Age of Stromatolites." During that period, stromatolites flourished in the primitive and uncrowded seas, forming important calcareous and dolomitic deposits all over the world. Silicified stromatolites from mid-Precambrian times (around 2 b yr ago) are found all over the world in association with unique deposits of Banded Iron Formations. These depostis, rich in ferrous ions, did not occur in later times and their formation may record the still-low oxygen content of the atmosphere or an interaction between blue-green algae and their surroundings.

This early development of intensive algal activity has had an important impact on the evolution of life conditions on earth. We know that the primary atmosphere was reducing and that blue-green algae evolved in these conditions; we also know that their resistance to lethal cosmic radiations allowed them to thrive at a time when no ozone shield protected the earth. It is very tempting to associate the first stromatolite-building communities with the

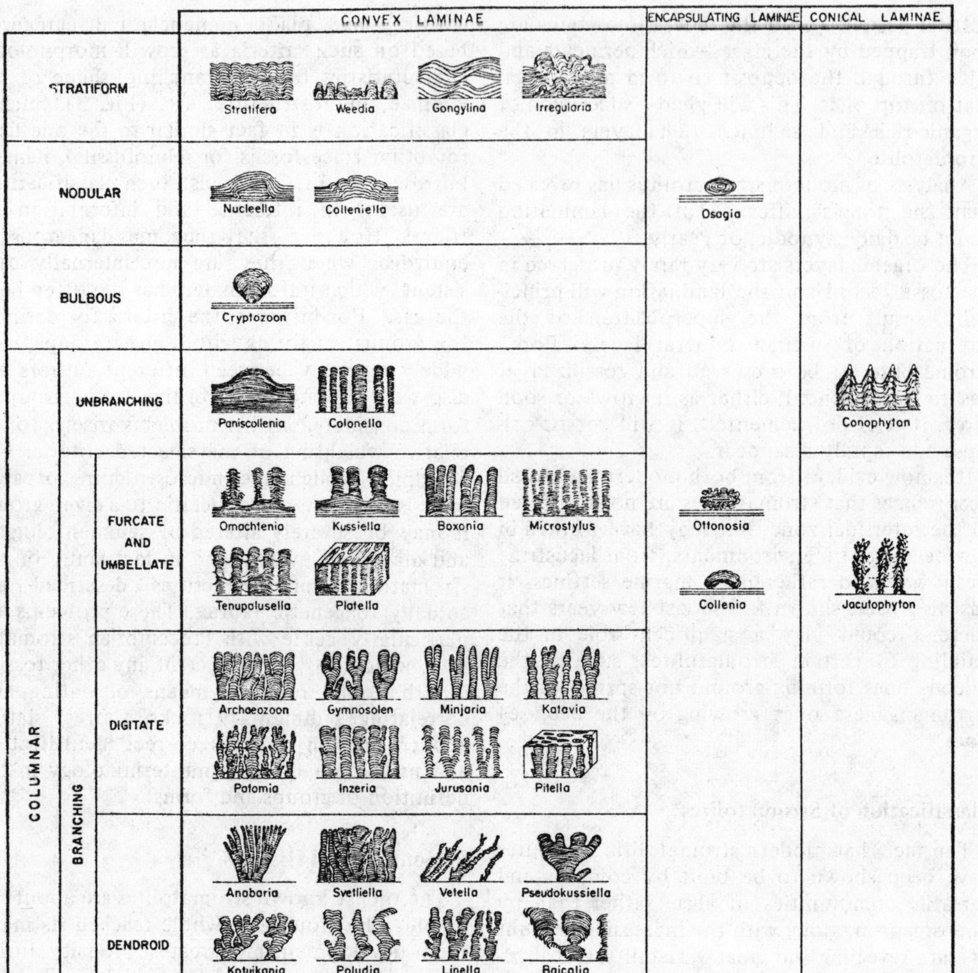

FIGURE 3. Morphology and general features of some stromatolite genera (from Hofmann, 1969).

beginning of photosynthesis on Earth. This is uncertain, however, as blue-green algae can have a heterotrophic metabolism and the earliest ones were undoubtedly heterotrophic. The photosynthetic function appears however to have been acquired rather early and blue-green algae started to evolve oxygen. This had two important consequences: first, oxygen progressively accumulated in the atmosphere, which became oxydant; second, part of this oxygen was ionized and contributed to the formation of an ozone layer, which shielded the earth from the lethal cosmic radiations. These new ecological conditions, imposed by 2 b yr of blue-green algal and bacterial activity, conditioned and oriented the differentiation of the first metazoans in the late Precambrian, and thus, animal life as we know it today.

Some 700 m yr ago, just before the dawn of the Paleozoic era, stromatolites suffered a decline marked by, among other things, a net decrease in diversity of columnar forms. This decline was strongly accentuated through early Paleozoic time and persists until today. Rock-building stromatolitic communities apparently were restricted due to the rise of the more complex eucaryotic algae and the metazoans in late Precambrian–early Paleozoic times. With the rise of invertebrates came more specialized and more sophisticated lime-depositing organisms, such as the stromatoporoids, the tabulate and rugose corals, the red algae, etc., which progressively displaced stromatolites. Apparently, stromatolites became more and more concentrated in the intertidal zone where the high resistance of blue-green algal mats to dessication, solar radiation, high salinities, etc. allowed them to live without competition.

The beginning of the Cenozoic era, some 65 m yr ago, marks the expansion of the calcareous red algae; they progressively invade the benthic marine environment from the intertidal zone down to the continental shelf and oceanic banks. These highly adapted and specialized encrusting algae, which have a true metabolized skeleton, completed the displacement of marine stromatolitic communities from the sea. Calcareous blue-green algae were then essentially confined to nonmarine environments where they continued to build important calcareous masses and reefs throughout the Tertiary.

Modern Stromatolitic Communities

Today, blue-green algae are still the most important lime-depositing algae in nonmarine environments; they are for example the main contributors to the formation of the extensive white calcareous muds that cover much of the Everglades in Florida. Similarly, in the Bahamas they form widespread stromatolitic deposits in seasonal marshes (Fig. 4) where periodic algal growth during the rainy season is associated with an *in situ* precipitation of fine crystals of calcite in the mesh of filaments. Stony encrustations, nodules, dams, banks, or "reefs" are also built by blue-green algae in alkaline streams and lakes all over the world. Finally, calcareous and siliceous laminated deposits are common in thermal springs in association with blue-green algae and/or bacteria. The siliceous microstromatolites growing in Yellowstone Park are interesting because they can give us important clues concerning the mode of formation of their Precambrian analogues.

One significant exception to the landward displacement of blue-green algae has been found in Shark Bay, Western Australia (Fig. 5). Here, in hypersaline waters that preclude the settlement of any competitors, various communities of blue-green algae are still building

FIGURE 4. Modern calcareous laminated domes built by nonmarine blue-green algae in the fresh water marshes of Andros Island, Bahamas.

rather voluminous calcareous masses in subtidal and intertidal environments. Besides the absence of competitors, the success of Shark Bay stromatolites results from a rapid calcification and cementation of the structures, a process not yet fully understood. With this exception, the majority of Holocene marine algal mats are confined to the intertidal and supratidal flats, where they contribute to the formation of laminated algal sediments. These deposits result from the periodical deposition of detrital particles on the algal mats by sediment-laden tidal or storm waters. These mats are not calcified and, when buried, are progressively reduced to dark organic laminae separating the sedimentary layers.

Uncalcified laminated blue-green algal colonies are also common in shallow lagoonal waters off Florida, the Bahamas, Bermuda, etc. These gelatinous biscuits evidently cannot be preserved and disappear at the end of their seasonal growth period.

Recent investigations of deep-sea manganese nodules have revealed their typical stromatolitic structure. They appear as stratiform, nodular, or small branching columnar forms showing a fine and regular lamination in section. Although many geochemists claim that these nodules are formed inorganically, microbiological studies reveal more and more that their growth is considerably enhanced by the metabolic activity of

FIGURE 5. Modern columnar stromatolites in the low intertidal zone near Carbla Point. Hamelin Pool, Shark Bay, Australia. (Photo kindly supplied by H. J. Hofmann.)

manganese-oxydizing communities of bacteria and fungi that coat the nodules (Monty, 1973).

Stromatolites and Stratigraphy

One of the most difficult problems in Precambrian stratigraphy has always been the relative dating and correlation of sedimentary rocks, because they lack true index fossils. Russian geologists have tried to get the best of what was available, i.e., stromatolites. They have devoted enormous efforts (1) to developing a methodology for studying, naming, and identifying Precambrian stromatolites; and (2) to investigating stromatolitic assemblages and attempting to base a stratigraphic scale on them. Their results have been understandably controversial. The invertebrates that are used in post-Precambrian stratigraphy are unique individuals resulting from unidirectional evolutionary phenomena, and no one would mistake a Cambrian trilobite for a Devonian one. Stromatolites, on the other hand, are biosedimentary structures built by complex and versatile algal *communities* and their general features are always dependent to some extent on environmental and diagenetic factors. Thus, identical stromatolites may be found in rocks that are not of the same age, but were formed under similar environmental conditions.

In spite of this, it does seem logical that blue-green assemblages would have changed their constitution with time during the long Precambrian eon as a result, for example, of the evolving atmospheric conditions. This might have originated a worldwide succession of algal communities that would have built particular stromatolites on which a stratigraphical scheme could be built. The Russian school has so far recognized four stromatolitic assemblages, on the basis of which they have drawn a stratigraphic scale for the late Precambrian. If they can date and correlate upper Precambrian rocks on a continental scale, it is likely that their scheme could also be used on an intercontinental scale, which would be a real achievement. Observations must be thoroughly extended over the entire Precambrian before we can have a clear view of the situation or define an unequivocal succession of assemblages through this vast period of time and thus to draw a scale with any confidence.

CLAUDE MONTY

References

All references cited here have extensive bibliographies.

Hofmann, H. J., 1969. Attributes of stromatolites, *Geol. Surv. Canada Paper 69-39*, 58p.

Hofmann, H. J., 1973. Stromatolites, characteristics and utility, *Earth-Sci. Rev., 9*, 339-373.

Monty, C. L. V., 1973. Precambrian background and Phanerozoic history of stromatolitic communities, an overview, *Soc. Geol. Belique Ann., 96*, 585-624.

Monty, C. L. V., 1977. Evolving concepts on the nature and the ecological significance of stromatolites, in E. Flügel, ed., *Fossil Algae*. Berlin: Springer Verlag, 15-35.

Schopf, J. W., 1970. Precambrian microorganisms and evolutionary events prior to the origin of vascular plants, *Biol. Rev., 45*, 319-353.

Walter, M., ed., 1976. *Stromatolites*. Amsterdam: Elsevier, 804p.

Cross-references: *Algae; Precambrian Life; Reefs and Other Carbonate Buildups.* Vol. III: *Algal Reefs.* Vol. VI: *Stromatolites.*

STROMATOPOROIDS

The stromatoporoids are an extinct group of marine invertebrates that secreted a layered skeleton of calcium carbonate and are common fossils in rocks of mid-Paleozoic age. Their soft parts are unknown. The hard tissue secreted by the animal, called the coenosteum, may be hemispherical, cabbage-shaped, pear-like, tabular, encrusting, or branching and tree-like. Coenostea may be several meters across but their average diameter is 10 cm. Globular and tabular stromatoporoids are commonly marked by growth bands a few millemeters thick called latilaminae. Stromatoporoids are the dominant fossils in Silurian and Devonian reef limestones.

Structure

Stromatoporoids have a complex internal structure that is studied under the microscope by means of thin sections cut perpendicular and tangential to the banding. Most coenostea are composed of laminae, which are parallel to the outer growth surface, and pillars, which are perpendicular to the laminae and occupy a radial position in hemispherical coenostea. Most specimens show astrorhizae—stellate systems of branching canals radiating from regularly spaced centers. The astrorhizal canals are grooves on the surface but are enclosed as the coenosteum grows and appear as tubes in the interior (Fig. 1). The surface of many species is marked by elevations on which the astrorhizae may be centered; in such species, the elevations are reflected in the interior by corresponding inflections of the laminae (Fig. 2B). In one large group of stromatoporoids, the pillars and laminae cannot be distinguished readily but the structure is a continuous, amalgamate network of tissue throughout the coenosteum (Fig. 2C). In most Ordovician stromatoporoids, laminae

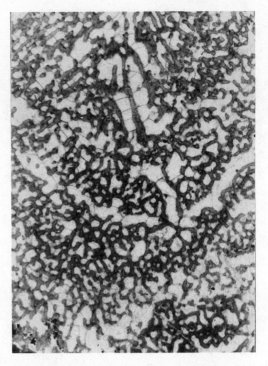

FIGURE 1. Astrorhizal canals in tangential section of *Steinerella gigantea* Schnorf X9; Cretaceous, Switzerland.

are not present and their place is taken by overlapping, upwardly convex cyst plates (Fig. 2D). The form of the laminae in Silurian and Devonian stromatoporoids varies from genus to genus. Laminae may be single layers of tissue that are imperforate, compact, or transversely fibrous or porous. They may have an axial light or dark zone dividing them into three layers. Some appear to be transversed by fine tubules. In the common genus *Actinostroma* the pillars periodically give off a set of about 6 horizontal processes, which join with others to form laminae that are open networks (Fig. 2A).

Pillars may be composed of the same material as the laminae and appear to be outgrowths of them or they may be independent. They may be rod-like or plate-like. They may pass through many laminae or they may be spool-shaped and confined to an interlaminar space (Fig. 2B). Short pillars may be superposed from one interlaminar space to the next.

The texture of microstructure of the hard tissue that makes up the pillars and laminae may be: (a) compact and speckled with opaque, irregular specks (1-5 μm across); (b) marked by opaque or light round spherules of larger size (15-50 μm)—these are called maculae or melanospheres if opaque and cellules if light;

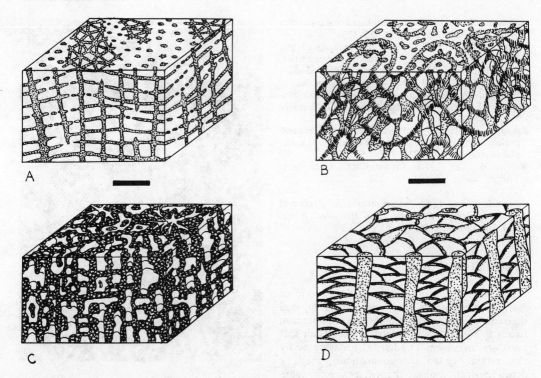

FIGURE 2. The structure of the stromatoporoids. Isometric blocks showing the appearance of four genera in vertical and tangential sections. The gross structure and microstructure are shown diagramatically. A, *Actinostroma;* B, *Anostylostroma;* C, *Stromatopora;* D, *Labechia.*

(c) finely fibrous and show trabecular microstructure like the hard tissue of scleractinian corals. Although the original microstructure of the stromatoporoid hard tissue is unknown, the textures in the calcite fossils appear to result from diagenetic modifications of trabecular and spherulitic aragonite skeletons in the process of inversion to calcite.

Classification

The classification of the stromatoporoids is based on the gross structure of the pillars, laminae, cysts, and amalgamate tissue, and also on the microstructure of these elements. No classification of stromatoporoids is widely accepted today, as students of the group do not agree on which features are valuable for classification. However, many classifications recognize the six major groups of Paleozoic stromatoporoids that are listed below without assignment as to rank in a hierarchical classification or any attempt to locate the original proposer of the subdivision:

1. Labechiids—composed basically of cysts and pillars
2. Clathrodictyids—simple laminae and nonsuperposed, simple pillars

3. Stromatoporellids—composite laminae and non-superposed pillars
4. Actinostromatids—pillars give off processes that join producing net-like laminae
5. Hermatostromatids—laminae and superposed pillars of various types
6. Stromatoporids—amalgamate tissue of cellular/melanospheric microstructure

Some paleontologists consider that similar fossils found in rocks of Permian to Cretaceous ages belong to the same group as the Paleozoic stromatoporoids. Others believe that absence of Late Paleozoic stromatoporoids and the consistently trabecular nature of the microstructure of Mesozoic forms make a close relationship of Mesozoic and Paleozoic groups improbable. Although the gap in the record is puzzling, the similarity of the two groups is so great that most paleontologists consider them both to be stromatoporoids but place them in different genera and higher taxa. The following informal divisions are recognized here:

7. Actinostromarids—gross structure much like actinostromatids
8. Milleporellids—amalgamate structures like the stromatoporids
9. Burgundiids—simple laminae and pillars comparable to the clathrodictyids

Geologic History

Stromatoporoids have been reported from Cambrian rocks of Siberia, but paleontologists differ on whether these fossils are stromatoporoids or archeocyathids. The first unequivocal stromatoporoids occur in rocks of early Middle Ordovician age, belong to the genus *Pseudostylodictyon*, and are composed of simple imbricating laminae with small upturns in the plates called denticles. During Middle and Late Ordovician time, the labechiids were the dominant stromatoporoids, building low mounds on the sea floor (e.g., Kapp, 1975). During Silurian time (e.g., Mori, 1968, 1970); laminate genera such as *Clathrodictyon* and amalgamate genera like *Stromatopora* became common, particularly in the patch reefs that characterized Middle Silurian rocks of the northern hemisphere. The actinostromatids were an important element in this fauna also. Paleozoic stromatoporoids were most abundant and diverse during Middle and early Late Devonian time (e.g., Kobluk, 1975) and thereafter decreased catastrophically in number and diversity. In the reefs of the Frasnian Stage (early Late Devonian) stromatoporids, hermatostromatids, stromatoporellids, and actinostromatids thrived. By the end of Devonian time, the labechiids, the most primitive of all the groups, returned to dominance, but above the basal Carboniferous beds, no trace of the group that dominated the mid-Paleozoic reef community is evident in the paleontologic record

The Mesozoic stromatoporoids are common in the reef limestones of the Jurassic and Early Cretaceous of southern Europe and the Middle East. The youngest stromatoporoid described comes from the Oligocene, but its affinity with the group is open to question.

Stromatoporoid reefs of the Paleozoic show ecological zonation. Certain species lived in the surf zone, others in deeper, stiller water of the forereef slope, and still others in the quiet waters of lagoons (Kapp, 1975; Kobluk, 1975). Branching species preferred the lagoonal environment; hemispherical or tabular species preferred more agitated waters. Stromatoporoids are not confined to reefal limestones but are found in well-bedded limestones and calcareous shales. The stromatoporoids from muddy environments commonly are small and encrust the hard parts of other animals.

Two viewpoints on the affinity of this extinct group to modern forms have persisted since the 19th century. One school has maintained that they are colonial coelenterates and their closest living relatives are hydrozoans like *Hydractinia*, an encrusting form from the North Atlantic. Those who support this hypothesis consider the astrorhizae are homologous to the hydrorhizae that connect the polyps of *Hydractinia* and provide interchange of nutrients between individuals of the colony.

Stromatopora

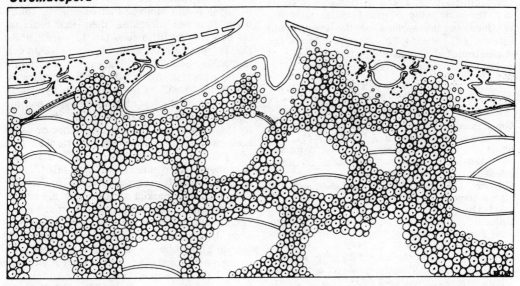

FIGURE 3. Cross section of a reconstruction of *Stromatopora* as an encrusting sponge (from Stearn, 1975). The hard tissue is spherulitic aragonite. The soft tissue is confined to the upper surface. Incurrent pores along the surface lead through connective tissue to flagellated chambers that empty into large astrorhizal canals.

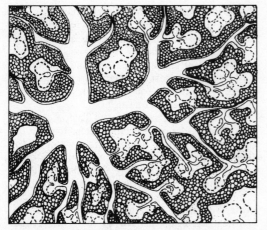

FIGURE 4. Reconstruction of the astrorhizal system in *Stromatopora* in a section parallel to the growing surface (from Stearn, 1975). Flagellated chambers empty into a stellate excurrent canal system that is just beneath the surface of the sponge.

Apart from the astrorhizae, which are not present on all specimens, the stromatoporoids show no sign of being colonial animals. The second viewpoint, that stromatoporoids are a form of encrusting sponge and that the astrorhizae are remnants of an excurrent canal system, has received support recently through the discovery of a class of encrusting calcareous sponges, the sclerosponges (Hartman and Goreau, 1970), that have remarkable similarities to stromatoporoids. Comparison with the living sclerosponges *Merlia* and *Astrosclera* suggests that the soft tissue of the fossil animals was confined to the surface of the coenosteum (Stearn, 1972, 1975). This interpretation of the stromatoporoid animal is illustrated in Figs. 3 and 4, a reconstruction of *Stromatopora*.

COLIN W. STEARN

References

Bogoyavlenskaya, O. V., 1969. Classification of stromatoporoids, *Paleontol. J.,* 1969, 457–471.

Flügel, E., and Flügel-Kahler, E., 1968. Stromatoporoidea (Hydrozoa Palaeozoica), *Fossilium Catalogus,* pt. 1, 115, 116, 1–681.

Galloway, J. J., 1957. Structure and classification of Stromatoporoidea, *Bull. Amer. Paleontology* 37, 345–470.

Galloway, J. J., and St. Jean, J., 1957. Middle Devonian Stromatoporoidea of Indiana, Kentucky, Ohio. *Bull. Amer. Paleontology,* 37, 24–208.

Hartman, W. D., and Goreau, T. E., 1970. Jamaican coralline sponges: Their morphology, ecology and fossil representatives, *Zool. Soc. London Symp.,* 25, 205–243.

Kapp, U. S., 1975. Paleoecology of Middle Ordovician stromatoporoid mounds in Vermont, *Lethaia,* 8, 195–207.

Kazmierczak, J., 1971. Morphogenesis and systematics of the Devonian Stromatoporoidea from Holy Cross Mountains, Poland, *Palaeontol. Polonica,* 26, 1–146.

Kobluk, D., 1975. Stromatoporoid paleoecology of the southeast margin of the Miette Carbonate Complex, Jasper Park, Alberta, *Bull. Canadian Petrol. Geol.* 23, 224–277.

Kobluk, D. R., 1978. Reef stromatoporoid morphologies as dynamic populations: Application of field data to a model and the reconstruction of an Upper Devonian reef, *Bull. Can. Petrol. Geo.,* 26, 218–236.

Lecompte, M., 1956. Stromatoporoidea, in R. C. Moore, ed., *Treatise on Invertebrate Paleontology,* pt. F, Coelenterata. Lawrence, Kansas: Geol. Soc. Amer. and Univ. Kansas Press. 107–144.

Mori, K., 1968, 1970. Stromatoporoids from the Silurian of Gotland. *Stockholm Contrib. Geol.,* 19, 1–100; 22, 1–112.

Nestor, K. E., 1964. Ordovician and Silurian stromatoporoids of Estonia, *Akad. Nauk. Estonian SSR Inst. Geol. Trudy,* 1–111.

Stearn, C. W., 1972. The relationship of the stromatoporoids to the sclerosponges, *Lethaia,* 5, 369–388.

Stearn, C. W., 1975. The stromatoporoid animal, *Lethaia,* 8, 89–100.

Stern, C. W., 1977. Studies of stromatoporoids by scanning electron microscopy, *Mém. Bur. Rech. Géol. Min.,* 89, 33–40.

Cross-references: *Coelenterata; Porifera; Reefs and Other Carbonate Buildups.*

SUBSTRATUM

The term *substratum* (pl. substrata) refers to the sediment at the sediment–water interface on and in which organisms live. As Purdy (1964) has noted, the substratum is a reflection of physical and biological environmental conditions. It is not surprising, then, that biofacies tend to parallel lithofacies (cf. Laporte, 1968), that is, that certain fossils are restricted to certain lithologies. Such a relationship is certainly to be expected for sessile benthic organisms, which are "tied" to the substratum by attachment. It is a little less obvious for vagile benthic organisms; but most of these are also "tied" to the substratum by food preference or other ecological factors. It is often overlooked that many nektic (free-swimming) and especially nektobenthic organisms are also "tied" to a certain substratum type by food preference. This fact has marked stratigraphic as well as paleontologic consequences, in that many organisms that were theoretically capable of wide distribution during life by virtue of their mobile life habits are still found as fossils only in certain lithologies. For example, Walker (1972) found certain Ordovician nautiloids to be restricted to a narrow range of lithologies and concluded that these represented in-life restrictions not postmortem transportation.

Thorson (1957) has documented especially

well the correlation between benthic communities and substratum type. Many early ecological workers (cf. Petersen, 1913) emphasized the substratum type as a controlling influence on community type, although later workers often have deemphisized substratum type relative to hydrologic factors as limiting organismic distributions. Still, Thorson (1957) has noted that modern communities in widely separated geographic regions but on similar substrata have similar taxonomic composition. He has referred to these as parallel communities. Walker and Laporte (1970) have shown that these *geographical parallel communities* have their counterparts in a temporal sense in stratigraphic sequences. Figure 1 is an example of these *temporal parallel communities.*

There is general agreement concerning the correlation between biotas and substrata— there is less agreement on the causes of this correlation. Basically, however, these causes fall into two categories: (1) properties of the sediment substratum itself that directly control organismic distributions, and (2) other environmental processes that are reflected in both substratum type *and* biotic distributions. An example of the first has been given by Rhoads and Young, 1970 (summarized and elaborated in Walker, 1974), in their discussion of the effects of mud-substratum fluidity, which is a reflection of sediment water content. High water content and concommitant high fluidity excludes epifaunal organisms or limits epifauna to those with special adaptations such as small size, high surface area to volume ratio, or "floatation" devices (flanges, spines, etc.; see Thayer, 1975). Walker and Alberstadt (1975) have detailed ecological successions on Ordovician mud substrata that began by the paving-over of soft mud by large, flat strophomenid brachiopods, which were adapted to floatation on these substrata by their shape. As these authors note, the large, flat shape has been independently developed at various times by other taxa as an adaptation to soft mud substrata (e.g., Cretaceous inoceramid bivalves). The water content, and thus the fluidity, of modern sediments is greatly increased by the stirring of infaunal deposit feeders (Rhoads, 1970). This mode of feeding tends to be more prevalent in fine-grained substrata (see below).

The difficulties in coping with a constantly shifting substratum is another example of direct control by substratum type. Walther (1894) long ago referred to sand flats as "the deserts of the sea" and Purdy (1964) has given an excellent example involving the oolite shoals of the Bahama banks (Fig. 2). Figure 3 shows the generalized consequences of increased sediment mobility.

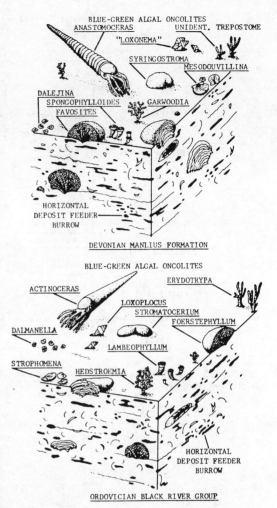

FIGURE 1. Reconstruction of subtidal communities from (below) Black River Group and (above) Manlius Formation (from Walker and Laporte, 1970).

Perhaps the best example of the second cause of substratum–biota correlation involves the relationship between sediment grain size and abundance of various feeding types. This relationship has been best documented in modern communities by Sanders (1956, 1958; for a recent paleontologic study, see Fürsich, 1976). The general relationship is summarized in Fig. 4. Suspension feeders are most abundant on sandy substrata because these occur in environments where currents are sufficient to deliver large quantities of water with its contained suspended food resource to the organism over time. These currents are also instrumental in removing fine sediment that might otherwise clog delicate feeding structures. Deposit feeders, on the other hand, tend to be most abundant on silty and clayey sub-

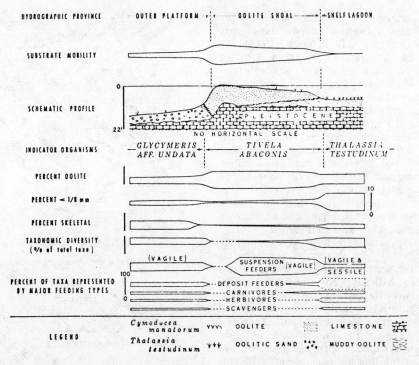

FIGURE 2. Relative taxonomic abundance of invertebrate feeding types on an idealized Bahamian oolite shoal (from Purdy, 1964).

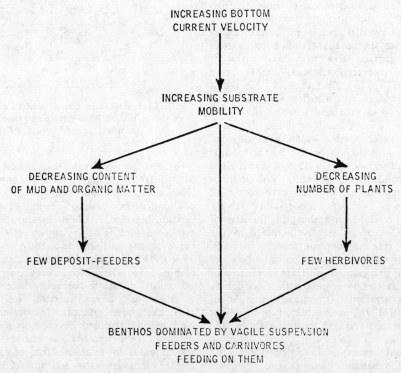

FIGURE 3. Simplified diagram of the ecological consequences of increasing substratum mobility (from Purdy, 1964).

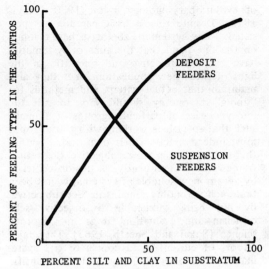

FIGURE 4. Generalized correlation between substratum grain size and the relative proportion of two benthic feeding types (Generalized from the work of Sanders, 1956, 1958.)

strata. These sediment sizes are generally deposited in quiet waters where organic detritus (with a specific gravity near that of sea water) is also deposited. Thus, the quantity of deposited organic detritus is high in fine-grained substrata and this abundant resource is exploited by deposit feeders. Therefore, the relative proportions of suspension and deposit feeders in an environment is a function of the interplay between water-exchange rate and the rate of deposition of organic detritus, both of which are functions of the hydrodynamic regime. In the sediment of the substratum, the hydrodynamic regime is reflected in part by grain size.

Finally, it should be noted that some substrata are themselves the direct or indirect product of biotic activity and nearly all substrata are modified to some degree by such activity. Reefs (q.v.) and algal stromatolites (q.v.) are prime examples of substrata produced by biotic activity, the first directly through skeletal build-up and the second indirectly through the sediment trapping and binding activity of algae. The accumulation of skeletal debris as sediment and the manipulation of sediment by burrowers (cf. Rhoads, 1970, 1974) are two ways in which the biota affects the substratum. Thus, substrata and the organisms that live on and in them form a feed-back loop in which the interrelationships are often complex.

KENNETH R. WALKER

References

Craig, G. Y., and Jones, N. S., 1966. Marine benthos, substrate and palaeoecology, *Palaeontology,* 9, 30–38.

Fürsich, E. E., 1976. Fauna-substrate relationships in the Corallian of England and Normandy, *Lethaia,* 9, 343–356.

Gray, J. S., 1974. Animal-sediment relationships, *Ann. Rev. Oceanogr. Marine Biol.,* 12, 223–261.

Laporte, L. F., 1968. Recent carbonate environments and their paleoecologic implications, in E. T. Drake, ed., *Evolution and Environment.* New Haven, Conn.: Yale Univ. Press, 229–258.

Petersen, C. G. J., 1913. Valuation of the sea II: The animal communities of the sea bottom and their importance for marine zoogeography, *Rept. Danish Biol. Station,* 21, 44p.

Purdy, E. G., 1964. Sediments as substrates, in J. Imbrie and N. D. Newell, eds., *Approaches to Paleoecology.* New York: Wiley, 238–271.

Rhoads, D. C., 1970. Mass properties, stability, and ecology of marine muds related to burrowing activity, in T. P. Crimes and J. C. Harper, eds., Trace fossils, *Geol. J. Spec. Issue,* 3, 391–406.

Rhoads, D. C., 1974. Organism-sediment relations on the muddy sea floor, *Ann. Rev. Oceanogr. Marine Biol.,* 12, 263–300.

Rhoads, D. C., and Young, D. K., 1970. The influence of deposit-feeding organisms on sediment stability and community trophic structure, *J. Marine Research,* 28, 150–178.

Sanders, H. L., 1956. Oceanography of Long Island Sound, 1952–1954, X. The biology of marine bottom communities, *Bull. Bingham Oceanogr. Coll.,* 15, 245–258.

Sanders, H. L., 1958. Benthic studies in Buzzards Bay, I. Animal-sediment relationships, *Limnol. Oceanogr.,* 3, 245–258.

Thayer, C. W., 1975. Morphologic adaptations of benthic invertebrates to soft substrata, *J. Marine Research,* 33, 177–189.

Thorsen, G., 1957. Bottom communities (sublittoral or shallow shelf), in J. W. Hedgpeth, ed., Treatise on marine ecology and paleoecology. Vol. 1, Ecology. *Geol. Soc. Amer. Mem.,* 76, 461–534.

Walker, K. R., 1972. Community ecology of the Middle Ordovician Black River Group of New York State, *Geol. Soc. Amer. Bull.,* 83, 2499–2524.

Walker, K. R., 1974. Mud substrata, in A. M. Ziegler et al., Principles of benthic community analysis, *Comp. Sed. Lab. Univ. Miami, Sedimenta,* vol. IV, 5.1–5.9.

Walker, K. R., and Alberstadt, L. P., 1975. Ecological succession as an aspect of structure in fossil communities, *Paleobiology,* 1, 238–257.

Walker, K. R., and Laporte, L. F., 1970. Congruent fossil communities from Ordovician and Devonian carbonates of New York State, *J. Paleontology,* 44, 928–944.

Walther, J., 1894. *Einleitung in de Geologie als historische Wissenschaft. 3, Lithogenesis der Gegenwart.* Jena: Fisher Verlag, 900p.

West, R. R., 1977. Organism-substrate relations: Terminology for ecology and paleoecology, *Lethaia,* 10, 71–82.

Cross references: *Actualistic Paleontology; Communities, Ancient; Dwarf Faunas; Paleoecology; Trace Fossils.*

SYSTEMATIC PHILOSOPHIES

The terms *systematics, classification,* and *taxonomy* all refer to the study of diversity of organisms, fossil and recent, and to the basis for their arrangement into recognized, ordered systems. In the literature, these terms have been used with unclear degrees of synonymy or unstated differences in connotation. To clarify this problem, Simpson (1961) and Mayr (1969:2) have defined (1) *systematics* as "the scientific study of the diversity of organisms and of any and all relationships among them;" (2) *classification* as the ordering of organisms into groups or sets on the basis of their relationships and the ranking of these sets within a hierarchical system of nomenclature; and (3) *taxonomy* as the theory and practice of classifying organisms. The terms classification and taxonomy thus closely coincide, the one emphasizing the practical aspects of naming and ranking of taxa, and the other working aspects of methods used to define and order these taxa. Systematics is a still more general concept encompassing all study of diversity and especially the scientific philosophical rationale for the methodology employed in doing taxonomy.

As Mayr (1969) notes, objective presentation of its principles is a late development in the history of systematics; the demeanor of the field has long been based on authority and tradition. Linnaeus, for example, has been titled "lawgiver of natural history" (Gregory, 1910, 38). Simpson (1945, 13, 217–218) has claimed that "good classification is conservative" and implied that in some cases one should avoid altering deeply rooted nomenclature even if it is rendered inconsistent by reinterpretation of material. Mayr (1969, 18) observes that "the young taxonomist learned as apprentice to a master." Systematic work has passed through periods of stagnation and disrepute when issues became dominated merely by legalism and the authoritarian aspects of classifying. Application of new subfields in biology and paleontology, or new discoveries in established ones, seemed to offer revival from these depressed periods. Systematics was spurred, for example, by the growth of 19th century comparative morphology and embryology after Darwin, and by development of the so-called "new systematics" through integration of a number of new fields

of evolutionary biology in the 1930s (Mayr, 1970). Despite this, a basic paradox has persisted. Many systematic zoologists have claimed on the one hand that the aims of systematics have become objective and scientific in the light of these new applications. Yet they also maintain that actual criteria and methods for "doing" taxonomy should vary to suit the idiosyncrasies of particular groups of animals, such that specialists working with those groups must indeed practice their own "arts," within the supposedly broader, objective theoretical framework. Consequently, a major criticism by recent new schools of systematics has been that such taxonomy remains in fact subjective; by not being vulnerable to objective self-criticism and evaluation, it is not rigorous science (Sokal and Sneath, 1963). The major import of current discussion is to give systematics a more explicitly stated, scientific philosophical basis.

Systematics Before Evolutionary Theory

Pre-Darwinian schools of systematic philosophy were largely typological in the sense that categories of animals as erected in a classification were characterized by a single unique attribute (either the ideal essence of the group, or else its *fundamentum divisionis,* the feature used to distinguish it from other groups of the same category). All members of a category must possess the defining character, and variance from it was considered trivial or ignored (in contrast to the modern preoccupation with intraspecific variation and, by extension, polythetic higher categories—see below). The roots of the typological outlook go back to Aristotle, and, through the scholastic scholars and Linnaeus, had a lasting influence on taxonomy through the 19th century. As a result of using single characters, and of the superficial (e.g., pre-phylogenetic) nature of those characters, many strange and incongruous groups were erected in early classifications. A well-known example is Linnaeus' "Bestiae" of mammals, lumping the pig with shrews, moles, opossums, and tree sloths. Under the typological scheme, widely different classifications could be advanced by changing the characters used to group animals (see Gregory, 1910 for a history of classifications of the mammals). Based on nonoverlapping sets of characters, such classifications were essentially diagnostic keys. Although predating the theory of evolution, they nevertheless influenced taxonomic thinking during and after the Darwinian revolution (see Simpson, 1961 for a brief summary).

Classical Evolutionary Taxonomy

Classical evolutionary taxonomy as propounded by Simpson (1961), Mayr (1963, 1969), Bock (1974 and references therein), and others, is firmly based on the Darwinian theory of evolution and the modern biological species concept. Given that evolution proceeds by means of natural selection, it is axiomatic to systematic work that resemblances among animals are the result of heritage from common ancestry, differences the result of descent with modification as a consequence of differential survival. The evolutionary and taxonomic unit is the biological species, characterized by (1) a gene pool kept distinct from others through reproductive isolation, (2) a more or less characteristic morphology, and (3) a unique ecological niche (Mayr, 1969). The origin of new types comes about only by gradual accumulation of small genetic changes over time, through a variety of mechansims described by population genetics. New types may arise by splitting of genetic lineages—speciation— or by transformation of single undividing lineages—phyletic gradualism (Eldredge and Gould, 1972). The important assumption systematics draws from evolutionary theory is that there is a "natural" classification of animals based on possession of common ancestry and community of descent (Simpson, 1961).

In practice, the taxonomist does not work with entire species, but classifies a series of specimens that are assumed to be a valid sample of the species' characteristic morphology. This, the phenotype, is presumed to reflect a corresponding genotype; the genotype is thus the true object of taxonomic work, although it is never directly observed by the taxonomist (Simpson, 1945). On the assumption that animals have a common origin, they are—by the process of reproduction—genetically continuous with this ancestry. Particularly in paleontological work, it has been a prevalent procedure to infer from a series of forms in stratigraphic succession a "dynamic flow in which the direction and rate of structural change can be directly observed, at least in part" (Simpson, 1945, 8). The specimens represent instantaneous samples from different time horizons in inferred evolutionary lineages. A large part of evolutionary taxonomic thought concerns the fate of lineages; Simpson and others have defined an extensive terminology to describe in a graphical conceptual manner the patterns of morphological change among lineages through time (Fig. 1). The explanation for these patterns involves ecological, genetic, and functional arguments which are best thought of as inferred evolutionary models based ultimately on morphological features and stratigraphic data of specimens.

Taxa are distinguished on the basis of (1) degree of overall similarity or dissimilarity— simple phenetic distance; (2) closeness of common ancestry; and (3) distinctness of their adaptations to particular adaptive zones. All three of these criteria are used to determine how useful specific characters might be toward determining relationships and erecting classifications. For example, morphologically the sea lions and true seals show close genealogical relationship to the bear and weasel-like carnivores, respectively. But they are grouped together and classified on a par with all the terrestrial carnivores (Suborder Pinnipedia, separate from Suborder Fissipedia) because they have ecologically (and therefore adaptively and morphologically) diverged far enough from their ancestral state to warrant equal taxonomic rank. In this case, one relies on the third criteria above to unite the aquatic carnivores into one taxon, as opposed to treating them separately and coordinately with the bear and weasel families among Fissipeds (Simpson, 1945).

In other cases, however, the first and second criteria become important, when the argument is made that characters with strong adaptive value are less useful in determining relationships than nonadaptive ones. The first type are likely to be acquired in distantly related groups by convergent evolution, while the second, being conservative, will indicate true relationship. In terms of the genetic model, different— unrelated—genotypes will respond to similar selection pressures of a given adaptive zone to yield convergent characters not indicative of true relationship; whereas similar genotypes when subjected to different selection pressures both yield divergent characters and also retain primitive unselected ones indicating their original close relationship. The outcome depends, of course, on the adaptive significance the taxonomist places on the characters investigated (see Thomson and Bossey, 1970 for an example of this line of reasoning applied to early tetrapod evolution).

Mayr (1965, 1969) has formulated this practice in another way by introducing a two-part definition of relationship: (1) genealogical, that is, recency of common ancestry, and (2) genetic, reflecting the proportion of total genotype held in common. The classic example of contradiction between these aspects occurs in the classification of birds and crocodiles. These share a common ancestor more immediate with each other than either does with other

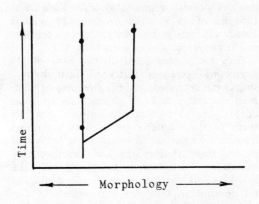

A Parallel evolution

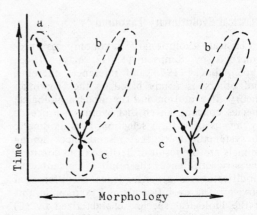

B Divergent evolution

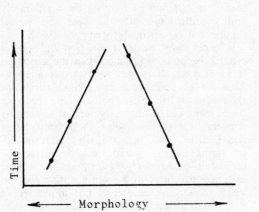

C Convergent evolution

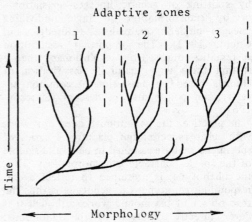

D Adaptive radiations

FIGURE 1. Patterns of phylogeny as seen by the classical evolutionary taxonomic school (based on Simpson, 1961). Dots represent specimens or samples, considered to be species, which are inferred to be points on continuous lineages. A, After diverging, two lineages continue through time at a fixed morphological distance. B, Two common classifications in the case of two diverging lineages. (In one case, a, b, and c are accorded equal rank in the hierarchy, taxon c being ancestral to a and b. In the second case, one of the offshoot lineages is considered too short lived to be accorded separate taxonomic status, and is subsumed in the ancestral taxon.) D, Successive species in two lineages appear more and more alike, but the similarities are not the result of common ancestry, rather the independent acquisition of convergent characters. D, Three lineages from a basal stock have entered new adaptive zones, leading to rapid increases in diversity, or adaptive radiations.

extant reptiles. Crocodiles are nevertheless classified with these reptiles because they have diverged so little from them and retain a large proportion of their genotype in common. Birds in contrast have a large proportion of their genotype newly evolved as a result of entering the aerial adaptive zone. Note the basic similarity between this approach and the phenetic distance criterion of numerical taxonomy (see below).

In classical evolutionary taxonomy, the adaptive-zone concept and its corollary of structural–functional types of major adaptive significance are important as an explanation for morphological "gaps" between taxa, and the oft-noted lack of transitional forms in the fossil and recent record. Any form that is quite different from its nearest genealogical relatives and appears to have entered a new adaptive zone is accorded equal or higher relative status in the classification hierarchy. Consequently, the species diversity of higher order categories may vary widely: monotypic orders accommodate isolated, very derived morphological types, while a family or tribe category may contain dozens of species, differing on a minor scale while conforming to a uniform structural plan. Taxa of all hierarchical rank are considered to be real entities in nature (Simpson, 1961), in that they represent a collective genotype that may or may not be fully known on the basis of material at hand. (Transitional lineages not breaking through to new adaptive zones are short-lived and unlikely to be found in the fossil record.) Simpson speaks of taxa at all hierarchical levels as capable of giving rise to taxa of any level equal to or above themselves. Thus a *monophyletic* taxon is derived from a single taxon of equal or lower rank; a *polyphyletic* taxon is derived from more than one ancestral taxon of higher rank (Fig. 2). If descendants are classified with their ancestors, the classification is vertical and emphasizes monophyletic groups. If contemporaneous descendants of one or several closely related ancestors are placed together to the exclusion of the ancestor(s), the classification is horizontal, and the descendent taxon is polyphyletic if it is of lower order than the ancestral one (Fig. 3). There are, however, no objective criteria for deciding how horizontal or vertical a classification should be (Simpson, 1961).

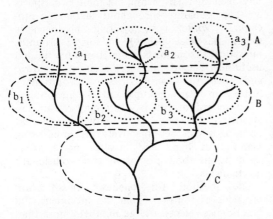

FIGURE 2. Monophyly and polyphyly in classical evolutionary taxonomy. Taxon A is polyphyletic. Taxa A and B are polyphyletic if B is of higher taxonomic rank than A, and C higher than B. Groups a_1, a_2, a_3, b_1, b_2, and b_3 are monophyletic.

Polyphyly has been employed extensively to classify confusingly diverse or poorly known groups of animals whose genealogical relationships could not be consistently discerned by simple character distribution. Instead of being classified on the basis of lineages, these arrays of forms are classified into *grades*—taxa that are polyphyletic and share only a common adaptive zone or general level of organization (see Schaeffer, 1967 for an application to fossil elasmobranchs). A grade is polythetic in the sense of Beckner (1959), that is, it is not uniquely defined by one character, nor does any one member of a grade possess all the characters used to define the grade.

In recent defenses of evolutionary taxonomy and in an attempt to make more explicit its working principles, Bock (1974, and references therein) has reemphasized the dual purpose of classification as seen by this school of systematics. That is, (1) reflection of the phylogenetic history of taxa and (2) an information retrieval/reference system to contain nomenclature, degree of morphological divergence and similarity, success of adaptive radiation, and other items of biological significance. To achieve these goals in practice, the taxonomist must chose characters whose homology among different groups can be discerned with minimum error, and use enough characters to counteract the probability that some of the supposed homologies will actually be non-homologies (i.e., convergences). According to Bock, classification is arrived at in the first instance by proposing and assessing homologies and it is then tested by investigating new characters using the same method. Not all observable characters are used, however, but only those deemed "useful" given the taxonomist's experience with the animals under study. Again, there are no standard, objective criteria for sorting out characters that can be used to define groups and indicate relationships from those that are not phylogenetically useful. Nevertheless, this must be done before reaching conclusions about classification, in order to avoid circular reasoning in which classification is based on characters whose homology among groups is in turn suggested by previous or existing classifications (Colless, 1967; Bock, 1974). The selection of certain characters for use in analysis has been termed phyletic weighting by Cain and Harrison (1960), who give some general, but still subjective, criteria for the basis of selection.

Numerical Taxonomy

Sokal and Sneath (1963) defined numerical taxonomy as "the numerical evaluation of the

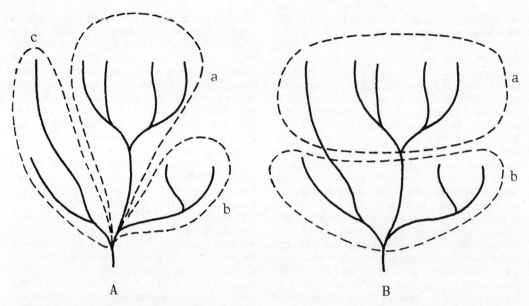

FIGURE 3. Horizontal vs. vertical classification. A, Vertical classification with taxa a, b, and c accorded equal rank. B, Horizontal classification: lineages in a are classified together excluding their separate ancestors in b.

affinity or similarity between taxonomic units and the ordering of these units into taxa on the basis of their affinities." Their controversial book *Principles of Numerical Taxonomy* (1963, extensively revised and expanded 1973) is both a critique of the classical evolutionary school and an exposition of numerical taxonomic methodology and of the variety of mathematical procedures in use. The computer, of course, was essential to the development of numerical taxonomy. In light of their criticism of the classical evolutionary school, Sokal and Sneath maintained that numerical methods would reestablish taxonomy as a repeatable and objective science.

In this system, classification is expressly not phylogenetic; it is based only on overall phenetic similarity, without reference to lineages, evolutionary origin of characters used, rates of convergence, divergence, etc. The exclusion of phylogenetic inferences stems from, among other things, the fragmentary nature of the fossil record, the need to treat even fossils phenetically, and the complexity of a classification scheme that could contain objectively both phenetic and phylogenetic information (Sokal and Sneath, 1973). It should be noted that these reasons are largely practical and not theoretical or methodological.

Ideas of phenetic, empirical taxonomy were first put forward by the 18th-century botanist Michel Adanson [1727–1806] and are the basis for the axioms of numerical taxonomy. These are that (1) taxa are to be based on as many

characters as possible thereby bearing maximum information content; (2) every character carries equal weight in the analysis; (3) affinity equals overall similarity; (4) various character correlations may be used to separate distinct taxa; and (5) taxonomy is strictly empirical and nonphylogenetic (Sokal and Sneath, 1973).

As with classical evolutionary taxonomy, the numerical school is in accord with modern evolutionary theory. Relationship is defined broadly to include phenetic resemblance and phylogenetic relationship, but these two aspects are kept separate throughout taxonomic procedure. While not rejecting phylogenetic systematics (see below) on logical grounds, numerical taxonomists claim that it has limited usefulness because phylogenies are not known, and with the complexities of evolutionary history in many groups and the incomplete fossil record of others, they may remain unattainable. Phenetic taxonomy is preferred because it can be objectively measured using material at hand and involves no *a priori* assumptions about phylogeny and evolutionary mechanisms.

The method for phenetic classification proceeds as follows. From the specimens, the taxonomist selects a large number (more than 40) of characters, by inspection or by measurement, which can be assigned numerical (code) values. Each operating taxonomic unit (OTU)— individual, population, species, or higher category depending on the level of taxonomy being done—is characterized by a set of the coded characters. Every pair of taxa are compared on the basis of all characters and the degree of similarity is recorded as the similarity coefficient. The outcome of a numerical analysis is always biased by the way in which characters are sampled and coded, and by the way in which the similarity coefficient is calculated (Ehrlich and Ehrlich, 1967; Sokal and Sneath, 1973). Highest values of the coefficient yield pairs of OTU's, the first level of grouping. Clustering of taxon pairs by computer (cluster analysis) based on the similarity coefficient matrix yields higher order groupings. This approach constitutes agglomerative clustering or classification "from below" as opposed to classification "from above" or dividing high-order groups into lower order ones, the common practice of pre-Darwinian typologists. The results are presented graphically in several ways. One of these is illustrated in Fig. 4, where the position of phenon lines on the verical scale gives the percent lower limit of affinity. The assignment of groups or phena to a particular rank in the hierarchy could be based on arbitrary divisions of the vertical scale (coded class marks at right of Fig. 4), and the phenetic scheme could thus be expressed in a Linnaean classification. Phenetic results can also be presented on scatter diagrams of three or more dimensions, contour

diagrams, and other graphs (Sneath and Sokal, 1973, 259–275).

Except for constructing keys and identifying specimens, straight phenetic classification has limited appeal to systematists, who are at some level concerned with phylogeny and the evolution of diversity. Numerical methods have therefore been adapted to arrive at phylogenetic dendrograms; in this case mathematical assumptions introduced during the procedure are to have phylogenetic significance. Of the total set of characters used, those considered "primitive" may be eliminated; the remaining "derived" characters are then analyzed in the usual manner. Two applications are outlined by Inger (1973). In the first, the so-called minimum-steps method produces a phylogenetic dendrogram in which the minimum number of evolutionary changes (mutations, for example) are involved in deriving all the derived characters from their primitive states (Fig. 5 and Table 1). The computer essentially inspects all possible cluster arrangements of the taxa and choses that dendrogram which requires the least number of changes, i.e., is the most parsimonious (see Camin and Sokal, 1965 for mathematical treatment of this method). The assumptions here are (1) that characters can be arranged in evolutionary sequences or morphoclines (this requires *a priori* knowledge of the phylogeny of each

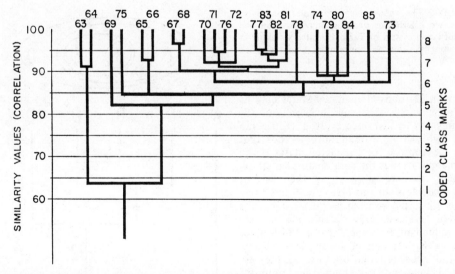

FIGURE 4. Dendrogram or diagram of relationships among 23 species of the bee subgenera *Chilosmia* and *Ashmeadiella* s. str. (from Sokal and Sneath, 1962). The relationships were obtained by the weighted pair-group method (WPGM). The ordinate is graduated in a Pearson product-moment correlation coefficient scale (coded by multiplying by 100). Numbers across the top of the figure are species code numbers. Horizontal lines across the dendrogram are phenon lines defining taxa at the minimum level of similarity at which the phenon line cuts the ordinate. Class intervals delimited by phenon lines along the similarity scale have had their class marks coded (on the right side of the dendrogram).

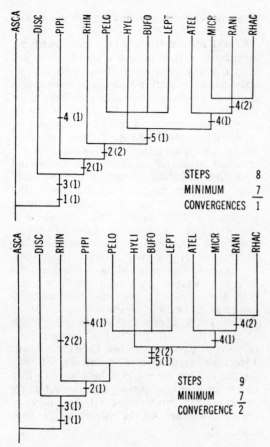

STEPS 8
MINIMUM 7
CONVERGENCES 1

STEPS 9
MINIMUM 7
CONVERGENCE 2

FIGURE 5. The minimum-steps method (from Inger, 1973). Two alternative dendrograms based on the character-taxon matrix of Table 1. Coded states are in parentheses next to the corresponding characters. See text.

TABLE 1. The Minimum Steps Method. (see also Fig. 5). Character-Taxon Matrix for 5 Characters in Several Frog Families.[a]

	Characters[b]				
	1	2	3	4	5
Ascaphidae	0	0	0	0	0
Discoglossidae	1	0	1	0	0
Pipidae	1	1	1	1	0
Rhinophrynidae	1	2	1	0	0
Microhylidae	1	2	1	2	1
Pelobatidae	1	2	1	0	1
Hylidae	1	2	1	0	1
Ranidae	1	2	1	2	1
Rhacophoridae	1	2	1	2	1
Bufonidae	1	2	1	0	1
Atelopodidae	1	2	1	1	1
Leptodactylidae	1	2	1	0	1
States	2	3	2	3	2
Min Steps	1	2	1	2	1 = 7

From Inger, 1973.
[a]Under each character are listed (by numerical code) the states in which it is found.

dendrograms are eliminated again on the basis of fewer steps (evolutionary changes) or fewer numbers of independent derivation (convergence), or on the assumption that convergences are more likely to occur in lineages within major groups than in those from widely different ones.

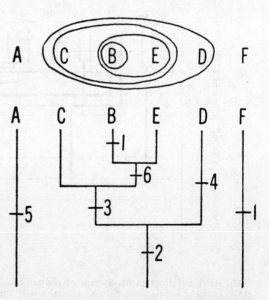

FIGURE 6. The combinatorial method (from Inger, 1973). Nested clusters (above) and dendrogram (below) based on character-taxon matrix in Table 2. All groups are monothetic and nonredundant. Note character 1 must be evolved twice.

character); (2) derived character states may be evolved in more than one lineage; and (3) they may not be lost once evolved. (Kluge and Farris, 1969, have worked out methods that allow for reversal of character states.) The second application is the combinatorial method, in which taxa are grouped on the basis of shared characters. In this case (Fig. 6 and Table 2), the clustering procedure followed by the computer systematically eliminates all so-called redundant groupings, that is, lower order groupings that are defined by the same character as the higher order groups in which they are nested (analogous to eliminating primitive characters in determining relationships by cladistic analysis—see below). The resulting dendrogram contains only monothetic and non-redundant groups. It is possible, of course, to have groups based on several characters, either unique or shared with other taxa. Alternative

TABLE 2. The Combinatorial Method: Types of Combinations (right)
Based on Hypothetical Character-Taxon Matrix (left)[a]

Taxa	States[b]						Taxa	States Shared	Nature of Group
	1	2	3	4	5	6			
A	0	0	0	0	1	0	AD	none	not monothetic
B	1	1	1	0	0	1	BCE	2, 3	monothetic, non-redundant
C	0	1	1	0	0	0	BC	2, 3	monothetic, redundant
D	0	1	0	1	0	0	CE	2, 3	monothetic, redundant
E	0	1	1	0	0	1	BE	2, 3, 6	monothetic, non-redundant
F	1	0	0	0	0	0	BCDE	2	monothetic, non-redundant

From Inger, 1973.
[a]See also Fig. 6.
[b]0 = character absent, 1 = character present.

Numerical cladistic analysis, using methods such as these, has been applied to constructing phylogenies of amino acid sequences in proteins (Fitch and Margoliash, 1967) based on immunological characters. A thorough literature review of numerical taxonomic work and theory is given in Sokal and Sneath (1973).

Phylogenetic Systematics (Cladistics)

Hennig's *Phylogenetic Systematics* (1966) was the first extensive formulation in English of the cladistic school of systematics. The book contains not only an exposition of cladistic methodology, but also an extensive discussion of the place of systematics in biology, the importance of the species concept, and the methods of taxonomy. Hennig's discussion was to serve as a general philosophical defense of cladistics and its priority among systematic schools. His methodology, variously amended, has been defended by Brundin (1966, 1972), Nelson (1973, and references therein), and others, and vigorously attacked by classical evolutionary taxonomists (see especially Mayr, 1974). The principle differences between cladistic methodology and that of the classical school concern (1) the definition of relationship; (2) the weighting of characters; (3) the degree to which evolutionary models are imputed into the method of analysis; and (4) the extent to which lineage relationships (ancestor-descendent relationships) are relied upon to arrive at phylogenies.

According to Hennig and later cladists, relationship for systematics should be restricted to the narrow definition of genealogical kinship among species and higher groups. This definition requires some species-splitting mechanism for the evolution of diversity, which, in its logical expression, is strictly hierarchical (Brundin, 1972, 108). Such kinship may be expressed in terms of sister-group relationships: for any three groups of organisms (regardless of relative size or content), two will be more closely related to each other—i.e., they will share a common ancestor more recently in the past—than either will be to the third. The sister-group relationship is expressed in a cladogram such as in Fig. 7A, where B and C are sister groups and the two groups A and (B + C) are also sister groups. The hypothesis stated here, that B and C are more closely related to each other than either is to A, is suggested by a distribution of characters such

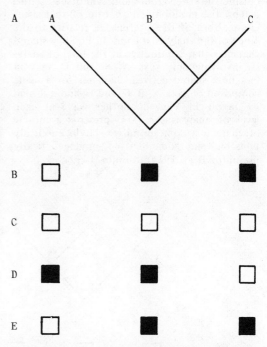

FIGURE 7. A, Cladogram showing a hypothesis of relationship among three taxa A, B, and C. B–E, Distributions of four hypothetical characters used to evaluate the dendrogram. Black has been determined to be derived or apomorphic; white, primitive or plesiomorphic. See text.

as in Fig. 7B: B and C share a (derived) character not possessed by A. An essential difference between cladistic and phenetic analysis is that it must be argued (somehow) that the character common to B and C is a shared *derived* character, i.e., it is unique to the sister group (B + C). If the shared character is derived, B and C constitute a clade, a strictly monophyletic taxon, whose ancestor possessed the character. If this is not established (i.e., the shared character is primitive or cannot be determined), the resemblance between B and C is phenetic only and may have no significance for discerning their genealogical relationship. A character distribution such as in Fig. 7C does not contradict the relationship shown in the cladogram, but is also equally compatible with either of the two other ways of relating A, B, and C. A distribution such as in Fig. 7D specifically contradicts the hypothesis expressed in the cladogram by suggesting that A and B, sharing a derived character, are more closely related to each other than either is to C. In cladistic terminology, shared derived characters are called synapomorphies, shared primitive characters are called symplesiomorphies. Symplesiomorphies are common not only to a proposed sister group but also to other taxa outside the sister group, and cannot be used to defend the monophyletic nature of that sister group. Both of these terms are relative to the scope of the analysis at hand. In Fig. 8, a shared character that is primitive at the level of analysis involving only taxa A, B, and C, may in fact be a shared derived character for a clade comprised of taxa A, B, C, and D, and not held by taxon E. Viewed another way: at each level of analysis, an OTU presents a mosaic of characters, some primitive—phylogenetically unusable—and some derived—phylogenetically useful (de Beer, 1954; Brundin, 1972).

In the cladistic system, all sister groups are monophyletic and every ancestral form is included as a member of the same taxon with all its descendents. Large clades may be distinguished from their sister groups by only a single derived character—irrespective of overall degree of divergence or morphological distance. The cladogram of itself implies no information about time, mechanism of speciation or phyletic evolution, rate of evolution, or adaptive significance of characters. This has important implications for paleontology since features in ancient forms have no *a priori* cause to be regarded as primitive simply on the basis of their stratigraphic position (Schaeffer et al., 1972). [Some element of time may be introduced when analyzing fossil forms (see Harper, 1976). The presence of two sister groups at a certain stratigraphic horizon implies a minimum (only) age of slightly older time for the existence of their common ancestor, and a series of such minimum ages can lead to statements on rates of phyletic branching.] Proposed phylogenies may be tested by checking them against additional character distributions— i.e., the same analysis is applied *seriatim* (Brundin, 1972, 111). Each character distribution will either (1) corroborate the original phylogeny as proposed (Fig. 7B); or (2) falsify it by contradicting the monophyletic nature of one or more clades (Fig. 7D); or (3) fail to do either of these (Fig. 7C).

Difficulties in cladistic practice arise in determining whether a character common to a group being analyzed is in fact primitive or derived. Classical evolutionary taxonomists have criticized the school on this point, arguing that by labelling a character as derived the investigator simply presupposes a phylogeny (Colless, 1969; Bock, 1974). Several methods have been suggested to determine the primitive

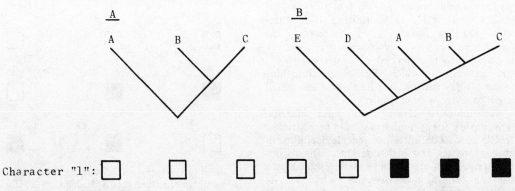

FIGURE 8. Relative nature of terms apomorphy and plesiomorphy. A single hypothetical character is present in Taxa A, B, and C and absent in D and E. In an analysis of A, B, and C only, it must be considered plesiomorphic for these groups (*A*). In a broader analysis, the same character is considered apomorphic for A, B, and C(*B*).

(plesiomorphic) or derived (apomorphic) nature of a character. Several of those discussed by Hennig (1966) include ancientness in the fossil record, developmental aspects, geographical distribution—approaches already standard in taxonomy. The method more commonly used by cladists after Hennig, however, is that of "out-group comparison." A character widely distributed among apparently unrelated groups is taken to be primitive for the smaller group in question. This does presume an existing phylogeny, but without involving all living things in the analysis (a logical regression *ad infinitum*), such an approach is practically necessary. Other difficulties encountered in cladistic analysis involve the choice of characters and the methods of recognizing homology. Cladistics of itself does not offer any striking advances here over traditional comparative morphology. It rather represents a logically consistent method for using characters considered to bear phylogenetic information. In some cases, e.g., relationships within species swarms, it may be next to impossible to determine unambiguously which characters are plesiomorphic, which apomorphic (Brundin, 1972, 110–111).

Hennig (1966) maintained that a taxonomy that is strictly phylogenetic provides the best overall reference system sought by systematics within the framework of evolution (but see the contrary opinion of Sneath and Sokal, 1973, 53–56). Cladists have argued that a hypothesis of relationships such as depicted in a cladogram is a sufficient presentation of the taxonomy in that no additional information can be given in a written classification not already present in the cladogram (Gaffney, 1975). If necessary, however, sister-group relations such as in Fig. 9A can be expressed uniquely in hierarchical manner as in Fig. 9B. Assigning absolute rank in the Linnaean sense (phylum, family, genus) to their hierarchical sets and subsets has been widely recognized as arbitrary. One is faced with the dilemma of either drastically increasing the number of hierarchical categories (one for each level of branching in the cladogram) and inventing new names for all sister groups; or of drastically revising the contents of established categories where these are paraphyletic. In addition, stability of nomenclature, such as advocated by Simpson (1945), cannot be expected from such cladistic classifications, since changes in hypotheses of relationship at one level must necessarily reorder the classification of all members of that particular sister group. Hennig (1966) has proposed ranking clades in a hierarchy on the basis of absolute time of branching. As yet, however, no satisfactory reconciliation between cladistic phylogenies and the Linnaean nomenclatorial heirarchy has been advanced.

STEVEN BARGHOORN

References

Becker, M., 1954. *The Biological Way of Thought.* New York: Columbia Univ. Press. 200p.

Bock, W. J., 1974. Philosophical foundations of classical evolutionary classification, *Systematic Zool.*, **22**, 375–392.

Brundin, L., 1966. Transatlantic relationships and their significance as evidenced by chironomid midges. I. Principles of phylogenetic systematics and phylogenetic reasoning, *Stockholm Kungl. Svenska Vetensk. Holl.*, **11**, 472p.

Brundin, L., 1972. Evolution, classification and causal biology, *Zool. Scripta*, **1**, 107–120.

Cain, A. J., and Harrison, G. A., 1960. Phyletic weighting, *Proc. Zool. Soc. London*, **135**, 1–30.

Camin, J. H., and Sokal, R. P., 1965. A method for

A

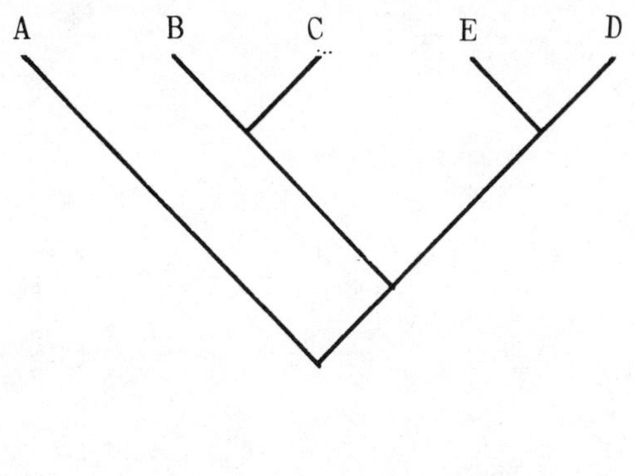

B

Group I (A-E)
- Subgroup I (A)
- Subgroup II (B-E)
  - Subgroup IIa (B-C)
    - Subgroup IIa-1 (B)
    - Subgroup IIa-2 (C)
  - Subgroup IIb (D-E)
    - Subgroup IIb-1 (D)
    - Subgroup IIb-2 (E)

FIGURE 9. Cladistic analysis and classification. See text.

deducing branching sequences in phylogeny, *Evolution,* **13**, 311–326.

Colless, D. H., 1969. The phylogenetic fallacy revisited, *Systematic Zool.,* **18**, 115–126.

Cracraft, J., and Eldredge N., eds., 1979. *Phylogenetic Analysis and Paleontology.* New York: Columbia Univ. Press.

DeBeer, G. A., 1954. *Archeopteryx* and evolution, *Adv. Sci.,* **42**, 1–11.

Ehrlich, P. R., and Ehrlich, A. H., 1967. The phenetic relationships of the butterflies. I. Adult taxonomy and the nonspecificity hypothesis, *Systematic Zool.,* **16**, 301–317.

Eldredge, N., and Gould, S. J., 1972. Punctuated equilibria: An alternative to phyletic gradualism, in T. J. M. Schopf, ed., *Models of Paleobiology.* San Francisco: Freeman, Cooper, 82–115.

Fitch, W. M., and Margoliash, E., 1967. Construction of phylogenetic trees, *Science,* **155**, 279–284.

Gaffney, E. S., 1975. Phylogeny and classification of the higher categories of turtles, *Amer. Mus. Nat. Hist. Bull.,* **155**, 387–436.

Goodman, M., and Moore, G. W., 1971. Immunodiffusion systematics of the primates. I. The Catarrhini, *Systematic Zool.,* **20**, 19–62.

Gregory, W. K., 1910. The orders of mammals, *Amer. Mus. Nat. Hist. Bull.,* **27**, 3–524.

Harper, C. W., Jr., 1976. Phylogenetic inference in paleontology, *J. Paleontology,* **50**, 180–193.

Hennig, W., 1966. *Phylogenetic Systematics.* Urbana: Univ. Illinois Press, 263p.

Inger, R. F., 1973. Numerical taxonomy, *Caldasia,* **11**, 7–28.

Kluge, A. G., and Farris, J. S., 1969. Quantitative phyletics and the evolution of anurans, *Systematic Zool.,* **18**, 1–32.

Mayr, E., 1963. *Animal Species and Evolution.* Cambridge Mass.: Harvard Univ. Press, 797p.

Mayr, E., 1965. Numerical phenetics and taxonomic theory, *Systematic Zool.,* **14**, 73–97.

Mayr, E., 1969. *Principles of Systematic Zoology.* New York: McGraw-Hill, 428p.

Mayr, E., 1970. *Populations, Species, and Evolution.* Cambridge, Mass.: Belknap Press, 453p.

Mayr, E., 1974. Cladistic analysis or cladistic classification? *Zeitschr. Zool. Syst. Evol.,* **12**, 94–128.

Nelson, G., 1973. Classification as an expression of phylogenetic relationship, *Systematic Zool.,* **22**, 344–359.

Schaeffer, B., 1967. Comments on elasmobranch evolution, in P. W. Gilbert, R. F. Matthewson, and O. P. Rall, eds., *Sharks, Skates and Rays.* Baltimore: Johns Hopkins Press, 3–35.

Schaeffer, B.; Hecht, M. K.; and Eldredge, N., 1972. Phylogeny and paleontology, in T. Dobzhansky, M. K. Hecht, and W. C. Steere, eds., *Evolutionary Biology,* vol. 6. New York: Appleton-Century-Crofts, 31–46.

Simpson, G. G., 1945. The principles of classification and a classification of mammals, *Amer. Mus. Nat. Hist. Bull.,* **85**, 1–350.

Simpson, G. G., 1961. *Principles of Animal Taxonomy.* New York: Columbia Univ. Press, 247p.

Sokal, R. R., and F. J. Rohlf, 1962. The comparison of dendograms by objective methods. *Taxon* **11**, 33–40.

Thomson, K., and Bossey, K. H., 1970. Adaptive trends and relationships in early Amphibia. *Forma et Functio,* **3**, 7–31.

Cross-references: *Biometrics in Paleontology; Evolution; Species Concept; Taxonomy, Numerical.*

T

TAPHONOMY

A most succinct and original definition of taphonomy is "the science of the laws of embedding" (Efremov, 1940). Efremov's original concept of the science included far more than the mere burial of organic relicts; he clearly meant the science to include studies of the many environmental phenomena that affect organic remains throughout their entire postmortem history. This broad usage of the term has been followed in many, but not all, subsequent writings.

Taphonomy is thus a very major part of the environmental aspects of paleontology (Fig. 1). Nearly all lifetime attributes of ancient organisms can be better understood with prior knowledge of the organisms' postmortem history; success in paleoecology, paleobiogeography, and evolutionary studies does often depend upon workers' ability to strip away the taphonomic overprint. Hence taphonomy is, most fundamentally, the discipline that allows scientists to provide "life after death" for organisms of the past.

Taphonomic studies have normally emphasized one or both of two general aspects: (1) the temporal sequence of taphonomic events, and/or (2) the general happenings or detailed processes that affect organic remains. Both of these approaches have their merits, yet it is obvious that knowledge of the processes surrounding the formation of fossil assemblages is most important to taphonomy. Unfortunately, our present understanding of many of these processes is inadequate. Both of the aforementioned study-aspects are included in the paper of Rolfe and Brett (1969), which contains an extensive bibliography of past work in taphonomy.

Temporal Sequence of Events

There are distinctive time intervals in the postmortem history of many organic remains. With the heritage of work by Germanic scientists in the early decades of the 20th century, Europeans have long recognized four important events in the total environmental history of a given fossil: birth, death, final burial, and discovery by a scientist (Fig. 1). The importance given to death, final burial, and discovery is reflected in some of the most widely used terms for fossil assemblages: *thanatocoenose,* for organisms that died together, *taphocoenose,* for remains that were buried together, and *oryctocoenose,* for remains that were found together in outcrop (Hecker, 1965; Sartenaer, 1959).

An excellent and detailed analysis of the temporal factors was provided by Clark and Kietzke (1967) in their study of Oligocene mammal collections from the South Dakota Badlands of North America. Clark and Kietzke wished to interpret the age structure, relative abundances, and other factors of their once-living Oligocene mammals. As a prerequisite for this interpretation they described, analyzed, and evaluated the sources of bias and error in their own collections. They separated seven groups of time-related factors that determined whether or not a particular mammal would become an identified specimen in their collections. These factors were associated with: (1) life attributes of the organisms, (2) their death, (3) the decomposition of bodies or corpses, (4) burial, (5) subsequent exposure, (6) collection techniques, and (7) preparation and curation methods (Fig. 2).

Many life attributes can strongly affect the subsequent preservation of remains. The lifetime distribution of organisms is often important, especially since there are some environments (such as uplands nonmarine settings) that are relatively rare in the accessible geologic record. The construction of skeletal tissues is another determinant; the presence, stable mineral content, and dense and fine-grained nature of hard parts all enhance the ultimate recovery of organic remains by scientists (Chave, 1964). Moderate to large size often increases the probability of recognition and collection of fossil remains, and this factor is but one reflection of the total population dynamics of the once-living organisms. Behavioral traits, including the formation of biogenic sedimentary structures, may improve the potential for preservation of information about a particular taxon. Autolysis can change the inorganic and organic chemistry of plant and animal tissues yielded to sediments and, for example, may decrease the potential for

Interval   Discipline

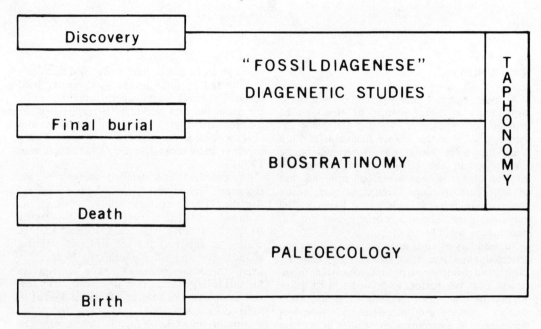

FIGURE 1. Most important events in the environmental history of fossils, and the disciplines concerned with the intervals between these events (from Lawrence, 1968).

finding certain arthropods preserved in the fossil record (Rolfe and Brett, 1969).

The cause and site of death both influence fossilization potential. For example, ingestion by predatory mammals can lead to the complete destruction of the endoskeletons of small prey mammals, while predatory borings drilled through invertebrates' exoskeletons can change patterns of postmortem transport (Lever and Thijssen, 1968). Although herbivores, overall, degrade the tissues passing through their tract, some plant remains may be preferentially preserved in the coprolites of herbivores or even carnivores (Jepsen, 1963). Organisms that die in cryptic settings (such as borings or burrows) are more likely to be preserved *in situ* and intact than are organisms, with a similar overall preservability, that die in more exposed environments.

Decomposition before burial produces the most striking and pervasive biases in the fossil record. Decay of soft parts, by microbial activity, is most important in this regard. Loss of soft tissues does eliminate the record of most plants and many invertebrates and also results in the initial dissociation of vertebrate remains (Weigelt, 1927; Simpson, 1960; Schäfer, 1972). Nonmicrobial organisms and physiocochemical

factors may likewise play a part in this decomposition. Clark (1967) provided a most useful summary of these events, and their timing, for vertebrate remains exposed on temperate steppes. The first two years of exposure are dominated by loss of flesh through vertebrate scavenging, insect activity, and microbial decay, with possibilities also for dehydration and mummification. The second through seventh years are characterized by the flaking of the remaining hard parts; rodents can contribute to this decomposition, but daily and annual temperature changes may be primarily responsible for the splitting and disintegration. Dissolutuion will begin in the fourth or fifth year, leading to final crumbling of the remains. Details of the early decomposition of human bodies have been analyzed numerous times (e.g., Pierce, 1949) and these findings must certainly apply to other terrestrial vertebrates. Microbes, burrowing and boring organisms, physical processes, and subsea chemical diagenesis combine to produce similar effects in the marine realm.

Burial factors can markedly affect the preservability of fossil remains. The rate of burial, nature of enclosing lithic matrix, and oxidation conditions at the burial site appear to be of

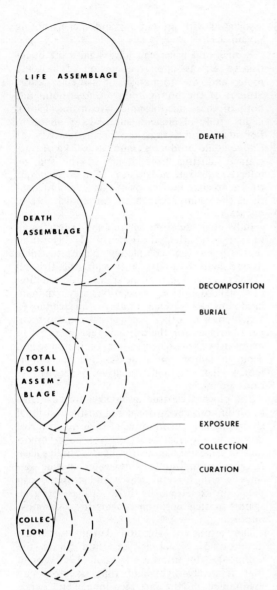

LIFE ASSEMBLAGE

DEATH

DEATH ASSEMBLAGE

DECOMPOSITION

BURIAL

TOTAL FOSSIL ASSEM- BLAGE

EXPOSURE

COLLECTION

CURATION

COLLEC- TION

FIGURE 2. General relationship between once-living communities and collections of fossils, showing the important sources of bias and error in the fossil collections (modified from Clark and Kietzke, 1967).

paramount importance here. Rapid burial is a classic criterion for preservation, because such burial effectively stops many of the previously active decompositional processes. Esentially instantaneous burial (especially by volcanic activity) has led to the preservation of unusual intact remains including the "buried forests" of Yellowstone National Park, Wyoming, USA (Dorf, 1964). The chemistry, mineral content, and particle size of the enclosing matrix can influence both chemical diagenesis and physical distortion of fossils

during compaction or tectonism. Fossil remains enclosed in finer-grained sediments, for example, are more likely to be strongly deformed during compaction than are those enclosed in similar-sized epiclastic sediments. Yet this same finer-grained matrix should enhance preservation during subsequent structural events and metamorphism (Bucher, 1953). Preservation is strongly enhanced by reducing conditions at the burial site because such conditions tend to eliminate the preburial decompositional agents. The exquisite preservation of ichthyosaurs from the Jurassic Posidonia Shales of Germany (Hofmann, 1958) attests to the role these settings have in adding to our knowledge of the fossil record. Burial itself may be a multistage event, with reexposure and additional preservational changes following an initial covering of the remains. Excellent examples of these latter complexities are in Seilacher (1971) and Boyd and Newell (1972).

Subsequent exposure of the remains may restart many of the decompositional events and can lead to the reworking of remains out of their original stratigraphic position. Care in collecting and curating duties will obviously enhance the value of documented fossil remains.

Like all generalizations, the preceding interpretations of the temporal factors nearly all suffer from known deviations. The factors are extremely complex and interacting. Together, they often produce very strong biases in the collections found in outcrops and in museums or laboratories. Indeed, these latter collections most commonly represent very fragmentary and blurred snapshots of the original populations and communities.

Processes and Scenarios

Lawrence (1968) recognized two general types of process-related scenarios that characterize the postmortem history of remains: (1) the nonpreservation or alteration of remains, and (2) the movement of organic relicts from lifetime positions, with or without major transport away from life settings. Alteration/ nonpreservation effects are commonly described in terms of whether the causative processes are primarily biological, physical, or chemical in nature; movement is ordinarily described with reference to the medium carrying out the processes—normally wind or water currents and waves. The net result of these processes is usually a major loss of information about once-living fossil communities.

Reorientation of remains, selective transport of individual taxa away from their lifetime or thanatic environments, and differential valve transport in bivalved organisms can be of

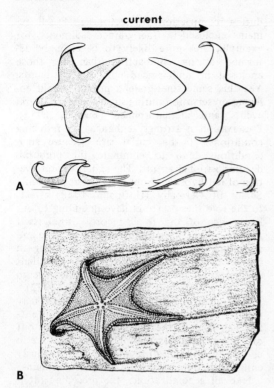

current

A

B

FIGURE 3. Fossil orientation (from Müller, 1963). A: Two possible current orientation patterns in dead starfishes, depending upon whether the oral (left) or aboral (right) surface is uppermost on the substrate. B: These patterns provided clues to the analysis of current regimes in Lower Devonian shales from Germany.

importance in studies of some nonmarine and some inshore marine fossil communities (Shotwell, 1955; Nagle, 1967). In addition, the orientation of individual fossils (Fig. 3) can often contribute to our knowledge of flow regimes in ancient sedimentary basins. These factors, however, appear to have little impact upon recognition and reconstruction of the widespread and abundant level-bottom marine communities of the past (Johnson, 1972). (See also Futterer, 1978.)

Biological destruction of skeletons may continue after the remains' burial but is most common in preburial depositional environments, and most fundamentally results from living organisms' search for food and shelter. Processes include chemical dissolution and mechanical disintegration by boring organisms, as well as possible solution or abrasion of skeletons during their passage through the gut of deposit-feeding animals. Ginsburg (1957) has summarized the results of these processes, while Carriker et al. (1969) have reviewed our knowledge of the diversity and processes

associated with present-day and ancient boring organisms.

Boring organisms can have significant quantitative effects upon the erosion of biogenic rocks and the formation of skeletal debris. Studies of the boring sponge *Cliona* point out both of these possible roles. Neumann (1966), in his study of biogenic subtidal carbonates in Bermuda, found that the boring activity of *Cliona* could produce as much as 6–7 kg of fine-grained detritus from 1 m^2 of the sponge-infested substrate in 100 days. This corresponds to an erosion rate of over 1 cm/yr, and confirms the sponges' capability for rapid substrate disintegration.

Individual skeletons on the sea bottom may be similarly affected. Hartman (1958) found that *Cliona* can completely penetrate adult oyster shells in as little as two months; Driscoll (1967a) suggested that the sponges' destruction of skeletal debris on the sea floor may eliminate hard substrate surfaces in areas, thus leading to a depletion of other species that require these hard surfaces for their attachment. Here, as in many other possible examples, taphonomic and ecologic factors must be considered concurrently when community analyses or reconstructions are made.

The physical destruction of skeletons can also occur in both depositional and postdepositional environments; yet in terms of their quantitative influence, physical processes are most important in depositional environments. Studies have largely emphasized shallow-water marine settings where forceful waves and currents can break up skeletons. In these settings, particle-against-particle abrasion appears to be the most important process.

The results of abrasion depend upon the duration of skeletons' exposure above the sediment-water interface. Driscoll (1970) found that relatively light and thin skeletons are maintained on the interface longer and hence are overall more subject to the effects of abrasion. Skeletal microarchitecture also exerts a major influence upon the rate of abrasive wear. Dense and fine-grained skeletons, and those with relatively little organic matrix, do appear to be most durable (Chave, 1964). Sediment substrates provide an additional major control over abrasion effects, since abrasive impact can occur both skeleton-against-skeleton and skeleton-against-substrate. Durability decreases as sediment grain size increases (Fig. 4) and as sediment sorting decreases (Driscoll and Weltin, 1973).

Driscoll (1970) summarized many of these findings by suggesting that two postmortem environments exist in many shallow-marine settings: one above the sediment–water inter-

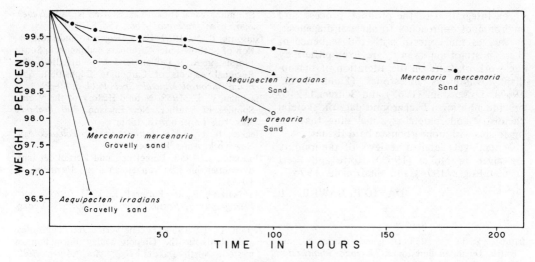

FIGURE 4. Weight reduction due to abrasion for different bivalve species on the beaches of Massachusetts, USA, showing that the lightest- and thinnest-valved species (*Mya*) abrades most rapidly on a percentage basis, and that abrasion rates increase as substrate grain size increases (from Driscoll, 1967b; copyright © 1967 by the Society of Economic Paleontologists and Mineralogists).

face where abrasion, boring, and chemical diagenesis may all be important postmortem processes, and another, below the interface, where chemical diagenesis may have paramount effects (Fig. 5).

Workers have long realized that skeletons of calcium carbonate are most widespread among the phyla represented by fossils (Lowenstam, 1963). Therefore analyses of the postmortem chemical and mineralogical changes in skeletons have been most intimately related to studies of carbonate diagenesis. Historical trends in these general diagenetic studies are reflected in work upon the calcareous fossils themselves.

A most important impetus to recent studies of skeletal diagenesis was provided by the seminal work of Bathurst (1958) on diagenetic fabrics in British Carboniferous limestones.

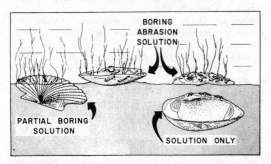

FIGURE 5. Important processes of carbonate skeleton destruction in shallow marine environments, indicating the relatively greater importance of chemical diagenesis *below* the sediment-water interface (from Driscoll, 1970; copyright © 1970 by the Society of Economic Paleontologists and Mineralogists).

Skeletal debris is important in many limestones, and the fabrics of these particles must be evaluated. Since skeletal fabric changes can only be deciphered with prior knowledge of the original skeletal architecture, this approach to diagenesis helped to spur work upon the original mineralogy and microfabrics of important taxa found in sedimentary rocks (Majewske, 1969; Bathurst, 1971; Horowitz and Potter, 1971).

Although alteration fabrics are widespread in fossils, these diagenetic products result from geologic processes that must also be understood for fossil remains. Following the ideas of Chave (1960), Land (1967) investigated the diagenesis of skeletal carbonates with focus upon the three most important mineral phases: calcite, the relatively unstable aragonite, and Mg-calcite. He recognized three basic processes in the stabilization of carbonate hard tissues: (1) "solid state" reactions, including the inversion of aragonite to calcite and the exsolution of Mg-calcite to yield calcite and dolomite; (2) partial or complete dissolution of the unstable phases; and (3) replacement of these latter phases by either dolomite or a noncarbonate mineral. Details of these general processes have been investigated by many workers, with a thorough summary provided by Bathurst (1971).

Environmental settings further influence these diagenetic changes in skeletal tissues. Subsea, subaerial, and subsurface environments can all produce distinctive diagenetic\products, whose recognition can tell us a great deal about the postdepositional environments through which the remains have passed (Purdy, 1968).

The integration of the product, process, and environment approaches to skeletal diagenesis, as well as the concern with the sequence of these postmortem events, are exemplified by the work of Land on the alteration of Bermudian fossiliferous limestones (Land et al., 1967). Lowenstam (1963) and Schraer (1970) provide additional background data on skeletal chemistry and mineralogy, including that for organisms with noncarbonate hard tissues.

General yet detailed reviews of taphonomy are given in Müller (1963), Rolfe and Brett (1969), Roger (1974), and MacDonald (1976).

DAVID R. LAWRENCE

References

Bathurst, R. G. C., 1958. Diagenetic fabrics in some British Dinantian limestones, *Liverpool Manchester Geol. J.*, **2**, 11-36.

Bathurst, R. G. C., 1971. *Carbonate Sediments and Their Diagenesis.* Amsterdam: Elsevier, 620p.

Boyd, D. W., and Newell, N. D., 1972. Taphonomy and diagenesis of a Permian fossil assemblage from Wyoming, *J. Paleontology*, **46**, 1-14.

Bucher, W. H., 1953. Fossils in metamorphic rocks: A review, *Geol. Soc. Amer. Bull.*, **58**, 843-918.

Carriker, M. R.; Smith, E. H.; and Wilce, R. T., eds., 1969. Penetration of calcium carbonate substrates by lower plants and invertebrates, *Amer. Zoologist*, **9**, 629-1020.

Chave, K. E., 1960. Carbonate skeletons to limestones: Problems, *Trans. N.Y. Acad. Sci.*, ser. II, **23**, 14-24.

Chave, K. E., 1964. Skeletal durability and preservation, in J. Imbrie and N. D. Newell, eds., *Approaches to Paleoecology*, New York: Wiley, 377-387.

Clark, J., 1967. Paleogeography of the Scenic Member of the Brule Formation, in Clark, et al., 1967, 75-110.

Clark, J., and Kietzke, K. K., 1967. Paleoecology of the Lower Nodular Zone, Brule Formation, in the Big Badlands of South Dakota, in Clark et al., 1967, 111-137.

Clark, J.; Beerbower, J. R.; and Kietzke, K. K., 1967. Oligocene Sedimentation, Stratigraphy, Paleoecology and Paleoclimatology in the Big Badlands of South Dakota, *Fieldiana, Geol. Mem. 5*, 158p.

Dorf, E., 1964. The petrified forests of Yellowstone Park, *Sci. American*, **210**, 107-114.

Driscoll, E. G., 1967a. Attached epifauna-substrate relations, *Limnol. Oceanogr.*, **12**, 633-641.

Driscoll, E. G., 1967b. Experimental field study of shell abrasion, *J. Sed. Petrology*, **37**, 1117-1123.

Driscoll, E. G., 1970, Selective bivalve shell destruction in marine environments, a field study, *J. Sed. Petrology*, **40**, 898-905.

Driscoll, E. G., and Weltin, T. P., 1973, Sedimentary parameters as factors in abrasive shell reduction, *Palaeogeography, Palaeoclimatology, Palaeoecology*, **13**, 275-288.

Efremov, I. A., 1940. Taphonomy; new branch of paleontology, *Pan-Amer. Geol.*, **74**, 81-93.

Futterer, E., 1978. Studient über die Einregelung, Anlagerung und Ein bettung biogener Hartteile im Strömungskanal (Fossil-Lagerstätten Nr. 44), *Neues. Jahrb. Geol. Paläontol. Abh.*, **156**, 87-131.

Ginsburg, R. N., 1957. Early Diagenesis and Lithification of Shallow-water Carbonate Sediments in South Florida, in R. J. LeBlanc and J. G. Breeding, eds., Regional Aspects of Carbonate Deposition, *Soc. Econ. Paleontol. Mineral., Spec. Publ. 5*, 80-99.

Hartman, W. D., 1958. Natural History of the Marine Sponges of Southern New England, *Bull. Peabody Mus. Yale Univ.*, no. 12, 155p.

Hecker, R. F., 1965. *Introduction to Paleoecology.* New York: Amer. Elsevier, 166p.

Hofmann, J., 1958. Einbettung und Zerfall der Ichthyosaurier im Lias von Holzmaden, *Meyniana*, **6**, 10-55.

Horowitz, A. S., and Potter, P. E., 1971. *Introductory Petrography of Fossils.* New York: Springer-Verlag, 302p.

Jepsen, G. L., 1963. Eocene vertebrates, coprolites, and plants in the Golden Valley Formation of western North Dakota, *Geol. Soc. Amer. Bull.*, **74**, 673-684.

Johnson, R. G., 1972. Conceptual Models of Benthic Marine Communities, in T. J. M. Schopf, ed., *Models in Paleobiology.* San Francisco: Freeman, Cooper, 148-159.

Land, L. S., 1967. Diagenesis of skeletal carbonates, *J. Sed. Petrology*, **37**, 914-930.

Land, L. S.; MacKenzie, F. T.; and Gould, S. J., 1967. Pleistocene history of Bermuda, *Geol. Soc. Amer. Bull.*, **78**, 993-1006.

Lawrence, D. R., 1968. Taphonomy and information losses in fossil communities, *Geol. Soc. Amer. Bull.*, **79**, 1315-1330.

Lever, J., and Thijssen, R., 1968. Sorting phenomena during the transport of shell valves on sandy beaches studied with the use of artifical valves, *Symp. Zool. Soc. London*, no. 22, 259-271.

Lowenstam, H. A., 1963. Biological problems relating to the composition and diagenesis of sediments, in T. W. Donnelly, ed., *The Earth Sciences—Problems and Progress in Current Research.* Chicago: Univ. Chicago Press, 137-195.

Macdonald, K. B., 1976. Paleocommunities: Toward some confidence limits, in R. W. Scott and R. R. West, eds. *Structure and Classification of Paleocommunities.* Stroudsburg; Pa.: Dowden, Hutchinson & Ross, 87-106.

Majewske, O. P., 1969. *Recognition of Invertebrate Fossil Fragments in Rocks and Thin Sections.* Leiden: Brill, 101p.

Müller, A. H., 1963. *Lehrbuch der Paläozoologie, vol. 1, Allgemeine Grundlagen.* Jena: Gustav Fischer Verlag, 387p.

Nagle, J. S., 1967. Wave and current orientation of shells, *J. Sed. Petrology*, **37**, 1124-1138.

Neumann, A. C., 1966. Observations on coastal erosion in Bermuda and measurements of the boring rate of the sponge Cliona lampa, *Limnol. Oceanogr.*, **11**, 92-108.

Pierce, W. D., 1949. Fossil arthropods of California. 17, The silphid burying beetles in the asphalt deposits, *S. Calif. Acad. Sci. Bull.*, **48**, 54-70.

Purdy, E. G., 1968. Carbonate diagenesis: An environmental survey, *Geol. Roman.*, **7**, 183-228.

Roger, J., 1974. *Paléontologie Générale.* Paris: Masson et Cie, 419p.

Rolfe, W. D. I., and Brett, D. W., 1969. Fossilization Processes, in G. Eglinton and M. T. J. Murphy, eds., *Organic Geochemistry: Methods and Results.* Berlin and New York: Springer-Verlag, 213-244.

Sartenaer, P., 1959. La plongée en scaphandre autonome au service de la Taphonomie, *Bull. Inst. Oceanogr., Monaco,* no. 1159, 14p.

Schäfer, W., 1972. *Ecology and Paleoecology of Marine Environments.* Chicago: Univ. Chicago Press, 568p.

Schraer, H., ed., 1970. *Biological Calcification: Cellular and Molecular Aspects.* New York: Appleton-Century-Crofts, 462p.

Seilacher, A., 1960. Strömungsanzeichen im Hunsrückschiefer, *Notizbl. Hess. Landesamt Bodenforsch. Wiesbaden,* 88, 88-106.

Seilacher, A., 1971. Preservational history of ceratite shells, *Palaeontology,* **14,** 13-21.

Shotwell, J. A., 1955. An approach to the paleoecology of mammals, *Ecology,* **36,** 327-337.

Simpson, G. G., 1960. The history of life, in S. Tax, ed., *Evolution After Darein,* vol. 1. Chicago: Univ. Chicago Press, 117-180.

Weigelt, J., 1927. *Rezente Wirbeltierleichen und ihre paläobiologische Bedeutung.* Leipzig: Max Weg, 227p.

Cross-references: *Actualistic Paleontology; Biostratinomy; Diagenesis of Fossils—Fossildiagenese; Fossils and Fossilization; Paleoecology, Terrestrial.* Vol. I: *Mass Mortality in the Sea.* Vol. VI: *Biostratinomy; Coquina, Criquina; Diagenesis.*

TASMANITIDS

Tasmanitids are a group of fossil unicellular organisms, spheroidal to ovoidal or bean-shaped, or biconvex to disc-shaped (this latter shape being generally or constantly produced by compression during diagenesis). Their color ranges from yellow to red-brown or almost black, depending on the degree of humic staining and on the diagenetic history of the enclosing sediment; the range of diameters (or longest axes) is from about 100 to 600 μm. The wall is thick; it is penetrated by pores of two kinds—micropores, radially arranged and large enough to be visible under the light microscope, and ultrapores, similarly arranged but very much smaller and only visible under the electron microscope (cf. Kjellström, 1968). These pores may pass partially or completely through the wall (Fig. 1).

The composition of the wall remains to be determined in detail. Wall (1962), on the basis of chemical tests, concluded that it consisted of a complex lipoid substance, with little or no cellulose. Kjellström (1968), from an examination of the infrared microspectra, concluded that long chains of alifatic saturated carbohydrates were present. Irregular openings or splits on one flank are of relatively common

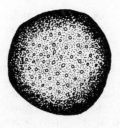 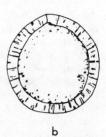

a   b

FIGURE 1. Tasmanitids (after Eisenack, 1958). a, *Tasmanites* sp., Early Ordovician, Baltic; exterior showing pores. b, *Tasmanites huronensis,* Late Silurian, Baltic; optic section. ca X135.

occurrence. The presence of a circular opening (a pylome) in the wall of some tasmanitids has been reported but remains to be confirmed. Similar, but probably unrelated, acritarchs have such an opening; and a mistake in identification appears likely.

P. Taugourdeau [1962] has recorded the finding of thick-walled circular spheres, perhaps tasmanitids, in linear tetrads, and A. Combaz [1967] found similar forms in loosely associated masses. However, these occurrences are exceptional: Tasmanitids, whether present in low number or in huge quantities in a particular stratum, normally occur singly.

History of Study

A full bibliographic review of the tasmanitids has been published by Muir and Sarjeant (1971).[1] These organisms were first reported by J. D. Hooker [1852] from the uppermost Silurian (Ludlovian) of England and described simply as "sphaeroidal bodies." Subsequently J. W. Dawson, in a series of papers and books published between 1863 and 1888, reported the presence of similar bodies in the Devonian and Carboniferous of Canada. Dawson originally believed them [1871] to be spore-cases (sporangia), naming them *Sporangites huronensis:* later [1884, 1886] he considered them to be the spores of a sporocarp, *Protosalvinia,* which he erroneously regarded as the sporangium of a plant with lepidodendroid affinities.

In the meantime, it had been discovered that they occurred in vast number in the Permian "white coal" (tasmanite) of Tasmania. T. S. Ralph [1865], in a paper published only in summary, identified them correctly as algae but proposed no name for them; it was not until a decade later that E. T. Newton proposed for them the name *Tasmanites punctatus.*

In 1905, P. F. Reinsch found similar forms in

[1] The references for works dated in square brackets may be found in full in Muir and Sarjeant, 1971.

the French Upper Cretaceous and gave them the informal name "palinosphères;" this never came into use.

In 1931, A. Eisenack discovered them in Silurian boulders in the North German drift; unaware of earlier studies, he at first named his forms *Bion solidum,* but later used that species as type for a new genus, *Leiosphaera.* A number of species of spherical acritarchs were subsequently placed in that genus and came to be popularly designated "leiospheres."

In 1941, R. Kräusel reconsidered Dawson's sporocarps and showed them to be small thallophytic plants, whose spores (trilete, produced from tetrads) evinced few common features with the species *huronensis.* He therefore transferred this species to Eisenack's genus *Leiosphaera.* Unaware of his work, Schopf, Wilson, and Bentall [1944] reviewed the taxonomy of *Sporangites* and *Tasmanites,* rejecting the former name. F. W. Sommer [1956], describing rich assemblages of these forms from the Devonian of Brazil, made the latter genus the type of a new algal family, the Tasmanaceae.

The confusion in nomenclature was finally sorted out by Eisenack (1958), who showed *Sporangites huronensis* and *Leiosphaera solida* to be synonyms, concluding that the correct name for this species was *Tasmanites huronensis.* One other species formerly attributed to *Leiosphaera* was considered sufficiently similar in morphology to be placed in *Tasmanites:* the remaining forms of dissimilar morphology were placed in a new genus, *Leiosphaeridia* (now regarded as a genus of sphaeromorphid acritarchs).

In 1963, Downie, Evitt, and Sarjeant attributed four other described genera to the family Tasmanaceae (*Crassosphaera, Pleurozonaria, Tytthodiscus* and *Zonosphaeridium*). Subsequent studies have greatly increased the number of genera and species (some 30 genera and approximately 110 species have now been described) and have greatly extended the known stratigraphical range.

Affinities

Despite the early suggestion by Ralph [1865] that they were algae, the view that these organisms were plant spores, originally put forward by J. W. Dawson [1871], was long maintained. Its final rejection can be said to have come through the work of Schopf, Wilson, and Bentall [1944], who concluded that *Tasmanites* was to be interpreted as an alga of unknown affinity.

In 1958, Eisenack classed these organisms, together with his leiospheres and a whole array of spiny microfossils, as "hystrichospheres,"

a group he then viewed as an independent order of microorganisms.

Wall (1962) demonstrated the close morphological and compositional similarity of these microorganisms to two living genera of unicellular algae, *Pachysphaera* and *Halosphaera.* His conclusion was that the former genus was in fact a living representative of the fossil *Tasmanites.* Studies by Evitt [1961] had in the meantime demonstrated that the most typical hystrichospheres were dinoflagellate cysts; and in 1963 Evitt proposed that the residue, of unproven affinity, should be termed "acritarchs." A classification for the acritarchs (q.v.) was formulated by Downie, Evitt, and Sarjeant [1963]: The views of Wall were accepted and the tasmanitids were excluded from the group Acritarcha, being instead placed along side *Pachysphaera* and *Halosphaera* as members of the green algae (Chlorophyceae). A subsequent reshuffling of the classification of the algae has occurred and these latter genera are now placed in the class Prasinophyceae; accordingly Downie (1967) placed the tasmanitids into this new class.

Further electron-microscope studies by Jux (1968, 1969) showed that, in the life cycle of *Pachysphaera marshalliae,* the motile alga is free-swimming and quite unlike *Tasmanites.* However, the alga encysts prior to reproduction; a thick-walled zoosporangium is formed, inside which the zoospores develop. This zoosporangium is spherical and penetrated by systems of pores, closely similar to those of fossil tasmanitids: and indeed, it was the zoosporangia of *Pachysphaera* and *Halosphaera* that Wall had examined. Ultimately, the zoosporangium bursts open; an inner sac, from which the zoospores are subsequently liberated, is then released and the abandoned outer membrane falls to the bottom. Since it has to resist fungal and bacterial attack during the period in which the zoospores are developing, the wall material is a macromolecular substance that is not readily subject to decay. For this reason, it has a high fossilization potential; hence the preservation of the abandoned zoosporangia—the tasmanitids.

Geological Distribution

The earliest tasmanitids yet definitely recognized are from the uppermost Cambrian, but similar spherical forms with thick walls occur as far back as the late Precambrian. Tasmanitids have been recorded intermittently throughout the Paleozoic, being immensely abundant at some horizons. In the Mesozoic and Cenozoic, records are fewer. It is clear that some ecological control determines their presence in, or absence from, particular sediments, but the nature of this control has not yet been ascertained.

Since tasmanitids are so resistant to decay, they frequently survive erosion to be redeposited in younger sediments. Indeed, Devonian tasmanitids were recorded by Johnson and Thomas [1884] in the water supply of the city of Chicago! In general, reworked forms can be recognized readily by their higher degree of humic staining; in absence of such staining, however, stratigraphic confusion is likely to arise in view of the constancy of their morphology throughout the period since the Cambrian.

WILLIAM A. S. SARJEANT

References

Downie, C., 1967. The geological history of the microplankton, *Rev. Palaeobot. Palynol.,* **1,** 269–281.

Eisenack, A., 1958. *Tasmanites* Newton 1875 und *Leiosphaeridia* n. g. als Gattungen der Hystrichosphaeridea, *Palaeontographica* B, **110,** 1-19.

Jux, U., 1968. Über den Feinbau der Wandung bei *Tasmanites* Newton, *Palaeontographica* B, **124,** 112-124.

Jux, U., 1969. Über den Feinbau der Zystenwandung von *Pachysphaera marshalliae* Parke, 1966, *Palaeontographica* B, **125,** 104-111.

Kjellström, G., 1968. Remarks on the chemistry and ultrastructure of the cell wall of some Palaeozoic leiospheres, *Geol. Fören. Forh., Stockholm,* **90,** 221-238.

Muir, M. D., and Sarjeant, W. A. S., 1971. An annotated bibliography of the Tasmanaceae and of related living forms (Algae: Prasinophyceae), in S. Jardiné, ed., *Microfossiles organiques du Paléozoique,* vol. 3. Paris: Editions C. N. R. S., 56-117.

Wall, D., 1962. Evidence from recent plankton regarding the biological affinities of *Tasmanites* Newton, 1875, and *Leiosphaeridia* Eisenack, 1958, *Geol. Mag.,* **99,** 36-62.

Cross-references: *Algae; Plankton; Protista.*

TAXONOMY, NUMERICAL

Use of the term *numerical taxonomy* in the 1960s was likely to invoke a strong response. Both paleontologists and neontologists were strongly polarized on the value of using phenetic taxonomy, in which classifications are based solely on organisms' similarity rather than on their inferred phylogenetic relationships (see, for example, Mayr, 1965). Numerical taxonomy was often equated with the phenetic school and the points of view propounded by R. R. Sokal and P. H. A. Sneath]1963] in their book *Principles of Numerical Taxonomy.*

Paleontologists, particularly, were in a quandry. The developers and strongest proponents of phenetic taxonomy were in the fields of entomology and microbiology, where useful fossils are rare. Yet paleontologists were rightly convinced that the chronistic dimension of their studies of the fossil record was a source of valuable information in addition to the phenetic information and inferred cladistic or ancestor-descendant relationships available for all organisms.

While the controversy raged, a generation of paleontologists appeared who were well versed in statistical analysis, multivariate methods, and the use of computers (see *Computer Applications in Paleontology*). As a result, much work was done that was clearly within the scope of numerical taxonomy without the authors' necessarily regarding themselves as numerical taxonomists (e.g., Cheetham, 1968; Gould, 1969; Spencer, 1970). In short, numerical taxonomy has become synonymous with numerical systematics, mathematical taxonomy, multivariate morphometrics, and taxometrics (see Sneath and Sokal, 1973). Taxonomists began using computer-dependent quantitative procedures when they showed promise of providing insight into the problems under study. At the same time, many of the same procedures were being applied in ecology and paleoecology, giving systematists and ecologists a new common ground.

Nevertheless, in spite of the growing convergence of numerical taxonomy as originally used and what might be called the new orthodox taxonomy, it has been the numerical taxonomists who have formulated the new principles on which much of the new taxonomy is coming to be based. Sneath and Sokal (1973) have listed these principles and have discussed them in detail, summarizing the "fundamental position of numerical taxonomy."

1. The greater the content of information in the taxa of a classification and the more characters on which it is based, the better a given classification will be.
2. A priori, every character is of equal weight in creating natural taxa.
3. Overall similarity between any two entities is a function of their individual similarities in each of the many characters in which they are being compared.
4. Distinct taxa can be recognized because correlations of characters differ in the groups of organisms under study.
5. Phylogenetic inferences can be made from the taxonomic structures of a group and from character correlations, given certain assumptions about evolutionary pathways and mechanisms.
6. Taxonomy is viewed and practiced as an empirical science.
7. Classifications are based on phenetic similarity.

They defined numerical taxonomy as "the grouping by numerical methods of taxonomic units on the basis of their character states."

The main uses of numerical taxonomy and related techniques have been in constructing classifications using cluster analysis, in ordinating groups of taxonomic units in order to get a clearer understanding of their similarities to each other, and in attempting to reconstruct phylogenies.

Cluster Analysis

The many methods of cluster analysis that have been developed are too numerous to discuss in detail. Williams (1971) has reviewed many of these methods, and Sneath and Sokal (1973) have discussed the following eight aspects of clustering methods that a user must understand about a computer program for cluster analysis:

Agglomerative vs. Divisive Methods. Agglomerative methods are the most commonly used methods of cluster analysis. They start with all the operational taxonomic units (OTU's) separated and then join them into clusters on the basis of the similarities of the units to each other and to previously formed clusters. Divisive methods, on the other hand, start with all OTU's in one set and employ various procedures to divide the set into subsets.

Hierarchic vs. Nonhierarchic Methods. Hierarchic methods of cluster analysis are the most familiar to us and are preferred over non-hierarchic methods for traditional biological taxonomy. Nonhierarchic methods include ordination, to be discussed in the next section; unrooted graphs, such as might result from modification of an ordination; and "clustering systems" (Lance and Williams, 1967), such as k-means clustering (MacQueen, 1967), which is useful for very large data sets because the number of clusters can be preset by the user. Sneath and Sokal (1973) have pointed out that nonhierarchic clustering "is preferred when emphasis is placed on faithful representation of the relationships among the OTU's rather than on a summarization of these relationships."

Nonoverlapping vs. Overlapping Methods. Nonoverlapping clusters are intuitively the most appealing because of our long experience with the Linnean System in which classifications are both hierarchic and nonoverlapping. However, Michener (1963) has suggested permitting OTU's to belong to more than one higher taxon. Such a scheme may result in less distortion of nomenclature and of phenetic relationships among OTU's. Jardine and Sibson (1971) and others have employed overlapping clusters.

Sequential vs. Simultaneous Methods. Ordination methods are simultaneous because all OTU's are considered simultaneously (see next section). However, most cluster analysis has been sequential, by which is meant "that a recursive sequence of operations is applied to the set of OTU's . . ." (Sneath and Sokal, 1973).

Local vs. Global Criteria. Some sequential agglomerative clustering methods give a good estimate of the similarities of OTU's within a cluster but distort similarities among OTU's in different clusters because of averaging that is done during clustering. Principal component analysis, on the other hand, may give a good estimate of the relationships between clusters but not a true representation of OTU's within clusters (Rohlf, 1970). The disparity of reliability between the two methods is expected to receive more attention by numerical taxonomists in the future.

Direct vs. Iterative Solutions. Direct clustering was regarded by Sneath and Sokal (1973) as clustering that is done "in a straightforward manner." The cluster that results is regarded as the solution to the clustering problem. Iterative clustering involves rearranging the members of clusters in order to maximize some criterion, such as the cophenetic correlation coefficient. Direct clustering has been used far more extensively than iterative clustering.

Weighted vs. Unweighted Clustering. Much has been written about the relative merits of weighting characters equally or according to their supposed "taxonomic importance." By weighted clustering, however, the numerical taxonomist refers to weighting that is done during the clustering procedure. For example, one commonly used clustering procedure, the weighted pair group method with arithmetic averages (WPGMA), gives joining branches equal weight, regardless of the number of OTU's that have already joined each branch. Thus, if a cluster of ten OTU's A, joined a single OTU, B, the two branches A and B, would be given equal weight in assigning the average similarity to a third cluster or OTU, C, added during the next step in the cluster analysis. Furthermore, if one dimension of a cluster is regarded as more important than another, so that the cluster is forced to grow in an elongate shape, that is another kind of weighting. Unweighted clustering regards all OTU's and all dimensions as being of equal importance and, hence, gives them equal weight. The unweighted pair group method with arithmetic averages (UPGMA) has been found empirically to give less distortion of similarity relationships, and theoretical justification has been given for this being so. The measure of lack of distortion that is usually used is the cophenetic correlation coefficient, and UPGMA clustering consistently gives a higher coefficient than other clustering procedures.

Nonadaptive vs. Adaptive Clustering. Nonadaptive clustering means that the clusters are computed according to a fixed method. Most cluster analysis has been of this type. Adaptive clustering implies that the clustering algorithms to be used are chosen according to the type of clusters that are encountered during the clustering procedure. The algorithms may be changed during clustering according to what the computer has "learned" about the data set.

Fig. 1 shows a phenogram computed from a taxonomic distance matrix by an agglomerative, hierarchic, nonoverlapping, sequential, direct, unweighted, nonadaptive clustering method that gives better reliability locally than globally (Kaesler, 1970). The method is UPGMA, one of the most commonly used clustering methods. Users of agglomerative, hierarchic clustering methods such as UPGMA and WPGMA have usually omitted any measure of the amount of distortion introduced during the clustering procedure. One of the most useful methods of measuring distortion is the coeffi-

cient of cophenetic correlation. This coefficient is a product-moment correlation coefficient computed between correspongding elements of the original similarity matrix and a matrix of similarities as shown by the phenogram. In the figure, the cophenetic correlation is 0.764, indicating a moderate but not serious amount of distortion.

Ordination

Whereas cluster analysis is intended for making classifications, ordination enables the user to visualize the taxonomic units he is studying with respect to each other as plotted in a space. If there are t OTU's being studied, each with n characters, the ordination shows the t OTU's plotted in a space with from 1 to n or $t - 1$ dimensions, which ever is less (Sneath and Sokal, 1973). Of course, it is not physically possible to plot points in a space of more than three dimensions. Therefore, mathematical means are used to reduce the number of dimensions in the space with as little loss of information or distortion of the relationships as possible. Five methods have been used most

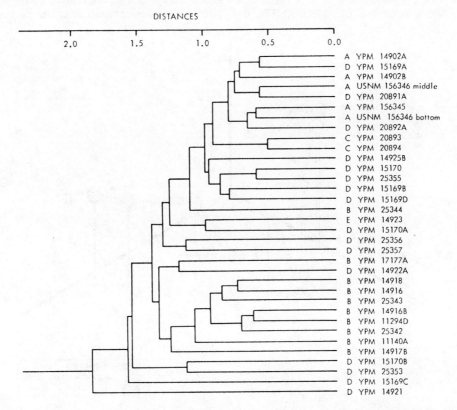

FIGURE 1. Phenogram prepared by an agglomerative, hierarchic clustering method (UPGMA) showing similarities among 34 specimens of the fusulinid genus *Pseudoschwagerina* from 5 species or subspecies (from Kaesler, 1970). YPM numbers refer to numbers given to the specimens by the Yale Peabody Museum of Natural History.

commonly: principal component analysis, principal coordinate analysis, multiple factor analysis, canonical variate analysis, and non-metric multidimensional scaling.

Principal Component Analysis. Mathematical treatments of principal component analysis have been given by many authors. Two of the clearest and most concise are by Davis (1973) and Sneath and Sokal (1973). Viewed geometrically, principal component analysis is a means of transforming the axes of the n-dimensional character or attribute space (A-space) in which the OTU's occur so that the first axis explains the maximum amount of variance, the second, orthogonal to the first, explains the maximum of what remains, and so forth. There are n such principal components (normalized eigenvectors), but often as few as two or three of them explain nearly all of the variation in the data.

The matrix that results from a principal component analysis gives the "directions of the set of k orthogonal axes in the A-space and are known as the *principal axes*" (Sneath and Sokal, 1973). In order to plot the new coordinate points of the OTU's, the equation $P = VX$ is used, "where P is the $k \times t$ matrix of coordinates of the t OTU's on the k principal axes (factors)" (Sneath and Sokal, 1973), V is the transposed $n \times k$ matrix of principal components, and X is the $n \times t$ matrix of data that has been standardized by rows. The coordinates in the matrix P may be used to plot the t OTU's in the new A-space. Since the axes are orthogonal, they may be selected for use independently of each other.

Fig. 2 (Rowell, 1970) is an example of a three-dimensional principal component ordination of ten species of trilobites. A fourth dimension familiar to paleontologists has been added by showing the biostratigraphic zones to which the species belong. The broken lines show a minimal-distance simply connected network based on average distance.

Principal Coordinate Analysis. This method, developed by Gower (1966), is a significant advance over principal component analysis because it permits one to compute principal components from a distance matrix. It is useful when the number of characters greatly exceeds the number of OTU's, since it is not necessary to deal with a large $n \times n$ matrix of correlations amoung characters. When Euclidean distances are used and the data contain no missing values, results of principal coordinate analysis and principal component analysis are proportional.

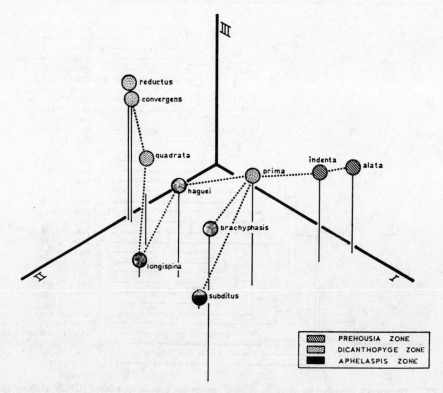

FIGURE 2. Principal component ordination showing relationships among ten species of trilobites as represented by the first three principal components (from Rowell, 1970).

Multiple Factor Analysis. Multiple factor analysis was used widely in geology during the 1960s, especially by sedimentologists. Its use in numerical taxonomy has declined since the early work by R. R. Sokal (see Rohlf and Sokal, 1962), but a few authors have made effective use of it (Gould, 1969; Gould and Littlejohn, 1973).

Multiple factor analysis involves computation of a factor matrix from a matrix of correlation coefficients by essentially the same procedure as was used for principal component analysis. The only difference is that the diagonal of the correlation matrix must contain estimates of the communality, which is "the proportion of the variance due to common factors in the study" (Sneath and Sokal, 1973). This is necessary since the goal of multiple factor analysis is to extract common factors, those that affect all OTU's to come extent, rather than special factors that affect only single OTU's. The correct value of the communality must be obtained by iteration. The factor matrix is then rotated according to various criteria to a position that gives *simple structure*. Simple structure has been somewhat loosely defined, so that different algorithms may give different solutions. Finally, the rotated orthogonal factors may be made oblique, indicating correlations amoung the factors. For a more thorough discussion, see Sneath and Sokal (1973).

Canonical Variate Analysis. This method is related to Mahalanobis' generalized distance and to multivariate discriminant analysis (Sneath and sokal, 1973). The lengths of the axes of the space in which the OTU's are to be plotted are distorted so that the hyperellipsoids containing the OTU's in the various groups become hyperspheres. The method has not been widely used, largely because it is statistical method and few sets of data meet the rigorous assumptions on which the method is based.

Fig. 3 (Spencer, 1970) shows a particularly

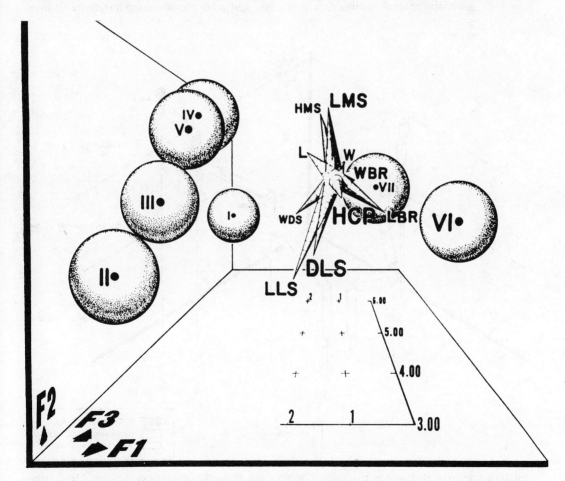

FIGURE 3. Canonical variate analysis resulting in ordination of brachiopods from seven stratigraphic units (from Spencer, 1970). Cones near the center show the influence of the measured characters.

effective use of the method in ordination. The cones near the center of the diagram represent the influence of the measured characters, and the spheres with Roman numerals are the unit spheres representing the brachiopods from each of seven localities sampled.

Nonmetric Multidimensional Scaling. Sneath and Sokal (1973) have emphasized the general nature of this method since it can operate on either similarity or dissimilarity matrices. It is well suited to much paleontological data because it will also accept rank orders of dissimilarity. The method attempts to find a monotonic relationship between the original dissimilarity matrix and the matrix of dissimilarities in a space of k dimensions, where k is very much smaller than t. If a monotonic relationship exists, the ordination is considered perfect (Sneath and Sokal, 1973). Otherwise, the *stress* that measures the degree of dissimilarity between the two is computed according to the method derived by Kruskal (1964). Nonmetric multidimensional scaling has not

been widely used in paleontology. Rowell et al. (1973) have made very effective use of the method, and we can expect that its use will grow as more paleontologists become aware of its advantages and familiar with its computation.

Reconstructing Phylogenies

Paleontologists have perhpas more reason than other biologists to be concerned with phylogeny, and the chronistic data available to them gives them an advantage over neontologists that more than compensates for the lack of genetical and embryological information about fossils. At the same time, they have felt less strongly than their neontological colleagues the need for numerical aid in reconstructing phylogenies. Nevertheless, numerical taxonomic methods can be of use in reconstructing phylogenies, particularly the methods referred to as numerical cladistics; and some paleontologists have made use of these methods. The philosophy and mathematics of the numerical cladistic

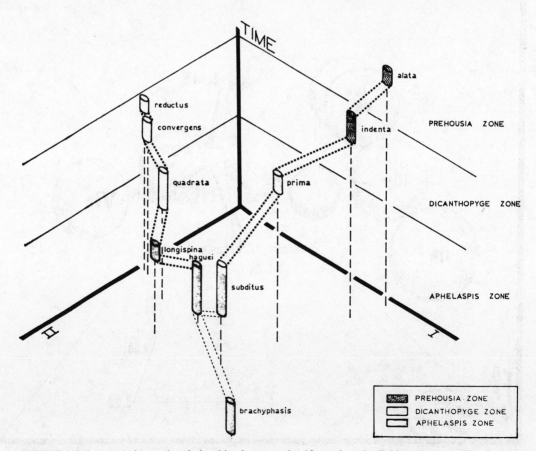

FIGURE 4. Inferred phylogenetic relationships between the 10 species of trilobites shown in Fig. 2 (from Rowell, 1970). The three dimensions include the first and second principal components and geologic time. The heavy dashed lines represent the minimal distance simply connected graph also shown in Fig. 2.

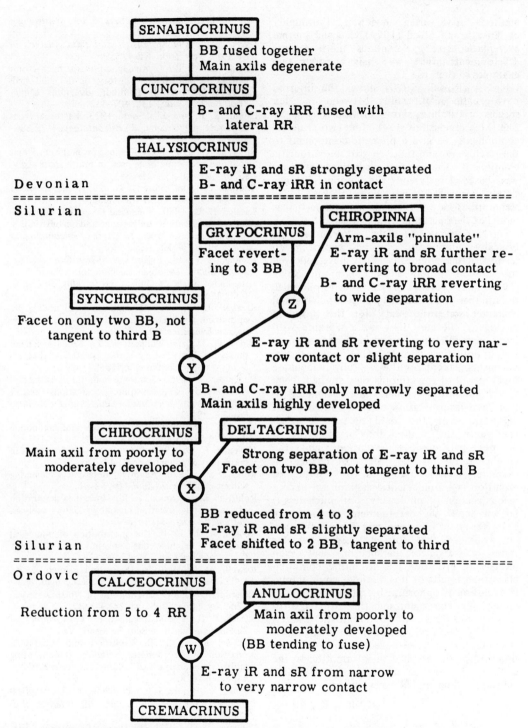

FIGURE 5. Cladogram with chronology made into postulated phylogenetic tree, showing changes in characters at each stage (from Kesling and Sigler, 1969).

methods have been reviewed thoroughly by Sneath and Sokal (1973), and the journal *Systematic Zoology* contains more recents developments. Here we shall consider two examples of their use.

Fig. 4 (Rowell, 1970) shows the inferred phylogenetic relationships between the ten species of trilobites shown previously (Fig. 2). The three dimensions are the first two principal components, to give a phenetic component to the phylogeny, and time, to give it a chronistic component. The inferred cladistic component was based largely on the minimal-distance simply connected graph shown in Fig. 2. No assumptions have been made about a particular model of the operation of evolution.

Fig. 5 (Kesling and Sigler, 1969) is a more typically numerical cladistic approach to reconstructing phylogeny. They have followed the methods proposed by Camin and Sokal (1965) and Wagner (1961) in order to minimize the number of evolutionary steps or changes in character states necessary for the proposed phylogeny. Because they were working with fossil material, they might have been able to test hypothetical ancestral OTU's against real organisms, except for the woefully inadequate fossil record of the crinoids they studied. They found no serious temporal inconsistencies, so that their interpretation of the phylogeny was consistent with the fossil record. A disadvantage of the method they used is that it assumes parsimonious evolution, and the number of evolutionary steps or changes in character states is minimized. Irreversibility of evolution is a fundamental assumption in any reconstruction of phylogeny, but much less is known about the development of successive character states in an evolving lineage. We have no assurance that for successive character states convergence, parallelism, and even reversal of evolution are not more commonplace than results of this method would imply. If evolution is opportunistic rather than deterministic, its course through time will be difficult to ascertain with deterministic models bound by assumptions about the changing of character states. Nevertheless, as an approximation and in the absence of information to the contrary, it may be unwarranted to assume anything but parsimonious evolution.

ROGER L. KAESLER

References

Camin, J. H., and Sokal, R. R., 1965. A method for deducing branching sequences in phylogeny, *Evolution,* 19, 311-326.

Cheetham, A. H., 1968. Morphology and systematics of the bryozoan genus *Metrarabdotos, Smithsonian Misc. Collect.,* 153, 1-121.

Davis, J. C., 1973. *Statistics and Data Analysis in Geology.* New York: Wiley, 550p.

Gould, S. J., 1969. An evolutionary microcosm: Pleistocene and Recent history of the land snail *P. (Poecilozonites)* in Bermuda. *Bull. Mus. Comp. Zool. Harvard Univ.,* 138, 407-532.

Gould, S. J., and Littlejohn, 1973. Factor analysis of caseid pelycosaurs, *J. Paleontology,* 47, 886-891.

Gower, J. C., 1966. Some distance properties of latent root and vector methods used in multivariate analysis, *Biometrika,* 53, 325-338.

Jardine, N., and Sibson, R., 1971. *Mathematical Taxonomy.* New York: Wiley, 286p.

Kaesler, R. L., 1970. Numerical taxonomy in paleontology: Classification, ordination, and reconstruction of phylogenies, *Proc. N. Amer. Paleontological Conv.,* pt. B, 84-100.

Kesling, R. V., and Sigler, J. P., 1969. *Cunctocrinus,* a new Middle Devonian calceoncrinid crinoid from the Silica Shale of Ohio, *Contrib. Mus. Paleontol. Univ. Michigan,* 22, 339-360.

Kruskal, J. B., 1964. Multidimensional scaling by optimizing goodness of fit to a nonmetric hypothesis, *Psychometrica,* 29, 1-27.

Lance, G. N., and Williams, W. T., 1967. A general theory of classificatory sorting strategies. II. Clustering systems, *Computer J.,* 10, 271-297.

MacQueen, J., 1967. Some methods for classification and analysis of multivariate observations, in *5th Berkely Symp. on Math. Stat. and Probability, Proc.,* 1, 281-297.

Mayr, E., 1965. Numerical phenetics and taxonomic theory, *Systematic Zool.,* 14, 73-97.

Michener, C. D., 1963. Some future developments in taxonomy, *Systematic Zool.,* 12, 151-172.

Rohlf, F. J., 1970. Adaptive hierarchical clustering schemes, *Systematic Zool.,* 19, 58-82.

Rohlf, F. J., and Sokal, R. R., 1962. The description of taxonomic relationships by factor analysis. *Systematic Zool.,* 11, 1-16.

Rowell, A. J., 1970. The contribution of numerical taxonomy to the genus concept, *Proc. N. Amer. Paleontological Conv.,* pt. C, 264-293.

Rowell, A. J.; McBride, D. J.; and Palmer, A. R., 1973. Quantitative study of Trempealeauian (latest Cambrian) trilobite distribution in North America, *Geol. Soc. Amer. Bull.,* 84, 3429-3442.

Sneath, P. H. A., and Sokal, R. R., 1973. *Numerical Taxonomy.* San Francisco: Freeman.

Spencer, R. S., 1970. Evolution and geographic variation of *Neochonetes granulifer* (Owen) using multivariate analysis of variance, *J. Paleontology,* 44, 1009-1028.

Wagner, W. H., Jr., 1961. Problems in the classification of ferns *Rec. Adv. Bot. 9th Internat. Bot. Congr.,* 1(9), 841-844.

Williams, W. T., 1971. Principles of clustering, *Ann. Rev. Ecol. Systematics,* 2, 303-326.

Cross-references: *Biometrics in Paleontology; Computer Applications in Paleontology; Systematic Philosophies.*

TENTACULITA

Tentaculita is a class name proposed by Bouček (1964) for enigmatic animal remains confined to the Paleozoic era. The fossils are elongate, straight, tapering calcareaous tubes closed at the small end (Fig. 1); in many species this closed end is a tiny bulbous chamber. The exterior commonly has multiple annulate ridges but may bear longitudinal striations or lirae. Although the earliest Tentaculita may be as old as Middle Cambrian, the Silurian–Devonian was their time of greatest abundance and diversity, and they may have become extinct by the end of the Devonian. They are most common in fine-grained sedimentary rocks, and locally may be so abundant as to be rock formers. More than 150 species have been described.

Classification of tentaculoids has been a zoological problem. They have been referred to numerous phyla, but most commonly to the Mollusca because of supposed homologies to the shell of Holocene specialized pelagic gastropods. Recently, Blind and Stürmer (1977) found soft-bodied fossile suggesting that at least some tentaculites were cepholopod

mollusks, while material studied by Towe (1978) exhibited shell structure similar to some brachiopods.

Several class names have been proposed for various combinations of tentaculoids and other enigmatic fossils; one of the more recent was Coniconchia, by G. P. Ljaschenko in 1955. Fisher (1962) recognized their uniqueness by placing them in a new class, Crinconarida, which he assigned to the Mollusca. Bouček removed them from this phylum because they possess a multilayered wall as well as other nonmolluscan features; he did not assign Tentaculita to another phylum but implied strongly that they may be related to Annelida. The limits of Tentaculita and the arrangement of their internal subdivisions differs from Crinconarida.

The Tentaculita are divided into five orders; only two are widespread, and these are best typified by *Tentaculites* and *Nowakia,* respectively. *Tentaculites* and allies are relatively long, thick-walled forms, considered part of the shallow-water benthos. Yoder and Erdtmann (1975) have suggested dimorphism in one species but this interesting notion has not been investigated further. By shortening the tube and reducing the wall to a thin layer, they may have given rise to the pelagic nowakioids. Representatives of the remaining three orders are relatively obscure fossils; all are considered benthic in habit.

Recent Russian work by G. P. Ljaschenko on benthic forms suggests that they may be useful as local guide fossils, though heretofore they have had little utility in stratigraphic work. The pelagic forms apparently evolved at a rapid rate, and because of their habit they may be valuable for intercontinental correlation (Churkin and Carter, 1970).

ELLIS L. YOCHELSON*

References

Blind, W., and Stürmer, W., 1977. *Viriatella fuchsi* Kutscher (Tentaculoidea) mit Sipho und Fangarmen, *Neues Jahrb. Geol. Paläontol. Mh.,* **1977,** 513–522.
Bouček, B., 1964. *The Tentaculites of Bohemia: Their Morphology, Taxonomy, Ecology, Phylogeny and Biostratigraphy.* Prague: Czech. Acad. Sci., 215p.
Churkin, M., Jr., and Carter, C., 1970. Devonian tentaculitids of east-central Alaska: Systematics and biostratigraphic significance, *J. Paleontology,* **44,** 51–68.
Fisher, D. W., 1962. Small conoidal shells of uncertain affinities, in R. C. Moore, ed., *Treatise on Inverte-*

FIGURE 1. *Tentaculites attenuatus* Hall from the Middle Devonian of Ontario, Canada; ca. X4.

*Publication authorized by the Director, U.S. Geological Survey.

brate Paleontology, pt. W, Miscellanea. Lawrence, Kansas: Geol. Soc. Amer. and Univ. Kansas Press, W98–W143.

Ljaschenko, G. P., 1959. *Konikonchii devona central-nych i vostocnych oblastej Russkoj platformy.* Leningrad: Gostoptechizdat, 220p.

Towe, K. M., 1978. *Tentaculites:* Evidence For a brachiopod affinity? *Science,* **201,** 626–628.

Yoder, R. L., and Erdtmann, B–D., 1975. *Tentaculites attenuatus* Hall and *T. bellulus* Hall: A redescription and interpretation of these species as dimorphs, *J. Paleontology,* **49,** 374–386.

Cross-references: *Cornulitidae; Mollusca.*

TINTINNIDS

In 1902, Lorenz described some minute organisms—90–100 μm in length—found in great numbers in the rocks of the Portlandian (Upper Jurassic) of Switzerland. He named them *Calpionella alpina,* considering them to be *incertae sedis* (of uncertain position). Although for a long time their taxonomic affinities were unknown, they were widely used by Alpine geologists as good guide fossils for the Portlandian. In the ensuing years, several additional species were described, on the basis of differences in the morphology of the shell, or "lorica."

In studying some Upper Jurassic and Lower Cretaceous sediments, this author found a variety whose lorica closely resembled those of a group of present-day pelagic Infusorians (Colom, 1934, 1939, 1948). Several new genera and species were described, some very similar to living forms, such as *Stenosemellopsis,* which is close to *Stenosemella; Tintinnopsella,* derived from *Tintinnopsis; Favelloides,* from *Favella,* etc.

In specimens from the modern plankton, the loricas of the Tintinnids are almost always found empty. Seen through the light microscope, their loricas appear as a simple, bright silhouette similar to those seen in thin sections of rocks preserved by calcite (Figs. 1 and 2). By these comparisons, it was found that all described fossil forms belong to the large group of loricate Infusorians of the present order Oligotricha, some of whose representatives belong to oceanic plankton (holoplankton), while others live in neritic waters (meso-plankton). They are also found in pollen samples from ancient peat bogs (Visscher, 1970). In the Upper Jurassic of Greece are specimens in various phases of sexual reproduction (conjugation), uniting the two loricas just as is seen in living forms (Colom, 1965).

The chemical composition of the lorica is still uncertain, some being gelatinous, while

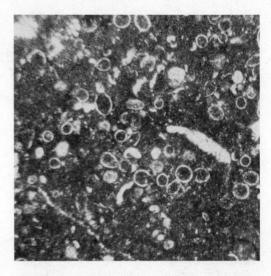

FIGURE 1. Thin section from the Portlandian (Upper Jurassic) of Mallorca, with abundant tintinnids: *Calpionella alpina* Lorenz, *Crassicollaria* sp.; X80.

others are pseudochitinous(?). In modern forms, the loricas that are most useful taxonimically are those (1) whose walls exhibit particular structures ("prismatic," "alveolar," etc.) when seen in section, and (2) that have a tendency to incorporate extraneous particles in the lorica, including quartz and other mineral grains, mica flakes, coccoliths, etc. According to recent analysis of the loricas of modern tintinnids, carried out in the *Instituto de Investigaciones Pesqueras* in Barcelona (R. Margalef,

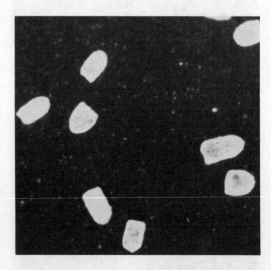

FIGURE 2. Fossil tintinnid loricas preserved in calcite but separated from the matrix by the oxygenated water method; seen with reflected light on a black background. (Micrograph taken from a preparation by Prof. Emberger, University of Bordeaux, X120.)

pers. comm.), the silica content may reach 50–60%.

These data are very interesting for the micropaleontologist, explaining the means by which the delicate loricas have been fossilized, with the calcite slowly replacing the aggregated particles that originally formed the walls of the loricas. The Jurassic and Lower Cretaceous species are seen only as thin silhouettes of calcite, which preserves more or less faithfully the true outline of the original form. In other species, the loricas have been preserved under the chitinous(?) layer, as seen in the Triassic tintinnids described by Visscher (1970) and in the species of Eicher (1965) from the Albian-Cenomanian of the Rocky Mountains. Another important feature in calcified fossil species, besides the general shape of the lorica, is the morphology of the oral collar. This feature, when preserved, is helpful in permitting comparisons between living and fossil species.

Tintinnids are today widely used by bio-stratigraphers, most notably in the pelagic sequences of the Mesozoic Tethys, where tintinnids reach their maximum abundance and diveristy (particularly in the Upper Jurassic-Lower Cretaceous; Fig. 3). Until 1960, only the Jurassic and Cretaceous species were known. But in 1962, Cuvillier and Saccal (1963) described a Paleozoic form, *Vautrinella lapparenti,* from the Upper Devonian of the northern Sahara; this was found later in France in equivalent strata. Some poorly preserved tintinnids have been described from the Silurian of Spain (Hermes, 1966; Fig. 4). Other records include those found in the Lower Carboniferous (marine) of northern Spain (Cuvillier and Barreyre, 1964), and from the Triassic New Red Sandstone (brackish water facies) of Ireland (Visscher, 1970), and the Triassic of Sicily (Agip Mineraria, 1959).

Colomiella, with its long cylindrical collar, characteristic of Aptian forms, was thought to be the termination of the Mesozoic Tethyan

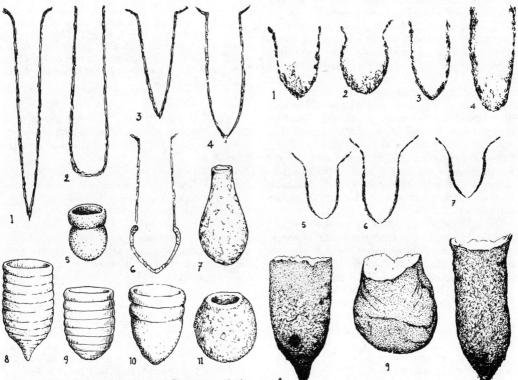

FIGURE 3. Upper Jurassic-Lower Cretaceous tintinnids. 1, *Salpingellina levantina* (Colom); 2, *Coxliellina berriasiensis* (Colom); 3, *Amphorella subacuta* (Colom; 4, *Tintinnopsella longa* (Colom); 5, *Codonella bojiga* Eicher; 6, *Colomiella mexicana* Bonet; 7, *Tintinnopsis ampullula* Eicher; 8, *Coxliella coloradoensis* Eicher; 9, *C. atricollium* Eicher; 10, *Dicloeopella borealis* Eicher; 11, *Tintinnopsis parovalis* Eicher. All ca. X112.

FIGURE 4. Paleozoic and lower Mesozoic tintinnids. 1–4, four different unnamed forms of Silurian tintinnids, from material of this age recognized by Hermes, ca. X100. 5–7, *Vautrinella lapparenti* Cuvillier and Sacal, from the Devonian of the Sahara, ca. X100. 8–10, isolated chitinous (?) loricas of Triassic tintinnids (after Visscher, 1970), ca. X140.

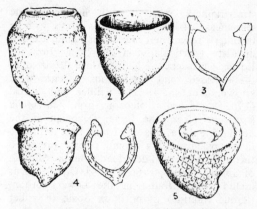

FIGURE 5. Eocene tintinnids (after Tappan and Loeblich, 1968). 1,3, *Remanellina eocoenica* Tappan and Loeblich; 2, *Yvonniellina companula* (LeCalvez); 4, *Tytthocorys coronula* Tappan and Loeblich; 5, *Yvonnniellina feugueri* (LeCalvez). All ca. X57.

line, but Eicher (1965) has demonstrated the presence of Albian and Cenomanian tintinnids in South Dakota, Colorado, and Wyoming, describing six species and one genus.

This suggested that other species might be found in the Tertiary, and Tappan and Loeblich (1968) have since confirmed this, describing several genera and species from the Eocene of Europe (Fig. 5). Tintinnids have yet to be found in Upper Cretaceous strata, however.

It is likely that new discoveries will reveal the presence of tintinnids in younger Cenozoic deposits as well. The fossil record suggests that the tintinnids appeared at the beginning of the Paleozoic, surviving into modern times, always maintaining a uniformity of morphological characteristics due, no doubt, to the constancy of the pelegic medium in which they lived for vast periods of time.

There remains a group of species whose relationship with the true tintinnids is somewhat uncertain. These forms, which are unusually large and possess thick walls of calcite (Fig. 6) were described by Radoĉić (1959, 1963) from the Upper Jurassic of Dalmatia. Farinacci (1963) considered them simply to be sections of shell of the bivalve *Bankia*, while Radoĉić believes them to be true tintinnids. Not having examined specimens of any of these species, this writer does not wish to affirm or negate this assignment.

Studies throughout the stratigraphic record have provided proof that tintinnids have played an important part in the history of the ancient seas of our planet, revealing a succession of species through the different epochs. In the Mesozoic of the Tethys region such microorganisms are almost always associated with radiolarian and *Nannoconus* facies, constituting in

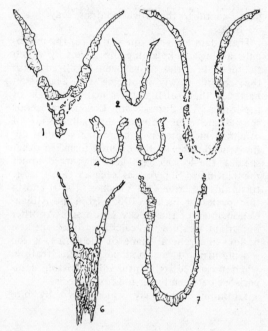

FIGURE 6. Possible tintinnids from the Upper Jurassic of Dalmatia (after Radoiĉić, 1959, 1963). 1, *Favelloides liliiformis* Radoiĉić; 2-3, *Campbelliella milesi* Radoiĉić; 4-5, *Hadziella zetae* Radoiĉić; 6, *Daturellina zetica* Radoiĉić; 7, *Tintinnopsella besici* Radoiĉić. All ca. X13.

general that basic pelagic element that forms the fine biogenic muds of the Tethys sea floor, from Mexico and Central America to the Himalayas.

It is possible that some of the species recently described from the Eocene are more closely related to order Arcellinidae (Family Arcellidae, etc.) than to the tintinnids, based on the appearance of the lorica. However, having no experience whatsoever with the Arcellinidae, the relationships of these Eocene forms with the true tintinnids are accepted here—but with some reservation.

G. COLOM*

References

Agip Mineraria, 1959. *Microfacies Italiane*. Milano: Agip Mineraria, Pl. XXII.

Colom, G., 1934. Estudios sobre las Calpionellas, *Bol. Real Soc. Española Hist. Nat.*, **34**, 379–388.

Colom, G., 1939. Tindinnidos fósiles (Infusorios Oligótricos), *Las Ciencias, Madrid*, 4(4), 1–11.

Colom, G., 1948. Fossil Tindinnids: Loricated Infusoria of the Order of the Oligotricha, *J. Paleontology*, **22**, 233–263.

Colom, G., 1965. Essais sur la biologie, la distribution

*Deceased

géographique et stratigraphique des tintinnoïdiens fossiles, *Eclogae Geol. Helvetiae,* **58,** 319-334.

Cuvillier, J., and Barreyre, M., 1964. Présence de tintinnoïdiens dans le viséen des Asturies, *Rev. Micropaléont.,* **7,** 80.

Cuvillier, J., and Barreyre, M., 1964. Présence de Tintinnoïdiens dans le Dévonien Supérieur du Sahara septentrional, *Rev. Micropaleontol.,* **6,** 73-75.

Eicher, D. L., 1965. Cretaceous tintinnids from the Western Interior of the United States, *Micropaleontology,* **11,** 449-456.

Farinacci, A., 1963. L'Organismo "C" Favre, 1927, appartiene alle Teredinidae? *Geol. Romana,* **2,** 151-176.

Hermes, J. J., 1966. Tintinnids from the Silurian of the Betic Cordilleras, Spain, *Rev. Micropaleontol.,* **8,** 211-214.

Loeblich, A. R., and Tappan, H., 1968. Annotated index to the genera, subgenera and suprageneric taxa of the ciliate Order Tintinnida, *J. Protozool.,* **15,** 185-192.

Lorenz, T., 1902. Geologische Studien in Grenzgebiete zwischen helvetischer und ostalpiner Fazies. II. Der Südliche Rhatikon, *Näturf. Gesell. Frieburg, Ber.,* **12,** 34-95.

Radoičić, R., 1959. Large Tintinnina: *Campbelliella* nov. gen. and *Daturellina* nov. gen., *Geol. Geofiz., Istraz. NR Srbije, Vesnick,* **17,** 79-86.

Radoicić, R., 1963. *Hadziella zetae* gen nov., sp. nov. Aberrant Tintinnina, *Srpskog Geol. Drustva, Zapisnici,* **1960-61,** 193-194.

Remane, J., 1978. Calpionellids, in B. U. Haq and A. Boersma, eds., *Introduction to Marine Micropaleontology.* New York: Elsevier, 161-170.

Tappan H., and Loeblich, A. R., Jr., 1968. Lorica composition of modern and fossil tintinnids (ciliate Protozoa), systematics, geologic distribution, and some new Tertiary taxa, *J. Paleontology,* **42,** 1378-1394.

Visscher, H., 1970. On the occurrence of chitinoid loricas of Tintinnida in an Early Triassic palynological assemblage from Kingscourt, Ireland, *Geol. Surv. Ireland, Bull.,* **1,** 61-64.

Cross-references: *Plankton; Protista.* Vol. I; *Zooplankton.*

TRACE FOSSILS

The main task of paleontology is to study the bodies or hard parts of animals and plants preserved as fossils. These remains are usually termed "body fossils" in order to distinguish them from a large number of other fossils, now collectively called "trace fossils." Tracks, trails, burrows, borings, and other such structures are assigned to the latter. These "biogenic structures," produced in loose sediments or in hard substrates, are the result of various activities of organisms—creeping, crawling, walking, burrowing, boring, feeding, resting, and so on. Trace fossils thus constitute an integral part

of the fossil record, and certain of them are equally valuable in studies of the sedimentary record.

Definitions

The term *trace fossil* was proposed by Simpson (1957), and the German equivalent *Spuren-Fossil* by Krejci-Graf (1932). These terms were intended to embrace all kinds of biological activities on, as well as within, a body of sediment at the time of its accumulation. The shortest definition, given by Haas (1954), was the single sentence: "Trace fossils are features or structures in sediments left by living organisms." Thus excluded are markings that do not reflect a behavioral function, such as those produced by rolling or drifting dead organisms. The German equivalent *Lebensspur* was proposed by Abel (1912, 1935); however, Abel's meaning included pathological phenomena, etc., and was used in a broader sense than is generally employed today. Trace fossils commonly have been assigned to the "problematica" (e.g., Caster, 1957), but this term is inappropriate because it refers primarily to body remains that cannot at present be assinged to a definite systematic position. The term "lebensspur" (without italics or capital letters) has now been accepted by many workers writing in English. Another synonym of trace fossil is the term *bioglyph,* proposed by the Russian Vassoievitch (1953); inorganic structures were called *mechanoglyphs,* the two together constituting *hieroglyphs.* Various other terms were listed by Häntzschel (1975).

A noteworthy shortcoming of most of the above definitions is that they tend to stress "biogenic sedimentary structures" to the virtual exclusion of "borings" and other distinctive traces, which are in fact trace fossils (Bromley, 1970). Borings, unlike burrows, are excavated into consolidated substrates, such as rock, shell, bone, or wood, and therefore are not sedimentary structures in a strict sense. Recent reviews and redefinition of basic concepts and their interrelationships (Fig. 1) are provided by Frey (1973) and Simpson (1975); these definitions are strengthened by virtue of the consensus. they represent among numerous ichnologists in several nations.

The science dealing with trace fossils is termed *ichnology* [W. Buckland, circa 1830] and comprises the entire field of work. *Neoichnology* is used for the study of present-day lebensspuren and *palichnology,* or *paleoichnology,* for ancient ones. Assemblages of trace fossils have been designated *ichnocoenoses* (L. S. Davitashvili, *in* Radwański and Roniewicz, 1970; Lessertisseur, 1955), thereby following

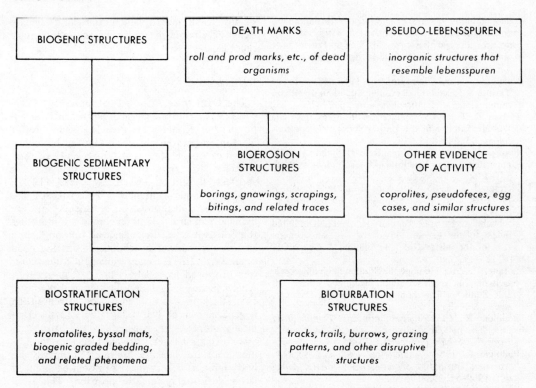

FIGURE 1. Major relationships among biogenic structures and of these to other phenomena (from Frey, 1971).

a concept widly adopted by paleoecologists. In a sense, this term is less problematical in ichnology than its counterparts are in general paleoecology—*biocoenosis, thanatocoenosis,* etc.—because trace fossils usually cannot be reworked or transported, or if so, are usually easily recognized as such. (See Dörjes and Hertweck, 1975.)

Occurrences

In general, trace fossils are more abundant in marine than in continental facies, and are more abundant in Phanerozoic than in Cryptozoic rocks. But trace fossils occur in virtually all kinds of clastic and biogenic sediments, of all geologic ages and sedimentary facies, from the Precambrian to the Holocene. Indeed, they are so prominent locally that they cannot reasonably be excluded from more traditional paleontological and sedimentological investigations.

Preservation

Although found in sediments deposited in all environments, especially clastic series, most lebensspuren are potentially transient structures compared with the hard parts of body fossils. Unlike shells, recent biogenic sedimentary structures are easily obliterated by tides,

currents, or substrate erosion. Thus, at first glance, these traces seem to have little chance of becoming fossilized. However, the multitude of trace fossils preserved in many different facies proves the possibility of their fossilization. Furthermore, where preserved, the trace fossils ordinarily remain *in situ;* they are therefore more reliable indicators of environment than are transportable shells. Finally, biogenic sedimentary structures (that escape mechanical destruction) tend to be enhanced by diagenetic processes that destroy skeletal remains in the same sediments; in certain rocks trace fossils thus provide the *only* record of ancient life.

Preservation of biogenic sedimentary structures depends on the features of the substrate in which the structures are imprinted: sediment grain size, content of water, plasticity, and other factors. Even more important is the site of the animals' activities, whether on the surface or inside the sediments. Trace fossils originating along the interface between a sand and an underlying mud layer have much better chances for preservation. Moreover, such traces ordinarily show morphological characteristics more distinctly than do surficial traces, many of which are indistinct or at least faded. This difference was especially emphasized by Seilacher (e.g., 1962), although considered less important by Osgood (1970, 292–294). Several different types of preservation have been distinguished, according to the position of trace fossils relative to the sediments (see Simpson, 1975; Hallam, 1975).

Also important in preservation is the particular way in which an animal builds its domicile, or moves through or across sediments (Frey, 1971, 101–102). In many cases, the walls of domiciles made by animals burrowing inside the sediment are somewhat consolidated by mucus (e.g., by polychaete worms). The form of these structures may thus be preserved before being filled with sediment from the surface, and such breaks in sediment fabrics act as loci for diagenetic processes that further enhance preservation.

Interpretation

During the early days of geological and paleontological science, the real nature of many trace fossils was not recognized. Many of them were believed to be algae. Their supposed relationship to recent genera of algae was expressed by the names given to them, e.g., the name *Fucoides* for a trace fossil that, to our present knowledge, has nothing to do with the recent marine alga *Fucus*. Genus names involving the suffix *-phycus* also reflect the original interpretation of the fossils as algae. Even some vertebrate tracks were formerly interpreted erroneously, such as the classic "bird tracks" from the Triassic sandstones of the Connecticut valley. Another example are trace fossils showing toe-like impressions that, for many decades, were thought to have been made by vertebrates, particularly by small saurians (Fig. 2). Later on, by Caster's careful investigations of the tracks of recent *Limulus* (see Caster, 1957, and references therein), they were proved to be made by limuloid animals.

Accurate interpretations of trace fossils are not possible without considering recent lebensspuren. They show that the traces made by animals of very different systematic position may be very similar or identical in shape, e.g., the functionally equivalent feeding burrows of the polychaete *Arenicola* and the enteropneust *Balanoglossus*. At the same time one individual animal can often produce very different kinds of traces, according to the site or kind of its activity. Thus, although certain broad generalities can be stated (Frey, 1971, 104–105; Osgood, 1975), the task of designating the specific producer of an invertebrate animal's track, trail, or burrow is ordinarily nearly impossible. Only among well-preserved vertebrate tracks does a morphological analysis

FIGURE 2. Crawling trace of *Limulus,* frequently interpreted erroneously by early workers as a vertebrate trackway (after Malz, 1964).

(by measurement, etc.) enable an experienced vertebrate paleontologist to reconstruct the animals responsible and to discern their more or less definite place in the zoological system. (See W. Soergel, *in* Abel, 1935; Baird, 1954; and Sarjeant, 1975, for examples.)

Classification

In terms of form and function, no conventional system of classification of living or fossil organisms is really applicable to the myriad of trace fossils. Vagaries in preservation of lebensspuren, as well as their great variety resulting from different activities of individual animals, renders every classification difficult to use. Moreover, the formation of lebensspuren, and to some extent their shapes, are controlled largely by the characteristics of the substrate in which they are imprinted. Therefore, a "system" having higher categories can be based only on a viewpoint concerned with the real nature of lebensspuren: their behavioral or *ethological* significance.

To date the classification given by Seilacher (1953) seems to be most suitable. Its applicability to both fossil and recent forms has been demonstrated often. Seilacher distinguished five major ethologic groups (Fig. 3): (1) dwelling structures (*domichnia*), the more or less per-

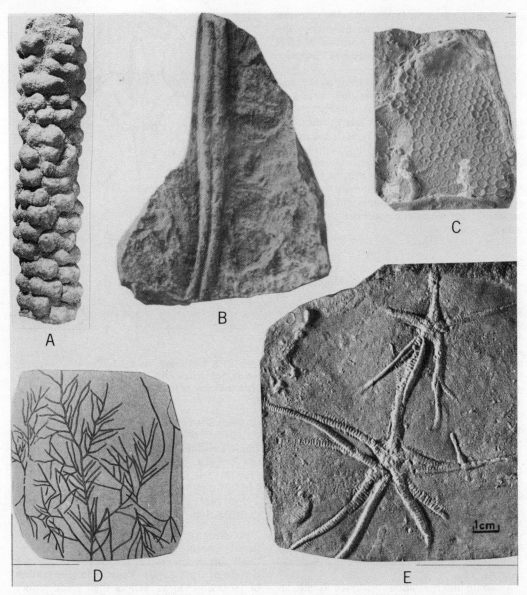

FIGURE 3. Examples of major ethological groups of trace fossils. A, *Ophiomorpha*, a dwelling structure, Upper Cretaceous, North Dakota, USA; and B, *Rouaultia*, a crawling trace, Ordovician, Portugal (from Häntzschel, 1975, courtesy of the Geological Society of America and the University of Kansas Press). C, *Paleodictyon*, a polygonal grazing trace, Lower Tertiary flysch, Poland (specimen in Geol. Inst. Univ. Hamburg). D, *Chondrites*, a plant-like feeding structure, Lower Tertiary flysch, Switzerland (from Heer, 1877). E, *Asteriacites*, the resting trace of a sea star, exhibiting lateral repetition, Lower Jurassic, Germany (from Häntzschel and Reineck, 1968).

manent domiciles of endobenthic animals; (2) feeding structures (*fodinichnia*), made within the substrate by sediment-ingesting animals; (3) grazing traces (*pascichnia*), produced by animals feeding or grazing on sediment surfaces, mostly in tight, winding or meandering patterns; (4) crawling traces (*repichnia*), various trackways, trails, furrows, and galeries made by the crawling movements of animals; and (5) resting traces (*cubichnia*), impressions made by resting or hiding animals, mostly epipsammonic species, the traces corresponding somewhat to the animals' shapes. (See Frey, 1971, 96-99, Table 3; Simpson, 1975.)

A few additional ethologic groups have been proposed by Müller (1962): *mordichnia*—biting and gnawing traces; *cursichnia*—walking traces;

natichnia—swimming traces; and *volichnia*—flying traces; as well as some new names to comprise larger groups, such as the *cibichnia* (for fodinichnia, mordichnia, and pascichnia, collectively) and *movichnia* (for all kinds of locomotion traces). Most such concepts are an unnecessary complication of Seilacher's simple, highly useful scheme. For this reason, the even more detailed classification by Vialov (1972) has attracted little use.

One significant kind of trace has not been stressed in many previous classifications, however: the *escape structure,* a lebensspur that relfects an animal's attempt to stay upon, or at a specific distance below, the substrate surface despite rapid aggradation or degradation (Fig. 4). Simpson (1975) has given this structure the formal designation *Fugichnion.* "Equilibrium structure" is a somewhat comparable informal term, but "escape structure" is already embedded in thought and some literature, and is especially appropriate if one realizes that an animal must sometimes escape downward as well as upward. Most escape structures are modified forms of feeding and dwelling structures. They are important paleoecologically because they indicate stresses on the animals involved, and sedimentologically because they provide a sensitive measure of deposition and erosion (see Howard, 1975).

Each of the above groups has characteristic features that are independent of the systematic position of the trace-making animals; indeed, the basic characteristics of each may be equally valid for animals having very different systematic positions.

Nomenclature

The binary system of nomenclature, traditionally applied to body fossils, is usually also applied to trace fossils. However, this custom is problematical in ichnology because here "genera" and "species" have an entirely different meaning. Moreover, present rules for systematics and nomenclature of body fossils, notably the *International Code of Zoological Nomenclature,* either do not admit trace fossils as valid taxa, or make inadequate provisions for them (Häntzschel and Kraus, 1972). Because of such inconsistencies, some workers have published only morphological descriptions, without designating the trace fossils by "generic" or "specific" names. This practice is recommended for morphologically nondescript or uncharacteristic forms. But it is important to recognize that unnamed fossils, even if distinctive, tend to escape the attention of other paleontologists because the names do not appear in faunal lists, in handbooks or textbooks, or in international compilations.

Due to the former confusion between trace fossils and plant fossils, and the corresponding belief that variations in trace-fossil morphology had relatively narrow limits, too many "genera" and "species" were distinguished, named, and described (see Häntzschel, 1975). Such practices have produced large numbers of useless taxa.

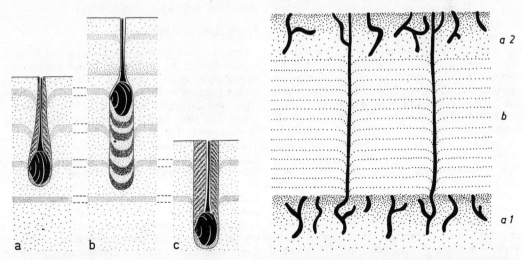

FIGURE 4. Examples of escape structures (from Reineck, 1958). *Left,* recent clam burrows: a, negligible deposition, growing animal gradually burrows deeper; b, rapid deposition, clam migrates upward, producing nested funnel-like laminae called "sitz marks;" c, substrate scour, clam migrates downward. *Right,* recent polychaete burrows: a1, negligible deposition, normal burrows; b, rapid deposition, worm makes escape burrows upward; a2, negligible deposition again.

As experience shows, however, new names are still being introduced. Inasmuch as possible, the new names ought to convey a clear conception of the object being named, especially with regard to its morphology, i.e., a name should be descriptive rather than interpretive, geographic, or patronymic (*Haentzschelinia* Vialov—representing a star-like feeding trace—is not a good example of a new name, although the old one for it, "*Spongia ottoi* Geinitz," is misleading). For the sake of consistency and stability, until the International Commission finally makes a ruling on the matter of trace fossils, it seems best to adopt the International Code of Zoological Nomenclature as if it already applied—albeit tacitly, keeping in mind the particular meaning of these "generic" and "specific" names. This practice should also include the use of italicized names for genera and species.

Some workers object to this "dual" system of nomenclature because they, where not already familiar with both trace fossil and body fossil genera, cannot distinguish the two by names alone. This ambiguity cannot be rectified among the many long-established genera, but Seilacher (1953) suggested an excellent label for subsequent generic names: the suffix *-ichnus* or *-ichnium*. Also useful are such suffixes as *-ichnites* and *-craterion*.

Ecological Significance

Regional investigations and comparisons of ichnocoenoses (particularly by Seilacher) showed a significant result: certain associations of trace fossils are independent of geologic age and have a long time range. They are found in specific facies, characterized by lithology and inorganic sedimentary structures obviously reflecting special ecological and sedimentological environments. These ichnocoenoses are typified by the frequency of occurrence, or conspicuous absence, of the different ethological types of traces (dwelling structures, feeding traces, crawling traces, etc.). The relation between these types is termed the "trace-" or "ichnospectrum" of an ichnocoenose (Seilacher, 1958). Such ichnocoenoses indicate a rich autochthonous life on and (or) within the sediment, and are a characteristic biological feature of the facies. They are therefore of great significance in paleoceological interpretations.

One of the most typical and best known ichnocoenoses is that of the flysch facies. Flysch sediments occurring in all geologic eras and regions were deposited as orogenic sediments within geosynclines, and, according to now widely accepted opinions, at great water depths. Frequently, the flysch is characterized by a lack of body fossils but an abundance of trace fossils, at least within certain strata (Fig. 5). Among trace fossils of the flysch, grazing traces are the most typical element. They have been found in a surprising variety of shapes, but their basic characters remain identical, or at least similar to one another, in most flysch sediments of the three geologic eras.

In such a way, Seilacher (1967) distinguished six "universal ichnofacies," the *Scoyenia*, *Glossifungites*, *Skolithos*, *Cruziana*, *Zoophycos*, and *Nereites*, each consisting of certain diagnostic assemblages of trace fossils and indicating different types of environmental conditions (see Frey, 1975a, Table 1). However, interpreting the exact environmental conditions represented by the different trace-fossil assemblages is not always simple. The real relationship between trace-making animals and various sedimentological facies is less a matter of such things as water depth, distance from shore, grain size, and composition of sediments than of such things as substrate stability, wave and (or) current energy, and suspended or deposited food supplies. In the way that ripple marks are good current indicators but poor depth indicators, trace fossils are good indicators of certain things but not others, especially not of distinct facies in the sense of *delta, bar,* or *shelf,* etc. Rather, as with inorganic sedimentary structures, one must first ask "What does this structure mean in terms of individual environmental parameters?" and then "In what facies are these particular environmental parameters apt to exist?" If this is done, trace fossils can be extremely useful environmental indicators (see numerous chapters in Frey, 1975b).

Stratigraphical Significance

Usually trace fossils represent good facies fossils. Owing to their origin and ecological significance, many similar or identical forms have a long time range through the eras. Such a wide vertical distribution has been proved often—e.g., the characteristic tracks of limulids, known from the Paleozoic to the present. Therefore, few trace fossils may serve as true guide fossils, although some are more diagnostic than others—e.g., the feeding burrow *Phycodes,* for the *Phycodes*-strata in the German Lower Ordovician, and the fecal pellet string *Tomaculum;* both have a wide horizontal distribution in European Ordovician strata. Certain trilobite crawling and resting traces are equally distinctive, such as *Cruziana* and *Rusophycus* (see Seilacher, 1970). Detailed morphological investigations of additional trace fossils may help disclose forms having such specific morphologic peculiarity and short vertical distribution that they may be used as guide fossils, at least for

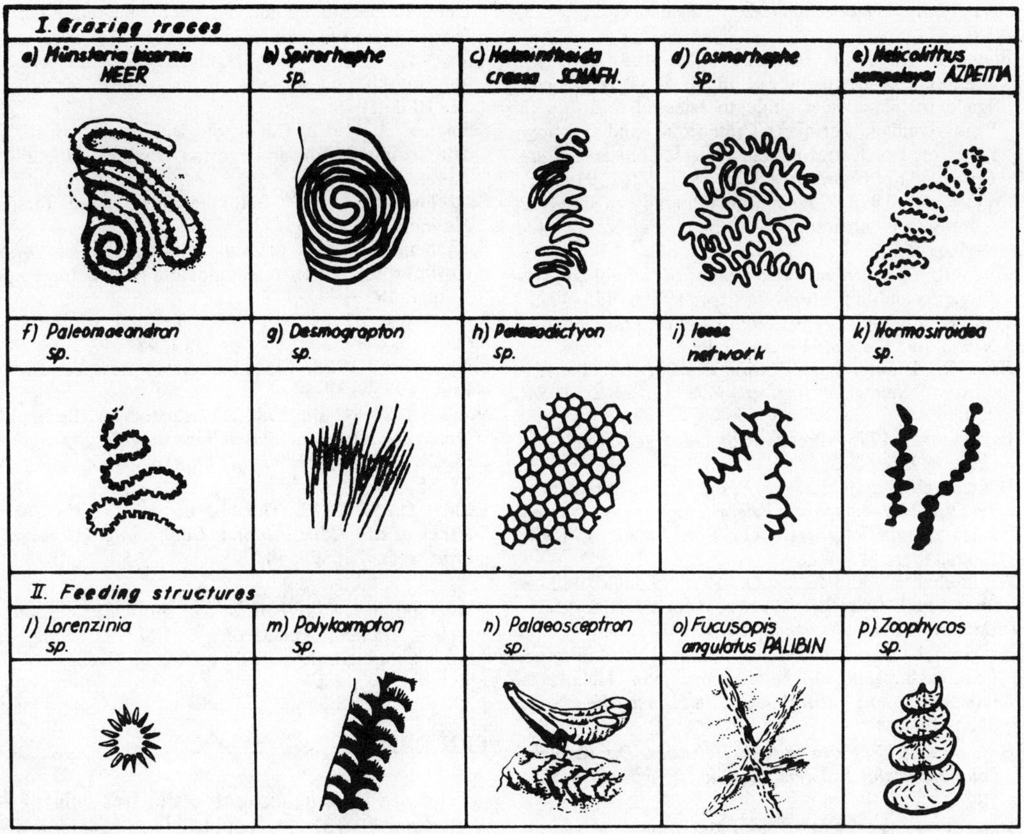

I. Grazing traces

a) *Münsteria bicornis* MEER	b) *Spirorhaphe* sp.	c) *Helminthoida crassa* SCHAFH.	d) *Cosmorhaphe* sp.	e) *Helicolithus sampelayoi* AZPEITIA
f) *Paleomaeandron* sp.	g) *Desmograpton* sp.	h) *Paleodictyon* sp.	i) loose network	k) *Hormosiroidea* sp.

II. Feeding structures

l) *Lorenzinia* sp.	m) *Polykampton* sp.	n) *Palaeosceptron* sp.	o) *Fucusopis angulosus* PALIBIN	p) *Zoophycos* sp.

FIGURE 5. Schematic reconstruction of an ichnocoenose from the psammites of the Cretaceous-Tertiary flysch (after Müller, 1963).

local time categories and (or) marker beds (Crimes, 1975). Some trace fossils are also good geopetals, or "way-up" criteria.

Summary

Trace fossils are valuable components of the fossil record, and frequently they are the only evidence for numerous soft-bodied animals that are not otherwise preserved. But trace fossils represent behavior of animals, not actual body parts, and therefore must be named and studied from a different standpoint than body fossils. Fortunately, this difficulty is compensated by the great paleoecological and sedimentological significance of numerous kinds of trace fossils. The field of ichnology is now well established as a viable facet of sedimentary geology and paleobiology.

WALTER HÄNTZSCHEL
ROBERT W. FREY*

*I was honored when Prof. Häntzschel asked me to coauthor this paper with him, and I now dedicate it to his memory. (RWF)

References

Abel, O., 1912. *Grundzüge der Paläobiologie der Wirbeltiere.* Stuttgart: Thieme, 708p.

Abel, O., 1935. *Vorzeitliche Lebensspuren.* Jena: Gustav Fischer, 644p.

Baird, D., 1954. *Chirotherium lulli,* a pseudosuchian reptile from New Jersey, *Bull. Mus. Comp. Zool. Harvard Univ.,* **3,** 165–192.

Basan, P. B., ed., 1978. Trace fossil concepts, *Soc. Econ. Paleontol. Mineral. Short Course,* **5,** 201p.

Bromley, R. G., 1970. Borings as trace fossils and *Entobia cretacea* Portlock, as an example, in Crimes and Harper, 1970, 49–90.

Caster, K. E., 1957. Problematica, *Geol. Soc. Amer. Mem. 67*(2), 1025–1032.

Crimes, T. P.. 1975. The stratigraphical significance of trace fossils, in Frey, 1975b, 109–130.

Crimes, T. P., and Harper, J. C., eds., 1970. *Trace Fossils.* Liverpool: Seel House Press, 547p.

Crimes, T. P., and Harper, J. C., eds., 1977. *Trace Fossils 2.* Liverpool: Seel House Press, 351p.

Dörjes, J., and Hertweck, G., 1975. Recent biocoenoses and ichnocoenoses in shallow-water marine environments, in Frey, 1975b, 459–491.

Ekdale, A. A., ed., 1978. Trace fossils and their importance in paleoenvironmental analysis, *Palaeo-*

geography, Palaeoclimatology, Palaeoecology, **23**, 167-323.

Frey, R. W., 1971. Ichnology–The study of fossil and Recent lebensspuren, in B. F. Perkins, ed., Trace fossils, a field guide to selected localities in Pennsylvanian, Permian, Cretaceous, and Tertiary rocks of Texas, and related papers, *Louisiana State Univ., School Geoscience Misc. Publ., 71-1*, 91-125.

Frey, R. W., 1973. Concepts in the study of biogenic sedimentary structures, *J. Sed. Petrology*, **43**, 6-19.

Frey, R. W., 1975a. The realm of ichnology, its strengths and limitations, in Frey, 1975b, 495-491.

Frey, R. W., ed., 1975b. *The Study of Trace Fossils*. New York: Springer-Verlag, 564p.

Haas, O., 1954. Zur Definition des Begriffs "Lebensspuren," *Neues Jahrb. Geol. Paläontol, Mh.,* **1954**, 379.

Hallam, A., 1975. Preservation of trace fossils, in Frey, 1975b, 55-63.

Häntzschel, W., 1975. Trace fossils and problematica, *Treatise on Invertebrate Paleontology*, pt. W, Miscellanea, suppl. 1. Kansas: Geol. Soc. Amer. and Univ. Kansas Press, W3-W269.

Häntzschel, W., and Kraus, O., 1972. Names based on trace fossils (ichnotaxa): request for a recommendation, *Bull. Zool. Nomencl.,* **29**, 137-141.

Häntzschel, W., and Reineck, H.-E., 1968. Fazies-Untersuchungen im Hettangium von Helmstedt (Niedersachsen). *Mitt. Geol. Staatsinst. Hamburg,* **37**, 5-39.

Heer, O., 1877. *Flora Fossilis Helvetiae. Die vorweltliche Flora der Schweiz.* Zurich : J. Wurster & Co., 182p.

Howard, J. D., 1975. The sedimentological significance of trace fossils, in Frey, 1975b, 131-146.

Krejci-Graf, K., 1932. Definition der Begriffe Marken, Spuren, Fährten, Bauten, Heiroglyphen und Fucoiden, *Senckenbergiana,* **14**, 19-39.

Lessertisseur, J., 1955. Traces fossiles d'activité animale et leur signification paléobiologique, *Soc. Geol. France, Mem.,* n.s., **74**, 1-150.

Malz, H., 1964. *Kouphichnium walchi,* die Geschichte einer Fährte und Tieres, *Natur und Museum,* **94**, 81-104.

Müller, A. H., 1962. Zur Ichnologie, Taxiologie und Ökologie fossiler Tiere, *Frieberger Forschungshefte* C, **151**, 5-49.

Müller, A. H., 1963. *Lehrbuch der Paläozoologie, vol. 1, Allgemeine Grundlagen.* Jena: Gustav Fischer Verlag, 387p.

Osgood, R. G., Jr., 1970. Trace fossils of the Cincinnati area, *Paleontogr. Amer.,* **6**, 281-444.

Osgood, R. G., Jr., 1975. The paleontological significance of trace fossils, in Frey, 1975b, 87-108.

Radwański, A., and Roniewicz, P., 1970. General remarks on the ichnocoenose concept, *Bull. Acad. Polonaise des Sciences, sér. sci. géol. et géogr.,* **18**, 51-56.

Reineck, H-E., 1958. Wühlbau-Gefüge in Abhängigkeit von Sediment-Umlagerungen *Senckenbergiana Lethaea,* **39**, 1-24.

Sarjeant, W. A. S., 1975. Tracks and impressions of vertebrates, in Frey, 1975b, 283-324.

Seilacher, A., 1953. Studien zur Palichnologie. I.

Über die Methoden der Palichnologie, *Neues Jahrb. Geol. Paläontol., Abh.,* **96**, 421-425.

Seilacher, A., 1958. Zur Ökologischen Charakteristik von Flysch und Molasse, *Eclogae Geol. Helvetiae,* **51**, 1062-1078.

Seilacher, A., 1962. Paleontological studies on turbidite sedimentation and erosion, *J. Geol.,* **70**, 227-234.

Seilacher, A., 1967. Bathymetry of trace fossils, *Marine Geol.,* **5**, 413-428.

Seilacher, A., 1970. *Cruziana* stratigraphy of "nonfossiliferous" Palaeozoic sandstones, in Crimes and Harper, 1970, 447-476.

Simpson, S., 1957. On the trace-fossil *Chondrites, Quart. J. Geol. Soc. London,* **112**, 475-499.

Simpson, S., 1975. Classification of trace fossils, in Frey, 1975b, 39-54.

Vassoievitch, N. B., 1953. O nekotorykh flishevykh (znakah). (On some flysch textures), *Trudy Lvovs. Geol. Obsh., Univ. Ivan Franko, Geol. ser.,* **3**, 17-85.

Vialov, O. S., 1972. The classification of the fossil traces of life, *24th Internat. Geol. Cong. (Montreal) Proc., sect.* 7, 639-644.

Cross-references: *Actualistic Paleontology; Communities, Ancient; Paleoecology.* Vol. VI: *Biogenic Sedimentary Structures; Bioturbation; Fucoid.*

TRILOBITA

Trilobita is an extinct Paleozoic class of Arthropoda (q.v.), which are invertebrate animals with jointed legs. (Living arthropods include insects, spiders, centipedes, and crustaceans.) Trilobites were most numerous and varied in the Cambrian period, progressively diminishing thereafter but persisting into the Middle Permian.

Morphology

Like other arthropods, the trilobite body was bilaterally symmetrical and divided into segments, the name referring to the characteristic three-fold longitudinal division into a central raised axis and lateral pleural regions. The outer shell or exoskeleton was made of a chitin-like material impregnated with calcium carbonate (see *Chitin*), and covered the dorsal (upper) surface of the body and extended a short distance inward on the under or ventral side. The body (Fig. 1) also shows a transverse division into three parts: cephalon, thorax, and pygidium. The anterior head or cephalon is formed by fusion of a number of segments. The more-or-less swollen midpart (glabella) was, in many forms, divided by transverse furrows; the compound eye was borne on the pleural region of the head, the genal spine projecting from the posterolateral corner. The eye lobe was present

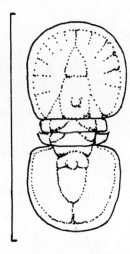

FIGURE 4. *Ptychagnostus* sp., Middle Cambrian, Utah, USA. Scale = 1 cm.

genera, and present in a few in which there is also a suture. In these genera there may be three thoracic segments. The axis is prominent and distinctly subdivided on the pygidium, and the pleural regions are not furrowed in many genera. These trilobites could enroll and the edges of cephalon and pygidium fitted closely. Extremely similar species of *Ptychagnostus* are present in the Middle Cambrian of Sweden, western USA, and Australia; and species of other genera have been shown to be widely distributed geographically yet to have limited range in time. They are thus valuable in intercontinental correlation. Only a few long-ranging and widely distributed genera persist into the Late Ordovician. This group embraces the order Agnostida of the *Treatise* (Moore, 1969).

Group 2. A relatively large cephalon, with its well-segmented glabella and long crescentic eye lobe, numerous thoracic segments, and a small pygidium typify group 2. *Redlichia* (Fig. 5) has facial sutures, which traverse the eye lobe; this type of trilobite is found in late Lower and early Middle Cambrian rocks of the Mediterranean region, Asia, and Australia. Contemporaneous in the Lower Cambrian is the western European and North American *Olenellus*, which lacks facial sutures. *Paradoxides* and its allies are typical of the Middle Cambrian of western Europe and eastern North America, and are distinguished by a forwardly expanding, well-furrowed glabella. The families in this group (Order Redlichiida of the *Treatise*) thus show a provincial distribution, but the distribution of other genera shows that no barriers impassable to any trilobite separated these provinces.

Group 3. The dominantly Lower and Middle Cambrian trilobites of group 3 are characterized

by a parallel-sided or forwardly expanding glabella which reaches the anterior border of the cephalon, 5–11 thoracic segments, medium to large pygidium, and spines on the pleurae. *Olenoides* (Figs. 2 and 3) represents a North American family of this worldwide group, which was termed Order Corynexochida in the *Treatise*. Growth stages are very like those known from group 4, and it is doubtful if the two should be separated.

Group 4. A large and varied assemblage of Middle and Upper Cambrian trilobites, group 4 contains twice the number of genera in groups 1–3. *Ptychoparia* (Fig. 6) from the Middle Cambrian, has the glabella narrowing forward and not reaching the anterior margin of the cephalon, the suture crossing the posterior margin of the cephalon inside the genal spine, fourteen thoracic segments and a pygidium that is considerably smaller than the cephalon. Other members of this group lack eyes, some have the glabella smooth or the entire cephalon unfurrowed, others have spines on the axis and pleurae, and some have the pygidium of a similar size to the cephalon. Many genera are widely distributed but short ranging in time; thus, subdivision and correlation of Upper Cambrian rocks in particular are based on genera of this group. It comprises families of Cambrian age placed in the Order Ptychopariida in the *Treatise*.

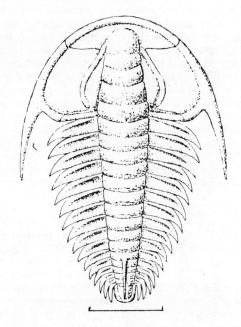

FIGURE 5. *Redlichia chinensis,* Middle Cambrian, China (from Moore, 1959, courtesy of the Geological Society of America and the University of Kansas Press). Scale = 1 cm.

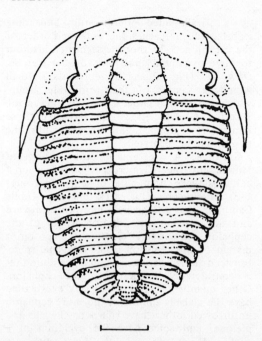

FIGURE 6. *Ptychoparia striata,* Middle Cambrian, Czechoslovakia. Scale = 1 cm.

Group 5. Group 5 embraces families that dominated during the Ordovician period; these families were included in the Order Ptychopariida in the *Treatise,* and may have been derived from ancestors here placed in group 4. No new families but new genera appear with diminishing frequency from the Silurian to the Middle Permian. Glabella may narrow forward and be furrowed, or may be parallel-sided or forwardly expanding and lack furrows. An eye lobe is present in most families, the suture running back to cross the posterior border inside the genal spine. The number of thoracic segments varies from 5 to >30, but 8–12 is characteristic. The pygidium may be smaller, equal to, or larger in size than the cephalon. Genal spines are commonly present, and pleural or axial spines may be developed. *Bumastus* (Fig. 7) belongs to an Ordovician and Silurian family in which both cephalon and pygidium are smooth and convex, and there are 8–10 thoracic segments. *Ampyx* (Fig. 8) typifies an Ordovician group in which the glabella expands forward, there are no eyes, and the suture is marginal. There are 5 or 6 thoracic segments, and the pygidium is smaller than the cephalon. *Warburgella* (Fig. 9), a Silurian genus, represents a group that appears in the Ordovician and persists into the Permian. Many new genera of the group evolved in the Devonian and Carboniferous periods and have been shown to be of considerable stratigraphical value.

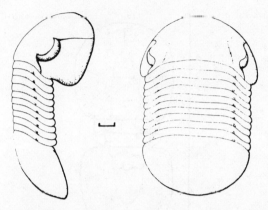

FIGURE 7. *Bumastus barriensis,* Middle Silurian, Great Britain. Scale = 1 cm.

Group 6. Characteristic of Ordovician to Devonian rocks is a group which, corresponding to the Order Phacopida of the *Treatise,* contains less than half the number of genera in group 5. It includes such genera as *Ceraurus, Ceraurinella* (Fig. 1), *Staurocephalus* (Fig. 10), *Calymene,* and *Phacops,* all of which have the suture running out from the eye lobe to cross the lateral margin of the cephalon. The mode of development seems to be similar throughout this varied group, in which the glabella may expand or narrow forward and be furrowed or not furrowed; the eye lobe in rare cases may be absent; the number of thoracic segments varies between 8 and approximately 20; and the pygidium is small to large. Spines and tubercles are commonly developed on the external surface. In *Phacops* and its allies, the eye is formed of

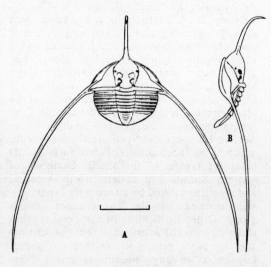

FIGURE 8. *Ampyx virginiensis,* Middle Ordovician, Virginia, USA (from Whittington, 1959). A, dorsal view; B, lateral view; scale = 1 cm.

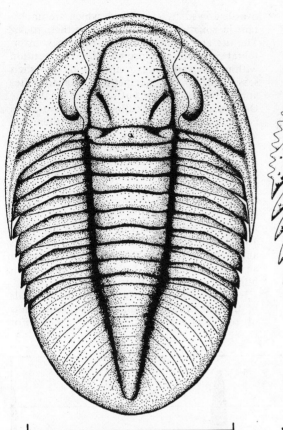

FIGURE 9. *Warburgella ludlowensis,* Upper Silurian, Great Britain (from Owens, 1973; copyright © 1973 by the Palaeontographical Society). Scale = 1 cm.

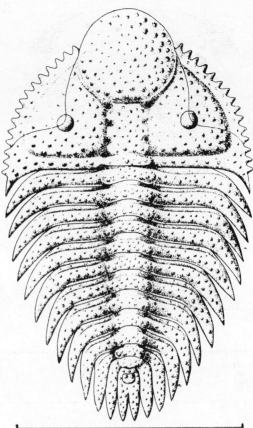

FIGURE 10. *Staurocephalus clavifrons,* Upper Ordovician, western Europe (from Kielan, 1957). Scale = 1 cm.

thick, biconvex lenses separated from one another by a layer of cuticle; this type of eye did not form a mosaic image; rather it was an aggregate of individual eyes, well able to detect movement and may have concentrated the light. Members of all the families are present in the Lower Ordovician, and several families persist into the Devonian where many new genera are evolved, but none has been demonstrated after the end of this period.

Group 7. Two different families of Ordovician to Devonian trilobites, typified by *Odontopleura* (Fig. 11) and *Lichas,* are here included in a single group, not separated as two Orders, Lichida and Odontopleurida, as in the *Treatise.* Their ancestry and relationships are problematical because both their developmental stages and adult morphology are different from those of other trilobite groups. The *Odontopleura* type has a furrowed glabella narrowing forward, eye lobes on raised areas or stalks, and a characteristically spinose exoskeleton. In *Lichas,* the large glabella is subdivided by longi-

tudinal furrows, the pygidium may be large, and the external surface is tuberculate and may be spinose.

Distribution

Most fossiliferous Cambrian rocks yield trilobites, and generally the remains of these animals far outnumber those of others. This dominance in numbers and kinds, and their rapid evolution, means that zonation and correlation of these rocks is based almost exclusively upon them. In the Ordovician the picture changes, for though new families arise, trilobites are not the preponderant fossils; they are outnumbered at many levels by brachiopods, cephalopods, and other mollusks, or by bryozoans. This is the case in many clastic or limey sediments; and, although the zonation of such strata in the Ordovician is based in part on trilobites, equally important are nautiloids and brachiopods. In the shale facies, trilobites are an important element in the faunas, but are

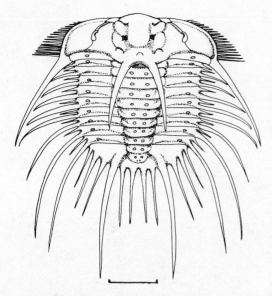

overshadowed by graptolites, which are important in zonation and correlation of these rocks. The rise and diversification of these new invertebrate groups intensified the competition within the varied environments of the shelf seas, and in this competition the trilobites appear to have slowly but surely lost. After the Ordovician, though existing groups showed a continuous evolution of new genera, there was a gradual decline and final extinction (Fig. 12). A factor in this gradual decline may have been that the trilobites were feeble swimmers and drifters, and were not armed with pincers and powerful jaws. Their defense from more active animals was to hide partly buried in the sediments or amid vegetation, or to enroll for protection of the soft parts of the body. Predators armed with pincers or jaws, such as eurypterids and nautiloids, or fish with jaws in Devonian and later times, may have preyed upon trilobites and so hastened their decline.

It has long been known that particular kinds of trilobites occur abundantly in certain kinds of rocks, rarely or not at all in others. For

FIGURE 12. Seven main groups of trilobites and their distribution in time. Black areas are roughly proportional to numbers of genera in each group (modified from Whittington, 1954). Solid line at right shows numbers of new subfamilies appearing through time, broken line is numbers of subfamilies dying out (after Stubblefield, 1959).

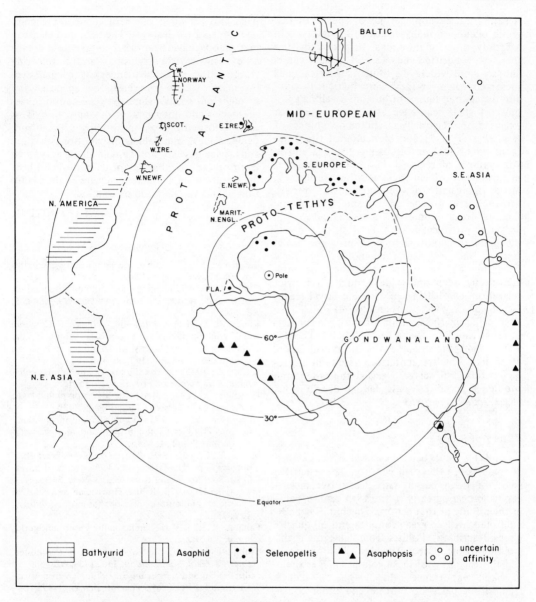

FIGURE 13. Distribution of five different faunal provinces of trilobites during the Lower Ordovician, southern hemisphere, polar projection (from Whittington and Hughes, 1972; used by permission of The Royal Society of London). Palaeogeography based on palaeomagnetic and other data, present-day outlines of land masses used for ease of understanding. Abbreviations for geographical areas: E.IRE., Southeastern Ireland, with England and Wales adjacent to it; E.NEWF., eastern Newfoundland; FLA., state of Florida, USA; MARIT.-N.ENGL., parts of the Maritime provinces of Canada and New England states of USA; SCOT., Scotland; W.IRE., northwestern Ireland; W.NEWF., western Newfoundland; W. NORWAY, western Norway.

example, in the Cambrian, *Ptychagnostus* (Fig. 4) and its allies are common in dark grey or black silty and shaly rocks formed in muddy waters, poor in oxygen, in which carbonaceous matter and pyrite accumulated to give the dark color. Similarly, certain of the varied trilobites contained in pure white limestones of Ordovician and Silurian age associated with reefs, are confined to such rocks. Thus, the local environment influenced the trilobite fauna we find preserved, though such faunas may occur on different continents in similar rocks at similar times. Other trilobites seem to show no such limitations, for example an Ordovician group characterized by broad axial and narrow pleural regions, very large eyes, and a small pygidium.

A similar succession in time of species of this group occurs on many continents, preserved in a variety of different rocks. It is thought that these species drifted and swam in surface waters and so were widely distributed, their remains preserved wherever they were carried. Regionally distributed faunas of bottom-dwelling trilobites of shelf seas have also been noted, for example, the regionally restricted occurences of *Redlichia* (Fig. 5) and olenellids of the Cambrian, mentioned in discussing group 2 above. An example of regional distribution of faunas in Lower Ordovician rocks is shown in Fig. 13. More than half the genera known in each province are unique to it, and while some families are in common, others are present in only one province. Such a distribution suggests independent evolution in each province. Isolation of one province from another may have been by an ocean (proto-Atlantic or mid-European) deep and wide enough to restrict migration by drifting of floating larval stages (see *Larvae*). Water temperature may also have been a factor, the location of the pole suggesting that the *Selenopeltis* fauna inhabited colder waters than, for example, the Bathyurid fauna. Maps like this are tentative, based on inadequate data, but serve to reveal the fascinating problems posed by past faunal distributions (see *Paleobiogeography*).

History of Study

Trilobites have been treasured by collectors of fossils since the 17th century. The scientific study of them began with descriptive monographs appearing early in the 19th century, and in the middle of that century Joachim Barrande published his two great volumes and 60 quarto plates describing trilobites from Paleozoic rocks around Prague, Czechoslovakia. The accuracy and thoroughness of these volumes makes them of basic importance. Monographs of faunas from other parts of Europe and other continents followed in increasing numbers. Charles D. Walcott, formerly Secretary of the Smithsonian Institution, and Professor Charles E. Beecher, Yale University, gave the first descriptions of discoveries that made known trilobite limbs. Fifty years ago, 200 genera of trilobites were known. Today the number is at least 12 times as great, and new kinds continue to be found on all continents. This is because of exploration, restudy of known areas of Paleozoic rocks, and better techniques in extracting and studying fossils. Particularly successful has been the etching of limestones from all Paleozoic systems in acid to release exoskeletons replaced by silica. Such specimens reveal in detail the inner and outer surface of the exoskeleton, and

have also provided size series, including many new, small growth stages. The light and electron microscope have been used to give new details of eye and cuticle structure. Careful collecting from stratigraphical sequences has revealed evolutionary trends in series of species, and brought out relationships between communities of species and environment of deposition. This new information on morphology, evolution, paleoecology, and faunal distributions in time and space reveals how much there is yet to learn of the paleobiology of trilobites, and how far we are from a coherent picture of the evolution of this varied group of animals.

H. B. WHITTINGTON

References

Bergström, J., 1973. Organization, life and systematics of trilobites, *Fossils and Strata*, **2**, 1–69.

Crimes, T. P., 1975. The production and preservation of trilobite resting and furrowing traces, *Lethaia*, **8**, 35–48.

Kielan, Z., 1955. A new trilobite of the genus *Ceraurus* and the significance of cephalic spines in the ontogeny and phylogeny of trilobites, *Acta Geol. Polonica*, **5**, 215–240. (In Polish)

Kielan, Z., 1957. On the trilobite family Staurocephahdae, *Acta Paleontol. Polonica*, **2**, 155–183.

Levi-Setti, R., 1975. *Trilobites. A Photographic Atlas.* Chicago: University of Chicago Press, 243p.

Martinsson, A., ed., 1975. Evolution and morphology of the Trilobita, Trilobitoidea, and Merostomata, *Fossils and Strata*, **4**, 468p.

Moore, R. C., ed., 1959. *Treatise on Invertebrate Paleontology*, pt. O, Arthropoda 1. Lawrence, Kansas: Geol. Soc. Amer. and Univ. Kansas Press, 560p.

Owens, R. M., 1973. British Ordovician and Silurian Proetidae (Trilobita), *Palaeontogr. Soc. Monogr.*, **127**, 98p.

Palmer, A. R., 1974. Search for the Cambrian world, *Amer. Sci.*, **62**, 216–224.

Stubblefield, C. J., 1959. Evolution in trilobites, *Quart. J. Geol. Soc. London*, **115**, 145–162.

Stürmer, W., and Bergstrom, J., 1973. New discoveries on trilobites by X-rays, *Palaont. Zeitschr.*, **47**, 104–141.

Whittington, H. B., 1954. Status of invertebrate paleontology, 1953.6. Arthropoda Trilobita, *Bull. Mus. Comp. Zool. Harvard Univ.*, **112**, 193–200.

Whittington, H. B., 1956. Silicified Middle Ordovician trilobites—the Odontopleuridae, *Bull. Mus. Comp. Zool. Harvard Univ.*, **114**, 153–288.

Whittington, H. B., 1959. Silicified Middle Ordovician trilobites-Remopleurididae, Trinucleidae, Raphiophoridae, Endymioniidae. *Bull. Mus. Comp. Zool. Harvard Univ.*, **121**, 371–496.

Whittington, H. B., and Hughes, C. P., 1972. Ordovician geography and faunal provinces deduced from trilobite distribution, *Phil. Trans. Roy. Soc. London, Ser. B*, **263**, 235–278.

Cross-references: *Arthropoda; Burgess Shale; Chitin; Larvae; Paleobiogeography.*

TROPHIC GROUPS

Study of feeding, or trophic, relationships is an area of ecological research that has a long history (Blegvad, 1915; Yonge, 1928, 1954; Lindeman, 1942; Jorgensen, 1966; Mann, 1969). Until recently, paleontologists have largely ignored the possibility of deducing feeding relationships among fossil taxa. Although food is not always a dominant limiting factor, it is a crucial aspect of any organism's life and is involved in the structuring of biotic communities. In recent years, some paleontologic research has been couched in trophic terms (Olson, 1966; Walker, 1972a,b; Rhoads et al., 1972; Scott, 1974, 1976; 1978; Thayer, 1974; Levinton and Bambach, 1975; Stanton and Dodd, 1976). Since the best-preserved fossil communities are composed of marine organisms, the present discussion emphasizes marine trophic relationships.

As Walker and Bambach (1974) have noted, the feeding process can be separated into four aspects: (1) types of available food, (2) location of food resources, (3) selection mechanisms, and (4) specific adaptations of various organisms for food acquisition. With regard to the first aspect, little in the way of organic material goes unused as a food resource. Indeed the materials used range from dissolved organic molecules, through particulate organic detritus of small size (both living and dead), to "particles" the size of whales. Three of these resources have received increasing attention because of their apparent importance in the benthic food web: (1) dissolved organic molecules—some organisms can subsist on this alone (cf. Stephen, 1967); (2) organic molecules adsorbed on sediment grains (cf. Anderson et al., 1958), and (3) organic detritus (Fox, 1950) which may be derived from terrigenous as well as marine sources.

Figure 1 summarizes the location and general types of food resources available in various parts of the marine environment. The most important aspect of the distribution is the concentration of food at and immediately below the sediment–water interface. In deeper waters,

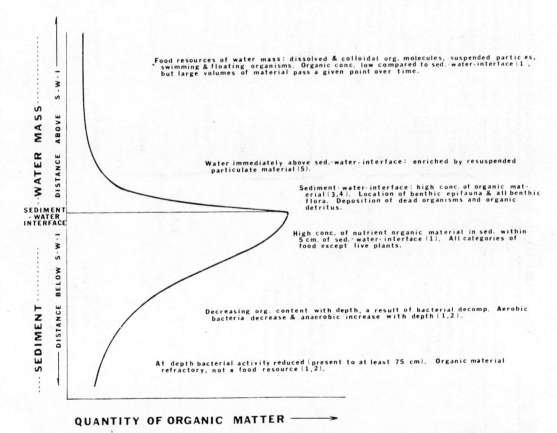

FIGURE 1. Location of food resources with respect to the sediment-water interface (from Walker and Bambach, 1974). Relative amounts of organic material at different levels from: 1, Zobell and Anderson, 1936; 2, Zobell, 1939; 3, Sverdrup et al., 1942; 4, Trask, 1939; 5, Marshall, 1970.

TABLE 1. Suggested Classification and Terminology of Feeding in Benthic Marine Invertebrates

Feeding Category (Trophic Group)	General Definition	Life Site	Location of Food Acquisition	Food Resources Used	Feeding Mechanisms	General Remarks
I. Suspension feeders	Remove food from suspension in the water mass without need to subdue or dismember particles.	Epifaunal	High— Water mass Low—	Swimming and floating organisms, dissolved and colloidal organic molecules, organic detritus less important.	Chiefly ciliary, mucoid and setous mechanisms for dealing with small particles, flagellate in Porifera.	Filter feeding is a subdivision of suspension feeding in which an organic filtration mechanism is involved in food acquisition. Suspension feeding organisms living in sediment may either move into water mass to feed (some sabellids) or draw water downward into burrow (some polychaets and molluscan bivalves).
		Infaunal (burrowers)		Swimming and floating organisms, resuspended organic detritus, dissolved and colloidal organic molecules.		
II. Deposit feeders	Remove food from sediment either selectively or nonselectively. Without need to subdue or dismember particles.	Epifaunal	Sediment-water interface In sediment (shallow)	Particulate organic detritus, living and dead smaller members of benthic flora and fauna, and organic-rich grains.	Non-selective: sediment swallowing in bulk. Selective: select some fraction of the sediment which tends to have a concentration of organic material. Selected either by size and/or specific gravity usually by ciliary activity.	Selectivity generally inefficient. Some epifaunal animals dig or probe the sediment for food.
		Infaunal (burrowers)	Sediment-water interface Shallow In sediment Deep	Same as above.	Chiefly selective by specific gravity or size fractionation using ciliary action.	Examples: tellinid and some nuculoid molluscan bivalves.
				All categories except live plants.	Non-selective: sediment swallowing in bulk. Selective: size fractionation or specific gravity separation by ciliary activity.	Selectivity generally inefficient. Some interstitial fauna. Other authors consider this a type of suspension feeding.
III. Browsers	Acquire food by scraping plant material from environmental surfaces or by chewing of rasping larger plants.	Epifaunal	Sediment-water interface	Benthic flora.	Scraping and seizing-chewing mechanisms for dealing with larger particles.	Many eat some organic detritus, so category grades into epifaunal deposit feeding.

IV. Carnivores	Capture live prey.	Epifaunal	Sediment-water interface	Benthic epifaunal meio- and macro-fauna.	Active: capture live prey after active search or pursuit and feed by seizing and either swallowing whole, chewing, or externally digesting prey. Passive: no search or pursuit. Feeds on organisms which pass sedentary location of predator. Use same mechanisms of feeding as active.	Usually sedentary or attached.
		Nekto-benthic	Sediment-water interface	Same	Active only, same mechanisms as above.	Very active swimming forms which depend on benthos for prey.
		Infaunal	In sediment	Benthic infaunal meio- and macro-fauna.	Active only, same mechanisms as above.	Usually active in shallow *and* deep deposit feeder zones.
V. Scavengers	Eat larger particles of dead organisms.	Epifaunal	Sediment-water interface	Dead, partially decayed organisms.	Seizing and either chewing, swallowing whole, or external digestion.	As particle size eaten decreases, category grades into deposit feeders. Many carnivores will eat recently dead but undecayed prey and, thus, this category also intergrades with carnivores.
		Infaunal	In sediment			
VI. Parasites	Fluids or tissues of host provide nutrition.	Same as host	Same as host	Mostly fluids and soft tissues.	Sucking of fluids, direct absorption.	Extremely rare in fossil record, for they are usually small and soft bodied in life.

From Walker and Bambach, 1974.

there is another zone of relative concentration in the photic zone where phytoplankton occur in abundance.

The properties of various food resources that allow selection by organisms of particular food types from among the plethora available essentially fall into three categories: (1) size of organic particles, (2) chemical composition, and (3) specific gravity. The size of particles ingested is often directly limited by mouth size and the "carrying capacity" of any sorting mechanism used in food acquisition. Many organisms are capable of distinguishing between particles of similar sizes based on chemical composition. Finally, the separation of organic particles and sediment grains seems frequently to be accomplished on the basis of differences in specific gravity.

Adaptations that have been developed by various groups of invertebrate organisms are summarized in Fig. 2. These feeding mechanisms are generally fairly constant within a major

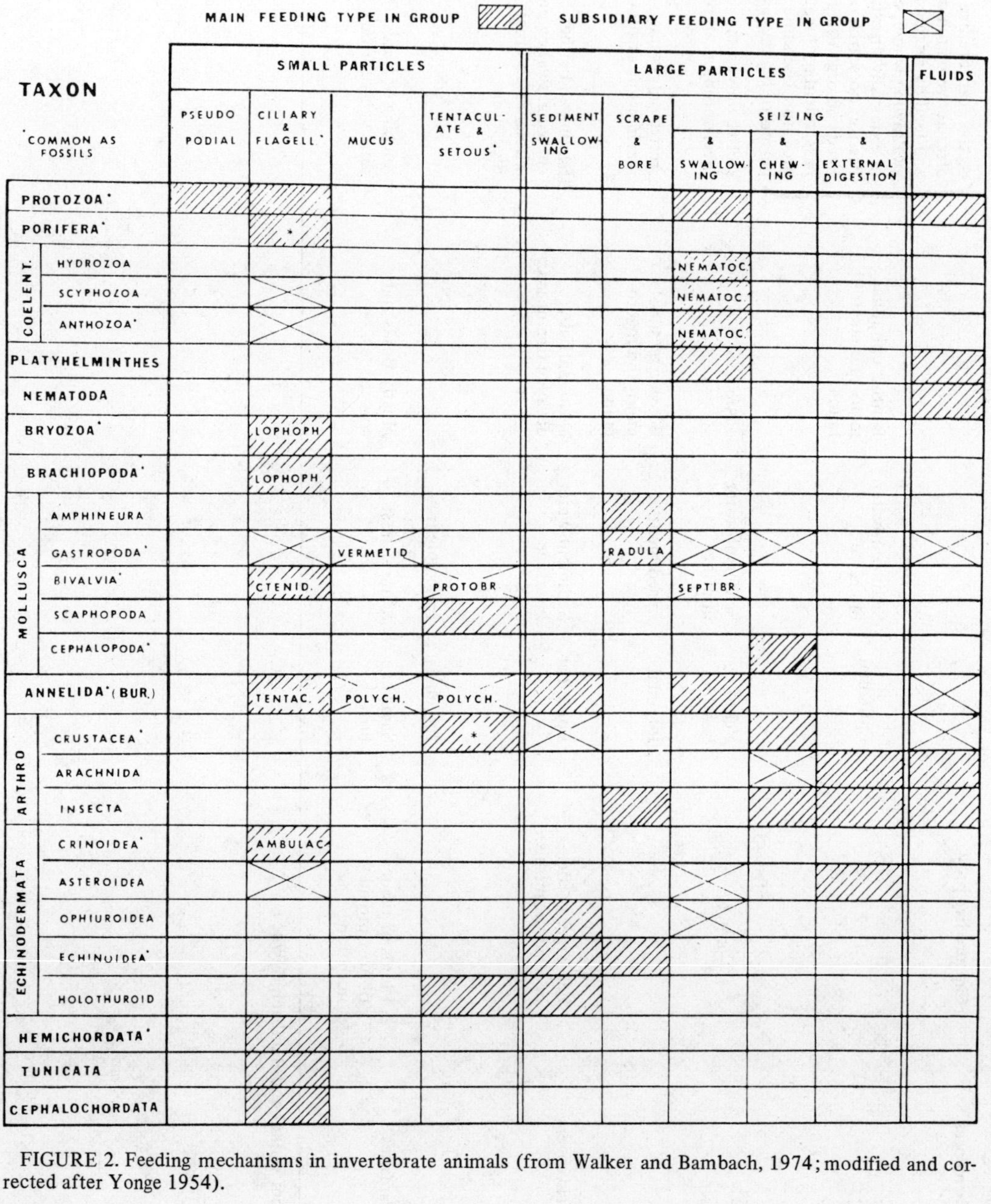

MAIN FEEDING TYPE IN GROUP ▨   SUBSIDIARY FEEDING TYPE IN GROUP ⊠

TAXON (COMMON AS FOSSILS)	SMALL PARTICLES				LARGE PARTICLES					FLUIDS
	PSEUDO-PODIAL	CILIARY & FLAGELL	MUCUS	TENTACULATE & SETOUS	SEDIMENT SWALLOWING	SCRAPE & BORE	SEIZING & SWALLOWING	SEIZING & CHEWING	SEIZING & EXTERNAL DIGESTION	
PROTOZOA*	main	main					main			main
PORIFERA*		main *								
COELENT. HYDROZOA							NEMATOC main			
COELENT. SCYPHOZOA		⊠					NEMATOC main			
COELENT. ANTHOZOA*		⊠					NEMATOC main			
PLATYHELMINTHES										main
NEMATODA										main
BRYOZOA*		LOPHOPH main								
BRACHIOPODA*		LOPHOPH main								
MOLLUSCA AMPHINEURA						main				
MOLLUSCA GASTROPODA*		⊠	VERMETID main			RADULA main	⊠	⊠	⊠	⊠
MOLLUSCA BIVALVIA*		CTENID. main		PROTOBR main			SEPTIBR			
MOLLUSCA SCAPHOPODA				main						
MOLLUSCA CEPHALOPODA*								main		
ANNELIDA*(BUR.)		TENTAC. main	POLYCH. main	POLYCH. main	main		main		⊠	
ARTHRO CRUSTACEA*				main *	⊠		⊠	main	main	
ARTHRO ARACHNIDA								⊠	main	
ARTHRO INSECTA							main	main		
ECHINODERMATA CRINOIDEA*		AMBULAC ⊠								
ECHINODERMATA ASTEROIDEA						⊠	⊠	main	main	
ECHINODERMATA OPHIUROIDEA					main		⊠			
ECHINODERMATA ECHINOIDEA*					main	main				
ECHINODERMATA HOLOTHUROID				main	main					
HEMICHORDATA*		main								
TUNICATA		main								
CEPHALOCHORDATA		main								

FIGURE 2. Feeding mechanisms in invertebrate animals (from Walker and Bambach, 1974; modified and corrected after Yonge 1954).

higher category, so that this figure can be used in interpreting fossil taxa. Of course, adaptive morphological evidence should be used and no group has sufficient constancy of trophic adaptation for the figure to be uncritically used as a key.

Finally, with these various aspects of the feeding process in mind, a classification of trophic groups may be erected. Table 1 is such a classification. Some workers have developed other trophic group schemes (see in particular, Turpaeva, 1957, Savilov, 1957, and Newell, 1970). The paleontologic advantage of viewing feeding relationships between trophic groups rather than between individual taxa is that these trophic groups have remained fairly constant since at least the late Precambrian while, of course, the taxa filling these feeding niches have changed with time. Thus, viewing trophic relationships in terms of these synthetic trophic groups allows comparison of trophic structure of communities of different ages.

KENNETH R. WALKER

References

Anderson, A. E.; Jonas, E. C.; and Dum, H. T., 1958. Alteration of clay minerals by digestive processes of marine organisms, *Science*, 127, 190–191.

Blegvad, H., 1915. Food and conditions of nourishment of invertebrate animals found on or in the sea bottom in Danish waters, *Rept. Danish Biol. Stat.*, 22, 5–49.

Fox, D. L., 1950. Comparative metabolism of organic detritus by inshore animals, *Ecology*, 31, 100–108.

Jorgensen, C. B., 1966. *Biology of Suspension Feeding.* Oxford: Pergamon Press, 357p.

Levinton, J. S., and Bambach, R. K. 1975. A comparative study of Silurian and Recent deposit-feeding bivalve communities, *Paleobiology*, 1, 97–124.

Lindeman, R. L., 1942. The trophic-dynamic aspect of ecology, *Ecology*, 23, 399–418.

Mann, K. H., 1969. The dynamics of aquatic ecosystems, *Ecol. Research*, 6, 1–81.

Marshall, N., 1970. Food transfer through the lower trophic levels of the benthic environment, in J. H. Steel, ed., *Marine Food Chains*. Berkeley: Univ. California Press, 52–66.

Newell, R. C., 1970. *Biology of Intertidal Animals.* New York: Elsevier, 555p.

Nikitin, B. N., 1957. *Trans. Inst. Okeanol., Marine Biol., Acad. Sci. USSR Press*, 20, 67–148. (Publ. in US by Amer. Inst. Biol. Sci., Washington, D.C.)

Olson, E. C., 1966. Community evolution and the origin of mammals, *Ecology*, 47, 291–302.

Rhoads, D. C.; Speden, I. G.; and Waage, K. M., 1972. Trophic group analysis of Upper Cretaceous (Maestrichtian) bivalve assemblages from South Dakota, *Bull. Amer. Assoc. Petrol. Geologists*, 56, 1100–1113.

Savilov, A. I., 1957. Biological aspects of the bottom fauna grouping of the North Okhotsk Sea, in Nikitin, 1957, 67–136.

Scott, R. W., 1974. Bay and shoreface benthic communities in the Lower Cretaceous, *Lethaia*, 7, 315–330.

Scott, R. W., 1976. Trophic classification of benthic communities, in R. W. Scott and R. R. West, eds., *Structure and Classification of Paleocommunities.* Stroudsburg, Pa.: Dowden, Hutchinson & Ross, 29–66.

Scott, R. W., 1978. Approaches to trophic analysis of paleocommunities, *Lethaia*, 11, 1–14.

Stanton, R. J., Jr., and Dodd, J. R., 1976. The application of trophic structures of fossil communities in paleoenvironmental reconstruction, *Lethaia*, 9, 327–342.

Stephen, G. C., 1967. Dissolved organic material as a nutritional source for marine and estuarine invertebrates, in G. M. Lauff, ed., Estuaries. *Amer. Assoc. Adv. Sci. Publ. 83*, 367–373.

Sverdrup, H. U., et al., 1942. *The Ocean.* Englewood Cliffs, N. J.: Prentice-Hall, 1087p.

Thayer, C. W., 1974. Marine paleoecology in the Upper Devonian of New York, *Lethaia*, 7, 121–155.

Trask, P. D., ed., 1939a. *Recent Marine Sediments.* Tulsa: Amer. Assoc. Petrol. Geologists, 419–453.

Trask, P. D., 1939b. Organic content of Recent marine sediments, in Trask, 1939a, 428–453.

Turpaeva, E. P., 1957. Food interrelationships of dominant species in marine benthic biocoenoses, in Nikitin, 1957, 137–148.

Walker, K. R., 1972a. Trophic analysis: A method for studying the function of ancient communities, *J. Paleontology*, 46, 82–93.

Walker, K. R., 1972b. Community ecology of the Middle Ordovician Black River Group of New York state, *Geol. Soc. Amer. Bull.*, 83, 2499–2524.

Walker, K. R., and Bambach, R. K., 1974. Feeding by benthic invertebrates: Classification and terminology for paleoecological analysis, *Lethaia*, 7, 67–78.

Yonge, C. M., 1928. Feeding mechanisms in invertebrates, *Biol. Rev.*, 3, 21–76.

Yonge, C. M., 1954. Food of invertebrates, *Tab. Biologicae*, 21, 25–68.

Zobell, C. E., 1939. Occurrence and activity of bacteria in marine sediments, in Trask, 1939a, 419–427.

Zobell, C. E., and Anderson, D. Q., 1936. Vertical distribution of bacteria in marine sediments, *Bull. Amer. Assoc. Petrol. Geologists*, 20, 258–269.

Cross-references: *Actualistic Paleontology; Communities, Ancient; Paleoecology.*

V

VERTEBRATA

Vertebrates, backboned animals, are members of Phylum Chordata (q.v.); other chordates are the acorn worms and pterobranchs (Hemichordata; q.v.), tunicates (Urochordata), and lancelets (Cephalochordata), collectively known as protochordates. Distinctive anatomical features of chordates are: an internal supporting rod (notochord) at some stage of life history—segmental vertebrae form around this in vertebrates and provide attachment for muscles; pharyngeal gill slits; and tubular dorsal nervous system.

Vertebrates have a closed circulatory system in which blood moves forward from a ventral heart, dorsally through aortic arches and gills, and posteriorly through an aorta that lies just ventral to the notochord or vertebrae. The blood contains red hemoglobin as an oxygen carrier; tissues have creatine phosphoric acid as an energy-releasing catalyst; embryonic development is bilaterally symmetrical (deuterostomatous); and the body cavity (coelom) arises during embryonic development by splitting of initially solid mesoderm (schizocoelous).

Primitive aquatic vertebrates, collectively referred to as fishes, belong to five classes: The jawless ostracoderms and cyclostomes (Agnatha), primitive armored fishes with jaws (Placodermi), cartilaginous fishes (Chondrichthyes), spiny fishes (Acanthodii), and bony fishes (Osteichthyes). The air-breathing, land-dwelling vertebrates or tetrapods comprise the classes Amphibia, Reptilia, Aves, and Mammalia (see Table 1).

Class Agnatha

The jawless vertebrates, Division Agnatha, include two families of living eel-like fishes, the parasitic lampreys and scavenging hag-fishes, and three or four groups of middle Paleozoic fossils with thin, bony exoskeletal "shells," the ostracoderms. The modern agnathans (cyclostomes) lack calcified skeletal tissues and until recently were unknown as fossils. Impressions of a fossil lamprey from the Pennsylvanian of Illinois (Bardack and Zangerl, 1967) show that this family is of great antiquity.

Agnathans differ from other vertebrates in the position of the branchial skeleton, external to the associated gill pouches, muscles, nerves, and branchial arteries, instead of medial to these, and the brain case lacks the embryonic elements called trabeculae. These and other differences suggest that the Agnatha, including the ancient ostracoderms, had diverged from the ancestry of other vertebrates well before the development of jaws, and that none of the known agnathans represent an ancestral group to the higher vertebrates.

Ostracoderms. Ostracoderms are known from broken fragments of bony plates in Early Ordovician deposits. More complete material is known from the Late Silurian and Devonian. They were extinct by the end of the Devonian.

Ostracoderms are grouped into three or four orders: Osteostraci, Anaspida, Heterostraci, and sometimes Thelodonti or Coelolepida.

The Osteostraci or cephalaspid-like ostracoderms had flattened bodies, sense organs on the top of the armored head, and a relatively short, muscular tail covered by scales. Their proportions suggest a bottom-dwelling habit. The presence of a thin, bony lining to all openings in the cartilaginous endocranium has enabled the Swedish paleontologist Dr. Erik A. Stensiö to demonstrate the relationship of cephalaspid ostracoderms to lampreys, which they resemble in the absence of jaws, number and arrangement of gill openings, median-dorsal opening of the nasohypophyseal duct, and presence of only two semicircular canals in the inner ear (in contrast to three in gnathostomes). Primitive Osteostraci such as *Tremataspis* lacked paired fins; pectoral fins evolved in some later members of the Osteostraci independently of their origin in jawed fishes.

The Anaspida differ from the Osteostraci in more fusiform bodies; small scales instead of a single bony shield over the head region; and larger, laterally directed eyes. Their internal anatomy is unknown but sense organs of the head are arranged as in cephalaspids. *Jamoytius,* a Late Silurian fossil, originally described as a protochordate, is now considered an anaspid.

The Heterostraci (Pteraspida) have several large bony plates surrounding the anterior part of the body, small laterally placed eyes, no dorsal nasohypophyseal opening, and an opercular plate covering the gills. They lack paired fins

TABLE 1. Classification of the Vertebrates

Class Agnatha, U.Camb.-Holo.
  Order Osteostraci, U.Sil.-M.Dev.
  Order Anaspida, M.Sil.-U.Dev.
  Order Heterostraci, U.Camb.-Dev.
  Order Coelolepida (Thelodonti), U.Sil.-L.Dev.
  Order Cyclostomata, Penn.-Holo.
Class Acanthodii, U.Sil.-L.Perm.
Class Placodermi, U.Sil.-U.Dev.
Class Chondrichthyes, M.Dev.-Holo.
  Subclass Elasmobranchii, M.Dev.-Holo.
  Subclass Holocephali, U.Dev.-Holo.
Class Osteichthyes, L.Dev.-Holo.
  Subclass Actinopterygii, L.Dev.-Holo.
    Infraclass Chondrostei, M.Dev.-Holo.
    Infraclass Holostei, U.Perm.-Holo.
    Infraclass Teleostei, U.Jur.-Holo.
  Subclass Sarcopterygii, L.Dev.-Holo.
    Order Dipnoi, L.Dev.-Holo.
    Order Crossopterygii, L.Dev.-Holo.
      Suborder Rhipidistia, L.Dev.-Perm.
      Suborder Coelacanthini, L.Dev.-Holo.
Class Amphibia, U.Dev.-Holo.
  Subclass Labryinthodonta, U.Dev.-U.Trias.
  Subclass Lepospondyli, L.Miss.-L.Perm.
  Subclass Lissamphibia, L.Trias.-Holo.
Class Reptilia, L.Penn.-Holo.
  Subclass Anapsida, L.Penn.-Holo.
    Order Cotylosauria, L.Penn.-U.Trias.
    Order Mesosauria, L. Perm.
    Order Chelonia, U.Trias.-Holo.
  Subclass Lepidosauria, U.Perm.-Holo.
  Subclass Archosauria, L.Trias.-Holo.
    Order Thecodontia, L.Trias.-U.Jur.
    Order Crocodilia, U.Trias.-Holo.
    Order Pterosauria, U.Trias.-U.Cret.
    Order Saurischia, U.Trias.-U.Cret.
    Order Ornithischia, U.Trais.-U.Cret.
  Subclass Synaptosauria, L.Trias.-U.Cret.
  Subclass Ichthyopterygia, M.Trias.-U.Cret.
  Subclass Synapsida, L.Penn.-L.Jur.
Class Aves, U.Jur.-Holo.
Class Mammalia, M.Trias.-Holo.

and swam by a hypocercal tail. Repetsky (1978) has recently reported heterostracan remains in rocks of Late Cambrian age.

Stensiö (1928) suggested that the Heterostraci were ancestral to hag-fishes (Myxinoidei), and divided the living cyclostomes into two independent subclasses, the Pteraspidomorphi (Heterostraci, Coelolepida, and Myxinoidei) and the Cephalaspidomorphi (Osteostraci, Anaspida, and Petromyzontida). This view has been questioned by Romer (1962) and others on the grounds that living cyclostomes show great morphological similarity among themselves and especially because they possess in common a unique feeding mechanism, the rasping tongue. Jarvik (1965) called attention to considerable differences in the detailed structure of the rasping organs of myxinoids and lampreys and

denied their homology, thus supporting Stensiö's opinion. Although numerous hypothetical restorations of the head structure of pteraspids have been published, none of these rest on a firm foundation of observation comparable to the basis for interpretation of cephalaspid morphology and relationships. Regardless of the degree of relationship, or lack of it, between lampreys and hag-fishes, no convincing evidence has been adduced to support the myxinoid-heterostracean grouping.

Indistinct impressions of jawless fishes and a variety of small scales from Late Silurian and Early Devonian rocks are often classified together as a fourth "Order," Coelolepida; the most adequately known remains of this group appear to be immature specimens of the other ostracoderm orders.

Gnathostomes—The Origin of Jaws

All other vetebrates have moveable jaws, which enable them to feed upon a far greater variety of food than is available to filter-feeders. With jaws and teeth came the ability to capture and eat relatively large organisms, and this permitted the fishes to grow to greater size. Thus the development of jaws stands as the first major evolutionary advance among vertebrate animals. It led to adaptive diversification in a far greater number of ways than ostracoderms or their descendants ever achieved.

Vertebrate jaws lie immediately in front of the jointed gill arches of gnathostome fishes and bear the same relationship to the muscles, nerves, and blood vessels as do the gill arches. Jaws and gill bars are serially homologous structures, and theory suggests that the gill supporting function came first, and later the most anterior gill arch came to be used for feeding and was modified into jaws. As pointed out under Agnatha, the jointed gill bars of gnathostomes are situated internally to the nerves, blood vessels, and muscles, and cannot be modifications of the branchial basket of agnathans.

Class Acanthodii

Acanthodians are primitive jawed fishes of the middle and Late Paleozoic whose relationships to modern fishes are as yet uncertain (see *Pisces*). Professor D. M. S. Watson of London found evidence of gill rakers along the hyoid arch of *Acanthodes* and inferred that acanthodians retained a full functional gill opening between the mandibular and hyoid arches. Their upper jaws, according to this view, could not have been supported against the braincase by a buttressing hyomandibular as are those of both sharks and bony fishes, and Watson believed that they, together with the armored placoderms, represented an extremely primitive stage in the evolution of jaws. The more recent discovery that *Acanthodes* has a normal gnathostome palate supported by the hyomandibular suggests that these fishes are merely such primitive members of the normal gnathostome group that we are unable to determine their relationships more closely.

Acanthodians are normal fusiform to elongate eel-like fishes known from Late Silurian to Early Permian time. They have large eyes; gills open by separate slits but these are covered ventrally by a partial operculum; fins are preceded by stout bony spines like those of sharks; as many as five pairs of spines may occur between the pectoral and pelvic fins. This arrangement is often cited as supporting the "fin-fold" theory of vertebrate appendages. The tail is heterocercal like those of sharks and primitive bony fishes. Scales are thick, small, rhombic, similar to those of the earliest bony fishes.

Class Placodermi

A varied assemblage of Late Silurian and Devonian fishes resemble ostracoderms in having well-developed bony armor over the front of the body, but differ from their jawless contemporaries in having well-developed jaws, large eyes, and generally a division of the armor into separate head and shoulder sections. In anatomical details of the braincase, they show enough similarity to sharks to suggest some relationship. Perhaps as a group they represent the bony forerunners of the cartilaginous fishes. But no good intermediates connect these fishes, so placoderms are placed in a separate class (see *Pisces*).

Arthrodires, the most typical placoderms, have heavy bony plates covering head and thoracic regions, separated by a joint in the neck region (whence the name, meaning *jointed neck*) that permitted the head to be raised and lowered with respect to the primitive shoulder girdle. Early actinolepids had stout immovable spines in position of the pectoral fins, and appear to be bottom dwellers. More typical arthrodires such as *Coccosteus,* or the giant *Dinichthys* [whose head and shoulder skeleton alone are 8 ft (2.4 m) long], had small pectoral spines and a moveable fin. These fishes were powerful predators of the Devonian seas.

Antiarchs also had armored heads, joined to the thorax in a different fashion from arthrodires. They differed in having more strictly bottom-dwelling adaptations, the eyes and other sense organs lying high on top of the head and the jaws being small and weak. The antiarch pectoral spine was movably jointed to the thoracic armor.

Several other groups of placoderms are known, including the ray-like rhenanid *Gemundina* and the ptyctodonts with crushing teeth.

Class Chondrichthyes

Fishes with cartilaginous skeletons, sharks, rays, and chimaeras, were once thought to be the most primitive jawed fishes because no bone is present in their skeleton, except in the bases of the teeth. Although it is not yet possible to point with certainty to a bony ancestor of these fishes, several lines of evidence suggest that the cartilaginous fishes have lost the ability to form bone in the course of their evolution, and retain the cartilaginous skeleton of embryos throughout life. Numerous similarities between the skulls of certain placoderms and those of sharks, and the rather sudden appearance of a variety of fossil sharks late in the Devonian, just as the arthrodires were on the wane, point to the placoderms as the most likely ancestor of the Chondrichthyes. Instead of being survivors of a primitive, ancestral group of fishes, the sharks and chimaeras now are regarded as a specialized branch that appear later (Late Devonian) than the bony fishes (Early Devonian).

Two distinct groups of cartilaginous fishes can be recognized from the outset. The Elasmobranchii—sharks and rays—have 5-7 pairs of gills, each opening separately to the outside. Their jaws are movably attached to the braincase and braced behind by the upper element of the hyoid arch, the hyomandibular. Teeth are numerous, replaced in rapid succession, and extremely varied in shape. The outside of the skin is covered with small tooth-like structures, placoid scales. The tail is usually heterocercal; but in rays, which are flattened, bottom-dwelling sharks with crushing teeth, the tail may become whip-like.

Chimaeras, Subclass Holocephali, have gills opening into a chamber covered by skin and having a single outside opening; the upper jaw cartilage is united with the braincase and not braced by a hyomandibular. Tooth succession is slow; the teeth are porous crushing plates. The skin is naked and the tail whip-like. Various crushing teeth from late Paleozoic rocks known as bradyodonts have been referred to the holocephali. Some of these, such as *Helodus* (whose skull is known) and the cochliodonts, may represent early members of this line.

More definite relatives of the chimaeras are *Squatinactis* from the Mississippian and *Squaloraja* and *Myriacanthus* of the Jurassic. The three modern families all date back to the Mesozoic, and are represented mainly by teeth and fin spines.

Ørvig has pointed out numerous resemblances between chimaeras and the placoderms known as ptyctodonts, and suggested that the Holocephali are modern survivors of the placoderms. The Jurassic chimaeroids, however, approach sharks to some extent, so the relationships between these groups remain questionable.

The earliest known sharks are Late Devonian cladodonts such as *Cladoselache*, which have broad-based fins that could function only as stabilizers. Schaeffer (1967) has emphasized the diversity of fin structure among cladodont and ctenacanth sharks of the late Paleozoic, and even more specialized fins are seen in the recently described Iniopterygii of the Mississippian. From the cladodonts evolved the typically Mesozoic hybodonts, fishes much like the living Port Jackson shark (*Heterodontus*) with more maneuverable fins attached by a narrow base. All these primitive sharks are amphistylic, that is, the upper jaw cartilage articulates with the postorbital process of the braincase as well as being supported by the hyomandibular.

In the Jurassic, hyostylic sharks evolved by the loss of the contact between palatoquadrate and braincase. The resulting greater flexibility of the mouth apparently permitted a great radiation of feeding types, and the modern shark genera rapidly evolved, most of them appearing in the Cretaceous. Rays are a specialized group of hyostylic sharks that have evolved crushing dentitions and greatly expanded pectoral fins, which form their sole swimming organ. They also appeared in the Cretaceous (Schaeffer, 1967).

Sharks and rays are dominantly marine fishes and have been highly successful in that habitat, although never as diverse as the bony fishes. A very few elasmobranchs have invaded fresh water; the Eocene ray *Xiphotrygon* from the Green River lake beds of Wyoming may be such; living freshwater sharks occur in lakes in Central America and in Thailand.

Class Osteichthyes

Bony fishes differ from sharks in the presence of bone in the endoskeleton and also in having bony scales instead of toothlike placoid scales. The gills open into a chamber covered by an operculum supported by the hyomandibular bone, and bounded behind by the large cleithral bone of the shoulder girdle. A swim bladder is developed from a pocket of the digestive tract.

Two subclasses of bony fishes can be recognized from earliest Devonian time, the ray-finned fishes or Actinopterygii, to which the vast majority of fishes belong, and the lobe-finned fishes or Sarcopterygii, which include the lung fishes (Dipnoi), coelacanths, and rhipidistians; the latter two are often grouped together as Crossopterygii. Rhipidistians are of particular interest because they include the ancestors of the tetrapods or land-dwelling vertebrates. Ray-finned fishes differ from lobefins in details of their scale histology, arrangement of sensory canals on the head and the related pattern of skull bones, and the course of evolution of the median and tail fins. Their fin skeletons consist of a series of supporting bony rays that articulate individually with internal supports (median pterygiophores and paired girdles), to which the fin musculature attaches.

Ray-finned fishes include nearly all the familiar types of fishes except sharks. Many problems concerning their relationships and evolution are still unsolved so the current classification is to a large extent "horizontal" rather than phylogenetic.

Subclass Actinopterygii: Chondrostei. Fishes of the infraclass Chondrostei lack ossified vertebral centra. Primitive members of this group have a single dorsal fin; heterocercal tail fin (in which the vertebral column and body muscles extend to the tip of the upper lobe of the tail, and the caudal fin is developed from ventral rays); numerous jointed or "soft" fin rays that always exceed the number of the endoskeletal fin supports; pelvic fins "abdominal," that is, adjacent to the vent; thick, usually rhombic, scales with a shiny outer layer of enamel-like ganoin; large eyes, small dorsally placed nostrils, and a large mouth with slender jaws. In all, the cheek region of the skull is covered by the large maxillary bone, firmly attached to the preoperculum.

The earliest genus, *Cheirolepis* (Middle Devonian) has tiny scales like the acanthodians; more typical palaeoniscoids of the later Paleozoic and Triassic have larger, nonoverlapping rhombic scales. Deep-bodied fishes with short mouths evolved by Pennsylvanian time (*Platysomus*). Surviving chondrosteans are the sturgeons and paddle-fishes, which retain the primitive heterocercal tail but have degenerate scales or naked skins and small mouths, and the African bichir (*Polypterus*) and its allies, which retain ganoid scales but have specialized fins. (These were long confused with the lobe-finned fishes under the name "fringe-finned ganoids.")

During the Triassic various paleoniscoid families evolved "advanced" characteristics such as reduced numbers of fin rays, more symmetrical (hemiheterocercal) tails, or loss of the posterior expansion of the maxillary that covered the cheek. Practically every combination of these features is found among the families sometimes grouped as subholosteans, which flourished during the Triassic. This independent evolution of adaptive improvements in the structure of various organ systems in different families has been termed *mosaic evolution.*

Holostei. The infraclass Holostei are fishes with incompletely ossified centra—actually vertebrae vary from thin bony rings to well-ossified cylinders within the group. The tail is nearly or quite symmetrical externally but retains vestiges of the primitive heterocercal condition in its skeleton. Fin rays are all soft but equal the internal supports in number. The maxillary no longer touches the preopercular, permitting considerable expansion of the jaw adductor muscle. At the front of the mouth, the premaxillary bone forms a larger part of the jaw margin. Scales vary from the rhombic ganoid type of the gar pike (*Lepisosteus*) to overlapping bony plates lacking or almost lacking any ganoin.

During the Mesozoic the holosteans displayed a wide variety of adaptations, from the streamlined and predaceous furids to deep bodied, nibbling semionotids such as *Dapedium* and pycnodonts (which survived into the Eocene), elongate, armored aspidorhynchids, swordfish-like pachycormids, and "progressive" pholidophoroids and leptolepids which approach the teleosts in structure. The living bowfin (*Amia*) and aberrant gar pikes (*Lepisosteus*) are survivors of this intermediate level of fish evolution.

Teleostei. Teleost fishes have well ossified vertebrae; symmetrical tails supported by modified haemal spines known as hypural bones; thin overlapping bony scales, which are oval (cycloid) in more primitive orders, or have a toothed exposed margin (ctenoid) in spiny-finned groups. The premaxillary bone forms the functional upper jaw and the toothless maxillary becomes modified into part of the apparatus for protruding the jaws.

A primitive level of teleost organization is shown by such fish as the salmon and trout, herrings, some eels, and certain deep-sea fishes, whose fin rays are jointed or soft (hence Malacopterygii), whose pelvic fins are adjacent to the vent (abdominal), and whose swim bladders remain connected to the digestive tract by an open duct (physostomous).

Formerly united in a single order Isospondyli, these are now considered to represent at least three major divisions of the teleost group. Tarpon (*Elops*), bonefish (*Albula*), and true eels (*Anguilla*) and their relatives form one of these primitive branches. They are well represented by fossils as far back as the Cretaceous. Of questionable relationship to these are the herrings and their allies (Clupeomorpha), a diverse and successful group likewise of Cretaceous origin. More divergent are the osteoglossomorphs, confined today to fresh waters of the southern hemisphere continents, but widespread in Eocene time in northern regions as well. Giant predaceous fishes of the Cretaceous such as *Portheus* have been referred to this division, though Bardack has questioned this.

The greatest diversity is presented by the branch known as the Protacanthopterygii, which includes salmon, trout and their allies among its more primitive members, and which leads to the more highly specialized ostariophysan and spiny finned fishes.

An important group of freshwater fishes—carp, catfish, the varied characins, gymnotid and electric eels of South America and Africa—have the first four vertebrae fused together and modified to form a series of ossicles (Webberian apparatus), which transmit vibrations from the swim bladder to the perilymph of the auditory labyrinth. These constitute the Order Ostariophysi; their fins commonly have strong leading spines but otherwise are soft rayed.

Several other orders represent diverse lines of teleost radiation, some of them foreshadowing some of the structures of the spiny finned teleosts or Acanthopterygii, which form the highest grade of evolution of the bony fishes. In the spiny finned order, the anterior dorsal fin and anterior part of the anal fin have solid, stout spines instead of jointed rays. Pelvic fins have migrated forward from the vent and attach to various anterior parts of the body—the shoulder girdle or even the chin—where they are more efficient at controlling swimming movements. The swim bladder loses its connection to the digestive tract and is completely closed (physoclystous). Scales are of the ctenoid type with toothed posterior margins.

Teleost fishes apparently underwent extremely rapid evolution in the Cretaceous following attainment of the protrusible mouth mechanism, for even the more advanced orders are represented in Cretaceous rocks. The fossil record is large but by no means sufficiently well known to permit construction of a detailed phylogeny.

Subclass Sarcopterygii. The Sarcopterygii appear in the Early Devonian, slightly before the first record of the Actinopterygii. The earliest genera have heterocercal tails, which evolve into symmetrical types by addition of a dorsal lobe instead of shortening the muscular axis. Musculature of the paired fins, and the median fins of crossopterygians, extend beyond the body to form a central lobe of the fin which has a jointed bony axis. Such fins are highly flexible and had the potential of developing into limbs suitable for walking on land.

Numerous anatomical distinctions between sarcopterygians and ray-finned fishes include the smaller eyes and larger olfactory organs and consequent platybasic form of the braincase; close articulation of the upper jaw to the braincase (autostylic) instead of suspension by an enlarged hyomandibular; irreconcilably different arrangement of the roofing bones of skull roof and cheek regions; two dorsal fins in the early members of the subclass, and "cosmoid" scales with a thick middle layer of dentine-like material lacking from the "ganoid" scales of early actinopterygians.

Dipnoi. Lung fish or Dipnoi first appear in Early Devonian marine deposits but had invaded fresh waters by Middle Devonian time and passed the rest of their history there. Their skulls have an enlarged median series of bony plates and tend to lose the more lateral elements. Only the very earliest possess teeth on the margins of the jaws; all the rest have distinctive palatal tooth plates with a fan-like arrangement of radiating ridges. The body form evolved from fusiform to elongate. By Permian time, the habit of burrowing in mud to survive dry seasons had been established, as shown by fossil burrows; the living African genus *Protopterus* and South American *Lepidosiren* retain this custom. The Australian *Epiceratodus* has undergone negligible morphological change from the structure of the Triassic *Ceratodus,* a fossil known long before its living relative was discovered.

Dipnoans are an excellent example of arrested evolution—adaptations acquired early in their history have enabled them to survive unchanged for 200 m yr (see *Living Fossil*).

Coelacanthini. Coelacanth fishes were thought until 1939 to have become extinct in the Cretaceous; their fossil record goes back to the Early Devonian. The order survives in the genus *Latimeria,* living today in a restricted area of the Indian Ocean. Like the dipnoans they early evolved a distinctive structure that has persisted since late Paleozoic time; they appear to be mainly marine rather than freshwater fishes, though certain Carboniferous and Triassic genera occur in probably freshwater deposits.

Coelacanths have large muscular fin lobes from which the membrane supporting rays emerge in a diverging radiate pattern. The tail is symmetrical with a projecting median lobe. Unlike other sarcopterygians, they lack internal nostrils, and the swim bladder, often preserved in fossils, lacks an open duct.

Rhipidistia. Rhipidistians are of particular interest as the immediate precursors of tetrapods or terrestrial vertebrates. Their ancestral position is shown by close similarity to the earliest tetrapods, the labyrinthodont amphibians, in the arrangement of the bones of the skull roof; cheek, and palate; the presence of similar palatal tusks; the labyrinthine infolding

of enamel and dentine of the teeth; vertebral centra composed of paired pleurocentra and ventral intercentra; and especially the arrangement of the basal rays of the paired fins, which are homologous to the bones of the tetrapod limbs and wrist or ankle. These various structures are best known in *Eusthenopteron* of the Late Devonian. Rhipidistians became extinct in the later Carboniferous, but their terrestrial descendants have occupied every niche of the lands, learned to fly through the air, and reinvaded the sea to successfully compete with fishes.

The question of whether tetrapods (air-breathing vertebrates with legs instead of fins) arose only once from a rhipidistian ancestor, or whether the tailed amphibia (Urodela) were descended from a different fish (a porolepiform) from the ancestor of the remaining tetrapods (an osteolepiform) has been debated inconclusively for many years. Recent summaries of these conflicting views may be found in Jarvik (1966) and Thomson (1967).

Class Amphibia

Frogs and toads, salamanders, and little-known worm-like burrowing tropical animals called caecilians are the modern survivors of the first vertebrates to acquire the ability to leave the water and slowly crawl over dry land. They are placed in the class Amphibia (q.v.), meaning double life, because their eggs are laid in water (or very moist soil) and develop into a fish-like larva, the tadpole, which later undergoes a metamorphosis in which its gills are lost, limbs are acquired, and it becomes able to leave the water, at least for short journeys on the land. Amphibians are less completely adapted structurally and physiologically for terrestrial life than reptiles and higher tetrapods; modern species survive by specialized behavior. In particular, they live in moist places or else come out in the open only at night to avoid excessive loss of water from their skins, which lack the impervious epidermis of reptiles.

Amphibians first appear in Late Devonian rocks of East Greenland. *Ichthyostega* (see *Vertebrate Paleontology*, Fig. 1) forms one of the finest examples provided by the fossil record of an intermediate form between two distinct grades of organization ranked as classes in zoological classification. Its skull is intermediate in proportions between those of crossopterygian fishes and later amphibians, and retains vestiges of the opercular bones that cover the gill chamber of fishes. It has fully developed tetrapod limbs and feet, but retains distinct remnants of the tail fin of its fish ancestors. The reason for the evolution of the tetrapod limb, which made possible the emergence of vertebrates from water onto land, is uncertain. An appealing theory that tetrapod ancestors inhabited a seasonally arid region where the streams and pools in which they lived sometimes dried up and eliminated those individuals that were unable to crawl over dry land to another pool imposes extremely stringent requirements for the earlier stages in development of the limb structure. A sufficiently efficient crawling mechanism to permit survival under these conditions would seem to require previous development of a strong limb which thus remains unexplained. The alternative seems to be evolution of limbs for crawling in water too shallow to swim, and then out on the shore, under humid conditions. This might have been done either in search for food, or very likely in search of safe places to deposit eggs.

During the latter part of the Paleozoic era, Amphibia were the principal terrestrial vertebrates and evolved into animals of relatively large size and diverse habits. Most of these had large bony heads, from which they are sometimes called stegocephalians (no longer regarded as a formal systematic group). Most had teeth with complex folds of enamel and dentine, from which the name Labyrinthodontia is derived. Labyrinthodonts survived to the end of the Triassic and are believed to have given rise to frogs and toads, if not all more recent amphibians. Certain Paleozoic amphibians, the microsaurs, nectrideans, aïstopods, and lysorophids, have spool-like vertebrae unlike those of labyrinthodonts and are placed in a separate subclass, the Lepospondyli. Modern salamanders have similar vertebral centra, and have been considered descendants of the lepospondyls. However, they also show many deep-seated similarities to the frogs and toads, which in turn appear to be derived from labyrinthodonts. At present it is uncertain which set of characters is fundamental and indicative of relationship and which has arisen independently in unrelated orders.

Class Reptilia

Reptiles (q.v.) are fully adapted to terrestrial life; both their scaly skins and their specialized excretory apparatus are designed to conserve water; and, most important, they have evolved an egg with a protecting shell and a series of internal membranes—the amnion (to enclose an aquatic environment for the embryo), the allantois (to provide for respiration and waste disposal), and the yolk sac (in which food is stored). The amniote egg can be laid on dry land, freeing reptiles from the necessity of returning to water to breed, and enabling them to occupy many terrestrial environments, including deserts.

Reptiles are generally considered to have descended from amphibian ancestors, but thus far the fossil record has failed to show the transitional steps. The early Permian *Seymouria* combines a very reptile-like postcranial skeleton with a primitive amphibian skull. *Diadectes,* long considered to be a cotylosaurian reptile related to pareiasaurs, procolophonids, and turtles, appears to be closer to *Seymouria* and is now included with the seymouriamorphs among the Amphibia, sometimes as a separate suborder (or even class), Batrachomorpha. *Gephyrostegus,* a Carboniferous relative of *Seymouria,* most closely approaches the structure expected in the ancestor of the reptiles. Neither reptiles nor fossils suitable to be reptilian ancestors have been found in prePennsylvanian deposits.

Modern reptiles, except turtles, Mesozoic dinosaurs and their kin, including birds, and the mammal-like reptiles of the Permian and Triassic, all can be derived from a Late Carboniferous and Permian stock known as the captorhinomorph cotylosaurs. The fossil record of these, especially the primitive family Romeriidae, extends well back into the Pennsylvanian and includes the earliest known reptiles, *Hylonomus, Cephalerpeton,* and *Romeriscus.* These animals lack the otic notch of labyrinthodonts and diadectids; and their stapes, instead of projecting upward toward the otic notch as in labyrinthodonts, extend downward from the braincase to the quadrate bone like the hyomandibular of fishes, from which it is derived. One may reasonably infer that captorhinomorphs had separated from the ancestors of the labyrinthodonts before the latter developed their characteristic middle-ear structure. An order of lepospondylous amphibians, the Microsauria, resembles captorhinomorph cotylosaurs in general skull shape and the absence of an otic notch, and also in the form of their vertebrae. But details of the stapes, double occipital condyles, the absence of vertebral intercentra, and a reduced number of toes on the feet all bar them from an ancestral position; and it seems likely that the superficial resemblances are merely results of adaptive convergence (Carroll, 1969).

Captorhinomorph cotylosaurs early gave rise to the synapsid branch to which the mammal-like reptiles of later Permian and Triassic rocks belong, and from which the mammals evolved early in the Mesozoic era. Less certainly, and with few satisfactory connecting links, they gave rise to diapsid reptiles—with two-arched skulls—both the lepidosaurs (scaly reptiles), including modern lizards and snakes, and the archosaurs (ruling reptiles), which include dinosaurs (q.v.), crocodiles, the flying ptero-dactyls, and ancestors of the birds. Relation-ships and ancestry of turtles, of the porpoise-like ichthyosaurs, and of the sea-serpent plesiosaurs are unsatisfactorily known; no fossils clearly connect them with particular early forms.

Modern reptiles and birds differ from mam-mals in the arrangement of the blood vessels connecting the heart and aorta as well as in other anatomical features such as having only a single instead of double condyle for the joint between skull and neck vertebrae. The warm-blooded birds and mammals thus represent the ends of two lines of development, which di-verged well back in Pennsylvanian time. (The first pelycosaur, *Protoclepsydrops,* is only one stage younger than *Romeriscus,* the oldest known reptile.) These two branches are known as the Sauropsida—culminating in birds and including all living reptiles—and Theropsida—culminating in mammals and including the varied mammal-like reptiles of the Permian and Triassic. Although this division is extremely important, it cannot be applied to many groups of extinct reptiles, for example, the ichthyo-saurs, whose arterial arrangements are not preserved in fossils. Consequently, paleontolo-gists prefer to classify reptiles by the arrange-ment of the bony arches of the skull to which the jaw muscles attach, even though these do not form an infallible guide to relationships.

Class Aves

Feathers are unique to birds; they provide insulation for the body—important for main-taining a high and uniform body temperature—and also form the firm airfoil surface for the wings that is necessary for flight. A four-chambered heart completely separates the systemic and pulmonary circulation streams as in mammals, but the aortic arch is on the right side as in reptiles and in contrast to mammals. Many other anatomical features are closer to those of crocodilians, the surviving order of archosaurian reptiles, than to other vertebrates. The rear limbs and feet are like those of thero-pod dinosaurs; the pelvis is modified to permit an unusually large egg to be laid and to provide strong support for the body on the rear legs. Parental care of the young is usual.

The attainment of the ability to fly opened a large new adaptive zone to birds, in which they soon outcompeted the flying reptiles (Pterosaurs; see *Reptilia*). *Archaeopteryx* of the Late Jurassic shows a mixture of conservative reptilian features such as a long tail, clawed fingers, and an unossified sternum, along with well-developed avian characteristics, especially feathers, and is one of the outstanding fossil connecting links between separate major sys-tematic classes.

Fossil birds are rare compared to remains of other classes because of the delicate light-weight construction of their bones and also because their flying habits and woodland habitat keep them away from sites favorable for fossil accu-mulation. Their fossil record consists largely of members of marsh, shore, and aquatic groups, and at certain sites such as the Rancho La Brea tar pits, of predators and carrion feeders. Sev-eral aquatic orders are known from the Creta-ceous, including the large nonflying diver, *Hesperornis,* which retained teeth in its jaws. Large ground-living birds that lost the ability to fly have evolved several times, including the Eocene *Diatryma* of North America, the middle Tertiary phororhachids of South America, the Pleistocene moas of New Zealand, and several still-living groups such as ostriches, rheas, emus, and the kiwi. Penguins represent a different adaptation that led to loss of flying ability, in this case, connected with marvelous dexterity at swimming. Fossil penguins are known as early as the Eocene; like the successful ground birds, they inhabit regions with few or no ter-restrial predatory mammals.

Class Mammalia

The warm-blooded, furry, generally quadru-pedal animals that nurse their young with maternal milk—mammals—evolved from the synapsid branch of the reptiles during the Trias-sic period. Paleontologists find it necessary to distinguish various fossil groups by characters of the skeleton and use the nature of the jaw bones and sound-conducting apparatus of the middle ear to separate mammals from reptiles. Mammals have a single pair of bones forming the lower jaw, which correspond to the tooth-bearing element of the complex reptile jaw. Be-hind this bone, reptiles have several accessory bones, one of which, the articular, forms a joint with the quadrate bone of the skull. The mam-malian mandible articulates with a different skull bone, the squamosal. Reptiles have a single rod-like bone, the stapes, to conduct sound vibrations from the ear drum to the organ of hearing in the inner ear. Mammals have a chain of three bones, the innermost of which can be shown to be homologous to the reptilian stapes, whereas the other two represent the quadrate and articular, which formed the rep-tilian jaw articulation! Details of the transition from the reptilian to mammalian structure of this part of the skull are now well known. Fos-sils with double articulations between skull and mandible, corresponding to the reptilian and the mammalian joints, have been described from the Middle Triassic of Argentina (*Probainognathus*) and Late Triassic of South Africa (*Diarthrognathus*). Depending on whether one defines mammals as having a single bone on each side of the lower jaw, or as having a jaw articulation between dentary and squamosal, or as having three auditory ossicles, various intermediate fossils would be arbitrarily assigned to different classes, Reptilia or Mammalia. Some evidence suggests that the structural changes just described took place in several related but genetically separate lines of advanced mammal-like reptiles, so that Class Mammalia, as so defined, is technically an evolutionary grade of polyphyletic origin. It also seems likely that some of the physiological traits characteristic of mammals were already present among the advanced therapsid reptiles, but no satisfactory criteria have yet been found for redefinition of Mammalia to include these forms and no others.

During the Jurassic and Cretaceous, mammals were of small size and limited diversity (see *Mammals, Mesozoic*). For nearly a century following their discovery, they were regarded as marsupials. A majority of the Mesozoic forms differed from both marsupials and placental mammals in the form and occlusal relationships of their cheek teeth, and also in the arrangement of bones and nerve openings in the braincase. The living monotremes (egg-laying mammals) of Australia differ from other mammals in the same details of cranial anatomy, and appear to be surviving members of the prototherian division of the Mammalia. Their many privitive features may have been retained since early Mesozoic time. The Mesozoic orders Triconodonta, Docodonta, and Multituberculata are often classified with the living monotremes as Prototheria.

Therian mammals, characterized by an occlusal pattern of interlocking triangular cheek teeth and the formation of the side wall of the braincase by the alisphenoid bone, appeared in the latest Triassic. The Mesozoic orders Symmetrodonta and Pantotheria, as well as the Cretaceous-to-Holocene marsupials and placentals, form this greater branch of the Mammalia. Following the extinction of dinosaurs at the end of the Cretaceous, they rapidly expanded into the unoccupied niches (the appearance of new types in the fossil record may also represent migration into areas of fossil accumulation), underwent rapid adaptive diversification, and began to produce animals of larger size. All orders for which the record is not deficient had appeared by the end of the Eocene, and the subsequent history is largely that of specialization of various adaptive lines, accompanied by elimination of many of the earlier "aberrant" groups. The rapid structural evolution of mammals through the Cenozoic and their repeated intercontinental migrations give them great importance in correlation of continental deposits.

JOSEPH T. GREGORY

References

Bardack, D., and Zangerl, R., 1967. First fossil lamprey: A record from the Pennsylvanian of Illinois, *Science, 162*, 1265–1267.

Carroll, R. L., 1969. Problems of the origin of Reptilia, *Biol. Rev., 44*, 393–432.

Clemens, W. A., 1970. Mesozoic mammalian evolution, *Ann. Rev. Ecol. Syst., 1*, 357–390.

Hecht, M. K. et al., 1977. *Major Patterns in Vertebrate Evolution.* N.Y.: Plenum Press, 908p.

Heilmann, G., 1927. *The Origin of Birds.* New York: Appleton, reprinted by Dover, 208p.

Hopson, J. A., and Crompton, A. W., 1969. Origin of mammals, *Evolutionary Biology, 3*, 15–72.

Howard, H., 1950. Fossil evidence of avian evolution, *Ibis, 92*, 1–21.

Jarvik, E., 1965. Die Raspelzunge der Cyclostomen und die pentadactyle Extremität der Tetrapoden als Beweise für monophyletische Herkunft, *Zool. Anz., 175*, 101–143.

Jarvik, E., 1966. Remarks on the structure of the snout in *Megalichthys* and certain other rhipidistid crossopterygians, *Arkiv Zool., 19*, 41–98.

Kurtén, B., 1972. *The Age of Mammals.* New York: Columbia Univ. Press, 250p.

McKenna, M. C., 1969. The origin and early differentiation of therian mammals, *New York Acad. Sci. Ann., 167*, 217–240.

Moy-Thomas, J. A., and Miles, R. S., 1971. *Palaeozoic Fishes.* Philadelphia, Toronto: Saunders, 259p.

Olson, E. C., et al., 1965. Symposium on evolution and relationships of the Amphibia, *Am. Zool., 5*, 263–334.

Ørvig. T., ed., 1968. *Current Problems of Lower Vertebrate Phylogeny.* Nobel Symposium, vol. 4, 1967, Stockholm/New York: Wiley, 540p.

Ostrom, J. H., 1974. *Archaeopteryx* and the origin of flight, *Quart. Rev. Biol., 49*, 27–47.

Repetsky, J. E., 1978. A fish from the Upper Cambrian of North America, *Science, 200*, 529–531.

Romer, A. S., 1946. The early evolution of fishes, *Quart. Rev. Biol., 21*, 33–69.

Romer, A. S., 1956. *The Osteology of the Reptiles.* Chicago: Univ. Chicago Press, 772p.

Romer, A. S., 1962. Vertebrate evolution. Reviews and comments, *Copeia*, 1962(1), 223–227.

Romer, A. S., 1966. *Vertebrate Paleontology*, 3rd ed. Chicago: Univ. Chicago Press, 468p.

Schaeffer, B., 1952. Rates of evolution in the coelacanth and dipnoan fishes, *Evolution, 6*, 101–111.

Schaeffer, B., 1967. Comments on elasmobranch evolution, in P. W. Gilbert, R. F. Mathewson, and D. P. Rall, eds., *Sharks, Skates and Rays.* Baltimore: Johns Hopkins Press, 3–35.

Schaeffer, B., 1968. The origin and basic radiation of the Osteichthyes, in Ørvig, 1968, 207–222.

Schaeffer, B., and Rosen, D. E., 1961. Major adaptive levels in the evolution of the Actinopterygian feeding mechanisms, *Am. Zool., 1*, 187–204.

Stahl, B. J., 1974. *Vertebrate History: Problems in Evolution.* New York: McGraw-Hill, 594p.

Stensiö, E. A., 1928. The Downtonian and Devonian vertebrates of Spitzbergen. Part I. Family Cephala-spidae, *Skr. Svalbard Nordishavet,* 12(1927), 1–391.

Thomson, K. S., 1967. Notes on the relationship of the rhipidistian fishes and the ancestry of the tetrapods, *J. Paleontology,* 41, 660–674.

Thomson, K. S., 1969. The biology of the lobe-finned fishes, *Biol. Rev.,* 44, 91–154.

Thomson, K. S., 1971. The adaptation and evolution of early fishes, *Quart. Rev. Biol.,* 46, 139–166.

Cross-references: *Agnatha; Amphibia; Aves; Bones and Teeth; Dinosaurs; Mammals, Mesozoic; Mammals, Placental; Marsupials; Multituberculates; Pisces; Reptilia; Vertebrate Paleontology.*

VERTEBRATE PALEONTOLOGY

Study of the fossil record of the history of animals with backbones is termed vertebrate paleontology. Vertebrates make up the major part of the phylum Chordata (q.v.) and include all fishes, amphibians, reptiles, birds, and mammals. (See entries for these classes.)

Vertebrate paleontology developed as a branch of zoology through the successful use of the methods of comparative anatomy for identification of vertebrate fossils in the 18th century, and through the brilliant demonstrations by George Cuvier [1796–1812] that the bones of mammoths, mastodons, and the peculiar ungulates whose skeletons were found in gypsum quarries near Paris could not be the remains of living animals but must represent extinct species. Cuvier also used stratigraphic data to demonstrate a succession of extinct faunas during the history of the earth.

Following the publication of Charles Darwin's *On the Origin of Species by Means of Natural Selection* in 1859, the same comparative method was applied by W. Kowalewsky, A. Gaudry, E. D. Cope, and O. C. Marsh to fossil mammals from successive stratigraphic levels to establish lines of descent among various families such as the horses. These early studies have become standard examples of the fossil record of evolutionary change. Many phyletic series of fossils were described in the late 19th and early 20th centuries, and additional details continue to accumulate rapidly, as phylogenetic systematics forms the central endeavor of vertebrate paleontologists.

Although Kowalewsky and Cope described the mechanical action of various tooth elements and configurations of limb bones of mammals, the study of functional morphology progressed slowly until about 1950. Earlier restorations of locomotor musculature of titanotheres, dino-saurs, and various primitive and mammal-like reptiles by W. K. Gregory, C. L. Camp, and A. S. Romer were directed more to problems of homology than of function. D. M. S. Watson attempted to analyze the swimming movements of marine reptiles. Problems of evolution of the jaw articulation and ear ossicles in the transition from reptile to mammal have been the subject of numerous investigations, especially by Watson and his students. Recently the mechanics of tooth occlusion, the posture and locomotion of dinosaurs, the configuration of tails and fins in fishes, the function of horns and massive skulls in herbivorous tetrapods, and the peculiarly enlarged nasal passages of duck-billed dinosaurs have all been subjected to mechanical and behavioral analyses.

G. G. Simpson introduced the use of statistical methods in the analysis of paleontological data in his studies of Tertiary mammals about 1937, and together with a number of zoologists, was instrumental in replacing the typological systematic procedures of 19th century paleontology with the population concept of species and other viewpoints of the "New Systematics." Paleontology since 1950 is much more involved in the dynamics of evolutionary biology than it was during the early 20th century.

The last quarter century has likewise witnessed attempts to study the composition, trophic structure, and evolution of animal communities, and to apply ecological principles to problems such as extinction and some of the major evolutionary transitions involved in the origins of new classes. I. A. Efremov formulated principles of taphonomy, which deals with the processes leading to the burial and preservation of fossils; J. A. Shotwell proposed criteria for inferring which members of a fossil assemblage belonged to the biotic community closest to the site of deposition; and E. C. Olson analyzed trophic structure of and changes in the earliest terrestrial communities (see *Paleoecology, Terrestrial* for references).

Vertebrate paleontology has never lacked connections with geology, although these are not so strong as its ties to zoology. Cuvier recognized a sequence of five vertebrate faunas, and his successors soon discovered additional steps in the development of terrestrial vertebrate life. Vertebrate fossils have regularly been included by stratigraphers in their summaries of the fossil content of various formations. In some instances, e.g., the presence or absence of dinosaurs as a criterion of the Cretaceous/Tertiary boundary, vertebrate fossils have been accorded considerable weight in stratigraphic definitions.

Systems of biostratigraphic zonation based

upon fossil mammals have been proposed for sequences of continental deposits that lacked marine invertebrate faunas; perhaps the earliest of these was proposed for the Tertiary of France by Paul Gervais in 1852. In North America, vertebrate faunas were equated with various formational units of the Great Plains and Rocky Mountains regions during the latter half of the 19th century, and the first attempts at their biostratrigraphic analysis were made by H. F. Osborn and W. D. Matthew between 1899 and 1910. A system of "North American Provincial Ages" for Cenozoic continental vertebrates introduced by H. E. Wood and others in 1941 is now widely used, with some modifications (see Tedford, 1970), and similar systems have been developed for South America by G. G. Simpson (replacing an earlier terminology used by F. Ameghino), and for Europe by L. Thaler. A biostratigraphic zonation of the continental Permian and Triassic Karroo System of South Africa was developed by Robert Broom early in the present century; recent work on Triassic vertebrate faunas in India and South America is laying the foundation for provincial biostratigraphic zonation of continental deposits in these areas. In other parts of the time scale, especially the Paleozoic and Cretaceous, vertebrate-bearing deposits are associated with fossiliferous marine strata sufficiently closely to permit application of the more universally used stratigraphic terms based upon marine invertebrates, and there has been no need for a separate scheme of zonation.

Dispersal of terrestrial or freshwater vertebrates from one continent to another has provided extremely useful data for intercontinental correlation. The first appearance in one region of an animal whose immediate ancestors are known on a different continent establishes the time of invasion closely, especially when this is confirmed by evidence of the spread of some other form in the opposite direction. Perhaps the most cited example of this method is the use of the migration of *Hipparion* from North America to Eurasia late in the Miocene for correlation of the North American deposits. Although the first appearance of *Hipparion* in Europe is no longer considered to mark the beginning of the Pliocene, it establishes the synchroneity of North American Clarendonian faunas with those of the Vallesian stage of Europe (Berggren and Van Couvering, 1974)— a correlation now supported also by potassium-argon dates of 11 to 12 m yr B.P.

In addition to the usual methods of correlation by identity of species or genera and by the proportions of taxa in common between assemblages, vertebrate paleontologists frequently use a method termed "stage of evolu-

tion," first proposed by A. Gaudry, as a basis for estimating the relative age of a deposit containing fossils that are unknown elsewhere, yet are related to or part of lineages whose histories of morphological change are known.

To summarize, vertebrate paleontology has contributed the concept of extinct species, which played an important role in modifying traditional ideas about the age of the earth and also posed the major problem that led to the theory of organic evolution. It has provided important examples of lineages, showing evolutionary changes through time that also can serve as controls on hypotheses about the causes and nature of evolution. Vertebrate fossils have also been useful for geologic correlation, particularly of nonmarine deposits and sometimes on an intercontinental scale, though their relative scarcity and sporadic occurrence greatly limits their application to geologic problems. Finally, they give important evidence of former geographic connections between the continental land masses (or the absence of such connections) which contribute to a better understanding of tectonic history.

Preservation of Vertebrate Fossils

Vertebrate fossils consist mainly of the hard skeletal parts, bones and teeth; but under unusual conditions, traces of softer body tissues may be preserved as well. Such rare fossils include the frozen mammoth and rhinoceros cadavers of Arctic Siberia, mummified ground sloths in caves of the southwestern USA and in Argentina. Remarkably preserved bodies of primitive mammals and other vertebrates occur in Eocene lignites (fraunkohle) of the Geiseltal near Halle an der Saale, Germany. The carbonized outlines of whole ichthyosaurs are found in the Early Jurassic black shales at Holzmaden, Germany. Impressions of feathers of the earliest known bird and of the delicate wing membranes of pterodactyls (flying reptiles) as well as of viscera of fishes are preserved in the Late Jurassic lithographic limestones of Solenhofen, Germany (q.v.). Ironstone nodules in the Pennsylvanian at Mazon Creek, Illinois (q.v.), preserve impressions of the body outlines and viscera of larval amphibians. Late Cretaceous siltstones in Montana and Alberta preserve impressions of the skin of mummified dinosaurs. The thin layer of perichondral bone lining all cavities in the skull of certain Early Devonian ostracoderms has preserved minute details of the internal anatomy of the earliest vertebrates.

Fossil eggs have been found at a number of localities; the earliest of these was laid by an unidentified reptile in the Early Permian of

Texas. Far better known are the nests of eggs of the small horned dinosaur, *Protoceratops,* found in Cretaceous sandstones of the Gobi desert, and the larger eggs of the sauropod dinosaur *Hypselosaurus* from the Late Cretaceous of southern France. Several kinds of fossil bird eggs have been found in Tertiary rocks.

Fossil footprints are found in many formations, especially those representing ancient tidal flats or lake margins. Many data concerning the posture and habits of extinct forms of life may be inferred from a study of such markings (Sarjeant, 1975). Fossil burrows reveal much about the mode of life of certain vertebrates that may be found inside. Sometimes fossil bone fragments are found bearing the marks of teeth, either of predators, or, more commonly, scavengers such as hyaenas or the characteristic parallel grooves left by gnawing rodents. Fossil excrement or coprolites (q.v.) containing fish scales or bone fragments may reveal something of the food of the ancient animals.

Techniques. Care must be taken to avoid destruction of fossil bones in removing them from the surrounding earth or rock matrix. Although some bones may be collected intact, most are cracked and will fall into fragments unless proper precautions are observed. Hardening the specimen with a plastic lacquer or thin glue, which will penetrate fissures and cement their walls (after drying) is often sufficient.

Upon finding a specimen in place, all fragments that have become detached and scattered, especially downslope from the outcropping bone, should be gathered and placed where they will not be trod upon or covered by debris from the excavation. Whenever possible the bone or bones should be first exposed from above only. After its extent has been determined, and the upper surface hardened, excavation continues around the sides until the specimen is supported on a narrow pedestal. If sturdy, it may then be hardened and removed, preferably with a bit of the underlying matrix adhering to it, to be cleaned away later. Each specimen should be labeled or numbered immediately, with the data recorded in field notes, before wrapping it for transportation to laboratory or museum.

Entire skeletons or much-fractured bones are best removed together with sufficient surrounding rock to keep the parts in proper position. Instead of completely excavating each bone, a trench is dug around the entire specimen after the upper surface of the skeleton has been exposed and hardened. Two or more layers of paper tissue are applied to all exposed bone surfaces, and an outer layer over the entire surface of the block. (This paper will adhere more closely if applied wet and tapped lightly with a whiskbroom). Strips of burlap are then dipped in a mixture of plaster of Paris and laid over the block and snugly against its sides, to form a firm supporting cast. A girdle around the outside of the block to secure the ends of the covering bandages is useful. Large blocks (over 10 kg) require two or more layers of plaster bandages. After the plaster is dry, the pedestal supporting the skeleton is broken or cut at sufficient depth to avoid injury to the specimen, and the block is carefully turned over. Excess matrix may be removed from the under side. Depending upon the means and distance of transportation to the laboratory, the block may be left open (and carried with the exposed side up!) or covered with another set of plaster bandages. The outside of the block should be numbered or marked with the locality.

In the laboratory, the plaster is cut through around the outside of the block and removed, generally from the under side. The specimen is than carefully exposed and hardened. A sand table is convenient both for supporting irregularly shaped blocks and especially for holding individual bones while pieces are being glued in place. A variety of tools including vibrators, dental drills, air scribes, and dental abrasive machines are commonly used, as well as hammer, chisel, scrapers, and needles, to remove the adhering rock with minimum damage to the fossils. In research collections, the bones are cleaned, separated, and placed in trays in drawers, with labels and usually with catalogue numbers written in ink upon each bone or fragment. Many techniques are used to mount specimens for display, some of which are described in the manuals by Camp and Hanna (1935) and by Kummel and Raup (1965).

In recent years, many small teeth and bones of lizards, rodents, and other small vertebrates have been recovered by washing large quantities (hundreds of kilograms to hundreds of tons) of poorly consolidated fossiliferous sediment on screens that retain the small bones and other coarse particles of sediment while permitting the enclosing sand and silt to be washed away. These concentrates are then dried and either "picked" in the field or placed in plastic bags, with labels, to be sent to the laboratory for final separation of the bones from the rock fragments and sorting. This technique has increased our knowledge of the smaller animals that lived along with the larger kinds whose remains are more conspicuous.

Where the matrix surrounding fossil bones is limestone, or rock with lime cement, it is sometimes possible to dissolve away the rock with dilute acid (acetic and formic acids are especially suitable as they are less likely to dissolve the bone itself). One side of fish skeletons may be imbedded in plastic and the unexposed, and often better preserved, side then prepared by dissolving away the rock with acid.

Other skeletons, particularly those in coal or mudstone nodules, may be studied by removing the remains of the bone from the enclosing rock and then preparing a latex cast of the mold left by the bones and teeth. Important information on vertebrate structure has also been obtained by grinding away thin layers of the specimen and making photographs, drawings, or acetate peels of the surface at closely spaced intervals. Reconstructions from such serial sections by the wax-plate technique of embryologists reveal the internal structures of skulls or other objects in great detail.

Fossil Vertebrates and Organic Evolution

Some of the best documented examples of evolutionary change through time are provided by fossil vertebrates. Perhaps the most striking of these are the intermediate connecting links between major vertebrate classes, of which *Ichthyostega* and *Archaeopteryx* are outstanding. These fossils retain vestiges of the distinctive features of the ancestral class along with the hallmarks of the derived group. More cogent to the demonstration of gradual evolutionary change are stratigraphically ordered sequences of fossils that show progressive steps in the development of adaptive morphological structures, presumable in response to selective forces. The numerous mammal-like reptiles of the Permian and Triassic periods reveal the sequence of individually minor modifications that converted the skeletal structures of primitive reptiles into those distinctive of mammals (Fig. 1), and include remains of animals so intermediate in structure between these classes that definition of the "boundary" between reptile and mammal becomes difficult. A brief, up-to-date account of the origin of mammals is given by Crompton and Jenkens (1973). Similar to these in their implications are less fully documented sequences of the lineages of various dinosaurs, marine reptiles, labyrinthodont amphibians, bony fishes, and many families of mammals. The ancestry of the horse is among the most completely known, the changes between fossils from successive stratigraphic levels often being no greater than the individual variation within a single population. Episodes of speciation or branching accompany the adaptive changes in the several horse lineages.

Another aspect of evolution, the divergence of differently adapted organisms from a common ancestor, is also shown by vertebrate fossils. The massive quadrupedal horned dinosaurs (Ceratopsia) can be traced back through *Protoceratops* and *Psittacosaurus* to relatively

primitive ornithopods similar to *Hypsilophodon* and the Triassic *Fabrosaurus*, animals that are equally suitable structural forerunners of the bipedal *Iguanodon* and varied duck-bill (hadrosaur) contemporaries of *Triceratops* and its relatives. Similarly among mammals, the ancestral lineages of present-day carnivores—dogs, bears, civets, cats, and the like—can be traced back separately to the Oligocene, but Eocene fossil carnivora appear about as close to one family as to another, and Paleocene carnivores are distinguishable with difficulty from the remains of contemporary herbivores. The distinctive dental patterns of modern and Tertiary mammals converge to a common type in the Cretaceous. These records thus strongly support the derivation of all mammals from a common ancestor.

A glance at the skeleton of *Ichthyostega* (Fig. 2) from the late Devonian of East Greenland reveals an animal with short, stout limbs, heavy body, and large head typical of late Paleozoic labyrinthodont amphibians. Closer examination of its skull (Fig. 3C) shows its proportions are intermediate between those of a rhipidistian fish such as *Eusthenopteron* (Fig. 3A) and typical labyrinthodonts (Fig. 3D), which have a longer snout and shorter posterior segment than fishes, nostrils low on the side of the snout much like the ancestral fish, and vestiges of the bones of the gill cover (opercular bones), which are large and important in fishes but absent from all other amphibians. Along the tail, a series of small bony rods, the fin rays, are likewise vestiges of the supports for the tail fin of fishes. *Ichthyostega* is clearly one step in the transition from fish to tetrapod; the many problems involved in the biological change from aquatic to terrestrial life are discussed by Stahl (1974) and Thomson (1969).

Archaeopteryx, the Late Jurassic bird (Fig. 4) now known from five skeletons, all from the same region in Bavaria, had the complex and highly distinctive feathers of birds, and was bird-like in the proportions of the bones of its forelimb, already modified to support the wing, and the structure of its rear feet. Unlike modern birds, the metacarpal bones of the hand were not united into a rigid wing support, and large, reptile-like claws were present on its three fingers. The skull was reptile-like with teeth in the jaws and a braincase relatively smaller than is typical of birds. A reptilian rather than avian pattern of hip bones, a long lizard-like bony tail, and the lack of a large breastbone conclude its reptilian similarities. It has frequently been said that, were it not for the feathers, the skeleton of *Archaeopteryx* would have been regarded as just another kind of fossil reptile; it has recently been shown that most of its anatomical

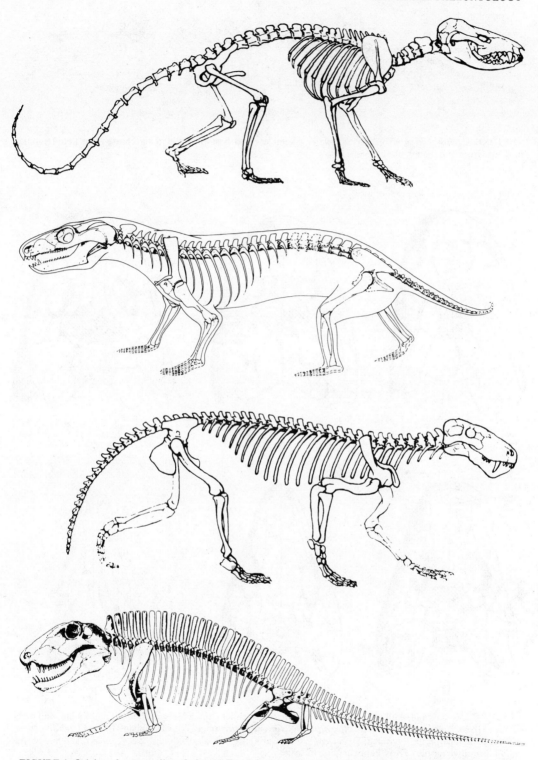

FIGURE 1. Origin of mammalian skeleton. From bottom to top: *Sphenacodon,* Early Permian pelycosaur (from Romer and Price, 1940); *Lycaenops,* Late Permian gorgonopsian (from Kuhn, 1951, after Colbert, 1948); *Massetognathus,* Triassic cynodont (from Jenkins, 1970); *Didelphys,* modern opposum (from Kuhn, 1951). Not to scale.

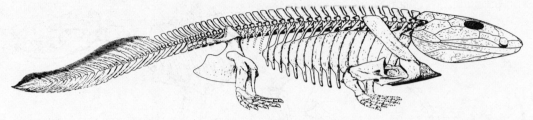

FIGURE 2. *Ichthyostega* (from Jarvik, 1955). Skeleton of earliest known land vertebrate from Late Devonian rocks of Greenland; length about 1 m.

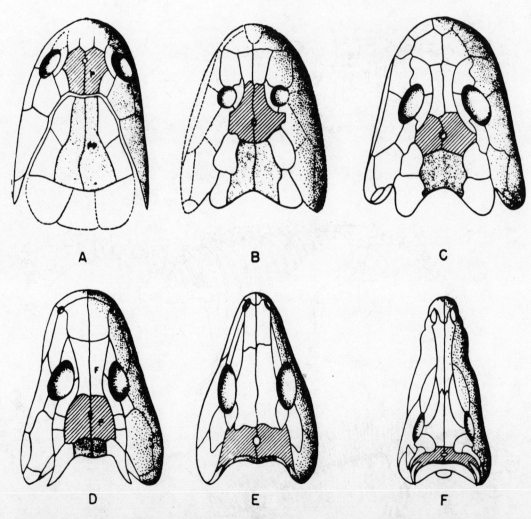

FIGURE 3. Changes in skull proportions from rhipidistian fishes through labyrinthodont amphibians and early reptiles (from Romer, 1956). A, *Eusthenopteron,* and B, *Elpistostege,* Late Devonian fishes; C, *Ichthyostega,* earliest amphibian, Late Devonian; D, *Palaeogyrinus,* Carboniferous labyrinthodont; E, *Romeria,* and F, *Sphenacodon,* primitive reptiles, Early Permian. Not to scale.

FIGURE 4. *Archaeopteryx* (from Swinton, 1958). Skeleton of long-tailed, toothed bird from the Late Jurassic of Germany; length about 40 cm. Reprinted by permission of the Trustees, British Museum (Natural History).

features except feathers were shared by the small, bipedal carnivorous dinosaurs, coelurosaurs, which now appear to be the group of reptiles from which birds were derived (Ostrom, 1974; for a different view of the closest reptilian relatives of birds, see Walker, 1972).

The story of the evolution of the modern horse from the small Eocene *Hyracotherium*, popularly known as eohippus, involves modification of the short, flexible, five-toed feet of primitive mammals into elongate single-toed structures specialized for rapid running, and development of a relatively large surface area and efficient grinding pattern on the cheek teeth, which later acquired high prismatic crowns to compensate for rapid wear by the harsh grasses (and adhering sand particles) which came to form their principal food (Fig. 5); the brain enlarged and became specialized for visual orientation. Details may be found in Simpson (1951).

Functional Morphology

Cuvier and Richard Owen stressed the ways in which the form and structure of each organ was adapted to perform a particular function.

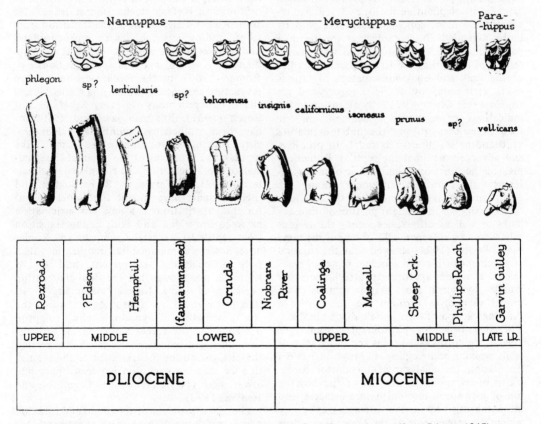

FIGURE 5. Evolution of high crowned grinding teeth in Late Tertiary horses (from Stirton, 1947).

Cuvier especially emphasized the necessity that all organs of an animal must function harmoniously together if the organism is to occupy its particular role in nature successfully. Early 19th century anatomists and paleontologists such as H. von Meyer inferred the modes of locomotion and food of extinct animals from the structure of their limbs, feet, and teeth.

Later in the 19th century, Kowalewsky and Cope showed that geologically younger fossils of various families of herbivorous mammals had more efficient teeth for grazing than their early Tertiary antecedents, and that the limb and foot bones and joints of ungulates likewise became progressively better constructed for running rapidly in the later fossils. Cope tried to explain the details of skeletal form through the effects of mechanical stresses on bones and teeth.

Analogy with living relatives or with the functions of structurally similar living animals has always been an important tool in the interpretation of fossils. But where no close analogue exists the method fails, and inferences must be based upon mechanical principles and anatomical analyses. Arguments about whether the huge sauropod dinosaur, *Diplodocus,* stood erect on its elephantine limbs, as Marsh had reconstructed it, or sprawled lizard-like with its limbs extended to the sides, as the German paleontologist G. Tornier insisted was characteristic of all reptiles, led to comparative studies of the limb and body musculature of various living vertebrates by W. K. Gregory and his students (see Olson, 1971). By attention to the indications of muscular attachment on the fossil bones they restored the limb musculature of titanotheres, dinosaurs, and both primitive and advanced mammal-like reptiles. Romer explained the functional significance of major differences in muscle arrangement among the several groups, and demonstrated the mechanical necessity for the upright posture of dinosaur limbs as well as differences among the various suborders (e.g., Romer, 1927). About the same time, Watson (1924) showed that the unusual arrangement of hip and shoulder bones of plesiosaurs provided appropriate attachments for muscles by which these marine reptiles could propel themselves through the water with a "rowing" action of their paddle-shaped limbs.

Starting from his photographic studies of Leland Stanford's race horses and other animals while walking and running, E. Muybridge laid a foundation for analysis of locomotor movements of tetrapods that has been refined by the use of high-speed motion picture cameras, and most recently by x-ray cinematography, by which the movements of the bones themselves can be closely monitored. Most functional analysis of fossil vertebrates during the past 25 years has been combined with related studies of the kinetics and behavior of corresponding actions of recent animals.

Functional problems that have particularly attracted the interest of vertebrate paleontologists include: The role of fins, and especially various forms of asymmetrical tails, in the swimming of fishes. Experimental analyses of swimming in living fishes by Harris (1963–1938) have been applied to the function of the tails of ostracoderms and ichthyosaurs (see Stahl, 1974, 155–159). Hildebrand (1974) and Gray (1968) have described the varied patterns of movement in running mammals, and Maynard-Smith and R. J. G. Savage have analyzed some of the underlying muscular mechanics. W. Auffenberg and C. Gans have dealt with locomotion of snakes and other limbless types. The flight capabilities and aerodynamics of the large flying reptile *Pteranodon* have been analyzed by Bramwell and Whitfield (1974) and by Stein (1975). Galton (1970) and Ostrom have reassessed the posture and locomotion of the bipedal dinosaurs and concluded that they held the back straight and at a rather low angle with the horizontal, in contrast to earlier restorations.

Studies of feeding mechanisms in fishes range from Romer's discussion of the transition from filter feeding invertebrates to protochordates and ostracoderms, and the eventual development of jaws and macrophagous feeding (e.g., Romer, 1967), to the mechanical analyses of jaw action and the evolution of jaw muscles and cheek bones in bony fishes by Schaeffer and Rosen (1961). Reptilian jaws and dentitions have been studied by a number of paleontologists, with most attention to the mammal-like reptiles and problems of the origin of the mammalian jaw articulation and complex dental patterns. D. M. S. Watson, F. R. Parrington, and their students have sought functional reasons for the substitution of a new jaw articulation between lower jaw and skull in the transition from reptile to mammal. Crompton and Romer have described mammal-like reptiles in which both articulations are present, and K. A. Kermack has shown that the dual articulation persisted in some Jurassic mammals. H. R. Barghusen and J. A. Hopson have restored the jaw musculature of cynodonts, and Crompton has analyzed the forces acting on the articulation during mastication. Kermack has related the changes during the transition to the acquisition of shearing occlusion between upper and lower teeth (for references see Crompton and Jenkins, 1973).

The allied problem of the origin and function of the multicuspid cheek teeth of mammals has likewise been attacked by Crompton and his students, using x-ray cinematography as well as analysis of wear facets on the cusps of rare Mesozoic mammalian teeth (Crompton and Jenkins, 1968). Most studies of mammalian teeth have stressed cusp homology, but P. M. Butler has made important analyses of the functions of various portions of the crowns of molars.

Numerous functional explanations have been offered for the varied crests on the skulls of duck-billed dinosaurs, including underwater feeding, olfaction, sound production, and intraspecific mating displays (e.g., Hopson, 1975). The function of horns of the horned dinosaurs, and of titanotheres, the massive skull caps of dinocephalians and pachycephalosaurs, and the enlarged antlers of the "Irish Elk" have been related to their probable roles in combat or display.

JOSEPH T. GREGORY

References

Bramwell, C. D., and Whitfield, G. R., 1974. Biomechanics of Pteranodon, *Phil. Trans. Roy. Soc. London, Ser. B.* **267**, 503–581.

Berggren, W. A., and Van Couvering, J. A., 1974. The Late Neogene, *Palaeography, Palaeoclimatology, Palaeoecology,* **16**, 1–216.

Camp, C. L., and Hanna, G. D., 1973. *Methods in Paleontology.* Berkeley: Univ. California Press, 153p.

Colbert, E. H., 1948. The mammal-like reptile *Lycaenops, Bull. Am. Mus. Nat. Hist.,* **89**, 353–504.

Crompton, A. W., and Jenkins, F. A., Jr., 1968. Molar occlusion in Late Triassic mammals, *Biol. Rev.,* **43**, 427–458.

Crompton, A. W., and Jenkins, F. A., Jr., 1973. Mammals from reptiles: A review of mammalian origins, *Ann. Rev. Earth Planetary Sci.,* **1**, 131–155.

Galton, P. M., 1970. The posture of hadrosaurian dinosaurs, *J. Paleontology,* **44**, 464–473.

Gray, J., 1968. *Animal Locomotion.* London: Weiden Field and Nicolson, 474p.

Harris, J. E., 1936–1938. The role of the fins in the equilibrium of the swimming fish, *J. Exp. Biol.,* **13**, 476–493; **15**, 32–47.

Hildebrand, M., 1974. *Analysis of Vertebrate Structure.* New York: Wiley, 710p.

Hopson, J. A., 1975. The evolution of cranial display structure in hadrosaurian dinosaurs, *Paleobiology,* **1**, 21–43.

Jarvik, E., 1955. The oldest tetrapods and their forerunners, *Sci. Monthly,* **80**, 141–154.

Jenkins, F. A., Jr., 1970 The Chañares (Argentina) Triassic reptile fauna VII. The postcranial skeleton of the traversodontid *Massetognathus pascuali* (Therapsida, Cynodontia), *Breviora, Mus. Comp. Zool. Harvard Univ.,* **352**, 28p.

Kuhn, E., 1951. *Geschichte der Wirbeltiere.* Zürich: Emil Rüegg, 168p.

Kummel, B., and Raup, D., 1965. *Handbook of Paleontological Techniques.* San Francisco: Freeman, 852p.

Olson, E. C., 1971. *Vertebrate Paleozoology.* New York: Wiley, 839p.

Ostrom, J. H., 1974. *Archaeopteryx* and the origin of flight, *Quart. Rev. Biol.,* **49**, 27–47.

Romer, A. S., 1927. The pelvic musculature of ornithischian dinosaurs, *Acta Zool.,* **8**, 225–275.

Romer, A. S., 1956. The early evolution of land vertebrates, *Proc. Amer. Phil. Soc.,* **100**, 157–167.

Romer, A. S., 1966. *Vertebrate Paleontology,* 3rd ed. Chicago: Univ. Chicago Press, 468p.

Romer, A. S., 1967. Major steps in vertebrate evolution, *Science,* **158**, 1629–1637.

Romer, A. S., and Price, L. I., 1940. Review of the Pelycosauria, *Geol. Soc. America Spec. Pap.,* **28**, 538p.

Sarjeant, W. A. S., 1975. Fossil tracks and impressions of vertebrates, in R. W. Frey, ed., *The Study of Trace Fossils.* New York: Springer-Verlag, 283–324.

Schaeffer, B., and Rosen, D. E., 1961. Major adaptive levels in the evolution of the actinopterygian feeding mechanism, *Amer. Zool.,* **1**, 187–204.

Simpson, G. G., 1951. *Horses.* New York: Oxford Univ. Press, 247p.

Stahl, B. J., 1974. *Vertebrate History: Problems in Evolution.* New York: McGraw-Hill, 594p.

Stein, R. S., 1975. Dynamic analysis of *Pteranodon igens:* A reptilian adaptation to flight, *J. Paleontology,* **49**, 534–548.

Stirton, R. A., 1947. Observations on evolutionary rates of hypsodonty, *Evolution,* **1**, 32–41.

Swinton, W. E., 1958. *Fossil Birds.* London: Brit. Mus. (Nat. Hist.), 63p.

Tedford, R. H., 1970. Principles and practices of mammalian geochronology in North America, *Proc. N. Amer. Paleontological Conv.,* pt. F, 666–703.

Thomson, K., 1969. The biology of the lobe finned fishes, *Biol. Rev.,* **44**, 91–154.

Walker, A. D., 1972. New light on the origins of birds and crocodiles, *Nature,* **237**, 257–263.

Watson, D. M. S., 1924. The elasmosaurid shoulder-girdle and forelimb, *Proc. Zool. Soc. London,* **1924**, 885–917.

Cross-references: *Agnatha; Amphibia; Aves; Bones and Teeth; Coprolite; Dinosaurs; Morphology, Functional; Mammals, Mesozoic; Mammals, Placental; Marsupials; Multituberculates; Paleobiogeography of Vertebrates; Paleoecology, Terrestrial; Paleopathology; Pisces; Reptilia; Sexual Dimorphism; Vertebrata.*

Z

ZOOXANTHELLAE

Zooxanthellae are the endosymbiotic dino-flagellates in Holocene reef-building corals. This association is generally believed to provide the physiological basis of the existence of coral reefs today. In view of the wide array of animals that have algal symbionts ("zoochlorellae" if the algae are green), and in view of the interest of paleontologists in reefs and reef-like communities, it is not surprising that several extinct groups have been offered as possible bearers of algal symbionts. The acceptance of such views carries considerable paleoecological implications.

The association between corals and dino-flagellates is in itself insufficient to yield reefs—strains of *Astrangia danae* off Woods Hole, Mass., for example, contain zooxanthellae. Nor are coral reef-like structures dependent on the association—witness the extensive banks of corals in deep, cold water (see *Reefs and other Carbonate Buildups*). Nevertheless, most of the calcium carbonate contributed by corals to tropical reefs has its origin in species with zooxanthellae. Within the limits of tropical temperatures, sufficient light (shallow water), and normal salinity, and in the presence of sufficiently oligotrophic conditions to require tight recycling of nutrients, hermatypic (reef-building) corals flourish. Nitrogenous wastes produced by the polyp are taken up by the alga, which in turn leaks large amounts (up to 75%) of the carbon it reduces back into the host's tissues. In the presence of light, the requirements for CO_2 and O_2 are also complementary.

The primary benefit to the animal bearing photosynthetic endosymbionts is energetic. A wide variety of animals, from *Paramecium bursaria*, to *Spongilla*, *Chlorohydra*, and the acoel *Convoluta*, carry algae but deposit no skeleton of calcium carbonate. Goreau (summarized, 1961) suggested that hermatypic corals actively took up Ca^{++} ions, which precipitated metabolically derived bicarbonate, yielding carbonic acid. The algae aided the reaction by taking up the acid. Goreau demonstrated that calcification proceeded at a far higher rate in the presence of light. Pearse and Muscatine (1971) demonstrated that the highest rate of calcification, at the tip of a branched coral, coexists with the lowest concentration of zooxanthellae, so the pleasing simplicity of the Goreau model has been subject to complications. Simkiss (1964) suggested that zooxanthellae absorbed phosphates that would otherwise poison the crystallization of calcium carbonate. Other difficulties involve the chemistry and role of the organic matrix, and the relation of carbon and oxygen isotope ratios in $CaCO_3$ to the action of zooxanthellae. Muscatine (1974) reviews the physiology of symbiosis, and gives references for the problems of relating calcification and zooxanthellae.

In view of the many phyla containing species that have algal symbionts, it is justified to believe that extinct organisms did as well; but to demonstrate this relation involves a daunting chain of circumstantial evidence. First, because the details of the chemical relationship between zooxanthellae and crystals of calcium carbonate remain obscure, the interpretation of fossil microstructure is uncertain. Whether any effect of zooxanthellae is diagnostic of their activity, and theirs alone, is unknown. Second, energetics is the *raison d'être* of endosymbiosis, not the deposition of calcium carbonate or the production of hard parts suitable for fossilization. What is crucial to the paleontologist was not necessarily quite as important to the extinct organism. Is it to be taken as intuitively obvious that the making of a corallum is an energetically costly task? No one has reported the cost to a coral of a gram of new skeleton, or what difference zooxanthellae make. Perhaps the involvement of algal symbionts in the formation of skeleton, as opposed to physiology, has been over-rated because of the disproportionate interest in the interpretation of the test. Neither calcium nor carbonate is in short supply on a reef—perhaps corals are excreting a waste product in the least inconvenient manner. The assumption that zooxanthellae allow the host to produce a more massive test contributes to the third difficulty—the lack of an appropriate gestalt for recognizing possible cases of endosymbiosis. Massive deposition of $CaCO_3$ as a result of zooxanthellae is perhaps not the best first approximation. *Tridacna* and the image of corals building a reef are what first come to mind when the question of zooxanthellae and calcification is raised. But the same process occurs between the delicate valves of *Corculum cardissa* (Kawaguti, 1950) as between the massive valves of *Tridacna*. It is worth recalling that the bulk of $CaCO_3$ on a reef comes not from corals but from fragmented algae, and that those hermatypic corals most dependent on zooplankton, or at least most adept at capturing them, are the most massive (Porter, 1974).

A couple of very general conditions are likely to hold for extinct animals that had algal symbionts. Because light is a controlling factor, organisms thought to be adapted to coping with vast amounts of extremely fine-grained sediments are unlikely candidates. The energetics of endosymbiosis are apparently favored when the supply of nutrients is extremely limited. Facies rich in reduced carbon are unlikely to contain animals that were the hosts of algal endosymbionts. That photosynthetic endosymbiosis occurs today in fresh water and tropical seas, and not in estuaries or cold, enriched marine water is perhaps a reflection of these ecological limits. But the most important factors relate to the reconstruction of functional morphology, especially signs of unusually extensive exposure of tissue to light.

The windows in the shell of *C. cardissa* come close to being ideal indicators of endosymbiosis. Vogel (1975) found a similar shell structure in a rudist bivalve, *Osculigera*. Kauffman and Sohl (1974) reported the impression of mantle vessels on the external side of the free valve of some rudists, and the elaboration of that valve in some forms into a seive plate. Taken with the reef-like nature of rudist localities and the massiveness of many forms, it is difficult to escape the conclusion that they were served by algal symbionts (see *Rudists*).

Two superfamilies of Permian strophomenid brachiopods, the lyttoniaceans, with dorsal valves transformed into something not unlike half-open Venetian blinds, and richthofeniaceans, cemented conical ventral valves with dorsal valves set on a knife-edge hinge well within, have been offered as probable hosts to algal symbionts, in a series of papers by M. J. S. Rudwick and R. Cowen (summary and bibliography in Cowen, 1975). Again, signs of extruded mantle tissue are found on the external side of the dorsal valve. The placement of spines and seives over the aperture of the ventral valve, and especially the resorption and redeposition of spines indicates large amounts of exposed mantle, at least on occasion. The evidence for symbiosis is part of an extraordinary interpretation of the functional anatomy of these brachiopods, which replaces ciliary feeding with rhythmic pumping of the dorsal valves. The resulting reefs, if such they are, have more in common with egg shells than poured concrete, but as argued above, this is not necessarily damaging. Rudwick's paradigmatic approach to functional anatomy is well served by these papers, and an able and provocative alternative is offered by Grant (1975).

Many recent foraminifera are invested with algal symbionts, and it is reasonable to theorize that many extinct ones were as well. Cowen (1970) credits D. J. Gobbett with the suggestion that some Permian forams contained symbiotic algae. Discoidal fusulinids with a sinuate cross section and subspherical fusulinids with equatorial aprons suggest basking in quiet waters, with an occasional gentle current flipping the shaded side to the light.

Finally, Sorauf (1974) suggests that regularly spaced growth lines on the underside of dissepiments in the tabulate coral *Favosites* are a reflection of a circadian rhythm resulting from the enhancement of calcification by endosymbiotic photosynthesis. As the polyp resides atop the completed dissepiment, the underside is not exposed to secondary calcification and retains a record of its fabrication. Sorauf found similar patterns in Holocene hermatypic corals, and less distinct growth lines in ahermatypes. As his specimen was silicified, he could not make further comparisons on the details of crystallization. Were *Favosites* not sometimes found in mud, subject to high levels of sedimentation (Philcox, 1971), or were there some sign of coenosteum, indicating elaboration of tissue for absorbing light and decreasing the possibility that *Favosites* was a sclerosponge rather than a coral (Hartman and Goreau, 1975), one might be less uncertain about this intriguing report.

STAN RACHOOTIN

References

Cowen, R., 1970. Analogies between the Recent bivalve *Tridacna* and the fossil brachiopods Lyttoniacea and Richthofeniacea, *Palaeogeography, Palaeoclimatology, Palaeoecology*, 8, 329–344.

Cowen, R., 1975. Flapping valves in brachiopods, *Lethaia*, 8, 23–29.

Goreau, T., 1961. Problems of growth and calcium deposition in reef corals, *Endeavour*, 20, 32–39.

Grant, R., 1975. Methods and conclusions in functional analysis: A reply, *Lethaia*, 8, 31–33.

Hartman, W., and Goreau, T., 1975. A Pacific tabulate sponge, living representative of a new order of sclerosponges, *Postilla*, 167, 21p. (See also Flügel 1976, *Lethaia*, 9, 405–419.)

Kauffman, E., and Sohl, N., 1974. Structure and evolution of Antillean Cretaceous rudist frameworks, *Verhandl. Naturf. Ges. Basel*, 84, 399–467.

Kawaguti, S., 1950. Observations on the heart cockle *Corculum cardissa* (L.) and its associated zooxanthellae, *Pac. Sci.,* **4,** 43–49.

Muscatine, L., 1974. Endosymbiosis of cnidarians and algae, in L. Muscatine and H. M. Lenhoff, eds., *Coelenterate Biology: Reviews and New Perspectives.* New York: Academic Press, 359–395.

Pearse, V., and Muscatine, L., 1971. Role of symbiotic algae (zooxanthellae) in coral calcification, *Biol. Bull.,* **141,** 350–363.

Philcox, M., 1971. Growth forms and role of colonial coelenterates in reefs of the Gower formation (Silurian), Iowa, *J. Paleontology,* **45,** 338–346.

Porter, J., 1974. Zooplankton feeding by the reef building coral *Montastrea cavernosa,* Proc. 2nd Internat. Coral Reef Symp., **1,** 111–125.

Simkiss, K., 1964. Phosphates as crystal poisons in calcification, *Biol. Rev.,* **39,** 487–505.

Sorauf, J., 1974. Growth lines on tabulae of *Favosites* (Silurian, Iowa), *J. Paleontology,* **48,** 553–555.

Vogel, K., 1975. Endosymbiotic algae in rudists? *Palaeogeography, Palaeoclimatology, Palaeoecology,* **17,** 327–332.

Cross-references: *Algae; Corals; Reefs and Other Carbonate Buildups; Rudists.*

INDEX

Boldface entries represent article titles; boldface page numbers indicate first page of main articles.

855